ASTRONOMY AND ASTROPHYSICS ABSTRACTS

A Publication of the Astronomisches Rechen-Institut Heidelberg
Member of the Abstracting Board of the International
Council of Scientific Unions

Volume 11
Literature 1974, Part 1

Edited by
S. Böhme · U. Esser · W. Fricke · U. Güntzel-Lingner
F. Henn · D. Krahn · H. Scholl · G. Zech

Springer-Verlag Berlin Heidelberg GmbH 1974

Astronomisches Rechen-Institut
Heidelberg
Director: Prof. Dr. W. Fricke

Astronomy and Astrophysics Abstracts
Editor-in-Chief: F. Henn

Astronomy and Astrophysics Abstracts
is prepared under the auspices
of the International Astronomical Union

ISBN 978-3-662-12294-5   ISBN 978-3-662-12292-1 (eBook)
DOI 10.1007/978-3-662-12292-1

© by Springer-Verlag Berlin Heidelberg 1974
Originally published by Astronomisches Rechen-Institut Heidelberg in 1974
Softcover reprint of the hardcover 1st edition 1974

Preface

Astronomy and Astrophysics Abstracts, which has appeared in semi-annual volumes since 1969, is devoted to the recording, summarizing and indexing of astronomical publications throughout the world. It is prepared under the auspices of the International Astronomical Union (according to a resolution adopted at the 14th General Assembly in 1970).

Astronomy and Astrophysics Abstracts aims to present a comprehensive documentation of literature in all fields of astronomy and astrophysics. Every effort will be made to ensure that the average time interval between the date of receipt of the original literature and publication of the abstracts will not exceed eight months. This time interval is near to that achieved by monthly abstracting journals, compared to which our system of accumulating abstracts for about six months offers the advantage of greater convenience for the user.

Volume 11 contains literature published in 1974 and received before August 15, 1974; some older literature which was received late and which is not recorded in earlier volumes is also included. Beginning with this volume some minor changes of our classification scheme have been made.

We acknowledge with thanks contributions to this volume by Dr. J. Bouška, who surveyed journals and publications in the Czech language and supplied us with abstracts in English, and by the Commonwealth Scientific and Industrial Research Organization (C.S.I.R.O.), Sydney, for providing titles and abstracts of papers on radio astronomy.

It is a pleasure to express our warmest thanks again to Miss Helga Ballmann, Mrs Monika Betz, Mrs Karola Gudé, Miss Lore Kiefert, and Mrs Ingrid Wolf, who typed the text of this volume on IBM 72 Composers and compiled the pages from abstract slips in a perfect form for offset reproduction, to Mrs Elisabeth Feigenbutz, for punching material for the author index.

Heidelberg, September 1974
                                                                    Siegfried Böhme
                                                                    Ute Esser
                                                                    Walter Fricke
                                                                    Ulrich Güntzel-Lingner
                                                                    Frieda Henn
                                                                    Dietlinde Krahn
                                                                    Hans Scholl
                                                                    Gert Zech

Contents

Preface
Introduction 1
Abbreviations 3

Periodicals, Proceedings, Books, Activities
    001   Periodicals 5
    002   Bibliographical Publications, News Notes from Current Journals 16
    003   Books 19
    004   History of Astronomy, Chronology 26
    005   Biography 29
    006   Personal Notes 30
    007   Obituaries 32
    008   Observatories, Institutes 34
    009   Notes on Observatories, Planetaria, and Exhibitions 50
    010   Societies, Associations, Organizations 51
    011   Reports on Colloquia, Congresses, Meetings, Symposia, and Expeditions . . 56
    012   Proceedings of Colloquia, Congresses, Meetings, and Symposia 58
    013   Reports on Astronomy in Various Countries and Particular Fields,
           International Cooperation 60
    014   Teaching in Astronomy 61
    015   Miscellanea 63

Applied Mathematics, Physics
    021   Mathematics, Computing, Data Processing 65
    022   Physical Papers Related to Astronomy and Astrophysics 68

Astronomical Instruments and Techniques
    031   Astronomical Optics, Methods of Observation and Reduction,
           Automation 74
    032   Astronomical Instruments 79
    033   Radio Telescopes and Equipment 82
    034   Astronomical Accessories (Spectrometers, Photometers, etc.) 87
    035   Clocks and Frequency Standards 92
    036   Photographic Auxiliaries 93

Positional Astronomy, Celestial Mechanics
- 041 Positional Astronomy, Astrometry, Star Catalogues and Atlases 94
- 042 Celestial Mechanics, Figure of Celestial Bodies 98
- 043 Astronomical Constants 105
- 044 Time, Rotation of the Earth 105
- 045 Latitude Determination, Polar Motion 107
- 046 Astronomical Geodesy, Satellite Geodesy, Navigation 109
- 047 Ephemerides, Almanacs, Calendars 111

Space Research
- 051 Extraterrestrial Research, Spaceflight Related to Astronomy and Astrophysics 112
- 052 Astrodynamics and Navigation of Space Vehicles 114
- 053 Lunar and Planetary Probes and Satellites 117
- 054 Artificial Earth Satellites 118
- 055 Observations of Earth Satellites, Lunar and Planetary Probes 120

Theoretical Astrophysics
- 061 General Theoretical Problems of Astrophysics, Gravitational Instability, Neutrino Astronomy, Infrared X-Ray, Gamma-Ray, Astronomy, Abundances and Origin of Elements 121
- 062 Hydrodynamics, Magnetohydrodynamics, Plasma 126
- 063 Radiative Transfer, Scattering 133
- 064 Stellar Atmospheres, Stellar Envelopes, Mass Loss 137
- 065 Star Formation, Stellar Structure and Evolution, Neutron Stars 143
- 066 Relativistic Astrophysics (without Cosmology), Background Radiation, Gravitation Theory 153

Sun
- 071 Solar Photosphere, Spectrum 162
- 072 Sunspots, Faculae, Solar Activity Cycles 167
- 073 Solar Chromosphere, Flares, Prominences 171
- 074 Solar Corona, Solar Wind 178
- 075 Solar Patrol 187
- 076 Solar UV, X Rays, Gamma Radiation 188
- 077 Solar Radio Radiation 191
- 078 Solar Cosmic Radiation 197
- 079 Solar Eclipses 200
- 080 Solar Atmosphere, Figure, Internal Constitution, Neutrinos, Magnetic Fields, Rotation, Miscellanea 202

Earth
- 081 Figure, Composition, and Gravity of the Earth 207
- 082 The Earth's Atmosphere Including Refraction, Scintillation, Extinction, Airglow, Site Testing 210
- 083 Ionosphere 217
- 084 Aurorae (220), Geomagnetic Field (223), Radiation Belts (229) 220
- 085 Solar-Terrestrial Relations 230

Planetary System
- 091 Physics of the Planetary System (Planetary Atmospheres, Figure, Interior, Magnetic Fields, Rotation, etc.) 231
- 092 Mercury 233
- 093 Venus 236
- 094 Moon: Dynamics (239), Global Properties (241), Local Properties (247) . . 239
- 095 Lunar Eclipses 254
- 096 Lunar Occultations 254
- 097 Mars (256), Mars Satellites (261) 256
- 098 Minor Planets 261
- 099 Jupiter (264), Jupiter Satellites (269) 264
- 100 Saturn (271), Saturn Satellites (272) 271
- 101 Uranus, Neptune, Pluto, Transplutonian Planets 274
- 102 Comets (Origin, Structure, Atmospheres, Dynamics) 275
- 103 Comets: Listed Objects 277
- 104 Meteors, Meteor Streams 286
- 105 Meteorites, Meteorite Craters 290
- 106 Interplanetary Matter, Interplanetary Magnetic Field, Zodiacal Light . . . 298
- 107 Cosmogony of the Planetary System 302

Stars
- 111 Stellar Parallaxes 304
- 112 Proper Motions, Radial Velocities, Space Motions 305
- 113 Stellar Magnitudes, Colors, Photometry 307
- 114 Stellar Spectra, Temperatures, Spectroscopy 311
- 115 Stellar Luminosities, Masses, Diameters, HR-Diagrams and Others 324
- 116 Stellar Magnetic Field, Figure, Rotation 326
- 117 Binary and Multiple Stars, Theory 328
- 118 Visual Double and Multiple Stars 331
- 119 Spectroscopic Binaries 333
- 120 Variable Stars: Catalogues, Ephemerides, Miscellanea 335
- 121 Eclipsing Variables 336
- 122 Physical Variables, Flare Stars, Pulsation Theory 344
- 123 Variable Stars: Lists of Observations, Individual Observations 353
- 124 Novae 355
- 125 Supernovae, Supernova Remnants 357
- 126 Low-luminosity Stars, Subdwarfs, White Dwarfs 362

Interstellar Matter, Gaseous Nebulae, Planetary Nebulae
- 131 Interstellar Matter, Polarization of Starlight (365),
  H I, H II Regions (376) 365
- 132 Emission Nebulae, Reflection Nebulae 381
- 133 Planetary Nebulae 384
- 134 Crab Nebula 386

Radio Sources, Quasars, Pulsars, Infrared, X-Ray, Gamma-Ray Sources, Cosmic Radiation
- 141 Radio Sources, Quasars (388), Pulsars (397), Infrared Sources (401) . . . 388
- 142 X-Ray, Gamma-Ray Sources 403
- 143 Cosmic Radiation 412

Stellar Systems
- 151 Kinematics and Dynamics of Stellar Systems 417
- 152 Stellar Associations 421
- 153 Galactic Clusters 422
- 154 Globular Clusters 425
- 155 Structure and Evolution of the Galaxy 428
- 156 Galactic Magnetic Field 435
- 157 Galactic Radio Radiation 436
- 158 Single and Multiple Galaxies (437), Peculiar Objects (446) 437
- 159 Magellanic Clouds 447
- 160 Clusters of Galaxies 448
- 161 Intergalactic Matter 450
- 162 Structure and Evolution of the Universe, Cosmology 451

Author Index 455
Subject Index 537

Introduction

Astronomical bibliographies

Astronomy and Astrophysics Abstracts begins documentation and abstracting as from the year 1969. For information on astronomical literature before this date consultation of one of the following bibliographies is suggested:
(1) J. J. de Lalande, Bibliographie Astronomique, Paris 1803 (this work covers the time from 480 B. C. to the year 1803, VIII + 966 pages).
(2) J. C. Houzeau, A. Lancaster, Bibliographie générale de l'astronomie, Volume I (in two parts), Bruxelles 1882, 1887, Volume II, Bruxelles 1889. The complete title of Volume II is "Bibliographie générale de l'astronomie ou catalogue méthodique des ouvrages, des mémoires et des observations astronomiques, publiés depuis l'origine de l'imprimerie jusqu'en 1880". A new edition of these volumes was prepared by D. W. Dewhirst in 1964.
(3) Bibliography of Astronomy, 1881 - 1898. The literature of this period was recorded on standard slips by the Observatoire Royal de Belgique. From the material (some 52.000 items) a microfilm version was produced by University Microfilms Limited, Tylers Green, High Wycombe, Buckinghamshire, England, in 1970.
(4) Astronomischer Jahresbericht, 1899 gegründet von Walter Wislicenus, herausgegeben vom Astronomischen Rechen-Institut in Heidelberg (formerly in Berlin), Verlag W. de Gruyter, Berlin. For the period from 1899 to 1968 sixty-eight volumes were published, each of which, in general, covers the literature of one year.
(5) Bulletin Signalétique – Section, Astronomie, Physique Spatiale, Géophysique. Published by Centre de Documentation du Centre National de la Recherche Scientifique, Paris. This publication is a continuation of "Bibliographie Mensuelle de l'Astronomie" founded in 1933 by the Société Astronomique de France. The publication is continued.
(6) Referativnyj Zhurnal. Founded in 1953 and published by Vsesoyuznyj Institut Naučnoj i Techničeskoj Informatsii, Akademiya Nauk, Moskva. The publication is continued.

Concept of Astronomy and Astrophysics Abstracts

This abstracting service aims to present a comprehensive documentation of the literature in all fields of astronomy and astrophysics. It appears in semi-annual volumes, two of which cover the literature of a calendar year. The half-yearly period of issue is regarded as an optimal period of time for summarizing papers into subject categories and for the presentation of abstracts as quickly as possible after the publication of the original literature. The time limits at which the documentation begins and ends for a volume are not sharply defined, except in the sense that all literature will be covered which was received by the editors within these limits.

Vol. 11 is devoted to the recording, summarizing and indexing of astronomical publications of the year 1974 received from January 1st, 1974 to August 15, 1974; it also records a number of papers issued before 1974 but received within the period of time.

The main characteristics of the concept of Astronomy and Astrophysics Abstracts may be summarized briefly.

(1) Titles of papers are given in the language of their authors whenever possible. If they are not in English but supplied with English translations they will be given in English. Abstracts are presented in English, French or German. Titles of papers in Russian are given in English.
(2) Authors' abstracts are used whenever possible. As a rule, popular articles were not abstracted; however their titles are usually given with the notation "Popular article".
(3) As a rule, each paper has been classified into one of 108 numbered subject categories and allocated a serial number within the category. In this way each item is numbered by six figures, the first three of which indicate the number of the category. Three further figures indicate the serial number within the category, which was allocated in the order of the receipt of the abstract. Reference to an abstract in Volume 1 is indicated by "01" before the number of the category; for example, 01.074.028, denotes Volume 1, category 074, abstract 028. Vol. 2 is indicated by "02", etc., Vol. 11 by "11".

A paper may have been classified into more than one category. Then its abstract has been allocated a number in one of the categories involved, and in the other category (or categories) the paper has been indicated by the title and a reference to the abstract number.

Papers whose authors are not named were treated like those with authors' names, with one exception: reports from correspondents of journals whose names were unknown were not numbered.

(4) There are categories which suggest the presentation of the material in subject groups. For instance, a subject group may be formed by all information received on the same solar eclipse, comet, nova, etc. The unsorted presentation of such material in a subject category would be inconvenient for the user, even if the individual comet, etc. were included in the subject index.

The following subject categories are subdivided into subject groups:

008 Observatories, Institutes. The publications of observatories and astronomical institutes are listed in alphabetical order of the towns of the institutions, each town forming a numbered subject group. For each publication a reference to an abstract number is made.
010 Societies, Associations, Organizations. The publications of each one form a subject group. The groups are presented in alphabetical order.
079 Solar eclipses. All publications related to one solar eclipse form a subject group.
103 Comets: Listed Objects. All publications related to the same comet form a numbered group.
124 Novae. All publications related to one nova form a subject group.
125 Supernovae. All publications related to one supernova form a subject group.

(5) Border fields of astronomy and astrophysics have been taken into account by presenting titles of papers occasionally without abstracts. The selection of papers for inclusion has been made according to the degree of relevance to astronomical research.

Transliteration of the Russian alphabet

The transliteration of the Russian alphabet in use in Astronomy and Astrophysics Abstracts is presented here.

А	а	a	Р	р	r
Б	б	b	С	с	s
В	в	v	Т	т	t
Г	г	g	У	у	u
Д	д	d	Ф	ф	f
Е	е	e	Х	х	kh
Ё	ё	e	Ц	ц	ts
Ж	ж	zh	Ч	ч	ch
З	з	z	Ш	ш	sh
И	и	i	Щ	щ	shch
Й	й	j	Ъ	ъ	''
К	к	k	Ы	ы	y
Л	л	l	Ь	ь	'
М	м	m	Э	э	eh
Н	н	n	Ю	ю	yu
О	о	o	Я	я	ya
П	п	p			

This transliteration was recommended by the Abstracting Board of the International Council of Scientific Unions in 1969. It is essentially the same as the transliteration proposed by the Academy of Sciences, Moscow, and used by the Referativnyj Zhurnal. It may be noted that the letters can be read and printed by usual data processing machines.
If the names of Russian authors in the literature are transliterated very different from this scheme we present the names in the form in which they are given in the references cited and in addition in round brackets according to our transliteration table.

Sources of information

The majority of sources of information for this volume are given in section **001 Periodicals** and in section **008 Observatories, Institutes**. The term "periodical" has been used in its widest sense for publications in a sequence of undetermined duration, even if the intervals of appearance are not regular. Section 001 records 291 periodicals with their full titles and with abbreviations which are in use in Astronomy and Astrophysics Abstracts. It may be noted that the titles of the periodicals are given in their original languages, and that Russian titles have been transliterated applying the transliteration given above. Section 008 records 132 periodicals; these are publication series of observatories and astronomical institutes which have not been included in section 001. The abbreviations of the titles of the periodicals have been given so that in most cases they permit recognition of the full title without recourse to the key in section 001. The steadily growing number of periodicals makes it necessary to use more extensive abbreviations and to abandon the use of very condensed ones.
Other abstracting journals have been consulted in order to examine the degree of completeness of our service. Occasionally, in particular in Physics Abstracts, Referativnyj Zhurnal, and Bulletin Signalétique abstracts of papers were found which had not come to our attention. In such cases Astronomy and Astrophysics Abstracts cites these papers, but also gives reference to the abstracting service which acted as the source.

Classification into a scheme of subject categories

The subdivision of astronomy and its border fields into subject categories is facilitated by the fact that the astronomical objects appear to be particularly well suited for the formation of categories. Sun, moon, earth, planets, comets, and meteorites, the various kinds of stars, galaxies, radio sources, quasars, and pulsars etc. suggest natural subdivisions. It may be assumed that such subdivisions can be maintained for long periods of time. Experience shows, however, that progress in research may imply changes in the classification scheme, in particular, in fields where the expansion of knowledge is explosive.
A few explanatory remarks may be in order on some of the subject categories. Section 002 includes short news notes whose titles and authors are given, but the authors of the notes have not been included in the author index. In section 003 books on astronomy and astrophysics and its border fields are listed which came to our notice from March 1974 to August 1974. References to book reviews are given if the review appeared quickly.
For completeness of documentation, personal notes (section 006) and obituaries (section 007) are listed. In section 012 (Proceedings of Colloquia, Congresses, Meetings, and Symposia) the proceedings etc. are listed with titles and editors. The individual papers are classified into their corresponding subject categories, but not included in the subject index. The main subjects of these symposia are cited in the index under section 012.
Errata to papers communicated by the authors are listed at the end of the corresponding subject categories.

Author index and subject index

The subject category and the serial number forming six figures for each abstract have been used as a means of reference in the author index and the subject index. These references are more precise than page references. They offer considerable advantages in indexing by means of data processing machines, and they are more convenient for the user.
The author index of this volume contains 7312 names. A complete reference comprises six figures, three for the subject category and three for the serial number within the category. In the case of more than one reference to abstracts in one category, the number of the category is given only once and not repeated in the immediately following references. The total number of papers (some do not give names of authors) recorded in this volume is about 6300.
We consider the subject index as only a first approximation to an optimal index covering all fields of astronomy and astrophysics and their border fields. The assigning of one or more key words to a paper is undoubtedly a difficult task. Some journals have started giving key words together with the titles of papers. These key words are chosen by the authors themselves and are in many cases identical with our designations of subject categories with no additional specification. In fact, in some cases it may be more useful to refer to a subject category as a whole than to an item number, in particular, if the total number of abstracts in a category is very small, and if more specific key words do not provide a proper description of the paper.
While each volume is scheduled to contain an author index and a subject index, the magnetic tapes containing the index information will be used to produce separate index volumes (authors and subjects) at intervals of a few years.
The text of the publication was typed on IBM 72 Composers in the editorial office, and it was given to the printer in a form ready for offset reproduction. For the preparation of the indexes a new sorting program has been developed. The introduction of small and capital letters in the layout caused some difficulties. Special programs had to code the capital letters into small ones. For the layout a TN chain for a 1403 IBM high speed printer was used. All the programs are written in PL/I. The computations are carried out on an IBM 360/44.

Abbreviations

AAS	American Astronomical Society	Geogr.	Geography, etc.
AAVSO	American Association of Variable Star Observers	Geophys.	Geophysics, etc.
		Ges.	Gesellschaft
Abh.	Abhandlungen	Glav.	Glavnyj (Main)
Abstr.	Abstract	Gos.	Gosudarstvennyj (State)
Abt.	Abteilung	HRD	Herzsprung-Russell diagram
Acad.	Academy, etc.	Hydrogr.	Hydrography, etc.
Accad.	Accademia	IAF	International Astronautical Federation
Adv.	Advances	IAU	International Astronomical Union
AG	Astronomische Gesellschaft	ICSU	International Council of Scientific Unions
AIAA	American Institute of Aeronautics and Astronautics	IEEE	Institute of Electrical and Electronics Engineers
AJB	Astronomischer Jahresbericht	Industr.	Industry, etc.
Akad.	Akademie	Inform.	Information
An.	Anales, etc.	Inst.	Institute, etc.
Ann.	Annals, etc.	Instn.	Institution
Arch.	Archiv, etc.	Ionosph.	Ionosphere, etc.
Ark.	Arkiv	Issled.	Issledovaniya (Research)
ASA	Astronomical Society of Australia	Ist.	Istituto
Asoc.	Asociación	Izv.	Izvestiya (News)
ASP	Astronomical Society of the Pacific	Jb.	Jahrbuch
Ass.	Association	JO	Journal des Observateurs
ASSA	Astronomical Society of Southern Africa	Journ.	Journal
Astrofis.	Astrofisica, etc.	Kl.	Klasse
Astrofiz.	Astrofizika, etc.	Lab.	Laboratory
Astron.	Astronomy, etc.	Mag.	Magazine
Astronaut.	Astronautics, etc.	Mat.	Matematica, etc.
Astrophys.	Astrophysics, etc.	Math.	Mathematics, etc.
ASV	Astronomical Society of Victoria	Mech.	Mechanics, etc.
ASWA	Astronomical Society of Western Australia	Med.	Mededelingen
Atmosph.	Atmosphere, etc.	Medd.	Meddelande, Meddelser
BA	Bulletin Astronomique	Mekhan.	Mekhanika, etc.
BAA	British Astronomical Association	Mém.	Mémoires
BAN	Bulletin of the Astronomical Institutes of the Netherlands	Mem.	Memoirs, Memorandum, etc.
		Meteorol.	Meteorology, etc.
Ber.	Berichte	MIT	Massachusetts Institute of Technology
BIH	Bureau International de l'Heure (Paris)	Mitt.	Mitteilungen
Bol.	Boletin	MVS Sonneberg	Mitteilungen über Veränderliche Sterne, Sonneberg
Boll.	Bolletino		
Bull.	Bulletin	Nachr.	Nachrichten
Byull.	Byulleten' (Bulletin)	NASA	National Aeronautics and Space Administration
Circ.	Circular		
Cl.	Classe	Nat.	Naturwissenschaftlich, etc.
Coll.	Collection	Naut.	Nautics, etc.
Commun.	Communication	NBS	National Bureau of Standards
Comun.	Comunicazioni	NRAO	National Radio Astronomy Observatory (Green Bank)
Contr.	Contributions, etc.		
COSPAR	Committee on Space Research	NRL	Naval Research Laboratory (Washington)
C.S.I.R.O.	Commonwealth Scientific Industrial Research Organization	Obs.	Observatory, etc.
		OSA	Optical Society of America
Dep.	Department	Oss.	Osservatorio, Osservazioni, etc.
Diss.	Dissertation	Ped.	Pedagogika, etc. (Pedagogics)
Div.	Division	Phil.	Philosophical
Dokl.	Doklady (Reports)	Phys.	Physics, etc.
ESO	European Southern Observatory	Planet.	Planetary
ESRO	European Space Research Organization	Priklad.	Prikladnoj (Applied)
Fis.	Fisica, etc.	Proc.	Proceedings
Fiz.	Fizika, etc.	Progr.	Progress, etc.
Fys.	Fysica, etc.	Pubbl.	Pubblicazioni
Géod.	Géodésie, etc.	Publ.	Publications
Geod.	Geodesy, etc.	Rap.	Raportoj
Geofis.	Geofisica, etc.	RAS	Royal Astronomical Society
Geofiz.	Geofizika, etc.	RAS Canada	Royal Astronomical Society of Canada
Geofys.	Geofysik, etc.	Rech.	Recherches
Geol.	Geology, etc.	Rend.	Rendiconti

Abbreviations

Rep.	Report	**Techn.**	Technics, etc.
Repr.	Reprint	**Tekhn.**	Tekhnika, etc.
Res.	Research	**Teor.**	Teoreticheskij
Rev.	Review, etc.	**Terr.**	Terrestrial, etc.
Ric.	Ricerche	**TH**	Technische Hochschule
Roy.	Royal, etc.	**Theor.**	Theoretical
SAF	Société Astronomique de France	**Tidssk.**	Tidsskrift
SAI	Società Astronomica Italiana	**Trans.**	Transactions
SAO	Smithsonian Astrophysical Observatory	**Trudy**	Trudy (Publications)
SAS	Société Astronomique de Suisse	**Tsentr.**	Tsentral'nyj (Central)
Sci.	Science, etc.	**Tsirk.**	Tsirkulyar (Circular)
Sect.	Section	**TU**	Technical University
Ser.	Series, etc.	**Uch. Zap.**	Uchenye Zapiski (Treatise)
S. I. R.	Service International Rapide des Latitudes	**Univ.**	University, etc.
Sitz.-Ber.	Sitzungsberichte	**URSI**	Union Radio Scientifique Internationale
Soc.	Society	**Verh.**	Verhandlungen
Soobshch.	Soobshcheniya (Communications)	**Veröff.**	Veröffentlichungen
Sternw.	Sternwarte	**Wet.**	Wetenschappen
Stud. Cerc.	Studii şi Cercetari	**Wiss.**	Wissenschaften, etc.
Supl.	Suplemento	**Zeitschr.**	Zeitschrift
Suppl.	Supplement	**ZfA**	Zeitschrift für Astrophysik
SuW	Sterne und Weltraum	**Zhurn.**	Zhurnal (Journal)

Periodicals, Proceedings, Books, Activities

001 Periodicals

AAS Photo-Bull.
AAS (American Astronomical Society) Photo-Bulletin. Published by the Working Group on Photographic Materials. Produced by Eastman Kodak Co., Rochester, N. Y.

Acad. Roy. Belgique, Bull. Cl. Sci.
Académie Royale de Belgique, Bulletin de la Classe des Sciences (Koninklijke Academie van België, Mededelingen van de Klasse der Wetenschappen). 5e Série. Palais des Académies, Bruxelles.

Acta Astron.
Acta Astronomica. Publisher: Komitet Astronomii, Polskiej Akademii Nauk, Warszawa - Kraków.

Acta Astronaut.
Acta Astronautica. Journal of the International Academy of Astronautics. Publisher: Pergamon Press Inc., Elmsford, New York, U.S.A.; Pergamon Press Ltd., Oxford, England.

Acta Astron. Sinica
Acta Astronomica Sinica. Published by Purple Mountain Observatory, Academia Sinica, Nanking, China.

Acta Cosmologica
Acta Cosmologica. Published by Obserwatorium Astronomiczne Uniwersytetu Jagiellońskiego, Kraków, Poland.

Acta Geophys. Sinica
Acta Geophysica Sinica. Published by Science Press, Peking, China.

Acta Phys. Austriaca
Acta Physica Austriaca. Publisher: Springer-Verlag, Wien.

Actas Acad. Nacional Cienc. Lima
Actas de la Academia Nacional de Ciencias Exactas, Fisicas y Naturales de Lima. Lima - Peru.

Acta Univ. Carolinae Math. Phys.
Acta Universitatis Carolinae, Mathematica et Physica. Administrace: Matematicko-fyzikální fakulta University Karlovy, Praha.

Adv. Astron. Astrophys.
Advances in Astronomy and Astrophysics. Publisher: Academic Press, New York – London.

AIAA Journ.
AIAA Journal. A Publication of the American Institute of Aeronautics and Astronautics devoted to Aerospace Research and Development. Published by the American Institute of Aeronautics and Astronautics, New York, N.Y

American Scient.
American Scientist. Society of Sigma Xi, New Haven, Conn.

Ann. Françaises Chronométrie Micromécanique
Annales Françaises de Chronométrie et de Micromécanique, publication annuelle de l'Observatoire de Besançon, du Centre Technique de l'Industrie Horlogère et de la Société Française de Chronométrie et de Micromécanique. Rédaction et administration: Observatoire de Besançon. Publiées avec le concours du Centre National de la Recherche Scientifique et des organismes corporatifs.

Ann. Géophys.
Annales de Géophysique. Revue Internationale trimestrielle, publiée par le Centre National de la Recherche Scientifique, Paris.

Ann. Obs. Astron. Météorol. Toulouse
Annales de l'Observatoire Astronomique et Météorologique de Toulouse. Publisher: Gauthier-Villars, Paris.

Ann. Physics
Annals of Physics. Publisher: Academic Press Inc., New York, N.Y.

Ann. Physik
Annalen der Physik. 7. Folge. Publisher: Johann Ambrosius Barth, Leipzig.

Ann. Physique
Annales de Physique. Publisher: Masson et Cie., Paris.

Ann. Soc. Sci. Bruxelles
Annales de la Société Scientifique de Bruxelles. Série I: Sciences Mathématiques, Astronomiques et Physiques. Published by Institut de Physique, Heverlé-Louvain.

Annual Rep. Astron. Inst. Greece
Annual Reports of the Astronomical Institutes of Greece. Published by the Greek National Committee for Astronomy. Academy of Athens, Research Center for Astronomy and Applied Mathematics.

Annual Rev. Astron. Astrophys.
Annual Review of Astronomy and Astrophysics. Publisher: Annual Reviews Inc., Palo Alto, California.

Ann. Univ.-Sternw. Wien
Annalen der Universitäts-Sternwarte Wien. In Kommission bei Ferd. Dümmlers Verlag, Bonn.

Anzeiger. Österreich. Akad. Wiss. Math.-Nat. Kl.
Anzeiger. Österreichische Akademie der Wissenschaften. Mathematisch-Naturwissenschaftliche Klasse. Publisher: Springer-Verlag, Wien.

Applied Optics
Applied Optics. A monthly publication of the Optical Society of America. Published for the Optical Society of America by the American Institute of Physics, New York, N. Y.

001 Periodicals

Arch. Sci. Genève
Archives des Sciences, éditées par la Société de Physique et d'Histoire Naturelle de Genève. Publisher: Imprimerie Kundig, Genève. Subscription address: Librairie Payot, Genève.

Ark. Astron.
Arkiv för Astronomi. Utgivet av Kungliga Svenska Vetenskapsakademien, Stockholm. Printed by Almqvist & Wiksell, Stockholm.

Ark. Geofys.
Arkiv för Geofysik. Kungliga Svenska Vetenskapsakademien, Stockholm. Printed by Almqvist & Wiksell, Stockholm.

Artificial Satellites
Artificial Satellites. Publication of Polish Scientific Institutions. Polish Academy of Sciences, National Committee of Geophysics and Geodesy, National Committee for Space Research, Warsaw. Publishing Office: Palac Kultury i Nauki, Warszawa.

Asoc. Argentina Astron. Bol.
Asociación Argentina de Astronomía. Boletin. Editor: Instituto Argentino de Radioastronomía, Provincia de Buenos Aires, Argentina. Printer: Talleres Gráficos "Renovación", La Plata, República Argentina.

Astrofizika
Astrofizika. Izdatel'stvo Akademii Nauk Armyanskoj SSR, Erevan. [An English translation is published in "Astrophysics".]

Astrofiz. Issled. Izv. Spets. Astrofiz. Obs.
Astrofizicheskie Issledovaniya. Izvestiya Spetsial'noj Astrofizicheskoj Observatorii. Akademiya Nauk SSSR. Publishers: Izdatel'stvo "Nauka", Leningradskoe Otdelenie, Leningrad.

Astron. Astrophys.
Astronomy and Astrophysics. A European Journal. Published by Springer-Verlag, Berlin – Heidelberg – New York.

Astron. Astrophys. Suppl. Ser.
Astronomy and Astrophysics. Supplement Series. A European Journal. Published by the Astronomical Institute Lausanne and Geneva Observatory, Switzerland, on behalf of the Board of Directors.

Astronautik
Astronautik. Organ der Hermann-Oberth-Gesellschaft e.V. Astronautik-Verlag, Druckerei H. Brandt, Delmenhorst (Germany).

Astron. in der Schule
Astronomie in der Schule. Zeitschrift für die Hand des Astronomielehrers. Herausgegeben vom Verlag Volk und Wissen, Berlin. Redaktion: Sternwarte Bautzen.

Astron. Journ.
The Astronomical Journal. Published for the American Astronomical Society by the American Institute of Physics, New York, N. Y. Editorial Office: Department of Astronomy, Columbia University, New York, N. Y.

Astron. Nachr.
Astronomische Nachrichten. Publisher: Akademie-Verlag, Berlin.

Astron. Tidssk.
Astronomisk Tidsskrift. Edited by Astronomisk Selskab, København; Norsk Astronomisk Selskap, Oslo; Svenska Astronomiska Sällskapet, Stockholm. Printed by John Griegs Boktrykkeri, Bergen.

Astron. Tsirk.
Astronomicheskij Tsirkulyar, izdavaemyj Byuro Astronomicheskikh Soobshchenij Akademii Nauk SSSR. Tipografiya Astrosoveta AN SSSR, Moskva.

Astron. Vestn.
Astronomicheskij Vestnik. Publishers: Izdatel'stvo "Nauka", Moskva.

Astron. Zhurn. Akad. Nauk SSSR
Astronomicheskij Zhurnal. Akademiya Nauk SSSR. Publishers: Izdatel'stvo "Nauka", Moskva. [An English translation is published in "Soviet Astronomy AJ"].

Astrophysics
Astrophysics. The Faraday Press cover-to-cover translation of Astrofizika. The Faraday Press, Inc., New York, N. Y.

Astrophys. Journ.
The Astrophysical Journal. Published in collaboration with the American Astronomical Society by the University of Chicago Press, Chicago, Illinois.

Astrophys. Journ. Suppl. Ser.
The Astrophysical Journal. Supplement Series. Published in collaboration with the American Astronomical Society by the University of Chicago Press, Chicago, Illinois.

Astrophys. Letters
Astrophysical Letters. An International *EXPRESS* Journal. Published monthly by Gordon and Breach Science Publishers Ltd., New York – London – Paris.

Astrophys. Norvegica
Astrophysica Norvegica. Edited by The Institute of Theoretical Astrophysics, University of Oslo (Det Norske Videnskaps-Akademi i Oslo). Universitets-forlaget, Oslo.

Astrophys. Space Sci.
Astrophysics and Space Science. An International Journal of Cosmic Physics. Published by D. Reidel Publishing Company, Dordrecht – Holland.

Atti Accad. Nazionale Lincei. Mem.
Atti della Accademia Nazionale dei Lincei. Serie Ottava. Memorie. Classe di Scienze fisiche, matematiche e naturali. Sezione I: Matematica, Meccanica, Astronomia, Geodesia e Geofisica. Published by Accademia Nazionale dei Lincei, Roma.

Atti Accad. Nazionale Lincei. Rend.
Atti della Accademia Nazionale dei Lincei. Serie Ottava. Rendiconti. Classe di Scienze fisiche, matematiche e naturali. Published by Accademia Nazionale dei Lincei, Roma.

Atti Soc. Astron. Italiana
Atti della Società Astronomica Italiana. Publisher: Tipografia Baccini & Chiappi, Firenze (Italy).

Australian Journ. Phys.
Australian Journal of Physics. Published by the Commonwealth Scientific and Industrial Research Organization, East Melbourne, Victoria.

001 Periodicals

Australian Journ. Phys. Astrophys. Suppl.
Australian Journal of Physics, Astrophysical Supplement. Published by Commonwealth Scientific and Industrial Research Organization, East Melbourne, Victoria.

BAV Rundbrief
BAV Rundbrief. Mitteilungsblatt der Berliner Arbeitsgemeinschaft für Veränderliche Sterne. Editor: BAV Berliner Arbeitsgemeinschaft für Veränderliche Sterne eV., Berlin.

BBSAG Bull.
Bedeckungsveränderlichen Beobachter der Schweizerischen Astronomischen Gesellschaft, [Swiss Astronomical Society's Eclipsing Variable Observers], Bulletin. To be obtained from R. Diethelm, Winterthur, Switzerland.

Bild der Wiss.
Bild der Wissenschaft. Zeitschrift über die Naturwissenschaften und die Technik in unserer Zeit. Publisher: Deutsche Verlagsanstalt, Stuttgart.

Bol. Inst. Mat., Astron., Fis. Univ. Nacional Córdoba
Boletin del Instituto de Matematica, Astronomia y Fisica, Universidad Nacional de Córdoba (R. A.).Dirección General de Publicaciones, Córdoba (Argentina).

Bol. Liga Latinoamericana Astron.
Boletin de la Liga Latinoamericana de Astronomia. Publicado por la Asociacion Argentina Amigos de la Astronomia, Buenos Aires, Argentina.

Boll. Geod. Sci. Affini
Bolletino di Geodesia e Scienze Affini. Pubblicazione dell'Istituto Geografico Militare, Firenze.

Boundary-Layer Meteorology
Boundary-Layer Meteorology. An International Journal of Physical and Biological Processes in the Atmospheric Boundary Layer. Published by D. Reidel Publishing Company, Dordrecht–Holland.

British Astron. Ass. Circ.
British Astronomical Association, Circular. Editorial Office: 97 Hawkswood Drive, Hailsham, Sussex.

Bull. American Astron. Soc.
Bulletin of the American Astronomical Society. Published for the American Astronomical Society by the American Institute of Physics Inc., New York, N. Y.

Bull. Astron. Inst. Czechoslovakia (BAC)
Bulletin of the Astronomical Institutes of Czechoslovakia. Published under the auspices of the Czechoslovak Academy of Sciences by Academia, Praha. Editor: Astronomical Institutes of the Czechoslovak Academy of Sciences, Praha.

Bull. Astron. Inst. Netherlands (BAN)
Bulletin of the Astronomical Institutes of the Netherlands. Publisher: North-Holland Publishing Company, Amsterdam. After Vol. 20 replaced by "Astronomy and Astrophysics".

Bull. Géod.
Bulletin Géodésique, being the Journal of the International Association of Geodesy. Nouvelle Série. Publié par le Bureau Central de l'Association Internationale de Géodésie, Paris.

Bull. Geograph. Survey Inst.
Bulletin of the Geographical Survey Institute. Published by the Geographical Survey Institute, Ministry of Construction, Tokyo, Japan.

Bull. Obs. Astron. Beograd
Bulletin de l'Observatoire Astronomique de Béograd. Editor: Observatoire Astronomique de Béograd. Printed by Naucna delo, Béograd.

Bull. Sci. Yougoslavie
Bulletin Scientifique. Conseil des Academies des Sciences et des Arts de la RSF de Yougoslavie. Section A: Sciences Naturelles, Techniques et Médicales. Redaction et Administration: Opatička ul. 18/II, Zagreb (Yougoslavie).

Bull. Signal.
Bulletin Signalétique. Section 120: Astronomie, Physique spatiale, Géophysique. Centre de Documentation du Centre Nationale de la Recherche Scientifique, Paris.

Bull. Signal.
Bulletin Signalétique. Bibliographie des Sciences de la Terre. Section 220, Cahier A: Minéralogie, Géochimie, Géologie extraterrestre. Centre de Documentation du C.N.R.S., Paris; Département Documentation du B.R. G.M., Orléans.

Bull. Soc. Roy. Sci. Liège
Bulletin de la Société Royale des Sciences de Liège. L'Université, Liège.

Byull. Abastuman. Astrofiz. Obs.
Abastumanskaya Astrofizicheskaya Observatoriya, Gora Kanobili. Byulleten'. Akademiya Nauk Gruzinskoj SSR. Publishers: Izdatel'stvo "Metsniereba", Tbilisi.

Byull. Stantsij Optichesk. Nablyud. Iskusstv. Sputnikov Zemli
Byulleten' Stantsij Opticheskogo Nablyudeniya Iskusstvennykh Sputnikov Zemli. Published by Astronomicheskij Sovet Akademii Nauk SSSR, Moskva.
Beginning with number 60 (1971) the title of the publication changed in Nablyudeniya Iskusstvennykh Nebesnykh Tel.

Canadian Journ. Phys.
Canadian Journal of Physics. Published by the National Research Council of Canada, Ottawa. Printed in Canada by the University of Toronto Press, Toronto, Ont.

Celestial Mechanics
Celestial Mechanics. An International Journal of Space Dynamics. Publishers: D. Reidel Publishing Company, Dordrecht–Holland.

Ciel et Terre
Ciel et Terre. Bulletin de la Société Belge d'Astronomie, de Météorologie et de Physique du Globe. Administration: Avenue Circulaire, 3, Bruxelles. Printed by Imprimerie R. Louis, Bruxelles.

Circ. d'Information
Circulaire d'Information. Union Astronomique Internationale. Commission des Etoiles Doubles. Address: Observatoire de Meudon, Meudon, France.

Coelum
Coelum. Periodico bimestrale per la Divulgazione dell' Astronomia. Editor: Osservatorio Astronomico Universitario di Bologna.

Comments Astrophys. Space Phys.
Comments on Astrophysics and Space Physics. A Journal of Critical Discussion of the Current Literature. Comments on Modern Physics: Part C. Publishers: Gordon and Breach Science Publishers, Inc., New York – London

Comptes Rendus Acad. Bulg. Sci.
Comptes Rendus de l'Académie bulgare des Sciences. (Doklady Bolgarskoj Akademii Nauk). Sofia.

Comptes Rendus Acad. Sci. Paris
Comptes Rendus hebdomadaires des Séances de l'Académie des Sciences, publié avec le concours du Centre National de la Recherche Scientifique. Imprimerie: Gauthier-Villars, Paris.

Contr. Atmosph. Phys.
Contributions to Atmospheric Physics – Beiträge zur Physik der Atmosphäre. Publisher: Friedrich Vieweg & Sohn, Braunschweig.

Cosmic Electrodynamics
Cosmic Electrodynamics. An International Journal devoted to Geophysical and Astrophysical Plasmas. Printed in The Netherlands by D. Reidel Publishing Company, Dordrecht–Holland.

COSPAR Inform. Bull.
COSPAR. Information Bulletin. Address: COSPAR Secretariat, Paris.

Deutsche Geod. Kommission Bayer. Akad. Wiss.
Deutsche Geodätische Kommission bei der Bayerischen Akademie der Wissenschaften. Reihe A: Höhere Geodäsie; Reihe B: Angewandte Geodäsie; Reihe C: Dissertationen; Reihe D: Tafelwerke; Reihe E: Geschichte und Entwicklung der Geodäsie. Published by Verlag der Bayerischen Akademie der Wissenschaften, München.

Documentat. Observateurs
Documentation des Observateurs. Rédaction: Station d'Astrophysique de Forcalquier.

Documentat. Observateurs Circ.
Documentation des Observateurs. Circulaire. Rédaction: Station d'Astrophysique de Forcalquier.

Dokl. Akad. Nauk
Doklady Akademii Nauk SSSR. Seriya Matematika, Fizika. Publishers: Izdatel'stvo "Nauka", Moskva.

Dunsink Obs. Publ.
Dunsink Observatory Publications. The Observatory of the School of Cosmic Physics, Dublin Institute for Advanced Studies, Dublin.

Earth Extraterr. Sci.
Earth and Extraterrestrial Sciences. Published by Gordon and Breach Science Publishers, London.

Earth Planet. Sci. Letters
Earth and Planetary Science Letters. A Letter Journal devoted to the Development in Time of the Earth and Planetary System. Publisher: North-Holland Publishing Company, Amsterdam.

El Universo
El Universo. Organo de la Sociedad Astronomica de Mexico, Mexico, D. F.

Endeavour
Endeavour. A review of the progress of science, published in four languages by Imperial Chemical Industries Limited, London.

ESO Bull.
European Southern Observatory, Bulletin. Edited by European Southern Observatory. Office of the Director: Hamburg.

ESO Techn. Rep.
European Southern Observatory, (ESO), Technical Report. Published by the European Southern Observatory Telescope Project Division, CERN, Geneva, Switzerland.

Gaz. Astron. Mém.
Gazette Astronomique. Mémoires van het Sterrenkundig Genootschap van Antwerpen, (de la Société d'Astronomie d'Anvers), Antwerpen. Printer: «De Voorzorg», A. Van Leuvenhaege, Antwerpen.

Geochim. Cosmochim. Acta
Geochimica et Cosmochimica Acta. Journal of the Geochemical Society. Publishing House: Pergamon Press, Ltd., Oxford.

Geodezja Kartografia
Geodezja i Kartografia. Komitet Geodezji Polskiej Akademii Nauk. Publisher: Państwowe Wydawnictwo Naukowe, Warszawa.

Geomagn. Aeronom.
Geomagnetizm i Aehronomiya. Akademiya Nauk SSSR. Izdatel'stvo "Nauka", Moskva [An English translation is published in "Geomagnetism and Aeronomy".]

Geophys. Journ.
The Geophysical Journal of the Royal Astronomical Society. Published for the Royal Astronomical Society by Blackwell Scientific Publications, Oxford – Edinburgh.

Geophys. Res. Letters
Geophysical Research Letters. Published monthly by the American Geophysical Union, Washington, D.C., U.S.A.

Gerlands Beiträge Geophys.
Gerlands Beiträge zur Geophysik. Publisher: Akademische Verlagsgesellschaft Geest & Portig K.-G., Leipzig.

Glasnik Mat.
Glasnik Matematicki. Published by the Society of Mathematicians and Physicists of the S. R. of Croatia. Publisher: Drustvo Matematicara i Fizicara S. R. Hrvatske, Zagreb.

Helvetica Phys. Acta
Helvetica Physica Acta. Schweizerische Physikalische Gesellschaft. Publisher: E. Birkhäuser, Basel.

IAU Circ.
International Astronomical Union, Circular. Central Bureau for Astronomical Telegrams, Smithsonian Astrophysical Observatory, Cambridge, Mass.

ICSU Bull.
ICSU Bulletin. International Council of Scientific Unions. Secretariat: 51, Bd de Montmorency, Paris, France.

Icarus
  Icarus. International Journal of Solar System Studies. Publisher: Academic Press, New York – London.

ICSU Bull.
  ICSU Bulletin. International Council of Scientific Unions. Secretariat: 7, Via Cornelio Celso, Rome, Italy.

IEEE Spectrum
  IEEE Spectrum. Published monthly by the Institute of Electrical and Electronics Engineers, Inc., New York, N. Y.

Inform. Bull. Southern Hemisphere
  Information Bulletin of the Southern Hemisphere. Editorial Office: Observatorio Astronómico, La Plata, Argentina.

Inform. Bull. Variable Stars
  Commission 27 of the I.A.U. Information Bulletin on Variable Stars. Konkoly Observatory, Budapest.

Infrared Physics
  An International Research Journal. Publisher: Pergamon Press Ltd., Oxford – London – New York.

International Journ. Theor. Phys.
  International Journal of Theoretical Physics. Publisher: Plenum Publishing Company, Donington House, London.

Irish Astron. Journ.
  The Irish Astronomical Journal. A Quarterly Publication under the auspices of the Observatories of Armagh and Dunsink. Subscription address: Managing Editor, Irish Astronomical Journal, Armagh Observatory, Northern Ireland.

Izv. Akad. Nauk Armyan. SSR
  Izvestiya Akademii Nauk Armyanskoj SSR. Fizika. Publisher: Izdatel'stvo AN Armyanskoj SSR, Erevan.

Izv. Glav. Astron. Obs. Pulkovo
  Izvestiya Glavnoj Astronomicheskoj Observatorii v Pulkove. Akademiya Nauk SSSR. Izdanie Glavnoj astronomicheskoj observatorii v Pulkove, Leningrad.

Izv. Komissii Fiz. Planet
  Izvestiya Komissii po Fizike Planet. Akademiya Nauk SSSR. Astronomicheskij Sovet. Moskva.

Izv. Krymskoj Astrofiz. Obs.
  Izvestiya Krymskoj Astrofizicheskoj Observatorii. Akademiya Nauk SSR. Publishers: Izdatel'stvo "Nauka", Moskva.

Jenaer Rundschau (Jena Review)
  Jenaer Rundschau (Jena Review). Publisher: VEB Verlag Technik, Berlin.

JETP Letters
  JETP Letters. A translation of JETP Pis'ma v Redaktsiyu of the Academy of Sciences in the USSR. Published semimonthly by the American Institute of Physics, Lancaster, Pennsylvania.

Journ. Astronaut. Sci.
  The Journal of the Astronautical Sciences. Published by the American Astronautical Society Inc., Baltimore, Md.

Journ. Astron. Soc. Victoria
  The Journal of the Astronomical Society of Victoria. Printed by D. Buscombe Printers, Glen Waverley, Victoria.

Journ. Astron. Soc. Western Australia
  The Journal of the Astronomical Society of Western Australia. Edited by the Astronomical Society of Western Australia, Perth, W. A.

Journ. Atmosph. Sci.
  Journal of the Atmospheric Sciences. Published by the American Meteorological Society, Boston, Mass.

Journ. Atmosph. Terr. Phys.
  Journal of Atmospheric and Terrestrial Physics. Publishers: Pergamon Press, Oxford – London – New York.

Journ. British Astron. Ass.
  Journal of the British Astronomical Association. Subscription address: British Astronomical Association, Burlington House, Piccadilly, London.

Journ. British Interplanet. Soc.
  Journal of the British Interplanetary Society. Printed in Great Britain by Unwin Brothers Ltd., The Gresham Press, Old Woking, Surrey, and published by The British Interplanetary Society, London.

Journ. Fluid Mechanics
  Journal of Fluid Mechanics. Published by Cambridge University Press, London – New York.

Journ. Geophys.
  Journal of Geophysics / Zeitschrift für Geophysik. Publisher: Springer-Verlag, Berlin–Heidelberg–New York

Journ. Geophys. Res.
  Journal of Geophysical Research. An International Scientific Publication. Published three times a month by the American Geophysical Union, Washington, D. C. First section: Space Physics; Second section: Physics and chemistry of the solid earth, planetology, geodesy; Third section: Oceans and atmospheres.

Journ. History Astron.
  Journal for the History of Astronomy. Publisher: Science History Publications Ltd., Cambridge, England. American Representative: Neale Watson Academic Publications, Inc., New York City, U.S.A.

Journ. Navigation
  The Journal of Navigation. Published quarterly by The Royal Institute of Navigation at the Royal Geographical Society, London.

Journ. Optical Soc. America
  Journal of the Optical Society of America. Publisher: American Institute of Physics, New York.

Journ. Phys. A. General Phys.
  Journal of Physics A. General Physics. Europhysics Journal. Published by the Institute of Physics and the Physical Society, London, England, in association with the American Institute of Physics, New York.

Journ. Physique
  Journal de Physique. Publication de la Société Française de Physique, Paris.

Journ. Plasma Phys.
  Journal of Plasma Physics. Publishers: Cambridge University Press, London.

Journ. Proc. Roy. Soc. New South Wales
Journal and Proceedings of the Royal Society of New South Wales. Publisher: Science House, Sydney, N.S.W. (Australia).

Journ. Quant. Spectrosc. Radiat. Transfer
Journal of Quantitative Spectroscopy & Radiative Transfer. Publisher: Pergamon Press, Oxford – New York.

Journ. Roy. Astron. Soc. Canada
The Journal of the Royal Astronomical Society of Canada, devoted to the advancement of astronomy and allied sciences. Printed by the University of Toronto Press, Toronto, Ontario, Canada.

Kometn. Tsirk. *Kiev*
Kometnyj Tsirkulyar. Gruppa po Issledovaniyu Komet Astrosoveta i Mezhduvedomstvennyj Geofizicheskij Komitet Akademii Nauk SSSR. Kievskij Universitet im. T. G. Shevchenko.

Komety i Meteory
Komety i Meteory. Akademiya Nauk Tadzhikskoj SSR. Astronomicheskij Sovet Akademii Nauk SSSR. Publishers: Izdatel'stvo "Donish", Dushanbe.

Kosmich. Issled.
Kosmicheskie Issledovaniya. Akademiya Nauk SSSR. Publishers: Izdatel'stvo "Nauka", Moskva.

Kozmos
Kozmos. Popular Astronomical Journal of the Slovak Central Observatory in Hurbanovo. Publisher: Slovenská ústredná hvezdáren v Hurbanove.

L'Astronomie
L'Astronomie et Bulletin de la Société Astronomique de France. Revue mensuelle. Rédaction: Société Astronomique de France, Paris.

L'Universo
L'Universo. Rivista dell'Instituto Geografico Militare. Direzione, Redazione e Amministrazione: Istituto Geografico Militare, Firenze.

Magnitnye Polya Solnech. Pyaten
Magnitnye Polya Solnechnykh Pyaten. (Supplements to Solnechnye Dannye. Byulleten' (*Solar Data*)). Publishers: Izdatel'stvo "Nauka", Leningrad.

Math. Rev.
Mathematical Reviews. Published by the American Mathematical Society, Providence, R. I.

Mem. Fac. Sci. Kyoto Univ.
Memoirs of the Faculty of Science, Kyoto University. Series of Physics, Astrophysics, Geophysics, and Chemistry. Printed by Yamashiro Printing Publishing Co. Ltd., Kamigyo, Kyoto.

Mem. Roy. Astron. Soc.
Memoirs of the Royal Astronomical Society. Published for the Royal Astronomical Society by Blackwell Scientific Publications, Oxford – Edinburgh.

Mem. Soc. Astron. Italiana
Memorie della Società Astronomica Italiana. Nuova Serie. Pubblicate sotto gli auspici del Consiglio Nazionale dell Ricerche. Publisher: Tipografia Baccini & Chiappi, Firenze.

Mercury
Mercury. The Journal of the Astronomical Society of the Pacific. Published by the Astronomical Society of the Pacific, San Francisco, California.

Messtechnik
Messtechnik (Zeitschrift für Instrumentenkunde). Publishers: Verlag Friedrich Vieweg & Sohn GmbH, Braunschweig.

Meteoritics
Meteoritics. The Journal of the Meteoritical Society. Published quarterly by The Meteoritical Society and Arizona State University Bureau of Publications. Editorial address: Center for Meteorite Studies, The Arizona State University, Tempe, Arizona.

Meteoritika
Akademiya Nauk SSSR. Komitet po Meteoritam. Publishers: Izdatel'stvo "Nauka", Moskva.

Minor Planet. Bull.
The Minor Planet Bulletin. Bulletin of the Minor Planets Section of the Association of Lunar and Planetary Observers. Editorial Office: R. G. Hodgson, Dordt College, Sioux Center, Iowa, U.S.A.

Mitt. Astron. Ges.
Mitteilungen der Astronomischen Gesellschaft, Hamburg. Printed by G. Braun, GmbH, Karlsruhe.

Monthly Notes Astron. Soc. Southern Africa
Monthly Notes of the Royal Astronomical Society of Southern Africa. Published by the Astronomical Society of Southern Africa, Royal Observatory, Cape Province, South Africa.

Monthly Notes International Polar Motion Service
Monthly Notes of the International Polar Motion Service. Published by the Central Bureau, International Latitude Observatory of Mizusawa, Mizusawa-shi, Iwate-ken, Japan.

Monthly Notices Roy. Astron. Soc.
Monthly Notices of the Royal Astronomical Society. Published for the Royal Astronomical Society by Blackwell Scientific Publications, Oxford – Edinburgh.

Moon
The Moon. An International Journal of Lunar Studies. Publisher: D. Reidel Publishing Company, Dordrecht – Holland.

MVS Sonneberg
Mitteilungen über Veränderliche Sterne. Edited by Sternwarte Sonneberg (Zentralinstitut für Astrophysik, Bereich Sternphysik) der Deutschen Akademie der Wissenschaften zu Berlin.

Nablyud. Iskusstv. Nebesn. Tel
Nablyudeniya Iskusstvennykh Nebesnykh Tel. Published by Astronomicheskij Sovet Akademii Nauk SSSR, Moskva.

Nachr. Akad. Wiss. Göttingen
Nachrichten der Akademie der Wissenschaften in Göttingen. II. Mathematisch-Physikalische Klasse. Vandenhoeck & Ruprecht, Göttingen.

Nachr. Karten-, Vermessungswesen
Nachrichten aus dem Karten- und Vermessungswesen. Editor: Institut für Angewandte Geodäsie (Abt. II des

Deutschen Geodätischen Forschungsinstituts). Published by Verlag des Instituts für Angewandte Geodäsie, Frankfurt a. M.

Nature
Nature. Editorial and Publishing Offices: Macmillan Journals Limited, 4 Little Essex Street, London; 711 National Press Building, Washington, D. C.

Nature, Phys. Sci.
Nature, Physical Science. Editorial and Publishing Offices: Macmillan Journals Limited, London – Washington.

Naturwissenschaften
Die Naturwissenschaften. Publisher: Springer-Verlag, Berlin – Heidelberg – New York.

Nauchn. Informatsii
Nauchnye Informatsii. Astronomicheskij Sovet Akademii Nauk SSSR, Moskva.

Nuovo Cimento
Il Nuovo Cimento. Rivista Internazionale e Organo della Società Italiana di Fisica, Series A, B. Publisher: Nicola Zanichelli, Editore, Bologna.

Nuovo Cimento Lettere
Lettere al Nuovo Cimento, a Cura della Società Italiana di Fisica. Editrice Compositori, Bologna.

Nuovo Cimento Rivista
Rivista del Nuovo Cimento a cura della Società Italiana di Fisica. Editrice Compositori, Bologna.

Nuovo Cimento Suppl.
Supplemento al Nuovo Cimento. Publisher: Nicola Zanichelli, Editore, Bologna.

Observations Artificial Earth Satellites
Observations of Artificial Satellites of the Earth (Nablyudeniya Iskusstvennykh Sputnikov Zemli). Magyar Tudományos Akadémia Csillagvizsgáló Intézete, Budapest.

Observatory
The Observatory. A Review of Astronomy. Publishers: The Editors of "The Observatory", Royal Greenwich Observatory, Herstmonceaux Castle, Hailsham, Sussex, England.

Optik
Optik. Zeitschrift für das gesamte Gebiet der Licht- und Elektronenoptik. Publishers: Wissenschaftliche Verlagsgesellschaft mbH., Stuttgart.

Orion Schaffhausen
Orion. Zeitschrift der Schweizerischen Astronomischen Gesellschaft (SAG). Bulletin de la Société Astronomique de Suisse (SAS). Administration: Generalsekretariat der SAG, Schaffhausen.

Österreich. Zeitschr. Vermessungswesen
Österreichische Zeitschrift für Vermessungswesen. Editor and Publisher: Österreichischer Verein für Vermessungswesen, Wien.

Peremennye Zvezdy, Byull.
Peremennye Zvezdy, Byulleten', izdavaemyj Astronomicheskim Sovetom Akademii Nauk SSSR. Published by Astronomicheskij Sovet Akademii Nauk SSSR, Moskva.

Peremennye Zvezdy, Prilozhenie
Peremennye Zvezdy, Prilozhenie (The Variable Stars, Supplement). Astronomicheskij Sovet Akademii Nauk SSSR, Moskva.

Phil. Mag.
The Philosophical Magazine. A Journal of Theoretical, Experimental and Applied Physics. Eighth Series. Publisher: Taylor & Francis, Ltd., London.

Phil. Trans. Roy. Soc. London
Philosophical Transactions of the Royal Society of London. Series A, Mathematical and Physical Sciences. Published by the Royal Society, London.

Phys. Abstr.
Physics Abstracts. Science Abstracts, Series A. An INSPEC Publication, published by The Institution of Electrical Engineers, London.

Phys. Ber.
Physikalische Berichte. Herausgegeben von der Deutschen Physikalischen Gesellschaft e. V.und von der Deutschen Akademie der Wissenschaften zu Berlin. Friedrich Vieweg & Sohn, Braunschweig.

Phys. Blätter
Physikalische Blätter. Physik-Verlag, Mosbach/Baden.

Phys. Bull.
Physics Bulletin. Published by the Institute of Physics and the Physical Society, London, England.

Phys. Earth Planet. Interiors
Physics of the Earth and Planetary Interiors. A journal devoted to observational and experimental studies of the Earth and Planetary interiors and their theoretical interpretation by the physical sciences. Publisher: North-Holland Publishing Company, Amsterdam, Netherlands.

Phys. Fluids
The Physics of Fluids. Published by the American Institute of Physics, New York, N.Y.

Phys. Letters
Physics Letters. Volumes A and B. Publisher: North-Holland Publishing Company, Amsterdam.

Phys. Rev. A
Physical Review A, General Physics. Published for the American Physical Society by the American Institute of Physics, Lancaster, Pa., and New York, N.Y.

Phys. Rev. B
Physical Review B, Solid State. Published for the American Physical Society by the American Institute of Physics, Lancaster, Pa., and New York, N. Y.

Phys. Rev. C
Physical Review C, Nuclear Physics. Published for the American Physical Society by the American Institute of Physics, Lancaster, Pa., and New York, N.Y.

Phys. Rev. D
Physical Review D, Particles and Fields. Published for the American Physical Society by the American Institute of Physics, Lancaster, Pa., and New York, N.Y.

Phys. Rev. Letters
Physical Review Letters. Published weekly by The Amer-

ican Physical Society, New York, N. Y.

Phys. Today
Physics Today. Published by the American Institute of Physics, New York, N.Y.

Physica
Physica. Publishers: North-Holland Publishing Company, Amsterdam, The Netherlands, on request of the Foundation "Physica", Utrecht.

Physica Scripta
Physica Scripta. (Formerly Arkiv för Fysik). Published by the Royal Swedish Academy of Sciences, Stockholm.

Planet. Space Sci.
Planetary and Space Science. Pergamon Press, Oxford – London – New York.

Plasma Physics
Plasma Physics. Publisher: Pergamon Press, Oxford, England.

Pokroky
Pokroky matematiky, fyziky a astronomie. Editor: Jednota čs. matematiků a fyziků. Publisher: Academia, Praha.

Postępy Astron.
Postępy Astronomii. Czasopismo Poświecone Upowszechnianiu Wiedzy Astronomicznej. Polskie Towarzystwo Astronomiczne, Warszawa. Printed in Poland by Pánstwowe Wydawnictwo Naukowe, Łódź.

Priroda
Priroda. Publishers: Izdatel'stvo "Nauka", Moskva.

Proc. Astron. Soc. Australia
Proceedings of the Astronomical Society of Australia. Published for the Society by Sydney University Press, Sydney.

Proc. Cambridge Phil. Soc.
Proceedings of the Cambridge Philosophical Society (Mathematical and Physical Sciences). Publishers: Cambridge University Press, London.

Proc. IEEE
Proceedings of the IEEE. Published monthly by the Institute of Electrical and Electronics Engineers, Inc., New York, N. Y.

Proc. Koninkl. Nederl. Akad. Wet.
Koninklijke Nederlandse Akademie van Wetenschappen. Proceedings. Series B, Physical Sciences. Publishers: North-Holland Publishing Company, Amsterdam.

Proc. National Acad. Sci. U.S.A.
Proceedings of the National Academy of Sciences of the United States of America. Published monthly by the National Academy of Sciences, Washington, D.C.

Proc. Roy. Soc. London
Proceedings of the Royal Society of London. Series A: Mathematical and Physical Sciences. Published by the Royal Society, London.

Progr. Theor. Phys. Japan
Progress of Theoretical Physics. Published for the Research Institute for Fundamental Physics and the Physical Society of Japan. Publication Office: Progress of Theoretical Physics, Yukawa Hall, Kyoto University, Kyoto, Japan.

Progr. Theor. Phys. Suppl.
Supplement of the Progress of Theoretical Physics. Published for the Research Institute for Fundamental Physics and The Physical Society of Japan. Publication Office: Progress of Theoretical Physics, Yukawa Hall, Kyoto University, Kyoto, Japan.

PTB Mitt.
PTB Mitteilungen. Amts- und Mitteilungsblatt der Physikalisch-Technischen Bundesanstalt, Braunschweig – Berlin.

Publ. Astron. Soc. Japan
Publications of the Astronomical Society of Japan. Published by the Astronomical Society of Japan. Office of the Society: Tokyo Astronomical Observatory, Mitaka, Tokyo. Agent: Maruzen Co. Ltd. (Export Department), Nihonbashi, Tokyo, Japan.

Publ. Astron. Soc. Pacific
Publications of the Astronomical Society of the Pacific. Published in Provo, Utah, by the Astronomical Society of the Pacific, San Francisco, California. Printed by Brigham Young University Press, Provo, Utah.

Publ. Roy. Obs. Edinburgh
Publications of the Royal Observatory, Edinburgh. Published by The Royal Observatory, Edinburgh, Scotland.

Publ. Tartu Astrofiz. Obs.
W. Struve nimelise Tartu Astrofüüsika Observatooriumi, Publikatsioonid. Eesti NSV Teaduste Akadeemia, Tartu.

Quarterly Journ. Roy. Astron. Soc.
Quarterly Journal of the Royal Astronomical Society. Published for the Royal Astronomical Society by Blackwell Scientific Publications, Oxford.

Radio Sci.
Radio Science. Published by the American Geophysical Union, Richmond, Virginia.

Referativ. Zhurn. 51. Astron.
Referativnyj Zhurnal. 51. Astronomiya. Vsesoyuznyj Institut Nachnoj i Tekhnicheskoj Informatsii. Moskva.

Referativ. Zhurn. 52. Geod. i Aehros"emka
Referativnyj Zhurnal. 52. Geodeziya i Aehros"emka. Vsesoyuznyj Institut Nauchnoj i Tekhnicheskoj Informatsii. Moskva.

Referativ. Zhurn. 62. Issled. kosm. prostranstva
Referativnyj Zhurnal. 62. Issledovanie Kosmicheskogo Prostranstva. Vsesoyuznyj Institut Nauchnoj i Tekhnicheskoj Informatsii. Moskva.

Rep. Progr. Phys.
Reports on Progress in Physics. Published by The Institute of Physics and the Physical Society, London.

Rev. Geophys. Space Phys.
Reviews of Geophysics and Space Physics (formerly Reviews of Geophysics). Published by the American Geophysical Union, Richmond, Virginia.

Revista Astron.
　Revista Astronomica. Organo de la Asociación Argentina Amigos de la Astronomia, Buenos Aires.

Rev. Modern Phys.
　Reviews of Modern Physics. Published for The American Physical Society by the American Institute of Physics, Lancaster, Pa., and New York, N.Y.

Rev. Sci. Instruments
　Reviews of Scientific Instruments. Published by the American Institute of Physics, Lancaster, Pa., and New York, N.Y.

Rev. Mexicana Astron. Astrofis.
　Revista Mexicana de Astronomia y Astrofisica. Dirección: Instituto de Astronomia, Universidad Nacional Autónoma de México, México, D.F.

Rezul'taty Nablyud. Sovet. Iskusstv. Sputnikov Zemli
　Rezul'taty Nablyudenij Sovetskikh Iskusstvennykh Sputnikov Zemli. Published by Astronomicheskij Sovet Akademii Nauk SSSR, Moskva. Replaced after No. 140 by Rezul'taty Nablyudenij Iskusstvennykh Sputnikov Zemli.

Rezul'taty Nablyud. Iskusstv. Sputnikov Zemli
　Rezul'taty Nablyudenij Iskusstvennykh Sputnikov Zemli. Published by Astronomicheskij Sovet Akademii Nauk SSSR, Ryazanskij Gosudarstvennyj Pedagogicheskij Institut, Ryazan'.

Říše hvězd
　Říše hvězd. Czechoslovak popular astronomical journal. Publisher: Orbis, Praha.

Roy. Astron. Soc. New Zealand Circ.
　Royal Astronomical Society of New Zealand, Variable Star Section, Circular. Publication Office: Greerton, Tauranga, New Zealand.

Rumanian Sci. Abstr.
　Rumanian Scientific Abstracts. Natural Sciences. Publishers: The Scientific Documentation Centre of the Academy of the Socialist Republic of Romania, București.

Sci. American
　Scientific American. Published monthly by Scientific American, Inc., New York, N.Y.

Science
　Science. American Association for the Advancement of Science, Washington, D.C.

Scient. Sinica
　Scientia Sinica. Edited by Editorial Committee of Scientia Sinica, Peking. Published by Science Press, Peking, China.

Sci. Progr. Découverte
　Science Progrès Découverte (formerly Science Progrès, La Nature). Revue publiée avec la participation du Palais de la Découverte. Published by Dunod, Editeur, Paris. Imprimerie Bayeusaine, Bayeux.

Sci. Rep. Tôhoku Univ.
　The Science Reports of the Tôhuku University. First Series (Physics, Chemistry, Astronomy). Published by the Faculty of Science, Tôhoku University, Sendai, Japan.

Sitzungsber. Akad. Wiss. Berlin
　Sitzungsberichte der Akademie der Wissenschaften der DDR. Klasse für Mathematik, Physik und Technik. Publisher: Akademie-Verlag, Berlin.

Sitz.-Ber. Bayer. Akad. Wiss.
　Bayerische Akademie der Wissenschaften. Mathematisch-Naturwissenschaftliche Klasse. Sitzungsberichte. Publisher: Verlag der Bayerischen Akademie der Wissenschaften, München.

Sitz.-Ber. Heidelberger Akad. Wiss.
　Sitzungsberichte der Heidelberger Akademie der Wissenschaften. Mathematisch-Naturwissenschaftliche Klasse. Publisher: Springer-Verlag, Heidelberg.

Sitz.-Ber. Österreich. Akad. Wiss.
　Sitzungsberichte. Österreichische Akademie der Wissenschaften. Mathematisch-Naturwissenschaftliche Klasse. Abteilung II: Mathematik, Astronomie, Meteorologie und Technik. Publisher: Springer-Verlag, Wien.

Sky Telescope
　Sky and Telescope. Published by Sky Publishing Corporation, Cambridge, Mass.

Smithsonian Contr. Astrophys.
　Smithsonian Contributions to Astrophysics. Smithsonian Institution Astrophysical Observatory, Cambridge, Mass. Printed by Smithsonian Institution Press, City of Washington. For sale by the Superintendent of Documents, U. S. Government Printing Office, Washington, D. C.

Smithsonian Year
　Smithsonian Year. Annual Report of the Smithsonian Institution, including the financial report of the Executive Committee of the Boards of Regents. Published by the Smithsonian Institution, Washington, D.C.

Solar Physics
　Solar Physics. A Journal for Solar Research and the Study of Solar Terrestrial Physics. Publishers: D. Reidel Publishing Company, Dordrecht–Holland.

Solnechnye Dannye Byull.
　Solnechnye Dannye. Byulleten'. *(Solar Data).* Publishers: Izdatel'stvo "Nauka", Leningradskoe Otdelenie, Leningrad.

Soobshch. Byurakan. Obs.
　Soobshcheniya Byurakanskoj Observatorii. Akademiya Nauk Armyanskoj SSR, Erevan.

Soobshch. Gos. Astron. Inst. Shternberg
　Soobshcheniya Gosudarstvennogo Astronomicheskogo Instituta im P.K. Shternberga. Publishers: Izdatel'stvo Moskovskogo Universiteta, Moskva.

Southern Stars
　Southern Stars. The Journal of the Royal Astronomical Society of New Zealand (Inc.). Address of the Society: P.O. Box 3181, Wellington C1, New Zealand.

Soviet Astron. AJ
　Soviet Astronomy AJ. A translation of the Astronomical Journal of the Academy of Sciences of the USSR. Published by the American Institute of Physics, Inc., New York, N.Y.

Spaceflight
　Spaceflight. A Publication of the British Interplanetary

Society. Printed by Eyre & Spottiswoode Limited at Grosvenor Press, Portsmouth, and published by the British Interplanetary Society, London.

Space Sci. Rev.
Space Science Reviews. Publishers: D. Reidel Publishing Company, Dordrecht–Holland.

Springer Tracts Modern Phys.
Springer Tracts in Modern Physics. (Ergebnisse der exakten Naturwissenschaften). Springer-Verlag, Berlin–Heidelberg–New York.

Sterne
Die Sterne. Zeitschrift für alle Gebiete der Himmelskunde. Johann Ambrosius Barth, Leipzig.

Sternenbote
Sternenbote. Monatsschrift für Österreichs Amateurastronomen. Publisher: Astronomisches Büro, Hermann Mucke, Wien.

Stockholms Obs. Ann.
Stockholms Observatoriums Annaler. Printed by Almquist & Wiksell, Stockholm.

Strolling Astronomer
The Strolling Astronomer. The Journal of The Association of Lunar and Planetary Observers, Publication Office: The Strolling Astronomer, Box 3 AZ, University Park, New Mexico.

Stud. Cerc. Astron.
Studii şi Cercetări de Astronomie. Editura Academiei Republicii Socialiste România. Editorial Office: Observatorul Astronomic, Bucureşti.

Stud. Geophys. Geod.
Studia geophysica et geodaetica. Published for the Geophysical Institute of the Czechoslovak Academy of Sciences by Academia, Praha.

Stud. Soc. Sci. Torunensis
Studia Societatis Scientiarum Torunensis, Toruń – Polonia. Sectio F (Astronomia).

Stud. Univ. Babeş-Bolyai
Studia Universitatis Babeş-Bolyai. Series Mathematica-Physica. Publishers: Intreprinderea Poligrafica, Cluj.

SuW
Sterne und Weltraum. Astronomische Monatsschrift. Publisher: Verlag Sterne und Weltraum Dr. Vehrenberg, Düsseldorf, Germany.

Tellus
Tellus, a bi-monthly Journal of Geophysics. Svenska Geofysiska Foreningen. Printed in Sweden by Almqvist & Wiksells Boktryckeri AB, Uppsala.

Trans. Astron. Obs. Yale Univ.
Transactions of the Astronomical Observatory of Yale University. Published by the Observatory, New Haven.

Trans. IAU
Transactions of the International Astronomical Union. Published and distributed for the IAU (UAI) by D. Reidel Publishing Company, Dordrecht–Holland/Boston–U.S.A.

Trans. Roy. Soc. Canada
Transactions of the Royal Society of Canada. Published by the Royal Society of Canada, National Research Building, Ottawa.

Trudy Astrofiz. Inst. Alma-Ata
Trudy Astrofizicheskogo Instituta, Alma-Ata. Akademiya Nauk Kazakhskoj SSR. Publishers: Izdatel'stvo "Nauka" Kazakhskoj SSR, Alma-Ata.

Trudy Glav. Astron. Obs. Pulkovo
Trudy Glavnoj Astronomicheskoj Observatorii v Pulkove. Akademiya Nauk SSSR. Izdanie Glavnoj astronomicheskoj observatorii v Pulkove, Leningrad.

Trudy Inst. Teor. Astron., *Leningrad*
Trudy Instituta Teoreticheskoj Astronomii. Akademiya Nauk SSSR. Publishers: Izdatel'stvo "Nauka", Leningrad.

Trudy Tashkent. Astron. Obs.
Trudy Tashkentskoj Astronomicheskoj Observatorii. Akademiya Nauk Uzbekskoj SSR. Publishers: Izdatel'stvo "FAN" Uzbekskoj SSR, Tashkent.

Tsirk. Astron. Inst. Tashkent
Tsirkulyar Astronomicheskogo Instituta. Akademiya Nauk Uzbekskoj SSR. Izdatel'stvo "FAN" Uzbekskoj SSR, Tashkent.

Tsirk. Astron. Obs. L'vov
Tsirkulyar. Astronomicheskaya Observatoriya. L'vovskij Ordena Lenina Gosudarstvennyj Universitet imeni Ivana Franko. Publisher: Izdatel'stvo L'vovskogo Universiteta, L'vov.

Umschau
Umschau in Wissenschaft und Technik. Umschau-Verlag Frankfurt a. M.

Urania Barcelona
Urania. Revista de Astronomia y Ciencias Afines. Organo de la Sociedad Astronómica de España y América, Barcelona; Unión Nacional de Astronomia y Ciencias Afines, Madrid.

Urania Kraków
Urania. Miesiecznik Polskiego Towarzystwa Miłośników Astronomii, Kraków. Publisher: Krakowska Drukarnia Prasowa, Kraków.

Vasiona
Vasiona. Revue d'Astronomie et d'Astronautique. Bulletin de la Société Astronomique "R. Bosković", Beograd.

Veröff. Astron. Rechen-Inst. Heidelberg
Veröffentlichungen des Astronomischen Rechen-Instituts Heidelberg. Verlag G. Braun, Karlsruhe.

Veröff. Sternw. Sonneberg
Deutsche Akademie der Wissenschaften zu Berlin. Institut für Sternphysik. Veröffentlichungen der Sternwarte in Sonneberg. Publisher: Akademie-Verlag, Berlin.

Vesmír
Vesmír. Přírodovědecky časopis Čs. akadmie věd. Publisher: Academia, Praha.

Vestn. Khar'kov. Univ.
Vestnik Khar'kovskogo Universiteta. Seriya Astronomicheskaya. Publishers: Izdatel'stvo Khar'kovskogo Universiteta, Khar'kov.

Vestn. Kiev. Univ.
Vestnik Kievskogo Universiteta. Seriya Astronomii. Publishers: Izdatel'stvo Kievskogo Universiteta, Kiev.

VJS Naturforsch. Ges. Zürich
Vierteljahresschrift der Naturforschenden Gesellschaft in Zürich. Printer and Publisher: Leeman AG, Zürich.

Weltraumfahrt
Weltraumfahrt, Raketentechnik. Publisher: Umschau-Verlag, Frankfurt a/Main.

Wiss. Zeitschr. Friedrich-Schiller Univ. Jena
Wissenschaftliche Zeitschrift der Friedrich-Schiller-Universität. Jena. Mathematisch-Naturwissenschaftliche Reihe. Edited by the Rektor der Friedrich-Schiller-Universität Jena.

Wiss. Zeitschr. Humboldt-Univ. Berlin
Wissenschaftliche Zeitschrift der Humboldt-Universität zu Berlin. Mathematisch-Naturwissenschaftliche Reihe. Edited by the Rektor der Humboldt-Universität, Berlin.

Yamamoto Circ.
Yamamoto Circular. Published by the Yamamoto Observatory, Kamitanakami – Kiryutyo, Otu, Siga-ken, Japan.

Zeitschr. Angew. Physik
Zeitschrift für Angewandte Physik. Publisher: Springer-Verlag, Berlin–Heidelberg–New York.

Zeitschr. Astrophys. (ZfA)
Zeitschrift für Astrophysik. Publisher: Springer-Verlag, Berlin–Heidelberg–New York. After Vol. 69 (1968) replaced by "Astronomy and Astrophysics".

Zeitschr. Naturforschung
Zeitschrift für Naturforschung. Europhysics Journal. Teil a: Astrophysik, Physik, Physikalische Chemie. Published by Verlag der Zeitschrift für Naturforschung, Tübingen, Germany.

Zeitschr. Physik
Zeitschrift für Physik. Publisher: Springer-Verlag, Berlin–Heidelberg–New York.

Zemlya i Vselennaya
Zemlya i Vselennaya. Astronomiya, Geofizika, Issledovaniya Kosmicheskogo Prostranstva. Nauchno-Populyarnyj Zhurnal Akademii Nauk SSSR. Publishers: Izdatel'stvo "Nauka", Moskva.

Zenit
Populair wetenschappelijk maandblad over sterrenkunde/weerkunde/ruimtevaart/ruimte-onderzoek/aanverwante wetenschappen en technieken. Bureau: Stichting De Koepel, Utrecht.

Zentralblatt Math. Grenzgebiete
Zentralblatt für Mathematik und ihre Grenzgebiete. Publisher: Springer-Verlag, Berlin–Heidelberg–New York.

Zvaigžņota Debess
Latvijas PSR Zinātņu Akadēmijas Radioastrofizikas Observatorijas Populārzinātnisks Gadalaiku Izdevums. Izdevnieciba "Zinātne", Riga.

002 Bibliographical Publications, News Notes from Current Journals

002.001 Science news.
Priroda, No. 1.74, p. 120 - 126 (1974). In Russian.
The Martian relief resembles the earth, p. 120; Meeting of scientists in Baku [XXIVth congress of the International Astronautical Federation, 1973, October 7 – 13], (*S. A. Nikitin*), p. 121 - 122.

002.002 Science news.
Priroda, No. 2.74, p. 107 - 118 (1974). In Russian.
Starts of space vehicles in the USSR (September – October 1973), p. 107 - 108; The flight of Intercosmos 10 (*S. A. Nikitin*), p. 108; Investigation of the circumlunar plasma, p. 108 - 109; Tidal friction – source of flares of novae? p. 109; The largest redshift, p. 109; Boomerang sounds the upper atmosphere, p. 116.

002.003 News from science and other informations.
Zemlya i Vselennaya, 1974, No. 1. In Russian.
Artificial aurorae, p. 32; Polar caps and dust storms on Mars, p. 42; Astonishing double stars (*M. S. Frolov*), p. 49; Riddling flares of gamma-radiation, p. 55; Powerful radio-burst in the galaxy 3C 120, p. 69.

002.004 Twenty years Referativnyj Zhurnal "Astronomiya". I. S. Shcherbina-Samojlova.
Tsirkulyar Vses. astron.-geod. o-va, 1973, No. 24, p. 94. In Russian.

002.005 Indexing astronomy.
J. D. Mulholland, with a comment from AIP by S. Schiminovich, A. W. K. Metzner.
Phys. Today, Vol. 27, No. 1, p. 9, 11, 13 (1974). – Letters.

002.006 Science news.
Priroda, No. 3.74, p. 103 - 114 (1974). In Russian.
Flight of Soyuz 13 finished, p. 103 - 104; Forward to Venus and Mercury, p. 104 - 105; Ejection of lunar rocks over 3000 km, p. 105; An outburst changes the matter of meteorites, p. 105 - 106; Consequences of the solar flare of 1972, p. 106.

002.007 Science news.
Priroda, No. 4.74, p. 107 - 118 (1974). In Russian.
Starts of space vehicles in the USSR (November – December 1973), p. 107 - 108; Explorer 50, p. 108; "Lunar" crater on earth, p. 108; Comet Kohoutek did not fulfil hopes (*A. N. Simonenko*), p. 108 - 109; New calculations of the motion of the satellites of Jupiter, p. 109; Deuterium discovered in space, p. 109.

002.008 News from science and other informations.
Zemlya i Vselennaya, 1974, No. 2. In Russian.
How much dust in the central regions of galaxies? p. 28; The nature of extragalactic X-ray sources, p. 35; A very long eclipse, p. 36 - 37; Disputes on Stephan's Quintet (*B. V. Komberg*), p. 44 - 45; Search for the Tashatkan rock (*V. V. Rychkov*), p. 66; In the objective – Jupiter, p. 80.

002.009 Astronomijas jaunumi.
Zvaigžņotā debess, 1973. gada rudens, p. 14 - 18.
Kas ir kvazāri? (*A. Balklavs*), p. 14 - 15; Jauna kvazāra 3C279 radionovērojumu interpretācija (*I. Šmelds*), p. 15 - 16; Vai aiz Plutona ir vēl kādas Saules sistēmas planetas? (*J. Francmanis*), p. 16 - 17; Oranžas krāsas ieži Mēness krāterī (*I. Daube*), p. 17; Kosmisko staru indikācija ... cilvēka galvā (*M. Paupere*), p. 17 - 18.

002.010 Annotations on the papers on geomagnetism and aeronomy in "News of higher educational establishments. Radiophysics", 1972, Vol. 15, Nos. 8 - 9.
Geomagn. Aeronom., Vol. 14, 387 - 388 (1974). In Russian.

002.011 Science news.
Priroda, No. 5.74, p. 102 - 115 (1974). In Russian.
New contribution to the science on planets, p. 102 - 103; Atmospheric "Explorer", p. 104; Redistribution of masses in the atmosphere and the rotation of the earth (*N. S. Sidorenko*), p. 104 - 105; New determination of the mass of a globular cluster, p. 105 - 106; Chemical elements on the lunar ground, p. 111.

002.012 The Crawford Collection of books and manuscripts on the history of astronomy, mathematics, etc., at the Royal Observatory, Edinburgh. E. G. Forbes.
British Journ. History Sci.,Vol. 6, 459 - 461 (1973).

002.013 Science news.
Priroda, No. 6.74, p. 90 - 103 (1974). In Russian.
Starts of space vehicles in the USSR (January – February 1974), p. 91; Data of Explorer 48, p. 92; The third crew aboard Skylab, p. 92 - 93; Search for "black holes", p. 93; Encounter of the earth with comets in the past, p. 93 - 94; Lunar tides and the magnetism of the moon, p. 94; The age of tektites – 35 mill. years, p. 97.

002.014 Astronomijas jaunumi.
Zvaigžņotā debess, 1973./74. gada ziema, p. 14 - 23.
Saules aktivitāte granulācijas ietvaros (*N. Cimahoviča*), p. 14 - 16; Jauni interesanti pulsāri (*A. Balklavs*), p. 17 - 18; Pirmie seismiskie novērojumi uz Mēness (*L. Roze*), p. 18 - 19; Zemes putekļu pavadoņi (*J. Francmanis*), p. 19 - 21; Infrasarkano staru astronomijas jaunumi (*A. Balklavs*), p. 21 - 23.

002.015 Astronomijas jaunumi.
Zvaigžņotā debess, 1974. gada pavasaris, p. 22 - 27.
Jauns pierādījums Saules vispārējā magnētiskā lauka eksistencei (*N. Cimahoviča*), p. 22 - 23; Saules radiostarojuma kvaziperiodisko fluktuāciju pētījumi speciālajā astrofizikas observatorijā (*G. Ozoliņš*), p. 23 - 24; Par Saules neitrino novērojumiem (*J. Francmanis*), p. 24 - 26; Saule un Zemes klimats (*N. Cimahoviča*), p. 26 - 27.

002.016 Bibliographie générale des marées terrestres. Supplement I 1972 - 1973. P. Melchior (Editor).
Ass. Internationale Géod., Centre International des Marées Terrestres, (Obs. Roy. Belgique, Bruxelles), p. 116 - 134 (1974).

002.017 News from science and other informations.
Zemlya i Vselennaya, 1974, No. 3. In Russian.
New solar telescope (*V. M. Mozhzherin*), p. 11 - 13; Expansion of supernova remnants, p. 20; Spot group of a new cycle? (*V. F. Chistyakov*), p. 20; The neutrino flux from cosmos, p. 56; Volcanism and the chemistry of the atmosphere, p. 77; The corona in Lyman α (*E. V. Ivanov*), p. 77; Craters on ... Venus, p. 77.

002.018 Chronicle.
Urania Kraków, Vol. 45, 14 - 16, 47 - 52 (1974).
In Polish.
(1) More about the comet Kohoutek, (*A. Pilski*); Thirteen new pulsars, (*B. Kuchowicz*); Age of the Sikhote-Alin meteorite, (*J. Pokrzywnicki*). (2) Interplanetary mission Pioneer 10, (*J. Pokrzywnicki*); Observations of cosmic gamma rays,

(B. Kuchowicz); The reddening of the quasar 3C 286, (B. Kuchowicz); Does the Barnard star possess satellites, (J. Pokrzywnicki); Lonar crater in India, (J. Pokrzywnicki); Mercury and the moon, (J. Pokrzywnicki).

002.019 **Chronicle.**
Urania Kraków, Vol. 45, 81 - 88, 117 - 120 (1974). In Polish.
(3) Observations of Seyfert galaxies in the optical domain, (T. Szymczak); The distance to the Crab nebula, (Z. Paprotny); An explosion in the galaxy NGC 5548, (Z. Paprotny); The farthest known object in the universe? (Z. Paprotny); The nearby star, (Z. Paprotny); A strange star in Aquila, (Z. Paprotny); The hypothesis on a lunar core, (J. Pokrzywnicki); Mascons and moonquakes, (J. Pokrzywnicki); Pluto's spin period, (J. Pokrzywnicki); A comet related to Perseids, (Z. Paprotny). (4) Dense interstellar clouds as source of cosmic gamma rays, (B. Kuchowicz); Lithium, berylium and boron in the peculiar star χ Cancri, (B. Kuchowicz); About changes in the period of rotation of the earth, (L. Zajdler); The evolution of planetoids, (Z. Paprotny); Was the Tunguska meteorite a black hole? (B. Kuchowicz).

002.020 **Chronicle.**
Urania Kraków, Vol. 45, 144 - 147, 176 - 183 (1974). In Polish.
(5) Great meteor above British Islands, (B. Lang); Further X-ray sources in clusters of galaxies, (B. Kuchowicz); New success of exobiology, (B. Kuchowicz); Missing link? (M. Różyczka); The orbits of Phobos and Deimos, (J. Pokrzywnicki); The chemistry of Titan, (Z. Paprotny). (6) Pioneer 10 investigates Jupiter, (T. Kwast); Results of comet Kohoutek investigations, (J. Pokrzywnicki); A crazy hypothesis: neutron stars bombarded by comets generate gamma-photons, (B. Kuchowicz); The day shortened by $0\overset{s}{.}001$ in January 1974, (I. Domiński); What was the title of Copernicus work? (H. Korpikiewicz).

002.021 **Astronomy and Astrophysics Abstracts. Vol. 10, Literature 1973, Part II.**
S. Böhme, W. Fricke, U. Güntzel-Lingner, F. Henn, D. Krahn, U. Scheffer, G. Zech (Editors).
Published for Astronomisches Rechen-Institut, Heidelberg by Springer-Verlag, Berlin–Heidelberg–New York. 10 + 661 pp. Price DM 86.00; (US $ 35.10) [Subscription price DM 68.80 (US $ 28.10)] (1974).

002.022 **General news.**
Minor Planet Bull., Vol. 1, 17 - 19 (1974).

002.023 **Astronomische Kurzberichte.**
Sternenbote, 17. Jahrgang, p. 10 - 18, 25 - 36, 51 - 59, 71 - 82, 96 - 104, 117 - 125 (1974).

002.024 **Centre de Données Stellaires. Information Bulletin No. 6.** J. Jung (Editor).
Compiled at Observatoire de Strasbourg, 38 pp. (1974).
The individual contributions are included in their corresponding subject categories – see abstracts 021.028, 021.029, 041.071 - 041.073, 113.041, 113.042.

002.025 **Centre de Données Stellaires. Information Bulletin No. 7.** J. Jung (Editor).
Compiled at Observatoire de Strasbourg, 59 pp. (1974).
The individual contributions are included in their corresponding subject categories – see abstracts 021.030, 021.031, 041.074, 114.164.

002.026 **A bibliography of the theory and application of the phase-lock principle.**
W. C. Lindsey, R. C. Tausworthe.
California Inst. Technol. JET Propulsion Lab. Techn. Rep. 32-1581, 91 pp. (1973). – A valuable list of 800 references. – *ACM*

002.027 **Nouvelles brèves.**
Ciel et Terre, Vol. 90, 74 - 78, 229 - 231 (1974).
L'éclipse rasante du 30 juin 1973; L'obliquité de Mars; La petite planète exceptionnelle 1973 NA; Les variations séculaires des orbites planétaires; La comète Clark (1973i); La comète Sandage (1973k); Redécouvertes de comètes périodiques; Une céphéide de très longue période; La comète Kohoutek (1973f); Nouvelle comète Gehrels (1973n); Désignation définitive des comètes de 1972.

002.028 **Rassegna delle riviste e notizie brevi.** P. Maffei.
Coelum, Vol. 42, 27 - 30, 73 - 82, 121 - 129 (1974).

002.029 **AFCRL bibliography for the first quarter of 1973.**
J. W. Salisbury (Editor); J. E. M. Adler, J. P. Dybwad, G. R. Hunt, C. J. Lenhoff, L. M. Logan, D. A. Long, J. W. Salisbury, L. F. Tolendino, K. P. Zinnow (Contributors).
Icarus, Vol. 21, 369 - 386 (1974).

002.030 **News and comments.** E. Öpik.
Irish Astron. Journ., Vol. 11, 23 - 30 (1973).
The Mediterranean a former desert; Two meteorite craters; The oldest rocks on earth; Start of life on earth; Dinosaurs; Microorganisms on Jupiter; Quaternary temperatures; Planetary missions; Interstellar meteoric particles; Mercury surface composition; Topography of Venus; Russian Venus probe; Surface conditions on Venus; Spectral reflectivity of Uranus; No lunar volcanoes; Liquid core in the moon; Moonquakes and lunar structure; Ages of some lunar craters.

002.031 **Astronomical notebook.** J. S. Griffith.
Journ. British Interplanet. Soc., Vol. 27, 66 - 72, 148 - 153, 226 - 231, 301 - 304, 385 - 388, 465 - 470, 540 - 547 (1974).
(1) Dust near the sun; Ice ages and life on Mars; Kohoutek and Skylab; Cometary nuclei; Meteoroid impacts on the earth/moon system; Mars; Titan's green house; Radii of satellites and asteroids; Missions to comets and asteroids; Other planetary systems; Peculiar A stars; The appearance of a star orbiting around a black hole; (2) Solar system chemistry; Planetesimal formation; The solar atmosphere; Earthquakes and the earth's magnetism; The light of the night sky; The atmosphere of Venus; The surface of the moon; The solar neutrino problem; The satellite TD-1A; Cosmic rays and pulsars; Helium polar caps on stars?; Optical atlas of galactic supernova remnants; The galactic nucleus; (3) The interior of Jupiter; Split tails of comets; Cosmic rays in interplanetary space; Rapidly pulsating star; New light on black holes; A variable galaxy; A group of interconnected galaxies; A miniquasar?; Radio astrometry; The origin of the universe; Gravitational lens; Detection of missing matter; Repulsive gravitation?; (4) Project Cyclops; Life – terrestrial and extraterrestrial; No planet of Barnard's star; Protostar and protoplanet?; Infrared emission from binaries; Runaway black holes; Faster than light; The evolution of quasars; (5) The early solar nebula; The fate of satellites of the moon; Saturn's rings; Saturn's gas ring?; X-ray sources as distance indicators?; Comets and neutron stars; Supermassive star explosions; The distance of the quasars; (6) The new solar cycle; The rocks of Venus; Particle size in the rings of Saturn; Energy sources in Jupiter; The origin of comets; Supernovae and neutron star formation; Binaries into supernovae?; An X-ray object in the Cygnus Loop; Globular cluster observations; The rare earths in Beta Coronae Borealis; Cosmology and radio source counts; Redshifts of radio galaxies; Intergalactic dust; (7) The atmosphere of Mercury; Ethane and acetylene in the atmosphere of Jupiter; Atmosphere on Ceres?; Solar-cycle variations in the solar wind; Origin of the

asteroids; Planetary engineering on Mars; Neutral hydrogen in galaxies; Large Magellanic Cloud; Isolated radiofrequency pulses; Interstellar titanium; Magnetic fields and the early stages of the universe; Extragalactic radio sources; Stephan's Quintet; Herbig-Haro objects; 'Sunspot' activity on flare stars; Black holes and binaries; Another black hole?; Cepheid variables; Collision of red giant and white dwarfs.

002.032 **Noted in the current journals.**
D. Morrison, N. D. Morrison.
Mercury, (Journ. Astron. Soc. Pacific), Vol. 2, No. 6, p. 13 - 15, 18 (1973); Vol. 3, No. 1, p. 10 - 13 (1974).
Vol. 2, No. 6: Fourth lunar science conference; First observations at 34 microns; The first large radio redshift; The dark markings on Mars and their variations. Vol. 3, No. 1: Radar and radio observations of the rings of Saturn; The internal energy source of Jupiter; Nuclear reactions in stars; Evidence against a planetary companion of Barnard's star; A possible stellar supernova remnant in the Cygnus Loop; New solar activity cycle under way.

002.033 **News: search & discovery.**
Phys. Today, Vol. 27, No. 1, p. 17 - 20; No. 2, p. 17 - 20; No. 3, p. 17 - 20; No. 4, p. 17 - 20; No. 5, p. 17 - 20; No. 6, p. 17 - 20; No. 7, p. 17 - 20 (1974).
(1) Material in rare meteorites may pre-date solar system, p. 17, 20; (2) Evidence accumulates for a black hole in Cygnus X-1, p. 17, 19 - 20; (3) Colliding-beam results upset quark advocates, p. 17, 20; (5) Nucleon-nucleon correlations in p—He^4? , p. 20; (6) Neutrino interactions may explain supernova explosions, p. 17 - 18; (7) Antineutrino pulses could signal birth of neutron star, p. 17, 20.

002.034 **Forschung und Technik.**
Phys. Blätter, 30. Jahrgang, p. 23 - 27, 64 - 72, 126 - 131; 175 - 180, 227 - 231, 272 - 275 (1974).
Neue große Rotverschiebung bei OQ 172, p. 27; Struktur und geologisch-morphologische Eigenschaften des Landegebietes von Luna 20, p. 68 - 70; Das Rätsel des galaktischen Deuteriums, p. 70 - 71; OH 471, ein Himmelskörper mit der extremen Rotverschiebung z = 3.40, p. 71 - 72; Polarisationsformen von Schwerewellen als Tests für Gravitationstheorien, p. 126 - 127; Titan-Atmosphäre könnte Gasring um Saturn erzeugen, p. 129 - 130; Groß-Teleskop kommt nach Spanien, p. 130 - 131; Neutronenstern Her X-1 und HZ Her — ein bedeckungsveränderliches Binärsystem, p. 175 - 179; Fünfte Schwefelverbindung im interstellaren Gas, p. 180; Komet Kohoutek (1973f) — eine Enttäuschung für Astronomen? (*L. Kohoutek, H. J. Wendker*), p. 227 - 229; Absorption in galaktischem Wasserstoff deutet auf großen Abstand eines Quasars, p. 229 - 230; Magnetsterne, p. 230 - 231; Präzession als Ursache geomagnetischer Schwankungen, p. 231; Mariner 10 an den Planeten Venus und Merkur (*H. W. Köhler*), p. 273 - 274.

002.035 **Science and the citizen.**
Sci. American, Vol. 230, No. 1, p. 50 - 52; No. 2, p. 42 - 45; No. 3, p. 44 - 46; No. 4, p. 48 - 51; No. 5, p. 59 - 61 (1974).
The water hole, No. 1, p. 52; Palomar in space, No. 2, p. 44; The hills of Mercury, No. 3, p. 44 - 45; Rime of the modern Mariner, No. 4, p. 48 - 49; Sidereal messengers, No. 5, p. 59; The quasar as a galaxy, No. 5, p. 60.

002.036 **Mitteilungen aus Wissenschaft und Literatur.**
Sterne, 50. Jahrgang, p. 60 - 64 (1974).

002.037 **Kurzberichte aus der Forschung.**
SuW, Vol. 13, 17 - 18, 50 - 52, 93 - 95, 128 - 131, 164 - 166, 201 - 204 (1974).
(1) Infrarotobjekte in Dunkelwolken; Infrarot- und Röntgenvariabilität von Cyg X-3; Zur Diskussion um die Rotverschiebung von Quasaren; Veränderliche Radiostrahlung eines Mirasternes; Uran auf der Venus; Europa entwickelt bemanntes Raumfluglabor; (2) Plasma-Analysator liefert erste Ergebnisse der Pioneer 10-Mission; Optische Eigenschaften der unteren Venusatmosphäre; Existieren schwarze Löcher tatsächlich? ; Krater auf der Venus ; (3) Photometrische Untersuchungen an Pluto; Ergebnisse aus Merkurdurchgangs-Beobachtungen; Planetoid 1932 HA Apollo; Wie werden die Nahaufnahmen von Merkur aussehen? ; Der Radio-Radius der Sonne; (4) Stärkere Argumente für nicht-kosmologische Rotverschiebungen? ; Über die Herkunft der Kometen; Beobachtungen des Kometen Kohoutek 1973f auf dem Calar Alto; Röntgenbilder von der Sonnenkorona; Neutraler Wasserstoff in frühen Galaxien; Helligkeitsschwankungen im Zodiakallicht; Die Birne; (5) H_2O^+ im Schweif und in der Koma von Komet Kohoutek 1973f (*K. Wurm*); Die optisch veränderliche Galaxie X Comae; Eine Gruppe von stark geröteten OB^+-Sternen in Norma; Interstellare Elementhäufigkeiten; (6) Intergalaktische Materie in Galaxienhaufen; Wissenschaftliche Aufgabe des 1. europäischen Satelliten auf geostationärer Umlaufbahn; Astronomie im fernen Infrarot; Situationsbericht Erderkundungssatellit ERTS-1.

002.038 **Bibliography.** Z. Kopal, M. Moutsoulas, J. W. Salisbury, F. B. Waranius (Editors).
The Moon, Vol. 9, 457 - 519; Vol. 10, 207 - 303 (1974).

002.039 **News and views.**
Nature, Vol. 247, 2 - 6, 127 - 130, 251 - 256, 333 - 340, 423 - 430; Vol. 248, 99 - 106, 191 - 198, 277 - 284, 377 - 384, 471 - 478, 549 - 556, 729 - 736; Vol. 249, 9 - 16, 105 - 112, 207 - 214, 305 - 312, 405 - 414, 509 - 516, 689 - 696, 795 - 802 (1974).
(Vol. 247): Satellites advance geodetic research, p. 4 - 5; Pioneer 10 exceeds expectations, p. 5; Geomagnetism and climatic change, p. 127; Impact of TV sensors on astronomical research, p. 251 - 252; X-ray sources and Occam's razor, p. 333; X-ray 'star' in Cygnus Loop, p. 339; Swings, roundabouts and Sco X-1, p. 339 - 340; Meteorites which 'bounce' off the earth, p. 423; Heliocentric theory in China, p. 425 - 426; Solar neutrino problem still unsolved, p. 427; Rhythms and the earth's rotation, p. 428 - 429; (Vol. 248): Even small meteoroids are fluffy, p. 99; Are accretion disks stable? , p. 101; Darwin glass related to tektite fall? , p. 101 - 102; When the lunar crust formed, p. 191; Are supernovae disruptive? , p. 192; Whys and wherefores of cosmic rays, p. 197; Where do meteorites come from? , p. 278 - 279; Sunspots affect Indian monsoon, p. 281; Limited progress with quantum gravity, p. 282 - 283; Earth spin theory queried by doubting Thomas, p. 383 - 384; Ghost neutrinos emerge from the mathematics, p. 471 - 472; Origin of asteroids, p. 550 - 551; Solar wind blows gusty at sunspot maximum, p. 555; Unpredictability of comets, p. 732; (Vol. 249): BL Lac identified with giant galaxy, p. 10; More about Mercury, p. 11 - 12; Slaved disk model for Hercules X-1, p. 107; Interplanetary dust on earth, p. 110 - 111; How special is the universe? (*P. C. W. Davies*), p. 208 - 209; Changing views of the moon's magnetism (*P. J. Smith*), p. 209 - 210; Super-rotation of the upper atmosphere (*D. W. Hughes*), p. 405 - 406; Einstein in crisis? (*P. C. W. Davies*), p. 510 - 511; Pioneering Jupiter meeting at RAS (*J. Gribbin*), p. 513; Collaboration fixes unique QSO spectra (*J. Gribbin*), p. 689; Solar geomagnetic disturbances and weather (*J. Gribbin*), p. 802.

002.040 **News notes.**
Sky Telescope, Vol. 47, 28, 83 - 84, 166 - 167, 240 - 241, 314, 372 - 373; Vol. 48, 22 - 23 (1974).
(1) Crab nebula's distance; A close pair of quasars; Communications test; (2) New solar cycle starts; Radio observation of Ceres; X Comae as a variable galaxy; The Leo I dwarf galaxy; The Jansky; (3) News about asteroids; About a new variable

star; Accretion onto black holes; Satellite view of an aurora; (4) Composition of Jupiter's atmosphere; Rotation of Vesta; New comet Bradfield (1974b); Massive stars and infrared sources; Massive visual binary; Deep-sky catalogue; (5) High-resolution planetary observation; Variable stars in globular clusters; Plate No. 100,000; (6) X-ray bright points on the sun; Radcliffe Observatory closes; Skylab is stabilized in orbit; An artificial lunar atmosphere; Trojan-type orbits in the earth-sun system; F-type supergiant stars; California astronomer; (Vol. 48, No. 1): Methylamine discovered in space; Pluto's rotation; Epsilon Eridani's unseen companion; Speeding up astrophotography; Supernova in NGC 4414; Water in comet Kohoutek.

003 Books (Astronomy and Astrophysics)

003.001 **Vistas in astronomy, Vol. 15.**
A. Beer (Editor).
Pergamon Press, Oxford–New York–Toronto–Sydney–Braunschweig. 7 + 187 pp. Price £ 9.00 (1973). – The individual contributions are included in their corresponding subject categories - see abstracts 015.014, 031.011, 034.014, 061.019, 064.019, 091.005, 102.008, 112.005, 151.020.

003.002 **Astrometriya i Astrofizika, Vypusk 20.**
E. P. Fedorov (Editor).
Respublikanskij Mezhvedomstvennyj Sbornik. Akademiya Nauk Ukrainskoj SSR, Glavnaya Astronomicheskaya Observatoriya. Izdatel'stvo "Naukova Dumka", Kiev. 112 pp. Price 1 Rbl. 3 Kop. (1973). In Russian. – The papers included are abstracted in their corresponding subject categories – see abstracts 031.016, 031.017, 041.010 - 041.012, 054.006, 064.021, 094.137, 114.046, 122.028.

003.003 **Ionosphere and solar-terrestrial relations.**
M. P. Rudina (Editor).
Akademiya Nauk Kazakhskoj SSR, Trudy Sektora Ionosfery, Tom 3. Izdatel'stvo "Nauka" Kazakhskoj SSR, Alma-Ata. 160 pp. Price 1 Rbl. 16 Kop. (1972). In Russian. – The individual contributions within the subject scope of Astronomy and Astrophysics Abstracts are included in their corresponding categories – see abstracts 034.021, 077.022, 083.021 - 083.027, 084.237, 143.012.

003.004 **Cosmical geophysics.**
A. Egeland, Ø. Holter, A. Omholt (Editors).
Universitetsforlaget, Oslo–Bergen–Tromsö, Norway. 360 pp., with a historic preamble by the editors, p. 11 - 17. Price $ 20.00 (1973). – Review in Icarus, Vol. 21, 211 - 212; 1974 (F. C. Michel). – The individual contributions within the subject scope of Astronomy and Astrophysics Abstracts are included in their corresponding categories – see abstracts 074.040, 080.019, 082.045, 082.062, 083.028, 084.016 - 084.018, 084.239 - 084.243, 143.016.

003.005 **Solar processes and their observations.**
Yu. M. Slonim (Editor).
Akademiya Nauk Uzbekskoj SSR, Astronomicheskij Institut. Izdatel'stvo "FAN" Uzbekskoj SSR, Tashkent. 126 pp. Price 1 Rbl. 3 Kop. (1973). In Russian. – The individual contributions are included in their corresponding subject categories – see abstracts 034.024, 034.025, 071.019, 072.019, 072.020, 073.036, 073.037.

003.006 **Annual Review of Earth and Planetary Sciences, Vol. 2.**
F. A. Donath, F. G. Stehli, G. W. Wetherill (Editors).
Annual Reviews Inc., Palo Alto, California. 8 + 478 pp. Price $ 12.50 (1974). – For the individual contributions within the subject scope of Astronomy and Astrophysics Abstracts – see abstracts 091.013, 094.126, 097.046, 105.105.

003.007 **Annuaire 1974 du Bureau des Longitudes.** Encyclopédie Physique et Spatiale.
Gauthier-Villars Éditeur, Paris. 16 + 701 + A8 + B6 + C14 + D25 + E116 pp. (1974). – Part 1: Éphémérides astronomiques; Part 2: Terre; Part 3: Système solaire; Part 4: Système de mesure: Part 5: Géographie de la France; Part 6: Supplément pour l'anné 1975; Part 7: Notices; Part 8: Tables analytiques.

003.008 **Investigation of the sun and red stars. 1.**
A. Balklavs (Editor).
Akademiya Nauk Latvijskoj SSR, Radioastrofizicheskaya Observatoriya. Izdatel'stvo "Zinatne", Riga. 72 pp. Price 19 Kop. (1974). In Russian. – The individual contributions are abstracted in their corresponding subject categories – see abstracts 034.049, 113.028 - 113.030.

003.009 **In honorem S. Placidis.** Special volume dedicated to Professor S. Plakidis on his 80th birthday.
D. Kotsakis (Editor), with speeches by J. Xanthakis, L. Carapiperis, I. Argyrakos, A. Makridis, D. Kotsakis and a reply by S. Plakidis (in Greek).
Published by C. Zisouli, Athens. 7 + 395 pp. (1974). – The individual contributions within the subject scope of Astronomy and Astrophysics Abstracts are included in their corresponding categories – see abstracts 004.066, 032.035, 032.036 034.057, 035.006, 041.032, 042.073, 042.074, 051.014, 066.091, 066.092, 077.059, 079.100, 085.012, 121.074, 141.091, 151.046, 151.047.

003.010 **The accuracy of orbits of comets and artificial earth satellites.**
Uchenye Zapiski Latvijskogo Gosudarstvennogo Universiteta im. Petra Stuchki, Vol. 175, Astron., vyp. (No.) 8, 79 pp. Price 36 Kop. (1973). In Russian. – The papers included are abstracted in their corresponding subject categories – see abstracts 052.038, 052.039, 055.010, 102.025.

003.011 **Astronomical Institute of the Academy of Sciences of the Uzbek SSR — 100 years.**
V. P. Shcheglov (Editor).
Akademiya Nauk Uzbekskoj SSR; Ordena Trudovogo Krasnogo Znameni Astronomicheskij Institut. Izdatel' stvo "Fan" Uzbekskoj SSR, Tashkent. 144 pp. Price 1 Rbl. 50 Kop. (1974). In Russian. — The individual papers are included in their corresponding subject categories — see abstracts 008.126 009.016, 041.065 - 041.067, 044.016, 046.016, 055.011, 075.007, 081.026, 082.109, 104.068, 122.109.

003.012 **Optimization of the reduction of observations of artificial earth satellites and time.**
Uchenye Zapiski Latvijskogo Gosudarstvennogo Universiteta im. Petra Stuchki, Vol. 202, Astron., vyp. (No.) 10, Riga. 84 pp. Price 40 Kop. (1974). In Russian.

003.013 **Transactions of the International Astronomical Union, Volume XV B: Proceedings of the Fifteenth General Assembly, Sydney 1973 and Extraordinary Assembly, Poland 1973.**
G. Contopoulos, A. Jappel (Editors).
D. Reidel Publishing Company, Dordrecht — Holland/Boston — U.S.A. 9 + 334 pp. Price Dfl. 120.00 (1974).
    The present volume gives a general picture of the Union's recent activity. It contains the report of the Executive Committee, the report of the General Assembly, including the Commissions, Meetings, a short report on the Extraordinary General Assembly and an Appendix with the Members and Commissions of the IAU and the approved names of Lunar and Martian features.

003.014 **Exploration of the universe.**   G. Abell.
Brief edition, Holt-Blond Limited, London. 14 + 483 pp. Price £ 6.75 (1973). — Review in Journ. History Astron.,Vol. 5, 142 (1974).

003.015 **Nicolaus Copernicus and his epoch.**
J. Adamczewski.
Charles Scribner's Sons, New York. 164 pp. Price $ 7.95 (1974). — Review in Sky Telescope, Vol. 48, 46 (1974).

003.016 **IR-theory and practice of infrared spectroscopy.**
N. L. Alpert, W. E. Keiser, H. A. Szymanski.
Plenum, Publishing Corporation, New York — London. 380 pp. Price $ 7.95 (1973).

003.017 **Pictorial guide to the moon.**   D. Alter.
Revised by J. H. Jackson.
Thomas Y. Crowell Company, New York. Third edition. 8 + 216 pp. Price $ 8.95 (1973).

003.018 **Pictorial astronomy.**   D. Alter, C. H. Cleminshaw, J. G. Phillips.
Thomas Y. Crowell Company, New York. 328 pp. Price $ 10.00 (1974).

003.019 **Philosophical problems of the science on the universe. Collected addresses, discourses and papers.**
V. A. Ambartsumyan.
AN ArmSSR, Erevan. 426 pp. Price 1 Rbl. 75 Kop. (1973). In Russian. — Review in Referativ. Zhurn. 51. Astron., 6.51. 62 (1974).

003.020 **Asimov on astronomy.**   I. Asimov.
Doubleday & Company, Inc., New York. 16 + 238 pp. Price $ 8.95 (1974).

003.021 **Jupiter: the largest planet.**   I. Asimov.
Lothrop, Lee, and Shepard Company, New York. 224 pp. Price $ 5.95 (1973). — Review in Strolling Astronomer, Vol. 24, 241; 1974 (P. K. Mackal).

003.022 **The tragedy of the moon.**   I. Asimov.
Doubleday & Company, Inc., New York. 16 + 220 pp. Price $ 6.95 (1973).

003.023 **J. E. Bodes Sternatlas 1782. Vorstellung der Gestirne auf XXXIV Kupfertafeln.**
H. Vehrenberg (Editor).
Facsimile reprint of the original edition. Treugesell-Verlag, Düsseldorf. 40 + 32 pp. Price DM 43.50 (1973). — Reviews in Orion Schaffhausen, 32. Jahrgang, p. 33; 1974 (E. Wiedemann); Sky Telescope, Vol. 47, 188 (1974); Sky Telescope, Vol. 47, 256 - 257; 1974 (J. Ashbrook); SuW, Vol. 13, 66, 68; 1974 (F. Schmeidler).

003.024 **The earth from space.**
J. Bodechtel, H.-G. Gierloff-Emden.
Arco Publishing Co., Inc., New York, N.Y. 176 pp. Price $ 16.95 (1974). — Review in Sky Telescope, Vol. 47, 257 (1974).

003.025 **Sternberg State Astronomical Institute. Brief history and description.**
L. N. Bondarenko, D. Ya. Martynov.
Moskovskij universitet, Moskva. 52 pp. Price 16 Kop. (1973). In Russian. — Review in Referativ. Zhurn. 51. Astron., 4.51. 32 (1974).

003.026 **Mars and the mind of man.**   R. Bradbury, A. C. Clarke, B. Murray, C. Sagan, W. Sullivan.
Harper & Row Publishers, New York — London — Mexico City — Sydney. 14 + 144 pp. Price $ 7.95 (1973). — Review in Science, Vol. 184, 663 - 664; 1974 (W. K. Hartmann).

003.027 **Experiments in the principles of space travel.**
F. M. Branley.
Thomas Y. Crowell Company, New York, N.Y. 113 pp. Price $ 4.50 (1973). — Review in Sky Telescope, Vol. 47, 50 (1974).

003.028 **The solar chromosphere.**
R. J. Bray, R. E. Loughhead.
International Astrophysics Series. Chapman and Hall, London. 19 + 384 pp. Price £ 9.00 (1974). — Contents: (1) Historical introduction, (2) Spicules and other fine structures at the solar limb, (3) The morphology and dynamics of the quiet chromosphere observed on the disk, (4) Physical conditions in the quiet chromosphere, (5) The fine structure of the active chromosphere, (6) The propagation and dissipation of waves in a compressible gravitationally-stratified atmosphere, (7) Theories of the heating of the chromosphere and of the origin of spicules. — Review in Monthly Notes Astron. Soc. Southern Africa, Vol. 33, 62 - 64; 1974 (A. H. Jarrett).

003.029 **Astrophysics: Part A: optical and infrared.**
N. P. Carleton (Editor).
Academic Press, Inc., New York. 608 pp. Price $ 43.50, £ 20.90 respectively (1974).

003.030 **Albert Einstein. Leben und Werk. Eine Biographie.**
R. W. Clark.
Translation of the English edition: Einstein, the life and times (1973) by M. Raeithel-Thaler.
Bechtle-Verlag, Esslingen. 15 + 507 pp. Price DM 48.00 (1974).

003.031 **Fundamental astronomy.**   F. W. Cole.
John Wiley & Sons Inc., New York — Toronto. 476 pp. Price $ 10.95 (1974). — Review in Sky Telescope, Vol. 47, 406 (1974).

003.032 **Fundamental astronomy: solar system and beyond.**

F. W. Cole.
John Wiley & Sons Inc., New York – Toronto. 496 pp. Price $ 14.95 (1974).

003.033 Dessinons, réalisons huit cadrans solaires.
M. Collenot.
To be purchased from M. Collenot, 26, rue de Neubourg, Evreux (France). 50 pp. Price F 22.00 (1973). – Review in L'Astronomie, 88ᵉ année, p. 113 (1974).

003.034 Physics of the earth and planets. A. H. Cook.
Macmillan, London – Basingstoke; Halsted (Wiley), New York. 10 + 316 pp. Price £ 8.50, $ 24.75 respectively (1973). – Reviews in Journ. British Interplanet. Soc., Vol. 27, 238 (1974); Journ. History Astron., Vol. 5, 142 (1974); Nature, Vol. 248, 262 - 263; 1974 (*D.C. Tozer*).

003.035 Il soggiorno di Nicolò Copernico in Italia.
B. Corrado.
Cappelli Editore, Bologna. 50 pp. Price L. 1200 (1973).

003.036 An introduction to experimental astronomy.
R. B. Culver.
W. H. Freeman & Company, San Francisco, California. 195 pp. Price $ 4.00 (1974). – Review in Sky Telescope, Vol. 47, 330 (1974).

003.037 Problems of modern astrophysics.
O. I. Dalgatov, V. G. Lapchinskij.
Knizhnoe izdatel'stvo, Makhachkala, Dagestan. 20 pp. Price 4 Kop. (1973). In Russian. – Review in Referativ. Zhurn. 51. Astron., 3.51.48 (1974).

003.038 Astrology: fact or fiction? K. J. Delano.
Our Sunday Visitor, Inc., Huntington, Indiana. 127 pp. Price $ 2.50, $ 1.95 respectively (1973). – Reviews in Sky Telescope, Vol. 47, 50 (1974); Strolling Astronomer, Vol. 24, 206; 1974 (*B. M. Frank*).

003.039 Raumfahrt in Stichworten. E. Elsner.
Verlag Ferdinand Hirt, Kiel – Wien. 272 pp. Price DM 38.00 (1973). – Review in SuW, Vol. 13, 139 - 140 (1974).

003.040 Problems of modern marine astronomy.
V. I. Ermakov.
Knigoizdat, Kaliningrad. 72 pp. Price 18 Kop. (1973). In Russian. – Review in Referativ. Zhurn. 51. Astron., 5.51.187 (1974).

003.041 The measurement of frequency and time interval.
L. Essen.
National Physical Laboratory, Teddington. 5 + 55 pp. Price £ 1.00 (1973). – Review in Nature, Vol. 249, 90; 1974 (*D. H. Sadler*).

003.042 The second fifteen years in space.
S. Ferdman (Editor).
Univelt, Inc., Tarzana, Calif. 201 pp. Price $ 15.00 (1973). Review in Sky Telescope, Vol. 47, 258 (1974).

003.043 Antonio, Arminio, Vittorio Nobile, astronomi all'Osservatorio di Capodimonte. C. N. Fiore.
Casa Editrice Aurelia, Roma. 92 pp. Price L. 2000 (1974). Review in Mem. Soc. Astron. Italiana, Nuova Ser., Vol. 44, 707 - 709; 1973/74 (*G. Abetti*).

003.044 UFOs. Interplanetary visitors. R. E. Fowler.
Exposition Press, Jericho, N.Y. 18 + 366 pp. Price $ 8.50 (1974).

003.045 An introduction to astronomy.
L. W. Fredrick, R. H. Baker.
Van Nostrand Reinhold Company Limited, London – New York – Melbourne – New Delhi – Toronto. 8th edition. 453 pp. Price $ 10.95 (1974). – Review in Sky Telescope, Vol. 48, 43 - 45; 1974 (*G. S. Mumford*).

003.046 Das Weltall. Eine moderne Kosmogonie.
C. Friedemann.
Urania-Verlag, Leipzig – Jena – Berlin. 3rd revised edition. 224 pp. Price M 6.80 (1973). – Reviews in Astron. in der Schule, 11. Jahrgang, p. 19; 1974 (*A. Muster*); SuW, Vol. 13, 32; 1974 (*J. Staude*).

003.047 Planets, stars and nebulae studied with photopolarimetry. T. Gehrels (Editor).
University of Arizona Press, Tucson, Az. 1133 pp. Price $ 27.50 (1974). – Review in Sky Telescope, Vol. 48, 46 (1974).

003.048 Dictionary of scientific biography. Vol. VIII: Jonathan Homer Lane – Pierre Joseph Macquer.
C. C. Gillispie (Editor).
Charles Scribner's Sons, New York. 12 + 624 pp. Price $ 35.00 (1973).

003.049 The analysis of tides. G. Godin.
Liverpool University Press, Liverpool, Great Britain. 264 pp. Price £ 14.00 (1973). – Review in Journ. Navigation London, Vol. 27, 129 - 131; 1974 (*N. J. Rock*).

003.050 From the black hole to the infinite universe.
D. Goldsmith, D. Levy.
Holden-Day, Inc., San Francisco. 6 + 330 pp. Price $ 6.95 (1974). – Review in Sky Telescope, Vol. 74, 258 (1974).

003.051 Introduction to the dynamics of planetary atmospheres. G. S. Golitsyn.
Gidrometeoizdat, Leningrad. 104 pp. Price 73 Kop. (1973). In Russian.

003.052 Spectral line broadening by plasmas. H. R. Griem.
Academic Press, Inc., New York. 422 pp. Price $ 31.50, £ 15.10 respectively (1974).

003.053 From Copernicus to "Copernicus". V. S. Gubarev.
Politizdat, Moskva. 128 pp. Price 23 Kop. (1973). In Russian. – Review in Priroda, No. 4.74, p. 126 (1974).

003.054 A revolution in the earth sciences: From continental drift to plate tectonics. A. Hallam.
Clarendon Press, Oxford. 7 + 127 pp. Price £ 4.00 (1973). Review in Endeavour, No. 118, Vol. 33, 50; 1974 (*D. McKenzie*).

003.055 Astrophysical concepts. M. Harwit.
John Wiley & Sons, New York – London – Sydney – Toronto. 561 pp. Price £ 6.95, $ 14.95 respectively (1973). Reviews in Astron. Nachr., Vol. 295, 201; 1974 (*K.-H. Schmidt*); Bull. Astron. Inst. Czechoslovakia, Vol. 25, 187; 1974 (*J. Palouš, V. Vanýsek*); Journ. British Astron. Ass., Vol. 84, 312; 1974 (*S. Mitton*); Journ. British Interplanet. Soc., Vol. 27, 238 (1974).

003.056 Cosmic ray physics: nuclear and astrophysical aspects. S. Hayakawa.
Translated from the English edition. Mir, Moskva. 701 pp. Price 3 Rbl. 62 Kop. (1973). In Russian. – Review in Referativ. Zhurn. 51. Astron., 5.51.651 (1974).

003.057 Der zweiten Erde auf der Spur. Signale aus unserem

Sonnensystem. H. Heuseler.
Deutsche Verlags-Anstalt, Stuttgart. 192 pp. Price DM 28.00 (1974). – Review in Orion, 32. Jahrgang, p. 133 - 134; 1974 (*E. Wiedemann*).

003.058 **Concepts of contemporary astronomy.** P. W. Hodge.
McGraw-Hill Book Company, New York. 547 pp. Price $ 9.95 (1974). – Review in Sky Telescope, Vol. 47, 406 (1974).

003.059 **Slide set for astronomy.** P. W. Hodge.
McGraw-Hill Book Company, New York. 255 slides, Price $ 225.00; Study guide $ 3.95 (1973).

003.060 **An observer's guide to the planet Mercury.** R. G. Hodgson.
Available from Rev. Hodgson, A.L.P.O., Mercury recorder and Director of Dordt College Observatory, Sioux Center, Iowa. Price $ 2.00. – Review in Strolling Astronomer, Vol. 24, 204 - 206; 1974 (*C. R. Chapman*).

003.061 **An introductory survey of the constellations.** R. G. Hodgson.
Dordt College Observatory, Sioux Center, Iowa. 34 pp. Price $ 1.50 (1973). – Review in Strolling Astronomer, Vol. 24, 242 - 243; 1974 (*B. M. Frank*).

003.062 **My father's watch: aspects of the physical world.** D. F. Holcomb, P. Morrison.
Prentice-Hall Inc., Englewood Cliffs, N. J. 390 pp. Price $ 11.50 (1974). – Review in Sky Telescope, Vol. 47, 406 (1974).

003.063 **National Maritime Museum, guide to the old Royal Observatory, Greenwich.** D. Howse.
H.M.S.O., London. 23 pp. Price 20 p. (1973). – Review in Observatory, Vol. 94, 143; 1974 (*D. Jones*).

003.064 **Nicolaus Copernicus. An essay on his life and work.** F. Hoyle.
Harper and Row Publishers, New York – London – Mexico City – Sydney. 12 + 94 pp. Price $ 5.95 (1973).

003.065 **The atmosphere of Titan.** D. M. Hunten (Editor).
National Aeronautics and Space Administration, Washington. NASA SP-340 (available from National Technical Information Service, Springfield, Va.), 177 pp. Price $ 5.25 (1974). – Review in Sky Telescope, Vol. 47, 406 (1974).

003.066 **Basic principles of plasma physics, a statistical approach.** S. Ichimaru.
W. A. Benjamin Inc., Reading, Mass. 324 pp. Price $ 19.50 hardcover; $ 12.50 paperback (1973).

003.067 **Astronomy: Fundamentals and frontiers.** R. Jastrow, M. H. Thompson.
John Wiley & Sons Inc., New York. Second edition. 519 pp. Price $ 14.50 (1974). – Review in Sky Telescope, Vol. 48, 46 (1974).

003.068 **Cosmology now.** L. John (Editor).
BBC Publications, London. 168 pp. Price £ 2.75 (1973). – Review in Nature, Vol. 247, 75; 1974 (*J. Gribbin*).

003.069 **Geography and geology of the planets (planetology).** Yu. A. Khodak.
Course of lections. Izdatel'stvo Moskovskogo gosudarstvennogo pedagogicheskogo instituta im. V. I. Lenina, Moskva, 134 pp. (1972). In Russian. – Review in Priroda, No. 2.74, p. 126; 1974 (*N. Kh. Platonov*).

003.070 **General theory of relativity.** C. W. Kilmister.
The Pergamon Group of Companies, Oxford – New York – Braunschweig. 365 pp. Price $ 11.50 hardcover, $ 6.50 paperback (1974).

003.071 **Studienbücher Physik: Astronomie.** K. Kolde.
Verlag Moritz Diesterweg/Otto Salle, Frankfurt/Main. 10 + 150 pp. Price DM 7.80 (1973). – Review in SuW, Vol. 13, 65 - 66; 1974 (*A. Kunert*).

003.072 **Mapping of the moon – past and present.** Z. Kopal, R. W. Carder.
D. Reidel Publishing Company, Dordrecht-Holland/Boston – U.S.A. Astrophysics and Space Science Library, Vol. 50, 8 + 237 pp. Price hfl 70.00 (1974). – Contents: (1) History of lunar mapping: 1600 – 1960; (2) Rotation and librations of the moon; (3) Selenographic coordinates; (4) Shape of the moon; (5) Relative elevations on the moon; (6) U.S. Air Force lunar mapping; (7) Lunar mapping at Lowell Observatory; (8) U.S. Air Force space support mapping; (9) U.S. Army lunar mapping; (10) U.S.S.R. lunar mapping; (11) National Geographic lunar mapping.

003.073 **Principles of plasma physics.** N. A. Krall, A. W. Trivelpiece.
McGraw-Hill Book Company, New York. 674 pp. Price $25,00 (1973). – Review in Phys. Today, Vol. 27, No. 3, p. 57, 59; 1974 (*J. L. Hirshfield*).

003.074 **Solar and lunar eclipses of the ancient near east from 3000 B.C. to 0 with maps.** M. Kudlek, E. H. Mickler.
Verlag Butzon & Bercker Kevelaer, Neukirchen-Vluyn, 8 + 119 pp. (1971). – Review in Journ. History Astron., Vol. 5, 136; 1974 (*G. S. Hawkins*).

003.075 **Der Almagest. Die Syntaxis Mathematica des Claudius Ptolemäus in arabisch-lateinischer Überlieferung.** P. Kunitzsch.
Otto Harrassowitz, Wiebaden. ISBN 3 447 01517 9, 16 + 384 pp. Price DM 160.00 (1974).

003.076 **Comets, meteorites and men.** P. Lancaster Brown.
Robert Hale & Company, London; Taplinger Publishing Company, New York. 255 pp. Price £3.20, $12.50 respectively (1973). – Reviews in Sky Telescope, Vol. 47, 406 (1974); Spaceflight, Vol. 16, 197; 1974 (*M. J. Anslow*).

003.077 **Sons of the blue planet.** L. Lebedev, B. Lyk'yanov, A. Romanov.
National Aeronautics and Space Administration, Washington. NASA TT-F-728 (available from National Technical Information Service, Springfield, Va.), 327 pp. Price $7.50 (1973). Review in Sky Telescope, Vol. 48, 46 (1974).

003.078 **Theoretical physics. Vol. 4: Quantum statistics and physical kinetics.** B. G. Levich.
North-Holland Publishing Company, Amsterdam – London – New York. 400 pp. Price Dfl. 47.00 (1973). – Review in Space Sci. Rev., Vol. 15, 542 - 543; 1974 (*N. G. van Kampen*).

003.079 **Astronomie selbst erlebt.** K. Lindner.
Urania-Verlag Leipzig – Jena – Berlin, Leipzig. 184 pp. Price M 12.80 (1973). – Review in Astron. in der Schule, 11. Jahrgang, p. 47; 1974 (*H. Bernhard*).

003.080 **The master of light.** D. M. Livingston.
Charles Scribner's Sons, New York, N. Y. 376 pp. Price $12.50 (1973). – Reviews in Sky Telescope, Vol. 47, 50 (1974); Vol. 47, 252 - 253; 1974 (*T. E. Bell*).

003.081 Physical science in the modern world. Part I: An overview of the physical universe; Part II: A closer look at the physical world; Part III: The twentieth century view of matter and energy. J. B. Marion.
Academic Press, Inc., New York – London. 720 pp. Price $12.95 approx. (1974).

003.082 Continental drift – the evolution of a concept. U. Marvin.
Smithsonian Institution Press, Washington. 256 pp. Price $12.50 (1973). – Review in Geochim. Cosmochim. Acta, Vol. 38, 655 - 657; 1974 (*E. R. Oxburgh*).

003.083 Our world in space. R. McCall, I. Asimov.
New York Graphic Society, Ltd., Greenwich, CT, U.S.A. 176 pp. Price $22.50 (1974). – Review in Sky Telescope, Vol. 48, 46 (1974).

003.084 A navigator's universe. The Libro de Cosmographia of 1538. P. de Medina.
Translated and with an introduction by U. Lamb. University of Chicago Press, Chicago – London. 10 + 224 pp. Price £8.35 (1972). – Essay review in Journ. History Astron., Vol. 5, 54 - 56; 1974 (*L. de Albuquerque*).

003.085 Astrospectroscopy – the language of the universe. O. A. Mel'nikov, V. S. Popov.
Znanie, Moskva. 62 pp. Price 10 Kop. (1973). In Russian.

003.086 The cosmology of Giordano Bruno.
P. H. Michel. Translated by R. E. W. Maddison. Hermann, Paris; Methuen & Co. Ltd., London; Cornell University Press, Ithaka, N. Y. 306 pp. Price £4.50 (1973). Reviews in Journ. History Astron., Vol. 5, 141 (1974); Nature, Vol. 249, 865 - 866; 1974 (*E. G. Forbes*).

003.087 Cosmic rays in the interplanetary space. L. I. Miroshnichenko.
Nauka, Moskva. 158 pp. Price 53 Kop. (1973). In Russian. – Review in Priroda, No. 6.74, p. 110 (1974). In Russian.

003.088 Yearbook of astronomy 1974. P. Moore (Editor).
Sidgwick and Jackson, London; W. W. Norton and Co., Inc., New York, N.Y. 204 pp. Price £2.25, $6.50 respectively (1973). – Reviews in Journ. History Astron., Vol. 5, 142 (1974); Strolling Astronomer, Vol. 24, 203; 1974 (*J. R. Smith*).

003.089 An introduction to celestial mechanics. F. R. Moulton.
Dover Publications, Inc., New York. 2nd edition. 436 pp. Price $3.00 (1972). – Review in Phys. Blätter, 30. Jahrgang, p. 191; 1974 (*H. J. Fahr*).

003.090 Grundzüge der Astronomie. B. Müller.
BSB B. G. Teubner Verlagsgesellschaft, Leipzig. 188 pp. Price M 8.90 (1973). – Review in Astron. in der Schule, 11. Jahrgang, p. 18 - 19; 1974 (*K.-G. Steinert*).

003.091 Bernhard Schmidt. P. Müürsepp.
Kirjastus 'Valgus', Tallinn, Estonia. 48 pp. Price 17 Kop. (1972). In Estonian. – Review in Journ. History Astron., Vol. 5, 65; 1974 (*E. Eelsalu*).

003.092 The special theory of relativity. H. Muirhead.
Halsted (Wiley), New York. 12 + 164 pp. Price $12.75 (1973).

003.093 Nicolaus Copernicus Gesamtausgabe. Band I: De Revolutionibus (Faksimile des Manuskriptes). Edited im Auftrage der Kommission für die Copernikus-Gesamtausgabe by H. M. Nobis.
Verlag Dr. H. A. Gerstenberg, Hildesheim. 14 + 56 + ca 213 pp. Price DM 140.00 (1974).

003.094 The backyard astronomer. A. E. Nourse.
Franklin Watts, Inc., New York. 118 pp. Price $7.95 (1973). – Reviews in Sky Telescope, Vol. 47, 120 (1974); Sky Telescope, Vol. 47, 328 - 329; 1974 (*W. R. Benton*).

003.095 L'univers et ses métamorphoses. R. Omnès.
Hermann, Paris. 184 pp. Price F 26.00 (1973).

003.096 Ideas from astronomy. L. W. Page.
Addison-Wesley Publishing Co. Inc., Reading, Mass.- London – Amsterdam – Canada. 250 pp. Price $3.36 (1973). Reviews in Sky Telescope, Vol. 47, 188 (1974); Vol. 47, 404 - 405; 1974 (*J. Sternig*).

003.097 The many-body problem. W. E. Parry.
Oxford University Press, New York. 217 pp. Price $20.00 (1973).

003.098 J. Kepler, Dissertatio cum Nuncio Sidereo, and Narratio de quattuor Jovis satellibus.
Latin text, translated with introduction and notes by E. Pasoli, G. Tabarroni. Bottega D'Erasmo, Turin. 12 + 159 pp. Price 12,000 Lire (1972). – Essay review in Journ. History Astron., Vol. 5, 58 - 60; 1974 (*W. R. Shea*).

003.099 Reflecting theory and natural science. T. Pavlov (Editor).
AN SSSR. In-t filos., Bolg. AN In-t filos. Sofiya, 'Nauka i izkustvo'. 348 pp. (1973). In Russian. – Review in Referativ. Zhurn. 51. Astron., 5.51.1 (1974).

003.100 Raumflugkörper. Ein Typenbuch. H. Pfaffe, P. Stache.
Transpress VEB Verlag für Verkehrswesen, Berlin. 2nd revised edition. 271 pp. Price M 16.80 (1973). – Review in Astron. in der Schule, 11. Jahrgang, p. 47; 1974 (*M. Schukowski*).

003.101 Interstellar communication. C. Ponnamperuma, A. G. W. Cameron.
Houghton Mifflin Company, Boston. 226 pp. Price $5.95 (1974). – Review in Sky Telescope, Vol. 48, 46 (1974).

003.102 Methods of experimental physics. Vol. 10: Physical principles of far-infrared radiation. C. L. Robinson.
Academic Press, Inc., New York. 460 pp. Price $29.00 (1973). Review in Phys. Today, Vol. 27, No. 1, p. 78 - 79; 1974 (*K. D. Moeller*).

003.103 Illustrated sources in history: Astronomy. C. A. Ronan.
David & Charles, Newton Abbot; Barnes & Noble Books, New York. 112 pp. Price £3.25 (1973). – Review in Journ. History Astron., Vol. 5, 65; 1974 (*A. Armitage*).

003.104 Astrophysics. W. K. Rose.
Holt, Rinehart and Winston Inc., New York. 14 + 287 pp. Price $14.00 (1973). – Review in Phys. Today, Vol. 27, No. 4, p. 93 - 94; 1974 (*R. C. Henry*).

003.105 La couronne solaire. J.-P. Rozelot, with a preface by J.-C. Pecker.
Gaston Doin et Cie, Paris. 144 pp. (1973). – Review in L'Astronomie, 88ᵉ année, p. 220; 1974 (*B. Morando*).

003.106 **Copernicus. Man and thought.** E. Rybka, P. Rybka. Translated from the Polish edition by Yu. Danilov, A. Bondarev. Mir, Moskva. 326 pp. Price 1 Rbl. 96 Kop. (1973). In Russian. – Review in Priroda, No. 5.74, p. 124 (1974).

003.107 **Filmstrip: Exploring the planets.** I. Ridpath, with drawings by D. A. Hardy. Hulton Educational Publications Ltd., Amersham. 35 frames (colour). Price £2.00 (1973). – Review in Journ. British Astron. Ass., Vol. 84, 227 - 228; 1974 (*E. A. Beet*).

003.108 **Gravitation waves in Einstein's theory of gravitation.** V. D. Sacharow. John Wiley & Sons Limited, Chichester, GB. 200 pp. Price £5.85 (1973). – Review in Phys. Blätter, 30. Jahrgang, p. 237; 1974 (*R. Meinhardt*).

003.109 **The cosmic connection – an extraterrestrial perspective.** C. Sagan. Anchor Press Doubleday, New York. 14 + 274 pp. Price $7.95 (1973). – Reviews in Journ. British Interplanet. Soc., Vol. 27, 313 - 314; 1974 (*A. R. Martin*); Science, Vol. 184, 663 - 664; 1974 (*W. K. Hartmann*); Sky Telescope, Vol. 47, 188 (1974).

003.110 **Cargèse lectures in physics. Vol. 6.** E. Schatzman (Editor). Gordon and Breach Science Publishers Ltd., London. 731 pp. Price £20.30 (1973). – Review in Solar Physics, Vol. 36, 239; 1974 (*C. de Jager*).

003.111 **Cosmic gas dynamics.** E. Schatzman, L. Biermann. John Wiley & Sons Inc., New York. 291 pp. Price $14.95 (1974).

003.112 **Physik der Sterne und der Sonne.** H. Scheffler, H. Elsässer. B.I. Wissenschaftsverlag, Bibliographisches Institut, Mannheim – Wien – Zürich. 535 pp. Price DM 48.00 (1974).

003.113 **Physik des Erdkörpers.** B. Schick, G. Schneider. Ferdinand Enke Verlag, Stuttgart. 267 pp. Price DM 59.00 (1973). – Review in Naturwissenschaften, 61. Jahrgang, p. 223; 1974 (*G. Angenheister*).

003.114 **Sonnenuhren.** H. Schumacher. Verlag Georg D. W. Callwey, München. 182 pp. Price DM 58.00 (1973). – Reviews in Sky Telescope, Vol. 47, 50 (1974); Vol. 47, 186 - 187; 1974 (*G. Lovi*); SuW, Vol. 13, 32 - 33; 1974 (*K. Schaifers*).

003.115 **Weisse Zwerge – Schwarze Löcher. Eine Einführung in die relativistische Astrophysik und Kosmologie.** R. U. Sexl, H. Sexl. Available from Institut für Theoretische Physik, Wien. 230 pp. Price öS 50.00 (1974). – Review in Sternenbote, 17. Jahrgang, p. 127 - 128 (1974).

003.116 **Galileo.** D. Shapere. University of Chicago Press, Chicago, Ill. 161 pp. Price $9.75 (1974). – Review in Sky Telescope, Vol. 48, 46 (1974).

003.117 **Kapitoly z astrofyziky.** M. Široká, J. Široký. Státní pedagogické nakladatelství, Praha. 171 pp. Price Kčs 9.50 (1973).

003.118 **Soviets in space.** P. L. Smolders. Taplinger Publishing Co., Inc., New York. 285 pp. Price $9.95 (1974). – Review in Sky Telescope, Vol. 47, 406 (1974).

003.119 **Himmelsmechanik, Band III. Allgemeine Störungen.** K. Stumpff under cooperation of J. Meffroy. VEB Deutscher Verlag der Wissenschaften, Berlin. Hochschulbücher für Physik, Band 41. 600 pp. Price DM 130.00 (1974).

003.120 **The legacy of Mikolaj Kopernik. One man's love affair with the universe.** C. J. Szymczak. Great Lakes Publishing Co., Milwaukee, Wis. 18 + 84 pp. Price $3.00 (1973).

003.121 **The moon.** J. H. Tatsch. Tatsch Associates, Sudbury, Mass. 338 pp. Price $10.00 (1974). – Review in Sky Telescope, Vol. 47, 406 (1974).

003.122 **Concise dictionary of physics.** J. Thewlis (Editor). Pergamon Press Ltd., Oxford. 8 + 366 pp. Price £ 5.50 (1973). Review in Journ. British Astron. Ass., Vol. 84, 307 - 308; 1974 (*C. A. Ronan*).

003.123 **A commentary on the Dresden Codex: a Maya hieroglyphic book.** J. E. S. Thompson. Memoirs of the American Philosophical Society, Philadelphia. 156 pp. Price $ 25.00 (1972). – Review in Journ. History Astron., Vol. 5, 137 - 138; 1974 (*D. Pingree*).

003.124 **Wave propagation.** I. Tolstoy. McGraw-Hill Book Company, New York. 466 pp. Price $ 18.50 (1973). – Review in Phys. Today, Vol. 27, No. 4, p. 85, 86 - 87; 1974 (*A. O. Williams, Jr.*).

003.125 **Science awakening II (the birth of astronomy).** B. L. van der Waerden. Noordhoff International Publishing, Leiden, Netherlands. 14 + 347 pp. Price Dfl 98.00 (1974).

003.126 **Atmospheric energetics.** J. Van Mieghem. Oxford University Press, New York. 306 + 9 pp. Price $ 24.00 (1973). – Review in Icarus, Vol. 21, 210 - 211; 1974 (*J. R. Holton*).

003.127 **Man explores the planets.** N. A. Varvarov. Mashinostroenie, Moskva. 191 pp. Price 44 Kop. (1973). In Russian. – Review in Referativ. Zhurn. 51. Astron., 3.51.49 (1974).

003.128 **The invisible universe.** G. L. Verschuur. Springer-Verlag New York Inc., New York. 190 pp. Price $ 5.90 (1974).

003.129 **Galactic and extra-galactic radio astronomy.** G. L. Verschuur, K. I. Kellermann (Editors). Springer-Verlag New York Inc., New York. 440 pp. Price $ 37.80 (1974).

003.130 **Sundials: their theory and construction.** A. E. Waugh. Dover Publications, Inc., New York. 228 pp. Price $ 3.50 (1973). – Review in Sky Telescope, Vol. 47, 399 - 401; 1974 (*J. A. Eddy*).

003.131 **A random walk in science.** R. L. Weber (Compiler). Crane, Russak & Company, Inc., New York. 206 pp. Price $ 12.50 (1973). – Review in Sky Telescope, Vol. 48, 46 (1974).

003.132 **The unpublished first version of Isaac Newton's Cambridge lectures on optics, 1670 - 1672.**
Introduction by D. T. Whiteside.
University Library, Cambridge, England. 10 + 129 pp. Price £ 10.00 (1973). − Review in Journ. History Astron., Vol. 5, 142 (1974).

003.133 **Kant's cosmogony** as in his essay on the retardation of the rotation of the earth and his natural history and theory of the heavens.
Translated by W. Hastie, with a new introduction by G. J. Whitrow.
Johnson Reprint Corporation, New York − London. Sources of Science, No. 133. 40 + 205 pp. (1970).

003.134 **The science of astronomy.** J. A. Woods, D. R. Hazard.
Harper & Row Publishers, New York − London − Mexico City − Sidney. 466 pp. Price $ 8.50 (1974). − Review in Sky Telescope, Vol. 48, 46 (1974).

003.135 **Progress in optics, Vol. XI.** E. Wolf (Editor).
Elsevier Publishing Company, Amsterdam − London − New York. 372 pp. Price Dfl. 100.00 (1973).

003.136 **Gravitational waves in Einstein's theory.** V. D. Zakharov.
Translated from Russian by R. N. Sen.
Israel Program for Scientific Translations, Jerusalem − London. Halsted Press (a division of John Wiley & Sons, Inc.), New York. 20 + 183 pp. (1973). − Contents: (1) Approximation methods for the investigation of gravitational waves; (2) The Cauchy problem for Einstein's equations; (3) Gravitational waves: substance of the problem; (4) Pirani's criterion; (5) Bel's criteria; (6) Lichnerowicz's criterion; (7) The Zel'manov criterion; (8) Other criteria for gravitational fields; (9) Propagation of gravitational waves; (10) Plane gravitational waves defined by an absolutely parallel vector field; (11) The asymptotic properties of fields of gravitational radiation; (12) Gravitational waves and chronometric invariants; (13) The problem of gravitational waves and physical experiment.

003.137 **Dust in the atmosphere and in circumterrestrial space.**
Trudy Simpoz. po mezhplanet. pyli rab. gruppy po optich. nestabil'n. zemn. atmosf. Astrosoveta AN SSSR, Pulkovo, dek. 1970 g. Nauka, Moskva. 212 pp. Price 1 Rbl. 63 Kop. (1973). In Russian.

003.138 **Comets.** R. Ash, I. Grant.
Ash and Grant, London. 48 pp. Price 60 p. (1973). Distributed by Universal-Tandem Publishing Company Limited, London. − Review in Journ. British Astron. Ass., Vol. 84, 147; 1974 (C. A. Ronan).

003.139 **J. C. Poggendorff: Biographisch-literarisches Handwörterbuch der exakten Naturwissenschaften.**
Edited by the Sächsischen Akademie der Wissenschaften zu Leipzig under the redaction of H. Salié.
Akademie-Verlag, Berlin. Band VIIb - Teil 5, 1. Lieferung, 160 pp. Price DM 24.00; VIIb - Teil 5, 2. Lieferung, 160 pp. Price DM 24.00 (1974).

003.140 **Bildatlas des Sonnensystems. Ferne Welten nah gesehen.** B. Stanek, L. Pesek.
Hallwag Verlag, Bern − Stuttgart. 202 pp. Price DM 58.00, Fr. 58.00 respectively (1974). − Reviews in Orion, 32. Jahrgang, p. 134; 1974 (E. Wiedemann); Umschau, 74. Jahrgang, p. 429; 1974 (D. Eschenbach).

003.141 **Besucher aus dem All. Das Geheimnis der unbekannten Flugobjekte.** A. Schneider.
Hermann Bauer Verlag, Freiburg. 364 pp. Price DM 28.00. Review in SuW, Vol. 13, 65; 1974 (J. Staude).

003.142 **The science of astronomy.**
Harper & Row, New York. 466 pp. Price $ 8.50 (1974).

003.143 **Advanced scanners and imaging systems for earth observations.**
National Aeronautics and Space Administration, Goddard Space Flight Center, NASA SP-335. [Available from U.S. Government Printing Office, Washington, D.C.]. 604 pp. Price $ 3.90 (1973). − Review in Sky Telescope, Vol. 47, 258 (1974).

003.144 **Apollo 17: preliminary science report.**
National Aeronautics and Space Administration, Washington. NASA SP-330, [may be ordered from the Superintendent of Documents, U.S. Government Printing Office, Washington, D.C.]. 680 pp. Price $ 7.95 (1973). − Review in Sky Telescope, Vol. 47, 406 (1974).

003.145 **Discoveries in the depths of space.** B. Bova.
Translated into Hebrew from the English edition 'The new astronomies' (1971) by Y. Kirsch.
Am Oved Publishers Ltd., Tel Aviv, Israel. 206 pp. Price IL 12.00 (1974).

003.146 **Woodmansterne colour slides-set MN07-Apollo 17.**
Woodmansterne Publication Ltd. Watford. Price £ 0.85 (1973). − Review in Journ. British Astron. Ass., Vol. 84, 308 - 309; 1974 (H. Miles).

003.147 **NASA-Zeiss-Farbdias.** Ser. 6 (Apollo 16), Ser. 7 (Apollo 17).
Astro-Bilderdienst SAG, Burgdorf. Price sFr. 19.00 (1974).

003.148 **A line to the sun.** (Film, 16-mm, colour, sound).
A. Sidi Productions, Leeds. Hire. £ 5.80 per showing. Purchase £ 195.00 (1973). − Review in Journ. British Astron. Ass., Vol. 84, 309 (1974).

003.149 **Leuchtglobus "Planet Erde".**
B. Diewerge, with an introduction by W. Gleissberg.
Columbus-Verlag, Paul Oestergaard, Berlin − Stuttgart. ∅ = 34 cm. Price DM 138.00. − Review in SuW, Vol. 13, 141 - 142; 1974 (A. Kunert).

003.150 **Simple astronomy.** I. Nicolson, D. Pottinger.
Nelson and Sons, London; Charles Scribner's Sons, New York. 64 pp. Price £ 1.95, $ 6.95 respectively (1973). Review in Journ. British Astron. Ass., Vol. 84, 147 - 148; 1974 (C. A. Ronan).

003.151 **Nicholas Copernicus: Complete works. I.**
Polish Academy of Science, Warsaw − Cracow; Macmillan, London. 25 + 439 pp. (1972). − Review in Science, Vol. 184, 660 - 661 (1974).

003.152 **Astronomy of star positions.** H. Eichhorn.
Frederick Ungar Publishing Co. Inc., New York, N.Y. 357 pp. Price $ 25.00 (1974).

004 History of Astronomy, Chronology

004.001 **On the establishment of astronomy at the Academy of Sciences.** B. V. Kukarkin, Z. L. Ponizovskij.
Priroda, No. 1.74, p. 44 - 49 (1974). In Russian.

004.002 **A visit to Tycho Brahe's observatory.** E. Simonsen.
Sky Telescope, Vol. 47, 86 - 88 (1974).

004.003 **Ireland's Birr Castle today.**
Sky Telescope, Vol. 47, 89 - 90 (1974).

004.004 **The astrolabe.** J. D. North.
Sci. American, Vol. 230, No. 1, p. 96 - 106 (1974).
This scientific instrument of the Middle Ages was used for both astronomical and terrestrial observations. It also served as an analogue computer, particularly for determining the local time.

004.005 **Keplerian planetary eggs, laid and unlaid, 1600 – 1605.** D. T. Whiteside.
Journ. History Astron., Vol. 5, 1 - 21 (1974).

004.006 **The foundation of the first Göttingen observatory: a study in politics and personalities.**
E. G. Forbes.
Journ. History Astron., Vol. 5, 22 - 29 (1974).

004.007 **The Kermario alignments.** A. Thom, A. S. Thom.
Journ. History Astron., Vol. 5, 30 - 47 (1974).

004.008 **The origin of the Gregorian civil calendar.**
N. Swerdlow.
Journ. History Astron., Vol. 5, 48 - 49 (1974).

004.009 **A megalithic lunar observatory in Islay.** A. Thom.
Journ. History Astron., Vol. 5, 50 - 51 (1974).

004.010 **Adriaan van Maanen on the significance of internal motions in spiral nebulae.** N. S. Hetherington.
Journ. History Astron., Vol. 5, 52 - 53 (1974).

004.011 **The Maskelyne manuscripts at the Royal Greenwich Observatory.** E. G. Forbes.
Journ. History Astron., Vol. 5, 67 - 69 (1974).

004.012 **Lunar-solar calendar in primeval Armenia.**
A. A. Martirosyan.
Vestn. obshchestv. nauk. AN ArmSSR, 1973, No. 7, p. 23 - 42. In Armenian. – Abstr. in Referativ. Zhurn. 51. Astron., 3.51.3 (1974).

004.013 **Abu-Rajkhan Biruni.** N. Abubakirov.
Nauka i zhizn', 1973, No. 9, p. 49 - 57. In Russian.

004.014 **Copernicus and his epochal discovery.**
Priroda i znanie (NRB), Vol. 26, No. 5, p. 2 - 7 (1973). In Bulgarian. – Abstr. in Referativ. Zhurn. 51. Astron., 3.51.14 (1974).

004.015 **The work of Nicolaus Copernicus.** F. Hoyle.
The planets today. Symposium 1973, (see 012.003), p. 105 - 114 (1974).

004.016 **Michelson and his interferometer.**
R. S. Shankland.
Phys. Today, Vol. 27, No. 4, p. 36 - 39, 41, 43, with a correction in Vol. 27, No. 5, p. 71 (1974).
Pioneering applications in such diverse fields as astronomy, atomic spectra and mensuration followed the initial disappointment over the failure to detect a luminiferous ether.

004.017 **100 years Astronomical Institute of the Uzbek Academy of Sciences.** V. P. Shcheglov.
Zemlya i Vselennaya, 1974, No. 2, p. 46 - 52. In Russian.

004.018 **Urgeschichtliche Ortung im Bergland.**
G. Innerebner.
SuW, Vol. 13, 59 - 60 (1974).

004.019 **Ein außergewöhnliches Osterdatum.**
W. Seggewiß.
SuW, Vol. 13, 92 - 93 (1974).

004.020 **The story of Groombridge 1830.** J. Ashbrook.
Sky Telescope, Vol. 47, 296 - 297 (1974).

004.021 **Nocturnal: Instrument voor het opnemen van de tijd.** W. Kastelein.
Zenit, Vol. 1, No. 4, p. 29 - 31 (1974).

004.022 **Koperniks un ciņa par heliocentrisko pasaules uzskatu.** U. Dzērvītis.
Zvaigžņota debess, 1973. gada rudens, p. 1 - 10.

004.023 **The history of the planet Vulcan.** B. Hellyer.
Journ. British Astron. Ass., Vol. 84, 192 - 193 (1974).

004.024 **The authenticity of Ptolemy's parallax data – Part II.** R. R. Newton.
Quarterly Journ. Roy. Astron. Soc., Vol. 15, 7 - 27 (1974).

004.025 **Copernikanisches und ptolemäisches Weltbild vom Standpunkt der allgemeinen Relativitätstheorie.**
G. Dautcourt.
Sterne, 50. Jahrgang, p. 30 - 34 (1974).

004.026 **On a little known paper of Rittenhaus on diffraction (To the hundred-seventhy-fifth anniversary of his death).**
Sb. statej. Sukhum. gos. ped. in–t. Sukhumi, 1973, p. 247 - 252. In Russian. – Abstr. in Referativ. Zhurn. 51. Astron., 4.51.34 (1974).

004.027 **La méthode de Kepler est-elle une non-méthode?**
J.-C. Pecker.
L'Astronomie, 88e année, p. 2 - 16 (1974).

004.028 **Retour sur Copernic, Kepler, Bessel et les parallaxes..**
J.-C. Pecker.
L'Astronomie, 88e année, p. 83 - 92 (1974).

004.029 **Jean-Baptiste Donati et la grande comète de 1858.**
J. Pernet.
L'Astronomie, 88e année, p. 93 - 98 (1974).

004.030 **Stonehenge.**
A. Thom, Ar. S. Thom, Al. S. Thom.
Journ. History Astron., Vol. 5, 71 - 90 (1974).

004.031 **Measurement of time in ancient and mediaeval Armenia.** B. E. Tumanian.
Journ. History Astron., Vol. 5, 91 - 98 (1974).

004.032 The nova of A.D. 1006 in European and Arab records. N. A. Porter.
Journ. History Astron., Vol. 5, 99 - 104 (1974).

004.033 Saturn and his anses. A. Van Helden.
Journ. History Astron., Vol. 5, 105 - 121 (1974).

004.034 Cosmological teaching in the seventeenth-century Scottish universities, Part 1. J. L. Russell.
Journ. History Astron., Vol. 5, 122 - 132 (1974).

004.035 Veränderliche Epizyklen von Āryabhata bis Copernicus. W. Petri.
Mitt. Astron. Ges., No. 34, p. 91 (1973/74). – Abstr. AG.

004.036 Copernicus' indflydelse på Tycho Brahe. K. P. Moesgaard.
Astron. Tidssk., Årg. 7, p. 1 - 16 (1974).

004.037 Har Eratosthenes mätt jordens storlek? P. Collinder.
Astron. Tidssk., Årg. 7, p. 49 - 58 (1974).

004.038 Der Beitrag von Immanuel Kant zur Entwicklung der Planetenkosmogonie. G. Jackisch.
Astron. in der Schule, 11. Jahrgang, p. 5 - 8 (1974).

004.039 Die philosophische Bedeutung von Kants kosmogonischen Vorstellungen. R. Wahsner.
Astron. in der Schule, 11. Jahrgang, p. 8 - 12 (1974).

004.040 Tobias Mayer und die Gründung der ersten Sternwarte zu Göttingen. E. G. Forbes.
SuW, Vol. 13, 191 - 196 (1974).

004.041 A computerized checklist of astrolabes. S. L. Gibbs.
Journ. History Astron., Vol. 5, 143 (1974).

004.042 Kritische Bemerkungen zu zwei Texten über die Sonnenbewegung. W. Hübner.
Centaurus, Vol. 17, 253 - 259 (1973).

The author checks the numerical contents of two manuscripts from the Catalogus codicum astrologorum Graecorum, one of the XIth century, written by Paulus, the other of the VIth century.

004.043 Theory and observation in medieval astronomy. B. R. Goldstein.
Isis, Vol. 63, 39 - 47 (1972).

This paper concerns examples of observations of astronomical phenomena not discussed in the 'Almagest' such as planetary brightness, transits, and comets that are found in medieval astronomical treatises in order to solve certain problems arising out of Ptolemaic astronomy.

004.044 500 Jahre Copernicus. W. Grossmann.
Zeitschr. Vermessungswesen, Vol. 98, 341 - 346 (1973).

004.045 Ibn Yūnus' very useful tables for reckoning time by the sun. D. A. King.
Arch. History Exact Sci., Vol. 10, 342 - 394 (1973).

This paper consists of a description and mathematical analysis of the main corpus of tables which was used in medieval Cairo for timekeeping by the sun and regulating the astronomically defined times of Muslim prayer. The study is based on seven Arabic manuscripts located in libraries in Cairo and Europe.

004.046 Die "Ägypter" und die "Chaldäer". B. L. van der Waerden.
Sitzungsber. Heidelberger Akad. Wiss., Math. - Nat. Kl. 1972, 5th paper, 31 pp. (1972).

004.047 On the ancient astronomy and mathematics. E. Maula.
Arkhimedes 1972, No. 1 - 2, p. 45 - 69 (1972). In Finnish.

004.048 Levi ben Gerson's lunar model. B. R. Goldstein.
Centaurus, Vol. 16, 257 - 284 (1972).

The author is concerned with the lunar tables from Levi ben Gerson's (1288 - 1344) astronomical treatise, which is preserved in four Hebrew manuscripts and in four manuscripts of the medieval Latin translation; the lunar tables are included in two Hebrew manuscripts only. Levi ben Gerson invented a new lunar model that eliminates some contradictions between the theory of Ptolemy and the results of the observation. In addition the author gives the translation of three fragments from the work of Levi ben Gerson.

004.049 Copernicanesimo e filosofia nel Rinascimento. R. Migliavacca.
Coelum, Vol. 42, 64 - 72 (1974).

004.050 Celebrating a quinquecentennial. O. Gingerich.
Science, Vol. 184, 660 - 663 (1974). – The author reviews some publications and symposia which have appeared on occasion of the 500th birthday of Copernicus.

004.051 The harmony of the spheres. B. R. Gaizauskas.
Journ. Roy. Astron. Soc. Canada, Vol. 68, 146 - 151 (1974).

004.052 Astroarchaeology and megalithic civilizations. C. Maxia, E. Proverbio.
Scientia – Rivista Sci., Anno 1972, (Sept./Oct.), 5 pp. = Pubbl. Stazione Astron. Internazionale Latitudine, Carloforte-Cagliari, Nuova Ser., No. 26.

004.053 Orientamenti astronomici di monumenti nuragici. C. Maxia, E. Proverbio.
Rend. Ist. Lombardo, Accad. Sci. Lettere, Cl. Sci. A, Vol. 107, 298 - 311 = Pubbl. Stazione Astron. Internazionale Latitudine Carloforte-Cagliari, Nuova Ser., No. 28 (1973).

004.054 Kants Kosmologie und der physische Teil des naturwissenschaftlichen Weltbildes. H.-J. Treder.
Sterne, Vol. 50, 65 - 73 (1974).

004.055 Die Kosmogonie Immanuel Kants. I. H. Lambrecht.
Sterne, Vol. 50, 74 - 81 (1974).

004.056 Kants Beitrag zur Frage der Verzögerung der Erdrotation. H.-J. Felber.
Sterne, Vol. 50, 82 - 90 (1974).

004.057 Frühe Spektralanalyse von Fraunhofer bis Kirchhoff. H.-U. Fuchs.
Orion, 32. Jahrgang, p. 98 - 102 (1974).

004.058 Encyclopaedist from Middle Asia. To the 1000th anniversary of Al-Biruni's birthday.
L. B. Alaev.
Byul. Komis. SSSR po delam YUNESKO, 1973, No. 3 (13), p. 26 - 27. In Russian. – Abstr. in Referativ. Zhurn. 51. Astron., 5.51.3 (1974).

004.059 Kant and modern concepts of the universe. G. M. Idlis.
Priroda, No. 6.74, p. 73 - 81 (1974). In Russian.

004.060 **Pleiades in Hesiod's works and days.** A. Fresa.
Reprinted from Atti Accad.Pontaniana, Nuova Ser., Vol. 22 = Ist. Univ. Navale, Astron. Generale e Sferica, Napoli, Seminario No. 7, 11 pp. (1973). In Italian.

In the introduction to "Agriculture", Hesiod presents an agricultural calender in connection with Pleiades. The author, considering Alcyone the most bright star of this asterism, calculates the dates of its seasonal phenomena: the heliacal rising and setting, and the cosmic setting. These dates have been calculated for the time 800 B.C. and for localities of Greece situated near the 39° and 40° parallel, namely as regards the epoch and region where Hesiod lived.

004.061 **La geometria del sistema planetario all'epoca di Copernico.** T. Nicolini.
Reprinted from Atti Accad.Pontaniana, Nuova Ser., Vol. 22 = Ist. Univ. Navale, Astron. Generale e Sferica, Napoli. Seminario No. 8, 24 pp. (1973).

004.062 **Difusión del sistema de Copérnico en el mundo.** A. Romañá.
Publ. Obs. Ebro, *Tortosa*, Miscelánea, No. 31, 31 pp. (1973).

004.063 **Astronomical alignment of the Big Horn Medicine Wheel.** J. A. Eddy.
Science, Vol. 184, 1035 - 1043 (1974).

Cairns of an unexplained Amerindian rock pattern appear to have been aligned to the summer solstice.

004.064 **Copernicus.** J. R. Ravetz.
Journ. British Astron. Ass., Vol. 84, 257 - 271 (1974). − Christmas lecture 1973.

004.065 **A hole in the sky.** J. Ashbrook.
Sky Telescope, Vol. 48, 9 - 10 (1974).

004.066 **Nicolaus Copernicus.** C. J. Macris.
In honorem S. Placidis, (see 003.009), p. 209 - 226 (1974). In Greek.

004.067 **1000th anniversary of Abu-Rajkhan Biruni's birthday (4. 9. 973 − 13. 12. 1048).**
M. S. Dimitrijević.
Vasiona, Vol. 22, 39 - 41 (1974). In Serbo-Croatian.

004.068 **Heliocentric theory in China. − In commemoration of the quincentenary of the birth of Nicolaus Copernicus.** T. Hsi, T. Yen, S. Po, C. Wang, C. Chen, M. Chen.
Scientia Sinica, Vol. 16, 364 - 376 (1973). − This paper has been read to a forum held by Academia Sinica and the Chinese Astronomical Society on June 22, 1973 to mark the quincentenary of the birth of Nicolaus Copernicus.

004.069 **La rivoluzione astronomica Copernicana.** W. Shea.
Mem Soc. Astron. Italiana, Nuova Ser., Vol. 44, 671 - 687 (1973/74).

004.070 **Einige physikalische und erkenntnistheoretische Aspekte der Arbeiten Isaac Newtons.**
G. Jackisch.
Astron. in der Schule, 11. Jahrgang, p. 53 - 56 (1974).

004.071 **Ancient Slaves and astronomy.** E. Javorka.
Kozmos, Vol. 5, 20 - 22, 52 - 53 (1974). In Slovak.

004.072 **What did the Copernicus anniversary bring us?** K. Ziołkowski.
Urania Kraków, Vol. 45, 2 - 5 (1974). In Polish.

004.073 **Was Copernicus a revolutionary?**
J. Zieleniewski.
Urania Kraków, Vol. 45, 10 - 13 (1974). In Polish.

004.074 **The library of Nicolaus Copernicus in Toruń.** L. Jarzębowski.
Urania Kraków, Vol. 45, 102 - 106 (1974). In Polish.

004.075 **Die arabischen Sternbilder des Südhimmels.** P. Kunitzsch.
Separate print from "Der Islam" [Walter de Gruyter, Berlin − New York], Vol. 51, No. 1, p. 37 - 54 (1974).

004.076 **Die Venustheorie von Vasistha in der Pancasiddhantika des Varaha Mihira.** P. B. Wirth.
Centaurus, Vol. 18, 29 - 43 (1974).

004.077 **Joseph-Nicolas Delisle (1688−1768).** N. I. Nevskaja.
Revue Histoire Sci., Vol. 26, 289 - 313 (1974).

This article examines the principal facets of the scientific activity of Joseph-Nicolas Delisle: history of astronomy; astronomy; celestial mechanics; astrophysics; meteorology and physics; geodesy; cartography and geography; oriental studies. The author emphasizes the influence of the French astronomer on the education of scientific personnel in Russia during the first half of the 18th century.

004.078 **Die Prä- und Post-Kopernikaner von Oberösterreich.** A. Adam.
Beiträge zur Kopernikusforschung, Katalog Oberösterreich. Landesmuseum No. 86, p. 4 - 6 (1973).

004.079 **Johannes von Gmunden − Georg von Peuerbach. Ihre geistigen Auswirkungen auf Oberösterreich, im besonderen auf die Klöster St. Florian, Kremsmünster und Wilhering.** H. Jung.
Beiträge zur Kopernikusforschung, Katalog Oberösterreich. Landesmuseum No. 86, p. 7 - 24 (1973).

004.080 **Der "Tractatus Cylindri" des Johannes von Gmunden.** Introduced, edited and translated by K. Ferrari d'Occhieppo, P. Uiblein.
Beiträge zur Kopernikusforschung, Katalog Oberösterreich. Landesmuseum No. 86, p. 25 - 85 (1973).

004.081 **The authenticity of Ptolemy's eclipse and star data.** R. R. Newton.
Quarterly Journ. Roy. Astron. Soc., Vol. 15, 107 - 121 (1974).

004.082 **'Long Meg and her nine daughters'.** K. Menzel.
Journ. Astron. Soc. Western Australia, Vol. 25, May, p. 3 - 9 (1974).

J. E. Bodes Sternatlas 1782. Vorstellung der Gestirne auf XXXIV Kupfertafeln. See Abstr. 003.023.

Il soggiorno di Nicolò Copernico in Italia. See Abstr. 003.035.

From Copernicus to "Copernicus". See Abstr. 003.053.

Der Almagest. Die Syntaxis Mathematica des Claudius Ptolomäus in arabisch-lateinischer Überlieferung. See Abstract 003.075.

J. Kepler, Dissertatio cum Nuncio Sidereo and Narratio de quattuor Jovis satellibus. See Abstr. 003.098.

Copernicus. Man and thought. See Abstr. 003.106.

Galileo.   See Abstr. 003.116.

The legacy of Míkolaj Koperník. One man's love affair with the universe.   See Abstr. 003.120.

A commentary on the Dresden Codex: a Maya hieroglyphic book.   See Abstr. 003.123.

The unpublished first version of Isaac Newton's Cambridge lectures on optics, 1670 - 1672.
See Abstr. 003.132.

Wilhelm Foersters chronologisches Werk.
See Abstr. 005.013.

Nikolaus Copernicus und sein Werk.
See Abstr. 005.017.

005   Biography

005.001   Something about a Connecticut amateur.
J. Ashbrook.
Sky Telescope, Vol. 47, 26 - 27 (1974).

005.002   20th-century astronomer.   D. P. Cruikshank.
Sky Telescope, Vol. 47, 159 - 164 (1974). — Concerning G. P. Kuiper.

005.003   Nikolas Copernicus (To the 500th anniversary of his birthday).   D. Ya. Martynov.
Vestn. Mosk. un-ta. Fiz. Astron., Vol. 14, 387 - 396 (1973). In Russian. — Abstr. in Referativ. Zhurn. 51. Astron., 3.51.12 (1974).

005.004   Nicolaus Copernicus (1473–1543).   J. R. Ravetz.
The planets today. Symposium 1973, (see 012.003), p. 5 - 9 (1974).

005.005   Auf den Spuren Karl Schwarzschilds.   H. Kienle.
SuW, Vol. 13, 79 - 82 (1974).

005.006   Zum 100. Geburtstag von Karl Schwarzschild (9. Oktober 1973).   N. Richter.
Jenaer Rundschau, (Jena Review), 18. Jahrgang, p. 315 - 317 (1973).

005.007   F. Zwicky, a special kind of astronomer.
C. Payne-Gaposchkin.
Sky Telescope, Vol. 47, 311 - 313 (1974).

005.008   M. Viljevs (1893 - 1919).   A. Andžāns.
Zvaigžņotā debess, 1973. gada rudens, p. 19 - 24.

005.009   Harlow Shapley.   B. J. Bok.
Quarterly Journ. Roy. Astron. Soc., Vol. 15, 51 - 55 (1974).

005.010   Herman Zanstra.   H. H. Plaskett.
Quarterly Journ. Roy. Astron. Soc., Vol. 15, 57 - 64 (1974).

005.011   Karl Schwarzschild, zu seinem 100. Geburtstag am 9. Oktober 1973.   N. Richter.
Sterne, 50. Jahrgang, p. 8 - 12 (1974).

005.012   Karl Schwarzschild und die Wechselbeziehungen zwischen Astronomie und Physik.   H.-J. Treder.
Sterne, 50. Jahrgang, p. 13 - 19 (1974).

005.013   Wilhelm Foersters chronologisches Werk.
P. Aufgebauer.
Sterne, 50. Jahrgang, p. 51 - 59 (1974).

005.014   Nicolaus Copernicus — great scientist and social man.   A. I. Khadzhiolov.
Priroda (Sofia), Vol. 22, 78 - 83 (1973). In Bulgarian. — Abstr. in Referativ. Zhurn. 51. Astron., 4.51.16 (1974).

005.015   Jean-Louis Pons, découveur de comètes, 1761 - 1831.   J.-P. Brunet, M.-J. Meynent.
L'Astronomie, 88e année, p. 24 - 27 (1974).

005.016   The prince of observers.   C. E. Worley.
Sky Telescope, Vol. 47, 370 - 372 (1974). — Concerning G. Van Biesbroeck's work mainly in double star measurement.

005.017   Nikolaus Copernicus und sein Werk.
F. Schmeidler.
Mitt. Astron. Ges., No. 34, p. 9 - 17 (1973/74). — Presented at the "Wissenschaftliche Tagung der Astron. Ges., Oberkochen, 1973 April 24 - 27".

005.018   Karl Schwarzschild und die Berliner Akademie der Wissenschaften.   C. Grau.
Astron. in der Schule, 11. Jahrgang, p. 33 - 36 (1974).

005.019   Fr. Walter Miller's variable star researches.
P. J. Treanor.
Ric. Astron., Specola Vaticana, *Castel Gandolfo*, Vol. 8, (No. 23), 469 - 474 (1973).

005.020   Johann Heinrich Mädler. Zum 100. Todestag des Astronomen am 14. März 1974.
D. Wattenberg.
Blick in das Weltall, Archenhold-Sternw., Berlin-Treptow, 22. Jahrgang, p. 25 - 31, 39 - 54 (1974).

005.021 **K. Doplers (1803–1853).** M. Zepe.
Zvaigžņotā debess, 1973./74. gada ziema, p. 37 - 39.

005.022 **Double star observer extraordinary.**
W. S. Finsen.
Sky Telescope, Vol. 48, 24 - 25 (1974). — Concerning W. H. van den Bos.

005.023 **V. V. Kavrajskij**, 1884, April 22 - 1954, February 26.

K. A. Zvonarev.
Astron. Zhurn. Akad. Nauk SSSR, Vol. 51, 677 (1974). In Russian. English translation in Soviet Astron. AJ, Vol. 18, No. 3.

Nicolaus Copernicus and his epoch.
See Abstr. 003.015.

Bernhard Schmidt.   See Abstr. 003.091.

006 Personal Notes

K. D. Abhyankar, Director of Nizamiah and Rangapur Observatories.
Observatory, Vol. 94, 32 (1974).

J. Albano, director of the La Plata Observatory, Argentina.
Inform. Bull. Southern Hemisph., No. 23, p. 38 (1973).

V. Ambarzumyan received the Cothenius-Medal.
Physikalische Blätter, 30. Jahrgang, p. 288 (1974).

N. P. Barabashov, 80th birthday.
V. I. Ezerskij, K. N. Kuz'menko, V. Kh. Pluzhnikov.
Zemlya i Vselennaya, 1974, No. 3, p. 58 - 61. In Russian.

L. Biermann received the medal of the Royal Astronomical Society, London.
Mercury, (Journ. Astron. Soc. Pacific), Vol. 3, No. 1, p. 14 (1974).

L. Biermann received the Gold Medal of the Royal Astronomical Society.
Physikalische Blätter, 30. Jahrgang, p. 288 (1974).

L. Biermann received the Gold Medal of the Royal Astronomical Society, London.
SuW, Vol. 13, 78 (1974).

K. Bullen received the medal of the Royal Astronomical Society, London.
Mercury, (Journ. Astron. Soc. Pacific), Vol. 3, No. 1, p. 14 (1974).

M. Burbidge resigns from UK astronomy position.
Phys. Today, Vol. 27, No. 1, p. 112 (1974).

S. J. Burnell received the Albert A. Michelson medal.
Observatory, Vol. 94, 32 (1974).

V. G. Fesenkov, 85th birthday.
A. I. Eremeeva, I. T. Zotkin.
Meteoritika, vyp. (No.) 33, p. 3 - 13 (1974). In Russian.

A. Hewish received the Albert A. Michelson medal.
Observatory, Vol. 94, 32 (1974).

A. Hunter, Director of the Royal Greenwich Observatory at Herstmonceux Castle.   J. E. Kennedy.
Journ. Roy. Astron. Soc. Canada, Vol. 68, 42 - 43 (1974).

A. Hunter, Director of the Royal Greenwich Observatory.
Observatory, Vol. 94, 31 (1974).

W. Iwanowska received the honory doctor degrees from Leicester, Manitoba and Toruń.   A. Woszczyk.
Urania Kraków, Vol. 45, 98 - 99 (1974). In Polish.

R. Kippenhahn received the Carus Medal.
SuW, Vol. 13, 4 (1974).

Z. Kopal, 60th birthday.   A. Rükl.
Říše hvězd, Vol. 55, 76 - 77 (1974). In Czech.

J. Meurers, 65th birthday.
Phys. Blätter, 30. Jahrgang, p. 96 (1974).

O. Obůrka, 65th birthday.
Z. Mikulášek, Z. Pokorný.
Říše hvězd, Vol. 55, 77 (1974). In Czech.

G. Perry received the medal of the Royal Astronomical Society, London.
Mercury, (Journ. Astron. Soc. Pacific), Vol. 3, No. 1, p. 14 (1974).

S. Plakidis, 80th birthday.
See Abstr. 003.009.

A. Sandage received the Cresson Medal.
Observatory, Vol. 94, 32 (1974).

R. F. Sisteró, interim director of the Córdoba

Observatory.
Inform. Bull. Southern Hemisph., No. 23, p. 38 (1973).

R. F. Sisteró, interim Director of the Observatorio de Cordoba.
Observatory, Vol. 94, 95 (1974).

C. P. Sonett, director of the Lunar and Planetary Laboratory at the University of Arizona.
Phys. Today, Vol. 27, No. 2, p. 84 (1974).

C. P. Sonett, director of a new department of planetary sciences at the University of Arizona.
Physics Today, Vol. 27, No. 3, p. 85 (1974).

L. Spitzer, Jr. received the Henry Draper medal.
Phys. Today, Vol. 27, No. 7, p. 57 - 58 (1974).

A. Unsöld received the Cothenius Medal.
SuW, Vol. 13, 4 (1974).

A. R. Upgren, director of the Van Vleck Observatory, Wesleyan University, Middletown, Connecticut.
Mercury, (Journ. Astron. Soc. Pacific), Vol. 2, No. 6, p. 9 (1973).

J. E. Wampler, director at Siding Spring.
Mercury, (Journ. Astron. Soc. Pacific), Vol. 3, No. 1, p.14 (1974).

D. Wattenberg, 65th birthday. D. B. Herrmann.
Blick in das Weltall, Archenhold-Sternw., Berlin-Treptow, 22. Jahrgang, p. 67 - 69 (1974).

F. L. Whipple received the Henry Medal of the Smithsonian Institution.
Phys. Today, Vol. 27, No. 1, p. 83 (1974).

J. P. Wild received the medal of the Royal Astronomical Society, London.
Mercury, (Journ. Astron. Soc. Pacific), Vol. 3, No. 1, p. 14 (1974).

P. Wild received the Herschel Medal of the Royal Astronomical Society.
Journ. Astron. Soc. Victoria, Vol. 27, 25 (1974).

007 Obituaries

C. G. Abbot, 1872 May 31 - 1973 December 17.
W. O. Roberts.
Phys. Today, Vol. 27, No. 5, p. 65, 67 (1974).

C. G. Abbot, 1872 - 1973.   J. Ashbrook.
Sky Telescope, Vol. 47, 75, 100 (1974).

I. S. Bowen, 1898 December 21 - 1973 February 6.
L. H. Aller.
Quarterly Journ. Roy. Astron. Soc., Vol. 15, 193 - 196 (1974).

N. Brice died 1974 January 31.
T. R. McDonough.
Icarus, Vol. 22, 233 - 234 (1974).

L. S. Copeland died 1973 November 14.
Sky Telescope, Vol. 47, 30 (1974).

S. Einarsson died 1974 March 25.
Sky Telescope, Vol. 47, 373 (1974).

M. Ewing died 1974 May 4.   C. L. Drake.
Phys. Today, Vol. 27, No. 7, p. 59 (1974).

V. C. A. Ferraro died 1974 January 3.
Nature, Vol. 248, 89 (1974).

P. W. Gast, 1930 September 11 - 1973 May 16.
D. W. Strangway.
The Moon, Vol. 9, 2 - 3 (1974).

G. E. Gjellestad, 1914 August 11 - 1972 January 11.
E. Jensen.
Quarterly Journ. Roy. Astron. Soc., Vol. 15, 196 (1974).

B. Hacar died 1974 March 9.   F. Konečný.
Říše hvězd, Vol. 55, 116 (1974). In Czech.

S. Herrick, died 1974 March 20.   J. Ashbrook.
Sky Telescope, Vol. 48, 3 (1974).

R. Hindmarsh died.
Observatory, Vol. 94, 94 (1974).

M. Kamienski, 1879 November 24 - 1973 April 18.
J. M. Witkowski.
Quarterly Journ. Roy. Astron. Soc., Vol. 15, 48 - 50 (1974).

A. Kaplan, 1901 March 7 - 1973 September 29.
G. Florsch.
L'Astronomie, 88e année, p. 178 - 180 (1974).

M. P. Karpowicz, 1913 September 5 – 1973 July 23.
K. Rudnicki.
Postępy Astron., Vol. 22, 55 - 56 (1974). In Polish.

M. Kneissl, 1907 - 1973.   F. Kobold.
Bull. Géod., Nouvelle Sér., Année 1974, No. 111, p. 5 - 8.

G. F. G. Knipe, 1916 August 16 - 1973 December 19.   J. Hers.
Monthly Notes Astron. Soc. Southern Africa, Vol. 33, 2 - 3 (1974).

G. P. Kuiper, 1905 – 1973.   C. Sagan.
Icarus, Vol. 22, 117 - 118 (1974).

G. P. Kuiper, 1905 December 7 - 1973 December 24.
Nature, Vol. 248, 539 - 540 (1974).

G. P. Kuiper died 1973 December 23.
Observatory, Vol. 94, 94 (1974).

G. P. Kuiper, 1905 December 7 – 1973 December 24.   E. A. Whitaker.
Phys. Today, Vol. 27, No. 3, p. 85, 87 (1974).

G. P. Kuiper died 1973 December 23.
Science, Vol. 183, 1323 (1974).

G. P. Kuiper died 1973 December 23.
Sky Telescope, Vol. 47, 83 (1974).

G. P. Kuiper died 1973.   C. Titulaer.
Zenit, Vol. 1, No. 1, p. 13 (1974).

A. P. Mackerras, 1899 August 28 - 1973 August 20.
H. Wood.
Journ. British Astron. Ass., Vol. 84, 198 - 199 (1974).

W. J. Miller, 1904–1973.
Coelum, Vol. 42, 136 (1974).

W. J. Miller died 1973 November 30.
Phys. Today, Vol. 27, No. 2, p. 91 (1974).

W. J. Miller died 1973 November 30.
Sky Telescope, Vol. 47, 28 (1974).

L. G. E. Morley, 1894 - 1973, July 29.
G. A. Eiby.
Southern Stars, Vol. 25, 101 - 103 (1973).

L. Orkisz, 1900 - 1973.   E. Rybka.
Urania Kraków, Vol. 45, 185 - 186 (1974). In Polish.

A. G. Pereguda, 1910 - 1973 June 25.
Izv. Krymskoj Astrofiz. Obs., Vol. 49, 117 (1974). In Russian.

M. Pstrzoch-Karpowicz, 1913 - 1973.
K. Rudnicki, L. Zajdler.
Urania Kraków, Vol. 45, 21 - 24 (1974). In Polish.

J. Sałabun died July 1973.
Astron. in der Schule, 11. Jahrgang, p. 67 (1974).

J. Sałabun, 1902 August 31 – 1973 July 13.
K. Rudnicki.
Postępy Astron., Vol. 22, 53 - 54 (1974). In Polish.

H. Shapley, 1885 November 2 – 1972 October 20.
K. Ferrari d'Occhieppo.
Almanach Österreich. Akad. Wiss., 123. Jahrgang, p. 315 - 321 (1973).

H. Simard, 1922 - 1973.
M. M. Thomson, M. Normandin.
Journ. Roy. Astron. Soc. Canada, Vol. 68, 53 - 54 (1974).

L. Simiand, 1913 - 1973.
L'Astronomie, 88e année, p. 33 (1974).

N. B. Slater, 1912 - 1973 January 31.
W. H. McCrea.

Quarterly Journ. Roy. Astron. Soc., Vol. 15, 65 - 67 (1974).

S. Tolansky, 1907 November 17 - 1973 March 4. M. R. C. McDowell.
Quarterly Journ. Roy. Astron. Soc., Vol. 15, 56 (1974).

G. Van Biesbroeck died 1974 February 23.
Circ. d'Inform. (U.A.I. Commission des Étoiles Doubles), Obs. Meudon, No. 62 (1974).

G. A. Van Biesbroeck died 1974 February 23.
Phys. Today, Vol. 27, No. 7, p. 59 (1974).

G. Van Biesbroeck died 1974 February 23.
Sky Telescope, Vol. 47, 215 (1974).

W. H. van den Bos died 1974 March 30.
Circ. d'Inform. (U.A.I. Commission des Étoiles Doubles), Obs. Meudon, No. 63 (1974).

W. H. van den Bos died 1974 March 30. W. S. Finsen.
Monthly Notes Astron. Soc. Southern Africa, Vol. 33, 60 - 61 (1974).

A. N. Vyssotsky died 1973 December 31.
Sky Telescope, Vol. 47, 166 (1974).

F. Zwicky, 1898–1974. F. Lot.
L'Astronomie, 88e année, p. 221 - 224 (1974).

F. Zwicky died 1974 February 8.
Observatory, Vol. 94, 94 (1974).

F. Zwicky died.
Orion Schaffhausen, 32. Jahrgang, p. 63 (1974).

F. Zwicky died 1974 February 8. P. Wild.
Orion Schaffhausen, 32. Jahrgang, p. 113 - 114 (1974).

F. Zwicky died 1974 February 8. H. Arp.
Phys. Today, Vol. 27, No. 6, p. 70 - 71 (1974).

F. Zwicky died 1974 February 8.
Sky Telescope, Vol. 47, 215 (1974).

F. Zwicky, 1898 February 14 - 1974 February 8.
SuW, Vol. 13, 77 - 78 (1974).

008 Observatories, Institutes

Reports, communications and publications of observatories and astronomical institutes are recorded in this section; included are numbered series of reprints. Whenever possible, the numbers of the abstracts referring to the publications are given. Observatories and institutes are listed in alphabetical order of their towns. In some cases observatory publications do not give the name of the town; the following list which gives names and towns of some institutions may serve as an aid in such cases.

Aarne Karjalainen Observatory	Oulu, Finland
Algonquin Radio Observatory	Lake Traverse, Ontario, Canada
Allegheny Observatory	Pittsburgh, Pennsylvania
Archenhold-Sternwarte	Berlin-Treptow, Germany
Arthur J. Dyer Observatory	Nashville, Tennessee
Astronomical Latitude Station, Polish Academy of Sciences	Borowiec, Poland
Bosscha Observatory	Lembang, Indonesia
Boyden Observatory	Bloemfontein, South Africa
Bureau International de l'Heure	Paris, France
Cajigal Observatory	Caracas, Venezuela
California Institute of Technology	Pasadena, California
Cape of Good Hope	Cape Town, South Africa
Carter Observatory	Wellington, New Zealand
Catalina Station	Tucson, Arizona
Cavendish Laboratory	Cambridge, England
Ceskoslovenská Akademie Ved Astronomický Ustav	Praha, Czechoslovakia
Chamberlin Observatory, University of Denver	Denver, Colorado
Commonwealth Observatory	Canberra, Australia
Corralitos Observatory	Las Cruces, New Mexico
David Dunlap Observatory, University of Toronto	Richmond Hill, Ontario
Dearborn Observatory	Evanston, Illinois
Department of Astronomy and Observatory, Univ. California	Los Angeles, California
Department of Astronomy, University of Texas	Austin, Texas
Division Radiophysics, C.S.I.R.O. University Grounds	Sydney, N.S.W., Australia
Dominion Astrophysical Observatory	Victoria, British Columbia
Dominion Observatory	Ottawa, Ontario
Dominion Radio Astrophysical Observatory	Penticton, British Columbia
Dudley Observatory	Albany, New York
Dunsink Observatory	Dublin, Ireland
Engelhardt Observatory	Kazan, U.S.S.R.
European Southern Observatory	Hamburg, Federal German Republic
Five College Observatories	Amherst, Massachusetts
Florida State University Radio Observatory	Tallahassee, Florida
Flower and Cook Observatories, University of Pennsylvania	Philadelphia, Pennsylvania
Fraunhofer Institut	Freiburg, Federal German Republic
Georgetown Observatory	Washington, D.C.
Goddard Space Flight Center	Greenbelt, Maryland
Goethe Link Observatory, University of Indiana	Bloomington, Indiana
Hale Observatories	Pasadena, California
Harvard College Observatory	Cambridge, Massachusetts
Harvard Radio Astronomy Station	Cambridge, Massachusetts
Haystack Observatory	Westford, Massachusetts
Heinrich-Hertz-Institut	Berlin, Germany
High Altitude Observatory, University of Colorado	Boulder, Colorado
Institute for Astronomy, University of Hawaii	Honolulu, Hawaii
Institute for Theoretical Astronomy (Institut Teoreticheskoj Astronomii)	Leningrad, U.S.S.R.
Institute of Theoretical Astrophysics, Blindern	Oslo, Norway
Inter-American Observatory	Cerro-Tololo, (La Serena), Chile
International Latitude Observatory	Mizusawa, Japan
Joint Institute for Laboratory Astrophysics (JILA)	Boulder, Colorado
Kandilli Observatory	Istanbul, Turkey
Kansas University Observatory	Lawrence, Kansas
Kapteyn Astronomical Laboratory	Groningen, Netherlands
Karl-Schwarzschild-Observatorium	Tautenburg, German Democratic Republic
Kenneth Mees Observatory	Rochester, New York
Kwasan Observatory	Kyoto, Japan
Lamont-Hussey Observatory	Bloemfontein, South Africa
Leander McCormick Observatory University of Virginia	Charlottesville, Virginia
Lee Observatory	Beirut, Lebanon
Leopold-Figl-Observatorium	Wien, Austria
Leuschner Observatory	Berkeley, California
Lick Observatory	Santa Cruz, (Mount Hamilton), California
Lindheimer Astronomical Research Center	Evanston, Illinois
Lockheed Solar Observatory	Saugus, California
Lohrmann-Observatorium für Geodätische Astronomie	Dresden, German Democratic Republic
Louisiana State University Observatory	Baton Rouge, Louisiana
Lowell Observatory	Flagstaff, Arizona
Lunar and Planetary Laboratory	Tucson, Arizona
Max-Planck-Institut für Astronomie	Heidelberg, Federal German Republic
Max-Planck-Institut für Phyik und Astrophysik	München, Federal German Republic
Max-Planck-Institut für Radioastronomie	Bonn, Federal German Republic
McDonald Observatory	Fort Davis, Texas
McMath Hulbert Observatory	Pontiac, Michigan
Michigan State University Observatory	East Lansing, Michigan
Molonglo Radio Observatory, University of Sydney	Sydney, New South Wales
Mount Cuba Observatory	Wilmington, Delaware
Mount John Observatory	Lake Tekapo, New Zealand
Mount Palomar Observatory	Pasadena, California
Mount Wilson Observatory	Pasadena, California
Mullard Radio Astronomy Observatory	Cambridge, England
Narrabri Observatory, University of Sydney	Sydney, New South Wales

Abstracts 11.008.001 - 11.008.006

National Bureau of Standards	Washington, D. C.	Sagamore Hill Radio Observatory	Bedford, Massachusetts
National Observatory, USA	Kitt Peak, Arizona	Saint-Michel, l'Observatoire	Haute Provence, France
National Radio Astronomy Observatory	Charlottesville, Virginia Green Bank, West Virginia Tucson, Arizona	San Fernando Observatory Smithsonian Astrophysical Observatory	El Segundo, California Cambridge, Massachusetts
New Mexico State University Observatory	Las Cruces, New Mexico	Specola Astronomica Vaticana Specola di Padova	Castel Gandolfo, Italy Asiago, Italy
Nizamiah Observatory	Hyderabad, India	Sproul Observatory	Swarthmore, Pennsylvania
Nuffield Radio Astronomy Laboratories, Jodrell Bank University of Manchester	Manchester, England	Sternberg Observatory Steward Observatory, University of Arizona	Moscow, U.S.S.R. Tucson, Arizona
Observatoire Royal de Belgique	Uccle, Belgium	United States Naval Observatory	Washington, D.C.
Observatorio de Cartuja	Granada, Spain	University of Florida, Radio Observatory	Gainesville, Florida
Observatorio del Ebro	Tortosa, Spain	University of Illinois Observatory	Urbana, Illinois
Observatorio Fabra	Barcelona, Spain	University of Michigan Observatories	Ann Arbor, Michigan
Observatory, University of Michigan	Ann Arbor, Michigan	University of South Florida Observatory	Tampa, Florida
Ohio State University Radio Observatory	Columbus, Ohio	Uttar Pradesh State Observatory	Naini Tal, India
Ole Roemer-Observatoriet	Aarhus, Denmark	Van Vleck Observatory	Middletown, Connecticut
Onsala Space Observatory	Gothenburg, Sweden	Wallace Observatory	Cambridge, Massachusetts
Owens Valley Radio Observatory	Big Pine, California	Warner and Swasey Observatory	Cleveland, Ohio
Perkins Observatory, Ohio State and Wesleyan Universities	Delaware, Ohio	Washburn Observatory West Melton Observatory	Madison, Wisconsin Christchurch, New Zealand
Purple Mountain Observatory	Nanking, China	Wilhelm-Förster Sternwarte	Berlin
Radcliffe Observatory	Pretoria, South Africa	Yale University Observatory	New Haven, Connecticut
Remeis-Sternwarte	Bamberg, Federal German Republic	Yerkes Observatory Zentralinstitut für Astrophysik, Sternwarte Babelsberg, (Fachbereich Kosmische Physik)	Williams Bay, Wisconsin Potsdam-Babelsberg, German Democratic Republic
Republic Observatory	Johannesburg, South Africa		
Rosemary Hill Observatory	Gainesville, Florida		
Royal Radar Establishment, Radio Astronomy Division	Malvern, England		

008.001 **Abastumani**

Chronicle.
Byull. Abastumansk. Astrofiz. Obs., No. 45, p. 159 - 162 (1974). In Russian.

Abastumanskaya Astrofizicheskaya Observatoriya, Gora Kanobili, Byulleten', No. 45 (I. F. Alaniya, 11.122.050; I. F. Alaniya, 11.122.051; N. L. Magalashvili, Ya. I. Kumsishvili, 11.121.043; N. L. Magalashvili, Ya. I. Kumsishvili, 11.121.044; V. A. Oshchepkov, 11.117.018; A. S. Tskhovrebadze, E. I. Tetruashvili, M. Sh. Gigolashvili, 11.073.035; A. Sh. Khatisov, G. N. Salukvadze, 11.097.201; O. R. Bolkvadze, R. I. Kiladze, 11.099.213; O. R. Bolkvadze, 11.096.013; A. Sh. Khatisov, 11.103.004; N. M. Martsvaladze, 11.082.046; T. G. Megrelishvili, 11.082.047; G. G. Mikirtumova, 11.082.048; G. G. Mikirtumova, G. V. Rozenberg, 11.082.049; V. M. Iskandarova, 11.082.050; D. F. Kharchilava, V. M. Iskandarova, 11.082.051; D. F. Kharchilava, V. M. Iskandarova, 11.082.052; T. A. Guseva, R. I. Kiladze, G. D. Matveev, A. Sh. Khatisov, 11.054.013).

008.002 **Albany**

Dudley Observatory, *Albany, New York,* Reports, No. 8 (A. G. D. Philip, 11.113.027).

008.003 **Alger**

Université d'Alger. Annales de l'Observatoire d'Alger, Tome 3, Fasc. 5 (A. Ghezloun, M. Benhocine, A. Marouf, J. Pham-Van, 11.041.020).

008.004 **Ames, Iowa**

Erwin W. Fick Observatory annual report 1972–1973, Iowa State University, Ames, Iowa. – Observatory report.   W. I. Beavers.
Bull. American Astron. Soc., Vol. 6, 30 - 32 (1974).

008.005 **Amherst**

Five College Astronomy Department: Amherst College, Amherst, Massachusetts; Hampshire College, Amherst, Massachusetts; Mount Holyoke College, South Hadley, Massachusetts; Smith College, Northampton, Massachusetts; University of Massachusetts, Amherst, Massachusetts. Observatory report.   T. Arny.
Bull. American Astron. Soc., Vol. 6, 32 - 36 (1974).

008.006 **Ann Arbor**

Department of Astronomy, The University of Michigan, Ann Arbor, Michigan. – Observatory report. W. A. Hiltner.
Bull. American Astron. Soc., Vol. 6, 121 - 124 (1974).

Publications of the Observatory of the University of Michigan, Vol. 12, No. 1 (F. Holden, 11.118.011).

008.007 Arcetri

Osservazioni e Memorie dell' Osservatorio Astrofisico di Arcetri, Fascicolo 99 (F. Mazzucconi, T. Grisendi, R. Baldini, G. Marcucci, 11.075.005), 101 (P. Arena, P. Patriarchi, 11.077.037), 102 (G. Ceppatelli, A. Righini, 11.082.075; R. Barletti, M. Meco, S. Paloschi, 11.082.076), 103 (P. Arena, P. Patriarchi, 11.077.038).

008.008 Asiago

The first year of observation with the Copernicus 182 cm telescope at Asiago. C. Barbieri, L. Rosino. Proc. ESO/SRC/CERN conference, (see 012.021), p. 199 - 200 (1974).

008.009 Baton Rouge

Louisiana State University Observatory, Baton Rouge, Louisiana. — Observatory report. A. U. Landolt. Bull. American Astron. Soc., Vol. 6, 107 - 109 (1974).

Contributions of the Louisiana State University Observatory, Nos. 71 (A. U. Landolt, K. L. Blondeau, 08.120.021), 72 (J. S. Drilling, A. U. Landolt, 08.121.100), 73 (H. E. Bond, 08.121.101), 74 (J. S. Drilling, 09.114.021), 75 (A. U. Landolt, 09.121.014), 76 (J. S. Drilling, 09.113.018), 77 (J. S. Drilling, P. Pesch, 09.113.019), 78 (N. A. Higginbotham, P. Lee, 09.114.117), 79 (A. U. Landolt, 09.122.076), 80 (H. E. Bond, 09.158.064), 81 (S. Brown, P. Lee, 09.133.038), 82 (H. E. Bond, A. G. D. Philip, 09.113.056), 83 (H. E. Bond, 10.114.049), 84 (A. U. Landolt, 10.122.101), 85 (A. U. Landolt, 10.113.091), 86 (A. U. Landolt, 10.122.103), 87 (A. U. Landolt, 10.113.087), 88 (P. Lee, R. Kenning, 10.133.008).

008.010 Bedford

Sagamore Hill Radio Observatory, Air Force Cambridge Research Laboratories, Bedford, Massachusetts. Observatory report. D. A. Guidice. Bull. American Astron. Soc., Vol. 6, 174 - 176 (1974).

008.011 Beirut

Lee Observatory, American University of Beirut, Lebanon. Monthly Bulletin, Astronomical Section, 1974 January — May (F. Bruin, H. Hourani, N. G. Bustati, 11.075.010).

008.012 Belo Horizonte

Instituto de Ciencias Exatas da Universidade Federal de Minas Gerais. (Institute of Exact Sciences of the Federal University of Minas Gerais). Inform. Bull. Southern Hemisph., No. 23, p. 16 (1973).

008.013 Beograd

Bulletin de l'Observatoire Astronomique de Belgrade, Vol. 29, F. 1, No. 125 (I. Pakvor, 11.032.055; I. Pakvor, 11.032.056; S. Petković, M. Kralj, 11.032.057; G. Teleki, 11.045.017; M. Jovanović, 11.032.058; M. Jovanović, D. Vesić, M. Lončarević, 11.044.021; M. Mijatov, 11.032.059; M. Djokić, 11.032.060; M. B. Protitch, 11.096.035; M. B. Protitch, M. Simić, 11.096.036; G. M. Popović, 11.118.027; D. M. Olević, 11.118.028; G. M. Popović, D. J. Zulević, D. M. Olević, 11.118.029; D. M. Olević, 11.118.030; V. Erceg, 11.118.031; G. M. Popović, 11.118.032; 11.096.037; J. Arsenijević, A. Kubičela, T. Angelov, 11.122.128; J. Arsenijević, A. Kubičela, 11.122.129; B. Popović, 11.042.088).

Publications de l'Observatoire Astronomique de Beograd, No. 18 (G. Teleki, 11.012.023).

008.014 Berkeley

University of California Observatory, Santa Cruz, California. I. Berkeley Campus. — Observatory report. Bull. American Astron. Soc., Vol. 6, 235 - 244 (1974).

008.015 Berlin

Inaugurazione del nuovo osservatorio di Berlino. See Abstr. 011.036.

Veröffentlichungen der Wilhelm-Foerster-Sternwarte Berlin, Nos. 34 (B. Wedel, 11.014.009), 35 (B. Wedel, 11.079.100).

008.016 Berlin-Adlershof

Heinrich-Hertz-Institut, Solare Beobachtungsergebnisse. Akademie der Wissenschaften der DDR, Zentralinstitut für Solar-Terrestrische Physik (Heinrich-Hertz-Institut), Berlin-Adlershof. HHI Solar Data, Vol. 25, 1974 January — March (E. A. Lauter, C.-U. Wagner, A. Böhme, F. Fürstenberg, D. Scholz, S. Böhm, 11.075.011).

008.017 Berlin-Treptow

Blick in das Weltall, Archenhold-Sternwarte Berlin-Treptow. Astronomische Veranstaltungen und Mitteilungen für Sternfreunde, 22. Jahrgang, Nos. 1–6.

008.018 Big Pine, California

Owens Valley Radio Observatory, California Institute of Technology, Big Pine, California. — Observatory report. G. J. Stanley. Bull. American Astron. Soc., Vol. 6, 162 - 166 (1974).

008.019 Bloemfontein

Boyden Observatory. Inform. Bull. Southern Hemisph., No. 23, p. 24 (1973).

008.020 Bonn

Mitteilungen der Astronomischen Institute Bonn, No. 141 (H. Dürbeck, 10.034.114).

Veröffentlichungen der Astronomischen Institute Bonn, Nos. 86 (W. Köhnlein, 11.094.162), 87 (F. Gieseking, 11.152.003).

008.021 Borowiec

Polish Academy of Sciences, Astronomical Latitude Station, Borowiec – Poland, Circular No. 128 (11.044.022).

008.022 Boulder

Joint Institute for Laboratory Astrophysics of the National Bureau of Standards and the University of Colorado, Boulder, Colorado. – Observatory report. C. J. Hansen.
Bull. American Astron. Soc., Vol. 6, 66 - 74 (1974).

008.023 Buenos Aires

Instituto de Astronomía y Física del Espacio. (Institute of Astronomy and Space Physics).
Inform. Bull. Southern Hemisph., No. 23, p. 6 (1973).

Instituto de Astronomía y Física del Espacio (IAFE), Buenos Aires, Argentina, Serie de Publicaciones Técnicas, No. 3 (C. E. Alarcón, M. Pupareli, 11.034.087).

Instituto de Astronomía y Física del Espacio, (IAFE), Buenos Aires, Argentina, Tirada Aparte Nos. 5 (A. E. Ringuelet, R. H. Méndez, 09.133.009), 6 (A. E. Ringuelet, R. H. Méndez, 09.133.010), 7 (J. Sahade, 11.117.039).

008.024 Cambridge, England

University of Cambridge, Institute of Astronomy. D. Lynden-Bell.
Quarterly Journ. Roy. Astron. Soc., Vol. 15, 169 - 181 (1974).

008.025 Cambridge, Massachusetts

Harvard College Observatory, Cambridge, Massachusetts. – Observatory report. A. Dalgarno.
Bull. American Astron. Soc., Vol. 6, 45 - 48 (1974).

Smithsonian Institution. Astrophysical Observatory. Research in Space Science. SAO Special Reports, Nos. 345 (M. R. Pearlman, D. Hogan, K. Goodwin, D. Kurtenbach, 11.082.073), 353 (E. M. Gaposchkin, 11.013.016), 356 (J. W. Slowey, 11.052.020), 357 (M. R. Pearlman, J. L. Bufton, D. Hogan, D. Kurtenbach, K. Goodwin, 11.082.074), 359 (R. L. Kurucz, 11.022.074).

Wallace Observatory Report, Massachusetts Institute of Technology, Cambridge, Massachusetts. Observatory report. T. B. McCord.
Bull. American Astron. Soc., Vol. 6, 185 - 186 (1974).

008.026 Cape Town

University of Cape Town. Department of Astronomy. B. Warner.
Monthly Notes Astron. Soc. Southern Africa, Vol. 33, 7 - 10 (1974).

008.027 Cardiff, Wales, England

University College, Cardiff, Astronomical Communications Nos. 1 (T. L. John, D. J. Morgan, D. Davies, 11.022.005), 2 (R. Y. Chiao, N. C. Wickramasinghe, 09.158.096), 3 (N. C. Wickramasinghe, 10.131.179), 4 (N. C. Wickramasinghe, K. Nandy, 11.131.016), 5 (N. C. Wickramasinghe, 11.131.076).

008.028 Carloforte

Rapporto annuale per l'anno 1972.
E. Proverbio.
Pubbl. Stazione Astron. Internazionale Latitudine, Carloforte-Cagliari, Nuova Ser., No. 30, 13 pp. (1973).

Circolari della Stazione Astronomica Internazionale di Latitudine, Carloforte–Cagliari, Ser. B (3), No. 3 (E. Proverbio, S. Uras, 11.045.006), Ser. B (4), No. 4 (E. Proverbio, S. Uras, 11.045.007), Ser. B (6), No. 8 (E. Proverbio, S. Uras, 11.045.008).

Pubblicazioni della Stazione Astronomica Internazionale di Latitudine, Carloforte–Cagliari, Nuova Serie, Nos. 13bis (E. Proverbio, F. Carta, F. Mazzoleni, 08.045.016), 18bis (E. Proverbio, F. Carta, F. Mazzoleni, 08.045.008), 21 (S. Mancuso, L. Milano, E. Proverbio, 08.041.005), 22 (E. Proverbio, A. Poma, 08.041.044), 22bis (E. Proverbio, A. Poma, 09.041.017), 24 (E. Proverbio, V. Quesada, 10.045.005), 24bis (E. Proverbio, V. Quesada, 09.045.016), 26 (C. Maxia, E. Proverbio, 11.004.052), 27 (E. Proverbio, V. Quesada, 11.045.009), 28 (C. Maxia, E. Proverbio, 11.004.053), 30 (E. Proverbio, 11.008.000), 31 (E. Proverbio, V. Quesada, A. Simoncini, 11.035.005).

008.029 Castel Gandolfo

Specola Vaticana. Annual report 1973: Report of the Astronomical Observatory; Report of the Astrophysical Laboratory. P. J. Treanor, J. Junkes.
Printed in Vatican City, 16 pp. (1974).

Ricerche Astronomiche, Specola Vaticana, Città del Vaticano, Vol. 8, Nos. 19 (M. F. McCarthy, 11.125.022), 20 (M. F. McCarthy, 11.125.023), 21 (M. F. McCarthy, G. Araya, 11.125.102), 22 (W. J. Miller, 11.122.079), 23 (P. J. Treanor, 11.005.019).

Specola Vaticana, Castel Gandolfo, Comunicazione, Nos. 58 (P. S. Osmer, J. E. Hesser, W. E. Kunkel, B. M. Lasker, M. F. McCarthy, A. U. Landolt, 07.125.107), 59 (M. F. McCarthy, 10.114.122), 60 (P. J. Treanor, 09.082.101), 61 (P. J. Treanor, 10.113.082), 62 (M. F. McCarthy, 11.125.044), 63 (M. F. McCarthy, 11.125.102).

Vatican Observatory Publications, Specola Vaticana, Città del Vaticano, Vol. 1, No. 5 (G. V. Coyne, T. A. Lee, E. De Graeve, 11.114.087).

008.030 Catania

Osservatorio Astrofisico di Catania, Pubblicazione, No. 152 (G. Godoli, V. Sciuto, R. A. Zappalà, E. Catinoto, G. Domina, S. Sciuto, G. Celeani, G. Sapienza, 11.075.004).

008.031 Cerro Tololo

Kitt Peak National Observatory, Tucson, Arizona; Cerro Tololo Inter-American Observatory, La Serena, Chile. Observatory reports. B. T. Lynds.
Bull. American Astron. Soc., Vol. 6, 75 - 104 (1974).

Observatorio Interamericano. (Cerro Tololo Interamerican Observatory).
Inform. Bull. Southern Hemisph., No. 23, p. 19 - 23 (1973).

008.032 Charlottesville

National Radio Astronomy Observatory, Charlottesville, Virginia; Green Bank, West Virginia; and Tucson, Arizona. – Observatory reports. D. S. Heeschen.
Bull. American Astron. Soc., Vol. 6, 144 - 153 (1974).

Publications of the Leander McCormick Observatory of the University of Virginia, Vol. 15, Part 7 (L. W. Fredrick, W. A. Gutsch, 11.158.122).

008.033 Chicago

The University of Chicago: Department of Astronomy and Astrophysics, Chicago, Illinois. The Yerkes Observatory, Williams Bay, Wisconsin 1972–1973. – Observatory reports. E. N. Parker, W. F. van Altena.
Bull. American Astron. Soc., Vol. 6, 15 - 21 (1974).

008.034 Cincinnati

Minor Planet Circulars (MPC), Nos 3603 - 3684 (P. Herget, 11.098.046).

008.035 Cleveland

Warner and Swasey Observatory, Case Western Reserve University, Cleveland, Ohio. – Observatory report. W. P. Bidelman.
Bull. American Astron. Soc., Vol. 6, 186 - 190 (1974).

Publications of the Warner and Swasey Observatory, Case Western Reserve University, Vol. 1, No. 3 (W. R. Kubinec, 11.155.039), 4 (C. B. Stephenson, 11.114.086).

Warner and Swasey Observatory, Case Western Reserve University, Reprints, Nos. 219 (J. S. Drilling, P. Pesch, 09.113.019), 233 (J. F. Dolan, 08.142.051), 234 (A. E. Metzger, J. F. Dolan, 08.142.052), 237 (P. A. Wehinger, B. Hidajat, 09.113.048), 238 (S. B. Parsons, E. Peytremann, 09.064.014), 239 (N. Sanduleak, C. B. Stephenson, 08.114.165), 240 (W.B. Weaver, 08.114.167), 241 (S. W. McCuskey, R. S. McMillan, 09.115.005), 243 (W. B. Weaver, S. A. Naftilan, 09.122.078), 244 (N. Sanduleak, C. B. Stephenson, 10.114.152), 245 (W. B. Weaver, 10.122.044), 246 (S. W. McCuskey, 10.155.038), 248 (C. E. Irvine, N. J. Irvine, 10.114.048), 256 (W. P. Bidelman, D. J. MacConnell, G. M. Frey, Jr., 10.114.130), 257 (C. B. Stephenson, 10.114.128).

008.036 Coimbra

Anais do Observatório Astronómico da Universidade de Coimbra. Fenómenos solares, Vol. 14 (11.075.012).

008.037 College Park

Astronomy Program, University of Maryland, College Park, Maryland. – Observatory report. F. J. Kerr.
Bull. American Astron. Soc., Vol. 6, 112 - 118 (1974).

008.038 Columbus

The Observatories of the Ohio State and Ohio Wesleyan Universities, Columbus and Delaware, Ohio. Observatory reports. A. Slettebak.
Bull. American Astron. Soc., Vol. 6, 156 - 161 (1974).

Contributions from the Perkins Observatory, Ohio State – Ohio Wesleyan Universities. Series I, II.
See Abstr. 008.040.

Ohio State University Radio Observatory, Columbus, Ohio. – Observatory report. J. D. Kraus.
Bull. American Astron. Soc., Vol. 6, 161 - 162 (1974).

008.039 Córdoba

Observatorio Astronómico. (Astronomical Observatory, National University of Córdoba). L. A. Milone.
Inform. Bull. Southern Hemisph., No. 23, p. 6 - 8 (1973).

Observatorio Astronómico (*Universidad Nacional de Córdoba, Argentina*), Tirada Aparte, Nos. 185 (R. F. Sisteró, 11.162.047), 208 (R. F. Sisteró, M. E. Castore de Sisteró, 09.121.055), 209 (R. F. Sisteró, 09.162.039), 210 (R. F. Sisteró, 10.162.028), 213 (E. Brandi, J. J. Clariá, 10.114.033).

008.040 Delaware

The Observatories of the Ohio State and Ohio Wesleyan Universities, Columbus and Delaware, Ohio. Observatory reports. A. Slettebak.
Bull. American Astron. Soc., Vol. 6, 156 - 161 (1974).

Contributions from the Perkins Observatory, Ohio State—Ohio Wesleyan Universities. Series I, Nos. 138 (E. G. Buerger, 09.133.012), 139 (L. H. Aller, S. J. Czyzak, E. Craine, J. B. Kaler, 09.133.034), 140 (P. C. Keenan, 10.115.018), 141 (R. F. Wing, 10.122.066), 142 (W. W. Morgan, P. C. Keenan, 10.114.044), 143 (P. C. Keenan, 10.114.119), 144 (S. J. Czyzak, J. J. Santiago, 10.131.064), 145 (E. R. Capriotti, 10.133.050), 146 (J. A. Graham, A. Slettebak, 10.113.072), 147 (S. J. Czyzak, L. H. Aller, 10.022.047), 148 (L. H. Aller, S. J. Czyzak, 10.133.039), 149 (M. V. Penston, R. F. Wing, 10.113.009), 150 (R. F. Wing, 10.114.141), 151 (R. F. Wing, G. W. Lockwood, 10.122.043), 152 (G. H. Newsom, S. O'Connor, R. C. M. Learner, 11.022.081), 153 (J. W. Warner, 10.158.102), 154 (L. H. Aller, S. J. Czyzak, E. Craine, J. B. Kaler, 09.133.034), 155 (R. F. Wing, J. Stock, 10.114.248).

Contributions from the Perkins Observatory, Ohio State – Ohio Wesleyan Universities. Series II, Nos. 32 (P. B. Boyce, N. M. White, R. Albrecht, A. Slettebak, 09.034.015), 33 (J. A. Graham, A. Slettebak, 09.114.127), 34 (G. W. Collins II, 10.115.003), 35 (R. F. Wing, 10.113.054), 36 (G. W. Collins II, P. F. Buerger, 11.063.002), 37 (G. O. Boeshaar, 11.133.003), 38 (M. G. Smith, R. F. Wing, 10.112.006).

008.041 Dushanbe

Chronicle.
Komety i Meteory, *Dushanbe*, No. 21, p. 49 (1972).
In Russian.

Byulleten' Instituta Astrofiziki, Akademiya Nauk Tadzhikskoj SSR, Nos. 59 (K. Kh. Saidov, O. F. Zolova, 11.104.040; O. Alimov, 11.083.032; O. Alimov, 11.083.033), 60 (O. P. Vasil'yanovskaya, 11.122.052; T. G. Nikulina, 11.122.053; N. N. Kiselev, 11.124.103; G. E. Erleksova, 11.122.054; G. E. Erleksova, V. Satyvaldiev, 11.032.018; T. G. Nikulina, 11.122.055; S. N. Zhurkina, 11.122.056; T. G. Nikulina, 11.122.057), 61 (S. G. Pomagaev, 11.151.034; Yu. V. Borisov, 11.122.058; K. Kh. Saidov, O. F. Zolova, 11.104.041; K. Kh. Saidov, 11.104.042; Yu. V. Borisov, 11.034.027), 62 (O. V. Dobrovol'skij, O. Mamadov, 11.103.120; D. Latipov, Sh. O. Isamutdinov, 11.104.043; K. Kh. Saidov, 11.104.044; K. Kh. Saidov, O. F. Zolova, 11.104.045; O. Alimov, D. Latipov, 11.083.034; A. G. Krylov, 11.054.014; N. N. Kiselev, 11.032.019; T. G. Nikulina, 11.122.059).

008.042 East Lansing

Michigan State University, East Lansing, Michigan. Observatory report. A. P. Linnell.
Bull. American Astron. Soc., Vol. 6, 124 - 126 (1974).

008.043 Edinburgh

Royal Observatory, Edinburgh. H. A. Brück.
Quarterly Journ. Roy. Astron. Soc., Vol. 15, 36 - 47 (1974).
Report for the year ending 1973 March 31.

Communications from the Royal Observatory, Edinburgh, Nos. 152 (K. Nandy, R. J. Dodd, R. D. Wolstencroft, 10.113.026), 155 (G. E. Bromage, K. Nandy, 10.114.159), 156 (K. Nandy, H. Seddon, 10.131.153), 157 (M. T. Brück, 11.132.002), 158 (V. C. Reddish, 11.065.116), 159 (K. Nandy, G. I. Thompson, C. M. Humphries, 11.113.006), 160 (W. McD. Napier, R. J. Dodd, 11.098.002), 161 (R. J. Dodd, 11.155.005), 162 (G. I. Thompson, C. M. Humphries, K. Nandy, 11.114.015), 165 (B. N. G. Guthrie, 11.160.004), 166 (R. D. Cannon, 10.013.015), 167 (W. A. Cormack, 11.034.077).

Publications of the Royal Observatory, Edinburgh, Vol. 9, No. 2 (R. J. Dodd, 11.112.028).

008.044 El Leoncito

Yale-Columbia Southern Observatory and U.S. Naval Observatory Southern Hemisphere Expedition.
Inform. Bull. Southern Hemisph., No. 23, p. 8 - 10 (1973).

008.045 El Segundo

The Aerospace Corporation, El Segundo, California: (I) Electronics Research Laboratory; (II) Space Physics Laboratory. – Observatory reports. G. A. Paulikas.
Bull. American Astron. Soc., Vol. 6, 1 - 6 (1974).
Research in astronomy is carried out by groups in the Electronics Research Laboratory and the Space Physics Laboratory. The Space Physics Laboratory operates the Aerospace Corporation's San Fernando Observatory and also conducts astronomical observations at X-ray and EUV wavelengths using rockets and satellites. The activities of these groups are described.

008.046 Evanston

Lindheimer Astronomical Research Center and Dearborn Observatory, Evanston, Illinois; Corralitos Observatory, Las Cruces, New Mexico. – Observatory reports.
J. A. Hynek.
Bull. American Astron. Soc., Vol. 6, 104 - 106 (1974).

008.047 Flagstaff

Lowell Observatory, Flagstaff, Arizona. – Observatory report. J. S. Hall.
Bull. American Astron. Soc., Vol. 6, 109 - 112 (1974).

008.048 Fort Davis

Atmospheric extinction at McDonald Observatory, 1960–68. See Abstr. 082.041.

008.049 Gainesville

University of Florida Observatories, Gainesville, Florida: (I) Rosemary Hill Observatory, F. B. Wood; (II) Radio Observatory, A. G. Smith. – Observatory reports.
Bull. American Astron. Soc., Vol. 6, 36 - 41 (1974).

008.050 Grahamstown

Rhodes University Physics Department.
Inform. Bull. Southern Hemisph., No. 23, p. 24 - 26 (1973).

008.051 Graz

Mitteilungen der Universitätssternwarte Graz,
Nos. 12 (H. F. Haupt, A. Schroll, 11.113.021), 16 (H. Haupt, A. Schroll, 11.098.042).

008.052 Green Bank

National Radio Astronomy Observatory, Charlottesville, Virginia; Green Bank, West Virginia; and Tucson, Arizona. – Observatory reports. D. S. Heeschen.
Bull. American Astron. Soc., Vol. 6, 144 - 153 (1974).

National Radio Astronomy Observatory, *Green Bank,* Reprints, Series A, Nos. 315 (G. E. Assousa, J. W. Erkes, 10.125.038), 316 (R. M. Hjellming, 10.142.081), 317 (K. I. Kellermann, I. I. K. Pauliny-Toth, 10.141.082), 318 (G. L. Verschuur, 10.155.048), 319 (S. D. Peterson, 10.158.094), 320 (D. R. W. Williams, 10.125.020), 321 (K. I. Kellermann, 10.141.063), 322 (B. G. Clark, 11.033.046), 323 (G. S. Shostak, 11.158.001), 324 (M. S. Roberts, 11.158.011), 325 (P. L. Baker, 11.131.043), 326 (P. A. R. Ade, J. D. G. Rather, P. E. Clegg, 11.080.005), 327 (B. Balick, R. H. Gammon, L. H. Doherty, 11.132.004), 328 (Y. Terzian, B. Balick, C. Bignell, 11.133.008), 329 (J. J. Condon, 11.141.038), 330 (J. F. C. Wardle, G. K. Miley, 11.141.005), 331 (K. H. Wesseling, J. P. Basart, J. L. Nance, 11.033.047), 332 (P. S. Yeung, R. L. Brown, 11.063.032), 333 (G. L. Verschuur, 11.131.508), 334 (M. R. Kundu, T. Velusamy, P. E. Hardee, 11.125.015), 335 (W. B. Burton, 11.155.012), 336 (B. E. Turner, B. Zuckerman, 11.131.009), 337 (H. M. Tovmassian, Y. Terzian, 10.158.136), 338 (A. R. Thompson, R. N. Bracewell, 11.033.006), 339 (G. S. Shostak, 11.160.015), 340 (R. L. Brown, J. Gómez-González, 11.131.053), 341 (D. C. Backer, J. R. Fisher, 11.141.319), 342 (P. C. Gregory, E. R. Seaquist, 11.122.021), 343 (Y. Terzian, B. Dennison, B. Balick, 11.131.521), 344 (A. H. Bridle, E. B. Fomalont, 11.141.043), 345 (I. Fejes, 11.131.526), 346 (T. D. Kinman, J. F. C. Wardle, E. K. Conklin, B. H. Andrew, G. A. Harvey, J. M. Macleod, W. J. Medd, 11.141.050), 347 (E. J. Grayzeck, F. J. Kerr, 11.131.532), 348 (R. N. Manchester, 11.156.002), 349 (G. L. Verschuur, 11.155.048), 350 (M. L. De Jong, 11.141.045).

National Radio Astronomy Observatory, *Green Bank,* Reprints, Series B, Nos. 421 (J. M. Moran, G. D. Papadopoulos, B. F. Burke, K. Y. Lo, P. R. Schwartz, D. L. Thacker, K. J. Johnston, S. H. Knowles, A. C. Reisz, I. I. Shapiro, 10.131.121), 422 (L. E. Snyder, D. Buhl, 10.132.023), 423 (K. Y. Lo, K. P. Bechis, 10.131.125), 424 (F. I. Shimabukuro, G. A. Chapman, E. B. Mayfield, S. Edelson, 09.077.068), 425 (R. L. Brown, B. Balick, 10.131.138), 426 (D. S. De Young, M. S. Roberts, W. C. Saslaw, 10.158.081), 427 (D. Buhl, L. E. Snyder, P. R. Schwartz, J. Edrich, 09.155.086), 428 (M. A. Kaftan-Kassim, 10.133.028), 429 (R. M. Hjellming, L. C. Blankenship, B. Balick, 09.141.040), 430 (R. Dube, E. J. Groth, L. Rudnick, D. T. Wilkinson, 10.155.016), 431 (L. E. Snyder, D. Buhl, 09.131.132), 432 (G. M. Tovmasyan, 10.160.022), 433 (R. N. Manchester, E. Tademaru, J. H. Taylor, G. R. Huguenin, 10.141.534), 434 (F. W. Peterson, W. A. Dent, 10.141.100), 435 (A. C. Reisz, I. I. Shapiro, J. M. Moran, G. D. Papadopoulos, B. F. Burke, K. Y. Lo, P. R. Schwartz, 10.131.224), 436 (S. von Hoerner, 10.141.101), 437 (M. Morris, B. Zuckerman, P. Palmer, B. E. Turner, 10.131.222), 438 (R. H. Becker, M. R. Kundu, 10.141.011), 439 (T. G. Phillips, K. B. Jefferts, P. G. Wannier, 10.131.211), 440 (B. E. Turner, 10.131.210), 441 (G. R. Knapp, W. K. Rose, F. J. Kerr, 10.154.022), 442 (M. R. Kundu, T. Gergely, 10.072.050), 443 (B. E. Turner, B. Zuckerman, P. Palmer, M. Morris, 10.131.207), 444 (R. C. Bignell, 10.125.054), 445 (N. C. Wickramasinghe, K. Nandy, 11.131.016), 446 (M. R. Kundu, T. Velusamy, R. H. Becker, 11.073.008).

008.053 Greenbelt

Goddard Space Flight Center, Greenbelt, Maryland, Documents X-590-73-340 (J. G. Marsh, B. C. Douglas, S. M. Klosko, 11.046.011), X-592-73-328 (D. P. Rubincam, 11.094.010), X-592-73-334 (M.A. Khan, 11.081.018), X-592-73-339 (F. J. Lerch, 11.046.012), X-660-73-383 (L. A. Fisk, B. Kozlovsky, R. Ramaty, 11.143.024), X-660-73-392 (F. B. McDonald, B. J. Teegarden, J. H. Trainor, W. R. Webber, 11.143.025), X-660-74-5 (H. T. Wang, R. Ramaty, 11.076.013), X-660-74-94 (R. Ramaty, B. Kozlovsky, 11.073.015), X-661-73-393 (R. E. Rothschild, E. A. Boldt, S. S. Holt, P. J. Serlemitsos, 11.142.060), X-661-74-11 (S. S. Holt, E. A. Boldt, R. E. Rothschild, P. J. Serlemitsos, 11.142.061), X-661-74-71 (R. D. Price, 11.143.061), X-921-74-144 (C. A. Wagner, B. C. Douglas, R. G. Williamson, 11.052.046), X-921-74-161 (D. E. Smith, R. Kolenkiewicz, R. W. Agreen, P. J. Dunn, 11.046.028).

Goddard Space Flight Center, Greenbelt, Maryland, Separate prints (J. F. Ormes, V. K. Balasubrahmanyan, 09.143.007), (D. E. Smith, R. Kolenkiewicz, P. J. Dunn, 11.081.009), (R. F. Silverberg, J. F. Ormes, V. K. Balasubrahmanyan, 10.143.049), (V. K. Balasubrahmanyan, J. F. Ormes, 10.143.051), R. Ramaty, C. C. Cheng, S. Tsuruta, 11.142.002), (S. S. Holt, E. A. Boldt, P. J. Serlemitsos, S. S. Murray, R. Giacconi, E. M. Kellogg, T. A. Matilsky, 11.142.043), (R. E. Rothschild, E. A. Boldt, S. S. Holt, P. J. Serlemitsos, 11.142.048).

008.054 Greenwich

Greenwich solutions.
Nature, Vol. 249, 297 - 298 (1974).

Royal Observatory Annals, [Royal Greenwich Observatory, Herstmonceux], No. 9 (K. C. Blackwell, M. E. Buontempo, 11.041.008).

H. M. Nautical Almanac Office, Royal Greenwich Observatory, Library Reprint, Nos. 302 (G. A. Wilkins, A. T. Sinclair, 11.091.028), 304 (A. T. Sinclair, 11.100.201), 305 (G. A. Wilkins, see 11.003.013, p. 77 - 79).

008.055 Groningen

Nederlandse Vereniging voor Weer- en Sterrenkunde. Observations of Variable Stars. Report (Kapteyn Astronomical Laboratory, Groningen – Netherlands), No. 25 (L. Plaut, H. Feijth, 11.123.019).

008.056 Hamburg

ESO – det faelleseuropaeiske astronomiske ob-

servatorium i Chiles Andesbjerge. R. M. West.
Astron. Tidssk., Årg. 7, p. 59 - 68 (1974).

The Messenger – El Mensajero, No. 1, with an address 'to all ESO staff members' by A. Blaauw.
Edited by European Southern Observatory (ESO), Hamburg. 6 pp. (1974).

This newsletter is planned in order to inform ESO personnel about what is going on in the various establishments of the organization. Furthermore it may serve to give the world outside some impression of what happens inside ESO.

Photos vom Südhimmel.
SuW, Vol. 13, 152 - 153 (1974).

The southern sky surveys – a review of the ESO Sky Survey Project. See Abstr. 041.023.

European Southern Observatory (ESO), Bulletin No. 10 (B. E. Westerlund, 11.082.084; R. M. West, 11.041.023).

European Southern Observatory (ESO), Technical Report Nos. 1 (W. Richter, D. Plathner, J. F. Rozeveld van der Ven, 11.032.043), 2 (R. N. Wilson, 11.032.044), 3 (R. N. Wilson, 11.031.060).

Deutsches Hydrographisches Institut, Hamburg.
Astronomische Zeit- und Breitenbestimmungen, Empfangszeiten von Zeitsignalen, 1973 July - September (11.044.023).

008.057 Hannover

Technische Universität, Hannover, Astronomische Station des Instituts für Theoretische Geodäsie, No. 9 (K. Pilowski, 11.032.029).

008.058 Hartebeestspoortdam

Leids Zuidelijk Station. (Southern Station of the Leiden Observatory, Holland).
Inform. Bull. Southern Hemisph., No. 23, p. 26 - 27 (1973).

008.059 Heidelberg

Astronomy and Astrophysics Abstracts, Vol. 10 (S. Böhme, W. Fricke, U. Güntzel-Lingner, F. Henn, D. Krahn, U. Scheffer, G. Zech, 11.002.021).

Astronomisches Rechen-Institut in Heidelberg, Mitteilungen, Serie A, Nos. 74 (W. Lohmann, 11.155.006), 75 (R. Wielen, 11.151.022).

Astronomisches Rechen-Institut in Heidelberg, Mitteilungen, Serie B, Nos. 39 (J. Schubart, 11.098.001), 40 (M. Miyamoto, 11.151.007), 41 (R. Wielen, 11.122.017).

008.060 Holmdel, New Jersey

Bell Laboratories, Crawford Hill Laboratory, Holmdel, New Jersey. – Observatory report. A. A. Penzias.
Bull. American Astron. Soc., Vol. 6, 14 - 15 (1974).

008.061 Honolulu

University of Hawaii, Institute for Astronomy, Honolulu, Hawaii. – Observatory report. J. T. Jefferies.
Bull. American Astron. Soc., Vol. 6, 48 - 56 (1974).

008.062 Iowa

The University of Iowa, Iowa City, Iowa. – Observatory report. J. S. Neff.
Bull. American Astron. Soc., Vol. 6, 63 - 66 (1974).

008.063 Jena

Mitteilungen der Universitäts-Sternwarte zu Jena, Nos. 98 (H. Lambrecht, 09.131.118), 111 (C. Friedemann, 07.122.007), 112 (W. Pfau, 07.131.019), 113 (J. Gürtler, 07.133.019), 114 (W. Pfau, 07.131.116), 117 (J. Dorschner, C. Friedemann, J. Gürtler, H. Oleak, K.-H. Schmidt, 08.141.082).

008.064 Kiev

Astrometriya i Astrofizika, Kiev, Vyp. (No.) 20 (E. P. Fedorov, 11.003.002).

008.065 Kitt Peak

Kitt Peak National Observatory, Tucson, Arizona; Cerro Tololo Inter-American Observatory, La Serena, Chile. Observatory report. B. T. Lynds.
Bull. American Astron. Soc., Vol. 6, 75 - 104 (1974).

"Stellar" research at the Kitt Peak National Observatory. See Abstr. 013.014.

008.066 Krim

Izvestiya Krymskoj Astrofizicheskoj Observatorii, Akademiya Nauk SSSR, Tom 49 (S. I. Gopasyuk, T. T. Tsap, 11.080.011; M. D. Gusejnov, 11.072.004; E. A. Baranovskij, 11.072.005; B. M. Vladimirskij, L. S. Levitskij, 11.072.006; L. S. Levitskij, 11.078.005; L. A. Eliseeva, L. I. Yurovskaya, 11.072.007; Yu. F. Yurovskij, 11.079.102; D. N. Rachkovskij, 11.063.010; T. S. Belyakina, E. S. Brodskaya, 11.121.020; A. N. Kulapova, N. I. Shakhovskaya, 11.122.014; R. E. Gershberg, N. I. Shakhovskaya, 11.122.015; E. A. Vitrichenko, V. A. Marsakov, P. N. Kholopov, G. S. Tsarevskij, 11.122.016; L. P. Metik, I. I. Pronik, 11.158.016; A. N. Abramenko, O. P. Gollandskij, V. V. Prokof'eva, 11.141.306; A. N. Abramenko, V. V. Prokof'eva, N. A. Ushakova, 11.097.015; G. M. Popov, V. I. Pronik, 11.031.008).

008.067 La Plata

Observatorio Astronómico. (Astronomical Observatory, National University of La Plata).

Inform. Bull. Southern Hemisph., No. 23, p. 10 (1973).

Separata Astronómica, Observatório Astronómico – La Plata – Argentina, Nos. 114 (C. A. Hernández, 07.121.013), 115 (J. M. Simon, L. R. de Novarini, R. Platzeck, 07.034.144), 116 (C. Jaschek, O. Ferrer, 07.119.015), 117 (M. Jaschek, E. Brandi, 08.114.023), 118 (C. A. Altavista, 08.042.036), 119 (O. H. Levato, 08.116.011), 120 (A. Feinstein, H. G. Marraco, I. Mirabel, 09.153.006), 121 (M. Jaschek, C. Jaschek, 09.114.057), 122 (A. E. Ringuelet, R. H. Méndez, 09.133.009), 123 (A. E. Ringuelet, R. H. Méndez, 09.133.010), 124 (O. Ferrer, C. Jaschek, 09.153.028), 125 (S. Malaroda, 09.114.175), 126 (J. C. Muzzio, 09.113.057), 127 (A. Feinstein, 10.113.068), 128 (C. Jaschek, M. Jaschek, 10.114.125), 129 (C. Jaschek, 10.114.144), 130 (E. Brandi, J. J. Clariá, 10.114.033).

Separata Geofísica, Observatório Astronómico – La Plata – Argentina, No. 16 (H. R. Affolter, O. Schneider, 08.084.213),

008.068 Las Cruces

Lindheimer Astronomical Research Center and Dearborn Observatory, Evanston, Illinois; Corralitos Observatory, Las Cruces, New Mexico. – Observatory reports. J. A. Hynek.
Bull. American Astron. Soc., Vol. 6, 104 - 106 (1974).

New Mexico State University, Department of Astronomy, Las Cruces, New Mexico. – Observatory report. W. L. Reitmeyer.
Bull. American Astron. Soc., Vol. 6, 153 - 156 (1974).

008.069 Lawrence, Kansas

Kansas University Observatory, Lawrence, Kansas. Observatory report. S. J. Shawl.
Bull. American Astron. Soc., Vol. 6, 74 - 75 (1974).

008.070 Lembang

Bandung Institute of Technology, Department of Science. **Publications of the Bosscha Observatory**, Nos. 7 (I. Radiman, B. Hidajat, 11.155.078), 8 (B. Hidajat, 11.065.124).

Contributions from the Bosscha Observatory, Bandung Institute of Technology, Department of Science, Nos. 46 (B. Hidajat, I. Radiman, 11.155.079), 47 (P. A. Wehinger, B. Hidajat, 09.113.048), 48 (R. W. Carlson, J. C. Bhattacharyya, B. A. Smith, T. V. Johnson, B. Hidajat, S. A. Smith, G. E. Taylor, B. O'Leary, R. T. Brinkmann, 10.099.039), 50 (W. Sutantyo, 10.117.023).

008.071 Leningrad

Ephemerides of minor planets for 1975, (G. A. Chebotarev, 11.098.043).

Trudy Astronomicheskoj Observatorii, (Transactions of the Astronomical Observatory), *Leningrad*, Vol. 30 (A. K. Kolesov, 11.091.022; A. S. Zentsova, 11.063.035; G. P. Apushkinskij, A. A. Shenogin, B. Ya. Losovskij, A. N. Tsyganov, M. T. Levchenko, 11.033.051; G. V. Khozov, T. N. Khudyakova, 11.113.036; M. K. Babadzhanyants, S. K. Vinokurov, V. A. Hagen-Thorn, E. V. Semenova, 11.158.091; V. A. Dombrovskij, T. A. Polyakova, 11.122.098; E. T. Belokon', O. S. Shulov, 11.122.099; V. A. Antonov, 11.151.044; M. A. Belozerova, 11.151.045; E. I. Timoshkova, K. V. Kholshevnikov, 11.042.068; R. P. Eremenko, E. N. Polyakhova, 11.054.020; V. B. Titov, 11.042.069; M. S. Zverev, 11.041.029; O. A. Mel'nikov, R. Kh. Salman-Zade, Yu. A. Solonskij, E. D. Khilov, 11.022.060).

008.072 London, Canada

The Observatories of the University of Western Ontario, London, Ontario, Canada. – Observatory report. W. H. Wehlau.
Bull. American Astron. Soc., Vol. 6, 197 - 199 (1974).

008.073 London, England

University College London, Mullard Space Science Laboratory. R. L. F. Boyd.
Quarterly Journ. Roy. Astron. Soc., Vol. 15, 182 - 192 (1974).
Report for the period 1972 September 1 – 1973 September 30.

008.074 Los Angeles

University of California Observatory, Santa Cruz, California. II. Los Angeles Campus. – Observatory report. G. O. Abell.
Bull. American Astron. Soc., Vol. 6, 244 - 248 (1974).

008.075 Louvain-La-Neuve

Institut d'Astronomie et de Géophysique, Georges Lemaître, Université Catholique de Louvain, Sciences I, Louvain-La-Neuve, Belgique. Contribution No. 13 (P. Paquet, M. Honorez, 11.021.023).

Publications de l'Institut d'Astronomie et de Géophysique, Georges Lemaître, Louvain-La-Neuve, Vol. 6, Nos. 1 (J. Kovalevsky, 11.042.050), 2 (O. Godart, 10.021.015), 3 (O. Godart, 10.151.033), 4 (A. Boury, 10.080.021), 5 (P. Melchior, 11.081.022).

008.076 Madison

Washburn Observatory, University of Wisconsin, Madison, Wisconsin. – Observatory report. R. C. Bless.
Bull. American Astron. Soc., Vol. 6, 190 - 195 (1974).

008.077 Madrid

Universidad Complutense de Madrid – Facultad de Ciencias. Seminario de Astronomia y Geodesia, Publicación, Nos. 72 (M. J. Fernández-Figueroa, 11.114.053), 73 (M. J.

Sevilla, 11.032.016), 74 (R. Parra, M. J. Sevilla, 11.054.017), 75 (M. Rego Fernández, M. J. Fernández-Figueroa, 11.114.091), 76 (E. Simonneau, 11.064.042).

008.078 Manchester

University of Manchester, Nuffield Radio Astronomy Laboratories, Jodrell Bank.   B. Lovell.
Quarterly Journ. Roy. Astron. Soc., Vol. 15, 28 - 35 (1974). Report for the year ending 1973 August 31.

008.079 Meudon

Radio astronomy on decameter wavelengths at Meudon and Nancay Observatories. (Report from Solar Institute).   See Abstr. 077.067.

008.080 Middletown

Van Vleck Observatory, Wesleyan University, Middletown, Connecticut. – Observatory report.
A. R. Upgren.
Bull. American Astron. Soc., Vol. 6, 182 - 183 (1974).

008.081 Minneapolis

University of Minnesota, Minneapolis, Minnesota, Separate print (W. J. Luyten, 11.112.013).

008.082 Mizusawa

Monthly Notes of the International Polar Motion Service, 1973 Nos. 11 - 12, 1974 Nos. 1 - 4 (11.045.018).

008.083 Mons

Centre Universitaire de l'État à Mons. Faculté des Sciences. Département d'Astrophysique. Communications. Nos. 28 (L. Houziaux, F. Ostan, 11.041.022), 39 (L. Houziaux, 11.114.092).

008.084 Moskva

Chronicle. [Dissertations at the Sternberg Institute]. Astron. Tsirk., No. 813, p. 8 (1974). In Russian.

Chronicle. [Dissertations at the Sternberg-Institute].
L. N. Bondarenko.
Astron. Tsirk., No. 815, p. 6 - 8 (1974). In Russian.

Sternberg State Astronomical Institute. Brief history and description.   See Abstr. 003.025.

Soobshcheniya Gosudarstvennogo Astronomicheskogo Instituta im. P. K. Shternberga. Izdatel'stvo Moskovskogo Universiteta, Nos. 183 (N. G. Bochkarev, 11.062.049; I. G. Reznikov, V. S. Strelnitskij, 11.022.051), 184 (Yu. N. Lipskij, V. V. Novikov, A. P. Popov, 11.114.096; N. I. Kozhevnikov, B. Kalzhanov, S. O. Obashev, 11.032.032; N. I. Kozhevnikov, 11.082.092; A. V. Bugaevskij, Yu. N. Lipskij, Yu. P. Pskovskij, M. M. Pospergelis, 11.074.071), 185 (D. Ya. Martynov, 11.121.070), 186 (G. G. Koman, 11.052.034; G. G. Pavlov, 11.042.054).

Trudy Gosudarstvennogo Astronomicheskogo Instituta im. P. K. Shternberga. Izdatel'stvo Moskovskogo Universiteta, Vol. 44 (N. M. Artyukhina, E. P. Kalinina, 11.112.014; L. P. Panteleeva, 11.112.015; V. V. Nesterov, Yu. I. Prodan, 11.046.015; T. V. Goryainova, 11.041.027; K. G. Steinert, 11.032.033).

008.085 München

Max-Planck-Institut für Physik und Astrophysik, Institut für Extraterrestrische Physik, Garching bei München, MPI-PAE/Extraterr. 95 (V. Schönfelder, G. Lichti, 11.142.064).

008.086 Naini Tal

Uttar Pradesh State Observatory, *Naini Tal*, Reprints, Nos. 56 (J. P. Chaturvedi, 07.115.005), 57 (G. S. D. Babu, S. D. Sinvhal, 08.116.007), 58 (M. C. Pande, G. C. Joshi, B. M. Tripathi, V. P. Gaur, 08.114.046), 59 (S. K. Gupta, A. K. Bhatnagar, 08.122.113), 60 (R. C. Kapoor, S. D. Sinvhal, 08.122. 112), 61 (R. C. Kapoor, 09.122.086), 62 (G. S. D. Babu, P. P. Saxena, 08.103.100), 63 (A. K. Bhatnagar, S. K. Gupta, 09.122.097), 64 (S. D. Sinvhal, C. D. Kandpal, H. S. Mahra, S. C. Joshi, J. B. Srivastava, 11.032.031), 65 (S. C. Joshi, 11.122.087), 66 (R. C. Kapoor, B. B. Sanwal, S. D. Sinvhal, 10.122.113), 67 (V. P. Gaur, M. C. Pande, B. M. Tripathi, 09.072.037), 68 (K. Sinha, 09.071.028), 69 (K. R. Bondal, G. C. Joshi, M. C. Pande, 09.114.111), 70 (M. C. Pande, G. C. Joshi, 09.114.112), 71 (P. P. Saxena, 10.082.124), 72 (M. C. Pande, V. P. Gaur, 10.071.008).

008.087 Nançay

Radio astronomy on decameter wavelengths at Meudon and Nançay Observatories. (Report from Solar Institute).   See Abstr. 077.067.

008.088 Napoli

Istituto Universitario Navale, Astronomia Generale e Sferica, Napoli. Pubblicazione, Nos. 6 (T. Nicolini, 11.032.030), 7 (E. Fichera, A. Pugliano, 11.046.014), 8 (E. Fichera, 11.045.012).

Istituto Universitario Navale, Astronomía Generale e Sferica, Napoli. Seminario Nos. 4 (E. Fichera, 11.052.029), 5 (E. Fichera, 11.052.030), 6 (T. Nicolini, 11.022.050), 7 (A. Fresa, 11.004.060), 8 (T. Nicolini, 11.004.061).

008.089 Nashville

Dyer Observatory, Vanderbilt University, Nashville, Tennessee. – Observatory report.   A. M. Heiser.

Bull. American Astron. Soc., Vol. 6, 28 - 30 (1974).

008.090 Neuchâtel

Rapport d'activité pour l'exercice 1973 et Rapport sur le Concours chronométrique 1973.
J. Bonanomi, G. Fischer.
Observatoire Cantonal de Neuchâtel, 25 pp. (1974).

Observatoire de Neuchâtel, Bulletin. Série B, 1973 July - December (11.044.026); Série D, 1973 July - December (11.044.027).

008.091 New Haven

Yale University Observatory, New Haven, Connecticut. — Observatory report.   P. Demarque.
Bull. American Astron. Soc., Vol. 6, 199 - 204 (1974).

008.092 Nice

La presse et l'Observatoire de Nice en 1972 — 1973.
G. Ringeard.
Bull. d'Information, Ass. Développement International Obs. Nice, No. 10, p. 83 - 84 (1973).

Rapport d'activité de l'Observatoire de Nice pour 1972. Équipes scientifiques; Équipes techniques; Publications de l'Observatoire de Nice.
Bull. d'Information, Ass. Développement International Obs. Nice, No. 10, p. 33 - 82, 85 - 90 (1973).

008.093 Oslo

Institute of Theoretical Astrophysics, Blindern — Oslo, Report, Nos. 38 (T. L. Hansen, 11.082.083), 39 (O. Engvold, Ø. Hauge, 11.071.036).

Institutt for Teoretisk Astrofysikk, Blindern — Oslo. Småtrykk, Nos. 79 (Ø. Hauge, 11.061.055), 81 (T. S. Ringnes, 11.097.202).

008.094 Ottawa

Contributions Astrophysics Branch, National Research Council of Canada, Ottawa, Ontario, NRC Nos. 13496 (M. B. Bell, D. N. Fort, 10.141.092), 13540 (B. H. Andrew, 10.141.094), 13672 (T. D. Kinman, J. F. C. Wardle, E. K. Conklin, B. H. Andrew, G. A. Harvey, J. M. Macleod, W. J. Medd, 11.141.050).

Contributions from the Earth Physics Branch, Ottawa, Canada, Nos. 416 (M. R. Dence, A. G. Plant, 09.094.629), 444 (D. H. Weichert, 11.066.147), 449 (J. C. Gupta, 11.083.054), 500 (M. K. Paul, 11.081.034).

008.095 Palo Alto

Lockheed Palo Alto Research Laboratory, Palo Alto, California. — Observatory report.   B. McCormac.
Bull. American Astron. Soc., Vol. 6, 106 - 107 (1974).

008.096 Paris

Bureau International de l'Heure. Rapport annuel pour 1973.   R. Michard (Editor).
Printing Office: Observatoire de Paris. 5 + A21 + B47 + C14 pp. (1974).
Contents: Methods of computation and explanations; Tables and figures; Time signals.

Bureau International de l'Heure, (B. I. H.), Circulaires B/C Nos. 213 - 219 (11.045.019); D86 - D92 (11.044.029).

008.097 Parque Pereyra Iraola

Instituto Argentino de Radioastronomía. (Argentine Radioastronomy Institute).
Inform. Bull. Southern Hemisph., No. 23, p. 11 - 12 (1973).

008.098 Pasadena

Hale Observatories, operated by Carnegie Institution of Washington and California Institute of Technology, Pasadena, California. Annual report of the director, 1972 - 1973.   H. W. Babcock, J. B. Oke.
Reprinted from Carnegie Institution, Washington, Year Book, Vol. 72, 97 - 164 (1972/73).

Current research programmes at the Hale Observatories.   See Abstr. 013.012.

008.099 Philadelphia

Flower and Cook Observatory, University of Pennsylvania, Philadelphia, Pennsylvania. — Observatory report.   B. S. P. Shen.
Bull. American Astron. Soc., Vol. 6, 41 - 44 (1974).

008.100 Pittsburgh

Allegheny Observatory, University of Pittsburgh, Pittsburgh, Pennsylvania. — Observatory report.
J. H. Kiewiet de Jonge.
Bull. American Astron. Soc., Vol. 6, 6 - 8 (1974).

008.101 Praha

Académie Tchécoslovaque des Sciences, Institut Astronomique, Station de l'Heure à Prague, Série 6, Nos. 8 - 9 (L. Webrová, V. Ptáček, 11.044.028).

008.102 Pretoria

200 years of observations at the Radcliffe Obser-

vatory. (A) at Oxford. (B) at Pretoria.   P. J. Andrews.
Inform. Bull. Southern Hemisph., No. 23, p. 2 - 5 (1973).

Radcliffe Observatory.
Inform. Bull. Southern Hemisph., No. 23, p. 27 - 33 (1973).

Radcliffe sells 74 inch telescope.
Monthly Notes Astron. Soc. Southern Africa, Vol. 33, 4 - 5 (1974).

Communications from the Radcliffe Observatory, Pretoria, Nos. 128 (P. R. Warren, 10.114.050), 129 (I. S. Glass, M. W. Feast, 10.113.030), 130 (P. J. Andrews, A. D. Thackeray, 10.118.013).

Radcliffe Observatory, *Pretoria,* **Reprints,** Nos. 127 (R. M. Catchpole, M. W. Feast, 10.114.031), 128 (M. W. Feast, T. Lloyd Evans, 10.114.032), 129 (M. W. Feast, I. S. Glass, 10.152.005), 130 (P. J. Andrews, T. Lloyd Evans, 10.154.008).

008.103   **Princeton**

Princeton University Observatory, Princeton, New Jersey. – Observatory report.   L. Spitzer, Jr.
Bull. American Astron. Soc., Vol. 6, 166 - 172 (1974).

008.104   **Pulkovo**

Trudy Glavnoj Astronomicheskoj Observatorii v Pulkove, Seriya 2, Vol. 81 (N. V. Fatchikhin, 11.112.006; N. M. Bronnikova, 11.112.007; A. V. Bolbochanu, 11.112.008).

008.105   **Richland, Washington**

Battelle Northwest Laboratories. Rattlesnake Mountain Observatory, Richland, Washington. – Observatory report.   R. A. Stokes.
Bull. American Astron. Soc., Vol. 6, 13 - 14 (1974).

008.106   **Richmond Hill**

David Dunlap Observatory, University of Toronto, Richmond Hill, Ontario, Canada. – Report 1972 July 1 – 1973 June 30.   D. A. MacRae.
Bull. American Astron. Soc., Vol. 6, 21 - 28 (1974).

David Dunlap Observatory, University of Toronto.
D. A. MacRae.
Quarterly Journ. Roy. Astron. Soc., Vol. 15, 146 - 166 (1974). Report for the year 1972 July 1 to 1973 June 30.

Communications from the David Dunlap Observatory, University of Toronto, Richmond Hill, Ontario, Canada, Nos. 381 (R. F. Garrison, 10.115.025), 382 (J. E. Winzer, 10.122.040), 383 (R. C. Bignell, 10.141.052), 384 (C. M. Coutts, 10.122.063), 385 (S. van den Bergh, 10.133.009), 386 (H. Sawyer Hogg, 10.122.049), 387 (D. L. DuPuy, 10.122.080), 388 (J. F. Heard, R. Hurkens, 10.119.015), 389 (R. F. Garrison, 10.114.120), 390 (N. R. Walborn, 10.155.069), 391 (R. C. Bignell, 10.141.131), 392 (S. Pineault, S. P. S. Anand, 10.062.030), 393 (H. B. Sawyer Hogg, appendix I of 10.120.001).

008.107   **Riga**

Hronika. Jāni Ikaunieki atceroties.   Ā. Alksne.
Zvaigžņotā debess, 1973./74. gada ziema, p. 58 - 59.

Uchenye Zapiski Latvijskogo Gosudarstvennogo Universiteta im. P. Stuchki, Vol. 175, Astron., vyp. (No.) 8, (11.003.010); Vol. 202, Astron., vyp. (No.) 10 (11.003.012).

008.108   **Rio de Janeiro**

Observatorio Nacional. (National Observatory).
Inform. Bull. Southern Hemisph., No. 23, p. 16 - 17 (1973).

008.109   **Rochester**

C. E. Kenneth Mees Observatory, Rochester, New York. – Observatory report.   S. Sharpless.
Bull. American Astron. Soc., Vol. 6, 118 - 121 (1974).

C. E. Kenneth Mees Observatory, University of Rochester, Rochester, N. Y., **Reprint,** Nos. 44 (C. Sturch, 11.113.014), 49 (L. Taff, M. P. Savedoff, 10.131.129), 50 (S. Sharpless, 11.008.000).

008.110   **Roma**

Monthly Bulletin. Osservatorio Astronomico di Roma, Nos. 186, 188 - 191 (M. Cimino, M. Torelli, F. Casamassima, V. Croce, 11.075.013).

Photographic Journal of the Sun, Osservatorio Astronomico di Roma, Nos. 75 - 80 (M. Cimino, 11.075.014).

008.111   **San Diego**

University of California Observatory, Santa Cruz, California. III. San Diego Campus. – Observatory report.
Bull. American Astron. Soc., Vol. 6, 248 - 252 (1974).

008.112   **San Fernando**

Memoria de las actividades en 1973.
Inst. y Obs. de Marina, San Fernando (Cadiz), España, 14 pp. (1974).

Instituto y Observatorio de Marina, San Fernando (Cadiz), España, Serie C, No. 75 (A. Orte, M. Sánchez, I. Vitini, J. B. Fernández, F. Parra, 11.044.011).

008.113   **San Juan**

Observatorio "Félix Aguilar". (Félix Aguilar" Astronomical Observatory, School of Engineering, National University of Cuyo).
Inform. Bull. Southern Hemisph., No. 23, p. 13 - 14 (1973).

008.114 San Miguel, Buenos Aires

Observatorio Nacional de Física Cósmica. (National Observatory of Cosmic Physics).
Inform. Bull. Southern Hemisph., No. 23, p. 14 (1973).

008.115 Santa Cruz

University of California Observatory, Santa Cruz, California. IV. Lick Observatory, D. E. Osterbrock; **V. Board of Studies in Astronomy and Astrophysics,** W. G. Mathews. Observatory reports.
Bull. American Astron. Soc., Vol. 6, 252 - 258 (1974).

Current research programs at the Lick Observatory. See Abstr. 013.013.

008.116 Santiago

Departamento de Astronomía, Universidad de Chile. (Astronomy Department, University of Chile).
Inform. Bull. Southern Hemisph., No. 23, p. 23 (1973).

008.117 São Paulo

Centro de Radio Astronomía e Astrofísica da Universidade Mackenzie. (Center of Radio Astronomy and Astrophysics, Mackenzie University).
Inform. Bull. Southern Hemisph., No. 23, p. 17 - 18 (1973).

008.118 Seattle, Washington

University of Washington, Astronomy Department, Seattle, Washington. – Observatory report. G. Wallerstein.
Bull. American Astron. Soc., Vol. 6, 195 - 197 (1974).

008.119 Sendai

Sendai Astronomiaj Raportoj, Nos. 127 (M. Takeuti, 10.122.093), 141 (Y. Shibata, M. Takeuti, 10.065.052), 142 (M. Seki, 10.131.094).

008.120 Skalnaté Pleso

Thirty years of the Astronomical Observatory Skalnaté Pleso. Ľ. Pajdušáková.
Kozmos, Vol. 5, 65 - 68 (1974). In Slovak.

008.121 Sonneberg

Akademie der Wissenschaften der DDR, Zentralinstitut für Astrophysik, Sternwarte Sonneberg. **Mitteilungen der Sternwarte zu Sonneberg,** No. 60 (G. A. Richter, 06.158.103; G. Jackisch, 07.158.020; S. Rössiger, 07.131.020; G. Jackisch, 08.114.106; W. Götz, 08.153.022; W. Götz, 08.153.023; S. Rössiger, W. Wenzel, 08.122.086; G. A. Richter, I. Meinunger, 08.113.032; W. Wenzel, 08.065.094; S. Rössiger, 09.113.036).

Akademie der Wissenschaften der DDR, Zentralinstitut für Astrophysik. **Veröffentlichungen der Sternwarte in Sonneberg,** Vol. 8, No. 3 (W. Götz, 11.153.018).

Zentralinstitut für Astrophysik. **Mitteilungen über Veränderliche Sterne,** Sonneberg, Vol. 6, No. 6 (R. Hudec, K. Juza, S. Rössiger, 11.113.045; L. Meinunger, 11.123.042; W. Wenzel, 11.133.019; H. Geßner, 11.123.043; M. Heß, 11.123.044; P. Ahnert, 11.123.045; M. Werner, 11.123.046; H. Geßner, 11.123.047; P. Ahnert, 11.121.089; H.-J. Jerominek, 11.123.048).

008.122 Stanford, California

Stanford Radio Astronomy Institute, Stanford University, Stanford, California. R. N. Bracewell.
Bull. American Astron. Soc., Vol. 6, 258 - 260 (1974).

008.123 Swarthmore

Sproul Observatory, Swarthmore, Pennsylvania. 1 July 1972 - 30 June 1973. – Observatory report.
S. L. Lippincott, W. D. Heintz.
Bull. American Astron. Soc., Vol. 6, 176 - 177 (1974).

Sproul Observatory, Swarthmore, Pennsylvania, **Reprints,** Nos. 217 (P. van de Kamp, 09.008.109), 218 (W. D. Heintz, 09.118.001), 219 (W. D. Heintz, 09.118.003), 220 (S. L. Lippincott, 09.117.024), 221 (W. D. Heintz, 09.118.010), 222 (J. L. Hershey, 09.041.020), 223 (S. L. Lippincott, 09.111.010), 224 (P. van de Kamp, 10.115.011), 225 (S. L. Lippincott, 10.115.013), 226 (W. D. Heintz, 10.117.011), 227 (W. D. Heintz, 10.111.005), 228 (J. L. Hershey, 10.117.026), 229 (P. van de Kamp, 10.118.022).

008.124 Sydney

CSIRO Division of Radiophysics. Research activities 1973. J. P. Wild.
Division Radiophys. CSIRO, Separate print, 38 pp. (1973).
Cosmic radio astronomy; interstellar radio spectroscopy *(J. L. Caswell, R. X. McGee);* mapping the radio Milky Way *(G. A. Day);* measuring cosmic magnetic fields *(J. A. Roberts);* supernova remnants and planetary nebulae *(D. K. Milne);* pulsars *(M. M. Komesaroff, J. G. Ables);* using the Culgoora radioheliograph for cosmic radio astronomy *(O. B. Slee);* the southern quasar hunt *(J. G. Bolton);* solar radio astronomy; radio bursts in the corona *(N. R. Labrum, S. F. Smerd);* experimental techniques and instrumentation – electronic techniques for radio astronomy at Parkes *(B. F. Cooper, J. W. Brooks);* the Parkes 1024 channel digital correlator *(J. G. Ables);* making the Parkes 64-m telescope capable of operating at all wavelengths greater than 1 cm *(D. E. Yabsley);* the importance of well designed primary feeds and their realization *(B. M. Thomas);* the 3-frequency heliograph *(K. V. Sheridan);* an electro-optical spectrograph *(T. W. Cole);* support for the Apollo missions to the moon *(J. G. Bolton);* radio navigation – development of a new generation aircraft landing system *(H. C. Minnett).*

Division of Radiophysics, C.S.I.R.O., Sydney, **Radiophysics Publication** RPP 1650 (N. Fourikis, 11.033.050), 1660 (W. K. Huchtmeier, R. A. Batchelor, 09.155.058), 1716

(U. J. Schwarz, D. J. Cole, D. Morris, 10.033.059).

Division of Radiophysics, C. S. I. R. O., Sydney,
Australia. Separate prints (J. V. Wall, R. D. Cannon, 11.141.
070; H. M. Tovmassian, 11.153.003; H. M. Tovmassian, E. T.
Shahbazian, 11.153.004; H. M. Tovmassian, E. T. Shahbazian,
S. E. Nersessian, 11.153.005; H. M. Tovmassian, E. T.
Shahbazian, S. E. Nersessian, 11.153.006; H. M. Tovmassian,
S. E. Nersessian, 11.153.007; J. V. Wall, D. J. Cole, 11.141.
023; W. K. Huchtmeier, R. A. Batchelor, 11.131.514; D. B.
Melrose, 11.074.047; D. B. Melrose, 11.074.048; A. J.
Shimmins, J. G. Bolton, 11.141.109).

Narrabri Observatory, University of Sydney.
R. Hanbury Brown.
Inform. Bull. Southern Hemisph., No. 23, p. 15 (1973).

008.125 Tartu

Tartu Astronoomia Observatoorium, Teated,
No. 44 (T. Nugis, 11.065.048; M. Ilmas, T. Nugis, 11.114.054;
T. Nugis, 11.114.055; T. Nugis, 11.114.056; T. Nugis, T.
Feklistova, 11.114.057), 45 (H. Eelsalu, 11.155.013), 46
(M. Jôeveer, 11.155.014; M. Jôeveer, 11.155.015; M. Jôeveer,
11.155.016).

008.126 Tashkent

Short history of the Tashkent Astronomical Observatory — Astronomical Institute of the Academy of Sciences of the Uzbek SSR (1873 - 1973). V. P. Shcheglov.
Astronomical Institute of the Academy of Sciences of the
Uzbek SSR — 100 years, (see 003.011), p. 6 - 21 (1974).
In Russian.

Chronicle.
Tsirk. Astron. Inst., *Tashkent,* No. 43 (390), p. 30 - 31 (1973).
In Russian.

**List of papers published by members of the
institute in 1972.**
Tsirk. Astron. Inst., *Tashkent,* No. 43 (390), p. 32 - 33 (1973).
In Russian.

**Astronomical Institute of the Academy of Sciences
of the Uzbek SSR — 100 years.** See Abstr. 003.011.

Tsirkulyar Astronomicheskogo Instituta, Akademiya Nauk Uzbekskoj SSR, Nos. 40 (387) (N. A. Omelina,
11.044.005; F. G. Mustaeva, 11.075.002; A. G. Rakhimov,
Eh. Rakhmatov, 11.103.100; O. S. Tursunov, 11.041.016),
41 (388) (A. A. Latypov, 11.112.009), 42 (389) (N. A.
Omelina, 11.044.005; F. G. Mustaeva, 11.075.002; A. A.
Latypov, 11.112.010; L. I. Bashtova, A. G. Rakhimov,
11.097.039), 43 (390) (N. A. Omelina, 11.044.005; F. G.
Mustaeva, 11.075.002; A. A. Latypov, 11.112.011; M. R.
Ehshmatov, Eh. Rakhmatov, 11.096.014), 44 (391) (M. F.
Bykov, 11.041.017).

008.127 Tautenburg

**Zentralinstitut für Astrophysik. Mitteilungen des
Karl-Schwarzschild-Observatoriums Tautenburg** der Akademie
der Wissenschaften der DDR, Nos. 66 (W. Högner, H. Löchel,
N. Richter, 09.102.011), 67 (N. Richter, 09.158.115), 68
(N. Richter, 11.005.006), 69 (H.-U. Sandig, 11.041.013), 70
(H. G. Beck, 11.032.010), 71 (F. Börngen, W. Högner, P.
Lochno, N. Richter, 11.132.007), 72 (F. Börngen, 11.032.
011).

008.128 Tokyo

Annals of the Tokyo Astronomical Observatory,
University of Tokyo, Second Series, Vol. 14, No. 1 (H. Kinoshita, H. Nakai, S. Aoki, 11.041.026; H. Kinoshita, G. Hori,
H. Nakai, 11.042.053), 2 (E. Hiei, M. Fukatsu, 11.073.092;
K. Kodaira, 11.114.171; F. Sato, K. Akabane, 11.131.555).

Tokyo Astronomical Bulletin, Tokyo Astronomical
Observatory, Second Series, Nos. 228 (H. Tabara, S. Kikuchi,
Y. Mikami, N. Kawano, N. Kawajiri, T. Ojima, M. Inoue, M.
Konno, K. Tomino, T. Daishido, 11.158.303), 229 (K.
Akabane, T. Miyaji, Y. Chikada, 11.033.049), 230 (K. Ichimura, Y. Shimizu, K. Okida, 11.122.092), 231 (T. Nakamura,
11.099.217), 232 (K. Akabane, Y. Chikada, K. Miyazawa,
11.033.094), 233 (M. Kiyokawa, S. Kikuchi, 11.121.097).

Time and Latitude Bulletins, Tokyo Astronomical
Observatory, Vol. 47, Nos. 8 - 12 (11.044.025.).

University of Tokyo, Tokyo Astronomical Observatory, Report Vol. 16, No. 4 (S. Hata, 11.074.070; K. Saito,
K. Tomita, 11.104.063; H. Shibasaki, 11.036.005; A. Tojo,
K. Saito, 11.082.091; K. Saito, S. Shinozawa, 11.103.122;
N. Sekiguchi, F. Miyamoto, 11.045.013).

Tokyo Astronomical Observatory, Reprints
Nos. 445 (K. Akabane, M. Morimoto, K. Nagane, K. Miyazawa,
T. Miyaji, H. Tabara, H. Hirabayashi, N. Kaifu, Y. Chikada,
11.131.024), 446 (T. Hirayama, Y. Nakagomi, 11.073.033),
447 (K. Nariai, 11.124.001), 448 (M. Saitō, 11.064.010), 449
(Y. Yamashita, 11.112.001), 450 (A. Takechi, 11.082.090),
451 (K. Nishi, 10.071.069), 452 (K. Akabane, H. Nakajima,
K. Ohki, F. Moriyama, T. Miyaji, 10.077.085).

Contributions from the Department of Astronomy,
University of Tokyo, Nos. 164 (T. Yoneyama, 10.062.004),
165 (M. Hirai, 11.114.095), 166 (M. Yuasa, 10.098.030),
167 (M. Kobayashi, Y. Osaki, 10.065.099), 168 (H. Yoshimura, 11.062.011), 169 (N. Kaifu, T. Iguchi, T. Kato, 11.155.
007), 170 (K. Nomoto, D. Sugimoto, 11.065.021), 171
(W. Unno, M.-K. Fujimoto, 11.065.022).

**Data Report of Hydrographic Observations. Series
of Astronomy and Geodesy,** Maritime Safety Agency, Tokyo,
Japan, No. 8 (T. Mori, Y. Ganeko, Y. Harada, M. Sasaki, M.
Yamaguti, 11.096.038).

008.129 Toledo

Ritter Astrophysical Research Center, The University of Toledo, Toledo, Ohio. — Observatory report.
A. N. Witt.
Bull. American Astron. Soc., Vol. 6, 172 - 174 (1974).

008.130 Torino

Contributi dell'Osservatorio Astronomico di Torino, (Pino Torinese), Nos. 70 (S. Vaghi, V. Zappalà, 10.008.113), 73 (S. Vaghi, V. Zappalà, 10.103.123), 74 (V. Banfi, 10.042. 105), 76 (S. Vaghi, V. Zappalà, 10.098.074), 77 (F. Scaltriti, 10.121.133).

Osservatorio Astronomico di Torino, Pino Torinese. Time Service, Bulletin No. 6 (C. Moranzino, 11.044.024).

Pubblicazioni Varie Fuori Serie dell'Osservatorio Astronomico di Torino (Pino Torinese), Nos. 54 (M. G. Fracastoro, 11.015.017), 55 (M. G. Fracastoro, 10.079.101), 56 (M. G. Fracastoro, Review of 10.003.101).

008.131 Tortosa

Publicaciones del Observatorio del Ebro, Miscelánea, Nos. 30 (A. Romañá, 10.051.010), 31 (A. Romañá, 11.004.062).

008.132 Toruń

Bulletin of the Astronomical Observatory of N. Copernicus University in Toruń, Nos. 50 (W. Iwanowska, 11.133.016; W. Iwanowska, 11.126.011; J. Krempeć, 11.113.032; S. Krawczyk, J. Krełowski, 11.122.091), 51 (S. Gąska, 11.098.019; S. Kasperczuk, L. Dybkowski, 11.098.020; L. Dybkowski, S. Kasperczuk, 11.098.021; F. Karmiński, 11.104.062).

008.133 Tucson

University of Arizona, Department of Planetary Sciences and Lunar Planetary Laboratory, Tucson, Arizona. Observatory report. C. P. Sonett.
Bull. American Astron. Soc., Vol. 6, 8 - 13 (1974).

National Radio Astronomy Observatory, Charlottesville, Virginia; Green Bank, West Virginia; and Tucson, Arizona. - Observatory reports. D. S. Heeschen.
Bull. American Astron. Soc., Vol. 6, 144 - 153 (1974).

008.134 Uccle

Bulletin Astronomique. (Astronomisch Bulletin). Observatoire Royal de Belgique (Koninklijke Sterrenwacht van België), Vol. 8, No. 2 (G. Evrard, R. Florée, C. Gonze, A. Koeckelenbergh, 11.075.006).

Observatoire Royal de Belgique (Koninklijke Sterrenwacht van België), Communications (Mededelingen), Série B, Nos. 81 (J. Dommanget, 10.041.020), 82 (P. Paquet, 10.055.030), 83 (A. Koeckelenbergh, 10.073.101), 84 (D. Djurovic, 11.044.009).

008.135 Uppsala

Uppsala Astronomiska Observatoriums Annaler, Band 5, No. 7 (B. Gustafsson, 11.064.044).

008.136 Utrecht

Utrechtse Sterrekundige Overdrukken, Sterrewacht "Sonnenborgh", Utrecht, Nos. 224 (J. P. de Grève, C. de Loore, C. de Jager, 08.064.039), 225 (K. S. de Boer, R. Hoekstra, K. A. van der Hucht, T. M. Kamperman, H. J. Lamers, S. R. Pottasch, 08.114.116), 228 (T. M. Kamperman, K. A. van der Hucht, H. J. Lamers, R. Hoekstra, 09.114.031), 229 (H. F. van Beek, 11.034.053), 230 (M. A. Raadu, M. Kuperus, 09.074.002), 231 (L. D. de Feiter, C. de Jager, 09.076.005), 232 (R. J. Rutten, 09.071.011), 233 (H. J. Lamers, K. A. van der Hucht, M. A. J. Snijders, N. Sakhibullin, 09.114.101), 234 (M. Kuperus, H. Rosenberg, 11.074.072), 235 (K. A. van der Hucht, H. J. Lamers, 09.119.011), 236 (E. P. J. van den Heuvel, C. de Loore, 09.142.102), 237 (C. Chiuderi, R. Giachetti, H. Rosenberg, 11.077.057), 238 (M. Burger, K. A. van der Hucht, H. J. Lamers, 09.114.155), 239 (E. P. J. van den Heuvel, 09.142.039), 240 (J. A. M. Bleeker, A. J. M. Deerenberg, J. Heise, K. Yamashita, Y. Tanaka, 09.142.029), 241 (C. Veth, T. de Graauw, J. C. Shelton, H. van de Stadt, 10.080.002), 242 (J. R. W. Heintze, 10.115.029), 243 (H. F. van Beek, L. D. de Feiter, 11.034.054), 244 (R. Hoekstra, K. A. van der Hucht, C. de Jager, T. Kamperman, H. J. Lamers, 11.114.107), 245 (C. Caroubalos, M. Pick, H. Rosenberg, C. Slottje, 10.077.007), 246 (L. D. de Feiter, 10.073.031), 247 (G. Hensberge, E. P. J. van den Heuvel, M. H. Paes de Barros, 10.119.011), 248 (E. P. J. van den Heuvel, C. de Loore, 10.142.051), 249 (J. Houtgast, A. van Sluiters, 10.084.245), 250 (C. de Jager, 10.010.015), 251 (L. D. de Feiter, 10.074.060), 252 (L. D. de Feiter, C. de Jager, 10.073.059), 253 (H. F. van Beek, L. D. de Feiter, 10.034.047), 254 (P. Hoyng, G. A. Stevens, 10.076.016), 256 (K. A. van der Hucht, H. J. Lamers, 11.012.012), 257 (A. D. Fokker, 10.077.039), 258 (A. D. Fokker, 10.077.040), 259 (A. D. Fokker, 10.010.041), 260 (T. M. Kamperman, K. A. van der Hucht, H. J. Lamers, R. Hoekstra, 09.114.049), 261 (M. Grewing, H. J. Lamers, C. M. Walmsley, C. Wulf-Mathies, 10.119.001), 262 (E. P. J. van den Heuvel, J. P. Ostriker, 10.142.049), 263 (T. de Graauw, H. van de Stadt, 10.094.121).

008.137 Valencia

Actividades del Observatorio Astronómico de la Facultad de Ciencias de Valencia: 1968–1973. A. López García.
Urania Barcelona, Año 58, Nos. 277 - 278, p. 237 - 247 (1973).

008.138 Victoria

The Dominian Astrophysical Observatory, Victoria, B.C. J. B. Hutchings.
Journ. Roy. Astron. Soc. Canada, Vol. 68, 109 - 110 (1974).

Publications of the Dominion Astrophysical Observatory, Victoria, B. C., Vol. 14, Nos. 8 (D. Crampton, A. Leir, F. Younger, 11.114.170), 9 (C. L. Morbey, 11.031.084).

008.139 Villanova, Pennsylvania

Villanova University, Department of Astronomy,

Villanova, Pennsylvania. – Observatory report.
E. F. Jenkins.
Bull. American Astron. Soc., Vol. 6, 183 - 185 (1974).

008.140 Vilnius

Chronicle.
Bull. Vilnius Astron. Obs., No. 36, p. 33 - 34 (1973).
In Russian.

Vilniaus Astronomijos Observatorijos Biuletenis (Bulletin of the Vilnius Astronomical Observatory), No. 36 (V. Straižys, 11.113.003; A. Bartkevičius, A. Gurklytė, G. Kavaliauskaitė, A. Kazlauskas, R. Kalytis, Z. Sviderskienė, J. Sperauskas, J. Sūdžius, V. Jasevičius, 11.113.004; A. Bartkevičius, 11.113.005).

008.141 Warsaw

Warsaw University Observatory and Astronomical Institute, Polish Academy of Sciences, Reprint Nos. 339 (M. Kozłowski, B. Paczyński, 09.065.152), 340 (S. M. Ruciński, 09.117.032), 341 (A. Żytkow, 09.064.071), 342 (G. Sitarski, 09.103.111), 352 (E. Ergma, B. Paczyński, 11.065.032), 353 (A. Kruszewski, 11.141.035), 354 (M. A. Abramowicz, 11.065.033).

008.142 Washington

U.S. Naval Observatory, Washington, D.C.
(1 July 1972 - 30 June 1973). – Observatory report.
K. A. Strand.
Bull. American Astron. Soc., Vol. 6, 177 - 182 (1974).

Astronomical Papers prepared for the use of the American Ephemeris and Nautical Almanac, Vol. 22, Part I (C. J. Cohen, E. C. Hubbard, C. Oesterwinter, 11.042.066).

U. S. Naval Observatory, Washington, D. C. Time Service Publications, Series 4, Nos. 361 - 386 (11.044.012); Series 7, Nos. 314 - 339 (11.044.013).

National Aeronautics and Space Administration, Washington, D.C. –Observatory reports. N. G. Roman.
Bull. American Astron. Soc., Vol. 6, 126 - 140 (1974). – Concerning the reports of Ames Research Center; Goddard Space Flight Center; Jet Propulsion Laboratory; Johnson Space Center; Langley Research Center; Marshall Space Flight Center.

National Bureau of Standards, Washington, D.C.
Observatory report. L. Hagan.
Bull. American Astron. Soc., Vol. 6, 140 - 144 (1974).

008.143 Westford, Massachusetts

Haystack Observatory, Northeast Radio Observatory Corporation, Westford, Massachusetts. – Observatory report. P. B. Sebring.
Bull. American Astron. Soc., Vol. 6, 56 - 60 (1974).

008.144 Wien

Astronomische Mitteilungen Wien, Nos. 10 (R. Dvorak, E. Göbel, 10.151.046), 11 (J. Hopmann, 11.155.076), 12 (K. Ferrari d'Occhieppo, H. Jenkner, 11.155.077), 13 (J. Hopmann, 11.118.025), 14 (J. Hopmann, 11.118.026).

008.145 Williams Bay

The University of Chicago: Department of Astronomy and Astrophysics, Chicago, Illinois. The Yerkes Observatory, Williams Bay, Wisconsin 1972–1973. – Observatory reports. E. N. Parker, W. F. van Altena.
Bull. American Astron. Soc., Vol. 6, 15 - 21 (1974).

008.146 Williamstown

Hopkins Observatory, Williams College, Williamstown, Massachusetts. – Observatory report. J. M. Pasachoff.
Bull. American Astron. Soc., Vol. 6, 60 - 61 (1974).

008.147 Wrocław

Wrocław Astronomical Observatory, Reprint Nos. 89 (J. Bem, T. Jastrzębski, 09.103.112), 90 (S. Płocieniak, B. Rompolt, 09.071.044), 91 (A. Opolski, 10.122.073), 92 (T. Ciurla, 10.122.132).

008.148 Yorktown Heights, New York

IBM Thomas J. Watson Research Center, Yorktown Heights, New York. – Observatory report.
M. C. Gutzwiller.
Bull. American Astron. Soc., Vol. 6, 61 - 62 (1974).

008.149 Zelenchukskaya

Chronicle. E. I. Popova.
Nauchn. Informatsii, vyp. (No.) 27, p. 108 - 111 (1973). In Russian.

Soobshcheniya Spetsial'noj Astrofizicheskoj Observatorii, Akademiya Nauk SSSR, vyp. (No.) 5 (S. V. Rublev, 11.064.033; V. V. Leushin, 11.114.074), 6 (L. I. Snezhko, 11.121.045; P. A. Fridman, 11.033.007; V. K. Dubrovich, 11.066.032), 7 (I. D. Karachenstev, 11.158.057).

008.150 Zürich

Tätigkeitsbericht der Eidgenössischen Sternwarte Zürich für das Jahr 1973. M. Waldmeier.
Zürich, 7 pp. (1974).

Astronomische Mitteilungen der Eidgenössischen Sternwarte Zürich, Nos. 324 (M. Waldmeier, 11.079.100), 325 (M. Waldmeier, S. E. Weber, 11.079.100), 326 (M. Waldmeier, 11.075.015), 327 (A. Zelenka, 11.080.063), 330 (M. Waldmeier, 11.074.114).

Quarterly Bulletin on Solar Activity (Zürich), Nos. 181 - 183 (M. Waldmeier, R. Howard, G. Olivieri, M. Bernot, H. Tanaka, 11.075.016).

009 Notes on Observatories, Planetaria, and Exhibitions

009.001 **Chile's mountain observatories revisited.** J. B. Irwin.
Sky Telescope, Vol. 47, 10 - 16 (1974). — Concerning Cerro Tololo, Las Campanas, La Silla resp. ESO.

009.002 **A Hong Kong observatory.** J. H. C. Liu.
Sky Telescope, Vol. 47, 221 - 226 (1974).

009.003 **People's Observatory in Lvov.** G. I. Pashchenko.
Zemlya i Vselennaya, 1974, No. 2, p. 79. In Russian.

009.004 **50 Jahre Zeiss-Planetarium.** H.-U. Keller.
SuW, Vol. 13, 17 (1974).

009.005 **Die Sternwarte des Coppernicus-Gymnasiums in Norderstedt.** A. Langkavel.
SuW, Vol. 13, 25 (1974).

009.006 **Rhodes University, Department of Physics.** E. E. Baart.
Monthly Notes Astron. Soc. Southern Africa, Vol. 33, 6 - 7 (1974).

009.007 **University of the Orange Free State. Boyden Observatory, Department of Astronomy.** A. H. Jarrett.
Monthly Notes Astron. Soc. Southern Africa, Vol. 33, 11 - 12 (1974).

009.008 **University of South Africa. Department of Mathematics and Astronomy.** J. Wolterbeek.
Monthly Notes Astron. Soc. Southern Africa, Vol. 33, 12 (1974).

009.009 **Jupiter mit seinen vier größten Monden. Ein neues Zusatzgerät aus Jena für Planetarien.** H. Letsch.
Jenaer Rundschau, (Jena Review), 18. Jahrgang, p. 333 - 335 (1973).

009.010 **The public observatory at Burke Baker Planetarium.** C. Sumners.
Sky Telescope, Vol. 47, 315 - 317 (1974).

009.011 **The planetarium: Yesterday, today, and tomorrow.** C. F. Hagar.
Mercury, (Journ. Astron. Soc. Pacific), Vol. 2, No. 6, p. 2 - 9 (1973).

009.012 **A public observatory.** B. Mitchell.
Journ. British Astron. Ass., Vol. 84, 123 - 125 (1974).

009.013 **Observatorijas un astronomi. Blagoveščenskas platuma laboratorija.** G. Šeptunovs.
Zvaigžņotā debess, 1973./74. gada ziema, p. 32 - 36.

009.014 **Observatorijas un astronomi.**
Zvaigžņotā debess, 1974. gada pavasaris, p. 34 - 43. Budapeštas observatorija (*J. Francmanis*), p. 34 - 39; Profesoram B. Voroncovam-Veljaminovam 70 gadi (*Z. Cirse*), p. 39 - 40; Padomju planetoloģijas pamatlicējs (*M. Grigorivs*), p. 40 - 41; F. Argelanderam — 175 (*A. Maslovskis*), p. 42 - 43.

009.015 **An inexpensive home-built planetarium projector.** S. B. Smith.
Sky Telescope, Vol. 48, 27 - 29 (1974).

009.016 **The Ulugh Begh International Latitude Station in Kitab.** A. M. Kalmykov.
Astronomical Institute of the Academy of Sciences of the Uzbek SSR — 100 years, (see 003.011), p. 52 - 58 (1974). In Russian.

009.017 **Astronomija na Čolinoj Kapi (Astronomy at Čolina Kapa).** M. Stupar.
Vasiona, Vol. 22, 12 - 14 (1974).

009.018 **L'attività dell'Istituto Geografico Militare nel 1973. Programma dei lavori previsti per l'anno 1974.** C. Revelli.
Boll. Geod. Sci. Affini, Anno 33, p. 129 - 136 (1974).

009.019 **Planetarium of space flights in Olsztyn.** S. Oszczak.
Urania Kraków, Vol. 45, 37 - 41 (1974). In Polish.

009.020 **Feierliche Eröffnung der Copernicus-Ausstellung in Wien am 25. Mai 1973.** K. Ferrari d'Occhieppo.
Almanach Österreich. Akad. Wiss., 123. Jahrgang, p. 589 - 597 (1973).

009.021 **Department of Natural Philosophy, University of Aberdeen.** R. V. Jones.
Quarterly Journ. Roy. Astron. Soc., Vol. 15, 167 - 168 (1974). Report for the year ending 1973 September 30.

009.022 **Annuaire de l'Institut Météorologique et Climatologique, Université de Thessaloniki. Annuaire No. 27—Spécial I.** G. C. Livadas.
Thessaloniki, 12 + 6 pp. (1973).

The author gives a description of the new building of the Institute of Meteorology and Climatology, as well as an analytical list of the installations and equipment of the meteorological station, which started operating on 1959 January 1.

009.023 **Aus der Arbeit der Volkssternwarten.**
Sterne, 50. Jahrgang, p. 122 - 123 (1974). – 20 Jahre Pionier- und Volkssternwarte Schneeberg (*B. Zill*).

009.024 **Planetarium, Modell des Sonnensystems.**
Astronomische Gesellschaft Burgdorf.
Astro-Bilderdienst SAG, Burgdorf. Price sFr. 15.00 (1974). Review in Orion Schaffhausen, 32. Jahrgang, p. 38 (1974).

009.025 **Skil-Craft-Planetarium** – Eine Sphärenkugel höchster Präzision. Mit einem ausführlichen Anleitungsbuch.
Kosmos-Service, Stuttgart, Price DM 45.00. – Review in SuW, Vol. 13, 64; 1974 (*G. D. Roth*).

Planetarium für Stuttgart.
SuW, Vol. 13, 149 (1974).

URSIES and the Bartol Coudé Observatory.
See Abstr. 034.064.

010 Societies, Associations, Organizations

010.001 **American Association of Variable Star Observers (AAVSO)**

No publication received.

010.002 **American Astronomical Society (AAS)**

The 142nd meeting of the American Astronomical Society, held 26–29 March 1974 at Lincoln, Nebraska.
Bull. American Astron. Soc., Vol. 6, 211 - 227 (1974).

Late-paper abstracts from the 141st meeting of the American Astronomical Society held 2–5 December 1973 at Tucson, Arizona.
Bull. American Astron. Soc., Vol. 6, 261 - 266 (1974).

Abstracts of papers presented at the High-Energy Astrophysics Division meeting held 6–8 December 1973 at Tucson, Arizona.
Bull. American Astron. Soc., Vol. 6, 269 - 282 (1974).

Abstracts of papers presented at the Solar Physics Division meeting held 9–11 January 1974 at Honolulu, Hawaii.
Bull. American Astron. Soc., Vol. 6, 284 - 298 (1974).

010.003 **Association of Lunar and Planetary Observers (ALPO)**

Announcements.
Strolling Astronomer, Vol. 24, 206 - 208, 253 - 256 (1974).

List of materials and services supplied by ALPO recorders. J. R. Smith.
Strolling Astronomer, Vol. 24, 195 - 198 (1974).

Minor Planets Section news.
Minor Planet Bull., Vol. 1, 19 - 20, 23, 29, 37 - 38 (1974).

010.004 **Astronomical Society of Australia (ASA)**

Society business: Report of the council May 1973.
Proc. Astron. Soc. Australia, Vol. 2, 228 (1973).

010.005 **Astronomical Society of Czechoslovakia**

No publication received.

010.006 **Astronomical Society of the Pacific (ASP)**

No publication received.

010.007 **Astronomical Society of Southern Africa (ASSA)**

Notices.
Monthly Notes Astron. Soc. Southern Africa, Vol. 33, 1 - 2, 31 - 32, 47 - 48, 59 (1974).

010.008 **Astronomical Society of Victoria (ASV)**

Society notes.
Journ. Astron. Soc. Victoria, Vol. 27, 30 - 31 (1974).

Variable Stars Section. T. B. Tregaskis.
Journ. Astron. Soc. Victoria, Vol. 27, 13 (1974).

Annual report, 1973.
A. E. Coombs, R. J. C. Lawrence, D. H. Walker, J. L. Perdrix, S. Chivers, J. Patchett, D. H. Whitehead, W. G. H. Tregear, B. S. Adcock.
Journ. Astron. Soc. Victoria, Vol. 27, 2 - 11, 12 - 13 (1974).
Included are reports on the activities of different sections of the Society.

010.009 **Astronomical Society of Western Australia (ASWA)**

Reports of proceedings – 254th – 258th ordinary meetings.
Journ. Astron. Soc. Western Australia, Vol. 25, January - June (1974).

Comet section A. S. W. A. F. de Jong.
Journ. Astron. Soc. Western Australia, Vol. 25, January, February (1974).

010.010 **Astronomische Gesellschaft (AG)**

Wissenschaftliche Tagung der Astronomischen Gesellschaft in Oberkochen vom 24. - 27. April 1973: Bericht über die Tagung. K. Schaifers.
Mitt. Astron. Ges., No. 34, p. 7 - 8 (1973/74).

010.011 **Astronomisk Selskab København**

No publication received.

010.012 **British Astronomical Association (BAA)**

Notices.
Journ. British Astron. Ass., Vol. 84, 82 - 83, 163 - 165, 242 - 246 (1974).

Meetings of the Association.
Journ. British Astron. Ass., Vol. 84, 88 - 94, 165 - 175, 247 - 255 (1974).

Lunar Section. P. Moore.
Journ. British Astron. Ass., Vol. 84, 200 - 202 (1974).

Solar Section. V. Barocas.
Journ. British Astron. Ass., Vol. 84, 129 - 135 (1974).

Variable Star Section. J. E. Isles.
Journ. British Astron. Ass., Vol. 84, 136 - 137, 203 - 208 (1974).

New members elected.
Journ. British Astron. Ass., Vol. 84, 152 - 159, 231 - 236 313 - 317 (1974).

Computing Section. C. Dinwoodie.
Journ. British Astron. Ass., Vol. 84, 209 - 216 (1974).

The annual general meeting of the Association held on Wednesday 1973 October 31 at 23 Savile Row, London W. 1.
Journ. British Astron Ass., Vol. 84, 84 - 88 (1974).

Residential course on observational astronomy at Horncastle, 1973 November.
T. P. Byatt, J. F. Ravest.
Journ. British Astron. Ass., Vol. 84, 95 (1974).

Education committee. E. A. Beet.
Journ. British Astron. Ass., Vol. 84, 255 - 256 (1974).

Lunar Section: Report of Section meeting.
P. Moore.
Journ. British Astron. Ass., Vol. 84, 280 - 282 (1974).

010.013 **British Interplanetary Society (BIS)**

Society news.
Spaceflight, Vol. 16, 77 (1974).

Society meetings.
Spaceflight, Vol. 16, 154 - 157 (1974).

010.014 **Committee on Space Research (COSPAR)**

Latest results of the investigations of the moon, Mars and Venus (16th COSPAR meeting, Konstanz, May 23 – June 6, 1973). O. Wołczek.
Postępy Astron. Vol. 22, 57 - 66 (1974). In Polish.

010.015 **European Space Research Organization (ESRO)**

European space looks forward to the 1980s.
A. Dattner.
Nature, Vol. 249, 398 - 399 (1974).

ESRO verjaarde in maart. Europa doet tien jaar aan ruimte-onderzoek. G. W. E. Beekman.
Zenit, Vol. 1, No. 4, p. 2 - 5 (1974).

010.016 **International Astronautical Federation (IAF)**

Paths and perspectives of development of aeronautics.
S. A. Nikitin, Yu. A. Shkolenko.
Zemlya i Vselennaya, 1974, No. 2, p. 53 - 58. In Russian.
XXIVth congress of the International Astronautical Federation in Baku, 1973, Oct. 7 - 13.

010.017 **International Astronomical Union (IAU)**

International Astronomical Union, Information Bulletin, Nos. 31, 32. 37 + 13 pp. (1974).
G. Contopoulos.
Contents: General assemblies; Executive committee; Commissions; IAU symposia and colloquia; Other scientific meetings; IAU publications; Other publications; International organizations; Membership.

Astronomi in Australia, 21 al 30 agosto a Sydney.
M. G. Fracastoro.
Coelum, Vol. 42, 35 - 37 (1974).

IAU symposium 65 "Exploration of the planetary system". W. M. Irvine.
Icarus, Vol. 21, 202 - 207 (1974). − Meeting review.

Fifteenth general assembly of the International Astronomical Union, Sydney, Australia, August 21 - 30, 1973. G. Hill.
Journ. Roy. Astron. Soc. Canada, Vol. 68, 89 - 95 (1974).

I symposia N. 56 e 57 dell'Unione Astronomica Internazionale. G. Godoli.
Mem. Soc. Astron. Italiana, Nuova Ser., Vol. 44, 691 - 693 (1973/74).

Il simposium N. 65 dell'Unione Astronomica Internazionale. C. Blanco.
Mem. Soc. Astron. Italiana, Nuova Ser., Vol. 44, 694 - 695 (1973/74).

Welttreffen der Astronomen in Australien und Polen. E. Wiedemann.
Orion Schaffhausen, 32. Jahrgang, p. 16 (1974).

Extraordinary meeting of the International Astronomical Union, Poland, Warsaw − Toruń − Krakow, September 4 - 12, 1973.
Postępy Astron., Vol. 22, 3 - 5 (1974). In Polish.

Stability of the solar system and of the small stellar systems. IAU symposium No. 62. K. Ziołkowski.
Postępy Astron., Vol. 22, 7 - 13 (1974). In Polish.

Confrontation of cosmological theories with observational data. IAU symposium No. 63.
A. Czerny, M. Heller, W. Zonn.
Postępy Astron., Vol. 22, 15 - 26 (1974). In Polish.

Gravitational radiation and gravitational collapse. IAU symposium No. 64. J. P. Lasota.
Postępy Astron., Vol. 22, 27 - 28 (1974). In Polish.

Late phases of stellar evolution. IAU symposium No. 66. B. Paczyński.
Postępy Astron., Vol. 22, 29 - 31 (1974). In Polish.

XV. Generalkonferenz der Internationalen Astronomischen Union (Kommission 31). G. Becker.
PTB Mitt., 84. Jahrgang, p. 123 (1974).

The report on the activity of the I.A.U. Commission 8 study group on astronomical refraction.
Publ. Obs. Astron. Beograd, No. 18, (see 012.023), p. 7 - 8 (1974).

Extraordinary general assembly of the International Astronomical Union. A. G. Masevich.
Zemlya i Vselennaya, 1974, No. 2, p. 58 - 59. In Russian. Poland, 1973, Sept. 4 - 12.

"Colloquia Copernicana". Z. K. Sokolovskaya.
Zemlya i Vselennaya, 1974, No. 2, p. 59 - 63. In Russian.

Symposium on cosmology. A. G. Doroshkevich.
Zemlya i Vselennaya, 1974, No. 2, p. 64 - 65. In Russian.

Starptautiskās astronomu savienības Ārkārtējā ģenerālā asambleja Polijas Tautas Republikā.
Z. Alksne, A. Balklavs, J. Francmanis.
Zvaigžņotā debess, 1974. gada pavasaris, p. 1 - 18.

Transactions of the International Astronomical Union, Volume XV B: Proceedings of the Fifteenth General Assembly, Sydney 1973 and Extraordinary Assembly, Poland 1973. See Abstr. 003.013.

A report on the activity of I.A.U. Commission No. 46 (Teaching of Astronomy). See Abstr. 014.051.

010.018 Meteoritical Society

Annual meeting−Meteoritical Society.
J. E. Kennedy, M. R. Dence.
Journ. Roy. Astron. Soc. Canada, Vol. 68, 41 - 42 (1974).
1973 August 26 - 31, Davos.

Abstracts of papers presented at the 36th annual meeting of the Meteoritical Society, August 26−31, 1973, Davos, Switzerland.
Meteoritics, Vol. 8, 323 - 469 (1973). − See Abstr. 082.025; 094.006; 094.130 - 094.136; 094.327 - 094.357; 105.018 - 105.085; 107.008.

010.019 Nederlandse Vereniging voor Weer- en Sterrenkunde

No publication received.

010.020 Polskie Towarzystwo Astronomiczne (PTA)

No publication received.

010.021 Polskie Towarzystwo Miłośników Astronomii (PTMA)

PTMA chronicle.
Urania Kraków, Vol. 45, 16 - 21, 52 - 55, 88 - 89, 123 - 124, 185 - 189 (1974). In Polish.

010.022 Royal Astronomical Society (RAS)

Meetings of the Society.
Observatory, Vol. 94, 1 - 4, 33 - 40, 97 - 108 (1974).

Meetings of the Society.
Quarterly Journ. Roy. Astron. Soc., Vol. 15, 1 - 6, 71 - 81 (1974).

Royal Astronomical Society Education Committee.
R. J. Tayler.
Quarterly Journ. Roy. Astron. Soc., Vol. 15, 68 (1974).

010.023 Royal Astronomical Society of Canada (RAS Canada)

No publication received.

010.024 Royal Astronomical Society of New Zealand (RAS New Zealand)

The Royal Astronomical Society of New Zealand (Inc.). 51st annual report of council for the year ended 1973 September 30.   P. A. Read, P. D. Cain.
Southern Stars, Vol. 25, 77 - 83 (1973).

Report of Variable Star Section.   F. M. Bateson.
Southern Stars, Vol. 25, 84 - 90 (1973). – Report for the year ended 30th September, 1973.

Report of the Treasurer.   B. E. Stonehouse.
Southern Stars, Vol. 25, 90 - 94 (1973). – Report for the year ended 30th September 1973.

51st Conference Abstracts.
Southern Stars, Vol. 25, 95 - 100 (1973). – See abstracts 011.022, 031.021 - 031.023, 066.026, 122.045.

010.025 Schweizerische Astronomische Gesellschaft (SAG)

Aus der SAG und den Sektionen.
Orion Schaffhausen, 32. Jahrgang, p. 32, 83 - 85, 88 (1974).

Jahresbericht 1973 der 'Astronomischen Arbeitsgruppe' der 'Naturforschenden Gesellschaft Schaffhausen'. H. Rohr.
Orion, 32. Jahrgang, p. 123 (1974).

Die SAG-Generalversammlung vom 4./5. Mai 1974 in Genf.   E. Wiedemann.
Orion, 32. Jahrgang, p. 124 (1974).

Jahresbericht des Zentralpräsidenten, für das Jahr 1973, erstattet an der Generalversammlung der SAG in Genf am 4. Mai 1974.   W. Studer.
Orion, 32. Jahrgang, p. 125 (1974).

Bericht des Generalsektretärs der SAG über seine Tätigkeit im Jahre 1973, erstattet an der Generalversammlung in Genf am 4. Mai 1974.   H. Rohr.
Orion, 32. Jahrgang, p. 126 - 127 (1974).

010.026 Sociedad Astronómica de México

No publication received.

010.027 Società Astronomica Italiana (SAI)

Discorso inaugurale alla XVI riunione della Società Astronomica Italiana.   G. Righini.
Mem. Soc. Astron. Italiana, Suppl. Vol. 44, p. S217 - S219 (1974).

Verbale dell'assemblea generale della Società Astronomica Italiana.
Mem. Soc. Astron. Italiana, Suppl. Vol. 44, p. S 317 - S 322 (1974). – Arcetri, 1972 October 15.

La XVII riunione della Società Astronomica Italiana. G. Righini.
Mem. Soc. Astron. Italiana, Suppl. Vol. 44, p. S 325 - S 326 (1974). – Bologna, 26–27–28 ottobre 1973.

Verbale dell'assemblea generale della Società Astronomica Italiana.
Mem. Soc. Astron. Italiana, Suppl. Vol. 44, p. S327 - S333 (1974). – Bologna, 1973 October 28.

010.028 Société Astronomique de France (SAF)

Les séances de la Société.   B. Clouet, L. Tartois.
L'Astronomie, 88e année, p. 29 - 32, 76 - 78, 111 - 112, 144 - 146, 187 - 190, 236 - 239 (1974).

Commission des Instruments et de la Photographie Astronomique.   A. Hamon.
L'Astronomie, 88e année, p. 42 - 56 (1974).

Commission des Surfaces Planétaires.
J. Dragesco.
L'Astronomie, 88e année, p. 103 - 106, 225 - 227 (1974).

Compte rendu de la Commission du Soleil.
M.-J. Martres.
L'Astronomie, 88e année, p. 228 - 234 (1974).

010.029 Société Astronomique "R. Bošković"

XXIst meeting of the Astronomical Society "Ruđer Bošković".   P. M. Đurković.
Vasiona, Vol. 22, 32 (1974). In Serbo-Croatian.

Odluka Upravnog odbora Astronomskog društva "Ruđer Bošković" (Resolution of the Executive Committee of the Astronomical Society "Ruđer Bošković").
Vasiona, Vol. 22, 32 (1974).

010.030 Société Chronométrique de France

No publication received.

010.031 Société Belge d'Astronomie, de Météorologie et de Physique du Globe

Séance mensuelle.   M. Ducuroir.
Ciel et Terre, Vol. 90, 79 - 80, 159, 233 - 236 (1974).

010.032 Svenska Astronomiska Sällskapet

No publication received.

010.033 VAGO (Astronomical-Geodetical Society of the USSR)

IIIrd plenary session of the VAGO Central Council of the Vth convocation.   V. A. Bronshtehn.
Tsirkulyar Vses. astron.-geod. o-va, 1973, No. 24, p. 6 - 7. In Russian. – Abstr. in Referativ. Zhurn. 51. Astron., 3.51.34 (1974).

Report on the activities of the Soviet Union Astro-

nomical-Geodetical Society in 1972.
Tsirkulyar Vses. astron.-geod. o-va, 1973, No. 24, p. 8 - 59.
In Russian. − Abstr. in Referativ. Zhurn. 51. Astron.,
3.51.35 (1974).

Financial report of VAGO for the year 1972.
Tsirkulyar Vses. astron.-geod. o-va, 1973, No. 24, p. 60 - 63.
In Russian.

Resolution of the inspection commission of VAGO
on the general and finance activities of the Central Council of
VAGO for the year 1972.
Tsirkulyar Vses. astron.-geod. o-va, 1973, No. 24, p. 64 - 65.
In Russian.

Resolution of the IIIrd plenary session of the
Central Council of VAGO of the Vth convocation, Tbilisi,
20 April, 1973.
Tsirkulyar Vses. astron.-geod. o-va, 1973, No. 24, p. 66 - 73.
In Russian.

Resolution of the first All-Union conference of
VAGO departments.
Tsirkulyar Vses. astron.-geod. o-va, 1973, No. 25, p. 67 - 70.
In Russian. − Abstr. in Referativ. Zhurn. 51. Astron., 6.51.46
(1974).

25 years of the Latvian department of the Astro-
nomical-Geodetical Society of the USSR (VAGO).
M. A. Dirikis, Ya. M. Kletnieks, Yu. L. Frantsman.
Zemlya i Vselennaya, 1974, No. 1, p. 70 - 73. In Russian.

First All-Union conference of presidents of educational and methodical sections of district branches of VAGO.
See Abstr. 014.014.

Outline of the work of the educational and methodical section of the Gor'kij district branch of VAGO.
See Abstr. 014.025.

Activity of the Chelyabinsk branch of VAGO in
supporting the high school.   See Abstr. 014.027.

010.034 Vereniging voor Sterrenkunde, België

No publication received.

010.035 **The Nantucket Maria Mitchell Association.**
Seventy-second annual report for the year ending
December 31, 1973.
The Nantucket Maria Mitchell Association, Vestal Street,
Nantucket, Massachusetts, 71 pp. (1974). − Included is the
annual report of the director of Maria Mitchell Observatories,
by *D. Hoffleit.*

010.036 **On the occasion of the 250th anniversary of the
Academy of Sciences of the USSR.**
Astron. vestn., Vol. 8, 65 - 69 (1974). In Russian.

010.037 **Rapport d'activité de l'ADION par le Secrétaire
Général et rapport financier.**   J.-C. Pecker.
Bull. d'Information, Ass. Développement International Obs.
Nice, No. 10, p. 19 - 29 (1973). − Report 1972.

010.038 **Institut de France. Académie des Sciences. Annuaire
pour 1974.**
Gauthier-Villars, Paris. 261 pp. (1973).
Cet annuaire a pour but d'exposer l'état de l'Académie
des sciences dans le présent et dans le passé. Il contient en outre
des indications sur les concours de ses prix et sur ses fondations.

010.039 **Société Royale d'Astronomie d'Anvers, [Koninklijk
Sterrenkundig Genootschap van Antwerpen].**
Cinquante-quatrième rapport 1973.   J. Storms.
Imprimerie: «La Prévoyance», Antwerpen. 18 + 10 pp. (1974).
In French and Flemish.

010.040 **Nachrichten der Vereinigung der Sternfreunde e.V.**
SuW, Vol. 13, 25, 99 - 101, 134 - 136, 171 - 174,
209 - 211 (1974).

010.041 **Attività dell'Associazione; Relazione morale dell'
A.A.B. per l'anno 1973.**   A. Betti.
Giornale Ass. Astrofili Bolognesi, *Bologna,* No. 33, p. 10 - 11;
No. 34, p. 14 (1974).

011 Reports on Colloquia, Congresses, Meetings, Symposia, and Expeditions

011.001 New mathematical methods in physics and problems in general relativity. K. Bleuler.
Phys. Blätter, 30. Jahrgang, p. 131 - 134 (1974). — Report on a meeting in Bonn, 1973 July 3 - 6.

011.002 First USSR conference on astronomy and geodesy, Tbilisi, 17–19 April 1973.
Tsirkulyar Vses. astron.-geod. o-va, 1973, No. 24, p. 3 - 5. In Russian.

011.003 Special study group N° V–16 of International Association of Geodesy. Information — and workingmeeting 4 - 7 September 1973 in Uppsala, Sweden. E. Tengström.
Bull. Géod., Nouvelle Sér., Année 1974, No. 111, p. 17 - 19.

011.004 Conference on the physics of cosmic rays, (December 12 - 15, 1972 Apatity, USSR). Z. Kobyliński.
Postępy Astron., Vol. 22, 66 - 67 (1974). In Polish.

011.005 Nuova teoria sull'origine del cosmo. C. Traversi.
L'Universo, Anno 53, p. 1326 - 1327 (1973). — Simposio in onore di Copernico, Miami 1973 dicembre.

011.006 APS and OSA meet in Washington. J. Golden.
Phys. Today, Vol. 27, No. 4, p. 46 - 49, 51, 53 (1974).
A varied program stressing the relation between physics and optics will be offered at the first joint meeting of the societies since 1941.

011.007 Großer Erfolg für das Symposium "Planetenforschung" vom 15. bis 16. 2. 1974 in Heidelberg. H. W. Köhler.
SuW, Vol. 13, 78 (1974).

011.008 Die Frühjahrstagung in Garching. J. Staude.
SuW, Vol. 13, 111, 113 (1974).

011.009 Lunar subjects on the conference for studying natural resources of the earth and on problems of cartography of planets from space photographs. G. A. Burba.
Kosmich. Issled., Vol. 12, 315 - 317 (1974). In Russian.

011.010 Kopernika jubilejai veltītie sviņīgie sarīkojumi. S. Francmane, M. Dīriķis, I. Daube.
Zvaigžņotā debess, 1973. gada rudens, p. 25 - 30.

011.011 Ģeodēzistu zinātniski praktiskā konference Jelgavā. E. Grāvītis.
Zvaigžņotā debess, 1973. gada rudens, p. 31 - 33.

011.012 Astronomu devums Latvijas Valsts universitātes XXXII zinātniskajai konferencei. L. Roze.
Zvaigžņotā debess, 1973. gada rudens, p. 33 - 35.

011.013 Cilvēks un kosmiskā vide. N. Cimahoviča.
Zvaigžņotā debess, 1973. gada rudens, p. 35 - 36.

011.014 XIXth meeting of the Argentine Astronomical Association, San Juan, September 26 - 29, 1973.
Inform. Bull. Southern Hemisph., No. 23, p. 35 - 37 (1973).

011.015 Working Seminar on Solar Physics, Instituto de Astronomía y Física del Espacio, Buenos Aires, Argentina, October 8 - 16, 1973.
Inform. Bull. Southern Hemisph., No. 23, p. 37 (1973).

011.016 All-Union conference on cosmic-ray physics (Kharkov, September 25 - 28, 1973). M. Kats.
Geomagn. Aeronom., Vol. 14, 386 (1974). In Russian.

011.017 Symposium zur Teleskopeinweihung, Berlin, 9–10 November 1973. A. Kunert.
SuW, Vol. 13, 151 (1974).

011.018 "Colloquia Copernicana" 1973 in Toruń. D. B. Herrmann.
Sterne, 50. Jahrgang, p. 47 - 50 (1974).

011.019 Astronomers meet administrators. E. Phillips.
Nature, Vol. 248, 273 - 274 (1974).

011.020 Scientific session of the Department of General Physics and Astronomy, USSR Academy of Sciences (April 25 - 26, 1973).
Uspekhi fiz. nauk, Vol. 111, 553 (1973). In Russian. — Abstr. in Referativ. Zhurn. 51. Astron., 4.51.42 (1974).

011.021 European seminary of the students of astronomy in Poland, 1973. P. Heinzel, M. Šolc.
Říše hvězd, Vol. 55, 29 - 33 (1974). In Czech.

011.022 I.A.U. symposium No. 59 on stellar instability and evolution. F. M. Bateson.
Southern Stars, Vol. 25, 97 - 98 (1973). — Conference abstract.

011.023 Astronomische Tagung in Strassburg. R. A. Naef.
Orion, 32. Jahrgang, p. 128 - 130 (1974). — 1974 March 29 - 31.

011.024 3. Frühjahrstagung der VdS in Würzburg. E. Wiedemann.
Orion, 32. Jahrgang, p. 130 - 131 (1974). — 1974 April 6.

011.025 Symposium on solar-terrestrial physics and the XIIIth session of COSPAR. L. I. Miroshnichenko.
Geofiz. byul. Mezhduved. geofiz. kom. pri Prezidiume AN SSSR, 1973, No. 25, p. 27 - 28. In Russian.

011.026 Acceleration and propagation of solar cosmic rays (seminar in Leningrad). L. I. Miroshnichenko.
Geofiz. byul. Mezhduved. geofiz. kom. pri Prezidiume AN SSSR, 1973, No. 26, p. 32 - 34. In Russian.

011.027 Konferences un sanāksmes.
Zvaigžņotā debess, 1973./74. gada ziema, p. 46 - 56. Pirmajā Vissavienības astronomijas un ģeodēzijas konferencē (*A. Alksnis, M. Dīriķis*), p. 46 - 52; Jauno astronomijas speciālistu vasaras skola Vītnā (*I. Šmelds*), p. 52 - 55; Saules sistēmas mazo ķermeņu pētījumu darba grupas apspriede Rīgā (*M. Dīriķis*), p. 55 - 56.

011.028 Konferences un sanāksmes.
Zvaigžņotā debess, 1974. gada pavasaris, p. 44 - 49.

Pirmā konference speciālajā astronomiskajā observatorijā (*J. Francmanis*), p. 44 - 47; Saules pētnieku sanāksme Karpatos (*M. Eliāss, G. Ozoliņš*), p. 47 - 49.

011.029 Summer School on "Particles and Fields in Space" in Balatonfüred, June 4 - 15, 1973.
Z. Kobyliński.
Postępy Astron., Vol. 22, 143 - 144 (1974). In Polish.

011.030 A conference on "Cosmic ray physics", Kharkov, September 25 - 28, 1973. Z. Kobyliński.
Postępy Astron., Vol. 22, 144 - 145 (1974). In Polish.

011.031 International symposium "Gravitational radiation and gravitational collapse".
L. M. Ozernoj, V. F. Shvartsman.
Zemlya i Vselennaya, 1974, No. 3, p. 65 - 68. In Russian. Warsaw, Sept. 1973.

011.032 Late stages of the evolution of stars.
A. V. Tutukov, Yu. L. Frantsman.
Zemlya i Vselennaya, 1974, No. 3, p. 68 - 70. — Symposium of the IAU in Warsaw, Sept. 1973.

011.033 Philosophical problems of astronomy and cosmic research (Report of the scientific conference dedicated to the 500th anniversary of N. Copernicus).
A. Eh. Voskobojnikov.
Nauch. dokl. vyssh. shkoly. Filos. n., 1973, No. 6, p. 146 - 152. In Russian. — Abstr. in Referativ. Zhurn. 51. Astron., 6.51.34 (1974).

011.034 On results of the All-Union symposium on plasma physics of the upper atmosphere and of the circumterrestrial cosmic space. Cosmophysical latitude expedition.
Yu. G. Shafer.
Geophys. byul. Mezhduved. geofiz. kom. pri Prezidiume AN SSSR, 1973, No. 26, p. 46 - 48. In Russian. — Abstr. in Referativ. Zhurn. 62. Issled. kosmich. prostranstva, 6.62.8 (1974).

011.035 Rezultati II i III meteorske ekspedicije (Results of the II. and III. meteor expedition).
K. Pavlovski.
Vasiona, Vol. 22, 14 - 18 (1974).

011.036 Inaugurazione del nuovo osservatorio di Berlino.
F. Zagar.
Mem. Soc. Astron. Italiana, Nuova Ser., Vol. 44, 696 - 697 (1973/74).

011.037 Septième symposium international sur les marées terrestres, (Sopron, 10–15 Septembre 1973).
R. Lecolazet.
Bull. Géod., Nouvelle Sér., Année 1974, No. 112, p. 137 - 143 (1974).

011.038 International colloquium on reference coordinate systems for earth dynamics. — Torun, Poland, 26–31 August 1974.
Bull. Géod., Nouvelle Sér., Année 1974, No. 112, p. 219 - 223 (1974).

011.039 25th international astronautical assembly.
R. Pešek.
Říše hvězd, Vol. 55, 41 - 46 (1974). In Czech.

011.040 Protokoll der 119. Sitzung der Schweiz. Geodätischen Kommission vom 2. Juni 1973 im Bernerhof in Bern mit Auszügen aus den Berichten über die Tätigkeit im Jahre 1972.
Société Helvétique des Sciences Naturelles, (Schweiz. Naturforschende Gesellschaft). Spross + Co., Kloten. 22 pp. (1974).

011.041 First international symposium: "The use of artificial satellites for geodesy and geodynamics", Ateny – Lagonissi, 14–21 May 1973.
W. Dobaczewska, B. Kolaczek, J. Śledziński.
Geodezja i Kartografia, Vol. 23, 157 - 163 (1974). In Polish.

011.042 Summaries of papers presented at the Royal Astronomical Society specialist discussion on "Astronomy as education" on 1973 November 9.
Observatory, Vol. 94, 109 - 112 (1974).

011.043 Royal Astronomical Society specialist discussion on "Propagation of cosmic rays in the Galaxy" 1974 January 11.
Observatory, Vol. 94, 112 - 116 (1974).

012 Proceedings of Colloquia, Congresses, Meetings, and Symposia

012.001 **Papers presented at the seventh annual general meeting of the Astronomical Society of Australia, held at Monash University on 16, 17 and 18 May 1973.**
Proc. Astron. Soc. Australia, Vol. 2, (No. 4), 170 - 227 (1973). The individual contributions are included in their corresponding subject categories — see abstracts 031.009, 045.001, 046.001, 062.018, 064.017, 077.004 - 077.009, 080.012, 099.026, 114.039, 132.005, 141.032, 141.033, 141.307, 142.023 - 142.025, 151.017, 151.018.

012.002 **Geophysical and geochemical exploration of the moon and planets.** Proceedings of a symposium held at the Lunar Science Institute, Houston, Texas, January 10 - 12, 1973.
D. W. Strangway (Editor).
The Moon, Vol. 9, Nos. 1/2, 245 pp. (1974). — The individual contributions are included in their corresponding subject categories — see abstracts 091.003, 094.002 - 094.004, 094.111 - 094.125, 094.318 - 094.323, 097.020 - 097.022, 097.047, 098.008, 107.005, 107.006.

012.003 **The planets today.** Contributions to the Royal Society's commemoration of the quincentenary of the birth of Nicolaus Copernicus, with a preface by H. Massey, W. H. McCrea.
The Royal Society, London = Proc. Roy. Soc. London, Ser. A, Vol. 336, 1 - 114 (1974). — The individual contributions are included in their corresponding subject categories — see 004.015, 005.004, 091.006, 094.128, 097.023, 099.027.

012.004 **Physics of dense matter.** International Astronomical Union, Symposium No. 53, held in Boulder, Colorado, U.S.A., 21 - 25 August 1972.
C. J. Hansen (Editor), assisted by L. H. Volsky, with introductory remarks by C. J. Hansen, W. E. Brittin, and concluding remarks by A. G. W. Cameron.
D. Reidel Publishing Company, Dordrecht–Holland/Boston–U.S.A. 10 + 327 pp. Price hfl 102.00 (1974). — The individual contributions are included in their corresponding subject categories — see abstracts 061.021 - 061.029, 065.037 - 065.043, 125.011, 126.006, 126.007, 141.036, 141.309, 141.310, 162.008.

012.005 **Proceedings of the conference on the numerical solution of ordinary differential equations, 19, 20 October 1972, The University of Texas at Austin.**
D. G. Bettis (Editor).
Lecture Notes in Mathematics, Vol. 362. Springer-Verlag, Berlin–Heidelberg–New York. 8 + 490 pp. Price DM 34.00 (1974). — The individual contributions within the subject scope of Astronomy and Astrophysics Abstracts are included in their corresponding categories — see abstracts 021.004 - 021.011, 042.034 - 042.036, 052.012, 052.013, 062.022, 151.021 - 151.028.

012.006 **Copernicus ... and modern dynamical astronomy.**
P. K. Seidelmann (Editor).
Celestial Mechanics, Vol. 9, 295 - 363 (1974). — The individual contributions are included in their corresponding subject categories — see abstracts 042.037, 042.038, 098.009, 102.010, 107.010, 151.031.

012.007 **Black holes. (Les astres occlus).** Les Houches, août 1972. Cours de l'École d'été de Physique théorique. C. DeWitt, B. S. DeWitt (Editors).
Gordon and Breach Science Publishers, New York–London–Paris. 12 + 552 pp. Price £ 13.50 (1973). — Review in Nature, Vol. 247, 494; 1974 (*G. W. Gibbons*). — The individual contributions are included in their corresponding subject categories — see abstracts 066.020 - 066.025, 142.038.

012.008 **Gravitational radiation and gravitational collapse.** International Astronomical Union, Symposium No. 64, held in Warsaw, Poland, 5 - 8 September 1973, Copernican Symposium.
C. Dewitt-Morette (Editor), with an introduction by C. Dewitt-Morette, A. Trautman, and a summary by J. A. Wheeler.
D. Reidel Publishing Company, Dordrecht–Holland/Boston–U.S.A. 16 + 224 pp. Price Dfl. 78.25 (1974). — The individual contributions are included in their corresponding subject categories — see abstracts 033.018 - 033.23, 065.076, 066.044 - 066.076, 117.025, 124.008, 141.321, 142.054, 142.056, 142.112, 143.020, 155.034, 162.020.

012.009 **Bericht über die wissenschaftliche Tagung der Astronomischen Gesellschaft in Oberkochen/Württbg., 24 - 27 April 1973.** K. Schaifers (Editor).
Mitt. Astron. Ges., No. 34, 170 pp. Price DM 20.00 (1973/74). The individual contributions are included in their corresponding subject categories — see abstracts 004.035, 005.017, 021.014, 031.026 - 031.028, 032.021, 034.031 - 034.034, 041.019, 061.046, 061.047, 064.039, 065.066 - 065.073, 065.081, 066.043, 072.025, 073.044, 073.045, 076.010, 077.028, 077.029, 080.025, 082.059, 098.017, 102.022, 103.100, 105.014, 117.023, 117.024, 118.009, 119.013, 122.067, 125.018, 131.073, 131.074, 131.537, 155.029, 155.030, 156.001.

012.010 **Galaxies and relativistic astrophysics.** Proceedings of the First European Astronomical Meeting, Athens, September 4 - 9, 1972. Vol. 3.
B. Barbanis, J. D. Hadjidemetriou (Editors), with introductory remarks by A. Blaauw.
Springer Verlag, Berlin–Heidelberg–New York. 12 + 247 pp. Price DM 126.00 (1974). — The individual contributions are included in their corresponding subject categories — see abstracts 062.045, 065.078, 065.079, 066.078, 066.079, 125.019, 125.102, 134.004, 134.005, 141.074, 141.322, 141.610, 151.040 - 151.043, 158.068 - 158.080, 159.002, 160.021.

012.011 **Proceedings of the symposium on the collection and analyses of astrophysical data,** Charlottesville, Virginia, USA, November 13–15, 1972.
Y. Biraud (Editor).
Astron. Astrophys., Suppl. Ser., Vol. 15, (No. 3), 321 - 534 (1974). — The individual contributions are included in their corresponding subject categories — see abstracts 021.015 - 021.022, 031.029 - 031.036, 033.025 - 033.045, 034.038, 071.027, 094.157, 141.075, 141.323.

012.012 **Reports of new ultraviolet observations,** presented at the Open Meeting of Working Group 3, Panel 3.A of COSPAR. Sixteenth Plenary Meeting 23 May–5 June 1973 Konstanz, F.R.G. K. A. van der Hucht, H. J. Lamers (Editors), with a preface by C. de Jager.
The Astronomical Institute at Utrecht, Internal report ROF 72. Space Res. Lab., Utrecht, Netherlands, 5 + 208 pp. = Utrechtse Sterrekundige Overdrukken, No. 256 (1974). — The individual contributions are included in their corresponding subject categories — see abstracts 032.034, 034.055, 082.094, 113.033, 113.034, 114.108 - 114.118, 115.013, 131.096 - 131.099, 132.027, 155.049.

012.013 **Atti della XVI Riunione e Giornate di Studio su**

'I nuclei delle galassie', Firenze, 13–14–15 ottobre 1972.
Mem. Soc. Astron. Italiana, Suppl. Vol. 44, p. S 217 - S 322 (1974). – The individual contributions are included in their corresponding subject categories – see abstracts 010.027, 141.090, 155.052, 158.095 - 158.097.

012.014 **New problems in astrometry.** International Astronomical Union, Symposium No. 61 held in Perth, Western Australia, 13–17 August, 1973.
W. Gliese, C. A. Murray, R. H. Tucker (Editors), with a general discussion, summary and conclusions by B. J. Bok.
D. Reidel Publishing Company, Dordrecht – Holland/Boston – U.S.A. 12 + 335 pp. Price hfl. 100.00 (1974). – The individual contributions are included in their corresponding subject categories – see abstracts 013.008, 031.051 - 031.055, 032.037 - 032.039, 034.059 - 034.061, 041.033 - 041.064, 043.005, 094.011, 111.004 - 111.008, 112.018 - 112.026, 141.092 - 141.096, 155.057.

012.015 **Supernovae and supernova remnants.** Proceedings of the International Conference on Supernovae held in Lecce, Italy, May 7–11, 1973.
C. B. Cosmovici, E. D'Anna, A. Borghesi (Editors), with a preface by C. B. Cosmovici, concluding remarks by L. Woltjer, a review of the conference by H. Arp.
Astrophysics and Space Science Library, Vol. 45. D. Reidel Publishing Company, Dordrecht – Holland/Boston – U.S.A. 15 + 387 pp. Price Dfl. 100.00 (1974). – The individual contributions are included in their corresponding subject categories – see abstracts 061.058, 125.029 - 125.063, 125.100 - 125.103, 134.006 - 134.008, 141.332, 158.099.

012.016 **Correlated interplanetary and magnetospheric observations.** Proceedings of the seventh ESLAB symposium held at Saulgau, W. Germany, 22–25 May, 1973.
D. E. Page (Editor).
Astrophysics and Space Science Library, Vol. 42, D. Reidel Publishing Company, Dordrecht – Holland/Boston – U.S.A. 14 + 662 pp. Price Dfl. 195.00 (1974). – The individual contributions are included in their corresponding subject categories – see abstracts 074.076 - 074.078, 076.017, 076.018, 078.023 - 078.035, 083.046, 084.031 - 084.033, 084.276 - 084.291, 085.013, 099.057, 106.033 - 106.036.

012.017 **Cosmic activity. Influence on science and technics.** XXIV international astronautical congress. Baku, 1973, October 7 - 13.
VINITI, Moskva. 448 pp. (1973). In Russian.

012.018 **Stars and the Milky Way system.** Proceedings of the First European Astronomical Meeting, Athens, September 4–9, 1972, Vol. 2.
L. N. Mavridis (Editor), with introductory remarks by A. Blaauw.
Springer-Verlag, Berlin – Heidelberg – New York. 13 + 368 pp. Price DM 138.00 (1974). – The individual contributions are included in their corresponding subject categories – see abstracts 031.057, 031.058, 034.064, 042.078, 042.079, 061.061, 061.062, 064.052, 091.025, 096.023, 098.023, 114.129, 114.130, 117.035, 117.036, 118.015, 121.077, 122.106 - 122.108, 124.103, 131.106 - 131.110, 131.546, 131.547, 141.614, 151.053, 155.059 - 155.066, 158.107.

012.019 **Giornate di studio sull'astronomia X spaziale. Il satellite EXOSAT,** with concluding remarks by B. Bertotti.
Mem. Soc. Astron. Italiana, Nuova Ser., Vol. 44, 481 - 640 (1973/74). – The individual contributions are included in their corresponding subject categories – see abstracts 021.024, 021.025, 034.072, 034.073, 054.022, 054.023, 061.069, 061.070, 114.151, 141.615, 142.095 - 142.099, 142.100.

012.020 **Analyse extraterrestrischen Materials.** Symposium, Wien 1973, Oktober 2 - 3.
W. Kiesl, H. Malissa, Jr. (Editors).
Springer-Verlag, Wien–New York. 9 + 326 pp. Price DM 56.00 (1974). – The volume is dedicated to Prof. Dr. F. Hecht on the occasion of his 70th birthday. – The individual contributions are included in their corresponding subject categories – see abstracts 022.069, 061.071, 061.072, 074.110, 094.174 - 094.177, 094.392 - 094.395, 105.149 - 105.156, 116.014, 131.122.

012.021 **Proceedings ESO/SRC/CERN conference on research programmes for the new large telescopes,** Geneva, 27–31 May 1974.
A. Reiz (Editor), with a preface and opening words by A. Blaauw, and two panel discussions.
European Southern Observatory Telescope Project Division, Geneva. 17 + 398 pp. (1974). – The individual contributions are included in their corresponding subject categories – see abstracts 008.000, 013.012 - 013.014, 031.067, 032.045 - 032.053, 036.009, 036.010, 065.114, 122.117, 131.125, 131.552, 141.111 - 141.115, 153.031, 154.023 - 154.026, 155.072, 158.117 - 158.119, 159.006, 159.007, 162.036.

012.022 **Proceedings of a symposium on "Celestial mechanics",** held in Kyoto, Japan, December 3–4, 1973.
G. Hori, M. Yuasa (Editors).
Dep. Astron., Univ. Kyoto, 73 pp. (1974). In Japanese. – The individual contributions are included in their corresponding subject categories – see abstracts 042.081 - 042.083, 044.019, 045.015, 098.026, 100.212, 151.058, 151.059.

012.023 **The present state and future of the astronomical refraction investigations.** Proceedings of the Study Group on Astronomical Refraction of the International Astronomical Union Commission 8.
G. Teleki (Editor).
Publ. Obs. Astron. Beograd, No. 18, 234 pp. (1974). – The individual contributions are included in their corresponding subject categories – see abstracts 010.017, 031.075 - 031.082, 046.025, 082.111 - 082.116.

012.024 **The reception of Copernicus' heliocentric theory.** Proceedings of a symposium organized by the 'Nicolas Copernicus Committee' of the International Union of the History and Philosophy of Science, Toruń, Poland, 1973.
J. Dobrzycki (Editor).
D. Reidel Publishing Company, Dordrecht – Holland/Boston – U.S.A. 368 pp. Price Dfl. 65.00, $ 24.00 respectively (1973).

012.025 **Molecules in the galactic environment.** Proceedings of a symposium, November 1971.
M. A. Gordon, L. E. Snyder (Editors).
Wiley-Interscience, John Wiley and Sons, New York – London – Sydney – Toronto. 18 + 475 pp. Price £ 9.50, $ 18.95 respectively (1973). – Review in Bull. Astron. Inst. Czechoslovakia, Vol. 25, 185 - 186; 1974 (*J. Kleczek*).

012.026 **Proceedings of the fourth lunar science conference.**
W. A. Gose (Editor).
The Pergamon Group of Companies, Oxford – New York – Braunschweig. 3336 pp. in 3 volumes. Price $ 100.00 (1973).

012.027 **Annales Françaises de Chronométrie et de Micromécanique.** Premiere partie: Communications présentées au Congrès de Chronométrie franco-allemand de Besançon les 15, 16 et 17 juin 1973. Deuxième partie: Communications présentées aux réunions d'automne de la

Société Chronométrique de France à Paris et mémoires scientifiques.
Publication annuelle de l'Observatoire de Besançon du Centre Technique de l'Industrie Horlogère et de la Société Chronométrique de France, 8ᵉ année, 165 pp. Price F 45.00 (1973).
For the individual contributions within the subject scope of Astronomy and Astrophysics Abstracts — see abstracts 035.008 - 035.018, 044.031.

012.028 **Investigation of the upper atmosphere of the earth. Transactions of the international symposium on space meteorology. Kiev, March 1972.**
A. I. Ivanovskij, G. A. Kokin (Editors).
Gidrometeoizdat, Moskva. 257 pp. Price 1 Rbl. 15 Kop. (1973). In Russian.

012.029 **From the history of aviation and aeronautics. Vypusk 17—18.**
Materials of the XIIIth international congress on the history of science, Moskva, 18—24 August, 1971. Moskva. 253 pp. Price 60 Kop. (1972). In Russian. — Review in Referativ. Zhurn. 62. Issled. kosmich. prostranstva, 3.62.4 (1974).

012.030 **Cospar life sciences and space research 11.**
Proceedings of a meeting, Madrid, May 1972.
P. H. A. Sneath (Editor).
Akademie Verlag, Berlin. 10 + 308 pp. Price DM 64.00 (1973).

012.031 **Communication with extraterrestrial intelligence.**
(Conf. Proc. Erevan, USSR, 5—11 September 1971).
C. Sagan (Editor).
The MIT Press, Cambridge, Mass. — London, GB. 34 + 428 pp. Price $ 10.00 (1973). — Reviews in Journ. British Interplanet. Soc., Vol. 27, 312 - 313; 1974 (*A. R. Martin*); Nature, Vol. 248, 459; 1974 (*I. Ridpath*).

012.032 **Sixth Texas symposium on relativistic astrophysics.**
D. J. Hegyi (Editor).
New York Academy of Sciences, New York, N.Y. 364 pp. Price $ 31.00 (1973).

012.033 **Symposium on Copernicus. Papers from a meeting, April 1973.**
Proceedings of the American Philosophical Society, Philadelphia, Vol. 117, No. 6 (pp. 413 - 552). Price $ 1.00 (1974).

012.034 **Gamma-ray astrophysics. Proceedings of a symposium, Greenbelt, Md., April 1972.**
F. W. Stecker, J. I. Trombka (Editors).
National Aeronautics and Space Administration, Washington, D.C., NASA SP-339, [available from the Superintendent of Documents, Washington, D.C.], 16 + 412 pp. Price $ 2.90 (1973).

013 Reports on Astronomy in Various Countries and Particular Fields, International Cooperation

013.001 **Astronomy in China today.**   G. K. Miley.
Sky Telescope, Vol. 47, 148 - 152 (1974).

013.002 **The Academy of Sciences and development of national astronomy.**   A. A. Mikhajlov.
Zemlya i Vselennaya, 1974, No. 2, p. 38 - 44. In Russian.

013.003 **The 250th anniversary of the Academy of Sciences and astronomy.**   P. G. Kulikovskij.
Astron. Zhurn. Akad. Nauk SSSR, Vol. 51, 225 - 232 (1974). In Russian. English translation in Soviet Astron. AJ, Vol. 18, No. 2.

013.004 **The state of astronomy in New Zealand.**
F. M. Bateson.
Southern Stars, Vol. 25, 71 - 76 (1973).

013.005 **Die astronomische Forschung in der Schweiz.**
E. Wiedemann.
Orion, 32. Jahrgang, p. 115 (1974).

013.006 **On interrelations between astronomical disciplines and on their system-theoretical description.**
H. Eelsalu.
ENSV Tead. Akad. Loodusuurijate Seltsi Aastaraamat, Ezhegodnik O-va estestvoispyt. AN EstSSR, Vol. 62, 88 - 97 (1973). In Estonian. — Abstr. in Referativ. Zhurn. 51. Astron., 5.51.2 (1974).

013.007 **The Astronomical Council of the Academy of Sciences of the USSR and the development of Soviet astronomy.**   G. S. Khromov.
Zemlya i Vselennaya, 1974, No. 3, p. 4 - 11. In Russian.

013.008 **Projet de centre de recherches astrométriques et géodynamiques en France.**   J. Kovalevsky.
IAU Symposium No. 61, (see 012.014), p. 263 - 267 (1974).

013.009 **Astronomy of the invisible.**
J. Milogradov-Turin.
Vasiona, Vol. 22, 1 - 7 (1974). In Serbo-Croatian.

013.010 **Lavori astronomici pubblicati su periodici Italiani nell'anno 1972.**   P. Maffei.
Mem. Soc. Astron. Italiana, Nuova Ser., Vol. 44, 713 - 716 (1973/74).

013.011 **Advances in astronomy in the year 1973.**
J. Grygar.

Říše hvězd, Vol. 55, 46 - 52, 65 - 72, 87 - 98 (1974). In Czech.

013.012 **Current research programmes at the Hale Observatories.** J. B. Oke.
Proc. ESO/SRC/CERN conference, (see 012.021), p. 29 - 34 (1974).

013.013 **Current research programs at the Lick Observatory.** M. F. Walker.
Proc. ESO/SRC/CERN conference, (see 012.021), p. 35 - 42 (1974).

013.014 **"Stellar" research at the Kitt Peak National Observatory.** A. A. Hoag.
Proc. ESO/SRC/CERN conference, (see 012.021), p. 43 - 52 (1974).

013.015 **Polish astronomy in China.** H. Lubomirski.
Urania Kraków, Vol. 45, 75 - 78 (1974). In Polish.

013.016 **1973 Smithsonian Standard Earth (III).**
E. M. Gaposchkin (Editor), with a historical introduction by C. A. Lundquist, F. L. Whipple.
Smithsonian Astrophys. Obs., *Cambridge, Mass.*, Special Report 353, 16 + 388 pp. (1973). – For the individual contributions see abstracts 046.026, 046.027, 052.045, 081.030, 081.031.

014 Teaching in Astronomy

014.001 **Zur Weiterbildung der Lehrer im Fach Astronomie ab Schuljahr 1973/74.** W. Wohlert.
Astron. in der Schule, 11. Jahrgang, p. 2 - 4 (1974).

014.002 **Effektive Wege zur Lehrplanerfüllung im Stoffgebiet "Astrophysik – Stellarastronomie".**
H.-P. Eckert.
Astron. in der Schule, 11. Jahrgang, p. 12 - 15 (1974).

014.003 **Astronomie in der Sekundarstufe I – Analyse einer Fragebogenaktion.** W. Winnenburg.
SuW, Vol. 13, 19 - 23 (1974).

014.004 **Kontaktstudium Astronomie in Heidelberg.** G. Groschopf.
SuW, Vol. 13, 23 (1974).

014.005 **Zur Vorbereitung des zweiten Beobachtungsabends.** A. Zenkert.
Astron. in der Schule, 11. Jahrgang, p. 15 - 18 (1974).

014.006 **Civic observatories.** R. J. Livesey, H. Valentine.
Journ. British Astron. Ass., Vol. 84, 194 - 195 (1974).

014.007 **Problemhafter Astronomieunterricht.** J. Stier.
Astron. in der Schule, 11. Jahrgang, p. 40 - 44 (1974).

014.008 **Schule und Astronomie.** M. I. Rossa.
SuW, Vol. 13, 208 (1974).

014.009 **Praktische Ratschläge beim Einsatz von Fernglas und Fernrohr.** B. Wedel.
Der Physikunterricht, Jahrgang 8, (No. 2), 43 - 56 = Veröff. Wilhelm-Foerster-Sternw., Berlin, No. 34 (1974).

014.010 **On the state of professional training of teachers in astronomy at pedagogical colleges of the USSR.**
M. M. Dagaev.
Tsirkulyar Vses. astron.-geod. o-va, 1973, No. 25, p. 47 - 50. In Russian. – Abstr. in Referativ. Zhurn. 51. Astron., 5.51.48 (1974).

014.011 **Some inferences from the work of the Physical-Astronomical Department of the A. M. Gor'kij Pedagogical Institute in Gor'kij.** A. V. Artem'ev.
Tsirkulyar Vses. astron.-geod. o-va, 1973, No. 25, p. 51 - 55. In Russian. – Abstr. in Referativ. Zhurn. 51. Astron., 5.51.49 (1974).

014.012 **Astronomy in the courses of geography at pedagogical colleges.** S. G. Kulagin, L. D. Kovbasyuk.
Tsirkulyar Vses. astron.-geod. o-va, 1973, No. 25, p. 56 - 57. In Russian. – Abstr. in Referativ. Zhurn. 51. Astron., 5.51.50 (1974).

014.013 **On the state of astronomical teaching at high schools according to inspections made by VAGO and real means of raising the efficiency of astronomical teaching.**
E. P. Levitan.
Tsirkulyar Vses. astron.-geod. o-va, 1973, No. 25, p. 13 - 20. In Russian. – Abstr. in Referativ. Zhurn. 51. Astron., 5.51.52 (1974).

014.014 **First All-Union conference of presidents of educational and methodical sections of district branches of VAGO.** A. V. Artem'ev.
Tsirkulyar Vses. astron.-geod. o-va, 1973, No. 25, p. 3 - 5. In Russian. – Abstr. in Referativ. Zhurn. 51. Astron., 5.51.53 (1974).

014.015 **Aims and purposes of educational and methodical sections of VAGO.** V. V. Radzievskij.
Tsirkulyar Vses. astron.-geod. o-va, 1973, No. 25, p. 5 - 12. In Russian. – Abstr. in Referativ. Zhurn. 51. Astron., 5.51.55 (1974).

014.016 **On teaching astronomy at high schools and colleges.** E. V. Sandakova.
Tsirkulyar Vses. astron.-geod. o-va, 1973, No. 25, p. 27 - 28. In Russian. − Abstr. in Referativ. Zhurn. 51. Astron., 5.51.56 (1974).

014.017 **On the structurization of astronomical knowledge and methodical problems of training.** E. P. Levitan.
Tsirkulyar Vses. astron.-geod. o-va, 1973, No. 24, p. 90 - 93. In Russian. − Abstr. in Referativ. Zhurn. 51. Astron., 5.51.57 (1974).

014.018 **Making more active the thinking of pupils in studying astronomy.** Yu. N. Klevenskij.
Tsirkulyar Vses. astron.-geod. o-va, 1973, No. 25, p. 42 - 44. In Russian. − Abstr. in Referativ. Zhurn. 51. Astron., 5.51.58 (1974).

014.019 **Analysis of high school astronomical educational literature.** Eh. V. Kononovich.
Tsirkulyar Vses. astron.-geod. o-va, 1973, No. 25, p. 29 - 34. In Russian. − Abstr. in Referativ. Zhurn. 51. Astron., 5.51.59 (1974).

014.020 **Some suggestions concerning the improvement of astronomical teaching at high schools and technical schools.** N. P. Khrisanov.
Tsirkulyar Vses. astron.-geod. o-va, 1973, No. 25, p. 66. In Russian. − Abstr. in Referativ. Zhurn. 51. Astron., 5.51.60 (1974).

014.021 **On the elements of cosmonautics and astronomy in teaching at schools.** P. N. Karnitskij.
Tsirkulyar Vses. astron.-geod. o-va, 1973, No. 25, p. 45 - 46. In Russian.

014.022 **On the presentation of the theme "Artificial celestial bodies" in the course of general astronomy.** B. V. Tumanyan.
Tsirkulyar Vses. astron.-geod. o-va, 1973, No. 25, p. 58 - 59. In Russian.

014.023 **Selected exercises in the fundamentals of stellar astronomy at pedagogical colleges.** M. T. Emel'yanenko.
Tsirkulyar Vses. astron.-geod. o-va, 1973, No. 25, p. 60 - 61. In Russian. − Abstr. in Referativ. Zhurn. 51. Astron., 5.51.64 (1974).

014.024 **On teaching astronomy at Leningrad high schools and colleges.** O. K. Ukhova.
Tsirkulyar Vses. astron.-geod. o-va, 1973, No. 25, p. 24 - 26. In Russian. − Abstr. in Referativ. Zhurn. 51. Astron., 5.51.65 (1974).

014.025 **Outline of the work of the educational and methodical section of the Gor'kij district branch of VAGO.** E. I. Melokhrino.
Tsirkulyar Vses. astron.-geod. o-va, 1973, No. 25, p. 21 - 23. In Russian. − Abstr. in Referativ. Zhurn. 51. Astron., 5.51.66 (1974).

014.026 **On the training of teachers in astronomy and raising to a higher level the teaching of the subject.** N. A. Istoshin.
Tsirkulyar Vses. astron.-geod. o-va, 1973, No. 25, p. 39 - 41. In Russian. − Abstr. in Referativ. Zhurn. 51. Astron., 5.51.67 (1974).

014.027 **Activity of the Chelyabinsk branch of VAGO in supporting the high school.** E. G. Ponomarev.
Tsirkulyar Vses. astron.-geod. o-va, 1973, No. 25, p. 64 - 65. In Russian. − Abstr. in Referativ. Zhurn. 51. Astron., 5.51.68 (1974).

014.028 **New visual methods and astronomical supplies and instruments.** E. G. Demidovich.
Tsirkulyar Vses. astron.-geod. o-va, 1973, No. 25, p. 62 - 63. In Russian. − Abstr. in Referativ. Zhurn. 51. Astron., 5.51.69 (1974).

014.029 **Astronomija skolā.**
Zvaigžņotā debess, 1973./74. gada ziema, p. 40 - 45. Pirmā skolēnu astronomijas olimpiāde (*J. Miezis, A. Asare*), p. 40 - 42; Apspriede par astronomijas mācīšanu (*A. Asare*), p. 43 - 45.

014.030 **L'astronomia nella scuola secondaria.** L. Stefanini.
Coelum, Vol. 42, 97 - 101 (1974).

014.031 **Astronomy in the training of teachers in geography.** B. A. Volynskij.
Zemlya i Vselennaya, 1974, No. 3, p. 74. In Russian.

014.032 **50 years of astronomical teaching at the Herzen Pedagogical College [in Leningrad].** O. K. Ukhova.
Teor. fiz. i astronomiya. Leningrad, 1973, p. 127 - 141. In Russian. − Abstr. in Referativ. Zhurn. 51. Astron., 6.51.47 (1974).

014.033 **On a new method of examining knowledge in astronomy of college students.** M. P. Chesnokova.
Teor. fiz. i astronomiya. Leningrad, 1973, p. 188 - 196. In Russian. − Abstr. in Referativ. Zhurn. 51. Astron., 6.51.48 (1974).

014.034 **On lessons in natural history in the fourth class.** K. Lupoj.
Nar. obrazovanie, 1974, No. 2, p. 51 - 52. In Russian. − Abstr. in Referativ. Zhurn. 51. Astron., 6.51.50 (1974).

014.035 **Telescopes for high school and college observatories.** V. A. Bronshtehn.
Tsirkulyar Vses. astron.-geod. o-va, 1973, No. 25, p. 35 - 38. In Russian. − Abstr. in Referativ. Zhurn. 51. Astron., 6.51.52 (1974).

014.036 **Main ideas on comets. I. Motion of a comet.** P. M. Đurković.
Vasiona, Vol. 22, 23 - 29 (1974). In Serbo-Croatian.

014.037 **Main ideas on comets. II. Physical peculiarities.** P. M. Đurković.
Vasiona, Vol. 22, 46 - 52 (1974). In Serbo-Croatian.

014.038 **Current trends in Ph.D. production and employment among astronomers.** D. Goldsmith.
Bull. American Astron. Soc., Vol. 6, 233 - 234 (1974).

014.039 **Begriffe im Astronomieunterricht.** M. Schukowski.
Astron. in der Schule, 11. Jahrgang, p. 56 - 59 (1974).

014.040 **Diafilm − ein effektives Anschauungsmittel im Astronomieunterricht.** E. P. Levitan.
Astron. in der Schule, 11. Jahrgang, p. 59, 62 (1974).

014.041 **Astronomy in school.** E. A. Beet.
Phys. Education, (*GB*), Vol. 8, 437 - 442 (1973).

014.042 **The place of astronomy in the study of physics.** R. A. R. Tricker.
Phys. Education, (*GB*), Vol. 8, 449 - 454 (1973).

014.043 **The mass of Jupiter.** A. Sutton.
Phys. Teacher, (*USA*), Vol. 11, 488 - 489 (1973).

014.044 **Teaching telescopes.** J. S. Reid.
Phys. Education, (*GB*), Vol. 9, 82 - 86 (1974).

014.045 **Photoelectric devices in astronomy and space research. I.** G. R. Carruthers.
Phys. Teacher, (*USA*), Vol. 12, 135 - 143 (1974).

014.046 **The horizontal sundial.** H. Erlichson.
American Journ. Phys., Vol. 42, 372 - 373 (1974).

014.047 **A teaching machine for elementary astronomy.** L. Houziaux.
Observatory, Vol. 94, 109 (1974). – Abstract.

014.048 **A philosophy for elementary practical classes.** D. Clarke.
Observatory, Vol. 94, 109 - 110 (1974). – Abstract.

014.049 **Vacation courses in astronomy for graduates and teachers.** E. A. Müller.
Observatory, Vol. 94, 110 (1974). – Abstract.

014.050 **Interdisciplinary astronomical education.** P. A. H. Seymour.
Observatory, Vol. 94, 110 - 111 (1974). – Abstract.

014.051 **A report on the activity of I.A.U. Commission No. 46 (Teaching of Astronomy).** D. McNally.
Observatory, Vol. 94, 111 - 112 (1974). – Abstract.

Cosmological teaching in the seventeenth-century Scottish universities, Part 1. See Abstr. 004.034.

015 Miscellanea

015.001 **Herschel and extra-terrestrial life.** C. A. Ronan.
Observatory, Vol. 94, 19 (1974). – Letter.

015.002 **On the possibility of applying cybernetic methods to search for highly organized systems in the universe.** B. N. Panovkin.
Izv. vyssh. ucheb. zavedenij. Radiofizika, Vol. 16, 1452 - 1454 (1973). In Russian. – Abstr. in Referativ. Zhurn. 62. Issled. kosmich. prostranstva, 3.62.478 (1974).

015.003 **A first look at the 1973 register.** B. F. Porter, S. F. Barisch, R. W. Sears.
Phys. Today, Vol. 27, No. 4, p. 23 - 27, 29, 31 - 33 (1974).
Trends emerge from the initial analysis of the APS–AIP register; over 50,000 respondents provided information for this assessment of physics and astronomy manpower.

015.004 **Astronomie auf Briefmarken.** E. Schmidt.
SuW, Vol. 13, 31 (1974).

015.005 **A new frontier: Do we still think primitively?** W. A. Elliott.
Spaceflight, Vol. 16, 179 - 180 (1974).

015.006 **Difficulty of interstellar radio communication.** D. R. Bates.
Nature, Vol. 248, 317 - 318 (1974). – Letter.

015.007 **Boodschappen uit het heelal.** G. van Wageningen.
Zenit, Vol. 1, No. 1, p. 7 - 8 (1974).

015.008 **Die Beiträge der Astronomie zum naturwissenschaftlichen Weltbild und ihre Popularisierung.** H.-J. Treder.
Sterne, 50. Jahrgang, p. 5 - 7 (1974).

015.009 **L'astronomie dans la philatélie.** N. Smagghe.
L'Astronomie, 88e année, p. 114 - 115 (1974).

015.010 **Comments on "Directed panspermia".** W. R. Chappell, R. R. Meglen, D. D. Runnells.
Icarus, Vol. 21, 513 - 515 (1974).
The authors point out that a more careful consideration of the elements chromium, nickel, and molybdenum does not support the conclusion of Crick and Orgel (1973).

015.011 **Sea-water and the origins of life.** T. H. Jukes.
Icarus, Vol. 21, 516 - 517 (1974).
Molybdenum is more abundant in sea-water than chromium or nickel, which does not support the panspermia hypothesis as proposed by Crick and Orgel.

015.012 **Reply: "Comments on 'Directed panspermia'" and "Seawater and the origin of life".** L. E. Orgel.
Icarus, Vol. 21, 518 (1974).

015.013 **'Stray' planets, their formation and the possibilities of CETI.** A. T. Lawton.
Spaceflight, Vol. 16, 188 - 189 (1974).

015.014 **Astronomy, the people and the governments.** J.-C. Pecker, G. Ringeard.
Vistas in astronomy, Vol. 15, (see 003.001), 1 - 12 (1973).

015.015 **Antimatter UFO's.** J. Epstein.
Phys. Today, Vol. 27, No. 3, p. 15 (1974).

015.016 Über einige Probleme des alternden Beobachters. W. W. Spangenberg.
Sterne, Vol. 50, 109 - 121 (1974).

015.017 Condizioni fisiche e possibilità di vita nel sistema solare. Conversazione del ciclo "Ipotesi su civiltà extraterrestri". M. G. Fracastoro.
Reprinted from Terzoprogramma, No. 1 = Pubbl. Varie Fuori Ser. Oss. Astron. Torino, (Pino Torinese), No. 54, 15 pp. (1973).

015.018 Legal measure units and astronomy. J. Bouška.
Říše hvězd, Vol. 55, 111 - 115 (1974). In Czech.

015.019 Could the inhabitants of the earth settle and live on Mars. J. Pokrzywnicki.
Urania Kraków, Vol. 45, 41 - 44 (1974). In Polish.

015.020 The beginning of comparative planetology. R. S. Young.
Space Life Sci., (Netherlands), Vol. 4, 505 - 515 (1973).

The study of the origin of life question is related to the comparative study of the planets in our solar system and in fact the universe as a whole. Data relevant to the origin of life is being accumulated from the earth, planets, stars and interstellar space.

015.021 Thermodynamics of thermal radiation from stars photoautotrophs and biospheres. A. Rueda.
Space Life Sci., (Netherlands), Vol. 4, 469 - 489 (1973).

Photoautotrophs are almost the exclusive providers of chemical free energy to the earth biosphere. Their importance in coadjuvating the evolutionary development of higher forms of life in other planets is briefly discussed from this point of view. A simple analysis based on the non-equilibrium thermodynamics of thermal radiation fields is performed. The analysis relates well known standard parameters of stars of the main sequence to the thermodynamic bounds on the free energy acquisition of planetary photosynthetic processes activated by the star radiation.

Possible consequences of nearby supernova explosions for atmospheric ozone and terrestrial life.
See Abstr. 082.093.

The outer solar system: perspectives for exobiology.
See Abstr. 091.027.

Interstellar molecules and the origin of life.
See Abstr. 131.126.

Applied Mathematics, Physics

021 Mathematics, Computing, Data Processing

021.001 **Runge-Kutta-Nyström methods.**
D. G. Bettis, K. Sepehnoori.
Bull. American Astron. Soc., Vol. 6, 208 (1974).– Abstr. AAS.

021.002 **Canonical forms for symplectic and Hamiltonian matrices.** A. J. Laub, K. Meyer.
Celestial Mechanics, Vol. 9, 213 - 238 (1974).
This paper gives a constructive method for finding canonical forms for symplectic and Hamiltonian matrices. No restrictions are made on the eigenvalues or their multiplicity. Real canonical forms are treated in detail.

021.003 **Use of punch card holders in the course of astronomy.** S. S. Peruanskij.
Vestn. vyssh. shkoly, 1973, No. 9, p. 25 - 26. In Russian. Abstr. in Referativ. Zhurn. 51. Astron., 3.51.42 (1974).

021.004 **Exploration methods for the solution of initial value problems and their practical realization.**
J. Stoer.
Proc. conference on numerical solution of ordinary differential equations, 1972, (see 012.005), p. 1 - 21 (1974).

021.005 **Changing stepsize in the integration of differential equations using modified divided differences.**
F. T. Krogh.
Proc. conference on numerical solution of ordinary differential equations, 1972, (see 012.005), p. 22 - 71 (1974).

021.006 **The order of differential equation methods.**
J. C. Butcher.
Proc. conference on numerical solution of ordinary differential equations, 1972, (see 012.005), p. 72 - 75 (1974).

021.007 **Equations of condition for high order Runge-Kutta-Nyström formulae.** D. G. Bettis.
Proc. conference on numerical solution of ordinary differential equations, 1972, (see 012.005), p. 76 - 91 (1974).

021.008 **On the non-equivalence of maximum polynomial degree Nordsieck-Gear and classical methods.**
R. Danchick.
Proc. conference on numerical solution of ordinary differential equations, 1972, (see 012.005), p. 92 - 106 (1974).

021.009 **Phase space analysis in numerical integration of ordinary differential equations.** B. E. Howard.
Proc. conference on numerical solution of ordinary differential equations, 1972, (see 012.005), p. 107 - 127 (1974).

021.010 **Multi-off-grid methods in multi-step integration of ordinary differential equations.** P. R. Beaudet.
Proc. conference on numerical solution of ordinary differential equations, 1972, (see 012.005), p. 128 - 148 (1974).

021.011 **Comparison of numerical integration techniques for orbital applications.** H. Moore.
Proc. conference on numerical solution of ordinary differential equations, 1972, (see 012.005), p. 149 - 166 (1974).

021.012 **Ein Prozeßrechner im Dienste der Sonnenphysik. Moderne Datenerfassung beim Fraunhofer Institut in Anacapri.** F.-L. Deubner.
SuW, Vol. 13, 47 - 50 (1974).

021.013 **Digitale Auswertung der Sonnenfinsternisaufnahmen vom 30. 6. 73.** R. Beck, H.-J. Pröll.
SuW, Vol. 13, 123 - 127 (1974).

021.014 **A computer program for the determination of orbits from three observations.**
J. Pfleiderer, E. Braunsfurth, H. J. Dorst, C. Durst.
Mitt. Astron. Ges., No. 34, p. 90 - 91 (1973/74). – Presented at the "Wissenschaftliche Tagung der Astron. Ges., Oberkochen, 1973 April 24 - 27".

021.015 **A directable modular approach to data processing.**
T. Cram.
Astron. Astrophys., Suppl. Ser., Vol. 15, 339 - 341 (1974).
Conference paper (see 012.011).

021.016 **A sensitive method for detecting dispersed radio emission.** J. H. Taylor.
Astron. Astrophys., Suppl. Ser., Vol. 15, 367 - 369 (1974).
Conference paper (see 012.011).

021.017 **Maximum entropy spectral analysis.** J. G. Ables.
Astron. Astrophys., Suppl. Ser., Vol. 15, 383 - 393 (1974). – Conference paper (see 012.011).

021.018 **Interactive processing of map data produced by supersynthesis radio telescopes.**
R. D. Ekers, R. J. Allen, J. R. Luyten.
Astron. Astrophys., Suppl. Ser., Vol. 15, 469 (1974). – Conference paper (see 012.011).

021.019 **The role of fast Fourier transform computers in astronomy.** J. L. Yen.
Astron. Astrophys., Suppl. Ser., Vol. 15, 483 - 484 (1974). Conference paper (see 012.011).

021.020 **Programming and control languages for automated laboratory experimentation.** J. Hudson.
Astron. Astrophys., Suppl. Ser., Vol. 15, 487 - 495 (1974). Conference paper (see 012.011).

021.021 **FORTH: a new way to program a mini-computer.**
C. H. Moore.
Astron. Astrophys., Suppl. Ser., Vol. 15, 497 - 511 (1974). Conference paper (see 012.011).

021.022 **A data acquisition and control system for the 46 m telescope at the Algonquin Park Radio Observatory.**
J. L. Wolfe, L. A. Higgs.
Astron. Astrophys., Suppl. Ser., Vol. 15, 513 - 515 (1974). Conference paper (see 012.011).

021.023 **Lissage des observations par la méthode de Whittaker-Robinson - Vondrak.** P. Paquet, M. Honorez.

Inst. Astron. Géophys., Georges Lemaître, Univ. Catholique Louvain, Belgique, Contr. No. 13, 24 pp. (1974).

021.024 Note sull'analisi dati. P. Mussio.
Mem. Soc. Astron. Italiana, Nuova Ser., Vol. 44, 629 - 633 (1973/74).

021.025 Organizzazione dell'analisi dati. G. Pizzichini.
Mem. Soc. Astron. Italiana, Nuova Ser., Vol. 44, 635 - 638 (1973/74).

021.026 Computer application in astronomy. K. Ziolkowski.
Informatyka, (*Poland*), Vol. 10, No. 1, p. 4 - 7 (1974). In Polish.

021.027 La distribution binomiale appliquée à l'examen des ensembles empiriques d'erreurs d'observations géodésiques. M. K. Szacherska.
Geodezja i Kartografia, Vol. 23, 75 - 91 (1974). In Polish.
La méthode de l'examen des ensembles empiriques d'erreurs et du choix de paramètres optimaux de la distribution binomiale, complétée par le programme BERNOULLI (ODRA-1204, ALGOL-1204), a été appliquée à l'analyse des ensembles d'observations géodésiques.

021.028 Concept of a data center. B. Hauck.
Centre de Données Stellaires, Inform. Bull. No. 6, p. 1 - 8 (1974).

021.029 Report on the Strasbourg Stellar Data Center presented at the IAU meeting at Sydney of the Working Group on Numerical Data. J. Jung.
Centre de Données Stellaires, Inform. Bull. No. 6, p. 9 - 13 (1974).

021.030 The role of data banks in astronomy experience with the Stellar Data Center in Strasbourg. J. Jung.
Centre de Données Stellaires, Inform. Bull. No. 7, p. 2 - 6 (1974).

021.031 Automated information retrieval at the Stellar Data Center. F. Ochsenbein.
Centre de Données Stellaires, Inform. Bull. No. 7, p. 7 - 37 (1974).

021.032 APL in astronomy. E. E. Mendoza V.
Rev. Mexicana Astron. Astrofis., Vol. 1, No. 1, p. 11 - 23 (1974).
This paper describes APL, an algorithmic language. Examples are given to demonstrate its versatility in the solving of astronomical problems.

021.033 Use of APL in Runge-Kutta methods. E. E. Mendoza V, S. Hacyan.
Rev. Mexicana Astron. Astrofis., Vol. 1, No. 1, p. 35 - 43 (1974).
This paper uses APL to describe algorithms of classical Runge-Kutta functions up to the eighth-order with and without integration step size control for solving systems of ordinary differential equations of the first order.

A brief note on searching for periodicities in data obtained at irregular intervals. See Abstr. 031.084.

NOD2 a general system of analysis for radioastronomy. See Abstr. 033.025.

The data handling of the Westerbork Synthesis Radio Telescope. See Abstr. 033.026.

Source fitting on the multibeam Molonglo Cross. See Abstr. 033.027.

The Parkes 1024 channel digital autocorrelator and its conversational mode software system. See Abstr. 033.029.

The 100 channel spectral line processor for Algonquin Radio Observatory. See Abstr. 033.030.

Aperture synthesis with a non-regular distribution of interferometer baselines. See Abstr. 033.031.

Phase error correction in multi-element radio interferometer by data processing. See Abstr. 033.033.

Interpolation of synthesis data and some effects of ringlobes. See Abstr. 033.036.

The Stanford five-element array: an instrument for real-time fan beam synthesis and rapid rotation-synthesis. See Abstr. 033.039.

A digital pulsar processor. See Abstr. 033.041.

On-site processing of data obtained with the Westerbork Synthesis Telescope. See Abstr. 033.044.

The FORTH program for spectral line observing on NRAO's 36 ft telescope. See Abstr. 033.045.

Quantization noise of correlation spectrometers. See Abstr. 034.038.

A general computer program for the application of the rigorous blockadjustment solution in photographic astrometry. See Abstr. 041.009.

Planetary perturbations in Newcomb's tables of the sun. See Abstr. 041.026.

A general computer program for the application of the rigorous block-adjustment solution in photographic astrometry. See Abstr. 041.062.

Astronomical data files at the U.S. Naval Observatory—ephemerides, star catalogues, and observations. See Abstr. 041.068.

Star catalog cross index. See Abstr. 041.069.

First complement to the list of star catalogues available in machine-readable form at the Astronomisches Rechen-Institut Heidelberg. See Abstr. 041.072.

Elements of the outer planets for one million years. See Abstr. 042.066.

Construction d'éphémérides condensées. See Abstr. 047.012.

Sequential filter design for precision orbit determination and physical constant refinement. See Abstr. 052.001.

Determination of the boundaries of shadow parts of artificial satellite orbits in the case of the conical earth shadow. II. See Abstr. 054.020.

On the construction and use of model stellar atmo-

spheres for stars of spectral types later than F0.
See Abstr. 064.044.

Analysis of solar granulation using laser-light.
See Abstr. 071.018.

Observation of the solar photospheric oscillations using a sodium optical resonance device and an optical spatial filtering. See Abstr. 071.027.

Obtaining normalized line profiles with an electronic digital computer. See Abstr. 114.058.

Magnetic tape of Bidelman's spectral data file.
See Abstr. 114.138.

The astronomical data file for the Celescope Catalog. See Abstr. 114.139.

Four stellar-diameter measurements by a new technique: amplitude interferometry. See Abstr. 115.001.

Métodos de cálculo de órbitas de sistemas dobles visuales; Programa para el cálculo de efemérides de estrellas dobles visuales. See Abstr. 118.005.

Modelling the light curves and determinations of periods of variable stars. See Abstr. 120.001.

Application of computers to determination of epochs by the Hertzsprung method. See Abstr. 120.002.

Use of electronic computers in studying periods of light variations of variable stars. See Abstr. 120.003.

The direct method of light curve solution of an eclipsing binary system with extended atmosphere. Calculation and application of weights. Computer programmes.
See Abstr. 121.033.

An almost complete survey of 21 cm line radiation for $|b| \geqslant 10°$. II. The accurate data on machine readable magnetic tape. See Abstr. 157.006.

022 Physical Papers Related to Astronomy and Astrophysics

022.001 A note on ionization equilibrium. D. L. Lafferty.
Astrophys. Journ., Vol. 187, 209 - 210 (1974).
The steady-state equations relating the population densities of ionization levels can be obtained from a more general condition than that of detailed balance. That the average value of the charge of the ionized fractions remains fixed is sufficient to ensure the proper recurrence relation.

022.002 The hyperfine Λ-doubling spectrum of sulfur hydride in the $^2\Pi_{3/2}$ state.
W. L. Meerts, A. Dymanus.
Astrophys. Journ., (Letters), Vol. 187, L45 - L46 (1974).
The hyperfine Λ-doubling spectrum of the five lowest rotational states of the $^2\Pi_{3/2}$, $v=0$ state is measured using the molecular-beam electric-resonance method. From the Stark splittings of the observed transitions the value of the electric dipole moment of SH is determined: $\mu = 0.7580(1)$ D.

022.003 Highly excited states of atoms in a magnetic field. R. F. O'Connell.
Astrophys. Journ., Vol. 187, 275 - 276 (1974).
The Zeeman effect in the Ba I absorption spectrum has been observed by Garton and Tomkins. One striking feature of the spectrum they obtained is the existence of a sequence of σ lines, extending across the zero-field series limit into the continuum, with a regular spacing of about $1.5\hbar\omega$, where ω is the cyclotron frequency. A semiclassical explanation is given for this result.

022.004 Theoretical analysis of the Al I absorption spectrum. C. D. Lin.
Astrophys. Journ., Vol. 187, 385 - 387 (1974).
A plot of oscillator strengths of the Al I $3s^23p\,^2P \to 3s^2nd\,^2D$ transitions and the adjoining continuum, by the quantum-defect method, shows a good fit of the recent Kohl and Parkinson continuum data with earlier discrete measurements.

022.005 The free-free transitions of Cl$^-$.
T. L. John, D. J. Morgan, D. Davies.
Monthly Notices Roy. Astron. Soc., Vol. 166, 9P - 10P (1974).
The free-free continuous absorption coefficient of Cl$^-$ is determined with wavefunctions calculated in the close coupling approximation neglecting exchange. The cross-section is shown to be an order of magnitude larger than the corresponding results for H$^-$, He$^-$ and C$^-$.

022.006 Opacity distribution functions and absorption in Schumann-Runge bands of molecular oxygen.
T.-M. Fang, S. C. Wofsy, A. Dalgarno.
Planet. Space Sci., Vol. 22, 413 - 425 (1974).
We discuss the application of opacity distribution functions to the calculation of the transmission of solar radiation in the atmosphere of the earth. We discuss also the selection of molecular constants and the construction of the O_2 cross sections in the Schumann-Runge bands and adjoining continua and we apply the results to the calculation of the photodissociation rates for a number of constituents of the mesosphere.

022.007 New laboratory measurements of the A $^2\Pi - X\,^2\Sigma$ and B $^2\Sigma - X\,^2\Sigma$ transitions of CaH.
L.-E. Berg, L. Klynning.
Astron. Astrophys., Suppl. Ser., Vol. 13, 325 - 344 (1974).
The absorption spectrum of calcium hydride in the region 6130–7125 Å has been photographed using a 10.685 m concave grating spectrograph. Wavelengths, vacuum wave numbers and identifications of 1425 lines from 10 bands belonging to the A $^2\Pi - X\,^2\Sigma$ and B $^2\Sigma - X\,^2\Sigma$ transitions are listed.

022.008 Molecular constants of the TiO molecule.
J. G. Phillips.
Astrophys. Journ., Suppl. Ser., No. 232, Vol. 26, 313 - 331 (1973).
Molecular constants for all known singlet and triplet states of TiO are presented. The triplet constants are based on extensive analyses at Berkeley of the three known triplet systems. The singlet constants are drawn in part from Berkeley analyses, and in part from analyses conducted elsewhere.

022.009 A summary of transition probabilities for atomic absorption lines formed in low-density clouds.
D. C. Morton, W. H. Smith.
Astrophys. Journ., Suppl. Ser., No. 233, Vol. 26, 333 - 363 (1973).
A table of wavelengths, statistical weights, and excitation energies has been prepared for 944 atomic spectral lines in 221 multiplets whose lower energy levels lie below 0.275 eV. Oscillator strengths were adopted for 635 lines in 155 multiplets from the available experimental and theoretical determinations. Radiation damping constants also were derived for most of these lines. This table contains the lines most likely to be observed in absorption in interstellar clouds, circumstellar shells, and the clouds in the direction of quasars where neither the particle density nor the radiation density is high enough to populate the higher levels. All ions of all elements from hydrogen to zinc have been included which have resonance lines longward of 912 Å, although a number of weaker lines of neutrals and first ions have been omitted.

022.010 On the extreme ultraviolet emission spectrum of Fe XIII. D. R. Flower, H. Nussbaumer.
Astron. Astrophys., Vol. 31, 353 - 360 (1974).
The comparison of the observed and calculated intensities of the ultra-violet emission lines of Fe XIII in the 200–300 Å range enables the electron density of the emitting region to be determined. The observed intensities date from a rocket flight on April 4, 1969. Consistent values of the electron density are derived from both the relative and the absolute line intensities. The necessary atomic data, principally radiative transition probabilities and electron collision cross-sections, have been calculated with full allowance for configuration interaction. The results of these atomic physics calculations are tabulated. Also presented are the computed intensities of the strongest emission lines of Fe XIII in the 200–400 Å range, which may serve for similar analyses of future observational data.

022.011 The Poynting-Robertson effect and Eddington limit for electrons scattering with hard photons.
G. R. Blumenthal.
Astrophys. Journ., Vol. 188, 121 - 129 = Lick Obs. Bull. No. 648 (1974).
The Klein-Nishina cross-section is used to calculate the mean force due to Compton scattering on electrons with arbitrary velocity. The Eddington-limit luminosity is then calculated as a function of photon energy and as a function of temperature for a source of hard radiation. The rate at which angular momentum is lost by gas orbiting a luminous point source is also calculated. Some implications for compact cosmic X-ray sources are discussed.

022.012 Autoionization spectra of beryllium (Be I and Be II) in the 110- to 140-eV energy range.
G. Mehlman, J. M. Esteva.
Astrophys. Journ., Vol. 188, 191 - 195 (1974).
A total of 35 new resonances have been identified in the absorption spectrum of beryllium, 20 of them belonging to

the neutral atom, the others to the singly ionized one. The upper levels involve excitation of one of the K-shell electrons and appear in most cases dominated by the core structure.

022.013 The spectra of S XIII and S XIV in the region 25–40 Å.
S. Goldsmith, L. Oren, L. Cohen.
Astrophys. Journ., Vol. 188, 197 - 200 (1974).

The spectra of S XIII and S XIV in the wavelength region 25–40 Å consist of transitions between levels of principal quantum numbers $n = 2$ to $n = 3$. Twenty new transitions were identified by the aid of various extrapolations. The methods of calculations are discussed in detail.

022.014 On the question of possible condensation of π^--mesons in nuclear matter.
G. P. Alodzhants (*Alojants*), D. M. Sedrakyan, E. V. Chubaryan.
Astrofizika, Vol. 9, 581 - 588 (1973). In Russian. English translation in Astrophysics, Vol. 9, No. 4.

022.015 Reinvestigation of rotational-line intensity factors in diatomic spectra.
E. E. Whiting, R. W. Nicholls.
Astrophys. Journ., Suppl. Ser., No. 235, Vol. 27, 1 - 19 (1974).

The theory of the intensity factors of rotational lines in diatomic molecular spectra is reviewed with an emphasis on removing obscurities and resolving ambiguities that exist.

022.016 Atomic and molecular properties of metals from artificial cloud experiments in the upper atmosphere.
S. Drapatz, L. Haser, K.-W. Michel.
Zeitschr. Naturforschung, Vol. 29a, 411 - 418 (1974).

Time and frequency resolved observations of metal vapor clouds in the upper atmosphere can be used to obtain information on atomic and molecular properties of the metals involved. Experimental techniques of evaporation, the methods of observation and some of the more important results are discussed for Sr, Ba and Eu, in particular those which cannot be derived by laboratory experiment, namely, radiative transitions and oxidation processes which involve metastable states. Especially one finds that the photoionization and oxidation processes of Eu and Ba are significantly different in contrast to theoretical predictions.

022.017 Phase-uncertainties of individual modes in nonlinear systems.   H. Wilhelmsson.
Phys. Scripta, Vol. 9, 61 - 66 (1974).

The purpose of the paper is to study the evolution and partition of initially small phase-uncertainties (or perturbations) of individual modes in nonlinear systems of waves. The characteristic times are estimated for the transfer of phase-uncertainties between the different interacting modes as well as for the growth (or dissipation) of phase-uncertainties of each mode in stable and nonlinearly unstable situations.

022.018 Additions to the analysis of vanadium V.
J. O. Ekberg.
Phys. Scripta, Vol. 9, 96 - 98 (1974).

This paper presents some additions to the analysis by Van Deurzen (1973). 42 lines have been classified as ground-term transitions and 6 lines as combinations between excited terms.

022.019 Direct measurements of lifetimes of low-lying excited electronic states in nitric oxide.
J. Brzozowski, N. Elander, P. Erman.
Phys. Scripta, Vol. 9, 99 - 103 (1974).

Lifetimes of 32 bands in NO originating from the $A^2\Sigma^+$, $B^2\Pi$, $B'^2\Delta$ and $D^2\Sigma^+$ states and a system reported by Migeotte have been directly measured, using the high frequency deflection technique. The results are compared to earlier direct measurements and the lifetimes are discussed in relation to various types of states.

022.020 Transition state calculations of oscillator strengths in the Xα local exchange approximation.
P. G. Ellis, O. Goscinski.
Phys. Scripta, Vol. 9, 104 - 108 (1974).

Oscillator strengths for the first four $s - p$ transitions in Li, Na and K have been obtained from single Hartree-Fock-Slater computations, employing half-integral occupation numbers for the transition orbitals. The suggested form of the transition moment leads to f-values in good agreement with experiment.

022.021 Neue Wegweiser in der Physik.   D. A. Bromley.
Umschau, 74. Jahrgang, p. 233 - 239 (1974).

022.022 Nouvelle analyse du spectre hertzien du radical hydroxyl.
J.-L. Destombes, C. Marlière, F. Rohart, J. Burie.
Comptes Rendus Acad. Sci. Paris, Sér. B, Vol. 278, 275 - 278 (1974).

Les transitions principales de dédoublement Λ correspondant à J = 13/2 sont identifiées.

022.023 The speed of light and the new meter.
Z. Bay, J. A. White.
Phys. Today, Vol. 27, No. 4, p. 9, 11 (1974).

022.024 On a solution of equations of statistical equilibrium for hydrogen.   V. M. Sobolev.
Solnechnye Dannye 1974 Byull., No. 1, p. 65 - 74 (1974). In Russian.

A system of equations of statistical equilibrium and ionization equilibrium of the hydrogen atom with 15 levels has been solved, the concentration of the hydrogen atoms in the normal state ranging from 10^{10} to 10^{14} and electronic temperatures from 5000° to 50000°. Relative populations of the levels with n = 2, 3, 4 ... 10 at these values of parameters are tabulated.

022.025 Excitation cross sections of some decay products of C_2H_2 by electron impact.
I. V. Sushanin, S. M. Kishko.
Astron. Zhurn. Akad. Nauk SSSR, Vol. 51, 447 - 448 (1974). In Russian. English translation in Soviet Astron. AJ, Vol. 18, No. 2. − Short note.

022.026 Analytic variational calculation of the ground-state binding energy of hydrogen in intermediate and intense magnetic fields.   L. W. Wilson.
Astrophys. Journ., Vol. 188, 349 - 352 (1974).

A "prescription" of simple functions is given for an analytic variational calculation of the binding energy of the hydrogen ground state. The calculation still yields a relatively smooth transition between intermediate and intense fields. An explicit calculation of the binding energy as $B \to \infty$ is given.

022.027 Transitions $2s^2 2p^k – 2s 2p^{k+1}$ of the F I, O I, and N I isoelectronic sequences.
G. A. Doschek, U. Feldman, R. D. Cowan, L. Cohen.
Astrophys. Journ., Vol. 188, 417 - 422 (1974).

Transitions of the type, $2s^2 2p^k – 2s 2p^{k+1}$, have been identified for the elements from titanium through nickel for ions of the fluorine, oxygen, and nitrogen isoelectronic sequences. Wavelengths, visual intensity estimates, and energies are given. The energy differences of levels of the ground configuration are compared with predictions based on semi-empirical equations derived by Edlén. Some of the lines of these isoelectronic sequences should be strong lines in solar-flare spectra.

022.028 **Oscillator strengths for ionized iron and manganese.**
K. S. de Boer, D. C. Morton, S. R. Pottasch, D. G. York.
Astron. Astrophys., Vol. 31, 405 - 408 (1974).

The observed strengths of interstellar absorption lines of Fe II and Mn II in the spectra of α Vir, β Cen, π Sco, and ζ Oph along with laboratory f values of some of these lines between 2343 and 2606 Å have been used to determine curves of growth for these ions and the f-values of 10 lines of Fe II and 3 lines of Mn II between 1055 and 1261 Å. The Fe and Mn abundances are derived.

022.029 **Reflectance of amorphous-cubic NH_3 frosts and amorphous-hexagonal H_2O frosts at 77K from 1400 to 3000 Å.**
J. G. Pipes, E. V. Browell, R. C. Anderson.
Icarus, Vol. 21, 283 - 291 (1974).

The directional hemispherical reflectance of ammonia and water frosts in the range from 1400 to 3000 Å was measured at 77K. Amorphous and cubic ammonia frosts and amorphous and hexagonal water frosts were studied.

022.030 **Étude spectroscopique d'un jet de plasma d'argon hors d'équilibre thermodynamique local.**
P. Ranson, J. Chapelle.
Journ. Quant. Spectrosc. Radiat. Transfer, Vol. 14, 1 - 18 (1974).

The authors have studied the properties of an argon plasma jet seeded with a few percent of hydrogen. At atmospheric pressure, significant deviations from local thermodynamic equilibrium are shown. The plasma parameters (electron density, temperature) have been determined from the broadening and the absolute intensity of hydrogen lines. The transition probabilities and the broadening parameters of some AI lines between 4000 and 6000 Å have been measured.

022.031 **On the electronic transition moment variation and the r-centroid approximation in interpretation of spectral intensities of diatomic molecules.** R. W. Nicholls.
Journ. Quant. Spectrosc. Radiat. Transfer, Vol. 14, 233 - 237 (1974). – Note.

022.032 **Velocity distortions of spectral lines with combined Doppler and collision broadening.**
H. K. Chen, S. S. Penner.
Journ. Quant. Spectrosc. Radiat. Transfer, Vol. 14, 239 - 241 (1974).

The authors present the results of theoretical studies of velocity distortions produced by directed motion in a conical source flow on lines with combined Doppler and collision broadening.

022.033 **Calculation of wavelengths and oscillator strengths for the isoelectronic sequence of Li.**
U. I. Safronova.
Journ. Quant. Spectrosc. Radiat. Transfer, Vol. 14, 251 - 257 (1974). In Russian.

Feynman diagram techniques have been applied to the calculation of wavelengths and oscillator strengths of the ground state and of a number of low-lying excited states for Li-like ions. Contributions have been calculated to the first order for the nonrelativistic energy, relativistic corrections and dipole matrix elements. The results were used for calculations of the wavelengths and oscillator strengths of the transitions $1s^2 2s - 1s^2 2p$, $1s^2 2s - 1s^2 3p$, $1s^2 2p - 1s^2 3s$, $1s^2 2p - 1s^2 3d$, $1s^2 3s - 1s^2 3p$, and $1s^2 3p - 1s^2 3d$ for Li-like ions. Results are compared with experimental data for the isoelectronic sequence of Li.

022.034 **The similarity law for the profile of a spectral line and an inelastic cross section.**
A. V. Demura, V. S. Lisitsa.
Journ. Quant. Spectrosc. Radiat. Transfer, Vol. 14, 273 - 286 (1974). In Russian.

Connection is made between the velocity dependence of the inelastic cross section and the profile of the broadened spectral line. The results obtained lead to the conclusion that line-broadening theory is useful in the calculations of inelastic cross sections, where the region of intermediate energies of the perturbing particles is included. Detailed calculations of the inelastic cross section have been made for the dipole interaction because exact solutions of the broadening problem are available for comparison in this case. Comparison is made with the results of other authors. The possibility of calculating the effects of field rotation is discussed.

022.035 **Measurement of intensities of multiplets in the $2\nu_3$-band of methane at low temperatures.**
S. Sarangi, P. Varanasi.
Journ. Quant. Spectrosc. Radiat. Transfer, Vol. 14, 365 - 376 (1974).

The integrated intensities of the multiplets $P(1)-P(10)$, $R(0)-R(9)$, and of the Q-branch in the $2\nu_3$-band of $^{12}CH_4$ have been measured at 102°K, 152°K, 202°K, 251°K, and 300°K. Comparison of the data with theoretical line strengths confirms, at all of the temperatures mentioned, the intensity anomalies observed by Margolis (1973) for lines in this band.

022.036 **Measurements of gf-values for Fe I lines.**
A. Gilbert, K. G. P. Sulzmann, S. S. Penner.
Journ. Quant. Spectrosc. Radiat. Transfer, Vol. 14, 455 - 478 (1974).

Observations have led to gf-values for selected low-excitation lines of multiplet 13 (6280.620 Å, 6358.690 Å, 6498.950 Å). gf-values were obtained of high-excitation lines belonging to multiplets 1032, 843 and 928 (5187.92 Å, 5242.50 Å, 5379.58 Å). Curves of growth for the strong, high-excitation lines have been used to determine half-widths for collision broadening of Fe by the gas mixture existing under equilibrium conditions.

022.037 **The f-value of the NO $\delta(0, 0)$ band by the line absorption method.**
M. Mandelman, T. Carrington.
Journ. Quant. Spectrosc. Radiat. Transfer, Vol. 14, 509 - 521 (1974).

The resonance-line absorption method has been applied to the isolated but unresolved (0, 0) band of the NO δ system, $C^2\Pi - X^2\Pi$, and yields $f = (2.2 \pm 0.3) \times 10^{-3}$. The result is compared with previous measurements obtained by four other independent methods.

022.038 **Experimentally determined oscillator strengths for molecular hydrogen—I. The Lyman and Werner bands above 900 Å.** W. Fabian, B. R. Lewis.
Journ. Quant. Spectrosc. Radiat. Transfer, Vol. 14, 523 - 535 (1974).

Photoelectric absorption measurements have been performed on molecular hydrogen under conditions where the line profile is dominated by Doppler broadening. A curve of growth analysis was used to relate the measured equivalent width to the line oscillator strength, and the appropriate band oscillator strength was calculated from this using Hönl–London factors. Oscillator strengths for eleven Lyman and six

Werner bands obtained in this way show close agreement with those deduced from inelastic electron scattering data.

022.039 Experimentally-determined oscillator strengths for molecular hydrogen—II. The Lyman and Werner bands below 900 Å, the B'-X and the D-X bands.
B. R. Lewis.
Journ. Quant. Spectrosc. Radiat. Transfer, Vol. 14, 537 - 546 (1974).

Results are presented of experimental oscillator-strength measurements for four Lyman, one Werner, two B'-X and two D-X bands of molecular hydrogen in the wavelength region between 850 and 900 Å.

022.040 The effect of electrolytes on sublimation of H_2O ice.
E. A. Kajmakov, V. I. Sharkov.
Komety i Meteory, *Dushanbe*, No. 20, p. 3 - 5 (1971).
In Russian.

022.041 The pressure of sublimation products over water ices.
E. A. Kajmakov, V. I. Sharkov.
Komety i Meteory, *Dushanbe*, No. 20, p. 6 - 8 (1971).
In Russian.

022.042 Effective Gaunt factors for electron impact excitation in multiply-charged nitrogen and oxygen ions.
J. Davis.
Journ. Quant. Spectrosc. Radiat. Transfer, Vol. 14, 549 - 554 (1974).

Electron impact excitation Gaunt factors and thermally averaged Gaunt factors are obtained for the resonance and several nonresonance transitions for various ionization stages of nitrogen and oxygen. In those cases where comparisons can be made, the agreement is good. In addition, a simplified expression for the Gaunt factor is suggested by the data.

022.043 Correlation effects in the theory of combined Doppler and pressure broadening—I. Classical theory.
J. Ward, J. Cooper, E. W. Smith.
Journ. Quant. Spectrosc. Radiat. Transfer, Vol. 14, 555 - 590 (1974).

The present work is concerned with the combined influence of radiator-perturber collisions and radiator translational motion in the context of foreign gas broadening of optical transitions in neutral radiators.

022.044 Line identifications in the beam-foil spectrum of boron. H. G. Berry, J. L. Subtil.
Phys. Scripta, Vol. 9, 217 - 220 (1974).

Many new lines between 2000 and 5500 Å from displaced terms in the B II spectrum are observed using the beam-foil technique. Possible transitions in the doubly-excited quartet system of B III are compared with previous observations in Be II. Other transitions in B II, B III, and B IV are also identified.

022.045 Beam-foil spectroscopy of highly ionized argon.
J. P. Buchet, M. C. Buchet-Poulizac, A. Denis,
J. Désesquelles, G. Do Cao.
Phys. Scripta, Vol. 9, 221 - 225 (1974).

The spectra of argon are investigated between 150 Å and 5500 Å using the beam-foil technique at 1.15 MeV/nucleon. More than thirty observed lines have been identified as hydrogenic transitions in Ar X–XV. In the UV range, most lines are due to 3–3 and $2s^2 2p^n - 2s 2p^{n+1}$ transitions.

022.046 A new analysis of the $A\,^2\Delta - X\,^2\Pi$ system of CH.
B. M. Krupp.
Astrophys. Journ., Vol. 189, 389 - 397 (1974).

A more precise formula of the Hill and Van Vleck type is developed for the rotational levels of the components of doublet states in diatomic molecules. An additional rotational constant H_ν is accurately included in the development. New values of the molecular constants of the $A\,^2\Delta - X\,^2\Pi$ system of CH are obtained using the smoothed data of Gerö and of Kiess and Broida.

022.047 Depolarization cross-sections of Fe^{13+} in collision with protons. F. Masnou-Seeuws.
Astron. Astrophys., Vol. 32, 229 - 230 (1974).

Cross-sections are computed for transitions between Zeeman sublevels in the ground state of Fe^{13+} colliding with protons. A semi-classical method is used as in the earlier work of Masnou-Seeuws and McCarroll (1972). The coupled equations are solved numerically without further approximation. The results can be useful for the study of depolarization of the green line $\lambda = 5303$ of the solar corona.

022.048 Dipole and quadrupole integrals for the C I, N I, and O I sequences. S. O. Kastner, C. Wade.
Astrophys. Journ., Suppl. Ser., No. 243, Vol. 27, 247 - 260 (1974).

The Coulomb approximation tables of Oertel and Shomo, together with binding-energy values obtained by a screening approximation, have been used to produce values of the dipole and quadrupole radial integrals needed in obtaining transition probabilities for ions of six, seven, and eight electrons.

022.049 Hot hydrogen atoms: initiators of reactions of interest in interstellar chemistry and evolution.
K. Hong, J.-H. Hong, R. S. Becker.
Science, Vol. 184, 984 - 987 (1974).

Photochemically generated hot hydrogen atoms initiate reactions with simple molecular substrates including methane to produce organic alcohols, amines, acids, amino acids, and other compounds. The typical quantum yields for the formation of amino acids are 2×10^{-5} to 4×10^{-5}. Hot hydrogen atoms may be important initiators of reactions in interstellar space and in planetary atmospheres.

022.050 Il significato della velocità della luce.
T. Nicolini.
Reprinted from Atti Accad. Pontaniana, Nuova Ser., Vol. 21= Ist. Univ. Navale, Astron. Generale e Sferica, Napoli. Seminario No. 6, 30 pp. (1973).

022.051 Line strengths for rotational transitions of water vapour. I. G. Reznikov, V. S. Strelnitskij.
Soobshch. Gos. Astron. Inst. Shternberga, No. 183, p. 40 - 49 (1973). In Russian.

Values of pure rotational line strengths of H_2O vapour up to $J = 12$ are given.

022.052 Lifetimes of some highly-excited terms in Cl IV– Cl IX.
S. N. Bhardwaj, H. G. Berry, T. Mossberg.
Phys. Scripta, Vol. 9, 331 - 334 (1974).

Lifetime measurements using the beam-foil technique at wavelengths between 2000 and 4000 Å and chlorine beam energies of 1–4 MeV are reported.

022.053 The spectrum of three times ionized titanium, Ti IV.
J. W. Swensson, B. Edlén.
Phys. Scripta, Vol. 9, 335 - 337 (1974).

The K I-like spectrum of titanium, Ti IV, has been observed and analysed.

022.054 Collective oscillations in many electron atoms. II. Scattering and slowing down. J. J. Monaghan.
Australian Journ. Phys., Vol. 27, 169 - 175 (1974).

The slowing down of fast charged particles by their interaction with many electron atoms is considered using the hydrodynamic version of the Thomas-Fermi model. The agreement obtained with experiment is excellent over a wide range of parameters but worsens as the velocity of the charged particle decreases.

022.055 **Experimental study of Stark broadening of the allowed and forbidden transitions $2\ ^3P$–$4\ ^3D$, $4\ ^3F$ (λ = 4471.5 Å, λ = 4470 Å) in neutral helium.**
H. W. Drawin, J. Ramette.
Zeitschr. Naturforschung, Vol. 29a, 838 - 850 (1974).
Measurements were made of the profile of the He I 4471.5 Å line and its forbidden component at λ = 4469.9 Å, emitted from an afterglow plasma of electron densities in the range 2.2×10^{14} to 1.5×10^{15} cm^{-3}. The various quantities which define position and form of the profile have been compared with the theoretical calculations of Griem and of Barnard et al.

022.056 **Hook-method measurements of gf-values for ultraviolet Fe I and Fe II lines on a shock tube.**
M. C. E. Huber.
Astrophys. Journ., Vol. 190, 237 - 240 (1974).
Transition probabilities for 14 lines of Fe II and 12 lines of Fe I in the wavelength region 2560–2737 Å were measured by use of a shock tube and the hook method. Absolute oscillator strengths for resonance lines of Fe I reported by Banfield and Huber (1973) were used to determine the number density of neutral iron in the shock-heated gas. With the assumption of thermal equilibrium, the density of singly ionized iron atoms in this gas was then computed from the measured temperature and pressure with the aid of the Saha equation.

022.057 **Electronic transitions of the LaO molecule.**
L. Schoonveld, S. Sundaram.
Astrophys. Journ., Suppl. Ser., No. 246, Vol. 27, 307 - 318 (1974).
For use in stellar absorption studies, a complete analysis of all available experimental data on the LaO molecule has been made. A valid and consistent set of vibrational constants has been derived for each of the known electronic states of the molecule. Potential energy curves (RKR and Morse) have been constructed. Franck-Condon intensity factors and R-centroids have been calculated.

022.058 **The fundamental rotation-vibration band of TiO.**
J. G. Phillips.
Astrophys. Journ., Suppl. Ser., No. 247, Vol. 27, 319 - 330 (1974).
The locations of lines in the fundamental rotation-vibration band of TiO are derived from (a) the molecular constants, and (b) a differencing process invoking the known structures of the bands of the γ-system.

022.059 **Laboratory microwave spectrum of ethylene oxide.**
C. Hirose.
Astrophys. Journ., (Letters), Vol. 189, L145 - L146 (1974).
Observed laboratory frequencies for H_2COCH_2 and the rotational and centrifugal distortion constants derived therefrom are listed.

022.060 **Spectra of diatomic molecules. I.**
O. A. Mel'nikov, R. Kh. Salman-Zade, Yu. A. Solonskij, E. D. Khilov.
Trudy Astron. Obs., Leningrad, Vol. 30 (= Uchenye Zapiski Leningr. Un-ta, No. 373 = Seriya Matem. Nauk, vyp. (No.) 50), p. 192 - 228 (1974). In Russian.
Questions of the systematics and theory of two-atomic molecular spectra of astrophysical interest are discussed.

022.061 **On the role of different excitation and ionization mechanisms of hydrogen.** V. M. Sobolev.
Solnechnye Dannye 1974 Byull., No. 3, p. 53 - 65 (1974). In Russian.
From the solution of the equation of statistical equilibrium the values of the Balmer decrement for different n_1 and T_e have been determined; the ratios of the number of radiation processes to that of collision ones were calculated for excitation and also for ionization. The role of different mechanisms was analyzed.

022.062 **Electron impact excitation of metastable helium-like ions.** J. A. Tully, J. M. P. Serrão.
Astron. Astrophys., Vol. 33, 187 - 189 (1974).
The Coulomb-Born approximation has been used to study collisional excitation of the transitions $2\ ^1S \to n\ ^1S, n\ ^1P$ and $2\ ^3S \to n\ ^3S, n\ ^3P$ in positive ions of the helium isoelectronic sequence. Scaled collision strengths are tabulated as functions of (a) the atomic number Z and (b) the incident electron energy in threshold units X_0. f-values for the optically allowed transitions have been calculated in the dipole-length approximation.

022.063 **Lifetime measurement of excited nickel I levels.**
U. Becker, L. H. Göbel, W.-D. Klotz.
Astron. Astrophys., Vol. 33, 241 - 247 (1974). In German.
Eighteen excited Ni I-levels have been investigated with respect to the lifetimes by means of the zero-field level-crossing method. The spectral lines under investigation have wavelengths in the region of 3000–4000 Å and are due to transitions between the excited odd configurations $3d^9\ 4p$ and $3d^8\ 4s4p$ and the ground configurations $3d^9\ 4s$ and $3d^8\ 4s^2$. The f-values are compared with those of other authors.

022.064 **The reliability of recent calculations of electron impact excitation cross sections for light ions.**
W. D. Robb.
Bull. American Astron. Soc., Vol. 6, 214 (1974). – Abstr. AAS.

022.065 **L-shell vacancy production cross sections in H^+ + Ar and Ar^+ + Ar collisions.** M. E. Rudd.
Bull. American Astron. Soc., Vol. 6, 214 (1974). – Abstr. AAS.

022.066 **Quasi-Landau spectrum of a hydrogen like atom in a high magnetic field.**
A. F. Starace.
Bull. American Astron. Soc., Vol. 6, 214 (1974). – Abstr. AAS.

022.067 **Ionization of molecules of astrophysical interest.**
G. N. Haddad, J. A. R. Samson.
Bull. American Astron. Soc., Vol. 6, 214 (1974). – Abstr. AAS.

022.068 **On the photoionization of carbon dioxide.**
J. L. Gardner, J. A. R. Samson.
Bull. American Astron. Soc., Vol. 6, 214 (1974). – Abstr. AAS.

022.069 **Die Anwendung der Elektronenmikrosonde zur Bestimmung der Wertigkeit und Koordination in Mineralien.** M. K. Pavićević.
Analyse extraterrestrischen Materials, (see 012.020), p. 289 - 306 (1974).

022.070 **The phase diagram of graphite grains.**
M. Takada, M. Hirai.
Publ. Astron. Soc. Japan, Vol. 26, 197 - 206 (1974).
The authors obtained a phase diagram between pure graphite grains and monoatomic carbon gas in the temperature range 1000–5000 K. The chemical potential of pure graphite grains was calculated with a platelet model of the elastic body in which the structural anisotropy of graphite was considered. This model is better than a Debye approximation in which three-dimensional isotropy such as used by Hoyle and Wick-

ramasinghe (1962) is assumed. The phase diagram calculated agrees fairly well with that obtained from experimental data.

022.071   Search of cosmic ray quarks at large zenith angles.
T. Kifune, Y. Hieda, S. Kurokawa, K. Tsunemoto, Y. Kimura, T. Nishikawa.
Journ. Phys. Soc. Japan, Vol. 36, 629 - 633 (1974).

022.072   X-ray ionization cross-sections, and ionization equilibrium equations modified by Auger transitions.
J. C. Weisheit.
Astrophys. Journ., Vol. 190, 735 - 740 (1974).

Approximate subshell cross-sections for ionization by photons of energy 5000 eV $\geq E \geq$ 100 eV have been computed for all ions of C, N, O, Ne, Mg, Al, Si, and S. Each subshell cross-section τ has been fitted to a simple power-law form, $\tau \propto E^{-\beta}$, and the fitting parameters are tabulated. In addition, the Auger transitions that follow inner-shell ionizations are considered in detail, and modified ionization equilibrium equations are derived for elements of nuclear charge $Z \leq 18$.

022.073   Energy spectrum of hydrogen-like atoms in a strong magnetic field.   G. L. Surmelian, R. F. O'Connell.
Astrophys. Journ., Vol. 190, 741 - 742 (1974).

An expression for the energy spectrum of hydrogen-like atoms in strong magnetic fields is derived in terms of the energy spectrum of hydrogen. The development of a parametric expression for the latter is thus motivated. The authors present such an expression for the ground-state energy of hydrogen in magnetic fields $B \simeq 10^7 - 10^{12}$ gauss, obtained by least-squares fittings of numerical results obtained previously from a variational calculation.

022.074   Semiempirical calculation of gf values, II: Fe I $(3d + 4s)^8 - (3d + 4s)^7$ 4p.   R. L. Kurucz.
Smithsonian Astrophys. Obs., *Cambridge, Mass.*, Special Report 359, 9 + 138 pp. (1974).

Slater parameters, eigenvalues, and principal eigenvector components are presented for the $(3d + 4s)^8$ and $(3d + 4s)^7$ 4p configurations of Fe I. Transition integrals calculated from scaled Thomas-Fermi-Dirac wavefunctions are listed. The complete gf transition array $(3d + 4s)^8 - (3d + 4s)^7$ 4p is presented, together with partial sums required for computing radiative, Stark, and van der Waals damping constants.

022.075   Empirical computation of collisional ionization rates of atoms and ions by electrons.
J. Cantó, E. Daltabuit.
Rev. Mexicana Astron. Astrofis., Vol. 1, No. 1, p. 5 - 9 (1974).

A two-parameter empirical formula is proposed to compute the electron collisional ionization cross sections for atoms and ions. An interpolation formula is given to compute one of the parameters when experimental information is missing, and the corresponding ionization rate is computed.

022.076   Microwave spectra of molecules of astrophysical interest. 4. Hydrogen sulfide.
P. Helminger, F. C. de Lucia, W. H. Kirchhoff.
National Bureau of Standards, Washington. 13 pp. (1973).

022.077   Microwave spectra of molecules of astrophysical interest. 5. Water vapor.
F. C. de Lucia, P. Helminger, W. H. Kirchhoff.
National Bureau of Standards, Washington. 12 pp. (1973).

022.078   Microwave spectra of molecules of astrophysical interest. 6. Carbonyl sulfide and hydrogen cyanide.
A. G. Maki.
National Bureau of Standards, Washington. 26 pp. (1973).

022.079   Microwave spectra of molecules of astrophysical interest. 7. Carbon monoxide, carbon monosulfide and silicon monoxide.   F. J. Lovas, P. N. Krupene.
National Bureau of Standards, Washington. 24 pp. (1973).

022.080   Microwave spectra of molecules of astrophysical interest. 8. Sulfur monoxide.   E. Tiemann.
National Bureau of Standards, Washington, 14 pp. (1973).

022.081   Re-examination of the spectrum of strontium: autoionization in the spectrum of neutral strontium.
G. H. Newsom, S. O'Connor, R. C. M. Learner.
Journ. Phys. B, Atomic Molecular Phys., Vol. 6, 2162 - 2176 = Contr. Perkins Obs., Ohio State–Ohio Wesleyan Univ., Ser. I, No. 152 (1973).

The spectrum of strontium has been re-examined within the range 2100 Å to 26000 Å. A large number of new lines have been observed and many of these are diffuse. This has led to the identification of new energy levels including the even parity autoionizing levels arising from the 4d5d configuration. A complete list is presented of Sr I and Sr II wavelengths observed within the range 7800 Å to 26000 Å together with approximate intensities.

New standards for light velocity and wavelength.
Phys. Today, Vol. 27, No. 1, p. 112 (1974).

Errata

022.901   Correction: 'He and Ne cross sections in natural Mg, Al, and Si targets and radionuclide cross sections in natural Si, Ca, Ti, and Fe targets bombarded with 14- to 45- MeV protons' [Journ. Geophys. Res., Vol. 78, 6428 - 6442 (1973)].   J. R. Walton, A. Yaniv, D. Heymann, D. Edgerley, M. W. Rowe.
Journ. Geophys. Res., Vol. 79, 314 - 315 (1974).

Astronomical Instruments and Techniques

031 Astronomical Optics, Methods of Observation and Reduction, Automation

031.001 A modified coherent average operator in the search for periodical variations of unknown period in low-level optical sources. G. Sedmak.
Astron. Astrophys., Vol. 30, 345 - 347 (1974).
This note shows a truly numerical variant of the classical coherent average technique, which may be useful in identifying periodical component(s) of unknown period(s) in low-level optical sources.

031.002 Least squares adjustment with relatively large observation errors, inaccurate initial approximations, or both. H. Eichhorn, W. G. Clary.
Monthly Notices Roy. Astron. Soc., Vol. 166, 425 - 432 (1974).
This paper concerns least squares adjustment problems in which the condition equations are non-linear in both the observations and the adjustment parameters ('unknowns'). The authors present a solution algorithm which takes into account the second (as well as the first) order derivatives in the adjustment residuals (observation errors) and the corrections to the initially available approximations to the adjustment parameters.

031.003 A new iterative technique for the deconvolution of real data. K. Taylor.
Astrophys. Space Sci., Vol. 26, 327 - 336 (1974).
A fast new deconvolution technique has been developed which employs a complex iterative scheme in order to deconvolve severely broadened noisy data. A great many computer-simulated experiments have been made using the iterative scheme in order to test its reliability and stability in many different situations.

031.004 Selective diffraction with possible application to solar research. Y. Öhman.
Solar Physics, Vol. 34, 253 - 258 (1974).
The writer describes diffraction phenomena in transmission gratings made as strips of alternating color. He discusses diffraction in slits formed in semitransparent material including liquid or gaseous layers and particularly when line- and band-absorption is present. He suggests that atmospheric seeing should be influenced by selective diffraction in telluric lines. He also discusses possible seeing effects in the solar atmosphere and prominences and hints at the possibility that irregular or semiregular selective diffraction may appear in the deep layers of sunspots possibly contributing to the scattering of light.

031.005 Measurements of velocity dispersions and Doppler shifts from digitized optical spectra.
S. M. Simkin.
Astron. Astrophys., Vol. 31, 129 - 136 (1974).
Techniques are discussed for measuring Doppler shifts and velocity broadening profiles of stars and galaxies from digitized optical spectra. They are based on methods of time series analysis and rely heavily for their feasibility on application of the Fast Fourier Transform algorithm. Two functions are described for measuring Doppler shifts. In addition, techniques are documented that allow one to measure velocity broadening in spectra. It is pointed out that the methods described here can be applied to a wide variety of problems.

031.006 On some properties of the instrumental diffraction halo. I. A. Prokof'eva.
Solnechnye Dannye 1973 Byull., No. 11, p. 62 - 70 (1974). In Russian.

031.007 On the peculiarities of an instrumental halo. I. A. Prokof'eva.
Solnechnye Dannye 1973 Byull., No. 12, p. 98 - 103 (1974). In Russian.

031.008 Simple fast speed mirror-lens system with spherical surfaces. G. M. Popov, V. I. Pronik.
Izv. Krymskoj Astrofiz. Obs., Vol. 49, 105 - 116 (1974). In Russian.

031.009 Very long baseline interferometry in the southern hemisphere.
D. S. Robertson, J. S. Gubbay, A. T. Legg.
Proc. Astron. Soc. Australia, Vol. 2, 184 - 187 (1973). — Invited paper — see 012.001.

031.010 Analysis of coordinate measurement errors of a N-shaped slit diaphragm due to manufacture errors of the latter. Yu. B. Arkhangel'skij, S. V. Korotkov, M. K. Lijgant, G. P. Chuguev.
Teoriya i proektir. vysokotochn. sistem upr. Leningrad, Nauka, 1973, p. 127 - 133. In Russian. — Abstr. in Referativ. Zhurn. 51. Astron., 3.51.194 (1974).

031.011 High-speed photometry. R. E. Nather.
Vistas in astronomy, Vol. 15, (see 003.001), 91 - 112 (1973).

031.012 Planetenphotographie mit Amateur-Teleskopen. E. Wiedemann.
Orion Schaffhausen, 32. Jahrgang, p. 24 (1974).

031.013 Sonnenbeobachtungen für den Amateur. M. Waldmeier.
Orion Schaffhausen, 32. Jahrgang, p. 43 - 44 (1974).

031.014 Sonnenphotographie. H. Treutner.
Orion Schaffhausen, 32. Jahrgang, p. 44 - 49 (1974).

031.015 Neue Möglichkeiten der Sonnenbeobachtung für den Amateur. J. Schaedler, G. Klaus.
Orion Schaffhausen, 32. Jahrgang, p. 50 - 51 (1974).

031.016 Autocorrelation functions of star image vibrations and their applications to the estimate of errors of astrometric observations. I. G. Kolchinskij.
Astrometriya i Astrofizika, *Kiev*, vyp. (No.) 20, (see 003.002), p. 19 - 39 (1973). In Russian.
The paper deals with autocorrelation functions $R(\tau)$ of

star image vibrations in telescopes obtained by measuring 53 trails. Formulas of the theory of random processes in continuous and discrete variations were applied to estimate the average dispersion of the micrometer reading per star. The obtained $R(\tau)$ were used for concrete calculations. The dispersion estimates are given at different realization duration and at different zenith distances. The optimum measurement intervals are determined in correspondence with Vilenkin's theory.

031.017 **Some results of an investigation of image vibrations of stars at small angular distances.**
M. V. Bratijchuk, I. I. Motrunich, I. V. Shvalagin.
Astrometriya i Astrofizika, *Kiev*, vyp. (No.) 20, (see 003.002), p. 40 - 46 (1973). In Russian.

031.018 **Kontrastübertragung und Auflösungsvermögen.**
L. D. Schmadel, M. A. M. van Venrooy.
SuW, Vol. 13, 53 - 55 (1974).

031.019 **Gedanken zu einer Erweiterung der Definition der Aufgabe der Photogrammetrie.** W. Rüger.
Jenaer Rundschau, (Jena Review), 18. Jahrgang, p. 342 - 344 (1973).

031.020 **A quantitative treatment of solar 'seeing', II. Microthermal measurements in the immediate vicinity of telescopes.** C. E. Coulman.
Solar Physics, Vol. 34, 491 - 506 (1974).

A previously described microthermal measurement technique is applied to diagnose the causes and locations of sources of harmful 'seeing' effects in solar telescopes. Examples of the investigation of a compact 30-cm refractor at Culgoora Observatory and the massive 60-in. McMath reflector at Kitt Peak Observatory are described.

031.021 **Comparison of Houghton correctors and Schmidt plates, Part I: history and design.** N. J. Rumsey.
Southern Stars, Vol. 25, 95 (1973). – Conference abstract.

031.022 **Comparison of Houghton correctors and Schmidt plates, Part II.** G. R. Nankivell.
Southern Stars, Vol. 25, 95 - 96 (1973). – Conference abstract.

031.023 **Methods for achieving large apertures for optical telescopes at relatively low cost.** N. J. Rumsey.
Southern Stars, Vol. 25, 97 (1973). – Conference abstract.

031.024 **Television photometry: the Mariner 9 experience.** A. T. Young.
Icarus, Vol. 21, 262 - 282 (1974).

Television photometry is compared to conventional techniques. Reduced data from the Mariner 9 cameras should, under optimum conditions, have been accurate to a few percent. However, a combination of unstable camera properties and various unfortunate circumstances produced serious nonlinearities (typically 20–40%) and other systematic errors. The means of estimating these errors are described in detail.

031.025 **A simple method of estimation and consideration of the loss in photon counters.**
A. S. Nikolow, W. Schöneich.
Astron. Nachr., Vol. 295, 159 - 161 (1974). In German.

The statistical pulse loss by the use of a photon counting photometer can be corrected on the base of the equation $n = Ne^{-N\tau}$. A simple method is given for determination of the parameter τ which can be realized on each photometer.

031.026 **Über Methoden zur Prüfung astronomischer Optik.** C. Kühne.
Mitt. Astron. Ges., No. 34, p. 43 - 64 (1973/74). – Presented at the "Wissenschaftliche Tagung der Astron. Ges., Oberkochen, 1973 April 24 - 27".

031.027 **Die Prüfung des Zentrierzustandes von Großteleskopen.** A. Behr.
Mitt. Astron. Ges., No. 34, p. 132 - 133 (1973/74). Presented at the "Wissenschaftliche Tagung der Astron. Ges., Oberkochen, 1973 April 24 - 27".

031.028 **Temperaturunabhängige Fernrohrobjektive.** H. Köhler.
Mitt. Astron. Ges., No. 34, p. 133 - 143 (1973/74). Presented at the "Wissenschaftliche Tagung der Astron. Ges., Oberkochen, 1973 April 24 - 27".

031.029 **Some remarks on the ideal automated observatory.** G. Westerhout.
Astron. Astrophys., Suppl. Ser., Vol. 15, 327 - 330 (1974). Conference paper (see 012.011).

031.030 **Automation and astronomy.** J. G. Davies.
Astron. Astrophys., Suppl. Ser., Vol. 15, 331 - 332 (1974). – Conference paper (see 012.011).

031.031 **Real-time compensation for autocorrelation clipper bias.** W. F. Davis.
Astron. Astrophys., Suppl. Ser., Vol. 15, 381 - 382 (1974). Conference paper (see 012.011).

031.032 **Phase correction in Fourier transform spectroscopy.** T. Hilgeman.
Astron. Astrophys., Suppl. Ser., Vol. 15, 399 - 400 (1974). Conference paper (see 012.011).

031.033 **A hybrid correlator for interferometric observations.** B. Anderson, P. Richards.
Astron. Astrophys., Suppl. Ser., Vol. 15, 405 - 406 (1974). Conference paper (see 012.011).

031.034 **Pattern recognition.** T. Orhaug.
Astron. Astrophys., Suppl. Ser., Vol. 15, 407 - 412 (1974). – Conference paper (see 012.011).

031.035 **Fourier transform supersynthesis.** S. Kenderdine.
Astron. Astrophys., Suppl. Ser., Vol. 15, 413 - 415 (1974). Conference paper (see 012.011).

031.036 **Speckle interferometry and possible extensions.** A. Labeyrie.
Astron. Astrophys., Suppl. Ser., Vol. 15, 463 (1974). – Conference paper (see 012.011).

031.037 **Nomenclature of polarized light: linear polarization.** D. Clarke.
Applied Optics, Vol. 13, 3 - 5 (1974).

031.038 **Comments on: Recent advances in astronomical optics.** R. A. Buchroeder.
Applied Optics, Vol. 13, 22 (1974). – Letter.

031.039 **Quasi-monochromatic calibration source for the wavelength range from 0.22 μm to 7.0 μm.**
J. Fried, D. Labs.
Applied Optics, Vol. 13, 197 - 199 (1974).

This paper describes an optical system (consisting of a calibrated radiation source, a special double monochromator, and a Cassegrain telescope) that produces a quasi-monochromatic beam with known radiation power at its exit. The wavelength range extends from 0.22 μm to 7.0 μm. Some important remarks concerning the calibration procedure of the monochromator (polarization, slitwidths, etc.) are given.

031.040 Nomenclature of polarized light: elliptical polarization. D. Clarke.
Applied Optics, Vol. 13, 222 - 224 (1974).

031.041 Method for obtaining a flat field with Rosin-Amon tilted plate correctors. A. Offner.
Applied Optics, Vol. 13, 242 (1974). — Letter.

031.042 Interferometric star tracking. A. B. DeCou.
Applied Optics, Vol. 13, 414 - 424 (1974).

A new star-tracking technique based on interferometry is described and analyzed in detail. A heuristic comparison is made with traditional star-tracking methods that demonstrates several advantages in the interferometric approach for very high accuracy systems. A detailed error analysis is performed on several versions of the system that use all solid-state detection. Applications of the new system in large orbiting astronomical observatories and deep space laser communications systems are also discussed.

031.043 Imaging of extended objects through a turbulent atmosphere. R. F. Lutomirski, H. T. Yura.
Applied Optics, Vol. 13, 431 - 437 (1974).

The extended Huygens-Fresnel principle is used to derive an explicit expression for the image plane illuminance distribution of an extended object with an arbitrary luminance emittance distribution. The combined simultaneous effects of attenuation, background luminance, and atmospheric turbulence are given. A quantitative comparison of these effects is made, and their contribution to the over-all loss in resolution is given. Explicit expressions are derived for both the atmospheric-optical system modulation transfer function and image plane modulation. Numerical results for the image plane modulation are presented for imaging both up and down along an atmospheric path under various viewing conditions.

031.044 Radiometric calibrations for the earth radiation budget experiment.
J. R. Hickey, A. R. Karoli.
Applied Optics, Vol. 13, 523 - 533 (1974).

The earth radiation budget experiment is scheduled to be flown aboard the NIMBUS F satellite that is to be launched in mid-1974. This experiment includes channels to measure solar radiation, earth-reflected radiation (albedo), and earth-emitted long-wave radiation. The calibration techniques and the standards employed are included in the topics of this paper. Problems associated with achieving the desired levels of confidence for a multifaceted, high accuracy satellite experiment are discussed and some results of preliminary calibrations are presented.

031.045 Ein neues lichtstarkes katadioptrisches System für die Astrophotographie. E. Wiedemann.
Orion, 32. Jahrgang, p. 116 - 117 (1974).

031.046 Planeten-Photographie mit Amateur-Teleskopen. H. Treutner.
Orion, 32. Jahrgang, p. 118 - 123 (1974).

031.047 A system for automatic reduction of data on rocket sounding of the atmosphere.
V. I. Kozlov, I. S. Moshnikov.
Issled. verkhn. atmosf. Zemli. Moskva, Gidrometeoizdat, 1973, p. 213 - 221. In Russian. — Abstr. in Referativ. Zhurn. 62. Issled. kosmich. prostranstva, 5.62.520 (1974).

031.048 Core-halo stellar images: a possibly physiological phenomenon. V. Icke.
Observatory, Vol. 94, 41 - 44 (1974).

It is shown that a stellar image as seen by a visual observer at a telescope appears to be sharper than the objective image, owing to the nonlinear response of the retina. Hence the results of visual observations cannot be used directly to promote the construction of enormous telescopes.

031.049 Device for the mean moment of star transit determination. M. Hłond.
Postępy Astron., Vol. 22, 133 - 135 (1974). In Polish.

Description of an electronic arrangement for the determination of the mean moment of a star transit is given.

031.050 Seeing: its cause and cure. A. T. Young.
Astrophys. Journ., Vol. 189, 587 - 604 (1974).

Following Fried, the statistics of images formed by an annular telescope aperture looking through the turbulent atmosphere are calculated; the deleterious effect of a central obscuration is much larger in the presence of seeing than for the telescope alone. A method of optimizing resolution through the atmosphere is explained in detail, with particular attention to the requirements for image-motion compensation. Some common misconceptions about image structure and motion, and various unsound proposals for beating the seeing, are criticized. For the range of turbulent spectra actually observed in the atmosphere, no large improvement in resolution can be obtained at infrared wavelengths. Furthermore, the optimum telescope aperture is rather small; the apparent advantages of large telescopes are due solely to a good signal-to-noise ratio, not good resolution.

031.051 Radio astrometry using connected-element interferometers. B. Elsmore.
IAU Symposium No. 61, (see 012.014), p. 111 - 117 (1974). Invited paper.

031.052 The present state and future of the astronomical refraction investigations. G. Teleki.
IAU Symposium No. 61, (see 012.014), p. 235 - 236 (1974). Abstract.

031.053 Environmental systematics and astronomical refraction, I. J. A. Hughes.
IAU Symposium No. 61, (see 012.014), p. 237 (1974). — Abstract.

031.054 Sur la mesure des distances par télémetrie laser. R. Bouigue.
IAU Symposium No. 61, (see 012.014), p. 275 (1974). — Abstract.

031.055 The influence of the reduction model on the systematic accuracy of adjusted parameters.
H. Eichhorn, A. E. Rust.
IAU Symposium No. 61, (see 012.014), p. 303 (1974). — Abstract.

031.056 On the diffraction of an instrumental halo on the slit. I. A. Prokof'eva.
Solnechnye Dannye 1974 Byull., No. 3, p. 106 - 112 (1974). In Russian.

031.057 High speed evaluation of photographic plates. V. C. Reddish.
Stars and the Milky Way system, Proc. 1972, (see 012.018), p. 249 - 254 (1974). — Invited lecture.

031.058 Analysis of electronographic images. M. R. S. Hawkins.
Stars and the Milky Way system, Proc. 1972, (see 012.018), p. 255 - 264 (1974).

031.059 **A binary star scanner utilizing digitalized photoelectric techniques.** D. R. Curott, B. Atwood.
Bull. American Astron. Soc., Vol. 6, 263 (1974). – Abstr. AAS.

031.060 **Test methods for secondary mirrors of Cassegrain telescopes with special reference to the ESO 3.6 m telescope.** R. N. Wilson.
European Southern Obs., Techn. Rep. No. 3, 6 + 38 pp. (1974).

This report is intended to give a resumé and critical analysis of all the methods known to the author which have been proposed for testing convex secondary mirrors. For the sake of completeness, methods which cannot be applied to the 3.6 m telescope have been included so that it can be made clear why they are inapplicable.

031.061 **An iterative technique for the rectification of observed distributions.** L. B. Lucy.
Astron. Journ., Vol. 79, 745 - 754 (1974).

An iterative technique is described for generating estimates to the solutions of rectification and deconvolution problems in statistical astronomy. The behavior of the technique is explored by applying it to problems whose solutions are known in the limit of infinite sample size, and excellent results are obtained after a few iterations. The astronomical use of the technique is illustrated by applying it to the problem of rectifying distributions of $v \sin i$ for aspect effect; calculations are also reported illustrating the technique's possible use for correcting radio-astronomical observations for beam-smoothing. Application to the problem of obtaining unbiased, smoothed histograms is also suggested.

031.062 **Results of registration of mean star transit moments.** M. Ogriņš.
Uch. Zap. Latv. Gos. Univ., Vol. 202, Astron., vyp. (No.) 10, (see 003.012), p. 3 - 11 (1974). In Russian.

A device for registration of mean star transit moments is described.

031.063 **About the accuracy of photoelectric registration of star transit moments.** M. Ogriņš.
Uch. Zap. Latv. Gos. Univ., Vol. 202, Astron., vyp. (No.) 10, (see 003.012), p. 12 - 29 (1974). In Russian.

031.064 **Complex program for automatic reduction of photographic observations of satellites.**
J. Balodis.
Uch. Zap. Latv. Gos. Univ., Vol. 202, Astron., vyp. (No.) 10, (see 003.012), p. 52 - 67 (1974). In Russian.

031.065 **Some thoughts on Newtonian telescope alignment.** F. L. Redburn.
Journ. Astron. Soc. Victoria, Vol. 27, 25 - 27 (1974).

031.066 **A method of demonstrating the nature of light pollution.** J. S. Neff.
Mercury, (Journ. Astron. Soc. Pacific), Vol. 3, No. 1, p. 8 - 9 (1974).

031.067 **The measurement and analysis of photographic images.** E. B. Newell.
Proc. ESO/SRC/CERN conference, (see 012.021), p. 347 - 355 (1974).

031.068 **How to construct an amateur telescope (11), (12), (13).** L. Newelski.
Urania Kraków, Vol. 45, 106 - 113, 140 - 144, 174 - 176 (1974). In Polish.

031.069 **Measurement of elastic properties of a primary mirror cell with the aid of holographic interferometry.** A. F. Fercher, E.-D. Knohl, H. Schurle.
Messtechnik [Zeitschr. Instrumentenkunde], Vol. 81, 289 - 293 (1973). In German.

031.070 **Speckle holography.** P. T. Gough, R. H. T. Bates.
Optica Acta, (*GB*), Vol. 21, 243 - 254 (1974).

The results of a simulation, using diffuse laser light and an optical bench, of a recent holographic interpretation of Labeyrie's speckle interferometry are reported. The approach applies to a wide class of multiple objects viewed through a turbulent medium.

031.071 **An automatic data-collecting equipment in astronomy.** J.-H. Bigay, D. Bubet.
Electron. Microelectron. Industr., (*France*), No. 180, p. 39 - 43 (1973).

The equipment installed at the high mountain observatory telescope at Lyon and incorporating miniaturized linear and digital integrated circuits is described. The photometric data are recorded on tape and analyzed by computer. The photomultiplier type photometer with its optics and electronics is illustrated diagrammatically, together with the digital clock, and the filter control system.

031.072 **Automatic correction of aberration in astro-optical systems.** D. Su, Y. Wang.
Acta Astron. Sinica, Vol. 15, 51 - 60 (1974). In Chinese.

031.073 **The personal equations of photoelectric transits.**
Photoelectric-transit Division, Shanghai Obs., Acad. Sinica.
Acta Astron. Sinica, Vol. 15, 73 - 78 (1974). In Chinese.

031.074 **Zelf een kijker bouwen (3). Vorm van de spiegel te meten met hulp van kunststerretje.** T. de Vries.
Zenit, Vol. 1, No. 6, p. 34 - 35 (1974).

031.075 **The present state and future of the astronomical refraction investigations.** G. Teleki.
Publ. Obs. Astron. Beograd, No. 18, (see 012.023), p. 9 - 16 (1974).

031.076 **Some notes on refraction in radio astronomy.** W. J. Altenhoff.
Publ. Obs. Astron. Beograd, No. 18, (see 012.023), p. 17 - 20 (1974).

031.077 **Some comments on astronomical refraction.** B. R. Bean, G. Teleki.
Publ. Obs. Astron. Beograd, No. 18, (see 012.023), p. 21 - 44 (1974).

031.078 **The short period terms in image motion.** R. Fukaya, H. Yasuda.
Publ. Obs. Astron. Beograd, No. 18, (see 012.023), p. 45 - 47 (1974).

031.079 **Report to the members of Study Group on Astronomical Refraction.** E. Høg.
Publ. Obs. Astron. Beograd, No. 18, (see 012.023), p. 61 - 62 (1974).

031.080 **Environmental systematics and astronomical refraction I.** J. A. Hughes.
Publ. Obs. Astron. Beograd, No. 18, (see 012.023), p. 63 - 81 (1974).

031.081 **Report to the Study Group on Astronomical Refraction.** I. G. Kolchinskij.
Publ. Obs. Astron. Beograd, No. 18, (see 012.023), p. 181 - 182 (1974). – Presented at conference on astronomical refraction, Pulkovo, 4–5 July, 1973.

031.082 **Physical and meteorological factors connected with astronomical refraction.** G. Teleki.
Publ. Obs. Astron. Beograd, No. 18, (see 012.023), p. 213 - 234 (1974).

031.083 **A spatial Fourier analyser.** A. Ogata, K. Adati.
Nagoya Univ. Inst. Plasma Phys. Techn. Rep. IPPJ-T-13, 15 pp. (1973).

031.084 **A brief note on searching for periodicities in data obtained at irregular intervals.** C. L. Morbey.
Publ. Dominion Astrophys. Obs., Victoria, Vol. 14, (No. 9), 185 - 189 (1973).

A simple method for extracting periods from data obtained at irregular intervals is described.

031.085 **Tips für die Astropraxis.**
SuW, Vol. 13, 96 - 97, 132 - 133, 169 - 170, 205 - 208 (1974).
(3) Ein binokularer Schiefspiegler (*H. O. von Seggern*), (4) Eine einfache Methode zur indirekten Kometennachführung (*P. Stättmayer*); (5) Arbeitsmöglichkeiten mit modernen Schmalfilmkameras (*H. Schumacher*); (6) Untersuchungen von Filmemulsionen (*P. Höbel*).

031.086 **Direct computation of the distribution function from the characteristic function.** A. F. Dos Santos.
Proc. IEEE, Vol. 62, 533 - 534 (1974).

A simple and exact method to compute the cumulative distribution function directly from the characteristic function is proposed. This method is completely insensitive to errors associated with the divergence of the integral defining the Fourier transform of the distribution function.

031.087 **Come e che cosa osservare.** L. Baldinelli.
Giornale Ass. Astrofili Bolognesi, *Bologna,* No. 33, p. 3 - 5; No. 34, p. 5 - 8 (1974).

The role of fast Fourier transform computers in astronomy. See Abstr. 021.019.

Programming and control languages for automated laboratory experimentation. See Abstr. 021.020.

Automated information retrieval at the Stellar Data Center. See Abstr. 021.031.

Inferring phase information from modulus information in two-dimensional aperture synthesis. See Abstr. 033.032.

Space-frequency synthesis. See Abstr. 033.037.

Computer control of the 100 m telescope. See Abstr. 033.043.

Modern instrumentation for optical astronomy. See Abstr. 034.023.

Systèmes des coordonnées célestes. See Abstr. 041.032.

Numerical experiments concerning the determination of the orbits of close visual doubles considering second order terms in the observing errors. See Abstr. 118.003.

Digital dispersion spectroscopy: a technique for the study of very rapid variability of celestial radio sources. See Abstr. 141.075.

032 Astronomical Instruments

032.001 **Small telescopes constructed in the USSR.**
A. D. Marlenskij.
Zemlya i Vselennaya, 1974, No. 1, p. 74 - 77. In Russian.

032.002 **An X-ray telescope sensitive at high energies.**
T. M. Palmieri.
Astrophys. Space Sci., Vol. 26, 431 - 445 (1974).
A telescope is described which is capable of producing images of point sources of X-rays without recourse to reflection optics. A mathematical approach to the operation and to the signal-to-noise properties of the telescope is presented. This is followed by several examples of its response and a discussion of detectors that could be used with the device.

032.003 **Instrumental polarization of the large extra-eclipsing coronograph of the IZMIRAN and the Pulkovo Observatory.** O. E. Den.
Solnechnye Dannye 1973 Byull., No. 12, p. 91 - 97 (1974). In Russian.

032.004 **Determination of the flexure on the basis of equatorial points to the north and south of the zenith.**
K. N. Kuz'menko, V. Kh. Pluzhnikov.
Vestn. Khar'kov. Univ. No. 99 (Ser. Astron. No. 8), p. 42 - 46 (1973). In Russian.

032.005 **The value of a revolution of the micrometer of the Kharkov meridian circle.** N. G. Zuev.
Vestn. Khar'kov. Univ. No. 99 (Ser. Astron. No. 8), p. 49 - 51 (1973). In Russian.

032.006 **Multi-unit large optical telescope with controllable mirror shape.** M. I. Gvozdev, N. A. Dimov, N. L. Zhernokleev, V. S. Zuev, P. D. Kalachev, Yu. L. Kokurin, T. I. Marchenko, E. P. Orlov, V. A. Sautkin.
Uspekhi fiz. nauk, Vol. 111, 558 - 560 (1973). In Russian. Abstr. in Referativ. Zhurn. 51. Astron., 3.51.81 (1974).

032.007 **Determination of the general term of the magnitude equation of the Tashkent normal astrograph.**
A. A. Latypov.
Izv. AN UzSSR. Ser. fiz.-mat. n., 1973, No. 4, p. 54 - 56. In Russian. — Abstr. in Referativ. Zhurn. 51. Astron., 3.51.189 (1974).

032.008 **New 60-cm telescope in Ostrowik.** K. Stępień.
Postępy Astron., Vol. 22, 49 - 51 (1974). In Polish.
A short history and a description of the new telescope recently put into operation at Ostrowik is given.

032.009 **A new wide-field triple lens paraboloid field corrector.** C. G. Wynne.
Monthly Notices Roy. Astron. Soc., Vol. 167, 189 - 197 (1974).
It is shown that, by a substantial change of lens shapes, a three lens corrector can be designed for a paraboloid, to give a very similar performance to the Ritchey-Chrétien triple corrector. Numerical examples are given.

032.010 **Prismatische Korrektionsplatte der Schmidt-Kamera 1340/2000/4000.** H. G. Beck.
Jenaer Rundschau, (Jena Review), 18. Jahrgang, p. 322 - 323 (1973).

032.011 **Die Reichweite der Tautenburger Objektivprismenaufnahmen.** F. Börngen.
Jenaer Rundschau, (Jena Review), 18. Jahrgang, Separatprint, 4 pp. (1973).

032.012 **Der astronomische Gerätebau in der UdSSR.**
N. N. Michel'son.
Jenaer Rundschau, (Jena Review), 18. Jahrgang, p. 336 - 341 (1973).

032.013 **A visit to the Soviet Union's 6-meter reflector.**
A. G. D. Philip.
Sky Telescope, Vol. 47, 290 - 295 (1974).

032.014 **The 72-inch "Copernicus telescope".**
C. Barbieri, L. Rosino, R. Stagni.
Sky Telescope, Vol. 47, 298 - 300 (1974).

032.015 **The lensless Schmidt camera for astrophotography.**
C. Ashcraft, with a note by R. E. Cox.
Sky Telescope, Vol. 47, 333 - 338 (1974).

032.016 **Cálculo de las constantes de distorsión y parámetros del disco obturador para cámaras balísticas.**
M. J. Sevilla.
Urania Barcelona, Año 58, Nos. 277 - 278, p. 72 - 83 (1973).

032.017 **Das neue 75-cm-Spiegelteleskop der Wilhelm-Foerster-Sternwarte.** B. Wedel.
SuW, Vol. 13, 167 - 168 (1974).

032.018 **Investigation of the photometric field error for the objective of the 16″-Zeiss astrograph.**
G. E. Erleksova, V. Satyvaldiev.
Byull. Inst. Astrofiz., *Dushanbe*, No. 60, p. 34 - 35 (1972). In Russian.

032.019 **Investigation of the optics of the AZT-8 telescope and the 40-cm astrograph.** N. N. Kiselev.
Byull. Inst. Astrofiz., *Dushanbe*, No. 62, p. 37 - 39 (1974). In Russian.

032.020 **A propos du télescope de Schmidt.** A. Brun.
L'Astronomie, 88e année, p. 107 - 108 (1974).

032.021 **Design study of a glass meridian circle. II: Building.**
E. Hoeg.
Mitt. Astron. Ges., No. 34, p. 150 - 155 (1973/74).
Presented at the "Wissenschaftliche Tagung der Astron. Ges., Oberkochen, 1973 April 24 - 27".

032.022 **On polarimetry in solar active regions. VI: Experimental compensation of telescopic phase retardation; influences on Zeeman polarimetry.**
E. Wiehr, M. Rossbach.
Solar Physics, Vol. 35, 343 - 349 (1974).
Phase retardation caused by the Grégory-coudé telescope at Locarno is determined empirically and compared with calculations. The praxis of compensation by means of two $\lambda/8$-plates (Bowen compensator) is described and tested.

032.023 **SCLERA: an astrometric telescope for experimental relativity.** J. R. Oleson, C. A. Zanoni, H. A. Hill, A. W. Healy, P. D. Clayton, D. L. Patz.
Applied Optics, Vol. 13, 206 - 211 (1974).
An $f/100$, 12.2-m focal length photoelectric telescope designed specifically for daytime astrometry of objects near the sun is now operative at its Tucson, Arizona, site. The design goal was to achieve accuracies of order 0.001 sec of arc in field position measurements of stars. To accomplish this,

many features reducing systematic and random errors are employed.

032.024 The study of flexure effects on the mobile parts of the great meridian instrument of the Bucharest Observatory and their influence on the accurate determination of stellar positions. I. Rusu.
Doctoral thesis, Bucharest Astronomical Observatory, Bucharest. 149 pp. (1974).

032.025 Ein vereinfachtes Protuberanzen-Fernrohr. G. Richter.
Sterne, Vol. 50, 105 - 108 (1974).

032.026 On the comparison of the screw values of two zenith telescopes of the Poltava Gravimetrical Observatory.
V. K. Budz'ko, T. B. Kurgan, R. I. Popova, A. P. Stehpa.
Vrashchenie i prilivn. deformatsii Zemli. Vyp. (No.) 5. Kiev, Nauk. dumka, 1973, p. 111 - 113. In Russian. – Abstr. in Referativ. Zhurn. 51. Astron., 5.51.197 (1974).

032.027 On a peculiarity of the T4 Wild instrument. B. N. D'yakov.
Geod. i kartografiya, 1973, No. 11, p. 22 - 23. In Russian. Abstr. in Referativ. Zhurn. 52. Geodeziya i Aehrosemka, 5.52.114 (1974).

032.028 Performance of Schmidt and field-corrected telescopes. P. Fellgett.
Monthly Notices Roy. Astron. Soc., Vol. 167, 83P - 85P (1974). – Short communication.

032.029 Erste geographische Ortsbestimmungen mit den Zenitkameras der astronomischen Station Hannover.
K. Pilowski.
Techn. Univ. Hannover, Astron. Station Inst. Theor. Geod., No. 9, 19 + 4 pp. = Ergänzung I zur Monographie 1973 „Zur Konstruktion und Theorie und zur Verwendung transportabler Zenitkameras für die geographische Ortsbestimmung und für die Fundamentale Positionsastronomie".
The first astronomic observations and their reduction of the determination of geographical position with two zenith cameras are discussed. The zenith cameras are a development at the Astronomische Station des Instituts für Theoretische Geodäsie der Technischen Universität Hannover.

032.030 Variazioni progressiva e annuale della flessione laterale del Telescopio Zenitale di Carloforte nell' intervallo 1900–1960. T. Nicolini.
Rend. Acad. Sci. Fis. Mat. Soc. Nazionale Sci., Lettere ed Arti Napoli, Ser. 4, Vol. 39, 155 - 160 = Ist. Univ. Navale Napoli, Astron. Generale e Sferica, Pubbl. No. 6 (1972).
The monthly means of the side-flexure of the visual Z. T. of the Carloforte Latitude Station, from 1900 to 1960, shows a progressive irregular fluctuation, with a superposed lesser annual oscillation.

032.031 The 104-cm telescope of Uttar Pradesh State Observatory. S. D. Sinvhal, C. D. Kandpal, H. S. Mahra, S. C. Joshi, J. B. Srivastava.
'Optical astronomy with moderate size telescopes', Symposium held at Centre of Advanced Study in Astronomy, Osmania Univ., Hyderabad, November 1969, p. 20 - 34 = Uttar Pradesh State Obs., *Naini Tal*, Repr. No. 64 (1972).

032.032 Distribution of thermal fluctuations of the air in the pavilion of a solar telescope.
N. I. Kozhevnikov, B. Kalzhanov, S. O. Obashev.
Soobshch. Gos. Astron. Inst. Shternberga, No. 184, p. 17 - 20 (1973). In Russian.

032.033 Investigation of the AFR-1 astrograph of the Moscow Observatory. K. G. Steinert (*Shtajnert*).
Trudy Gos. Astron. Inst. Shternberga, Vol. 44, 195 - 202 (1973). In Russian.

032.034 The UV sky-survey telescope S2/68 in the TD-1 A satellite. A. Boksenberg, R. G. Evans, R. G. Fowler, S. K. Gardner, L. Houziaux, C. M. Humphries, C. Jamar, D. Macau, J. P. Macau, D. Malaise, A. Monfils, K. Nandy, G. I. Thompson, R. Wilson, H. Wroe.
Astron. Inst. Utrecht, Internal Rep. ROF 72, (see 012.012), p. C8.1 (1974). – Abstract.

032.035 The new 48-inch Cassegrain coudé telescope. D. Kotsakis.
In honorem S. Placidis, (see 003.009), p. 161 - 176 (1974). In Greek.

032.036 The 25-inch Newall refractor of the astronomical station in Pentele (Athens). C. Banos.
In honorem S. Placidis, (see 003.009), p. 277 - 284 (1974). In Greek.

032.037 Modern developments of the meridian circle. E. Høg.
IAU Symposium No. 61, (see 012.014), p. 243 - 255 (1974). Invited paper.

032.038 Instrumental parameters of the U.S. Naval Observatory's automatic transit circle (ATC).
B. L. Klock, F. S. Gauss.
IAU Symposium No. 61, (see 012.014), p. 257 - 259 (1974).

032.039 The results of investigations of the Pulkovo horizontal meridian circle (HMC).
G. I. Pinigin, L. A. Sukharev, G. M. Timashkova.
IAU Symposium No. 61, (see 012.014), p. 261 - 262 (1974).

032.040 An automated photometric telescope.
S. J. Hill, A. P. Linnell.
Bull. American Astron. Soc., Vol. 6, 218 (1974). – Abstr. AAS.

032.041 A visit to the Russian 6-meter reflecting telescope at Zelenchukskaya. A. G. D. Philip.
Bull. American Astron. Soc., Vol. 6, 261 (1974). – Abstr. AAS.

032.042 Report on the Kitt Peak vacuum telescope.
W. Livingston, J. Harvey, K. Pierce, D. Schrage, C. Slaughter.
Bull. American Astron. Soc., Vol. 6, 291 (1974). – Abstr. AAS.

032.043 The mechanical design of the 3.6 metre telescope.
W. Richter, D. Plathner, J. F. Roozeveld van der Ven.
European Southern Obs., Techn. Rep. No. 1, 2 + 46 pp. (1974).
This report describes those parts of the telescope which are definitively designed. Most of them are being currently executed.

032.044 The optics of the ESO 3.6 metre telescope. R. N. Wilson.
European Southern Obs., Techn. Rep. No. 2, 3 + 33 + 20 pp. (1974).
It is the purpose of this report to give a resumé of the astronomical/technical arguments leading to the choice of the telescope parameters; to discuss in some detail the optical design considerations which fixed the final form of the mirrors; and to give an account of the manufacture and test results with final values of the parameters.

032.045 Large telescope astronomy. J. L. Greenstein.
Proc. ESO/SRC/CERN conference, (see 012.021),

p. 11 - 25 (1974).

032.046 Use of large optical telescopes at infrared wavelengths. E. E. Becklin.
Proc. ESO/SRC/CERN conference, (see 012.021), p. 57 - 70 (1974).

032.047 Preliminary results of the Baranne white pupil image tube nebular spectrograph.
A. Baranne, N. Carrozzi, G. Comte, G. Courtès, J. M. Deharveng, R. Duflot, G. Monnet, A. Pellet, with the technical assistance of G. di Biaggio.
Proc. ESO/SRC/CERN conference, (see 012.021), p. 231 - 237 (1974).

032.048 Applications of Schmidt telescopes – with emphasis on direct and spectral surveys. R. M. West.
Proc. ESO/SRC/CERN conference, (see 012.021), p. 321 - 341 (1974).

032.049 The U.K. 48-inch Schmidt telescope and some current research programmes. R. D. Cannon.
Proc. ESO/SRC/CERN conference, (see 012.021), p. 343 - 345 (1974).

032.050 How to use large optical telescopes to the fullest extent. G. R. Burbidge.
Proc. ESO/SRC/CERN conference, (see 012.021), p. 363 - 369 (1974).

032.051 Considerations for allocation of telescope time. R. R. Shobbrook.
Proc. ESO/SRC/CERN conference, (see 012.021), p. 371 - 372 (1974).

032.052 Scheduling telescopes with very many users. K. C. Freeman.
Proc. ESO/SRC/CERN conference, (see 012.021), p. 373 - 374 (1974).

032.053 Organization of telescope use. P. A. Strittmatter.
Proc. ESO/SRC/CERN conference, (see 012.021), p. 375 - 379 (1974).

032.054 Design of an X-ray telescope system for an orbiting astronomical observatory.
J. A. Bowles, T. J. Patrick, P. H. Sheather, A. M. Eiband.
Journ. Phys. E, Sci. Instruments, Vol. 7, 183 - 191 (1974).
Describes the design and construction of a package of X-ray telescopes launched in the NASA satellite Copernicus (OAO-3).

032.055 The meridian marks of the large transit instrument of the Belgrade Observatory. I. Pakvor.
Bull. Obs. Astron. Belgrade, Vol. 29, F. 1, No. 125, p. 3 (1971/72). – Presented at the IAU meeting, Commission 8, Brighton 1970.

032.056 Micrometer of the large transit instrument of the Belgrade Observatory, 1971. I. Pakvor.
Bull. Obs. Astron. Belgrade, Vol. 29, F. 1, No. 125, p. 4 - 6 (1971/72).

032.057 Automatic device (A. D.) for contact-micrometer on the transit instrument.
S. Petković, M. Kralj.
Bull. Obs. Astron. Belgrade, Vol. 29, F. 1, No. 125, p. 7 - 8 (1971/72).

032.058 Sur l'influence des irrégularités des tourillons de l'instrument de passages sur la valeur observée de Cp.
M. Jovanović.
Bull. Obs. Astron. Belgrade, Vol. 29, F. 1, No. 125, p. 13 - 14 (1971/72).

032.059 Examination of the flexure of Belgrade meridian circle by the daytime. M. Mijatov.
Bull. Obs. Astron. Belgrade, Vol. 29, F. 1, No. 125, p. 19 - 25 (1971/72).

032.060 Analysis of the determinations of the corrections of the micrometer screw value of the Belgrade zenith telescope. M. Djokić.
Bull. Obs. Astron. Belgrade, Vol. 29, F. 1, No. 125, p. 25 - 29 (1971/72).

032.061 Beiträge zur Untersuchung des Babelsberger Meridiankreises. J. Liebert.
Akad. Wiss. DDR, Forschungsbereich Kosmische Phys., Veröff. Zentralinst. Phys. Erde, No. 27, 101 pp. (1973).
For the purpose of testing the efficiency of the Pistor & Martins meridian circle (189/2640 mm), the pivots, the graduated circle, the micrometer and the levels were examined, and an extensive series of observations of fundamental stars were carried out.

"Stellar" research at the Kitt Peak National Observatory. See Abstr. 013.014.

A quantitative treatment of solar 'seeing', II. Microthermal measurements in the immediate vicinity of telescopes. See Abstr. 031.020.

Test methods for secondary mirrors of Cassegrain telescopes with special reference to the ESO 3.6 m telescope. See Abstr. 031.060.

URSIES and the Bartol Coudé Observatory. See Abstr. 034.064.

Second Greenwich catalogue of stars for 1950.0. See Abstr. 041.008.

Ground-based television patrol of Martian cloud formation in 1969 and 1971–1972. See Abstr. 097.024.

Survey of faint H II regions in the nearest galaxies by means of the large reducteur focal of the 193 cm telescope of the Haute Provence Observatory. See Abstr. 131.552.

Untersuchungen über die Sternassoziation Cyg T 1. See Abstr. 152.003.

Errata

032.901 Erratum: 'General analysis of aplanatic Cassegrain, Gregorian, and Schwarzschild telescopes' [Applied Optics, Vol. 11, 2817 - 2832 (1972)]. W. B. Wetherell.
Applied Optics, Vol. 13, 242 (1974).

032.902 Addendum: 'The ultra-violet sky-survey telescope in the TD-1A satellite' [Monthly Notices Roy. Astron. Soc., Vol. 163, 291 - 322 (1973)].
A. Boksenberg, R. G. Evans, R. G. Fowler, I. S. K. Gardner, L. Houziaux, C. M. Humphries, C. Jamar, D. Macau, D. Malaise, A. Monfils, K. Nandy, G. I. Thompson, R. Wilson, H. Wroe, J. P. Macau.
Monthly Notices Roy. Astron. Soc., Vol. 167, 663 (1974).
See abstract 10.032.014.

033 Radio Telescopes and Equipment

033.001 **The Synthesis Radio Telescope at Westerbork. General lay-out and mechanical aspects.**
J. W. M. Baars, B. G. Hooghoudt.
Astron. Astrophys., Vol. 31, 323 - 331 (1974).
This paper describes the design specifications and the general lay-out of the antennas and the baseline. Geodetic and astrometric methods of measurement are described, which have been used to obtain an overall accuracy of 1 mm in the baseline. The drive system and pointing accuracies are briefly discussed. The characteristics of the reflectors are evaluated, both on the basis of measurements on the construction and aperture efficiency determinations.

033.002 **The Synthesis Radio Telescope at Westerbork. The 21 cm continuum receiver system.**
J. L. Casse, C. A. Muller.
Astron. Astrophys., Vol. 31, 333 - 338 (1974).
The SRT operates as a set of 20 simultaneous interferometers. The paper describes the first receiver system. It is a correlation receiver capable of measuring the complete polarization state of the received radiation.

033.003 **A radio telescope of whole earth's size.**
L. Matveenko.
Nauka i zhizn', 1973, No. 10, p. 25 - 32. In Russian. — Abstr. in Referativ. Zhurn. 51. Astron., 3.51.63 (1974).

033.004 **Investigation of a device for compensating wind disturbances of a radio telescope antenna.**
G. G. Basistov.
Teoriya i proektir. vysokotochn. sistem upr. Leningrad, Nauka, 1973, p. 29 - 34. In Russian. — Abstr. in Referativ. Zhurn. 51. Astron., 3.51.99 (1974).

033.005 **Long delayed echoes: the search for a solution.**
A. T. Lawton, S. J. Newton.
Spaceflight, Vol. 16, 181 - 187, 195 (1974).

033.006 **Interpolation and Fourier transformation of fringe visibilities.** A. R. Thompson, R. N. Bracewell.
Astron. Journ., Vol. 79, 11 - 24 (1974).
This paper investigates some of the methods for computing maps of radio brightness from observations of complex fringe visibility obtained with radio telescopes using earth rotation synthesis. In particular the authors are interested in data obtained with east–west arrays such as those at Cambridge, Westerbork, and Stanford, although most of the conclusions reached are more generally applicable. As an illustrative example they use the Stanford array.

033.007 **Recurrent algorithms of restoration of a radio astronomical image.** P. A. Fridman.
Soobshch. Spets. Astrofiz. Obs. AN SSSR, *Zelenchukskaya*, vyp. (No.) 6, p. 19 - 26 (1972). In Russian.
On the basis of a mean-squares criterion of quality, recurrent algorithms of restoration are found of brightness temperature distribution smoothed by the radio telescope beam in the presence of noise.

033.008 **Preliminary results of measurements with an interferometer with retransmitting of meter wave range ($\lambda = 3.5$ m).**
V. V. Vitkevich, G. I. Dobysh, V. S. Artyukh, R. D. Dagkesamanskij, B. K. Izvekov, S. A. Sukhodol'skij, V. A. Frolov.
Izv. vyssh. ucheb. zavedenij. Radiofizika, Vol. 16, 1320 - 1324 (1973). In Russian. — Abstr. in Referativ. Zhurn. 51. Astron., 4.51.115 (1974).

033.009 **Electric parameters and guiding system of the RT-25×2 radio telescope.**
A. G. Kislyakov, V. I. Chernyshev, A. A. Nosov, Yu. P. Shandra.
Izv. vyssh. ucheb. zavedenij. Radiofizika, Vol. 16, 1409 - 1417 (1973). In Russian. — Abstr. in Referativ. Zhurn. 51. Astron., 4.51.116 (1974).

033.010 **Directivity and characteristics of the RT-25×2 radio telescope.**
A. G. Kislyakov, V. I. Turchin, A. L. Fogel', V. I. Chernyshev.
Izv. vyssh. ucheb. zavedenij. Radiofizika, Vol. 16, 1418 - 1424 (1973). In Russian. — Abstr. in Referativ. Zhurn. 51. Astron., 4.51.117 (1974).

033.011 **Raising the sensitivity of the RT-25×2 radio telescope of the Crimean Astrophysical Observatory at the 3-cm wavelength range for using the radio interferometer with a maximum resolving power.**
L. D. Bakhrakh, M. I. Grigor'eva, V. A. Efanov, L. R. Kogan, V. I. Kostenko, L. I. Matveenko, G. S. Misezhnikov, I. G. Moiseev, Yu. G. Monin, M. M. Mukhina, V. B. Shtejnshlejger.
Izv. vyssh. ucheb. zavedenij. Radiofizika, Vol. 16, 1425 - 1428 (1973). In Russian. — Abstr. in Referativ. Zhurn. 51. Astron., 4.51.118 (1974).

033.012 **Investigation and practical application of radio astronomical alignment to the large Pulkovo telescope.**
G. B. Gel'frejkh, O. A. Golubchina.
Izv. vyssh. ucheb. zavedenij. Radiofizika, Vol. 16, 1429 - 1433 (1973). In Russian. — Abstr. in Referativ. Zhurn. 51. Astron., 4.51.119 (1974).

033.013 **Study of ionospheric effects on the characteristics of the UTR-2 radio telescope.**
A. V. Men', A. V. Antonov.
Izv. vyssh. ucheb. zavedenij. Radiofizika, Vol. 16, 1434 - 1438 (1973). In Russian. — Abstr. in Referativ. Zhurn. 51. Astron., 4.51.120 (1974).

033.014 **Radiometer of 3 to 3.8 mm range with an input commutator.**
V. N. Voronov, R. Kirakosyan.
Izv. vyssh. ucheb. zavedenij. Radiofizika, Vol. 16, 1439 - 1441 (1973). In Russian. — Abstr. in Referativ. Zhurn. 51. Astron., 4.51.121 (1974).

033.015 **A 4-cm range radiometer with a parametric input amplifier and liquid nitrogen cooling.**
A. B. Berlin, D. V. Korol'kov, G. M. Timofeeva, A. S. Berlin, N. T. Tishchenkov, O. A. Arsen'eva, R. I. Vergasov.
Izv. vyssh. ucheb. zavedenij. Radiofizika, Vol. 16, 1444 - 1447 (1973). In Russian. — Abstr. in Referativ. Zhurn. 51. Astron., 4.51.122 (1974).

033.016 **Autonomous automatic radio telescope control as an extremum problem.**
N. M. Aleksandrovskij, P. V. Ermuratskij, B. N. Sevryukov.
Izv. vyssh. ucheb. zavedenij. Ehlektromekhanika, 1973, No. 10, p. 1151 - 1156. In Russian. — Abstr. in Referativ. Zhurn. 51. Astron., 4.51.123 (1974).

033.017 **Meteorological features affecting large aperture synthesis radio telescopes.** R. N. Hardy.
Nature, Vol. 249, 431 - 433 (1974).
The first few months of observing with the 5-km teles-

cope at the Mullard Radio Astronomy Observatory, University of Cambridge, showed that there were frequent significant effects from intermediate systems, with scales of 10 to 100 km or more. An investigation was undertaken to relate these periods of anomalous phase changes – 'events' – to meteorological phenomena and the results are summarised here.

033.018 The prospects for high sensitivity gravitational antennae. V. B. Braginsky.
IAU Symposium No. 64, (see 012.008), p. 28 - 34 (1974).

033.019 Gravitational radiation detector magnetic tapes from Rochester and Maryland. M. Lee, J. Weber.
IAU Symposium No. 64, (see 012.008), p. 35 (1974). – Abstract.

033.020 Observations with wide-band gravitational radiation detectors. R. W. P. Drever, J. Hough, R. Bland, G. W. Lessnoff.
IAU Symposium No. 64, (see 012.008), p. 37 (1974). – Abstract.

033.021 The use of cryogenic techniques to achieve high sensitivity in gravitational wave detectors.
S. P. Boughn, W. M. Fairbank, M. S. McAshan, H. J. Paik, R. C. Taber, T. P. Bernat, D. G. Blair, W. O. Hamilton.
IAU Symposium No. 64, (see 012.008), p. 40 - 51 (1974).

033.022 Optimization of gravitational burst detectors using piezoelectric transducers. D. Maeder.
IAU Symposium No. 64, (see 012.008), p. 52 (1974). – Abstract.

033.023 Electromagnetic detectors of gravitational waves.
V. B. Braginsky, L. P. Grishchuk, A. G. Doroshkievich, Ya. B. Zel'dovich, I. D. Novikov, M. V. Sazhin.
IAU Symposium No. 64, (see 012.008), p. 54 - 58 (1974).

033.024 A solar radio spectrograph for the frequency range 70–810 MHz. A. Tlamicha.
Bull. Astron. Inst. Czechoslovakia, Vol. 25, 163 - 167 (1974).
The equipment for obtaining solar radio spectra of bursts on metric- and deci-metric wavelengths is described.

033.025 NOD2 a general system of analysis for radioastronomy. C. G. T. Haslam.
Astron. Astrophys., Suppl. Ser., Vol. 15, 333 - 338 (1974). Conference paper (see 012.011).

033.026 The data handling of the Westerbork Synthesis Radio Telescope. H. W. van Someren Gréve.
Astron. Astrophys., Suppl. Ser., Vol. 15, 343 - 352 (1974). Conference paper (see 012.011).

033.027 Source fitting on the multibeam Molonglo Cross. D. F. Crawford.
Astron. Astrophys., Suppl. Ser., Vol. 15, 353 - 354 (1974). Conference paper (see 012.011).

033.028 A digital correlation spectrometer for supersynthesis.
F. K. Bowers, T. L. Landecker, D. A. Whyte.
Astron. Astrophys., Suppl. Ser., Vol. 15, 395 - 396 (1974). Conference paper (see 012.011).

033.029 The Parkes 1024 channel digital autocorrelator and its conversational mode software system.
J. G. Ables, J. W. Brooks, B. F. C. Cooper, A. J. Hunt, G. G. Moorey.
Astron. Astrophys., Suppl. Ser., Vol. 15, 397 (1974). – Conference paper (see 012.011).

033.030 The 100 channel spectral line processor for Algonquin Radio Observatory.
L. A. Higgs, C. W. McLeish, T. G. O'Neill.
Astron. Astrophys., Suppl. Ser., Vol. 15, 401 - 403 (1974). Conference paper (see 012.011).

033.031 Aperture synthesis with a non-regular distribution of interferometer baselines. J. A. Högbom.
Astron. Astrophys., Suppl. Ser., Vol. 15, 417 - 426 (1974). Conference paper (see 012.011).

033.032 Inferring phase information from modulus information in two-dimensional aperture synthesis.
P. J. Napier, R. H. T. Bates.
Astron. Astrophys., Suppl. Ser., Vol. 15, 427 - 430 (1974). Conference paper (see 012.011).

033.033 Phase error correction in multi-element radio interferometer by data processing. M. Ishiguro.
Astron. Astrophys., Suppl. Ser., Vol. 15, 431 - 443 (1974). Conference paper (see 012.011).

033.034 On changing the synthesized beam pattern within a source distribution map. Y. L. Chow, G. A. Pelletier.
Astron. Astrophys., Suppl. Ser., Vol. 15, 445 - 448 (1974). Conference paper (see 012.011).

033.035 Very large array configurations for the observation of rapidly varying sources. W. K. Klemperer.
Astron. Astrophys., Suppl. Ser., Vol. 15, 449 - 451 (1974). Conference paper (see 012.011).

033.036 Interpolation of synthesis data and some effects of ringlobes. R. N. Bracewell, A. R. Thompson.
Astron. Astrophys., Suppl. Ser., Vol. 15, 453 (1974). – Conference paper (see 012.011).

033.037 Space-frequency synthesis. J. N. Douglas.
Astron. Astrophys., Suppl. Ser., Vol. 15, 455 (1974). – Conference paper (see 012.011).

033.038 The Texas small-diameter source catalogue.
F. N. Bash, J. N. Douglas.
Astron. Astrophys., Suppl. Ser., Vol. 15, 457 - 461 (1974). Conference paper (see 012.011).

033.039 The Stanford five-element array: an instrument for real-time fan beam synthesis and rapid rotation-synthesis. R. S. Colvin, L. R. D'Addario.
Astron. Astrophys., Suppl. Ser., Vol. 15, 465 - 468 (1974). Conference paper (see 012.011).

033.040 A real-time image forming system for the Arecibo sphere. J. J. Condon, J. M. Durdin, C. Hazard, D. L. Jauncey, L. M. Lalonde, M. J. Yerbury.
Astron. Astrophys., Suppl. Ser., Vol. 15, 471 - 473 (1974). Conference paper (see 012.011).

033.041 A digital pulsar processor. V. Boriakoff.
Astron. Astrophys., Suppl. Ser., Vol. 15, 479 - 481 (1974). – Conference paper (see 012.011).

033.042 Total power statistical confusion surveys.
J. J. Condon, D. L. Jauncey.
Astron. Astrophys., Suppl. Ser., Vol. 15, 485 - 486 (1974). Conference paper (see 012.011).

033.043 Computer control of the 100 m telescope.
P. Stumpff, J. Schraml.
Astron. Astrophys., Suppl. Ser., Vol. 15, 517 - 523 (1974).

Conference paper (see 012.011).

033.044 On-site processing of data obtained with the Westerbork Synthesis Telescope. E. Raimond.
Astron. Astrophys., Suppl. Ser., Vol. 15, 525 - 527 (1974).
Conference paper (see 012.011).

033.045 The FORTH program for spectral line observing on NRAO's 36 ft telescope.
C. H. Moore, E. D. Rather.
Astron. Astrophys., Suppl. Ser., Vol. 15, 529 - 534 (1974).
Conference paper (see 012.011).

033.046 The effect of digitization errors on detection of weak signals in noise. B. G. Clark.
Proc. IEEE, Vol. 61, 1654 - 1655 = National Radio Astron. Obs., *Green Bank,* Repr. Ser. A, No. 322 (1973).

A common method of detecting weak signals in the presence of noise is to digitize samples of the signal and average the samples to reduce the effect of noise. It is shown that errors in the digitizer can affect this process even if they are much smaller than the noise variance.

033.047 Simultaneous interferometer phase and water vapor measurements.
K. H. Wesseling, J. P. Basart, J. L. Nance.
Radio Sci., *(USA)*, Vol. 9, 349 - 353 = National Radio Astron. Obs., *Green Bank,* Repr. Ser. A, No. 331 (1974).

Phase fluctuations on the three-element, 11-cm wavelength interferometer at the National Radio Astronomy Observatory were compared with the difference in the atmospheric water vapor content over the respective antennas along a line-of-sight path to a celestial radio source.

033.048 Control of the position of the hour axis of a solar radio telescope with automatic guide.
A. A. Drozdovskij, A. Kurbanov.
Dokl. AN TadzhSSR, Vol. 16, No. 10, p. 13 - 15 (1973). In Russian. – Abstr. in Referativ. Zhurn. 51. Astron., 5.51.135 (1974).

033.049 A reflector type beam switching for millimeter wave telescope, and its application to a search for interstellar molecule of CH_3NH_2.
K. Akabane, T. Miyaji, Y. Chikada.
Tokyo Astron. Bull., Second Ser., No. 229, p. 2639 - 2647 (1974).

A rotating reflector covering the primary feed of Cassegrain system has been used for a beam switching the 6 meter millimeter wave radio telescope of the Tokyo Astronomical Observatory. Performance of the switching system and some observational results with it to find an interstellar molecular line of CH_3NH_2 are presented.

033.050 An 18 cm turnstile junction as a polarisation splitter.
N. Fourikis.
Proc. Instn. Radio Electron. Engineers Australia, Vol. 34, 403 - 405 = Division Radiophys. CSIRO, Sydney, Radiophys. Publ. RPP 1650 (1973).

A six-port turnstile junction designed for a radio astronomy application to operate in the 1612, 1666 and 1720 MHz bands is described. The matching structure is fully described and experimental results are given. The junction is used with the Parkes radio telescope.

033.051 The use of highly sensitive wide-band amplifiers in the millimeter region for radio astronomical observations. G. P. Apushkinskij, A. A. Shenogin, B. Ya. Losovskij, A. N. Tsyganov, M. T. Levchenko.
Trudy Astron. Obs., *Leningrad,* Vol. 30 (= Uchenye Zapiski Leningr. Un-ta, No. 373 = Seriya Matem. Nauk, vyp. (No.) 50), p. 29 - 48 (1974). In Russian.

The use of TWT's as low noise wide-band high-frequency amplifiers is considered. Peculiarities of TWT's in the millimeter range are discussed. The 8-millimeter range TWT receiver and the wide-band atmospheric noise compensation device are described. The receiver was used to observe the planet Mars and low radio brightness regions on the sun. The results of these observations are presented.

033.052 Numerical evaluation of electromagnetic fields due to dipole antennas in the presence of stratified media. L. Tsang, R. Brown, J. A. Kong, G. Simmons.
Journ. Geophys. Res., Vol. 79, 2077 - 2080 (1974).

Two numerical methods are used to evaluate the integrals that express the em fields due to dipole antennas radiating in the presence of a stratified medium. In contrast to previous analytical methods that applied only to two-layer cases the numerical methods can be used for any arbitrary number of layers with general properties.

033.053 A radio telescope for measurements of the slowly varying component of solar radio emission at 3.2 cm wavelength.
Yu. B. Vedeneev, M. S. Durasova, G. A. Lavrinov, T. S. Podstrigach, N. M. Prytkov, O. I. Yudin.
Solnechnye Dannye 1974 Byull., No. 3, p. 94 - 97 (1974). In Russian.

033.054 The Synthesis Radio Telescope at Westerbork. Principles of operation, performance and data reduction.
J. A. Högbom, W. N. Brouw.
Astron. Astrophys., Vol. 33, 289 - 301 (1974).

A description is given of the main properties of the Westerbork Synthesis Radio Telescope as an instrumental system for radio astronomical observations. Sufficient basic aperture synthesis theory is included to derive the main characteristics of the instrument such as the field of view, the synthesized beam, the grating responses, the sidelobe structure and the sensitivity under different operating conditions. The main steps in the standard data processing from the receiver outputs to the computed sky brightness distribution are described as well as some of the commonly used additional programs that are available for the further processing of the maps.

033.055 Results from a swept-frequency ionospheric impedance probe. E. K. Miller, H. F. Schulte, Jr.
Planet. Space Sci., Vol. 22, 1017 - 1030 (1974).

Experimental results are presented for the impedance of a rocket-borne dipole antenna immersed in the ionospheric plasma. The dependence of several interesting impedance artifacts upon the antenna position relative to the earth's magnetic field and rocket motion through the ionospheric plasma are shown.

033.056 On optimum synthetic linear arrays with application to radioastronomy.
F. Biraud, E. J. Blum, J. C. Ribes.
IEEE Trans. Antennas Propagation, Vol. AP-22, 108 - 109 (1974).

033.057 NRC's supersynthesis interferometer – radio waves of the galaxies.
Sci. Dimension, *(Canada)*, Vol. 5, No. 6, p. 24 - 28 (1973).

Describes a system which consists of two radio telescopes mounted on an east-west track some 300 metres long. The signals received by the two moveable 8.5-metre parabolic reflectors are combined to produce the effective resolution of a 300-metre paraboloid.

033.058 On the measurement of antenna beamwidth using extraterrestrial radio sources. W. L. Stutzman.

IEEE Trans. Antennas Propagation, Vol. AP-22, 493 - 495 (1974).

The effect of source angular size on the measurement of antenna half-power beamwidth using extraterrestrial sources is analyzed for a circularly symmetric Gaussian antenna pattern and a circular uniform disk source. Results are presented in the form of universal curves which can be used conveniently in practical applications.

033.059 **A new wideband, fully steerable, decametric array at Clark Lake.** W. C. Erickson, J. R. Fisher.
Radio Sci., (USA), Vol. 9, 387 - 401 (1974).

033.060 **Experimental determination of attenuation in corrugated circular waveguides.**
A. D. Olver, P. J. B. Clarricoats, S. L. Chong.
Electronics Letters, Vol. 9, 424 - 426 (1973).

033.061 **A microwave circulator that's smaller than a quarter.** R. Knerr.
Bell Lab. Rec., Vol. 51, No. 3, p. 79 - 84 (1973).

033.062 **A varactor-tuned Q-band Gunn oscillator.**
C. S. Aitchison, R. Davies.
International Journ. Electronics, Vol. 35, No. 1, p. 105 - 108 (1973).

033.063 **The corrugated horn as an antenna range standard.** R. Caldecott, C. A. Mentzer, L. Peters, Jr.
IEEE Trans. Antennas Propagation, Vol. AP-21, 562 - 564 (1973).

The authors describe the properties of both rectangular and conical corrugated horns, and consider their application for use in standard gain measurements. – *BMT*

033.064 **Astigmatism in reflector antennas.**
J. R. Cogdell, J. H. Davis.
IEEE Trans. Antennas Propagation, Vol. AP-21, 565 - 567 (1973).

The characteristics of the astigmatic phase error in large parabolic aerials are described. – *BMT*

033.065 **Some errors in the calculation of the radiation patterns of reflector antennas, using Kirchhoff integration.** J. Dijk, E. J. Maanders.
Electronics Letters, Vol. 9, 510 - 512 (1973).

The far sidelobes calculated using Kirchhoff integration are unreliable and lead to estimates of noise temperature that are too high. – *DNC*

033.066 **Correlator performance and the squared coherency.** R. F. Webber, W. H. Delashmit.
Electronics Letters, Vol. 9, 496 - 497 (1973).

033.067 **A new wideband, fully steerable, decametric array at Clark Lake.** W. C. Erickson, J. R. Fisher.
College Park, Maryland, Univ. Maryland, Astron. Program, Clark Lake Radio Obs., 27 pp. (1973).

Description and progress report on the Clark Lake 3.0 km × 1.8 km 'T' array, operating frequency is 15 MHz to 125 MHz. – *JGA*

033.068 **Antennas employing conical dielectric horns. Part 1. Propagation and radiation characteristics of dielectric cones.** P. J. B. Clarricoats, C. E. R. C. Salema.
Proc. Instn. Electr. Engineers, (GB), Vol. 120, 741 - 749 (1973).

033.069 **Antennas employing conical dielectric horns. Part 2. The Cassegrain antenna.**
P. J. B. Clarricoats, C. E. R. C. Salema.
Proc. Instn. Electr. Engineers, (GB), Vol. 120, 750 - 756 (1973).

033.070 **Antenna with multibeam applications.**
R. Claydon.
Electronics Letters, Vol. 9, 517 - 518 (1973).

033.071 **The circular synthetic radar.** R. H. MacPhie.
IEEE Trans. Aerospace Electronic Systems, Vol. AES-9, 608 - 611 (1973).

033.072 **Eigenvalues of spherical hybrid modes in corrugated conical horns.**
M. S. Narasimhan, M. S. Govindarajan.
Proc. Instn. Electr. Engineers, (GB), Vol. 120, 965 - 967 (1973).

033.073 **Short axial length broad-band horns.** J. L. Kerr.
IEEE Trans. Antennas Propagation, Vol. AP-21, 710 - 714 (1973).

Description of short horns giving about 10 : 1 bandwidth and approximately constant half-power beamwidth of 30 deg. *DNC*

033.074 **A new method for calculating correction factors for near-field gain measurements.**
A. C. Ludwig, R. A. Norman.
IEEE Trans. Antennas Propagation, Vol. AP-21, 623 - 628 (1973).

Using spherical wave expansions of measured far-field patterns, near field gains are predicted. – *DNC*

033.075 **Broad-band antenna array with application to radio astronomy.**
C. L. Rufenach, W. M. Cronyn, K. L. Neal.
IEEE Trans. Antennas Propagation, Vol. AP-21, 697 - 700 (1973).

Low frequency broadband array using zig-zag log periodic elements is fed by an open wire line matrix with time-delay gradients to obtain frequency independent beam positons.

033.076 **Some data for the design of low-crosspolarisation feeds.** S. I. Ghobrial.
Electronics Letters, Vol. 9, 465 - 466 (1973).

Curves show the variation in dimensions of a five-dipole array as a function of the semi-angular aperture of paraboloidal reflectors. – *ACM*

033.077 **Modified diffraction coefficients for focusing reflectors.** G. L. James, G. T. Poulton.
Electronics Letters, Vol. 9, 537 - 538 (1973).

The calculated near-field of a parabolic reflector under plane-wave illumination shows good agreement with the physical-optics approach. – *ACM*

033.078 **Frequency-selective surfaces for multiple-frequency antennas.** G. H. Schennum.
Microwave Journ., Vol. 16, 55 - 57, 76 (1973).

Curves developed from empirical data are given for rapid design of FSS reflectors. Actual scatter and transparency radiation patterns are presented for an FSS hyperbolic sub-reflector. – *ACM*

033.079 **Horn design saves weight without performance loss.** H. S. Jones, Jr.
Microwaves, Vol. 12, No. 10, p. 72 - 73, 75 - 76 (1973).

Some design details for pyramidal and sectoral horns using copper-plated dielectric-foam construction. Advantages cited are light weight and ease of including corrugated surfaces. *DNC*

033.080 **An integrated wide-band varactor-tuned Gunn oscillator.** R. A. Gough, B. H. Newton.
IEEE Trans. Electron. Devices, Vol. ED-20, 863 - 865 (1973).

033.081 **High-power frequency multiplier using MIS varactors.** F. Schumacher.
IEEE Trans. Microwave Theory Techn., Vol. MTT-21, 648 - 649 (1973).

A 5.5 W pulsed power at 5.4 GHz frequency doubler is described, efficiency 55%. –JBS

033.082 **Crosspolarisation with Cassegrainian and front-fed reflectors.** P. J. Wood.
Electronics Letters, Vol. 9, 597 - 598 (1973).

033.083 **Probability distribution for the resultant intensity due to random interference of two correlated M-vectors.** S. Okui.
International Journ. Electronics, Vol. 35, 623 - 626 (1973).

033.084 **Nomograms for estimating clear-sky zenith atmospheric attenuation in range 80 - 130 GHz.**
C. J. Gibbins.
Electronics Letters, Vol. 9, 605 - 607 (1973).

033.085 **Offset-reflector antennas with offset feeds.**
A. W. Rudge.
Electronics Letters, Vol. 9, 611 - 613 (1973).

Some theoretical and experimental results for offset feeds with offset-reflector antennas. – DNC

033.086 **Modular design of millimeter antennas with some guides for combining the basic ingredients.**
T. C. Cahill, G. Gill, H. Syrogos.
Microwave Journ., Vol. 16, No. 11, p. 53 - 58 (1973).

This article includes descriptions of corrugated feeds, polarizers, transducers and transitions. – BMT

033.087 **On the optimum radiometer.** J. E. Ohlson.
Radio Sci., (USA), Vol. 8, 841 - 844 (1973).

033.088 **A monolithic silicon wide-band amplifier from dc to 1 GHz.**
J. B. Coughlin, R. J. H. Gelsing, P. J. W. Jochems, H. J. M. van der Laak.
IEEE Journ. Solid-State Circuits, Vol. SC-8, 414 - 419 (1973).

033.089 **Radiation patterns of pyramidal dielectric waveguides.**
N. Brooking, P. J. B. Clarricoats, A. D. Olver.
Electronics Letters, Vol. 10, 33 - 34 (1974).

033.090 **Phase shifter with high amplitude accuracy.**
G. Demmel, D. Lukoschus.
Electronics Letters, Vol. 10, 22 - 23 (1974).

033.091 **Antenna input impedance computation.**
D. Abeyasekere.
Proc. Instn. Radio Electronics Engineers Australia, Vol. 35, 5 - 10 (1974).

A method of analyzing wire antennas and computing the driving point impedance based on the Richmond-Thiele equation is given. – DJC

033.092 **A reference noise standard for millimeter waves.**
W. C. Daywitt.
IEEE Trans. Microwave Theory Techn., Vol. MTT-21, 845 - 847 (1973).

033.093 **Two-port dipolar switch.** J. Helszajn.
Electronics Letters, Vol. 10, 46 - 47 (1974).

033.094 **A beam switching of phase-shifter type for millimeter-wave telescope.**
K. Akabane, Y. Chikada, K. Miyazawa.
Tokyo Astron. Bull., Second Ser., No. 232, p. 2675 - 2685 (1974).

A rotating phase-shifter is installed between the primay feed and the secondary reflector of the Cassegrain system of the 6 meter millimeter-wave telescope at the Tokyo Astronomical Observatory as a device for beam switching. Performance of the beam switching and its application to some millimeter-wave observation are presented.

033.095 **Aperture antenna with non-Gaussian phase errors.**
P. Beckmann.
Proc. IEEE, Vol. 62, 532 - 533 (1974).

The letter derives the mean radiation pattern of an aperture with random phase errors of the excitation when the errors are a stationary random, but not necessarily normal, function with given probability distribution and correlation coefficient. The case of exponential errors is investigated as an example.

A sensitive method for detecting dispersed radio emission. See Abstr. 021.016.

A data acquisition and control system for the 46 m telescope at the Algonquin Park Radio Observatory. See Abstr. 021.022.

Radio astrometry using connected-element interferometers. See Abstr. 031.051.

An iterative technique for the rectification of observed distributions. See Abstr. 031.061.

Some notes on refraction in radio astronomy. See Abstr. 031.076.

The Synthesis Radio Telescope at Westerbork: the 80-channel filter spectrometer. See Abstr. 034.002.

Solar radio emissions. See Abstr. 077.059.

034 Astronomical Accessories (Spectrometers, Photometers, etc.)

034.001 The orbiting stellar ultraviolet spectrophotometer S59 in ESRO's TD-1A satellite.
C. de Jager, R. Hoekstra, K. A. van der Hucht, T. M. Kamperman, H. J. Lamers, A. Hammerschlag, W. Werner, J. G. Emming.
Astrophys. Space Sci., Vol. 26, 207 - 262 (1974).

The ultraviolet stellar spectrophotometer S59 of the Utrecht Astronomical Institute uses the stabilization properties of the ESRO TD-1A satellite. The optical, mechanical and electronic properties of the instrument and its tracking system are described in some detail, as well as the optical and technical performance in laboratory tests and in orbit. Some results obtained during the first half year of operation are briefly described.

034.002 The Synthesis Radio Telescope at Westerbork: the 80-channel filter spectrometer.
R. J. Allen, J. P. Hamaker, K. J. Wellington.
Astron. Astrophys., Vol. 31, 71 - 78 (1974).

This paper describes the design and operation of a multichannel filter spectrometer which has been constructed for the study of radio spectral lines with the Westerbork Synthesis Radio Telescope. In order to reduce the delay between conception and realization of the project, the spectrometer was designed to make extensive use of the previously constructed low-frequency parts of the continuum receiver and to require only minor changes in the on-line and off-line computer programs. The essential characteristics of this continuum polarization receiving system are described, along with the modifications necessary to convert that system to a spectrometer. Tables of the overall system characteristics at 21 cm and 6 cm wavelength are given for the standard interferometer configurations used at Westerbork. The normal methods of observation and calibration are described and the resulting stability, sensitivity, and dynamic range of the instrument are discussed.

034.003 A grating spectrograph for a college observatory.
D. J. Schroeder.
Sky Telescope, Vol. 47, 96 - 99 (1974).

034.004 An instrumental profile of the double-diffraction spectrophotometer with digital registration.
E. K. Kokhan, N. I. Pechinskaya.
Solnechnye Dannye 1973 Byull., No. 12, p. 84 - 90 (1974). In Russian.

034.005 An instrument for investigation of small electron fluxes with energies $E_e \geqslant 40$ keV aboard the Prognoz satellites. M. A. Zel'dovich, T. G. Zubieva, M. A. Kovalenskaya, Yu. I. Logachev, V. N. Lutsenko, N. F. Pisarenko, I. A. Savenko, I. P. Shestopalov.
Kosmich. Issled., Vol. 12, 143 - 145 (1974). In Russian. Brief information.

034.006 Comparison of the instruments UIM-21 and KIM-3.
P. P. Pavlenko.
Vestn. Khar'kov. Univ. No. 99 (Ser. Astron. No. 8), p. 36 - 42 (1973). In Russian.

034.007 A new kind of spectrohelioscope for observing solar prominences. C. L. Stong.
Sci. American, Vol. 230, No. 3, p. 110 - 115 (1974).

034.008 Structural analysis of a two-circuit optical tracking system. A. F. Burov, V. F. Gumen, R. N. Loparev, V. A. Myasnikov, A. A. Parshin, O. V. Popov.
Teoriya i proektir. vysokotochn. sistem upr. Leningrad, Nauka, 1973, p. 5 - 13. In Russian. – Abstr. in Referativ. Zhurn. 51. Astron., 3.51.235 (1974).

034.009 Optimization of characteristics of a regime to search for a signal from a light source.
R. N. Loparev, L. P. Khomenko.
Teoriya i proektir. vysokotochn. sistem upr. Leningrad, Nauka, 1973, p. 22 - 25. In Russian. – Abstr. in Referativ. Zhurn. 51. Astron., 3.51.236 (1974).

034.010 Discrete systems of light beam self correction in ray guides. Z. N. Kuteva, R. K. Makarova.
Teoriya i proektir. vysokotochn. sistem upr. Leningrad, Nauka, 1973, p. 34 - 39. In Russian. – Abstr. in Referativ. Zhurn. 51. Astron., 3.51.237 (1974).

034.011 Method for computing the threshold sensibility of photomultipliers.
L. N. Tikhomirova, S. V. Korotkov, V. T. Pivovarov.
Teoriya i proektir. vysokotochn. sistem upr. Leningrad, Nauka, 1973, p. 43 - 47. In Russian. – Abstr. in Referativ. Zhurn. 51. Astron., 3.51.238 (1974).

034.012 Computation of the photocurrent and of the signal to noise ratio of photomultipliers used in photoelectric tracking systems.
L. N. Tikhomirova, M. E. Zhandarov, V. T. Pivovarov.
Teoriya i proektir. vysokotochn. sistem upr. Leningrad, Nauka, 1973, p. 48 - 51. In Russian. – Abstr. in Referativ. Zhurn. 51. Astron., 3.51.239 (1974).

034.013 Television digital sensor for tracking point light sources.
V. G. Vorob'ev, R. N. Loparev, O. V. Popov.
Teoriya i proektir. vysokotochn. sistem upr. Leningrad, Nauka, 1973, p. 76 - 82. In Russian. – Abstr. in Referativ. Zhurn. 51. Astron., 3.51.241 (1974).

034.014 Image tubes in astronomy. J. D. McGee.
Vistas in astronomy, Vol. 15, (see 003.001), 61 - 89 (1973).

034.015 Das Filmen von Sonnenprotuberanzen.
H. Brägger, E. Moser.
Orion Schaffhausen, 32. Jahrgang, p. 54 - 59 (1974).

034.016 Ein extrem kurz gebautes Spektrohelioskop.
F. Veio, H. Treutner.
Orion Schaffhausen, 32. Jahrgang, p. 62 (1974).

034.017 A scanning device attached to the horizontal solar telescope. P. A. Golub, G. A. Korkotyan.
Solnechnye Dannye 1974 Byull., No. 1, p. 100 - 103 (1974). In Russian.

The block scheme of an automatic scanning device for the horizontal solar telescope of the Pulkovo Observatory and two programs of its operation are described.

034.018 A semiconducting electron spectrometer for investigations in space.
N. V. Alekseev, P. V. Vakulov, L. M. Kovrygina, Yu. V. Mineev, B. Ya. Shcherbovskij.
Kosmich. Issled., Vol. 12, 301 - 304 (1974). In Russian. Brief information.

034.019 Honderd maal korter belichten met elektronografie.
T. de Groot.
Zenit, Vol. 1, No. 3, p. 24 - 25 (1974).

034.020 **Charged particle measurements on board the Intercosmos 5 satellite made by the PG-1-A apparatus.**
S. Fischer, V. Veselý, M. Šícha, L. Láska, M. Novák, Z. Němeček, J. Studnička, V. Hrachová-Řezáčová, M. Tichý, J. Dubinský.
Bull. Astron. Inst. Czechoslovakia, Vol. 25, 69 - 74 (1974).

The electronic PG-1-A apparatus, which was placed on the earth-satellite Intercosmos 5, is described. An example of the measurements is presented.

034.021 **Apparatus for the observation of fast fluctuations of the solar radio radiation flux.**
A. A. Bezotosnyj, O. G. Gontarev.
Ionosphere and solar-terrestrial relations, (see 003.003), p. 129 - 131 (1972). In Russian.

034.022 **A portable driven mount for 35 mm cameras.**
D. G. Daniels.
Journ. British Astron. Ass., Vol. 84, 189 - 191 (1974).

034.023 **Modern instrumentation for optical astronomy.**
F. J. Ahern.
Journ. Roy. Astron. Soc. Canada, Vol. 68, 99 - 107 (1974).

034.024 **Investigation of the ASP-20 spectrograph by means of a laser.** I. Sattarov.
Solar processes and their observations, (see 003.005), p. 89 - 106 (1973). In Russian.

The results of a detailed test of the solar spectrograph ASP-20 installed at the horizontal solar telescope of the Astronomical Institute of the Academy of Sciences of the UzSSR are presented.

034.025 **The transmission of interference-polarization filters for the Hα and K Ca$^+$ lines.** Z. B. Korobova.
Solar processes and their observations, (see 003.005), p. 107 - 113 (1973). In Russian.

034.026 **Apparatus for the investigation of energy and charge spectra of solar cosmic rays at the satellites "Prognoz".** N. N. Volodichev, A. I. Vorob'ev, N. L. Grigorov, G. Ya. Kolesov, O. B. Likin, G. S. Malkiehl', E. I. Morozova, N. F. Pisarenko, I. A. Savenko, A. A. Suslov.
Geomagn. Aeronom., Vol. 14, 218 - 223 (1974). In Russian.

034.027 **The ASP-9 scanning electrospectrophotometer.**
Yu. V. Borisov.
Byull. Inst. Astrofiz., *Dushanbe*, No. 61, p. 31 - 32 (1972). In Russian.

034.028 **Sensitivity variations of silicon vidicons.**
A. M. Title.
Solar Physics, Vol. 35, 233 - 237 (1974).

Local peak to peak sensitivity fluctuations of Si vidicon targets that can exceed 100 % in the red have been observed. These fluctuations are due to Fabry-Pérot interference in the target.

034.029 **Pupil imagery in astronomical spectrographs.**
C. F. W. Harmer.
Monthly Notices Roy. Astron. Soc., Vol. 167, 311 - 318 (1974).

Increased use of astronomical spectrographs for relatively wide field applications is shown to emphasize the need for instruments to be designed as part of a system, i.e. matched to their parent telescopes. This study first suggests and then demonstrates that consideration of pupil imagery is important in making telescope and spectrograph a properly integrated unit.

034.030 **A new stabilizer of the anodic continuous current of a photomultiplier.**
E. Tătaru, M. Tătaru.
Studia Univ. Babeș-Bolyai, Ser. Phys., Anul 19, Fasc. 1, p. 13 - 17 (1974). In Romanian.

034.031 **Ortsabhängigkeit der Empfindlichkeit von kleinen Bleisulfid-Detektoren.**
U. Fahrbach, W. Hofmann, D. Lemke.
Mitt. Astron. Ges., No. 34, p. 144 (1973/74). − Abstr. AG.

034.032 **Heterodyn-Spektrometer für den Submillimeterbereich.** W. Reinert, G. V. Schultz.
Mitt. Astron. Ges., No. 34, p. 144 - 146 (1973/74).
Presented at the "Wissenschaftliche Tagung der Astron. Ges., Oberkochen, 1973 April 24 - 27".

034.033 **Eigenschaften von Bildverstärkern mit "proximity-focus".** E. Kreysa, G. V. Schultz.
Mitt. Astron. Ges., No. 34, p. 146 - 147 (1973/74).
Presented at the "Wissenschaftliche Tagung der Astron. Ges., Oberkochen, 1973 April 24 - 27".

034.034 **Der Echelle-Spektrograph des Figl-Observatoriums für Astrophysik (FOA).** W. W. Weiss.
Mitt. Astron. Ges., No. 34, p. 147 - 150 (1973/74).
Presented at the "Wissenschaftliche Tagung der Astron. Ges., Oberkochen, 1973 April 24 - 27".

034.035 **The GSFC EUV and X-ray spectroheliograph on OSO-7.** J. H. Underwood, W. M. Neupert.
Solar Physics, Vol. 35, 241 - 258 (1974).

The Goddard Space Flight Center instrument carried on the pointed section of the OSO-7 satellite is described. Representative results from each of the subsystems are presented.

034.036 **On polarimetry in solar active regions. VII: A new Zeeman polarimeter and its advantages as compared to other designs.** E. Wiehr.
Solar Physics, Vol. 35, 351 - 359 (1974).

A new modulation procedure for Zeeman polarimeters is described and tested. The azimuth rotation by means of two steady $\lambda/4$-plates, combined with the common EOLM, has several advantages as compared to 'two-EOLM-polarimeters'. As a first application of the new polarimeter, the linear polarization is measured in a sunspot penumbra.

034.037 **Fotoelektroniske billedrør anvendt i astronomien.**
R. F. Nielsen.
Astron. Tidssk., Årg. 7, p. 17 - 23 (1974).

034.038 **Quantization noise of correlation spectrometers.**
F. K. Bowers, R. J. Klingler.
Astron. Astrophys., Suppl. Ser., Vol. 15, 373 - 380 (1974). Conference paper (see 012.011).

034.039 **Moderate resolution VUV rocket spectrograph.**
A. W. Mantz, B. L. Sowers, J. J. Lange.
Applied Optics, Vol. 13, 193 - 196 (1974).

A moderate resolution rocket spectrograph is described for use in the vacuum ultraviolet (VUV) spectral region from 1050 Å to 2000 Å. The spectrograph incorporates a Carruthers electronographic camera, a plane grating, and focusing optics to provide a well defined field of view for an extended atmospheric source. A detailed description of the spectrograph and sample spectra of the fourth positive band of CO from 1200 Å to 1800 Å are given.

034.040 **White light extended source shearing interferometer.**
J. C. Wyant.
Applied Optics, Vol. 13, 200 - 202 (1974).

A grating lateral shear interferometer is described that

can be used with a white light source. The use of the interferometer with certain types of extended sources is also demonstrated.

034.041 **Field widened Michelson spectrometer with no moving parts.** J. G. Hirschberg.
Applied Optics, Vol. 13, 233 - 234 (1974). − Letter.

034.042 **Far-infrared filters utilizing small particle scattering and antireflection coatings.**
K. R. Armstrong, F. J. Low.
Applied Optics, Vol. 13, 425 - 430 (1974).

High transmission, low pass scatter filters for blocking at wavelengths from 3.5 µm to 50 µm and single-layer antireflection coatings for optical materials used in the 25-µm–300-µm region of the spectrum are described. The application of both techniques to liquid-He-cooled filters permits the construction of efficient low pass and medium-width band pass filters for use throughout the far infrared.

034.043 **Photon-counting polarizing radiometer.**
K. L. Coulson, R. L. Walraven, G. I. Weigt, L. B. Soohoo.
Applied Optics, Vol. 13, 497 - 498 (1974).

Photon counting is used in this instrument to obtain high precision in the measurement of the intensity, degree of polarization, and orientation of the plane of polarization of light in the natural environment. The dual-channel configuration, with four interference filters in each channel, provides the possibility of measurements in eight narrow spectral intervals in the wavelength range 0.32–0.90 µm.

034.044 **Extreme ultraviolet spectrometer for space research.**
S. Kumar, F. Paresce, S. Bowyer, M. Lampton.
Applied Optics, Vol. 13, 575 - 580 (1974).

A normal incidence spectrometer for use in the wavelength region from 200 Å to 1270 Å has been developed. The design and calibration of the instrument are described in detail. The spectrometer can be employed to detect extreme uv radiation at a minimum flux level of 1 rayleigh with a spectral resolution of 40 Å and a spatial resolution of 6°. Data on the extreme uv night sky spectrum between 780 Å and 1270 Å obtained with this instrument on a recent flight are presented.

034.045 **Astronomical demonstration of an infrared upconverter.**
T. R. Gurski, H. W. Epps, S. P. Maran.
Nature, Vol. 249, 638 - 639 (1974). − Letter.

034.046 **A superheterodyne spectrum analyzer for the 1.1 - − 1.7 - mm wavelength range.**
Yu. Yu. Kulikov, L. I. Fedoseev, A. A. Shvetsov, V. R. Amirkhanyan.
Izv. vyssh. ucheb. zavedenij. Radiofizika, Vol. 16, 1442 - 1443 (1973). In Russian. − Abstr. in Referativ. Zhurn. 51. Astron., 5.51.121 (1974).

034.047 **The transmission of polarization interference filters for Hα and K Ca⁺ lines.** Z. B. Korobova.
Solnech. protsessy i ikh nablyudeniya. Tashkent, "Fan", 1973, p. 107 - 113. In Russian. − Abstr. in Referativ. Zhurn. 51. Astron., 5.51.245 (1974).

034.048 **Test of the ASP-20 solar spectrograph by use of a laser.** I. Sattarov.
Solnech. protsessy i ikh nablyudeniya. Tashkent, "Fan", 1973, p. 89 - 106. In Russian. − Abstr. in Referativ. Zhurn. 51. Astron., 5.51.246 (1974).

034.049 **On the dynamic range of photoelectric multipliers in the regime of photon counting.**
G. Spulgis, I. Šmelds.
Investigation of the sun and red stars, (see 003.008), p. 63 - 69 (1974). In Russian.

034.050 **A coherence interferometer for direct measurement of the atmospheric transfer function.**
J. C. Dainty, R. J. Scaddan.
Monthly Notices Roy. Astron. Soc., Vol. 167, 69P - 73P (1974).

An interferometer for measuring the atmospheric transfer function has been built and tested. The authors briefly describe the theory and design of the instrument and also present some preliminary results.

034.051 **Photometer and spectrometer for solar X-ray radiation.** M. Hŕond.
Postępy Astron., Vol. 22, 109 - 123 (1974). In Polish.

A discussion of problems concerning the design and construction of solar X-ray photometers and spectrometers for soft-range radiation measurements is given.

034.052 **Methods of balancing self-control of iris-photometers.** R. Schielicke.
Astron. Nachr., Vol. 295, 195 - 200 = Mitt. Univ.-Sternw. Jena, No. 121 (1974). In German.

Following a critical review of automatic output techniques of equality information in iris-type photometers put into practice previously, new possibilities are discussed. One of these methods, which is based on digital techniques and is easy to realize, resolves differences between the intensities of the comparison and measuring beam of less than 10^{-3}.

034.053 **Development and performance of a solar hard X-ray spectrometer.** H. F. van Beek.
Diss. Wiskunde en Natuurwetenschappen, Univ. Utrecht, 88 pp. = Utrechtse Sterrekundige Overdrukken No. 229 (1973).

034.054 **First results of the solar hard X-ray spectrometer on board the ESRO TD-1A satellite.**
H. F. van Beek, L. D. de Feiter.
Space Research XIII, p. 777 - 780 = Utrechtse Sterrekundige Overdrukken No. 243 (1973).

034.055 **Présentation préliminaire du tir de l'expérience Janus.** G. Courtès.
Astron. Inst. Utrecht, Internal Rep. ROF 72, (see 012.012), p. C16.1 - C16.2 (1974).

034.056 **A self-operating meteor camera.**
B. Mayer.
Sky Telescope, Vol. 48, 54 - 56 (1974).

034.057 **The Richardson spectrograph for the 25-inch Newall refractor.** P. G. Laskarides.
In honorem S. Placidis, (see 003.009), p. 189 - 208 (1974). In Greek.

034.058 **The computer-interactive photometer. An adaptive device.** G. A. De Biase, G. Sedmak.
Astron. Astrophys., Vol. 33, 1 - 8 (1974).

The note describes the structure and the performance of an astronomical photoelectric photometer capable of adaptive measurements. The system is made by integrating the photometer with a digital computer on line to it. The computer is programmed so as to generate an adaptive photometer. The theoretical performance is determined and a tentative machine is studied by comparing the results obtained from numerical simulations with theoretical predictions. The results show the clear superiority of the adaptive system over any other non-adaptive use of the given photometer.

034.059 **Upgrading of the Lick-Gaertner automatic measuring system.** L. B. Robinson, S. Vasilevskis.
IAU Symposium No. 61, (see 012.014), p. 295 - 298 (1974).

034.060 **The COSMOS plate measuring machine.**
N. M. Pratt.
IAU Symposium No. 61, (see 012.014), p. 299 - 300 (1974).

034.061 **A new high speed measuring machine for astrometric plates.** P. M. Routly.
IAU Symposium No. 61, (see 012.014), p. 301 (1974). – Abstract.

034.062 **Device for measurement of thermal radiation fluxes.**
O. A. Gerashchenko, S. A. Sazhina.
Kosmich. issledovaniya na Ukraine. Resp. mezhved. sb., 1973, vyp. (No.) 3, p. 62 - 66. In Russian. – Abstr. in Referativ. Zhurn. 62. Issled. kosmich. prostranstva, 6.62.109 (1974).

034.063 **A new chopper design for astronomical infrared photometry.**
U. Fahrbach, K. Haussecker, D. Lemke.
Astron. Astrophys., Vol. 33, 265 - 267 (1974).
This chopper design is applicable as a focal plane modulator, as a chopping secondary or as a chopping primary of a telescope. The modulation function can be made close to a square wave with fast change between the two end positions. Frequency and throw of the chopper can be changed during operation. The basic design of the chopper driver is that of a servo circuit controlling both position and speed of the modulator.

034.064 **URSIES and the Bartol Coudé Observatory.**
A. A. Wyller.
Stars and the Milky Way system, Proc. 1972, (see 012.018), p. 265 - 272 (1974).

034.065 **Optimization of a "phoswich"-type X-ray detector in the energy range 10–150 keV.**
A. Scheepmaker.
Bull. American Astron. Soc., Vol. 6, 272 (1974). – Abstr. AAS.

034.066 **A large double scatter detector for γ-rays and neutrons.** D. Herzo, R. Koga, R. S. White.
Bull. American Astron. Soc., Vol. 6, 276 (1974). – Abstr. AAS.

034.067 **A multi-channel coronal spectrophotometer.**
D. A. Landman, F. Q. Orrall, R. Zane.
Bull. American Astron. Soc., Vol. 6, 290 (1974). – Abstr. AAS.

034.068 **Results from the Lockheed magnetograph.**
R. C. Smithson.
Bull. American Astron. Soc., Vol. 6, 294 (1974). – Abstr. AAS.

034.069 **Improvement of birefringent filters.** A. M. Title.
Bull. American Astron. Soc., Vol. 6, 295 (1974).
Abstr. AAS.

034.070 **Drift in interference filters.**
A. M. Title, T. P. Pope.
Bull. American Astron. Soc., Vol. 6, 295 (1974). – Abstr. AAS.

034.071 **A 512-channel solar magnetograph.**
D. Trumbo, J. Harvey, W. Livingston, C. Slaughter.
Bull. American Astron. Soc., Vol. 6, 296 (1974). – Abstr. AAS.

034.072 **Strumentazione del satellite EXOSAT.**
H. Horstman.
Mem Soc. Astron. Italiana, Nuova Ser., Vol. 44, 507 - 518 (1973/74).

034.073 **Programmi sperimentali di astronomia in raggi X.**
G. Spada.
Mem. Soc. Astron. Italiana, Nuova Ser., Vol. 44, 519 - 538 (1973/74).

034.074 **A device for measuring convex hyperbolic mirrors.**
R. Fangor.
Urania Kraków, Vol. 45, 79 - 81 (1974). In Polish.

034.075 **Far infrared filters for solar observation.**
W. G. Mankin.
Infrared Phys., Vol. 13, 333 - 336 (1973).
Reststrahlen filters have long been used for isolating bands of radiation in the far infrared because they combine high transmission with very good rejection of the visible and near infrared. A simple arrangement for holding one to four crystals to form such a filter is presented.

034.076 **Speckle interferometry through small multiple apertures. Michelson stellar interferometer and aperture synthesis in optics.** F. Roddier.
Optics Commun., (Netherlands), Vol. 10, 103 - 105 (1974).
The use of a Michelson stellar interferometer or multiple-hole interferometers is considered as a special case of speckle interferometry. Formulas are given for the power spectrum of the interference pattern produced by the multiple apertures.

034.077 **An altazimuth control system.** W. A. Cormack.
Journ. Phys. E, Sci. Instruments, Vol. 7, 280 - 282 = Commun. Roy. Obs. Edinburgh No. 167 (1974).
A programmable calculator is used for on-line control of a large altazimuth mounted stellar Michelson interferometer.

034.078 **Crystals for stellar spectrometers.**
N. G. Alexandropoulos, G. G. Cohen.
Applied Spectrosc., (USA), Vol. 28, 155 - 164 (1974).
The principles of crystal evaluation as they apply to the special problems of X-ray astronomy are presented. A number of crystals were evaluated, and the energy dependence of the diffraction properties are reported.

034.079 **A sun glint heliostat for atmospheric spectroscopy.**
A. E. S. Green, R. A. Sutherland, G. Ganguli.
Rev. Sci. Instruments, Vol. 45, 60 - 63 (1974).

034.080 **The use of Josephson effect for the detection of far infrared and microwave radiation.** J. Warman.
Rev. Mexicana Astron. Astrofis., Vol. 1, No. 1, p. 55 - 66 (1974).
The theoretical and experimental bases are presented for the application of the Josephson effect for the detection of electromagnetic radiation of astronomical objects in the far infrared, millimeter and centimeter wavelengths.

034.081 **A folded Schmidt camera for a television multi-channel sensor.**
L. E. Celaya, F. Cobos, C. Firmani.
Rev. Mexicana Astron. Astrofis., Vol. 1, No. 1, p. 75 - 79 (1974).

034.082 **A simple polarimetric system for the Lick Observatory image-tube scanner.** K. H. Nordsieck.
Publ. Astron. Soc. Pacific, Vol. 86, 324 - 329 = Lick Obs. Bull., No. 656 (1974).
The insertion of a simple, stationary, three-element filter in front of the Cassegrain spectrograph and image-tube scanner of the Lick 120-inch telescope allows simultaneous measurement of the Stokes parameters as a function of wavelength across at least 2000 Å.

034.083 **Improving the research capability of small spectrographs.** D. R. Hollars, H. J. Reitsema.
Publ. Astron. Soc. Pacific, Vol. 86, 330 - 333 (1974).

The authors have improved the research capability of their Cassegrain spectrograph by (1) altering the grating mounts to permit their use at both positive and negative angles, and (2) using baked IIIa-J in addition to IIa-O emulsion for spectra.

034.084 **The use of the EBS television camera tube for the acquisition of faint objects.**
T. D. Kinman, M. Green.
Publ. Astron. Soc. Pacific, Vol. 86, 334 - 335 (1974).

The performance of a low-light-level television system using an EBS camera tube is given and compared with a similar camera using an SEC tube.

034.085 **A computer-controlled photometer and polarimeter.**
T. D. Kinman, C. T. Mahaffey.
Publ. Astron. Soc. Pacific, Vol. 86, 336 - 339 (1974).

A system using a computer-controlled photometer and polarimeter is described and examples of its performance are given.

034.086 **Characteristics of photomultipliers suitable for recording near infrared radiation.**
G. A. Terez, E. I. Terez.
Astron. Tsirk., No. 818, p. 2 - 5 (1974). In Russian.

034.087 **Detector adaptivo de proporción.**
C. E. Alarcón, M. Pupareli.
Inst. Astron. Fís. Espacio, (IAFE), *Buenos Aires, Argentina*, Ser. Publ. Técnicas, No. 3 (1973).

Radiometric calibrations for the earth radiation budget experiment. See Abstr. 031.044.

Preliminary results of the Baranne white pupil image tube nebular spectrograph. See 032.047.

Infrared techniques. See Abstr. 061.062.

Sources of gravity waves. See Abstr. 066.013.

The Yerkes Observatory photoelectric parallax scanner. See Abstr. 111.007.

TV investigations of ultra-short period optical radiation sources. See Abstr. 141.306.

035 Clocks and Frequency Standards

035.001 **Sonnenuhr für mittlere Ortszeit in Dortmund-Wellinghofen.**
Umschau, 74. Jahrgang, p. 127 - 128 (1974).

035.002 **Le cadran lunaire.** L. Janin.
Orion Schaffhausen, 32. Jahrgang, p. 3 - 11 (1974).

035.003 **The Boughton Monchelsea sundial.** G. E. Taylor.
Journ. British Astron. Ass., Vol. 84, 184 - 188 (1974).

035.004 **A portable electronic clock for timing grazing occultations.** J. Hers.
Monthly Notes Astron. Soc. Southern Africa, Vol. 33, 33 - 38 (1974).

035.005 **Comparaisons de phase avec émission Loran-C et variation de l'index de réfraction.**
E. Proverbio, V. Quesada, A. Simoncini.
Bull. Annuel Soc. Suisse Chronométrie, Lab. Suisse Recherches Horlogères, Vol. 6, Fasc. 4, 455 - 459 = Pubbl. Stazione Astron. Internazionale Latitudine, Carloforte-Cagliari, Nuova Ser., No. 31 (1973).

035.006 **Atomic frequency standards.** M. Zikides.
In honorem S. Placidis, (see 003.009), p. 93 - 104 (1974). In Greek.

035.007 **Das Nationale Zeitnormal der Physikalisch-Technischen Bundesanstalt.** G. Becker.
Separate print from Kleinheubacher Berichte 17, 14 pp. (1974).
The National Time Standard of the Physikalisch-Technische Bundesanstalt (PTB) consists of the primary time and frequency cesium standard CS1 controlling a group of 7 cesium atomic clocks using additional atomic reference standards. The role of MEZ (PTB) as well as plans concerning the establishment of a "German Time Service" is described. Details of the construction of CS1 as well as problems concerning the realization of the second with the standard are discussed. Finally the "coded time information" is described introduced in June 1973 transmitted by the German standard frequency and time signal transmitter DCF77 (77.5 kHz) disseminating MEZ (PTB).

035.008 **Recherches actuelles sur les quartz à l'École Nationale Supérieure de Chronométrie et de Micromécanique.** P. Mesnage.
Ann. Françaises Chronométrie Micromécanique, 8e année, (see 012.027), 7 - 9 (1973).

035.009 **Étude d'un schéma équivalent du résonateur à quartz rendant compte de ses propriétés non-linéaires.** J.-J. Gagnepain.
Ann. Françaises Chronométrie Micromécanique, 8e année, (see 012.027), 10 - 21 (1973).

035.010 **Étude topographique par rayons X de cristaux piézoélectriques.** R. Besson, G. Marduel.
Ann. Françaises Chronométrie Micromécanique, 8e année, (see 012.027), 22 - 29 (1973).

035.011 **Mesure automatique des paramètres des résonateurs à quartz.** C. Pegeot, J. J. Gagnepain.
Ann. Françaises Chronométrie Micromécanique, 8e année, (see 012.027), 30 - 35 (1973).

035.012 **Subminiature quartz tuning fork for uses in watches and clocks.** J. H. Staudte.
Ann. Françaises Chronométrie Micromécanique, 8e année, (see 012.027), 36 - 44, 45 - 53 (1973). – In English and French.

035.013 **Les quartz d'horlogerie.**
J. Jouannic, C. Guerin, P. Plaud.
Ann. Françaises Chronométrie Micromécanique, 8e année, (see 012.027), 54 - 77 (1973).

035.014 **Présentation de la montre à quartz Montrélec.** P. Vovelle.
Ann. Françaises Chronométrie Micromécanique, 8e année, (see 012.027), 95 - 96 (1973). – Ce texte constitue un résumé de la communication qui sera présentée au Congrès Franco-Allemand de Chronométrie en Juin 1973.

035.015 **Remarques sur les contrôles chronométriques des montres à quartz.** A. Remond, M. Vincent.
Ann. Françaises Chronométrie Micromécanique, 8e année, (see 012.027), 110 - 118 (1973).

035.016 **Nouvelles piles électriques pour applications horlogères.** G. Lehmann.
Ann. Françaises Chronométrie Micromécanique, 8e année, (see 012.027), 125 - 135 (1973).

035.017 **La chronométrie dans l'espace.** B. Guinot.
Ann. Françaises Chronométrie Micromécanique, 8e année, (see 012.027), 136 - 143 (1973). – Communication présentée à la Société Chronométrique de France lors de l'Assemblée Générale du 23 Novembre 1972.

035.018 **Quelques améliorations à la technique de la thermocompression.** J.-L. Milan, B. Mazodier.
Ann. Françaises Chronométrie Micromécanique, 8e année, (see 012.027), 147 - 155 (1973).

Dessinons, réalisons huit cadrans solaires.
See Abstr. 003.033.

Sonnenuhren. See Abstr. 003.114.

Sundials: their theory and construction.
See Abstr. 003.130.

Nocturnal: Instrument voor het opnemen van de tijd. See Abstr. 004.021.

036 Photographic Auxiliaries

036.001 A film for obtaining isophotic contours.
R. Beck, W. Hünecke.
Sky Telescope, Vol. 47, 274 - 276 (1974).

036.002 Effects of moisture, oxygen, baking, and water-hypersensitization on the sensitivity of Kodak spectroscopic plates, types IIIa-J, IV-N, and I-N.
T. A. Babcock, P. M. Ferguson, T. H. James.
Astron. Journ., Vol. 79, 92 - 98 (1974).

The individual effects of oxygen and moisture on the photographic sensitivity of a model silver halide emulsion and of Kodak spectroscopic films and plates, types IIIa-J, IV-N, and I-N were examined. The relation of the effects of oxygen and moisture to techniques such as baking of IIIa-J films or hypersensitizing IV-N plates is also discussed. A technique of exposing an infrared plate, such as Kodak spectroscopic plates, type IV-N, or I-N in a high-humidity, oxygen-free environment, appears to be an efficient means of achieving hypersensitization without the fog increases or nonuniformities caused by liquid water hypersensitization.

036.003 Methode zur Trennung schwächster Flächenhelligkeiten von der allgemeinen Hintergrundschwärzung fotografischer Himmelsaufnahmen und deren Äquidensitometrie. W. Högner.
Sterne, 50. Jahrgang, p. 35 - 38 (1974).

036.004 Speeding up astronomical photography.
P. M. Corben, V. C. Reddish, M. E. Sim.
Nature, Vol. 249, 22 - 24, with a correction in Vol. 249, 396 (1974). – Letter.

036.005 On the characteristic of Kodak Tri X film for different developers. H. Shibasaki.
Tokyo Astron. Obs., Report No. 63, Vol. 16, 714 - 725 (1974). In Japanese.

036.006 Hypersensitization of Kodak type 103a-O plates by nitrogen baking. R. L. Scott, A. G. Smith.
Astron. Journ., Vol. 79, 656 - 658 = Contr. Rosemary Hill Obs., Univ. Florida, *Gainesville, Florida*, No. 41 (1974).

It is shown that the speed of Kodak type 103a-O spectroscopic plates can be increased by factors of from two to three by baking in a controlled nitrogen atmosphere. Extensive sky tests have shown that baking introduces a statistically significant but practically unimportant loss in photometric precision.

036.007 Intensification of photographic images by neutron irradiation. P. M. Perry.
Bull. American Astron. Soc., Vol. 6, 218 (1974). – Abstr. AAS.

036.008 Investigation of the distortion of an aerophotofilm emulsion layer. K. Šteins, J. Balodis.
Uch. Zap. Latv. Gos. Univ., Vol. 202, Astron., vyp. (No.), 10, (see 003.012), p. 30 - 51 (1974). In Russian.

036.009 Photographic recording at large telescopes.
A. G. Millikan.
Proc. ESO/SRC/CERN conference, (see 012.021), p. 309 - 319 (1974).

036.010 Plates sensitization by μs preflashing.
A. Braccesi, F. Fusipecci.
Proc. ESO/SRC/CERN conference, (see 012.021), p. 357 - 358 (1974).

036.011 Kiezen van de juiste film zeer belangrijk voor maanfotografie. M. A. M. van Venrooij.
Zenit, Vol. 1, No. 6, p. 39 - 40 (1974).

036.012 Comparison of IIIa-J Kodak plates with A-500 RP and A-700 RP films.
I. I. Brejdo, O. M. Mikhajlova.
Astron. Tsirk., No. 815, p. 5 - 6 (1974). In Russian.

036.013 Testing of astronomical films on the telescope AST-452. G. V. Novikova.
Astron. Tsirk., No. 817, p. 7 - 8 (1974). In Russian.

036.014 On noise of photographic emulsions.
A. B. Palej.
Astron. Tsirk., No. 818, p. 5 - 7 (1974). In Russian.

Sonnenphotographie. See Abstr. 031.014.

Neue Möglichkeiten der Sonnenbeobachtung für den Amateur. See Abstr. 031.015.

Planeten-Photographie mit Amateur-Teleskopen. See Abstr. 031.046.

Positional Astronomy. Celestial Mechanics

041 Positional Astronomy, Astrometry, Star Catalogues and Atlases

041.001 **On the construction of new astronomical refraction tables.** A. I. Nefed'eva.
Astron. Zhurn. Akad. Nauk SSSR, Vol. 51, 208 - 215 (1974). In Russian. English translation in Soviet Astron. AJ, Vol. 18, No. 1.

Based on averaged monthly aerological data the values of astronomical refraction were calculated for three summer and three winter months. Computations were made for stations of the USSR disposed over $10°$ intervals of latitude and longitude. These values of astronomical refraction were compared with the values taken from the Pulkovo refraction tables.

041.002 **On the "system" of the meridian circle of the Kharkov Astronomical Observatory in the determination of declinations in the years 1948–53.** K. N. Kuz'menko, N. S. Olifer, V. Kh. Pluzhnikov.
Vestn. Khar'kov. Univ. No. 99 (Ser. Astron. No. 8), p. 46 - 48 (1973). In Russian.

041.003 **Finding the errors of the form $\Delta\alpha_\delta$ of coordinates of stars located symmetrically to the zenith.** N. G. Zuev.
Vestn. Khar'kov. Univ. No. 99 (Ser. Astron. No. 8), p. 52 - 58 (1973). In Russian.

041.004 **Algorithm for calculation of mean and apparent positions of stars.** K. N. Derkach.
Vestn. Khar'kov. Univ. No. 99 (Ser. Astron. No. 8), p. 58 - 64 (1973). In Russian.

041.005 **Planetary ephemerides based on radar, transit circle, and Mariner 9 data.** M. S. W. Keesey.
Bull. American Astron. Soc., Vol. 6, 206 (1974).– Abstr.AAS.

041.006 **The use of radar-ranging in planetary ephemeris development.** E. M. Standish, Jr.
Bull. American Astron. Soc., Vol. 6, 206 (1974).– Abstr. AAS.

041.007 **Determination of the influence of the terrestrial refraction on the celestial refraction near the horizon.** V. V. Kirichuk.
Geod., kartogr. i aehrofotosemka. Resp. mezhved. nauch.-tekhn. sb., 1973, vyp. (No.) 18, p. 21 - 23. In Russian. Abstr. in Referativ. Zhurn. 51. Astron., 3.51.179; 52. Geod. Aehrosemka, 3.52.94 (1974).

041.008 **Second Greenwich catalogue of stars for 1950.0.** K. C. Blackwell, M. E. Buontempo, with an appendix on 'Cooke transit circle collimation investigation' by K. C. Blackwell.
Roy. Obs. Ann., [Roy. Greenwich Obs., Herstmonceux], No. 9, 4 + 66 pp. (1973).

This catalogue of observed right ascension of 375 FK3 stars is the outcome of trial observations made at Greenwich with the Cooke reversible transit circle from 1937 September 29 to 1940 September 5. The observations, based on the system of the FK3, have been reduced without proper motion to the equinox of 1950.0. Part I contains the mean result for all observations of each star. Part II contains the separate results above and below pole for each of the 77 circumpolar stars contained in Part I.

041.009 **A general computer program for the application of the rigorous blockadjustment solution in photographic astrometry.** C. de Vegt, H. Ebner.
Monthly Notices Roy. Astron. Soc., Vol. 167, 169 - 181 (1974).

The evaluation of stellar positions from overlapping plates by rigorous blockadjustment methods leads to a very large system of normal equations containing both, stellar positions and plate constants as unknowns. These general normal equations can be reduced to a subsystem with banded-bordered coefficient matrix, containing only the plate constants as unknowns. The general structure of a computer program, now under development at the Hamburg Observatory, is described in detail.

041.010 **Declinations of 899 stars in the FK4 system.** V. V. Konin.
Astrometriya i Astrofizika, *Kiev*, vyp. (No.) 20, (see 003.002), p. 69 - 96 (1973). In Russian.

The processing of stellar work of FKSZ and the Poltava zenith program by the differential method in 1945–1949 on the Odessa meridian circle is described. The performed results are given as a catalogue of stellar declinations in the system FK4. Corrections of declinations of reference stars are given and a comparison of the obtained declinations with fundamental systems is made.

041.011 **Declinations of 194 stars of the program of the Pulkovo wide-angle zenith telescope in the FK4 system.** L. A. Klepikova.
Astrometriya i Astrofizika, *Kiev*, vyp. (No.) 20, (see 003.002), p. 97 - 105 (1973). In Russian.

The paper deals with declinations of 194 stars of the Pulkovo zenith zone in the FK4 system. The declinations are deduced from observations obtained on the Odessa meridian circle in 1965–1968. The mean-square error of one observation in δ is $\pm 0''.36$. The mean epoch of observation of the catalogue is 1965.89.

041.012 **Declinations of 96 additional stars of the program of the Pulkovo zenith telescope.** V. V. Konin.
Astrometriya i Astrofizika, *Kiev*, vyp. (No.) 20, (see 003.002), p. 106 - 109 (1973). In Russian.

The paper deals with the results of differential determinations of additional stellar declinations of the Pulkovo zenith telescope program. The results are obtained from the author's observations on the meridian circle of the Odessa Astronomical Observatory in 1949–1950.

041.013 **Astrometrische Versuche mit der Tautenburger Schmidt-Kamera.** H.-U. Sandig.
Jenaer Rundschau, (Jena Review), 18. Jahrgang, p. 318 - 321 (1973).

041.014 **On the systematic and accidental errors of the Yale Catalogues in the declination zones from $-0°$ to $-20°$.** E. V. Khrutskaya.

Astron. Tsirk., No. 809, p. 3 - 6 (1974). In Russian.

041.015 **On the determination of errors $\Delta\alpha_\delta$ of the Fundamental Catalogue from meridian observations.**
N. G. Zuev.
Astron. Tsirk., No. 810, p. 6 - 8 (1974). In Russian.

041.016 **Observations of the sun, Mercury and Venus in right ascension in 1963 in Tashkent.** O. S. Tursunov.
Tsirk. Astron. Inst., *Tashkent*, No. 40 (387), p. 17 - 22 (1973). In Russian.

041.017 **Systematic and random differences TMK –FK4 in the right ascensions of stars obtained on the basis of observations with the Tashkent meridian circle.**
M. F. Bykov.
Tsirk. Astron. Inst., *Tashkent*, No. 44 (391), 32 pp. (1973). In Russian.

041.018 **Überprüfung des Smithsonian Astrophysical Observatory (SAO)-Atlas auf Vollständigkeit bis $9^m_.0$.**
P. Wolff.
Sterne, 50. Jahrgang, p. 39 - 46 (1974).

041.019 **Vergleich einiger Positionskataloge mit dem AGK 3.**
W. Dieckvoß, C. de Vegt.
Mitt. Astron. Ges., No. 34, p. 90 (1973/74). – Abstr. AG.

041.020 **Résultats des observations faites à Alger avec l'astrolabe impersonnel A. Danjon OPL 8. Temps et latitude 1972.**
A. Ghezloun, M. Benhocine, A. Marouf, J. Pham-Van.
Ann. Obs. Astron. Alger, Vol. 3, Fasc. 5, 19 pp. (1974).

041.021 **The secular motion of the earth's poles and its influence on astronomical coordinates.**
G. M. Kaganovskij.
Geod. i kartografiya, 1973, No. 11, p. 9 - 15. In Russian. Abstr. in Referativ. Zhurn. 51. Astron., 5.51.186; 52. Geodeziya i Aehrosemka, 5.52.116 (1974).

041.022 **Coordonnées écliptiques 1970,0 des étoiles du Star Catalog du Smithsonian Astrophysical Observatory.**
L. Houziaux, F. Ostan.
Centre Univ. État Mons, Fac. Sci., Dép. Astrophys., Commun. No. 28, 5 pp. (1973).
Le présent catalogue donne les coordonnées écliptiques 1970,0 des 258.997 étoiles du Star Catalog du Smithsonian Astrophysical Observatory (1966). Il a été constitué à partir du catalogue en coordonnées écliptiques 1970,0 reprenant par longitude écliptique croissante les objets du Star Catalog (Houziaux, 1970).

041.023 **The southern sky surveys – a review of the ESO Sky Survey Project.** R. M. West.
European Southern Obs., Bull. No. 10, p. 25 - 40 (1974).

041.024 **Positional star catalogues. Part I.** P. Rybka.
Postępy Astron., Vol. 22, 81 - 89 (1974). In Polish.
In the first part of the review of positional star catalogues, the basic problems connected with these catalogues are described. After a short historical outline of the old catalogues the principles of modern catalogues as well as their errors are described.

041.025 **Positional star catalogues. Part II.** P. Rybka.
Postępy Astron., Vol. 22, 91 - 107 (1974). In Polish.
A chronological description of fundamental catalogues is given. The interrelation between the catalogues and their characteristics are discussed.

041.026 **Planetary perturbations in Newcomb's tables of the sun.** H. Kinoshita, H. Nakai, S. Aoki.
Ann. Tokyo Astron. Obs., Second Ser., Vol. 14, (No. 1), 1 - 13 (1974).
The authors developed a computer program, which calculates the solar coordinates, based on Newcomb's trigonometric expressions for the solar motion given in the tables of the sun (1895).

041.027 **Ts2 [Z2] catalogue of corrections to the right ascensions of stars of the Time Service programme.**
T. V. Goryainova.
Trudy Gos. Astron. Inst. Shternberga, Vol. 44, 185 - 194 (1973). In Russian.
This paper contains the catalogue Ts2 of corrections to the right ascensions of 245 stars of the Time Service programme. The catalogue is based on the results of the observations by the Time Service from January 1963 to April 1964.

041.028 **Les conjonctions de Vénus avec diverses régions du ciel.** J. Dommanget.
Ciel et Terre, Vol. 90, 179 - 186 (1974).

041.029 **On the instrumental system of coordinates in meridian astrometry.** M. S. Zverev.
Trudy Astron. Obs., *Leningrad*, Vol. 30 (= Uchenye Zapiski Leningr. Un-ta, No. 373 = Seriya Matem. Nauk, vyp. (No.) 50), p. 177 - 191 (1974). In Russian.
The instrumental system of coordinates is to be represented as the systematic differences between observational results and a fundamental catalogue. To improve the R.A. system of the fundamental catalogue with respect to the $\Delta\alpha_\delta$ errors, the results of differential observations reduced by a quasi-absolute method can be used, as well as the absolute star catalogues. For controlling and a more strict relating to the pole it is recommended to observe the Polarissima stars. An improvement of the $\Delta\alpha_\alpha$ system can be obtained by long-lasting series of observations of stars in a wide equatorial zone.

041.030 **An analysis of the comparison of the General Catalogue of the USRR Time Services and the FK4.**
P. M. Afanas'eva, V. L. Gorshkov.
Astron. Zhurn. Akad. Nauk SSSR, Vol. 51, 652 - 657 (1974). In Russian. English translation in Soviet Astron. AJ, Vol. 18, No. 3.
Tables of systematic differences of the right ascensions and proper motions of common stars of the KCB and FK4 are presented and analyzed.

041.031 **Influence of the instrumental system on a photographic catalogue system.**
D. N. Ponomarev, N. B. Frolova.
Astron. Zhurn. Akad. Nauk SSSR, Vol. 51, 675 - 676 (1974). In Russian. English translation in Soviet Astron. AJ, Vol. 18, No. 3.
The results of comparisons of new Moscow zenith star catalogues with AGK3R, AGK2 and AGK3 catalogues are given. It is shown that any photographic catalogue system is affected very much by some causes of instrumental, meteorological and methodical character.

041.032 **Systèmes des coordonnées célestes.**
D. Vlachos, with a preface by I. Argyrakos.
In honorem S Placidis, (see 003.009), p. 33 - 52 (1974). In Greek.

041.033 **New impetus to astrometry.** W. Fricke.
IAU Symposium No. 61, (see 012.014), p. 1 - 4 (1974). – Inaugural address.

041.034 **Report on IAU colloquium No. 20, 'Meridian astro-**

nomy'. G. van Herk.
IAU Symposium No. 61, (see 012.014), p. 7 - 21 (1974). — Invited paper.

041.035 Plans for the improvement and extension of the FK4. W. Fricke.
IAU Symposium No. 61, (see 012.014), p. 23 - 30 (1974). Invited paper.

041.036 Observations relevant to the improvement of the fundamental system. W. Gliese.
IAU Symposium No. 61, (see 012.014), p. 31 - 32 (1974).

041.037 Report on the compilation of the PFKSZ-2 catalogue.
Ya. S. Yatskiv (*Ia. S. Iatskiv*), A. N. Kur'yanova (*Kurianova*), D. D. Polozhentsev, M. S. Zverev.
IAU Symposium No. 61, (see 012.014), p. 33 - 34 (1974).

041.038 On the improvement of the α_δ system of the FK4 catalogue.
C. Anguita, G. Carrasco, P. Loyola, A. A. Naumova, A. A. Nemiro, D. D. Polozhentsev, T. A. Polozhentseva, V. N. Shishkina, R. Taibo, G. M. Timashkova, M. P. Varin, V. A. Varina, M. S. Zverev.
IAU Symposium No. 61, (see 012.014), p. 35 (1974). — Abstract.

041.039 Absolute determination of right ascensions of stars at high geographical latitude during the polar night.
G. M. Petrov.
IAU Symposium No. 61, (see 012.014), p. 37 - 38 (1974).

041.040 On stars with the most favourable background in observational history. G. van Herk.
IAU Symposium No. 61, (see 012.014), p. 39 - 40 (1974).

041.041 Southern reference star program: progress report. J. L. Schombert.
IAU Symposium No. 61, (see 012.014), p. 41 (1974).

041.042 Investigation of magnitude effect in the AGK3R catalogue. J. L. Schombert, T. E. Corbin.
IAU Symposium No. 61, (see 012.014), p. 43 - 55 (1974).

041.043 Mouvements propres des étoiles de repère du catalogue photographique. R. Bouigue.
IAU Symposium No. 61, (see 012.014), p. 57 (1974). — Abstract.

041.044 Observations in Chile and results related to southern hemisphere systems. C. Anguita.
IAU Symposium No. 61, (see 012.014), p. 63 - 72 (1974). Invited paper.

041.045 Preliminary results of relative declination observations made with the Repsold meridian circle at Cerro Calán. G. Carrasco.
IAU Symposium No. 61, (see 012.014), p. 73 - 78 (1974).

041.046 Observations by the Hamburg SRS-expedition to Perth. E. Høg, I. Nikoloff.
IAU Symposium No. 61, (see 012.014), p. 79 - 84 (1974).

041.047 On progress and methods of reduction of the Melbourne observations of fundamental stars.
K. N. Tavastsherna.
IAU Symposium No. 61, (see 012.014), p. 85 (1974). — Abstract.

041.048 Contribution des astrolabes à la détermination du système de référence dans l'hémisphère sud.
G. Billaud, S. Débarbat.
IAU Symposium No. 61, (see 012.014), p. 87 - 90 (1974).

041.049 Some urgent programmes of meridian observations especially for the southern hemisphere.
M. S. Zverev.
IAU Symposium No. 61, (see 012.014), p. 91 - 94 (1974). Invited paper.

041.050 Report on the Cape participation in the southern reference star project. R. H. Tucker.
IAU Symposium No. 61, (see 012.014), p. 95 (1974).

041.051 Report on the Cape photographic survey.
S. V. M. Clube, W. Nicholson.
IAU Symposium No. 61, (see 012.014), p. 97 (1974).

041.052 Potentialities of Yale astrometric materials.
E. D. Hoffleit.
IAU Symposium No. 61, (see 012.014), p. 99 - 103 (1974).

041.053 Thoughts on the future of photographic catalogues.
H. Wood.
IAU Symposium No. 61, (see 012.014), p. 105 - 108 (1974).

041.054 Radio and optical astrometry. C. M. Wade.
IAU Symposium No. 61, (see 012.014), p. 133 - 139 (1974). — Invited paper.

041.055 Accurate optical positions using the Palomar Sky Atlas. R. W. Hunstead.
IAU Symposium No. 61, (see 012.014), p. 175 - 179 (1974).

041.056 Reduction of Schmidt photographs with the aid of reference marks on the curved plate.
P. Lacroute.
IAU Symposium No. 61, (see 012.014), p. 181 (1974). — Abstract.

041.057 On the project of a new fourfold coverage of the northern hemisphere. C. De Vegt.
IAU Symposium No. 61, (see 012.014), p. 209 - 215 (1974).

041.058 Les effets de la réfraction atmosphérique sur les coordonnées tangentielles en astrométrie.
J. Dommanget.
IAU Symposium No. 61, (see 012.014), p. 239 (1974). — Abstract.

041.059 Prospects of space astrometry.
P. Bacchus, P. Lacroute.
IAU Symposium No. 61, (see 012.014), p. 277 - 282 (1974).

041.060 Astrometry with the large space telescope.
W. F. Van Altena, O. G. Franz, L. W. Fredrick.
IAU Symposium No. 61, (see 012.014), p. 283 - 293 (1974).

041.061 Star positions from overlapping plates in the Cape zone $-40°$ to $-52°$. S. V. M. Clube.
IAU Symposium No. 61, (see 012.014), p. 305 - 306 (1974).

041.062 A general computer program for the application of the rigorous block-adjustment solution in photographic astrometry. C. De Vegt, H. Ebner.
IAU Symposium No. 61, (see 012.014), p. 307 - 308 (1974).

041.063 La réduction des clichés astrographiques de champ quelconque. J. Dommanget.
IAU Symposium No. 61, (see 012.014), p. 309 (1974). — Abstract.

041.064 Precise reduction to the apparent places of stars.
S. Yumi, K. Hurukawa, T. Hirayama.
IAU Symposium No. 61, (see 012.014), p. 319 (1974).

041.065 Meridian astrometry. M. F. Bykov.
Astronomical Institute of the Academy of Sciences of the Uzbek SSR – 100 years, (see 003.011), p. 59 - 68 (1974). In Russian.

041.066 Photographic astrometry.
A. A. Latypov, A. G. Rakhimov.
Astronomical Institute of the Academy of Sciences of the Uzbek SSR – 100 years, (see 003.011), p. 69 - 79 (1974). In Russian.

041.067 Theoretical astronomy, theory of instruments and celestial mechanics. P. P. Loginov.
Astronomical Institute of the Academy of Sciences of the Uzbek SSR – 100 years, (see 003.011), p. 80 - 87 (1974). In Russian.

041.068 Astronomical data files at the U.S. Naval Observatory–ephemerides, star catalogues, and observations.
A. D. Fiala.
Bull. American Astron. Soc., Vol. 6, 217 - 218 (1974).

041.069 Star catalog cross index. J. Mead.
Bull. American Astron. Soc., Vol. 6, 218 (1974).

041.070 International Information Bureau on Astronomical Ephemerides.
Bull. B.I.I.E.A. (IAU–COSPAR), Paris. Information cards, Nos. 78 - 93 (1973). – Notes on data files available at the Astronomisches Rechen-Institut, Heidelberg.

041.071 The bibliographical star index.
R. Cayrel, J. Jung, A. Valbousquet.
Centre de Données Stellaires, Inform. Bull. No. 6, p. 24 - 31 (1974).

041.072 First complement to the list of star catalogues available in machine-readable form at the Astronomisches Rechen-Institut Heidelberg. T. Lederle.
Centre de Données Stellaires, Inform. Bull. No. 6, p. 32 - 37 (1974).

041.073 The Astrographic Catalog.
P. Lacroute, A. Valbousquet.
Centre de Données Stellaires, Inform. Bull. No. 6, p. 38 (1974).

041.074 Star catalogs and files available at the Stellar Data Center. J. Jung, M. Guilbaut.
Centre de Données Stellaires, Inform. Bull. No. 7, p. 40 - 58 (1974).

041.075 Catalogue of declinations of 200 stars of the Pulkovo Latitudinal Program in the FK4 system.
N. P. Krasnenko.
Astron. Tsirk., No. 816, p. 6 - 7 (1974). In Russian.

041.076 Color star atlas. P. Moore.
Crown Publishers Inc., New York, N.Y. 112 pp. Price $ 7.95 (1973). – Review in Sky Telescope, Vol. 47, 188 (1974).

041.077 Drehbare Sternkarte "Zodiak". J. Schultz.
Edited by the Math. Astron. Sektion am Goetheanum, Dornach (Switzerland). 10th edition. Price DM 31.00. Review in SuW, Vol. 13, 64 - 65; 1974 (*G. D. Roth*).

Autocorrelation functions of star image vibrations and their applications to the estimate of errors of astrometric observations. See Abstr. 031.016.

Device for the mean moment of star transit determination. See Abstr. 031.049.

Radio astrometry using connected-element interferometers. See Abstr. 031.051.

The influence of the reduction model on the systematic accuracy of adjusted parameters. See Abstr. 031.055.

Environmental systematics and astronomical refraction I. See Abstr. 031.080.

The study of flexure effects on the mobile parts of the great meridian instrument of the Bucharest Observatory and their influence on the accurate determination of stellar positions. See Abstr. 032.024.

Accurate positions of Mars and its satellite Deimos from photographic observations in Tashkent. See Abstr. 097.039.

The U.S. Naval Observatory parallax program. See Abstr. 111.004.

Stellar proper motions with reference to galaxies. See Abstr. 112.005.

Absolute proper motions of 14600 stars in 85 areas of the northern sky as obtained from galaxies with the Pulkovo normal astrograph. See Abstr. 112.006.

Catalogue of relative proper motions of 1474 stars in the α Cyg region. See Abstr. 112.007.

Catalogue of absolute proper motions of stars with respect to galaxies in Kapteyn's Selected Area 32. See Abstr. 112.008.

Proper motion and parallax programmes for large telescopes. See Abstr. 112.018.

Remarks on the definition and determination of proper motions. See Abstr. 112.020.

Proper motions with respect to galaxies. See Abstr. 112.021.

Report on the Yale-Columbia southern proper motion program. See Abstr. 112.022.

The errors of absolute photographic proper motions of stars relative to galaxies. See Abstr. 112.023.

Progress report on the Herstmonceux programme of measurement of proper motions in the northern Selected Areas. See Abstr. 112.024.

On the relationship between the apparent magnitudes given in several catalogues and the *UBV* system. See Abstr. 113.020.

Positions of extragalactic radio sources from very long baseline interferometry. See Abstr. 141.092.

Position solution of compact radio sources using long coherence VLBI. See Abstr. 141.093.

The Texas radio astrometric survey. See 141.094.

042 Celestial Mechanics, Figure of Celestial Bodies

042.001 Long periodic terms in the solar system.
P. Bretagnon.
Astron. Astrophys., Vol. 30, 141 - 154 (1974). In French.

The author has studied the long-period variations of the eight planets of the solar system (Pluto is excluded). He first calculated the Lagrangian solution. He then introduced the long-period terms of fourth order in excentricities and inclinations in the disturbing function. In a second approximation he took into account the contribution of the short-period terms which provide the perturbations of the first order with respect to the masses.

042.002 Vertical stability of periodic orbits in the restricted problem. II. Hill's case. M. Hénon.
Astron. Astrophys., Vol. 30, 317 - 321 (1974).

The author studies the stability of plane periodic orbits with respect to perturbations perpendicular to the plane in Hill's limiting case of the restricted three-body problem. This, combined with earlier results on horizontal stability, allows a determination of the three-dimensional stability of the orbits.

042.003 On the theory of motion in the field of a rotationally symmetric potential. T. A. Agekyan.
Dokl. AN SSSR, Ser. Mat. Fiz., Vol. 214, 783 - 786 (1974). In Russian.

042.004 A new class of periodic motions of a solid body with one fixed point in a Newtonian force field.
V. G. Demin, F. I. Kiselev.
Dokl. AN SSSR, Ser. Mat. Fiz., Vol. 214, 997 - 998 (1974). In Russian.

042.005 The problem of four fixed centres and its application to the theory of the motion of heavenly bodies.
I. S. Kozlov.
Astron. Zhurn. Akad. Nauk SSSR, Vol. 51, 191 - 198 (1974). In Russian. English translation in Soviet Astron. AJ, Vol. 18, No. 1.

042.006 On the stability of a particular solution of the problem of motion of three homogeneous spheroids.
V. V. Vidyakin.
Astron. Zhurn. Akad. Nauk SSSR, Vol. 51, 199 - 207 (1974). In Russian. English translation in Soviet Astron. AJ, Vol. 18, No. 1.

The problem of stability in the Lyapunov sense of a particular solution close to the Lagrangian triangular solution in the problem of three material points and that of the motion of three spheroids has been investigated. It has been proved that for sufficiently small values of the eccentricities of the meridional section of the interacting spheroids and under certain conditions applied to the semi-major axes and the masses of spheroids, the solution under consideration is stable.

042.007 Approximate axial alignment times for spinning bodies. D. C. McAdoo, J. A. Burns.
Icarus, Vol. 21, 86 - 93 (1974).

Using a different approximate scheme for each of two separate body shapes, the authors compute for torque-free rotating quasi-rigid bodies the rate of damping of the nutation angle due to internal energy dissipation. The first method calculates the alignment rate as well as the free precession period of a Maxwell body; it modifies, and thereby makes clearer, a similar calculation by Gerstenkorn (1967). The second method deals with an elongated body and models it by symmetrically placing masses along a massless rod, where all energy dissipation occurs through the hysteresis loss of the strain energy stored during the precession motion. The results agree well with those of another approximate solution (Burns and Safronov, 1973) and numerically check that solution for the alignment times of asteroids.

042.008 Normal forms for real linear Hamiltonian systems with purely imaginary eigenvalues.
N. Burgoyne, R. Cushman.
Celestial Mechanics, Vol. 8, 435 - 443 (1974).

This note gives a concise algorithm for computing a normal form for a real linear Hamiltonian differential equation which has purely imaginary eigenvalues. This algorithm is then applied to the differential equation which comes from the quadratic terms of the Hamiltonian of the restricted three body problem at a Lagrange equilateral triangle equilibrium point.

042.009 Escape or retention in the three-body problem.
J. S. Griffith, R. D. North.
Celestial Mechanics, Vol. 8, 473 - 479 (1974).

Sufficient conditions for escape and for retention in the three-body problem are derived which, for escape, are more stringent than those previously derived and, for retention, avoid the difficulty of containing the inverse of the mass of the escaping particle.

042.010 Sur les solutions particulières du problème généralisé des trois corps solides.
G. N. Duboshin (*Doubochine*).
Celestial Mechanics, Vol. 8, 495 - 521 (1974).

The generalized problem of translatory-rotatory motion of three rigid bodies whose elementary particles act upon each other according to arbitrary laws of forces along the straight line joining them is discussed.

042.011 The Ideal Resonance Problem: a comparison of the solutions expressed in terms of mean elements and in terms of initial conditions. A. H. Jupp.
Celestial Mechanics, Vol. 8, 523 - 530 (1974).

In an earlier publication (Jupp, 1972), a solution of the Ideal Resonance Problem is exhibited explicitly in terms of the 'mean' elements; to second order in the small parameter in the case of libration, and to first order in the case of deep circulation. Both representations possess a singularity when the 'mean' modulus of the Jacobi elliptic functions is unity. Is is shown here that, provided particular coefficients associated with the problem satisfy specific relations, the singularity is removed, and the resulting solution is applicable throughout the deep resonance region. The solution is then expressed in terms of general initial conditions. Again, in general, the solution has a singularity associated closely with the limiting motion, and the circulation part of the solution is restricted to deep circulation. It is shown that when the previously-mentioned coefficients satisfy particular constraints, the singularity is removed. In addition, with the same constraints, the deep-circulation solution is applicable throughout the circulation region.

042.012 On the free rotation of a rigid body. A. H. Jupp.
Celestial Mechanics, Vol. 9, 3 - 20 (1974).

The general solution of the Ideal Resonance Problem is employed to formulate a second-order formal series solution of the problem of a freely-rotating rigid body which has two of its principal moments of inertia differing by a small quantity This solution is firstly expressed in terms of the mean elements, and then in terms of the initial conditions. The latter solution is global in nature being applicable over the whole phase plane.

042.013 Periodic solutions near a resonant equilibrium of a Hamiltonian system. D. S. Schmidt.
Celestial Mechanics, Vol. 9, 81 - 103 (1974).

The author considers a conservative Hamiltonian system with two degrees of freedom where the linearized system consists of two harmonic oscillators whose frequencies ω_1 and ω_2 are commensurable or close to being so. Near the equilibrium of such a system he proves the existence of natural families of periodic orbits and analyzes their relation to the natural families which follow from Lyapunov's theorem. He gives a complete description of all families and provides an easy existence proof at the same time.

042.014 A second-order global solution of the Ideal Resonance Problem. B. Garfinkel, C. A. Williams.
Celestial Mechanics, Vol. 9, 105 - 125 (1974).

The Ideal Resonance Problem, as formulated in 1966 (Astron. Journ., Vol. 71, 657 - 669 (1966)), is defined by the Hamiltonian $F = B(y) + 2\mu^2 A(y) \sin^2 x$, $\mu \ll 1$. Following the procedure adopted in the construction of a first-order global solution from earlier papers, the authors derive a second-order solution from the von Zeipel-Bohlin recursive algorithm. The singularities inherent in the Bohlin expansion in powers of μ have been suppressed by means of the regularizing function constructed in Celestial Mechanics, Vol. 5, 189 - 203 (1972), and the singularities in the coefficients at $AB'' = 0$ have been removed by the normalization technique.

042.015 Spin-orbit coupling: a unified theory of orbital and rotational resonances.
P. J. Hamill, L. Blitzer.
Celestial Mechanics, Vol. 9, 127 - 146 (1974).

The dynamics of the spin-orbit interaction of a sphere and a rotating asymmetrical rigid body are examined. No restrictions are imposed on the masses, on the orientation of the rotation axis to the orbit plane, or on the orbit eccentricity.

042.016 On the equivalence of Deprit's and Hori's perturbation theories. M. Rapaport.
Astron. Astrophys., Vol. 31, 79 - 82 (1974). In French.

042.017 Adiabatic pulsations and convective instability of uniformly rotating gaseous masses.
R. K. Kochhar, S. K. Trehan.
Astrophys. Space Sci., Vol. 26, 271 - 287 (1974).

Third-order virial equations are used to investigate the oscillations and the stability of the sequence of uniformly rotating compressible Maclaurin spheroids, referred to in an inertial frame. It is seen that in the case of the oscillations belonging to the third harmonics, the frequency spectrum of the Maclaurin sequence referred to in an inertial frame is distinct from the spectrum of the Maclaurin sequence considered stationary in a rotating frame of reference.

042.018 Rotational dynamics of deformable celestial bodies. I.: Tidal deformations. J. N. Tokis.
Astrophys. Space Sci., Vol. 26, 447 - 476 (1974).

Euler's equations for a deformable body are set up in an inertial (or fixed) coordinate system without any restriction on the stress tensor. Application of these equations is made for a simple viscous fluid body. Then, the Eulerian equations are formulated explicitly for three-dimensional rotation of self-gravitating compressible celestial bodies of arbitrary structure, and the viscosity of their material is treated as an arbitrary function of spatial coordinates, with special respect to a description of the effects of tidal deformation in a close pair of such bodies.

042.019 Rotational dynamics of deformable celestial bodies. II: Tidal friction and axial rotation. J. N. Tokis.
Astrophys. Space Sci., Vol. 26, 477 - 495 (1974).

In a previous paper the Eulerian differential equations, which govern three-dimensional rotation of celestial bodies consisting of fluid material, have been set up with special respect to a description of effects of viscous friction exhibited in a binary system consisting of a close pair of such bodies. In order to study their solution, in the present investigation the author departs from one differential equation which governs the simple rotation about an axis perpendicular to the orbital plane of the system. A solution of this differential equation is given using modified Bessel functions for the case of constant orbital elements. Methods are discussed for numerical integration of this equation when the rotating body is spherical, with constant viscosity at the surface of the body, or spheroidal with constant viscosity throughout the whole body; these methods have been extended to the case in which the orbital elements of the system are given very approximately by explicit functions of the time.

042.020 Operations with Fourier series in celestial mechanics. J. Chapront, M. Chapront, J. L. Simon.
Astron. Astrophys., Vol. 31, 151 - 155 (1974). In French.

The authors describe three different methods of handling Fourier series on computers which have been studied and applied to several problems of celestial mechanics at the Bureau des Longitudes (e.g. to planetary theories and the lunar main problem).

042.021 Numerical study of discrete plane area-preserving mappings. F. Rannou.
Astron. Astrophys., Vol. 31, 289 - 301 (1974).

Dynamical systems with two degrees of freedom can be reduced to the study of a two-dimensional mapping. The author considers discrete mappings operating only on integers. This allows an exact numerical study, without round-off errors. Any point belongs to a finite cycle. These mappings are strictly one-to-one. He defines, for comparison, a "random mapping" and gives its principal properties.

042.022 Mean elements of the principal planets.
P. K. Seidelmann, L. E. Doggett, M. R. DeLuccia.
Bull. American Astron. Soc., Vol. 6, 205 - 206 (1974). Abstr. AAS.

042.023 Quadrature solution of the classical two-body problem if the gravitational constant G varies inversely as the age of the universe. J. P. Vinti.
Bull. American Astron. Soc., Vol. 6, 206 (1974). – Abstr. AAS.

042.024 L_2 points in the sun, solar nebula, Jupiter, asteroid problem. M. Lecar.
Bull. American Astron. Soc., Vol. 6, 206 (1974). – Abstr. AAS.

042.025 A note on velocity-related series expansions in the two-body problem. R. A. Broucke.
Bull. American Astron. Soc., Vol. 6, 207 (1974). – Abstr. AAS.

042.026 On the improvement of algebraic manipulation systems for celestial mechanics. W. H. Jefferys.
Bull. American Astron. Soc., Vol. 6, 208 (1974). – Abstr. AAS.

042.027 Advantages of the Encke equations compared to regularized Cowell equations. E. Everhart.
Bull. American Astron. Soc., Vol. 6, 208 - 209 (1974).

042.028 On the comparison of numerical theories of motion. J. D. Mulholland.
Bull. American Astron. Soc., Vol. 6, 209 (1974). – Abstr. AAS.

042.029 **Espace des phases dans le problème plan des trois corps.** B. Elmabsout.
Comptes Rendus Acad. Sci. Paris, Sér. A, Vol. 278, 459 - 461 (1974).

Classification topologique des sous-variétés d'énergie h et de moment angulaire ω donnés.

042.030 **Analysis of the restricted circular twice-averaged three-body problem in the case of close orbits.**
M. L. Lidov, S. L. Ziglin.
Celestial Mechanics, Vol. 9, 151 - 173 (1974).

This paper shows a new limiting class of solution of the restricted circular twice-averaged three-body problem and gives its qualitative investigation. The space orbits of the 'asteroids' everywhere close to the circular orbit of the body with the least finite mass are considered. Certain qualitative singularities of the behavior of solutions connected with the critical inclination (Krasinsky, 1972) are noted. The paper also shows an analogous class of solutions of the twice-averaged problem, dealing with perturbations of the satellite orbit by a 'mascon'.

042.031 **The angle of escape in the three-body problem.**
D. G. Saari.
Celestial Mechanics, Vol. 9, 175 - 181 (1974).

In the three-body problem an upper bound is found for the angle defined by the invariable plane and the position vector of an escaping particle.

042.032 **About regions of convergence of expansions of differential equations of the three-dimensional restricted three-body problem in the vicinity of the collinear libration points.** G. I. Shirmin.
Celestial Mechanics, Vol. 9, 183 - 190 (1974).

The analysis of regions of convergence of expansions of the right-hand sides of the differential equations of motion in the vicinity of the collinear libration points in the circular restricted three-body problem in powers of coordinates of the infinitesimal body due to Moulton is shown to be erroneous, and his results are corrected. The generalisation of Moulton's results to analogous expansions of the equations in the elliptic problem of three bodies made by R. W. Farquhar (Farquhar, 1968) is shown to be groundless.

042.033 **On the analytical structure of partial derivatives of the Hamilton functions in the circular restricted three-body problem.** E. A. Grebenikov, M. N. Kiosa.
In-t teor. i eksperim. fiz. ITEhF-56. Moskva. 16 pp. Price 6 Kop. (1973). In Russian. − Abstr. in Referativ. Zhurn. 51. Astron., 3.51.110 (1974).

042.034 **Examples of transformations improving the numerical accuracy of the integration of differential equations.** J. Baumgarte, E. Stiefel.
Proc. conference on numerical solution of ordinary differential equations, 1972, (see 012.005), p. 207 - 236 (1974).

042.035 **Use of Green's functions in the numerical solution of two-point boundary value problems.**
L. J. Gallaher, I. E. Perlin.
Proc. conference on numerical solution of ordinary differential equations, 1972, (see 012.005), p. 374 - 407 (1974).

042.036 **Multirevolution methods for orbit integration.**
O. Graf.
Proc. conference on numerical solution of ordinary differential equations, 1972, (see 012.005), p. 471 - 490 (1974).

042.037 **Planetary theories.** J. M. A. Danby.
Celestial Mechanics, Vol. 9, 297 - 302 (1974).
Presented at a colloquium on 'Copernicus ... and modern dynamical astronomy', Washington, 1972 December.

The present condition of planetary theory, i.e., the computation and comprehension of the motion of the planets, is reviewed. Account is taken of the new and anticipated demands of observational accuracy, new mathematical methods in perturbation theory, and the use of the high-speed computer to perform algebraic as well as numerical work.

042.038 **The problem of three bodies.** V. Szebehely.
Celestial Mechanics, Vol. 9, 359 - 363 (1974).
Presented at a colloquium on 'Copernicus ... and modern dynamical astronomy', Washington, 1972 December.

The gravitational problem of three bodies is presented in the general case, without restrictions on the distances and masses of the participating bodies. Recent advances are discussed and the consequences of the Laplacean instability in stellar dynamics are described.

042.039 **Numerical investigation of the planar restricted three-body problem. I. Periodic orbits of the second generation in the Sun-Jupiter system.** V. V. Markellos.
Celestial Mechanics, Vol. 9, 365 - 380 (1974).

A method is described for the determination of families of periodic orbits, of the restricted problem of three bodies, as branchings of a given family of stable periodic orbits. Poincaré's method of successive crossings of a surface of section is applied for a value of the mass parameter corresponding to the Sun-Jupiter case of the restricted problem. New families are found, of the type of direct asteroids, having long periods and closing in space after many revolutions of the third body about the sun.

042.040 **Sufficient conditions for hyperbolic-elliptic escape and for 'ejection without escape' in the three-body problem.** C. Marchal.
Celestial Mechanics, Vol. 9, 381 - 393 (1974).

The three-body problem is decomposed into the motion of the binary m_1, m_2 and the motion of the third mass m_3 with respect to the center of masses of m_1 and m_2. Some limitations are assumed on the motion of the binary m_1, m_2 which imply limits on the motion of m_3. Sufficient criteria are demonstrated for the hyperbolic-elliptic escape and for the 'ejection without escape'. The new criteria are compared with those given in the literature.

042.041 **Effects of the mass ratio on the existence of retrograde satellites in the circular plane restricted problem.** D. Benest.
Astron. Astrophys., Vol. 32, 39 - 46 (1974).

The extended region of stability for retrograde satellites is investigated when the mass of the planet increases. A preliminary study is made of the characteristic and stability properties of the simple-periodic orbits. Then the general case of non-periodic orbits is considered. Contrary to expectation, it is found that the band of stability continues to exist even when the simple-periodic family has become partly unstable for the mass ratio $\mu > 0.0477$.... This is explained by the presence of a family of stable double-periodic orbits, which is described and tabulated.

042.042 **Second order perturbations for Jupiter and Saturn. Comparison with Le Verrier.**
J. L. Simon, J. Chapront.
Astron. Astrophys., Vol. 32, 51 - 64 (1974). In French.

The authors present a construction of the perturbations for the Keplerian elements of Jupiter and Saturn. First and second order effects with respect to the disturbing masses are taken into account. Results are compared in detail with those of Le Verrier, then on the whole with a numerical integration. A general discussion is done to explain the deviations between the two theories and the numerical integration.

042.043 Periodic Trojan-type orbits in the earth–sun system.
P. R. Weissman, G. W. Wetherill.
Astron. Journ., Vol. 79, 404 - 412 (1974).

Periodic orbits about the triangular equilibrium points are found for the planar restricted three-body problem using the earth–sun system. The maximum semimajor axis for tadpole orbits ranges from the infinitesimal orbit at 1.000 a.u. to the near-limiting orbit at 1.00285 a.u. Horseshoe orbits are found for 1.0029 to 1.0080 a.u., larger horseshoes being unstable because of close approaches to the earth. The limit of stability for nonperiodic orbits is found to occur for maximum semimajor axes near 1.0020 a.u. Near-periodic tadpole orbits appear to be stable against perturbations by Jupiter and Venus for periods of at least 10^4 yr. The possibility that minor planets actually exist in such orbits is considered.

042.044 On the accurate calculation of second order perturbations in the motion of celestial bodies.
Yu. V. Plakhov.
Izv. vyssh. ucheb. zavedenij. Geod. i aehrofotosemka, 1973, No. 5, p. 45 - 48. In Russian. – Abstr. in Referativ. Zhurn. 62. Issled. kosmich. prostranstva, 4.62.352 (1974).

042.045 Periodic solutions of a disturbed elliptic restricted three-body problem. V. Matas.
Bull. Astron. Inst. Czechoslovakia, Vol. 25, 129 - 134 (1974).

Existence of periodic motions of an infinitesimal body has been proved in the vicinity of the libration points of the elliptic restricted three-body problem "in spite of" the assumption of a perturbation due to the presence of the gravitational and radiative fields of a fourth (finite-mass) body and a resisting medium.

042.046 On non-existence of periodic solutions of the variational equations with respect to the collinear libration points of the elliptic restricted three-body problem.
V. Matas.
Bull. Astron. Inst. Czechoslovakia, Vol. 25, 135 - 138 (1974).

Based on a previously published (1973) separation of the variational equations with respect to the collinear libration points of the elliptic restricted three-body problem into a system of Hill equations it is shown that these variational equations have no non-trivial periodic solution. A numerical analysis has been used to support this qualitative result and to give stability conclusions concerned with the solutions of the variational equations with respect to the collinear libration points of the planar restricted three-body problem.

042.047 On non-existence of periodic solutions of the variational equations with respect to the triangular libration points of the planar elliptic restricted three-body problem. V. Matas.
Bull. Astron. Inst. Czechoslovakia, Vol. 25, 139 - 142 (1974).

Presumed a restrictive inequality between functions of e and μ, it is shown that the variational equations with respect to the triangular libration points of the planar elliptic restricted three-body problem have no non-trivial periodic solutions. Some stability conclusions regarding the general solution of these equations are given.

042.048 Asymptotic expansions in the perturbed two-body problem. F. Verhulst.
Diss. Rijksuniversiteit Utrecht, Utrecht. 119 pp. (1973).

042.049 A stable manifold theorem for degenerate fixed points with applications to celestial mechanics.
R. McGehee.
Journ. Diff. Equations, Vol. 14, 70 - 88 (1973).

An orbit of the classical three-body problem is called parabolic if two of the particles remain bounded while the third approaches infinity with zero velocity. It is proved for three special cases that the set of parabolic orbits forms a real analytic submanifold of the phase space.

042.050 Problèmes de mécanique céleste. J. Kovalevsky.
Rev. Questions Scientifiques, Vol. 144, 189 - 210 = Publ. Inst. Astron. Géophys., Georges Lemaître, Louvain, Vol. 6, No. 1 (1973).

This paper is based on a lecture the author has given as a visiting professor at the chair of Lemaître at Leuven. Its interest lies in a mainly verbal survey of a number of developments in celestial mechanics which makes the paper useful as an introduction to the subject for many scientists working outside this field. Two problems are considered in more detail: the dynamics of stellar systems and the theory of planetary systems, especially the three-body problem.

042.051 Construction of explicit solutions of the three-body problem by the method of semi-analytic programming. V. A. Prikhod'ko.
In-t teor. i ehksperim. fiz. ITEhF-60. Moskva. 31 pp. Price 11 Kop. (1973). In Russian. – Abstr. in Referativ. Zhurn. 51. Astron., 5.51.146 (1974).

042.052 Construction of equations of stable programmed motion of a mass point. A. F. Vishnyakov.
Probl. mekh. upravlyaemogo dvizheniya. Vyp. (No.) 3. Perm', 1973, p. 59 - 64. In Russian. – Abstr. in Referativ. Zhurn. 51. Astron., 5.51.149 (1974).

042.053 Modified Jacobi polynomial and its applications to expansions of disturbing functions.
H. Kinoshita, G. Hori, H. Nakai.
Ann. Tokyo Astron. Obs., Second Ser., Vol. 14, (No. 1), 14 - 35 (1974).

The authors modify a general linear transformation of the n-th degree spherical harmonics into a convenient form for applications in celestial mechanics. Using this transformation they can easily obtain general expansions of disturbing functions expressed by the combination of the spherical harmonics. They compute the explicit expressions for the modified Jacobi polynomials $Q_n^{(m, m')}(z)$, which appear in the coefficient of the linear transformation, up to $(n, m, m') = (10, 10, 10)$, and show them in this paper.

042.054 Application of Jacobi's polynomials to series expansion of the perturbing function. G. G. Pavlov.
Soobshch. Gos. Astron. Inst. Shternberga, No. 186, p. 46 - 58 (1973). In Russian.

042.055 Die Bewegung in der Nähe der Dreieckspunkte des elliptischen eingeschränkten Dreikörperproblems II.
J. Tschauner.
Celestial Mechanics, Vol. 9, 419 - 435 (1974).

The explicit separation of the motion close to the equilateral points of the elliptic restricted problem into two 2nd-order components presented in a previous paper is used for stability investigations of this motion. By means of simple approximate solutions for the components there are computed analytical curves of constant basic frequency and of stability boundaries with high accuracy in the plane of parameters g and e. The results of other authors are corrected or made more precise.

042.056 Second-order matching in the restricted three-body problem (small μ).
J. V. Breakwell, L. M. Perko.
Celestial Mechanics, Vol. 9, 437 - 450 (1974).

Elliptic orbits around the large primary are matched to hyperbolas, osculating at closest approach, around the small primary of the circular restricted three-body problem. The distance of closest approach to the small primary is assumed to be of the same order as the mass-ratio μ of small to large primary. The dependence of the hyperbola on initial conditions for the elliptic orbit is carried to second order jointly in μ and in the variations of the initial conditions, which are three-dimensional rather than two-dimensional.

042.057 **Sur les solutions Lagrangiennes du problème des trois corps solides avec la loi de Weber.**
G. N. Duboshin (*Doubochine*).
Celestial Mechanics, Vol. 9, 451 - 463 (1974).

The author considers the problem of translational-rotational motion of three solid bodies, for which the elementary particles attract each other according to different Weber's laws for each pair of bodies. This problem represents a special case of the generalized problem of three solids considered in a previous work (Duboshin, 1974) and it gives an example of the verification of the existence conditions for the Lagrangian solutions.

042.058 **The dynamical decay of unstable 4-body systems.**
R. S. Harrington.
Celestial Mechanics, Vol. 9, 465 - 470 (1974).

Unstable 4-body systems with negative energy can ultimately decay to (1) a binary plus two single stars, (2) two separate binaries, or (3) a stable triple plus a single star. One hundred random 2-dimensional and one hundred random 3-dimensional 4-body systems have been numerically integrated to determine the statistics of the end products. Of the final stable triples and binaries, 19 % were triples, which agrees well with observational estimates of the ratio of triples to binaries. The results were essentially the same for 2- and 3-dimensional systems.

042.059 **On the stability of Laplace's solutions of the unrestricted three body problem.** A. L. Kunitsyn.
Celestial Mechanics, Vol. 9, 471 - 481 (1974).

The instability criterion of a nonlinear mechanical system neutral to the first approximation is formulated for the internal resonance case which is characterized by the existence of commensurabilities between the frequencies of the system. The criterion derived is used for determining the regions of instability of Laplace's constant triangular solutions of the unrestricted three-body problem. It is shown that in the region where necessary Routh-Joukovsky stability conditions are satisfied, there may exist eight resonance-unstable sets of the masses of the three bodies.

042.060 **On the differential geometry of trajectory deviation with particular emphasis on Keplerian motion.**
E. L. Liipola.
Celestial Mechanics, Vol. 9, 483 - 489 (1974).

The Pontryagin/Lawden scalar product of the deviation phase vector and the adjoint phase vector may be identified with Lagrange's reciprocal formula for two variant motions if the acceleration field is conservative. Hence for two slightly different trajectories with common end-points, the terminal velocity differences must have equal scalar product with Lawden's primer vectors. The final velocity difference is orthogonal to the constant time of flight locus for isoenergetic motions from a common initial point. A Keplerian trajectory behaves like a rigid curve as far as radial deviation is concerned in the case of a small change of direction of the initial velocity vector.

042.061 **Étude du choc de deux petites masses en présence d'un troisième corps de masse prépondérante.**
M. Irigoyen.
Celestial Mechanics, Vol. 9, 491 - 506 (1974).

The analytical study of the evolution in the rectilinear problem of three bodies leads to consider the collision between two bodies, M_2 and M_3, in the presence of the third body, M_1. This problem, which seems to be difficult to approach in the general case, can partly be solved if the masses of M_2 and M_3 are equal and can be neglected in regard to M_1. In this particular case of the general problem, the mechanical study of a collision between M_2 and M_3, leads to two distinct types of collisions: 'instantaneous collisions', and 'collisions with repetition'.

042.062 **A grid search for families of periodic orbits in the restricted problem of three bodies.**
V. V. Markellos, W. Black, P. E. Moran.
Celestial Mechanics, Vol. 9, 507 - 512 (1974).

A method is described for the numerical determination of families of periodic orbits in the planar restricted problem of three bodies. The families are sought in their representation as curves in a two-dimensional space of parameters. A grid search is applied to the study of the evolution of satellite motion when the mass parameter μ is varied. Only that part of the space of parameters is investigated for which the relative energy constant takes values larger than that corresponding to the inner Lagrangian point L_2. Critical values of the mass parameter are determined for which new families of simple or double periodic orbits appear inside the closed ovals of zero velocity.

042.063 **Recherche du terme de $n$ième ordre des développements de l'astrodynamique.** E. Fichera.
Celestial Mechanics, Vol. 9, 513 - 516 (1974).

General expressions for the functions $E = f(M)$, $r/a = f(M)$ and $a/r = f(M)$ in the two-body problem are obtained as functions of eccentricity e without using Bessel functions.

042.064 **Normal forms for Hamiltonian systems.**
K. R. Meyer.
Celestial Mechanics, Vol. 9, 517 - 522 (1974).

A generalization of Birkhoff's normal form for a Hamiltonian developed about a critical point is given. The normal form is characterized intrinsically.

042.065 **Reduction of the two-dimensional elliptic restricted problem of three bodies to Hill's equation.**
B. Érdi.
Astron. Journ., Vol. 79, 653 - 655 (1974).

The differential equations of motion of the two-dimensional elliptic restricted problem are investigated. It is shown that a formal solution in the form of a power series of the eccentricity can be constructed by solving Hill's equation for a given periodic solution of the circular problem.

042.066 **Elements of the outer planets for one million years.**
C. J. Cohen, E. C. Hubbard, C. Oesterwinter.
Astron. Papers, U. S. Naval Obs., *Washington, D. C.*, Vol. 22, (Part 1), 1 - 92 (1973). [For sale by the Superintendent of Documents, U. S. Government Printing Office, Washington, D. C., Price $2.20].

The coordinates of the outer planets were computed by numerical integration for one million years centered at the present time. The results are presented in the form of element plots and a comparison is made with general theory. A discussion of the accuracy of the integration is also given.

042.067 **The equilibrium of a rotating body of arbitrary density.** P. Lanzano.
Astrophys. Space Sci., Vol. 29, 161 - 178 (1974).

This paper extends Clairaut's theory of rotational equilibrium to third order terms in a small parameter and is meant to be a sequel to a 1962 publication by the author bearing on the same topic.

042.068 **On effects of the sun and moon on the motion of a satellite of a planet.**
E. I. Timoshkova, K. V. Kholshevnikov.
Trudy Astron. Obs., *Leningrad*, Vol. 30 (= Uchenye Zapiski

Leningr. Un-ta, No. 373 = Seriya Matem. Nauk, vyp. (No.) 50), p. 141 - 156 (1974). In Russian.

The secular and long-period perturbations in the elements of the Eulerian orbit of a satellite caused by gravitation of a third body (moon, sun) are obtained.

042.069 The expansion of the perturbation function for relations of the major semiaxes close to unity. V. B. Titov.

Trudy Astron. Obs., *Leningrad*, Vol. 30 (= Uchenye Zapiski Leningr. Un-ta, No. 373 = Seriya Matem. Nauk, vyp. (No.) 50), p. 164 - 176 (1974). In Russian.

The expansion of the perturbating function of the problem of three bodies in powers of $\alpha - \lambda$ is derived, α being the ratio of the major semiaxes of perturbed and perturbing bodies and λ an arbitrary number. For the three obtained expansions the convergence is studied. One can improve the convergence of the expansion by suitable choice of λ.

042.070 On the stability of the triangular libration points in the Sun – Jupiter system. A. P. Markeev.

Astron. Zhurn. Akad. Nauk SSSR, Vol. 51. 627 - 634 (1974). In Russian. English translation in Soviet Astron. AJ, Vol. 18, No. 3.

The stability of the triangular Lagrangian solution for the restricted elliptic three-body problem is investigated. The values of e and μ are taken for the case Sun – Jupiter. It has been proved that in the case of the plane three-body problem stability for most of initial conditions takes place. As to the three-dimensional case the libration points are stable taking into account the terms up to the fourth order in the normal form of the Hamiltonian.

042.071 A form of equations of perturbed motion in canonical elements. S. N. Yashkin.

Astron. Zhurn. Akad. Nauk SSSR, Vol. 51. 635 - 640 (1974). In Russian. English translation in Soviet Astron. AJ, Vol. 18, No. 3.

Formulae are derived which take into account first-order perturbations in canonical elements for orbits of small eccentricity. A method is worked out to calculate the coefficients dependent on the elements L, G and H in the terms of the expansion of the disturbing function.

042.072 Expansion of the coordinates of the elliptic motion in power series of the eccentricity. K. Ahmad.

Astron. Zhurn. Akad. Nauk SSSR, Vol. 51, 641 - 645 (1974). In Russian. English translation in Soviet Astron. AJ, Vol. 18, No. 3.

042.073 The applicability of a new integral in the restricted three-body problem II. G. Bozis.

In honorem S. Placidis, (see 003.009), p. 301 - 316 (1974).

042.074 A note on the analytic continuation of periodic orbits. J. D. Hadjidemetriou.

In honorem S. Placidis, (see 003.009), p. 385 - 394 (1974).

042.075 On increasing the accuracy of second order perturbation computations in the motion of celestial bodies. Yu. V. Plakhov.

Izv. vyssh. ucheb. zavedenij. Geod. i aehrofotosemka, 1973, No. 5, p. 45 - 48. In Russian. – Abstr. in Referativ. Zhurn. 51. Astron., 6.51.144 (1974).

042.076 On the representation of periodic functions of many arguments by means of Fourier polynomials using computers. S. G. Kapralov.

Probl. mekh. upravlyaemogo dvizheniya. Vyp. (No.) 3. Perm', 1973, p. 104 - 110. In Russian. – Abstr. in Referativ. Zhurn. 51. Astron., 6.51.146 (1974).

042.077 Numerical exploration of the capture problem. G. Horedt.

Acta Astron., Vol. 24, 207 - 213 (1974).

A large mass M is considered and a small mass m_1 moving initially circularly around M. The author investigates single, close, single encounters between m_1 and the initially hyperbolic orbits of (1) massless particles, (2) small masses with a mass m_2 comparable to m_1. Generally, the capture probability of the particles or of m_2 into an elliptic orbit with small semimajor axis increases if m_1 increases ($m_1 \ll M$). It increases also if the velocity at infinity of the particles or of m_2 as well as the radii of m_1, m_2 decrease.

042.078 The problem of three bodies. V. Szebehely.

Stars and the Milky Way system, Proc. 1972, (see 012.018), p. 273 - 278 (1974). – Invited lecture.

042.079 Periodic orbits of the restricted problem for various values of the mass-ratio.

G. Contopoulos, M. Zikides.

Stars and the Milky Way system, Proc. 1972, (see 012.018), p. 279 - 282 (1974).

042.080 Stability regions for quasiperiodic motion in the restricted problem of three bodies.

W. H. Jefferys.

Astron. Journ., Vol. 79, 710 - 721 (1974).

Surfaces of section, plotted in configuration space, have been computed for the motion of the massless particle in the restricted problem of three bodies. Nine mass ratios and a wide variety of Jacobi constants were investigated; over four thousand orbits were computed, about half of which were finally used. The plots of surface of section have been reduced to plots of stability regions, following a method due to Henon. Sample surfaces of section are also given.

042.081 Pericentric and apocentric velocities. Y. Hatanaka.

Proc. symposium "Celestial mechanics", Kyoto 1973, (see 012.022), p. 20 - 29 (1974). In Japanese.

042.082 Extensions du développement de Lagrange. T. Inoue.

Proc. symposium "Celestial mechanics", Kyoto 1973, (see 012.022), p. 30 - 38 (1974). In Japanese.

042.083 Estimation of an inequality in the development of planetary disturbing functions. G. Hori.

Proc. symposium "Celestial mechanics", Kyoto 1973, (see 012.022), p. 39 - 42 (1974). In Japanese.

042.084 Symmetry transformations of the classical Kepler problem. H. H. Rogers.

Journ. math. Phys., Vol. 14, 1125 - 1129 (1973).

The non-relativistic problem of Kepler is studied from a geometrical point of view. The finite symmetry transformations can be viewed as an automorphism of the original orbit followed by a rigid rotation of the original orbit into the new orbit.

042.085 Once more the Kepler problem. F. A. E. Pirani.

Nuovo Cimento B, Ser. 11, Vol. 19B, 189 - 207 (1974).

Kustaanheimo's spinor linearization of the Kepler problem is written in an $SL_{2,c}$-covariant formalism. Two canonical versions are constructed; in one of them the Newtonian time is a dynamical variable. In the other version, the group of linear transformations which permutes orbits is isomorphic to $O_{4,2}$, and the intersection of this group with the symplectic group is isomorphic to O_4.

042.086 On the motion of the planets. E. Balint.

Fiz. Szemle, (*Hungary*), Vol. 24, No. 1, p. 17 - 24 (1974). In Hungarian.

Analytical solutions to the Newtonian equations of motion of the planets are derived to the first approximation using the equations of unperturbed heliocentral motion. Perturbations of the second, third and fourth orders of the outer planets are examined with the aid of Poincaré and Lagrange-Laplace theories. Early results of the secular perturbations of the planets are discussed.

042.087 **Toroidal figures of equilibrium.** C.-Y. Wong.
Astrophys. Journ., Vol. 190, 675 - 694 (1974).

A rotating mass in the form of a torus allows an effective balance between the gravitational attraction and the force due to centrifugal acceleration, when the angular momentum is very large. Attention is therefore fixed on a toroidal mass idealized for simplicity to be homogeneous, incompressible, and rotating uniformly about the central symmetry axis. The toroidal figures of equilibrium are determined in a self-consistent manner for various values of the angular momentum. A hypothesis of toroidal formation and breakup in galaxies and pre-main-sequence stars is postulated and compared with observations.

042.088 **Lagrange-koeficientoj f, g, h por la perturbata movigo.** B. Popović.
Bull. Obs. Astron. Belgrade, Vol. 29, F. 1, No. 125, p. 60 - 64 (1971/72).

An introduction to celestial mechanics.
See Abstr. 003.089.

Himmelsmechanik, Band III. Allgemeine Störungen.
See Abstr. 003.119.

Comparison of numerical integration techniques for orbital applications. See Abstr. 021.011.

Theoretical astronomy, theory of instruments and celestial mechanics. See Abstr. 041.067.

Relativistic spinor regularization of the astrodynamical problem of two bodies. See Abstr. 052.045.

On the equilibrium figures of an ideal rotating liquid in the post-Newtonian approximation of general relativity. I: Equilibrium conditions. See Abstr. 062.020.

On the equilibrium figures of an ideal rotating liquid in the post-Newtonian approximation of general relativity. II: Maclaurin's P-ellipsoid. See Abstr. 062.021.

A case of metastability for slowly rotating, supermassive objects. See Abstr. 065.027.

The effect of gravitational radiation-reaction on the evolution of the Riemann S-type ellipsoids.
See Abstr. 066.016.

Mögliche Zusammenhänge zwischen Apsidendrehungen bei Planetenbahnen und der Ausbreitungsgeschwindigkeit der Gravitation. See Abstr. 066.077.

Stability of asteroid orbits passing close to Jupiter. See Abstr. 098.007.

On the secular perturbations of asteroids.
See Abstr. 098.026.

An interaction effect between oblateness perturbation and mutual perturbation in the Enceladus-Dione system. See Abstr. 100.212.

On the long-term motion of Pluto.
See Abstr. 101.001.

043 Astronomical Constants

043.001 On the constant of gravitation.
V. A. Krat, I. L. Gerlovin.
Astrophys. Space Sci., Vol. 26, 521 - 522 (1974). – Letter.

043.002 A determination of the rate of change of G.
T. C. Van Flandern.
Bull. American Astron. Soc., Vol. 6, 206 (1974). – Abstr. AAS.

043.003 On the gravitational constant.
V. A. Krat, I. L. Gerlovin.
Doklady Akad. Nauk SSSR, Ser. Mat. Fiz., Vol. 215, 305 - 306 (1974). In Russian.

043.004 Definition of the meter. J. D. Mulholland.
Science, Vol. 183, 1248 - 1249 (1974).

043.005 Accuracy estimation of astronomical constants from long baseline interferometer observations.
H. G. Walter.
IAU Symposium No. 61, (see 012.014), p. 131 (1974). – Abstract.

Quadrature solution of the classical two-body problem if the gravitational constant G varies inversely as the age of the universe. See Abstr. 042.023.

Mean elements of the principal planets. See Abstr. 091.010.

Unterscheidbarkeit der galaktischen Rotation von der Präzession. Ein "Copernicanisches" Problem des 20. Jahrhunderts. See Abstr. 155.075.

Limits on variation of G from clusters of galaxies. See Abstr. 160.002.

044 Time, Rotation of the Earth

044.001 Sur une variation brusque de la durée du jour en janvier 1974. M. Feissel, B. Guinot.
Comptes Rendus Acad. Sci. Paris, Sér. B, Vol. 278, 293 - 294 (1974).

Une réduction de près de 0.001 s de la durée du jour est survenue dans les premières semaines de l'année 1974. Cette perturbation est la plus abrupte qu'on ait jamais mesurée.

044.002 Sur la recherche d'une nutation presque diurne de l'axe de rotation terrestre dans la terre et des erreurs sur les termes principaux de nutation dans l'espace.
N. Capitaine.
Comptes Rendus Acad. Sci. Paris, Sér. B, Vol. 278, 355 - 358 (1974).

044.003 Neue Anpassung der mittleren Weltzeit UTC (coordinated universal time) an die mittlere Erdzeit (UT 2).
Orion Schaffhausen, 32. Jahrgang, p. 11 (1974).

044.004 Round-off error of universal time. G. P. Pilnik.
Astron. Zhurn. Akad. Nauk SSSR, Vol. 51, 431 - 437 (1974). In Russian. English translation in Soviet Astron. AJ, Vol. 18, No. 2.

It is shown that for tidal and nutation studies the astronomical time observations must be made with an accuracy up to $0.^{s}0001$. The need for applying a plumb-line variation correction is discussed.

044.005 Determination of time (TU1). N. A. Omelina.
Tsirk. Astron. Inst., *Tashkent,* Nos. 40 (387), 42 (389), 43 (390) (1973). In Russian.

044.006 The application of ancient astronomy to the study of time. R. R. Newton.
Endeavour, No. 118, Vol. 33, 34 - 39 (1974).

The measurement of time has been an important aspect of astronomy since its earliest beginnings. By the analysis of astronomical data that have survived since ancient times, we are able to study the deviations of solar time from a uniformly flowing time base. The results have important implications for modern astronomy in the study of the non-gravitational accelerations of the earth and moon.

044.007 Les résidus des déterminations horaires à l'Observatoire de Bucarest pour la période 1958–1966.
M. Ciobanu, G. Oprescu.
Stud. Cerc. Astron., Vol. 19, 35 - 51 (1974).

044.008 Raccordement des échelles de temps utilisant les ondes ultra-courtes.
V. Stavinschi, R. Dorobanțu.
Stud. Cerc. Astron., Vol. 19, 53 - 66 (1974). In Romanian.

On présente les résultats obtenus en utilisant les impulsions des cadres contenues dans le signal complexe de télévision, pour raccorder les échelles de temps entre l'Observatoire Astronomique de Prague et l'Observatoire Astronomique de Bucarest. Les mesures ont été faites avec un chronomètre électronique Rochar; les écarts ne dépassent pas 1 µs.

044.009 Sur un terme harmonique de 122 jours dans la rotation de la terre et dans le mouvement du pôle.
D. Djurovic.
Ciel et Terre, Vol. 90, 45 - 60 (1974).

044.010 Sur la rotation de la terre et l'époque éocambrienne.
A. Dauvillier.
Ciel et Terre, Vol. 90, 136 - 142 (1974).

L'auteur montre comment le problème du ralentissement de la vitesse de rotation de la terre dans le passé, qui demeure indéterminé du point de vue astronomique, peut être précisé par des considérations géophysiques.

044.011 Rotación de la tierra año 1972. Resultados obtenidos en San Fernando con el Astrolabio Impersonal Danjon OPL No. 37. I. − Tiempo y latitud. II. − Observaciones de Saturno.   A. Orte, M. Sánchez, I. Vitini, J. B. Fernández, F. Parra.
Published by Inst. y Obs. de Marina, San Fernando (Cadiz), Ser. C, No. 75, 30 pp. (1974).

044.012 Daily phase values.
U. S. Naval Obs., Washington, D. C., Time Service Publ., Ser. 4, Nos. 361 - 386 (1974). − 1974 January 2 − June 26.

044.013 Preliminary times and coordinates of the pole.
U. S. Naval Obs., Washington, D. C. Time Service Publ., Ser. 7, Nos. 314 - 339 (1974). − 1974 January 3 − June 27.

044.014 On the amplitude modulation of waves in the variations of the velocity of the diurnal rotation of the earth.   A. A. Korsun', N. S. Sidorenkov.
Astron. Zhurn. Akad. Nauk SSSR, Vol. 51, 658 - 663 (1974). In Russian. English translation in Soviet Astron. AJ, Vol. 18, No. 3.

A spectral analysis of characteristics of the velocity of the earth's rotation shows some harmonic components. The possibility of their explanation by an amplitude modulation of seasonal oscillations is discussed. The qualitative conclusions are confirmed by calculations of amplitude variations for the yearly, half-yearly and quarterly waves from month to month for the period 1956.5–1972.8.

044.015 The yearly variations of the siderial day duration as a result of light ray deviation in the gravitational field of the sun.   B. V. Bondarev.
Astron. Zhurn. Akad. Nauk SSSR, Vol. 51, 664 - 665 (1974). In Russian. English translation in Soviet Astron. AJ, Vol. 18, No. 3.

The influence of the gravitational field of the sun on the angular distance between a star and the vernal equinox point is considered. A formula describing the change of the sidereal day duration in dependence on the position of the earth in its orbit is derived.

044.016 Time measurement.   B. V. Yasevich.
Astronomical Institute of the Academy of Sciences of the Uzbek SSR − 100 years, (see 003.011), p. 43 - 51 (1974). In Russian.

044.017 Time what is it exactly?   V. Protić.
Vasiona, Vol. 22, 33 - 39 (1974). In Serbo-Croatian.

044.018 Corrections to Czechoslovak time signals.
V. Ptáček.
Říše hvězd, Vol. 55, 19, 38, 58, 78, 101, 118, 138 (1974). In Czech. − 1973 October - 1974 April.

044.019 Theory of the rotation of the earth. I.
H. Kinoshita.
Proc. symposium "Celestial mechanics", Kyoto 1973, (see 012.022), p. 67 - 71 (1974). In Japanese.

044.020 Analysis of the periodicity of the irregular rotation of the earth.
S. Luo, S. Liang, S. Ye, S. Yan, Y. Li.
Acta Astron. Sinica, Vol. 15, 79 - 85 (1974). In Chinese.

A periodogram analysis is applied to the 1820–1970 data of the earth's rotation to search for hidden periods ranged from a few years to several decades. The observed variation of the earth's rotation in the above-mentioned time interval of 150 years has been fitted by a combination of 12 periods. Most of these periods are very close to the periods of sunspots, the moon, and the planets.

044.021 Détermination astronomique de l'heure.
M. Jovanović, D. Vesić, M. Lončarević.
Bull. Obs. Astron. Belgrade, Vol. 29, F. 1, No. 125, p. 15 - 19 (1971/72).

044.022 Time and latitude service.
Polish Acad. Sci., Astron. Latitude Station, Borowiec, Circ. No. 128, 12 pp. (1974). − 1973 October–December.

044.023 Astronomische Zeit- und Breitenbestimmungen. Empfangszeiten von Zeitsignalen.
Edited by Deutsches Hydrographisches Institut, Hamburg. 6 pp. (1973). − 1973 July - September.

044.024 Time service.   C. Moranzino (Editor).
Oss. Astron. Torino (Pino Torinese), Bull. No. 6, 15 pp. (1973). − Results of the time determinations 1973 September - December.

044.025 International Time and Latitude Service at the Tokyo Astronomical Observatory during 1973.
Tokyo Astron. Obs., Time and Latitude Bull., Vol. 47, Nos. 8 - 12, p. 49 - 82 (1973). − Results of the time determinations 1973 August - December.

044.026 Détermination astronomique de l'heure et de la latitude.
Obs. Neuchâtel, Bull. (B), 1973 July - December (1974).

044.027 L'heure astronomique définitive de l'Observatoire de Neuchâtel.
Obs. Neuchâtel, Bull. (D), 1973 July - December (1974).

044.028 Détermination astronomique de l'heure et heures demi-définitives de réception des signaux horaires.
L. Webrová, V. Ptáček.
Acad. Tchécoslov. Sci. Inst. Astron., Station de l'Heure, Prague, Sér. 6, Nos. 8 - 9, 14 + 15 pp. (1974). − 1973 March - June.

044.029 Universal time and coordinates of the pole; Emission time of time signals; Coordinated universal time; Independent local atomic time scales AT (i); Informations.
Bureau International de l'Heure, (B. I. H.), Paris, Circ. D86 - D92 (1974). − 1974 January - July.

044.030 Astronomische Zeit- und Breitenbestimmungen, Empfangszeiten von Zeitsignalen, Präzisionszeitvergleiche.
Akad. Wiss. DDR, Zentralinst. Phys. Erde, Abt. Geod. Astron., Jahrgang 1973, Nos. 3 - 4 (1974). − 1973 May–August.

044.031 Note sur la liaison transatlantique par Loran-C.

M. Granveaud, B. Mazodier.
Ann. Françaises Chronométrie Micromécanique, 8ᵉ année, (see 012.027), 144 - 146 (1973).

The generation of TAI (Temps Atomique International) is carried out by using the Loran-C links between seven laboratories. The autocovariance coefficients of the various pairs of local atomic times are computed. Some anomalies relative to the atlantic link are made apparent.

044.032 **The historical acceleration of the earth.**
R. R. Newton.
Geophys. Survey, (*Netherlands*), Vol. 1, No. 2, p. 123 - 145 (1973).

The measurement of frequency and time interval. See Abstr. 003.041.

Kants Beitrag zur Frage der Verzögerung der Erdrotation. See Abstr. 004.056.

Comparaisons de phase avec émission Loran-C et variation de l'index de réfraction. See Abstr. 035.005.

Das Nationale Zeitnormal der Physikalisch-Technischen Bundesanstalt. See Abstr. 035.007.

Résultats des observations faites à Alger avec l'astrolabe impersonnel A. Danjon OPL 8. Temps et latitude 1972. See Abstr. 041.020.

Tensor of inertia of the atmosphere, annual variations of its components and variations of the earth's rotation. See Abstr. 082.077.

Rotation of solid bodies in the solar system. See Abstr. 091.001.

045 Latitude Determination, Polar Motion

045.001 **On the regularity of motions of the pole.**
H. J. M. Abraham.
Proc. Astron. Soc. Australia, Vol. 2, 203 (1973). — Presented at the annual general meeting of the Astron. Soc. Australia — see 012.001.

045.002 **Determination of the latitude from observations of stars symmetric to the prime vertical.**
V. V. Kirichuk.
Geod., kartogr. i aehrofotosemka. Resp. mezhved. nauch.-tekhn. sb., 1973, vyp. (No.) 18, p. 120 - 122. In Russian.
Abstr. in Referativ. Zhurn. 51. Astron., 3.51.180; 52. Geod. Aehrosemka, 3.52.92 (1974).

045.003 **Palaeomagnetic results and late Precambrian glaciations.**
M. W. McElhinny, J. W. Giddings, B. J. J. Embleton.
Nature, Vol. 248, 557 - 561 (1974).

A combined late Precambrian to early Palaeozoic polar wander path, which differs from that previously proposed for Africa and South America, is derived for Gondwanaland. The widespread distribution of late Precambrian glaciations can be explained in terms of this polar migration.

045.004 **Chandler wobble and viscosity in the earth's core.**
J. Verhoogen.
Nature, Vol. 249, 334 - 335 (1974).

Jeffreys, and Munk and MacDonald, have stated that viscosity of the core cannot account for the observed damping of the Chandler wobble. This statement may be in error.

045.005 **Resonance effects in polar motion measurable by radio interferometry and laser ranging.**
M. A. Graber.
Journ. Geophys. Res., Vol. 79, 1709 - 1710 (1974).

The resonant effect of the liquid core on polar motion can be measured with the greatest probability of success by determining the resonance-induced amplification of the luni-solar nutations with frequencies near resonance compared to the calculated rigid body amplitudes. The changes in amplitude of the nutations are expected to be less than 0.15 arc sec. The techniques of radio interferometry and laser ranging would allow the observation of these amplitudes.

045.006 **Results of latitude observations for the year 1971 with VZT.** E. Proverbio, S. Uras.
Circ. Stazione Astron. Internazionale Latitudine, Carloforte−Cagliari, Ser. B (3), No. 3, 195 pp. (1972).

045.007 **Results of latitude observations for the year 1972 with VZT.** E. Proverbio, S. Uras.
Circ. Stazione Astron. Internazionale Latitudine, Carloforte−Cagliari, Ser. B (4), No. 4, 239 pp. (1973).

045.008 **Statistical analysis of latitude observations in 1972.** E. Proverbio, S. Uras.
Circ. Stazione Astron. Internazionale Latitudine, Carloforte−Cagliari, Ser. B (6), No. 8, 46 pp. (1973).

The purpose of this paper is to determine the amounts of different random errors: instrumental, personal and refractional, that characterize the latitude observations obtained by means of a conventional visual zenith telescope at the Carloforte International Astronomical Station.

045.009 **Velocity and direction of plate displacements by latitude observations.**
E. Proverbio, V. Quesada.
Ann. Geofis., Vol. 26, 289 - 302 = Pubbl. Stazione Astron. Internazionale Latitudine, Carloforte-Cagliari, Nuova Ser., No. 27 (1973).

The results obtained by the authors (1972), providing completely independent evidence on the plate tectonics hypothesis, are discussed. The choice of the rotational pole position of the plates is carefully analyzed.

045.010 **Some features of the "wind effect" at night and day-time latitude observations.**
N. A. Popov, O. V. Chuprunova.
Vrashchenie i prilivn. deformatsii Zemli. Vyp. (No.) 5. Kiev, Nauk. dumka, 1973, p. 106 - 111. In Russian. — Abstr. in Referativ. Zhurn. 51. Astron., 5.51.185 (1974).

045.011 **Zemes polu svārstības.** L. Rihlova.
Zvaigžņotā debess, 1973./74. gada ziema, p. 1 - 4.

045.012 **Il polo d'inerzia iniziale (1900.0–1907.0).**
E. Fichera.
Rend. Acad. Sci. Fis. Mat. Soc. Nazionale Sci., Lettere ed Arti Napoli, Ser. 4, Vol. 40, 58 - 66 = Ist. Univ. Navale Napoli, Astron. Generale e Sferica, Pubbl. No. 8 (1973).

045.013 **On the "anomalous polar motion".**
N. Sekiguchi, F. Miyamoto.
Tokyo Astron. Obs., Report No. 63, Vol. 16, 907 - 934 (1974). In Japanese.

045.014 **The secular variation of longitudes and plate tectonic motion.** E. Proverbio, V. Quesada.
Bull. Géod., Nouvelle Sér., Année 1974, No. 112, p. 187 - 212 (1974).

045.015 **What does change the Chandler motion?**
S. Takagi.
Proc. symposium "Celestial mechanics", Kyoto 1973, (see 012.022), p. 72 - 73 (1974). In Japanese.

045.016 **Coordinates of the earth's instantaneous pole from 1825.0 to 1897.9.** Z. Li, H. Xu, B. Lin.
Acta Astron. Sinica, Vol. 15, 86 - 92 (1974). In Chinese.

045.017 **Refractional pairs of the International Latitude Service and meteorological elements.**
G. Teleki.
Bull. Obs. Astron. Belgrade, Vol. 29, F. 1, No. 125, p. 9 - 12 (1971/72).

045.018 **Monthly Notes of the International Polar Motion Service.**
IPMS Monthly Notes, International Latitude Obs. Mizusawa (Japan). 1973 Nos. 11 - 12, p. 87 - 102; 1974 Nos. 1 - 4, p. 1 - 31 (1974). — Announces the values of latitudes observed at the collaborating stations during 1973 November - December; 1974 January - April.

045.019 **Coordonnées du pôle instantané rapportées à l'origine conventionnelle internationale et corrections de longitude TU1 –TU0, à 0h TU.**
Bureau International de l'Heure, (B. I. H.), Paris, Circ. B/C Nos. 213 - 219 (1974). — 1974 January - July. — Valeurs interpolées et extrapolées.

The Ulugh Begh International Latitude Station in Kitab. See Abstr. 009.016.

Résultats des observations faites à Alger avec l'astrolabe impersonnel A. Danjon OPL 8. Temps et latitude 1972. See Abstr. 041.020.

The secular motion of the earth's poles and its influence on astronomical coordinates. See Abstr. 041.021.

Time and latitude service. See Abstr. 044.022.

Astronomische Zeit- und Breitenbestimmungen. Empfangszeiten von Zeitsignalen. 1973 July - September. See Abstr. 044.023.

International Time and Latitude Service at the Tokyo Astronomical Observatory during 1973. See Abstr. 044.025.

Détermination astronomique de l'heure et de la latitude. See Abstr. 044.026.

Astronomische Zeit- und Breitenbestimmungen, Empfangszeiten von Zeitsignalen, Präzisionszeitvergleiche. See Abstr. 044.030.

Geographical coordinates and their variability. See Abstr. 046.016.

Dynamic techniques for studies of secular variations in position from ranging to satellites. See Abstr. 046.028.

046 Astronomical Geodesy, Satellite Geodesy, Navigation

046.001 **National mapping's astro-geodetic complex.**
M. McK. Luck, M. J. Miller, P. J. Morgan.
Proc. Astron. Soc. Australia, Vol. 2, 203 - 206 (1973). – Presented at the annual general meeting of the Astron. Soc. Australia – see 012.001.

046.002 **Astronomische Ortsbestimmung mit Hilfe des Computers.** M. Frick, M. Henkel.
Orion Schaffhausen, 32. Jahrgang, p. 12 - 13 (1974).

046.003 **Astronomische Ortsbestimmung bei Ortswechsel.**
M. Frick.
Orion Schaffhausen, 32. Jahrgang, p. 70 - 71 (1974).

046.004 **Precision clocks as a basis for a navigation system.**
S. S. D. Jones.
Journ. Navigation London, Vol. 27, 33 - 40 (1974).

046.005 **Polaris and all that.** N. Goodman.
Journ. British Astron. Ass., Vol. 84, 96 - 122 (1974). Presidential address.

046.006 **Modèle de la composition des erreurs de mesurages géodésiques.** M. K. Szacherska.
Geod. i Kartografia, Vol. 23, 21 - 51 (1974). In Polish.

046.007 **Satellite geodesy with cameras and lasers.**
D. G. King-Hele.
Endeavour, No. 118, Vol. 33, 3 - 10 (1974).

The launching of satellites gave a great impetus to geodesy, one of the oldest of the sciences: during the 1960s the shape of the earth became known correct to a few metres, thanks largely to camera observations of satellites. With the advent of laser measurements, now accurate to 50 centimetres and perhaps soon to 10 centimetres, a previously undreamt-of precision will be attained and geodesy will in the 1970s be well placed to make a strong impact on the other earth sciences.

046.008 **Quelques applications du cercle de simultanéité dans la triangulation par satellites.**
C. Popovici, M. Cîrşmaru.
Stud. Cerc. Astron., Vol. 19, 3 - 15 (1974). In Romanian.

046.009 **Certaines applications des observations simultanées optiques et Doppler dans la géodésie géométrique spatiale.** A. Dinescu.
Stud. Cerc. Astron., Vol. 19, 17 - 24 (1974).

046.010 **Errors of integration in the variational equations for the simple layer model in satellite geodesy.**
H. Fröhlich, K. R. Koch.
Mitt. Inst. Theor. Geod., Univ. Bonn, No. 25, 3 + 33 pp.(1974). In German.

For the representation of the earth's gravity field in satellite geodesy by the potential of a simple layer the derivatives of the satellite positions with respect to the unknown density values of the simple layer are needed. They are obtained by the integration of the variational equations. To compute their coefficients, integrals over surface elements into which the surface of the earth is divided are solved numerically. The errors of this integration in case of a spherical earth are investigated in the variational equations.

046.011 **Analyses of precision reduced optical observations from the International Satellite Geodesy Experiment (ISAGEX).**
J. G. Marsh, B. C. Douglas, S. M. Klosko.
Goddard Space Flight Center, Greenbelt, Maryland, GSFC Document X-590-73-340, 6 + 18 pp. (1973).

The authors have previously reported the results of a dynamical solution which combined ISAGEX laser data with the optical and laser tracking data recorded during the National Geodetic Satellite Program (NGSP) and the CNES/SAO 1968 Observing Program for the simultaneous recovery of co-ordinates for over 70 tracking stations (Marsh, Douglas, Klosko, 1973). The present paper represents a continuation of the analyses of the ISAGEX data using the optical data presently available.

046.012 **Geometrical geodesy techniques in Goddard earth models.** F. J. Lerch.
Goddard Space Flight Center, Greenbelt, Maryland, GSFC Document X-592-73-339, Preprint, 5 + 35 pp. (1974). – Presented at the International Association of Geodesy Symposium on 'Computational methods in geometrical geodesy', Oxford, United Kingdom, September 2–8, 1973.

The method for combining geometrical data with satellite dynamical and gravimetry data for the solution of geopotential and station location parameters is discussed. A linear regression model is formulated for combining the data, based upon the statistical technique of weighted least squares. Reduced normal equations, independent of satellite and instrumental parameters, are derived for the solution of the geodetic parameters. Exterior standards for the evaluation of the solution and for the scale of the earth's figure are discussed.

046.013 **The use of ABC tables in preparing for star observations.** P. A. Thompson.
Journ. Navigation, London, Vol. 27, 259 - 261 (1974).

046.014 **Determination of astronomical co-ordinates in the trigonometric points of the Italian triangulation. Position lines method.** E. Fichera, A. Pugliano.
Ist. Univ. Navale Napoli, Astron. Generale e Sferica, Pubbl. No. 7, 20 pp. (1973). In Italian.

046.015 **Observations of the latitude of the Moscow Observatory made on the ZTL-180 during the period 1958 - 1963.** V. V. Nesterov, Yu. I. Prodan.
Trudy Gos. Astron. Inst. Shternberga, Vol. 44, 147 - 184 (1973). In Russian.

The authors describe the series of the latitude observations. All observed latitudes (about 13 thousand) and the declination system are given.

046.016 **Geographical coordinates and their variability.**
V. P. Shcheglov.
Astronomical Institute of the Academy of Sciences of the Uzbek SSR – 100 years, (see 003.011), p. 22 - 42 (1974). In Russian.

046.017 **Datum shifts of the Adindan-African 12th parallel survey-datum.** G. Obenson.
Bull. Géod., Nouvelle Sér., Année 1974, No. 112, p. 145 - 148 (1974).

046.018 **On the search for reliable criteria of the accuracy of precise levelling based on statistical considerations of the discrepancies.** A. M. Wassef.
Bull. Géod., Nouvelle Sér., Année 1974, No. 112, p. 149 - 163 (1974).

046.019 **La mesure de la base-étalon de Mata das Virtudes.**
T. Honkasalo.

Inst. Geográfico Cadastral, *Lisboa*, Caderno Técnico Informação No. 33, 25 + 41 pp. (1973).

046.020 Coordinates of observatories.
IAU Circ., No. 2645 (1974).

046.021 Die Arbeiten des Sonderforschungsbereiches 78 Satellitengeodäsie der Technischen Universität München im Jahre 1973. Presented by R. Sigl.
Veröff. Bayer. Kommission Internationale Erdmessung, Bayer. Akad. Wiss., München. Astron.-Geod. Arbeiten, No. 32, 120 pp. (1974).
This publication gives the report on the two projects on satellite geodesy at the Technical University of Munich for 1973. Besides the reports on activities some papers on research work are presented. The following contributions are within the subject scope of Astronomy and Astrophysics Abstracts: Bestimmung einer Referenzbahn mit Hilfe der Mehrzielmethode, *G. Dichtl*, p. 29 - 31; Bestimmung von Referenzbahnen künstlicher Erdsatelliten als Randwertaufgabe, *K. H. Ilk*, p. 32 - 34; Bahnbestimmung und SHORT-ARC Verfahren zur Bestimmung von Stationskoordinaten, *H. Ludwig*, p. 42 - 44; Die Bezugssysteme der Satellitengeodäsie und ihre Relativbewegungen, *E. Nagel*, p. 45 - 47; Arbeiten zur Bahn- und Feldparameterbestimmung, *C. Reigber*, p. 48 - 50; Die Bestimmung der Koeffizienten der Sampling Funktionen, *H. F. Schmidt*, p. 51 - 59; Bestimmung periodischer Lösungen eines Bewegungsproblems mit Hilfe von Entwicklungen nach orthonormierten Eigenfunktionen, *M. Schneider*, p. 60 - 63; Numerische Erfahrungen bei der Berechnung von Bahnparametern für die räumliche Glättung bei Mini-Arcs, *W. Ehrnsperger*, p. 102 - 106.

046.022 Auswertung von Altimetermessungen für das Modell der einfachen Schicht des Erdschwerefeldes in der Satellitengeodäsie. W. Benning.
Deutsche Geod. Kommission, Bayer. Akad. Wiss. München, Ser. C, No. 200, 79 pp. (1974). — Diss. Rheinisch. Friedrich-Wilhelms-Univ. Bonn.

046.023 Astronomische Ortsbestimmung.
K. Bretterbauer.
Sternenbote, 17. Jahrgang, p. 86 - 96 (1974).

046.024 Méthodes d'élimination des déformations d'un image causées par la déformation du matériel photographique. A. Bujakiewicz.
Geodezja i Kartografia, Vol. 23, 141 - 155 (1974). In Polish.

046.025 Review of refraction effects of the atmosphere on geodetic measurements to celestial bodies.
C. H. Joshi, I. I. Mueller.
Publ. Obs. Astron. Beograd, No. 18, (see 012.023), p. 83 - 157 (1974).

046.026 SAO network: instrumentation and data reduction.
M. R. Pearlman, J. M. Thorp, C. R. H. Tsiang, D. A. Arnold, C. G. Lehr, J. Wohn.
Smithsonian Astrophys. Obs., *Cambridge, Mass.*, Special Rep. 353, (see 013.016), p. 13 - 84 (1973).
The SAO optical satellite-tracking network that supported the National Geodetic Satellite Program is described. Particular attention is given to the instrumentation of the lasers, the Baker-Nunn cameras, and the station timing systems in use during the program. Network operations and data-reduction techniques are also discussed, along with a history of network site locations.

046.027 Determination of station coordinates.
E. M. Gaposchkin, J. Latimer, G. Veis.
Smithsonian Astrophys. Obs., *Cambridge, Mass.*, Special Rep. 353, (see 013.016), p. 309 - 388 (1973).
The analysis of satellite data combined with surface measurements has resulted in the determination of the coordinates of 90 satellite-tracking sites.

046.028 Dynamic techniques for studies of secular variations in position from ranging to satellites.
D. E. Smith, R. Kolenkiewicz, R. W. Agreen, P. J. Dunn.
Goddard Space Flight Center, Greenbelt, Maryland, GSFC Document X-921-74-161, Preprint, 6 + 30 pp. (1974).

046.029 Astronomische Ortsbestimmungen in den Antarktisstationen Wostok und Mirny. J. Liebert.
Reprinted from Vermessungstechnik, 21. Jahrgang, p. 381 - 382 = Mitt. Zentralinst. Phys. Erde, *Potsdam*, No. 324 (1973).

046.030 Geodäsie und Physik der Erde. Aktuelle geowissenschaftliche Probleme im Mittelpunkt der Diskussionen eines internationalen Symposiums in Potsdam.
E. Buschmann.
Reprinted from Vermessungstechnik, 21. Jahrgang, p. 457 - 463 = Mitt. Zentralinst. Phys. Erde, *Potsdam*, No. 355 (1973).

1973 Smithsonian Standard Earth (III).
See Abstr. 013.016.

La distribution binomiale appliquée à l'examen des ensembles empiriques d'erreurs d'observations géodésiques.
See Abstr. 021.027.

Astronomisch-geodätische Lotabweichungen als Beitrag zu den geophysikalischen Untersuchungen im Nördlinger Ries. See Abstr. 105.104.

047 Ephemerides, Almanacs, Calendars

047.001 **Astronomiskais Kalendārs 1974. Gadam.**
M. Dīriķis (Editor).
Latvijas PSR Zinātņu Akadēmija; Radioastrofizikas Observatorija; Vissavienības Astronomijas un Ģeodēzijas Biedrības Latvijas Nodaļa. Izdevniecība "Zinātne", Riga. 172 pp. Price 38 Kop. (1973).

047.002 **Events of 1974 in the graphic time table.**
Sky Telescope, Vol. 47, 35 - 37 (1974).

047.003 **Kalendāra reforma.** I. Daube.
Zvaigžņotā debess, 1973. gada rudens, p. 37 - 38.

047.004 **Elementi astronomici per il calendario dell'anno 1974.**
Boll. Geod. Sci. Affini, Anno 33, p. 105 - 119 (1974). Calcolati dall'Osservatorio Astrofisico di Arcetri — Firenze.

047.005 **A thumbnail almanac for the moon.**
O. L. Harvey.
Sky Telescope, Vol. 47, 384 (1974).

047.006 **The Air Almanac 1974, September - December.**
Her Majesty's Stationery Office, London; United States Naval Observatory, Washington. 246 + A84 + F4 pp. Price £ 2.00 (1974).

047.007 **Anuário Astronômico 1974.**
Published by Instituto Astronômico e Geofísico, Universidade de São Paulo, São Paulo, Brasil. 11 + 112 + 181*+ 2 pp. (1973).

047.008 **Philippine Astronomical Handbook 1974.**
Prepared under the supervision of S. V. Inciong.
Republic of the Philippines — Department of National Defense Philippine Atmospheric, Geophysical and Astronomical Services Administration (Weather Bureau), Quezon City. 11 + 58 pp. (1973).

047.009 **The Astronomical Ephemeris for the year 1975.**
Issued by Her Majesty's Nautical Almanac Office, London; Nautical Almanac Office, United States, Naval Observatory, Washington. Her Majesty's Stationery Office, London. 8 + 568 pp. and an appendix with the physical ephemerides of Mars and Saturn, 9s pp. Price £ 3.50 net (1973).

047.010 **Connaissance des Temps ou des mouvements célestes pour l'an 1975 à l'usage des astronomes et des navigateurs.**
Publiée par le Bureau des Longitudes under the supervision of B. Morando.
Gauthier-Villars Éditeur, Paris. 42 + 493 + A145 pp. Price F 220.00 (1974).

047.011 **Interesting astronomical phenomena in 1974.**
D. Bonov.
Priroda (NRB), Vol. 22, No. 6, p. 3 - 6 (1973). In Bulgarian.

047.012 **Construction d'éphémérides condensées.**
A. Deprit.
Comptes Rendus Acad. Sci. Paris, Sér. B, Vol. 278, 1055 - 1057 (1974).
Les éphémérides traditionnelles de l'astronomie peuvent être représentées en séries de polynomes de Chebyshev qui sont l'approximation polynomiale la meilleure au sens de Chebyshev sur un intervalle de temps donné. Sous cette forme les éphémérides deviennent condensées: elles donnent lieu à des algorithmes simples et universels à programmer pour des mini-calculateurs.

047.013 **Astronomical Yearbook of the USSR for the year 1977.** V. K. Abalakin (Editor).
Institut Teoreticheskoj Astronomii Akademii Nauk SSSR. Izdatel'stvo "Nauka", Leningradskoe Otdelenie, Leningrad. 719 pp. Price 7 Rbl. 25 Kop. (1974). In Russian.

047.014 **Tables of Sunrise, Sunset, Twilight, Moonrise, and Moonset 1974.**
Prepared under the supervision of S. V. Inciong.
Republic of the Philippines — Department of National Defense, Philippine Atmospheric, Geophysical and Astronomical Services Administration (Weather Bureau), Quezon City. 11 + 57 pp. (1973).

047.015 **1974 wieder "Paradoxe Ostern".** H. Haupt.
Sternenbote, 17. Jahrgang, p. 66 - 68 (1974).

047.016 **Das Himmelsjahr 1974.** M. Gerstenberger.
Kosmos-Verlag, Stuttgart. 110 pp. Price DM 7.80 (1973).

047.017 **Almanaque Nautico 1975.**
Published by Instituto y Observatorio de Marina, San Fernando (Cádiz). Printed in Spain by Imprenta del Observatorio de Marina, San Fernando. 416 + 30 pp. (1974).

047.018 **An astronomical calendar for 1974.**
G. Ottewell.
Furman University Physics Dept., Greenville, S. C., 27 pp. Price $ 2.00 (1973).

047.019 **Nautical astronomical yearbook for 1974.**
I. D. Zhongolovich.
In-t teor. astron. AN SSSR. Gidrogr. upr. M-va oborony, Moskva. 444 pp. Price 4 Rbl. (1972). In Russian.

047.020 **Nautisches Jahrbuch oder Ephemeriden und Tafeln für das Jahr 1975,** zur Bestimmung der Zeit, Länge und Breite zur See nach astronomischen Beobachtungen.
Edited by "Deutsches Hydrographisches Institut", Hamburg. 124. Jahrgang, 3 + 43 + 365 + 30 pp. (1974).

Annuaire 1974 du Bureau des Longitudes.
See Abstr. 003.007.

Planetary perturbations in Newcomb's tables of the sun. See Abstr. 041.026.

Space Research

051 Extraterrestrial Research, Spaceflight Related to Astronomy and Astrophysics

051.001 Flight-time minimisation for an energy-limited flyby star probe. C. Powell.
Journ. British Interplanet. Soc., Vol. 27, 267 - 272 (1974).
Numerical results are given for a mission to Barnard's star. It is found that to achieve a flight time under 40 years, an exhaust velocity above 8000 km/sec and a mass/power ratio under 0.5 kg/MW are required.

051.002 Interstellar communication — antenna or artifact? A. T. Lawton.
Journ. British Interplanet. Soc., Vol. 27, 286 - 294 (1974).

051.003 The large space telescope. D. Baker.
Spaceflight, Vol. 16, 7 - 8 (1974).

051.004 American prospects in space. R. N. Watts, Jr.
Sky Telescope, Vol. 47, 308 - 310 (1974).

051.005 Outer space research — some results, problems, prospects. R. Z. Sagdeev, Yu. I. Zajtsev.
Priroda, No. 5.74, p. 4 - 13 (1974). In Russian.

051.006 Extra-atmospheric observations of the sun. V. A. Krat.
Priroda, No. 5.74, p. 34 - 41 (1974). In Russian.

051.007 On the feasibility of applying the methods of cybernetics to the detection of high-intelligence systems in the universe. B. N. Panovkin.
Izv. vyssh. ucheb. zavedenij. Radiofizika, Vol. 16, 1452 - 1454 (1973). In Russian. — Abstr. in Referativ. Zhurn. 51. Astron., 4.51.5 (1974).

051.008 New results of space research. A. A. Blagonravov, Yu. A. Surkov.
Session of COSPAR in the Federal Republic of Germany.
Vestn. AN SSSR, 1973, No. 12, p. 69 - 72. In Russian. Abstr. in Referativ. Zhurn. 62. Issled. kosmich. prostranstva, 5.62.7 (1974).

051.009 Some astrophysical and selenodetic results of the reduction of photographs from an automatic interplanetary station of the Zond series.
Yu. N. Lipskij, Zh. F. Rodionova, V. I. Chikmachev, M. F. Shabanov, V. V. Shevchenko.
Izv. vyssh. ucheb. zavedenij. Geod. i aehrofotosemka, 1973, No. 5, p. 53 - 63. In Russian. — Abstr. in Referativ. Zhurn. 62. Issled. kosmich. prostranstva, 5.62.163 (1974).

051.010 Kosmosa apgūšana.
Zvaigžņotā debess, 1973./74. gada ziema, p. 24 - 31. Jauns eksperiments kosmosā, p. 24; Atkal uz "sarkano" planētu (*I. Kovaļs*), p. 24 - 27; "Skylab" (*E. Mūkins*), p. 28 - 31.

051.011 Kosmosa apgūšana.
Zvaigžņotā debess, 1974. gada pavasaris, p. 28 - 33. Kosmonautika un zinātniski tehniskais progress, p. 28 - 31; "Skylab" (*E. Mūkins*), p. 31 - 33.

051.012 Cosmos and problems of public instruction. O. M. Belotserkovskij.
Zemlya i Vselennaya, 1974, No. 3, p. 50 - 56. In Russian.

051.013 A space potpourri. R. N. Watts, Jr.
Sky Telescope, Vol. 48, 25 - 26 (1974).

051.014 Principles and applications of astronautics. M. Moutsoulas.
In honorem S. Placidis, (see 003.009), p. 237 - 250 (1974). In Greek.

051.015 Soviet-American space experiment. K. D. Bushuev.
Vestn. AN SSSR, 1974, No. 1, p. 59 - 66. In Russian.
Abstr. in Referativ. Zhurn. 62. Issled. kosmich. prostranstva, 6.62.10 (1974).

051.016 Reduction methods for the scientific information of the Proton space probes. V. V. Akimov, V. V. Beletskij, V. V. Golubkov, G. N. Zlotin, S. I. Karmanov, I. N. Kiknadze, V. E. Nesterov, V. M. Pokras, V. L. Prokhin, I. D. Rapoport, I. G. Khatskevich.
Izuchenie kosmich. luchej na iskusstven. sputnikakh Zemli. Moskva, Nauka, 1973, p. 138 - 168. In Russian. — Abstr. in Referativ. Zhurn. 62. Issled. kosmich. prostranstva, 6.62.116 (1974).

051.017 Solar physics investigations on Skylab. E. M. Reeves.
Bull. American Astron. Soc., Vol. 6, 225 - 226 (1974). Abstr. AAS.

051.018 Astronautics in the year 1973. M. Grün, P. Koubský.
Říše hvězd, Vol. 55, 105 - 109 (1974). In Czech.

051.019 How to travel by kettle in space. A. Drożyner.
Urania Kraków, Vol. 45, 169 - 174 (1974). In Polish.

051.020 'Explorer-50' interplanetary monitoring platform.
Telecommun. Journ., (*Switzerland*), Vol. 41, No. 1, p. 10 - 13 (1974).

051.021 Large space vehicles — platforms for second generation in-situ wake observations. U. Samir.
Journ. Astronaut. Sci., Vol. 20, 347 - 369 (1973).
The use of large space vehicles such as space station, and space shuttles as platforms for wake observations is examined. Several experimental approaches are outlined and discussed. Examples of specific experiments that can be put on board the shuttle/module platform are presented and discussed.

051.022 Balloon flight of the COS-B experiment. B. G. Taylor.

ESRO/ELDO Bull., (*France*), No. 23, p. 9 - 11 (1973).

051.023 **EXOSAT and 'black holes'.** R. D. Andresen.
ESRO/ELDO Bull., (*France*), No. 23, p. 12 - 15 (1973).

051.024 **Scientific objectives of space exploration.** H. Mark.
IEEE Trans. Nuclear Sci., Vol. NS 21, No. 1, p. 34 - 37 (1974).

A general review of the work being done on: the solar wind and its interaction with the earth's and other planets' magnetic fields; lunar and planetary exploration; stellar and galactic exploration including radio and X-ray astronomy.

051.025 **Kosmosforschung heute, Marslandung als Forschungsunternehmen der 70er Jahre.**
A. L. Mangelsen.
Universitas, [Wiss. Verlagsgesellschaft, Stuttgart], Vol. 29, No. 1, p. 11 - 18 (1974).

051.026 **Astronautica.**
Coelum, Vol. 42, 31 - 35, 82 - 83 (1974).

051.027 **Space report.**
Journ. British Interplanet. Soc., Vol. 27, 73 - 79, 154 - 158, 232 - 235, 305 - 309, 389 - 396, 471 - 477, 548 - 557 (1974).

051.028 **Space report.**
Spaceflight, Vol. 16, 35 - 37, 65 - 68, 76, 105 - 110, 119, 149 - 151, 190 - 193, 195, 213 - 217, 219 (1974).
Earth benefits from Skylab, p. 35 - 36; Geological map from Soyuz, p. 65; Rare earth organism, p. 65 - 66; CETI signals, p. 66; 'Black hole' confirmed, p. 67; Close-up on Jupiter, p. 67 - 68; A galactic Sputnik? p. 105; Probe from Barnard's star, p. 105 - 106; Acid clouds of Venus, p. 106 - 107; Jupiter fly-by: first results, p. 107 - 108; CETI meeting in Copenhagen, p. 150; Radar probes Mercury, p. 150 - 151; Water molecules in Kohoutek, p. 151; Earth-like molecules from space, p. 151; CETI investigations in the U.S.S.R., p. 190 - 191; Orion 2 results, p. 190 - 191; Mariner reveals 'new Venus', p. 193, 195; Mars probes off target, p. 213 - 214; Large infrared telescope, p. 216 - 217; Pioneer–Venus spacecraft, p. 217; Film on Mars, p. 219.

The second fifteen years in space.
See Abstr. 003.042.

Soviets in space. See Abstr. 003.118.

La chronométrie dans l'espace.
See Abstr. 035.017.

052 Astrodynamics and Navigation of Space Vehicles

052.001 Sequential filter design for precision orbit determination and physical constant refinement.
D. W. Curkendall, C. T. Leondes.
Celestial Mechanics, Vol. 8, 481 - 494 (1974).

Earth-based spacecraft tracking data have historically been processed with classical least squares filtering techniques both for navigation purposes and for physical constant determination. In this paper, sequential estimation is motivated for application to deep space navigation and celestial mechanics experiment purposes and a computational structure which retains many of the advantages of the extant least squares techniques is discussed.

052.002 Dynamics of flexible space vehicles with active attitude control. P. C. Hughes.
Celestial Mechanics, Vol. 9, 21 - 39 (1974).

The implications of flexible appendages on the attitude dynamics of a space vehicle are examined in general terms. Two families of natural vibration modes, referred to as 'constrained' and 'unconstrained', are discussed and the relationships between them derived. The incorporation of either set of modes into a simulation of the general attitude motion (under the influence of perturbing torques and control torques) is explained. The influence of rotors on these results is also explored.

052.003 Special perturbations employing osculating reference states.
G. H. Born, E. J. Christensen, L. K. Seversike.
Celestial Mechanics, Vol. 9, 41 - 53 (1974).

The concept of employing osculating reference position and velocity vectors in the numerical integration of the equations of motion of a satellite is examined. The choice of the reference point is shown to have a significant effect upon numerical efficiency and the class of trajectories described by the differential equations of motion. A number of formulations, including a universal one, are considered. Comparisons of the numerical characteristics of these techniques with those of the Encke method are presented.

052.004 Regularization and the artificial earth satellite problem. D. G. Saari.
Celestial Mechanics, Vol. 9, 55 - 72 (1974).

Some of the properties of singularities of a system of differential equations, which includes several formulations of the artificial earth satellite problem, are derived. Using them, it is shown that this problem cannot be regularized by using the current commonly used ideas and definitions of regularization.

052.005 The gradient of the scalar field of variations of the kinematic parameters of Keplerian trajectories.
V. A. Modestov.
Kosmich. Issled., Vol. 12, 134 - 138 (1974). In Russian.
Brief information.

052.006 Precession in spin-orbit coupling. L. Blitzer.
Bull. American Astron. Soc., Vol. 6, 205 (1974).
Abstr. AAS.

052.007 Lunar perturbations of artificial satellites of the earth. G. E. O. Giacaglia.
Celestial Mechanics, Vol. 9, 239 - 267 (1974).

The author obtains the disturbing function for the lunar perturbations using ecliptic elements for the moon and equatorial elements for the satellite. Secular, long-period, and short-period perturbations are then computed, with the expressions kept in closed form in both inclination and eccentricity of the satellite. Alternative expressions for short-period perturbations of high satellites are also given, assuming small values of the eccentricity.

052.008 Graphisches Konstruktionsschema zur Bestimmung der Bewegungsparameter eines drallstabilisierten Flugkörpers mit Nutationsdämpfung unter dem Einfluß von Lagekorrekturimpulsen. K. Becker.
Celestial Mechanics, Vol. 9, 269 - 289 (1974).

It is shown that the motion of a spinning body with nutation damping due to a series of torque pulses can be completely derived from a regular polygon determined by the ratio of inertias, the integral of a single pulse, the momentum and a constant characterizing damping. Simple rules are shown for the application of the scheme on pulse frequencies which are multiples or fractions of spin frequency.

052.009 Methods of determination and prediction of the orbits of earth satellites in the presence of errors in the mathematical interpretation of their motion.
A. I. Nazarenko, L. G. Markova.
Prikl. zadachi kosmich. ballistiki. Moskva, Nauka, 1973, p. 36 - 68. In Russian. – Abstr. in Referativ. Zhurn. 51. Astron., 3.51.150 (1974).

052.010 Applied problems of cosmic ballistics.
AN SSSR. In-t kosmich. issled. Nauka, Moskva. 140 pp. Price 62 Kop. (1973). In Russian.

052.011 On coupled motion of space vehicles with solar sails. N. D. Dzhumanaliev, M. I. Kiselev.
Trudy Kirg. un-ta. Ser. fiz. n., 1973, vyp. (No.) 2, p. 143 - 147. In Russian. – Abstr. in Referativ. Zhurn. 62. Issled. kosmich. prostranstva, 3.62.316 (1974).

052.012 Calculation of precision satellite orbits with non-singular elements (VOP formulation).
C. E. Velez, P. J. Cefola, A. C. Long, K. S. Nimitz.
Proc. conference on numerical solution of ordinary differential equations, 1972, (see 012.005), p. 183 - 206 (1974).

052.013 Computation of solar perturbations with Poisson series. R. Broucke.
Proc. conference on numerical solution of ordinary differential equations, 1972, (see 012.005), p. 237 - 259 (1974).

052.014 The determination of the satellite orbit of Mariner 9.
G. H. Born, E. J. Christensen, A. J. Ferrari, J. F. Jordan, S. J. Reinbold.
Celestial Mechanics, Vol. 9, 395 - 414 (1974).

This paper presents a comprehensive analysis of the Mars orbital phase of the Mariner 9 trajectory as determined from earth based radio data. Both the method and accuracy of the orbit determination process are reviewed. Analysis is presented to show the effects of Mars gravity model and node in the plane of the sky errors on the accuracy of orbit determination. In addition the long term evolution of the orbit from insertion through the first 500 revolutions is presented, and decomposed into effects from the Mars gravity field, n-body perturbations, and solar radiation pressure.

052.015 Integrable cases in the problem of orbital evolution of a satellite under the joint influence of an outer body and the noncentral field of a planet.
M. L. Lidov, M. V. Yarskaya.
Kosmich. Issled., Vol. 12, 155 - 170 (1974). In Russian.

052.016 **On the validity of estimates of the parameters of motion of space objects.** P. E. Ehl'yasberg.
Kosmich. Issled., Vol. 12, 171 - 178 (1974). In Russian.

052.017 **On rotational motions of a free body relative to non-principal axes of inertia in a Newtonian gravitational field.** V. A. Grobov, N. M. Zavrazhina.
Prikl. mekhanika, Vol. 9, No. 11, p. 69 - 73 (1973). In Russian Abstr. in Referativ. Zhurn. 62. Issled. kosmich. prostranstva, 4.62.358 (1974).

052.018 **Über die Navigation von Raumsonden.**
H. F. Fliegel, D. A. O'Handley, J. W. Zielenbach.
SuW, Vol. 13, 197 - 200 (1974).

052.019 **Satellite artificiel. Solution approchée au premier ordre.** C. Delmas.
Anais Acad. Brasil. Ci., Vol. 44, 25 - 31 (1972).

052.020 **Radiation-pressure and air-drag effects on the orbit of the balloon satellite 1963 30D.** J. W. Slowey.
Smithsonian Astrophys. Obs., *Cambridge, Mass.*, Special Report 356, 9 + 93 pp. (1974).

Computed orbits of the balloon satellite 1963 30D are given every 2 days over an interval of 456 days near the beginning of the satellite's lifetime and an interval of 824 days near the end of its lifetime. The effects of radiation pressure on the satellite are examined in some detail. It is found that the variations in all the elements can be represented by use of a single parameter to specify the effect of diffuse reflection from the satellite's surface, and that this parameter remains constant, or nearly so, during the entire 7-year lifetime.

052.021 **Analytical study of the solar radiation influence upon the orbit of a satellite.** N. N. Motorina.
Izv. AN MoldSSR. Ser. fiz.-tekhn. i mat. n., 1973, No. 3, p. 33 - 40. In Russian. – Abstr. in Referativ. Zhurn. 51. Astron., 5.51.158 (1974).

052.022 **The motion of an electrically charged satellite in the central gravitational and magnetic fields of the earth.**
E. B. Bibik.
Probl. mekh. upravlyaemogo dvizheniya. Vyp. (No.) 3. Perm', 1973, p. 33 - 39. In Russian. – Abstr. in Referativ. Zhurn. 62. Issled. kosmich. prostranstva, 5.62.364 (1974).

052.023 **On the influence of light pressure on the motion of an artificial earth satellite.** N. N. Motorina.
Probl. mekh. upravlyaemogo dvizheniya. Vyp. (No.) 3. Perm', 1973, p. 143 - 147. In Russian. – Abstr. in Referativ. Zhurn. 62. Issled. kosmich. prostranstva, 5.62.365 (1974).

052.024 **Elementary theory of some perturbations of orbits of artificial satellites.** S. V. Izmajlov.
Teor. fiz. i astronomiya. Leningrad, 1973, p. 142 - 150. In Russian. – Abstr. in Referativ. Zhurn. 62. Issled. kosmich. prostranstva, 5.62.367 (1974).

052.025 **On the conservation of conditionally periodic motions in the disturbed classical problem of two fixed centres.** I. F. Kiselev.
Probl. mekh. upravlyaemogo dvizheniya. Vyp. (No.) 3. Perm', 1973, p. 111 - 115. In Russian. – Abstr. in Referativ. Zhurn. 62. Issled. kosmich. prostranstva, 5.62.368 (1974).

052.026 **Numerically analytical method for calculation of the evolution of near artificial earth satellites.**
A. A. Krasovskij, E. I. Bushuev, Eh. P. Kompaniets, A. I. Vasil'eva.
Kosmich. issledovaniya na Ukraine. Resp. mezhved. sb. 1973, vyp. (No.) 2, p. 83 - 91. In Russian. – Abstr. in Referativ. Zhurn. 62. Issled. kosmich. prostranstva, 5.62.372 (1974).

052.027 **Influence of the ellipticity of an orbit on the track.**
V. L. Kalachev.
Probl. mekh. upravlyaemogo dvizheniya. Vyp. (No.) 3. Perm', 1973, p. 97 - 103. In Russian. – Abstr. in Referativ. Zhurn. 62. Issled. kosmich. prostranstva, 5.62.373 (1974).

052.028 **Relativistic spinor regularization of the astrodynamical problem of two bodies.** P. E. Kustaanheimo.
Commentationes Phys.-Math., Vol. 44, No. 1, 5 pp. (1974).

The motion of two mass points in a three-dimensional space under their mutual Newtonian attraction is described in a Lorentz-invariant way by regularized harmonic oscillations of the two eigen-spinors of the relativistic distance motor between the mass points. Seven simple motor integrals of the spinor differential equations of the motion and four identities between these integrals are deduced.

052.029 **Appunti introduttivi alla navigazione spaziale. Parte prima. Analisi delle ipotesi Newtoniane.**
E. Fichera.
Ist. Univ. Navale, Astron. Generale e Sferica, Napoli. Seminario No. 4, 55 pp. (1973).

052.030 **Appunti introduttivi alla navigazione spaziale. Parte seconda. Lo sviluppo di Lagrange in astrodinamica.** E. Fichera.
Ist. Univ. Navale, Astron. Generale e Sferica, Napoli. Seminario No. 5, 55 pp. (1973).

052.031 **Determination of the orbit of an artificial earth satellite from measurements of the distance, radial velocity and angular coordinate.** Yu. S. Savrasov.
Kosmich. Issled., Vol. 12, 337 - 345 (1974). In Russian.

052.032 **Optimal motion of a space vehicle in the neighbourhood of a Keplerian ellipse with arbitrary eccentricity.** V. I. Lashkin.
Kosmich. Issled., Vol. 12, 353 - 359 (1974). In Russian.

052.033 **Some tasks of astrodynamics.** G. N. Duboshin.
Zemlya i Vselennaya, 1974, No. 3, 62 - 65. In Russian.

052.034 **Intermediate orbits of artificial lunar satellites.**
G. G. Koman.
Soobshch. Gos. Astron. Inst. Shternberga, No. 186, p. 3 - 45 (1973). In Russian.

052.035 **Elementary theory of some perturbations of orbits of artificial satellites.** S. V. Izmajlov.
Teor. fiz. i astronomiya. Leningrad, 1973, p. 142 - 150. In Russian. – Abstr. in Referativ. Zhurn. 51. Astron., 6.51.151 (1974).

052.036 **Equations of the perturbed motion of elliptical orbits with large eccentricities.** Eh. A. Borisov.
Izv. vyssh. ucheb. zavedenij. Geod. i aehrofotosemka, 1973, No. 6, p. 59 - 67. In Russian. – Abstr. in Referativ. Zhurn. 51. Astron., 6.51.152; 62. Issled. kosmich. prostranstva, 6.62.230 (1974).

052.037 **Determination of the orbits of artificial earth satellites.** L. Sekhnal.
Nauka i chelovechestvo. 1974. Moskva, Znanie, 1973, p. 221 - 227. In Russian. – Abstr. in Referativ. Zhurn. 51. Astron., 6.51.162; 62. Issled. kosmich. prostranstva, 6.62.231 (1974).

052.038 **On the accuracy of the determination of satellite orbits.** K. Šteins, L. Laucenieks.

Uch. Zap. Latv. Gos. Univ., Vol. 175, Astron., vyp. (No.) 8, (see 003.010), p. 3 - 16 (1973). In Russian.

052.039 **A method of applying a mobile barrier for predictions of satellite appearing.**
K. Šteins, L. Laucenieks.
Uch. Zap. Latv. Gos. Univ., Vol. 175, Astron., vyp. (No.) 8, (see 003.010), p. 17 - 37 (1973). In Russian.

052.040 **About the calculation of normal and synchronous positions of satellites.** J. Žagars.
Uch. Zap. Latv. Gos. Univ., Vol. 202, Astron., vyp. (No.) 10, (see 003.012), p. 68 - 74 (1974). In Russian.

052.041 **Numerical averaging in orbit prediction.**
C. Uphoff.
AIAA Journ., Vol. 11, 1512 - 1516 (1973).

This paper is a presentation of some useful aspects of a numerical averaging technique that has been applied with considerable success to the orbit prediction problem for artificial satellites. A particularly useful set of averaged equations is presented along with a description of an efficient algorithm for their solution. Examples of the advantages to be obtained by use of the method are included.

052.042 **Integrals for optimal flight over a spherical earth.**
H. G. Moyer.
AIAA Journ., Vol. 11, 1441 - 1443 (1973).

Integrals of the Euler-Lagrange differential equations that govern the optimal trajectories of a glider, airplane, or rocket obeying point dynamics are found.

052.043 **Mission design and navigation for a 1977–1978 Venus swingby/Mercury orbiter.**
S. K. Asnin, D. G. Roos.
Journ. Spacecraft and Rockets, (*USA*), Vol. 10, 631 - 637 (1973).

Analyses characteristics of the interplanetary trajectory and determines midcourse delta velocity (ΔV) requirements. The paper also examines constraints which influence the design of the orbiter phase of the mission, proposes an orbit selection rationale, and defines a strategy for mapping the Mercury surface.

052.044 **Round trip Mars missions using looping trajectories in the 1980–2000 time period.** J. F. Kibler.
Journ. Spacecraft and Rockets, (*USA*), Vol. 10, 686 - 688 (1973).

052.045 **Satellite dynamics.** E. M. Gaposchkin.
Smithsonian Astrophys. Obs., *Cambridge, Mass.*, Special Rep. 353, (see 013.016), p. 85 - 192 (1973).

The perturbations of an artificial close-earth satellite are developed in analytical form. Gravitational perturbations due to the geopotential, the sun, the moon, the body tide, and the ocean tides are treated; and nongravitational perturbations due to atmospheric drag and radiation pressure are developed. Also discussed are applications of the development for orbit determination and computation.

052.046 **The ROAD program.**
C. A. Wagner, B. C. Douglas, R. G. Williamson.
Goddard Space Flight Center, Greenbelt, Maryland, GSFC Document X-921-74-144, 5 + 39 pp. (1974).

The philosophy, history, operation, calibration of and some analyses with the ROAD (Rapid Orbit Analysis and Determination) program are described. This semi-numeric trajectory program integrates and analyses mean element variations for earth orbits with great efficiency. Through its use, extensive zonal, resonant harmonic and earth tidal determinations have been made at Goddard Space Flight Center since 1969.

Use of Green's functions in the numerical solution of two-point boundary value problems. See Abstr. 042.035.

Multirevolution methods for orbit integration. See Abstr. 042.036.

Modified Jacobi polynomial and its applications to expansions of disturbing functions. See Abstr. 042.053.

Bestimmungsgleichungen für Resonanzparameter der Ordnung 13 aus der Analyse von Bahnen der Satelliten GEOS B, BEC und D1D. See Abstr. 081.029.

Errata

052.901 **Errata: 'Effects of gravity-gradient torque on the rotational motion of a triaxial satellite in a precessing elliptic orbit'** [Celestial Mechanics, Vol. 6, 127 - 150 (1972)]. J. E. Cochran.
Celestial Mechanics, Vol. 9, 534 (1974).

053 Lunar and Planetary Probes and Satellites

053.001 **News of Mariner 10 and Skylab.** R. N. Watts, Jr.
Sky Telescope, Vol. 47, 19 - 23 (1974).

053.002 **Pioneer 10 mission to Jupiter.** P. H. Abelson.
Science, Vol. 183, 261 (1974).

053.003 **Pioneer 10.** C. F. Hall.
Science, Vol. 183, 301 - 302 (1974). − Report.

053.004 **Analysis of the transfer trajectories using Jupiter's gravitational field.**
B. Ts. Bakhshiyan, N. G. Khavenson, P. E. Ehl'yasberg.
Prikl. zadachi kosmich. ballistiki. Moskva, Nauka, 1973, p. 3 - 10. In Russian. − Abstr. in Referativ. Zhurn. 51. Astron., 3.51.154; 62. Issled. kosmich. prostranstva, 3.62.322 (1974).

053.005 **Titan as a gravitational brake.**
V. Vinge, A. Sorkin.
Journ. British Interplanet. Soc., Vol. 27, 129 - 131 (1974).

It is demonstrated that launch opportunities exist later in this decade for a Saturn orbiter mission using Titan's gravitational field in place of an on-board retro-rocket. The feasibility of gravitational braking for other outer planet missions is briefly considered.

053.006 **Pioneer 10 passes Jupiter.** R. N. Watts, Jr.
Sky Telescope, Vol. 47, 23 (1974).

053.007 **Experimental flight "Apollo − Soyuz".**
A. A. Leonov.
Zemlya i Vselennaya, 1974, No. 2, p. 13 - 17. In Russian.

053.008 **Die Sonnensonde Helios und ihre Experimente.**
C. Leinert, H. Link, E. Pitz.
SuW, Vol. 13, 86 - 91 (1974).

053.009 **Children of the dawn. Part 2. The case for asteroidal missions.** A. D. Farmer.
Spaceflight, Vol. 16, 9 - 19, 37 (1974).

053.010 **The 1973 Mariner mission to Venus and Mercury.**
A. J. Hooke.
Spaceflight, Vol. 16, 25 - 34, 46 - 54 (1974).

053.011 **Report from Jupiter. Part 1.** D. Baker.
Spaceflight, Vol. 16, 140 - 144 (1974).

053.012 **Schroot op de maan: 38 toestellen.** T. de Vries.
Zenit, Vol. 1, No. 1, p. 34 - 35 (1974).

053.013 **Pioneer 10 Jupiter al ver voorbij.** T. de Vries.
Zenit, Vol. 1, No. 4, p. 20 (1974).

053.014 **Mariner 10 Venus encounter.** J. A. Dunne.
Science, Vol. 183, 1289 - 1291 (1974).

The Mariner 10 spacecraft encountered Venus at 1701 G.M.T. on 5 February 1974. The preplanned encounter science sequence was executed satisfactorily, accomplishing all objectives despite a number of spacecraft problems that had occurred in the early phases of the flight. Seven experiments were conducted, including observations of the solar wind interaction region, extreme ultraviolet and infrared emissions, radio occultation, and imaging.

053.015 **The off-line data processing system for the solar probe Helios A from the experimenters stand-point.**
I. Mistrik.
Journ. British Interplanet. Soc., Vol. 27, 401 - 409 (1974).

A description of the objectives of the solar probe Helios-A is given.

053.016 **Perturbations of Pioneer 6 telemetry signal during solar occultation.** A. A. Chastel, J. F. Heyvaerts.
Nature, Vol. 249, 21 - 22 (1974).

Attention has been drawn recently to unexplained perturbations in the telemetry signal of Pioneer 6 (2,300 MHz) during solar occultation. Various explanations of those phenomena have been attempted and are discussed here. It is shown that low frequency waves are the only ones that may give rise to the perturbations. The authors emphasize that if such processes are responsible for the perturbations of the signal, it is simple to explain why these perturbations are correlated with solar flares.

053.017 **Averaging the probability of contact of lunar stations.**
G. V. Alferov.
Probl. mekh. upravlyaemogo dvizheniya. Vyp. (No.) 3. Perm', 1973, p. 10 - 17. In Russian. − Abstr. in Referativ. Zhurn. 62. Issled. kosmich. prostranstva, 5.62.375 (1974).

053.018 **Starts of space vehicles in the USSR (January–April 1973).** M. S. Dimitrijević.
Vasiona, Vol. 22, 11 - 12 (1974). In Serbo-Croatian.

053.019 **'Mariner-10'.**
Telecommun. Journ., (Switzerland), Vol. 41, No. 1, p. 14 - 18 (1974).

The spacecraft launched on Nov. 3, 1973 for exploration of Mercury is described with reference to its construction, communications, tracking and attitude control, power supply and scientific objectives.

053.020 **The Apollo-Soyuz programme.** H. W. Kohler.
Techn. Rundschau, (Switzerland), Vol. 66, No. 10, p. 41 - 45 (1974). In German.

Australia's Skylark to coordinate with Skylab.
IEEE Spectrum, Vol. 11, No. 1, p. 20 (1974).

Around the solar system in 1,800 days.
Nature, Vol. 248, 544 (1974).

Russia's Mars probes.
Nature, Vol. 248, 632 (1974).

Pioneer 11 retargeted.
Sky Telescope, Vol. 47, 287, 307 (1974).

Pioneer 11 dichter langs Jupiter.
Zenit, Vol. 1, No. 4, p. 20 (1974).

Television photometry: the Mariner 9 experience.
See Abstr. 031.024.

Intermediate orbits of artificial lunar satellites.
See Abstr. 052.034.

Probabilities for the peak flux and fluence of energetic solar protons incident on interplanetary spacecraft.
See Abstr. 078.043.

The planet Mercury as viewed by Mariner 10.
See Abstr. 092.020.

Mariner 10 Mercury encounter.
See Abstr. 092.033.

The flyby of Jupiter.   See Abstr. 099.010.

Pioneer 10 am Jupiter vorbeigeflogen. Erste Ergebnisse.   See Abstr. 099.034.

054   Artificial Earth Satellites

054.001 **Soyuz 13 manned mission.**   R. N. Watts, Jr.
Sky Telescope, Vol. 47, 83 (1974).

054.002 **Skylab mission completed.**   R. N. Watts, Jr.
Sky Telescope, Vol. 47, 227 - 230 (1974).

054.003 **On some peculiarities of the application of a one vector algorithm for determination of the orientation of artificial earth satellites.**   V. N. Zigunov, N. E. Kalenov, A. E. Ordanovich, A. A. Rostovskaya.
Kosmich. Issled., Vol. 12, 45 - 54 (1974). In Russian.

054.004 **Time base of the satellite laser radar at Ondřejov Observatory.**   P. Navara.
Bull. Astron. Inst. Czechoslovakia, Vol. 25, 46 - 51 (1974). In Russian.

The measuring of the distances of earth satellites with an accuracy better than ±10 m, as in the case of the satellite laser radar, requires that the time of measurement be known with an accuracy better than ±1 ms. The main contents of the paper is the description of how the time base of the satellite laser radar was determined at the Ondřejov Observatory with an accuracy relative to the UTC better than ±100 µs.

054.005 **The orbit of 1963-30D from 1968 to 1971.**
A. N. Winterbottom, D. G. King-Hele.
Journ. British Interplanet. Soc., Vol. 27, 132 - 143 (1974).

054.006 **Frequency spectrum of the image vibrations of artificial earth satellites.**
M. V. Bratijchuk, I. I. Motrunich, V. P. Ryabov, I. V. Shvalagin.
Astrometriya i Astrofizika, Kiev, vyp. (No.) 20, (see 003.002), p. 47 - 54 (1973). In Russian.

054.007 **Großer Erfolg der 2. bemannten Skylab-Mission.**
H. W. Köhler.
SuW, Vol. 13, 5 - 8 (1974).

054.008 **The evolution of Skylab.**   D. Dooling.
Spaceflight, Vol. 16, 20 - 24 (1974).

054.009 **Skylab: 59 days in space. Part 1, 2.**   D. Baker.
Spaceflight, Vol. 16, 55 - 62, 170 - 173 (1974).

054.010 **Analytical estimates of the accuracy of determination of the parameters of motion of artificial earth satellites in the reduction of unequally accurate measurements** with correlation errors.   L. F. Porfir'ev, V. V. Smirnov.
Kosmich. Issled., Vol. 12, 299 - 301 (1974). In Russian. Brief information.

054.011 **Photométrie photoélectrique du satellite Echo 2.**
L. Neužil.
Bull. Astron. Inst. Czechoslovakia, Vol. 25, 75 - 78 (1974).

Sur les courbes photométriques d'éclipses du satellite artificiel Echo 2 l'auteur a trouvé l'influence probable de la couche absorbante élevée.

054.012 **The fourth flight of the Soviet stratospheric solar observatory.**
V. A. Krat, L. Z. Dul'kin, M. A. Validov, I. Ya. Vakhrameev, V. N. Karpinskij, Yu. S. Muzalevskij, R. P. Nikolaev, B. A. Protsenko, V. M. Sobolev, Z. N. Tabakova, Yu. L. Shakhbazyan.
Astron. Tsirk., No. 807, p. 1 - 3 (1974). In Russian.

054.013 **An investigation of the accuracy of positional observations of distant artificial space objects by means of television methods.**
T. A. Guseva, R. I. Kiladze, G. D. Matveev, A. Sh. Khatisov.
Byull. Abastumansk. Astrofiz. Obs., No. 45, p. 145 - 158 (1974). In Russian.

Faintness and great angular velocity of distant artificial space objects require television methods to be applied for measuring their spherical coordinates. Televisual method (direct reading) and telephotography (from television screen) are most frequently used. Accuracy of these methods applied at Abastumani Astrophysical Observatory has been investigated.

054.014 **The errors of measurements of artificial earth satellite tracks recorded on films.**   A. G. Krylov.
Byull. Inst. Astrofiz., Dushanbe, No. 62, p. 33 - 36 (1974). In Russian.

054.015 **The micrometeoroid detector aboard the satellite Prospero.**   D. K. Bedford, H. W. Bryan.
Journ. British Interplanet. Soc., Vol. 27, 459 - 464 (1974).

054.016 **Skylab: 59 days in space. Part three.**
D. Baker.
Spaceflight, Vol. 16, 206 - 209 (1974).

054.017 **Cálculo de efemérides y previsiones de pasos de satélites geodésicos.**   R. Parra, M. J. Sevilla.
Revista Real Acad. Ci. Exactas, Fís. Nat., Madrid, Vol. 67,

505 - 538 = Univ. Complutense Madrid, Fac. Ci., Seminario Astron. Geod., Publ. No. 74 (1973).

054.018 **Orientation of scientific instruments of the Prognoz artificial earth satellites. Method and apparatus.**
V. S. Aleksandrov, S. I. Babichenko, S. I. Karmanov, I. P. Karpinskij, A. I. Sigunov, A. E. Stefanovich.
Kosmich. Issled., Vol. 12, 440 - 446 (1974). In Russian.

054.019 **Satellites artificiels: Observations de périodes photométriques, 1971–1973.** J. Meeus.
Ciel et Terre, Vol. 90, 201 - 214 (1974).

054.020 **Determination of the boundaries of shadow parts of artificial satellite orbits in the case of the conical earth shadow. II.** R. P. Eremenko, E. N. Polyakhova.
Trudy Astron. Obs., *Leningrad*, Vol. 30 (= Uchenye Zapiski Leningr. Un-ta, No. 373 = Seriya Matem. Nauk, vyp. (No.) 50), p. 156 - 164 (1974). In Russian.

The algorithm for determination of the coefficients in the canonical shadow equation and for the exact solution of this equation is given for the case of an artificial earth satellite. The algorithm is presented as ALGOL-60 procedure. The procedure permits to determine the polar coordinates (radius-vector and true anomaly) of the boundary points of the shadow part of the orbit.

054.021 **Some results from Skylab's solar experiments.**
Sky Telescope, Vol. 48, 11 - 17 (1974).

054.022 **Introduzione storica al progetto EXOSAT (Helos).** C. Dilworth.
Mem. Soc. Astron. Italiana, Nuova Ser., Vol. 44, 481 - 482 (1973/74).

054.023 **Il progetto «EXOSAT» dell'Organizzazione Europea per la Ricerca Spaziale (E.S.R.O.).** V. Manno.
Mem. Soc. Astron. Italiana, Nuova Ser., Vol. 44, 483 - 505 (1973/74).

054.024 **Skylab – the first space laboratory.** T. Kwast.
Urania Kraków, Vol. 45, 72 - 75 (1974). In Polish.

054.025 **Attitude control for the Netherlands astronomical satellite (ANS).** P. van Otterloo.
Philips Techn. Rev., (*Netherlands*), Vol. 33, No. 6, p. 162 - 176 (1973).

054.026 **Satellite digest.** Compiled by G. Falworth.
Spaceflight, Vol. 16, 38, 75 - 76, 119, 152 - 153, 194 - 195, 218 - 219 (1974). — A monthly listing of all known artificial satellites and spacecraft.

NASA plans 26 space launches during 1974; all will be unmanned.
IEEE Spectrum, Vol. 11, No. 2, p. 22 (1974).

News of Mariner 10 and Skylab.
See Abstr. 053.001.

055 Observations of Earth Satellites, Lunar and Planetary Probes

055.001 **Optical observations of distant artificial cosmic objects.**
E. S. Agapov, V. F. Anisimov, P. P. Dobronravin, V. M. Mozhzherin, V. K. Prokof'ev, N. S. Chernykh.
Priroda, No. 2.74, p. 10 - 17 (1974). In Russian.

055.002 **Simultaneous observation solutions for NASA–MOTS and SPEOPT station positions on the North American Datum.** J. S. Reece, J. G. Marsh.
Bull. Géod., Nouvelle Sér., Année 1974, No. 111, p. 53 - 71.
Simultaneous observations of the GEOS-I and II flashing lamps by the NASA MOTS and SPEOPT cameras on the North American Datum (NAD) have been analyzed using geometrical techniques to provide an adjustment for the station coordinates. Two separate adjustments have been obtained.

055.003 **Poljot 1 – 1963-43-1. Visual observations. Equatorial coordinates (1950.0) July - October 1971.**
Rezul'taty Nablyud. Iskusstv. Sputnikov Zemli, vyp. (No.) 14 (154), 69 pp. (1973). In Russian.

055.004 **Explorer 32 – 1966-44-1. Poljot 1 – 1963-43-1. Visual observations. Equatorial coordinates (1950.0) February - November 1972; April - December 1972.**
Rezul'taty Nablyud. Iskusstv. Sputnikov Zemli, vyp. (No.) 15 (155), 70 pp. (1973). In Russian.

055.005 **Poljot 1 – 1963-43-1. Explorer 19 – 1963-53-1. Visual observations. Horizontal coordinates. March - November 1972; March - October 1972.**
Rezul'taty Nablyud. Iskusstv. Sputnikov Zemli, vyp. (No.) 16 (156), 56 pp. (1973). In Russian.

055.006 **Explorer 19 – 1963-53-1. Explorer 32 – 1966-44-1. Oreol 1971-119-1. Visual observations. Horizontal coordinates March - November 1972.**
Rezul'taty Nablyud. Iskusstv. Sputnikov Zemli, vyp. (No.) 17 (157), 55 pp. (1973). In Russian.

055.007 **Explorer 19 – 1963-53-1. Explorer 39 – 1968-66-1. Oreol 1 – 1971-119-1. Visual observations. Equatorial coordinates (1950.0) February - March 1972; February - September 1972; March - December 1972.**
Rezul'taty Nablyud. Iskusstv. Sputnikov Zemli, vyp. (No.) 18 (158), 62 pp. (1973). In Russian.

055.008 **Explorer 19 – 1963-53-1. Visual observations. Equatorial coordinates (1950.0) April - December 1972.**
Rezul'taty Nablyud. Iskusstv. Sputnikov Zemli, vyp. (No.) 19 (159), 60 pp. (1973). In Russian.

055.009 **Explorer 19 – 1963-53-1. Visual observations. Horizontal coordinates (January - June 1973).**
Rezul'taty Nablyud. Iskusstv. Sputnikov Zemli, vyp. (No.) 20 (160), 60 pp. (1973). In Russian.

055.010 **On the accuracy of photographic observations at the Riga satellite tracking station.** L. Laucenieks.
Uch. Zap. Latv. Gos. Univ., Vol. 175, Astron., vyp. (No.) 8, (see 003.010), p. 62 - 71 (1973). In Russian.

055.011 **Observations of artificial earth satellites.**
Yu. M. Ivanov, A. Kadyrov.
Astronomical Institute of the Academy of Sciences of the Uzbek SSR – 100 years, (see 003.011), p. 139 - 140 (1974). In Russian.

055.012 **Visual observations of artificial earth satellites in Finland 1973.**
Prepared under the supervision of A. Tuominen, with an introduction by P. Järvi.
Observations of Satellites, No. 14, (published by the Finnish Meteorological Institute, Helsinki, Finland), 10 + 121 pp. (1974).

055.013 **Artificial space objects (Vyp. (No.) 8). 1971, Jan. 1 - 1971, Dec. 31.**
G. A. Lejkin (Editor), compiled by V. V. Bazykin.
Nablyud. Iskusstv. Nebesn. Tel, No. 64, 84 pp. (1973). In Russian.

055.014 **Photographische Positionsbestimmung künstlicher Erdsatelliten mit einer Tracking-Kamera.**
K.-H. Marek.
Akad. Wiss. DDR, Forschungsbereich Kosmische Phys., Veröff. Zentralinst. Phys. Erde, No. 15, 133 pp. (1973).
The present paper deals with the theoretical extension of the approximation of topocentric satellite orbits by small circles and leads this principle to a technological ability. The theoretical and practical foundation of the observation technique for satellite-observation devices of the 2nd generation are developed. They have been confirmed experimentally during the test of camera SBG, the recently most modern device of this kind.

055.015 **Die Genauigkeit von Satellitenbeobachtungen mit der Kamera SBG.** L. Stange.
Reprinted from Vermessungstechnik, 20. Jahrgang, No. 12 = Mitt. Zentralinst. Phys. Erde, *Potsdam*, No. 279, 2 pp. (1972).

Theoretical Astrophysics

061 General Theoretical Problems of Astrophysics, Gravitational Instability, Neutrino Astronomy, Infrared, X-Ray, Gamma-Ray Astronomy, Abundances and Origin of Elements

061.001 **Absorption of infrared radiation by electrons in the field of a neutral hydrogen atom.** J. R. Stallcop.
Astrophys. Journ., Vol. 187, 179 - 183 (1974).

The determination of the absorption of infrared radiation by free-free transitions of the negative hydrogen ion has been extended to higher temperatures. A simple analytical expression for the absorption coefficient has been derived; it is accurate for a wide range of high temperatures.

061.002 **Free-free absorption of infrared radiation in collisions of electrons with neutral rare-gas atoms.**
J. R. Stallcop.
Astron. Astrophys., Vol. 30, 293 - 296 (1974).

A relationship between the inverse bremsstrahlung absorption cross section and the electron neutral momentum transfer cross section has been utilized to determine the infrared free-free continuum absorption coefficient for the negative ions of helium, neon, argon, krypton, and xenon. Analytical expressions for the absorption coefficient have also been developed. From the results of this calculation, one can determine the absorption coefficient per unit electron density per neutral atom for temperatures in the range $2500°K - 25000°K$. The results are compared with those from tabulations of previous calculations and those computed from theoretical values of the phase shifts for the elastic scattering of electrons by neutral atoms.

061.003 **Fluctuations in the X-ray background.**
P. A. G. Scheuer.
Monthly Notices Roy. Astron. Soc., Vol. 166, 329 - 337 (1974).

The fluctuations in the X-ray background at high galactic latitudes give an upper limit to the counts of extragalactic X-ray sources near the intensity level corresponding to one source per beam width. The general mathematical relation between the source counts and the probability distribution of the resulting contribution to the fluctuations is known, and is here computed for various extrapolations of the observed source counts to fainter sources. The observed limits to background fluctuations are consistent with a uniform distribution of sources in space.

061.004 **Nucleosynthesis in supernova outbursts and the chemical composition of the envelopes of neutron stars.** G. S. Bisnovatyj-Kogan, V. M. Chechetkin.
Astrophys. Space Sci., Vol. 26, 3 - 24, 25 - 46 (1974). In Russian and English.

The formation of chemical elements in the envelopes of neutron stars is considered at the densities $\varrho = 10^7$ to 10^{13} g cm^{-3}. The formation of the heavy nuclei in supernovae explosions is considered briefly. Rough estimates are made for the differences in chemical composition of ejected matter during the explosions of stars of different masses and supernovae of different types.

061.005 **Similarity solution for unsteady accretion flow.**
S. Sakashita.
Astrophys. Space Sci., Vol. 26, 183 - 188 (1974).

Similarity solution for unsteady accretion flow in a gravitional field of a point mass is obtained. Characteristic features of the flow pattern are discussed. It is shown that shock waves appeared in the accretion flow propagate outward as $r_s \propto t^{2/3}$.

061.006 **Induced pair production and opacity due to blackbody radiation.**
S. A. Bonometto, F. Lucchin, P. Marcolungo.
Astron. Astrophys., Vol. 31, 41 - 46 (1974).

The authors present a general scheme for studying the interaction between very high energy particles and black-body radiation. The case of electromagnetic interactions is studied in detail, and the fundamental role of the reaction $\gamma e \rightarrow 3e$ (induced pair production) as a limiting factor for the transparency of the metagalaxy to very high energy photons and electrons is stressed.

061.007 **Production of superheavy nuclei by multiple capture of neutrons.**
W. M. Howard, J. R. Nix.
Nature, Vol. 247, 17 - 20 (1974).

A new calculation of fission barriers and neutron separation energies for heavy neutron-rich nuclei indicates that it is unlikely that superheavy nuclei can be produced either in the astrophysical r process or in man-made nuclear explosions.

061.008 **Heavy ion transfer reactions in nuclear astrophysical processes.** T. W. Conlon.
Nature, Vol. 247, 268 - 269, with corrections, Vol. 249, 96, 786 (1974).

The purpose of this note is to identify for the heavy ion systems involved which reactions are likely to have a non-negligible rate.

061.009 **Astrophysical fuel-coolant interactions.**
D. E. T. F. Ashby, D. J. Buchanan, R. S. Peckover.
Nature, Vol. 247, 272 - 273 (1973).

If a hot liquid comes into contact with a cold vaporisable liquid a violent explosion may occur; such interactions are often called fuel-coolant interactions (FCIs). In this letter the authors apply FCI theory to the problem of matter and antimatter interacting on an astronomical scale.

061.010 **γ-ray observations from the OSO-3 satellite.**
G. D. Badhwar, M. F. Kaplon, D. A. Valentine.
Astrophys. Space Sci., Vol. 27, 147 - 155 (1974).

The result on γ-rays obtained from the analysis of 5800 orbits of data from the University of Rochester telescope on board the OSO-3 satellite are presented. For γ-rays of energy greater than 100 MeV, an upper limit of 2.3×10^{-4} cm^{-2} s^{-1} std has been placed on the diffuse (assumed isotropic) flux. An upper limit to the flux from the sun is set at 3.2×10^{-5} and 2.4×10^{-5} cm^{-2} s^{-1} for energies greater than 50 MeV and 100 MeV, respectively.

061.011 **A new rocket measurement of the diffuse X-ray background.**
E. Horstman-Moretti, F. Fuligni, H. M. Horstman, D. Brini.
Astrophys. Space Sci., Vol. 27, 195 - 201 (1974).

A rocket measurement of the X-ray cosmic diffuse background is presented. The data were obtained with detectors of different apertures. The result is compared to a previous rocket measurement where a screened detector had been used to evaluate the instrumental background. Results are in substantial agreement.

061.012 **Variations in the primordial helium abundance.**
G. R. Gisler, E. R. Harrison, M. J. Rees.
Monthly Notices Roy. Astron. Soc., Vol. 166, 663 - 671 (1974).

The authors consider how the cosmic helium abundance might be affected by primordial irregularities. It is shown that temperature fluctuations cannot account for variations of Y in the manner proposed by Silk & Shapiro (1971) and Silk (1972). The authors then show that curvature variations, which are associated with the temperature fluctuations and were ignored by Silk & Shapiro, are in general capable of producing a modest variation in the cosmic helium abundance.

061.013 **The escape of energetic charged particles from trapping magnetic structures.** G. A. Stevens.
Astron. Astrophys., Vol. 31, 157 - 163 (1974).

The diffusive motion of charged particles in a magnetic field is derived from Liouville's equation supplemented with a term describing a relaxation to momentum isotropy. The accessory relaxation time is estimated for collisional scattering and for scattering due to magnetic field fluctuations. It is concluded that the most probable escape mechanism of energetic protons from the sun is a direct injection into interplanetary space from an open magnetic configuration.

061.014 **Some applications of the relationship between emissivity and absorptivity of a medium.**
V. V. Sobolev.
Astrofizika, Vol. 9, 515 - 524 (1973). In Russian. English translation in Astrophysics, Vol. 9, No. 4.

The relationship between emissivity and absorptivity of a medium which has been found earlier (1973) is applied to three particular cases: 1) semi-infinite medium, 2) plane layer, 3) homogeneous sphere. In each case the emissivity is given in terms of absorptivity, and vice versa.

061.015 **Cosmic far ultraviolet background.**
A. Davidsen, S. Bowyer, M. Lampton.
Nature, Vol. 247, 513 - 516 (1974).

Observations of a diffuse ultraviolet background at high galactic latitude might reveal the existence of an intergalactic medium. Although existing results provide only weak constraints, current techniques are sufficient to set interesting limits on its density and temperature.

061.016 **The age of the elements.** D. N. Schramm.
Sci. American, Vol. 230, No. 1, p. 69 - 77 (1974).

Study of the formation and spontaneous decay of radioactive nuclei can reveal when certain of the elements were created. From this information one can infer the age of the universe.

061.017 **On the formation and unfolding of pulse height distributions.** P. Hoyng, G. A. Stevens.
Astrophys. Space Sci., Vol. 27, 307 - 321 (1974).

The present paper deals with the question how one can reconstruct original photon or particle spectra from measured pulse height distributions. Finally, a numerical example is given.

061.018 **Some problems of physics and astrophysics.**
V. L. Ginzburg.
Fizika segodnya i zavtra. Leningrad, Nauka, 1973, p. 5 - 60. In Russian. − Abstr. in Referativ. Zhurn. 51. Astron., 3.51.68 (1974).

061.019 **Neutrinos in the universe.** T. de Graaf.
Vistas in astronomy, Vol. 15, (see 003.001), 161 - 181 (1973).

061.020 **Origin of the chemical elements and stellar evolution.** R. L. Sears.
Journ. Roy. Astron. Soc. Canada, Vol. 68, 1 - 12 (1974).
Presented at a symposium on "Chemical evolution in the universe", organized by the Royal Society of Canada, Queen's University, Kingston, Ontario, June 6, 1973.

061.021 **The equation of state of matter at sub-nuclear density.** J. W. Negele.
IAU Symposium No. 53, (see 012.004), p. 1 - 25 (1974).

061.022 **Equation of state at densities greater than nuclear density.** H. A. Bethe.
IAU Symposium No. 53, (see 012.004), p. 27 - 46 (1974).

061.023 **Variational method for dense systems.**
V. R. Pandharipande.
IAU Symposium No. 53, (see 012.004), p. 47 - 65 (1974).

061.024 **'Statistical bootstrap' equation of state for cold ultra-dense matter.** J. C. Wheeler.
IAU Symposium No. 53, (see 012.004), p. 77 - 91 (1974).

061.025 **A simple equation of state of matter at super-nuclear densities.** Y. C. Leung, C. G. Wang.
IAU Symposium No. 53, (see 012.004), p. 93 - 104 (1974).

061.026 **Pions in neutron star matter.** R. F. Sawyer.
IAU Symposium No. 53, (see 012.004), p. 105 - 110 (1974).

061.027 **Hypercollapsed nuclear matter.** Y. Ne'eman.
IAU Symposium No. 53, (see 012.004), p. 111 - 115 (1974).

061.028 **Phase diagram of a charged Bose gas.**
J. P. Hansen, B. Jancovici, D. Schiff.
IAU Symposium No. 53, (see 012.004), p. 991 (1974). Abstract.

061.029 **Superluminal sound and ferromagnetic transition in the Zeldovich model.** G. Kalman, S. T. Lai.
IAU Symposium No. 53, (see 012.004), p. 169 - 182 (1974).

061.030 **Powerful γ-ray bursts − a new astronomical discovery.**
O. F. Prilutskij, I. L. Rozental, V. V. Usov.
Priroda, No. 3.74, p. 93 - 95 (1974). In Russian.

061.031 **Flash-up of cosmic γ radiation from observations aboard AES Cosmos 461.**
E. P. Mazets, S. V. Golenetskij, V. N. Il'inskij.
Priroda, No. 3.74, p. 96 - 97 (1974). In Russian.

061.032 **Multiple inverse Compton scattering and the diffuse X-ray component.** A. Bui-Van, K. Hurley.
Astrophys. Journ., (Letters), Vol. 188, L51 - L53 (1974).

It is shown that, when multiple scattering is taken into account the spectrum of inverse Compton produced X-radiation is different from the single-scatter spectrum. In particular, the γ-ray fluxes in the 1–10 MeV region can increase by a factor of about 3 over the single-scatter spectrum. The significance of this result with respect to the diffuse cosmic X-ray flux is discussed.

061.033 Diffuse cosmic gamma-radiation above 10 MeV.
G. H. Share, R. L. Kinzer, N. Seeman.
Astrophys. Journ., Vol. 187, 511 - 519 (1974).

An investigation of the diffuse component of cosmic γ-radiation above 10 MeV was performed with an emulsion spark-chamber telescope carried to a depth of 2.5 g cm^{-2} at a cutoff rigidity of ~11.5 GV. An upper limit to the integral flux above 10 MeV was obtained from an extrapolation to the top of the atmosphere. Upper limits to the extraterrestrial intensity were also derived from a measurement of the γ-ray spectrum at 2.5 g cm^{-2}. These and other measurements raise questions concerning the existence of an additional component of diffuse γ-rays suggested above 1 MeV.

061.034 Cosmic gamma-ray bursts from relativistic dust grains. J. E. Grindlay, G. G. Fazio.
Astrophys. Journ., (Letters), Vol. 187, L93 - L96 (1974).

A cosmic-ray theory, in which all the observed energy originally resides in one "particle," is developed to account for the principal observed features of cosmic γ-ray bursts. In particular, two burst events that could be associated with the 1969 glitch of the Crab pulsar are discussed.

061.035 Cosmic gamma-ray bursts from directed stellar flares. K. Brecher, P. Morrison.
Astrophys. Journ., (Letters), Vol. 187, L97 - L99 (1974).

The authors suggest that the recently observed nonsolar γ-ray bursts arise in directed stellar flares, hypothetical events akin to giant transient coronal streamers. Directed beams of inverse-Compton scattered photons could produce the time structure, spectrum, and intensity observed.

061.036 Kosmische Magnetfelder. M. Reinhardt.
Naturwissenschaften, 61. Jahrgang, p. 143 - 149 (1974).

After a general review of cosmic magnetic fields, there is a discussion of the methods used in astronomy to observe magnetic fields, with a short catalogue of detected cosmic magnetic fields. Finally, the two theories of the origin of these fields are considered: dynamo-mechanisms, and the primordial hypothesis.

061.037 Possible cosmic dust origin of terrestrial plutonium-244. K. Sakamoto.
Nature, Vol. 248, 130 - 132 (1974). – Letter.

061.038 Stoten gammastraling uit het heelal.
C. de Jager.
Zenit, Vol. 1, No. 3, p. 13 - 17 (1974).

061.039 Non-linear Compton magneto-bremsstrahlung.
P. Stewart.
Astron. Astrophys., Vol. 32, 13 - 15 (1974).

The spectrum radiated by a charged cloud is calculated for the case of a pulse of intense circularly polarized electromagnetic radiation which is switched on adiabatically and which propagates parallel to a magnetostatic field; the cloud is initially at rest. The results are applied to the Crab nebula.

061.040 Thick-target measurement of the (p, γ) stellar reaction rates on the nuclides ^{12}C, ^{29}Si, ^{46}Ti, ^{47}Ti, and ^{56}Fe.
N. A. Roughton, M. J. Fritts, R. J. Peterson, C. S. Zaidins, C. J. Hansen.
Astrophys. Journ., Vol. 188, 595 - 601 (1974).

Stellar reaction rates for (p, γ) on ^{12}C, ^{29}Si, ^{46}Ti, ^{47}Ti, and ^{56}Fe have been obtained from thick-target measurements. The experimental and theoretical techniques are discussed, and a comparison is made to stellar rates derived from a statistical nuclear model. The capture reaction on ^{56}Fe is discussed in some detail because of the presence of a strong nonstatistical state – an isobaric analog resonance – in the energy range of interest. Some discussion is included regarding differences between experimental and theoretical cross-sections.

061.041 Neutrino reactions in strong magnetic field.
V. Canuto, C. Chiuderi, C. K. Chou, L. Fassio-Canuto.
Astrophys. Space Sci., Vol. 28, 145 - 161 (1974).

The authors present the energy losses due to several neutrinos processes: (1) synchrotron neutrinos, (2) pair annihilation neutrinos, (3) plasmon neutrinos, and (4) photoneutrinos in the presence of a superstrong magnetic field. Numerical results are tabulated and illustrated for several values of densities and temperatures.

061.042 On resonant interactions of atmospheric waves.
K. B. Dysthe, C. Jurén, L. Stenflo.
Phys. Scripta, Vol. 9, 226 - 228 (1974).

A nonlinear interaction between acoustic- and internal gravity waves in an inhomogeneous atmosphere is considered. Due to wave motion from a dense to a less dense medium, we conclude that the importance of the non-linear effect is increasing with height. The coupling coefficients are calculated, and the problem concerning resonance conditions is considered. It is shown that a transfer of energy between the waves can take place more rapidly than in the case of a wave interaction in a homogeneous medium.

061.043 Electron-hydrogen photoattachment as a source of ultraviolet absorption. G. W. F. Drake.
Astrophys. Journ., Vol. 189, 161 - 163 (1974).

The absorption coefficient is calculated for the photoattachment process in which a hydrogen atom, an electron, and a photon combine to form the $2p^2\,^3P$ state of H$^-$. The process is surprisingly efficient and should be considered as a source of continuous ultraviolet absorption in stellar atmospheres or laboratory plasmas with high electron densities.

061.044 L'astronomie infrarouge. J.-C. Pecker.
L'Astronomie, 88e année, p. 57 - 75 (1974).

061.045 Highly charged ions in astrophysics.
A. H. Gabriel.
Phys. Scripta, Vol. 9, 306 - 312 (1974).

The study of highly ionized atoms in astrophysical sources is reviewed with particular reference to the spectroscopy of the sun. The range of spectroscopic observations is surveyed in terms of the general properties of the solar atmosphere. Such spectra contain important information on the physical properties of the emitting regions. Some of the methods are described by which spectral intensities can be related, through the processes of ionization recombination and excitation, to the temperature and density structure of the source.

061.046 Ferromagnetismus in Sternen hoher Dichte.
H. P. Gail, J. Schmid-Burgk.
Mitt. Astron. Ges., No. 34, p. 80 (1973/74). – Abstr. AG.

061.047 Gibt es auch stationäre $\alpha\omega$-Dynamos?
W. Deinzer, H.-U. v. Kusserow, M. Stix.
Mitt. Astron. Ges., No. 34, p. 155 - 158 (1973/74).
Presented at the "Wissenschaftliche Tagung der Astron. Ges., Oberkochen, 1973 April 24 - 27".

061.048 Molecular hydrogen in X-ray astronomy.

B. Margon.
Nature, Vol. 249, 24 - 25 (1974).

There is a serious discrepancy between determinations of the column density of interstellar matter in the direction of the Crab nebula. Virtually every soft X-ray observation with sufficient energy resolution to derive this column density yields a result substantially in excess of determinations based on 21-cm absorption. It seems likely that this situation also applies to many or all galactic X-ray sources. The author suggests that interstellar molecular hydrogen is responsible for a significant component of the necessary opacity.

061.049 **Multiplex methods and advantages in X-ray astronomy.** T. M. Palmieri.
Astrophys. Space Sci., Vol. 28, 277 - 287 (1974).

The multiplex advantage is defined and applied in considering various techniques presently used in X-ray astronomy. It is concluded that the multiplex advantage will be useful in evaluating future techniques to be used in X-ray astronomy.

061.050 **Interpretation of double structure in the celestial γ-ray bursts.** J. V. Jelley.
Nature, Vol. 249, 747 - 748 (1974).

The author makes the suggestion that the sources of the γ-ray bursts may be identified with the collapse of a rotating magnetised star, following a model of the type computed by LeBlanc and Wilson (1970). The main objective of this suggestion is that if acceptable theoretically it could readily account for the double pulses.

061.051 **Theories of the geomagnetic and solar dynamos.** D. Gubbins.
Rev. Geophys. Space Phys., Vol. 12, 137 - 154 (1974).

A review of recent advances in dynamo theory is presented in relation to the problems of the generation of the earth's and the sun's magnetic field. Some relevant modern observations and estimates of physical quantities are discussed, and the lack of knowledge about the dynamical state of the earth's core is emphasized. Most attention is given to the low-conductivity limit because this allows many of the theories to be exhibited in their simplest form. Dynamical dynamo models, driven either by an applied body force or by thermal convection, are described.

061.052 **Instabilities and nonlinear processes in geophysics and astrophysics.** A. Hasegawa.
Rev. Geophys. Space Phys., Vol. 12, 273 - 280 (1974).

A review of some nonlinear effects related to plasma instabilities is presented. Discussions are included on (1) quasi-linear diffusion, in which a comparison between the effects of electromagnetic and electrostatic turbulence is given; (2) anomalous resistivity, in which a general derivation of the anomalous resistivity is presented and compared with some concrete examples; (3) wave-wave and wave-particle interactions, in which a general theory of parametric instabilities and quenching is presented together with the effects of the nonlinear Landau damping on stabilization of linear instabilities; and (4) modulational instability, in which a general derivation of the instability conditions and the consequence of the instability are discussed.

061.053 **Outburst of cosmic gamma-radiation according to observations aboard the artificial earth satellite Cosmos 461.**
E. P. Mazets, S. V. Golenetskij, V. N. Il'inskij.
Pis'ma v ZhurnEhTF, Vol. 19, 126 - 128 (1974). In Russian. Abstr. in Referativ. Zhurn. 62. Issled. kosmich. prostranstva, 5.62.200 (1974).

061.054 **Die diffuse kosmische Gamma-Strahlung.**
V. Schönfelder.
Naturwissenschaften, 61. Jahrgang, p. 250 - 256 (1974).

Observation of diffuse cosmic gamma radiation can provide information that relates to much earlier ages of our universe than that obtainable from studies of other regions of the electromagnetic spectrum. The first part of this work briefly describes the experimental methods used and the measurements obtained; the second half discusses the origin of the diffuse gamma radiation.

061.055 **Om grunnstoffenes opprinnelse og alder.**
Ø. Hauge.
Naturen 1973, No. 3, p. 107 - 117 = Inst. Teor. Astrofys., Blindern – Oslo, Småtrykk No. 79 (1973).

061.056 **Gamma rays from black holes.**
G. H. Dahlbacka, G. F. Chapline, T. A. Weaver.
Nature, Vol. 250, 36 - 37 (1974).

As a result of electron ion decoupling, the compression of interstellar matter falling onto an isolated black hole results in ion temperatures greater than 100 MeV as the matter approaches the Schwarzschild radius. Here, experimental pion production cross sections are used to calculate the rate of production of γ rays from this hot gas. A characteristic γ-ray spectrum is produced, which peaks at 18 MeV regardless of the mass of the black hole or the interstellar density.

061.057 **X-ray transition radiation formed in molecular clouds.**
G. G. Bakhshyan, G. M. Garibyan, C. Yang.
Astrofizika, Vol. 10, 93 - 98 (1974). In Russian. – English translation in Astrophysics, Vol. 10, No. 1.

The X-ray transition radiation formed by ultrarelativistic charged particles on gas molecules is investigated. A comparison of this radiation with the bremsstrahlung shows that there is a frequency region in which the cross section of transition radiation is larger than that of bremsstrahlung. The X-ray transition radiation formed by cosmic electrons in a molecular cloud of the type of Ori-A is calculated.

061.058 **Search for celestial X-ray bursts.**
T. L. Cline, U. D. Desai.
Astrophys. Space Sci. Library, Vol. 45, (see 012.015), 261 - 265 (1974).

061.059 **Gamma-ray astronomy: the last observational frontier.** J. Gribbin.
Nature, Vol. 250, 92 - 93 (1974).

A recent symposium provided an insight not only into the present status of gamma-ray astronomy, but also into the conventions of scientific symposia.

061.060 **Polytropic sheets, cylinders and spheres with negative index.** Y. P. Viala, G. Horedt.
Astron. Astrophys., Vol. 33, 195 - 202 (1974).

The principal properties of plane-symmetric, cylindrical and spherical polytropes with negative polytropic index are derived.

061.061 **Infrared astronomy.** J. Borgman.
Stars and the Milky Way system, Proc. 1972, (see 012.018), p. 188 - 208 (1974). – General lecture.

061.062 **Infrared techniques.** M. J. Smyth.
Stars and the Milky Way system, Proc. 1972, (see 012.018), p. 232 - 242 (1974). – Invited lecture.

061.063 **Low energy X-ray experiments on the third Small Astronomy Satellite.**
D. R. Hearn, H. V. Bradt, G. W. Clark, W. H. G. Lewin, H. W. Schnopper, S. Rappaport, W. F. Mayer, J. E. McClintock, J. A. Richardson.

Bull. American Astron. Soc., Vol. 6, 272 (1974). – Abstr. AAS.

061.064 **High energy γ-ray astronomy with SAS-II.**
D. A. Kniffen.
Bull. American Astron. Soc., Vol. 6, 272 - 273 (1974). Abstr. AAS.

061.065 **The importance of spallation induced radioactivity for X- and gamma-ray astronomy.** C. S. Dyer.
Bull. American Astron. Soc., Vol. 6, 275 (1974). – Abstr. AAS.

061.066 **The multiplex advantage in X-ray astronomy.**
T. M. Palmieri.
Bull. American Astron. Soc., Vol. 6, 277 (1974). – Abstr. AAS.

061.067 **The Fokker-Planck approximation to charged-particle transport in an irregular magnetic field.**
J. R. Jokipii.
Bull. American Astron. Soc., Vol. 6, 277 (1974). – Abstr. AAS.

061.068 **Shock induced nucleosynthesis of deuterium and other light elements.**
R. I. Epstein, D. N. Schramm. W. D. Arnett.
Bull. American Astron. Soc., Vol. 6, 281 (1974). – Abstr. AAS.

061.069 **Panoramica dello stato attuale dell'astronomia X.**
G. C. Perola.
Mem. Soc. Astron. Italiana, Nuova Ser., Vol. 44, 539 - 552 (1973/74).

061.070 **Osservazioni nel campo della gamma astronomia.**
G. Sironi.
Mem. Soc. Astron. Italiana, Nuova Ser., Vol. 44, 617 - 623 (1973/74).

061.071 **Zur Kosmochemie von Kohlenstoff-14.**
R. Boeckl.
Analyse extraterrestrischen Materials, (see 012.020), p. 75 - 83 (1974).

061.072 **Die Bedeutung der Spurenelemente für die Kosmochemie.** W. Kiesl.
Analyse extraterrestrischen Materials, (see 012.020), p. 109 - 123 (1974).

061.073 **Neutrinos and stellar structure.** V. Vanýsek.
Vesmír, Vol. 53, 16 - 17 (1974). In Czech.

061.074 **Bootstrapped electron pair production in extragalactic sources.** S. A. Bonometto, F. Lucchin.
Nuovo Cimento B, Ser. 11, Vol. 18B, 225 - 236 (1973).
Possible observational peculiarities due to the occurrence of electron pair production in extra-galactic sources are studied; the case of steep spectra is outlined; a possible enhancement of the pair production reaction due to its very occurrence is also proven for a set of models built on the basis of the data observed for 3C 273 and NGC 1275.

061.075 **High energy astronomy.** J. Labeyrie.
Rheinisch-Westfael. Akad. Wiss. Vorträge, No. 229, p. 39 - 79 (1973). In French.
A review of developments, mainly in the 1960's and to date, covering: observational techniques; principle results already obtained on solar particles; composition and origin of galactic cosmic particles; cosmic electrons; cosmic X- and γ-rays.

061.076 **Neutrino radiation in spherically-symmetric gravitational fields. I. The energy-momentum tensor for the one-neutrino and many-particle neutrino field.**
J. B. Griffiths.
General Relativity and Gravitation, (GB), Vol. 4, 361 - 370 (1973).
The physical situation of a star emitting neutrinos is considered. Some difficulties in the classical theory are mentioned, and a more detailed approach to the properties of neutrino radiation in general relativity is given.

061.077 **Problems and achievements of nuclear astrophysics. III. Slow neutron capture in stars.** B. Kuchowicz.
Postępy Fiz., Vol. 24, 669 - 690 (1973). In Polish.
Recent ideas on the mechanism of nucleosynthesis of the majority of heavy nuclides are outlined. Arguments based on theory and on correlations between neutron capture cross sections and nuclidic abundances point in favour of slow neutron capture as the mechanism which is responsible for the production of the chemical elements between the iron group and bismuth.

061.078 **Astrophysical upper bounds on neutrino-nucleon cross-section at energy $E \geqslant 3 \times 10^{17}$ eV.**
V. S. Berezinsky, A. Yu. Smirnov.
Phys. Letters B, (Netherlands), Vol. 48B, 269 - 272 (1974).
The possibility of experiments with cosmic neutrinos of energy $E > 3 \times 10^{17}$ eV is discussed. An upper bound on neutrino-nucleon cross-section is obtained from the zenith angular distribution of extensive air showers under some particular astrophysical assumptions.

061.079 **High temperature nuclear astrophysics.**
W. A. Fowler.
Quarterly Journ. Roy. Astron. Soc., Vol. 15, 82 - 106 (1974).
The George Darwin Lecture delivered on 1973 December 14.

061.080 **On the transport of charged particles in turbulent fields: comparison of an exact solution with the quasilinear approximation.** J. R. Jokipii, I. Lerche.
Plasma Phys., (GB), Vol. 15, 619 - 625 (1973).

Problems of modern astrophysics.
See Abstr. 003.037.

Astrophysics. See Abstr. 003.104.

Use of large optical telescopes at infrared wavelengths. See Abstr. 032.046.

Gravitational instability of a rotating plasma.
See Abstr. 062.059.

Nuclear goblins and cosmic gamma ray bursts.
See Abstr. 065.056.

Finite nuclear size effects on neutrino-pair bremsstrahlung in neutron stars. See Abstr. 065.102.

Cosmic gamma-ray emission and comet collisions with compact stars. See Abstr. 102.017.

On the opacity of the interstellar medium to ultrasoft X-rays and extreme-ultraviolet radiation.
See Abstr. 131.040.

Cosmic gamma-ray burst detected with an instrument on board the OGO-5 satellite. See Abstr. 142.011.

Galactic arm structure and gamma-ray astronomy.
See Abstr. 155.026.

High-energy X-rays from the Perseus cluster.
See Abstr. 160.006.

062 Hydrodynamics, Magnetohydrodynamics, Plasma

062.001 On a criterion for the onset of dynamical instability by a nonaxisymmetric mode of oscillation along a sequence of differentially rotating configurations.
S. Chandrasekhar.
Astrophys. Journ., Vol. 187, 169 - 174 (1974).

A variational base is derived for locating the point of onset of dynamical instability along a sequence of differentially rotating configurations. The underlying idea is to seek the conditions for the existence of quasi-stationary, nonaxisymmetric modes of deformation in frames of reference rotating with various preassigned angular velocities.

062.002 Spectral and polarization properties of radiation generated by plasma turbulence.
C. Chiuderi, P. Veltri.
Astron. Astrophys., Vol. 30, 265 - 270 (1974).

The authors have studied a system which consists of a cold plasma containing an ultrarelativistic distribution of electrons. The response of such a system to a periodic electromagnetic disturbance is first discussed. The problem is then specialized to the case of a longitudinal stationary plasma wave. The emission and propagation of radiation in such a system is analysed in detail by a method which is analogous to the one used in synchrotron radiation theory. Explicit results are given for a power-law distribution of relativistic electrons, and they are compared to synchrotron emission under the same conditions. It is shown that the frequency dependence and polarization are quite similar in the two cases.

062.003 Dynamical stability of stationary dynamo magnetic fields.   E. H. Levy.
Astrophys. Journ., Vol. 187, 361 - 367 (1974).

The effect of the magnetic field stress on the dynamical stability of stationary dynamo magnetic fields is investigated. It is found that half of the stationary states, which arise in the kinematic approach to hydromagnetic dynamo theory, are dynamically unstable and thus will not appear in nature. The rest are stable.

062.004 Laser-plasma spectra of highly ionized fluorine.
U. Feldman, G. A. Doschek, D. J. Nagel, W. E. Behring, R. D. Cowan.
Astrophys. Journ., Vol. 187, 417 - 420 (1974).

Lines between 11.3 Å and 17.2 Å of lithium-like, helium-like, and hydrogen-like fluorine have been observed in spectra of laser-produced plasmas. These lines include nine members of the Lyman series of F IX; eight members of the principal series of F VIII; and satellite lines arising from doubly excited configurations of F VII and F VIII. A wavelength list of fluorine lines is given, and physical conditions in the plasma are discussed.

062.005 Development of initial disturbances in a scattering, radiating, absorbing viscous heat-conducting gas.
V. A. Prokof'ev.
Dokl. AN SSSR, Ser. Mat. Fiz., Vol. 214, 1284 - 1287 (1974). In Russian.

062.006 Spherical harmonic analysis of the Navier-Stokes equation in magnetofluid dynamics.   M. C. Frazer.
Phys. Earth Planet. Interiors, Vol. 8, 75 - 82 (1974).

The equation of motion (Navier-Stokes equation) for a uniformly rotating, compressible, magnetic, viscous fluid is analyzed in terms of infinite series of spherical surface harmonics. Differential equations are obtained for the radial functions of the poloidal and toroidal harmonics of the velocity, corresponding to those obtained by Bullard and Gellman for the magnetic field from the electromagnetic induction equation.

062.007 Longitudinal photons in plasmas.
M. A. Gintsburg.
Astron. Zhurn. Akad. Nauk SSSR, Vol. 51, 218 - 221 (1974). In Russian. English translation in Soviet Astron. AJ, Vol. 18, No. 1. – Short note.

062.008 Rotating self-gravitating fluid spheroid with a magnetic field.   S. K. Trehan, M. Singh.
Astrophys. Space Sci., Vol. 26, 167 - 171 (1974).

The equilibrium of a self-gravitating fluid spheroid is examined in the presence of a rotation and a poloidal magnetic field. It is shown that 'true equilibrium' allows only rigid rotation for a spheroid of a small eccentricity.

062.009 Rayleigh-Taylor instability of a viscous compressible plasma of variable density.   P. K. Bhatia.
Astrophys. Space Sci., Vol. 26, 319 - 325 (1974).

A study has been made of the problem of the Rayleigh-Taylor instability of a hydromagnetic plasma of varying density to investigate the influence of the simultaneous presence of the effects of compressibility and viscosity. The solution is shown to be characterized by a variational principle. Based on the variational principle proper solutions have been obtained for a semi-infinite plasma, in which the density has a one-dimensional gradient along the direction of a uniform vertical magnetic field, confined between two planes.

062.010 The stability of a velocity shear in the presence of an inhomogeneous magnetic field.   P. R. Wilson.
Astrophys. Space Sci., Vol. 26, 363 - 369 (1974).

The stability of a velocity shear in the presence of a parallel but non-uniform magnetic field is considered in general terms. Two special cases are then investigated; (1) the well known case of a plane interface at which a discontinuity in the magnetic field coincides with the velocity shear; (2) an axially symmetric flow in which discontinuities in the magnetic and velocity fields occur at a cylindrical surface whose axis is parallel to the flow.

062.011 Global fluid motions in a uniformly rotating convective spherical shell studied as a model of the solar convection zone by a variational method.   H. Yoshimura.
Publ. Astron. Soc. Japan, Vol. 26, 9 - 51 (1974).

The behavior of the global nonaxisymmetric fluid motions in a uniformly rotating spherical shell of Boussinesq fluids with superadiabatic temperature gradient simulating the solar convection zone is investigated by a variational method from slow-rotation regime to fast-rotation regime, assuming that the horizontal scale of the motions is much larger than the vertical dimension. The global fluid motions in the spherical shell are classified into the convection mode and the Rossby mode according to the behavior of the motions.

062.012 Hydromagnetic stability of a self-gravitating plasma in the presence of Hall currents.   P. K. Bhatia.
Publ. Astron. Soc. Japan, Vol. 26, 109 - 116 (1974).

The hydromagnetic stability of a self-gravitating incompressible plasma of varying density has been investigated to include the effects of Hall currents. A variable horizontal magnetic field, stratified vertically, is assumed to be present in the system. A variational principle is shown to characterize the problem. Proper solutions have been obtained for a semi-infinite plasma in which the density has a one-dimensional (exponential) vertical stratification. It is found that the Hall currents have a destabilizing influence.

062.013 Scattering of Alfvén waves by random density fluctuations. G. C. Valley.
Astrophys. Journ., Vol. 188, 181 - 189 (1974).

The scattering of coherent Alfvén waves by random density fluctuations is considered, using both the method of first-order smoothing and the first Born approximation. The method of first-order smoothing provides dispersion relations for the propagation of Alfvén waves in an infinite medium in which the density is a spatially homogeneous random function. The Born approximation is used to solve the problem of the scattering of a coherent Alfvén wave by a patch of density fluctuations and to calculate the mean square scattered dilatation (divergence of velocity) and vorticity and the ratio of average total energy scattered into magnetosonic modes to that scattered into Alfvén modes. Applications of the results to hydromagnetic wave propagation in the solar wind, scattering of Alfvén waves by tangential discontinuities, and the problem of heating the solar wind are given.

062.014 Gleichzeitige Bestimmung von Elektronenkonzentration und Elektronenstoßzahl in einem örtlichen langsam veränderlichen Magnetoplasma (Teil II). W. Muschler.
Zeitschr. Naturforschung, Vol. 29a, 75 - 83 (1974).

A preceding paper was concerned with the principles of a wave propagation experiment. In this paper supplementary numerical calculations are carried out using a plasma model that corresponds to conditions observable in the earth's lower day time ionosphere and that is characterized by exclusively positive gradients of electron condensation N_e. Furthermore, a modification of the original principle is discussed renouncing measurements of refractive index n and using only those of wave polarization ρ. Numerical calculations on error propagation demonstrate a significant superiority of the modified version.

062.015 High-beta plasmas confined in poloidal magnetic fields. B. Lehnert, T. Hellsten, R. Raggi.
Phys. Scripta, Vol. 9, 53 - 60 (1974).

Experiments on rotating plasmas provide methods for studying part of the important problems of plasma equilibrium and stability at high beta values. This report describes investigations on high-density plasmas being performed by means of probe and Rogowski coil recordings.

062.016 The minimum-power effect of a magnetized plasma. B. Lehnert, J. Bergström, M. Bureš, S. Holmberg, E. Tennfors.
Phys. Scripta, Vol. 9, 109 - 118 (1974).

An experimental and theoretical analysis is performed of a fully ionized, magnetized, quasi-steady plasma being separated from a surrounding neutral gas blanket by partially ionized boundary regions: For rotating as well as high-frequency heated plasmas there exists a sharply defined minimum heating power P and temperature T below which a steady fully ionized state cannot be maintained. The theoretically predicted behaviour of the plasma is roughly in agreement with the experiments in the power range close to the minimum value.

062.017 Étude hydrodynamique et thermodynamique d'un fluide anisotrope relativiste. M. Cissoko.
Comptes Rendus Acad. Sci. Paris, Sér. A, Vol. 278, 463 - 466 (1974).

062.018 A plasma hypothesis for anomalous OH emission. D. B. Melrose.
Proc. Astron. Soc. Australia, Vol. 2, 206 - 208 (1973). − Presented at the annual general meeting of the Astron. Soc. Australia − see 012.001.

062.019 Non-uniformly rotating, self-gravitating, compressible masses with internal meridian circulation. I. W. Roxburgh.
Astrophys. Space Sci., Vol. 27, 425 - 435 (1974).

The structure of self-gravitating, inviscid, compressible fluids is investigated assuming a polytropic relation between pressure and density. A class of solutions with non-uniform rotation and internal meridian circulation are presented and the stream lines of the flow calculated using a perturbation technique.

062.020 On the equilibrium figures of an ideal rotating liquid in the post-Newtonian approximation of general relativity. I: Equilibrium conditions.
K. A. Pyragas, N. P. Bondarenko, O. V. Kravtsov.
Astrophys. Space Sci., Vol. 27, 437 - 452 (1974).

Distributions of velocities and condition of equilibrium for a perfect rotating fluid are obtained by Lyapunov's method in the general relativistic post-Newtonian approximation. The Born-type rigid rotation is discussed.

062.021 On the equilibrium figures of an ideal rotating liquid in the post-Newtonian approximation of general relativity. II: Maclaurin's P-ellipsoid.
N. P. Bondarenko, K. A. Pyragas.
Astrophys. Space Sci., Vol. 27, 453 - 466 (1974).

Maclaurin's P-ellipsoid, which is an equilibrium figure in the post-Newtonian approximation of general relativity, is constructed in the neighbourhood of Maclaurin's classical ellipsoid. Its shape and rotation velocity are investigated.

062.022 Shooting-splitting method for sensitive two-point boundary value problems.
P. J. Firnett, B. A. Troesch.
Proc. conference on numerical solution of ordinary differential equations, 1972, (see 012.005), p. 408 - 433 (1974).

062.023 The transfer of polarized radiation in a magnetoactive plasma.
V. V. Zheleznyakov, E. V. Suvorov, V. E. Shaposhnikov.
Astron. Zhurn. Akad. Nauk SSSR, Vol. 51, 243 - 251 (1974). In Russian. English translation in Soviet Astron. AJ, Vol. 18, No. 2.

The transfer equation is obtained for the radiation polarization tensor in a medium with arbitrary anisotropy and nonorthogonality of normal wave polarization. The possibility of spatial oscillations in the radiation intensity is pointed out, as well as the oscillations in the frequency spectrum of radiation from a plasma slab. In conclusion the mutual correspondence and applicability limits of the transfer equations for the polarization tensor obtained earlier by Kavabata, Zheleznyakov, Sazonov and Tsytovich are discussed.

062.024 On the magneto-dipole emission of an atom in a plasma. E. B. Klejman, I. M. Ojringel.
Astron. Zhurn. Akad. Nauk SSSR, Vol. 51, 448 - 450 (1974). In Russian. English translation in Soviet Astron. AJ, Vol. 18, No. 2. − Short note.

062.025 Physics of compact nonthermal sources. I. Theory of radiation processes.
T. W. Jones, S. L. O'Dell, W. A. Stein.
Astrophys. Journ., Vol. 188, 353 - 368 (1974).

The canonical nonthermal spectrum, taken to be incoherent electron-synchrotron radiation from a quasi-static source, is investigated allowing for the possibility of anisotropic electron distributions and highly ordered magnetic fields. From considerations of circular polarization, synchrotron self-absorption, self-Compton radiation, and Compton scattering off the microwave background, a system of restrictions upon the magnetic field and angular size of the source is parametrized in terms of the electron anisotropy and field geometry. The production and maintenance of small pitch angles for relativistic electrons is investigated.

062.026 **Rotating magnetosphere: far-field solutions.** F. C. Michel.
Astrophys. Journ., Vol. 187, 585 - 588 (1974).

General principles are established for the solution of the distant magnetic field structure about a rotating axisymmetrically magnetized object surrounded by plasma of negligible inertia. These principles are applied to the interesting case of a dipole magnetic field (as well as the monopole case) and can readily be extended to any other axisymmetric multipole. Analytic solutions are shown, and the author thereby computes relevant physical quantities (charge density, magnetic field intensity, torque, etc.).

062.027 **On the passage of radiation through inhomogeneous, moving media. I. The plane, differentially sheared medium.** I. Lerche.
Astrophys. Journ., Vol. 187, 589 - 596 (1974).

The author sets up the equations which describe the transport of rays and waves in differentially sheared media. For media of constant phase refractive index, he illustrates how a thin pencil of waves entering almost normal to the surface is distorted and refracted as it passes through the slab. The results are of interest to the problem of transport of radiation through pulsar magnetospheres, which corotate near the underlying neutron star, and which differentially shear near the so-called speed-of-light circle.

062.028 **On the passage of radiation through inhomogeneous, moving media. II. The rotating, differentially sheared medium.** I. Lerche.
Astrophys. Journ., Vol. 187, 597 - 607 (1974).

Using an idealized model, the author sets up and discusses the transport of radiation (under the geometrical optics approximation) through a rotating medium where the rotational speed may be comparable to the speed of light. In an appendix he gives the results of a calculation for wave propagation using a simple mechanical model of such a medium. The results obtained may be of interest in connection with the problem of radiation transport through pulsar magnetospheres.

062.029 **On the dynamo theory of cosmic magnetic fields. II. Representation of vector fields as sum of a poloidal and a toroidal part.** K.-H. Rädler.
Astron. Nachr., Vol. 295, 73 - 84 (1974). In German.

This paper deals with mathematical fundamentals of the foregoing one on the treatment of spherical dynamo models. A representation of arbitrary vector fields in the three-dimensional euclidian space as a sum of a poloidal and a toroidal part is founded. In this way well-known representations of axisymmetric or not necessarily axisymmetric but solenoidal fields are generalized. Possibilities of further generalization are investigated.

062.030 **On the correlation tensor of second rank for an inhomogeneous turbulent velocity field.** K.-H. Rädler.
Astron. Nachr., Vol. 295, 85 - 92 (1974). In German.

This paper deals with the correlation tensor of the second rank for an inhomogeneous turbulent velocity field. It is supposed that all the deviations from a homogeneous isotropic and reflection symmetric turbulence may be described only by the gradient of turbulence intensity. On the basis of symmetry considerations, the general form of the correlation tensor is derived and discussed. Furthermore, its special form for a solenoidal field is derived and discussed. Finally, some remarks are made on the correlation tensor for other kinds of inhomogeneous turbulence.

062.031 **On the Reynolds stresses in mean-field hydrodynamics. I. Incompressible homogeneous isotropic turbulence.** F. Krause, G. Rüdiger.
Astron. Nachr., Vol. 295, 93 - 99 (1974).

The authors consider an incompressible fluid undergoing turbulent motions. Using the methods of mean-field electrodynamics the relations between the Reynolds stresses and the mean velocity field are deduced in linear approximation. The homogeneous isotropic case is treated in more detail and the authors show that the effects can be described by a turbulent viscosity which proved to be positive in any case.

062.032 **Hydrodynamic instability of electromagnetic and plasma waves at gyrofrequency harmonics.** E. Ya. Zlotnik.
Izv. vyssh. ucheb. zavedenij. Radiofizika, Vol. 16, 1652 - 1659 (1973). In Russian. − Abstr. in Referativ. Zhurn. 51. Astron., 5.51.440 (1974).

062.033 **A relaxation theory of plasma-broadened He II lines.** J. T. O'Brien, C. F. Hooper, Jr.
Journ. Quant. Spectrosc. Radiat. Transfer, Vol. 14, 479 - 496 (1974).

The relaxation theory of plasma line broadening developed by Smith and Hooper, and extended by Smith, is applied to a charged radiator. The 304 Å Lyman alpha line of He II is chosen as an example. The fact that the radiator is charged complicates the treatment of electron perturbations; two different approaches are included.

062.034 **Dispersion and stability of waves in plasmas in the presence of a Coriolis force.** F. Verheest.
Astrophys. Space Sci., Vol. 28, 91 - 99 (1974).

The dispersion law for the propagation of waves in cold magnetized plasmas is derived for arbitrary directions of the rotation axis with respect to the static magnetic field. The waves are shown to be stable, not only in the case of a cold plasma, but in any plasma case which yields hermitian mobility tensors. An interesting special case is when the rotation and magnetization axes are parallel.

062.035 **Phase-bunching and other non-linear processes occurring in gyroresonant wave-particle interactions.** R. Gendrin.
Astrophys. Space Sci., Vol. 28, 245 - 266 (1974).

The author considers the movement of individual electrons in a magnetized plasma in which a monochromatic wave is propagating in the whistler mode. He derives simple expressions which give the displacement of the electrons as a function of time, the phase angle that their velocity vector makes with the magnetic component of the wave, their pitch angle and energy changes. A useful formula is obtained which gives the velocity range over which particles remain trapped inside the wave, as a function of the wave intensity and of the initial phase angle of the particle. From the derived expressions, it is possible to compute the phase-bunching effect which occurs approximately at one trapping wavelength behind the leading edge of the interaction region. The author deduces also the total amount of energy which is taken from (or given to) the wave by magnetospheric electrons in both cases of naturally existing or artificially injected particles. It is shown that these non-linear amplification processes can lead to very large VLF amplitude in the magnetosphere.

062.036 **On the equilibrium and stability of rotating high-beta plasmas.** B. Lehnert.
Phys. Scripta, Vol. 9, 229 - 236 (1974).

In this paper two problems are treated. The first concerns the equilibrium state of a rotating plasma and the associated radial expansion of the plasma body and its "frozen-in" magnetic field, as produced by the centrifugal force. The second problem concerns stability.

062.037 The contribution to nonlinear frequency shifts from diffusion in velocity space.
L. Falk, O. D. Kocherga.
Phys. Scripta, Vol. 9, 237 - 240 (1974).

The resonant interaction of four longitudinal plasma waves is investigated taking into account the influence of the waves on the background distribution function. This effect gives an essential contribution to the nonlinear frequency shift. Langmuir waves in a cold plasma are considered as an example.

062.038 Controlled experiments from the space shuttle.
F. L. Scarf.
Space Sci. Rev., Vol. 15, 933 - 950 (1974).

A study has been carried out to identify the controlled space physics and plasma physics experiments that can be performed from the space shuttle on sortie missions of 7 to 30-day duration.

062.039 Sur les ondes de choc en magnétohydrodynamique anisotrope relativiste. M. Cissoko.
Comptes Rendus Acad. Sci. Paris, Sér. A, Vol. 278, 1233 - 1236 (1974).

Propagation des ondes de choc dans un fluide anisotrope relativiste, non dissipatif et de conductivité infinie. Système complet d'équations de choc satisfaites par un fluide anisotrope polytropique.

062.040 Rayonnement cohérent d'ondes électrostatiques quasi-perpendiculaires dû à un faisceau d'électrons.
J. Lavergnat, R. Pellat, R. P. Singh.
Comptes Rendus Acad. Sci. Paris, Sér. B, Vol. 278, 763 - 765 (1974).

L'émission cohérente d'un faisceau d'électrons dans les modes de Bernstein est étudiée théoriquement et les résultats obtenus sont comparés à une expérience.

062.041 Onde de neutralisation d'un faisceau d'électrons injectés dans un plasma.
J. Lavergnat, R. Pellat.
Comptes Rendus Acad. Sci. Paris, Sér. B, Vol. 278, 827 - 829 (1974).

On calcule l'onde de neutralisation accompagnant un faisceau d'électrons injectés dans un plasma et on montre son influence sur l'émission d'ondes cohérentes.

062.042 Le tenseur de la conductivité électrique d'un plasma magnétoactif. M. Cristea.
Studia Univ. Babeş–Bolyai, Ser. Phys., Anul 19, Fasc. 1, p. 45 - 58 (1974). In Romanian.

On calcule le tenseur de la conductivité électrique pour un plasma magnétoactif, homogène, infini, faiblement ionisé. Le terme de collision de l'équation cinétique a la forme simplifiée proposée par Bhatnagar, Gross et Krook. Finalement on étudie quelques cas particuliers.

062.043 Radiation field of a uniformly moving charge in an anisotropic medium.
G. P. Sastry, S. Datta Majumdar.
Zeitschr. Naturforschung, Vol. 29a, 687 - 692 (1974).

Fourier integrals are set up for the field of a point charge moving uniformly in an arbitrary direction in a uniaxial medium anisotropic in ϵ only. The integrals break up into several parts two of which yield the ordinary and extraordinary cones with uniform azimuthal potential distribution. The remaining integrals neither contribute to the energy radiated nor affect the size and the shape of the cones, but merely distort the field within the cones. The integrals are evaluated exactly in the non-dispersive case and closed expressions for the potential are obtained.

062.044 M.H.D. en astrophysique. J.-C. Pecker.
L'Astronomie, 88ᵉ année, p. 153 - 154 (1974).

062.045 On some properties of relativistic MHD flows.
I. Lukačević.
Galaxies and relativistic astrophysics, Proc. 1972, (see 012.010), p. 183 - 189 (1974).

062.046 On the diffraction of Alfvén waves in a magnetoactive plasma.
N. S. Bellyustin, V. V. Tamojkin.
Izv. vyssh. ucheb. zavedenij. Radiofizika, Vol. 16, 1467 - 1474 (1973). In Russian. – Abstr. in Referativ. Zhurn. 62. Issled. kosmich. prostranstva, 5.62.317 (1974).

062.047 Hydromagnetische Wellen im zähen, anisotrop leitfähigen ionosphärischen Plasma.
H.-R. Lehmann, R. Treumann, D. Johanning.
Gerlands Beiträge Geophys., Vol. 83, 113 - 120 (1974).

The dispersion relation of mhd-waves in a viscous and anisotropic conducting plasma is derived. Solutions are given for the two special cases of parallel and perpendicular propagation. In the latter case an undamped fast magnetosonic wave propagates, that is accompanied by a damped viscous disturbance.

062.048 The conditions of evolution of rotational shock waves of an anisotropic pressure plasma and their cosmical applications. K. G. Ivanov.
Geomagn. Aeronom., Vol. 14, 538 - 540 (1974). In Russian. Brief information.

062.049 Calculation of heating, ionization and excitation processes of plasma by fast electrons and soft X-rays and cosmic rays. N. G. Bochkarev.
Soobshch. Gos. Astron. Inst. Shternberga, No. 183, p. 3 - 39 (1973). In Russian.

The cascade process of electron deceleration is calculated in detail for the case of an electron decelerating from the initial energy to thermal velocities in plasma consisting of H I atoms, free electrons and small quantities of He I and He II. Tables of numerical values of coefficients for calculating rates of medium heating, ionizations and excitations of some H I, He I and He II levels by soft X-rays having different spectra and by cosmic rays having energy of 2 MeV/nucleon and 6.4 MeV/nucleon are given.

062.050 Influence of an external field on wave-wave interaction coefficients. L. Stenflo.
Phys. Scripta, Vol. 9, 341 - 342 (1974).

This paper considers the resonant generation of an electrostatic plasma wave by means of two other waves. It is shown that the coupling coefficient is changed into a certain time-dependent function when an external periodic electric field is applied.

062.051 The propagation of compressional Alfvén waves in nonuniform plasmas.
R. Morrow, M. H. Brennan.
Australian Journ. Phys., Vol. 27, 181 - 194 (1974).

A theoretical treatment is presented for the propagation of Alfvén waves in a plasma. It includes the effects of resistivity, ion-neutral collisions, the ion cyclotron frequency and radial nonuniformities in ion density, neutral particle density and temperature. The theory is applied to plasmas with conducting and nonconducting walls and the results are compared with those of experiments conducted in the afterglow of a shock-produced plasma.

062.052 Gyromagnetic absorption at the fundamental.
D. B. Melrose.

Australian Journ. Phys., Vol. 27, 279 - 283 (1974).

Using a conventional procedure to treat gyromagnetic absorption in the centre of the line at the fundamental, it is found that there is a small range of angles ($\sim 5°$ in the solar corona) where the damping is weak. The procedure used by Gershman (1960) is shown to be nonphysical in that the time-irreversible properties of the waves are calculated using the time-reversible response of the plasma, and vice versa.

062.053 **On the hydromagnetic Kelvin-Helmholtz instability between compressible fluids.** K. M. Srivastava.
Zeitschr. Naturforschung, Vol. 29a, 888 - 892 (1974).

The author has discussed the effect of gravity on the hydromagnetic Kelvin-Helmholtz instability of a plane interface between compressible, inviscid, infinitely conducting fluids. The stability of the interface is investigated including gravity. The solar plasma and the magnetospheric medium are supposed to be of equal density and to carry a uniform magnetic field in the direction of streaming.

062.054 **On the Reynolds stresses in mean-field hydrodynamics. II. Two-dimensional turbulence and the problem of negative viscosity.** F. Krause, G. Rüdiger.
Astron. Nachr., Vol. 295, 185 - 193 (1974).

The methods proposed in a foregoing paper are used for the derivation of the eddy viscosity of a two-dimensional homogeneous isotropic turbulence. In contrast to three-dimensional isotropic turbulence it is shown that for the two-dimensional case the eddy viscosity (1) has no definite sign, (2) tends to zero if the molecular viscosity tends to zero, (3) is negative for special cases. However, the modes with large wave numbers decay in any case as is shown by investigating a dispersion relation.

062.055 **A note on summing series of Bessel functions occurring in certain plasma astrophysical situations.**
I. Lerche.
Astrophys. Journ., Vol. 190, 165 - 166 (1974).

The author gives the generic method of summing certain series of Bessel functions which occur in plasma astrophysical situations like the solar wind and the Crab nebula. He believes the closed analytic expression may aid in numerical computations of these sums.

062.056 **Polarization of inverse plasmon scattering.**
R. A. Windsor, P. J. Kellogg.
Astrophys. Journ., Vol. 190, 167 - 173 (1974).

The scattering of electrostatic plasma waves by a flux of ultrarelativistic electrons passing through a plasma gives rise to a radiation spectrum which is similar to a synchrotron radiation spectrum. This mechanism, first considered by Gailitis and Tsytovich, is analogous to inverse Compton scattering, and the authors have named it inverse plasmon scattering. In an attempt to distinguish between these mechanisms, they have calculated the polarization level expected from inverse plasmon scattering. This attempt to differentiate between the two mechanisms by polarization effects has been unsuccessful.

062.057 **Dissipation of magnetic flux and magnetic energy in partially ionized gases.** J. C. Byrne, R. R. Burman.
Astrophys. Space Sci., Vol. 29, 179 - 189 (1974).

Macroscopic equations of motion are used to derive several forms of the generalized Ohm's law for partially ionized ternary gases in magnetic fields. The rate of dissipation of magnetic energy in partially ionized gases has been claimed by Cowling (1956) and by Kaplan and Pikelner (1970) to be enhanced by the magnetic field H when the current density j is not parallel to H. Here it is shown that the rate of dissipation of electromagnetic energy need not be enhanced, although the rate of heating of the gas and the rate of decrease of magnetic flux through any surface moving everywhere with the neutral component of the gas are enhanced when j is not parallel to H.

062.058 **Effects of an ambient medium on the emission, absorption and scattering of waves by atoms and molecules.** D. B. Melrose.
Astrophys. Space Sci., Vol. 29, 211 - 219 (1974).

A theory describing the interaction between atoms or molecules and waves in an arbitrary mode in an arbitrary ambient medium is developed. Rules for generalizing formulae describing processes for waves in vacuo to include the effects of a medium are stated and the illustrative examples of multipole radiation, the photo-electric effect and Rayleigh and Raman scattering are given. A possible astrophysical application of resonant scattering by molecules of electron plasma waves is discussed briefly.

062.059 **Gravitational instability of a rotating plasma.**
R. C. Sharma.
Astrophys. Space Sci., Vol. 29, L1 - L4 (1974).

The gravitational instability of an infinite homogeneous self-gravitating rotating plasma in the presence of a uniform vertical magnetic field has been studied to include the FLR effects. It has been found that the Jeans' criterion of instability remains unaffected even if rotation and FLR effects are included. The particular cases of the effect of FLR and rotation on the waves propagated along and perpendicular to the magnetic field have been discussed.

062.060 **Simulation experiment of the current dissipation and plasma acceleration in the neutral sheet.**
N. Ohyabu, S. Okamura, N. Kawashima.
Journ. Geophys. Res., Vol. 79, 1977 - 1979 (1974). – Brief report.

062.061 **Convective energy transport in the presence of a toroidal magnetic flux tube.** Yu. V. Vandakurov.
Astron. Zhurn. Akad. Nauk SSSR, Vol. 51, 672 - 674 (1974). In Russian. English translation in Soviet Astron. AJ, Vol. 18, No. 3.

The equilibrium of an axisymmetric magnetic flux tube in a convective zone is considered in weak-field approximation. The specific magnetic energy is supposed to be large compared with the convective energy. It is shown that perturbations of the entropy gradient due to the presence of the tube appear to exist in some adjacent non-magnetic zone spreading up to the symmetry axis.

062.062 **The stability of differential rotation with non-axisymmetric perturbations. I. The sufficient conditions.** C.-H. Sung.
Astron. Astrophys., Vol. 33, 99 - 104 (1974).

A normal-mode analysis is used to establish some sufficient conditions for stability of a differentially rotating, compressible flow between two coaxial cylinders subject to non-axisymmetric perturbations.

062.063 **Propagation of transverse waves in a radially expanding plasma.**
C. V. Solodyna, J. W. Belcher.
Journ. Geophys. Res., Vol. 79, 2297 - 2301 (1974).

The authors have investigated the properties of low-frequency transverse waves in an expanding plasma. The wave vector k, the background magnetic field B, and the streaming velocity of the plasma V are all assumed to lie along the radial direction. The authors present expressions for the radial dependence of the amplitude and phase of left and right circularly polarized waves, correct to first order in the wave frequency divided by the proton cyclotron frequency. Differences in the phase velocities of these two circular polarizations result in a Faraday rotation that can be substantial for typical interplanetary conditions near 1 AU. The authors also consider the implications of these results for a realistic solar wind model with the interplanetary magnetic field along the spiral direction.

062.064 Large-amplitude hydromagnetic waves.
A. Barnes, J. V. Hollweg.
Journ. Geophys. Res., Vol. 79, 2302 - 2318 (1974).

The authors examine several aspects of the theory of large-amplitude hydromagnetic waves and their behavior in the interplanetary medium. They consider the characteristic modes of the full MHD equations and their modification by collisionless and finite-frequency effects. They use a second-order analysis to study fluctuations that are not characteristic modes. The results are used to discuss several possible mechanisms by which hydromagnetic waves may heat the solar wind.

062.065 Wave-wave coupling in multiple-ion plasmas.
N. Brice.
Journ. Geophys. Res., Vol. 79, 2519 - 2520 (1974).

The purpose of this note is to point out that the presence of more than one ion in a plasma opens up several possibilities for wave-wave coupling that are not available in a single-ion plasma.

062.066 On the electrodynamic equilibrium of a space charge region around a rotating neutron star with an aligned magnetic field. W. G. Pilipp.
Astrophys. Journ., Vol. 190, 391 - 401 (1974).

The problem of the electrodynamic equilibrium around a rigidly rotating, perfectly conducting neutron star with an azimuthally symmetric magnetic field aligned with the rotation axis is considered. Starting with the assumption that a space charge cloud including the neutron star can adjust to such a shape confined to a finite volume that it is in electrodynamic equilibrium, it is shown that the magnetic field as well as the electric field outside the space charge region will vanish identically. Thus, there is an electrodynamic equilibrium everywhere in space. Furthermore, it is shown that either there is no magnetic field at all outside the neutron star or the space charge region must extend at least to the boundary of the light cylinder.

062.067 Acceleration of thermal particles in collapsing magnetic regions. R. H. Levine.
Astrophys. Journ., Vol. 190, 447 - 456 (1974).

The acceleration of thermal particles in the vicinity of a magnetic neutral sheet is discussed. This process is considered as the archetypal low-intensity process at a newly formed neutral sheet. The collapse of the field toward the neutral sheet produces acceleration by secularly increasing the velocity component perpendicular to the sheet. The average energy increase for a thermal distribution of particles is calculated both with and without the inclusion of Coulomb losses by accelerated particles to particles in the accelerating region. Emphasis is placed on the conditions necessary for acceleration in the presence of Coulomb losses. The applicability of this approach to regions which collapse very fast is discussed.

062.068 A new method for improving the accuracy of numerical hydrodynamics calculations.
L. D. Cloutman.
Bull. American Astron. Soc., Vol. 6, 217 (1974). – Abstr. AAS.

062.069 Resistive diffusion of force-free magnetic fields in a passive medium. B. C. Low.
Bull. American Astron. Soc., Vol. 6, 264 - 265 (1974). Abstr. AAS.

062.070 Dynamical perturbations of warm, incompressible fluid disks. B. F. Schutz, Jr., J. M. Bardeen.
Bull. American Astron. Soc., Vol. 6, 279 - 280 (1974).
Abstr. AAS.

062.071 X-ray satellite lines of hydrogenlike and heliumlike ions. U. Feldman, G. A. Doschek, D. J. Nagel, R. D. Cowan, R. R. Whitlock.
Bull. American Astron. Soc., Vol. 6, 286 (1974). – Abstr. AAS.

062.072 On stability of the flow of stratified rotating superposed fluids. N. Rudraiah, M. Shanthakumar.
Publ. Astron. Soc. Japan, Vol. 26, 221 - 239 (1974).

The effect of Coriolis force on the stability of two superposed fluids is examined under the assumption of ideal flow. The analysis reveals only one type of instability in contrast to two types of instability observed by Sontowski, Seidel, and Ames (1969) in the absence of Coriolis force. Thus the effect of Coriolis force on the superposed flow is to make the flow more stable. In particular, the special case of an upper stratified fluid rotating and a lower non-rotating one is examined in detail, and the authors find two separate and different types of instability. These general stability results are applied to a particular problem of the Kelvin-Helmholtz instability with rotation, and the authors find that the results are in agreement with those of Chandrasekhar (1961). They determine the rate of growth of initial instabilities, which depends on the density stratification and the angular velocity of the rotating gas.

062.073 Cosmic magnetohydrodynamics. D. L. Moss.
Phys. Bull., (GB), Vol. 24, 668 - 670 (1973).

The dynamic problem of the earth's core, the Galaxy and Ap stars is discussed.

062.074 The interpretation of resonance interaction of charged particles with an elliptically polarized plasma wave. S. S. Sazhin.
Geomagn. Aeronom., Vol. 14, 543 - 545 (1974). In Russian. Brief information.

062.075 On the analogy between a wave packet of quasi-monochromatic plasma waves and a particle in a gravitational field [stellar atmosphere]. V. A. Petrzilka.
Czech. Journ. Phys. B, Vol. B 24, 522 - 526 (1974).

In a cold collisionless plasma, the energy-mass relation for a wave packet propagating radially in a Schwarzschild gravitational field is derived. The quasi-particle description of the wave packet yields no new red-shift formula.

062.076 Index of refraction of plasma in motion.
M. Elitzur.
Astrophys. Journ., Vol. 190, 673 - 674 (1974).

Index of refraction is calculated for plasma which streams in a specified direction with relativistic velocities. When applied to emission from pulsars, the model indicates that the low-frequency cutoff observed in the spectrum of some slowly rotating pulsars cannot be due to plasma effects. It is suggested that it is the result of self-absorption by the emitting bunches.

062.077 Energy conversion between longitudinal and transverse waves by mode-mode coupling in a relativistic plasma. K. Y. Fu.
Plasma Phys., (GB), Vol. 15, 1221 - 1233 (1973).

062.078 Electromagnetic instabilities in non-uniform anisotropic plasmas. B. Buti, G. S. Lakhina.
Journ. Plasma Phys., (GB), Vol. 10, 249 - 263 (1973).

062.079 Source flows in magnetohydrodynamics.
C. P. Yu, A. L. Doe.
Phys. Fluids, (USA), Vol. 16, 696 - 698 (1973).

062.080 Steady-state solutions for relativistically strong elec-

tromagnetic waves in plasmas. C. E. Max.
Phys. Fluids, (*USA*), Vol. 16, 1277 - 1288 (1973).

062.081 **Stability of isothermal magnetohydrodynamic shock waves.** N. Bel, C. Laury-Micoulaut.
Journ. Plasma Phys., (*GB*), Vol. 10, 301 - 316 (1973).

062.082 **Parametric instability of a relativistically strong electromagnetic wave.** C. E. Max.
Phys. Fluids, (*USA*), Vol. 16, 1480 - 1489 (1973).

062.083 **Plasma instability at an X-type magnetic neutral point.**
P. J. Baum, A. Bratenahl, M. Kao, R. S. White.
Phys. Fluids, (*USA*), Vol. 16, 1501 - 1504 (1973).

Spectral line broadening by plasmas.
See Abstr. 003.052.

Basic principles of plasma physics, a statistical approach. See Abstr. 003.066.

Principles of plasma physics. See Abstr. 003.073.

063 Radiative Transfer, Scattering

063.001 **Monte Carlo simulation of Voigt distribution in photon diffusion problems.** J.-S. Lee.
Astrophys. Journ., Vol. 187, 159 - 161 (1974).

A simple and exact formula is presented for randomly selecting the emission or absorption frequency for a scattering photon from a Voigt distribution. Theoretical proof of the validity of this formula is given, and numerical values of the generated frequency distribution are compared with tabulated values of the Voigt function.

063.002 **The transfer of circularly polarized radiation.** G. W. Collins II, P. F. Buerger.
Astrophys. Journ., Vol. 187, 163 - 167 (1974).

The equations of radiative transfer for circularly polarized radiation arising through the mechanism proposed by Kemp are described. A comparison of numerical results with the observations of Grw + 70° 8247 yields a value for the average magnetic field in the atmosphere of this white dwarf star of 1.7×10^7 gauss. Discussion of the expected accuracy and applicability of the results leads to the conclusion that the Kemp theory is not inconsistent with the observations.

063.003 **The influence of blends upon the formation of spectral lines.** B. W. Lites.
Astron. Astrophys., Vol. 30, 297 - 300 (1974).

The effect of blending of normal stellar absorption lines is investigated in order to determine the error introduced when such blending is neglected in line transfer calculations. Solutions of the radiative transfer equations in some typical blended solar absorption lines demonstrate that, in cases where the line under consideration has only weaker blends confined to wavelengths outside the immediate line core, the neglect of the blending in the radiative transfer computations is indeed a valid approximation.

063.004 **On the finiteness of a discrete spectrum of the characteristic equation of the radiative transfer theory.** I. A. Feldman.
Dokl. AN SSSR, Ser. Mat. Fiz., Vol. 214, 1280 - 1283 (1974). In Russian.

063.005 **Radiative transfer in spherically symmetric systems – III. Fundamentals of line formation.**
P. B. Kunasz, D. G. Hummer.
Monthly Notices Roy. Astron. Soc., Vol. 166, 19 - 55 (1974).

A generalization of the variable Eddington factor method is presented that makes possible the solution of line formation problems in extended spherical atmospheres whose constitutive properties depend on radius in an arbitrary way. Extensive numerical results for Doppler broadening in models with power law opacities (n = 0, 2, 3) are presented and interpreted. The single-flight escape probability is derived for a general opacity law. The effects of dilution are considered and some properties of the infinite radius, finite optical depth models are inferred. An appendix contains the solution of the line transfer problem for a homogeneous sphere by the kernel-approximation method.

063.006 **Radiative transfer in spherically symmetric systems–IV. Solution of the line transfer problem with radial velocity fields.** P. B. Kunasz, D. G. Hummer.
Monthly Notices Roy. Astron. Soc., Vol. 166, 57 - 78 (1974).

A numerical procedure is presented for solving the line transfer problem with complete redistribution in spherically symmetric atmospheres in which the radial velocity is an arbitrary function of radius, limited by practical considerations to maximum velocities a few times the mean thermal velocity. Numerical results are discussed for three sequences of models, two with linear velocity laws and one with constant velocities, in which the effect of the transverse velocity gradient is demonstrated.

063.007 **The scattering of resonance-line radiation in the limit of large optical depth – II. Reflection and transmission of radiation incident upon a slab.**
J. P. Harrington.
Monthly Notices Roy. Astron. Soc., Vol. 166, 373 - 382 (1974).

A solution developed previously is extended to the case of a slab illuminated by monochromatic radiation. The frequency distribution of the reflected and transmitted radiation is expressed as an infinite series, and some approximate closed expressions are obtained. Application of the results to the problem of escape of photons from an expanding shell of great optical depth is discussed.

063.008 **Light scattering in an inhomogeneous atmosphere.** V. V. Sobolev.
Astron. Zhurn. Akad. Nauk SSSR, Vol. 51, 50 - 55 (1974). In Russian. English translation in Soviet Astron. AJ, Vol. 18, No. 1.

The problem of diffuse reflection and transmission of radiation by an isotropically scattering atmosphere is considered. The albedo of single scattering is first assumed to be an arbitrary function of optical depth, and then is specialized to be the step function. An atmosphere composed of two layers is considered in detail. Three methods are given of the solution of the problem for the two-layer atmosphere composed of a semi-infinite medium and a layer of finite optical thickness on it.

063.009 **Omnidirectional induced Compton scattering by relativistic electrons.**
V. M. Charugin, Yu. P. Ochelkov.
Astrophys. Space Sci., Vol. 26, 337 - 344, 345 - 352 (1974). In English and Russian.

The authors consider the induced Compton scattering of isotropic radiation by relativistic electrons and calculate the rate of relativistic electron heating and the distortions of the sources spectra. They restrict themselves to Thomson scattering approximation.

063.010 **A system of radiative transfer equations in a strong magnetic field.** D. N. Rachkovskij.
Izv. Krymskoj Astrofiz. Obs., Vol. 49, 51 - 57 (1974). In Russian.

Absorption coefficients of a quantum oscillator in a strong magnetic field are derived. A system of radiative transfer equations is formed for the Stokes parameters.

063.011 **Complete linearization of the integral equations in radiative transfer.** W. Kalkofen.
Astrophys. Journ., Vol. 188, 105 - 119 (1974).

The transfer equation is linearized and a procedure presented for constructing integral operators that give the effect of density perturbations on the specific intensity, the mean intensity, the net rate of photoexcitation, and the equations of statistical equilibrium. An iteration method based on the linearized integral equation is applied to the solution of the equations of radiative transfer and of statistical equilibrium for atomic models with two levels (one line) and three levels (three lines). The features of this method are compared with those of the complete linearization methods of Auer and Mihalas and of Skumanich and Domenico.

063.012 **Line formation in turbulent media.**
H. P. Gail, E. Hundt, W. H. Kegel, J. Schmid-Burgk, G. Traving.
Astron. Astrophys., Vol. 32, 65 - 72 (1974).

The theory of line formation is generalized in order to take into account the influence of stochastic velocity fields. A Fokker-Planck equation is derived for the combined probability $P(x; v, I)$ to find at x the velocity v and the intensity I. This equation is discussed for the case that the velocity distribution is gaussian. Solutions of the differential equation obtained in this way are discussed.

063.013 **Line formation in turbulent media: mathematics of profile computation.** J. Schmid-Burgk.
Astron. Astrophys., Vol. 32, 73 - 78 (1974).

A transfer equation is derived for computing the profiles of spectral lines formed in a turbulent medium. The transfer equation contains two coefficients, as factors to the source term to the intensity, respectively. Formulae for rapid calculation of these coefficients are given. In praticular, an analytical correction term to the usual expression of microturbulence theory is derived for the case of a small but finite ratio of photon mean free path to correlation length of the stochastic velocity field.

063.014 **Radiation reaction in non-linear Compton effect.**
S. Bonometto, L. Scrascia.
Astron. Astrophys., Vol. 32, 115 - 116 (1974).

The authors discuss the conditions that must hold to allow the radiation reaction to be neglected when studying the motion of particles (namely electrons) in a strong electromagnetic wave.

063.015 **Polarization of multiple scattered light in planetary atmospheres from solution of the coupled linear integral equations derived for mixed Rayleigh-isotropic scattering.** J. B. Kumer.
Journ. Quant. Spectrosc. Radiat. Transfer, Vol. 14, 165 - 187 (1974).

A set of coupled integral equations describing nonconservative multiple scattering for a mixed isotropic and Rayleigh single scattering phase function in inhomogeneous, plane-parallel planetary atmospheres is derived. The equations are applicable for frequency redistributions. Solution of the equations permits one to calculate the intensity and degree of polarization in an arbitrary direction outside or inside the plane-parallel scattering medium. The equations are readily adaptable to more complicated geometries. Solutions for several cases are presented to demonstrate the versatility and validity of the method. Tables of functions that one might utilize to obtain solutions in the escape function approximation are also given.

063.016 **Two-dimensional radiative equilibrium: A semi-infinite medium subjected to a finite strip of radiation.** W. F. Breig, A. L. Crosbie.
Journ. Quant. Spectrosc. Radiat. Transfer, Vol. 14, 189 - 209 (1974).

Exact numerical results are presented for the emissive power and radiative flux at the boundary of a two-dimensional, absorbing-emitting, semi-infinite medium bounded by (1) a strip of collimated radiation and (2) a constant temperature black strip.

063.017 **Non-gray radiative transfer.**
Y. Yener, M. N. Özişik, C. E. Siewert.
Journ. Quant. Spectrosc. Radiat. Transfer, Vol. 14, 259 - 271 (1974).

The normal-mode-expansion technique is used to establish the solution of the Milne problem basic to a generalized equation of radiative transfer. The non-gray model used includes the effects of absorption, scattering and losses due to photo-electric ionizations and collisions of the second kind. Accurate numerical results are presented for such physical quantities as the extrapolation distance, the integrated Planck function and the angular distribution of the exit intensity for selected values of the basic parameters.

063.018 **A comparison of differential and integral equations of radiative transfer.** W. Kalkofen.
Journ. Quant. Spectrosc. Radiat. Transfer, Vol. 14, 309 - 316 (1974).

The equations of radiative transfer and of statistical equilibrium of a two-level atom are solved by means of differential and integral equations for a one-dimensional medium. The numerical solutions are compared to the analytic solution. It is found that the integral equation for piecewise quadratic source functions gives more accurate results than does the differential equation.

063.019 **Radiation transport with anisotropic scattering.**
A. Razani.
Journ. Quant. Spectrosc. Radiat. Transfer, Vol. 14, 339 - 349 (1974).

Reflection and transmission of radiation with anisotropic scattering is considered, using an invariant-imbedding formulation of the transport equation. A modified transport approximation is proposed in which singly-scattered radiation is taken into account exactly and higher-order scatterings are obtained from the transport approximation. The results derived from the modified transport approximation are in good agreement with other calculations for the cases considered.

063.020 **Utilisation des approximants de Padé pour l'étude des largeurs équivalentes des raies formées en atmosphère diffusante.** Y. Fouquart.
Journ. Quant. Spectrosc. Radiat. Transfer, Vol. 14, 497 - 508 (1974).

In order to obtain the distribution of the photon optical paths, $p(\lambda)$, for a cloudy atmosphere, it is proposed to invert the Laplace transform by means of Pade approximants. The stability of the method is tested and some examples of $p(\lambda)$ are presented. The method is applied to the determination of the equivalent widths of spectral lines formed in a scattering atmosphere. An analytical expression is proposed for the Lorentz lines: the optical path corresponding to the weak lines is determined in the same manner as for the strong lines. Some examples of curves of growth are presented.

063.021 **Reflection function from a double layer – a new approach.** R. Bellman, R. Vasudevan.
Astrophys. Space Sci., Vol. 28, 115 - 128 (1974).

To improve the accuracy of calculations for the reflection and transmission functions, doubling techniques are in use. The central theme of this method is to derive the total reflection function when two portions of a medium are adjoined together. The method developed in this article splits the total reflected beam into two parts, one relating to that flux which suffers no scattering at all in the first portion of the medium and the second portion that undergoes at least one scattering in the first portion. Order-of-scattering analysis is also carried out.

063.022 **Sobolev's Φ-function of radiative transfer in planar and spherical scattering media.**
P. T. Y. Poon, S. Ueno.
Astrophys. Space Sci., Vol. 28, 233 - 244 (1974).

The part of Sobolev's Φ-function in plane-parallel and spherical, isotropically scattering atmospheres with internal source distribution is investigated from analytical and numerical aspects. With the aid of invariant imbedding, the authors computed Sobolev's Φ-function of Milne's integral equation for the planar case by solving the Cauchy system for the auxiliary function and Chandrasekhar's X- and Y-functions. The corresponding Φ-function for the spherical case is readily obtained from that for the planar case.

063.023 Efficient stream distributions in radiative transfer theory. C. Whitney.
Journ. Quant. Spectrosc. Radiat. Transfer, Vol. 14, 591 - 611 (1974).

This paper discusses a new, computationally-efficient method for approximating the integro-differential equation of radiative transfer with a finite set of coupled differential equations for discrete streams. The method uses recommended spatial distributions of streams that are quite different from those typically used in that they are based on the symmetry of several regular Platonic solids.

063.024 A note on the exponential kernel approximation. B. F. Armaly, T. T. Lam.
Journ. Quant. Spectrosc. Radiat. Transfer, Vol. 14, 651 - 656 (1974).

Integral equations, which appear in the study of radiative transfer through a semi-infinite, emitting-absorbing, and isotropically scattering medium, are investigated. Closed form, approximate solutions, which were obtained by using an exponential kernel approximation, are presented and compared with exact solutions.

063.025 A modification to the conventional equation of transfer needed to insure consistency with the equilibrium diffusion description. G. C. Pomraning.
Journ. Quant. Spectrosc. Radiat. Transfer, Vol. 14, 657 - 660 (1974).

It is shown that relativistic terms, normally neglected in the equation of transfer, are necessary for the equation to reduce to the proper diffusion limit for radiation hydrodynamic problems. Simplified relativistic terms are suggested which also lead to the correct diffusion limit.

063.026 Gravitational Stokes parameters. A. M. Anile, R. A. Breuer.
Astrophys. Journ., Vol. 189, 39 - 49 (1974).

The electromagnetic and gravitational Stokes parameters are defined in the general theory of relativity. The general-relativistic equation of radiative transfer for polarized radiation is then derived in terms of the Stokes parameters, for both high-frequency electromagnetic and gravitational waves. The authors conclude by generalizing the concept of Stokes parameters for the most general class of metric theories of gravity, where six (instead of two) independent states of polarization are present.

063.027 Algebraic recurrence relations for the finite-order scattering and transmission functions. P. T. Y. Poon, S. Ueno.
Journ. Quant. Spectrosc. Radiat. Transfer, Vol. 14, 85 - 92 (1974).

It is shown how to find algebraic recurrence relations for the finite-order scattering and transmission functions in terms of finite-order X- and Y-functions without referring to an initial-value method. These recurrence relations are suitable for the numerical computation of the finite-order scattering and transmission functions by use of a digital computer. A numerical example of the finite-order reflection functions for optical thickness 0.2 with albedo = 1 is listed. An example of the cumulative reflection functions for optical thickness 0.2 with three different albedos is presented.

063.028 The velocity-dependent source function in radiative transfer theory. C. J. Cannon, L. E. Cram.
Journ. Quant. Spectrosc. Radiat. Transfer, Vol. 14, 93 - 99 (1974).

The authors consider the effect of velocity fields upon the transfer of line radiation by two-level atoms. They show that a simultaneous solution of the radiation transfer equation and the time-dependent rate equations leads to an equation for the source function which contains the Lagrangian derivative. They discuss a physical interpretation of the derivative term and present a method for solving this type of problem. They exhibit calculations which show that, for quite reasonable velocity fields, large errors can be produced if the derivative terms in the rate equations are neglected.

063.029 Radiative transfer in a moving medium. C. Magnan.
Journ. Quant. Spectrosc. Radiat. Transfer, Vol. 14, 123 - 139 (1974).

The author illustrates first, for a one-dimensional medium, how to generalize the concept of the probability of quantum exit to the case of a moving atmosphere. The transfer of radiation takes place in a two-level atom with total redistribution in frequency at each scattering. The values of the reflection coefficient and of the source function at the surface of a semi-infinite medium with constant properties are given. The results are then extended to the three-dimensional case.

063.030 Solutions of the equation of transfer for a medium bounded by a perfect specular reflector in terms of those for a perfect absorber. G. W. Kattawar.
Journ. Quant. Spectrosc. Radiat. Transfer, Vol. 14, 157 - 158 (1974). − Note.

063.031 Comparison of the photon diffusion model and Kubelka-Munk equation with the exact solution of the radiative transport equation. L. F. Gate.
Applied Optics, Vol. 13, 236 - 238 (1974). − Letter.

063.032 Green's function approach to inverse Compton scattering. P. S. Yeung, R. L. Brown.
Phys. Rev. D, Particles and Fields, Vol. 8, 4286 - 4297 = National Radio Astron. Obs., *Green Bank,* Repr. Ser. A, No. 332 (1973).

The emission, absorption, and transfer of radiation produced through the Compton interaction of relativistic electrons with microwave photons in a homogeneous source is considered. It is shown that the intensity of Compton-scattered radiation can be conveniently expressed in terms of a Green's function, and this enables the radiation-transfer problem for arbitrarily many scatterings to be formalized in terms of a series of appropriately defined scattering vertex functions and absorption propagators. A rigorous discussion of the cross section for inverse Compton scattering is presented which is valid both in the extreme relativistic ($\gamma \gg 1$) and weakly relativistic limits ($\gamma \simeq 1$).

063.033 The effect of velocity gradients on multi-level atom non-LTE line source functions. I. M. Vardavas, C. J. Cannon.
Australian Journ. Phys., Vol. 27, 157 - 167 (1974).

Macroscopic velocity gradients are featured in calculations of several different line source functions for a model multi-level atom in non-LTE. The physical processes giving rise to these source functions, and their corresponding line profiles, are discussed in detail. In particular, three quite distinct coupling mechanisms between the various levels in the atom are considered and the results so obtained enable an easy interpretation of non-LTE line source functions to be made for any multi-level atom.

063.034 On the resolvent of Milne's integral equation for a spherical, isotropically-scattering medium. S. Ueno.

Journ. Quant. Spectrosc. Radiat. Transfer, Vol. 14, 245 - 249 (1974).

Recently, determination of the resolvent kernel of Milne's integral equation for a spherical, isotropically-scattering medium with internal sources has been made by several authors. In this paper, it is shown how to compute the resolvent kernel of the above Milne's equation in terms of the modified Sobolev's Φ-function, which is reduced to the angular integration of the source function in the diffuse radiation field by a finite slab.

063.035 Cooling and decrease of ionization in a medium without energy supply.   A. S. Zentsova.
Trudy Astron. Obs., *Leningrad*, Vol. 30 (= Uchenye Zapiski Leningr. Un-ta, No. 373 = Seriya Matem. Nauk, vyp. (No.) 50), p. 25 - 28 (1974). In Russian.

The decrease of temperature and degree of ionization is studied in a medium without energy supply. The cooling is assumed to be caused by radiative losses due to recombination and line radiation of O II, N II and Fe II.

063.036 The relation between emission and absorption coefficients for arbitrary radiation processes.
L. Oster.
Astron. Astrophys., Vol. 33, 151 - 152 (1974).

A universally applicable rule that allows the computation of absorption from emission coefficients and vice versa is presented and the synchrotron case treated as an example.

063.037 Radiative transfer in an extended axisymmetric aspherical object.   R. S. Kandel.
Bull. American Astron. Soc., Vol. 6, 215 (1974). – Abstr. AAS

063.038 Markov description of stationary turbulence at large Reynolds number.
T. Y. Yueh, Y. T. Li, T. S. Kuan, P. Chang.
Scientia Sinica, Vol. 17, 181 - 192 (1974).

063.039 Diffuse reflection of time-dependent parallel rays by a semi-infinite atmosphere.   M. Matsumoto.
Publ. Astron. Soc. Japan, Vol. 26, 241 - 253 (1974).

The exact solution of the time-dependent equation of transfer for the emergent intensity from a semi-infinite, plane-parallel, non-emitting, and homogeneous atmosphere scattering radiation anisotropically is obtained, based on the assumption that the phase function can be expanded in Legendre polynomials consisting of finite terms. Some numerical values of the reflected intensity with two parameters having dimensions of time, the mean free time and mean duration of temporal capture, are presented and discussed in the case of isotropic scattering.

063.040 Green's-function approach to inverse Compton scattering.   P. S. Yeung, R. L. Brown.
Phys. Rev. D, Particles and Fields, Vol. 8, 4286 - 4297 (1973).

A general description is presented for the transfer of Compton scattered radiation in a plasma by expressing the intensity of the scattered radiation in terms of a Green's function. The spectrum of once and twice scattered photons is explicitly calculated and displayed.

063.041 Variational principles for the radiative transfer equation of single scattering.   D. Anderson.
Journ. Inst. Math. and its Applications, London, Vol. 12, 55 - 62 (1973).

063.042 Numerical solution of the radiative transfer equation in spherical shells.   A. Peraiah, I. P. Grant.
Journ. Inst. Math. and its Applications, London, Vol. 12, 75 - 90 (1973).

063.043 Utilisation de la méthode des harmoniques sphériques dans les calculs de transfert radiatif. Extension au cas de couches diffusantes d'absorption variable.
J. L. Deuze, C. Devaux, M. Herman.
Nouvelle Rev. Optique, Vol. 4, 307 - 314 (1973).

Effects of an ambient medium on the emission, absorption and scattering of waves by atoms and molecules. See Abstr. 062.058.

On the use of mean escape probabilities to solve transfer problems in nebulae.   See Abstr. 132.026.

064 Stellar Atmospheres, Stellar Envelopes, Mass Loss

064.001 A model envelope for the shell star 1 Delphini.
J. M. Marlborough, A. P. Cowley.
Astrophys. Journ., Vol. 187, 99 - 105 (1974).

The stellar-wind model for the distribution of the circumstellar material is shown to account satisfactorily for the appearance of the Hα line. The resulting steady-state rate of mass loss for 1 Del is $1.5 \times 10^{-8} M_\odot$ per year. The circumstellar envelope is shown to extend to at least 30 stellar radii and probably more than 100 stellar radii above the surface of the star.

064.002 The composition and evolutionary status of the helium-rich stars.   P. S. Osmer, D. M. Peterson.
Astrophys. Journ., Vol. 187, 117 - 129 (1974).

A spectroscopic investigation of recently discovered helium-rich B-type stars shows that they form a well-defined group with the characteristics: (1) helium abundances comparable to hydrogen by number; (2) temperatures indicative of B0–B2 stars; (3) gravities appropriate to the zero-age main sequence; (4) generally low rotational velocities; (5) order-of-magnitude overabundances of oxygen and nitrogen; but (6) basically normal abundances of carbon, silicon, magnesium, and aluminum.

064.003 A comparison of the straight-mean, harmonic-mean, and multiple-picket approximations for the line opacities in cool model atmospheres.   D. F. Carbon.
Astrophys. Journ., Vol. 187, 135 - 145 (1974).

Three standard approximations—the straight mean, the harmonic mean, and the multiple picket—for the line opacity in a stellar atmosphere are compared in several numerical experiments. First, the approximations are used to determine the integrated fluxes and flux derivatives in two blanketed spectral intervals. Model atmospheres calculated with the straight mean and with the multiple picket are compared and found to differ significantly in their temperature structures. The models based on multiple pickets are found to be sensitive to changes in the turbulence parameter and the isotopic abundances. The limitations of the multiple-picket approximation are outlined.

064.004 Formation of the luminosity-sensitive O I multiplet at 7774 Å.
H. R. Johnson, R. W. Milkey, L. W. Ramsey.
Astrophys. Journ., Vol. 187, 147 - 150 (1974).

The authors have calculated line profiles and equivalent widths for the 7774 Å multiplet of O I for model atmospheres in the range $T_{eff} = 6000° - 6500°$ K for surface gravities in the range $\log g = 4.0-1.0$. They show that if the line formation problem is treated correctly the derived value of the microturbulence decreases considerably. They have obtained high-resolution observations of the central intensities of these lines in Canopus (F0 Ib) to test whether the very deep profiles theoretically predicted actually occur. The results clearly indicate large departures from LTE.

064.005 The CNO abundances of the CH stars in Omega Centauri.   R. A. Bell, R. J. Dickens.
Monthly Notices Roy. Astron. Soc., Vol. 166, 89 - 99 (1974).

The Swan bands in the spectra of the ω Cen CH stars RGO 55 and RGO 70 have been analysed, using synthetic spectra calculations, and yield $C^{12}/C^{13} \sim 10$ and $[C/H] \sim -0.8$. The nitrogen abundance, [N/H], obtained from bands of the blue and red CN systems, is approximately solar for both of the stars. The UBV colours of the CH star models show ultraviolet excesses of ~ 0.3 mag when compared with normal giant models. The weak appearance of the Ca I 4226 line in CH star spectra is shown to be caused to a large extent by the great strength of the CH and CN features.

064.006 The radiation from photospheres with dominant role of scattering in the opacity.
R. A. Syunyaev, N. I. Shakura.
Astron. Zhurn. Akad. Nauk SSSR, Vol. 51, 102 - 106 (1974). In Russian. English translation in Soviet Astron. AJ, Vol. 18, No. 1.

The radiation flux emerging from an optically thick medium with dominant role of electron scattering in the opacity is not described by Stephan–Boltzmann's law. Formulas connecting the flux of radiative energy with the temperature of radiating plasma for isothermic atmospheres with either homogeneous, exponential or power distributions of the matter density have been derived.

064.007 Detectability and nature of CNO anomalies in early-type main-sequence stars.
B. Baschek, M. Scholz.
Astron. Astrophys., Vol. 30, 395 - 402 (1974).

Using model atmosphere techniques, the authors calculate selected spectral features in near-main-sequence stars in the range O 9 – B 5 for eight different element mixtures which are chosen to characterize the anomalous CNO line strengths discovered by Walborn. They discuss when CNO abundance anomalies can be detected at classification dispersion through anomalous line strengths or through inconsistencies from different classification criteria. A comparison of the results with the predictions by Paczyński (1973) does not give much support to the hypothesis that meridional circulation causes the N type anomaly in hot main-sequence stars.

064.008 Diffusion processes in the envelopes of main-sequence A stars: model variations due to helium depletion.   G. Vauclair, S. Vauclair, A. Pamjatnikh.
Astron. Astrophys., Vol. 31, 63 - 70 (1974).

Diffusion of helium is computed in the sub-photospheric radiative part of the envelopes of $2 M_\odot$ and $1.5 M_\odot$ population I stars. The influence of the helium abundance variation on the structure of the envelopes is taken into account. Consequences of these computations on the interpretation of peculiar A stars, Am stars and on the variability of δ Scuti stars are discussed.

064.009 Magnetic braking by a stellar wind–IV. The effect of different poloidal field structures.
I. Okamoto.
Monthly Notices Roy. Astron. Soc., Vol. 166, 683 - 701 (1974).

The theory of magnetic braking by a stellar wind is extended to cover a variety of assumed poloidal fields, all symmetric about the rotation axis, all dipolar angular structure, but with radial structures varying from the curl-free at one extreme to the quasi-monopolar at the other. The rates of mass-loss $\dot{M}$ and angular momentum-loss $\dot{J}$ are computed as functions of four non-dimensional parameters: l, the ratio of gravitational to thermal energy density, ζ, the ratio of magnetic to thermal density, κ, the ratio of the centrifugal force to gravity, all measured at the coronal base r_s; and a parameter λ that describes the structure of the poloidal field.

064.010 A magnetohydrodynamic model of the circumstellar envelope of Be stars.   M. Saitō.
Publ. Astron. Soc. Japan, Vol. 26, 103 - 108 = Tokyo Astron. Obs. Repr. No. 448 (1974).

The author shows that steady-state solutions corresponding to the super-Alfvénic and sub-Alfvénic solutions of the stellar wind flow give good envelope models. In these models the magnetic field slightly transfers angular momentum. Thus matter is supported against the star's gravitational force mainly by the centrifugal forces near the equatorial surface and by the magnetic forces in the outer region.

064.011 **Model atmospheres for C type stars.**
F. Querci, M. Querci, T. Tsuji.
Astron. Astrophys., Vol. 31, 265 - 282 (1974).

Model atmospheres with abundances that could be representative of C type stars have been calculated for effective temperatures T_e = 4500, 4200, 3800 and 3400°K and gravities g=0.1, 1.0 and 10 cgs units. They are based on the classical hypothesis and take into account radiative and turbulent pressures. The authors compare their results with other late-type star models which represent the molecular opacity by some kind of mean, and show how the resulting $T(\tau)$ relations can change. Their surface temperatures are clearly lower. The calculated emergent fluxes are in reasonable agreement with observation except in the visual region.

064.012 **Polarization by rotationally distorted electron scattering atmospheres.**
J. P. Cassinelli, B. M. Haisch.
Astrophys. Journ., Vol. 188, 101 - 104 (1974).

The polarization to be expected from rotationally distorted early-type stars is determined using results of calculations of the transfer of linearly polarized light in extended atmospheres. The factors controlling the direction of the polarization are discussed. Plausible explanations are offered for the anomalous wavelength dependence of both the degree of polarization in the Wolf-Rayet star HD 50896 and of the direction of polarization in the Be star HD 45677.

064.013 **Stellar spectral synthesis in the ultraviolet.**
R. L. Kurucz.
Astrophys. Journ., (Letters), Vol. 188, L21 - L22 (1974).

A computer program for spectral synthesis has been developed, based on a list of data for atomic lines described by Kurucz, Peytremann, and Avrett. To demonstrate the usefulness of this program, the author presents a sample calculation of the region around the C IV resonance doublet at 155 nm, which is a feature in rocket and satellite spectra of B stars.

064.014 **Dimensions of the zones of ionized gas around the sources of strong ultraviolet radiation. I. Physics of the problem and principles of calculation.** J.-C. Pecker.
Astrofizika, Vol. 9, 525 - 548 (1973). In French. English translation in Astrophysics, Vol. 9, No. 4.

A very simplified study shows that around a source of UV radiation (black body at temperature T), He III, He II, H II and H I regions succeed to each other in this order, provided T is not too high. At higher temperatures, on the contrary, helium can be ionized when hydrogen is already neutral. The bases for a complete study of the problem are given; the effects of the photons of the three spectral regions $\lambda < 228$ Å, 228 Å $< \lambda < 504$ Å, 504 Å $< \lambda < 912$ Å are treated separately; the secondary photons are taken into consideration and tables allowing the computations are constructed.

064.015 **Transfer of rotational momentum in stellar envelopes by means of a magnetic field.** V. N. Morozov.
Astrofizika, Vol. 9, 567 - 579 (1973). In Russian. English translation in Astrophysics, Vol. 9, No. 4.

Equations describing the change of the angular velocity of rotation within spherical or cylindrical stellar envelopes with time in presence of a magnetic field are derived from the equations of magnetohydrodynamics. Analytical solutions of these equations are used to find the typical time of transfer of rotational momentum from the star to the envelope. The results are applied to explain phenomena of the Be-stars.

064.016 **Electrical field in atmospheres of early type stars caused by pressure gradient.**
M. Kopecký, P. Kotrč.
Bull. Astron. Inst. Czechoslovakia, Vol. 25, 64 - 66 (1974).

Calculations of the electric field strength caused by a pressure gradient in some of Mihalas's models of atmospheres of early type stars have been made.

064.017 **Relaxation oscillations in red-giant envelopes and the symbiotic stars.** P. R. Wood.
Proc. Astron. Soc. Australia, Vol. 2, 198 - 200 (1973). – Presented at the annual general meeting of the Astron. Soc. Australia – see 012.001.

064.018 **A graphical method for estimating overshoot in convective stellar atmospheres.** C. Chen.
Astrophys. Space Sci., Vol. 27, 303 - 305 (1974).

A graphical method for estimating convective overshoot in stellar atmospheres is proposed. Applying the method to the solar atmosphere, the author finds that a convective element which starts at a depth of about 1000 km below the top of convection zone can penetrate to a height about 300 km above it.

064.019 **Velocity fields in stellar atmospheres and the concept of microturbulence.**
G. Worrall, A. M. Wilson.
Vistas in astronomy, Vol. 15, (see 003.001), 39 - 60 (1973).

064.020 **Gravity-darkening in stars for general rotation laws.**
R. C. Smith, R. Worley.
Monthly Notices Roy. Astron. Soc., Vol. 167, 199 - 213 (1974).

The authors have shown that gravity-darkening in differentially rotating radiative atmospheres may depart strongly from the usually-assumed von Zeipel law. Two examples are given for circulation-free rotation laws, which show that the dependence of emergent flux on latitude can be quite strong. This could have important effects on close binary light curves.

064.021 **Determination of titanium oxide and atomic oxygen abundances in the atmospheres of M-type stars.**
A. V. Shavrina.
Astrometriya i Astrofizika, Kiev, vyp. (No.) 20, (see 003.002), p. 10 - 12 (1973). In Russian.

The abundance of titanium oxide in the atmospheres of five M-stars (α Sco, δ^2 Lyr, R Lyr, α Her A and β Peg) was determined from the profile of the (0.1) band head of the TiO γ-system. The atomic oxygen concentration is obtained from the dissociation equilibrium equation for α Sco, R Lyr, α Her A and β Peg, for which reliable values of TiI abundances are available.

064.022 **On the application of Christy's modified opacity formula.** A. Pamyatnykh.
Nauch. Informatsii, vyp. (No.) 27, p. 99 - 107 (1973). In Russian.

The problem of application of opacity tables for computations of the structure of stellar envelopes and for investigation of stellar instability is discussed. Christy's (1966) formula for the chemical composition X=0.7, Z=0.02 is slightly modified in order to employ it in deep layers of envelopes.

064.023 **On the problem of matter outflow from Wolf-Rayet stars.** Kh. F. Khaliullin.
Astron. Zhurn. Akad. Nauk SSSR, Vol. 51, 395 - 402 (1974). In Russian. English translation in Soviet Astron. AJ, Vol. 18, No. 2.

On the basis of the O − C graph, constructed from available photometric observations of the eclipsing binary V444 Cyg, an estimate of the intensity of the mass loss of the WR-component is made for several models of the mass loss.

064.024 A line-blanketed model stellar atmosphere of Sirius. J. W. Fowler.
Astrophys. Journ., Vol. 188, 295 - 307 (1974).

The method of artificial absorption edges has been applied to a grid of model atmospheres to include the effects of hydrogen and metal line blanketing on the atmospheric structure. A best-fit model of Sirius is obtained which is free of the discrepancies inherent in unblanketed or partially blanketed models. All major spectral features calculated from the model agree with the corresponding observed features to within the observational uncertainty. Some additional properties of the models are discussed.

064.025 Transfer of resonance-line radiation in differentially expanding atmospheres. III. Formation of P Cygni-type lines by a doublet line or two partially "blended" lines. T. G. Hewitt, P. D. Noerdlinger.
Astrophys. Journ., Vol. 188, 315 - 326 (1974).

The authors examine the transfer of resonance radiation in a spherically symmetric expanding stellar or quasi-stellar atmosphere in the case that the rest position of one line lies in the blueward absorption feature of another, under simplifying assumptions. They find that for most optical depths and most reasonable choices of the velocity-radius relation, an emission peak is seen only for the redward component; only for small optical depths and wide doublet separation is a clear emission peak of the blue component formed within the absorption feature of the redward component. They have also treated cases where emission in the atmosphere is added to the scattered radiation. Using these results, they establish an upper limit to the N V optical depth in the expanding atmosphere of the QSO PHL 5200, and show that the outflowing stellar winds in several hot supergiants in Orion probably involve a more graduate acceleration and less mass loss than heretofore postulated. Gradual acceleration is also indicated in ζ Pup and ξ Per.

064.026 On atmosphere models of late type stars. V. E. Panchuk.
Astron. Tsirk., No. 799, p. 4 - 6 (1973). In Russian.

064.027 Fe I line formation in solar-type giants and dwarfs. B. W. Lites, C. R. Cowley.
Astron. Astrophys., Vol. 31, 361 - 369 (1974).

The formation of Fe I lines is investigated in stellar models of solar T_{eff} but surface gravities ranging from $\log g = 4$ to $\log g = 2$. The results of these kinetic equilibrium computations demonstrate that the departures from LTE in the Fe I lines generally cause only small differences from equivalent widths computed under the LTE assumption.

064.028 The energy requirements of microturbulence. A. G. Hearn.
Astron. Astrophys., Vol. 31, 415 - 418 (1974).

The flux of energy required to maintain microturbulence against viscous dissipation has been calculated assuming that it results from a progressive longitudinal wave. This has been applied to the sun, a metallic line star 63 Tauri with a large microturbulent velocity and a B type star.

064.029 X-ray heating in HZ Her. P. A. Strittmatter.
Astron. Astrophys., Vol. 32, 7 -12 (1974).

The problem of X-ray heating of a stellar atmosphere is treated using a double grey body approximation and the results are applied to HZ Her.

064.030 The temperature and velocity distribution in Wolf-Rayet stars. C. J. Cannon.
Astron. Astrophys., Vol. 32, 79 - 83 (1974).

A methodology for determining the qualitative features of the physical parameters dominating the emission line spectra observed from Wolf-Rayet stars is discussed.

064.031 The effect of redistribution on the emission peaks from chromospheric-type stellar atmospheres. C. J. Cannon, I. M. Vardavas.
Astron. Astrophys., Vol. 32, 85 - 88 (1974).

The equation of transfer for spectral line radiation is solved for a physically idealistic model problem involving departures from complete redistribution. In particular, the authors consider a two-level atom one-dimensional atmosphere exhibiting a chromospheric-type temperature rise, and compute emergent absorption profiles having emission peaks similar to those observed in the core of Ca II and Mg II H and K lines.

064.032 Solid particles and stellar mass loss. J. D. Fix, D. R. Alexander.
Astrophys. Journ., (Letters), Vol. 188, L91 - L92 (1974).

Extensive model atmosphere calculations for cool giants and supergiants show that solid particles do not cause mass loss in such stars.

064.033 On Zanstra's theorem and its applications. S. V. Rublev.
Soobshch. Spets. Astrofiz. Obs. AN SSSR, Zelenchukskaya, vyp. (No.) 5, p. 3 - 17 (1972). In Russian.

A strict derivation is given of Zanstra's theorem for any hydrogen envelope with a radiative population of atomic levels. A derivation is given of a generalized formula allowing for the escape of Lyman quants from the envelope (in lines and beyond the series limit). A strict derivation is described of a modified equation of the "hydrogen method" that enables the "core" temperature to be estimated from the intensity (equivalent width) of any bright line observed in the spectrum of any stellar envelope.

064.034 Un modèle simple d'atmosphère stellaire. A. Heck, J. Manfroid.
Orion, 32. Jahrgang, p. 95 - 98 (1974).

064.035 A model atmosphere analysis of the metal-rich giant Iota Draconis. P. M. Williams.
Monthly Notices Roy. Astron. Soc., Vol. 167, 359 - 367 (1974).

Using model atmospheres and high dispersion (2.2 Å mm^{-1}) spectrograms observed with the Isaac Newton telescope, the K2 III giant ι Draconis was found to have twice the metal content of the solar composition giant ϵ Virginis. Using a least squares procedure, the microturbulence was found to be intermediate (1.4 km s^{-1}) between those of ϵ Vir and the sun. The temperature and surface gravity solved for simultaneously using colours, Fe I lines, the absolute magnitude and the Fe and Ti ionization equilibria. Using these parameters, the abundances of Sc, Ti, V, Cr, Mn, Fe, Co, Ni, Y and Zr were derived.

064.036 Some observational implications of extended static O-star model atmospheres. D. Mihalas, D. G. Hummer.
Astrophys. Journ., (Letters), Vol. 189, L39 - L43 (1974).

Some results and observational implications are presented for the first extended spherical non-LTE model atmospheres in hydrostatic and radiative equilibrium. These models all correspond to a star with $M = 60 \, M_\odot$, $L = 1.25 \times 10^6 \, L_\odot$, and $R = 24 \, R_\odot$, with an effective temperature $T_{eff} \simeq 39{,}500 \, °K$ and surface gravity $\log g \simeq 3.45$ (spectral type near O6). It has been possible to obtain models very close to the limit at which the radiation force balances the gravity. Hydrogen and helium ($Y = 0.1$) constitute the gas; six hydrogen lines are treated explicitly. These models show Lα in emission, the lower Balmer lines in absorption, the Balmer jump in absorption, and both infrared and ultraviolet excesses relative to the visual. Continuum jumps and gradients, Strömgren-system colors, and equivalent widths of Hα, Hβ, and Hγ are tabulated and discussed briefly.

064.037 **The effect of sphericity on stellar continuous energy distributions.** J. I. Castor.
Astrophys. Journ., Vol. 189, 273 - 283 (1974).

Model stellar atmospheres are constructed with the object of explaining the anomaly observed in a number of Wolf-Rayet stars that the line spectrum indicates a higher temperature than does the continuous spectrum. The models assume radiative and hydrostatic equilibrium, and local thermodynamic equilibrium. The spherical symmetry of the atmosphere is taken fully into account. It is found that the extension of the atmosphere, and the atmospheric density, depend critically upon the ratio of luminosity to mass for the model. Approximate analytic results are obtained which give a quantitative expression for this dependence. The principal result regarding the continuous spectrum is that, for the extended models, the brightness temperature is a monotonically increasing function of frequency, over the observed range.

064.038 **Thermal-convective instability of a stellar atmosphere with Hall effects.** R. C. Sharma.
Astron. Astrophys., Vol. 32, 223 - 225 (1974).

Thermal-convective instability of a stellar atmosphere has been considered to include Hall effects in the presence of a uniform vertical magnetic field. The criterion for monotonic instability has been found to be unchanged by the presence of Hall effects.

064.039 **Non-LTE-effects in atmospheres of A0 Ia-supergiants.** R. P. Kudritzki.
Mitt. Astron. Ges., No. 34, p. 85 - 86 (1973/74). – Presented at the "Wissenschaftliche Tagung der Astron. Ges., Oberkochen, 1973 April 24 - 27".

064.040 **The interpretation of emission line profiles from envelopes of Be stars. I. The radiation field.**
S. Kříž.
Bull. Astron. Inst. Czechoslovakia, Vol. 25, 143 - 152 (1974).

The paper presents a solution of the equation of line radiative transfer for a moving envelope of a Be star. Formulae are derived for the mean intensity of the radiation and for the emission line profile. Illustrative computations of line profiles are presented. The possibility of determining the physical conditions of the envelope directly from the observed emission line profiles is discussed.

064.041 **The enhancement of ultraviolet line opacities in the atmospheres of peculiar A stars.**
D. S. Leckrone, J. W. Fowler, S. J. Adelman.
Astron. Astrophys., Vol. 32, 237 - 244 (1974).

In the investigation of the photometric properties of Ap stars, as observed both by OAO-2 in the ultraviolet and by ground-based instruments, we have calculated line-blanketed model atmospheres in the effective temperature range $11\,000°$ to $15\,000°$K in which metal line opacities are arbitrarily enhanced by various factors. The enhanced opacity atmospheres predict flux distributions qualitatively similar to those observed for Ap stars from the vacuum ultraviolet through the visible.

064.042 **Problemas en la determinación de abundancias de elementos en las estrellas en condiciones de equilibrio termodinámico local y alejadas del equilibrio termodinámico local.** E. Simonneau.
Revista Real Acad. Ci. Exactas, Fís. Nat., Madrid, Vol. 67, 659 - 666 = Univ. Complutense Madrid, Fac. Ci., Seminario Astron. Geod., Publ. No. 76 (1973).

064.043 **On convection in stellar atmospheres.**
Å. Nordlund.
Astron. Astrophys., Vol. 32, 407 - 422 (1974).

Comparisons between different convection theories are made, and the reasons for the qualitative differences between them are considered. It is found that these differences can be understood and that, when the basic equations are handled carefully, the results of different theories are not very different.

064.044 **On the construction and use of model stellar atmospheres for stars of spectral types later than F0.**
B. Gustafsson.
Uppsala Astron. Obs. Ann., Vol. 5, No. 7, 16 pp. (1974).

A brief discussion is given on the properties of Feautrier-type methods when applied to the problem of constructing LTE model atmospheres with convection. Some recent determinations of metal abundances and other physical parameters of F dwarfs and G-K giants, with the aid of photoelectric photometry in very narrow bands and model-atmosphere analysis, are discussed.

064.045 **The thermal-convective instability in a stellar atmosphere with rotation and Hall effects.**
R. C. Sharma.
Zeitschr. Naturforschung, Vol. 29a, 867 - 869 (1974).

The problem of thermal-convective instability of a stellar atmosphere is considered to include the effects due to rotation and Hall currents in the presence of a uniform vertical magnetic field. The criterion for monotonic instability is found to be unchanged by the presence of rotation and Hall effects.

064.046 **Polarization of radiation from stellar envelopes.**
A. Z. Dolginov, N. A. Silant'ev.
Astron. Zhurn. Akad. Nauk SSSR, Vol. 51, 489 - 501 (1974). In Russian. English translation in Soviet Astron. AJ, Vol. 18, No. 3.

Explicit analytical formulae for the polarization of optical radiation scattered in an optically thin star envelope of arbitrary form are obtained. The scattering by randomly oriented or axially oriented particles small in comparison to radiation wavelength is considered. Approximate formulae are given for the Stokes parameters of radiation scattered by medium size particles in the star envelope.

064.047 **Emission Balmer decrement and electron density of stellar chromospheres.** R. E. Gershberg.
Astron. Zhurn. Akad. Nauk SSSR, Vol. 51, 552 - 559 (1974). In Russian. English translation in Soviet Astron. AJ, Vol. 18, No. 3.

The Balmer decrement shock theory for an optically thick medium with a gradient of velocity of internal motions is used to analyze the quiet state spectra of UV Cet-type flare stars. Spectra of 9 flare stars have been obtained with the 2.6-m Shajn reflector. Observed emission Balmer decrements are satisfactorily represented by this theory. From the representation the electron densities of stellar chromospheres are evaluated to be about $\sim 3 \times 10^{12}$ cm^{-3} and the probabilities of the Ly α-quantum escape are near to 10^{-6}. The application of the theory to the solar chromosphere permits to represent the observed Lyman decrement at the center of the quiet solar disc, to estimate the electron density, optical and geometrical thicknesses of the layer which radiates effectively the Lyα-quanta.

064.048 **Some consequences of a rapid Doppler width variation on the formation of stellar chromospheric lines.** C. Magnan.
Astron. Astrophys., Vol. 33, 139 - 141 (1974).

A simple formula related to the optical depth and to the width of a line formed in a chromospheric situation is derived. Some applications are indicated especially concerning the Wilson-Bappu relation.

064.049 **Line-blanketing and model stellar atmospheres. I. Statistical method and calculation of a grid of models.** E. Peytremann.
Astron. Astrophys., Vol. 33, 203 - 214 (1974).

The author describes a statistical method for the computation of line-blanketed model stellar atmospheres. By comparison with the solar spectrum, he adjusted a basic set of data (Corliss and Bozman, 1962) so that he was able to reproduce, on the average, the solar spectrum. A check of the method is performed by calculating a blanketed solar model, which is compared to the Harvard Smithsonian Reference Atmosphere. He then computed a grid of blanketed stellar atmospheres in the range $5000 \leq T_{\text{eff}} \leq 8500°K$, $2 \leq \log g \leq 4.5$. He discusses some of the results in terms of backwarming and surface cooling.

064.050 **Absolute fluxes of K chromospheric emission in main sequence stars.**
C. Blanco, S. Catalano, E. Marilli, M. Rodonò.
Astron. Astrophys., Vol. 33, 257 - 264 (1974).

From new observations of the K emission line of Ca II and additional data from various sources, absolute fluxes from the chromospheres of main sequence stars have been estimated.

064.051 **A fine analysis of the helium-rich star HD 184927.** N. A. Higginbotham, P. Lee.
Astron. Astrophys., Vol. 33, 277 - 288 = Contr. Louisiana State Univ. Obs., *Baton Rouge*, No. 95 (1974).

An LTE fine analysis of the helium-rich star HD 184927 has been performed by using a grid of flux-constant model atmospheres. Observational material includes both high and moderate dispersion spectrograms covering the spectral region λ 3600 Å – λ 6200 Å and moderate resolution spectral scans. The final model has been adopted to match the following criteria: (1) equal hydrogen abundance from both the hydrogen and helium equivalent widths, (2) ionization equilibria of Si II/Si III and S II/S III, (3) Hγ and Hδ profiles, (4) He I, $\lambda\lambda$ 4471, 4388, 4143 profiles, and (5) the continuous flux distribution (Balmer discontinuity). A micro-turbulent velocity has been determined from the metallic lines. This model is parameterized by: $T_{\text{eff}} = 22500°K$, $n_{\text{He}} = 0.50$, $\log g = 3.80$, $v_t = 6$ km/s. Mass, radius, luminosity, and distance are determined to be $M = 12.4\,M_\odot$, $R = 7.3\,R_\odot$, $\log(L/L_\odot) = 4.1$, and $r = 1260$ pc.

064.052 **Fine analytic abundance determination of Magellanic Cloud A-supergiants and its importance for the discrimination of theories for the chemical evolution of the Galaxy.** B. Wolf.
Stars and the Milky Way system, Proc. 1972, (see 012.018), p. 178 - 181 (1974).

064.053 **X-rays from hot, dense coronas.** N. M. Hoffman.
Bull. American Astron. Soc., Vol. 6, 213 (1974). Abstr. AAS.

064.054 **Continuum formation regions of hot expanding atmospheres.** J. P. Cassinelli, L. W. Hartmann.
Bull. American Astron. Soc., Vol. 6, 215 (1974). – Abstr. AAS.

064.055 **The formation of the Hϵ emission feature in late-type stars.** T. R. Ayres, J. L. Linsky.
Bull. American Astron. Soc., Vol. 6, 226 (1974). – Abstr. AAS.

064.056 **A new derivation of the diffusion equation.** C. R. Cowley.
Bull. American Astron. Soc., Vol. 6, 265 - 266 (1974). Abstr. AAS.

064.057 **Ionization equilibria in the atmospheres of late M-giant stars.** J. R. Auman, J. E. J. Woodrow.
Bull. American Astron. Soc., Vol. 6, 266 (1974). – Abstr. AAS.

064.058 **Stellar winds and related phenomena in surrounding nebulae.** S. B. Pikel'ner.
Comments Astrophys. Space Phys., Vol. 5, 151 - 158 (1973).

064.059 **Radiation pressure and the lithium-beryllium problem.** E. Schatzman, S. Vauclair.
Comments Astrophys. Space Phys., Vol. 5, 159 - 167 (1973).

064.060 **Self-similar homothermal flow of self-gravitating gas behind shock wave [in stellar atmospheres].**
S. C. Purohit.
Journ. Phys. Soc. Japan, Vol. 36, 288 - 292 (1974).

Self-similar homothermal flows of self-gravitating gas behind the spherical shock wave propagating in a nonuniform atmosphere at rest are investigated. Using a similarity variable the solutions corresponding to infinite star have been constructed. Both numerical and particular analytical solutions are obtained. These solutions are compared with the corresponding solutions of the adiabatic flow.

064.061 **Rechnungen zum molekularen Reaktions- und Kondensationsgleichgewicht, zur monochromatischen Opazität und zum Strahlungstransport in ausgedehnten, kühlen zirkumstellaren Hüllen.** D. Sieber.
Thesis Univ. Bonn (1973).

The phase diagram of graphite grains. See Abstr. 022.070.

Semiempirical calculation of gf values, II: Fe I $(3d + 4s)^8 - (3d + 4s)^7\,4p$. See Abstr. 022.074.

Electron-hydrogen photoattachment as a source of ultraviolet absorption. See Abstr. 061.043.

On the analogy between a wave packet of quasi-monochromatic plasma waves and a particle in a gravitational field [stellar atmosphere]. See Abstr. 062.075.

Radiative transfer in spherically symmetric systems – III. Fundamentals of line formation.
See Abstr. 063.005.

The effect of velocity gradients on multi-level atom non-LTE line source functions. See Abstr. 063.033.

Radiative transfer in an extended axisymmetric aspherical object. See Abstr. 063.037.

On the surface conditions for very massive stars. See Abstr. 065.021.

Effect of mass loss by stellar winds on the pre-main sequence stage of stellar evolution. See Abstr. 065.036.

Models of asymptotic-giant-branch stars. See Abstr. 065.127.

Observed departures from LTE in stellar Fe I lines. I. The sun. See Abstr. 071.043.

Effect of convection on radial velocities.
See Abstr. 112.027.

The calibration of *uby* photometry.
See Abstr. 113.017.

On the strength of Hα in the O-stars.
See Abstr. 114.016.

Spectral energy distributions of T Tauri stars.
See Abstr. 114.028.

Remarques sur les spectres continus théoriques d'étoiles O et B.   See Abstr. 114.037.

Spectrophotometry of stars of the Pleiades cluster. II. Absorption lines.   See Abstr. 114.046.

Spectroscopic studies of O-type stars. IV. Lines in the red region.   See Abstr. 114.050.

Lines of H I and He II in the spectra of Wolf-Rayet stars.   See Abstr. 114.054.

The carbon isotope ratio in Arcturus.
See Abstr. 114.089.

Spectroscopic observation of the carbon stars Y Canum Venaticorum and U Hydrae in the one-micron region.
See Abstr. 114.153.

Strong-line K stars. I. Photometry.
See Abstr. 114.173.

Strong-line K stars. II. Chemical abundances.
See Abstr. 114.174.

Theoretical aspects of light variations of magnetic stars.   See Abstr. 116.006.

Magnetic accretion and atmospheric peculiarities in early type stars.   See Abstr. 116.010.

Chemical compositions of cool helium and carbon white dwarfs.   See Abstr. 126.004.

The effects of rapid, differential rotation on the spectra of white dwarfs.   See Abstr. 126.017.

Further evidence for an interstellar source of night-time He I 584 Å radiation.   See Abstr. 131.036.

Acceleration of QSO clouds by radiation pressure.
See Abstr. 141.001.

Errata

064.901   Erratum: 'Theoretical effect of various broadening parameters on ultraviolet line profiles' [Astron. Astrophys., Vol. 17, 76 - 82 (1972)].   E. Peytremann. Astron. Astrophys., Vol. 30, 482 (1974).

065 Star Formation, Stellar Structure and Evolution, Neutron Stars

065.001 **Destruction of ^{14}N by ^{14}N$(e^-, \nu)^{14}$C$(\alpha, \gamma)^{18}$O in degenerate matter.** R. Mitalas.
Astrophys. Journ., Vol. 187, 155 - 158 (1974).

The destruction of ^{14}N by ^{14}N$(e^-, \nu)^{14}$C$(\alpha, \gamma)^{18}$O reactions in electron degenerate matter, where at densities $\sim 1 \times 10^6$ g cm^{-3} the electron Fermi energy is comparable to the ^{14}N electron-capture threshold energy of 156 keV, is investigated. The reaction is found to be unimportant in population II red-giant interiors.

065.002 **On the origin of the blue halo stars.**
A. V. Sweigart, J. G. Mengel, P. Demarque.
Astron. Astrophys., Vol. 30, 13 - 19 (1974).

The existence in the halo population of UV-bright stars lying above the horizontal branch in luminosity and blueward of the asymptotic branch is interpreted in terms of the final evolution of three stars with masses of 0.453, 0.51 and 0.60 $M_\odot$. The present computations cannot account for the subdwarf B stars. Difficulties associated with several interpretations of the subdwarf B stars are discussed.

065.003 **Low velocity M dwarfs and star formation.**
P. Biermann.
Astron. Astrophys., Vol. 30, 31 - 36 (1974).

Making the plausible assumption that the recently discovered high number of low velocity M dwarfs is local neither in space nor in time, two interesting consequences can be inferred: a) The velocity dispersion of Pop I stars is a property largely given to them at formation. b) The amplitude of the gravitational field of spiral structure changes only on time scales comparable with the lifetime of the Galaxy. The arguments leading to these conclusions give the basis for a discussion of the three known sites of star formation, OB associations, T Tauri associations and open clusters; it is speculated that clusters are formed from clouds and OB associations from intercloud medium.

065.004 **The theoretical low mass main sequence.**
A. S. Grossman, D. Hays, H. C. Graboske, Jr.
Astron. Astrophys., Vol. 30, 95 - 103 (1974).

The authors have investigated the structure of stars on the theoretical low mass main sequence for the mass range $0.085 - 0.5\, M_\odot$, spectral range M 0 to M 8. A chemical composition of $X = 0.68$, $Z = 0.03$ and l/H_p ratio of 1 were chosen. Constitutive physics for the lower main sequence requires interaction effects for the thermodynamic properties, intermediate electron screening effects for nuclear reactions, and a detailed treatment of all reactions in the p-p cycle. The resulting models show good agreement with the observational main sequence.

065.005 **On the influence of the opacity values on static stellar models. II. Main sequence stars.**
I. Parsian, S. Refsdal, R. Stabell.
Astron. Astrophys., Vol. 30, 275 - 280 (1974).

The method which was described in the first paper of this series is here applied to zero-age main-sequence stars of 0.5, 1, 1.8, 5, 9, and 15 $M_\odot$. Particularly the influence of opacity changes on luminosity and effective temperature is discussed. It is found that any small increase in the opacities anywhere in a main-sequence star will cause a decrease in luminosity and effective temperature.

065.006 **Secular stability of a 0.5 $M_\odot$ and a 1.1 $M_\odot$ star during hydrogen shell burning phases.**
M. Gabriel, A. Noels.
Astron. Astrophys., Vol. 30, 339 - 343 (1974).

Evolutionary models of 1.1 $M_\odot$ and .5 $M_\odot$ star in the hydrogen shell burning phase are tested for secular stability. Contrary to the suggestion of Balton and Eggleton, but in agreement with the discussion of Refsdal and Weigert, no instability is found.

065.007 **Multiple solutions and secular stability of a 7 $M_\odot$ star with core helium and shell hydrogen burning.**
D. Lauterborn, R. A. Siquig.
Astrophys. Journ., Vol. 187, 299 - 302 (1974).

A star of $M = 7\, M_\odot$ and $l/H_p = 0.5$, with a helium-burning core and a hydrogen-burning shell source, is shown to permit three equilibrium solutions over a certain range of core masses. Two of these solutions are secularly stable, while the third solution is unstable.

065.008 **Low-temperature photoneutron sources for stellar nucleosynthesis.** T. G. Harrison, T. W. Edwards.
Astrophys. Journ., Vol. 187, 303 - 311 (1974).

This paper investigates a neutron-source mechanism for stellar nucleosynthesis which consists of (p, γ) reactions followed by (γ, n) reactions between various nuclei involved in the proton-proton and CNO-cycle hydrogen-burning processes. Several photoneutron cycles including ^{13}C$(p, \gamma)^{14}$N and ^{14}N$(p, \gamma)^{15}$O followed by ^{13}C$(\gamma, n)^{12}$C are shown to be active in CNO burning regions of stars both on and off the main sequence. In addition to these primary photoneutron sources, several other photoneutron-source mechanisms are also investigated but are shown to be comparatively less active under similar environmental conditions.

065.009 **A photoneutron mechanism for the production of technetium-99 in the interior of evolved stars.**
T. W. Edwards, T. G. Harrison.
Astrophys. Journ., Vol. 187, 313 - 319 (1974).

A photoneutron mechanism for the production of ^{99}Te and possibly other radioactive isotopes is considered. This process operates coincidentally with the CNO cycle for hydrogen burning.

065.010 **Is the present stellar birthrate really determined by gas density?** D. C. Black, S. A. Kellman.
Astrophys. Space Sci., Vol. 26, 107 - 109 (1974).

Observed supernovae rates in Sb and Sc galaxies, and a recent re-examination of the mean gas density in these galactic types, implies that if the 'clumpiness' of gas in the disks of Sb and Sc galaxies is similar, the gas density is not the primary factor in determining the overall present stellar birthrate.

065.011 **Gravity modes in composite polytropic stars.**
M. Goossens, P. Smeyers.
Astrophys. Space Sci., Vol. 26, 137 - 151 (1974).

The authors study gravity modes of composite polytropic stars which consist of two convectively stable zones separated by a convectively unstable zone. In addition to the unstable gravity modes associated with the intermediate zone, they distinguish two types of stable gravity modes, one type being mainly associated with the core, the other one being mainly associated with the envelope. They find also some accidental 'resonances' between the core and the envelope.

065.012 **Multiple solutions of the equations of stellar structure. III. Secular stability of E models.**
D. Lauterborn, R. A. Siquig.
Astron. Astrophys., Vol. 30, 385 - 389 (1974).

It is shown that the secular behavior of stellar models with core helium and shell hydrogen burning can be inferred from their envelope structure only; the reaction of the core can be simulated with sufficient accuracy by appropriate boundary conditions at the bottom of the envelope. A large number of E model sequences is then tested for secular stability; it is found that in all cases the intermediate solution is secularly unstable.

065.013 **On circumstellar shells of protostellar origin.**
N. Berruyer.
Astron. Astrophys., Vol. 30, 403 - 409 (1974).

Numerical models of protostars indicate that the time scale for contraction of the envelope is greater than the contraction time of the core. The development of the H II region is calculated taking into account the protostellar origin of the circumstellar medium. The importance of the initial conditions in the protostellar medium is also discussed.

065.014 **The evolution of a massive protostar.**
I. Appenzeller, W. Tscharnuter.
Astron. Astrophys., Vol. 30, 423 - 430 (1974).

The hydrodynamic evolution of a massive protostar has been calculated starting from a homogeneous gas and dust cloud of 60 $M_\odot$ and an initial density of 10^{-19} g cm^{-3}.

065.015 **On the uniqueness of the solutions for stellar evolution.** H. Kähler, A. Weigert.
Astron. Astrophys., Vol. 30, 431 - 439 (1974).

The uniqueness of stellar evolution is investigated for the following usual approximations: (1) evolution of equilibrium models, (2) evolution on "thermal" time scales. For the second case, the concept of linear series is extended to non-equilibrium models, and non-static atmospheres have to be discussed. The connection between uniqueness and the stellar stability problem is also discussed.

065.016 **Asymptotic formulation of a differentially rotating star with small dissipation.** C. Sung.
Astrophys. Space Sci., Vol. 26, 305 - 317 (1974).

A system of nonlinear differential equations is formulated for a differentially rotating star with a small radiative dissipation. Certain dimensionless parameters are introduced, such as ϵ_1 to indicate the degree of small, but not necessarily infinitesimal, departure from the steady, axisymmetric state, and ϵ_2 to indicate the localization of perturbations etc. As examples of application, the Goldreich-Schubert and the Solberg-Hoiland instabilities are discussed.

065.017 **Neutrino pair emission by a magnetized stellar plasma.** N. J. Morgenstern Horing, C. Acquista.
Astrophys. Space Sci., Vol. 26, 387 - 390 (1974).

The decay of a plasmon into two neutrinos in the presence of an intense magnetic field has been studied by Canuto et al. (1970). They suggest that one of the principal longitudinal plasmon modes, which occurs only in magnetized plasmas, would cause certain magnetic stars to cool more rapidly than their unmagnetized counterparts. The authors show here that this mechanism is inoperative since the plasmon mode involved cannot be excited in the direction parallel to the magnetic field as considered by Canuto et al. Moreover, for $\omega_c/\omega_p \ll 1$, they show that the other principal longitudinal plasmon mode considered earlier by Adams et al. (1963) dominates the plasmon-neutrino decay cooling of magnetic stars.

065.018 **Off-center detonations in evolved stellar cores.**
J.-R. Buchler, J. C. Wheeler, Z. K. Barkat.
Astrophys. Space Sci., Vol. 26, 391 - 401 (1974).

Recent work on late stages of evolution has raised the possibility that the stars in the mass range from 4–8 $M_\odot$ may ignite carbon violently off-center. Hydrodynamic evolution has been carried out for models with various central densities and for temperature profiles both with and without an inversion. It has been found that, provided ignition leads to a detonation, both types of presupernova models have essentially the same final post detonation behavior. In particular, an off-center detonation is unable to prevent a complete disruption of the star if ignition occurs at low density.

065.019 **Magnetic braking during star formation I.**
J. Gillis, L. Mestel, R. B. Paris.
Astrophys. Space Sci., Vol. 27, 167 - 194 (1974).

The authors study the transport of angular momentum from a cool massive gas cloud by Alfvén waves travelling along the distorted magnetic field linking the cloud with the hot galactic background. The efficiency of braking is never so great as to keep the cloud even roughly corotating with the background.

065.020 **Pulsational instability towards non-radial oscillations in homogeneous stars of small mass.**
A. Noels, A. Boury, R. Scuflaire, M. Gabriel.
Astron. Astrophys., Vol. 31, 185 - 188 (1974).

The pulsational stability of a 0.5 $M_\odot$ star composed of a radiative core and an extended convective envelope is studied. The star is found strongly unstable because of the high temperature sensitivity of the He^3 reaction. This instability should affect only a very small range of masses.

065.021 **On the surface conditions for very massive stars.**
K. Nomoto, D. Sugimoto.
Publ. Astron. Soc. Japan., Vol. 26, 129 - 136 (1974).

The influence of atmospheric structure on internal stellar structure is studied for stars of masses $10^2 - 10^4 M_\odot$. Taking into account the effect of radiation pressure at vanishing optical depth, a correct photospheric boundary condition is obtained. When it is applied to the internal stellar structure with electron scattering opacity, there exist solutions in hydrostatic equilibrium, which have finite stellar radii, even for stars with dominant contribution of radiation pressure over gas pressure. This hydrostatic equilibrium is stationary in the sense that there is no steady mass loss from the star.

065.022 **Super-massive stars composed of matter and anti-matter.** W. Unno, M.-K. Fujimoto.
Publ. Astron. Soc. Japan, Vol. 26, 137 - 157 (1974).

Structure and stability are studied for super-massive stars composed of matter and anti-matter. The structure of the model is approximated by the polytrope of index 3, and the length scale of the model is fixed by the balance between the energy generation due to annihilation and the surface luminosity. Two types of matter-antimatter configurations, viz., the bubble model and the core-envelope model are considered. In both cases, models of $10^6 - 10^7 M_\odot$ give luminosity, total energy, and life time in order-of-magnitude agreement with quasars. Inconsistencies, however, arise, owing to the inefficieny of convective energy transport; the possible consequences are discussed.

065.023 **Measurement and theoretical analysis of some reaction rates of interest in silicon burning.**
A. J. Howard, H. B. Jensen, M. Rios, W. A. Fowler, B. A. Zimmerman.
Astrophys. Journ., Vol. 188, 131 - 139 (1974).

Total cross-sections for the ^{27}Al$(\alpha, n)^{30}$P, ^{35}Cl$(\alpha, n)^{38}$K, ^{46}Ti$(\alpha, n)^{49}$Cr, ^{40}Ca$(\alpha, p)^{43}$Sc, and ^{45}Sc$(p, n)^{45}$Ti reactions have been measured in the energy range of importance for nucleosynthesis in stellar silicon burning. Stellar reaction rates, $N_A \langle \sigma v \rangle$, have been deduced from the observed reaction cross-sections and have been represented approximately by theoretically reasonable analytic functions of the temperature. The effect of excited states in target and residual nuclei is

065.024 **Constraints on the evolutionary history of stars showing s-processed material.** B. M. Schlesinger.
Astrophys. Journ., Vol. 188, 141 - 143 = Publ. Goethe Link Obs., Indiana Univ., Bloomington, No. 156 (1974).

In some N carbon stars whose spectra show that other products of the s-process are present, technetium is absent, probably because it has decayed. The half-life of the technetium isotope produced by the s-process is 2×10^5 years; any production of technetium in these stars must therefore have occurred at least 5×10^5 years ago. Stars of mass greater than $\sim 9\,M_\odot$ are red supergiants for 2×10^5 years or fewer after they exhaust core helium. It follows that unless the stars discussed here have masses less than about $9\,M_\odot$, some production of heavy elements by the s-process occurs before core helium has been exhausted.

065.025 **Thermal pulses in helium shell-burning stars. III.** R. A. Gingold, D. J. Faulkner.
Astrophys. Journ., Vol. 188, 145 - 148 (1974).

The thermal and pulsational stability of a $0.7\,M_\odot$ pre-white-dwarf helium shell-burning star previously evolved by Wood and Faulkner has been examined and the nature of the instability found by them is confirmed.

065.026 **ORELA neutron capture and stellar nucleosynthesis.** B. J. Allen, R. L. Macklin.
Atomic Energy Australia, Vol. 16, No. 2, p. 14 - 21 (1973).

065.027 **A case of metastability for slowly rotating, supermassive objects.** H. Dedic, J.-L. Tassoul.
Astrophys. Journ., Vol. 188, 173 - 179 (1974).

This paper deals with finite-amplitude axisymmetric pulsations of slowly rotating, supermassive objects. Fully nonlinear isentropic motions are described by means of Lagrangian variables. The scalar virial theorem in the first post-Newtonian approximation of general relativity provides an approximate description of the oscillations of uniformly rotating bodies. A criterion for unconditional stability toward pseudo-radial pulsations is given in a practical form.

065.028 **Convection in A stars: convection suppressed in magnetic A stars?** D. J. Mullan.
Irish Astron. Journ., Vol. 11, 32 - 82 (1973).

This is a discussion of how convection in A stars is expected to be affected by the presence of magnetic fields. The paper is divided into two parts. Part I contains a non-technical account of how magnetic fields in sunspots and in stars can suppress convection, and how various features of magnetic A stars can be explained in the light of suppressed convection. Part II provides a more theoretical account of suppression of convection using recently computed models of convection zones in main sequence stars. This part of the article is also meant to serve as a review of the field. New results and speculations are summarized in the final section of Part II.

065.029 **Electron capture in carbon dwarf supernovae.** T. J. Mazurek, J. W. Truran, A. G. W. Cameron.
Astrophys. Space Sci., Vol. 27, 261 - 291 (1974).

The rates of electron capture on heavier elements under the extreme conditions predicted for dwarf star supernovae have been computed. An estimate of the maximum possible value of such rates is also given. The distribution of nuclei in nuclear statistical equilibrium has been calculated for the range of expected supernovae conditions. These nuclide abundance distributions are then used to compute nuclear equilibrium thermodynamic properties. The effects of the electron capture on such equilibrium matter are discussed. The results of supernova numerical hydrodynamics incorporating the computed equilibrium properties and the influence of electron capture are presented.

065.030 **A note on anisotropic convection and the rotation of stellar convective zones.** I. W. Roxburgh.
Astrophys. Space Sci., Vol. 27, 419 - 424 (1974).

The interaction of rotation and convection produces a latitude dependent anisotropic turbulent viscosity. When this friction is dominant, equilibrium of a convective outer layer of a star is achieved by an equatorial acceleration and a two-zone circulation, towards the equator at low surface latitudes.

065.031 **The photon-neutrino coupling theory and stellar models.** O. Vilhu.
Astrophys. Space Sci., Vol. 27, 513 - 515 (1974).

The effect of the photon-neutrino coupling theory, developed by Bandyopadhyay (1968), on stellar structure is studied by constructing a chemically homogeneous model of one solar mass in thermal equilibrium, and taking into account the neutrinos according to the theory. As anticipated, the effect is large and leads to great difficulties in the understanding of stellar constitution.

065.032 **Carbon burning with convective URCA neutrinos.** E. Ergma, B. Paczyński.
Acta Astron., Vol. 24, 1 - 16 (1974).

Results of simplified model computations for highly degenerate carbon-oxygen cores of intermediate mass stars are presented. Core evolution towards carbon ignition is followed for a variety of chemical compositions. Carbon burning with the convective URCA process on Na^{23} and Ne^{21} taken into account is studied for the initial carbon content of $(X_{12})_0 = 0.5$. The uncertainty of the results is discussed.

065.033 **Theory of level surfaces inside relativistic, rotating stars. II.** M. A. Abramowicz.
Acta Astron., Vol. 24, 45 - 53 (1974).

Some theorems connected with the properties of the level surfaces of pressure and angular velocity are derived and used to a discussion of the problem of thermodynamical equilibrium of a differentially rotating, relativistic star.

065.034 **On slowly rotating homogeneous masses in general relativity.** S. Chandrasekhar, J. C. Miller.
Monthly Notices Roy. Astron. Soc., Vol. 167, 63 - 79 (1974).

The present paper is devoted to a study of slowly rotating homogeneous masses in which the energy density is a constant. The structure of such configurations is determined with the aid of equations derived by Hartle in the exact framework of general relativity. These configurations have a natural limit in that the static, non-rotating, configurations must have radii (R) exceeding 9/8 times the Schwarzschild radius (R_S). The derived structures, for varying R/R_S, are illustrated by a series of graphs.

065.035 **On the analogy between neutron star models and isothermal gas spheres and their general relativistic instability.** S. Yabushita.
Monthly Notices Roy. Astron. Soc., Vol. 167, 95 - 102 (1974).

Investigation is made of the structure of gaseous spheres such that their equation of state is $p = q\rho$, where p and ρ are pressure and energy-density, respectively and q is a constant, and the relation is obtained between the central density and the mass of the gaseous spheres such that they exert the same pressure at the boundary. Assuming that the adiabatic index γ is related to q by the relation $\gamma = 1+q$, equation for radial oscillation about equilibrium is derived. By making use of the neutral mode solution, the degree of compression needed to make the sphere dynamically unstable is calculated.

065.036 **Effect of mass loss by stellar winds on the pre-main sequence stage of stellar evolution.**
S. Cuperman, A. Sternlieb.
Monthly Notices Roy. Astron. Soc., Vol. 167, 183 - 187 (1974).

Modified contraction times and total mass losses for some pre-main sequence stars which lose mass by stellar winds, have been calculated. In this, time varying rates of mass loss as deduced from the statistics of T-Tauri stars have been used. The results are compared with those corresponding to constant mass evolution.

065.037 **Properties and synthesis of heavy nuclei and properties of neutron star matter.** J. R. Buchler.
IAU Symposium No. 53, (see 012.004), p. 67 - 75 (1974).

065.038 **Quantum crystals in neutron stars.**
V. Canuto, S. M. Chitre.
IAU Symposium No. 53, (see 012.004), p. 133 - 150 (1974).

065.039 **Superfluidity in neutron stars.** G. Greenstein.
IAU Symposium No. 53, (see 012.004), p. 151 - 165 (1974).

065.040 **Neutron star structure from pulsar observations.**
D. Pines, J. Shaham, M. A. Ruderman.
IAU Symposium No. 53, (see 012.004), p. 189 - 207 (1974).

065.041 **Cooling of dense stars.** S. Tsuruta.
IAU Symposium No. 53, (see 012.004), p. 209 - 225 (1974).

065.042 **Hadron star models.**
J. M. Cohen, G. Börner.
IAU Symposium No. 53, (see 012.004), p. 237 - 250 (1974).

065.043 **Differential rotation in degenerate stars.**
H. M. Van Horn.
IAU Symposium No. 53, (see 012.004), p. 251 - 264 (1974).

065.044 **A self-similar gravitational contraction of protostars composed of molecular hydrogen.**
I. G. Kolesnik, D. K. Nadezhin.
Astron. Zhurn. Akad. Nauk SSSR, Vol. 51, 382 - 387 (1974). In Russian. English translation in Soviet Astron. AJ, Vol. 18, No. 2.

For a molecular hydrogen protostar a self-similar solution was found at the stage of gravitational contraction with volume energy losses.

065.045 **On the critical luminosity at the accretion and on shell sources of energy.** N. I. Shakura.
Astron. Zhurn. Akad. Nauk SSSR, Vol. 51, 441 - 443 (1974). In Russian. English translation in Soviet Astron. AJ, Vol. 18, No. 2.

The critical accretion on a neutron star in spherical case limited by the value of the critical Eddington luminosity is calculated. The influence of the shell sources of energy on the picture of matter infall is considered.

065.046 **The occurrence of a nonspherical thermal instability in red giant stars.** D. O. Richstone.
Astrophys. Journ., Vol. 188, 327 - 333 (1974).

The nonspherical thermal instability has been investigated for shell-burning stars, with special attention to the character of the problem at the boundaries of the shell. Approximate stability conditions are derived, and an analogy with the spherical instability (shell flash) is drawn. A number of models containing hydrogen- or helium-burning shells have been tested by numerical methods for stability against nonspherical thermal perturbations.

065.047 **Carbon and eruptive stars: surface enrichment of lithium, carbon, nitrogen, and ^{13}C by deep mixing.**
I.-J. Sackmann, R. L. Smith, K. H. Despain.
Astrophys. Journ., Vol. 187, 555 - 574 (1974).

It appears possible to form carbon stars from stars of both low ($\sim 1\,M_\odot$) and intermediate ($\sim 5\,M_\odot$) masses and from both population types. A single helium shell flash followed by deep mixing suffices for the former case, but on the order of 100 such flashes are required for the latter. The Cameron-Fowler mechanism for the production of high lithium abundances in late-type stars is shown to work satisfactorily, producing ^{7}Li/H up to about 10^{-7} in a single flash. Eruptive stars may result when the deep convective envelopes reach into helium-burning layers. Large amounts of energy can be liberated in a very short time; e.g., the binding energy of the envelope, 10^{48} ergs, can be exceeded in $\sim 10^6$ sec. Many of the gross features of the R Coronae Borealis stars can be understood by erupting carbon star models. For an unrepeated deep mixing, an upper limit can be placed on the mass of a carbon star, namely, $\sim 2\,M_\odot$. Repeated deep mixing can explain the observations known for S, Ba II, and C stars when they are considered to be intermediate-mass objects.

065.048 **On a nebular model for Wolf-Rayet stars.**
T. Nugis.
Tartu Astr. Obs.,Teated No. 44, p. 3 - 17 (1973). In Russian.

Some considerations are given to prove that the nebular model of Beals is well acceptable for explaining the phenomenon of Wolf-Rayet stars. The cause of the absence of any detectable phase shift which ought to be introduced by the transit-time effect in radial velocity curves of the binary V 444 Cyg (OB+WN 5) is discussed.

065.049 **Detailed study of the fate of the isotopic (σn) correlations in possible s-process conditions.**
M. Arnould.
Astron. Astrophys., Vol. 31, 371 - 380 (1974).

The present work examines the restrictions which could be put on the physical conditions which have possibly prevailed during the s-process(es) responsible for the solar system s-elements synthesis, and particularly on typical mean neutron irradiations.

065.050 **Diffusion, He-weak stars, and ^{3}He in 3 Cen A.**
S. Vauclair, G. Michaud, Y. Charland.
Astron. Astrophys., Vol. 31, 381 - 390 (1974).

Microscopic diffusion below the helium convection zone in A–B type stars leads to an underabundance of helium by a factor of thirty or so and to a relative overabundance of ^{3}He. Saturation of the radiation field is shown to put a limit on the helium that can be supported by the radiation field to at most 1% by number of the hydrogen abundance. The most important process is the photoionization of He II.

065.051 **Static criteria for stability of arbitrarily rotating stars.** G. S. Bisnovatyi-Kogan, S. I. Blinnikov.
Astron. Astrophys., Vol. 31, 391 - 404 (1974).

Static criteria for the hydrodynamical stability of stars are discussed. New formulations of static criteria, which are suitable for studying the stability of nonrotating and rapidly rotating isentropic stars are proposed. The practical use of the method is illustrated by numerical examples for some stellar models. The possibilities of generalization to the case of general relativity and frozen-in toroidal magnetic fields are also discussed, together with the difficulties thereby introduced.

065.052 **On stellar birthrates and age distributions.**
B. M. Tinsley.
Astron. Astrophys., Vol. 31, 463 - 465 (1974).

A general relationship is derived between the age distribution of A–F stars and the time-dependence of their birthrate.

Two recently derived age distributions, each for several hundred nearby evolving stars, are shown to be consistent with a birthrate that has decreased over the past 7 billion years, with e-folding time about 5 billion years.

065.053 Aflevering 1 - zwaartekracht en sterevolutie.
J. Stollman, P. Hut.
Zenit, Vol. 1, No. 4, p. 21 - 23 (1974).

065.054 Compounds in neutron-star crusts.
T. A. Witten, Jr.
Astrophys. Journ., Vol. 188, 615 - 625 (1974).

The outer crust of a neutron star is thought to consist of crystalline matter with bare nuclei embedded in a relativistic electron gas making up the lattice. The author gives evidence that chemical compounds are stable in this environment. He also presents rough dispersion curves for the HeFe compound and zero-point energies for two related compounds, and generalizes two theorems about Coulomb lattice vibrations to the case of compounds.

065.055 On the numbers, birthrates and final states of moderate- and high-mass stars.
J. P. Ostriker, D. O. Richstone, T. X. Thuan.
Astrophys. Journ., (Letters), Vol. 188, L87 - L89 (1974).

Using counts of stars in the solar neighborhood and conventional stellar-lifetime calculations, the authors find death rates of stars in the mass ranges ($4 \leq M/M_\odot \leq 8$) and ($8 < M/M_\odot$) and, assuming a steady state, show that the current death rate of high-mass stars is insufficient (by a factor ~5) to account for the estimated birthrate of pulsars and that the probable death rate of moderate-mass stars integrated over the age of the Galaxy is too high (by a factor >60) to allow them to produce and eject $1.4 M_\odot$ of iron-peak elements each.

065.056 Nuclear goblins and cosmic gamma ray bursts.
F. Zwicky.
Astrophys. Space Sci., Vol. 28, 111 - 114 (1974).

Among the cosmic and the macroscopic bodies composed of nuclear matter, the neutron stars and the nuclear goblins represent typical configurations. Nuclear goblins are stable only under large external pressure. When propelled to the surface of their parent stars, and there exploding, they produce flares. When ejected from their parent stars and exploding in interstellar space they will give rise to short bursts of gamma rays with energies centered around 0.78 MeV.

065.057 The effect of a toroidal magnetic field on the non-radial oscillations of polytropes.
M. J. Miketinac.
Astrophys. Space Sci., Vol. 28, 193 - 203 (1974).

Frequencies of non-radial oscillation of polytropic models of stars, belonging to spherical harmonics of orders l = 1, 2 and 3, are evaluated in a 'second approximation' by a variational method. Equilibrium configurations in the presence of toroidal magnetic fields are obtained numerically without any restriction on the field strength. The value of the ratio of the specific heats is assumed to be equal to 5/3 and only two polytropic indices, n = 1.5 and 3.0, are considered.

065.058 Effects of flattening on the stability and pulsations of rotating neutron stars.
Yu. L. Vartanyan, A. M. Kechiyants, A. V. Ovsepyan.
Dokl. AN ArmSSR, Vol. 56, No. 5, p. 290 - 294 (1973). In Russian. – Abstr. in Referativ. Zhurn. 51. Astron., 4.51.516 (1974).

065.059 Pulsational stability of stars in thermal imbalance. II. An energy approach.
W. R. Davey, J. P. Cox.
Astrophys. Journ., Vol. 189, 113 - 124 (1974).

A general expression is derived for the rate of change of the total pulsation energy (kinetic plus potential) of a star in thermal imbalance. It is shown that no intrinsically second-order terms appear for the case of infinitesimal oscillations. Quantitative agreement with Simon and with Kato and Unno is obtained in certain special cases. It is shown, however, that this energy approach cannot be the correct approach to the thermal imbalance problem. Moreover, this approach leads to disagreement with a result due to Unno and also with the results of the "small perturbation" approach used by Cox, Hansen, and Davey.

065.060 Do helium-shell flashes cause extensive mixing in low-mass stars? A. V. Sweigart.
Astrophys. Journ., Vol. 189, 289 - 291 (1974).

The effects due to the mixing of hydrogen into the convective zone produced by a helium-shell flash have been investigated for a $0.7 M_\odot$ population II star. A deep penetration of the envelope convection does not occur. Extensive mixing seems unlikely during the evolutionary phase considered.

065.061 A comparison of variable and nonvariable stars in the cepheid strip.
E. G. Schmidt, J. D. Rosendhal, C. P. Jewsbury.
Astrophys. Journ., Vol. 189, 293 - 302 (1974).

Spectra of three cepheids (RX Aur, X Cyg, and T Mon) and 11 nonvariable supergiants are analyzed. Additional evidence is presented for the presence of nonvariables within the cepheid strip. Temperatures determined from Hα line profiles and broad-band photometry were used together with model atmospheres and line strengths from the spectra to determine the gravities, turbulent velocities, and element abundances.

065.062 Post-Newtonian neutron stars.
R. V. Wagoner, R. C. Malone.
Astrophys. Journ., (Letters), Vol. 189, L75 - L78 (1974).

Models of neutron stars are constructed within the framework of the parametrized post-Newtonian approximation of metric theories of gravity. It is found that the observed masses of both the neutron star and the ordinary star in a binary system can differ considerably from the general-relativistic predictions.

065.063 Absorption of high energy heavy nuclei and γ rays at the surface of hot neutron stars.
T. N. Rengarajan.
Nature, Vol. 248, 569 - 571 (1974).

The author points out that high energy heavy nuclei and γ rays may also be absorbed because of their interactions with the blackbody photons emitted from the surface of the hot neutron star.

065.064 Rotation in late stages of stellar evolution.
J. Hardorp.
Astron. Astrophys., Vol. 32, 133 - 136 (1974).

Empirical arguments favor the view that the inner core of a star exchanges angular momentum with the envelope during evolution.

065.065 Secular stability of stars with central helium burning.
M. Gabriel, S. Refsdal, H. Ritter.
Astron. Astrophys., Vol. 32, 217 - 221 (1974).

A complete secular stability analysis, including a search for complex eigenvalues, has been performed for central He-burning stars of $7 M_\odot$ and $9 M_\odot$, and with l/H_p = 1.5. For the $9 M_\odot$ star the authors find that three static solutions with central He-burning are possible for certain chemical profiles, and that two of these solutions are secularly stable, whereas one is unstable. When only one static solution with central He-burning is allowed for a given chemical profile, the solution is always found to be secularly stable. This occurs often for

9 $M_\odot$ and always for 7 $M_\odot$.

065.066 **Stars with He-cores.** S. Refsdal.
Mitt. Astron. Ges., No. 34, p. 65 - 74 (1973/74).
Presented at the "Wissenschaftliche Tagung der Astron. Ges., Oberkochen, 1973 April 24 - 27".

065.067 **Exploding supermassive stars.**
I. Appenzeller, W. Tscharnuter.
Mitt. Astron. Ges., No. 34, p. 75 (1973/74).– Abstr. AG.

065.068 **Kollaps und Explosion eines rotierenden supermassiven Sterns.** J. Schmidt.
Mitt. Astron. Ges., No. 34, p. 75 - 78 (1973/74). – Presented at the "Wissenschaftliche Tagung der Astron. Ges., Oberkochen, 1973 April 24 - 27".

065.069 **Dynamical phases of supermassive stars.** K. Fricke.
Mitt. Astron. Ges., No. 34, p. 79 (1973/74).
Abstr. AG.

065.070 **Explosive nucleosynthesis in supermassive stars.**
K. Fricke, J. M. Audouze.
Mitt. Astron. Ges., No. 34, p. 79 (1973/74). – Abstr. AG.

065.071 **On the influence of opacity values on zero-age main-sequence models.**
I. Parsian, S. Refsdal, R. Stabell.
Mitt. Astron. Ges., No. 34, p. 81 (1973/74). – Abstr. AG.

065.072 **Ist die zeitliche Entwicklung eines Sterns eindeutig bestimmt?** H. Kähler, A. Weigert.
Mitt. Astron. Ges., No. 34, p. 84 (1973/74). – Abstr. AG.

065.073 **Berechnung des Emissionsspektrums einer Protosternhülle im Infrarot-Bereich.** D. Sieber.
Mitt. Astron. Ges., No. 34, p. 92 - 95 (1973/74). – Presented at the "Wissenschaftliche Tagung der Astron. Ges., Oberkochen, 1973 April 24 - 27".

065.074 **On the secular stability of models with an isothermal core.** M. Gabriel, M. L. Roth.
Astron. Astrophys., Vol. 32, 309 - 315 (1974).
The secular stability of models with an isothermal core is discussed. The determinant of secular stability for $\sigma = 0$ is related to quantities defined as far as possible at the surface of the core and at the bottom of the envelope. This relation is used to discuss the validity of 2 intuitive criteria for secular stability.

065.075 **γ-ray lines from accreting neutron stars.**
C. Reina, A. Treves, M. Tarenghi.
Astron. Astrophys., Vol. 32, 317 - 320 (1974).
The gravitational energy of an ion at the surface of neutron star is ~ 100 MeV/nucl. Therefore infalling ions may induce nuclear reactions in the external layer of the star. The γ-ray line emission produced by this process in accreting X-ray sources is evaluated in terms of the X-ray flux, assuming that the infalling matter and the stellar atmosphere have universal abundances.

065.076 **Surface composition of neutron stars that are accreting matter.** S. Tsuruta, R. Ramaty, G. Börner.
IAU Symposium No. 64, (see 012.008), p. 186 (1974). – Abstract.

065.077 **The structure of late type stars and the Hayashi track.** F. Caputo, V. Castellani, F. A. D'Antona.
Astrophys. Space Sci., Vol. 28, 303 - 324 (1974).
A topological approach in the HR diagram is given for the occurrence and the characteristics of convective envelopes in population II stars. The location of the Hayashi track and of the red giants branch are both investigated for various assumed masses of the stars and chemical compositions of the convective layers. An analysis is performed on the influence of the assumed mixing length. The observational consequences of a possible mass loss is discussed for both the red giants and the asymptotic branches. Possible causes for separation between these two branches are briefly examined. The case of the globular cluster ω Cen is investigated, in order to give some suggestions for a large dispersion in colour observed among the giants belonging to this cluster.

065.078 **The stability of stellar masses in general relativity.**
S. Chandrasekhar.
Galaxies and relativistic astrophysics, Proc. 1972, (see 012.010), p. 162 - 165 (1974). – Invited lecture.

065.079 **Stability of non-radial vibrational modes of relativistic neutron stars.** P. Cazzola, L. Lucaroni.
Galaxies and relativistic astrophysics, Proc. 1972, (see 012.010), p. 181 - 182 (1974).

065.080 **Generation of a stellar magnetic field by convection in the outer regions.**
L. Eh. Gurevich, E. D. Ehjedel'man.
Magnit. gidrodinamika, 1973, No. 4, p. 3 - 11. In Russian. Abstr. in Referativ. Zhurn. 51. Astron., 5.51.492 (1974).

065.081 **Die Abhängigkeit der Schönberg-Chandrasekhar-Grenze vom Wasserstoff-Profil.** M. L. Roth.
Mitt. Astron. Ges., No. 34, p. 82 - 83 (1973/74). – Presented at the "Wissenschaftliche Tagung der Astron. Ges., Oberkochen, 1973 April 24 - 27".

065.082 **The decay of magnetic fields in stars.**
G. A. E. Wright.
Monthly Notices Roy. Astron. Soc., Vol. 167, 527 - 537 (1974).
Assuming that the magnetic stars are in a state of radiative equilibrium with no circulation, it was shown in earlier work that field configurations are possible with surface fields far higher than those observed. It is shown in the present work that normal Ohmic decay could not have dissipated these fields within the stellar lifetime, leading to the conclusion that either the decay is accelerated by the action of instabilities or the star has lost most of its 'primeval' flux before reaching the main-sequence.

065.083 **Supergiant stars as chemical composition indicators.**
G. Bertelli, C. Chiosi.
Astron. Astrophys., Vol. 32, 399 - 406 (1974).
A new calibration relation in terms of the metal content Z is found for evolutionary models of massive stars in the post main sequence phases. It turns out that the mean difference between red and blue visual magnitudes of supergiant stars linearly depends on the metal content only. This relation is applied to investigate a possible variation of the heavy elements abundance of very young stars in galaxies of different luminosity and morphological type.

065.084 **Origin of neutron star magnetic fields.**
E. H. Levy, W. K. Rose.
Nature, Vol. 250, 40 - 41 (1974).
The uniformity of neutron star magnetic moments, as inferred from the observed properties of pulsars, seems to indicate that the magnetic fields are produced with large spatial scales as well as with uniform intensities. The authors suggest that convection during the carbon-burning stage (before collapse to a neutron star) in a massive, rapidly and differentially rotating, degenerate stellar core, leads to the production of a large scale magnetic field within the core itself, through

the action of a hydromagnetic dynamo.

065.085 Neutron star matter and neutron star models.
H. Heintzmann, W. Hillebrandt, M. F. El Eid, E. R. Hilf.
Zeitschr. Naturforschung, Vol. 29a, 933 - 946 (1974).

Various methods to study the ground state of neutron star matter are compared and the corresponding neutron star models are contrasted with each other.

065.086 The Doppler splitting of spectral lines in pulsating stars. A. J. Skalafuris.
Astrophys. Journ., Vol. 190, 91 - 93 (1974).

Uniformly accelerated hydrodynamic flow, with radiation pressure and dissipation, is considered. It is shown that during cepheid contractions only shock waves above a minimum strength can be generated by the star's dynamic gravity; hence strong lines are first to show the velocity discontinuity. Population I stars have small radial variations and cannot reach this minimum shock strength, hence can at most reveal only an initial set of spectral lines.

065.087 s-process nucleosynthesis in massive stars: core helium burning.
R. G. Couch, A. B. Schmiedekamp, W. D. Arnett.
Astrophys. Journ., Vol. 190, 95 - 100 (1974).

The authors have investigated the nucleosynthesis which takes place via neutron capture during core helium burning in massive stars ($M \gtrsim 15\ M_\odot$). The neutrons are produced by the ^{22}Ne$(\alpha, n)^{25}$Mg reaction. It was found that complete conversion of ^{22}Ne to ^{25}Mg can take place only in the most massive stars ($M \gtrsim 30\ M_\odot$). This study indicates that it is particularly important to obtain experimental data on the neutron-capture cross-sections of ^{22}Ne, ^{25}Mg, and ^{58}Fe.

065.088 Horizontal-branch evolution with semiconvection. I. Interior evolution. A. V. Sweigart, P. G. Gross.
Astrophys. Journ., Vol. 190, 101 - 107 (1974).

The evolution of a 0.66 $M_\odot$ star has been followed through the horizontal-branch phase with a full treatment of both convective overshooting and semiconvection. The dependence of the interior structure on the degree of superadiabaticity within the semiconvective zone is determined. A discussion of the input physics and numerical procedure is included.

065.089 Are the UV stars nuclear-powered? J. G. Hills.
Astrophys. Journ., Vol. 190, 109 - 115 (1974).

The UV stars (hot pre-white dwarfs) may be fueled by nuclear energy rather than merely by gravitational contraction as has usually been assumed. Evolutionary models of white dwarfs without nuclear burning are used to determine their contribution to the observed far-ultraviolet excess in galaxies and their frequency in stellar systems.

065.090 Relativistic stellar stability: preferred-frame effects. W.-T. Ni.
Astrophys. Journ., Vol. 190, 131 - 136 (1974).

In a previous paper, the PPN (parametrized post-Newtonian) formalism was used to analyze relativistic influences on stellar stability in nearly all metric theories of gravity. That analysis omitted all "preferred-frame" terms. In this paper, possible preferred-frame effects on stellar stability are examined and no new instabilities are found.

065.091 Rotating superfluid in neutron stars.
M. A. Ruderman, P. G. Sutherland.
Astrophys. Journ., Vol. 190, 137 - 139 (1974).

A rotating perfect fluid subject to axisymmetric, azimuthal, time-independent forces is shown to have a uniform angular acceleration on each coaxial cylindrical surface. In a rotating superfluid, uniform rotation is mimicked by a dense corotating paraxial array of quantized vortices. Despite forces which may vary greatly along individual vortex lines, the vortices move radially but remain parallel to the rotation axis. This resistance to bending suppresses the development of superfluid turbulence in the interiors of rapidly rotating neutron stars.

065.092 Photon opacity in surfaces of magnetic neutron stars.
J. Lodenquai, V. Canuto, M. Ruderman, S. Tsuruta.
Astrophys. Journ., Vol. 190, 141 - 152 (1974).

Approximate expressions are derived for free-free, bound-free, and Thomson cross-sections of photons by gaseous matter in the presence of superstrong magnetic fields. For photons in modes whose electric field polarization is perpendicular to this magnetic field, the cross-section is reduced by approximately the squared ratio of the photon frequency to the electron cyclotron frequency if this ratio is small.

065.093 An excitation mechanism for pulsations in Beta Cephei stars. Y. Osaki.
Astrophys. Journ., Vol. 189, 469 - 477 (1974).

A mechanism which might be responsible for the excitation of pulsations in β Cephei stars is proposed in which an eigenmode of the nonradial oscillations of the whole star resonates with the overstable convection in the rapidly spinning core of a massive star. The equations for adiabatic, nonradial oscillations have been solved for a realistic stellar model, and it is shown that the eigenfunction of the f-mode has large amplitudes only in the convective core and near the surface, which is a situation favorable to the operation of this mechanism.

065.094 An interpretation of the puzzling observations of FG Sagittae. I.-J. Christy-Sackmann, K. H. Despain.
Astrophys. Journ., Vol. 189, 523 - 530 (1974).

The three major phenomena observed in FG Sge—namely, (1) the existence of a planetary nebula, (2) the large increase in apparent brightness with the extremely fast, continued increase in spectral type, and (3) the observed increases in surface abundances with their respective time scales of ~6000 years, ~100 years, and ~1 year—can all be consistently explained in terms of helium-shell flashes, with the last flash taking place on the order of 10^2 years ago.

065.095 Nucleosynthesis of lithium in low-energy flares. R. Canal.
Astrophys. Journ., Vol. 189, 531 - 533 (1974).

New values for the rates of Li formation in low-energy flares occurring on stellar surfaces are reported.

065.096 Dynamical phases of rotating supermassive stars. K. J. Fricke.
Astrophys. Journ., Vol. 189, 535 - 542 (1974).

The dynamical evolution of supermassive stars including angular momentum has been calculated using a simple approximation procedure. The results indicate that—regardless of the amount of angular momentum—explosions do not occur as long as metals are initially absent. For normal population I composition, gigantic explosions are found involving masses from 4×10^5 to $2 \times 10^8\ M_\odot$, and explosion energies between 10^{56} and 10^{60} ergs which are also typical for the most violent local events observed in the universe.

065.097 Differences between the evolutionary tracks of young stars in the Galaxy and in the Magellanic Clouds. G. L. Hagen, S. van den Bergh.
Astrophys. Journ., (Letters), Vol. 189, L103 - L104 (1974).

The evolutionary tracks of massive young stars in the Galaxy are found to differ systematically from those of similar stars in the Small Magellanic Cloud. Evolving stars in the Large Cloud have characteristics that are intermediate between those

065.098 **On the semiconvective zone in the core-helium-burning stars.** H. Saio.
Astrophys. Space Sci., Vol. 29, 41 - 49 (1974).

The validity of current ideas to set up the semiconvective zone of stellar models in core-helium-burning phase has been considered mainly from the view point of time-scales related to the mechanism. The time-scale of outward motion of a convective shell has been estimated as being longer than that of evolution. From this point of view, it might be said that the models constructed with semiconvective zone are not appropriate for the core-helium-burning stars.

065.099 **Neutron stars and comets.** I. S. Shklovskij.
Astron. Zhurn. Akad. Nauk SSSR, Vol. 51, 665 - 667 (1974). In Russian. English translation in Soviet Astron. AJ, Vol. 18, No. 3.

Soft gamma-ray bursts can be generated during accretion of cometary material by neutron stars and white dwarfs.

065.100 **Population type and evolutionary state of carbon stars.** G. Barbaro, N. Dallaporta.
Astron. Astrophys., Vol. 33, 21 - 32 (1974).

In order to gain some insight into the role of carbon stars in the general pattern of stellar evolution, a phenomenological and theoretical analysis whose aim is to obtain their population types and evolutionary phases is conducted on the assumption that their spectra are due to mixing of the envelope with layers which have undergone nuclear processing.

065.101 **Models for nuclei of planetary nebulae and ultraviolet stars.**
J. I. Katz, R. C. Malone, E. E. Salpeter.
Astrophys. Journ., Vol. 190, 359 - 363 (1974).

A series of stellar models was evolved, all with a total mass of $0.65 M_\odot$, an initial carbon-oxygen core of mass $0.60 M_\odot$, an intermediate helium mantle, and an outer hydrogen-rich envelope, with mass varying from case to case. Although the most hydrogen-rich cases resulted in red giants, cases with less than $0.01 M_\odot$ in the hydrogen envelope evolved at high surface temperature. The early stages of development of these models are similar to observed central stars of planetary nebulae. The later stages still have a high luminosity; the relevance to "ultraviolet stars" is discussed.

065.102 **Finite nuclear size effects on neutrino-pair bremsstrahlung in neutron stars.** E. Flowers.
Astrophys. Journ., Vol. 190, 381 - 383 (1974).

The author considers the effects on neutrino-pair bremsstrahlung emission by electrons of the finite size of the charge distributions scattering the electrons. The neutrino bremsstrahlung emission from electrons scattering from a static lattice including finite size effects is evaluated. Two different descriptions of the nuclei present in neutron-star matter are considered.

065.103 **Correlation effects on the energy shifts of excited nucleons in neutron-star matter.** A. Nandy.
Astrophys. Journ., Vol. 190, 385 - 390 (1974).

The effects of correlations among the hadrons due to the presence of a universal repulsive core are considered and an upper limit found to the change in mass shifts obtained by Sawyer. This is seen to be rather large, and indicates that correlation effects have to be included in calculations of this type. The significance of these effects on the equation of state is also discussed.

065.104 **More about the nucleosynthesis of the nuclei between carbon and neon.**
M. Arnould, W. Beelen.
Astron. Astrophys., Vol. 33, 215 - 230 (1974).

Using Arnett's (1969) model of explosive nucleosynthesis, and as an extension of the work of Howard et al. (1971), the authors study the explosive processing of four compositionally different types of layers possibly surrounding an evolved C-O stellar core.

065.105 **The stability of stars in thermal imbalance.**
M. L. Aizenman, J. P. Cox.
Bull. American Astron. Soc., Vol. 6, 211 (1974). — Abstr. AAS.

065.106 **Electron-ion relaxation in a detonation.**
J. E. Littleton, J.-R. Buchler.
Bull. American Astron. Soc., Vol. 6, 211 - 212 (1974). Abstr. AAS.

065.107 **Are stellar surface heavy-element abundances systematically enhanced?** P. C. Joss.
Bull. American Astron. Soc., Vol. 6, 216 - 217 (1974). Abstr. AAS.

065.108 **Vela X1: neutron star or black hole?**
G. Wallerstein, D. R. Mikkelsen.
Bull. American Astron. Soc., Vol. 6, 276 (1974). — Abstr. AAS.

065.109 **Effects of differential rotation on magnetic fields in rapidly rotating degenerate dwarfs.** K. Brecher.
Bull. American Astron. Soc., Vol. 6, 281 (1974). — Abstr. AAS.

065.110 **On the e-process: its components and their neutron excesses.**
K. L. Hainebach, D. Clayton, W. D. Arnett, S. E. Woosley.
Bull. American Astron. Soc., Vol. 6, 281 (1974). — Abstr. AAS.

065.111 **s-process nucleosynthesis in massive stars.**
R. G. Couch, A. B. Schmiedekamp, W. D. Arnett.
Bull. American Astron. Soc., Vol. 6, 281 (1974). — Abstr. AAS.

065.112 **Peculiar stars and the r-process.**
J. B. Blake, D. N. Schramm, B. Kuchowicz.
Bull. American Astron. Soc., Vol. 6, 282 (1974). — Abstr. AAS.

065.113 **The evolution of a $15 M_\odot$ star.**
C. Chiosi, E. Nasi.
Mem. Soc. Astron. Italiana, Nuova Ser., Vol. 44, 665 - 670 (1973/74).

A star of $15 M_\odot$ with initial chemical composition given by $X = 0.700$ and $Z = 0.02$ has been evolved up to the core He-exhaustion stage. The present results are compared with other models of the same mass but different initial chemical composition.

065.114 **Problems about stars and stellar evolution: questions to observers and suggestion for the program of observations with the 360 cm of E.S.O.** E. Schatzman.
Proc. ESO/SRC/CERN conference, (see 012.021), p. 143 - 153 (1974).

065.115 **Propagation of strong shocks in self gravitating, conducting gases in stellar models.**
B. G. Verma, B. Prasad.
Acta Phys. Acad. Sci. Hungaricae, Vol. 34, 239 - 248 (1973).

065.116 **Mechanisms of star formation.** V. C. Reddish.
Phys. Bull., (GB), Vol. 24, 661 - 664 = Commun. Roy. Obs. Edinburgh, No. 158 (1973).

There is no generally accepted theory of the mechanism of star formation. Some of the theories which have been put forward are summarised. The cloudlike structure of interstellar gas, fragmentation of the clouds and condensation to form stars and the controlling factors involved are discussed.

065.117 **Neutron stars.** I. W. Roxburgh.
Phys. Bull., (GB), Vol. 24, 664 - 667 (1973).

The mass-radius and mass-density relations for models of white dwarfs and neutron stars are considered. The particle content of neutron stars of increasing density and the chemical structure is discussed and related to models of neutron stars. The possible origin of neutron stars is discussed.

065.118 **Proton superfluidity in neutron-star matter.**
T. Takatsuka.
Progr. Theor. Phys., (Japan), Vol. 50, 1754 - 1755 (1973).

065.119 **Nuclear reactions induced by macro-micro interactions in super-dense stars.** P. G. Amte.
Indian Journ. Phys., Vol. 47, 637 - 638 (1973).

Baryon stars with densities of the order of 10^{15} to 10^{16} gm/cm^3 can be considered to be macroscopic systems in which quasiparticles exist as excitations. There is a possibility of such quasiparticles inducing nuclear reactions by their interaction with a microparticle in the material of the star.

065.120 **Maximum mass of a neutron star.**
C. E. Rhoades, R. Ruffini.
Phys. Rev. Letters, Vol. 32, 324 - 327 (1974).

The extremal principle of selecting an equation of state that produces maximum critical mass on the basis of Einstein's theory of general relativity, Le Chatelier's principle and the principle of causality determines the maximum mass of a neutron star in its equilibrium conformation to be $3.2 M_\odot$.

065.121 **Synthesis of very heavy elements in supermassive stars.** T. Ohnishi.
Progr. Theor. Phys. (Japan), Vol. 51, 123 - 133 (1974).

The synthetic processes of very heavy elements taking place in a supermassive star expanding from a very high temperature are investigated. p-process-like nuclear reactions would occur and the proton-rich heavy nuclides in the mass region $150 \lesssim A \lesssim 215$ may be produced in this type of stars.

065.122 **Microscopic calculations of liquid and solid neutron star matter.**
S. Chakravarty, M. D. Miller, Chia-Wei Woo.
Nuclear Phys. A, (Netherlands), Vol. A220, 233 - 240 (1974).

As the first step to a microscopic determination of the solidification density of neutron star matter, variational calculations are performed for both liquid and solid phases using a very simple model potential.

065.123 **Effect of nucleon correlations on pion condensation in neutron stars.** R. Rajaraman.
Phys. Letters B, (Netherlands), Vol. 48B, 179 - 182 (1974).

It is pointed out that nucleon-nucleon correlations play an important role in the pion condensation problem. Their effect is estimated by a simple analytical method.

065.124 **Evolution of stars toward the main sequence.**
B. Hidajat.
Proc. ITB, Vol. 7, 119 - 130 = Bandung Inst. Technol., Dep. Sci., Publ. Bosscha Obs., Lembang, No. 8 (1973). – Review paper.

065.125 **Spherically symmetric radiating star.** P. Goyal.
Current Sci., (India), Vol. 42, 674 (1973).

The interior of a radiating star is filled with a mixture of matter and radiation. A spherically symmetric metric has been obtained which describes the interior of a radiating star of variable radius $r = R(t)$.

065.126 **Gravitierende Neutronen – Gleichgewicht und mikroskopische Stabilität.** W. Hillebrandt.
Thesis Univ. Köln, 43 pp. (1973).

065.127 **Models of asymptotic-giant-branch stars.**
P. R. Wood.
Astrophys. Journ., Vol. 190, 609 - 630 (1974).

The static structure and dynamical behavior of the envelopes of four $0.9\, M_\odot$ asymptotic-giant-branch stars are described. It is found that in a model of luminosity $\log L/L_\odot = 3.42$ the envelope pulsates steadily in the first-overtone mode. The full-amplitude pulsational properties of this model agree well with those of a Mira variable of the same period. Three more-luminous models ($\log L/L_\odot = 3.60$, 3.85, and 4.14) all pulsate in the fundamental mode while simultaneously undergoing violent relaxation oscillations. Mass loss occurs from the models of luminosity $\log L/L_\odot = 3.60$ and 4.14. A distinct outward-moving shell which forms in the two most luminous models suggests a connection with planetary-nebula ejection.

Toroidal figures of equilibrium. See Abstr. 042.087

Nucleosynthesis in supernova outbursts and the chemical composition of the envelopes of neutron stars.
See Abstr. 061.004.

Origin of the chemical elements and stellar evolution. See Abstr. 061.020.

Equation of state at densities greater than nuclear density. See Abstr. 061.022.

A simple equation of state of matter at super-nuclear densities. See Abstr. 061.025.

Pions in neutron star matter.
See Abstr. 061.026.

Thick-target measurement of the (p, γ) stellar reaction rates on the nuclides ^{12}C, ^{29}Si, ^{46}Ti, ^{47}Ti, and ^{56}Fe.
See Abstr. 061.040.

Ferromagnetismus in Sternen hoher Dichte.
See Abstr. 061.046.

Neutrinos and stellar structure.
See Abstr. 061.073.

The stability of differential rotation with non-axisymmetric perturbations. I. The sufficient conditions.
See Abstr. 062.062.

On the electrodynamic equilibrium of a space charge region around a rotating neutron star with an aligned magnetic field. See Abstr. 062.066.

Rapidly rotating stars, disks, and black holes.
See Abstr. 066.023.

On the stability of axisymmetric systems to axisymmetric perturbations in general relativity. V. Differentially rotating configurations. See Abstr. 066.104.

Problems in gravitational collapse.
See Abstr. 066.108.

Black hole-neutron star collisions.
See Abstr. 066.112.

The black hole in astrophysics: the origin of the concept and its role. See Abstr. 066.127.

Resolution of the praseodymium abundance anomaly in the Ba II stars. See Abstr. 114.160.

The linear polarization of 53 Camelopardalis.
See Abstr. 116.009.

On secular stability of tidally distorted stars of arbitrary structure.   See Abstr. 117.005.

Evolution of primary components of massive close binary systems.   See Abstr. 117.010.

Evolution of close binaries and Wolf-Rayet stars.
See Abstr. 117.011.

Evolution of massive close binaries.
See Abstr. 117.012.

Evolution of a close binary with a relativistic component.   See Abstr. 117.013.

Evolution of the secondary component of a close binary system.   See Abstr. 117.014.

Variable stars and stellar evolution.
See Abstr. 122.106.

Theoretical aspects of mixed-mode variable stars.
See Abstr. 122.111.

Neutron stars in supernova remnants.
See Abstr. 125.006.

Supernovae and neutron stars.
See Abstr. 125.011.

Ap-Sterne und Supernovae.   See Abstr. 125.018.

The evolution of rotating white dwarfs with outflow of matter.   See Abstr. 126.012.

Der kosmische Staub und seine Rolle bei der Sternentstehung.   See Abstr. 131.044.

Molecules as probes of the interstellar matter.
See Abstr. 131.106.

Evolution of central stars of planetary nebulae towards the crystallizing white dwarf stage.
See Abstr. 133.004.

Matter in superstrong magnetic fields.
See Abstr. 141.309.

Pulsar observations and neutron star models.
See Abstr. 141.310.

Free precession of neutron stars.
See Abstr. 141.320.

On the value of the magnetic field at the surface of a neutron star – X-ray pulsar.   See Abstr. 142.027.

π° – decay gamma rays from neutron stars.
See Abstr. 142.093.

Origin of cosmic rays, atomic nuclei and pulsars in explosions of massive stars.   See Abstr. 143.020.

Stellar evolution near the main sequence: on some systematic differences between cluster sequences and model calculations.   See Abstr. 153.017.

Statistische Untersuchungen zur Entwicklung junger Sternhaufen und ihrer Mitglieder.   See Abstr. 153.018.

Sensitivity of the star formation rate to the interstellar gas abundance of heavy elements.
See Abstr. 155.025.

Late stages of stellar evolution in the light of elliptical galaxies.   See Abstr. 158.100.

Some recent results from galactic and stellar evolution theory.   See Abstr. 158.121.

Errata

065.901   Errata: 'Origin of cosmic rays, atomic nuclei, and pulsars in explosions of massive stars'. [Astrophys. Journ., (*Letters*), Vol. 184, L47 - L51 (1973)].
W. D. Arnett, D. N. Schramm.
Astrophys. Journ., (*Letters*), Vol. 187, L47 (1974).

066 Relativistic Astrophysics (without Cosmology), Background Radiation, Gravitation Theory

066.001 **Black-hole physics: some effects of gravity on the radiation emission.**
F. de Felice, L. Nobili, M. Calvani.
Astron. Astrophys., Vol. 30, 111 - 118 (1974).

The effects of the gravitational dragging on the electromagnetic radiation emitted by particles which move on right bound orbits around a Kerr black hole, are found to change the photon frequency shift $1 + z$ from values $\lesssim 1$ to values $\gg 1$ with a periodicity of $\sim 10^{-4}\ M/M_\odot$ s. The energy of the emitted radiation, as measured by a distant observer, turns out to be confined in pulses with the same periodicity.

066.002 **A measurement of solar gravitational microwave deflection with the Westerbork Synthesis Telescope.**
K. W. Weiler, R. D. Ekers, E. Raimond, K. J. Wellington.
Astron. Astrophys., Vol. 30, 241 - 248 (1974).

A determination of the deflection of radiowaves at a frequency of 4995 MHz in the solar gravitational field was performed in October 1972 by measuring the relative positions of 3 C 273 and 3 C 279 during the occultation of 3 C 279 by the sun. Unique observing and analysis techniques reduced the atmospheric and instrumental instabilities to a level previously unobtainable in experiments of this type. At this accuracy, the effects of the bending in the solar corona were significant and required modeling. The measured gravitational deflection was then 0.96 ± 0.05 times the value predicted by general relativity.

066.003 **A search for isolated radiofrequency pulses.**
G. R. Huguenin, E. L. Moore.
Astrophys. Journ., (Letters), Vol. 187, L57 - L58 (1974).

The authors have searched for isolated radiofrequency pulses of extraterrestrial origin, relying on their dispersed arrival time to distinguish them from terrestrial interference. Simultaneous observations at widely separated stations produced no coincident events, although several isolated events showing the expected signature were seen at each station.

066.004 **Radiation of gravitational waves from a cluster of overdense stars.** Ya. B. Zel'dovich, A. G. Polnarev.
Astron. Zhurn. Akad. Nauk SSSR, Vol. 51, 30 - 40 (1974).
In Russian. English translation in Soviet Astron. AJ, Vol. 18, No. 1.

The article is concerned with the gravitational radiation mechanism due to processes in the galactic center, when a collapsed object passes another. Estimates were obtained on the number of events per year, when a gravitational pulse can be observed with a given spectrum.

066.005 **On the problem of circular orbits in the Kerr metric.** Z. Kh. Kurmakaev.
Astron. Zhurn. Akad. Nauk SSSR, Vol. 51, 187 - 190 (1974).
In Russian. English translation in Soviet Astron. AJ, Vol. 18, No. 1.

Constants and siderial periods in coordinate and proper time of the circular motion for a particle in the equatorial plane of a rotating body are determined. Some quantitative estimates are given.

066.006 **Small-scale temperature variations of the cosmic background radiation at 11.1 cm.**
K. S. Stankevich.
Astron. Zhurn. Akad. Nauk SSSR, Vol. 51, 216 - 218 (1974).
In Russian. English translation in Soviet Astron. AJ, Vol. 18, No. 1.

The results of measurements of small-scale spatial variations of the cosmic background at wavelength 11.1 cm using the 210-ft radio telescope at Parkes are presented. The temperature of fluctuations in the angular scale of $8' < \theta < 19'$ is equal to $4.2 \times 10^{-3}\ °K$.

066.007 **Generalization of Einstein's principle of equivalence so as to embrace the field equations of gravitation.** O. Klein.
Phys. Scripta, Vol. 9, 69 - 72 (1974).

It is shown that the apparent impossibility of deriving Einstein's field equations of gravitation in a way similar to that used by him in deriving the equations of particle motion and electromagnetism may be overcome by using the quantal vacuum with its fluctuations instead of the vacuum of ordinary physics with its absolute emptiness.

066.008 **Airborne measurement of the temperature of the cosmic microwave background at 3.3 mm.**
P. E. Boynton, R. A. Stokes.
Nature, Vol. 247, 528 - 530 (1974).

A 3.3 mm Dicke radiometer was carried above the tropopause on five flights in May and July of 1971; but only one flight was successful. During the course of that flight, the authors made 21 measurements of the background intensity at various zenith angles and report here a minimum variance weighted average of those points indicating a thermodynamic temperature of (2.48 ± 0.54) K.

066.009 **Upper bound on the electric charge of a black hole.**
W. T. Zaumen.
Nature, Vol. 247, 530 - 531 (1974).

An expression for an upper bound on the black hole's charge can be derived by considering a loss of charge by pair production. The discussion is limited to negatively charged black holes so that the ingoing particles are positrons. For positively charged black holes, one interchanges electrons and positrons.

066.010 **Action d'un champ de gravitation statique sur un milieu dispersif isotrope.** É. Argence.
Comptes Rendus Acad. Sci. Paris, Sér. B, Vol. 278, 319 - 322 (1974).

Analogie entre la propagation d'une onde dans un plasma isotrope soumis à l'action d'un champ de gravitation et la propagation de l'onde associée à un corpuscule, d'après un résultat de Louis de Broglie.

066.011 **Gravitational stability of an ultrarelativistic centrally symmetric collapse.**
V. K. Pinus, A. L. Frenkel'.
Sib. in-t zemn. magn. ionosfery i rasprostr. radiovoln. Sib. otd. AN SSSR. Irkutsk. 19 pp. (1973). In Russian. — Abstr. in Referativ. Zhurn. 51. Astron., 3.51.840 (1974).

066.012 **Zur Kenntnis der "Schwarzen Löcher".**
E. Wiedemann.
Orion Schaffhausen, 32. Jahrgang, p. 14 - 15 (1974). — Auszug aus dem NASA-Report 73–251.

066.013 **Sources of gravity waves.** T. J. Sejnowski.
Phys. Today, Vol. 27, No. 1, p. 40 - 43, 45, 47 - 48 (1974).

Although gravitational signals from supernovae and black holes are estimated to be below the noise level of present detectors, they may soon be observed with new generation gravity telescopes.

066.014 Is backward motion in time possible?
A. A. Grib.
Priroda, No. 4.74, p. 24 - 32 (1974). In Russian.

066.015 Kinetic theory of growth and damping of small excitations in expanding gravitating matter.
V. V. Seliverstov.
Astron. Zhurn. Akad. Nauk SSSR, Vol. 51, 293 - 299 (1974). In Russian. English translation in Soviet Astron. AJ, Vol. 18, No. 2.

Based on the kinetic equation with self-consistent Newtonian gravitation field growth and damping of the amplitudes of inhomogeneities of the density of matter in a spatially homogeneous expanding universe has been investigated.

066.016 The effect of gravitational radiation-reaction on the evolution of the Riemann S-type ellipsoids.
B. D. Miller.
Astrophys. Journ., Vol. 187, 609 - 620 (1974).

It is shown that the $2\,^1/_2$-post-Newtonian formalism for describing gravitational radiation-reaction is equivalent to the formalism developed by Burke and Thorne, if one is neglecting lower-order post-Newtonian corrections. The equations of motion governing the evolution by radiation-reaction of the Jacobi ellipsoid and other Riemann S-type ellipsoids are integrated for several initial configurations. Evolution proceeds toward a nonradiating state which in some cases is a Maclaurin spheroid and in other cases a Dedekind ellipsoid.

066.017 A relativity eclipse experiment refurbished.
B. S. DeWitt, R. A. Matzner, A. H. Mikesell.
Sky Telescope, Vol. 47, 301 - 306 (1974).

066.018 On the post-Newtonian approximation of theories of gravity. U. Kasper, D.-E. Liebscher.
Astron. Nachr., Vol. 295, 11 - 17 (1974).

The physical meaning of the parameters of the post-Newtonian approximation is considered from a theoretical point of view. It is shown that the post-Newtonian approximation of any Lorentz-covariant theory, which implies Poisson equations for every metric coefficient in the linear approximation, is the member of a three-parametric family of post-Newtonian metrics. Some tetrad theories are cited.

066.019 A possible variation in the rate of passage of time.
K. D. Barker.
Astron. Astrophys., Vol. 31, 461 (1974).

It is suggested that an acceleration of the rate of passage of time may be responsible for the galactic red-shifts. The mechanism is an acceleration of atomic frequencies in the laboratory whilst signals from distant sources are in transit.

066.020 The event horizon.   S. W. Hawking.
Black holes. Les Houches 1972, (see 012.007), p. 1 - 55 (1973).

066.021 Black hole equilibrium states.   B. Carter.
Black holes. Les Houches 1972, (see 012.007), p. 57 - 214 (1973).

066.022 Timelike and null geodesics in the Kerr metric.
J. M. Bardeen.
Black holes. Les Houches 1972, (see 012.007), p. 215 - 239 (1973).

066.023 Rapidly rotating stars, disks, and black holes.
J. M. Bardeen.
Black holes. Les Houches 1972, (see 012.007), p. 241 - 289 (1973).

066.024 Astrophysics of black holes.
I. D. Novikov, K. S. Thorne.
Black holes. Les Houches 1972, (see 012.007), p. 343 - 450 (1973).

066.025 On the energetics of black holes.   R. Ruffini.
Black holes. Les Houches 1972, (see 012.007), p. 451 - 546, and 19 appendices, p. R1 - R176 (1973).

The author summarizes most of the known results on mass limits for neutron stars, radiation (both electromagnetic and gravitational) emitted by single objects falling onto black holes, and general theory of the energetics of black holes. In the appendices some earlier papers, partly in collaboration with other authors, are presented.

066.026 Self-gravitation, rotation and all that... .
R. J. Hosking.
Southern Stars, Vol. 25, 99 - 100 (1973). — Conference abstract.

066.027 Black hole explosions?   S. W. Hawking.
Nature, Vol. 248, 30 - 31 (1974).

It seems that any black hole will create and emit particles such as neutrinos or photons at just the rate that one would expect if the black hole was a body with a temperature of $(\kappa/2\pi)\,(\hbar/2k) \approx 10^{-6}\,(M_\odot/M)K$ where κ is the surface gravity of the black hole.

066.028 Anisotropic spheres in general relativity.
R. L. Bowers, E. P. T. Liang.
Astrophys. Journ., Vol. 188, 657 - 665 (1974).

The authors consider spherically symmetric static distributions of matter which are assumed to be locally anisotropic. The equations of hydrostatic equilibrium for such systems are derived and investigated. The authors first consider an incompressible model with a highly idealized form of anisotropy which allows the structure equations to be integrated analytically. The resulting maximum mass and surface redshift (SRS) are compared with the corresponding isotropic model. In order to eliminate model-dependent effects, the authors then study the questions of the maximum SRS by generalizing Bondi's analysis of isotropic spheres in general relativity (Bondi 1964) to include anisotropic stresses.

066.029 Some remarks about relativistic line profiles.
D. Gerbal, M. Prud'Homme.
Journ. Quant. Spectrosc. Radiat. Transfer, Vol. 14, 351 - 356 (1974).

The authors give, by analogy with relativistic kinetic theory, a fully relativistic definition of line profiles. Some examples are given. Relativistic broadening of lines produces asymmetries and blue-shifting. The authors have also calculated the relativistic Voigt profile.

066.030 The accretion of matter by a collapsing star in the presence of a magnetic field.
G. S. Bisnovatyi-Kogan, A. A. Ruzmaikin.
Astrophys. Space Sci., Vol. 28, 31 - 44, 45 - 59 (1974). In Russian and English.

The exact nonstationary solution for the variation of the magnetic field in the Schwarzschild metric with a given spherically symmetric flow is obtained. On the assumption of equipartition between the magnetic and kinetic energies of a falling gas in the relativistic case, estimates of the stationary field and the intensity of synchrotron radiation are presented.

066.031 Tetrad field equations and a generalized Friedmann equation.   B. O. J. Tupper.

Astrophys. Space Sci., Vol. 28, 225 - 231 (1974).

The tetrad field equations of general relativity discussed in previous articles are applied to Robertson-Walker cosmological models. A generalized Friedmann equation is derived and some of its consequences are discussed.

066.032 **Generation of an electromagnetic wave by a plane gravitational wave in a constant magnetic field.**
V. K. Dubrovich.
Soobshch. Spets. Astrofiz. Obs. AN SSSR, *Zelenchukskaya*, vyp. (No.) 6, p. 27 - 36 (1972). In Russian.

Interaction is considered between a flat gravitational wave and a continuous homogeneous magnetic field in two cases: 1) in the vacuum, 2) in the medium with $\epsilon \neq 1$, $\sigma \neq 0$. On the basis of solutions of Einstein–Maxwell equations it is shown that in both cases an electromagnetic wave is generated having a frequency equal to that of an incident gravitational wave.

066.033 **Accretion onto black holes: the emergent radiation spectrum. III. Rotating (Kerr) black holes.**
S. L. Shapiro.
Astrophys. Journ., Vol. 189, 343 - 351 (1974).

Steady-state, spherically symmetric accretion of matter onto rotating black holes is examined for simple polytropic gases. The total luminosity and frequency spectrum of the radiation emitted by interstellar hydrogen accreting onto a rotating black hole and observed above 1 keV are computed.

066.034 **Tetrads and the gravitational-inertial field.**
G. E. Marsh.
Australian Journ. Phys., Vol. 27, 131 - 133 (1974).

The tetrad formulation of general relativity allows a nontensorial decomposition of the gravitational field into two components which have been thought to represent the permanent and inertial parts. It is shown here that this division does not hold for arbitrary motions in a flat space-time, and therefore cannot be expected to hold in more general spaces.

066.035 **Gravitational deflection of polarised radiation.**
M. Harwit, R. V. E. Lovelace, B. Dennison, D. L. Jauncey, J. Broderick.
Nature, Vol. 249, 230 - 233 (1974).

An upper limit is determined for the difference in the deflection of beams of orthogonally polarised radiation passing through the sun's gravitational field. A null result is anticipated by present theories of gravitation but this prediction has never been tested.

066.036 **Relative-distance Machian theories.**
J. B. Barbour.
Nature, Vol. 249, 328 - 329 (1974).

Mach's principle, in essence, requires that the dynamical law of the universe be expressed ultimately in terms only of the relative distances between the observable entities in the universe. The author proposes a general framework for constructing theories that satisfy this postulate automatically. A simple model shows how the Newtonian world picture can be satisfactorily 'Machianised'.

066.037 **Gravity-induced electric polarisation near the Schwarzschild limit.** W. Davidson, H. J. Efinger.
Nature, Vol. 249, 431 (1974).

Here the authors point out that strong gravitational fields, such as those encountered in collapsed stellar objects, may produce observable electromagnetic effects arising directly from the gravitational mass of electric charges.

066.038 **Géodésiques et événements horizons en théorie pentadimensionnelle.** J.-P. Duruisseau.
Comptes Rendus Acad. Sci. Paris, Sér. A, Vol. 278, 1229 - 1232 (1974).

L'auteur étudie les trajectoires des particules d'épreuves neutres et des rayons lumineux dans cértains cas particuliers. Ces résultats sont comparés à ceux obtenus pour la solution de Schwarzschild.

066.039 **Les décalages spectraux «anormaux».**
L. Gouguenheim.
L'Astronomie, 88e année, p. 17 - 23 (1974).

066.040 **Doppler effect in general relativity.** I.-M. Ganea.
Stud. Cerc. Astron., Vol. 19, 99 - 107 (1974).

In this paper a method due to Synge and Avez is applied to the case of a Schwarzschild field; the method consists in calculating the Doppler effect as in special relativity, by replacing the real source by a fictitious one deduced by parallel transport of source velocity from the source to the observer, along the null geodesic. The author gives the exact solution of the equations of parallel transport along the radial null geodesic, and uses this solution to find in this way the formulae of gravitational redshift for a source at rest, and of Doppler shift for a source in radial free fall.

066.041 **Observational background of Treder's gravitation theory in cosmology and the role of viscosity.**
H. Oleak.
Astron. Nachr., Vol. 295, 107 - 121 (1974).

Starting from solutions of Treder's field equations for a homogeneous and isotropic universe filled with incoherent matter relations between observational quantities are derived and discussed. The Friedmann time is of the order of Hubble age, the deceleration parameter may principally possess any positive or negative value, hence it is also possible to explain any m, z-relation completely in the frame of this theory. A decision between Einstein's theory and Treder's by cosmological observations is impossible. Considering a realistic medium which consists of particles and radiation, energy dissipation originating from bulk viscosity (due to the interaction of particles and quanta) prevents cosmological singularities. This process generates large amounts of rest-mass-free energy in the phase of maximum contraction and therefore could as well explain the large luminosity of compact single objects (quasars?).

066.042 **Fluctuations in the cosmic microwave background arising from low-frequency gravitational radiation.**
G. Dautcourt.
Astron. Nachr., Vol. 295, 123 - 131 (1974).

Intense low-frequency intergalactic gravitational radiation with wave lengths λ smaller than the Hubble distance $\lambda_H \cong 3000 \, (100/H_0)$ Mpc but not exceedingly small compared to λ_H, generates anisotropies in the microwave background radiation. Available data on large-scale microwave fluctuations do not exclude appreciable amounts of gravitational background radiation in the Megaparsec wave band. A more sensitive test is provided by a second far-field contribution, which has a small angular scale. Its amplitude depends strongly on the ratio of the (present) rest mass density to the Hubble constant, if a cosmological origin of the blackbody radiation is assumed. In a low-density universe, pre-galactic Compton scattering of the blackbody radiation is not able to reduce the fluctuations caused by the low-frequency gravitational wave field.

066.043 **Astrophysikalische Anwendungen des allgemein relativistischen Zweikörperproblems.**
M. Reinhardt, A. Rosenblum.
Mitt. Astron. Ges., No. 34, p. 79 (1973/74). – Abstr. AG.

066.044 **Mechanisms for the emission and absorption of gravitational radiation.** C. W. Misner.
IAU Symposium No. 64, (see 012.008), p. 3 - 15 (1974).

066.045 The method of virtual quanta and gravitational radiation. R. A. Matzner, Y. Nutku.
IAU Symposium No. 64, (see 012.008), p. 16 (1974). — Abstract.

066.046 Detection of gravitational radiation. J. A. Tyson.
IAU Symposium No. 64, (see 012.008), p. 17 - 27 (1974).

066.047 On the evaluation of the Munich-Frascati Weber-type experiment. P. Kafka.
IAU Symposium No. 64, (see 012.008), p. 38 (1974). — Abstract.

066.048 Meudon gravitational radiation detection experiment.
S. Bonazzola, M. Chevreton, J. Thierry-Mieg.
IAU Symposium No. 64, (see 012.008), p. 39 (1974). — Abstract.

066.049 Analysis of gravitational-wave detection experiments. D. M. Eardley, D. L. Lee, A. P. Lightman, R. V. Wagoner, C. M. Will.
IAU Symposium No. 64, (see 012.008), p. 53 (1974). — Abstract.

066.050 Interaction of gravitational radiation with a uniformly magnetized sphere.
V. de Sabbata, P. Fortini, C. Gualdi, L. Fortini Baroni.
IAU Symposium No. 64, (see 012.008), p. 59 (1974). — Abstract.

066.051 Gravitational radiation by ultrarelativistic bodies. P. J. Westervelt.
IAU Symposium No. 64, (see 012.008), p. 60 (1974). — Abstract.

066.052 The stability of relativistic systems. S. Chandrasekhar.
IAU Symposium No. 64, (see 012.008), p. 63 - 81 (1974).

066.053 Gravitational collapse. R. Penrose.
IAU Symposium No. 64, (see 012.008), p. 82 - 91 (1974).

066.054 Perturbations of a rotating black hole. S. A. Teukolsky.
IAU Symposium No. 64, (see 012.008), p. 92 (1974). — Abstract.

066.055 Recent work on Kerr stability and superradiant wave scattering. W. H. Press.
IAU Symposium No. 64, (see 012.008), p. 93 (1974). — Abstract.

066.056 Amplification of waves reflected from Kerr black holes. A. A. Starobinsky.
IAU Symposium No. 64, (see 012.008), p. 94 (1974). — Abstract.

066.057 Scalar waves in the exterior of a Schwarzschild black hole. S. Persides.
IAU Symposium No. 64, (see 012.008), p. 95 (1974). — Abstract.

066.058 Electromagnetic waves in the exterior of a Schwarzschild black hole.
H. Stephani, E. Herlt.
IAU Symposium No. 64, (see 012.008), p. 96 (1974). — Abstract.

066.059 On the description of high-frequency gravitational waves. M. A. H. MacCallum.
IAU Symposium No. 64, (see 012.008), p. 98 (1974). — Abstract.

066.060 Alternative approach to infinity. P. G. Bergmann.
IAU Symposium No. 64, (see 012.008), p. 99 (1974). Abstract.

066.061 The geodetic interval in a Riemannian space-time in the second post-Minkowskian approximation. R. W. John.
IAU Symposium No. 64, (see 012.008), p. 100 (1974). — Abstract.

066.062 A new general covariant approach to the general relativistic two-body problem. A. Rosenblum.
IAU Symposium No. 64, (see 012.008), p. 102 (1974). — Abstract.

066.063 Gravitational deviation reaction. T. J. Sejnowski.
IAU Symposium No. 64, (see 012.008), p. 103 (1974). — Abstract.

066.064 Tidal tensor and the emission and absorption of gravitational radiation. T. J. Sejnowski.
IAU Symposium No. 64, (see 012.008), p. 104 (1974). — Abstract.

066.065 Complex Maxwell and Einstein fields. E. T. Newman.
IAU Symposium No. 64, (see 012.008), p. 105 (1974). — Abstract.

066.066 On black and white holes. M. A. Markov.
IAU Symposium No. 64, (see 012.008), p. 106 - 131 (1974).

066.067 Properties of black holes relevant to their observation. J. M. Bardeen.
IAU Symposium No. 64, (see 012.008), p. 132 - 144 (1974).

066.068 On the problem of detection of isolated black holes. V. F. Shvartsman.
IAU Symposium No. 64, (see 012.008), p. 183 (1974). — Abstract.

066.069 Black holes in the early universe. B. J. Carr, S. W. Hawking.
IAU Symposium No. 64, (see 012.008), p. 184 (1974). — Abstract.

066.070 Quantum aspects of accretion onto black holes in the early universe. S. W. Hawking, G. W. Gibbons.
IAU Symposium No. 64, (see 012.008), p. 185 (1974). — Abstract.

066.071 A class of solutions of Einstein-Maxwell equations with the cosmological constant. J. F. Plebański.
IAU Symposium No. 64, (see 012.008), p. 188 - 190 (1974).

066.072 New solutions of Einstein equations representing spinning masses. H. Sato, A. Tomimatsu.
IAU Symposium No. 64, (see 012.008), p. 191 (1974). — Abstract.

066.073 A new solution of the Einstein-Maxwell equations for a system with mass, magnetic moment, charge, and angular momentum. L. Witten.
IAU Symposium No. 64, (see 012.008), p. 192 (1974). — Abstract.

066.074 **Accretion of matter onto black holes.**
R. A. Sunyaev.
IAU Symposium No. 64, (see 012.008), p. 193 (1974). – Abstract.

066.075 **Accretion onto relativistic objects.** M. J. Rees.
IAU Symposium No. 64, (see 012.008), p. 194 - 212 (1974).

066.076 **What information can be extracted from radio data about the existence of supermassive black holes?**
L. M. Ozernoy.
IAU Symposium No. 64, (see 012.008), p. 214 - 215 (1974).

066.077 **Mögliche Zusammenhänge zwischen Apsidendrehungen bei Planetenbahnen und der Ausbreitungsgeschwindigkeit der Gravitation.** K. Molsen.
Astronautik, Jahrgang 11, p. 1 - 5 (1974).

If one assumes the existence of a general isotropic gravitational radiation with an unknown, but finite velocity of propagation, the term of aberration has to be introduced into the theory of gravitation. The equations of motion changed according to this assumption lead to a small secular motion of the line of apsides. By a comparison with the observed values of the perihelion motion of the inner planets a lower bound of the gravitational radiation velocity can be estimated.

066.078 **Remarks on Schwarzschild black holes.**
G. C. McVittie.
Galaxies and relativistic astrophysics, Proc. 1972, (see 012.010), p. 166 - 173 (1974).

066.079 **Classical fields in the vicinity of a Schwarzschild black hole.** S. Persides.
Galaxies and relativistic astrophysics, Proc. 1972, (see 012.010), p. 174 - 180 (1974).

066.080 **f gravity and gravitational singularities.**
C. Sivaram, K. P. Sinha, E. A. Lord.
Nature, Vol. 249, 640 - 641 (1974). – Letter.

066.081 **Comments on two recent measurements of the solar gravitational red-shift.** J. L. Snider.
Solar Physics, Vol. 36, 233 - 234 (1974). – Research note.

066.082 **On the possibility of a physical interpretation of static polycentric solutions of Einstein's equations.**
S. L. Galkin.
Vestn. Mosk. un-ta. Fiz., astron., Vol. 14, 542 - 546 (1973). In Russian. – Abstr. in Referativ. Zhurn. 51. Astron., 5.51.766 (1974).

066.083 **The Schwarzschild peculiar sphere.**
R. F. Polishchuk.
Vestn. Mosk. un-ta. Fiz., astron., Vol. 14, 710 - 715 (1973). In Russian. – Abstr. in Referativ. Zhurn. 51. Astron., 5.51.769 (1974).

066.084 **Time delay of the radar echo as a time defect of desynchronization of a relativistic particle.**
A. E. Levashev, G. Eh. Susurin.
Izv. AN BSSR. Ser. fiz.-mat. n., 1973, No. 6, p. 63 - 69. In Russian. – Abstr. in Referativ. Zhurn. 51. Astron., 5.51.788 (1974).

066.085 **On the possibility of measurement of the dependence of the gravitational constant on time.**
V. B. Braginskij, V. L. Ginzburg.
Dokl. Akad. Nauk SSSR. Ser. Mat. Fiz., Vol. 216, 300 - 302 (1974). In Russian.

066.086 **Was Einstein aware of the Michelson-Morley experiment?** V. J. Joshi.
Observatory, Vol. 94, 81 (1974). – Letter.

066.087 **Observed deflection of light by the sun as a function of solar distance.**
P. Merat, J. C. Pecker, J. P. Vigier, W. Yourgrau.
Astron. Astrophys., Vol. 32, 471 - 475 (1974).

A detailed analysis using modern computing methods of all known measurements of the deflection of light near the solar limb shows a) the validity of Einstein's prediction beyond a certain distance of the sun, b) the possible existence of a new effect in the close vicinity of the solar limb which can be interpreted as resulting from a dispersive medium emitted by the solar surface.

066.088 **Do black holes really explode?**
P. C. W. Davies, J. G. Taylor.
Nature, Vol. 250, 37 - 38 (1974). – Letter.

066.089 **Gravitation and long-range weak interactions in neutrino fields. Ideas on a theory of solar neutrinos.**
H.-J. Treder.
Astron. Nachr., Vol. 295, 169 - 184 (1974). In German.

The non-evidence of the solar-neutrino current by the experiments of Davis et al. postulates a fundamental revision of the theory of weak interactions and of its relations to gravitation theory. The present paper is based on Pauli's hypothesis about the connection between weak and gravitational interactions.

066.090 **The mass-angular momentum-diagram and the black hole limit.** P. Brosche.
Astrophys. Space Sci., Vol. 29, L7 - L8 (1974).

The ratio angular momentum/mass squared of a wide variety of astronomical objects lies about 10^3 times above the value $G/c = 2.2 \times 10^{-18}$ cm^2 g^{-1} s^{-1} for extremely fast rotating black holes.

066.091 **Black holes as sources of gravity waves.**
D. Dionysiou.
In honorem S. Placidis, (see 003.009), p. 63 - 73 (1974). In Greek.

066.092 **Black holes and the structure of space-time.**
S. Persides.
In honorem S. Placidis, (see 003.009), p. 325 - 340 (1974).

066.093 **Aflevering 2 – zwaartekracht.**
J. Stollman, P. Hut.
Zenit, Vol. 1, No. 5, p. 18 - 19 (1974).

066.094 **The damping law of external fields of a collapsing body.**
A. Z. Patashinskij, V. K. Pinus, A. A. Khar'kov.
Zhurn. ehksperim. i teor. fiz., Vol. 66, 393 - 405 (1974). In Russian. – Abstr. in Referativ. Zhurn. 51. Astron., 6.51.704 (1974).

066.095 **Non-static spherically symmetric solutions of Einstein's equations.** M. P. Korkina.
Ukr. fiz. zhurn., Vol. 19, No. 1, p. 40 - 43 (1974). In Russian. Abstr. in Referativ. Zhurn. 51. Astron., 6.51.705 (1974).

066.096 **The global structure of non-singular metrics of Taub-Newman-Unti-Tamburino.** V. A. Ruban.
Gravitatsiya i teoriya otnositel'n. Vyp. (No.) 9. Kazan', Kazan. un-t, 1973, p. 38 - 59. In Russian. – Abstr. in Referativ. Zhurn. 51. Astron., 6.51.706 (1974).

066.097 **Microgeon with spin.** A. Ya. Burinskij.

Zhurn. ehksperim. i teor. fiz., Vol. 66, 406 - 411 (1974). In Russian. — Abstr. in Referativ. Zhurn. 51. Astron., 6.51.707 (1974).

066.098 Mapping test body paths of motion in a Schwarzschild field. R. S. Singatullin.
Gravitatsiya i teoriya otnositel'n. Vyp. (No.) 9. Kazan', Kazan. un-t, 1973, p. 60 - 66. In Russian. — Abstr. in Referativ. Zhurn. 51. Astron., 6.51.708 (1974).

066.099 On the stability of circular orbits in the gravitational field of two fixed centres.
N. F. Kamaletdinova, K. A. Piragas.
Gravitatsiya i teoriya otnositel'n. Vyp. (No.) 9. Kazan', Kazan. un-t, 1973, p. 115 - 126. In Russian. — Abstr. in Referativ. Zhurn. 51. Astron., 6.51.709 (1974).

066.100 Energy of a relativistic particle in general relativity.
M. P. Korkina, V. D. Gladush.
Ukr. fiz. zhurn., Vol. 19, No. 1, p. 82 - 84 (1974). In Russian. Abstr. in Referativ. Zhurn. 51. Astron., 6.51.710 (1974).

066.101 On the problem of harmonic coordinates.
Z. A. Shtejngrad.
Gravitatsiya i teoriya otnositel'n. Vyp. (No.) 9. Kazan', Kazan. un-t, 1973, p. 105 - 108. In Russian. — Abstr. in Referativ. Zhurn. 51. Astron., 6.51.724 (1974).

066.102 Interaction of a relativistic oscillator with the field of the earth and gravitational waves.
V. I. Bashkov.
Gravitatsiya i teoriya otnositel'n. Vyp. (No.) 9. Kazan', Kazan. un-t, 1973, p. 3 - 8. In Russian. — Abstr. in Referativ. Zhurn. 51. Astron., 6.51.730 (1974).

066.103 Oscillator in general relativity.
V. I. Golikov, M. D. Sherman.
Gravitatsiya i teoriya otnositel'n. Vyp. (No.) 9. Kazan', Kazan. un-t, 1973, p. 9 - 21. In Russian. — Abstr. in Referativ. Zhurn. 51. Astron., 6.51.731 (1974).

066.104 On the stability of axisymmetric systems to axisymmetric perturbations in general relativity. V. Differentially rotating configurations. C. M. Will.
Astrophys. Journ., Vol. 190, 403 - 410 (1974).
The author extends the theory developed in Papers I and II to allow for the effects of differential rotation. He derives a variational expression for the frequency of axisymmetric oscillations of relativistic differentially rotating configurations, and applies this formula to the quasi-radial modes of oscillation of slowly, differentially rotating stellar models. By appropriate redefinitions of variables, the theory is also applied to stationary nonaxisymmetric deformations of differentially rotating configurations.

066.105 The space-time of axisymmetric gravitating masses.
K. Y. Fu.
Astrophys. Journ., Vol. 190, 411 - 415 (1974).
Axisymmetric space-time associated with astrophysical bodies of axial symmetry, which probably results from the presence of a strong magnetic field, is discussed. The static solution shows no Schwarzschild singularity which exists for the case of spherical symmetry. The nonstationary solution indicates that initial anisotropic stresses cannot prevent the body from eventual collapse "three dimensionally". The limiting character of the space-time for the dustlike matter has been investigated in detail.

066.106 Radiation pressure in general relativity.
D. W. Marks.
Bull. American Astron. Soc., Vol. 6, 217 (1974). — Abstr. AAS.

066.107 A search for optical and radio bursts from collapsing objects. E. O'Mongain, T. C. Weekes.
Bull. American Astron. Soc., Vol. 6, 270 (1974). — Abstr. AAS.

066.108 Problems in gravitational collapse.
D. N. Schramm, W. D. Arnett.
Bull. American Astron. Soc., Vol. 6, 279 (1974). — Abstr. AAS.

066.109 Conservation of baryon and lepton number for black holes. W. T. Zaumen.
Bull. American Astron. Soc., Vol. 6, 279 (1974). — Abstr. AAS.

066.110 Pion gamma ray emission from black holes.
G. H. Dahlbacka, G. F. Chapline, T. A. Weaver.
Bull. American Astron. Soc., Vol. 6, 279 (1974). — Abstr. AAS.

066.111 Axisymmetric stability of Kerr black holes.
J. L. Friedman, B. F. Schutz, Jr.
Bull. American Astron. Soc., Vol. 6, 280 (1974). — Abstr. AAS.

066.112 Black hole-neutron star collisions.
J. Lattimer, D. Schramm.
Bull. American Astron. Soc., Vol. 6, 280 - 281 (1974). Abstr. AAS.

066.113 Few words on the isotropic radio emission that some call also relic background radiation.
M. Różyczka.
Urania Kraków, Vol. 45, 66 - 72 (1974). In Polish.

066.114 Eine analytische Formulierung der Mach-Einstein-Doktrin und der Relativität der Trägheit, basierend auf Riemanns Verallgemeinerung des Lagrangeschen Theorems.
H.-J. Treder.
Istituto Nazionale di Alta Matematica, Symposia Mathematica, Vol. 12, 111 - 138 (1973).

066.115 Über die mögliche lineare Form von Lorentz-kovarianten Gravitationstheorien. H.-J. Treder.
Ann. Physik, 7. Ser., Vol. 31, 1 - 17 (1974).

066.116 The motion of black holes. G. W. Gibbons.
Commun. Math. Phys., (*Germany*), Vol. 35, 13 - 23 (1974).
The motion of axisymmetric, non rotating black holes is discussed using the properties of Weyl solutions. A new exact solution is obtained representing a black hole chased by a negative mass particle, both objects being uniformly accelerated and all solutions representing a single black hole tidally distorted by an external static, axisymmetric gravitational field are obtained.

066.117 Gravitational collapse and black holes.
J. Taylor.
Phys. Bull., (*GB*), Vol. 24, 654 - 656 (1973).
The collapse of a heavy star and the features produced are discussed. The process of collapse is visualised by drawing the light cones at various points in space-time as the collapse ensues. Evidence for the existence of black holes is also discussed.

066.118 Two axisymmetric black holes cannot be in static equilibrium. H. M. Z. Hagen, H. J. Seifert.
International Journ. Theor. Phys., (*GB*), Vol. 8, 443 - 450 (1973).

066.119 Strong gravity and collapse. A. R. Prasanna.
Phys. Letters A, (*Netherlands*), Vol. 46A, 169 - 170 (1973).
The author has found that in the mode proposed by Salam et al. for arresting gravitational collapse, stars with mass

of the order of a solar mass and less do not cross their event horizon.

066.120 Axisymmetric stability of Kerr black holes.
J. L. Friedman, B. F. Schutz.
Phys. Rev. Letters, Vol. 32, 243 - 245 (1974).

A three-sequence proof is presented of the stability of all axisymmetric modes in Kerr rotating black holes.

066.121 Exact solutions of Einstein-conformal scalar equations. J. D. Bekenstein.
Ann. Physics, Vol. 82, 535 - 547 (1974).

A theorem is presented by means of which one can generate two Einstein conformal scalar solutions from a single Einstein-ordinary scalar solution. Two families of spherically symmetric static Einstein-conformal scalar solutions are obtained. A family of static spherically symmetric Einstein-Maxwell-conformal scalar solutions, which have black-hole geometries but are not genuine black holes are exhibited. Finally, all the Robertson-Walker cosmological models which contain both incoherent radiation and a homogeneous conformal scalar field are presented.

066.122 Cross-sections of a cylindrical antenna for gravitational waves. E. D'Anna, G. Pizzella, D. Trevese.
Nuovo Cimento Lettere, Ser. 2, Vol. 9, 231 - 234 (1974).

The possibility is considered that, if gravitational waves exist, they might generate, in a cylindrical antenna, longitudinal vibrations other than axial. The cross-section of a gravitational wave with a cylinder vibrating in such a radial model is calculated and a comparison made with that for axial vibrations.

066.123 Gedanken experiments to destroy a black hole.
R. Wald.
Ann. Physics, Vol. 82, 548 - 556 (1974).

066.124 Point charge in the vicinity of a Kerr black hole.
J. M. Cohen, L. S. Kegeles, C. V. Vishveshwara, R. M. Wald.
Ann. Physics, Vol. 82, 597 - 603 (1974).

The expression for the electromagnetic field of a point charge at rest on the symmetry axis near a rotating Kerr black hole is derived. Unlike the Schwarzschild case the charge is found to give rise to magnetic fields as seen by a stationary or locally nonrotating observer.

066.125 On the solution of the equations governing the coupled emission of gravitational and electromagnetic radiation [from black holes].
M. Johnston, R. Ruffini, M. Peterson.
Nuovo Cimento Lettere, Ser. 2, Vol. 9, 217 - 223 (1974).

Some analytic properties of the system of equations governing the coupled emission of gravitational and electromagnetic radiation are examined and an integration technique exemplified. The consequences for gravitational collapse phenomena are discussed.

066.126 Gravitational self-lens effect.
F. Winterberg, W. G. Phillips.
Phys. Rev. D, Particles and Fields, Vol. 8, 3329 - 3337 (1973).

This effect of a star on itself will be significant provided the star is near its Schwarzschild singularity and it is found that there are two important differences from the nonrelativistic case. The image of the star is magnified, by about 2.6 in the singularity limit, and a nonuniform light intensity will be observed over the stellar disc.

066.127 The black hole in astrophysics: the origin of the concept and its role. S. Chandrasekhar.
Contemporary Phys., (GB), Vol. 15, 1 - 24 (1974).

The formation of black holes is discussed with particular reference to the Schwarzschild black hole and the Kerr black hole. Observational evidence for the occurrence of neutron stars and black holes in close short-period binaries is also analysed.

066.128 Lines of force of a point charge near a Schwarzschild black hole. R. S. Hanni, R. Ruffini.
Phys. Rev. D, Particles and Fields, Vol. 8, 3259 - 3265 (1973).

These are investigated using Maxwell's equations for a curved space, the charge being at rest. A definition of lines of force is given in this situation, they are computed and plotted for the charge at different distances from the centre of the black hole. The transition from a Schwarzschild to a Reissner-Nordstrom black hole is discussed.

066.129 The tests of general relativity and scalar fields.
B. O. J. Tupper.
Nuovo Cimento B, Ser. 11, Vol. 19 B, 135 - 148 (1974).

Conditions are found under which a space-time will give results in agreement with the Schwarzschild values for the three classical tests of general relativity and the radar reflection experiment. The solutions of the field equations are shown to be conformal to the solutions of the Brans-Dicke field equations. The field equations in the presence of matter are also discussed.

066.130 Electromagnetic fields in general relativity: a constructive procedure. J. M. Cohen, L. S. Kegeles.
Phys. Letters A, (Netherlands), Vol. 47 A, 261 - 262 (1974).

A new procedure for obtaining explicit solutions to Maxwell's equations in curved spaces is presented. The formulation includes astrophysically important cosmological models, neutron star and black hole space-times.

066.131 Mach's principle for rotation. L. Pietronero.
Nuovo Cimento B, Ser. 11, Vol. 20B, 144 - 148 (1974).

066.132 On a phenomenological modification of Einstein's gravitational Lagrangian. H. Nariai.
Progr. Theor. Phys. (Japan), Vol. 51, 613 - 629 (1974).

A phenomenological approach to the modification of Einstein's gravitational Lagrangian is proposed in order to arrive at a regular isotropic model-universe being stable with respect to the excitation of gravitational and rotational waves in such a way that the modification does not seriously disturb the success of general relativity for the solar gravitational field.

066.133 Asymptotics of some relativistic Markov processes.
R. M. Dudley.
Proc. National Acad. Sci. USA, Vol. 70, 3551 - 3555 (1973).

Markov processes in special-relativistic position-velocity phase space are proved to have converging velocities as $t \to +\infty$ under some mild assumptions on transition probabilities. This offers a possible mechanism to explain the recession of galaxies.

066.134 Coherent neutrino scattering and stellar collapse.
J. R. Wilson.
Phys. Rev. Letters, Vol. 32, 849 - 852 (1974).

The hydrodynamic behaviour of stellar collapse, including neutrino-nuclear scattering inferred from neutral current theory is investigated for a model with a 1.49 $M_\odot$ Fe core surrounded by a thin Si-O layer and a low density C envelope.

066.135 Geodesic synchrotron radiation [from a black hole].
R. A. Breuer, P. L. Chrzanowski, H. G. Hughes III, C. W. Misner.
Phys. Rev. D, Particles and Fields, Vol. 8, 4309 - 4319 (1973).

Techniques are presented for the calculation of the high

frequency radiation emitted by a particle moving in a circular orbit around a black hole described by the Schwarzschild metric. The computation is useful in the case of highly relativistic particles since the high energy radiation is the dominant contribution.

066.136 An attempt to measure the far infrared spectrum of the cosmic background radiation. H. P. Gush.
Canad. Journ. Phys., Vol. 52, 554 - 561 (1974).

066.137 More on the Thomas precession in special relativity. S. Aranoff.
Nuovo Cimento Lettere, Ser. 2, Vol. 9, 603 - 606 (1974).

066.138 Some wave solutions in general relativity. L. K. Patel.
Proc. Cambridge Phil. Soc., Vol. 75, 261 - 267 (1974).

A general scheme for the derivation of wave solutions in general relativity is developed. Some solutions describing the flow of gravitational waves are discussed. Singular electromagnetic fields corresponding to one particular solution are also discussed.

066.139 Apparent horizons in the two-black-hole problem. A. Cadez.
Ann. Physics, Vol. 83, 449 - 457 (1974).

The apparent horizon of the two-black-hole problem on the time-symmetric spacelike hypersurface is studied. Its area is computed as a function of the separation parameter. The critical value of the separation parameter for which the two black holes merge is computed.

066.140 Relativity time-delay experiments utilizing 'Mariner' spacecraft. P. B. Esposito, J. C. Anderson.
Telecommun. Journ., (Switzerland), Vol. 41, 299 - 306 (1974).

Mariner spacecraft have been utilized as scientific instruments in the experimental determination of the relativistic time delay. Analysis of the Mariner data confirms the prediction of Einstein's theory of general relativity with a 3% uncertainty.

066.141 The wave tails and nonspherical perturbations of relativistic gravitational collapse.
A. Z. Patashinsky, V. K. Pinus, A. A. Kharkov.
Nuovo Cimento Lettere, Ser. 2, Vol. 9, 701 - 703 (1974).

066.142 Beat frequency oscillations near charged black holes and other electrovacuum geometries.
U. H. Gerlach.
Phys. Rev. Letters, Vol. 32, 1023 - 1025 (1974).

The normal modes of the perturbation of a charged black hole are determined in the high frequency approximation. This perturbation takes the form of an interaction between gravitational and e.m. modes which beat against each other in their propagation through space-time.

066.143 General relativity, unitarian Einstein theory, expansion and age of the universe. M. Pierucci.
Nuovo Cimento B, Ser. 11, Vol. 21 B, 69 - 96 (1974).

The inconsistency of some reserves of which the unitarian Einstein theory till now was the object, is shown. Hence, by two quite different relativistic methods, almost perfectly coincident results for the age of the universe are obtained. At last it is proved that the beginning of the expansion is not a repeatable event.

066.144 Black holes in static electrovac space-times. H. Muller Zum Hagen, D. C. Robinson, H. J. Seifert.
General Relativity and Gravitation, (GB), Vol. 5, 61 - 72 (1974).

The theorem of Israel which characterizes the Reissner-Nordstrom solutions as the only well behaved asymptotically flat electrovac spaces with a simple regular horizon is extended by weakening the assumptions. The possibilities of non-static or non-conservative electromagnetic fields in a static space-time are discussed and excluded by physical arguments.

066.145 Zwarte gaten, III. Gekromde ruimte. J. Stollman, P. Hut.
Zenit, Vol. 1, No. 6, p. 20 - 22 (1974).

066.146 Registration of signals from astronomical sources in the kHz region. A. I. Gusak, A. I. Kuznetsov.
Astron. Tsirk., No. 815, p. 3 - 4 (1974). In Russian.

066.147 An attempt to detect gravitational waves with the Yellowknife seismic array. D. H. Weichert.
Geophys. Journ. Roy. Astron., Soc., Vol. 35, 337 - 342 = Contr. Earth Phys. Branch, *Ottawa, Canada*, No. 444 (1973).

The Yellowknife array of 19 vertical short-period seismometers was used to search for small coherent signals in the frequency range from about 1 to 8 Hz. Consistent peaks are observed at several frequencies, but terrestrial or instrumental origins have been identified for most components. None of the observed components shows correlation with sidereal time. There is, therefore, no evidence that gravitational waves have been detected.

From the black hole to the infinite universe.
See Abstr. 003.050.

Gravitation waves in Einstein's theory of gravitation.
See Abstr. 003.108.

A determination of the rate of change of G.
See Abstr. 043.002.

EXOSAT and 'black holes'.
See Abstr. 051.023.

Relativistic spinor regularization of the astrodynamical problem of two bodies. See Abstr. 052.028.

Hypercollapsed nuclear matter.
See Abstr. 061.027.

Gamma rays from black holes.
See Abstr. 061.056.

On some properties of relativistic MHD flows.
See Abstr. 062.045.

Gravitational Stokes parameters.
See Abstr. 063.026.

Theory of level surfaces inside relativistic, rotating stars. II. See Abstr. 065.033.

On slowly rotating homogeneous masses in general relativity. See Abstr. 065.034.

On the analogy between neutron star models and isothermal gas spheres and their general relativistic instability. See Abstr. 065.035.

Hadron star models. See Abstr. 065.042.

Post-Newtonian neutron stars. See Abstr. 065.062.

The stability of stellar masses in general relativity. See Abstr. 065.078.

Stability of non-radial vibrational modes of relativistic neutron stars.   See Abstr. 065.079.

Vela X1: neutron star or black hole?   See Abstr. 065.108.

Search for "black holes" in close binary systems.   See Abstr. 121.045.

Are BL Lac-type objects nearby black holes?   See Abstr. 122.103.

The role of gravitational radiation in the evolution of dwarf novae.   See Abstr. 124.008.

Cosmic background radiation at 1.32 millimeters.   See Abstr. 131.133.

Black holes in binary systems: Instability of disk accretion.   See Abstr. 142.003.

2U 1700−37: Another black hole?   See Abstr. 142.010.

2U 0900−40: a black hole?   See Abstr. 142.020.

Collapsed objects and galactic X-ray sources.   See Abstr. 142.023.

Black holes and X-ray binaries.   See Abstr. 142.025.

The synthesis of close-binary light curves. VI. X-ray and collapsar binaries.   See Abstr. 142.028.

Upper limit on 2.5-second pulsations from Hercules X-1.   See Abstr. 142.029.

Soft X-ray variability of binary X-ray stars.   See Abstr. 142.030.

Millisecond temporal structure in Cygnus X-1.   See Abstr. 142.048.

Millisecond temporal structure in Cyg X-1.   See Abstr. 142.060.

Spectroscopic observations of the optical companion to Centaurus X-3.   See Abstr. 142.069.

The principle of correspondence in cosmology. I. The isotropic cosmos as Newtonian limit of relativistic theories of gravitation.   See Abstr. 162.013.

The effect of a linear rotational perturbation on the isotropy of the cosmic background radiation.   See Abstr. 162.022.

Is the cosmological constant really constant?   See Abstr. 162.032.

Sun

071 Solar Photosphere, Spectrum

071.001 **The nature of microturbulence in the solar photosphere.** A. M. Wilson, F. J. Guidry.
Monthly Notices Roy. Astron. Soc., Vol. 166, 219 - 233 (1974).

This paper contains two main results: (1) The apparent centre-limb increase of Doppler width in the cores of the sodium D lines is not caused by the frequency dependence of the lines source function. (2) The increase in Doppler width in these lines as we approach the limb is probably due to the inhomogeneous structure of the upper photosphere.

071.002 **Non-LTE analysis of the infrared O I-triplets in the solar spectrum.** E. Sedlmayr.
Astron. Astrophys., Vol. 31, 23 - 35 (1974).

All former LTE analyses of the individual lines of the solar infrared $\lambda 7773$ triplet always led to discrepancies between the theoretical results and the observations, especially in the centre to limb variation. In this paper the formation of these lines is studied in a multilevel NLTE approach.

071.003 **A preliminary theoretical line-blanketed model solar photosphere.** R. L. Kurucz.
Solar Physics, Vol. 34, 17 - 23 (1974).

A preliminary theoretical solar model is presented that produces closer agreement with observation than has been heretofore possible. The qualitative advantages and shortcomings of this model are discussed and projected improvements are outlined.

071.004 **On the possibility of detecting abundance inhomogeneities resulting from spallation reactions in the solar photosphere.** L. Hultqvist.
Solar Physics, Vol. 34, 25 - 32 (1974).

The possibility of detecting abundance inhomogeneities in the photosphere of the sun is discussed. These inhomogeneities are postulated to be a result of nuclear reactions caused by high energy protons from solar flares. From our present knowledge of flares and of nuclear reaction cross sections it is found that temporary inhomogeneities in Li-abundance of spectroscopically detectable amounts should not be excluded. Preliminary observations have given negative results.

071.005 **On the solar bismuth content.** Ø. Hauge.
Solar Physics, Vol. 34, 33 - 35 (1974). — Research note.

071.006 **Line broadening calculations for some infrared solar Fraunhofer lines.**
G. Deridder, W. van Rensbergen.
Solar Physics, Vol. 34, 77 - 90 (1974).

Theoretically calculated damping constants for the solar atmosphere at different optical depths, are compared with empirical values for three infrared multiplets of C I, O I and Si I resp. A test-calculation was also performed for the NaD-doublet broadened by a gas of He particles, for which many theoretical and experimental values are known.

071.007 **The influence of a photospheric spectrum of turbulence on the profiles of weak Fraunhofer lines.**
C. de Jager.
Solar Physics, Vol. 34, 91 - 103 (1974).

The author assumes that the motion field in the solar photosphere is described by a spectrum of turbulence, defined by suitably chosen parameters. For various values of the spectral parameters he computes average (i.e. averaged over a sufficiently large part of the photosphere) profiles of weak Fraunhofer lines. The resulting profiles which represent the distribution function of line-of-sight velocity components as modified by the transfer of radiation through the atmosphere, are thereupon still broadened by a function representing the influence of the distribution function of the granulation cell sizes.

071.008 **A comparison of EUV spectroheliograms and photospheric magnetograms.**
J. B. Gurman, G. L. Withbroe, J. W. Harvey.
Solar Physics, Vol. 34, 105 - 111 (1974).

The authors explore the relationship between intensity and field strength for a coronal line, Mg X $\lambda 625$, through use of data from the Harvard College Observatory (HCO) experiment on OSO-6 and the 40-channel magnetograph of the McMath solar telescope at Kitt Peak National Observatory (KPNO). From these data they infer distinctly different relationships between the strength of the photospheric magnetic field and density in the overlying corona in quiet and in active regions.

071.009 **Broadband polarization measurements on the quiet sun's disk near λ 5834.**
D. L. Mickey, F. Q. Orrall.
Astron. Astrophys., Vol. 31, 179 - 183 (1974).

The complete state of polarization of radiation from the quiet sun's disk has been measured as a function of distance from the limb in a 20 Å wide wavelength band centered at λ 5834 — a region relatively free of Fraunhofer lines. These measurements were made using a scanning polarimeter-photometer at Mt. Haleakala (Orrall, 1971). The results are presented.

071.010 **The contrast of granulation in different regions of the continuum.** I. V. Yudina.
Solnechnye Dannye 1973 Byull., No. 11, p. 71 - 72 (1974). In Russian.

The values of mean square contrasts of granulation have been compared for the blue and yellow regions of the continuum obtained by the photoelectric method. The contrast in the blue region was found to be higher than that in the yellow one, their ratio being 1.2. It is shown that the values of the contrasts in the minimum of solar activity are lower than those in the maximum.

071.011 **Telluric lines as radial velocity standards.**
A. T. Young.
Observatory, Vol. 94, 22 - 23 (1974). — Letter.

071.012 **Interaction of magnetic ropes of the solar photosphere.** A. A. Solov'ev.
Solnechnye Dannye 1974 Byull., No. 1, p. 78 - 83 (1974). In Russian.

The theory suggested previously was used to investigate

the interaction of two insulated magnetic ropes in the solar photosphere without external perturbations. The life-time of the rope fine structure has been estimated.

071.013 The determination of the velocity amplitude of the total photospheric motion field. I. The use of the weak and moderately weak lines.
N. N. Kondrashova, E. A. Gurtovenko.
Solar Physics, Vol. 34, 291 - 298 (1974).

By the analysis of the profiles of 20 weak lines observed at five centre-to-limb positions on the solar disk, radial and tangential components of the velocity amplitude of the photospheric motion field are derived in the range of optical depth $-3.0 \leqslant \lg \tau_5 \leqslant +0.5$.

071.014 Studies of granular velocities. V: The height dependence of the granular Doppler shifts.
W. Mattig, H. Schlebbe.
Solar Physics, Vol. 34, 299 - 301 = Mitt. Fraunhofer Inst., Freiburg, No. 124 (1974). − Research note.

071.015 Temperature fluctuations in solar granulation.
M. Lévy.
Astron. Astrophys., Vol. 31, 451 - 458 (1974).
The relaxation time of temperature fluctuations is analyzed by means of a hydrodynamic model for convective cells. Account is taken of radiative transfer. The instantaneous relaxation time at the origin ($\tau_T = 1/\beta$) is obtained in terms of various parameters: granulation scale, penetration coefficient, velocity, the amplitude of temperature fluctuations. Results for the disk-centre contrast and the variation in centre-to-limb contrast of granules are compared to those of Edmonds so as to scale the mean values of the parameters used to describe temperature and density perturbations.

071.016 Ballik-Ramsey band in the solar spectrum.
M. C. Pande, K. Sinha.
Bull. Astron. Inst. Czechoslovakia, Vol. 25, 83 - 84 (1974).

Equivalent width calculations for a line 1.83 μ of the 0−0 vibrational band of the Ballik-Ramsey band system has been carried out in two photospheric models — BCA and HSRA. Centre-to-limb variation of the above line is given for BCA. Calculations show that this band may be observable in the photospheric spectrum.

071.017 Formation of C_2 molecules in the solar atmosphere.
K. Sinha, M. C. Pande.
Bull. Astron. Inst. Czechoslovakia, Vol. 25, 84 - 86 (1974).

It is shown that while the triplet states of Swan bands are populated through an LTE path, the singlet states of Phillips bands are not. The problems regarding the observations of the various bands of C_2 molecules in the solar photospheric and umbral spectra are discussed in the light of this result.

071.018 Analysis of solar granulation using laser-light.
W. Schlosser, W. Klinkmann.
Astron. Astrophys., Vol. 32, 29 - 37 (1974). In German.

Laser-optical analysis of photographs of solar granulation fields gives both qualitative and quantitative information on basic parameters of the granulation in a quick and convenient manner. The general approach is to analyse the Fourier transform of the granulation pattern. This is achieved by illuminating slides of granulation fields with slightly converging laser-light. The resulting diffraction in the focal plane is the square of the Fourier-transform of the intensity distribution in the slide.

071.019 Investigation of the seeing from the photometric profile of a Mercury projection on the solar disk.
I. Sattarov.

Solar processes and their observations, (see 003.005), p. 114 - 122 (1973). In Russian.

Spectrograms of the transit of Mercury obtained on May 9, 1970 at the Tashkent horizontal solar telescope have been used to study the role of light scattering and blurring in spectrophotometric investigations of solar surface details. The observed profile of Mercury's shadow was compared with the calculated one assuming that the blurring function has different values. Dependence of light scattering and blurring on wavelengths was investigated.

071.020 Center-to-limb profiles of the aluminum autoionization lines in the solar spectrum.
H. C. McAllister.
Solar Physics, Vol. 35, 3 - 10 (1974).

The 1929 to 1941 Å region of the solar spectrum has been measured, using spectrograms obtained with a rocket-borne high resolution spectrograph. The center-to-limb behavior of the Al I autoionization lines of the $3s^2\, 3p\ ^2P° - 3s\, 3p^2\ ^2S$ doublet, at a spatial resolution of approximately three minutes of arc, is reported. The results provide additional data relating to the photospheric-chromospheric transition region.

071.021 Computation and observation of Zeeman multiplet polarization in Fraunhofer lines. II: Computation of Stokes parameter profiles. A. Wittmann.
Solar Physics, Vol. 35, 11 - 29 (1974).

A self-contained summary of the generalized Unno theory of LTE line formation in solar magnetic fields and its application to the numerical computation of Stokes parameter profiles is given. Within this context, computational details of general interest are described and numerical results for sunspot fields are given. Finally, a new method of computing the height of line formation is presented.

071.022 Temperature effects on measurements of photospheric magnetic fields.
B. Caccin, R. Falciani, A. Donati-Falchi.
Solar Physics, Vol. 35, 31 - 36 (1974).

The effects of line profile variations, due to the temperature enhancements connected with photospheric magnetographic measurements are evaluated in a quantitative way: taking into account these results, the differences found by many authors between the values of the field obtained with different Fraunhofer lines can be reduced within 10−30 %.

071.023 One- and multi-component models of the upper photosphere based on molecular spectra. II: CN (1, 1) of the CN violet system.
G. H. Mount, J. L. Linsky.
Solar Physics, Vol. 35, 259 - 276 (1974).

The authors have obtained center-to-limb photoelectric spectra of the CN(1, 1) B-X bandhead region λ 3868−3872 Å at Kitt Peak National Observatory. From these spectra and a detailed analysis of the formation of the CN(1, 1) spectrum they derive a best-fit upper photospheric model differing from the HSRA which is consistent with their previous CN (0, 0) λ 3883 spectra. They derive a solar carbon abundance of log A_c = 8.30 ± 0.10 compared to the HSRA value of log A_c = 8.55 ± 0.10. In addition they specify the regions of formation for the CN(0, 0) λ 3883.35 and CN(1, 1) λ 3871.38 bandheads at disc center and limb.

071.024 Temperature variations in the solar photosphere. II: A search for equator-to-pole differences in photospheric temperature.
R. Falciani, M. Rigutti, G. Roberti.
Solar Physics, Vol. 35, 277 - 280 (1974).

From the variations in equivalent width of a selected set of Fraunhofer lines, the authors derived the variation of effective temperature and microturbulence velocity between the equator and the regions at heliographic latitude of $\simeq 72°$.

071.025 **The solar abundance of germanium.**
J. E. Ross, L. H. Aller.
Solar Physics, Vol. 35, 281 - 286 (1974).

The solar abundance of germanium, deduced from two relatively unblended Ge I lines, $\lambda 3039.06$ and $\lambda 3269.50$ is found to be $\log N(\text{Ge}) = 3.50 \pm 0.05$ on the scale $\log N(\text{H}) = 12.00$ in good agreement with Cameron's recent solar system abundance $\log N(\text{Ge}) = 3.56$.

071.026 **Short period variation of the photospheric magnetic field.** I. K. Csada.
Solar Physics, Vol. 35, 325 - 330 (1974).

The photospheric magnetic data recorded from August 12, 1959 to September 29, 1967 and averaged over Bartels rotation periods are treated as zonal terms of the solar magnetic field which is expanded in a series of spherical harmonics. Numerical analysis of the reduced data gives seven periods. Three of these seem to be essential in the superposed variation of the solar magnetic field. The amplitudes and phase angles of the periodic terms in question are determined.

071.027 **Observation of the solar photospheric oscillations using a sodium optical resonance device and an optical spatial filtering.** E. Fossat.
Astron. Astrophys., Suppl. Ser., Vol. 15, 475 - 477 (1974). Conference paper (see 012.011).

071.028 **Studies of granular velocities. IV: Statistical analysis of granular Doppler-shifts.** W. Mattig, A. Nesis.
Solar Physics, Vol. 36, 3 - 9 = Mitt. Fraunhofer Inst., *Freiburg*, No. 123 (1974).

The authors present a statistical analysis of local Doppler shifts measured in the line $\lambda 6302.5$ of Fe I.

071.029 **The solar abundance of beryllium.**
J. E. Ross, L. H. Aller.
Solar Physics, Vol. 36, 11 - 19 (1974).

The solar abundance of beryllium is deduced from high-resolution Kitt Peak observations of the $\lambda 3130.43$ and $\lambda 3131.08$ lines of Be II interpreted by the method of spectrum synthesis. The authors find $\{Be\} \equiv \log N(\text{Be})/N(\text{H}) + 12 = 1.08$.

071.030 **The solar abundance of thulium.**
J. E. Ross, L. H. Aller.
Solar Physics, Vol. 36, 21 - 23 (1974).

The solar spectrum contains one relatively unblended line $\lambda 3131.258$ Tm II which yields a thulium abundance of log $N(\text{Tm})/N(\text{H}) + 12 = \{Tm\} = 0.80 \pm 0.10$, with the Corliss and Bozman f-value. A recent beam-foil experiment suggests that the thulium abundance may be reduced to $\{Tm\} = 0.30$.

071.031 **The [Ca II] $\lambda 7323.90$ line in the solar spectrum.**
R. W. Day.
Solar Physics, Vol. 36, 25 - 27 (1974). – Research note.

071.032 **An interpretation of the correlation in the intensity fluctuations in H and K of Ca II and b_1 of Mg I.**
K. R. Sivaraman.
Solar Physics, Vol. 36, 49 - 50 (1974). – Research note.

071.033 **Differential rotation of solar filaments.**
D. L. Glackin.
Solar Physics, Vol. 36, 51 - 60 (1974).

The latitudinal component of solar differential rotation and the possibility of a radial component are discussed and compared to the observed rotational velocities of solar filaments. It is shown that filaments in closer proximity to active regions usually exhibit no differential rotation, while those far from active regions generally show it clearly. Comparison with Mt. Wilson photospheric Doppler measurements shows that filaments rotate faster than the general photosphere and that, as is well known, the spot rate exceeds that for the general photosphere.

071.034 **Polarisation de la lumière, au bord du disque solaire, dans le proche infra-rouge.** J. L. Leroy.
Solar Physics, Vol. 36, 81 - 84 (1974).

The polarization of light observed near the sun's limb steeply goes down from the blue to the red part of the spectrum. New measurements performed through near-infrared filters have shown that this behaviour holds up to 8000 Å. On the other hand, the author has found that the proportion of polarized light remains nearly constant between 8000 and 10000 Å.

071.035 **Bright photospheric areas surrounding sunspot groups at 5700 Å.** R. A. Miller.
Solar Physics, Vol. 36, 91 - 99 (1974).

Sunspot groups at various central angles were chosen to study what appears to be an enhanced surrounding area, even at small central angles. The continuous band of observation was 200 Å wide, centered at 5700 Å. Four years of sunspot pictures from March 1957 to March 1961, were inspected and no instance of the non-appearance of the enhancement was noted.

071.036 **Elemental abundances, isotope ratios and molecular compounds in the solar atmosphere.**
O. Engvold, Ø. Hauge.
Inst. Theor. Astrophys., Blindern – Oslo, Rep. No. 39, 25 pp. (1974).

Solar abundances of chemical elements and solar isotope ratios are tabulated. The solar abundances of 67 elements are known. In addition, the upper limits for the abundances of 5 elements are listed. Isotope ratio investigations have been carried out for 16 elements. A separate list presents 21 molecules which have been identified in the solar atmosphere and another 15 which may be termed "possibly found".

071.037 **On the equivalent widths of spectral lines in the elements of solar granulation. II.** V. M. Sobolev.
Solnechnye Dannye 1974 Byull., No. 2, p. 55 - 62 (1974). In Russian.

The equivalent widths in three granules and the neighbouring intergranular regions of the Fe I, Fe II, Cr I, Sc II, Ni I, Ca I, Ti II, V I, Y II and CH lines are given. The mean equivalent widths of these lines, their mean ratios for granules and intergranular regions and also the mean differences for four granules have been calculated. The heights of absorbing layers are evaluated. From the Fe I lines the ratio of the mean temperatures of excitation in the intergranular regions and granules was found to be equal to 0.9. The values of electronic pressure for one granule and intergranular region are given.

071.038 **On the motion of solar plasma in photospheric granules.** V. A. Krat, A. A. Shpitalnaya.
Solnechnye Dannye 1974 Byull., No. 2, p. 63 - 69 (1974). In Russian.

The fluctuations of radial velocities of the photospheric elements (granules and intergranular regions) on the spectrograms of the solar disc centre obtained on July 30, 1970 by means of the Stratospheric Solar Observatory are measured. The resulting values of the velocities lead to the conclusion that the negative velocities correlate with ganules only statistically.

071.039 **Distribution of the brightness amplitude in photospheric granulation.**

L. M. Pravduyk, V. N. Karpinskij, A. V. Andrejko.
Solnechnye Dannye 1974 Byull., No. 2, p. 70 - 88 (1974). In Russian.

The photographs of granulation at different parts of the solar disk (from center to limb) taken during the flights of the Stratospheric Solar Observatory on July 30, 1970 and June 20, 1973 have been reduced. A square area with a side equal to 27000 km was taken as a single sample. A histogram of the frequency distribution of relative brightness fluctuations was constructed.

071.040 **Photoelectric photometry of the solar granulation in two regions of the continuum.** I. V. Yudina.
Solnechnye Dannye 1974 Byull., No. 2, p. 107 - 116 (1974). In Russian.

A direct photoelectric photometry of the photosphere was made simultaneously in two regions of the continuum. The results are compared with those of Vasil'eva et al. (1967). It is shown that errors due to atmospheric dispersion must be avoided.

071.041 **Hydraulic concentration of magnetic fields in the solar photosphere. I. Turbulent pumping.**
E. N. Parker.
Astrophys. Journ., Vol. 189, 563 - 568 (1974).

The author points out a simple hydraulic mechanism—turbulent pumping—that appears to account for the observed concentration of fields. The mechanism is treated by using the simple mixing-length theory of convection and turbulence.

071.042 **Hydraulic concentration of magnetic fields in the solar photosphere. II. Bernoulli effect.**
E. N. Parker.
Astrophys. Journ., Vol. 190, 429 - 436 (1974).

The magnetic filaments in the solar photosphere are subject to vigorous kneading and massaging by the convective turbulence at, and beneath, the visible surface. It is shown that the Bernoulli effect of the consequent surging of fluid up and down along the filaments is a major factor in concentrating the magnetic pressure of the filament.

071.043 **Observed departures from LTE in stellar Fe I lines. I. The sun.** M. A. Smith.
Astrophys. Journ., Vol. 190, 481 - 486 (1974).

The recent calculation of the excitation and ionization equilibrium of the Fe I atom in the sun by Athay and Lites has made possible a simple division of observed Fe I transitions into three groups, so that a convenient comparison of the predicted departures of each of these groups can be made with observed solar lines. The observed data, taken from the Michigan photoelectric solar atlas, consist of core depths and line strengths for lines measured at $\mu = 1$ and 0.3. The result of comparison of this non-LTE model with observations is presented.

071.044 **On the asymmetry of selected Fraunhofer lines, II.**
R. I. Kostik, T. V. Orlova.
Solar Physics, Vol. 36, 279 - 285 (1974).

The asymmetry of 11 absorption lines of neutral iron was determined from observations made with the double-pass system on the horizontal solar telescope ASU-5. An attempt was made to interpret this asymmetry in terms of progressive sound waves. The value of asymmetry computed theoretically was shown to be on average only 20 % of the observed value.

071.045 **One- and multi-component models of the upper photosphere based on molecular spectra. III: CH (0, 0) $\lambda 3144$ of the CH C-X system.**
G. H. Mount, J. L. Linsky.
Solar Physics, Vol. 36, 287 - 298 (1974).

The authors have obtained accurate center-to-limb photoelectric spectra of the CH (0, 0) C-X bandhead region $\lambda 3143-3148$ Å at Kitt Peak National Observatory. From these spectra and a detailed analysis of the formation of the CH (0, 0) spectrum they demonstrate that the best-fit upper photospheric model derived from their previous analyses of CN (0, 0) and CN (1, 1) spectra adequately explains the CH C-X observations. In addition they derive a solar carbon abundance of $\log A_c = 8.30 \pm 0.20$ compared to the HSRA value of $\log A_c = 8.55$.

071.046 **On the energy distribution in wavenumber spectra of the granular velocity field.** F.-L. Deubner.
Solar Physics, Vol. 36, 299 - 301 = Mitt. Fraunhofer Inst., Freiburg, No. 126 (1974).

Wavenumber spectra of spatial velocity fluctuations measured in selected high resolution solar photospheric spectrograms are presented. The statistical stability of the power estimates of particular 'peaks' in these spectra is discussed.

071.047 **Atomic-beam study of the 5-min solar wavelength oscillations.**
J. L. Snider, J. P. Eisenstein, G. R. Otten.
Solar Physics, Vol. 36, 303 - 312 (1974).

Long records of the wavelength oscillations of the solar 7699 Å potassium line have been obtained with an atomic-beam resonance-scattering apparatus taken to Kitt Peak. Data are presented giving the dependence of rms oscillation amplitude on the size of the area observed and power spectra of the oscillations are shown. The results are compared with recent work by others.

071.048 **The origin of the solar five-minute oscillation.**
S. Musman.
Solar Physics, Vol. 36, 313 - 319 (1974).

Two absorption lines formed in the lower photosphere were used to study simultaneous velocity and intensity fluctuations. No significant correspondence was found between the locations of granules and those of oscillation, even when a time lag was included. This result supports the explanation of the origin of the oscillations as a self-excited sound wave rather than the local response to a granule excitation.

071.049 **The response of an isothermal atmosphere to pressure fluctuations at its base and the five-minute oscillations in the solar photosphere.** R. L. Moore.
Solar Physics, Vol. 36, 321 - 337 (1974).

The steady-state vertical-velocity response of an isothermal atmosphere to pressure fluctuations of arbitrary period and horizontal wavelength at its base is derived in the approximation of dissipationless polytropic motion in the atmosphere. The correct behavior of the response is presented in some detail. Comparison of the response of the model, for the case of isothermal oscillations, with observed features of the photospheric oscillations indicates that, in addition to the evanescent photospheric oscillations which occur at the compression-wave propagation cut-off frequencies and which have horizontal wavelengths $\gtrsim 3000$ km, in the lower photosphere there are also smaller-scale evanescent oscillations which have horizontal wavelengths $\lesssim 1000$ km, periods ranging from 200 to 400 s, amplitudes comparable to that of the larger-scale oscillations, and in which the phase of the vertical velocity oscillation leads the phase of the pressure oscillation.

071.050 **A photoelectric study of absorption of the continuous spectrum by Fraunhofer lines. VII. Intensities of continuous solar emission.**
P. P. Kozak, I. V. Baranovskij, L. Yu. Ostroverkhaya.
Solnechnye Dannye 1974 Byull., No. 3, p. 98 - 105 (1974). In Russian.

The values of the integral of the monochromatic intensity of the continuous spectrum have been estimated for the solar disk center for a series of λ- and τ-values, with or without accounting for blanketing effect and also at different angular distances from the center of the solar disc. The monochromatic fluxes for the same λ and τ values have also been determined.

071.051 **Excitation temperatures of molecules and models of the solar atmosphere.** A. I. Khlystov.
Solnechnye Dannye 1974 Byull., No. 4, p. 89 - 95 (1974).
In Russian.
It is shown that the excitation temperatures of molecules cannot be used for a determination of the physical conditions in the solar atmosphere.

071.052 **A non-L.T.E. analysis of the solar Mg I spectrum.** R. C. Altrock, R. C. Canfield.
Bull. American Astron. Soc., Vol. 6, 219 - 220 (1974). Abstr. AAS.

071.053 **Partial redistribution effects in the solar magnesium II resonance lines.** R. W. Milkey, D. M. Mihalas.
Bull. American Astron. Soc., Vol. 6, 220 (1974). — Abstr. AAS.

071.054 **The solar chromium abundance as a case study in stellar abundance determination reliability.**
G. J. Garwood, J. C. Evans.
Bull. American Astron. Soc., Vol. 6, 220 (1974). — Abstr. AAS.

071.055 **A two-dimensional analysis of intensity fluctuations in MgI 4571 Å on the solar disk.**
R. C. Altrock, C. J. Cannon.
Bull. American Astron. Soc., Vol. 6, 284 (1974). — Abstr. AAS.

071.056 **The width of the solar 584 Å line of neutral helium.**
W. E. Behring, G. A. Doschek, U. Feldman, L. Cohen, J. Houston.
Bull. American Astron. Soc., Vol. 6, 284 (1974). — Abstr. AAS.

071.057 **A phenomenological study of high-resolution granulation photographs.**
B. J. La Bonte, G. W. Simon, R. B. Dunn.
Bull. American Astron. Soc., Vol. 6, 285 (1974). — Abstr. AAS.

071.058 **Preliminary results using a MgI b-line filter to photograph the photospheric network.** G. A. Chapman.
Bull. American Astron. Soc., Vol. 6, 285 (1974). — Abstr. AAS.

071.059 **A statistical study of ephemeral active regions in 1970 and 1973.** K. Harvey, J. Harvey.
Bull. American Astron. Soc., Vol. 6, 288 (1974). — Abstr. AAS.

071.060 **Magnetic flux measurements in the photosphere.** Ro. Howard.
Bull. American Astron. Soc., Vol. 6, 289 (1974). — Abstr. AAS.

071.061 **Time variations of the magnetic and velocity fields of the quiet photosphere.** D. K. Lynch.
Bull. American Astron. Soc., Vol. 6, 291 (1974). — Abstr. AAS.

071.062 **The response of an isothermal atmosphere to pressure fluctuations at its base and the five-minute oscillations in the solar photosphere.** R. L. Moore.
Bull. American Astron. Soc., Vol. 6, 292 (1974). — Abstr. AAS.

071.063 **The generation of the solar five-minute oscillation.** S. Musman.
Bull. American Astron. Soc., Vol. 6, 292 (1974). — Abstr. AAS.

071.064 **Extrapolation of photospheric magnetic fields into the corona.**
G. Poletto, A. Krieger, J. K. Silk, A. Timothy, G. Vaiana.
Bull. American Astron. Soc., Vol. 6, 292 - 293 (1974). Abstr. AAS.

071.065 **Solar sector boundary configuration from comparison of synoptic charts of the photospheric magnetic field with the observed interplanetary field.**
P. H. Scherrer, J. M. Wilcox, Ro. Howard.
Bull. American Astron. Soc., Vol. 6, 293 (1974). — Abstr. AAS.

071.066 **The widths of the solar He I and He II lines at 584, 537, and 304 Å.**
G. A. Doschek, W. E. Behring, U. Feldman.
Astrophys. Journal, (*Letters*), Vol. 190, L141 - L142 (1974).
The authors report direct measurements from a rocket spectrograph of the widths of the solar He I lines, $1s^2 - 1s2p$ and $1s^2 - 1s3p$, at 584 and 537 Å. They also report the width of the solar resonance line of He II at 304 Å. Line profiles corrected for the nonlinear film response are presented for the 584 and 304 Å lines.

Extra-atmospheric observations of the sun.
See Abstr. 051.006.

Highly charged ions in astrophysics.
See Abstr. 061.045.

Line-blanketing and model stellar atmospheres. I. Statistical method and calculation of a grid of models.
See Abstr. 064.049.

Some remarks on line weakenings in photospheric faculae. See Abstr. 072.021.

Chromospheric activity associated with moving photospheric magnetic fields. See Abstr. 073.038.

Chromospheric activity associated with moving photospheric magnetic field. See Abstr. 073.083.

Comparison of X-ray coronal structures with the dynamics and morphology of photospheric activity.
See Abstr. 076.006.

The relationship between the slowly varying component of solar radio emission and large scale photospheric magnetic field patterns. See Abstr. 077.032.

Extraterrestrial solar spectrum, 3000–6100 Å at 1-Å intervals. See Abstr. 080.030.

Dynamics of the solar magnetic field. IV. Examples of force-free magnetic-field evolution in response to photospheric motions. See Abstr. 080.062.

072 Sunspots, Faculae, Solar Activity Cycles

072.001 **The theory of finite-amplitude overstability and its application to periodic motions in sunspots.**
R. Van der Borght.
Monthly Notices Roy. Astron. Soc., Vol. 166, 191 - 202 (1974).

A theory is developed for the study of overstable motions in a fluid layer, within the Boussinesq approximation. The results are then applied to observed periodic motions in the umbra of sunspots.

072.002 **Remarks concerning Ward's 'The latitudinal motion of sunspots and solar meridional circulations'.** J. Tuominen.
Solar Physics, Vol. 34, 15 - 16 (1974). – Research note.

072.003 **On the manifestation of the north-south asymmetry of sunspot activity in the geomagnetic indices.**
N. N. Petrova.
Solnechnye Dannye 1973 Byull., No. 11, p. 73 - 79 (1974).
In Russian.

The catalogue of M-indices of geomagnetic activity during 1884–1968 was used for the study of the dependence of geomagnetic disturbances on the north-south asymmetry of sunspot activity. The form of the dependence tends to change according to the phase and to odd or even number of the 11-year cycle.

072.004 **On the inhomogeneity of the magnetic field in sunspot umbrae.** M. D. Gusejnov (*M. J. Huseynov*).
Izv. Krymskoj Astrofiz. Obs., Vol. 49, 15 - 24 (1974). In Russian.

The magnetic field intensities in a few sunspots were measured from 78 polarization spectrograms obtained with the solar tower telescope of the Crimean Astrophysical Observatory in 1957, July 1 - 15. It is concluded that the field in the umbra is not homogeneous and apparently consists of three principal components: a) bright umbral granules associated with a comparatively small field; b) the continuous background with a moderate field and c) dark umbral knots displaying the strongest field.

072.005 **On the determination of the temperature and density in a sunspot and in the chromosphere above the sunspot.** E. A. Baranovskij.
Izv. Krymskoj Astrofiz. Obs., Vol. 49, 25 - 30 (1974). In Russian.

For the determination of the temperature and density the observed and calculated profiles of the lines H_α and K Ca II are compared. The density in the sunspot and the chromosphere above the sunspot is found to be 30 times less as compared to undisturbed regions.

072.006 **North-south asymmetry in solar activity distribution and solar wind structure in the meridional plane of the sun.** B. M. Vladimirskij, L. S. Levitskij.
Izv. Krymskoj Astrofiz. Obs., Vol. 49, 31 - 35 (1974). In Russian.

The zone of avoidance of solar protons impact was located at the solar equatorial plane for 1938 - 1963 data. It is shown that this zone of avoidance for 1964 - 1969 was situated at heliolatitudes $-2°$ to $-5°$.

072.007 **The connection of sources of noise storms with the magnetic flux of spots.**
L. A. Eliseeva, L. I. Yurovskaya.
Izv. Krymskoj Astrofiz. Obs., Vol. 49, 42 - 45 (1974). In Russian.

From comparison of the magnetic flux of spots during their passage across the solar disk with the emission of noise storm it is shown that the source of radio emission in the corona is appearing in the interval of rise of the magnetic flux at the photosphere.

072.008 **Spectral lines suitable for sunspot investigations.**
J. Buurman.
Astron. Astrophys., Suppl. Ser., Vol. 15, 35 - 46 (1974).

Spectrograms from a large umbra ($\lambda\lambda$ 5000–6900 Å) have been obtained in both directions of circular polarization simultaneously. The observational data and their reduction is described. The spectrograms from the dark core of the umbra were used to select spectral lines with and without magnetic splitting, which are suitable for sunspot investigations. Tables of lines of various splitting patterns are presented. Some important profiles are reproduced and discussed. Some peculiar V I profiles are described. A new method is discussed to adjust the spectrograms of both polarization directions to the same wavelength scale by means of umbral molecular lines.

072.009 **The regularity in changes of the length of the solar spot cycles.** W. Szymański.
Postępy Astron., Vol. 22, 45 - 47 (1974). In Polish.

It seems that the changes in the length of the solar spot cycles constantly decrease.

072.010 **On the character of spatial-temporal relationships of different solar indices.**
Yu. I. Vitinskij, B. M. Rubashev.
Solnechnye Dannye 1974 Byull., No. 1, p. 84 - 91 (1974).
In Russian.

The coefficients of correlation between the values of the flare index for the given solar rotation and, correspondingly, those of the total area of sunspots, the magnetic field intensity, total area of Ca II plages, total area of photospheric faculae, total length of hydrogen filaments for the previous solar rotation have been determined in $40° \times 10°$ fixed regions during 1966–1970.

072.011 **On the problem of classification of light bridges in sunspots.** Kh. I. Abdusamatov.
Solnechnye Dannye 1974 Byull., No. 1, p. 91 - 94 (1974).
In Russian.

The classification of light bridges in sunspots by Korobova (1966) is considered. It is shown that all the bridges are genetically connected with parental sunspots and are the results of fragmentation of the magnetic force tube with the rope structure.

072.012 **The characteristics of the fluctuations of Wolf numbers near the maximum epoch of the 20th solar cycle.** Yu. I. Vitinskij.
Solnechnye Dannye 1974 Byull., No. 1, p. 103 - 105 (1974).
In Russian.

Values of the fluctuation index and lists of its positive and negative strong fluctuations for 1970–1972 and values of absolute and relative indices of "perturbations" for 1967–1972 are given. Some peculiarities of these characteristics are discussed.

072.013 **Sonnenfleckentätigkeit und interstellare Materie.**
A. Mahn.
SuW, Vol. 13, 44 - 46 (1974).

072.014 **Sunspot models with Alfvén wave emission.**
D. J. Mullan.

Astrophys. Journ., Vol. 187, 621 - 631 (1974).

Sunspot models have been computed on the assumption that the missing flux is transported by undissipated Alfvén waves. In order to estimate the flux of these waves, the author proposes an extension of Öpik's cellular model of convection to include the effects of a vertical magnetic field on horizontal gas flow. He derives the depth-dependence of the effective temperature T_e in a spot, and finds that it increases nonmonotonically from a low value at the surface to the solar value at depth z.

072.015 **The width of solar active zones.**
N. I. Kozhevnikov.
Astron. Tsirk., No. 802, p. 3 - 5 (1973). In Russian.

072.016 **Number of sunspot groups formed between 1962 and 1964 and their average lifetime.**
M. Kopecký, F. Kopecká.
Bull. Astron. Inst. Czechoslovakia, Vol. 25, 79 - 81 (1974).

The paper complements the values of the number of formed sunspot groups and their average lifetimes, already published earlier for the period from 1874 to 1961.

072.017 **Vibration-rotation bands of HF, HBr and HI in the sunspot spectrum.** M. C. Pande, V. P. Gaur.
Bull. Astron. Inst. Czechoslovakia, Vol. 25, 81 - 82 (1974).

The calculated equivalent widths of $R(7)$, $R(9)$ and $R(13)$ lines of $1-2$ vibration-rotation band of HF for Henoux's (1969) sunspot model suggest the presence of this band in the sunspot spectrum. The reasons for the absence of the vibration-rotation bands of HBr and HI in the sunspot spectrum are discussed.

072.018 **Investigation of an external dynamical influence on the solar cycle.** V. S. Prokudina.
Astron. Tsirk., No. 804, p. 3 - 5 (1973). In Russian.

072.019 **Proper motions of spots in multicentral groups.**
Z. B. Korobova, V. M. Tishchenko, A. A. Tskhaj, V. V. Baturin.
Solar processes and their observations, (see 003.005), p. 57 - 76 (1973). In Russian.

Proper motions in 25 multicentral groups (10 bipolar groups, 10 irregular multipolar spots and 5 complex groups) are investigated. It is shown that rotational motions prevail in groups of any magnetic class. No correlation between the vortex directions and the hemisphere have been found. The direction of the rectilinear motion of intermediate spots is dependent on their magnetic polarity. The rectilinear and rotational motion velocities are found to be $\approx$ 100 m/sec.

072.020 **The connection between sporadic chromospheric processes and the evolution of sunspot groups.**
K. F. Kuleshova.
Solar processes and their observations, (see 003.005), p. 77 - 80 (1973). In Russian.

An attempt is made to divide sporadic chromospheric processes according to the phases of a spot group evolution. It is shown that processes with rapid development such as surges are arising on the descending phase of development of an active region and are probably connected with magnetic field decay. On the contrary processes with slow evolution (loops-type prominences) are arising on the ascending phase and are probably connected with penetration of the field into chromosphere and corona.

072.021 **Some remarks on line weakenings in photospheric faculae.**
B. Caccin, R. Falciani, A. Donati-Falchi.
Solar Physics, Vol. 35, 41 - 45 (1974).

A revision of Schmahl's facular model is used for comparison with the HSRA as average photosphere. On its basis many of the available data, in the visible spectrum, can be explained quantitatively.

072.022 **The depth of sunspots.** T. Prokakis.
Solar Physics, Vol. 35, 105 - 110 (1974).

Good quality photographs of many regular sunspots were obtained in three spectral regions, with the use of narrow band filters. Isodensity contours were used for measurements of the umbra and penumbra size along different axes of sunspots. A parameter based on the morphology of the whole sunspot was defined for a better study of the Wilson effect.

072.023 **The cooling of a sunspot. IV: Reply to D. J. Mullan.**
P. R. Wilson, with a comment by D. J. Mullan.
Solar Physics, Vol. 35, 111 - 121 (1974).

072.024 **Sur les premiers signes de la reprise de l'activité solaire.** S. Koutchmy, C. Bareau, G. Stellmacher.
Comptes Rendus Acad. Sci. Paris, Sér. B, Vol. 278, 873 - 876 (1974).

Grâce à l'étude d'un spectre de la couronne solaire obtenu au cours de l'éclipse du 30 juin 1973 au Tchad, des raies interdites d'éléments fortement ionisés (Fe XIV, Ni XIII et XVII, Ar X) ont pu être mises en évidence au voisinage du pôle sud. La présence de ces raies, caractéristiques des régions coronales actives, comme la morphologie de la couronne solaire dans cette région montre que le nouveau cycle de 11 ans de l'activité solaire peut être en fait détecté au moins 3 ans avant son commencement.

072.025 **Vorläufige Epochen der Maxima und Minima des achtzigjährigen Sonnenfleckenzyklus.**
R. Hartmann.
Mitt. Astron. Ges., No. 34, p. 116 - 122 (1973/74).
Presented at the "Wissenschaftliche Tagung der Astron. Ges., Oberkochen, 1973 April 24 - 27".

072.026 **Large sunspot groups in the years 1955–1964.**
M. Kopecký, P. Kotrč.
Bull. Astron. Inst. Czechoslovakia, Vol. 25, 171 - 180 (1974).

On the basis of Greenwich observational data, a catalogue of large sunspot groups with an average area of more than 500 millionth of the solar disk, observed in the years 1955–1964, has been compiled and compared with a similar catalogue of Gnevysheva, compiled from Pulkovo sunspot catalogues. The time pattern of the occurrence of large sunspot groups in the 19th cycle of solar activity is discussed.

072.027 **Temporal variations of the magnetic field in sunspots.**
R. B. Schultz, O. R. White.
Solar Physics, Vol. 35, 309 - 316 (1974).

The authors obtained simultaneous spectra with a spatial resolution of $1/2''$ and a temporal resolution of 15 s in Hα, Ca II-K, Ca II 8542 Å, and three Fe I lines of the sunspot group responsible for the large flares of August, 1972 (McMath No. 11976). A time series taken 1972, August 3 in the Fe I 6173 Å Zeeman sensitive line was analyzed for oscillations of field strength and the angle between the field and the line of sight, and for changes of the field associated with the Ca II-K umbral flashes discovered by Beckers and Tallant (1969). The results place restrictions on magnetic modes of energy transport between the photospheric layers and the chromospheric layers where the umbral flashes are observed.

072.028 **Umbral intensities of large sunspots.**
G. Ekmann, P. Maltby.
Solar Physics, Vol. 35, 317 - 322 (1974).

Observations of the intensities of sunspot umbrae at the Oslo Solar Observatory during 5 yr are described. The results for 5 large sunspots observed in 6 (alternatively 7) wavelength regions are presented.

072.029 Comments on Wilson's model of cooling of a sunspot. M. H. Gokhale.
Solar Physics, Vol. 35, 323 - 324 (1974). — Research note.

072.030 Computation and observation of Zeeman multiplet polarization in Fraunhofer lines. III: Magnetic field structure of spot Mt. Wilson 18488. A. Wittmann.
Solar Physics, Vol. 36, 29 - 44 (1974).

From an analysis of Stokes parameter profiles measured in a unipolar sunspot, properties of its large-scale magnetic field structure have been derived. With due allowance for our present knowledge about the Wilson effect and the height of line formation, an expression for the field strength distribution $H(r, z)$ has been obtained. The field azimuth has been determined across a sunspot diameter on several days with due allowance for the Macaluso-Corbino effect, and significant azimuthal components H_x have been found especially in the umbra.

072.031 The temperature of penumbral filaments.
O. Kjeldseth Moe, P. Maltby.
Solar Physics, Vol. 36, 101 - 108 (1974).

The intensity of individual penumbral filaments has recently been measured at the Pic-du-Midi Observatory as well as from observations obtained during the third flight of the Soviet Stratospheric Solar Station. The authors have used the results of these measurements to calculate the corresponding average penumbral intensity as function of wavelength. The calculated average intensity is compared with the average intensity observed at the Oslo Solar Observatory. The run of temperature versus optical depth is given for bright and dark penumbral filaments. The variation of gas pressure with geometrical depth is discussed.

072.032 Models for different sunspot umbrae.
O. Kjeldseth Moe, P. Maltby.
Solar Physics, Vol. 36, 109 - 114 (1974).

The recently detected intensity difference between individual, large sunspots in the infrared spectral region is considered. The authors show that the intensity difference may be explained by a temperature difference of 140–160 K in the upper atmosphere keeping the temperature nearly unchanged below an optical depth approximately equal to unity. The change in temperature in the upper layers alters the observability of the deeper layers and the corresponding intensity.

072.033 The spectrum of low-frequency oscillations of the magnetic field of sunspots and low-frequency modulation of radio emission of solar active regions.
Eh. I. Mogilevskij, V. N. Obridko, B. D. Shel'ting.
Izv. vyssh. ucheb. zavedenij. Radiofizika, Vol. 16, 1357 - 1361 (1973). In Russian. — Abstr. in Referativ. Zhurn. 51. Astron., 5.51.427 (1974).

072.034 On the contrast of sunspots. G. S. Minasyants.
Solnechnye Dannye 1974 Byull., No. 2, p. 89 - 97 (1974). In Russian.

A photometry of high-resolution photographs of 72 sunspots and pores ($S_{umbra} \geq 4$ millionths of the solar hemisphere) in the λ 5190 Å region and 28 sunspots and pores ($S_{umbra} = 5$ millionths of the solar hemisphere) in the λ 6400 Å region has been made. An analysis of the contrasts has not shown an increase of sunspot brightness with a decrease of S_{umbra}.

072.035 Faculae and the solar oblateness. II. R. H. Dicke.
Astrophys. Journ., Vol. 190, 187 - 198 (1974).

Chapman and Ingersoll (1973) have suggested that the presence of a normally distributed error in their facular function may require a large increase in the estimate of the photospheric facular contribution to the solar oblateness signal. This effect is evaluated by using the Chapman-Ingersoll statistical assumptions and additional facular data from earlier studies. With the further assumption that there is no common systematic bias between the two facular signals, the above effect is found to be negligible, representing an additional 2 - 3 percent correction to the oblateness signal. The effects of systematic biases of the facular data are also examined.

072.036 The fine structure in the latitudinal zones on the sun. N. I. Kozhevnikov.
Astron. Zhurn. Akad. Nauk SSSR, Vol. 51, 582 - 587 (1974). In Russian. English translation in Soviet Astron. AJ, Vol. 18, No. 3.

The results of studying the distribution of sun spots living one day in heliographic latitude on the solar disk are given. This distribution was found to have nearly the form of periodic function with a period of about 2°. The investigation of other factors characterizing the activity of sunspots, living one day, showed the presence of a period of about 4°.

072.037 The nature of the sunspot phenomenon. I: Solutions of the heat transport equation.
E. N. Parker.
Solar Physics, Vol. 36, 249 - 274 (1974).

The author concludes that the inhibition of convection cannot be the cause of a sunspot. He suggests, instead, that a sunspot is a region of enhanced, rather than inhibited, energy transport and emissivity. The magnetic field of the sunspot causes a dynamical overstability in the outer thousand km of the convective zone, generating copious fluxes of hydromagnetic waves, which propagate rapidly out of the region along the magnetic field. He suggests that this heat engine is so efficient as to convert at least three fourths of the heat flux into waves.

072.038 Sub- and superhydrostatic equilibrium in sunspots. W. Mattig.
Solar Physics, Vol. 36, 275 - 277 = Mitt. Fraunhofer Inst., Freiburg, No. 127 (1974).

072.039 An inhomogeneous sunspot model. I. The effect of inhomogeneity on the evaluation of "empirical" models. V. N. Obridko.
Solnechnye Dannye 1974 Byull., No. 3, p. 73 - 76 (1974). In Russian.

It is shown that if the sunspot umbra spectrum is obtained with a resolution not high enough for separate observations of hot and cold components emission and the measured equivalent widths of total emissions are interpreted for obtaining a homogeneous model, then the "empirical" homogeneous model will have significant deviation from hydrostatics.

072.040 On the spatial orientation of light bridges in sunspots. Kh. I. Abdusamatov, I. Ferro.
Solnechnye Dannye 1974 Byull., No. 3, p. 77 - 82 (1974). In Russian.

The origin of light bridges in sunspots is considered to be the result of internal and external influences on the structure of the sunspot magnetic field. From the study of 198 bridges it has been found that they are mostly located along the solar meridian.

072.041 An inhomogeneous sunspot model. II. Spectral line intensification in a spot. V. N. Obridko.
Solnechnye Dannye 1974 Byull., No. 4, p. 72 - 75 (1974). In Russian.

It is shown that line intensification in a spot can be explained within the limits of a two-component model, each of the components being in hydrostatical equilibrium.

072.042 **On proper motions of the umbrae in a large sunspot group in August 1972.** Z. B. Korobova.
Solnechnye Dannye 1974 Byull., No. 4, p. 76 - 79 (1974). In Russian.

It is shown that the proper motions of the sunspots of preceding and following polarity in the large spot group on August 1972 have their own peculiarities which led to systematic relative shift of the regions with opposite polarity.

072.043 **On the size of a bipolar group at photospheric and chromospheric levels.**
G. I. Kornienko, Eh. P. Surkov.
Solnechnye Dannye 1974 Byull., No. 4, p. 80 - 88 (1974). In Russian.

A comparison of the distances between sunspots of a bipolar group at photospheric and chromospheric levels was made from photoheliograms and filtergrams in the Hα line. It was found that the distance between the group components at the upper (chromospheric) layer is larger than that at the lower (photospheric) layer. This phenomenon is supposed to be connected with variations of the sizes of large photometric inhomogeneities and fluctuations of velocity with height.

072.044 **On the problem of annual variations of solar activity. I. Dependence of the seasonal variation on the phase of the 11-year cycle.**
G. Ya. Vasil'eva, D. A. Kuznetsov, A. A. Shpital'naya, N. S. Petrova.
Solnechnye Dannye 1974 Byull., No. 4, p. 96 - 110 (1974). In Russian.

Wolf numbers W for 1749–1972 and sunspot areas Sp for 1878–1972 were considered. A component systematic with respect to the earth's position on the orbit, has been detected by averaging the W and Sp indices for a series of 11-year cycles from different samples. This seasonal component reveals a dependence on the phase of a cycle and is different for the even and odd cycles. An increase of the activity in July-August for the maximum epoch corresponds to its decrease for the minimum epoch. The variations of solar activity during the year in the minimum epoch and at the descending branch are symmetric relative to the projection of the direction of the galactic magnetic field in the vicinity of the solar system on the ecliptic.

072.045 **On the annual variation of strong fluctuations of Wolf numbers.** Yu. I. Vitinskij.
Solnechnye Dannye 1974 Byull., No. 4, p. 111 - 118 (1974). In Russian.

The annual variation of the number of strong fluctuations of the Wolf numbers and large sunspot groups has been detected at the descending branch of the 11-year cycles. This variation is shown to manifest itself more distinctly in even 11-year cycles, especially for positive fluctuations. A distinct maximum of the number of positive fluctuations has been found for odd cycles in September and even cycles in February–March. In conclusion some suggestions on the probable nature of solar cycles are discussed.

072.046 **Gyro-resonance emission transverse to the magnetic field of a large spot group.**
W. Graf, D. M. Rust.
Bull. American Astron. Soc., Vol. 6, 287 (1974). – Abstr. AAS.

072.047 **A possible link between sunspot magnetism and solar sector magnetism.** L. Svalgaard.
Bull. American Astron. Soc., Vol. 6, 294 - 295 (1974). Abstr. AAS.

072.048 **Relationships between MMF trajectories and sunspot penumbral filament structure.**
D. Vrabec, G. A. Chapman.
Bull. American Astron. Soc., Vol. 6, 296 (1974). – Abstr. AAS.

072.049 **A morphological study of a solar active region in August 1972. I. The morphology of the fine structure of sunspots.**
Solar Physics Division, Yunnan Obs., Acad. Sinica.
Acta Astron. Sinica, Vol. 15, 24 - 33 (1974). In Chinese.

Some photographs of the white-light fine structure of a large sunspot group in August 1972 have been obtained at Yunnan Observatory. The results of a detailed study of the daily variations of the morphology of the sunspot group are presented.

072.050 **Nieuwe zonnecyclus dient zich aan.** C. de Jager.
Zenit, Vol. 1, No. 6, p. 6 - 7, 9 (1974).

072.051 **Berechnung und Messung der Stokesparameter in solaren Zeeman-Multipletts.** A. Wittmann.
Thesis Univ. Göttingen, 78 pp. (1973).

On polarimetry in solar active regions. VII: A new Zeeman polarimeter and its advantages as compared to other designs. See Abstr. 034.036.

The stability of a velocity shear in the presence of an inhomogeneous magnetic field. See Abstr. 062.010.

Dispersion and stability of waves in plasmas in the presence of a Coriolis force. See Abstr. 062.034.

Convection in A stars: convection suppressed in magnetic A stars? See Abstr. 065.028.

Formation of C_2 molecules in the solar atmosphere. See Abstr. 071.017.

Bright photospheric areas surrounding sunspot groups at 5700 Å. See Abstr. 071.035.

On chromospheric phenomena accompanying birth and decay of sunspots. See Abstr. 073.064.

Eleven-years inversion of the green corona emission. See Abstr. 074.030.

The solar wind velocity in the eleven-year cycle No. 20 and the solar radar cross-section. See Abstr. 074.045.

The position of a type I storm source in the magnetic field of an active region. See Abstr. 077.005.

The radio emission spectrum of the quiet sun and its connection with the solar activity cycle. See Abstr. 077.050.

A correlative study of ssc's, interplanetary shocks, and solar activity. See Abstr. 106.031.

073 Solar Chromosphere, Flares, Prominences

073.001 **A theoretical and experimental study of Fe XIX to Fe XXIV solar-flare spectra and isoelectronic spectra in sulfur.** B. C. Fawcett, R. D. Cowan, R. W. Hayes.
Astrophys. Journ., Vol. 187, 377 - 383 (1974).

Atomic self-consistent field calculations are used to predict wavelengths and oscillator strengths of the $2p^n - 2p^{n-1}3d$ transitions in Fe XIX to Fe XXIV. The validity of these calculations is checked through comparison with new experimental data in the isoelectronic sequences. In particular, sulfur classifications are correlated and new S IX identifications reported. Data of value to the analysis of improved solar-flare spectrograms are presented.

073.002 **Determination of the heating time in the upper solar chromosphere.**
M. A. Livshits, O. G. Badalyan.
Astron. Zhurn. Akad. Nauk SSSR, Vol. 51, 139 - 147 (1974). In Russian. English translation in Soviet Astron. AJ, Vol. 18, No. 1.

Heat spreading is considered in a gas column with boundary temperatures 10^6 and 10^4 °K and with hydrostatically changing density independent on time. Numerical computations have been carried out by the method of finite differences. Density variation in time has been taken into account by simultaneous solution of the equations of heat conductivity and hydrostatic equilibrium. It has been shown that some increase of the density in the upper part of the column due to "evaporation" does not affect considerably the initial phases of heat disturbance spreading. In addition, a formal mass estimate of the gas flowing out of the chromosphere to the corona has been obtained.

073.003 **A possible mechanism of formation of surges and sprays.** Lyu Van Lyong.
Astron. Zhurn. Akad. Nauk SSSR, Vol. 51, 148 - 159 (1974). In Russian. English translation in Soviet Astron. AJ, Vol. 18, No. 1.

073.004 **Stationary equations of hydrogen excitation and ionization in prominences.**
Ts. Chultem, N. A. Yakovkin.
Solar Physics, Vol. 34, 133 - 150 (1974).

The statistical equilibrium equations for the continuum and first 10 levels of a hydrogen atom show that the radiation of a bright prominence (the brightness of the Hα line has attained 56 mÅ of the disc centre spectrum) is completely due to scattering of the sun radiation. The parameters are shown to depend greatly on the prominence optical thickness in the lines of the first subordinate series of a hydrogen atom. In the course of determination all the parameters and 100 interconnected integral equations of the radiation diffusion have been thickness-averaged; the population of levels has been calculated by observations using the self-absorption factors.

073.005 **Fine structure of a solar active region at 3.7 and 11.1 cm wavelengths.**
M. R. Kundu, R. H. Becker, T. Velusamy.
Solar Physics, Vol. 34, 185 - 188 (1974). – Research note.

073.006 **Theoretical chromospheric flare spectra. I. Hydrogen equilibrium for the kinematic flare-shock model of Nakagawa et al. (1973).**
R. C. Canfield, R. G. Athay.
Solar Physics, Vol. 34, 193 - 206 (1974).

The authors simultaneously solve the equations of radiative transfer and statistical equilibrium for a model hydrogen atom including Lyman-α, Lyman-β, Balmer-α and the Lyman, Balmer and Paschen continua. The model atmospheres they use are the results of Nakagawa et al. (1973) for a kinematic model of the chromospheric solar flare. They find that the models adequately predict the total intensity of Bα, its wing broadening, the presence of a red-shifted wing, the maximum electron density, the total line-of-sight second-level population and the narrowness in height of the Bα emitting region. The authors find that Nakagawa et al. seriously overestimate the radiative loss function, which will have a large effect on their models.

073.007 **On a method to determine the effective extension, visibility of the Balmer continuum, and the effective mass in solar flares.**
R. E. Guseinov, L. B. Tzirulnik, A. H. Babayev.
Solar Physics, Vol. 34, 207 - 215 (1974).

A method to determine the emission measure and hence the effective extension of solar flares is described. The conditions for the Balmer continuum to be visible in disk flares is also considered. The concept of the effective mass is introduced and it is shown that this mass is a little less than that of a plasma filling the whole spectroscopic volume and including thin and dense filaments and the intervals between them as well.

073.008 **Fine structure of a solar flare region at 3.7 and 11.1 cm wavelengths.**
M. R. Kundu, T. Velusamy, R. H. Becker.
Solar Physics, Vol. 34, 217 - 222 (1974).

On June 9, 1973, a flare associated burst was observed with the NRAO 3-element interferometer at 3.7 and 11.1 cm wavelength. The burst was of 'gradual rise and fall' type.

073.009 **Effect of solar corona conditions on flare particle propagation.** G. Cherki, J. P. Mercier, A. Raviart, L. Treguer, D. Maccagni, F. Perotti, G. Villa.
Solar Physics, Vol. 34, 223 - 229 (1974).

Data on high energy electrons and protons in different energy windows are analyzed and compared for two solar flares which occurred at 37 W solar longitude on the 25th February 1969 and the 29th March 1970. While the data for the first of these flares can be interpreted in the framework of a diffusion model with suitable values of the parallel diffusion coefficient, in order to explain the time behaviour of the different particles after the second event, we are led to suppose that the coronal magnetic fields are such that particles of different rigidity are ejected at different longitudes and that there is no good magnetic connection of the earth with the flare region.

073.010 **The support of prominences formed in neutral sheets.** M. Kuperus, M. A. Raadu.
Astron. Astrophys., Vol. 31, 189 - 193 (1974).

It is shown that prominence filaments formed in a magnetically neutral sheet in the corona can be supported against gravity by magnetic forces due to induced currents in the photospheric boundary. The supporting force is inversely proportional to the height above the photosphere and the filaments seem stable against small perturbations. The adopted magnetic field configuration allows for a prominence eruption. However, it is suggested that the so called disparition brusque may only occur if either the prominence loses weight by mass loss or through the influence of large scale changes in the photospheric magnetic field.

073.011 **Analysis of the spectra of two essentially different prominences.** V. N. Zujkov.

Solnechnye Dannye 1973 Byull., No. 11, p. 57 - 61 (1974). In Russian.

The profiles of some spectral lines of a bright prominence (1958) and a prominence with middle brightness (1959) have been investigated. The main difference was found to be in the following: the prominence of 1958 was characterized by higher values of the kinetic and excitation temperatures than that of 1959.

073.012 **Absolute spectrophotometry of the chromosphere and prominences in the H_8 and He 3888.65 lines from photographs of the total eclipse on March 7, 1970.**
V. M. Sobolev, G. F. Vyalshin.
Solnechnye Dannye 1973 Byull., No. 12, p. 63 - 72 (1974). In Russian.

A spectrophotometric reduction of the spectrograms of the chromosphere and prominences, taken during the solar eclipse on March 7, 1970 in Mexico, was made. For 4 heights and 10 tracings of the chromosphere the total intensities of the H_8 and He 3888.65 lines (in absolute units), the ratio of their intensities and the halfwidths of their profiles have been obtained. For four prominences an analog operation was made. The intensity gradients of the lines in the chromosphere have been calculated.

073.013 **On the problem of self-reversal of features of the $H\alpha$ line in flares.** N. D. Kostyuk.
Solnechnye Dannye 1973 Byull., No. 12, p. 72 - 76 (1974). In Russian.

The self-reversal of features of the $H\alpha$ line in solar flares has been interpreted on the basis of Ivanov's investigations of the transfer of resonance radiation in media with exponential distribution of the sources of radiation. An interpretation of all known features of the self-reversed $H\alpha$ line is given. Some parameters of the surface region of a flare have been estimated on the basis of spectroscopic data of the flare on August 4, 1972.

073.014 **Chromospheric flares during the 19th solar cycle.**
Yu. N. Dolginova.
Solnechnye Dannye 1973 Byull., No. 12, p. 76 - 84 (1974). In Russian.

Some characteristics of flare activity for all the flares of importance ≥ 2 and proton flares have been determined.

073.015 **H^2, H^3, He^3 production in solar flares.**
R. Ramaty, B. Kozlovsky.
Goddard Space Flight Center, Greenbelt, Maryland, GSFC Document X-660-74-94, Preprint, 26 + 10 pp. (1974).

The authors evaluate in detail the production of deuterium, tritium, and helium-3 from nuclear reactions of accelerated charged particles with the ambient solar atmosphere. They use updated cross sections and kinematics, extend their calculations to very low energies (~ 0.1 MeV/nucleon), and calculate the angular distribution of the secondary particles. Then they compare the calculations with data on accelerated isotopes from solar flares. Finally, they provide an explanation for He^3-rich events in terms of the angular distributions of secondary isotopes, and make predictions on the flux of 2.2 MeV gamma rays from such flares.

073.016 **Observations of the chromospheric network: Initial results from the Apollo Telescope Mount.**
E. M. Reeves, P. V. Foukal, M. C. E. Huber, R. W. Noyes, E. J. Schmahl, J. G. Timothy, J. E. Vernazza, G. L. Withbroe.
Astrophys. Journ., (Letters), Vol. 188, L27 - L29 (1974).

A preliminary analysis of early data taken by the HCO spectrometer on *Skylab* shows that the solar chromospheric network can be clearly seen with varying contrast in the extreme-ultraviolet emission characteristic of temperatures between $10^4\,°K$ (the Lyman continuum) and $3 \times 10^5\,°K$ (O VI). In the emission of Mg X, a coronal line formed at about $1.5 \times 10^6\,°K$, the network is generally unrecognizable.

073.017 **Entwicklung von Aktivitätsgebieten auf der Sonne.**
K. O. Kiepenheuer.
Umschau, 74. Jahrgang, p. 190 (1974).

073.018 **On Lyman emission of solar flares.**
L. Křivský, L. N. Kurochka.
Bull. Astron. Inst. Czechoslovakia, Vol. 25, 52 - 61 (1974).

A review is given of the present measurements in the range of the $L\alpha$ line and in the UV continuum at the time of solar flares. The relation of the flash intensities, recorded in these ranges, to the developmental stages of the flare are discussed. An estimate of the temperature of the flare region, radiating in $L\alpha$ and of the concentration of hydrogen atoms in their fundamental state is made.

073.019 **Development and spatial structure of proton flares near the limb, and coronal phenomena. VII. Change of activity prior to the proton flare of September 1, 1971.**
L. Křivský.
Bull. Astron. Inst. Czechoslovakia, Vol. 25, 62 - 64 (1974).

It was again found in the case of the proton flare of Sept. 1, 1971, which occurred beyond the western limb of the solar disk, that the flare activity of the active region increased several days prior to the occurrence of the proton flare.

073.020 **Algorithm and program of treatment of results of magnetic field measurements in solar prominences with a computer by the photographic method.**
V. G. Miller, V. N. Shmulevskij, V. S. Bashkirtsev, G. Ya. Smol'kov.
Issled. po geomagnetizmu, aehron. i fiz. Solntsa. Vyp. (No.) 28. Moskva, Nauka, 1973, p. 128 - 136. In Russian. – Abstr. in Referativ. Zhurn. 51. Astron., 3.51.463 (1974).

073.021 **Models of the solar chromosphere.** J. Paciorek.
Postępy Astron., Vol. 22, 33 - 43 (1974). In Polish.

Three empirical models of the chromosphere are confronted and discussed. The accuracy of their agreement with observational data is considered. Some attempts on the interpretation of the latest observations of the line and continuous spectra in the frame of non-homogeneous models are also discussed.

073.022 **Der Protuberanzenaufstieg vom 25. März 1967.**
G. Klaus.
Orion Schaffhausen, 32. Jahrgang, p. 52 - 54 (1974).

073.023 **Hydrogen emission in prominences.**
Ts. Chultem, N. A. Yakovkin.
Solnechnye Dannye 1974 Byull., No. 1, p. 74 - 77 (1974). In Russian.

The mechanisms of hydrogen excitation in prominences have been considered. A plane-parallel layer in the picture plane was used as model of a prominence. The numbers of elementary processes resulting in population and de-excitation have been calculated and the balance conditions have been constructed. The energy emitted and absorbed by the prominence, optical depths, the source functions and probabilities of quantum survival have also been estimated.

073.024 **Solar proton flares and the sectorial structure of the interplanetary magnetic field.**
V. N. Obridko, S. M. Mansurov, L. G. Mansurova.
Geomagn. Aeronom., Vol. 14, 3 - 7 (1974). In Russian.

073.025 **The nature of point-like sources of spectral lines, continuum and X-ray emission.** S. B. Pikel'ner.
Astron. Zhurn. Akad. Nauk SSSR, Vol. 51, 233 - 242 (1974).

In Russian. English translation in Soviet Astron. AJ, Vol. 18, No. 2.

The point-like sources of linear emission (moustaches) appear as the result of reconnection of magnetic lines in the process of annihilation of an old field of active regions with a new magnetic flux. This annihilation proceeds in a small region of the chromosphere and the main part of the magnetic energy is transformed into kinetic energy of the gas. The gas is heated up to $10^6 - 10^7$ K and it explains the bright points in the X-ray image of the sun, and the weak Mg XII lines in the UV spectrum of the corona. The conditions in the gas are calculated from an analysis of the Hα line and of a weak streak of continuous emission often associated with moustaches. The stationarity conditions for populations of three hydrogen energy levels and continuum let calculate these populations if n_e and T are given. The continuum emission is explained with $n_e = 10^{13} - 3 \times 10^{13}$ cm^{-3}, T = 7500 - 10000 K.

073.026 **On spatial variations in the intensity of chromospheric Hα.** K. B. Gebbie, R. Steinitz.
Astrophys. Journ., Vol. 188, 399 - 406 (1974).

The authors investigate the formation of patterns in Hα spectroheliograms and filtergrams. Introducing a source-sink-control diagram, they conclude that the Hα line source function in the quiet solar chromosphere is indirectly controlled by the photospheric radiation fields in the Balmer and Paschen continua. They demonstrate that in producing the observed patterns, horizontal spatial variations in the shape of the absorption profile are extremely effective compared to changes in the source and sink terms. Applying this mechanism, the authors compute asymptotic values for the contrasts and visibilities in chromospheric Hα.

073.027 **Determination of electron temperature of quiescent prominences by hydrogen lines.**
N. N. Morozhenko.
Solar Physics, Vol. 34, 303 - 308 (1974).

In order to determine the electron temperature T_e of quiescent prominences the author has used the equation of hydrogen ionization equilibrium and the condition of detailed balance in Lα. Calculations have been carried out for 21 prominences of different brightness. The values of the electron temperatures have a scattering within the observation and processing error range. It is suggested that the electron temperatures of all the prominences are equal and are close to 7300 K.

073.028 **Non-thermal line broadening in the solar chromosphere.** R. A. E. Fosbury.
Solar Physics, Vol. 34, 309 - 311 (1974). – Research note.

073.029 **On intensity ratios of helium and hydrogen lines in quiescent prominences.** N. N. Morozhenko.
Solar Physics, Vol. 34, 313 - 322 (1974).

In present paper the author makes an attempt to explain the change of the helium relative intensities in prominences of different brightness as a result of the change of the degree of excitation and ionization of helium. This change occurs both because of decrease of electron density in the outer layers, and because of the diffuse penetration of UV radiation field inside a homogeneous prominence.

073.030 **Theoretical model of flares and prominences. I: Evaporating flare model.** T. Hirayama.
Solar Physics, Vol. 34, 323 - 338 (1974).

This paper presents a theoretical model of the solar flare based on particle evaporation from the chromosphere and predicts various observed quantities obtained in Hα flare, EUV emissions, soft X-ray and the flare associated solar wind.

073.031 **Theoretical chromospheric flare spectra. II: Hydrogen equilibrium for Brown's (1973) models for heating by non-thermal electrons.** R. C. Canfield.
Solar Physics, Vol. 34, 339 - 348 (1974).

The author obtains simultaneous solutions of the equations of radiative transfer and statistical equilibrium for hydrogen excitation and ionization. The model atom includes Lα, Lβ, Balmer-α and the Lyman, Balmer and Paschen continua. The model atmospheres are those of Brown (1973) representing the effects of heating by non-thermal electrons on the Harvard-Smithsonian Reference Atmosphere.

073.032 **A correlation between time-overlapping solar flares and the release of energetic particles.**
G. M. Simnett.
Solar Physics, Vol. 34, 377 - 391 (1974).

The origin of a large co-rotating solar particle event in August, 1970, is discussed. Proton data from spacecraft at five widely separated heliocentric longitudes are used to identify two distinct release points which are over 100° apart in solar longitude. Optical flare data show a high incidence of time-overlapping flares between plage regions close to the two release points, indicating a good connection between them. Unusual X-ray and radio emissions are also observed from these regions.

073.033 **Observations of prominences in He II with a new 25 cm coronagraph.**
T. Hirayama, Y. Nakagomi.
Publ. Astron. Soc. Japan, Vol. 26, 53 - 56 = Tokyo Astron. Obs. Repr. No. 446 (1974).

Spectroscopic observations of quiescent prominences obtained with a new coronagraph at Norikura are reported. A brief description of the coronagraph is also given.

073.034 **On self-heating of flare ejections.**
V. P. Vasil'ev, V. I. Kucherov, V. I. Lapshin.
Astron. Tsirk., No. 799, p. 1 - 2 (1973). In Russian.

073.035 **Complex observations of the chromospheric flare on August 4, 1972.**
A. S. Tskhovrebadze, E. I. Tetruashvili, M. Sh. Gigolashvili.
Byull. Abastumansk. Astrofiz. Obs., No. 45, p. 63 - 70 (1974). In Russian.

The results of spectral observations and radial velocity measurements are presented.

073.036 **Solar flares and processes in an active region.**
Yu. M. Slonim.
Solar processes and their observations, (see 003.005), p. 3 - 56 (1973). In Russian.

On the basis of 30-year observations the connection between flares and phenomena on photospheric, chromospheric and coronal layers of an active region is considered. Spatial localization of the flares is also investigated.

073.037 **On a model of the solar chromosphere.**
A. Sadikov.
Solar processes and their observations, (see 003.005), p. 80 - 88 (1973). In Russian.

Some modern ideas of the structural and physical properties of the chromosphere are summarized. It is supposed to consist of cold and hot spicules with photospheric and coronal origin respectively.

073.038 **Chromospheric activity associated with moving photospheric magnetic fields.**
J.-R. Roy, A. G. Michalitsanos.
Solar Physics, Vol. 35, 47 - 54 (1974).

With the aid of Hα and Ca II K filtergrams and magnetograms of region McMath 12417 on 3, 4 and 5 July 1973, we have followed the evolution of a moving rim of positive magnetic flux 50″ long in an area dominated by negative flux.

Chromospheric activity in the form of brightenings and small surges was associated with this moving flux; a concentration of activity is observed at the locations where magnetic fields of opposite sign meet together.

073.039 **High-resolution photography of the solar chromosphere. XII: An attempt to measure vertical velocities of Hα bright mottles beyond the limb.**
R. E. Loughhead.
Solar Physics, Vol. 35, 55 - 61 (1974).

Measurements have been made of the heights of seven bright mottles photographed beyond the limb on sequences of broad-band Hα filtergrams. They reveal no evidence of a systematic upward or downward motion persisting throughout the lifetime of a mottle, and in no case does the observed change in height exceed 1" (725 km) over the period of observation.

073.040 **The galloping chromosphere.** C. Sawyer.
Solar Physics, Vol. 35, 63 - 81 (1974).

The author describes the qualities of the galloping chromosphere, seen as horizontally-moving waves, shows how some of the characteristics change from center to limb and from quiet areas to active regions, compares this aspect to that seen at other wavelengths and by other workers, and shows how observed periods depend on magnetic fields in the upper chromosphere.

073.041 **The formation of solar quiescent prominences by condensation.** E. Hildner.
Solar Physics, Vol. 35, 123 - 136 (1974).

The author models the formation of solar quiescent prominences by solving numerically the non-linear, time-dependent, magnetohydrodynamic equations governing the condensation of the corona. A two-dimensional geometry is used. Gravitational and magnetic fields are included, but thermal conduction is neglected. He concludes that: (1) condensation of coronal material due to thermal instability is possible if thermal conduction is inhibited; (2) hydrodynamical processes determine, in large part, the rate of condensation; (3) condensation can occur on a time scale compatible with the observed times of formation of quiescent prominences.

073.042 **Resistive diffusion of force-free magnetic fields in a passive medium. III. Acceleration of flare particles.**
B. C. Low.
Astrophys. Journ., Vol. 189, 353 - 358 (1974).

The author has previously suggested that solar flares are triggered by the resistive diffusion of force-free magnetic fields in a passive medium. He considers a one-dimensional model in which an increasingly large electric field is induced by a rapidly evolving magnetic field. He estimates, in the case of solar flares, the energies to which protons and electrons may be directly accelerated by such an induced electric field.

073.043 **Direct observational evidence for the propagation and dissipation of energy in the chromosphere.**
S.-Y. Liu.
Astrophys. Journ., Vol. 189, 359 - 365 (1974).

In studying a time sequence of the Ca II K-line spectrograms of high spatial and time resolutions taken in a quiet region at the center of the disk, the author finds intensity perturbations that propagate from the far wing toward the line core. This perturbation is interpreted as a local heating in the chromosphere due to an upward propagating disturbance generated in the lower layers. The author has estimated the mechanical flux and the excess radiative flux of one prominent feature for discussion purposes.

073.044 **Strahlungsdämpfung akustischer Wellen beim Austritt aus den optischen dicken Schichten der Sonnenatmosphäre.** P. Ulmschneider.
Mitt. Astron. Ges., No. 34, p. 111 - 114 (1973/74).
Presented at the "Wissenschaftliche Tagung der Astron. Ges., Oberkochen, 1973 April 24 - 27".

073.045 **Entstehung solarer Aktivitätsgebiete.** R. Born.
Mitt. Astron. Ges., No. 34, p. 115 (1973/74).
Abstr. AG.

073.046 **On the structure and the motion of a spicule.**
W. Unno, E. Ribes, I. Appenzeller.
Solar Physics, Vol. 35, 287 - 308 (1974).

A stationary two-dimensional isothermal flow parallel to the magnetic lines of force is studied in connection with the hydrodynamic support of a spicule. Observed large extension into the corona and high velocities can be explained consistently if the effective kinetic temperature within a spicule could be about 10^4 K in the chromospheric region and increase to about 2.5×10^4 K or more in the coronal region. In a special simple case, an analytic solution of equations of motion is obtained and is used for explaining why the pressure in a spicule can be higher than the normal surrounding pressure in upper levels. Observational implications are briefly discussed.

073.047 **On peculiarities of the H and K Ca II lines of quiescent prominences.** N. N. Morozhenko.
Solar Physics, Vol. 35, 395 - 400 (1974).

A study of the metallic lines in bright quiescent prominences indicates that the optical thickness in the K line of Ca II may reach values as high as 10^3. This is about 10 times larger than the optical thickness in the Hα line and may explain some peculiarities of the H and K lines in solar prominences.

073.048 **The loop prominence of 11 August 1972: a coronal continuum event.** R. R. Fisher.
Solar Physics, Vol. 35, 401 - 408 (1974).

Observations of a loop prominence formed after the flare of 11 August 1972 are discussed. Estimates of electron density are obtained from (a) the line ratio of the Ca XV forbidden lines and (b) a Thompson scattering model. Both methods give an approximate value of $n_e = 10^{11} \text{cm}^{-3}$. This density was high enough to render the loop structures visible as continuum features, corresponding to the Ca XV structures as seen in the plane of the sky.

073.049 **An eruptive prominence of June 10, 1973.**
O. Engvold, B. M. Rustad.
Solar Physics, Vol. 35, 409 - 417 (1974).

An eruptive prominence of June 10, 1973, showing ascending and expanding motions, has been recorded spectroscopically at Oslo Solar Observatory. The Ca II H line yields broadening velocities ranging from 5 km s^{-1} to 20 km s^{-1}. An extreme ratio of 1/30 was measured for line-broadening to expansion velocity. Flare activity succeeded the start of the observed disparition brusque phase of the prominence. The eruption is related to coronal disturbances recorded approximately 1 hr later by ATM, Skylab.

073.050 **The continuum of the extreme limb and the chromosphere at the 1970 eclipse.**
H. Kurokawa, K. Nakayama, T. Tsubaki, M. Kanno.
Solar Physics, Vol. 36, 69 - 79 (1974).

The flash spectra of the partial sun and the chromosphere were obtained at the total solar eclipse on 7 March, 1970. The authors studied the distributions of the surface brightness of the continuum at six wavelengths in the visual region to compare them with previous observations and the existing model atmospheres. The intensity distribution in the low chromosphere was also examined.

073.051 **High n emission and absorption lines of the sun.**
A. Greve.

Solar Physics, Vol. 36, 85 - 90 (1974).

Due to the expected peaks of the b_n curves of ions in the chromosphere-corona transition region and the corona, high n transition lines change from emission to absorption in the wavelength region from $\approx 1/2$ mm for elements in a low state of ionization, to ≈ 2 mm for elements in a high state of ionization. The observational consequences are discussed.

073.052 **On helium-like $1s2l-1snl'$ transitions in solar flare spectra.**
S. O. Kastner, W. M. Neupert, M. Swartz.
Solar Physics, Vol. 36, 121 - 128 (1974).

Expected wavelengths and intensities are computed for $1s2l-1snl'$ transitions in helium-like ions of the abundant elements from oxygen to iron, under coronal conditions. Probable observations of some of these lines, in the spectra of solar flares, are discussed and attention is called to a possible reversal of singlet and triplet intensities as compared to laboratory observations.

073.053 **Cooling of solar flare plasmas.**
W. T. Zaumen, L. W. Acton.
Solar Physics, Vol. 36, 139 - 144 (1974).

A simple model for the cooling of solar flare plasmas is considered. This model predicts that an increase in emission measure with decreasing temperature is a general feature of a cooling flare. The results are compared to solar flare data.

073.054 **A probable mean transit time of the flare-generated disturbances.** I. D. Niţă.
Solar Physics, Vol. 36, 145 - 149 (1974).

Using the correlation method the author has obtained 39 to 42 h as the most probable value of the mean transit time of the solar disturbances producing magnetic storms with sudden commencement; this involves a mean propagation speed of about 1000 km s^{-1} inside 1 AU.

073.055 **Correlation of a flare-wave and type II burst.**
K. L. Harvey, S. F. Martin, A. C. Riddle.
Solar Physics, Vol. 36, 151 - 155 (1974).

The authors have studied the relation of a flare-induced wave and the type II and III radio bursts associated with the 26 April 1969, 2258 UT flare. Their observations suggest the flare-wave and type II bursts were produced by a common source.

073.056 **An attempt of experimental discovery of a correlation of chromospheric height variations with fluctuations of the 3-cm solar radio emission flux.**
N. S. Kaverin, V. V. Pakhomov, Yu. V. Platov.
Izv. vyssh. ucheb. zavedenij. Radiofizika, Vol. 16, 1354 - 1356 (1973). In Russian. – Abstr. in Referativ. Zhurn. 51. Astron., 5.51.389 (1974).

073.057 **The overheating instability of a conductive gas flow in a transversal magnetic field with $Re_m \gg 1$ and a possible explanation of the nature of solar chromospheric flares.** V. S. Sokolov.
Izv. Sib. otd. AN SSSR, 1973, No. 13, ser. tekhn. n., vyp. (No.) 3, p. 86 - 96. In Russian. – Abstr. in Referativ. Zhurn. 51. Astron., 5.51.436 (1974).

073.058 **Solar flares as sources of energetic particles.**
L. D. De Feiter.
Space Sci. Rev., Vol. 16, 3 - 43 (1974).

A review is presented of the present status of our knowledge of solar flare phenomena with special emphasis on the production of suprathermal particles and their solar effects.

073.059 **R. A. Kopp, "Energy balance in the chromosphere-corona transition region".** E. E. Dubov.
Astron. Zhurn. Akad. Nauk SSSR, Vol. 51, 671 - 672 (1974). In Russian. English translation in Soviet Astron. AJ, Vol. 18, No. 3.

It is shown that the dependence found by Kopp of the EUV emitting fraction of the solar surface on height does not contradict the supposition that the transition zone ($5.5 < \lg T < 6.3$) emission originates above the inner part of supergranules. This means that in contrary to Kopp the conductive energy flux from the corona found earlier is true, and this conductive energy flux is most essential for the dependence of temperature on height in the transition zone.

073.060 **Solar-flare and laboratory plasma phenomena.**
T. N. Lee.
Astrophys. Journ., Vol. 190, 467 - 479 (1974).

Systematic comparisons between solar flares and a laboratory discharge indicate that a number of similarities and interesting contrasts exist between the two in the following areas: the thermal and nonthermal (suprathermal) X-ray continuum spectra, the emission line spectra (> 6 keV) of iron, the spectral and temporal behavior of microwave and X-radiations, and the plasma ejection.

073.061 **Comments on the role of conduction in optical flare heating.** J. C. Brown.
Solar Physics, Vol. 36, 371 - 374 (1974). – Research note.

073.062 **Spectrograph, filtergraph and magnetograph observations of the two-ribbon flare of 29 July, 1973.**
A. G. Michalitsanos, P. Kupferman.
Solar Physics, Vol. 36, 403 - 416 (1974).

The authors present high resolution detailed observations of the class 3N two-ribbon flare of 1973, July 29 (McMath 12461), which was associated with the disappearance of a large filament ('disparition brusque'). They conclude that the model of infall-impact of Hyder (1967) is not consistent with their filtergraph and spectrograph observations.

073.063 **Determination of physical conditions in an eruptive prominence.** M. N. Ivannikova.
Solnechnye Dannye 1974 Byull., No. 3, p. 65 - 72 (1974). In Russian.

The parameters characterizing the physical state of an active prominence are given. The brightnesses of the H, Sr II, Ti II and Ca II lines calculated with these parameters coincide with the observed ones.

073.064 **On chromospheric phenomena accompanying birth and decay of sunspots.**
G. I. Kornienko, Eh. P. Surkov.
Solnechnye Dannye 1974 Byull., No. 3, p. 82 - 88 (1974). In Russian.

Photoheliograms are compared with filtergrams obtained in the center ($\Delta \lambda = 0$) and symmetric wings ($\Delta \lambda = \pm 0.5$ Å) of the Hα line. The regularities in the emission region in the upper and middle chromosphere with respect to each other and dark areas, visible only at $\Delta \lambda = \pm 0.5$ Å, for birth and decay of sunspots are described.

073.065 **The eruptive prominence of August 21, 1973 observed from Skylab in the white light corona and in the He II 304 Å chromosphere.** A. I. Poland, J. D. Bohlin, G. E. Brueckner, J. D. Purcell, V. E. Scherrer, N. R. Sheeley, R. Tousey.
Bull. American Astron. Soc., Vol. 6, 219 (1974). – Abstr. AAS.

073.066 **Theoretical helium I emission line intensities for quiescent prominences.**
J. N. Heasley, D. Mihalas, A. I. Poland.
Bull. American Astron. Soc., Vol. 6, 219 (1974). – Abstr. AAS.

073.067 **The acceleration and magnetic fields of surges.**
J. M. Malville, E. Tandberg-Hanssen.
Bull. American Astron. Soc., Vol. 6, 219 (1974). – Abstr. AAS.

073.068 **On the formation of chromospheric Hα.**
R. Steinitz, K. B. Gebbie, V. Bar.
Bull. American Astron. Soc., Vol. 6, 264 (1974). – Abstr. AAS.

073.069 **Scatter free propagation of charged particles along diverging magnetic field lines.** J. A. Earl.
Bull. American Astron. Soc., Vol. 6, 278 (1974). – Abstr. AAS.

073.070 **Flares as impulsive flux transfer events: laboratory observations of the reconnection process.**
P. J. Baum, A. Bratenahl.
Bull. American Astron. Soc., Vol. 6, 284 (1974). – Abstr. AAS.

073.071 **The 1175 Å to 1900 Å ultraviolet spectrum of solar flares.**
G. E. Brueckner, J. D. Bohlin, O. K. Moe, K. R. Nicolas, J. D. Purcell, V. E. Scherrer, N. R. Sheeley, Jr., R. Tousey.
Bull. American Astron. Soc., Vol. 6, 285 (1974). – Abstr. AAS.

073.072 **Theoretical hydrogen spectra of chromospheric flares.** R. C. Canfield.
Bull. American Astron. Soc., Vol. 6, 285 (1974). – Abstr. AAS.

073.073 **Solar line profiles in range 200 Å – 700 Å.**
G. Cushman, G. Godden, W. A. Rense.
Bull. American Astron. Soc., Vol. 6, 285 (1974). – Abstr. AAS.

073.074 **Solar flare emission lines of highly-ionized iron and nickel.**
G. A. Doschek, U. Feldman, R. D. Cowan, L. Cohen.
Bull. American Astron. Soc., Vol. 6, 286 (1974). – Abstr. AAS.

073.075 **Comparison of impulsive optical and radio emission features of an energetic subflare.**
V. Gaizauskas, L. W. Avery.
Bull. American Astron. Soc., Vol. 6, 287 (1974). – Abstr. AAS.

073.076 **Can the dissipation of high frequency sound waves in the low chromosphere produce the temperature rise?** S. D. Jordan.
Bull. American Astron. Soc., Vol. 6, 289 (1974). – Abstr. AAS.

073.077 **Non-thermal processes in large solar flares.**
R. P. Lin, H. S. Hudson, T. W. Jones.
Bull. American Astron. Soc., Vol. 6, 290 (1974). – Abstr. AAS.

073.078 **Preliminary interpretation of diode array simultaneous observations of He I and Ca II line profiles in collaboration with ATM.**
J. L. Linsky, R. B. Dunn, D. M. Rust.
Bull. American Astron. Soc., Vol. 6, 290 (1974). – Abstr. AAS.

073.079 **An empirical interpretation for the time evolution of the Ca II K line.**
S.-Y. Liu, A. Skumanich.
Bull. American Astron. Soc., Vol. 6, 290 - 291 (1974). Abstr. AAS.

073.080 **The spectrum and vertical structure of the August 7, 1972 white light flare.**
M. E. Machado, D. M. Rust.
Bull. American Astron. Soc., Vol. 6, 291 (1974). – Abstr. AAS.

073.081 **The spectacular solar disturbance of 28 October 1972.**
M. K. McCabe, A. C. Riddle, R. T. Hansen.
Bull. American Astron. Soc., Vol. 6, 291 (1974). – Abstr. AAS.

073.082 **Spectrograph, filtergraph and magnetograph observations of the two ribbon flare of 29 July 1973.**
A. G. Michalitsanos, P. Kupferman.
Bull. American Astron. Soc., Vol. 6, 291 (1974). – Abstr. AAS.

073.083 **Chromospheric activity associated with moving photospheric magnetic field.**
J. R. Roy, A. G. Michalitsanos.
Bull. American Astron. Soc., Vol. 6, 293 (1974). – Abstr. AAS.

073.084 **The galloping chromosphere.** C. Sawyer.
Bull. American Astron. Soc., Vol. 6, 293 (1974). Abstr. AAS.

073.085 **Hα observations of the near-limb flare of 29 June 1973.** S. A. Schoolman, H. E. Ramsey.
Bull. American Astron. Soc., Vol. 6, 293 - 294 (1974). Abstr. AAS.

073.086 **An experimental model of solar flares in the corona.**
J. K. Silk, S. Kahler, A. Krieger, A. Timothy, G. Vaiana, R. Pallavicini.
Bull. American Astron. Soc., Vol. 6, 294 (1974). – Abstr. AAS.

073.087 **Rocket spectroheliogram observations of the heights of formation and sizes of bright features in the transition zone.**
G. W. Simon, P. H. Seagraves, R. Tousey, R. W. Noyes.
Bull. American Astron. Soc., Vol. 6, 294 (1974). – Abstr. AAS.

073.088 **The screw pinch and the solar flare.**
D. S. Spicer, C. C. Cheng.
Bull. American Astron. Soc., Vol. 6, 294 (1974). – Abstr. AAS.

073.089 **Motion of prominences through the corona.**
E. Tandberg-Hanssen, R. T. Hansen, A. C. Riddle.
Bull. American Astron. Soc., Vol. 6, 295 (1974). – Abstr. AAS.

073.090 **Spicules and fibrils in Hα and K.** H. Zirin.
Bull. American Astron. Soc., Vol. 6, 298 (1974). Abstr. AAS.

073.091 **Plasma instabilities in solar flares and their laboratory analogies.** L. Křivský.
Kozmos, Vol. 5, 45 - 46, 51, 76 - 78 (1974). In Czech.

073.092 **Photometric atlas of emission lines of the solar chromosphere between 3599 Å and 4017 Å.**
E. Hiei, M. Fukatsu.
Ann. Tokyo Astron. Obs., Second Ser., Vol. 14, (No. 2), 37 - 84 (1974).

A photometric atlas of chromospheric emission lines from 3599 Å to 4017 Å is made from a flash spectrogram taken at the 1966 total solar eclipse in Peru. The absolute intensity for the chromospheric lines and continua of the whole chromosphere is given in the photometric atlas.

073.093 **On the excitation and ionization of metal atoms in prominences.** K. S. Tavastsherna.
Astron. Tsirk., No. 812, p. 7 - 8 (1974). In Russian.

The solar chromosphere. See Abstr. 003.028.

Das Filmen von Sonnenprotuberanzen. See Abstr. 034.015.

Development and performance of a solar hard X-ray spectrometer. See Abstr. 034.053.

Perturbations of Pioneer 6 telemetry signal during solar occultation. See Abstr. 053.016.

Some results from Skylab's solar experiments. See Abstr. 054.021.

Resistive diffusion of force-free magnetic fields in a passive medium. See Abstr. 062.069.

Emission Balmer decrement and electron density of stellar chromospheres. See Abstr. 064.047.

On the possibility of detecting abundance inhomogeneities resulting from spallation reactions in the solar photosphere. See Abstr. 071.004.

On the determination of the temperature and density in a sunspot and in the chromosphere above the sunspot. See Abstr. 072.005.

On the character of spatial-temporal relationships of different solar indices. See Abstr. 072.010.

The connection between sporadic chromospheric processes and the evolution of sunspot groups. See Abstr. 072.020.

A dynamical model for the chromosphere-corona transition region. See Abstr. 074.010.

Coronal disturbances. II. The fast rearrangement of coronal magnetic fields. See Abstr. 074.028.

Excess heating of corona and chromosphere above magnetic regions by non-linear Alfvén waves. See Abstr. 074.052.

A discussion of interplanetary postshock flows with two examples. See Abstr. 074.081.

On the structure of the quiet corona near quiescent prominences. See Abstr. 074.089.

Measurement of the X-ray emission of the solar atmosphere during a period of low activity. See Abstr. 076.002.

Spatial distribution of soft X-ray and EUV emission associated with a chromospheric flare of importance 1b on August 2, 1972. See Abstr. 076.005.

Hard solar flare X-ray bursts on 8 December 1970. See Abstr. 076.011.

Ionization equilibrium in soft X-ray-emitting solar flares. See Abstr. 076.020.

An X-ray flare from Skylab: results and interpretation. See Abstr. 076.024.

Force-free magnetic fields in the X-ray flare of 16 June 1973. See Abstr. 076.031.

Electron density from line ratios in the XUV spectrum of a solar flare. See Abstr. 076.036.

Coupling of microwaves at a selected solar active centre. See Abstr. 077.002.

Fine structure of the radio spectra of early phase of plasma instability during the solar flare August 4, 1972. See Abstr. 077.021.

A search for 5 min periodic structure in solar 2 cm emission. See Abstr. 077.025.

On the role of the magnetic configuration of flares for production of type III solar radio bursts. See Abstr. 077.026.

A clue to the trigger for both the type III solar radio burst and the solar flare. See Abstr. 077.064.

Long base line interferometry of the sun at 3.7 and 11.1 cm wavelengths. See Abstr. 077.070.

Low-energy solar cosmic rays and chromospheric flares. See Abstr. 078.005.

Schwere solare kosmische Strahlung. See Abstr. 078.042.

Observation of transiron solar-flare nuclei in an Apollo 16 command-module window. See Abstr. 078.045.

A search for the footpoints of solar magnetic fields. See Abstr. 080.026.

Photoelectric drift scans II: time pulse evaluation, limb profiles, and the solar diameter. See Abstr. 080.031.

On the possibility of parametric amplification of Alfvén waves in the solar atmosphere. See Abstr. 080.038.

Dynamics of the solar magnetic field. III. Location of solar-flare excitation and the velocity field determined from magnetograms. See Abstr. 080.061.

Some statistical characteristics of sudden ionospheric disturbances and distribution of geoactive chromospheric flares over the solar disc. See Abstr. 083.006.

On the relationship between shock wave and magnetic bottle associated with the solar flare of 4 November 1968. See Abstr. 084.204.

Interplanetary shock waves generated by solar flares. See Abstr. 106.004.

Interplanetary magnetic field, solar flares and geomagnetic disturbances. See Abstr. 106.013.

Coherent propagation of charged-particle bunches in random magnetic fields. See Abstr. 106.014.

074 Solar Corona, Solar Wind

074.001 Interpretation of columnar content measurements of the solar-wind turbulence. P. S. Callahan.
Astrophys. Journ., Vol. 187, 185 - 190 (1974).

The temporal spectrum of measurements of the integrated electron columnar content between the earth and a distant source are considered. The relation between this spectrum and the three-dimensional wavenumber spectrum of the solar-wind density is derived under the assumption that the correlation scale is the smallest scale of interest. The specific case of round-trip measurements to a spacecraft is investigated for a power-law solar-wind density spectrum. The resulting spectrum has a power-law envelope one power less steep than the solar-wind density spectrum with several deep minima superposed.

074.002 Observations of the line 7892 Å of Fe XI by means of Pic-du-Midi Observatory's coronagraph.
J. C. Noens, J. P. Rozelot.
Astron. Astrophys., Vol. 30, 81 - 85 (1974). In French.

Very few observations of the forbidden line 7892 Å of Fe XI have been made up to now. The authors describe here some experimental results: isophotes maps, gradients, etc... A complete theoretical study of Fe XI has been made in a separate paper, the results of which are taken here to compare with observational results. Concerning the line of sight, various assumptions are made which are discussed at each step of the calculus. Finally, the authors obtain a method which allows the computations of the characteristic sizes, the vertical density gradient and the electron density at the bottom of the studied structures.

074.003 Effects of turbulence anisotropy on propagation and electromagnetic radiation of particle streams in the solar corona. J. Heyvaerts, G. Verdier de Genouillac.
Astron. Astrophys., Vol. 30, 211 - 222 (1974).

This paper studies the evolution of a plasma-beam instability in the presence of non-linear effects. These consist of induced ll-scattering of Langmuir waves on the polarization clouds of ions; Langmuir waves are considered to have all possible angular patterns. This problem is set in relation to type III solar radio-bursts, hence the parameters used are those presumed to be characteristic of coronal conditions. The authors investigate the influence of the turbulence level and anisotropy on the electromagnetic radiation resulting from it by the two conversion processes usually invoked for type III bursts.

074.004 Determination of Faraday rotation occurring between the burst-source in the solar corona and the earth. R. V. Bhonsle, S. K. Mattoo.
Astron. Astrophys., Vol. 30, 301 - 303 (1974).

The total Faraday rotation suffered by type III solar radio bursts, as measured by means of the two-bandwidth (7.5 and 12.5 kHz) time-sharing polarimeter at 35 MHz, has been found to be of the order of 10^3 radians. It is suggested that such low values of the Faraday rotation can be qualitatively explained on the basis of generation of type III bursts primarily at the second harmonic of the local plasma frequency.

074.005 The outer solar corona as observed from Skylab: Preliminary results. R. M. MacQueen, J. A. Eddy, J. T. Gosling, E. Hildner, R. H. Munro, G. A. Newkirk, Jr., A. I. Poland, C. L. Ross.
Astrophys. Journ., (Letters), Vol. 187, L85 - L88 (1974).

The white-light coronagraph experiment has made frequent, periodic observations of the solar corona from $1.5\,R_\odot$ to $6.0\,R_\odot$ during the Skylab mission, and these observations will permit the determination of the three-dimensional extent of coronal forms. There are several time scales on which visual changes in coronal structures occur, ranging from approximately one-half rotation to less than hours. A number of events corresponding to the shortest time scale — coronal transients — cause major restructuring of the corona.

074.006 The solar corona polarization at the total eclipse on September 22, 1968.
Yu. N. Lipskij, A. V. Bugaevskij, Yu. P. Pskovskij.
Astron. Zhurn. Akad. Nauk SSSR, Vol. 51, 160 - 166 (1974). In Russian. English translation in Soviet Astron. AJ, Vol. 18, No. 1.

The average polarization in all position angles for given distances from the limb is in good agreement with theoretical data. A considerable diminution of the degree of polarization was found near prominences and in a flash. The average orientation of the plane of polarization is radial, but it deviates from that in the vicinity of prominences.

074.007 Wave-trains in the solar wind. II: Comments on the propagation of Alfvén waves in the quiet interplanetary medium. A. K. Richter, D. J. Olbers.
Astrophys. Space Sci., Vol. 26, 95 - 105 (1974).

The equations of motion of all relevant parameters of Alfvén waves propagating from the sun outwardly into the expanding interplanetary medium are discussed for the case of a quiet, ideal, isotropic, one-fluid solar wind plasma.

074.008 The coronal radiance in the intermediate infrared.
W. G. Mankin, R. M. MacQueen, R. H. Lee.
Astron. Astrophys., Vol. 31, 17 - 21 (1974).

An observed value of the spectral mean of the coronal radiance in the intermediate infrared ($7.5-13\,\mu$) of $9 \pm 5 \times 10^{-7}$ watt-cm^{-2} sterad$^{-1} \mu^{-1}$ at $4\,R_\odot$ is presented, with the observed change in radiance with elongation angle over the range $3.5-12.5\,R_\odot$. The observations are consistent with the model of silicate emission proposed by Kaiser (1970), but a two component model is required to reconcile these observations and the measurements by Peterson (1971) at wavelengths of $2.2\,\mu$ and less.

074.009 Rigid and differential rotation of the solar corona.
E. Antonucci, L. Svalgaard.
Solar Physics, Vol. 34, 3 - 10 (1974).

The rotation of the solar corona has been studied using recurrence properties of the green coronal line (5303 Å) for the interval 1947–1970. Short-lived coronal activity is found to show the same differential rotation as short-lived photospheric magnetic field features. Long-lived recurrences show rigid rotation in the latitude interval $\pm 57°.5$. It is proposed that at least part of the variability of rotational properties of the solar atmosphere may be understood as a consequence of coexistence of differential and rigid solar rotation.

074.010 A dynamical model for the chromosphere-corona transition region. C. Chiuderi, I. Riani.
Solar Physics, Vol. 34, 113 - 124 (1974).

A dynamical, homogeneous model of the chromosphere-corona transition region and of the lower corona is presented, based on the hydrodynamical equations and on a semi-empirical relation deduced from radio observations. The model is shown to be in agreement with radio and UV observations and with the particle flux given by solar wind measurements. A comparison with the analogous static model shows that dynamical effects are very small.

074.011 **The profile and polarization of the coronal Lα line.** J. M. Beckers, E. Chipman.
Solar Physics, Vol. 34, 151 - 161 (1974).

The authors calculate the profile and polarization of the Lα line in the solar corona. Coronal temperature variation, solar wind and other non-thermal motions have been taken into account. Because of the relatively low atomic weight of hydrogen the profile of the Lα line is a sensitive indicator of the coronal temperature. The line polarization contains relatively little information except for strong magnetic fields (> 70 G).

074.012 **The identification of Fe IX and Ni XI in the solar corona.** L. Å. Svensson, J. O. Ekberg, B. Edlén.
Solar Physics, Vol. 34, 173 - 179 (1974).

The levels of the configuration $3s^2 3p^5 3d$ of Fe IX have been experimentally determined from their combinations with $3s3p^6 3d$ ^3D, ^1D in the region 300–400 Å. Wavelengths can now be accurately predicted for all transitions within $3s^2 3p^5 3d$, and eight of these can be identified with coronal lines from 2042 to 4585 Å. Also, identifications of solar lines from 171 to 245 Å with electric-dipole and magnetic-quadrupole transitions to the ground state, $3s^2 3p^6$ ^1S, are confirmed and extended. Solar identifications with corresponding transitions in Ni XI, both within $3s^2 3p^5 3d$ and to the ground state, are proposed on the basis of a short extrapolation.

074.013 **On the determination of coronal temperature from the decay of type III radio bursts.** A. C. Riddle.
Solar Physics, Vol. 34, 181 - 184 (1974). – Research note.

074.014 **The influence of non-uniform solar wind expansion on the angular momentum loss from the sun.**
E. R. Priest, G. W. Pneuman.
Solar Physics, Vol. 34, 231 - 241 (1974).

The influence on the rate of angular momentum loss from the sun of magnetic geometries which are not spherically symmetric is estimated. Departures from spherical symmetry are expected to influence significantly the loss rate by two effects – the presence of closed magnetic field regions with no loss and also the variability in the radial distance to the Alfvénic point, as stressed by Mestel (1968).

074.015 **On the bulk velocity of the solar wind α-particles.**
G. Moreno, F. Palmiotto.
Solar Physics, Vol. 34, 243 - 245 (1974). – Research note.

074.016 **Solar wind direction from HEOS 1 and Explorer 33 satellites.** A. Egidi, C. Signorini.
Solar Physics, Vol. 34, 247 - 251 (1974).

A description is given of the analysis of the solar wind angular data from the satellite HEOS 1 and a comparison is made with the data obtained by Explorer 33, in the period of time from January 1967 to April 1970. The observed flow direction shows a temporal behaviour which suggests the possibility of a long period regular variation.

074.017 **First order latitude effects in the solar wind.**
C. R. Winge, Jr., P. J. Coleman, Jr.
Planet. Space Sci., Vol. 22, 439 - 463 (1974).

The purpose of this paper is to demonstrate that even with spherically symmetric boundary conditions at the base of the corona, the magnetic field introduces a small latitude effect in the solar wind which should be measurably significant at large distances.

074.018 **The coronal aureola in the time of total solar eclipse.** O. Koutchmy, S. Koutchmy.
Astron. Astrophys., Suppl. Ser., Vol. 13, 295 - 303 (1974). In French.

The authors propose a computational method allowing to obtain the intensity of the coronal aureola from the measures of the solar aureola. The method is applied to 2 very different observations performed at the solar eclipses at Sept. 22, 1968 and March 7, 1970. In addition, a scattering function is given for some typical cases observed showing a large variety in values. The authors suggest to use this type of function in order to describe the scattered light superimposed on the lunar earthshine and the sky background.

074.019 **Evidence of solar-cycle variations in the solar wind.**
D. S. Intriligator.
Astrophys. Journ., (Letters), Vol. 188, L23 - L26 (1974).

Solar-wind observations are presented from 1965 July through 1971 June that show the first evidence for long-term variations in the solar wind associated with changes in the solar cycle. The observations indicate that the frequency of high-speed streams in the solar wind and their duration vary over the solar cycle.

074.020 **Heat transport in the solar wind.** E. V. Mishin.
Astrophys. Space Sci., Vol. 27, 351 - 366, 367 - 382 (1974). In Russian and English.

Heat transport is considered both for quiet and disturbed solar winds. It is shown that heat may be transferred during solar flares by sharp fronted thermal wave pulses. The effects of ionosonic turbulence on heat transport in a quiet solar wind are also investigated.

074.021 **Solution of one-fluid model equations with short range retarding magnetic forces for the quiet solar wind.** S. Cuperman, A. Harten.
Astrophys. Space Sci., Vol. 27, 383 - 387 (1974).

The discrepancy of the low predicted versus the observed coronal particle densities is investigated by considering radial magnetic forces acting at the base of the corona in the one-fluid model equations with anomalous thermal conductivity for the quiet solar wind.

074.022 **The solar wind and its influence on planetary atmospheres.** I. E. Turchinovich.
XXVI Gertsenovsk. chteniya. Teor. fiz. i astron. Nauch. dokl. Leningrad, 1973, p. 120 - 126. In Russian. – Abstr. in Referativ. Zhurn. 51.Astron., 3.51.455 (1974).

074.023 **Ergebnisse von zwei Jahren Korona-Forschung mit dem Radio-Heliographen von Culgoora.**
H. Urbarz.
Orion Schaffhausen, 32. Jahrgang, p. 63 - 69 (1974).

074.024 **Detection of tangential and contact discontinuities from measurements in the solar wind.**
K. G. Ivanov.
Geomagn. Aeronom., Vol. 14, 8 - 12 (1974). In Russian.

074.025 **Relative coronal abundances derived from X-ray observations. I. Sodium, magnesium, aluminum, silicon, sulfur, and argon.**
A. B. C. Walker, Jr., H. R. Rugge, K. Weiss.
Astrophys. Journ., Vol. 188, 423 - 440 (1974).

The authors have observed the coronal spectrum between 3.5 and 8.5 Å for several levels of coronal activity, and determined absolute fluxes for 79 lines and for the continuum. Identifications have been determined for the majority of the lines observed, and, in particular, for the resonance lines of Mg XI, Mg XII, Al XII, Al XIII, Si XIII, Si XIV, S XV, and S XVI. The measured fluxes of these lines have been used to construct a model of the temperature dependence of the coronal emission measure between 1.5 and 10×10^6 °K. This model is used to determine the relative abundances of Mg, Al, Si, and S. The relative abundances of Ar and Na were also computed, using the observed intensities of weak lines of

074.026 **Comments on 'Neutralization and stabilization of particle streams in the corona and type III radio bursts' by Dean F. Smith.** D. B. Melrose.
Solar Physics, Vol. 34, 421 - 425 (1974).

Electron streams which excite type III bursts experience a negligible energy loss in accelerating a return current. Furthermore the plasma oscillations generated as the return current turns on and off are too weak to be of significance as a source of the observed radiation.

074.027 **The coronal transient of 16 June 1972.** M. Koomen, R. Howard, R. Hansen, S. Hansen.
Solar Physics, Vol. 34, 447 - 452 (1974).

On 16 June 1972, the Naval Research Laboratory's coronagraph aboard OSO-7 tracked a huge coronal cloud moving outward from the sun. Concurrent observations of the inner corona made by the High Altitude Observatory at Mauna Loa showed bifurcation of the underlying coronal structure. Together, these observations can be interpreted as evidence for the stretching of the closed fields into a 'magnetic bottle', extending to at least eight radii from the center of the sun.

074.028 **Coronal disturbances. II. The fast rearrangement of coronal magnetic fields.**
W. J. Wagner, R. T. Hansen, S. F. Hansen.
Solar Physics, Vol. 34, 453 - 459 (1974).

A fast coronal transient event was observed simultaneously on 17 February 1972 by the Sacramento Peak Observatory 6-in. λ5303 filter coronagraph and the High Altitude Observatory K-coronameter. The authors interpret the rapid opening of green line structure cospatial with the disappearance of a white light streamer as material motion of iron ions and electrons. Together with the subsequent two-fold increase in K-corona brightness in an adjacent region, this is taken as evidence of a transference of electrons to a new streamer in a realignment of magnetic flux tubes accompanying a flare.

074.029 **Coronal magnetic structure at a solar sector boundary.** J. M. Wilcox, L. Svalgaard.
Solar Physics, Vol. 34, 461 - 470 (1974).

The persistent large-scale coronal magnetic structure associated with a sector boundary appears to consist of a magnetic arcade loop structure extending from one solar polar region to the other in approximately the north–south direction. This structure was inferred from computed coronal magnetic field maps for days on which a stable magnetic sector boundary was near central meridian, based on an interplanetary sector boundary observed to recur during much of 1968 and 1969.

074.030 **Eleven-years inversion of the green corona emission.** E. Antonucci.
Solar Physics, Vol. 34, 471 - 476 (1974).

A cross-correlation analysis of coronal green line intensity (5303 Å) and interplanetary magnetic field polarity for the period 1947–1970 shows that the coronal features are organized in a constant pattern with respect to the 4-sector structure through the solar cycle. A sudden inversion of the coronal pattern with respect to the sector structure takes place at the solar minima. The high emission regions of the green corona are located near the solar magnetic sector boundaries having polarities (−, +), (+, −), (−, +) during cycles 18, 19, 20 respectively in the northern hemisphere, and (+, −), (−, +), (+, −) in the southern hemisphere.

074.031 **Effects of heavy ions on electron temperatures in the solar corona and solar wind.** M. P. Nakada.
Journ. Geophys. Res., Vol. 79, 36 - 39 (1974).

The effects of the reduction in the thermal conductivity due to heavy ions on electron temperatures in the solar corona and solar wind are examined. Large enhancements of heavy ions in the corona appear to be necessary to give appreciable changes in the thermal gradient of the electrons. These enhancements, if they should occur, may contribute to the understanding of some low values of solar wind temperature measurements at 1 AU.

074.032 **Perturbation of the solar wind by the lunar atmosphere.** M. K. Wallis.
Journ. Geophys. Res., Vol. 79, 275 - 279 (1974).

The effects of atmospheric ions on the solar wind flow past the moon may be represented by source terms in a steady flow. The sources appropriate to ions of large Larmor radius are found to differ qualitatively from those given earlier (Siscoe and Mukherjee, 1972). The transverse force is the major perturbing effect and gives limits on the lunar atmospheric density at some 5 times higher than those previously derived.

074.033 **Further remarks on plasma instabilities produced by ions born in the solar wind.**
C. S. Wu, R. E. Hartle.
Journ. Geophys. Res., Vol. 79, 283 - 285 (1974). − Letter.

074.034 **The flow past Mars and Venus by the solar wind. General laws.** O. L. Vajsberg, A. V. Bogdanov.
Kosmich. Issled., Vol. 12, 279 - 284 (1974). In Russian.

074.035 **Shift and profile of the Mg 5184 Å line in the F-corona during the solar eclipse of 1973, June 30.**
Yu. A. Kozlov, S. B. Novikov, P. V. Shcheglov.
Astron. Tsirk., No. 808, p. 1 - 2 (1974). In Russian.

074.036 **On the Hα-emission in the integrated spectrum of the sky and corona during the total solar eclipse of 1972, July 10.** Yu. D. Mateshvili, T. I. Toroshelidze.
Astron. Tsirk., No. 808, p. 2 - 4 (1974). In Russian.

074.037 **Acceleration of the solar wind by the interplanetary magnetic field.** A. Barnes.
Astrophys. Journ., Vol. 188, 645 - 648 (1974).

The effect of the azimuthly symmetric (spiral-model) interplanetary magnetic field on radial solar-wind flow is reexamined. It is found that when the azimuthal motion of the plasma due to this field is considered, there is no significant radial acceleration of the plasma by the magnetic field.

074.038 **The adiabatic cooling of the protons in the solar wind: the case where the interplanetary magnetic-field is of spiral form.** M. Eyni, A. S. Kaufman.
Astrophys. Space Sci., Vol. 28, 177 - 183 (1974).

This is a sequel to the paper by Eyni and Kaufman (1971) in which the coordinate systems with respect to the magnetic direction and solar radial-direction were assumed coincident. In the present account, it is shown that the theoretical expression for the proton cooling derived in that paper is still applicable in the case of the spiral magnetic-field provided the appropriate transformation of kinetic temperature from the magnetic direction to the radial direction as derived here is used. A detailed comparison is made between the present model and other models based on the spiral form of the magnetic field; the agreement is satisfactory.

074.039 **Comments on paper 'The solar wind cycle, the sunspot cycle and the corona' by J. Hirshberg.**
L. Diodato, G. Moreno.
Astrophys. Space Sci., Vol. 28, L7 - L10 (1974). − Letter.

074.040 **The solar wind.** C.-G. Fälthammar.
Cosmical geophysics, (see 003.004), p. 93 - 102 (1973).

074.041 **The quiet corona: temperature and temperature gradient.**
S. J. Bame, J. R. Asbridge, W. C. Feldman, P. D. Kearney.
Solar Physics, Vol. 35, 137 - 152 (1974).

A study of the lower corona thermal properties was made using the best examples of solar wind heavy ion spectra obtained with Vela 5 and 6 plasma analyzers at times of quiet solar wind (low speed, low temperature). The multiple Si and Fe ion species peaks in the spectra were fit with solutions of the ionization equilibrium equations to determine 'freezing in' temperatures for the various species over a range of heliocentric distances.

074.042 **On the observation of scattered radio emission from sources in the solar corona.** A. C. Riddle.
Solar Physics, Vol. 35, 153 - 169 (1974).

The effects of scattering and refraction on radio waves in the solar corona are considered for several different coronal models. By considering a source near the plasma level in a spherically symmetric corona and in a streamer enhancement superimposed on a spherically symmetric corona we obtain results relating to bursts of types I, II and III.

074.043 **The coronal disturbance of 1972, August 12.**
A. C. Riddle, E. Tandberg-Hanssen, R. T. Hansen.
Solar Physics, Vol. 35, 171 - 179 (1974).

On 1972, August 12, there occurred a flare spray observed optically by both flare patrol and coronagraph instruments. A subsequent moving type IV radio burst was recorded by several instruments and distinct changes in the coronal brightness at $1.6\ R_\odot$ were measured. These observations combine to form one of the most complete sequences of measurements yet recorded covering the range from the chromosphere to about $6\ R_\odot$. The separate measurements are discussed and we show that they can be combined to form a relatively simple physical picture of the whole event.

074.044 **Coronal magnetic field structure derived from two-frequency radioheliograph observations.**
K. Kai, K. V. Sheridan.
Solar Physics, Vol. 35, 181 - 192 (1974).

An exceptional variety of positions and polarizations was found for two type I storms and numerous sporadic bursts observed during 15 consecutive days with the Culgoora radioheliograph at 80 and 160 MHz. The radio data are combined with optical data to derive a model of the coronal magnetic field structure for a complex of active regions; in this model the type I sources are located on multiple strong-field loops, and the sources of the sporadic bursts on widely diverging weak-field arches. It is also shown that this model can explain the seemingly erratic placement of the storm sources and the appearance of the sporadic sources round a large arc.

074.045 **The solar wind velocity in the eleven-year cycle No. 20 and the solar radar cross-section.**
S. Pintér.
Solar Physics, Vol. 35, 225 - 232 (1974).

The annual average values of the solar wind velocity over the period 1962–1972 were investigated on the basis of data obtained from different space probes. The comparison of the pattern of the annual average solar wind velocities observed by the Vela and Pioneer 6 satellites indicates that the pattern presented by Gosling et al. (1971) is realistic. The long-range trend in the solar wind velocity during the 11-year cycle is governed by the number and intensity of irregularities occurring in the corona.

074.046 **Solar coronal line profiles in the extreme-ultraviolet.** U. Feldman, W. E. Behring.
Astrophys. Journ., (*Letters*), Vol. 189, L45 - L46 (1974).

The authors report the first direct measurements of the widths of the lines emitted by the solar corona from 170 to 370 Å.

074.047 **Mode coupling in the solar corona. I. Coupling near the plasma level.** D. B. Melrose.
Australian Journ. Phys., Vol. 27, 31 - 42 (1974).

The theory of mode coupling in the radiation from solar type I storms is extended to treat coupling when the frequency f is near the plasma frequency f_p. The implications on the handedness of solar radio radiation are discussed. The possibility of depolarization due to mode coupling is considered briefly.

074.048 **Mode coupling in the solar corona. II. Oblique incidence.** D. B. Melrose.
Australian Journ. Phys., Vol. 27, 43 - 52 (1974).

Previous discussions of mode coupling at a quasi-transverse region have assumed vertical incidence and have thus invoked magnetic structures which violate div B = 0. A new method is developed here for calculating the coupling coefficients for oblique incidence so that coupling at a quasi-transverse region can be treated without invoking nonphysical magnetic structures. The present method is used to generalize the results of Cohen to allow oblique incidence.

074.049 **Role of collisions in the polarization rate of the forbidden emission lines of the solar corona. I. Depolarization by proton impact. Application to the green line of Fe XIV and to the infrared lines of Fe XIII.**
S. Sahal-Bréchot.
Astron. Astrophys., Vol. 32, 147 - 154 (1974).

The semi-classical perturbation theory with cut off is compared with the exact results of Masnou Seeuws and McCarroll (1972) and is shown to be sufficient for calculations of fine structure cross sections of positive ions. Formulae are provided for calculating cross sections between Zeeman sublevels and the depolarization rates by proton impact of the green line of Fe XIV and of the infrared lines of Fe XIII are computed.

074.050 **Photométrie photographique de la couronne solaire, observée au cours de l'éclipse totale du 10 juillet 1972.**
S. Koutchmy, N. I. Dzyubenko (*Dzubenko*), A. T. Nesmyanovich (*Nesmjanovich*), S. K. Vsekhsvyatskij (*Vsekhsvjatsky*).
Solar Physics, Vol. 35, 369 - 375 (1974).

A series of plates of the solar corona were obtained during the total solar eclipse of July 10th, 1972 near Anadyr (U.S.S.R.) using a standard eclipse coronagraph. A neutral radial filter was specially conceived to enable the study of the corona up to $5\ R_\odot$. The brightness of the corona $(K + F)$ is given along the N–S and E–W axes and also the total brightness recorded along the tangential directions, at selected radial distances.

074.051 **Curvature and surface distribution of the polar rays in the solar corona on 12 November 1966.**
J. G. Kirk, J. S. Newby.
Solar Physics, Vol. 35, 377 - 384 (1974).

Following a technique developed by Saito, the positions and orientations of the polar rays visible in a photograph of the total solar eclipse of 12 November 1966 were examined to infer the length of an equivalent bar magnet representing the sun's external magnetic field and the true distribution of polar rays from the two-dimensional projection seen in the photograph. The results are in good accord with expectation from Saito's analysis.

074.052 **Excess heating of corona and chromosphere above magnetic regions by non-linear Alfvén waves.**
Y. Uchida, O. Kaburaki.
Solar Physics, Vol. 35, 451 - 466 (1974).

Excess heating of the active region solar atmosphere is interpreted by the decay of MHD slow-mode waves produced in the corona through the non-linear coupling of Alfvén waves supplied from subphotospheric layers. It is stressed that the Alfvén-mode waves may be very efficiently generated directly in the convection layer under the photosphere in magnetic regions, and that such magnetic regions, at the same time, provide the 'transparent windows' for Alfvén waves in regard to the Joule and frictional dissipations in the photospheric and subphotospheric layers.

074.053 **A note on the solution of the saturation flux limited solar wind equations.** I. W. Roxburgh.
Solar Physics, Vol. 35, 481 - 487 (1974).

The solution curves of the differential equations determining the behaviour of the solar wind are calculated for the case where the heat flux has its maximum value $3/2\, nkTv_{th}$. All the supersonic solutions are asymptotically adiabatic, $T \sim r^{-4/3}$.

074.054 **The helium component of solar wind velocity streams.**
J. Hirshberg, J. R. Asbridge, D. E. Robbins.
Journ. Geophys. Res., Vol. 79, 934 - 938 (1974).

Systematic variations of the properties of the helium constituent of the solar wind in the velocity streams are described. It is found that the helium abundance n_α/n_p varies by about a factor of 2 as the stream is crossed. The velocity of the helium differs from that of the hydrogen by a few kilometers per second throughout much of the stream structure. This velocity difference is greatest immediately after the proton density peak passes, the helium velocity being typically 20 km/s faster than the protons at that position in the stream.

074.055 **Shapes of strong shock fronts in an inhomogeneous solar wind.** M. A. Heinemann, G. L. Siscoe.
Journ. Geophys. Res., Vol. 79, 1349 - 1355 (1974).

The shapes expected for solar-flare-produced strong shock fronts in the solar wind have been calculated, large-scale variations in the ambient medium being taken into account. The shock fronts were assumed to be spherically symmetric near the sun. It has been shown that for reasonable ambient solar wind conditions the mean and the standard deviation of the east–west shock normal angle are in agreement with experimental observations including shocks of all strengths.

074.056 **Alfvénic acceleration of solar wind helium and related phenomena. 1. Theory.** J. V. Hollweg.
Journ. Geophys. Res., Vol. 79, 1357 - 1363 (1974).

The author presents a new physical mechanism by which helium nuclei can be preferentially accelerated by Alfvén waves in the solar wind. The mechanism is described in detail.

074.057 **Some effects of diffusion on the flow of ions in the solar wind.** M. P. Nakada.
Journ. Geophys. Res., Vol. 79, 1364 - 1367 (1974).

The purpose of this study is to extend the work of Geiss et al. and to examine certain diffusive processes and their potential contribution to equalizing flow velocities of the ions in the solar wind.

074.058 **Transverse Alfvén waves in the solar wind: Arbitrary k, v_0, B_0, and $|\delta B|$.** J. V. Hollweg.
Journ. Geophys. Res., Vol. 79, 1539 - 1541 (1974).

Using a simple analysis based on energy conservation, we derive expressions for the spatial variation of the amplitudes of transverse Alfvén waves in the solar wind. We make no assumptions about the solar wind geometry or the directions of propagation, and we do not require that the wave amplitudes be small.

074.059 **Solar wind torque as an inhibitor of terrestrial rotation.** C. O. Hines.
Journ. Geophys. Res., Vol. 79, 1543 - 1545 (1974).

Coleman (1971) has suggested that the solar wind may exert sufficient torque on the terrestrial magnetosphere to effect observable changes in the earth's rate of rotation. Hirshberg (1972) has examined the torque available from the solar wind and has concluded that it is too small for the purpose. It is argued here that Hirshberg's development and conclusion could be undermined by a variant on her model, but that other considerations lead ultimately back to her conclusion. These considerations concern the rate of rotation that the polar ionosphere would achieve under the driving effect of the solar wind torque and the resistive effect of viscous coupling through the neutral atmosphere from the ground.

074.060 **Comments on 'Solar polar spin-down', by Kenneth Schatten.** P. A. Gilman.
Solar Physics, Vol. 36, 61 - 64 (1974). – Research note.

074.061 **Green corona and solar sector structure.**
E. Antonucci, L. Svalgaard.
Solar Physics, Vol. 36, 115 - 120 (1974).

Analysis of the green line corona for the interval 1947–1970 suggests the existence of large-scale organization of the emission. The green line emission at high northern latitudes is correlated with the emission at high southern latitudes 6, 15 and 24 days later, while the low latitude green corona seems to be correlated on both sides of the equator with no time lag. These coronal features are recurrent with a 27-day period at all latitudes between ±60°, and the authors associate these large-scale structures with the solar magnetic sector structure. The high correlation between northern and southern high-latitude emission at 15 days time lag is explained as a signature of a two-sector structure, while four sectors are associated with the 6 and 24 day peaks.

074.062 **The structure of the middle corona from observations at 80 and 160 MHz.**
G. A. Dulk, K. V. Sheridan.
Solar Physics, Vol. 36, 191 - 202 (1974).

Maps of the brightness distribution of the 'quiet sun' at 80 and 160 MHz reveal the presence of features both brighter and darker than average. The 'dark' regions are well correlated with dark regions on UV maps; the authors deduce that they result from 'coronal holes'. The 'bright' regions are associated with quiescent filaments and not plages or bright regions on microwave or UV maps; the authors deduce that they result from 'coronal helmets'.

074.063 **Observations of coronal disturbances from 1 to 9$R_\odot$. I: First event of 1973 January 11.**
R. T. Stewart, M. K. McCabe, M. J. Koomen, R. T. Hansen, G. A. Dulk.
Solar Physics, Vol. 36, 203 - 217 (1974). – This paper was presented at IAU Symp. No. 57 on 'Coronal Disturbances', Surfers Paradise, Australia, 7–11 September 1973.

Hα, white-light and radio observations of a coronal disturbance on 1973 January 11 commencing at about 00^h36^m UT show that a piston-driven shock wave propagated outwards through the corona to heights of at least 9 $R_\odot$. Probably most of the expelled coronal gas originated in a coronal enhancement in the lower corona. Shock-wave parameters are evaluated and a model of the disturbance outlined.

074.064 **Observations of coronal disturbances from 1 to 9 $R_\odot$. II: Second event of 1973 January 11.**

R. T. Stewart, R. A. Howard, F. Hansen, T. Gergely, M. Kundu.
Solar Physics, Vol. 36, 219 - 231 (1974).

Observations of a coronal disturbance on 1973 January 11 commencing at 18^h01^m UT are described. The event is homologous with an earlier disturbance from the same region of the corona. The observations suggest that a cloud of coronal gas containing ~4×10^{39} electrons propagated outwards to $\geqslant 5\ R_\odot$ behind a piston-driven shock wave travelling at a velocity of 800 to 1200 km s^{-1}.

074.065 **Coronal Holes – eine neue Erscheinung der Sonnenkorona.** G. Scholz.
Sterne, Vol. 50, 101 - 104 (1974).

074.066 **Coronal λ 5303 intensity, geomagnetic activity and solar sources of high-speed plasma streams.**
A. Gulbrandsen.
Planet. Space Sci., Vol. 22, 841 - 857 (1974).

Investigations during the last 30 yr of the relationship between the green emission line corona and geomagnetism have yielded contradictory results. The papers on this subject can be separated into 2 groups; (a) papers reporting a negative correlation between green line intensity and geomagnetic activity, and (b) papers reporting a positive correlation. The negative correlation seems to be the one better supported by solar wind theory and observations. It implies that solar active regions are the base of low expansion speeds, whereas quasi-stationary high-speed solar plasma streams originate from regions of low density and open magnetic field structures. In the present paper the author has re-examined some of the analyses described in the group (b) papers.

074.067 **The solar wind as a three-dimensional flow.**
M. Sroczyńska.
Postępy Astron., Vol. 22, 125 - 132 (1974). In Polish.

A method describing deviations of the solar wind from spherical symmetry (due for instance to solar rotation or to the inhomogeneous distribution of active centres on the disk) is given. It was assumed that the deviations are small and that each of the sources causing a deviation has axial symmetry.

074.068 **Intensity gradients and electron density in the solar corona.** J. P. Rozelot, J. C. Noëns.
Astron. Astrophys., Vol. 32, 453 - 455 (1974). In French.

The height gradient of coronal emission lines enables to determine the electron density in the solar corona. The authors show a dependence of the gradient on the intensity and give a method to obtain three important parameters: vertical density gradient, longitudinal extension and density at the bottom of the studied enhancement. Two upper limits owing to the limitations of the method are found. An application of the method is given.

074.069 **A study of the fast moving active phenomena in the solar corona.**
Ts. S. Khetsuriani, Eh. I. Tetruashvili.
Solnechnye Dannye 1974 Byull., No. 2, p. 98 - 106 (1974). In Russian.

Active coronal knots at the height of about 1 have been observed with the coronograph of the Abastumanı Astrophysical Observatory. During about 1 hour the intensities of these knots vary by 100% and even more. Radial velocities of some tens km/sec are sometimes observed. The mean electron concentration in these knots is $\bar{n}_e \approx 10^{10}$ cm^{-3}.

074.070 **On the variations of the solar coronal brightness and electron density according to solar activity.**
S. Hata.
Tokyo Astron. Obs., Report No. 63, Vol. 16, 673 - 698 (1974). In Japanese.

074.071 **Polarization of the inner corona at the total solar eclipse on September 22, 1968.**
A. V. Bugaevskij, Yu. N. Lipskij, Yu. P. Pskovskij, M. M. Pospergelis.
Soobshch. Gos. Astron. Inst. Shternberga, No. 184, p. 31 - 45 (1973). In Russian.

During the total solar eclipse on the 22nd of September, 1968 near Sary-Shagan 12 photographs of the solar corona were obtained through polaroids. The authors calculated the degree of polarization and the orientation of the plane of polarization up to a distance from the limb to $0.20-0.25\ R_\odot$; the accuracy of the results is 3–4 %. The average polarization in all position angles for the given distances from the limb is in good agreement with theoretical data. A considerable diminution of the degree of polarization was found near prominences and in a flash. The average orientation of the plane of polarization is radial, but it deviates from that in vicinity of prominences.

074.072 **Onderzoek naar magnetohydrodynamische en plasma instabiliteiten in de zonnekorona.**
M. Kuperus, H. Rosenberg.
Nederl. Tijdschr. Natuurk., Vol. 39, No. 9, p. 130 - 133 = Utrechtse Sterrekundige Overdrukken No. 234 (1973).

074.073 **Upper limits for the solar wind He$^+$ content at 1 AU.**
W. C. Feldman, J. R. Asbridge, S. J. Bame, P. D. Kearney.
Journ. Geophys. Res., Vol. 79, 1808 - 1812 (1974).

An upper limit for the steady state solar wind He$^+$ concentration is estimated from four Vela 5 and 6 heavy ion spectrums. The upper limit value for $N(\text{He}^+)/N(\text{H})$ of 2.5×10^{-6} so obtained is at least an order of magnitude less than that predicted from presently accepted models of the local interstellar medium. An explanation of the discrepancy is given.

074.074 **On the account of nonlinear stabilization of beams when interpreting polarization X-ray observations.**
M. A. Livshits, V. M. Tomozov.
Astron. Zhurn. Akad. Nauk SSSR, Vol. 51, 560 - 564 (1974). In Russian. English translation in Soviet Astron. AJ, Vol. 18, No. 3.

It is shown that electron beams with energy $E = 15$ keV stabilized by nonlinear processes can cover in the solar corona a path comparable with the free path length of an individual accelerated electron in a plasma. Polarization of X-ray radiation is due to electron beams with small pitch angles which scatter at plasma waves propagating along the magnetic field.

074.075 **The screw-like coronal ray on September 22, 1968.**
A. T. Nesmyanovich, N. I. Dzyubenko, Yu. A. Khomenko, O. S. Popov.
Astron. Zhurn. Akad. Nauk SSSR, Vol. 51, 577 - 581 (1974). In Russian. English translation in Soviet Astron. AJ, Vol. 18, No. 3.

The results of a detailed study of the structure of the north-eastern active region of the corona on September 22, 1968 are given. The coronal structure was apparently formed as a result of active processes on the photosphere – chromospheric level. The structural peculiarities of the ray are explained by propagation of a unstable helical wave along the ray. It is caused by the rotation of the magnetic field of a spot during the chromospheric flare.

074.076 **The solar wind and magnetospheric dynamics.**
C. T. Russell.
Astrophys. Space Sci. Library, Vol. 42, (see 012.016), 3 - 47 (1974).

074.077 **Pioneer solar plasma and magnetic field measurements in interplanetary space during August 2–17,**

1972. J. D. Mihalov, D. S. Colburn, H. R. Collard, B. F. Smith, C. P. Sonett, J. H. Wolfe.
Astrophys. Space Sci. Library, Vol. 42, (see 012.016), 545 - 553 (1974).

074.078 **Relation of coronal magnetic structure to the interplanetary proton events of August 2–9, 1972.**
E. C. Roelof, J. A. Lezniak, W. R. Webber, F. B. McDonald, B. J. Teegarden, J. H. Trainor.
Astrophys. Space Sci. Library, Vol. 42, (see 012.016), 563 - 571 (1974).

074.079 **The interaction between an impact-produced neutral gas cloud and the solar wind at the lunar surface.**
R. A. Lindeman, R. R. Vondrak, J. W. Freeman, C. W. Snyder.
Journ. Geophys. Res., Vol. 79, 2287 - 2296 (1974).

On April 15, 1970, the Apollo 13 S-IVB stage impacted the nighttime lunar surface approximately 140 km west of the Apollo 12 Alsep site and 410 km west of the dawn terminator. Beginning 20 s after impact the Suprathermal Ion Detector Experiment and the Solar Wind Spectrometer observed a large flux of positive ions and electrons. Two separate streams of ions were observed. An examination of the data shows that collisions between neutral molecules and hot electrons (50 eV) were probably an important ionization mechanism in the impact-produced neutral gas cloud.

074.080 **The solar wind He^{2+} to H^+ temperature ratio.**
W. C. Feldman, J. R. Asbridge, S. J. Bame.
Journ. Geophys. Res., Vol. 79, 2319 - 2323 (1974).

The Los Alamos Scientific Laboratory Imp 6 solar wind ion plasma data measured between March 18, 1971, and May 30, 1972, were searched for a dependence of the alpha particle to proton temperature ratio T_a/T_p on the ratio of the solar wind expansion scale time to the Coulomb collision energy transfer time τ_e/τ_c. For flow conditions characterized by nearly Maxwellian ion velocity distributions, T_a/T_p is observed to vary inversely with τ_e/τ_c: when τ_e/τ_c is small, $T_a/T_p \simeq 4$, and when $\tau_e/\tau_c \gtrsim 1$, $T_a/T_p \gtrsim 2$.

074.081 **A discussion of interplanetary postshock flows with two examples.**
K. W. Ogilvie, L. F. Burlaga.
Journ. Geophys. Res., Vol. 79, 2324 - 2330 (1974).

Plasma and magnetometer observations of two types of flare-associated shock flows are described and compared with present models. One type represents a class of flows in which the shock is followed by a stream and separated from it by a region in which density, temperature, and speed decrease monotonically. The other type is characterized by a complex region between the shock and the following stream, which has many discontinuities and fluctuations but in which there is no increase in helium concentration.

074.082 **The reliability of the daily quick look solar wind velocities as indicators of interplanetary activity.**
D. S. Intriligator.
Journ. Geophys. Res., Vol. 79, 2491 - 2492 (1974).

It is shown that the daily quick look solar wind velocities based on the Ames Research Center solar wind plasma data obtained on the Pioneer spacecraft are surprisingly reliable indicators of interplanetary activity.

074.083 **A new theory of coronal heating.** R. H. Levine.
Astrophys. Journ., Vol. 190, 457 - 466 (1974).

This paper examines an alternative to the heating of the corona by shocks or waves. It is argued that the corona is interspersed with magnetic neutral surfaces which collapse and accelerate particles to a few times their thermal velocity in the manner described in the preceding paper. The accelerated particles travel through the corona and lose their increased energy through Coulomb collisions. It is shown that this mechanism can provide the energy balance needed by the corona and that such acceleration can account for regions of enhanced or decreased heating in the corona.

074.084 **Nonspherical magnetized solar wind.**
M. Sroczyńska.
Acta Astron., Vol. 24, 165 - 188 (1974).

Stationary, axisymmetric, three-dimensional models of a magnetized solar wind are studied by standard small-perturbation technique. The zero-order model is represented by a spherically symmetric Parker's solution with polytropic formula instead of the energy equation. For certain boundary conditions imposed upon the flow it is possible to obtain models of the solar wind exhibiting a tendency to blow faster near the poles than near the equator in agreement with observational data. A method of describing the contribution to nonspherical deviations of the solar wind due to solar active centers is also proposed.

074.085 **Observations coronographiques avant et après les éclipses du 10 Juillet 1972 et du 30 Juin 1973.**
P. Poulain.
Solar Physics, Vol. 36, 339 - 342 (1974).

The author presents the results of coronagraphic observations that he has made at Pic-du-Midi Observatory on the occasion of the total solar eclipses of July 10th, 1972 and June 30th, 1973.

074.086 **The Fe XIV brightness measurements: 30 June 1973.** R. R. Fisher.
Solar Physics, Vol. 36, 343 - 344 (1974). – Research note.

074.087 **Representations of coronal magnetic fields including currents.** R. H. Levine, M. D. Altschuler.
Solar Physics, Vol. 36, 345 - 350 (1974).

Coronal electric currents are superposed on the calculated large-scale current-free (potential) magnetic field of the solar corona and the new magnetic configurations are mapped. The results indicate that relatively large coronal electric currents are required before significant topological deviations from the potential magnetic field configuration can be noticed.

074.088 **Dynamics of structural formations in the solar wind.**
G. F. Krymskij, I. A. Transkij.
Raspredelenie galakt. kosmich. luchej i dinamika struktur. obrazovanij v solnechn. vetre. Yakutsk, 1973, p. 154 - 197. In Russian. – Abstr. in Referativ. Zhurn. 62. Issled. kosmich. prostranstva, 6.62.156 (1974).

074.089 **On the structure of the quiet corona near quiescent prominences.** B. Fort, M. J. Martres.
Astron. Astrophys., Vol. 33, 249 - 255 (1974).

Monochromatic images of the solar corona in the light of the red and green lines of Fe X and Fe XIV respectively, obtained at the Pic-du-Midi Observatory, enable the geometrical structure of the quiet corona to be determined in the region of quiescent prominences, observed in Hα. The electron density of the emitting region is estimated from the results of a photometric analysis of the red and green line images. Some of the implications of the existence of "sheets of matter" in the quiet corona are discussed with particular reference to current ideas on coronal cavities.

074.090 **The evolution of the coronal structure of an active region.** A. Krieger, M. Gerassimenko, R. Petrasso, A. Timothy, G. Vaiana.
Bull. American Astron. Soc., Vol. 6, 265 (1974). – Abstr. AAS.

074.091 **Coronal loops associated with active filaments.**
J. Davis, R. Chase, A. Krieger, A. Timothy, G.

Vaiana.
Bull. American Astron. Soc., Vol. 6, 265 (1974). — Abstr. AAS.

074.092 Rigid and differential rotation of the solar corona.
E. Antonucci, L. Svalgaard.
Bull. American Astron. Soc., Vol. 6, 284 (1974). — Abstr. AAS.

074.093 Coronal loop structures associated with active filaments. J. Davis, R. Chase, A. Krieger, R. Simon, A. Timothy, G. Vaiana.
Bull. American Astron. Soc., Vol. 6, 286 (1974). — Abstr. AAS.

074.094 Coronal bright points.
L. Golub, R. Chase, A. Krieger, J. K. Silk, A. Timothy, G. Vaiana.
Bull. American Astron. Soc., Vol. 6, 287 (1974). — Abstr. AAS.

074.095 Some properties of major disruptions in the solar corona. R. T. Hansen, C. J. Garcia.
Bull. American Astron. Soc., Vol. 6, 287 - 288 (1974). Abstr. AAS.

074.096 Synoptic coronagraph observations around the 30 June 1973 eclipse.
S. F. Hansen, R. T. Hansen, Ru. Howard, M. Koomen.
Bull. American Astron. Soc., Vol. 6, 288 (1974). — Abstr. AAS.

074.097 Coronal rotational and flow motions measured at the 1970 and 1973 total eclipses.
J. Harvey, W. Livingston, L. Doe.
Bull. American Astron. Soc., Vol. 6, 288 (1974). — Abstr. AAS.

074.098 Solar wind magnetic field structure from radioburst analysis. E. G. Howard.
Bull. American Astron. Soc., Vol. 6, 288 - 289 (1974). Abstr. AAS.

074.099 Polarimetry of the outer corona from a jet aircraft during the June 1973 solar eclipse.
C. F. Keller, J. E. Tabor.
Bull. American Astron. Soc., Vol. 6, 289 - 290 (1974). Abstr. AAS.

074.100 Temporal behavior of the coronal structure of active regions. A. Krieger, L. Golub, J. K. Silk, A. Timothy, G. Vaiana, D. Webb.
Bull. American Astron. Soc., Vol. 6, 290 (1974). — Abstr. AAS.

074.101 Solar coronal holes as sources of recurrent geomagnetic disturbances. W. M. Neupert, V. Pizzo.
Bull. American Astron. Soc., Vol. 6, 292 (1974). — Abstr. AAS.

074.102 Near infrared photography of the outer corona during the 30 June 1973 solar eclipse.
L. J. Radziemski, C. G. Lilliequist.
Bull. American Astron. Soc., Vol. 6, 293 (1974). — Abstr. AAS.

074.103 A preliminary study of coronal structures by means of time-lapse photography.
N. R. Sheeley, J. D. Bohlin, G. E. Brueckner, J. D. Purcell, V. Scherrer, R. Tousey.
Bull. American Astron. Soc., Vol. 6, 294 (1974). — Abstr. AAS.

074.104 The evolution of coronal holes.
A. Timothy, A. Krieger, R. Petrasso, J. K. Silk, G. Vaiana.
Bull. American Astron. Soc., Vol. 6, 295 (1974). — Abstr. AAS.

074.105 The frequency and nature of coronal transient events observed by OSO-7.
R. Tousey, R. A. Howard, M. J. Koomen.
Bull. American Astron. Soc., Vol. 6, 295 - 296 (1974). Abstr. AAS.

074.106 Temperature and density of the coronal portion of an active region.
G. Vaiana, M. Gerassimenko, A. Krieger, A. Timothy, M. Landini, B. C. Monsignori-Fossi.
Bull. American Astron. Soc., Vol. 6, 296 (1974). — Abstr. AAS.

074.107 The fast rearrangement of coronal magnetic fields.
W. J. Wagner, R. T. Hansen, S. F. Hansen.
Bull. American Astron. Soc., Vol. 6, 296 (1974). — Abstr. AAS.

074.108 The temperature structure of a coronal active region.
A. B. C. Walker, Jr., E. B. Mayfield, D. L. McKenzie, J. H. Underwood.
Bull. American Astron. Soc., Vol. 6, 296 - 297 (1974). Abstr. AAS.

074.109 Coronal magnetic structure at a solar sector boundary. J. M. Wilcox, L. Svalgaard.
Bull. American Astron. Soc., Vol. 6, 297 (1974). — Abstr. AAS.

074.110 Interstellare Edelgasanteile im Sonnenwind (He, Ne, Ar). H. J. Fahr.
Analyse extraterrestrischen Materials, (see 012.020), p. 37 - 52 (1974).

074.111 Aligned magnetohydrodynamic solution for solar wind flow past the earth's magnetosphere.
J. R. Spreiter, A. W. Rizzi.
Acta Astronaut., Vol. 1, 15 - 35 (1974).
  Exact numerical solutions of the magnetohydrodynamic equations for a perfect dissipationless gas with aligned magnetic field are given for conditions representative of steady supersonic solar wind flow past an axisymmetric model of the earth's magnetosphere. The results are consistent with direct observations in space.

074.112 Polarigraphic observations of the inner corona and a condensation region in the solar corona.
T. Cao, J. You, Z. Wang, F. Hu.
Acta Astron. Sinica, Vol. 15, 11 - 24 (1974). In Chinese.
  This paper presents an analysis of the polarization characteristics of the radiation of the inner corona and its condensation region. Photographs of the corona were taken during the total solar eclipse of 22 September, 1968 in the western part of Sinkiang.

074.113 Are coronal holes M-regions? C. Jordan.
Observatory, Vol. 94, 141 (1974). — Letter.

074.114 Shape and structure of the corona at the solar eclipse of June 30, 1973. M. Waldmeier.
Astron. Mitt. Eidgenössisch. Sternw. Zürich, No. 330, 18 pp. (1974).

074.115 K corona and magnetic sector boundaries.
S. F. Hansen, C. Sawyer, R. T. Hansen.
Geophys. Res. Letters, Vol. 1, No. 1, p. 13 - 16 (1974).

La couronne solaire. See Abstr. 003.105.

On the extreme ultraviolet emission spectrum of Fe XIII. See Abstr. 022.010.

Depolarization cross-sections of Fe^{13+} in collision with protons. See Abstr. 022.047.

A film for obtaining isophotic contours. See Abstr. 036.001.

Some results from Skylab's solar experiments. See Abstr. 054.021.

Scattering of Alfvén waves by random density fluctuations. See Abstr. 062.013.

Gyromagnetic absorption at the fundamental. See Abstr. 062.052.

On the hydromagnetic Kelvin-Helmholtz instability between compressible fluids. See Abstr. 062.053.

Propagation of transverse waves in a radially expanding plasma. See Abstr. 062.063.

Large-amplitude hydromagnetic waves. See Abstr. 062.064.

Electromagnetic instabilities in non-uniform anisotropic plasmas. See Abstr. 062.078.

A comparison of EUV spectroheliograms and photospheric magnetograms. See Abstr. 071.008.

Extrapolation of photospheric magnetic fields into the corona. See Abstr. 071.064.

North-south asymmetry in solar activity distribution and solar wind structure in the meridional plane of the sun. See Abstr. 072.006.

Effect of solar corona conditions on flare particle propagation. See Abstr. 073.009.

Development and spatial structure of proton flares near the limb, and coronal phenomena. VII. Change of activity prior to the proton flare of September 1, 1971. See Abstr. 073.019.

The formation of solar quiescent prominences by condensation. See Abstr. 073.041.

The loop prominence of 11 August 1972: a coronal continuum event. See Abstr. 073.048.

High n emission and absorption lines of the sun. See Abstr. 073.051.

R. A. Kopp, "Energy balance in the chromosphere-corona transition region". See Abstr. 073.059.

An experimental model of solar flares in the corona. See Abstr. 073.086.

Rocket spectroheliogram observations of the heights of formation and sizes of bright features in the transition zone. See Abstr. 073.087.

Motion of prominences through the corona. See Abstr. 073.089.

Measurement of the X-ray emission of the solar atmosphere during a period of low activity. See Abstr. 076.002.

The relevance of bipolar type I storm structures to the theory of mode coupling in the solar corona. See Abstr. 077.004.

Decameter type IV bursts associated with coronal transients. See Abstr. 077.018.

Landau damping of type III bursts by the interplanetary electrons of high energy. See Abstr. 077.027.

Decameter type IV bursts associated with coronal transients. See Abstr. 077.069.

Millimeter emission related to coronal magnetic fields. See Abstr. 077.077.

Phenomenological presentation of the path of propagation of electromagnetic impulses in the ionosphere and solar corona. See Abstr. 083.004.

A theoretical relation between DST and the solar wind merging electric field. See Abstr. 083.053.

Motions of the bow shock induced by interplanetary disturbances. See Abstr. 084.224.

Relationship between interplanetary plasma parameters and geomagnetic *Dst*. See Abstr. 084.227.

Investigation of the association of magnetopause instability with interplanetary sector structure. See Abstr. 084.234.

Atmospheric ion wakes of Venus and Mars in the solar wind. See Abstr. 093.037.

A model for the steady interaction of the solar wind with the moon. See Abstr. 094.103.

On the interaction of the solar wind with the lunar limb. See Abstr. 094.167.

Solar wind nitrogen in lunar material. See Abstr. 094.393.

Magnetic fields of Mars and Venus: solar wind interactions. See Abstr. 097.020.

Comets, solar wind and the D/H ratio. See Abstr. 102.015.

Wechselwirkungen zwischen Sonnenwind und Komet Bennett. See Abstr. 103.100.

Correlation of interplanetary scintillation and spacecraft plasma density measurements. See Abstr. 106.038.

075 Solar Patrol

075.001 **Solar activity during 1972.** H. Hill.
Journ. British Astron. Ass., Vol. 84, 129 - 135 (1974). – Report of Solar Section of the British Astron. Association.

075.002 **Solar activity.** F. G. Mustaeva.
Tsirk. Astron. Inst., *Tashkent*, Nos. 40 (387), 42 (389), 43 (390) (1973). In Russian. – 1972 July - 1972 December.

075.003 **Nombres relatifs moyens pour l'année 1972.** M. Waldmeier.
L'Astronomie, 88ᵉ année, p. 82 (1974).

075.004 **Solar observations made at Catania Astrophysical Observatory during 1973.** G. Godoli, V. Sciuto, R. A. Zappalà, E. Catinoto, G. Domina, S. Sciuto, G. Celeani, G. Sapienza.
Oss. Astrofis. Catania, Pubbl. No. 152, 91 pp. (1974).
Sunspots; Hα and K faculae; Hα flares; Hα quiescent prominences; K quiescent prominences; Hα active prominences on disc and at limb; Hα disc and limb patrol hours.

075.005 **Osservazioni eseguite nell' Osservatorio Astrofisico di Arcetri durante il biennio 1968–1969.**
F. Mazzucconi, T. Grisendi, R. Baldini, G. Marcucci.
Osservazioni Mem. Oss. Astrofis. Arcetri, Fasc. 99, 201 pp. (1973).
Macchie; Flocculi Hα$_3$; Flocculi K$_{2,3,2}$; Brillamenti; Filamenti Hα$_3$; DSD; BSL; Periodo di osservazione al filtro monocromatico; Ore di esecuzione filtrogrammi.

075.006 **Observations photosphériques et chromosphériques solaires faites à Uccle en 1971.**
G. Evrard, R. Florée, C. Gonze, A. Koeckelenbergh.
Bull. Astron. Obs. Roy. Belgique, Vol. 8, 68 - 128 (1973).

075.007 **Investigations of the sun.** Yu. M. Slonim.
Astronomical Institute of the Academy of Sciences of the Uzbek SSR – 100 years, (see 003.011), p. 88 - 112 (1974). In Russian.

075.008 **Cinematographic observations for ATM and their comparison with some ATM results.**
H. Zirin, J. Holt, G. E. Brueckner, J. D. Bohlin, J. D. Purcell, V. E. Scherrer, N. R. Sheeley, R. Tousey.
Bull. American Astron. Soc., Vol. 6, 298 (1974). – Abstr. AAS.

075.009 **Solar activity in 1973.** J. Mergentaler.
Urania Kraków, Vol. 45, 149 - 150 (1974). In Polish.

075.010 **Solar photospheric observations.**
F. Bruin, H. Hourani, N. G. Bustati.
Lee Obs., American Univ. Beirut, Monthly Bull. Astron. Section, 1974 January – May (1974).
Sunspot relative numbers; Heliographic mean position and classification of the sunspot groups; Number of facular zones.

075.011 **Solare Beobachtungsergebnisse (Solar Data).**
E. A. Lauter, C.-U. Wagner, A. Böhme, F. Fürstenberg, D. Scholz, S. Böhm.
Zentralinst. für Solar-Terrestrische Physik (Heinrich-Hertz-Inst.), Akad. Wiss. DDR, HHI Solar Data, Vol. 25, January – March (1974). – Solar radio emission.

075.012 **Fenómenos solares (1942).**
Anais Obs. Astron. Univ. Coimbra, Vol. 14, 105 pp. (1974).

075.013 **Solar phenomena.** M. Cimino, M. Torelli, F. Casamassima, V. Croce.
Oss. Astron. Roma, Monthly Bull. Nos. 186, 188 - 191 (1973/74). – 1973 October, December - 1974 March: Daily total areas of sunspot-groups; Heliographic position, classification and area of sunspot-groups; Longitudinal sunspot magnetic fields; Hours of K-line cinematographic patrol; Hours of H$_\alpha$ cinematographic patrol; S.C.N.A. and S.E.A.; Explanation.

075.014 **Daily Hα chromosphere pictures, daily K$_{232}$ chromosphere pictures, daily white light photosphere pictures.** M. Cimino (Editor).
Photographic Journ. of the Sun, Oss. Astron. Roma, Nos. 75 - 80 (1973). – 1973 June 29 – December 9. – Rotations 1603 – 1608.

075.015 **Sunspot relative numbers for 1973.** M. Waldmeier.
Astron. Mitt. Eidgenössisch. Sternw. Zürich, No. 326, 10 pp. (1974).
The paper gives the daily values of the sunspot relative numbers for the whole disc, for the central zone and the daily number of sunspot groups.

075.016 **Sunspots (sunspot relative-numbers and sunspot-areas); Synoptic charts of solar magnetic fields (Mount Wilson Observatory); Eruptions chromosphériques brillantes; Intensité de la couronne solaire; Solar radio emission.**
M. Waldmeier, R. Howard, G. Olivieri, M. Bernot, H. Tanaka.
Quarterly Bull. Solar Activity (published by Eidgen. Sternw. Zürich), Nos. 181 - 183, p. 1 - 131 (1974). – 1973 January – September.

075.017 **Daily maps of the sun and geophysical graphs.**
Solnechnye Dannye 1973 Byull., No. 11, p. 1 - 56; No. 12, p. 1 - 56; 1974, No. 1, p. 1 - 64; No. 2, p. 1 - 54; No. 3, p. 1 - 52; No. 4, p. 1 - 71. In Russian.

075.018 **Magnetic fields of sunspots.**
Prilozhenie k Byulletenyu "Solnechnye Dannye", 1973, Nos. 10 - 12; 1974, Nos. 1 - 3. In Russian.

075.019 **Indices of geomagnetic activity.**
Journ. Atmosph. Terr. Phys., Vol. 36, 191 - 192, 559 - 560, 911 - 912 (1974). – 1973 August – 1974 January.

075.020 **Solar and solar system activity.**
R. J. J. Langton, J. R. Smith.
Journ. British Astron. Ass., Vol. 84, 138 - 140, 217 - 219, 297 - 299 (1974). – 1973 September – 1974 February.

075.021 **Geomagnetic and solar data.**
J. V. Lincoln (Editor).
Journ. Geophys. Res., Vol. 79, 316, 688, 1134, 1586, 2006, 2555 (1974). – 1973 September – 1974 February.

075.022 **L'activité solaire.** M.-J. Martres.
L'Astronomie, 88ᵉ année, p. 34 - 36, 79, 109 - 110, 149 - 150, 191, 235 (1974). – Rotations 1602 - 1609.

075.023 **Sunspot numbers.**
Sky Telescope, Vol. 47, 64, 135, 205, 276, 350, 417; Vol. 48, 60 (1974).

076 Solar UV, X Rays, Gamma Radiation

076.001 **The spatial distribution of Lyman-α on the sun.**
D. K. Prinz.
Astrophys. Journ., Vol. 187, 369 - 375 (1974).

One of the high spatial resolution, Lα line spectroheliograms obtained by the Naval Research Laboratory during a rocket flight on 1972 July 10 has been analyzed to show the distribution of Lα on the solar disk. The intensity distribution across selected cells, filaments, bright "points", and active regions is presented.

076.002 **Measurement of the X-ray emission of the solar atmosphere during a period of low activity.**
R. S. Wolff.
Solar Physics, Vol. 34, 163 - 172 (1974).

A large-area high-sensitivity X-ray spectrometer has been constructed and used to measure the 1.8–5.3 Å X-ray emission of the sun under quiescent conditions. The instrument utilizes Bragg reflection from mosaic graphite crystals. The data indicate that the X-ray emission can best be accounted for by a multitemperature model of the solar atmosphere in which both the over-all corona and active regions contribute to the X-ray spectrum. Theoretical calculations of the X-ray flux of a hot, optically thin plasma have been used to estimate the solar conditions at the time when the measurements were made.

076.003 **Interpretation of solar hard X-ray burst polarisation measurements.**
J. C. Brown, A. N. McClymont, I. S. McLean.
Nature, Vol. 247, 448 - 449 (1974).

The authors point out that the results as presented by Tindo et al. are in error since their calibration was based on the assumption that the hard X-ray flux from the flare itself tends to zero polarisation in its final stages of decay. They further consider how much this calibration error may affect interpretation of the Intercosmos results in terms of flare particles and suggest how further theoretical work, combined with results from a laboratory calibrated polarimeter, can yield information on the spatial location of hard X-ray flares as well as on their true polarisation.

076.004 **Observations of the solar spectrum in the region 150 Å to 870 Å emitted from the disk and above the limb.**
J. G. Firth, F. F. Freeman, A. H. Gabriel, B. B. Jones, C. Jordan, C. R. Negus, D. B. Shenton, R. F. Turner.
Monthly Notices Roy. Astron. Soc., Vol. 166, 543 - 560 (1974).

Photographic spectra have been obtained in the grazing-incidence region from a position on the quiet solar disk and from a region just above the visible limb. The payload, which was launched on a sun-stabilized Skylark rocket, contained three grazing-incidence spectrographs, each illuminated by a two-component grazing-incidence telescope mirror. The present paper presents a full identification of the spectra, which includes a number of intersystem transitions, notably in iron ions.

076.005 **Spatial distribution of soft X-ray and EUV emission associated with a chromospheric flare of importance 1b on August 2, 1972.**
W. M. Neupert, R. J. Thomas, R. D. Chapman.
Solar Physics, Vol. 34, 349 - 375 (1974).

Soft X-ray and extreme ultraviolet spectroheliographs carried by the OSO-7 have been used to record the development of XUV emission associated with a flare of importance 1b on August 2, 1972. Spatial resolution was 20″ and spectral resolution was adequate to select emission lines originating within well-defined ranges of electron temperature between 5×10^4 and 30×10^6 K. The observations are presented and the implications of these data on current flare models are discussed.

076.006 **Comparison of X-ray coronal structures with the dynamics and morphology of photospheric activity.**
V. V. Kasinskij, V. M. Tomozov.
Astron. Tsirk., No. 806, p. 1 - 3 (1974). In Russian.

076.007 **About forecasting of relative changes of intensity of short-wave radiation of the sun in the 11-year cycle.**
G. S. Ivanov-Kholodnyj.
Geomagn. Aeronom., Vol. 14, 188 - 191 (1974). In Russian.

076.008 **Observations of solar X-ray bursts in the energy range 5–15 keV.**
D. W. Datlowe, H. S. Hudson, L. E. Peterson.
Solar Physics, Vol. 35, 193 - 206 (1974).

The authors present the results of the first study of a large sample of separate bursts, 197 events associated with sub-flares and a few importance 1 events. The observations were made by a proportional counter on the satellite OSO-7 from October 1971 to June 1972. From these observations they show that the growth of the thermal energy in the flare plasma throughout the burst can be due to the heating of new cool material.

076.009 **Solar X-ray bright points.**
L. Golub, A. S. Krieger, J. K. Silk, A. F. Timothy, G. S. Vaiana.
Astrophys. Journ., (Letters), Vol. 189, L93 - L97 (1974).

The purpose of this Letter is to provide new quantitative and statistical data about bright points obtained by a preliminary examination of photographs from the X-ray spectrographic telescope aboard Skylab.

076.010 **Das solare UV-Spektrum in der Umgebung der Mg II-Resonanzlinien.**
A. Greve, C. D. McKeith, N. E. McKeith.
Mitt. Astron. Ges., No. 34, p. 107 - 109 (1973/74).
Presented at the "Wissenschaftliche Tagung der Astron. Ges., Oberkochen, 1973 April 24 - 27".

076.011 **Hard solar flare X-ray bursts on 8 December 1970.**
K. A. Anderson, W. A. Mahoney.
Solar Physics, Vol. 35, 419 - 430 (1974).

Hard X-ray bursts have been observed from two 1B flares located in the same sunspot region and separated in time by about 70 min on 8 December 1970. The bursts are composed by many 'flashes' of 2 to 20 seconds duration. Power spectrum analysis reveals no strong periodicities although a significant peak appears at 7.5 s.

076.012 **On anisotropy of solar hard X-ray emission.**
G. Pizzichini, A. Spizzichino, G. R. Vespignani.
Solar Physics, Vol. 35, 431 - 439 (1974).

A number of solar X-ray events above 10 keV and 20 keV were compiled in order to test for evidence of anisotropic emission. The results are not definite, although the two samples show apparently different behaviours.

076.013 **Neutron propagation and 2.2 MeV gamma-ray line production in the solar atmosphere.**
H. T. Wang, R. Ramaty.
Goddard Space Flight Center, Greenbelt, Maryland, GSFC

Document X-660-74-5, Preprint, 18 pp. (1974).

The authors have calculated the 2.2 MeV gamma-ray line intensity from the sun using a Monte-Carlo method for neutron propagation in the solar atmosphere. They provide detailed results on the total gamma-ray yield per neutron and on the time profile of the 2.2 MeV line from an instantaneous and monoenergetic neutron source. The parameters which have the most significant effects on the line intensity are the energies of the neutrons, the position of the neutron source on the sun, and the abundance of He^3 in the photosphere.

076.014 **A correction to McAllister's 1960 atlas of the solar spectrum.** E. G. Chipman.
Solar Physics, Vol. 36, 45 - 48 (1974). — Research note.

076.015 **The spectral dependence of solar soft X-ray flux values obtained by SOLRAD 9.**
K. P. Dere, D. M. Horan, R. W. Kreplin.
Journ. Atmosph. Terr. Phys., Vol. 36, 989 - 994 (1974).

The determination of solar energy flux values from ionization chamber experiments is discussed. Procedures and calculations necessary for the correction of SOLRAD 9 0.5–3 Å and 1–8 Å data and for the determination of the flare plasma temperature and emission measure are presented.

076.016 **The short-wave solar radiation spectrum at different activity levels.**
G. S. Ivanov-Kholodnyj, V. V. Firsov.
Geomagn. Aeronom., Vol. 14, 393 - 398 (1974). In Russian.

076.017 **Gamma ray observations during the August 1972 solar activity.**
E. L. Chupp, D. J. Forrest, A. N. Suri.
Astrophys. Space Sci. Library, Vol. 42, (see 012.016), 519 - 531 (1974).

076.018 **Time profiles and photon spectra of solar hard X-rays.** H. F. Van Beek, L. D. de Feiter, C. de Jager.
Astrophys. Space Sci. Library, Vol. 42, (see 012.016), 533 - 543 (1974).

076.019 **Is there a real need for doubling the solar EUV fluxes?** S. S. Prasad, D. R. Furman.
Journ. Geophys. Res., Vol. 79, 2463 - 2468 (1974).

The authors have examined the need for doubling the solar EUV flux, which has recently been suggested in the literature. They used two independent approaches: (1) comparison between electron production and loss using ion composition and density data and (2) comparison between observed and calculated electron fluxes. Their finding is that the reasonings so far advanced for doubling the solar EUV flux are not compelling.

076.020 **Ionization equilibrium in soft X-ray-emitting solar flares.**
K. J. H. Phillips, W. M. Neupert, R. J. Thomas.
Solar Physics, Vol. 36, 383 - 401 (1974).

Studies of the flare-produced line feature at 1.9 Å due to highly ionized iron show that it is emitted in conditions closely approximating steady-state ionization equilibrium. Calculations of the line flux per unit emission measure from time-dependent and steady-state ionization equilibria are compared with observed values during four flares in particular. Only for electron densities $N_e \lesssim 10^{10}$ cm^{-3} do the time-dependent equilibrium values give as good an approximation to the observed values as the steady-state equilibrium. This lower limit is compared with values of N_e derived from analyses of the temperature decline in each of these events, and with estimates of N_e given by other workers.

076.021 **A multithermal analysis of solar X-ray emission.**
K. P. Dere, D. M. Horan, R. W. Kreplin.
Solar Physics, Vol. 36, 459 - 472 (1974).

Data during quiet and flaring periods are analyzed and the general behavior of the differential emission measure during flares is presented. This analysis is based on experimental measurements of the efficiencies of the SOLRAD detectors.

076.022 **Hard solar flare X-ray bursts on 8 December 1970.**
K. A. Anderson, W. A. Mahoney.
Bull. American Astron. Soc., Vol. 6, 264 (1974). — Abstr. AAS.

076.023 **Solar X-ray bright points.**
A. Timothy, L. Golub, A. Krieger, J. K. Silk, G. Vaiana.
Bull. American Astron. Soc., Vol. 6, 265 (1974). — Abstr. AAS.

076.024 **An X-ray flare from Skylab: results and interpretation.**
G. S. Vaiana, S. Kahler, A. Krieger, R. Pallavicini, J. K. Silk.
Bull. American Astron. Soc., Vol. 6, 265 (1974). — Abstr. AAS.

076.025 **Coronal information from EUV disk spectral line intensities.** D. E. Billings, M. Alvarez.
Bull. American Astron. Soc., Vol. 6, 284 - 285 (1974). Abstr. AAS.

076.026 **Preliminary results of X-ray observations from an ATM support rocket.**
R. C. Catura, L. W. Acton, W. T. Zaumen.
Bull. American Astron. Soc., Vol. 6, 285 (1974). — Abstr. AAS.

076.027 **Observations of solar X-ray bursts in the energy range 5–15 keV.** D. Datlowe, H. Hudson.
Bull. American Astron. Soc., Vol. 6, 285 - 286 (1974). Abstr. AAS.

076.028 **SFD observations of EUV bursts during the ATM/Skylab missions.** R. F. Donnelly.
Bull. American Astron. Soc., Vol. 6, 286 (1974). — Abstr. AAS.

076.029 **Solar X-ray features and events.**
T. J. Janssens, G. A. Chapman, A. C. De Loach, D. L. McKenzie, J. E. Milligan, J. H. Underwood.
Bull. American Astron. Soc., Vol. 6, 289 (1974). — Abstr. AAS.

076.030 **Center-to-limb variation of impulsive solar X-ray bursts.** S. R. Kane.
Bull. American Astron. Soc., Vol. 6, 289 (1974). — Abstr. AAS.

076.031 **Force-free magnetic fields in the X-ray flare of 16 June 1973.** R. X. Meyer, E. B. Mayfield.
Bull. American Astron. Soc., Vol. 6, 291 (1974). — Abstr. AAS.

076.032 **The effect of partial frequency redistribution on the formation of the wings of Lyman-α.**
R. W. Milkey, D. Mihalas.
Bull. American Astron. Soc., Vol. 6, 291 - 292 (1974). Abstr. AAS.

076.033 **An atlas of the solar ultraviolet spectrum between 2220 and 2990 Å.**
E. F. Milone, W. P. Schneider, S. G. Tilford, R. Tousey.
Bull. American Astron. Soc., Vol. 6, 292 (1974). — Abstr. AAS.

076.034 **ATM observations of the time dependent intensity fluctuations in the extreme ultraviolet.**
J. E. Vernazza, P. K. Foukal, M. C. E. Huber, R. W. Noyes, E. M. Reeves, E. J. Schmahl, J. G. Timothy, G. L. Withbroe.
Bull. American Astron. Soc., Vol. 6, 296 (1974). — Abstr. AAS.

076.035 **Longitude dependence of hard and soft solar X-rays.**

J. Vorpahl.
Bull. American Astron. Soc., Vol. 6, 296 (1974). – Abstr. AAS.

076.036 Electron density from line ratios in the XUV spectrum of a solar flare.
K. G. Widing, J. D. Purcell, R. Tousey.
Bull. American Astron. Soc., Vol. 6, 297 (1974). – Abstr. AAS.

076.037 Extreme ultraviolet solar observations from the Harvard ATM experiment.
G. L. Withbroe, P. K. Foukal, M. C. E. Huber, R. W. Noyes, E. M. Reeves, E. J. Schmahl, J. G. Timothy, J. E. Vernazza.
Bull. American Astron. Soc., Vol. 6, 297 - 298 (1974). – Abstr. AAS.

076.038 The Lockheed OSO-I experiment: instrument capabilities and observing plans.
C. J. Wolfson, L. W. Acton, R. C. Catura.
Bull. American Astron. Soc., Vol. 6, 298 (1974). – Abstr. AAS.

076.039 Polarization of solar X-ray bursts.
S. L. Mandel'shtam.
Acta Phys. Acad. Sci. Hungaricae, Vol. 35, 47 - 50 (1974). In Russian.

The mechanism responsible for production of the recently discovered polarization of the solar X-ray bursts is considered theoretically in terms of interactions of a directional flux of accelerated electrons with the electrons and ions in a plasma.

076.040 Search for high energy gamma rays from the sun.
C. Y. Kim.
Canadian Journ. Phys., Vol. 52, 197 - 201 (1974).

An attempt to measure the flux of high energy solar gamma rays was made by comparing the angular distribution of the intensity from the direction of the sun and that of the intensity from the symmetrical direction about the zenith using two stacks of oriented nuclear emulsions flown by balloon on May 16, 1970, from Sioux Falls, South Dakota. An upper limit to the flux of solar gamma rays was estimated to be 4.3×10^{-4} photons cm^{-2} s^{-1} in the energy region above 10 MeV. Three importance-2 flares and five importance-1 flares were reported during the period of the balloon flight.

On the extreme ultraviolet emission spectrum of Fe XIII. See Abstr. 022.010.

Development and performance of a solar hard X-ray spectrometer. See Abstr. 034.053.

Programmi sperimentali di astronomia in raggi X. See Abstr. 034.073.

γ-ray observations from the OSO-3 satellite. See Abstr. 061.010.

X-ray satellite lines of hydrogenlike and heliumlike ions. See Abstr. 062.071.

A comparison of EUV spectroheliograms and photospheric magnetograms. See Abstr. 071.008.

Relative coronal abundances derived from X-ray observations. I. Sodium, magnesium, aluminum, silicon, sulfur, and argon. See Abstr. 074.025.

Solar coronal line profiles in the extreme-ultraviolet. See Abstr. 074.046.

On the account of nonlinear stabilization of beams when interpreting polarization X-ray observations. See Abstr. 074.074.

Coronal loops associated with active filaments. See Abstr. 074.091.

Non-relativistic solar electrons. See Abstr. 078.021.

Extraterrestrial solar spectrum, 3000–6100 Å at 1-Å intervals. See Abstr. 080.030.

Neutron propagation and 2.2 MeV gamma-ray line production in the solar atmosphere. See Abstr. 080.032.

Electron loss processes in the lower D-region during the decay of solar X-ray flare events. See Abstr. 083.020.

Local hydrogen gas and the background Lyman-alpha pattern. See Abstr. 106.009.

077 Solar Radio Radiation

077.001 Fine structure of the sun at 1.3 cm wavelength.
M. R. Kundu, T. Velusamy.
Solar Physics, Vol. 34, 125 - 131 (1974).

The two-element interferometer at Hat Creek Observatory was used at 1.3 cm wavelength to study the fine structure of the radio emissive regions on the sun. Observations of the quiet sun at 1.3 cm show sudden changes in the fringe amplitude and phase, lasting for typically about 5 – 8 min. Assuming that these events are identical in nature, a plot of peak amplitude vs the projected baseline at the time of the event suggests emission from a region of angular size of about 10".

077.002 Coupling of microwaves at a selected solar active centre. E. Scalise, Jr., P. Kaufmann.
Solar Physics, Vol. 34, 189 - 191 (1974). — Research note.

077.003 On the temporal structure of solar radio bursts of types I and III. G. F. Eliseev, P. V. Panov.
Solnechnye Dannye 1973 Byull., No. 12, p. 104 - 107 (1974). In Russian.

Records of bursts of types I and III with time resolution of 1 sec have been treated. The characteristic periods of the intensity variations are shown to be 2 sec for the bursts of type I and 3–4 sec for those of type III. In supposition of the correspondence between temporal and space scales it is concluded on a difference of characteristic sizes in the regions of the bursts generation.

077.004 The relevance of bipolar type I storm structures to the theory of mode coupling in the solar corona.
D. B. Melrose.
Proc. Astron. Soc. Australia, Vol. 2, 208 - 211 (1973). — Presented at the annual general meeting of the Astron. Soc. Australia — see 012.001.

077.005 The position of a type I storm source in the magnetic field of an active region.
G. A. Dulk, G. J. Nelson.
Proc. Astron. Soc. Australia, Vol. 2, 211 - 214 (1973). — Presented at the annual general meeting of the Astron. Soc. Australia — see 012.001.

077.006 A theory of type I solar radio bursts. W. N.-C. Sy.
Proc. Astron. Soc. Australia, Vol. 2, 215 - 217 (1973). — Presented at the annual general meeting of the Astron. Soc. Australia — see 012.001.

077.007 Evolutionary relation between type I storm activity and the S-component at centimetre wavelengths.
K. Kai, H. Sekiguchi.
Proc. Astron. Soc. Australia, Vol. 2, 217 - 219 (1973). — Presented at the annual general meeting of the Astron. Soc. Australia — see 012.001.

077.008 Unusual absorption of a solar type II burst by 'shadow' type III bursts. K. Kai.
Proc. Astron. Soc. Australia, Vol. 2, 219 - 222 (1973). — Presented at the annual general meeting of the Astron. Soc. Australia — see 012.001.

077.009 A moving radio burst on the limb of the sun observed at 80 and 160 MHz. D. J. McLean.
Proc. Astron. Soc. Australia, Vol. 2, 222 - 226 (1973). — Presented at the annual general meeting of the Astron. Soc. Australia — see 012.001.

077.010 On quasi-periodic components with periods from 30 to 60 min in the spectra of fluctuations of solar radio emission at 3 cm wavelength.
V. I. Aleshin, M. M. Kobrin, A. I. Korshunov.
Izv. vyssh. ucheb. zavedenij. Radiofizika, Vol. 16, 747 - 753 (1973). In Russian. — Abstr. in Referativ. Zhurn. 51. Astron., 3.51.444 (1974).

077.011 A new height estimate of local sources of the slowly varying component of solar radio emission at 3.2 cm wavelength.
V. N. Borovik, G. B. Gel'frejkh, B. I. Lubyshev.
Izv. vyssh. ucheb. zavedenij. Radiofizika, Vol. 16, 731 - 736 (1973). In Russian. — Abstr. in Referativ. Zhurn. 51.Astron., 3.51.467 (1974).

077.012 A 10.5 year period for the slowly varying component of the solar radio flux. A. E. Covington.
Journ. Roy. Astron. Soc. Canada, Vol. 68, 31 - 36 (1974).

A basic period of 10.5 years has been found from the Ottawa 10.7 cm daily radio flux observations and is identical to that found in recent studies of the more extensive sunspot numbers. Assuming this period, it is predicted that the next radio minimum will occur in January 1975 with an error of ± one solar rotation.

077.013 Generation of type III radio bursts by electron streams with large time of injection.
V. V. Zajtsev, M. V. Kunilov, N. A. Mityakov, V. O. Rapoport.
Astron. Zhurn. Akad. Nauk SSSR, Vol. 51, 252 - 260 (1974). In Russian. English translation in Soviet Astron. AJ, Vol. 18, No. 2.

Analysis is made on the generation of type III bursts by electron streams in the case when the time of injection of hot electrons from the flare region is considerably larger than the time of existence of a burst at the given fixed frequency. The plasma wave generation occurs due to escaping of fast electrons at the forward front of the stream. A satisfactory agreement of calculated curves of the time variation of type III bursts with experimental ones for meter and decameter wavelengths is observed.

077.014 On the nature of type I bursts in the sporadic solar radiation. G. M. Vereshkov.
Astron. Zhurn. Akad. Nauk SSSR, Vol. 51, 261 - 269 (1974). In Russian. English translation in Soviet Astron. AJ, Vol. 18, No. 2.

The phenomenon of the source of type I is the result of conversion of Langmuir and acoustic waves to electromagnetic waves. Langmuirs waves are generated by the instability on the front of Alfvén waves. The conversion takes place when packets of Alfvén and acoustic waves are crossed.

077.015 Towards a theory for type III solar radio bursts. II: The radiation source. D. F. Smith.
Solar Physics, Vol. 34, 393 - 411 (1974).

The mechanisms for the transformation of plasma waves into radiation near the fundamental and second harmonic of the plasma frequency are reviewed and equations are given for both the emission and absorption coefficients for these mechanisms. These results are applied to construct models of the radiation source for type III solar radio bursts both at high frequencies where the fundamental is dominant and at low frequencies where the second harmonic is dominant using two model plasma wave spectra, one being one-dimensional, the other isotropic.

077.016 **Kilometer-wave type III burst: Harmonic emission revealed by direction and time of arrival.**
H. Alvarez, F. T. Haddock, W. H. Potter.
Solar Physics, Vol. 34, 413 - 420 (1974).

A type III solar burst was observed at seven frequencies between 3.5 MHz and 80 kHz by the Michigan experiment aboard the IMP-6 satellite. From the data the authors determine burst direction-of-arrival as well as time-of-arrival. They predict these quantities using simple models whose parameters are varied to obtain a good fit to the observations. They find that between 3.5 MHz and 230 kHz the observed radiation was emitted at the fundamental of the local plasma frequency while below 230 kHz it was emitted at the second harmonic. The exciter particles that produced the burst onset and burst peak have velocities of 0.27 and 0.12, respectively, in units of the velocity of light.

077.017 **On a suggested explanation for fine structures observed in some wide-band bursts.** W. N.-C. Sy.
Solar Physics, Vol. 34, 427 - 431 (1974).

Plasma radiation arising from non-linear coupling of upper hybrid plasma waves and Bernstein waves is examined as a possible explanation (Rosenberg 1972) for the fine structures observed in some wide-band solar radio bursts. This radiation is found to be weakly polarized and therefore it cannot account for the fine structures, which are strongly polarized. The difficulties encountered in explaining several bands of radiation at comparable intensities are indicated.

077.018 **Decameter type IV bursts associated with coronal transients.** T. E. Gergely, M. R. Kundu.
Solar Physics, Vol. 34, 433 - 446 (1974).

The characteristics of four moving type IV bursts, observed with the 65–20 MHz swept-frequency interferometer of the Clark Lake Radio Observatory are discussed. All four bursts were associated with depletions in the electron content of the white light corona.

077.019 **Motion of the sources for type II and type IV radio bursts and flare-associated interplanetary disturbances.** K. Sakurai, J. K. Chao.
Journ. Geophys. Res., Vol. 79, 661 - 664 (1974).

Shock waves are indirectly observed as the source of type II radio bursts, whereas magnetic bottles are identified as the source of moving metric type IV radio bursts. The difference between the expansion speeds of these waves and bottles is examined during their generation and propagation near the flare regions. It is shown that, although generated in the explosive phase of flares, the bottles behave quite differently from the waves and that the bottles are generally much slower than the waves. It is shown that the transit times of disturbances between the sun and the earth give information about the deceleration of shock waves to their local speeds observed near the earth's orbit. The relationship among magnetic bottles, shock waves near the sun, and flare-associated disturbances in interplanetary space is briefly discussed.

077.020 **The excitation and decay of low frequency type III solar radio bursts.**
H. M. Bradford, V. A. Hughes.
Astron. Astrophys., Vol. 31, 419 - 429 (1974).

The authors have analyzed the dynamic spectra of a number of type III solar radio bursts obtained from Alouette I satellite records in the nominal frequency range of 2 MHz to 8 MHz. An objective new method of analysis was employed which utilized the measured moments of the shapes of the bursts up to third order. The received flux was represented as a convolution of functions which mainly represent the excitation and exponential decay of the plasma waves. Deconvolution with the aid of the measured moments yielded an upper limit on the decay time and a measure of the duration of the excitation.

077.021 **Fine structure of the radio spectra of early phase of plasma instability during the solar flare August 4, 1972.** L. Křivský, A. Tlamicha.
Bull. Astron. Inst. Czechoslovakia, Vol. 25, 126 - 127 (1974). Short communication.

077.022 **Measurement of the fluctuations of the solar radio radiation flux at 3 cm.**
A. A. Bezotosnyj, O. G. Gontarev, E. F. Rizov.
Ionosphere and solar-terrestrial relations, (see 003.003), p. 132 - 136 (1972). In Russian.

077.023 **Fluctuations of solar radio emission at 200 MHz observed at IZMIRAN on August, 4, 1972.**
A. A. Gnezdilov.
Astron. Tsirk., No. 807, p. 3 - 5 (1974). In Russian.

077.024 **Polarization fine structure in solar radio bursts of type III on short meter wavelengths.** C. Slottje.
Astron. Astrophys., Vol. 32, 107 - 110 (1974).

Accurately calibrated measurements with the 60-Channels Solar Radiospectrograph, now operated at Dwingeloo between 315 and 200 MHz, led to the discovery of fine structure in the time profile of the degree of polarization of some type III bursts. A more sophisticated model enables to deduce from the observed polarization profile and frequency drift rate, consistent values for the magnetic field strength in the source, the density gradient of the ambient plasma and the exciter speed.

077.025 **A search for 5 min periodic structure in solar 2 cm emission.** D. D. Sentman, S. D. Shawhan.
Solar Physics, Vol. 35, 83 - 103 (1974).

Two hundred and eighty-five hours of solar data obtained from the University of Iowa 2 cm radiometer during 1968–1969 were analyzed for evidence of periodic structure related to the 5 min periodic chromospheric oscillations detected in optical line emission. A power spectral analysis of the data failed to show any statistically significant periodic activity in the frequency range 1–15 MHz (periods of 1–16 min) for data organized according to solar activity in Hα, soft solar X-rays (2–12 Å), and several microwave frequencies (3–15 GHz). A small shift in power from low to higher frequencies in the power spectrum of the 2 cm data was found to be correlated with Hα and X-ray activity. Consistent statistical analyses of previous works reporting evidence for oscillations at microwave and extreme-ultraviolet frequencies indicate that confidence in these previous results is marginal.

077.026 **On the role of the magnetic configuration of flares for production of type III solar radio bursts.**
F. Axisa.
Solar Physics, Vol. 35, 207 - 224 (1974).

This paper investigates the possibility that the particular location of flare production sites in an active region is intimately connected to the production of type III radio bursts as well as centimetric and hard X-ray events.

077.027 **Landau damping of type III bursts by the interplanetary electrons of high energy.** M. G. Aubier.
Astron. Astrophys., Vol. 32, 141 - 146 (1974).

From the in situ measurements of electron flux the author derives the solar wind electron velocity distribution at 1 A.U. This distribution is very different from a maxwellian, even during solar quietest times. He extrapolates these results to regions closer to the sun. The conclusions depend on the unknown variation of the high energy tail of the electron distribution with distance from the sun, but it is likely that due to this tail, the Landau damping is extremely efficient at altitudes above about 2 $R_\odot$.

077.028 Die ersten Sonnenbeobachtungen mit dem Effelsberger 100-m-Teleskop bei 2.8 cm Wellenlänge.
E. Fürst, W. Hirth.
Mitt. Astron. Ges., No. 34, p. 102 - 106 (1973/74).
Presented at the "Wissenschaftliche Tagung der Astron. Ges., Oberkochen, 1973 April 24 - 27".

077.029 Zur Typ III-Burstaktivität solarer Aktivitätsregionen.
H. Urbarz.
Mitt. Astron. Ges., No. 34, p. 109 - 110 (1973/74).
Presented at the "Wissenschaftliche Tagung der Astron. Ges., Oberkochen, 1973 April 24 - 27".

077.030 Solar radio flux measurements on wavelengths of 37 cm, 56 cm and 115 cm.
A. Tlamicha, O. Kepka.
Bull. Astron. Inst. Czechoslovakia, Vol. 25, 160 - 162 (1974).

The reconstruction of the solar radio receivers at the Ondřejov Observatory is described. The determination of the value of the total flux from the spectral curve and the daily values of the solar flux in the period February 1973–November 1973 are presented.

077.031 Quasi-periodic pulsations at decimetric wavelengths during the solar radio events of August 4, 1972.
A. Tlamicha.
Bull. Astron. Inst. Czechoslovakia, Vol. 25, 168 - 170 (1974).

Observations of quasi-periodic pulsations during the large solar activity of August 4, 1972 made with the solar radio spectrograph at the Ondřejov Observatory are presented. The photometrical method has been used to analyse the periodicity of the pulsations of the bursts. The histogram shows a principal period of 0.3 sec.

077.032 The relationship between the slowly varying component of solar radio emission and large scale photospheric magnetic field patterns.
P. H. Scherrer, M. El-Raey.
Solar Physics, Vol. 35, 361 - 368 (1974).

Daily solar radio flux observations have been examined for a relationship to the large-scale photospheric magnetic field structure. Interplanetary magnetic field sector boundaries were used to indicate boundaries between photospheric field regions of opposite polarity. An enhancement in emission was found about four days before the boundary central meridian passage. Most of the effect came from emission near toward-to-away type boundaries. A higher level of emission appears to be associated with toward field regions than with away field regions.

077.033 A relationship between the brightness temperatures for type III bursts. D. B. Melrose.
Solar Physics, Vol. 35, 441 - 450 (1974).

It is widely accepted that the emission in type III solar radio bursts is produced by Langmuir waves, emission at the fundamental frequency arising when these waves are scattered by thermal ions and emission at the second harmonic when two Langmuir waves coalesce. In this paper the theory of these processes is used to derive an inequality on the energy density in Langmuir waves and an inequality on the expected brightness temperature at the second harmonic. Furthermore, the theory is used to derive a functional relation between the brightness temperature at the plasma frequency and the brightness temperature at the second harmonic for given plasma frequency.

077.034 Evidence of solar bursts directivity at 169 MHz from simultaneous ground based and deep space observations (STEREO-1 preliminary results).
C. Caroubalos, J. L. Steinberg.
Astron. Astrophys., Vol. 32, 245 - 253 (1974).

The STEREO-1 experiment was designed to detect and measure the directivity of solar burst radiation at 169 MHz by simultaneous observations from Nançay (France) and the Soviet planetary probe Mars-3. The equipments are described, the data reduction procedure is analyzed in detail and the errors involved evaluated. A preliminary account of the observations is given.

077.035 The directivity of type III bursts.
C. Caroubalos, M. Poquérusse, J. L. Steinberg.
Astron. Astrophys., Vol. 32, 255 - 267 (1974).

Simultaneous observations of type III bursts at 169 MHz from the earth and the Soviet planetary probe Mars-3 are reported. The authors compare the time profiles of the same event recorded on the Mars-3 probe and on the earth. Emphasis is placed on the observation of a limb source on November 14, 1971; a detailed analysis of these events is carried out and it is shown that some of them are pairs of fundamental and harmonic components when most of the others are harmonics. The time profile and the directivity measurements of disc and limb bursts are compared to theoretical predictions from ray-tracing computations in an inhomogeneous, spherical corona.

077.036 Solar bursts at $\lambda = 2$ cm on July 31, 1972.
J. P. Castelli, D. A. Guidice, D. J. Forrest, R. R. Babcock.
Journ. Geophys. Res., Vol. 79, 889 - 894 (1974).

On July 31, 1972, McMath plage 11976 was the site of many moderate-intensity solar radio bursts with unusual radio spectra, emission being limited to only short-centimeter (and possibly millimeter) wavelengths. This same plage region was responsible for the historic solar geophysical events of early August 1972. Most of the July 31 bursts observed were impulsive, with time profiles similar to those of the X-ray emission in the 7.5- to 15-keV and 15- to 30-keV ranges. From the radio spectral characteristics a number of concepts concerning conditions within the burst region are considered and found to present difficulties. Some general conclusions on the burst region properties are reached.

077.037 Solar radio emission. Outstanding events at 9285, 2830 and 1420 MHz, January–December 1970.
P. Arena, P. Patriarchi.
Osservazioni Mem. Oss. Astrofis. Arcetri, Fasc. 101, 49 pp. (1973).

077.038 Solar radio emission. Outstanding events at 9285, 2830 and 1420 MHz, January–December 1971.
P. Arena, P. Patriarchi.
Osservazioni Mem. Oss. Astrofis. Arcetri, Fasc. 103, 29 pp. (1974).

077.039 A coherent radiation mechanism for type IV dm radio bursts. J. Kuijpers.
Solar Physics, Vol. 36, 157 - 169 (1974).

An interpretation is presented of the decimetric type IV continuum with fine structure on March 6, 1972 and of the corresponding source region, in terms of Čerenkov plasma radiation and alternatively of synchrotron radiation, both in case of coherent and incoherent generation.

077.040 Solar longitude dependence of some characteristics of type III radio bursts from metric to hectometric wavelengths. K. Sakurai.
Solar Physics, Vol. 36, 171 - 178 (1974).

Using the observed data for metric and hectometric type III radio bursts, the dependence of burst characteristics on the solar longitude has been examined over a wide frequency range. It is found that there exists an east–west asymmetry for the extension of metric type III bursts into hectometric wavelength range.

077.041 Meter and decameter wavelength positions of solar bursts of July 31–August 7, 1972.
M. R. Kundu, W. C. Erickson.
Solar Physics, Vol. 36, 179 - 189 (1974).

The positional analysis of solar bursts at meter and decameter wavelengths observed during the period July 31–August 7, 1972 is presented. Most of the activity during this period was associated with the active regions McMath 11976 and 11970. Except near the CMP of region 11976, two regions of continuum emission were observed – one a relatively smooth continuum and the other a continuum superimposed with many type III's and other fine structure.

077.042 Preliminary results of the spectrum of fluctuations of solar radio emission with nearly continuous covering of the daily interval. A. B. Berlin, G. B. Gel'frejkh, V. G. Zandanov, A. N. Korzhavin, L. E. Treskova.
Izv. vyssh. ucheb. zavedenij. Radiofizika, Vol. 16, 1366 - 1368 (1973). In Russian. – Abstr. in Referativ. Zhurn. 51. Astron., 5.51.390 (1974).

077.043 A darkening ring in the solar radio brightness distribution at 3 cm wavelength (polar region).
I. I. Berulis, A. S. Grebinskij, O. V. Korobchuk, N. G. Franchuk, L. V. Yasnov.
Izv. vyssh. ucheb. zavedenij. Radiofizika, Vol. 16, 1375 - 1378 (1973). In Russian. – Abstr. in Referativ. Zhurn. 51. Astron., 5.51.396 (1974).

077.044 The problem of reconstruction of radio brightness distribution on the sun.
S. D. Kremenetskij, L. M. Risover, G. Ya. Smol'kov.
Izv. vyssh. ucheb. zavedenij. Radiofizika, Vol. 16, 1369 - 1374 (1973). In Russian. – Abstr. in Referativ. Zhurn. 51. Astron., 5.51.441 (1974).

077.045 Analysis of solar radio emission fluctuations at 3 cm with two radio telescopes spaced on 1500 km.
M. M. Kobrin, V. V. Pakhomov, M. S. Durasova, B. V. Timofeev, N. A. Prokof'eva, E. I. Lebedev, G. A. Lavrinov.
Izv. vyssh. ucheb. zavedenij. Radiofizika, Vol. 16, 1350 - 1353 (1973). In Russian. – Abstr. in Referativ. Zhurn. 51. Astron., 5.51.446 (1974).

077.046 The spectral index and fluctuations of solar radio emission at ~ 3 cm wavelength.
I. I. Berulis, A. P. Molchanov, V. P. Olyanyuk, I. E. Pogodin, O. Ya. Pudov, N. G. Franchuk, L. V. Yasnov.
Izv. vyssh. ucheb. zavedenij. Radiofizika, Vol. 16, 1362 - 1365 (1973). In Russian. – Abstr. in Referativ. Zhurn. 51. Astron., 5.51.447 (1974).

077.047 Regions with lower radio brightness on the sun from observations in the 8-millimeter wavelength range.
G. P. Apushkinskij, A. N. Tsyganov.
Izv. vyssh. ucheb. zavedenij. Radiofizika, Vol. 16, 1379 - 1382 (1973). In Russian. – Abstr. in Referativ. Zhurn. 51. Astron., 5.51.448 (1974).

077.048 Statistical investigation of characteristics of the slowly varying component of solar radio emission at 9 cm wavelength. Sh. B. Akhmedov.
Izv. vyssh. ucheb. zavedenij. Radiofizika, Vol. 16, 1388 - 1394 (1973). In Russian. – Abstr. in Referativ. Zhurn. 51. Astron., 5.51.449 (1974).

077.049 Study of the determined component of the total flux of solar radio emission.
S. A. Andrianov, L. V. Yasnov.
Izv. vyssh. ucheb. zavedenij. Radiofizika, Vol. 16, 1383 - 1387 (1973). In Russian. – Abstr. in Referativ. Zhurn. 51. Astron., 5.51.460 (1974).

077.050 The radio emission spectrum of the quiet sun and its connection with the solar activity cycle.
G. F. Eliseev, L. F. Lazareva.
Izv. vyssh. ucheb. zavedenij. Radiofizika, Vol. 16, 1395 - 1397 (1973). In Russian. – Abstr. in Referativ. Zhurn. 51. Astron., 5.51.461 (1974).

077.051 Type III bursts.
Solar Radio Group Utrecht.
Space Sci. Rev., Vol. 16, 45 - 89 (1974).

One of the purposes of this paper is to stress the fact that the solar type III radio burst is not the simple phenomenon it is sometimes suggested to be. Therefore the authors present its various properties derived from observations of the past 20 years. A few theories invoked to describe the basic features are reviewed.

077.052 Type-III radio bursts and their interpretation.
D. F. Smith.
Space Sci. Rev., Vol. 16, 91 - 144 (1974).

The observations of type-III solar radio bursts are briefly reviewed to set requirements on a model for their interpretation. The most important of these requirements is that the source must be an electron stream which is in a state of continuous quasilinear relaxation and which initially must have a nearly monotonically decreasing velocity distribution. Progress on a model for the plasma wave source is reviewed and it is concluded that no existing models are adequate. Progress on the radiation source is considered both in the absence and presence of a magnetic field. Calculations of scattering of radiation in a random medium are reviewed. It is concluded that these are adequate at high and low frequencies, but have not been carried out properly at intermediate frequencies where amplification of the fundamental may still be present.

077.053 Satellite observations of type III solar radio bursts at low frequencies.
J. Fainberg, R. G. Stone.
Space Sci. Rev., Vol. 16, 145 - 188 (1974).

Type III solar radio bursts have been observed from 10 MHz to 10 kHz by satellite experiments above the terrestrial plasmasphere. Solar radio emission in this frequency range results from excitation of the interplanetary plasma by energetic particles propagating outward along open field lines over distances from $5 R_\odot$ to at least 1 AU from the sun. This review summarizes the morphology, characteristics and analysis of individual as well as storms of bursts.

077.054 The mechanism responsible for 'shadow' type III solar radio bursts. I. Absorption due to Langmuir turbulence. D. B. Melrose.
Australian Journ. Phys., Vol. 27, 259 - 269 (1974).

The hypothesis is advanced that for 'shadow' type III solar radio events the absorption mechanism involves Langmuir turbulence, such absorption being the inverse of either fundamental or second harmonic plasma emission. The theory for both absorption processes is developed and applied to shadow type III events.

077.055 The mechanism responsible for 'shadow' type III solar radio bursts. II. Absorption due to ion sound turbulence. D. B. Melrose.
Australian Journ. Phys., Vol. 27, 271 - 277 (1974).

The hypothesis is explored that ion sound turbulence generated by the exciting agency for type III bursts is responsible for shadow type III events. The possible absorption mechanisms are listed: the most favourable are the coalescence of transverse waves and ion sound waves into Langmuir waves or the decay of transverse waves into Langmuir waves and ion sound waves.

077.056 **Stabilization of electron streams in type III solar radio bursts.**
K. Papadopoulos, M. L. Goldstein, R. A. Smith.
Astrophys. Journ., Vol. 190, 175 - 185 (1974).

The authors show that the electron streams that give rise to type III solar radio bursts are stable and will not be decelerated while propagating out of the solar corona.

077.057 **Nonlinear wave coupling in type IV solar radio bursts.** C. Chiuderi, R. Giachetti, H. Rosenberg.
Utrechtse Sterrekundige Overdrukken, No. 237, 21 pp. (1973).

In order to explain a fine structure of parallel ridges in stationary type IV continua, the emission due to the coupling of electrostatic upper hybrid waves and Bernstein waves at the sum frequency of the upper hybrid and harmonics of the gyrofrequency has been calculated. If the energy density of these electrostatic waves is more than 10 times the thermal fluctuation energy density then the observed zebra pattern can be emitted by a region with diameter of $\sim 10^3$ km.

077.058 **On the connection of fluctuations of noise storm continuum and burst activity of the sun in cm and dm ranges.** A. A. Gnezdilov.
Astron. Zhurn. Akad. Nauk SSSR, Vol. 51, 565 - 570 (1974). In Russian. English Translation in Soviet Astron. AJ, Vol. 18, No. 3.

A comparison between continuum fluctuation power spectra for three noise storms and burst activity within the range 9000 - 650 MHz was performed. A correlation between additional fluctuation occurrence and the increase of burst number as well as the growth of burst maximum was revealed.

077.059 **Solar radio emissions.** T. Prokakis.
In honorem S. Placidis, (see 003.009), p. 341 - 354 (1974). In Greek.

077.060 **High resolution interferometry of the sun at 3.7 cm wavelength.** K. R. Lang.
Solar Physics, Vol. 36, 351 - 367 (1974).

Interferometric observations of the sun at a wavelength $\lambda = 3.7$ cm and an effective angular resolution of $7''$ and $15''$ are presented.

077.061 **1.0 arc second structure on the sun at 3.71 cm wavelength.** R. W. Hobbs, S. D. Jordan, W. J. Webster, Jr., S. P. Maran, H. M. Caulk.
Solar Physics, Vol. 36, 369 - 370 (1974). – Research note.

077.062 **The slowly varying component of solar meter wavelength radiation: a non-thermal radio source.**
A. C. Riddle.
Solar Physics, Vol. 36, 375 - 381 (1974).

In this study, in which scattering is included for the first time, it is shown that scattering may lead to lower emission from density enhancements rather than higher emission as predicted by models in which refraction alone is considered. This strongly suggests that the emission observed at meter wavelengths is of non-thermal origin.

077.063 **Type II radio bursts and particle acceleration.**
Z. Švestka, L. Fritzová-Švestková.
Solar Physics, Vol. 36, 417 - 431 = Mitt. Fraunhofer Inst., *Freiburg*, No. 125 (1974).

328 particle events recorded during 30 months from January 1, 1966 to June 30, 1968 are compared with the occurrence of 166 type II radio bursts during the same period.

The results of this comparison give a convincing evidence that proton acceleration to higher energies in flares is closely connected with the type II burst occurrence. A detailed analysis indicates that we may need even three different acceleration mechanisms in flares.

077.064 **A clue to the trigger for both the type III solar radio burst and the solar flare.**
E. R. Priest, J. Heyvaerts.
Solar Physics, Vol. 36, 433 - 442 (1974).

The purpose of this note is to give a theoretical model for so-called 'neutral line absorbing features', which are situated in the solar chromosphere or lower corona and appear to be associated with solar type III radio bursts.

077.065 **On the third harmonic in solar radio bursts.**
V. V. Zheleznyakov, E. Ya. Zlotnik.
Solar Physics, Vol. 36, 443 - 449 (1974).

Possible mechanisms of the third harmonic generation in solar radio bursts are investigated. It is shown that the most essential is the coupling of plasma waves in a source with electromagnetic radiation at the second harmonic. This phenomenon may be responsible for the occurrence of the third harmonic in the type V event on 25 October 1972 recorded by Benz, as well as in the U-burst on 3 September 1957 reported by Haddock and Takakura.

077.066 **The third harmonic of type III solar radio bursts.**
T. Takakura, S. Yousef.
Solar Physics, Vol. 36, 451 - 458 (1974).

Two examples of pairs of 'J-shaped' type III bursts with 2:3 frequency ratio are given. It is suggested that these could be the second and the third harmonics of undetected fundamental radiation. The 80 MHz heliograph source of the third harmonic showed an apparent brightness temperature of 10^9 K, while the order of 10^{16} K seems to be required theoretically. This may imply that the apparent radio source was composed of many unresolved small sources of much higher brightness temperature.

077.067 **Radio astronomy on decameter wavelengths at Meudon and Nançay Observatories.** (Report from Solar Institute). A. Boischot.
Solar Physics, Vol. 36, 517 - 522 (1974).

The note describes the instruments which were operational at the end of 1973, and gives a brief survey of the main topics which have been studied up to now.

077.068 **On an outstanding event on the sun from observations at 4.25 mm on May 3, 1973.**
F. A. Grigoryan, A. G. Kislyakov, N. R. Khachatryan, V. I. Chernyshev.
Solnechnye Dannye 1974 Byull., No. 3, p. 89 - 93 (1974). In Russian.

The main characteristics of the outstanding event observed at 4.25 mm wavelength on May 3, 1973, are given. An attempt is made for obtaining the instantaneous spectrum of the radio emission intensity of the solar flare region connected with it.

077.069 **Decameter type IV bursts associated with coronal transients.** T. E. Gergely, M. R. Kundu.
Bull. American Astron. Soc., Vol. 6, 218 - 219 (1974). Abstr. AAS.

077.070 **Long base line interferometry of the sun at 3.7 and 11.1 cm wavelengths.**
M. R. Kundu, C. Alissandrakis, R. H. Becker.
Bull. American Astron. Soc., Vol. 6, 219 (1974). –Abstr. AAS.

077.071 **Characteristics of type-III bursts observed at km-λ**

wavelengths from IMP-6.
H. Alvarez, F. T. Haddock, W. H. Potter.
Bull. American Astron. Soc., Vol. 6, 284 (1974). – Abstr. AAS.

077.072 Polarization spectra of type III radio bursts.
J. C. Dodge.
Bull. American Astron. Soc., Vol. 6, 286 (1974). – Abstr. AAS.

077.073 High-resolution radio maps of solar active regions at a wavelength of 2.8 centimeters.
C. J. Grebenkemper, K. M. Price.
Bull. American Astron. Soc., Vol. 6, 287 (1974). – Abstr. AAS.

077.074 Brightness temperature of the sun at 8.6 millimeters – presence of an absorption feature?
W. Henze, M. Bleiweiss, F. Wefer.
Bull. American Astron. Soc., Vol. 6, 288 (1974). – Abstr. AAS.

077.075 The slowly varying component of solar meter wavelength radiation; a non-thermal radio source.
A. C. Riddle.
Bull. American Astron. Soc., Vol. 6, 293 (1974). – Abstr. AAS.

077.076 Relations between quiet sun radio observations and models of the solar atmosphere.
F. I. Shimabukuro.
Bull. American Astron. Soc., Vol. 6, 294 (1974). – Abstr. AAS.

077.077 Millimeter emission related to coronal magnetic fields. K. P. White III.
Bull. American Astron. Soc., Vol. 6, 297 (1974). – Abstr. AAS.

077.078 3.3-millimeter observations of the June 30, 1973 total solar eclipse.
W. J. Wilson, F. I. Shimabukuro, T. T. Mori.
Bull. American Astron. Soc., Vol. 6, 297 (1974). – Abstr. AAS.

077.079 Observations of solar radio emission at 146 MHz. Instrumental and data analysis. X. Liu, X. He.
Acta Astron. Sinica, Vol. 15, 61 - 72 (1974). In Chinese.

The 16-dish E-W interferometer of the Mi-Yun Station, Peking Observatory (working frequency 146 MHz) is described. Measured instrumental parameters are compared with theoretical ones. The observations of the quiet sun and the slowly varying component are described. The authors found that the apparent diameter of the meter-wave sun had not varied during the descending phase of the solar cycle. The results of measurements of type-I sources are analyzed. According to the rate of occurrence of type-I bursts, the sources are classified into four classes.

077.080 A statistical study of 5480 microwave solar radio events. F. L. Wefer.
Pennsylvania State Univ. Coll. Sci. Dep. Astron. Radio Astron. Obs. Sci. Rep. No. 026, 302 pp. (1973).

Uses the 5480 solar microwave events observed at Psurao at 10.7 GHz, 2.7 GHz and 960 MHz to draw statistical conclusions concerning burst characteristics and relationship to solar activity. – RDR

Fine structure of a solar active region at 3.7 and 11.1 cm wavelengths. See Abstr. 073.005.

Fine structure of a solar flare region at 3.7 and 11.1 cm wavelengths. See Abstr. 073.008.

Correlation of a flare-wave and type II burst. See Abstr. 073.055.

An attempt of experimental discovery of a correlation of chromospheric height variations with fluctuations of the 3-cm solar radio emission flux. See Abstr. 073.056.

Effects of turbulence anisotropy on propagation and electromagnetic radiation of particle streams in the solar corona. See Abstr. 074.003.

Determination of Faraday rotation occurring between the burst-source in the solar corona and the earth. See Abstr. 074.004.

On the determination of coronal temperature from the decay of type III radio bursts. See Abstr. 074.013.

Ergebnisse von zwei Jahren Korona-Forschung mit dem Radio-Heliographen von Culgoora. See Abstr. 074.023.

Comments on 'Neutralization and stabilization of particle streams in the corona and type III radio bursts' by Dean F. Smith. See Abstr. 074.026.

On the observation of scattered radio emission from sources in the solar corona. See Abstr. 074.042.

The coronal disturbance of 1972, August 12. See Abstr. 074.043.

Coronal magnetic field structure derived from two-frequency radioheliograph observations. See Abstr. 074.044.

Mode coupling in the solar corona. I. Coupling near the plasma level. See Abstr. 074.047.

Solar wind magnetic field structure from radioburst analysis. See Abstr. 074.098.

Solare Beobachtungsergebnisse (Solar Data). See Abstr. 075.011.

Non-relativistic solar electrons. See Abstr. 078.021.

Errata

077.901 Erratum: 'Kilometer-wave type III burst: harmonic emission revealed by direction and time of arrival' [Solar Physics, Vol. 34, 413 - 420 (1974)].
H. Alvarez, F. T. Haddock, W. H. Potter.
Solar Physics, Vol. 36, 534 (1974).

077.902 Errata: 'Polarization observations of solar active regions at 7.875 GHz' [Bull. American Astron. Soc., Vol. 5, 433 (1973)].
M. D. Papagiannis, K. K. Arora, J. Kogut, R. M. Straka.
Bull. American Astron. Soc., Vol. 6, 268 (1974). – Abstr. AAS.

078 Solar Cosmic Radiation

078.001 **Solar cosmic-ray acceleration by a plasma instability.**
S. P. Gary.
Astrophys. Journ., Vol. 187, 195 - 196 (1974).

Charged particles which are trapped in the potential troughs of an electrostatic plasma wave can be accelerated if the phase velocity of the wave increases with position. Energetic protons from solar flares may be accelerated in this way by the beam-plasma instability.

078.002 **An experiment for investigation of electrons of solar origin aboard the artificial satellites Prognoz.**
N. L. Grigorov, V. G. Kurt, Yu. I. Logachev, V. N. Lutsenko, O. B. Likin, N. S. Nikolaeva, N. F. Pisarenko, I. A. Savenko, N. Yu. Svechnikov, I. P. Shestopalov.
Kosmich. Issled., Vol. 12, 67 - 73 (1974). In Russian.

078.003 **Measurements of the proton and α-components of plasma aboard Prognoz during the period of high solar activity (August 1972).** V. V. Temnyj, A. A. Zertsalov, O. L. Vajsberg, Yu. E. Berezin.
Kosmich. Issled., Vol. 12, 74 - 79 (1974). In Russian.

078.004 **Tracing of high-latitude magnetic field lines by solar particles.** G. A. Paulikas.
Rev. Geophys. Space Phys., Vol. 12, 117 - 128 (1974).

Recent measurements of solar particles in the energy interval between hundreds of keV and a few MeV have shown that a direct connection exists between a portion of the high-latitude geomagnetic field and the interplanetary magnetic field. The purpose of this paper is to review the status of knowledge regarding this connection.

078.005 **Low-energy solar cosmic rays and chromospheric flares.** L. S. Levitskij.
Izv. Krymskoj Astrofiz. Obs., Vol. 49, 36 - 41 (1974). In Russian.

The frequency distribution of the amplitudes of PCA events is obtained for the years 1965 - 1969 (new solar cycle). A connection between solar proton flux and solar flare importance was studied. It is concluded that the generation of solar cosmic rays with energy ≥ 10 MeV takes place for all flares with importance ≥ 1. There are no physical differences between proton and non-proton flares. The longitudinal distribution of proton flares and large radio bursts at $\lambda 3 - 10$ cm was investigated for the new solar cycle.

078.006 **Cosmic-ray flares in Norilsk on January 24 and September 1, 1971.** L. I. Dorman, A. V. Palamarchuk, V. P. Karpov, Yu. A. Kurchenko.
Geomagn. Aeronom., Vol. 14, 151 - 152 (1974). In Russian. Brief information.

078.007 **Implications of the reported low energy electron gradients.** J. A. Lezniak, W. R. Webber.
Solar Physics, Vol. 34, 477 - 489 (1974).

A recently reported measurement of a small electron gradient in the energy range 1.9–8.4 MeV by Webber et al. (1973) is interpreted in terms of a large local value of the scattering mean free path for these particles. The possibility that the scattering mean free path may be large throughout the modulation region is then investigated under the assumption of an azimuthally symmetric modulation region of 5 AU extent, the applicability of the diffusion-convection-adiabatic energy loss transport equation, and a galactic origin for the low energy electrons. The implications for the solar modulation of electrons and the interstellar electron spectrum are discussed.

078.008 **Variability of intensity ratios of H to He and He to $Z \geq 3$ ions in solar energetic particle events.**
J. A. Van Allen, P. Venkatarangan, D. Venkatesan.
Journ. Geophys. Res., Vol. 79, 1 - 8 (1974).

By means of data from the University of Iowa solid state detector on Explorer 35, a study has been made of the intensity ratios of H to He and He to ions with atomic numbers $Z \geq 3$ in the sub-MeV per nucleon specific kinetic energy range for the following solar energetic particle events: June 9, 1968; November 18, 1968; April 11, 1969; May 13, 1969; and, by reference to a previous paper, the events of February 25 and March 30, 1969. The principal finding is that the intensity ratios H/He and He/($Z \geq 3$) in the sub-MeV per nucleon specific kinetic energy range vary markedly from event to event and more especially during the time history of individual events.

078.009 **The solar proton diffusion mean free path and the anisotropic particle event of November 18, 1968.**
J. J. Quenby, G. E. Morfill, A. C. Durney.
Journ. Geophys. Res., Vol. 79, 9 - 16 (1974).

Neutron monitor impact zones for the November 18, 1968, flare event are examined. The peak interplanetary intensity, deduced by using the asymptotic acceptance cones, agrees well with the interplanetary field direction. The purpose of this paper is to discuss the 100% anisotropic propagation seen initially in the event in terms of the observed power spectrum of interplanetary field fluctuations.

078.010 **Studies of modulation and solar particle events using neutron monitor multiplicities.** D. W. Kent.
Journ. Geophys. Res., Vol. 79, 271 - 274 (1974).

078.011 **Generation of charged particles on the sun on July 22, 1972.**
V. P. Grigor'eva, M. V. Kudryavtsev, V. G. Kurt, Yu. I. Logachev, A. S. Melioranskij, N. F. Pisarenko, I. A. Savenko, V. M. Shamolin, I. P. Shestopalov.
Kosmich. Issled., Vol. 12, 213 - 218 (1974). In Russian.

078.012 **Definition of the free transport path and intensity of primary protons for different phases of solar cosmic ray flares.**
E. V. Gorchakov, G. A. Timofeev, T. I. Morozova.
Geomagn. Aeronom., Vol. 14, 192 - 195 (1974). In Russian.

078.013 **Propagation of energetic solar particles in wedge-shaped streams.** Z. B. Rojkhvarger.
Geomagn. Aeronom., Vol. 14, 201 - 206 (1974). In Russian.

078.014 **Interplanetary acceleration of solar cosmic rays to relativistic energy.**
M. A. Pomerantz, S. P. Duggal.
Journ. Geophys. Res., Vol. 79, 913 - 919 (1974).

Although the August 1972 cosmic ray storm (i.e., a rapid succession of Forbush decreases) was the greatest in more than 3 decades of continuous observations, halving the flux of galactic cosmic rays above 1 GeV, its basic characteristics were not unusual despite its unprecedented magnitude and complexity. However, the associated ground level event of August 4, representing the arrival of relativistic solar particles, displayed abnormal features. It is suggested that the observed nucleonic intensity enhancement was a consequence of the acceleration of ambient lower-energy solar protons to relativistic energies by reflection between two shocks moving with respect to each other in the interplanetary medium.

078.015 **Solar proton flux enhancements at auroral latitudes.** I. B. McDiarmid, J. R. Burrows, M. D. Wilson.
Journ. Geophys. Res., Vol. 79, 1099 - 1103 (1974). — Brief report.

078.016 **Mechanism of the diurnal anisotropy of cosmic radiation.** R. P. Kane.
Journ. Geophys. Res., Vol. 79, 1321 - 1331 (1974).
The observed diurnal vector of cosmic ray neutron intensity at Deep River for the period 1965—1968 is resolved into a radially outward convection vector and a diffusion vector, days of Forbush decrease being omitted. The diffusion vector deviates from the interplanetary field direction by more than 1 hour in 60% of the cases and by more than 2 hours in 35% of the cases. The author proposes to examine in detail the behavior of the diffusive component vis-à-vis the orientation of the interplanetary magnetic field.

078.017 **A two-satellite study of low-energy protons over the polar cap during the event of November 18, 1968.** R. J. Hynds, G. Morfill, R. Rampling.
Journ. Geophys. Res., Vol. 79, 1332 - 1344 (1974).
A study of low-energy solar proton intensity profiles over the northern polar cap has been made for the November 18 event. Use of data from two satellites crossing the polar cap at closely related universal times has permitted separation of spatial and temporal variations in the intensity profiles early in the event. The resulting spatial profiles have been related to the structure and length of the magnetotail and have been considered in terms of current theories of formation of the magnetotail.

078.018 **On the access of solar protons to the synchronous altitude region.** J. B. Blake, E. F. Martina, G. A. Paulikas.
Journ. Geophys. Res., Vol. 79, 1345 - 1348 (1974).
Observations of 10- to 90-MeV solar protons were made near the synchronous altitude around the time of a sudden commencement at 0037 UT on August 9, 1972, by using sensors aboard spacecraft 1970-93A and 1972-10A.

078.019 **Energetic particles from the sun.** K. Sakurai.
Astrophys. Space Sci., Vol. 28, 375 - 519 (1974).
This paper discusses solar cosmic ray phenomena and related topics from the solar physical point of view. Basic physics of the solar atmosphere and solar flare phenomena are considered in some detail. The propagation of solar cosmic rays in the interplanetary space is discussed briefly by referring to the observed magnetic properties of this space. Some problems related to the physics of galactic cosmic rays are discussed.

078.020 **Multiple generation of cosmic radiation on the sun on August 4, 1972 according to data of measurements aboard Prognoz.** N. N. Volodichev, N. L. Grigorov, G. Ya. Kolesov, O. B. Likin, E. I. Morozov, N. F. Pisarenko, I. A. Savenko, A. A. Suslov, V. M. Ustinovshchikov.
Kosmich. Issled., Vol. 12, 483 - 485 (1974). In Russian. Brief information.

078.021 **Non-relativistic solar electrons.** R. P. Lin.
Space Sci. Rev., Vol. 16, 189 - 256 (1974).
This review summarizes both the direct spacecraft observations of non-relativistic solar electrons, and observations of the X-ray and radio emission generated by these particles at the sun and in the interplanetary medium.

078.022 **Relativistic electron events in interplanetary space.** G. M. Simnett.
Space Sci. Rev., Vol. 16, 257 - 323 (1974).
This review is concerned with relativistic electron events observed in interplanetary space. The different types of event are identified and illustrated. The relationships between solar X-ray and radio emissions and relativistic electrons are examined, and the relevance of the observations to solar flare acceleration models is discussed. A statistical analysis of electron spectra, the electron/proton ratio and propagation from the flare site to the earth is presented. A model is outlined which can account for the release of electrons from the sun in a manner consistent with observations of energetic solar particles and electromagnetic solar radiation.

078.023 **Observations of solar particle propagation.** L. J. Lanzerotti.
Astrophys. Space Sci. Library, Vol. 42, (see 012.016), 345 - 379 (1974).

078.024 **The value of multi-spacecraft measurements.** M. A. Shea, D. F. Smart.
Astrophys. Space Sci. Library, Vol. 42, (see 012.016), 381 - 396 (1974).

078.025 **Solar particle access into the inner magnetosphere.** G. Morfill.
Astrophys. Space Sci. Library, Vol. 42, (see 012.016), 399 - 417 (1974).

078.026 **Energetic solar particle access into the geomagnetic tail.** M. Scholer.
Astrophys. Space Sci. Library, Vol. 42, (see 012.016), 419 - 431 (1974).

078.027 **On a model of propagation of 5—300 MeV protons in the earth magnetic cavity. The role of neutral and plasma sheets.** R. Gall, S. Bravo.
Astrophys. Space Sci. Library, Vol. 42, (see 012.016), 433 - 447 (1974).

078.028 **Comparisons of solar proton trajectory computations using model magnetospheres.** G. R. Thomas, D. M. Willis, R. J. Pratt.
Astrophys. Space Sci. Library, Vol. 42, (see 012.016), 449 - 461 (1974).

078.029 **Solar electron access to the magnetosphere.** A. L. Vampola.
Astrophys. Space Sci. Library, Vol. 42, (see 012.016), 463 - 478 (1974).

078.030 **Simultaneous observations of proton and alpha fluxes during the period 29 October—3 November 1972.** L. Katz, P. L. Rothwell, G. K. Yates, B. Sellers, F. A. Hanser.
Astrophys. Space Sci. Library, Vol. 42, (see 012.016), 479 - 483 (1974).

078.031 **Measurements of solar protons in the near earth magnetotail.** V. Domingo, D. E. Page, K.-P. Wenzel.
Astrophys. Space Sci. Library, Vol. 42, (see 012.016), 507 - 516 (1974).

078.032 **MeV electrons, protons and alpha particles observed August 2—12, 1972.** D. E. Page, V. Domingo, K.-P. Wenzel.
Astrophys. Space Sci. Library, Vol. 42, (see 012.016), 573 - 585 (1974).

078.033 **Solar particle observations during the August 1972 event.** L. J. Lanzerotti, C. G. Maclennan.
Astrophys. Space Sci. Library, Vol. 42, (see 012.016), 587 - 596 (1974).

078.034 **Interplanetary particle fluxes observed by OV5–6 satellite.**
G. K. Yates, L. Katz, B. Sellers, F. A. Hanser.
Astrophys. Space Sci. Library, Vol. 42, (see 012.016), 597 - 602 (1974).

078.035 **Temporal variations of solar particle spectra (August 4–7, 1972).** G. Kremser, G. Pfotzer, E. Kirsch, H. Specht, W. Riedler, K. Zirm.
Astrophys. Space Sci. Library, Vol. 42, (see 012.016), 603 - 607 (1974).

078.036 **The response of the Deep River neutron monitor to an anisotropic flux of solar cosmic rays.**
M. A. Shea, D. F. Smart.
Journ. Geophys. Res., Vol. 79, 2487 - 2490 (1974).
Brief report.

078.037 **Propagation of flare protons in the solar atmosphere.**
R. Reinhard, G. Wibberenz.
Solar Physics, Vol. 36, 473 - 494 (1974).

The velocity dispersion for a large number of solar proton events is analyzed in the energy regime of 10–60 MeV. It is found for all events that the time from the flare to particle maximum is well represented by a sum of two components. The first component which is energy independent describes the propagation in the solar atmosphere, the second component describes the propagation in the interplanetary medium. The additional study of time intensity profiles, onset times, and space probe observations reveals that the propagation in the solar atmosphere consists of three processes: (1) a rapid transport process in the initial phase after the event fills up a "fast propagation region", (2) a large-scale drift process which is energy independent, and simultaneously (3) a diffusion process which yields the general broadening of the intensity time profiles for eastern hemisphere events.

078.038 **Observations of H and He isotopes in solar cosmic rays.**
G. J. Hurford, R. A. Mewaldt, E. C. Stone, R. E. Vogt.
Bull. American Astron. Soc., Vol. 6, 270 (1974). – Abstr. AAS.

078.039 **Production of solar neutrons.**
B. Kozlovsky, R. Ramaty.
Bull. American Astron. Soc., Vol. 6, 270 - 271 (1974). Abstr. AAS.

078.040 **Interpretation of observed charge states of low-energy solar particles.** J. R. Jokipii, A. J. Owens.
Bull. American Astron. Soc., Vol. 6, 271 (1974). – Abstr. AAS.

078.041 **Observations of low energy hydrogen and helium nuclei during solar quiet times.**
R. A. Mewaldt, G. J. Hurford, E. C. Stone, R. E. Vogt.
Bull. American Astron. Soc., Vol. 6, 274 (1974). – Abstr. AAS.

078.042 **Schwere solare kosmische Strahlung.**
D. Hovestadt.
Naturwissenschaften, 61. Jahrgang, p. 281 - 287 (1974).

During solar flares heavy solar cosmic rays are observed along with protons, alpha particles, and electrons. New experimental techniques have recently made the low-energy part ($E < 8$ MeV/nucleon) of the energy spectrum accessible to observation. At these low energies the heavier elements (Fe) are more abundant than the lighter elements (C, O), which contrasts with the relative abundance of the elements on the sun and in high-energy solar cosmic rays. Heavy elements generally seem to be highly ionized, even at low energies ($E < 1$ MeV/nucleon).

078.043 **Probabilities for the peak flux and fluence of energetic solar protons incident on interplanetary spacecraft.** N. Divine.
Jet Propulsion Lab. Techn. Rev., Vol. 3, No. 2, p. 37 - 44 (1973).

For future long-term interplanetary and planetary missions, techniques have been developed which use solar particle event data from 1956 through 1970 to predict the probability of exceeding any value of peak proton intensity or mission proton fluence. Dependences on proton energy, heliocentric distance, and phase of the solar cycle are included. The techniques are described and applied to the Mariner Jupiter/Saturn 1977 mission.

078.044 **Some characteristics of solar proton activity during the 20th cycle.**
Solar Activity Division, Peking Obs., Acad. Sinica.
Acta Astron. Sinica, Vol. 15, 34 - 42 (1974). In Chinese.

Some characteristics of the solar proton events and proton-active regions during the period 1964 Oct.–1972 Dec. of the 20th cycle are discussed.

078.045 **Observation of transiron solar-flare nuclei in an Apollo 16 command-module window.** E. K. Shirk.
Astrophys. Journ., Vol. 190, 695 - 702 (1974).

Transiron nuclei from the 1972 April 18 solar flare have been detected in one of the Apollo 16 command-module windows. Measurements of 39 events with $Z \geq 32$ and of 12 events with $Z \geq 44$ have been completed.

Apparatus for the investigation of energy and charge spectra of solar cosmic rays at the satellites "Prognoz".
See Abstr. 034.026.

High energy astronomy. See Abstr. 061.075.

Development and spatial structure of proton flares near the limb, and coronal phenomena. VII. Change of activity prior to the proton flare of September 1, 1971.
See Abstr. 073.019.

A correlation between time-overlapping solar flares and the release of energetic particles. See Abstr. 073.032.

Solar flares as sources of energetic particles.
See Abstr. 073.058.

Relation of coronal magnetic structure to the interplanetary proton events of August 2–9, 1972.
See Abstr. 074.078.

Type II radio bursts and particle acceleration.
See Abstr. 077.063.

Solar proton entry to the magnetosphere on 18 November 1968 and 25 February 1969 – I. Interpretation of satellite data using trajectory computations in a model magnetosphere. See Abstr. 084.260.

Solar proton entry to the magnetosphere on 18 November 1968 and 25 February 1969 – II. Comparison of trajectory computations in two model magnetospheres.
See Abstr. 084.261.

A study of magnetospheric boundaries by the simultaneous observations of solar and magnetospheric particles.
See Abstr. 084.282.

Variations of the trapped particle population and of cutoff and pitch angle distribution of simultaneously observed magnetospheric solar protons during substorm activity.
See Abstr. 084.406.

Cosmic ray ionization rates in the planetary atmospheres. See Abstr. 091.008.

Coherent propagation of charged-particle bunches in random magnetic fields. See Abstr. 106.014.

Piston shock waves in the interplanetary medium and Forbush effects. See Abstr. 106.028.

An interpretation of the observed oxygen and nitrogen enhancements in low energy cosmic rays. See Abstr. 143.024.

The anomalous abundance of cosmic ray nitrogen and oxygen nuclei at low energies. See Abstr. 143.025.

079 Solar Eclipses

079.001 **Finsternisse 1974.** L. D. Schmadel.
SuW, Vol. 13, 69 (1974).

079.002 **Solar-eclipse photography for the amateur.**
Journ. Astron. Soc. Western Australia, Vol. 25, February, p. 4 - 8 (1974).

079.100 **Solar eclipse 1973 June 30**

Die Expedition zur Beobachtung der totalen Sonnenfinsternis vom 30. Juni 1973. M. Waldmeier.
Astron. Mitt. Eidgenössisch. Sternw. Zürich, No. 324, 22 pp. (1973).
The main interest concerned the exploration of the corona; cameras of 5, 36, 50, 120, 150, 165 and 800 cm focal length have been used. The photographs have been taken partly in natural, partly in polarized light. A special program concerned the investigation of the sky-light at large distances from the sun. Furthermore, around second contact 12 pictures of the flash-spectrum have been obtained. With all instruments together, 155 photographs have been taken during totality and several additional ones during the partial phase. Most of the photographs serve for investigation of the brightness distribution, polarization, shape and structure of the corona.

Der Temperaturverlauf während der Sonnenfinsternis vom 30. Juni 1973. M. Waldmeier, S. E. Weber.
Astron. Mitt. Eidgenössisch. Sternw. Zürich, No. 325, 7 pp. (1973).

Solformørkelsen den 30. juni 1973 set fra Kenya.
K. Strandbaek.
Astron. Tidssk., Årg. 7, p. 33 - 35 (1974).

Solförmörkelse i Mauritanien.
T. Hanson, B. Stenholm, S. Söderhjelm.
Astron. Tidssk., Årg. 7, p. 35 - 39 (1974).

Observations of the total solar eclipse of June 30, 1973 according to the Soviet-French cooperation programme.
S. K. Vsekhsvyatskij, N. I. Dzyubenko, A. T. Nesmyanovich, O. S. Popov, G. A. Rubo, S. Koutchmy, C. Bareau, J. Begot, J. Fagot.
Astron. Tsirk., No. 800, p. 1 - 3 (1973). In Russian.

Observations of the solar eclipse of June 30, 1973 in Mauritania. G. N. Salukvadze.
Astron. Tsirk., No. 801, p. 7 - 8 (1973). In Russian.

L'éclipse solaire du 30 juin 1973.
G. W. Curtis.
Bull. d'Information, Ass. Développement International Obs. Nice, No. 10, p. 3 - 12 (1973).

80 minutes de totalité. P. Lena.
Bull. d'Information, Ass. Développement International Obs. Nice, No. 10, p. 13 - 16 (1973).

Observations au Kenya de l'éclipse totale de soleil du 30 juin 1973. D. Cardoen, H. Debehogne, A. Dreze, G. Roland, P. Simon, E. van Hemelrijck.
Ciel et Terre, Vol. 90, 187 - 195 (1974).

Simultaneous observation of TEC from north and south hemispheres during the 30th June 1973 solar eclipse.
D. A. Matsoukas, M. A. Anastassiadis.
In honorem S. Placidis, (see 003.009), p. 227 - 236 (1974).

Observation of the total solar eclipse of 30 June 1973. V. Rušín.
Kozmos, Vol. 5, 69 - 70 (1974). In Slovak.

The only Polish expedition for the total solar eclipse on June 30, 1973. J. M. Kreiner.
Postępy Astron., Vol. 22, 137 - 140 (1974). In Polish.
A short report on a students' expedition for the total solar eclipse on June 30, 1973 is given.

The moon's shadow at a solar eclipse.
V. Ashcraft.
Sky Telescope, Vol. 47, 61 (1974).

Die Finsternis der Tuareg. F. Dorst.
SuW, Vol. 13, 29 - 30 (1974).

Cracow–Sahara astronomical expedition.
M. Mazur.

Urania Kraków, Vol. 45, 5 - 6 (1974). In Polish.

Solar eclipse on June 30th, 1973. H. Brancewicz.
Urania Kraków, Vol. 45, 7 - 9 (1974). In Polish.

Supersonic astronomy. A. Marks.
Urania Kraków, Vol. 45, 99 - 102 (1974). In Polish.

Die totale Sonnenfinsternis vom 30. 6. 1973. Bericht über die Expedition der Wilhelm-Foerster-Sternwarte nach Mauretanien.
B. Wedel, with a contribution by D. Zucht on 'Die sich verfinsternde Landschaft'.
Veröff. Wilhelm-Foerster-Sternw., Berlin, No. 35, 33 pp. (1974).

The eclipse in the western Sahara.
S. B. Novikov, Yu. V. Platov.
Zemlya i Vselennaya, 1974, No. 1, p. 56 - 61. In Russian.

De totale zonsverduistering van 30 juni 1973.
D. Cardoen.
Zenit, Vol. 1, No. 4, p. 5 (1974).

Digitale Auswertung der Sonnenfinsternisaufnahmen vom 30. 6. 73. See Abstr. 021.013.

A relativity eclipse experiment refurbished.
See Abstr. 066.017.

Shape and structure of the corona at the solar eclipse of June 30, 1973. See Abstr. 074.114.

3.3-millimeter observations of the June 30, 1973 total solar eclipse. See Abstr. 077.078.

079.101 Solar eclipse 1973 December 24

Die Ringförmige Sonnenfinsternis vom 24. Dezember 1973. F. Dorst.
Orion Schaffhausen, 32. Jahrgang, p. 59 - 60 (1974).

Report on December's solar eclipse.
Sky Telescope, Vol. 47, 131 - 132 (1974).

Notes on December's annular eclipse.
Sky Telescope, Vol. 47, 198 - 201 (1974).

Sonnenfinsternis in Algerien. F. Dorst.
SuW, Vol. 13, 102 (1974).

079.102 Solar eclipse 1971 February 25

Observations of the solar eclipse on 25 February 1971 in Crimea at 10 cm wavelength. Yu. F. Yurovskij.
Izv. Krymskoj Astrofiz. Obs., Vol. 49, 46 - 50 (1974). In Russian.

079.103 Solar eclipse 1972 July 10

On the Hα-emission in the integrated spectrum of the sky and corona during the total solar eclipse of 1972, July 10. See Abstr. 074.036.

079.104 Solar eclipse 1966 November 12

Curvature and surface distribution of the polar rays in the solar corona on 12 November 1966.
See Abstr. 074.051.

079.105 Solar eclipse 1974 June 20

Solar eclipse 1974 June 20th, refinements of official predictions. F. Leroux.
Journ. Astron. Soc. Western Australia, Vol. 25, April(1974).

080 Solar Atmosphere, Figure, Internal Constitution, Neutrinos, Magnetic Fields, Rotation, Miscellanea

080.001 **Direct observation of temperature amplitude of solar 300-second oscillations.**
H. S. Hudson, C. A. Lindsey.
Astrophys. Journ., (*Letters*), Vol. 187, L35 - L36 (1974).

The 300-second oscillations form the dominant source of variability of the solar infrared continuum. The authors have observed them at 20 μ with an amplitude $\Delta T_{rms} = 3.0°$ K over an area with an effective diameter of 33".

080.002 **Possible interpretation of an anomalous redshift observed on the 2292 MHz line emitted by Pioneer-6 in the close vicinity of the solar limb.**
P. Merat, J.-C. Pecker, J.-P Vigier.
Astron. Astrophys., Vol. 30, 167 - 174 (1974).

An analysis of Goldstein's observations shows an anomalous redshift of the central frequency of the 2292 MHz band emitted by Pioneer-6 during its occultation by the sun. This shift, symmetrical with respect to the sun's center, does not correspond to any presently known physical effect. It can be quantitatively interpreted through an interaction between incident transverse photons and light neutral bosons emitted by the sun, an interaction of the type proposed earlier by the authors.

080.003 **A self-limiting inversion technique for centre to limb data.** J. N. Holt.
Astron. Astrophys., Vol. 30, 185 - 188 (1974).

A recently developed numerical technique for inverting a single absorption line profile is adapted to the inversion of centre to limb data. Its performance is compared with that of the Prony algorithm for a set of synthetic observations. The results of the centre to limb inversions of the mean sodium D_2 line at line-centre and at 2.0 Å from line-centre are compared to those from the single line inversion of the same line at disc centre.

080.004 **Mass transport induced by time-dependent oscillations of finite amplitude in a stratified perfect gas.**
J. Naze Tjötta, S. Tjötta.
Astron. Astrophys., Vol. 30, 249 - 258 (1974).

The mass transport induced by time-dependent oscillations of finite amplitude in a stratified gas is considered. The mass transport is given by the Lagrangian mean velocity, calculated to the second order in the Mach number of the oscillations. The authors find no mass transport in a non-dissipative model. Taking into account dissipation, however, the theory leads to non-zero vertical drift and horizontal flow. The vertical drift becomes zero in an incompressible model. The general theory is applied to study the flow in the solar atmosphere.

080.005 **Limits to solar limb darkening at a wavelength of 1.4 millimeters derived from antenna-beam parameters.** P. A. R. Ade, J. D. G. Rather, P. E. Clegg.
Astrophys. Journ., Vol. 187, 389 - 392 (1974).

Careful studies have been made to derive the performance characteristics of the NRAO 11-m dish in the 0.8–1.6 mm wavelength range. It is shown that incomplete reduction of data previously obtained with the same instrument has led certain authors to the erroneous conclusion that limb darkening exists at short millimeter wavelengths. Within the available resolution of the dish the quiet-sun brightness distribution is rectangular, with some indication of polar darkening.

080.006 **An explanation of the solar limb shift.**
M. H. Hart.
Astrophys. Journ., Vol. 187, 393 - 401 (1974).

Absorption lines in the solar spectrum are shifted as a result of collisions of the absorbing atoms with neutral hydrogen atoms. Computations using the Lennard-Jones potential and standard photospheric models correctly predict the magnitude and direction of the limb shift, as well as the functional forms of the observed variations of line shift and damping constant with distance from the center of the disk.

080.007 **On differential rotation.** H. Köhler.
Solar Physics, Vol. 34, 11 - 14 (1974).

In a first order approximation the influence of meridional circulations in a spherical shell on the radial dependence of the angular velocity is studied.

080.008 **The formation of Mg I 4571 Å in the solar atmosphere. IV: Empirical vs synthetic analyses.**
R. C. Altrock.
Solar Physics, Vol. 34, 37 - 56 (1974).

A comparison is made of synthetic and empirical analyses for the 4571 Å line of Mg I. First, several different inversion techniques are applied to synthetic line profiles. The results show that at least some of these techniques are able to correctly reproduce the input atmosphere to a reasonable degree. Secondly, these same techniques are applied to equivalent observational data. In this case some of the techniques yield results that can be shown to be of comparable quality to the synthetic analysis.

080.009 **Non-LTE profiles of the Al I autoionization lines.**
G. D. Finn, J. T. Jefferies.
Solar Physics, Vol. 34, 57 - 75 (1974).

A non-LTE formulation is given for the transfer of radiation in the autoionizing lines of neutral aluminum at λ1932 and λ1936 through both the Bilderberg and Harvard-Smithsonian model atmospheres. Numerical solutions for the common source function of these lines and their theoretical line profiles are calculated and compared with the corresponding LTE profiles. The results show that the non-LTE profiles provide a better match with the observations.

080.010 **Magnetic field annihilation. Petschek's mechanism.**
A. A. Solov'ev.
Solnechnye Dannye 1973 Byull., No. 12, p. 57 - 62 (1974). In Russian.

The well-known Petschek mechanism of magnetic field annihilation is considered. The basic equation of the problem is given, the condition of the "freezing" field in the wave region of the boundary layer being taken into account. The rate of annihilation is shown to depend strongly on the geometry of the boundary layer.

080.011 **Properties of the magnetic field evolution in a quiet region on the sun.** S. I. Gopasyuk, T. T. Tsap.
Izv. Krymskoj Astrofiz. Obs., Vol. 49, 3 - 14 (1974). In Russian.

The magnetic field evolution in a quiet region on the sun is studied. The records of the magnetic fields were made with a magnetograph in the lines Fe I λ5250 Å and Ca I λ6103 Å. The observational properties of the magnetic field variation can be explained by rising of the field to the surface of the sun and descending off this level.

080.012 **Solar neutrino problem.** A. J. R. Prentice.
Proc. Astron. Soc. Australia, Vol. 2, 226 - 227

(1973). – Presented at the annual general meeting of the Astron. Soc. Australia – see 012.001.

080.013 Nonadiabatic pulsations of a solar model.
V. I. Aleshin.
Astron. Zhurn. Akad. Nauk SSSR, Vol. 51, 445 - 447 (1974). In Russian. English translation in Soviet Astron. AJ, Vol. 18, No. 2.

It is shown that the most powerful solar flares can give rise to radial pulsations of the sun with period of 50 min. The time of attenuation of these oscillations is of about one month.

080.014 Solar models with low neutrino fluxes.
R. K. Ulrich.
Astrophys. Journ., Vol. 188, 369 - 378 (1974).

Models of the sun which currently produce all their energy from nuclear reactions are examined in very general terms. It is concluded that a counting rate of 0.5 SNU is a firm lower bound for such models. Limitations on the rapidly rotating core model of Demarque, Mengel, and Sweigart and the central magnetic field model of Chitre, Ezer, and Stothers are discussed. Modification of all the nuclear-reaction-rate cross-section factors by 2 standard deviations in such a way as to reduce the neutrino fluxes yields a model with a counting rate of 1.7 SNU.

080.015 A model of the solar convection zone.
H. C. Spruit.
Solar Physics, Vol. 34, 277 - 290 (1974).

A model of the convection zone is presented which matches an empirical model atmosphere (HSRA) and an interior model. A mixing length formalism containing four adjustable parameters is used. Thermodynamical considerations provide limits on two of these parameters. The average temperature-pressure relation depends on two or three combinations of the four parameters. It is shown that the mean temperature-pressure relation is fixed well by these data.

080.016 Internal rotation of the sun and the solar neutrino flux. I. W. Roxburgh.
Nature, Vol. 248, 209 - 211 (1974).

The low upper limit of 1 SNU on the observed neutrino flux from the sun obtained by Davis (1972) has proved an embarrassment to stellar physicists. The essential difficulty has been to produce a model with a low enough central temperature that can still produce the observed luminosity of the sun with an age of 4.7×10^9 yr. Arguments for a rapid central rotation, causing a low neutrino flux, are elaborated.

080.017 Is the sun an astrometric binary? N. Gunn.
Journ. British Astron. Ass., Vol. 84, 126 - 128 (1974).

080.018 Maxwellian relative energies and solar neutrinos.
D. D. Clayton.
Nature, Vol. 249, 131 (1974).

The solar neutrino discrepancy is regarded as very serious. The author wishes to add to the list of possible ad hoc explanations; namely that the scarcity of ^{8}B neutrinos reflects a departure of the distribution of relative kinetic energies from the Maxwellian distribution.

080.019 Solar radiation and particle emission.
E. Jensen.
Cosmical geophysics, (see 003.004), p. 19 - 43 (1973).

080.020 Polfelder der Sonne. M. Rossbach, H. Wöhl.
SuW, Vol. 13, 157 - 160 (1974).

080.021 Magnetic fine structure and the solar magnetic monopole. C. Sawyer.
Solar Physics, Vol. 35, 37 - 40 (1974). – Research note.

080.022 Solar neutrinos and the behavior of the Fermi coupling constant. A. Finzi.
Astrophys. Journ., Vol. 189, 157 - 160 (1974).

The solar neutrino puzzle can be solved if the Fermi coupling constant g is assumed to vary with the gravitational potential. The number $\Sigma(\phi\sigma)$ of expected absorption events is lower than Davis's upper limit of 1 SNU if at the center of the sun g exceeds its standard value by a factor of at least 1.3.

080.023 The oblateness of the sun.
R. H. Dicke, H. M. Goldenberg.
Astrophys. Journ., Suppl. Ser., No. 241, Vol. 27, 131 - 182 (1974).

The solar oblateness observations of 1966 are analyzed. Included are measures of the vertical and diagonal components of the oblateness. These components represent the fractional contractions of the solar disk along the north-south and the northeast-southwest diameters, respectively. Also included are the annular-ring measures of the equatorial brightening of the solar disk. The $25\overset{d}{.}67$ periodicity of the solar oblateness is studied. The periodic autocorrelation function of the oblateness fluctuation, the periodic fluctuation component in the data, and the cross-correlation function of the data with the periodic component are evaluated. The periodic fluctuation component exhibits sharp positive peaks separated by 12 days. The values for the oblateness obtained from the diagonal and vertical components are consistent with each other and yield a mean value of $\Delta r/r = 4.51 \pm 0.34 \times 10^{-5}$. The solar quadrupole moment consistent with this value, $J = 2.47 \pm 0.23 \times 10^{-5}$, implies a correction of $2\overset{''}{.}98 \pm 0\overset{''}{.}27$ per century to the classical excess motion of Mercury's perihelion.

080.024 The oblateness of the sun and relativity.
R. H. Dicke.
Science, Vol. 184, 419 - 429 (1974).

080.025 Statistische Analyse granularer Dopplerverschiebungen. W. Mattig, A. Nesis.
Mitt. Astron. Ges., No. 34, p. 114 - 115 (1973/74). – Abstr. AG.

080.026 A search for the footpoints of solar magnetic fields.
G. W. Simon, J. B. Zirker.
Solar Physics, Vol. 35, 331 - 342 (1974).

High-resolution measurements of magnetic fields have been made in quiet and active regions in order to determine whether the photospheric fine-structures ('crinkles'), recently photographed by Dunn (1972), coincide with the footpoints of strong, compact fields.

080.027 On the nature of plasma arcs in solar active regions.
R. Gajewski.
Solar Physics, Vol. 35, 385 - 394 (1974).

A mechanism is proposed explaining the structures consisting of plasma arcs, as observed in X-ray photographs of solar active regions. It is suggested that the width of the arcs corresponds to the cut-off wavelength of a Rayleigh-Taylor instability which develops due to a difference in density between the plasma in the arcs and the plasma in the surrounding region. The transverse component of the magnetic field necessary to stabilize the instability at a wavelength corresponding to the width of the arcs is estimated to be of the order of 0.1 G.

080.028 Heated solar atmosphere: a one-fluid model.
E. Leer.
Solar Physics, Vol. 35, 467 - 480 (1974).

A one-fluid model of the solar atmosphere is considered. The corona is heated by waves propagating out from the sun, and profiles for temperature, flow speed and number density are obtained. For a relatively quiet sun the inwards heat flux in the inner corona is constant in $T \lesssim 5-6 \times 10^5$ K and the temperature maximum is reached for $r - R_\odot = 0.4-0.5\, R_\odot$. The number density in the inner corona decreases with increasing particle flux.

080.029 Le mystère des neutrinos solaires. E. Schatzman.
L'Astronomie, 88ᵉ année, p. 155 - 171 (1974).

080.030 Extraterrestrial solar spectrum, 3000–6100 Å at 1-Å intervals. M. P. Thekaekara.
Applied Optics, Vol. 13, 518 - 522 (1974).

Standard values of the solar constant and extraterrestrial solar spectrum are reviewed. A detailed spectrum obtained from solar scans with a Perkin-Elmer, Model 112 monochromator was found to give sufficient detail. A normalization program was developed to make the Perkin-Elmer curve agree with the standard curve. Values of extraterrestrial solar spectral irradiance at 1-Å intervals in the range 3000–6100 Å have been derived. The results are presented in tabular form and as spectral charts.

080.031 Photoelectric drift scans II: time pulse evaluation, limb profiles, and the solar diameter.
A. Wittmann.
Solar Physics, Vol. 36, 65 - 68 (1974).

The method of computer controlled photoelectric drift scans has been improved by virtue of a new timing technique, allowing for an accuracy of 4 ms or equivalently 50 km on the sun. With this technique, the author's previous result for the solar semidiameter has been confirmed: $R = 960.277''$. The extreme solar limb intensity profile at 5012 Å has been derived from drift scans. A comparison with the computed profile has been made, and parameters for the base of the chromosphere have been derived.

080.032 Neutron propagation and 2.2 MeV gamma-ray line production in the solar atmosphere.
H. T. Wang, R. Ramaty.
Solar Physics, Vol. 36, 129 - 137 (1974).

The authors have calculated the 2.2 MeV gamma-ray line intensity from the sun using a Monte Carlo method for neutron propagation in the solar atmosphere. They provide detailed results on the total gamma-ray yield per neutron and on the time profile of the 2.2 MeV line from an instantaneous and monoenergetic neutron source.

080.033 A Gaussian spread function for the solar aureole. L. Staveland.
Solar Physics, Vol. 36, 235 - 238 (1974).

An analytical expression for the limb profile and the aureole is derived for a Gaussian spread function with $b \geq 2'$. For smaller values of the spread parameter b the expression given by Staveland (1972) gives a more correct result. Mullan's (1973) expression gives too low intensities in the range $\pm 2b$ around the limb.

080.034 The region of motion of a gravitating charged particle in a spiral magnetic field. A. V. Artem'ev.
Pyl' v atmosf. i okolozemn. kosmich. prostranstve. Moskva, Nauka, 1973, p. 30 - 37. In Russian. – Abstr. in Referativ. Zhurn. 51. Astron., 5.51.406 (1974).

080.035 The differential rotation of the solar surface.
P. J. Gierasch.
Astrophys. Journ., Vol. 190, 199 - 210 (1974).

The large-scale flow in the solar convection zone is discussed. The objective is to deduce from observation the principal physical balances in the governing equations. It is deduced that the zonal flow is geostrophic, the meridional flow is controlled by friction, and diffusive heating balances advective cooling due to vertical motion. A detailed calculation of the latitude profile of surface angular velocity is performed, based on approximate equations containing only the principal balances, and agrees well with observation. It is demonstrated that the amplitude of the flow may be determined by the constraint that the net equatorward angular momentum flux vanish. There is contradiction with the observation of no equator-pole flux variation, but this may be due to oversimplification of equations and boundary conditions.

080.036 On the sun's differential rotation: its maintenance by large-scale meridional motions in the convection zone. B. R. Durney.
Astrophys. Journ., Vol. 190, 211 - 221 (1974).

It is shown that if the observed differential rotation of the sun is generated in the lower, Boussinesq part of the convection zone, a large pole-equator difference in flux would also have to be present. The sun's angular velocity is evaluated as a function of depth and latitude under the assumption that the main effect of the interaction of rotation with convection is the generation of a small pole-equator difference in temperature in the lower part of the convection zone, which drives a meridional motion over the entire convection zone. The pole-equator difference in flux associated with this meridional circulation is negligible.

080.037 The solar temperature reversal and convective motions. A. J. Skalafuris.
Astrophys. Space Sci., Vol. 29, L9 - L17 (1974).

The author contends that simple hydrodynamic flow, properly analyzed with radiation as a source or sink of energy, has inherently in it a structure consistent with: (a) the solar temperature reversal, (b) the upper chromospheric temperature discontinuity, (c) convective cellular motions.

080.038 On the possibility of parametric amplification of Alfvén waves in the solar atmosphere.
N. S. Petrukhin.
Astron. Zhurn. Akad. Nauk SSSR, Vol. 51, 571 - 576 (1974). In Russian. English translation in Soviet Astron. AJ, Vol. 18, No. 3.

The excitation of Alfvén waves in a homogeneous flat acoustic resonator in a weak magnetic field perpendicular to the boundaries has been considered. It has been shown that parametric amplification of standing Alfvén waves is possible in the region of the density loop of free oscillations. The possibility of applying the results of the investigation to the undisturbed lower solar chromosphere is discussed.

080.039 Thermonuclear reactions and the solar neutrino radiation.
V. M. Gol'dman, G. A. Zisman, R. Ya. Shaulov.
Teor. fiz. i astronomiya. Leningrad, 1973, p. 96 - 104. In Russian. – Abstr. in Referativ. Zhurn. 51. Astron., 6.51.450 (1974).

080.040 Dynamics of the solar magnetic field. I. Method of examination of force-free magnetic fields.
Y. Nakagawa.
Astrophys. Journ., Vol. 190, 437 - 440 (1974).

In the solar atmosphere, prevailing physical conditions indicate that below the photospheric level the magnetic field is subject to passive change due to fluid motions, while above the photospheric level the magnetic field must satisfy closely the force-free conditions. On the basis of such understandings a practical method of studying the evolution of a general force-free magnetic field in the solar atmosphere utilizing magnetograph observations is described. The details of the formulations are presented, together with discussions of their physical significance.

080.041 **Dynamics of the solar magnetic field. II. The energy spectrum of large-scale solar magnetic fields.**
Y. Nakagawa, R. H. Levine.
Astrophys. Journ., Vol. 190, 441 - 446 (1974).

The energy spectrum of large-scale solar magnetic fields is obtained as a function of the circumferential wavenumber k along the heliographic latitudinal circle from the full-disk magnetograms of the Kitt Peak National Observatory covering one solar rotation. It is shown that the spectra in the higher latitudinal zones are different from those in the lower latitudinal zones.

080.042 **Differential rotation and sector structure of solar magnetic fields.** J. O. Stenflo.
Solar Physics, Vol. 36, 495 - 515 (1974).

The differential rotation and sector structure of solar magnetic fields has been studied using digitized data on photospheric magnetic fields recorded at the Mount Wilson Observatory during the period August 1959–May 1970.

080.043 **The solar neutrino puzzle.** J. Barnothy.
Bull. American Astron. Soc., Vol. 6, 271 (1974). Abstr. AAS.

080.044 **Helium inhomogeneities and the solar neutrino problem.**
P. Demarque, J. G. Mengel, A. V. Sweigart.
Bull. American Astron. Soc., Vol. 6, 271 (1974). – Abstr. AAS.

080.045 **A full-disk neutral-line analysis of active solar regions.** J. D. Farley.
Bull. American Astron. Soc., Vol. 6, 286 (1974). – Abstr. AAS.

080.046 **The small-scale structure of solar velocity fields.**
E. N. Frazier.
Bull. American Astron. Soc., Vol. 6, 287 (1974). – Abstr. AAS.

080.047 **Horizontal solar streamlines from Doppler line-of-sight velocity measurements.** R. G. Hendl.
Bull. American Astron. Soc., Vol. 6, 288 (1974). – Abstr. AAS.

080.048 **Infrared continuum observations of 300-sec oscillations.** H. S. Hudson.
Bull. American Astron. Soc., Vol. 6, 289 (1974). – Abstr. AAS.

080.049 **Non-LTE line formation in the presence of intermediate scale velocity fields.** R. A. Shine.
Bull. American Astron. Soc., Vol. 6, 294 (1974). – Abstr. AAS.

080.050 **An abundance analysis by center and limb profile fitting.** L. Testerman.
Bull. American Astron. Soc., Vol. 6, 295 (1974). – Abstr. AAS.

080.051 **Where do active regions grow?**
F. Ward, R. F. Carnevale.
Bull. American Astron. Soc., Vol. 6, 297 (1974). – Abstr. AAS.

080.052 **On the local equatorial deceleration of the solar atmosphere.**
G. Belvedere, L. Paternò.
Mem. Soc. Astron. Italiana, Nuova Ser., Vol. 44, 689 - 690 (1973/74). – Letter.

080.053 **Solar observation.** A. E. Coombs.
Journ. Astron. Soc. Victoria, Vol. 27, 18 - 24 (1974). – Presidential address delivered at annual general meeting, 1974 February 14.

080.054 **Solar neutrinos.** Z. Mikulášek.
Říše hvězd, Vol. 55, 4 - 7 (1974). In Czech.

080.055 **Solar-neutrino related M1 transition in ^{6}Li.**
C. Werntz, H. Uberall.
Zeitschr. Physik, Vol. 265, 405 - 410 (1973).

A two-cluster model is devised for the hypothetical 0^+, $T = 1$ level at 11.5 MeV in ^{6}Be which was proposed to explain the low value of the observed flux of high-energy solar neutrinos.

080.056 **The mystery of the missing neutrinos [from the sun].**
L. Ryder.
Phys. Education, (GB), Vol. 8, 484 - 485 (1973).

080.057 **Gentle neutrino scattering and the Davis experiment.**
E. Lubkin.
Phys. Letters A, (Netherlands), Vol. 46A, 431 - 432 (1974).

Energy loss from many very soft elastic scatterings of neutrinos could remove the energy from the neutrinos from the sun. This might be detectable through gross heating of cryogenic preparations by currently available neutrino beams.

080.058 **Recent theoretical interpretations of the solar five minute period oscillation.** A. G. Michalitsanos.
Earth Extraterr. Sci., Vol. 2, 125 - 138 (1973).

The properties of the theoretical models which have been proposed are discussed in context with recent observations of the 'wiggly line structure'. Models including the effects of magnetic fields on the 5-minute period oscillations in regions of moderate field strength of several hundred gauss are also reviewed.

080.059 **The enigma of solar neutrinos.** R. Omnes.
Recherche, (France), Vol. 5, No. 41, p. 15 - 21 (1974). In French.

The author outlines the five solar equations, and discusses several possible explanations of the discrepancies in the light of Davis's solar neutrino experiment.

080.060 **Der Neutrinofluß der Sonne bei differentieller Rotation und starkem Magnetfeld.** D. Bartenwerfer.
Thesis Univ. Göttingen, 56 pp. (1972).

080.061 **Dynamics of the solar magnetic field. III. Location of solar-flare excitation and the velocity field determined from magnetograms.** R. H. Levine, Y. Nakagawa.
Astrophys. Journ., Vol. 190, 703 - 709 (1974).

The velocity field comparable with the relative motions of sunspots is determined from consecutive magnetograms obtained near the disk center of the sun during 1972 August 2 - 4. The velocity fields show relative motions of sunspots in agreement with those deduced from white-light photographs. Further it is shown that the most likely positions of initial flare activity correspond to locations where a local maximum in the rate of strain of the velocity field coincides with either zero gradient or zero value of the vertical magnetic field. The detailed manner of determination is described together with discussions of the physical significance of the results.

080.062 **Dynamics of the solar magnetic field. IV. Examples of force-free magnetic-field evolution in response to photospheric motions.** Y. Nakagawa, K. Tanaka.
Astrophys. Journ., Vol. 190, 711 - 714 (1974).

The evolution of force-free magnetic fields in the solar atmosphere in response to photospheric motions is examined. With linear dipole magnetic fields it is shown that for a photospheric motion which increases general shear, the magnetic field evolves to produce open field lines leading to the formation of a 'magnetic bubble', and that for a motion which reduces general shear the evolution is characterized by the presence of flat field lines. The details of the analysis and the results of physical significance are discussed.

080.063 Die Asymmetrie des Hα-Absorptionskoeffizienten und die Entstehungshöhe von Hα in der Sonnenatmosphäre. A. Zelenka.
Astron. Mitt. Eigenössisch. Sternw. Zürich, No. 327, 121 pp. (1974).

The fine structure splitting of the Hα transition array results in a 0.14 Å separation between the most intensive, and in a 0.20 Å separation between the extreme components. These values are comparable with the Doppler width $\Delta\lambda_D$ representative of the layers where the solar Hα profile originates. Thus, in calculating the line absorption coefficient, one should not use the commonly considered Stark splitting, which is due to the quasi-static ion field, before it grows larger than the fine structure splitting. A representation is proposed which drives the Holtsmark profiles smoothly toward the fine structure profiles. An attempt is also made to substitute the Stark splitting in weak fields for pure fine structure splitting. Both variants yield an asymmetric absorption coefficient. The new absorption coefficients are tested by computing Hα profiles for the solar atmosphere.

080.064 Extraterrestrial values of solar radiation at the geographical latitude 45°49'N. I. Penzar.
Geofiz. zavod. Prirodoslovno-mat. fak. Sveučilišta, Zagreb – Zb. meteorol. i hidrol. radova, 4, 1973, p. 3 - 17, Beograd. Abstr. in Bull. Sci. Yougoslavie, Section A, Vol. 19, 39 (1974).

Selective diffraction with possible application to solar research. See Abstr. 031.004.

Theories of the geomagnetic and solar dynamos. See Abstr. 061.051.

Global fluid motions in a uniformly rotating convective spherical shell studied as a model of the solar convection zone by a variational method. See Abstr. 062.011.

On the equilibrium and stability of rotating high-beta plasmas. See Abstr. 062.036.

The energy requirements of microturbulence. See Abstr. 064.028.

Gravitation and long-range weak interactions in neutrino fields. Ideas on a theory of solar neutrinos. See Abstr. 066.089.

Computation and observation of Zeeman multiplet polarization in Fraunhofer lines. II: Computation of Stokes parameter profiles. See Abstr. 071.021.

Excitation temperatures of molecules and models of the solar atmosphere. See Abstr. 071.051.

Faculae and the solar oblateness. II. See Abstr. 072.035.

Strahlungsdämpfung akustischer Wellen beim Austritt aus den optischen dicken Schichten der Sonnenatmosphäre. See Abstr. 073.044.

Green corona and solar sector structure. See Abstr. 074.061.

Neutron propagation and 2.2 MeV gamma-ray line production in the solar atmosphere. See Abstr. 076.013.

Relations between quiet sun radio observations and models of the solar atmosphere. See Abstr. 077.076.

Absolute brightness temperature measurements at 2.1-mm wavelength. See Abstr. 091.011.

Earth

081 Figure, Composition, and Gravity of the Earth

081.001 Analysis of the orbit of Cosmos 387 (1970-111A) near 15th-order resonance. D. G. King-Hele.
Planet. Space Sci., Vol. 22, 509 - 524 (1974).

The variation of orbital inclination while the satellite was experiencing 15th-order resonance, as given by these 19 orbits and 55 U.S. Navy orbits, has been analysed to obtain equations accurate to 4 per cent for the geopotential coefficients of order 15 and odd degree (15, 17, 19...). These equations have subsequently been used (with others) in determining individual coefficients of order 15 and odd degree.

081.002 Green's function and tidal prediction. D. J. Webb.
Rev. Geophys. Space Phys., Vol. 12, 103 - 116 (1974).

This paper is concerned with applying Green's function techniques to the theory of the tides. Relationships between the mathematics of the Green's function and the physics of the ocean are developed, and the Green's function is then used to formally solve the tidal equation. By making assumptions about the smoothness of a related quantity, the response function, one can obtain a simple equation for tidal prediction.

081.003 General relationships among sound speeds. I. New experimental information. D. H. Chung.
Phys. Earth Planet. Interiors, Vol. 8, 113 - 120 (1974).

New ultrasonic data on some crystal structures important in mantle mineralogy are presented for compounds of Mg_2GeO_4, Fe_2GeO_4 and Fe_2SiO_4, both in olivine and spinel structures, and of GeO_2, SnO_2 and SiO_2 in the rutile structure, as their elastic parameters were determined by a powder-matrix method. With these new elasticity data, a power law for the velocity-density relationship arising from lattice dynamics is examined. The power law is useful in the study of the elasticity and constitution of the elasticity and constitution of the deep interiors of the earth and other planets.

081.004 Apparent polar wander relative to Australia during the Precambrian. R. A. Facer.
Earth Planet. Sci. Letters, Vol. 22, 44 - 50 (1974).

Palaeomagnetic data for Australian Precambrian rocks allow the preparation of a tentative apparent polar wander curve. Comparison of the Australian Precambrian apparent polar wander curve with proposed curves for Europe, Siberia, North America and Africa provides further evidence for the suggestion of relative continental movement during the Precambrian.

081.005 The absolute value of the geopotential. J. A. O'Keefe III.
Bull. Géod., Nouvelle Sér., Année 1974, No. 111, p. 81 - 84.

It is suggested that it would be worthwhile to determine the absolute value of the geopotential on the geopotential surface which corresponds to mean sea level. This number would replace the earth's semi-major axis as the parameter which fixes the earth's size. Fixing this number involves knowing the geopotential for a point on the orbit of a satellite whose true gravitational potential is also known.

081.006 La représentation du potentiel terrestre par masses ponctuelles. G. Balmino.
Bull. Géod., Nouvelle Sér., Année 1974, No. 111, p. 85 - 108.

Un modèle de représentation du potentiel terrestre par 126 masses ponctuelles de profondeurs situées entre 1000 et 1500 km a été construit à partir de la Standard Earth II du S.A.O. et utilisé avec succès pour la représentation du géoide, des anomalies de gravité, ainsi qu'en calcul d'orbites par correction différentielle utilisant des observations réelles de satellites artificiels. Les propriétés de décroissance rapide de ces fonctions sont mises en évidence, et leur utilisation envisagée à l'analyse scientifique de mesures altimétriques.

081.007 Geophysics — an overview. A. F. Spilhaus, Jr.
Phys. Today, Vol. 27, No. 3, p. 23 - 26 (1974).

081.008 Refining geoid heights. B. P. Day.
Nature, Vol. 249, 197 (1974). — Letter.

081.009 Development in geophysics. R. A. Lyttleton.
Astrophys. Space Sci., Vol. 28, L1 - L5 (1974). Letter.

081.010 Distribution of the normal gravity and expression of the parameters of the earth ellipsoid by expansion in ellipsoidal functions. L. A. Savrov.
Vestn. Mosk. un-ta. Fiz., astron., Vol. 14, 422 - 425 (1973). In Russian. — Abstr. in Referativ. Zhurn. 52. Geodeziya i Aehrosemka, 4.52.61 (1974).

081.011 On some questions of the problem of reductions of gravity. S. V. Evseev.
Geofiz. sb. AN USSR, 1973, vyp. (No.) 55, p. 3 - 8. In Russian. — Abstr. in Referativ. Zhurn. 52. Geodeziya i Aehrosemka, 4.52.64 (1974).

081.012 Application of the collocation method to the accuracy estimation of the earth gravity field determination. J. B. Zieliński.
Geod. i Kartografia, Vol. 23, 3 - 20 (1974). In Polish.

In the paper the formulae are developed which enable to estimate the accuracy of the earth gravity field determination from satellite orbit. The collocation method is briefly described.

081.013 Plate movement and continental magmatism. J. C. Briden, I. G. Gass.
Nature, Vol. 248, 650 - 653 (1974).

Extensive magmatic and metamorphic events have occurred within the earth's continental crust far from plate margins. During the last 800 million years Africa has only been host to such events during the breaks in the systematic motion of the African plate. This suggests that sublithospheric heat sources have little effect on a moving plate and can only produce recognisable effects in the upper crust if focused for long periods on the same part of the lithosphere.

081.014 Years of peak astronomical tides. D. E. Cartwright.
Nature, Vol. 248, 656 - 657 (1974). — Letter.

081.015 Comment on 'The direct mapping of gravity anomalies using Doppler tracking between a satellite pair' by Gary C. Comfort.   C. R. Schwarz, F. Morrison.
Journ. Geophys. Res., Vol. 79, 1233 - 1234, with a reply by G. C. Comfort, p. 1235 (1974).

081.016 Premières mesures d'albédo du nivéo-éolien sur terre.   A. Cailleux.
Comptes Rendus Acad. Sci. Paris, Sér. B, Vol. 278, 989 - 991 (1974).

081.017 La terre: une planète qui ne tourne pas rond.   K. Lambeck, F. Barlier.
Sci. Progrès Découverte, No. 3446, p. 40 - 47 (1972).

081.018 **Plumes in the mantle.**   M. A. Khan.
Goddard Space Flight Center, Greenbelt, Maryland, GSFC Document X-592-73-334, Preprint, 6 + 21 pp. (1973).
Free air and isostatic gravity anomalies for the purposes of geophysical interpretation are presented.

081.019 **Fifteenth-order harmonics in the geopotential.**
D. G. King-Hele, D. M. C. Walker, R. H. Gooding.
Nature, Vol. 249, 748 - 750 (1974). — Letter.

081.020 **Marées terrestres.**   P. Melchior (Editor).
Bull. d'Informations, (Obs. Roy. Belgique, Bruxelles), No. 68, p. 3754 - 3838 (1974).

081.021 **Analysis of 27 satellite orbits to determine odd zonal harmonics in the geopotential.**
D. G. King-Hele, G. E. Cook.
Planet. Space Sci., Vol. 22, 645 - 672 (1974).
The odd zonal harmonics in the geopotential are the terms independent of longitude and antisymmetric about the equator: they define the 'pear-shape' effect. The coefficients $J_3, J_5, J_7, \ldots$ of these harmonics have been evaluated by analyzing the variations in eccentricity of 27 orbits covering a wide range of inclinations. With the new set of values of the odd harmonics $J_3 - J_{17}$ the pear-shape tendency of the earth amounts to 44.7 m at the poles, instead of the previous 40 m, though the new geoid is within 1 m of the old at latitudes away from the poles.

081.022 **Harmonic analysis of earth tides.**   P. Melchior.
Methods Computational Phys., [Academic Press, Inc., New York - London], Vol. 13, 271 - 341 = Publ. Inst. Astron. Géophys., Georges Lemaître, Louvain, Vol. 6, No. 5 (1973).

081.023 **On the representation of the actual gravity field of the earth on the normal ellipsoidal field.**
A. Marussi.
Geophys. Journ. Roy. Astron. Soc., Vol. 37, 347 - 352 (1974).
An attempt is made to establish a one-to-one correspondence between points of the actual gravity field of the earth and the normal ellipsoidal one both referred to geographic co-ordinates and to potential, which permits the univocal definition of anomalies and deflections, and the use of the simple metric tensor of the normal field for their analytical manipulation. The fundamental equation of physical geodesy, as well as the integrability conditions for deflections and anomalies are derived.

081.024 **The scalar equations of infinitesimal elastic-gravitational motion for a rotating, slightly elliptical earth.**   M. L. Smith.
Geophys. Journ. Roy. Astron. Soc., Vol. 37, 491 - 526 (1974).
The purpose of this paper is to provide a theoretical formulation for describing and computing the infinitesimal normal modes of a rotating, slightly elliptical earth which is everywhere in hydrostatic equilibrium and has an isotropic elastic constitutive relation. The results of this paper, which are essentially generalizations of the theory of the free oscillations of a spherically symmetric earth, should enable us to investigate more carefully the elastic-gravitational normal mode spectrum of the rotating earth and, in turn, to assess the validity of this model in describing observed motions of the real earth.

081.025 **On the breakup of tectonic plates by polar wandering.**   H.-S. Liu.
Journ. Geophys. Res., Vol. 79, 2568 - 2572 (1974).
The observed boundary system of the major tectonic plates on the surface of the earth lends fresh support to the hypothesis of polar wandering. In this paper a dynamic model of the outer shell of the earth under the influence of polar shift is developed.

081.026 **Gravimetry.**   V. P. Shcheglov.
Astronomical Institute of the Academy of Sciences of the Uzbek SSR — 100 years, (see 003.011), p. 135 - 138 (1974). In Russian.

081.027 **Possibilities of determination of the local anomalies of the earth's gravitational field from satellite observations.**   J. Žagars, N. Žagars.
Uch. Zap. Latv. Gos. Univ., Vol. 202, Astron., vyp.(No.) 10, (see 003.012), p. 75 - 83 (1974). In Russian.

081.028 **Contribution à l'amélioration du potentiel terrestre. Choix d'une représentation — techniques nouvelles de détermination.**   G. Balmino.
Bull. Groupe Recherches Géod. Spatiale, No. 12, 1 + 71 pp. (1974).
A representation of the earth's potential by point masses has been derived from the S. A. O. Standard Earth II model (1969), and successfully used for computing geoid heights and gravity anomalies as well as satellite orbits by differential correction using observations. Besides, the author studied in detail the importance of new experiments, such as having a drag-free satellite in a low orbit, satellite to satellite tracking and satellite altimetry, for improving our knowledge of the earth's gravity field. Satellite altimetry in particular has been further studied since it can also be used successfully for determining satellite orbits.

081.029 **Bestimmungsgleichungen für Resonanzparameter der Ordnung 13 aus der Analyse von Bahnen der Satelliten GEOS B, BEC und D1D.**   C. Reigber.
Deutsche Geod. Kommission, Bayer. Akad. Wiss. München, Ser. C, No. 198, 120 pp. (1974). — Thesis for the habilitation, Techn. Univ. München.
With the aid of optical and laser data for a total of 8 arcs of the GEOS B (680021)-, BEC (650321)- and D1D (670141)- satellites 110 observation equations for the 13th order resonant harmonics have been derived by analyzing the resonance effect in the mean longitude, the inclination and the eccentricity. The necessary principles of the analysis are described in detail. The 110 equations have been solved for a preliminary 10 coefficients-solution up to the (18.13)-terms.

081.030 **Estimate of gravity anomalies.**
M. R. Williamson, E. M. Gaposchkin.
Smithsonian Astrophys. Obs., *Cambridge, Mass.*, Special Rep. 353, (see 013.016), p. 193 - 228 (1973).
The method of obtaining $5° \times 5°$ mean gravity anomalies from $1° \times 1°$ mean free-air gravimetry data is discussed, and various estimate procedures are considered. The assumption of the stationarity of the gravity data is also investigated. We

conclude that a simplified estimate procedure is the best one for obtaining the 5°× 5° mean anomalies.

081.031 **Determination of the geopotential.**
E. M. Gaposchkin, M. R. Williamson, Y. Kozai, G. Mendes.
Smithsonian Astrophys. Obs., *Cambridge, Mass.*, Special Rep. 353, (see 013.016), p. 229 - 308 (1973).

Laser and optical satellite-tracking data are combined with surface-gravity data to determine spherical harmonics representing the geopotential to 18th degree and order.

081.032 **Die Niveauflächen der Erde nach der Integralgleichung für das gravimetrische Zusatzglied und anderen Verfahren.** K. Arnold.
Akad. Wiss. DDR, Forschungsbereich Kosmische Phys., Veröff. Zentralinst. Phys. Erde, No. 19, 61 pp. (1973).

081.033 **Current estimates of mean earth ellipsoid parameters.** R. H. Rapp.
Geophys. Res. Letters, Vol. 1, No. 1, p. 35 - 38 (1974).

This paper examines new data and considers an adjustment technique to determine the parameters and their standard deviations of the mean earth ellipsoid.

081.034 **A method of evaluating the truncation error coefficients for geoidal height.** M. K. Paul.
Bull. Géod., Nouvelle Sér., Année 1973, No. 110, p. 413 - 425 = Contr. Earth Phys. Branch, *Ottawa, Canada*, No. 500 (1973).

Bibliographie générale des marées terrestres. Supplement I: 1972 - 1973. See Abstr. 002.016.

The earth from space. See Abstr. 003.024.

A revolution in the earth sciences: From continental drift to plate tectonics. See Abstr. 003.054.

Physik des Erdkörpers. See Abstr. 003.113.

Modified Jacobi polynomial and its applications to expansions of disturbing functions. See Abstr. 042.053.

Distribution of gold and rhenium between nickel-iron and silicate melts: implications for the abundance of siderophile elements on the earth and moon.
See Abstr. 094.375.

082 The Earth's Atmosphere Including Refraction, Scintillation, Extinction, Airglow, Site Testing

082.001 **Photographing the rising or setting sun.**
M. B. Stewart.
Sky Telescope, Vol. 47, 127 - 129 (1974).

082.002 **The light of the night sky.** R. G. Roosen.
Sky Telescope, Vol. 47, 231 - 234 (1974).

082.003 **Measurement of nitric oxide in the stratosphere between 17.4 and 22.9 km.**
B. A. Ridley, H. I. Schiff, A. W. Shaw, L. R. Megill, L. Bates, C. Howlett, H. Levaux, T. E. Ashenfelter.
Planet. Space Sci., Vol. 22, 19 - 24 (1974).

082.004 **Comments on the paper: 'Diurnal, annual and solar cycle variations of hydroxyl and sodium nightglow intensities in the Europe-Africa sector'.**
G. Moreels, R. L. Gattinger, A. V. Jones.
Planet. Space Sci., Vol. 22, 344 - 346 (1974). – Research note.

082.005 **Analysis of the orbit of 1970–65D, Cosmos 359 rocket.** D. M. C. Walker.
Planet. Space Sci., Vol. 22, 391 - 402 (1974).

2600 observations from the Hewitt cameras at Malvern and Edinburgh were available for 10 of the 42 orbits of Cosmos 359 rocket. Ten values of density scale height, at heights between 185 and 261 km, have been determined from analysis of the variations in perigee height. The average atmospheric rotation rate, for heights near 220 km, is found to be 1.04 rev/day.

082.006 **Air density at heights near 200 km from the orbit of 1970-65D.** D. M. C. Walker.
Planet. Space Sci., Vol. 22, 403 - 411 (1974).

Analysis of the orbit of 1970-65D has yielded 146 values of air density, which establish the variations in density between August 1970 and September 1971, at heights of 180–230 km. After corrections for day-to-night and solar activity variations and allowance for the expected increases in density at times of geomagnetic disturbance, a picture of the semi-annual variation emerges. The maximum density (November and April) was 1.5 times the minimum density (January and July) at heights near 225 km.

082.007 **The interpretation of 6300 Å airglow observations of ionospheric irregularities.**
P. L. Dyson, P. A. Hopgood.
Planet. Space Sci., Vol. 22, 495 - 497 (1974). – Research note.

082.008 **A measurement of the ozone concentration from 65 to 75 km at night.**
C. G. De Jonckheere, D. E. Miller.
Planet. Space Sci., Vol. 22, 497 - 499 (1974).

The concentration of ozone between 65 and 75 km has been determined from measurements of the attenuation of moonlight at 2570 Å made from a moon-pointing rocket payload. The results support earlier rocket measurements, but are in marked disagreement with some recent data obtained by Hays and Roble using the stellar occultation technique.

082.009 **Superrotation of upper atmosphere by global deposition of meteoroids.** V. Mitra.
Planet. Space Sci., Vol. 22, 559 - 568 (1974).

A new theory of the superrotation of upper atmosphere is worked out on the basis of global deposition of meteoroids assuming that a certain constant influx of meteoroids is continually falling upon the earth's atmosphere. It is found that a global deposition of 34 tons/day of meteoric material is required to account for the observed superrotation which agrees with the recent estimates on meteoric mass influx on the earth.

082.010 **The effect of displaced geomagnetic and geographic poles on the thermospheric neutral winds.**
R. G. Roble, R. E. Dickinson.
Planet. Space Sci., Vol. 22, 623 - 631 (1974).

082.011 **Association between the mid-winter stratospheric circulation and the first observation in the following summer of noctilucent clouds.** A. F. D. Scott.
Nature, Vol. 247, 269 - 271 (1974). – Letter.

082.012 **On the ozone at the basis of the thermosphere.**
A. I. Semenov.
Kosmich. Issled., Vol. 12, 146 (1974). In Russian. – Brief information.

082.013 **Staubschichten in der Atmosphäre.** F. Rössler.
Umschau, 74. Jahrgang, p. 89 - 90 (1974).

The scattered sunlight was measured as a function of height using rockets and balloons. A quantitative analysis allowed to determine the number of aerosols.

082.014 **Über den Energiehaushalt der Erdatmosphäre.**
F. Kasten.
Phys. Blätter, 30. Jahrgang, p. 53 - 63 (1974).

082.015 **Analyse de spectres stratosphériques de NO et de NO_2.**
J.-C. Fontanella, A. Girard, L. Gramont, N. Louisnard.
Comptes Rendus Acad. Sci. Paris, Sér. B, Vol. 278, 181 - 184 (1974).

082.016 **Flux de chaleur sensible et de vapeur d'eau associés à un «saut convectif» atmosphérique.**
B. Guillemet, R. Rosset, P. Mascart, H. Isaka.
Comptes Rendus Acad. Sci. Paris, Sér. B, Vol. 278, 231 - 233 (1974).

Les flux de chaleur et de vapeur d'eau présentent, au niveau du front convectif, un caractère intermittent lié au feuilletage de la couche limite planétaire.

082.017 **Prédétermination de l'intensité du rayonnement solaire direct par ciel clair et sous faible teneur de l'atmosphère en aérosols.** J. Fléchon, I. Touré.
Comptes Rendus Acad. Sci. Paris, Sér. B, Vol. 278, 365 - 367 (1974).

Les auteurs proposent une méthode de prédétermination du rayonnement solaire direct par ciel clair et sous faible teneur de l'atmosphère en aérosols. Le coefficient de trouble d'Ångström β étant inférieur à 0.1.

082.018 **Astronomical seeing and meteorological air mass analysis.** B. McInnes, M. Hartley, T. T. Gough.
Observatory, Vol. 94, 14 - 17 (1974).

The main purpose of this communication is to draw the attention of astronomers to the meteorological concept of air-mass source regions.

082.019 **Nucleation processes and aerosol chemistry.**
A. W. Castleman, Jr.

Space Sci. Rev., Vol. 15, 547 - 589 (1974).

As the result of many observable optical phenomena, the occasional existence of upper atmospheric aerosols has been known since the beginning of the century. Nevertheless, it is only during the last two decades that their persistent nature and extent of global distribution have become recognized. This review is addressed to the chemistry of upper atmospheric aerosols with particular attention to the chemical reactions and nucleation mechanisms responsible for their formation.

082.020 **Upon estimation of effective concentration of neutral atmosphere minor constituents NO and N on the basis of experimental values $[NO^+]/[O_2^+]$.**
V. V. Koshelev, S. I. Belinskaya.
Journ. Atmosph. Terr. Phys., Vol. 36, 315 - 323 (1974).

Presented are the estimations on effective concentration of neutral atmosphere minor constituents NO and N within a height range of 100–200 km for daytime and nighttime conditions. The estimations were made by using experimental values $[NO^+]/[O_2^+]$. In the final result the analysis of possible errors, caused by uncertainties in the data used, shows that such an error in estimation of effective concentration actually depends on how accurate is our knowledge of dissociative recombination rate constants of ions O_2^+ and NO^+.

082.021 **Fast variations of hydroxyl night airglow emission.**
M. V. Shagaev.
Journ. Atmosph. Terr. Phys., Vol. 36, 367 - 371 (1974).

The device is described which permits the photographing of hydroxyl emission spectra within exposures beginning with 0.5 min. Relationships between rotational temperatures and the intensities of hydroxyl bands 9-5 and 4-1 are given, as well as vibrational temperature variations.

082.022 **Vertical propagation of tides at meteor heights.**
J. L. Fellous, A. Spizzichino, M. Glass, M. Massebeuf.
Journ. Atmosph. Terr. Phys., Vol. 36, 385 - 396 (1974).

The vertical propagation of atmospheric tides has been studied from meteor trail observations over Garchy (France). A set of experimental data is described. It is found that higher-order modes of the semi-diurnal tide, and evanescent modes of the diurnal tide are often present, in particular during stratospheric warming events.

082.023 **Simultaneous measurements of the OH* nightglow in the (9-4), (8-3) and (5-1) bands and effects of quenching.** G. Fiocco, G. Visconti.
Journ. Atmosph. Terr. Phys., Vol. 36, 583 - 590 (1974).

Simultaneous measurements of the OH* nightglow emission in the (9-4), (8-3) and (5-1) bands have been carried out with two coupled tilting interference filter photometers in the period from May to December 1971. The nocturnal variation of the relative intensities is interpreted, with the help of existing mesospheric models, to ascertain the importance of the quenching processes.

082.024 **Numerical studies of oxygen-hydrogen constituents in the mesosphere and thermosphere.–Effect of changing chemical rate coefficients.**
M. R. Bowman, L. Thomas.
Journ. Atmosph. Terr. Phys., Vol. 36, 657 - 665 (1974).

An attempt is made to resolve discrepancies between observed concentrations of certain neutral constituents and those obtained previously from time-dependent solutions of the diffusion and continuity equations for an oxygen-hydrogen atmosphere. Particular attention is paid to the effects of revisions in chemical rate coefficients, but changes of eddy diffusion coefficient are also considered.

082.025 **New evidence concerning the origin of atmospheric He-3.**
F. Bühler, W. I. Axford, H. J. A. Chivers, K. Marti.
Meteoritics, Vol. 8, 334 (1973). – Abstr. Meteoritical Soc.

082.026 **On the transparency of the earth's atmosphere in the infrared region up to 2.5 μ obtained from solar observations at Pulkovo.** N. F. Kuprevich, A. Kh. Kurmaeva.
Solnechnye Dannye 1974 Byull., No. 1, p. 95 - 100 (1974). In Russian.

082.027 **The role of reaction of radiative association of atomic oxygen in the earth's atmosphere.**
B. A. Mirtov.
Geomagn. Aeronom., Vol. 14, 78 - 82 (1974). In Russian.

082.028 **About vertical intermixing in the meteor zone of the atmosphere.** I. A. Delov.
Geomagn. Aeronom., Vol. 14, 170 - 172 (1974). In Russian. Brief information.

082.029 **Observations of He I 584 Å nighttime radiation; evidence for an interstellar source of neutral helium.**
F. Paresce, S. Bowyer, S. Kumar.
Astrophys. Journ., Vol. 187, 633 - 639 (1974).

The intensity and spatial variations of the He I 584 Å radiation in the night sky were measured to an altitude of 264 km from a sounding rocket launched from Thumba, India. The data obtained are presented in the form of an all-sky map. Since the peak flux levels observed are approximately two orders of magnitude larger than that expected from geocoronal resonant scattering, the authors have compared their results with those expected from two interplanetary sources of neutral helium: the penetration of the local interstellar medium into the solar neighborhood and the dust deionization of solar-wind alpha particles. The interstellar helium source is compatible with the data to within the uncertainties of the models.

082.030 **On the distribution of He^+ in the plasmasphere from observations of resonantly scattered He II 304-Å radiation.** F. Paresce, C. S. Bowyer, S. Kumar.
Journ. Geophys. Res., Vol. 79, 174 - 178 (1974).

The intensity of the He II 304-Å radiation in the night sky was measured from a sounding rocket launched from White Sands Missile Range on June 9, 1972. The data obtained on this flight are compared with predictions of theoretical models of the He^+ altitude distribution within the plasmasphere.

082.031 **Empirical model of global thermospheric temperature and composition based on data from the OGO 6 quadrupole mass spectrometer.** A. E. Hedin, H. G. Mayr, C. A. Reber, N. W. Spencer, G. R. Carignan.
Journ. Geophys. Res., Vol. 79, 215 - 225 (1974).

An empirical global model for magnetically quiet conditions has been derived from longitudinally averaged N_2, O, and He densities by means of an expansion in spherical harmonics. The data were obtained by the OGO 6 neutral mass spectrometer and cover the altitude range 400–600 km for the period June 27, 1969, to May 13, 1971.

082.032 **Observations of the extreme ultraviolet nightglow.**
G. R. Riegler, G. P. Garmire.
Journ. Geophys. Res., Vol. 79, 226 - 232 (1974).

Measurements of extreme ultraviolet radiation from the night sky were made with a set of Parylene, aluminum, and tin filters behind a grazing incidence concentrator aboard an Aerobee 170 rocket. If the aluminum and tin filter data are interpreted as being due to only the He I 584-Å and He II 304-Å lines, the nightglow flux at 25° zenith angle and 270°

azimuth angle at 180-km altitude was $I(584) = 4.4 \pm 1.9$ R and $I(304) = 1.3 \pm 0.5$ R.

082.033 Evolution with solar activity of the atomic hydrogen density at 100 kilometers of altitude.
A. Vidal-Madjar, J. E. Blamont, B. Phissamay.
Journ. Geophys. Res., Vol. 79, 233 - 241 (1974).

The purpose of this paper is to extend the analysis of the data on the solar Lyman alpha line shape already published by Vidal-Madjar et al. (1973) by including two new factors: the use of a more precise exospheric temperature model, the Jacchia (1971) evaluation including the geomagnetic heating, and the analysis of 1 more year of data (1971) extending our observations toward lower solar activity conditions.

082.034 The stratospheric dust event of October 1971. F. E. Volz.
Journ. Geophys. Res., Vol. 79, 479 - 482 (1974). — Brief report.

082.035 Global characteristics in the diurnal variations of the thermospheric temperature and composition.
H. G. Mayr, A. E. Hedin, C. A. Reber, G. R. Carignan.
Journ. Geophys. Res., Vol. 79, 619 - 628 (1974).

082.036 $CO_2(001)$ and N_2 vibrational temperatures in the $50 \lesssim z \lesssim 130$ km altitude range.
J. B. Kumer, T. C. James.
Journ. Geophys. Res., Vol. 79, 638 - 648 (1974).

082.037 Metastable helium in the earth's upper atmosphere.
R. D. Rundel, R. F. Stebbings.
Journ. Geophys. Res., Vol. 79, 681 - 684 (1974). — Letter.

082.038 Preliminary data on the astropoint Akarhar.
V. N. Petrakov, V. G. Popov, V. V. Chichmar, S. P. Yatsenko.
Astron. Tsirk., No. 799, p. 6 - 8 (1973). In Russian.

082.039 The multifariousness of components of the seeing parameters. N. I. Kozhevnikov.
Astron. Tsirk., No. 802, p. 5 - 7 (1973). In Russian.

082.040 Dynamical analytical model of the upper atmosphere of the earth. M. A. Degtyarev.
Kosmich. Issled., Vol. 12, 311 - 312 (1974). In Russian. Brief information.

082.041 Atmospheric extinction at McDonald Observatory, 1960–68.
G. de Vaucouleurs, R. J. Angione.
Publ. Astron. Soc. Pacific, Vol. 86, 104 - 115 (1974).

Atmospheric extinction coefficients measured in the U, B, V bands at McDonald Observatory on 280 nights between December 1960 and October 1968 have been analyzed to derive the effects of seasonal variations, volcanic eruptions, and possibly secular trends. The spectral dependence of the various effects are interpreted in terms of a variable tropospheric dust content with a λ^{-1} extinction law, and fluctuating amount of roughly neutral volcanic smoke averaging 0.03 mag since March 1963.

082.042 Determination of nitric oxide concentrations from eclipse variations of ion concentrations.
W. L. Oliver.
Journ. Atmosph. Terr. Phys., Vol. 36, 801 - 810 (1974).

It is shown how nitric oxide concentrations in the upper atmosphere may be derived from the observation of changes in the $[NO^+]/[O_2^+]$ concentration ratio and electron concentration and temperature. Experimental data obtained during the 1970 March 7 eclipse at Wallops Island are analyzed to give nitric oxide concentration profiles above 130 km. A value of 2×10^7 cm^{-3} is obtained for the concentration at 130 km.

082.043 Vergleichende Messungen des astronomischen Seeing in Griechenland, Spanien, Südwestafrika und Chile. K. Birkle.
Diss. Naturwiss. Gesamtfakultät Ruprecht-Karl-Univ. Heidelberg. 5 + 62 pp. (1973).

082.044 Observation of $O^+(^2P^0 - ^2D^0)$ $\lambda 7,319$ Å emissions in the twilight and night airglow.
R. W. Carlson, K. Suzuki.
Nature, Vol. 248, 400 - 401 (1974). — Letter.

082.045 The neutral atmosphere. E. Hesstvedt.
Cosmical geophysics, (see 003.004), p. 63 - 71 (1973).

082.046 Effects of lunar tides on the $H\alpha$ emission intensity of the upper atmosphere. N. M. Martsvaladze.
Byull. Abastumansk. Astrofiz. Obs., No. 45, p. 87 - 92 (1974). In Russian.

Effects of lunar tides on the intensity of $H\alpha$ emission have been drawn from spectrographic (1958–1970) and interferometric (1968–1970) observations carried out in Abastumani. The intensity variations due to lunar tides have their minimum and maximum at new moon and at the quadratures, respectively.

082.047 Variations of scattered twilight intensity at the maximum of solar activity. T. G. Megrelishvili.
Byull. Abastumansk. Astrofiz. Obs., No. 45, p. 93 - 108 (1974). In Russian.

Seasonal, diurnal and altitude variations of scattered twilight intensity have been investigated at Abastumani Astrophysical Observatory. These investigations are based on spectral observations. The intensity variation is compared with solar activity and some characteristics of the lower atmosphere. It is stated that temporal and spatial intensity variations of scattered light reflect solar activity, seasonal solar radiation flux with a wavelength 10.7 cm, annual variation of total ozone amount as well as characteristic temperature and relative humidity of the layers near the troposphere. Semi-annual variations of the scattered light intensity have been revealed at the height of 70–100 km.

082.048 High-altitude dependence of the scattering coefficient and air turbidity in the mesosphere according to twilight probing data. G. G. Mikirtumova.
Byull. Abastumansk. Astrofiz. Obs., No. 45, p. 109 - 120 (1974). In Russian.

Scattering coefficient calculations and comparison of two independent methods of its determination have been made on the basis of measured sky brightness obtained within several years. High-altitude dependences of scattering coefficients and turbidity values are plotted.

082.049 Correlation analysis of polarimetric data of the twilight sky: The mesospheric aerosol layer.
G. G. Mikirtumova, G. V. Rozenberg.
Byull. Abastumansk. Astrofiz. Obs., No. 45, p. 121 - 128 (1974). In Russian.

Some 135 series of polarimetric measurements of twilight sky in the zenith have been analyzed. Standard autocorrelation matrices, intrinsic functions and intrinsic numbers were computed. As a result of the analysis a mesospheric aerosol layer was insulated at the level of 40–60 km.

082.050 Some results of measurements of the vertical ozone distribution in Abastumani from observations of 1971. V. M. Iskandarova.

Byull. Abastumansk. Astrofiz. Obs., No. 45, p. 129 - 134 (1974). In Russian.

082.051 **Variations of the total ozone amount with reference to the axis of the jet stream over the territory of Georgia.** D. F. Kharchilava, V. M. Iskandarova.
Byull. Abastumansk. Astrofiz. Obs., No. 45, p. 135 - 140 (1974). In Russian.

082.052 **The influence of direction and velocity of the wind on the ozone profile of the jet stream over the territory of Georgia.** D. F. Kharchilava, V. M. Iskandarova.
Byull. Abastumansk. Astrofiz. Obs., No. 45, p. 141 - 144 (1974). In Russian.

082.053 **About the problem of helium distribution in the upper atmosphere.**
L. A. Antonova, G. S. Ivanov-Kholodnyj.
Geomagn. Aeronom., Vol. 14, 287 - 292 (1974). In Russian.

082.054 **Excitation of the atmospheric luminescence with ionization by solar radiation.**
V. A. Krasnopolskij.
Geomagn. Aeronom., Vol. 14, 293 - 299 (1974). In Russian.

082.055 **Discussion on possible connections between the parameters of the meteor zone and the lower atmosphere.** K. A. Karimov, A. A. Shnejder.
Komety i Meteory, *Dushanbe*, No. 22, p. 46 - 51 (1973). In Russian.

The paper presents some aspects of ionosphere-troposphere coupling effects in September–October 1971. A method is given of determining the average temperature in the meteor zone.

082.056 **On a statistical analysis of average daily fluctuations of the upper atmosphere's parameter.**
V. E. Chertoprud.
Prikl. zadachi kosmich. ballistiki. Moskva, Nauka, 1973, p. 68 - 93. In Russian. – Abstr. in Referativ. Zhurn. 51. Astron., 4.51.493 (1974).

082.057 **Semi-annual variations of the atmosphere's density at altitudes of 200 - 300 km.** A. M. Alferov.
Prikl. zadachi kosmich. ballistiki. Moskva, Nauka, 1973, p. 107 - 119. In Russian. – Abstr. in Referativ. Zhurn. 51. Astron., 4.51.494 (1974).

082.058 **Biological modulation of the earth's atmosphere.** L. Margulis, J. E. Lovelock.
Icarus, Vol. 21, 471 - 489 (1974).
The authors review the evidence that the earth's atmosphere is regulated by life on the surface so that the probability of growth of the entire biosphere is maximized.

082.059 **Über den Fortgang des JOSO-Projektes.**
K. O. Kiepenheuer.
Mitt. Astron. Ges., No. 34, p. 101 - 102 (1973/74).
Presented at the "Wissenschaftliche Tagung der Astron. Ges., Oberkochen, 1973 April 24 - 27".

082.060 **Polarization of the airglow emission λ6300 in an artificially heated ionosphere.**
J. W. Chamberlain.
Journ. Geophys. Res., Vol. 79, 1239 - 1241 (1974).

082.061 **Mid-latitude artificial noctilucent clouds initiated by high-altitude rockets.**
B. Benech, J. Dessens.
Journ. Geophys. Res., Vol. 79, 1299 - 1301 (1974).

082.062 **Airglow.** G. Kvifte.
Cosmical geophysics, (see 003.004), p. 81 - 92 (1973).

082.063 **Contamination of the O I ($^3P_2$-$^1D_2$) emission line by the (9-3) band of OH $X^2\Pi$ in high-resolution measurements of the night sky.** G. Hernandez.
Journ. Geophys. Res., Vol. 79, 1119 - 1123 (1974). – Brief report.

082.064 **Twilight airglow. 2. N_2^+ emission at 3914 Å.**
W. E. Sharp.
Journ. Geophys. Res., Vol. 79, 1569 - 1571 (1974). Brief report.

082.065 **First satellite observations of the He^+ 304-Å radiation and its interpretation.**
C. S. Weller, R. R. Meier.
Journ. Geophys. Res., Vol. 79, 1572 - 1574 (1974). Brief report.

082.066 **Extreme ultraviolet observations of the latitudinal variation of helium.**
R. R. Meier, C. S. Weller.
Journ. Geophys. Res., Vol. 79, 1575 - 1578 (1974). Brief report.

082.067 **Lysende natskyer 1973.** J. Ø. Olesen.
Astron. Tidssk., Årg. 7, p. 39 - 40 (1974).

082.068 **Circumsolar sky radiation and turbidity of the atmosphere.** A. Ångström.
Applied Optics, Vol. 13, 474 - 477 (1974).
A statistical treatment of field measurements carried out by the Astrophysical Observatory of the Smithsonian Institution is presented. The brightness of a band of sky 10° wide, concentric with the sun, has been determined. The primary object was to obtain an idea of the integral scattering of the sun's radiation by the atmospheric aerosol. Results have been used to determine long-periodic changes in the scattering properties of the atmosphere and their relation to other phenomena.

082.069 **Some measurements of the attenuation of solar radiation during BOMEX.**
A. J. Drummond, G. D. Robinson.
Applied Optics, Vol. 13, 487 - 492 (1974).
Measurements of the upward and downward flux of solar radiation in two broad wavelength bands defined by glass filters were made at height between about 300 m and 12000 m in the neighborhood of Barbados during the BOMEX experiment in the summer of 1969. These were examined on selected occasions to determine the attenuation – absorption and upward scattering – that could not be attributed to atmospheric gases and must be considered to be caused by particles.

082.070 **Monte Carlo studies of the sky radiation at twilight.**
W. G. Blättner, H. G. Horak, D. G. Collins, M. B. Wells.
Applied Optics, Vol. 13, 534 - 547 (1974).
Results are presented in several wavelengths of the sky intensity and polarization along the solar vertical at twilight, as calculated using the backward Monte Carlo method applied to spherical shell atmospheres. Molecular scattering with anisotropy of molecules, ozone absorption, and refraction are taken into account.

082.071 **Analyse de spectres stratosphériques de NO_2 et HNO_3.** J.-C. Fontanella, A. Girard, L. Gramont, N. Louisnard.

Comptes Rendus Acad. Sci. Paris, Sér. B, Vol. 278, 995 - 998 (1974).

082.072 On the limits of noctilucent clouds within the earth's shadow. N. I. Novozhilov.
Astron. vestn., Vol. 8, 106 - 111 (1974). In Russian.

On the basis of observations performed at the stations of Canadian Centre in 1968 it was established that in 60–70% of the cases the quantity of noctilucent clouds exceeds optimum values. Evaluation of the refraction value capable of causing such a phenomenon is given. A conclusion about evolution of clouds caused directly by mesopause illumination by straight solar rays is also cited.

082.073 A meteorological report for the Mt. Hopkins Observatory: 1968–1971.
M. R. Pearlman, D. Hogan, K. Goodwin, D. Kurtenbach.
Smithsonian Astrophys. Obs., *Cambridge, Mass.*, Special Report 345, 5 + 53 pp. (1972).

082.074 SAO/NASA joint investigation of astronomical viewing quality at Mt. Hopkins Observatory: 1969–1971. M. R. Pearlman, J. L. Bufton, D. Hogan, D. Kurtenbach, K. Goodwin.
Smithsonian Astrophys. Obs., *Cambridge, Mass.*, Special Report 357, 11 + 83 + A6 + B37 + C7 pp. (1974).

Quantitative measurements of the astronomical seeing conditions have been made with a stellar-image monitor system at the Mt. Hopkins Observatory in Arizona. Correlations between seeing quality and local meteorological conditions are investigated. The theoretical basis for the relationship of atmospheric turbulence to optical effects is discussed in some detail, along with a description of the equipment used in the experiment. General site-testing comments and applications of the seeing-test results are also included.

082.075 JOSO site testing campaigns in Lampedusa and Lampione. G. Ceppatelli, A. Righini.
Osservazioni Mem. Oss. Astrofis. Arcetri, Fasc. 102, p. 3 - 52 (1973).

082.076 The micrometeorological instruments for the JOSO 1971 site-testing campaign in Lampione.
R. Barletti, M. Meco, S. Paloschi.
Osservazioni Mem. Oss. Astrofis. Arcetri, Fasc. 102, p. 53 - 64 (1973).

082.077 Tensor of inertia of the atmosphere, annual variations of its components and variations of the earth's rotation. N. S. Sidorenkov.
Izv. AN SSSR. Fiz. atmosf. i okeana, Vol. 9, 333 - 351 (1973). In Russian. − Abstr. in Referativ. Zhurn. 51. Astron., 5.51.183 (1974).

082.078 Some new aspects on the superrotation of the thermosphere. P. W. Blum, I. Harris.
Journ. Atmosph. Terr. Phys., Vol. 36, 967 - 978 (1974).

The motion of the thermosphere with a rotational velocity between 10 and 20 per cent in excess of the earth's rotational velocity has been deduced by King-Hele and his co-workers from the change of the inclination of satellite orbits. To date no completely satisfactorily explanation of the observations has been presented. In this paper it is shown that in the thermosphere there exists a small diurnal mean driving force in the eastward direction. This force has not previously been considered in analyses of superrotation. A critical review of the observations and a theoretical analysis that takes account of both equinox and solstice conditions is presented.

082.079 Correlation of 1.65 and 2.15 μ airglow emissions.
L. M. Kieffaber.
Journ. Atmosph. Terr. Phys., Vol. 36, 1079 - 1086 (1974).

082.080 OH airglow fluctuations at 1.65 and 2.15 μ.
L. M. Kieffaber.
Journ. Atmosph. Terr. Phys., Vol. 36, 1087 - 1091 (1974).

Analyses of fluctuations in airglow intensity at two wavelengths during the past two years are presented.

082.081 The 6300 Å O(^{1}D) airglow and dissociative recombination.
V. B. Wickwar, L. L. Cogger, H. C. Carlson.
Planet. Space Sci., Vol. 22, 709 - 724 (1974).

Measurements of night-time 6300 Å airglow intensities at the Arecibo Observatory have been compared with dissociative recombination calculations based on electron densities derived from simultaneous incoherent backscatter measurements. The agreement indicates that the nightglow can be fully accounted for by dissociative recombination.

082.082 Irradiation: a note. C. H. Cotter.
Journ. Navigation, London, Vol. 27, 261 - 263, with a comment by D. H. Sadler, p. 263 - 265 (1974).

082.083 Comparison of solar seeing and air temperature fluctuations at Oslo Solar Observatory.
T. L. Hansen.
Inst. Theor. Astrophys. Blindern − Oslo, Rep. No. 38, 14 pp. (1973).

The temperature fluctuations in air have been measured, mainly at a height of 12 m. The fluctuations are compared with both the diurnal solar seeing pattern as well as with simultaneous image quality measurements.

082.084 Meteorological observations on La Silla in 1971 and 1972. B. E. Westerlund.
European Southern Obs., Bull. No. 10, p. 5 - 23 (1974).

082.085 On radiative transfer in the 15-mkm band of CO_2 in the upper atmosphere of the earth.
S. R. Drejson, R. O. Manujlova, G. M. Shved.
Kosmich. Issled., Vol. 12, 396 - 401 (1974). In Russian.

082.086 On the concentration of nitric oxide at altitudes more than 100 km. V. A. Krasnopol'skij.
Geomagn. Aeronom., Vol. 14, 487 - 491 (1974). In Russian.

082.087 Day-time variation of the electron concentration of the upper atmosphere.
B. A. Mirtov, A. G. Starkova.
Geomagn. Aeronom., Vol. 14, 556 - 558 (1974). In Russian. Brief information.

082.088 Sunshine duration on Mt Olympus − Greece.
G. C. Livadas, V. A. Semertzidis.
Sci. Ann., Fac. Phys. Math., Univ. Thessaloniki, Vol. 13, 251 - 269 = Meteorologika, Publ. Meteorol. Inst. Univ. Thessaloniki, No. 28 (1973).

082.089 Sunshine duration in Neos Marmaras − Chalkidiki.
G. C. Livadas, P. J. Pennas.
Sci. Ann., Fac. Phys. Math., Univ. Thessaloniki, Vol. 13, 279 - 300 = Meteorologika, Publ. Meteorol. Inst. Univ. Thessaloniki, No. 29 (1973).

082.090 Night airglow intensities on August 9/10, 1972.
A. Takechi.
Rep. Ionosph. Space Res. Japan, Vol. 27, 183 - 185 = Tokyo Astron. Obs., Repr. No. 450 (1973).

082.091 The fish-eye camera observations of the brightness and polarization of the light from blue sky.

A. Tojo, K. Saito.
Tokyo Astron. Obs., Report No. 63, Vol. 16, 726 - 759 (1974). In Japanese.

082.092 Component of seeing conditions near the earth's surface. N. I. Kozhevnikov.
Soobshch. Gos. Astron. Inst. Shternberga, No. 184, p. 21 - 30 (1973). In Russian.

082.093 Possible consequences of nearby supernova explosions for atmospheric ozone and terrestrial life.
M. A. Ruderman.
Science, Vol. 184, 1079 - 1081 (1974).

Hard X-ray pulses or increased cosmic radiation originating in nearby supernova explosions may be capable of temporarily removing most of the earth's atmospheric ozone cover even when direct radiation effects at the earth's surface are negligible. Consequently, terrestrial life may be subject to relatively huge solar ultraviolet fluxes every few hundred million years.

082.094 Space observations of night-sky brightness.
A. B. Severny, E. I. Terez, A. M. Zvereva.
Astron. Inst. Utrecht, Internal Rep. ROF 72, (see 012.012), p. C25.1 (1974). − Abstract.

082.095 Observations of the green flash.
D. di Cicco, G. Ripley, P. Travers.
Sky Telescope, Vol. 48, 61 - 63 (1974).

082.096 The role of eddy turbulence in the development of self-consistent models of the lower and upper thermospheres. S. Chandra, A. K. Sinha.
Journ. Geophys. Res., Vol. 79, 1916 - 1922 (1974).

Numerical solutions of mutually coupled time-dependent equations of continuity, momentum, and energy balance are presented to illustrate the effect of eddy turbulence on the neutral composition and temperature of the lower and upper thermospheres. From the illustrative examples comprising parametric changes in the eddy diffusion coefficient the specific roles of eddy turbulence in the development of theoretical models of the thermosphere are discussed.

082.097 Variations in thermospheric composition: a model based on mass spectrometer and satellite drag data.
L. G. Jacchia.
Journ. Geophys. Res., Vol. 79, 1923 - 1927 (1974).

The seasonal-latitudinal and the diurnal variations of composition observed by mass spectrometers on the Ogo 6 satellite are represented by two simple empirical formulas. The formulas are of a very general nature and predict the behavior of these variations at all heights and for all levels of solar activity; they yield a satisfactory representation of the corresponding variations in total density, as derived from satellite drag.

082.098 Spatial and temporal behavior of atomic oxygen determined by Ogo 6 airglow observations.
T. M. Donahue, B. Guenther, R. J. Thomas.
Journ. Geophys. Res., Vol. 79, 1959 - 1964 (1974).

Maps are produced of the atomic oxygen density near 97 km showing a strong variation in latitude, longitude, universal time, and time of year. These densities are deduced from atomic oxygen green nightglow observations carried out from Ogo 6. Meridional wind patterns needed to support the asymmetries observed in local oxygen production and loss rates are deduced.

082.099 The measurement of O_2 number density by absorption of Lyman α. L. G. Smith, K. L. Miller.
Journ. Geophys. Res., Vol. 79, 1965 - 1968 (1974).

082.100 Diurnal and annual variations of astronomical refraction. A. I. Nefed'eva.
Astron. Zhurn. Akad. Nauk SSSR, Vol. 51, 646 - 651 (1974). In Russian. English translation in Soviet Astron. AJ, Vol. 18, No. 3.

Computations of the astronomical refraction at the Engelhardt Observatory have been made from aerologic observational data. It has been found that the diurnal and annual variations of the astronomical refraction are due to variations of the law of decrease of the atmosphere's density with height during the day and the year.

082.101 Berechnung des atmosphärischen Streulichts für Milchstraße und Zodiakallicht als extraterrestrische Quellen. H. J. Staude.
Diss. Naturwiss. Gesamtfakultät Ruprecht-Karl-Univ. Heidelberg. 8 + 69 + 31 pp. (1974).

082.102 Heating of the high-latitude thermosphere during magnetically quiet periods.
C. A. Reber, A. E. Hedin.
Journ. Geophys. Res., Vol. 79, 2457 - 2461 (1974).

A persistent mid- to high-latitude heating phenomenon is observed in both hemispheres in data from the Ogo 6 quadrupole mass spectrometer. The phenomenon is evidenced by an increase in N_2 density (indicative of a thermospheric temperature rise) and a depletion in helium (indicating an upwelling of air).

082.103 Oxygen recombination in the tropical nightglow.
P. S. Julienne, J. Davis, E. Oran.
Journ. Geophys. Res., Vol. 79, 2540 - 2543 (1974).

In order to explain O I lines in the tropical nightglow, complete and reasonably accurate recombination and emission rate coefficients are used to construct a detailed radiative cascade model of a recombining oxygen plasma. Effective recombination coefficients are given for all low quantum number O I levels for the two limiting cases of zero and infinite optical depth of the resonance lines.

082.104 Satellite observations of the equatorial Mg II dayglow intensity distribution.
J.-C. Gérard, A. Monfils.
Journ. Geophys. Res., Vol. 79, 2544 - 2550 (1974). − Brief report.

082.105 Spectrophotometry of the daylight sky and some questions of interpretation of measurements of artificial earth satellites.
G. Sh. Livshits, L. M. Musorina, Eh. L. Tem.
Pyl' v atmosf. i okolozemn. kosmich. prostranstve. Moskva, Nauka, 1973, p. 151 - 157. In Russian. − Abstr. in Referativ. Zhurn. 62. Issled. kosmich. prostranstva, 6.62.106 (1974).

082.106 Distribution of the electron concentration at heights between 100 and 210 km from data of direct measurements. S. I. Belinskaya, V. V. Koshelev.
Issled. po geomagnetizmu, aehron. i fiz. Solntsa. Vyp. (No.) 29. Moskva, Nauka, 1973, p. 121 - 126. In Russian. − Abstr. in Referativ. Zhurn. 62. Issled. kosmich. prostranstva, 6.62.210 (1974).

082.107 A representation of Jacchia's thermospheric models in spherical harmonic functions.
P. Blum, I. Harris.
Planet. Space Sci., Vol. 22, 955 - 960 (1974).

The Jacchia models are represented in terms of spherical harmonic functions. This representation has the advantage of ease of comparison with other global theoretical and empiric models that use this mathematical form. Furthermore it is analytic, continuous and has continuous derivatives all over the globe. An example of a similar representation for the total mass density at a particular height and level of solar activity is given as well.

082.108 **On the influence of the aerosols for satellite measurements of ozone at the thermospheric base.**
G. V. Rosenberg, A. I. Semenov, N. N. Shefov.
Planet. Space Sci., Vol. 22, 1035 - 1036 (1974).

The dust layer at an altitude of about 80 km causes the attenuation of the light of the stars during the OAO-2 satellite observations near the horizon. This is compared with absorption due to ozone near to 2400 Å in the spectral range, using ozone concentration measured by the OAO-2 satellite.

082.109 **Astroclimatic investigations.**
V. S. Shevchenko, V. G. Khetselius.
Astronomical Institute of the Academy of Sciences of the Uzbek SSR — 100 years, (see 003.011), p. 122 - 129 (1974). In Russian.

082.110 **On the distribution of water vapor in strato- and mesospheres measured by means of space vehicles.**
K. Ya. Kondratyev, A. A. Buznikov, V. P. Kozlov, A. G. Pokrovsky.
Acta Astronaut., Vol. 1, 125 - 133 (1974).

The possibilities of constructing a rational model of the water vapor vertical profile in the stratosphere and mesosphere from spectral measurements of transmittance from space vehicles are investigated. The main requirement to such a model is the agreement of its parameters with the measurement capabilities of the corresponding satellite instruments. The authors show that such an agreement can be obtained by making use of the Riemann information metric in the space of the states of the object measured.

082.111 **On the theory of astronomical refraction.**
B. Garfinkel.
Publ. Obs. Astron. Beograd, No. 18, (see 012.023), p. 49 - 50 (1974).

082.112 **Some kinematical effects of the atmospheric motion on the astronomical refraction.** T. Goto.
Publ. Obs. Astron. Beograd, No. 18, (see 012.023), p. 51 - 59 (1974).

082.113 **Air density distribution near the astronomical observing house.**
C. Kakuta, G. Teleki, T. Goto.
Publ. Obs. Astron. Beograd, No. 18, (see 012.023), p. 159 - 180 (1974).

082.114 **Recherche de sites astrométriques.** F. Laclare.
Publ. Obs. Astron. Beograd, No. 18, (see 012.023), p. 183 - 188 (1974).

082.115 **On some investigations of astronomical refraction.**
A. I. Nefed'eva.
Publ. Obs. Astron. Beograd, No. 18, (see 012.023), p. 189 - 192 (1974).

082.116 **On the computation of the astronomical refraction based on the exponential representation of refractive indices.** C. Sugawa, N. Kikuchi.
Publ. Obs. Astron. Beograd, No. 18, (see 012.023), p. 193 - 212 (1974).

082.117 **Hα nightglow observations in the antisolar direction.**
N. M. Martsvaladze, L. M. Fishkova, P. V. Shcheglov.
Astron. Tsirk., No. 820, p. 3 - 4 (1974). In Russian.

082.118 **Studying star image vibration by the photographic method under day conditions.**
A. G. Zhestkov, B. I. Kozarenko, L. P. Sedova.
Astron. Tsirk., No. 820, p. 4 - 5 (1974). In Russian.

Investigation of the upper atmosphere of the earth. Transactions of the international symposium on space meteorology. Kiev, March 1972. See Abstr. 012.028.

Opacity distribution functions and absorption in Schumann-Runge bands of molecular oxygen.
See Abstr. 022.006.

Some comments on astronomical refraction.
See Abstr. 031.077.

Report to the members of Study Group on Astronomical Refraction. See Abstr. 031.079.

Physical and meteorological factors connected with astronomical refraction. See Abstr. 031.082.

On the construction of new astronomical refraction tables. See Abstr. 041.001.

Determination of the influence of the terrestrial refraction on the celestial refraction near the horizon.
See Abstr. 041.007.

Review of refraction effects of the atmosphere on geodetic measurements to celestial bodies.
See Abstr. 046.025.

A lower limit to Jeans' escape rate.
See Abstr. 091.009.

Stray light suppression in optical space experiments.
See Abstr. 106.019.

083 Ionosphere

083.001 Exospheric temperature and composition from satellite beacon measurements. J. E. Titheridge.
Planet. Space Sci., Vol. 22, 209 - 222 (1974).

083.002 A model of ionospheric F-2 region storms in middle latitudes. K. Davies.
Planet. Space Sci., Vol. 22, 237 - 253 (1974).

083.003 Investigation of integral characteristics of the earth's ionosphere with the method of a dispersion interferometer during the flight of Luna 19. M. B. Vasil'ev, A. S. Vyshlov, M. A. Kolosov, A. P. Mestehrton, N. A. Savich, V. A. Samovol, L. N. Samoznaev, A. I. Sidorenko, D. Ya. Shtern.
Kosmich. Issled., Vol. 12, 87 - 91 (1974). In Russian.

083.004 Phenomenological presentation of the path of propagation of electromagnetic impulses in the ionosphere and solar corona. V. V. Ivanov, L. S. Chudnovskij.
Kosmich. Issled., Vol. 12, 109 - 114 (1974). In Russian.

083.005 On the theory of nocturnal ELF emissions. M. Y. Yu, R. S. B. Ong.
Phys. Scripta, Vol. 9, 67 - 68 (1974).
Continuous nocturnal sub-ELF emissions are explained in terms of the excitation of drift waves in the ionosphere. A simplified model of the ionosphere is used in the calculations. The results are in good agreement with observations.

083.006 Some statistical characteristics of sudden ionospheric disturbances and distribution of geoactive chromospheric flares over the solar disc.
V. V. Belikovich, E. A. Benediktov.
Izv. vyssh. ucheb. zavedenij. Radiofizika, Vol. 16, 1133 - 1137 (1973). In Russian. — Abstr. in Referativ. Zhurn. 51. Astron., 3.51.491 (1974).

083.007 Exospheric models of the topside ionosphere. J. Lemaire, M. Scherer.
Space Sci. Rev., Vol. 15, 591 - 640 (1974).
The historical evolution of the study of escape of light gases from planetary atmospheres is delineated, and the application of kinetic theory to the ionosphere is discussed.

083.008 Dst and SD variations of electron content at low latitude.
Y.-N. Huang, K. Najita, T. H. Roelofs, P. C. Yuen.
Journ. Atmosph. Terr. Phys., Vol. 36, 9 - 28 (1974).

083.009 Power spectra of ionospheric scintillations. D. G. Singleton.
Journ. Atmosph. Terr. Phys., Vol. 36, 113 - 133 (1974).
The diffraction theory of interplanetary scintillations, which allows a connection to be established between the spatial spectrum of electron-density fluctuations in the irregular medium and the power spectrum of intensity scintillations observed on the ground, is reviewed and applied to the case of the ionosphere.

083.010 Monte Carlo simulation of a model ionosphere—III. Photoelectron and escape electron spectra.
K. Schlegel.
Journ. Atmosph. Terr. Phys., Vol. 36, 183 - 187 (1974).
Photoelectron spectra in different altitudes and escape electron spectra are calculated from a Monte Carlo simulation of a model ionosphere described in two preceding papers. The results are in good accordance with the data obtained by other authors with experiments or calculations.

083.011 A numerical description of the winter anomaly in ionospheric absorption for a sunspot cycle.
J. Röttger, H. Schwentek.
Journ. Atmosph. Terr. Phys., Vol. 36, 363 - 366 (1974). Short paper.

083.012 About the choice of the best approximation of the N(h)-profile and its derivative for calculating effective heights of the ionosphere. T. L. Gulyaeva.
Geomagn. Aeronom., Vol. 14, 48 - 52 (1974). In Russian.

083.013 Regular absorption of cosmic radio radiation in the near-polar ionosphere. V. M. Lukashkin.
Geomagn. Aeronom., Vol. 14, 168 - 170 (1974). In Russian. Brief information.

083.014 Studies of ionospheric storms using a simple model. K. Davies.
Journ. Geophys. Res., Vol. 79, 605 - 613 (1974).

083.015 Behavior of the ionospheric F region during the great solar flare of August 7, 1972.
M. Mendillo, J. A. Klobuchar, R. B. Fritz, A. V. da Rosa, L. Kersley, K. C. Yeh, B. J. Flaherty, S. Rangaswamy, P. E. Schmid, J. V. Evans, J. P. Schödel, D. A. Matsoukas, J. R. Koster, A. R. Webster, P. Chin.
Journ. Geophys. Res., Vol. 79, 665 - 672 (1974). — Brief report.

083.016 Sporadic source of ionization in the D- and E-regions of the night ionosphere of mean latitudes.
V. F. Tulinov, V. M. Fejgin, V. A. Lipovetskij, Yu. M. Zhuchenko.
Kosmich. Issled., Vol. 12, 219 - 225 (1974). In Russian.

083.017 Investigation of the statistical characteristics of intensity fluctuations of energetic electrons aboard Oreol. V. A. Kuz'mina.
Kosmich. Issled., Vol. 12, 226 - 234 (1974). In Russian.

083.018 An experimental and theoretical study of the D-region—I. Mid-latitude D-region electron density profiles from the radio wave interaction experiment.
A. J. Ferraro, H. S. Lee, J. N. Rowe, A. P. Mitra.
Journ. Atmosph. Terr. Phys., Vol. 36, 741 - 754 (1974).

083.019 An experimental and theoretical study of the D-region—II. A semi-empirical model for mid-latitude D-region. J. N. Rowe, A. P. Mitra, A. J. Ferraro, H. S. Lee.
Journ. Atmosph. Terr. Phys., Vol. 36, 755 - 785 (1974).

083.020 Electron loss processes in the lower D-region during the decay of solar X-ray flare events.
G. E. Perona.
Journ. Atmosph. Terr. Phys., Vol. 36, 897 - 902 (1974).
Data concerning the recovery of the lower D-region during the decay of solar X-ray events, are examined. A simple discussion of such results in terms of a two positive-ion model, has shown possible ranges for some significant parameters of the D-region, in agreement with previous experimental determinations.

083.021 On the distribution function of electrons from velocities in the ionosphere.

B. V. Troitskij, V. I. Drobzhev.
Ionosphere and solar-terrestrial relations, (see 003.003), p. 3 - 5 (1972). In Russian.

083.022 Solar-lunar tidal phenomena and motions in the F-region.
M. P. Rudina, N. F. Solonitsyna, G. I. Gordienko.
Ionosphere and solar-terrestrial relations, (see 003.003), p. 20 - 28 (1972). In Russian.

083.023 Temperature fluctuations in the F-region of the ionosphere.
N. F. Solonitsyna, M. P. Rudina, G. I. Gordienko.
Ionosphere and solar-terrestrial relations, (see 003.003), p. 29 - 32 (1972). In Russian.

083.024 On the increase of the electron density in the F-region after proton flares.
N. F. Solonitsyna, M. P. Rudina, G. I. Gordienko.
Ionosphere and solar-terrestrial relations, (see 003.003), p. 33 - 37 (1972). In Russian.

083.025 Some morphological peculiarities of the F2-region.
P. E. Kozina.
Ionosphere and solar-terrestrial relations, (see 003.003), p. 38 - 41 (1972). In Russian.

083.026 Time-dependent variations of the recombination coefficient and of the profile of electron density in the lower ionosphere. I. D. Kozin, G. I. Gerasimov.
Ionosphere and solar-terrestrial relations, (see 003.003), p. 66 - 70 (1972). In Russian.

083.027 On the vertical motion of the lower ionosphere during a solar eclipse. B. A. Turkeeva.
Ionosphere and solar-terrestrial relations, (see 003.003), p. 74 - 77 (1972). In Russian.

083.028 Formation of the ionosphere. B. Landmark.
Cosmical geophysics, (see 003.004), p. 73 - 80 (1973).

083.029 Magnetospheric convection and the high-latitude F_2 ionosphere. W. C. Knudsen.
Journ. Geophys. Res., Vol. 79, 1046 - 1055 (1974).

083.030 Model of the ionosphere taking into account plasma motion along geomagnetic force lines.
I. A. Krinberg, V. A. Kuzmin, G. I. Gershengorn.
Geomagn. Aeronom., Vol. 14, 224 - 230 (1974). In Russian.

083.031 The atmospheric circulation at the altitudes of registration of drifts of meteor trails and ionospheric irregularities. P. B. Babadzhanov, L. N. Rubtsov, B. G. Solovej, R. P. Chebotarev.
Komety i Meteory, *Dushanbe*, No. 22, p. 20 - 29 (1973). In Russian.

083.032 On the correlation between the electron concentration of the E-region and the height of the sun.
O. Alimov.
Byull. Inst. Astrofiz., *Dushanbe*, No. 59, p. 28 - 30 (1971). In Russian.

083.033 Results of an investigation of the E-region of the ionosphere over Dushanbe in 1964. O. Alimov.
Byull. Inst. Astrofiz., *Dushanbe*, No. 59, p. 31 - 33 (1971). In Russian.

083.034 Some regularities of the behaviour of the night-time E-layer by data of mean latitude stations.

O. Alimov, D. Latipov.
Byull. Inst. Astrofiz., *Dushanbe*, No. 62, p. 26 - 32 (1974). In Russian.

A positive correlation is established between the number of appearances of the night-time E-layer and the number of sporadic meteors.

083.035 Ionisation ledges in the equatorial ionosphere.
R. Raghavarao, M. R. Sivaraman.
Nature, Vol. 249, 331 - 332 (1974).

The authors describe observations of cusps associated with echoes reflected from a feature aligned with the field below the anomaly field line and suggest how the ledge of ionisation in the equatorial topside ionosphere may be formed.

083.036 Enregistrement des perturbations ionosphériques SEA à l'Observatoire de Bucarest.
V. Dinulescu.
Stud. Cerc. Astron., Vol. 19, 31 - 33 (1974). In Romanian.

On donne les moments pour le commencement, le maximum et la fin des effets SEA (Sudden Enhancement of Atmospherics) observés à Bucarest pendant le mois de mai 1973, en comparaison des observations effectuées à l'Observatoire de Úpice (Tchécoslovaquie).

083.037 In situ measurements of the spectral characteristics of F region ionospheric irregularities.
P. L. Dyson, J. P. McClure, W. B. Hanson.
Journ. Geophys. Res., Vol. 79, 1497 - 1502 (1974).

083.038 Wavelength dependence of radio scintillation: ionosphere and interplanetary irregularities.
C. L. Rufenach.
Journ. Geophys. Res., Vol. 79, 1562 - 1566 (1974).
Brief report.

083.039 Empirical model of the ion structure of the region between 140 and 200 km.
A. K. Buravtsev, A. V. Vinitskij, V. V. Koshelev, L. A. Shchepkin.
Issled. po geomagnetizmu, aehron. i fiz. Solntsa. Vyp. (No.) 29. Moskva, Nauka, 1973, p.127 - 134. In Russian. – Abstr. in Referativ. Zhurn. 62. Issled. kosmich. prostranstva, 5.62. 332 (1974).

083.040 Electron density and temperature changes in the equatorial ionosphere during magnetic storms.
T. B. Jones, K. Davies.
Journ. Atmosph. Terr. Phys., Vol. 36, 1071 - 1078 (1974).

083.041 Characteristic parameters of the ionospheric-magnetospheric medium. I. V. Kovalevskij.
Kosmich. Issled., Vol. 12, 402 - 418 (1974). In Russian.

083.042 Altitude distribution of H^+ and O^+ ions in the F-region of the ionosphere with regard to the carry away effect (numerical example).
Eh. I. Ginzburg, V. F. Kim.
Geomagn. Aeronom., Vol. 14, 441 - 444 (1974). In Russian.

083.043 The seasonal anomaly in the electron concentration of maximum of the F2-region and of the external ionosphere in the period of high solar activity.
M. N. Fatkullin, N. M. Boenkova, A. D. Legen'ka.
Geomagn. Aeronom., Vol. 14, 546 - 548 (1974). In Russian. Brief information.

083.044 Parametric amplification of propagating electron plasma waves in the ionosphere.
D. Arnush, B. D. Fried, C. F. Kennel.
Journ. Geophys. Res., Vol. 79, 1885 - 1893 (1974).

The authors compute the linear parametric convective amplification of electron plasma waves by an O mode pump propagating vertically into a vertically stratified ionosphere. The upshifted plasma line compares favorably with Arecibo observations, center frequency, spectral width, and possibly amplitude when the radar is aimed in a northerly direction. The dependence of the computed enhanced plasma line shape upon density scale length, electron-to-ion temperature ratio, pump power, density gradient scale length, and backscatter radar incidence angle is investigated.

083.045 **Measurement of total cross sections for electron recombination with NO^+ and O_2^+ using ion storage techniques.** F. L. Walls, G. H. Dunn.
Journ. Geophys. Res., Vol. 79, 1911 - 1915 (1974).

Total recombination cross sections for NO^+ as a function of electron energy from 0.045 to 4 eV are reported. From these data, rate coefficients are calculated for electron temperatures extending up to 40000°K.

083.046 **Polar riometer absorption data August 1–16, 1972.** P. Stauning.
Astrophys. Space Sci. Library, Vol. 42, (see 012.016), 609 - 623 (1974).

083.047 **The interrelationship between the >130 keV electron trapping boundary, the VHF radar backscatter, and the visual aurora.** G. J. Romick, W. L. Ecklund, R. A. Greenwald, B. B. Balsley, W. L. Imhof.
Journ. Geophys. Res., Vol. 79, 2439 - 2443 (1974).

A coordinated program between satellite, VHF radar, and optical auroral observations was carried out in Alaska during the spring of 1972. On 18 days in February, March, and April the positions of the >130-keV isotropic and cutoff trapping boundaries in the evening sector of the oval were determined on satellite 1971-89a. The analysis of these data with respect to substorm phase and position of the aurora shows that the motion of the trapping boundary is a dynamic feature of the substorm that can be monitored and approximately located by the VHF radar echoes.

083.048 **Esro 4 gas analyzer results. 2. Direct measurements of changes in the neutral composition during an ionospheric storm.** G. W. Prölss, U. von Zahn.
Journ. Geophys. Res., Vol. 79, 2535 - 2539 (1974).

Direct measurements of the neutral composition at ionospheric heights with the gas analyzer aboard Esro 4 reveal large decreases in the atomic oxygen to molecular nitrogen ratio during the ionospheric storm in late February 1973.

083.049 **Atmospheric expansion from Joule heating.** H. F. Bates.
Planet. Space Sci., Vol. 22, 925 - 937 (1974).

An expression for the vertical velocity of the neutral atmosphere in the F-region is derived for Joule heating by the electric field that drives the auroral electrojet.

083.050 **Concerning the electron temperature discrepancy between in situ and remote probes in the E-region.** R. J. D'Arcy, J. Sayers.
Planet. Space Sci., Vol. 22, 961 - 966 (1974).

Rocket borne Langmuir probe measurements of electron temperature in the E-region are examined in relation to recent laboratory investigations of surface drift effects which can lead to erroneously high and time-dependent electron temperature measurements. The rocket data is consistent with the laboratory expectations thus supporting the suggested importance of surface effects in rocket measurements and in relation to the E-region discrepancy with simultaneous incoherent radar scatter measurements.

083.051 **The effects of including the Coriolis force on Joule dissipation in the upper atmosphere.** M. G. Heaps.
Planet. Space Sci., Vol. 22, 1031 - 1035 (1974).

Including the Coriolis term in the equations of motion affects the velocity-dependent Joule dissipation. The effect is most apparent in the night-time ionosphere where Joule dissipation, after an initial decrease, returns to approximately its initial value.

083.052 **Fluorescent ion jets for studying the ionosphere and magnetosphere.** K. W. Michel.
Acta Astronaut., Vol. 1, 37 - 69 (1974).

Fast ion jets, the fluorescent light of which can be recorded on ground, become of increasing importance in rocket-release studies of the structure and dynamics of the ionosphere and magnetosphere. The physical principles for the production of such jets of high intensity by means of shaped charges are outlined. Simple expressions are derived to describe the basic features during the expansion of such jets in the collision dominated and collisionless regime of the ionosphere.

083.053 **A theoretical relation between DST and the solar wind merging electric field.** G. Siscoe, N. Crooker.
Geophys. Res. Letters, Vol. 1, No. 1, p. 17 - 20 (1974).

083.054 **Lunar ionospheric and oceanic dynamo variations.** J. C. Gupta.
Gerlands Beiträge Geophys., Vol. 83, 1 - 15 = Contr. Earth Phys. Branch, *Ottawa, Canada,* No. 449 (1974).

Lunar harmonic coefficients obtained from the analysis of long series of geomagnetic data from a few stations are used to study the daily variation, according to several geophysical variables, of the lunar ionospheric (L_I) and oceanic (L_O) dynamo parts of L_2.

Results from a swept-frequency ionospheric impedance probe. See Abstr. 033.055.

Controlled experiments from the space shuttle. See Abstr. 062.038.

Rayonnement cohérent d'ondes électrostatiques quasi-perpendiculaires dû à un faisceau d'électrons. See Abstr. 062.040.

Hydromagnetische Wellen im zähen, anisotrop leitfähigen ionosphärischen Plasma. See Abstr. 062.047.

The effect of displaced geomagnetic and geographic poles on the thermospheric neutral winds. See Abstr. 082.010.

Upon estimation of effective concentration of neutral atmosphere minor constituents NO and N on the basis of experimental values $[NO^+]/[O_2^+]$. See Abstr. 082.020.

On the distribution of He^+ in the plasmasphere from observations of resonantly scattered He II 304-Å radiation. See Abstr. 082.030.

Global characteristics in the diurnal variations of the thermospheric temperature and composition. See Abstr. 082.035.

Metastable helium in the earth's upper atmosphere. See Abstr. 082.037.

Oxygen recombination in the tropical nightglow. See Abstr. 082.103.

Electric field measurements at the geostationary position. See Abstr. 084.289.

Some results of studies of meteor phenomena and the sporadic E-layer. See Abstr. 104.043.

High-energy cosmic ray intensity increase of non-solar origin and the unusual Forbush decrease of August 1972. See Abstr. 143.033.

084 Aurorae, Geomagnetic Field, Radiation Belts

Aurorae

084.001 **Auroral displays near the 'foot' of the field line of the ATS-5 satellite.**
S.-I. Akasofu, S. DeForest, C. McIlwain.
Planet. Space Sci., Vol. 22, 25 - 40 (1974).

084.002 **The OI λ1304 and λ1356 emissions in aurorae.**
D. J. Strickland, M. H. Rees.
Planet. Space Sci., Vol. 22, 465 - 481 (1974).
In this work the authors start with the production rates of excited O atoms derived from electron energy degradation considerations. They then predict altitude-intensity profiles for λ1304 and λ1356 and compare these to the observed profiles of Miller et al. (1968) and of Peek (1970). This approach provides insight into some aspects of the radiative transfer theory (e.g. the appropriate line frequency profile) and enables to comment on the accuracy of currently available excitation cross sections.

084.003 **Interferometric measurements of the λ7319 Å doublet emissions of OII.**
J. W. Meriwether, P. B. Hays, K. D. McWatters, A. F. Nagy.
Planet. Space Sci., Vol. 22, 636 - 638 (1974). — Research note.

084.004 **Stable auroral red arcs.** R. J. Hoch.
Rev. Geophys. Space Phys., Vol. 11, 935 - 949 (1973).
Stable auroral red (SAR) arcs are diffuse, persistent, practically monochromatic (λλ6300–6364Å) auroral forms peculiar to mid-latitude regions of earth. There is strong evidence that the source of energy is the ring current. Suggested mechanisms by which energy is transferred from the ring current to the electrons in the SAR arc region are (1) heat flow, (2) transfer of ring current proton kinetic energy to hydromagnetic waves, which are in turn damped by the electrons in the SAR arc region, and (3) direct influx of energetic electrons into the SAR arc region. Which of these mechanisms predominates is still not resolved at this time.

084.005 **New route for the quenching of $N_2(A^3\Sigma_u^+)$ in the aurora?** D. J. Malcolme-Lawes.
Nature, Vol. 247, 540 - 541 (1974). — Letter.

084.006 **Movement processes of auroral structures.**
P. Czechowsky.
Journ. Atmosph. Terr. Phys., Vol. 36, 61 - 77 (1974).

084.007 **Auroral conjugacy between Kerguelen Island and north-west U.S.S.R.** M. Fehrenbach, G. Weill, V. K. Roldugin, G. V. Starkov, E. A. Vasilkova.
Journ. Atmosph. Terr. Phys., Vol. 36, 407 - 416 (1974).

084.008 **Auroral photography from a satellite.**
E. H. Rogers, D. F. Nelson, R. C. Savage.
Science, Vol. 183, 951 - 952 (1974).
Photographs taken from satellites show the form, location, and intensity of the aurora from a new perspective. They provide an effective way of monitoring auroral activity on a worldwide basis and are likely to become one of the major tools in the effort to understand this phenomenon.

084.009 **Auroral electron fluxes parallel to the geomagnetic field lines.** J. M. Bosqued, G. Cardona, H. Rème.
Journ. Geophys. Res., Vol. 79, 98 - 104 (1974).

084.010 **Rocket measurements of $O_2(^1\Delta_g)$ emissions in the auroral zone.**
K. D. Baker, R. H. Bishop, L. R. Megill.
Journ. Geophys. Res., Vol. 79, 243 - 248 (1974).

084.011 **Atmospheric spreading of protons in auroral arcs.**
G. E. Iglesias, R. R. Vondrak.
Journ. Geophys. Res., Vol. 79, 280 - 282 (1974). — Brief report.

084.012 **Simplified formula for the determination of the altitude of an auroral arc from a single station.**
B. S. Dandekar.
Journ. Atmosph. Terr. Phys., Vol. 36, 829 - 834 (1974).
Rees' formula based on a graphical method of determining an altitude of an auroral arc from a single station, is presented as a set of two simultaneous equations. It is shown that solving these two simultaneous equations is a method more convenient than Rees' graphical method.

084.013 **On the coruscation of auroral particle flux at College, Alaska.** F. T. Berkey.
Journ. Atmosph. Terr. Phys., Vol. 36, 881 - 887 (1974).

084.014 **Satelliet kan poollicht van boven bekijken.**
C. de Jager.
Zenit, Vol. 1, No. 2, p. 2 -3 (1974).

084.015 **Energy and pitch angle distributions for auroral ions using the current sheet acceleration model.**
E. F. Jaeger, T. W. Speiser.
Astrophys. Space Sci., Vol. 28, 129 - 144 (1974).
Using a dipole plus tail magnetic field model, H^+, He^{++}, and O_{16}^{+6} ions are followed numerically, backward in time, from

an output plane perpendicular to the axis of the geomagnetic tail, to their point of entrance to the magnetosphere as solar wind particles in the magnetosheath. Using Liouville's theorem, and varying initial conditions at the output plane, the distribution function is found as a function of energy and pitch angle at the output plane. These results are then mapped to the auroral ionosphere using guiding center theory. Comparisons of the results with experimental observations are presented.

084.016 **Auroral particles.** B. Hultqvist.
Cosmical geophysics, (see 003.004), p. 161 - 179 (1973).

084.017 **Auroral morphology.** A. Omholt.
Cosmical geophysics, (see 003.004), p. 203 - 210 (1973).

084.018 **Particle precipitation: ionization and excitation.** A. Omholt.
Cosmical geophysics, (see 003.004), p. 221 - 235 (1973).

084.019 **Infrared processes in the auroral zone.**
R. H. Bishop, A. W. Shaw, R. Y. Han, L. R. Megill.
Journ. Geophys. Res., Vol. 79, 1729 - 1736 (1974).
Recently auroral observations have indicated the strong enhancement of certain infrared bands during periods of auroral activity. A theoretical model has been created in order to study the infrared processes associated with the aurora. The quantitative results of this study depend upon a thorough knowledge of the rates of all processes involved.

084.020 **Sound wave modulation of auroral X-rays.**
A. F. Wickersham, Jr.
Nature, Vol. 249, 538 - 540 (1974).
A uniquely intense pulsation of auroral-associated X-rays, followed by a solar cosmic flare in September 1966 and was detected by scintillation counters on a balloon platform at an altitude of about 32 km. The X-rays were strongly modulated with a period of 4–6 s. All the features observed can be interpreted as an interaction between atmospheric sound waves and a uniform flux of X-rays.

084.021 **Polar cap auroral electron fluxes observed with Isis 1.**
J. D. Winningham, W. J. Heikkila.
Journ. Geophys. Res., Vol. 79, 949 - 957 (1974).

084.022 **The expansive phase of magnetospheric substorms. 1. Development of the auroral electrojets and auroral arc configuration during a substorm.**
J. L. Kisabeth, G. Rostoker.
Journ. Geophys. Res., Vol. 79, 972 - 984 (1974).

084.023 **Satellite observations of auroral substorms.**
C. P. Pike, J. A. Whalen.
Journ. Geophys. Res., Vol. 79, 985 - 1000 (1974).
Satellite photographs of auroras obtained from a polar-orbiting U.S. Air Force satellite were used to study auroral substorms in the premidnight time sector during and after a geomagnetic storm period. The photographs provide a unique means of seeing in detail the interrelationship between discrete, continuous/diffuse, and polar cap auroras during a period of frequent occurrence of substorms.

084.024 **On determining the major SAR arc properties from a few measurable parameters.**
G. Hernandez, R. G. Roble.
Journ. Geophys. Res., Vol. 79, 1057 - 1064 (1974).

084.025 **Auroral substorms observed from above the north polar region by a satellite.**
A. L. Snyder, S.-I. Akasofu, T. N. Davis.
Journ. Geophys. Res., Vol. 79, 1393 - 1402 (1974).

084.026 **Auroral particle precipitation and Birkeland currents.** R. L. Arnoldy.
Rev. Geophys. Space Phys., Vol. 12, 217 - 231 (1974).
Correlated measurements that provide information on Birkeland (field aligned) currents are reviewed. Because of the obvious importance of field-aligned electric fields with regard to Birkeland currents, the last section of the paper is devoted to a presentation of recent data on the observations of field-aligned auroral electron fluxes and monoenergetic peaks in the spectrum.

084.027 **Auroral activity during 1972.** M. Hallissey.
Observatory, Vol. 94, 93 - 94 (1974).

084.028 **Backscatter results from Lindau – I. Observations of radio-auroras.**
P. Czechowsky, W. Dieminger, H. Kochan.
Journ. Atmosph. Terr. Phys., Vol. 36, 955 - 966 (1974).

084.029 **Photometric measurements of H-beta in the aurora.**
G. J. Romick, A. E. Belon, W. J. Stringer.
Planet. Space Sci., Vol. 22, 725 - 733 (1974).

084.030 **Proton precipitation and the $H\beta$ emission in a post-breakup auroral glow.** F. Söraas, H. R. Lindalen, K. Måseide, A. Egeland, T. A. Sten, D. S. Evans.
Journ. Geophys. Res., Vol. 79, 1851 - 1859 (1974).
Measurements of the hydrogen auroral emissions have been made by rocket-borne photometers simultaneously with observations of the energy spectra and pitch angle distributions of the precipitating energetic particles. The observed $H\beta$ profiles are compared with the profiles derived from the measured primary proton beam and obtained by using Eather's (1967) values for the pertinent excitation cross sections. Good agreement between measured and computed profiles is obtained.

084.031 **The magnetospheric electric field.**
J. G. Roederer.
Astrophys. Space Sci. Library, Vol. 42, (see 012.016), 257 - 261 (1974).

084.032 **Magnetospheric plasma flow and the nature of the magnetospheric boundary layer.** E. W. Hones, Jr.
Astrophys. Space Sci. Library, Vol. 42, (see 012.016), 263 - 275 (1974).

084.033 **Particles and plasmas in the earth's magnetotail at $60 R_E$.** K. A. Anderson, L. M. Chase, R. P. Lin, R. E. McGuire, J. E. McCoy.
Astrophys. Space Sci. Library, Vol. 42, (see 012.016), 297 - 316 (1974).

084.034 **Indications of a longitudinal component in auroral phenomena.** H. C. Stenbaek-Nielsen.
Journ. Geophys. Res., Vol. 79, 2521 - 2523 (1974).
A number of independent observations indicate the presence of a longitudinal component in auroral activity arising from asymmetries in the earth's internal magnetic field.

084.035 **Observations of the conjugate SAR arcs of September 28 - 30, 1967.** E. I. Reed, J. E. Blamont.
Journ. Geophys. Res., Vol. 79, 2524 - 2525 (1974).
Brief report.

084.036 **The aurora and the magnetosphere: The Chapman Memorial Lecture.** S.-I. Akasofu.
Planet. Space Sci., Vol. 22, 885 - 923 (1974).

The main part of the paper reviews recent progress in magnetospheric physics, in particular, in understanding the magnetospheric substorm. A number of magnetospheric phenomena can now be understood by viewing the solar wind-magnetosphere interaction as an MHD dynamo; auroral phenomena are powered by the dynamo. We have also succeeded in identifying magnetospheric responses to variations of the north–south and east–west components of the interplanetary magnetic field. The magnetospheric substorm is entirely different from the responses of the magnetosphere to the southward component of the interplanetary magnetic field. It may be associated with the formation of a neutral line within the plasma sheet and with an enhanced reconnection along the line.

084.037 **Fluctuations of electron precipitation to the dayside auroral zone modulated by compression and expansion of the magnetosphere.** T. Saito, F. Takahashi, A. Morioka, M. Kuwashima.
Planet. Space Sci., Vol. 22, 939 - 953 (1974).

Concurrent variations of cosmic noise absorption (CNA) fluctuations and geomagnetic fluctuations are classified into type 1 (substorm-type), type 2 (Pc5-type), and type 3 which is the object of the present study. Type 3 apparently has peculiar characteristics in that CNA fluctuations at a certain auroral-zone station show a pronounced positive correlation with magnetic fluctuations at distant low-latitude stations.

084.038 **A rocket instrument for imaging the aurora by means of high speed photometric scanning.**
I. W. H. Robertson, D. D. Wallis, C. D. Anger, D. W. Johnson.
Planet. Space Sci., Vol. 22, 1003 - 1015 (1974).

A high resolution rocket-borne mechanical flying spot scanner has evolved from similar ground-based instruments. The spinning motion of the rocket together with that of a motor-driven prism are used to scan a narrow-angle photometer systematically and repetitively over the full 4π sr region around the rocket. Results illustrating the capabilities of the instrument and data analysis system have been obtained from rocket AKD-VB-27 which was launched from the Churchill Research Range, 20 January 1971.

The interrelationship between the >130 keV electron trapping boundary, the VHF radar backscatter, and the visual aurora. See Abstr. 083.047.

Atmospheric expansion from Joule heating. See Abstr. 083.049.

Interplanetary magnetic field direction and auroral zone disturbances. See Abstr. 106.035.

Geomagnetic Field

084.201 The role of fluctuations in the interplanetary magnetic field in determining the magnitude of substorm activity. H. B. Garrett.
Planet. Space Sci., Vol. 22, 111 - 119 (1974).

084.202 Geomagnetic effects of interplanetary sector structure. A. Moldovanu.
Planet. Space Sci., Vol. 22, 193 - 208 (1974).

084.203 Some features of field line resonances in the magnetosphere. D. J. Southwood.
Planet. Space Sci., Vol. 22, 483 - 491 (1974).

084.204 On the relationship between shock wave and magnetic bottle associated with the solar flare of 4 November 1968. K. Sakurai, J. K. Chao.
Planet. Space Sci., Vol. 22, 493 - 494 (1974).

This note discusses the relation between shock wave and magnetic bottle, which were both associated with the solar proton flare of 4 November 1968. In particular, the formation and development of this wave and bottle are described.

084.205 Solar cycle dependence of periodic variations in geomagnetic K_p-index.
S. Abdel-Wahab, A. Goned.
Planet.Space Sci., Vol. 22, 537 - 544 (1974).

The geomagnetic K_p-index data for the 1932 - 1969 period have been investigated by means of a modified power spectrum technique on the basis of overlapping 2-yr intervals. The observed 27-, and 13.5-day periodicities show an obvious solar cycle dependence through the whole period concerned.

084.206 The compressed geomagnetic field as a function of dipole tilt. J. Y. Choe, D. B. Beard.
Planet. Space Sci., Vol. 22, 595 - 608 (1974).

084.207 The near earth magnetic field of the magnetotail current. J. Y. Choe, D. B. Beard.
Planet. Space Sci., Vol. 22, 609 - 615 (1974).

084.208 Response of dayside thermosphere to an intense geomagnetic storm. R. R. Allan.
Nature, Vol. 247, 23 - 25 (1974). – Letter.

084.209 Wheather and the earth's magnetic field. J. W. King.
Nature, Vol. 247, 131 - 134 (1974).

A comparison of meteorological pressures and the strength of the geomagnetic field suggests a possible controlling influence of the field on the longitudinal variation of the average pressure in the troposphere at high latitudes.

084.210 The earth's magnetosphere. J. G. Roederer.
Science, Vol. 183, 37 - 46 (1974).

The outer limits of man's environment provide a readily available "astrophysical plasma laboratory".

084.211 On the problem of geoeffectiveness of solar active regions. V. P. Kuleshova, E. V. Lavrova.
Solnechnye Dannye 1973 Byull., No. 11, p. 80 - 85 (1974). In Russian.

A possibility is considered of forecasting the geomagnetic activity on the basis of the data on transits of the active regions across the central meridian and also of the Kleczek flare index. Three types of regions have been detected according to the criterion of their flare activity. It is shown that the Kleczek integral index cannot be used to forecast the geomagnetic activity, and the effect of neighbouring active regions is to be taken into account.

084.212 A leading shock wave and the α-component in the transition region from measurements aboard the high-latitude satellite Prognoz.
O. L. Vajsberg, A. A. Zertsalov, V. V. Temnyj, Yu. E. Berezin.
Kosmich. Issled., Vol. 12, 80 - 86 (1974). In Russian.

084.213 Partial ring current models for worldwide geomagnetic disturbances.
N. Fukushima, Y. Kamide.
Rev. Geophys. Space Phys., Vol. 11, 795 - 853 (1973).

Direct observations of particles and magnetic field by means of satellites in the magnetosphere have given some supporting evidence for the existence of partial ring current in the magnetosphere. In this article a review is made of the analysis of low-latitude geomagnetic variation during magnetic storms or substorms, and the contribution of the partial ring current system to low-latitude geomagnetic variations is discussed.

084.214 Geomagnetic secular variation, 1962.5 to 1967.5.
S. R. C. Malin, A. D. Clark.
Geophys. Journ. Roy. Astron. Soc., Vol. 36, 11 - 20 (1974).

Mean values of observatory geomagnetic data for the intervals 1960.0 to 1965.0 and 1965.0 to 1970.0 are subjected to spherical harmonic analysis to obtain coefficients of the main magnetic field. Westward drift and movements of the centred and eccentric dipole are discussed, and the value of secular acceleration as an aid to prediction is assessed. The correlation between changes in westward drift and in the length of the day is re-examined and found to be of doubtful significance.

084.215 Effects of a time-dependent ring current on field lines of a magnetic dipole. C. S. Wang, J. S. Kim.
Geophys. Journ. Roy. Astron. Soc., Vol. 36, 43 - 56 (1974).

An analytical expression for the magnetic lines of force for a system consisting of a magnetic dipole and the symmetric ring current in free space was first derived in terms of the intensity of the ring current. Then, with the aid of the magnetic flux conservation theorem, an expression relating the latitudinal displacement of the magnetic lines of force and the change of the ring current intensity was derived, and the space and time variations of the magnetic lines of force with a change in the intensity of the ring current were examined. Several model calculations were made and their applications to geophysical phenomena were suggested.

084.216 Spherical harmonic analyses of the geomagnetic field for eight epochs between 1600 and 1910.
D. R. Barraclough.
Geophys. Journ. Roy. Astron. Soc., Vol. 36, 497 - 513 (1974).

The data for epochs between 1600 and 1910 in the catalogue of Veinberg and Shibaev have been analysed and two sets of spherical harmonic models of the geomagnetic field have been derived. The dipolar nature of the field in the Pacific region is investigated and the positions of the geomagnetic poles are derived. The variation of the position of the eccentric dipole with time is studied.

084.217 Palaeomagnetic evidence for the transitional behaviour of the geomagnetic field.
P. Dagley, E. Lawley.
Geophys. Journ. Roy. Astron. Soc., Vol. 36, 577 - 598 (1974).

A definition of transition intervals based on the definition of normal, intermediate and reversed pole positions proposed by Wilson, Dagley and McCormack is used to select data for comparison with two models of the transitional geomagnetic field. The data for twenty-three transition intervals reported by various authors are reviewed.

084.218 Spatial power spectrum of the main geomagnetic field, and extrapolation to the core.
F. J. Lowes.
Geophys. Journ. Roy. Astron. Soc., Vol. 36, 717 - 730 (1974).
The spatial 'power' spectrum of the main geomagnetic field has been estimated for harmonics up to n = 500. It is shown to consist of two components, long wavelengths being dominated by fields originating in the core, and short wavelengths by fields originating in the crust; the cross-over occurs at $n \geqslant 11$, a wavelength $\leqslant 3600$ km.

084.219 Variation semi-annuelle de l'activité magnétique et vitesse du vent solaire. P.-N. Mayaud.
Comptes Rendus Acad. Sci. Paris, Sér. B, Vol. 278, 139 - 142 (1974).
A partir d'une série de 100 années d'indices quantitatifs de l'activité magnétique, on établit l'existence d'une dépendance entre la phase de la variation semi-annuelle de l'activité magnétique et la vitesse moyenne du vent solaire, dépendance induite par l'effet d'aberration du vent solaire.

084.220 Large-scale auroral-zone electron precipitation event, briefly interrupted during a negative magnetic impulse. T. Pytte, J. Bjordal, K. Brønstad, I. Singstad, J. Stadsnes, H.Trefall, S. Ullaland, R. R. Brown, R. H. Karas.
Journ. Atmosph. Terr. Phys., Vol. 36, 29 - 42 (1974).
Multiple balloon recordings of bremsstrahlung X-rays from a large scale auroral-zone electron precipitation event are presented. Additional riometer recordings show that it extended from noon, via dusk, to midnight. The X-ray observations show electron precipitation over a range of L-values from $\simeq 5.5$ to 7.5.

084.221 Lunar effects in the counter electrojet near the magnetic equator. R. G. Rastogi.
Journ. Atmosph. Terr. Phys., Vol. 36, 167 - 170 (1974). Short paper.

084.222 About the eleven-year periodicity in the horizontal component of the earth's magnetic field.
Yu. R. Rivin.
Geomagn. Aeronom., Vol. 14, 119 - 123 (1974). In Russian.

084.223 A statistical method of defining intensity of the ancient geomagnetic field. E. N. Tarkhov.
Geomagn. Aeronom., Vol. 14, 144 - 147 (1974). In Russian.

084.224 Motions of the bow shock induced by interplanetary disturbances. H. J. Völk, R.-D. Auer.
Journ. Geophys. Res., Vol. 79, 40 - 48 (1974).

084.225 Association of DP and DR fields with the interplanetary magnetic field variation. Y. Kamide.
Journ. Geophys. Res., Vol. 79, 49 - 55 (1974).
The purpose of this paper is to show that there is a quantitative difference between substorms occurring in association with southward-directed interplanetary magnetic fields and those occurring in association with northward-directed fields.

084.226 High-latitude electric fields and the three-dimensional interaction between the interplanetary and terrestrial magnetic fields. F. S. Mozer, W. D. Gonzalez, F. Bogott, M. C. Kelley, S. Schutz.
Journ. Geophys. Res., Vol. 79, 56 - 63 (1974).

084.227 Relationship between interplanetary plasma parameters and geomagnetic Dst. R. P. Kane.
Journ. Geophys. Res., Vol. 79, 64 - 72 (1974).

The relationship between interplanetary plasma parameters and geomagnetic Dst is studied for individual storms and also on a statistical basis using hourly and 3-hourly values. It is shown that a turning of B_z from northward to southward acts only as a trigger for letting solar wind into the magnetosphere. Whereas this produces quick changes in the polar region, the mid-latitude Dst develops a few hours later, generally if the solar wind density N is high. Solar wind velocity seems to play a comparatively minor role.

084.228 A comparison of the geomagnetic ap index with fluctuations in B and V: Mariner 5.
D. E. Jones, J. R. Ballif, J. G. Melville.
Journ. Geophys. Res., Vol. 79, 286 - 288 (1974). – Letter.

084.229 Review on the usage of magnetospheric models.
R. Gall, A. Orozco.
Journ. Geophys. Res., Vol. 79, 293 - 296 (1974). – Letter.

084.230 Origin of the semiannual variation of geomagnetic Kp indices. T. Murayama.
Journ. Geophys. Res., Vol. 79, 297 - 300 (1974). – Letter.

084.231 30- to 100-keV protons upstream from the earth's bow shock.
R. P. Lin, C.-I. Meng, K. A. Anderson.
Journ. Geophys. Res., Vol. 79, 489 - 498 (1974).
In this report the authors present observations of protons upstream from the bow shock in the energy range 29-100 keV. These protons are found to exist continuously on interplanetary field lines that connect to the bow shock and may be related to the lower-energy protons and the waves that appear on these same field lines.

084.232 Observations of the internal structure of the magnetopause.
M. Neugebauer, C. T. Russell, E. J. Smith.
Journ. Geophys. Res., Vol. 79, 499 - 510 (1974).

084.233 Association of plasma sheet thinning with neutral line formation in the magnetotail.
A. Nishida, E. W. Hones, Jr.
Journ. Geophys. Res., Vol. 79, 535 - 547 (1974).

084.234 Investigation of the association of magnetopause instability with interplanetary sector structure.
B. R. Boller, H. L. Stolov.
Journ. Geophys. Res., Vol. 79, 673 (1974). – Letter.

084.235 Measurements of low-energy protons aboard Prognoz. A. A. Bednyakov, Yu. I. Logachev, S. P. Ryumin, I. A. Savenko, S. K. Stolboushkin, I. P. Shestopalov.
Kosmich. Issled., Vol. 12, 312 - 314 (1974). In Russian. Brief information.

084.236 The development of the auroral absorption substorm. O. Fjordheim, K. Henriksen.
Journ. Atmosph. Terr. Phys., Vol. 36, 811 - 820 (1974).

084.237 The latitude dependence of the seasonal course of the S_q-variations from data of Soviet magnetic observatories. V. N. Pogrebnoj.
Ionosphere and solar-terrestrial relations, (see 003.003), p. 107 - 110 (1972). In Russian.

084.238 Rotation of the geomagnetic field.
F. J. Lowes, with a reply by S. R. C. Malin, I. Saunders.
Nature, Vol. 248, 402 - 405 (1974). – Letter.

084.239 The geomagnetic field. F. Eleman.

Cosmical geophysics, (see 003.004), p. 45 - 62 (1973).

084.240 The magnetosphere. L. P. Block.
Cosmical geophysics, (see 003.004), p. 103 - 119 (1973).

084.241 Motion of charged particles in the magnetosphere. C.-G. Fälthammar.
Cosmical geophysics, (see 003.004), p. 121 - 142 (1973).

084.242 Particle observations in the magnetosphere. F. Søraas.
Cosmical geophysics, (see 003.004), p. 143 - 160 (1973).

084.243 Perturbations of the geomagnetic field. B. Hultqvist.
Cosmical geophysics, (see 003.004), p. 193 - 201 (1973).

084.244 Interaction of MGD waves with the magnetosphere. A. Best, H.-R. Lehmann, R. Treumann.
Geomagn. Aeronom., Vol. 14, 300 - 308 (1974). In Russian.

084.245 Evolution of the main geomagnetic field. N. V. Adam, N. P. Benkova, G. I. Kolomijtseva, T. N. Cherevko.
Geomagn. Aeronom., Vol. 14, 350 - 354 (1974). In Russian.

084.246 An investigation of wave-particle interactions and particle dynamics using electron beams injected from sounding rockets. J. R. Winckler.
Space Sci. Rev., Vol. 15, 751 - 780 (1974). — Presented at the Workshop on Controlled Magnetospheric Experiments, Kyoto, Japan, Sept., 1973.

Electrons with energy up to 40 keV have been injected into semi-trapped orbits from sounding rockets at Wallops Island, Virginia, and at Fort Churchill, Manitoba, Canada. It was possible to detect conjugate echoes at Wallops and possibly at Churchill, and to study the distribution of the echoes in space, time and energy. By combining observations of many echoes, a composite picture can be obtained of the beam patterns. Atmospheric scattering at the conjugate point of Wallops island has been extensively studied. Models of the magnetic fields are used to predict the bounce displacement, and by comparison with the observations to evaluate other effects such as electric field drift integrated over a complete bounce period. The process of rocket neutralization, beam-beam interactions, and electromagnetic radiation from the beams has been studied.

084.247 Controlled vlf wave injection experiments in the magnetosphere. R. A. Helliwell.
Space Sci. Rev., Vol. 15, 781 - 802 (1974).

Whistler-mode waves injected into the magnetosphere from ground sources (e.g., lightning discharge, vlf transmitters) are used to probe the distribution of ions and electrons in the magnetosphere. They also cause wave growth (vlf emissions) and precipitation of electrons.

084.248 Laboratory experiments of magnetospheric interest. C.-G. Fälthammar.
Space Sci. Rev., Vol. 15, 803 - 825 (1974).

Space-related laboratory experiments can play an important role as a complement to observations and active experiments in the magnetosphere. Excluding laboratory experiments for mere developing or testing of techniques for space experiments, we may distinguish between two major types: (1) partial scale model experiments and (2) experiments for clarifying basic plasma physical processes known or expected to be important in the magnetosphere. The limitations and potentialities of both types are discussed and examples of experiments are given.

084.249 Laboratory experiments directed toward the investigation of magnetospheric phenomena. I. M. Podgorny, E. M. Dubinin.
Space Sci. Rev., Vol. 15, 827 - 840 (1974).

The penetration of fast electrons (~ 5 keV) into an artificial magnetosphere and their precipitation on the terrella surface is investigated. These fast electrons act as 'radioactive tracers' allowing the experimental determination of the global picture of plasma flow around the magnetosphere and its intrusion into the latter.

084.250 Magnetosphere dynamics with artificial plasma clouds. J. M. Cornwall.
Space Sci. Rev., Vol. 15, 841 - 860 (1974).

Addition of cold plasma to the magnetosphere outside the plasmasphere can enhance both ion and electron electromagnetic cyclotron (EMC) instabilities. Calculations have been made of total ion-EMC amplification on a single pass through a lithium cloud. The dynamics of the cold lithium cloud have been studied in detail. Some remarks are made about the effects of added cold plasma on the Post-Rosenbluth electrostatic mode.

084.251 Stimulation of vlf amplification in the magnetosphere. H. B. Liemohn.
Space Sci. Rev., Vol. 15, 861 - 889 (1974).

Selected cases of plasma cloud and beam injections are reviewed quantitatively based on the linear theory for the cyclotron-resonance interaction. Only the interaction of vlf waves (3–30 kHz) with hot electrons (0.1–100 keV) is treated.

084.252 Potential experiments with superconducting magnets in earth orbit. W. P. Olson.
Space Sci. Rev., Vol. 15, 899 - 904 (1974).

Several experiments that can be performed in earth orbit with a superconducting magnet are discussed. They are divided into 2 classes, pure plasma physics experiments that can be performed in near earth orbit and planetary magnetosphere simulation experiments that are best conducted in weak background fields distant from the earth.

084.253 The French-Soviet 'ARAKS' experiment. R. Gendrin.
Space Sci. Rev., Vol. 15, 905 - 931 (1974).

In 1975, a rocket borne electron gun experiment will be achieved in Kerguelen Islands (South Indian Ocean), as a result of a cooperation between Soviet and French scientists. The gun will inject into the magnetosphere large currents (0.5 and 1 A) of high energy electrons (15 and 27 keV) with different initial pitch angles ($\sim 0°, 70, 140°$). The author describes the experiment and gives the results of some preliminary computations which have been made to predict the amplitude of the expected phenomena. A discussion is made of the respective ability for electron beam injection and cold plasma injection to artificially induce strong particle precipitation.

084.254 On the local time dependence of the bow shock wave structure. J. V. Olson, R. E. Holzer.
Journ. Geophys. Res., Vol. 79, 939 - 947 (1974).

084.255 On the cause of geomagnetic storms. C. T. Russell, R. L. McPherron, R. K. Burton.
Journ. Geophys. Res., Vol. 79, 1105 - 1109 (1974). — Brief report.

084.256 A model of the open magnetosphere. J. R. Kan, S.-I. Akasofu.

Journ. Geophys. Res., Vol. 79, 1379 - 1384 (1974).

084.257 **Prolonged tailward flow of plasma in the thinned plasma sheet observed at $r \approx 18 R_E$ during substorms.**
E. W. Hones, Jr., A. T. Y. Lui, S. J. Bame, S. Singer.
Journ. Geophys. Res., Vol. 79, 1385 - 1392 (1974).

084.258 **A model of the geomagnetic field for 1970.**
L. Hurwitz, E. B. Fabiano, N. W. Peddie.
Journ. Geophys. Res., Vol. 79, 1716 - 1717 (1974).
  A mathematical description of the geomagnetic field for 1970 has been derived by using 360,000 land, sea, and airborne measurements accumulated since 1939. The model is expressed by 168 spherical harmonic coefficients $g_n{}^m$, $h_n{}^m$ to $m = n = 12$. It was used as the basis for the 1970 edition of the U. S. World Charts of Magnetic Variation.

084.259 **Secular variations of the geomagnetic field and the classification of magnetic anomalies.**
Yu. P. Bulashevich.
Dokl. Akad. Nauk SSSR. Ser. Mat. Fiz., Vol. 216, 313 - 315 (1974). In Russian.

084.260 **Solar proton entry to the magnetosphere on 18 November 1968 and 25 February 1969 – I. Interpretation of satellite data using trajectory computations in a model magnetosphere.**
G. R. Thomas, D. M. Willis, R. J. Pratt.
Journ. Atmosph. Terr. Phys., Vol. 36, 995 - 1017 (1974).
  Latitudinal distributions of high-energy (~ 100 MeV) solar protons measured over the earth's polar caps by low-altitude polar-orbiting satellites are described. The data, which refer to the initial phases of the 18 November 1968 and 25 February 1969 solar events, are related to interplanetary particle and magnetic-field measurements by means of trajectory computations using a model magnetosphere.

084.261 **Solar proton entry to the magnetosphere on 18 November 1968 and 25 February 1969 – II. Comparison of trajectory computations in two model magnetospheres.**
D. M. Willis, G. R. Thomas, R. J. Pratt.
Journ. Atmosph. Terr. Phys., Vol. 36, 1019 - 1035 (1974).
  Solar-proton data recorded over the earth's polar caps during the initial phases of the 18 November 1968 and 25 February 1969 solar events are analyzed by computing proton trajectories in two different model magnetospheres, and the results obtained using these two models are compared.

084.262 **Simplified representations of the magnetopause boundary surface for a quantitative model of the magnetosphere.** G. R. Thomas, D. M. Willis, R. J. Pratt.
Journ. Atmosph. Terr. Phys., Vol. 36, 1037 - 1044 (1974).

084.263 **The motion and magnetic structure of the plasma sheet near $30 R_E$.** S. B. Bowling, R. A. Wolf.
Planet. Space Sci., Vol. 22, 673 - 686 (1974).

084.264 **Electric field and currents connected with Y-component of interplanetary magnetic field.**
S. V. Leontyev, W. B. Lyatsky.
Planet. Space Sci., Vol. 22, 811 - 819 (1974).

084.265 **On a dynamical system of coordinates in a geomagnetic trap.** V. V. Bogdanov, V. D. Pletnev.
Kosmich. Issled., Vol. 12, 380 - 386 (1974). In Russian.

084.266 **Investigation of plasma in the earth's magnetosphere and in the interplanetary space on the satellites of the Prognoz series.** V. V. Bezrukikh, A. P. Belyashin, G. I. Volkov, K. I. Gringauz, L. I. Denshchikova, G. N. Zastenker, V. G. Kaptsov, V. F. Kopylov, V. S. Mokrov, L. S. Musatov, U. N. Pozdnov, A. P. Remizov, M. Z. Khokhlov.
Geomagn. Aeronom., Vol. 14, 399 - 406 (1974). In Russian.

084.267 **Propagation of hydromagnetic waves in a three-dimensional magnetosphere. I.**
L. L. Van'yan, A. S. Lipatov.
Geomagn. Aeronom., Vol. 14, 496 - 501 (1974). In Russian.

084.268 **Division of anomalies of the variable geomagnetic field according to the results of a spherical analysis.**
M. N. Berdichevskij, M. S. Zhdanov.
Geomagn. Aeronom., Vol. 14, 506 - 511 (1974). In Russian.

084.269 **Analytical description of the geomagnetic field of past epochs and determination of the spectrum of magnetic waves in the earth's core. II.** S. I. Braginskij.
Geomagn. Aeronom., Vol. 14, 522 - 529 (1974). In Russian.

084.270 **On the existence of large regional anomalies in the geomagnetic field.**
N. G. Berlyand, V. S. Tsirel'.
Geomagn. Aeronom., Vol. 14, 530 - 537 (1974). In Russian.

084.271 **Magnetotail variations associated with the southward interplanetary magnetic field.**
C.-I. Meng, D. S. Colburn.
Journ. Geophys. Res., Vol. 79, 1831 - 1835 (1974).
  Ten weeks of simultaneous fine time resolution data of the interplanetary magnetic field (IMF), the solar wind parameters observed by the Explorer 33 and Vela 3 satellites, the magnetic field, and the particle fluxes in the magnetotail from the Imp 3 satellite are examined together with auroral zone magnetograms monitoring substorm activity to study the magnetotail variations associated with the latitudinal changes of the IMF.

084.272 **Energetic O^+ ions in the magnetosphere.**
R. D. Sharp, R. G. Johnson, E. G. Shelley, K. K. Harris.
Journ. Geophys. Res., Vol. 79, 1844 - 1850 (1974).
  Observations of energetic heavy ions are reported on seven occasions during three magnetic storms in 1969. Measurements were made at 1, 3, and 9 keV. Data from an orbit on March 24, 1969, are examined in some detail. The latitude distribution of these ion fluxes is compared to simultaneous satellite and ground-based measurements of other phenomena.

084.273 **Detached plasma regions in the magnetosphere.**
C. R. Chappell.
Journ. Geophys. Res., Vol. 79, 1861 - 1870 (1974).
  This paper presents the results of a study of detached plasma regions using cases from more than a year of cold plasma data measured by the Lockheed light ion mass spectrometer on Ogo 5. This data set includes all local times and covers radial distances from the top side ionosphere out to the magnetopause. The sensitivity of the spectrometer allows the measurement of plasma densities from about 0.1 ion/cm^3 to 10^6 ions/cm^3.

084.274 **Plasma sheet at lunar distance during magnetospheric substorms.**
F. J. Rich, D. L. Reasoner, W. J. Burke, E. W. Hones, Jr.
Journ. Geophys. Res., Vol. 79, 1981 - 1984 (1974).
  Evidence is described from the charged particle lunar environment experiment (CPLEE) to support the thesis that the plasma sheet thins at the onset of a substorm and expands during substorm recovery in the region $-60 R_E \leq X_{se} \leq -15 R_E$.

084.275 **Current flow in the magnetosphere during magnetospheric substorms.** G. Rostoker.

Journ. Geophys. Res., Vol. 79, 1994 - 1998 (1974).

A new model current system describing changes in real current flow in the magnetosphere associated with magnetospheric substorm activity is presented in this report. The physically new concept on which this current system is based is the lack of downflowing current in the night side magnetosphere during substorm expansion phase activity.

084.276 **The magnetospheric boundary.**
D. B. Beard, J. Y. Choe.
Astrophys. Space Sci. Library, Vol. 42, (see 012.016), 97 - 114 (1974).

084.277 **Magnetospheric boundaries and fields.**
W. P. Olson, K. A. Pfitzer.
Astrophys. Space Sci. Library, Vol. 42, (see 012.016), 115 - 129 (1974).

084.278 **A self consistent theory of the magnetotail.**
D. B. Beard, M. K. Bird.
Astrophys. Space Sci. Library, Vol. 42, (see 012.016), 131 - 137 (1974).

084.279 **OGO-5 observations of the magnetopause.**
C. T. Russell, M. Neugebauer, M. G. Kivelson.
Astrophys. Space Sci. Library, Vol. 42, (see 012.016), 139 - 157 (1974).

084.280 **Energetic electrons at the magnetopause.**
V. Domingo, D. E. Page, K.-P. Wenzel.
Astrophys. Space Sci. Library, Vol. 42, (see 012.016), 159 - 165 (1974).

084.281 **Polar magnetopause distance.**
A. Bahnsen, I. B. Iversen, E. Ungstrup.
Astrophys. Space Sci. Library, Vol. 42, (see 012.016), 167 - 171 (1974).

084.282 **A study of magnetospheric boundaries by the simultaneous observations of solar and magnetospheric particles.** J. B. Blake, A. L. Vampola.
Astrophys. Space Sci. Library, Vol. 42, (see 012.016), 173 - 186 (1974).

084.283 **The earth's bow shock fine structure.**
V. Formisano.
Astrophys. Space Sci. Library, Vol. 42, (see 012.016), 187 - 223 (1974).

084.284 **Proton fluxes upstream from the earth's bow shock.**
K. A. Anderson, R. P. Lin, C.-I. Meng.
Astrophys. Space Sci. Library, Vol. 42, (see 012.016), 225 - 227 (1974).

084.285 **HEOS 2 in relation to earlier polar cusp measurements.** K.-P. Wenzel.
Astrophys. Space Sci. Library, Vol. 42, (see 012.016), 229 - 235 (1974).

084.286 **ULF measurements at the polar cusp.**
A. Bahnsen, N. D'Angelo, E. Ungstrup.
Astrophys. Space Sci. Library, Vol. 42, (see 012.016), 237 - 247 (1974).

084.287 **Plasma observations in the high-latitude magnetosphere.** G. Paschmann, H. Grünwaldt, M. D. Montgomery, H. Rosenbauer, N. Sckopke.
Astrophys. Space Sci. Library, Vol. 42, (see 012.016), 249 - 253 (1974).

084.288 **The convergence of fact and theory on magnetospheric convection.** C. R. Chappell.
Astrophys. Space Sci. Library, Vol. 42, (see 012.016), 277 - 295 (1974).

084.289 **Electric field measurements at the geostationary position.** A. Pedersen.
Astrophys. Space Sci. Library, Vol. 42, (see 012.016), 337 - 342 (1974).

084.290 **Initial observations of magnetospheric boundaries by Explorer 45 (S^3).** T. A. Fritz, P. H. Smith, D. J. Williams, R. A. Hoffman, L. J. Cahill, Jr.
Astrophys. Space Sci. Library, Vol. 42, (see 012.016), 485 - 506 (1974).

084.291 **The design and development of a space laboratory to conduct magnetospheric and plasma research.**
A. Rosen.
Astrophys. Space Sci. Library, Vol. 42, (see 012.016), 649 - 662 (1974).

084.292 **Electron Echo experiment 1: Comparison of observed and theoretical motion of artificially injected electrons in the magnetosphere.**
R. W. McEntire, R. A. Hendrickson, J. R. Winckler.
Journ. Geophys. Res., Vol. 79, 2343 - 2354 (1974).

084.293 **Thermal and suprathermal plasma densities in the outer magnetosphere.** D. A. Gurnett, L. A. Frank.
Journ. Geophys. Res., Vol. 79, 2355 - 2361 (1974).

By using the low-frequency cutoff of electromagnetic noise trapped in the magnetosphere at frequencies above the local plasma frequency it is now possible to make very accurate, ±1%, electron density measurements in the low density region between the magnetopause and plasmapause. This technique for measuring the total plasma density has been used, together with measurements of the suprathermal proton intensities with the Lepedea instrumentation on the Imp 6 spacecraft, to determine the thermal proton densities in the region between the plasmapause and magnetopause.

084.294 **Near-earth magnetic disturbance in total field at high latitudes. 1. Summary of data from Ogo 2, 4, and 6.** R. A. Langel.
Journ. Geophys. Res., Vol. 79, 2363 - 2371 (1974).

Variations in the total (i.e., scalar) magnetic field data from the polar orbiting Ogo 2, 4, and 6 spacecraft (altitudes 400–1510 km) are summarized for invariant latitudes above 55°. Data from all degrees of magnetic disturbance are included. The data are presented in terms of the quantity ΔB (i.e., measured field magnitude minus the field magnitude from a spherical harmonic model of the quiet field).

084.295 **Near-earth magnetic disturbance in total field at high latitudes. 2. Interpretation of data from Ogo 2, 4, and 6.** R. A. Langel.
Journ. Geophys. Res., Vol. 79, 2373 - 2392 (1974).

Variations in the scalar magnetic field (ΔB) from the polar-orbiting Ogo 2, 4, and 6 spacecraft, with supporting vector magnetic field data from surface observatories, are analyzed at dipole latitudes above 55°. Data from all degrees of magnetic disturbances are included, the emphasis being on periods when $Kp = 2-$ to $3+$.

084.296 **Outline of a magnetospheric theory.**
W. J. Heikkila.
Journ. Geophys. Res., Vol. 79, 2496 - 2500 (1974).

Magnetospheric and auroral energy dissipation can be described in terms of electric fields and currents. The author suggests that the energy is created by an MHD generator in front of the magnetopause that derives its energy from the slowing down of the solar wind plasma. It drives a westward current in front of the eastward magnetopause surface current.

084.297 Synchrotron loss rates of energetic magnetospheric electrons. J. C. Baker.
Journ. Geophys. Res., Vol. 79, 2503 - 2505 (1974).

Loss rates due to synchrotron radiation have been calculated for electrons in the earth's magnetosphere with energies in the range 2–10 MeV. These loss rates are compared with measured rates and are shown to be of little importance for the earth. For Jupiter's magnetosphere the effect is much more pronounced.

084.298 The International Magnetospheric Study. J. G. Roederer.
Acta Astronaut., Vol. 1, 1 - 14 (1974).

The author first gives a brief qualitative description of the present state of knowledge of magnetospheric physics and then focuses on some of the research programs proposed for the International Magnetospheric Study 1976–78.

084.299 Paläomagnetismus. H. Heinz.
Sternenbote, 17. Jahrgang, p. 42 - 51 (1974).

084.300 On the heliographic latitude dependence of the interplanetary magnetic field as deduced from the 22-year cycle of geomagnetic activity. C. T. Russell.
Geophys. Res. Letters, Vol. 1, No. 1, p. 11 - 12 (1974).

The geomagnetic activity index, Ci, exhibits a double sunspot cycle in which 11-year periods alternately have more, then less, than average activity. The amplitude of the cycle-to-cycle variation has varied markedly since 1885, being large from 1885 to 1907, small from 1908 to 1948 and large again from 1949 to 1970. If we assume that the 22-year variation is caused by reversals every sunspot cycle of the heliographic latitude dependence of the dominant polarity of the interplanetary magnetic field, the behavior of the Ci index implies that the amplitude of the heliographic latitude dependence varies with time.

084.301 Determination of the external geomagnetic field from intensity measurements. G. E. Backus.
Geophys. Res. Letters, Vol. 1, No. 1, p. 21 - 22 (1974).

084.302 Editing and evaluating digitally recorded geomagnetic components at Canadian observatories.
J. M. Delaurier, E. I. Loomer, G. Jansen van Beek, A. Nandi.
Publ. Earth Phys. Branch, Dep. of Energy, Mines and Resources, Ottawa, Canada, Vol. 44, (No. 9), 235 - 242 (1974).

Automatic magnetic observatory systems (AMOS) incorporating a fluxgate sensor and a proton precession magnetometer are in operation at 10 Canadian observatories. Values of the field (H, D, Z, F) are recorded each minute on digital magnetic tape. Computer programs edit the data and reduce AMOS values to absolute observatory piers. Comparison of AMOS mean hourly values with similar values scaled from RUSKA magnetograms, and other tests, support the adoption of AMOS systems as primary recorders at Canadian observatories.

084.303 Compact bias coil systems for geomagnetic measurements. P. H. Serson.
Publ. Earth Phys. Branch, Dep. of Energy, Mines and Resources, Ottawa, Canada, Vol. 44, (No. 10), 243 - 247 (1974).

Measurements of the geomagnetic components with the proton magnetometer require a coil system for producing a homogeneous bias field in a direction which can be determined accurately. The theoretical advantages of various symmetrical arrangements of coils are reviewed, and the precision of mechanical construction necessary to achieve the theoretical performance is estimated.

084.304 An automatic magnetic observatory system. F. Andersen.
Publ. Earth Phys. Branch, Dep. of Energy, Mines and Resources, Ottawa, Canada, Vol. 44, (No. 11), 249 - 259 (1974).

A digitally recording magnetometer system is designed to replace the standard photographic magnetographs at geomagnetic observatories. The instrument can operate for up to 80 days without attention. Provision is made for immunity to brief power failures. The performance of the system is estimated from a comparison of hourly mean values with those of the St. John's Magnetic Observatory.

Erweitertes Modell der Magnetosphäre.
Umschau, 74. Jahrgang, p. 25 - 26 (1974).

A new concept of an open magnetosphere, in which geomagnetic field lines cross the boundary of the magnetosphere and connect directly with interplanetary magnetic-field lines, has become widely accepted and is described here.

Theories of the geomagnetic and solar dynamos.
See Abstr. 061.051.

Phase-bunching and other non-linear processes occurring in gyroresonant wave-particle interactions.
See Abstr. 062.035.

Controlled experiments from the space shuttle.
See Abstr. 062.038.

On the manifestation of the north-south asymmetry of sunspot activity in the geomagnetic indices.
See Abstr. 072.003.

A probable mean transit time of the flare-generated disturbances. See Abstr. 073.054.

Coronal λ 5303 intensity, geomagnetic activity and solar sources of high-speed plasma streams.
See Abstr. 074.066.

The solar wind and magnetospheric dynamics.
See Abstr. 074.076.

Solar coronal holes as sources of recurrent geomagnetic disturbances. See Abstr. 074.101.

A two-satellite study of low-energy protons over the polar cap during the event of November 18, 1968.
See Abstr. 078.017.

Solar particle access into the inner magnetosphere.
See Abstr. 078.025.

Energetic solar particle access into the geomagnetic tail. See Abstr. 078.026.

On a model of propagation of 5–300 MeV protons in the earth magnetic cavity. The role of neutral and plasma sheets. See Abstr. 078.027.

Comparisons of solar proton trajectory computations using model magnetospheres. See Abstr. 078.028.

Solar electron access to the magnetosphere.
See Abstr. 078.029.

Measurements of solar protons in the near earth magnetotail. See Abstr. 078.031.

Fluorescent ion jets for studying the ionosphere and magnetosphere. See Abstr. 083.052.

Indications of a longitudinal component in auroral phenomena. See Abstr. 084.034.

The aurora and the magnetosphere: The Chapman Memorial Lecture. See Abstr. 084.036.

Fluctuations of electron precipitation to the dayside auroral zone modulated by compression and expansion of the magnetosphere. See Abstr. 084.037.

Interplanetary control of magnetospheric dynamics. See Abstr. 106.033.

The relation between the azimuthal component of the interplanetary magnetic field and the geomagnetic field in the polar caps. See Abstr. 106.034.

Radiation Belts

084.401 Pitch angle effect on synchrotron radiation from electrons in the earth's magnetosphere.
S. Y. Peng, C. S. Wang, J. S. Kim.
Journ. Geophys. Res., Vol. 79, 138 - 141 (1974).

084.402 Geomagnetically trapped alpha particles. 3. Low-altitude outer zone alpha-proton comparisons.
J. F. Fennell, J. B. Blake, G. A. Paulikas.
Journ. Geophys. Res., Vol. 79, 521 - 528 (1974).

The outer zone population of trapped α particles and protons was measured off the equator in the L interval $2 \leqslant L \leqslant 4$ during 1969. The α particle spectra are best represented by a power law with spectral indices between 2.4 and 4.0 over the energy range $0.85 \leqslant E_\alpha \leqslant 9.0$ MeV. The α particle and proton spectra have the same slope at the same total energy.

084.403 Variations of the pitch-angle distribution of protons in the outer regions of the radiation belt.
A. S. Kovtyukh, M. I. Panasyuk, Eh. N. Sosnovets.
Kosmich. Issled., Vol. 12, 235 - 240 (1974). In Russian.

084.404 A study of equatorial inner belt protons from 2 to 200 MeV. E. S. Claflin, R. S. White.
Journ. Geophys. Res., Vol. 79, 959 - 965 (1974).

084.405 Particle saturation of the outer zone: a nonlinear model. M. Schulz.
Astrophys. Space Sci., Vol. 29, 233 - 242 (1974).

Properties of the steady state and transient behavior of geomagnetically trapped radiation are analyzed by means of phenomenological equations that concisely summarize the operative dynamical processes. The equations provide for a realistic coupling between electromagnetic wave energy, particle intensity, and pitch-angle anisotropy in the context of the outer zone.

084.406 Variations of the trapped particle population and of cutoff and pitch angle distribution of simultaneously observed magnetospheric solar protons during substorm activity. B. Häusler, M. Scholer, D. Hovestadt.
Journ. Geophys. Res., Vol. 79, 1819 - 1824 (1974).

It is the purpose of the present paper to discuss the influence of substorm activity on the solar proton behavior by using more simultaneous information about the substorm development, trapping boundary changes, and solar proton pitch angle distribution. The authors use low-altitude (Azur) and equatorial (ATS 5) data and examine in detail one substorm that occurred on March 29–30, 1970, during the presence of solar protons. The study shows that there exists an indication for strong pitch angle scattering of solar particles in the pseudotrapping region. Also observed is a strong dependence of the cutoff slope (solar proton flux versus latitude) on trapping boundary variations that occurred during different substorm phases.

084.407 Measurement of effective sections of the non-elastic interaction of protons with carbon and hydrogen nuclei in the energy range from 20 to 600 GeV aboard the space stations Proton 1, 2, 3. N. L. Grigorov, V. E. Nesterov, I. D. Rapoport, I. A. Savenko, G. A. Skuridin.
Izuchenie kosmich. luchej na iskusstven. sputnikakh Zemli. Moskva, Nauka, 1973, p. 10 - 33. In Russian. – Abstr. in Referativ. Zhurn. 62. Issled. kosmich. prostranstva, 6.62.177 (1974).

084.408 Investigation of high-energy electrons in cosmic space aboard Proton 1 and Proton 2.
N. L. Grigorov, L. F. Kalinkin, E. I. Kogan-Laskina, I. A. Savenko.
Izuchenie kosmich. luchej na iskusstven. sputnikakh Zemli. Moskva, Nauka, 1973, p. 104 - 111. In Russian. – Abstr. in Referativ. Zhurn. 62. Issled. kosmich. prostranstva, 6.62.178 (1974).

084.409 Investigation of high-energy electrons in the stratosphere. V. A. Bezus, A. M. Gal'per, N. L. Grigorov, V. V. Dmitrenko, L. F. Kalinkin, V. G. Kirillov-Ugryumov, B. I. Luchkov, A. S. Melioranskij, I. A. Savenko, Eh. M. Shermanzon.
Izuchenie kosmich. luchej na iskusstven. sputnikakh Zemli. Moskva, Nauka, 1973, p. 112 - 125. In Russian. – Abstr. in Referativ. Zhurn. 62. Issled. kosmich. prostranstva, 6.62.179 (1974).

Errata

084.901 Correction: 'Energy and momentum theorems in magnetospheric processes' [Rev. Geophys. Space Phys., Vol. 11, 289 - 353 (1973)]. G. L. Siscoe.
Rev. Geophys. Space Phys., Vol. 12, 135 (1974).

085 Solar-Terrestrial Relations

085.001 **Solar flares and the earth's rotation.**
N. S. Sidorenkov.
Priroda, No. 2.74, p. 104 - 105 (1974). In Russian.

085.002 **Transequatorial v.h.f. transmissions and solar-related phenomena.** M. P. Heeran, E. H. Carman.
Australian Journ. Phys., Vol. 26, 797 - 804 (1973).

Radio and ionospheric data are analysed to determine the influences of solar-geophysical phenomena on transequatorial v.h.f. transmissions along a European-Southern African circuit. Results over six years show a close dependence on sunspot number. The observed correlation with sudden ionospheric disturbances indicates periodic solar-dependent defocusing of transequatorial signals by the ionosphere, while the combined effects of neutral winds and the position of the magnetic equator appear to control the seasonal behaviour of the transmissions.

085.003 **Reliability of short-period forecasts of weather within a 27-day period.**
A. N. Lyubarskij, B. M. Rubashev.
Solnechnye Dannye 1973 Byull., No. 12, p. 108 - 112 (1974). In Russian.

085.004 **On the information content of ionospheric solar flare effect observations–I. Experimental evidence of solar and atmospheric control of the flare effects.**
K.-H. Ohle, R. Knuth, G. Entzian, J. Taubenheim.
Journ. Atmosph. Terr. Phys., Vol. 36, 513 - 524 (1974).

The purpose of this paper is to check some types of Sudden Ionospheric Disturbances (SID) in order to clarify their respective efficiencies to the incoming excessive solar X-radiation and to give some background for further statistical investigations.

085.005 **On the information content of ionospheric solar flare effect observations–II. Some model considerations on the interpretation of solar flare effects in the ionosphere.** J. Taubenheim, G. Entzian, R. Knuth, K.-H. Ohle.
Journ. Atmosph. Terr. Phys., Vol. 36, 525 - 535 (1974).

Analysis of records of LF sudden field anomaly (SFA) effects, together with the SOLRAD-9 records of the corresponding solar X-ray flare, shows that the quasi-phase height interpretation of the SFA effects is justified and that relatively simple model assumptions are able to provide a quantitative interpretation of these effects and their dependence on solar zenith angle as described in an earlier paper (1974).

085.006 **On the global relationship between the solar (corpuscular) activity and variations of atmospheric pressure.** V. E. Chertoprud, N. B. Mulyukova.
Astron. Tsirk., No. 804, p. 5 - 8 (1973). In Russian.

085.007 **Stratospheric ozone depletion and solar ultraviolet radiation on earth.** P. Cutchis.
Science, Vol. 184, 13 - 19 (1974).

Increased ultraviolet radiation and some consequent biological effects can be calculated.

085.008 **Lightning incidence in Britain and the solar cycle.**
M. F. Stringfellow.
Nature, Vol. 249, 332 - 333 (1974). – Letter.

085.009 **Estimate of the accuracy of forcasting short-period variations of the atmospheric density.**
M. I. Vojskovskij, I. I. Volkov, N. I. Gryazev, B. V. Kugaenko, V. M. Sinitsyn, P. E. Ehl'yasberg.
Prikl. zadachi kosmich. ballistiki. Moskva, Nauka, 1973, p. 93-107. In Russian. – Abstr. in Referativ. Zhurn. 51. Astron., 5.51.477 (1974).

085.010 **Geophysical consequences of solar flares in August 1972 according to observations at the Sverdlovsk Observatory.** Yu. P. Bulashevich, S. I. Ayubasheva, N. A. Ivanov, V. I. Utkin, V. A. Shapiro.
Dokl. Akad. Nauk SSSR. Ser. Mat. Fiz., Vol. 216, 528 - 531 (1974). In Russian.

085.011 **The annual distribution of atmospheric energy on a planetary scale.** J. P. Peixóto, A. H. Oort.
Journ. Geophys. Res., Vol. 79, 2149 - 2159 (1974).

The earth's climate, both near the surface and in the upper air, is of course largely determined by the atmospheric response to the seasonally variable solar input. The goal of obtaining a better understanding of this response and of other related processes has motivated the present study of the distribution of energy and the annual cycle in the atmospheric energetics.

085.012 **The influence of sunspots on rainfalls, thunderstorms and Etesians in Athens.**
L. N. Carapiperis.
In honorem S. Placidis, (see 003.009), p. 121 - 138 (1974).

085.013 **Solar activity and the weather.** L. Svalgaard.
Astrophys. Space Sci. Library, Vol. 42, (see 012.016), 627 - 639 (1974).

085.014 **Faraday rotation studies in Africa during the solar eclipse of June 30, 1973.**
A. N. Hunter, B. K. Holman, D. G. Feldgate, R. Kelleher.
Nature, Vol. 250, 205 - 206 (1974).

Four stations were set up in southern and eastern Africa to observe variations of total electron content during the eclipse of the sun on June 30, 1973. The authors conclude that if there were any travelling ionospheric disturbances, induced by the eclipse, they produced variations of less than 1% in the total electron content. Thus, it seems that the disturbances were very much smaller than had been anticipated.

085.015 **Can all clima periods and their disturbances be explained by the motion of planets?** R. Liese.
Deutsche Gewässerkundliche Mitt., Sonderheft 1973, p. 12 - 20. In German.

A general investigation on the passage of two planets in a certain direction shows that all climatic cycles as well as their periodical non-occurrence can be interpreted on the basis of such planetary passages.

085.016 **Solar magnetic fields and their influence on the earth.** J. M. Wilcox.
Naval Res. Rev., Arlington, (USA), Vol. 26, No. 8, p. 16 - 24 (1973).

085.017 **Special Committee on Solar-Terrestrial Physics: The International Magnetospheric Study (IMS) 1976–78.** H. Friedman.
ICSU Bull. No. 31, p. 27 - 34 (1973).

085.018 **On the influence of the solar (corpuscular) activity on variations of the atmospheric pressure.**
V. E. Chertoprud.
Astron. Tsirk., No. 820, p. 1 - 3 (1974). In Russian.

Planetary System

091 Physics of the Planetary System (Planetary Atmospheres, Figure, Interior, Magnetic Fields, Rotation, etc.)

091.001 **Rotation of solid bodies in the solar system.** S. J. Peale.
Rev. Geophys. Space Phys., Vol. 11, 767 - 793 (1973).

The effects of elastic distortion, nonprincipal axis rotation, precessing orbits, and internal dissipation on the rotation of a solid solar system body, which is in the gravitational field of an exterior body, are relatively easily analyzed by a Hamiltonian theory developed here. Examples of applications include the Chandler wobble, wobble of the moon, spin-orbit coupling, generalized Cassini laws, and tidal evolution.

091.002 **Quantization of the solar system.** J. M. Barnothy.
Bull. American Astron. Soc., Vol. 6, 209 (1974). Abstr. AAS.

091.003 **On the thermal evolution of the terrestrial planets.** P. E. Fricker, R. T. Reynolds, A. L. Summers.
The Moon, Vol. 9, 211 - 218 (1974). — Communication presented at the Lunar Science Institute Conference (see 012.002).

Physical and chemical constraints for such different planetary objects as the earth, the moon and meteorite parent bodies can best be satisfied by thermal history models having high initial temperatures. On the basis of thermal calculations it is suggested that the evolution of the other terrestrial planets (Mars, Venus and Mercury) was also characterized by high initial temperatures.

091.004 **Physical properties of the natural satellites.** D. Morrison, D. P. Cruikshank.
Space Sci. Rev., Vol. 15, 641 - 739 (1974).

This paper reviews the physical nature of the satellites of the planets, excluding the moon but including the rings of Saturn. Emphasis is placed on the best studied objects: Titan, Phobos and Deimos, the four Galilean satellites and the rings of Saturn.

091.005 **Photopolarimetry of planets and stars.** T. Gehrels.
Vistas in astronomy, Vol. 15, (see 003.001), 113 - 129 (1973).

091.006 **The dynamics of the planets and their satellites.** G. A. Wilkins, A. T. Sinclair.
The planets today. Symposium 1973, (see 012.003), p. 85 - 104 (1974).

091.007 **Inversion of gravity data for giant planets.** W. B. Hubbard.
Icarus, Vol. 21, 157 - 165 (1974).

In the present paper, using the polytrope of index one as a standard, the author will demonstrate the physical significance of the gravitational moments J_4 and J_6 for the giant planets, and derive approximate inversion relations which permit the calculation of certain interior properties, given observed values of J_2, J_4, and J_6.

091.008 **Cosmic ray ionization rates in the planetary atmospheres.** P. Velinov.
Journ. Atmosph. Terr. Phys., Vol. 36, 359 - 362 (1974).

Formulas are obtained for calculating the electron production rate of the solar cosmic rays in planetary atmospheres. This method is simpler to use than the expression developed by Dubach and Barker.

091.009 **A lower limit to Jeans' escape rate.** P. G. Gross.
Monthly Notices Roy. Astron. Soc., Vol. 167, 215 - 219 (1974).

A new and mathematically simple approximation to the classical, thermal escape rate of planetary atmospheres is presented. Using this and Jeans' escape rate, a permissible range for hydrogen escaping from the earth's atmosphere is derived. Several Monte Carlo and other theoretical calculations are found to fall within this permitted range.

091.010 **Mean elements of the principal planets.** P. K. Seidelmann, L. E. Doggett, M. R. DeLuccia.
Astron. Journ., Vol. 79, 57 - 60 (1974).

Mean elements of the principal planets for the epoch and the ecliptic and equinox of 1950.0 have been determined primarily as a basis for generating theories of the planets.

091.011 **Absolute brightness temperature measurements at 2.1-mm wavelength.** B. L. Ulich.
Icarus, Vol. 21, 254 - 261 (1974).

Absolute measurements of the brightness temperatures of the sun, new moon, Venus, Mars, Jupiter, Saturn, and Uranus, and of the flux density of DR21 at 2.1-mm wavelength are reported. Relative measurements at 3.5-mm wavelength are also presented.

091.012 **On the accuracy of the determination of the parameters of the atmospheres of planets from radio eclipsing phase measurements.**
V. A. Danilin, A. N. Kazantsev, A. V. Plotnikov.
Izv. vyssh. ucheb. zavedenij. Radiofizika, Vol. 16, 1405 - 1408 (1973). In Russian. - Abstr. in Referativ. Zhurn. 62. Issled. kosmich. prostranstva, 4.62.174 (1974).

091.013 **Satellites and magnetospheres of the outer planets.** D. A. Mendis, W. I. Axford.
Annual Rev. Earth Planet. Sci., Vol. 2, (see 003.006), 419 - 474 (1974).

091.014 **Planetologia comparata.** R. M. Nuccio.
Coelum, Vol. 42, 1 - 12, 53 - 63 (1974).

091.015 **The optical properties of Venus and the Jovian planets. II. Methods and results of calculations of the intensity of radiation diffusely reflected from semi-infinite homogeneous atmospheres.** J. M. Dlugach, E. G. Yanovitskij.
Icarus, Vol. 22, 66 - 81 (1974).

An efficient method is proposed for computation of the intensity of radiation diffusely reflected from a semi-infinite homogeneous atmosphere with arbitrary phase function. The authors present the results of calculations of the brightness distribution over the disk of a planet when the phase angle $\alpha = 0$, as well as spherical and geometrical albedos and other auxiliary functions. The curves of dependence of center-of-disk reflection coefficient and geometrical albedo of a planet upon spherical albedo were found to differ slightly for different phase functions. Using these dependences and results of spectrophotometric observations of the reflection coefficient for the center of disk of Saturn and geometrical albedo of Uranus, the integral (radiometric) albedo of these planets is estimated (0.50 ± 0.03 and 0.37 ± 0.05, respectively). The corresponding effective temperature is $T_e = 76 \pm 1°$ for Saturn and $T_e = 57° \pm 1°$ for Uranus.

091.016 **Nuclear reactions in the matter of the solar system.**
A. K. Lavrukhina, R. I. Kuznetsova, Eh. Rupp.
Meteoritika, vyp. (No.) 33, p. 14 - 22 (1974). In Russian.

091.017 **The $H_2O - CO_2$ system and planetary atmospheres.**
Yu. V. Alekhin, V. A. Zharikov, I. V. Zakirov.
Geokhimiya. Mineralogiya. Petrografiya. Tom 7. (Itogi nauki i tekhn. VINITI AN SSSR). Moskva, 1973, p. 5 - 78. In Russian. – Abstr. in Referativ. Zhurn. 51. Astron., 5.51.208 (1974).

091.018 **Nonstationary solution of the diffusion equation in a gravitational field with consideration of the transfer of particles and their disappearance.**
V. V. Makeev, V. M. Polyakov, V. V. Rybin.
Izv. vyssh. ucheb. zavedenij. Radiofizika, Vol. 16, 1660 - 1670 (1973). In Russian. – Abstr. in Referativ. Zhurn. 51. Astron., 5.51.220 (1974).

091.019 **On the accuracy of determination of the parameters of planetary atmospheres from radio phase eclipse observations.**
V. A. Danilin, A. N. Kazantsev, A. V. Plotnikov.
Izv. vyssh. ucheb. zavedenij. Radiofizika, Vol. 16, 1405 - 1408 (1973). In Russian.

091.020 **Deuterium enrichment of metallic hydrogen.**
W. B. Hubbard.
Astrophys. Journ., Vol. 190, 223 - 224 (1974).

If there is a first-order phase transition between molecular and metallic hydrogen in the liquid state, a simple statistical-mechanics calculation shows that the metallic phase will be slightly enriched in deuterium. The observed atmospheric deuterium abundance in the giant planets will be affected by at most 15 percent.

091.021 **An exact expression for the temperature structure of a homogeneous planetary atmosphere containing isotropic scatterers.** B. R. Barkstrom.
Astrophys. Journ., Vol. 190, 225 - 235 (1974).

An exact, analytic expression for the temperature structure in a radiative-equilibrium planetary atmosphere is derived, allowing isotropic scattering of the incident solar radiation. The equation of transfer for the solar radiation is solved by using the singular eigenfunction method. Simple expressions are derived for a number of interesting quantities, such as the solar flux, the albedo at a given latitude, and the Bond albedo. The temperature structure is found by using the Green's function for the thermal transport equation, with the solar-flux divergence as a source term. The expression for the temperature structure is evaluated using a single scattering albedo of 0.99 for the solar radiation with various ratios of solar to thermal extinction coefficients.

091.022 **Azimuth-dependent diffuse reflection of light from a semi-infinite planetary atmosphere.**
A. K. Kolesov.
Trudy Astron. Obs., *Leningrad*, Vol. 30 (= Uchenye Zapiski Leningr. Un-ta, No. 373 = Seriya Matem. Nauk, vyp. (No.) 50), p. 3 - 25 (1974). In Russian.

Sobolev's exact theory of anisotropic light scattering is used to solve the problem of diffuse reflection from a semi-infinite planetary atmosphere. The reflection function is expanded in terms of $\cos m\varphi$, where φ is the azimuth angle.

091.023 **Deconvolution in planetary photometry.**
K. Lumme.
Astron. Astrophys., Vol. 33, 39 - 41 (1974).

A method is proposed by which spreading in planetary images can be corrected. A solution of the deconvolution problem is given which can be used when the object is clearly larger than the half-width of the spreading function.

091.024 **Formation of coupled spectral lines in a planetary atmosphere.** J. W. Chamberlain, L. Wallace.
Astrophys. Journ., Vol. 190, 487 - 495 (1974).

Exact solutions have been obtained previously for the emergent intensities of two or more spectral lines that are coupled together by sharing a common upper level. Here, simple approximate solutions for the ith component are presented in terms of $H(\varpi, x_i)$, the H-functions of isotropic scattering for a weighted mean albedo (ϖ) for the whole multiplet. A table facilitating the calculation of $H(\varpi, x > 1)$ is given. For the case of a moderately inhomogeneous atmosphere an approximate solution is given in terms of $H(\varpi_e, x_i)$, the H-functions for an equivalent homogeneous atmosphere with an effective mean albedo of ϖ_e.

091.025 **Commensurable mean motions as a tool in solar system dynamical studies.** A. E. Roy.
Stars and the Milky Way system, Proc. 1972, (see 012.018), p. 290 (1974). – Abstract.

091.026 **The escape of light gases from planetary atmospheres.** D. M. Hunten.
Journ. Atmosph. Sci., Vol. 30, 1481 - 1494 (1973).

A recent description for the atmosphere of Titan is applied to other objects. A range of conditions is found in which the escape flux is determined by diffusion, and a simple expression is given for the flux. Models of the present earth, Mars and Venus are considered. Illustrative models are given of primitive atmospheres, some of which show hydrodynamic blowoff of H_2.

091.027 **The outer solar system: perspectives for exobiology.**
T. Owen.
Origins of Life, (*Netherlands*), Vol. 5, 41 - 55 (1974).

Summarizes current knowledge about the composition and structure of outer planet atmospheres with special emphasis on Jupiter, Saturn and Titan, in relation to the production of organic molecules of interest to studies of the origin of life.

091.028 **The dynamics of the planets and their satellites.**
G. A. Wilkins, A. T. Sinclair.
Proc. Roy. Soc. London, Ser. A, Vol. 336, 85 - 104 = H. M. Nautical Almanac Office, Roy. Greenwich Obs., Library Repr. No. 302 (1974).

The dynamical system of the planets and satellites shows many regularities that appear to have arisen during the evolution of the system. The study of some of the subsystems suggests that tidal friction has played an important role in bringing them into states in which the direct gravitational perturbations have become sufficiently strong to maintain the regular relationships between the orbits. In other cases it appears to be necessary to invoke the action of collision processes in order to find a satisfactory explanation of the observed regularities.

091.029 **Puzzles of the planets.** R. A. Lyttleton.
Australian Sci. Technol., Vol. 10, No. 10, p. 7 - 11 (1973).

091.030 **Effet de la polarisation sur la luminance du rayonnement diffusé par une atmosphère planétaire.**
J. Lenoble.
Nouvelle Rev. Optique, Vol. 4, 301 - 305 (1973).

Physics of the earth and planets.
See Abstr. 003.034.

Introduction to the dynamics of planetary atmospheres. See Abstr. 003.051.

Geography and geology of the planets (planetology).
See Abstr. 003.069.

Man explores the planets. See Abstr. 003.127.

Hot hydrogen atoms: initiators of reactions of interest in interstellar chemistry and evolution.
See Abstr. 022.049.

Mean elements of the principal planets.
See Abstr. 042.022.

Elements of the outer planets for one million years.
See Abstr. 042.066.

On stability of the flow of stratified rotating superposed fluids. See Abstr. 062.072.

Detailed study of the flate of the isotopic (σn) correlations in possible s-process conditions.
See Abstr. 065.049.

The solar wind and its influence on planetary atmospheres. See Abstr. 074.022.

General relationships among sound speeds. I. New experimental information. See Abstr. 081.003.

The physics of lunar and planetary surfaces.
See Abstr. 094.183.

Winds and the occultation experiment.
See Abstr. 097.058.

High-resolution Fourier spectra of stars and planets.
See Abstr. 114.101.

092 Mercury

092.001 **A new upper limit for an atmosphere of CO_2, CO on Mercury.** U. Fink, H. P. Larson, R. F. Poppen.
Astrophys. Journ., Vol. 187, 407 - 415 (1974).

High-resolution infrared spectra of Mercury ($1.9-2.7\mu$) obtained with the original "Connes" interferometer at the Steward Observatory 90-inch telescope have provided a very sensitive test for the possible presence of a CO_2, CO atmosphere. An improved upper limit of 0.12 cm-atm has been set for CO_2, and a new upper limit of 0.05 cm-atm has been set for CO. Upper limits of similar magnitude can be established for CH_4 and NH_3. Implications for the possible evolution of an atmosphere on Mercury are discussed.

092.002 **A well-observed transit of Mercury.**
J. Ashbrook.
Sky Telescope, Vol. 47, 4 - 9 (1974).

092.003 **Mercury transit: some late reports.**
J. Ashbrook.
Sky Telescope, Vol. 47, 202 - 204 (1974).

092.004 **Albedo variations on the surface of Mercury.**
L. Wilson.
Planet. Space Sci., Vol. 22, 99 - 109 (1974).

Photoelectric photometric (slit) scans of Mercury have been obtained and combined with a map of the surface markings to yield relative normal albedoes over about one quarter of the planet's total surface at a wavelength of 0.45 microns. Maximum albedo ratios at a resolution of one fifth of the planetary diameter are not less than 2 to 1 and probably near 2.5 to 1. The corresponding average lunar value is 2.3 to 1.

092.005 **Numerical simulation of the Mercury transit black drop phenomenon.** A. Wittmann.
Astron. Astrophys., Vol. 31, 239 - 243 (1974). – Research note.

092.006 **Le passage de Mercure, vu de Lausanne, le 10 novembre 1973.** M. Roud.
Orion Schaffhausen, 32. Jahrgang, p. 22 (1974).

092.007 **Passage de Mercure, le 10 novembre 1973 vu à Orcines (F63).** J. Dragesco.
Orion Schaffhausen, 32. Jahrgang, p. 23 (1974).

092.008 **Beobachtung des Merkur-Durchgangs vor der Sonne am 10. November 1973 in Locarno-Monti.**
W. Sandner, H. Bernhard.
Orion Schaffhausen, 32. Jahrgang, p. 28 - 30 (1974).

092.009 **Der Merkurdurchgang vom 10. November 1973.**
A. Wittmann, H. Wöhl.
SuW, Vol. 13, 41 - 43 (1974).

092.010 **Observations of the Mercury transition across the solar disk on November 10, 1973.** I. F. Nikulin.
Astron. Tsirk., No. 802, p. 2 - 3 (1973). In Russian.

092.011 **Surface features on Mercury.**
S. Zohar, R. M. Goldstein.
Astron. Journ., Vol. 79, 85 - 91 (1974).

A high-resolution study of the surface of Mercury with a 2.388 GHz radar reveals the existence of hills and valleys with heights (depths) of about 1 km. There is also evidence supporting the existence of craters with diameters of about 50 km and depths of about 700 m.

092.012 **Observation of the transit of Mercury over the solar disc on November 10, 1973.** Yu. A. Medvedev.
Astron. Tsirk., No. 806, p. 7 - 8 (1974). In Russian.

092.013 **The atmosphere of Mercury from observations during its transit across the solar disc on November 10, 1973.** N. A. Kozyrev.
Astron. Tsirk., No. 808, p. 5 - 6 (1974). In Russian.

092.014 **On the atmosphere of Mercury.** R. I. Kiladze.
Astron. Tsirk., No. 811, p. 7 - 8 (1974). In Russian.

092.015 **Mercury: Does its atmosphere contain water?** G. E. Thomas.
Science, Vol. 183, 1197 - 1198 (1974).
The atmosphere of Mercury, like that of the moon, is maintained in an extremely tenuous minimum state by weak solar wind accretion and radioactive decay processes, and depleted by strong removal mechanisms. Unlike the moon, it has a high daytime surface temperature that promotes the production of water vapor, which may be the dominant atmospheric constituent derived from solar wind protons.

092.016 **New results of investigations on Mercury and Venus.** A. Bonov.
Mat. i fizika (NRB), Vol. 16, No. 5, p. 55 - 58 (1973). In Bulgarian.

092.017 **Mariner 10 pictures of Mercury: First results.** B. C. Murray, M. J. S. Belton, G. E. Danielson, M. E. Davies, D. Gault, B. Hapke, B. O'Leary, R. G. Strom, V. Suomi, N. Trask.
Science, Vol. 184, 459 - 461 (1974).
Mercury has a heavily cratered surface containing basins up to at least 1300 kilometers diameter flooded with mare-like material. Many features are closely similar to those on the moon, but significant structural differences exist. Major chemical differentiation before termination of accretion is implied.

092.018 **Atmosphere of Mercury.** I. P. Williams.
Nature, Vol. 249, 234 (1974).
It had generally been expected that because of the high temperatures existing on Mercury and also because of the low surface gravity, any atmosphere originally present would quickly evaporate. All this ignores the solar wind which carries mass into the atmosphere. In reality the problem becomes a boundary layer problem in which the magnetic field of Mercury plays a part, but one can see simply that the main effect is to generate an atmosphere.

092.019 **Transit of Mercury across the solar disk.** V. A. Efanov, I. G. Moiseev, A. B. Severny.
Nature, Vol. 249, 330 - 331 (1974). – Letter.

092.020 **The planet Mercury as viewed by Mariner 10.** R. G. Strom.
Sky Telescope, Vol. 47, 360 - 369 (1974).

092.021 **Le passage de Mercure devant le soleil du 10 novembre 1973 observé par nos sociétaires.** P. de La Cotardière.
L'Astronomie, 88ᵉ année, p. 181 - 186 (1974).

092.022 **Erste Ergebnisse der Merkur-Erkundung.** H. Zimmer.
SuW, Vol. 13, 186 - 190 (1974).

092.023 **Observation of the transit of Mercury of 10 November 1973.** L. Hric.
Kozmos, Vol. 5, 56 (1974). In Slovak.

092.024 **Passaggio di Mercurio sul disco solare del 10 novembre 1973.** L. Pansecchi, Banducci, Mallegni, Mazzoni.
Coelum, Vol. 42, 38 - 40 (1974).

092.025 **Merkur-Durchgang, 1973 XI. 10^d.** W. Sandner.
Sterne, Vol. 50, 122 (1974).

092.026 **The possible atmosphere of Mercury.** J. H. Robinson.
Journ. British Astron. Ass., Vol. 84, 278 (1974).

092.027 **An 'extension' to the transit of Mercury, 1973 November 10.** H. Hill.
Journ. British Astron. Ass., Vol. 84, 279 - 280 (1974).

092.028 **Report on the planet Mercury, 1972 February to 1974 February.** J. H. Robinson.
Journ. British Astron. Ass., Vol. 84, 283 - 286 (1974).

092.029 **The nature of the subsurface of Mercury from microwave observations at several wavelengths.** J. N. Cuzzi.
Astrophys. Journ., Vol. 189, 577 - 586 (1974).
Recent microwave observations of Mercury at wavelengths spanning the range 0.3–18 cm are interpreted in the light of a new and refined model as being entirely consistent with a lunar-like regolith of particulate geological material in a vacuum.

092.030 **Prolaz Merkura ispred Sunca 10. 11. 1973 (Transit of Mercury across the sun on 10. 11. 1973).** G. Kren.
Vasiona, Vol. 22, 41 - 43 (1974).

092.031 **Possible histories of the obliquity of Mercury.** S. J. Peale.
Astron. Journ., Vol. 79, 722 - 744 (1974).
The possible evolutionary paths of the spin axis of Mercury are determined as a function of the permanent deformation and initial conditions. Transition into and subsequent evolution within the (3/2) spin resonance are also considered. From this analysis, the evolution of the moon and satellites of the major planets is also inferred.

092.032 **Mercury: more surprises in the second assessment.** W. D. Metz.
Science, Vol. 185, 132 (1974).

092.033 **Mariner 10 Mercury encounter.** J. A. Dunne.
Science, Vol. 185, 141 - 142 (1974). – Report.

092.034 **Preliminary infrared radiometry of the night side of Mercury from Mariner 10.** S. C. Chase, E. D. Miner, D. Morrison, G. Münch, G. Neugebauer, M. Schroeder.
Science, Vol. 185, 142 - 145 (1974).
The infrared radiometer experiment on Mariner 10 was designed to measure the surface brightness temperature along a near-equatorial track from mid-afternoon across the night side to mid-morning, local time. In this preliminary report the authors discuss only the data from the night side – 19 hours to 05 hours, local time.

092.035 **Observations at Mercury encounter by the plasma science experiment on Mariner 10.** K. W. Ogilvie, J. D. Scudder, R. E. Hartle, G. L. Siscoe, H. S. Bridge, A. J. Lazarus, J. R. Asbridge, S. J. Bame, C. M. Yeates.
Science, Vol. 185, 145 - 151 (1974).
An unexpectedly strong interaction between the solar wind and Mercury was detected by the plasma science experiment

when Mariner 10 encountered Mercury on 29 March 1974. This report presents preliminary results from the rearward-looking (anti-solar) electrostatic analyzer which forms part of the plasma science experiment on Mariner 10.

092.036 **Magnetic field observations near Mercury: preliminary results from Mariner 10.** N. F. Ness, K. W. Behannon, R. P. Lepping, Y. C. Whang, K. H. Schatten.
Science, Vol. 185, 151 - 160 (1974).

Results from a preliminary analysis of "quick-look" data obtained by the NASA-GSFC magnetic field experiment during Mercury encounter on 29 March 1974 are summarized in this report. There is substantial evidence in this initial assessment of the results to support the preliminary conclusion that an intrinsic planetary magnetic field exists.

092.037 **Electrons and protons accelerated in Mercury's magnetic field.**
J. A. Simpson, J. H. Eraker, J. E. Lamport, P. H. Walpole.
Science, Vol. 185, 160 - 166 (1974).

This is a preliminary report of measurements from the Mariner 10 spacecraft, which show the presence at Mercury of large and impulsive fluxes of electrons with energies >170 keV and protons with energies >500 keV distributed over regions comparable in size to the planet itself.

092.038 **Mercury's atmosphere from Mariner 10: preliminary results.**
A. L. Broadfoot, S. Kumar, M. J. S. Belton, M. B. McElroy.
Science, Vol. 185, 166 - 169 (1974).

Analysis of data obtained by the ultraviolet experiment on Mariner 10 indicates that Mercury is surrounded by a thin atmosphere consisting in part of helium. The total surface pressure of the atmosphere is less than about 2×10^{-9} millibar. Upper limits are set for the abundance of various gases, including hydrogen, oxygen, carbon, argon, neon, and xenon. The wavelength dependence of Mercury's surface albedo is similar to that of the moon over a broad range of wavelengths from 500 to 1600 angstroms.

092.039 **Mercury's surface: preliminary description and interpretation from Mariner 10 pictures.**
B. C. Murray, M. J. S. Belton, G. E. Danielson, M. E. Davies, D. E. Gault, B. Hapke, B. O'Leary, R. G. Strom, V. Suomi, N. Trask.
Science, Vol. 185, 169 - 179 (1974).

The surface morphology and optical properties of Mercury resemble those of the moon in remarkable detail and record a very similar sequence of events. Chemical and mineralogical similarity of the outer layers of Mercury and the moon is implied. No evidence of atmospheric modification of landforms has been found.

092.040 **Mercury: results on mass, radius, ionosphere, and atmosphere from Mariner 10 dual-frequency radio signals.** H. T. Howard, G. L. Tyler, P. B. Esposito, J. D. Anderson, R. D. Reasenberg, I. I. Shapiro, G. Fjeldbo, A. J. Kliore, G. S. Levy, D. L. Brunn, R. Dickinson, R. E. Edelson, W. L. Martin, R. B. Postal, B. Seidel, T. T. Sesplaukis, D. L. Shirley, C. T. Stelzried, D. N. Sweetnam, G. E. Wood, A. I. Zygielbaum.
Science, Vol. 185, 179 - 180 (1974).

As Mariner 10 flew by Mercury on 29 March 1974, dual-frequency radio transmissions from the spacecraft were monitored on earth. Mariner 10's trajectory allowed Mercury's mass to be estimated from these data and, because the spacecraft was occulted by the planet, also afforded investigators the opportunity to measure the radius of Mercury and to detect any possible atmosphere or ionosphere. The results of the preliminary analysis of the radio data are presented.

092.041 **Surface temperature and emissivity of Mercury.** O. L. Hansen.
Astrophys. Journ., Vol. 190, 715 - 717 (1974).

A new method, which is independent of flux calibration and bolometric emissivity, has been used to measure the surface temperature of the illuminated side of Mercury and to deduce the surface emissivity.

Observation of transit of Mercury – 1973 November 11.
Journ. Astron. Soc. Victoria, Vol. 27, 11 (1974).

'Snapshots' of planet Mercury show surface dotted with hills.
IEEE Spectrum, Vol. 11, No. 3, p. 26 (1974).

An observer's guide to the planet Mercury.
See Abstr. 003.060.

Investigation of the seeing from the photometric profile of a Mercury projection on the solar disk.
See Abstr. 071.019.

Brightness temperatures of Venus, Saturn and Mercury at 3.87 mm wavelength. See Abstr. 093.001.

093 Venus

093.001 **Brightness temperatures of Venus, Saturn and Mercury at 3.87 mm wavelength.**
V. N. Voronov, A. G. Kisljakov, A. V. Troitsky.
Astron. vestn., Vol. 8, 17 - 19 (1974). In Russian.
The following brightness temperature values were obtained: Venus $(387 \pm 50)°K$, Saturn $(115 \pm 15)°K$, Mercury $(335 \pm 44)°K$.

093.002 **Investigations of the Venus atmosphere with Soviet automatic stations.**
Yu. A. Surkov, B. M. Andrejchikov.
Zemlya i Vselennaya, 1974, No. 1, p. 33 - 37. In Russian.

093.003 **On the status of the rotation of Venus.**
R. A. Lyttleton.
Astrophys. Space Sci., Vol. 26, 497 - 511 (1974).
Owing to its extremely slow rotation, Venus must be regarded as a triaxial body with differences of all three principal moments of inertia comparable in magnitude. Equations for the rotatory motion are set up in a form suitable for numerical solution by machine-calculations. Results obtained suggest that for the actual planet the direction of the rotation axis may move almost randomly between the two hemispheres defined by the orbital plane and thus that the present direction near the south celestial pole of the orbit may be only a temporary situation. Order-of-magnitude considerations based on the equations of motion suggest that a timescale of order 10^7 to 10^8 yr may on average be required for large changes in direction of the rotation axis to take place.

093.004 **The lower ionosphere of Venus.**
J. Dubach, R. C. Whitten, J. S. Sims.
Planet. Space Sci., Vol. 22, 525 - 536 (1974).
Galactic cosmic ray bombardment provides a permanent background ionosphere in planetary atmospheres. A transport technique is used to compute the cosmic ray ionization rate profile in a model of the Venusian atmosphere at altitudes between 55 and 100 km. These ionization rates are then applied to a model of ion chemistry to predict equilibrium electron and ion density profiles. Ionization rates for typical solar flare proton events are available from earlier calculations and have been included.

093.005 **Venus clouds: structure and composition.**
A. T. Young.
Science, Vol. 183, 407 - 409 (1974).
The clouds of Venus consist of a fine sulfuric acid aerosol similar to that found in the earth's stratosphere. The acid aerosol on Venus appears to be uniformly mixed with the gas, at least in the visible layers, and possibly down to the cloud base.

093.006 **The energy flux and fluctuations of decimeter radio waves at the descent of Venera 7 and Venera 8 to the surface of the planet.** O. I. Yakovlev, A. I. Efimov, T. S. Timofeeva, K. M. Shvachkin.
Kosmich. Issled., Vol. 12, 100 - 103 (1974). In Russian.

093.007 **On the propagation of electromagnetic waves in the Venus atmosphere.**
V. S. Avduevskij, A. I. Nojkina, M. M. Skotnikov.
Kosmich. Issled., Vol. 12, 104 - 108 (1974). In Russian.

093.008 **Spectroscopic observations of spatial and temporal variations on Venus.**
A. T. Young, A. Woszczyk, L. G. Young.
Acta Astron., Vol. 24, 55 - 68 (1974).
Details of the Table Mountain spectroscopic patrol of Venus in Sept.-Oct. 1972 are given. The data indicate systematic variation over the disc, with more CO_2 absorption near the terminator than at the limb, and slightly more in the southern than in the northern hemisphere. The semiregular 4-day variation is confirmed.

093.009 **Lateral inhomogeneities in the Venus atmosphere: analysis of thermal infrared maps.**
A. P. Ingersoll, G. S. Orton.
Icarus, Vol. 21, 121 - 128 (1974).
The thermal infrared maps of Venus published by Murray, Wildey, and Westphal (1963) and Westphal, Wildey, and Murray (1965) have been analyzed systematically in order to separate the observed intensity into a limb-darkening component and a solar-associated component representing fixed patterns of intensity corotating with the earth and sun, respectively. Interesting new results are obtained for the solar-associated component.

093.010 **The Regulus occultation light curve and the real atmosphere of Venus.**
J. Veverka, L. Wasserman.
Icarus, Vol. 21, 196 - 198 (1974).
An inversion of the light curve observed during the July 7, 1959, occultation of Regulus by Venus leads to the conclusion that the light curve cannot be reconciled with models of the Venus atmosphere based on spacecraft observations.

093.011 **Circulation and dust content of the Venus atmosphere according to wind velocity measurements on the Venera 8 automatic interplanetary station.**
V. V. Kerzhanovich, M. Ya. Marov.
Doklady Akad. Nauk SSSR, Ser. Mat. Fiz., Vol. 215, 554 - 557 (1974). In Russian.

093.012 **Radiative transfer within the mesospheres of Venus and Mars.** V. Ramanathan, R. D. Cess.
Astrophys. Journ., Vol. 188, 407 - 416 (1974).
A simplified formulation is presented for radiative transfer within the mesosphere of a carbon dioxide atmosphere. The analysis, when applied to predict the global mean radiative-equilibrium temperature profile for Venus, gives excellent agreement with the more detailed calculations of Dickinson. With respect to Mars, a temperature minimum is found to occur at a pressure of 30 μb, consistent with the suggestion of a temperature minimum near 20 μb by the Mariner 9 television experiment. Qualitative consideration is also given to dynamics induced by diurnal changes in solar heating.

093.013 **Comparison of the far ultraviolet spectra of Venus and Mars.** H. W. Moos.
Journ. Geophys. Res., Vol. 79, 685 - 687 (1974).
Both Venus and Mars have largely CO_2 atmospheres and as a result should be similar above the turbopause. Thus it is of some interest to compare the far ultraviolet spectra of the two planets, both for similarities and for differences. This letter describes a comparison between the Mariner 9 ultraviolet spectra of Mars (Barth et al., 1972) and a spectrum of the total disc of Venus obtained from a sounding rocket (Rottman and Moos, 1973).

093.014 **Estimate of the temperature of the stratosphere of Venus from data of the automatic station Venera 8.**
Z. P. Cheremukhina, S. F. Morozov, N. F. Borodin.
Kosmich. Issled., Vol. 12, 264 - 271 (1974). In Russian.

093.015 On the interpretation of measurements of the illumination intensity in the Venus atmosphere.
N. L. Lukashevich, M. Ya. Marov, E. M. Fejgel'son.
Kosmich. Issled., Vol. 12, 272 - 278 (1974). In Russian.

093.016 Mariner 10 enthüllt Venus. W. Petri.
Umschau, 74. Jahrgang, p. 278 - 279 (1974).
Outstanding preliminary results are presented.

093.017 Water vapor in Venus determined by airborne observations of the 8200 Å band.
T. R. Gull, C. R. O'Dell, R. A. R. Parker.
Icarus, Vol. 21, 213 - 218 (1974).

The region of the 8200 Å band of H_2O was studied in spectra of Venus obtained with an echelle grating spectrograph operated at an altitude of 14.6 km in the NASA Learjet research aircraft. Taking advantage of low foreground absorption, observing at a time of velocity quadrature, differential spectroscopy with respect to lunar spectra, and spectrum averaging, we establish a value of H_2O of $3 \pm 20\,\mu$ for the total path over the entire disk.

093.018 Radar map of Venus at 3.8 cm wavelength.
A. E. E. Rogers, R. P. Ingalls, G. H. Pettengill.
Icarus, Vol. 21, 237 - 241 (1974).

Radar echoes from Venus have been mapped at a wavelength of 3.8 cm from approximately 270° to 10° in longitude and −50° to +50° in latitude during the inferior conjunctions of 1969 and 1972. Observations made in April 1969 and again in June 1972 both show the same regions of high reflectivity as well as several large regions of low reflectivity.

093.019 Die Venusatmosphäre und Mariner 10.
H. Zimmer.
SuW, Vol. 13, 160 - 163 (1974).

093.020 Preliminary infrared radiometry of Venus from Mariner 10.   S. C. Chase, E. D. Miner,
D. Morrison, G. Münch, G. Neugebauer.
Science, Vol. 183, 1291 - 1292 (1974).

The intensity of emission at 45 micrometers, measured with high spatial resolution along a single crossing of the Venus disk, is presented.

093.021 Observations at Venus encounter by the plasma science experiment on Mariner 10.
H. S. Bridge, A. J. Lazarus, J. D. Scudder, K. W. Ogilvie, R. E. Hartle, J. R. Asbridge, S. J. Bame, W. C. Feldman, G. L. Siscoe.
Science, Vol. 183, 1293 - 1296 (1974).

Preliminary results from the rearward-looking electrostatic analyzer of the plasma science experiment during the Mariner 10 encounter with Venus are described. They show that the solar-wind interaction with the planet probably involves a bow shock rather than an extended exosphere. An observed reduction in the flux of electrons with energies greater than 100 electron volts is interpreted as evidence for some direct interaction with the exosphere.

093.022 Venus: Mass, gravity field, atmosphere, and ionosphere as measured by the Mariner 10 dual-frequency radio system.   H. T. Howard, G. L. Tyler, G. Fjeldbo, A. J. Kliore, G. S. Levy, D. L. Brunn, R. Dickinson, R. E. Edelson, W. L. Martin, R. B. Postal, B. Seidel, T. T. Sesplaukis, D. L. Shirley, C. T. Stelzried, D. N. Sweetnam, A. I. Zygielbaum, P. B. Esposito, J. D. Anderson, I. I. Shapiro, R. D. Reasenberg.
Science, Vol. 183, 1297 - 1301 (1974).

The unique properties of the Mariner 10 radio system and the preliminary scientific results obtained from the analysis of the radio signals are described.

093.023 Magnetic field observations near Venus: Preliminary results from Mariner 10.   N. F. Ness,
K. W. Behannon, R. P. Lepping, Y. C. Whang, K. H. Schatten.
Science, Vol. 183, 1301 - 1306 (1974).

Results from a preliminary analysis of a limited data set are summarized in this report. A detached bow shock wave that develops as the super Alfvénic solar wind interacts with the Venusian atmosphere and far downstream disturbances associated with the solar wind wake have been observed.

093.024 Venus: Atmospheric motion and structure from Mariner 10 pictures.   B. C. Murray,
M. J. S. Belton, G. E. Danielson, M. E. Davies, D. Gault, B. Hapke, B. O'Leary, R. G. Strom, V. Suomi, N. Trask.
Science, Vol. 183, 1307 - 1315 (1974).

The Mariner 10 television cameras imaged the planet Venus in the visible and near ultraviolet for a period of 8 days at resolutions ranging from 100 meters to 130 kilometers. The general pattern of the atmospheric circulation in the upper tropospheric/lower stratospheric region is displayed in the pictures.

093.025 Ultraviolet observations of Venus from Mariner 10: Preliminary results.
A. L. Broadfoot, S. Kumar, M. J. S. Belton, M. B. McElroy.
Science, Vol. 183, 1315 - 1318 (1974).

An objective grating spectrometer on Mariner 10 has measured airglow in the wavelength range 200 to 1700 angstroms. The data reveal the presence of significant concentrations of hydrogen, helium, carbon, and oxygen atoms in the upper atmosphere of Venus. A preliminary analysis of the hydrogen data indicates an exospheric temperature of 400°K.

093.026 Search by Mariner 10 for electrons and protons accelerated in association with Venus.
J. A. Simpson, J. H. Eraker, J. E. Lamport, P. H. Walpole.
Science, Vol. 183, 1318 - 1321 (1974).

The University of Chicago instruments on board the Mariner 10 spacecraft bound for Mercury have measured energy spectra and fluxes of electrons from 0.18 to 30 million electron volts and protons from 0.5 to 68 million electron volts along the plasma wake and in the bow shock regions associated with Venus. Unusually quiet solar conditions and improved instrumentation made it possible to search for much lower fluxes of protons and electrons in similar energy regions as compared to earlier Mariner missions to Venus. No evidence for electrons or protons was found. The importance of these null results for determining the necessary and sufficient conditions for particle acceleration is discussed with respect to magnetometer evidence that Venus does not have a magnetosphere.

093.027 Internal structure of Venus.   R. A. Lyttleton.
Nature, Vol. 248, 624 (1974). − Letter.

093.028 Radar backscattering from Venus at oblique incidence at a wavelength of 70 cm.
T. Hagfors, D. B. Campbell.
Astron. Journ., Vol. 79, 493 - 502 (1974).

By illuminating Venus with circularly polarized radiation at 70-cm wavelength and observing the backscattered echo, the polarized circular diffuse scattering laws were measured for incidence angles up to 90°. It was found that at large incidence angles both scattering laws have a $\cos^n \theta$ dependence with n very close to 1.7. In addition, the degree of linear polarization of the received echo was measured as a function of the incidence angle along the apparent rotation axis. This was found to be very much smaller than for the moon, indicating very little subsurface reflection, and to have an angular dependence not increasing monotonically with θ. A model is proposed to explain this effect.

093.029 Spectral reflectance systematics for mixtures of powdered hypersthene, labradorite, and ilmenite.
D. B. Nash, J. E. Conel.
Journ. Geophys. Res., Vol. 79, 1615 - 1621 (1974).

The authors determine what systematic variations exist in the optical properties of a mineral mixture when the mineral concentration is varied from one end-member to another within a family of mineral phases. In this way they can determine what spectral property or combination of properties, if any, is uniquely diagnostic of mineral type and concentration in an unknown material. Their objective is to be able to interpret better in terms of mineral composition the spectra of the lunar surface obtained by remote techniques over wide areas of the lunar surface not sampled directly during the Apollo missions.

093.030 Snelle wolken hoog in de dampkring van Venus.
O. Namba.
Zenit, Vol. 1, No. 5, p. 38 - 40 (1974).

093.031 Atmosphere of Venus: implications of Venera 8 sunlight measurements.
A. A. Lacis, J. E. Hansen.
Science, Vol. 184, 979 - 982 (1974).

Venera 8 measurements of solar illumination within the atmosphere of Venus are quantitatively analyzed by using a multilayer model atmosphere. The analysis shows that there are at least three different scattering layers in the atmosphere of Venus and the total cloud optical thickness is $\gtrsim 10$.

093.032 The nature of planet Venus. G. N. Katerfeld.
Fiz.-mat. spisanie, Vol. 16, No. 2, p. 91 - 115 (1973). In Bulgarian.

093.033 Auswertung planetarer Daten der Venus-Sonden.
F. Frölich.
Gerlands Beiträge Geophys., Vol. 83, 95 - 100 = Mitt. Zentralinst. Phys. Erde, No. 356 (1974).

The characteristical data, hitherto available, of the atmosphere of Venus are studied in connection with questions on the causes of the formation and effects of such an extremely different situation relative to the terrestrial conditions.

093.034 Measurements of the wind velocity in the Venus atmosphere with the automatic interplanetary station Venera 8. B. N. Andreev, V. T. Guslyakov, V. V. Kerzhanovich, Yu. M. Kruglov, V. P. Lysov, M. Ya. Marov, L. V. Onishchenko, M. K. Rozhdestvenskij, V. P. Sorokin, Yu. N. Shnygin.
Kosmich. Issled., Vol. 12, 419 - 429 (1974). In Russian.

093.035 Evaluation of water vapor and ammonia content in the lower Venus atmosphere from radar measurements. T. V. Smirnova, A. D. Kuzmin.
Astron. Zhurn. Akad. Nauk SSSR, Vol. 51, 607 - 610 (1974). In Russian. English translation in Soviet Astron. AJ, Vol. 18, No. 3.

A comparison of radar measurement data of Venus and calculated data for various models of the Venus atmosphere was carried out. An evaluation of the water vapor and ammonia content in the lower Venus atmosphere was obtained. A new processing of radio interferometric measurements of Venus was made taking into account microwave absorption in the Venus atmosphere. The dielectric constant of the Venus surface was obtained to be $\epsilon = 3.6 \pm 1.0$.

093.036 Venus: an ionospheric model with an exospheric temperature of 350° K.
S. Kumar, D. M. Hunten.
Journ. Geophys. Res., Vol. 79, 2529 - 2532 (1974).

The electron densities from Mariner 5 are fitted by an ionosphere based on a neutral composition (at the homopause) containing 1% O and 2 ppm H_2. The corresponding exospheric temperature is 350°K. Ion-neutral reactions of the H_2 give a large source of nonthermal H atoms that may explain the dual scale height of Lyman α.

093.037 Atmospheric ion wakes of Venus and Mars in the solar wind.
P. A. Cloutier, R. E. Daniell, Jr., D. M. Butler.
Planet. Space Sci., Vol. 22, 967 - 990 (1974).

A model has been developed for the currents induced in the ionospheres of Venus and Mars by the flowing magnetized solar wind in a previous paper (Cloutier and Daniell, 1973). The altitudes of the ionopauses on both planets, determined from the electrodynamical models of the previous paper, are used to calculate the total rates of atmospheric mass loss to the solar wind for Venus and Mars. These loss rates are compared to the rates calculated by Michel (1971) based upon the limit of mass loading of the solar wind flow determined from hydrodynamic constraints. The distributions of planetary ions in the downstream wakes of Venus and Mars are calculated, and the interpretation of ion spectrometer measurements from close planetary encounters is discussed.

093.038 Venera 8: measurements of temperature, pressure and wind velocity on the illuminated side of Venus.
M. Ya. Marov, V. S. Avduevsky, V. V. Kerzhanovich, M. K. Rozhdestvensky, N. F. Borodin, O. L. Ryabov.
Journ. Atmosph. Sci., Vol. 30, 1210 - 1214 (1973).

093.039 Venera 8: measurements of solar illumination through the atmosphere of Venus.
V. S. Avduevsky, M. Ya. Marov, B. E. Moshkin, A. P. Ekonomov.
Journ. Atmosph. Sci., Vol. 30, 1215 - 1218 (1973).

Measurements on the flux of downward solar radiation through the atmosphere of Venus and at the planetary surface are reported.

093.040 Rotation of the upper atmosphere of Venus.
C. B. Leovy.
Journ. Atmosph. Sci., Vol. 30, 1218 - 1220 (1973).

It is suggested that the rapid rotation of the upper atmosphere of Venus indicates that there is an equatorial thermal bulge corresponding to a vertically averaged temperature variation of about 3K.

093.041 Photochemistry of the Venus atmosphere.
M. B. McElroy, Nien Dak Sze. Yuk Ling Yung.
Journ. Atmosph. Sci., Vol. 30, 1437 - 1447 (1973).

Carbon monoxide, produced in the Venus atmosphere by photolysis of CO_2, is removed mainly by reaction with OH. A detailed model, which accounts for the photochemical stability of Venus CO_2 is presented and discussed.

093.042 NASA scientists study acid clouds of Venus.
S. Miller, N. Panagakos.
NASA News Release, No. 73—268, p. 1 - 6 (1973).

093.043 Venus ionosphere: an interpretation of Mariner 10 observations. S. J. Bauer, R. E. Hartle.
Geophys. Res. Letters, Vol. 1, No. 1, p. 7 - 10 (1974).

The dayside ionosphere of Venus observed by Mariner 10 may be understood in terms of a dynamic interaction with the solar wind which results in a compressed topside above an "F_2 ledge" consisting of O^+ and a dynamically unaffected F_1 layer corresponding to a neutral temperature $T \approx 380°$ K and consisting of O_2^+ and CO_2^+. The top of the upper ledge appears to be an ionopause caused by solar wind scavenging of He^+, representing a solar-wind obstacle consistent with the bow shock observations.

Venus observed by Mariner.
Sky Telescope, Vol. 47, 235 - 240 (1974).

Les conjonctions de Vénus avec diverses régions du ciel. See Abstr. 041.028.

Mariner 10 Venus encounter. See Abstr. 053.014.

New results of investigations on Mercury and Venus. See Abstr. 092.016.

Evaluation of intensity of resonance line emission of inert gases in planetary atmospheres. See Abstr. 097.001.

Magnetic fields of Mars and Venus: solar wind interactions. See Abstr. 097.020.

Mars and Venus. See Abstr. 097.023.

News of plasma envelopes of Mars and Venus. See Abstr. 097.027.

The effect of radiative transfer on shear-flow instability in the atmospheres on Mars and Venus. See Abstr. 097.045.

The ionospheres of Mars and Venus. See Abstr. 097.056.

Cloud base levels for Jupiter and Venus and the heteromolecular nucleation theory. See Abstr. 099.028.

094 Moon: Dynamics, Global Properties, Local Properties

Moon, Dynamics

094.001 Radiographic inspection of the circumlunar space with the help of Luna 19.   M. B. Vasil'ev, V. A. Vinogradov, A. S. Vyshlov, O. G. Ivanovskij, M. A. Kolosov, N. A. Savich, V. A. Samovol, L. N. Samoznaev, A. I. Sidorenko, A. I. Shejkhet, D. Ya. Shtern.
Kosmich. Issled., Vol. 12, 115 - 121 (1974). In Russian.

094.002 Effect of a giant impact on the thermal evolution of the moon.   J. Arkani-Hamed.
The Moon, Vol. 9, 183 - 209 (1974). — Communication presented at the Lunar Science Institute Conference (see 012.002).

094.003 The formation of the moon.   J. A. O'Keefe III.
The Moon, Vol. 9, 219 - 225 (1974). — Communication presented at the Lunar Science Institute Conference (see 012.002).
Supporting evidence for the fission hypothesis for the origin of the moon is offered. Two dynamical objections to fission are shown to be surmountable under certain apparently plausible conditions.

094.004 Evolution of the earth-moon system.   H. Alfvén, G. Arrhenius.
The Moon, Vol. 9, 235 - 239 (1974). — Communication presented at the Lunar Science Institute Conference (see 012.002).

094.005 Lunar orbit determination in the presence of unmodeled accelerations.
D. S. Ingram, B. D. Tapley.
Celestial Mechanics, Vol. 9, 191 - 211 (1974).
A technique for estimating the state of an artificial satellite in the presence of unmodeled accelerations is presented. The unmodeled acceleration is approximated by a first-order Gauss-Markov sequence which can be separated into a timewise correlated component and a purely random component. Using this approximation, a sequential procedure for estimating the position, velocity, and the unmodeled acceleration is developed. The method is evaluated by reducing range-rate observations obtained by tracking the Apollo 10 and 11 spacecraft during the lunar orbit phase of the mission. Numerical results are presented which show that the observation residual pattern lies within the observation noise standard deviation.

094.006 Early thermal evolution of the moon, III. Formation of a liquid sulfide core.   M. Hallam.
Meteoritics, Vol. 8, 375 - 376 (1973). — Abstr. Meteoritical Soc.

094.007 On the question of the absence of free libration. K. S. Shakirov.
Astron. Tsirk., No. 809, p. 6 - 7 (1974). In Russian.

094.008 The evolution of the moon.
M. N. Toksöz, D. H. Johnston.
Icarus, Vol. 21, 389 - 414 (1974).
The authors present a report on the status of the current knowledge of the thermal evolution of the moon based on both their own calculations and other published works. They first list the data and constraints that are of major importance to the models. Then they describe the basic inputs, computational techniques, and temperature models. In the final section they compare the thermal evolution of the moon with that of the earth and Mars.

094.009 La lune. Un premier bilan du programme d'exploration lunaire.   R. Dejaiffe.
Ciel et Terre, Vol. 90, 1 - 44 (1974).

094.010 The early history of the lunar inclination.
D. P. Rubincam.
Goddard Space Flight Center, Greenbelt, Maryland, GSFC Document X-592-73-328, 11 + 167 pp. (1973).
The effect of tidal friction on the inclination of the lunar

orbit to the earth's equator for earth-moon distances of less than 10 earth radii is examined. The results obtained bear on a conclusion drawn by Gerstenkorn and others which has been raised as a fatal objection to the fission hypothesis of lunar origin, namely, that the present nonzero inclination of the moon's orbit to the ecliptic implies a steep inclination of the moon's orbit to the earth's equatorial plane in the early history of the earth-moon system. This conclusion is shown to be valid only for particular rheological models of the earth.

094.011 **Que pourra-t-on déduire des mesures de distance terre-lune par laser?** J. Kovalevsky.
IAU Symposium No. 61, (see 012.014), p. 269 - 274 (1974).

094.012 **On the moon's orbit.** T. Kwast.
Urania Kraków, Vol. 45, 136 - 140 (1974). In Polish.

The tragedy of the moon.
See Abstr. 003.022.

The moon. See Abstr. 003.121.

Rotation of solid bodies in the solar system.
See Abstr. 091.001.

Moon, Global Properties

094.101 Degassing of the moon. II. Degassing process now and in the past. J. Classen.
Astron. vestn., Vol. 8, 29 - 34 (1974). In Russian.

The observed concentration of the transient lunar phenomena near lunar craters can be interpreted as result of fracturation of lunar ground with accumulation of gases in holes that would be formed. The dependence of the degassing process upon the evolution of lunar surface and interior is discussed. The effect of the magma cooling rate on the degassing process is discussed also.

094.102 Comparison of dynamical and geometrical shape of the moon. I. V. Gavrilov, G. T. Yanovitskaya.
Phys. Earth Planet. Interiors, Vol. 8, 102 - 104 (1974).

The dynamical shape of the moon is described by means of the parameters of its gravity field and principal moments of inertia. These data are derived from the lunar artificial satellites (Luna 10, Orbiters 1–4), and astronomical measurements of the physical librations. The geometrical shape of the moon is determined from measurements of absolute heights of the lunar surface. The differences of dynamical and geometrical figures of the moon are analysed, and their possible interpretation is discussed.

094.103 A model for the steady interaction of the solar wind with the moon. P. J. Catto.
Astrophys. Space Sci., Vol. 26, 47 - 94 (1974).

The steady state interaction of the solar wind with the moon is modeled as a uniform, magnetized, quasi-neutral, collisionless, hypersonic, and hyper-Alfvénic flow of an electron-proton plasma past a perfectly ion absorbing, non-magnetized sphere. For the temperature of the electrons T much less than that of the ions T_i, steady state equations are derived self-consistently from the Vlasov and Maxwell equations and by expanding in a small parameter ϵ that measures the smallness of ∇B terms compared to a dominant term retained. A partial numerical solution is presented and discussed for the limit in which ϵ is much less than $\beta=$(ion pressure/magnetic pressure). In addition, a simple technique is presented whereby the steady state equations can be approximately extended to cases in which $T \gtrsim T_i$ for arbitrary ϵ/β.

094.104 Stereophotographs of the moon. Yu. K. Lyubimov.
Zemlya i Vselennaya, 1974, No. 1, p. 77 - 78. In Russian.

094.105 New names on the moon. J. Ashbrook.
Sky Telescope, Vol. 47, 170 - 171 (1974).

094.106 General colorimetric structure of the lunar surface. N. N. Evsyukov.
Vestn. Khar'kov. Univ. No. 99 (Ser. Astron. No. 8) p. 26 - 32 (1973). In Russian.

094.107 Statistical colour distribution on the lunar surface. N. N. Evsyukov.
Vestn. Khar'kov. Univ. No. 99 (Ser. Astron. No. 8), p. 32 - 36 (1973). In Russian.

094.108 Lunar seismicity, structure, and tectonics. D. R. Lammlein, G. V. Latham, J. Dorman, Y. Nakamura, M. Ewing.
Rev. Geophys. Space Phys., Vol. 12, 1 - 21 (1974).

Analysis of seismic signals from nine man-made impacts has led to a model of shallow lunar structure. Data relevant to the structure, state, and dynamics of the deep lunar interior derive from deep moonquakes, and the interpretation of these data is the subject of this paper.

094.109 Lunar magnetism. M. Fuller.
Rev. Geophys. Space Phys., Vol. 12, 23 - 70 (1974).

In this review, material presented up to and including the abstract of the Fourth Lunar Science Conference is covered: essentially these are the results from the Apollo 11, 12, 14, and 15 missions and some preliminary work for 16 and 17.

094.110 Stand der Mondforschung nach Apollo. R. Meißner.
Umschau, 74. Jahrgang, p. 74 - 80 (1974).

Samples from different lunar areas are analyzed and explained in terms of origin and structure. Data about the interior of the moon are used in order to derive physical parameters of the moon's body and compare its evolution with that of the earth.

094.111 Lunar velocity structure and compositional and thermal inferences. M. N. Toksöz, F. Press, A. M. Dainty, K. R. Anderson.
The Moon, Vol. 9, 31 - 42 (1974). – Communication presented at the Lunar Science Institute Conference (see 012.002).

Seismic data from the Apollo Passive Seismic Network stations are analyzed to determine the velocity structure and to infer the composition and physical properties of the lunar interior. Data from artificial impacts (S-IVB booster and LM ascent stage) cover a distance range of 70 – 1100 km. Travel times and amplitudes, as well as theoretical seismograms, are used to derive a velocity model for the outer 150 km of the moon.

094.112 A magnetic dynamo in the moon? E. H. Levy.
The Moon, Vol. 9, 49 - 56 (1974). – Communication presented at the Lunar Science Institute Conference (see 012.002).

The existence of fossil lunar magnetism has caused speculation that the moon had, at one time, an internally produced dynamo magnetic field. Quantitative analysis of this idea, constrained by the largest iron lunar core compatible with observations, implies that the moon would have had to rotate faster than its breakup angular velocity in order to support a dynamo magnetic field.

094.113 Temperature dependence of electrical conductivity and lunar temperatures. G. R. Olhoeft, A. L. Frisillo, D. W. Strangway, H. Sharpe.
The Moon, Vol. 9, 79 - 87 (1974). – Communication presented at the Lunar Science Institute Conference (see 012.002).

There are several pitfalls inherent in the extrapolation of lunar temperatures from laboratory measurements of electrical conductivity. These include the choice of representative material for the luanr interior, appropriate environmental conditions (pressure, fugacity, etc.) and the various measurement difficulties.

094.114 Preliminary results of the Apollo 17 infrared scanning radiometer. W. W. Mendell, F. J. Low.
The Moon, Vol. 9, 97 - 103 (1974). – Communication presented at the Lunar Science Institute Conference (see 012.002).

094.115 Lunar gravity via the Apollo 15 and 16 subsatellites. W. L. Sjogren, R. N. Wimberly, W. R. Wollenhaupt.
The Moon, Vol. 9, 115 - 128 (1974). – Communication presented at the Lunar Science Institute Conference (see 012.002).

Dense Doppler tracking coverage of the Apollo 15 and 16 subsatellites over ten and eighteen day periods when periapsis altitudes were 15 - 50 km has provided detailed gravity mapping of the lunar frontside. Many new gravity features are revealed including one that does not correlate with any visible topographic structure.

094.116 **Radon emanation from the moon, spatial and temporal variability.**
P. Gorenstein, L. Golub, P. Bjorkholm.
The Moon, Vol. 9, 129 - 140 (1974). — Communication presented at the Lunar Science Institute Conference (see 012.002).

094.117 **Density within the moon and implications for lunar composition.** S. C. Solomon.
The Moon, Vol. 9, 147 - 166 (1974). — Communication presented at the Lunar Science Institute Conference (see 012.002).

Density models for the moon, including the effects of temperature and pressure, can satisfy the mass and moment of inertia of the moon and the presence of a low density crust indicated by the seismic refraction results only if the lunar mantle is chemically or mineralogically inhomogeneous. Using $C/MR^2 = 0.400$ and the recent seismic evidence suggesting a thin, high density zone beneath the crust and a partially molten 'core', successful density models can be found for a range of temperature profiles.

094.118 **Problems associated with estimating the relative impact rates on Mars and the moon.**
G. W. Wetherill.
The Moon, Vol. 9, 227 - 231 (1974). — Communication presented at the Lunar Science Institute Conference (see 012.002).

094.119 **Moonquakes and lunar tectonism.**
G. Latham, D. Lammlein.
The Moon, Vol. 9, 233 (1974). — Communication presented at the Lunar Science Institute Conference (see 012.002). Abstract.

094.120 **Deep interior of the moon derived from seismic data.**
Y. Nakamura, D. Lammlein, G. Latham, M. Ewing, J. Dorman.
The Moon, Vol. 9, 233 (1974). — Communication presented at the Lunar Science Institute Conference (see 012.002). Abstract.

094.121 **Seismic data on lunar meteoroid impacts.**
J. Dorman, D. Lammlein, Y. Nakamura, G. Latham, M. Ewing.
The Moon, Vol. 9, 233 - 234 (1974). — Communication presented at the Lunar Science Institute Conference (see 012.002). — Abstract.

094.122 **Elasticity and constitution of the lunar crust and mantle: laboratory studies.** D. H. Chung.
The Moon, Vol. 9, 234 (1974). — Communication presented at the Lunar Science Institute Conference (see 012.002). Abstract.

094.123 **Lunar magnetic fields.** C. T. Russell, P. J. Coleman, Jr., B. R. Lichtenstein, G. Schubert, L. R. Sharp.
The Moon, Vol. 9, 234 - 235 (1974). — Communication presented at the Lunar Science Institute Conference (see 012.002). — Abstract.

094.124 **Apollo 15 and 16 gammy ray results.**
A. E. Metzger, R. C. Reedy, J. I. Trombka, L. E. Peterson, J. R. Arnold.
The Moon, Vol. 9, 242 - 243 (1974). — Communication presented at the Lunar Science Institute Conference (see 012.002). — Abstract.

094.125 **Bistatic-radar inference of lunar surface properties.**
G. L. Tyler, H. T. Howard.
The Moon, Vol. 9, 244 - 245 (1974). — Communication presented at the Lunar Science Institute Conference (see 012.002). — Abstract.

094.126 **Geophysical data and the interior of the moon.**
M. N. Toksöz.
Annual Rev. Earth Planet. Sci., Vol. 2, (see 003.006), 151 - 177 (1974).

094.127 **Topographic photographs of the lunar surface from Soviet automatic space vehicles.**
B. N. Rodionov, B. V. Nepoklonov, V. V. Kiselev, A. S. Selivanov, V. M. Govorov, V. V. Zasetskij, O. G. Ivanov, F. I. Babkov, R. M. Mikhajlov, E. A. Reshetov, M. V. Abramova, A. A. Bol'shoj, K. K. Davidovskij, V. G. Somal'.
Geod. i kartografiya, 1973, No. 10, p. 29 - 41. In Russian. Abstr. in Referativ. Zhurn. 62. Issled. kosmich. prostranstva, 3.62.465 (1974).

094.128 **Some aspects of the physics of the moon.**
S. K. Runcorn.
The planets today. Symposium 1973, (see 012.003), p. 11 - 33 (1974).

094.129 **The interior of the moon.** D. L. Anderson.
Phys. Today, Vol. 27, No. 3, p. 44 - 49 (1974).

094.130 **The lunar crust: a product of heterogenous accretion or differentiation of a homogeneous moon?**
R. Brett.
Meteoritics, Vol. 8, 331 - 333 (1973). — Abstr. Meteoritical Soc.

094.131 **Thermoluminescence effects in lunar materials and meteorites.** S. A. Durrani.
Meteoritics, Vol. 8, 359 - 360 (1973). — Abstr. Meteoritical Soc.

094.132 **An indication for variation of meteoroid flux based on exposure ages of individual lunar microcraters.**
J. B. Hartung, D. Storzer.
Meteoritics, Vol. 8, 376 - 377 (1973). — Abstr. Meteoritical Soc.

094.133 **Opposing partial melting models and the genesis of some lunar magmas.**
N. J. Hubbard, C.-Y. Shih.
Meteoritics, Vol. 8, 384 - 386 (1973). — Abstr. Meteoritical Soc.

094.134 **Lunar differentiation model.** J. A. Philpotts.
Meteoritics, Vol. 8, 419 - 420 (1973). — Abstr. Meteoritical Soc.

094.135 **The bulk chemistry of the moon: a two-component model.** H. Wänke, H. Palme.
Meteoritics, Vol. 8, 451 (1973). — Abstr. Meteoritical Soc.

094.136 **The moon's feldspathic crust: early differentiation or heterogeneous accretion?** J. A. Wood.
Meteoritics, Vol. 8, 467 - 468 (1973). — Abstr. Meteoritical Soc.

094.137 **Absolute altitudes of the moon's marginal zone in the system of selenodetic reference points.**
A. S. Duma.
Astrometriya i Astrofizika, *Kiev*, vyp. (No.) 20, (see 003.002),

p. 55 - 68 (1973). In Russian.

The paper deals with 960 absolute altitudes of points within the system of compiled catalogue of 2580 points on the moon. The method to determine altitudes is presented; material of observations and the procedure of measuring and reduction are described.

094.138 Automatic observations on the lunar surface for the investigation of the rotation of the moon.
M. D. Polanuer.
Astron. Zhurn. Akad. Nauk SSSR, Vol. 51, 425 - 430 (1974). In Russian. English translation in Soviet Astron. AJ, Vol. 18, No. 2.

A method of the utilization of the results of position measurements of stars passing through the field of view of a rotating astronomical telescope established on the lunar surface for the determination of physical libration constants of the moon is proposed.

094.139 Possible lunar ring dikes.
W. S. Cameron, J. L. Padgett.
The Moon, Vol. 9, 249 - 294 (1974).

Features exhibiting one or more of the following four criteria were included as lunar analogs to terrestrial ring dikes: (1) inner ridge(s) approximately concentric with the crater wall, (2) inner rill(s) approximately concentric with the crater wall, (3) outer ridge(s) and/or rill(s) approximately concentric with the crater wall, and (4) interior and exterior slopes of the crater wall approximately equal (implying extrusion of lava along a ring fracture). Features exhibiting each of the four criteria were found and some had combinations of two or more including rills merging into ridges. The initial search showed a distribution of the possible lunar ring dikes that was non-random and strongly associated with the margins of the maria, further implying that they are volcanic features. This relation was upheld when extended by the recent survey.

094.140 The ultraviolet reflectivity of the moon.
J. H. Carver, B. H. Horton, R. S. O'Brien, G. G. O'Connor.
The Moon, Vol. 9, 295 - 303 (1974).

The ultraviolet flux from the entire lunar disk has been measured in a series of rocket flights from Woomera at several wavelength bands in the range 2400–2900 Å and also at the wavelength of the hydrogen Lα line (1216 Å). Comparison of these measurements with other observations shows that between the visible and middle ultraviolet part of the spectrum, the lunar albedo decreases sharply towards shorter wavelengths falling to (0.7 ± 0.1) percent at 2400 Å which is a factor of ten less than the visible albedo. The measured albedo at 1216 Å is (0.3 ± 0.1) percent.

094.141 Zonal comparison of modern selenodetic reference systems.
A. A. Gurshtein, A. A. Konopikhin, G. A. Burba.
The Moon, Vol. 9, 305 - 321 (1974).

An account is given of the results of a comparison of existing basic selenodetic systems in the equatorial zone of the moon together with plan and altitude data, which have been provided by means of a specially worked out method, based on the use of the LAC charts of the moon (scale 1 : 1000000), and which does not require the presence of common catalogued reference points. As a result of the comparison a set of better points has been obtained forming a catalogue which may be referred to as LPL. The selection was made on the basis of magnitude and character of both the systematic and random errors.

094.142 Correlation between region of enhanced mare flooding on the lunar farside and Apollo orbital results.
V. Gornitz.
The Moon, Vol. 9, 323 - 325 (1974). – Research note.

094.143 Multivariate analyses of crater parameters and the classification of craters.
B. S. Siegal, J. C. Griffiths.
The Moon, Vol. 9, 397 - 413 (1974).

Multivariate analyses were performed on certain linear dimensions of six genetic types of craters. A total of 320 craters, consisting of laboratory fluidization craters, craters formed by chemical and nuclear explosives, terrestrial maars and other volcanic craters, and terrestrial meteorite impact craters, authenticated and probable, were analyzed in the first data set in terms of their mean rim crest diameter (D_r), mean interior relief (R_i), rim height (R_e), and mean exterior rim width (W_e). The second data set contained an additional 91 terrestrial craters of which 19 were of experimental percussive impact and 28 of volcanic collapse origin, and which was analyzed in terms of D_r, R_i, and R_e. Principal component analyses were performed on the six genetic types of craters.

094.144 Observations of electrons at the lunar surface.
B. E. Goldstein.
Journ. Geophys. Res., Vol. 79, 23 - 35 (1974).

Observations of electrons at the Apollo 12 and 15 sites by the ALSEP Solar Wind Spectrometer experiments showed qualitative differences. In the geomagnetic tail the Apollo 15 instrument provided measurements of lunar photoelectron flux from 5 to 40 eV; at 20 eV the flux was 2×10^6 el/cm^2 sec eV ster. The estimated height-integrated conductivity of the photoelectron layer is 10^{-4} to 10^{-5} ohm^{-1}. A theoretical model, ignoring the magnetic field, indicates that most solar wind electrons should reach the lunar surface; a lunar surface potential at the subsolar point of +5 to −3 V is calculated and compared with the Apollo 15 site observations.

094.145 Apollo-experiment brengt aan het licht: toch een atmosfeer rondom de maan. C. de Jager.
Zenit, Vol. 1, No. 1, p. 14 - 17 (1974).

094.146 Geometric albedo of the moon in the ultraviolet.
A. N. Bentley, C. G. De Jonckheere, D. E. Miller.
Astron. Journ., Vol. 79, 401 - 403 (1974).

A measurement of the lunar irradiance in five passbands in the near ultraviolet has been made with a rocket-borne photometer. The results show that the geometric albedo of the moon decreases from 5.6% at 3270 Å to 4.3% at 2570 Å. These values conform to the steady downward trend in geometric albedo towards shorter wavelengths indicated by the ground-based UBV photometric data, but are at least a factor of two higher than earlier rocket measurements at wavelengths shorter than 3000 Å.

094.147 Internal origin for lunar rilled craters and the maria?
J. L. Whitford-Stark.
Nature, Vol. 248, 573 - 575 (1974).

If the rilles in large rilled craters are the same age as the enclosing crater, the distribution of the craters precludes them from being anything other than of internal origin and, as a consequence, the maria and their basins must also be of internal origin.

094.148 Lunar magnetism and an early cold moon.
D. W. Strangway, H. A. Sharpe.
Nature, Vol. 249, 227 - 230 (1974).

The authors consider whether an early hot outside, cold inside moon, which evolved into the present cold outside, hot inside moon, could quantitatively account for the evidence of an ancient magnetic field.

094.149 Creation of an artificial lunar atmopshere.
R. R. Vondrak.

Nature, Vol. 248, 657 - 659 (1974).

The tenuous nature of the lunar atmosphere is maintained by rapid loss of gases released at the lunar surface. But would such rapid loss still occur if the density of the lunar atmosphere were greatly increased from its present value? The author seeks to answer this question by evaluating quantitatively the atmospheric loss mechanisms operating in the lunar environment. One conclusion is that if the density of the lunar atmosphere is increased, a point can be reached where loss occurs so slowly that it is negligible over human time scales.

094.150 **The lunar atmosphere.**
R. R. Hodges, Jr., J. H. Hoffman, F. S. Johnson.
Icarus, Vol. 21, 415 - 426 (1974).

The atmosphere of the moon is exceedingly tenuous and appears to consist mainly of noble gases. The solar wind impinges on the lunar surface, supplying detectable amounts of helium, neon and ^{36}Ar. Influxes of solar wind protons and carbon and nitrogen ions are significant. Radiogenic ^{40}Ar and ^{222}Rn produced within the moon have been detected.

094.151 **Moonquake predetermination and tides.**
W. B. Chapman, B. M. Middlehurst, A. L. Frisillo.
Icarus, Vol. 21, 427 - 436 (1974).

A pattern in moonquakes, which correlates with the monthly tidal cycle, also correlates with a phase shifted pattern of a 7-month tidal cycle. The lead of approximately 2 months in moonquake occurrence can be explained if local tidal forces are combined with a moonquake-driving force. This force, assumed to result from the 6-yr physical libration in latitude, would cause N–S sliding of an outer layer across a solid layer within a decoupled core in a lunar model.

094.152 **Neutral and ion exosphere models for lunar hydrogen and helium.** R. E. Hartle, G. E. Thomas.
Journ. Geophys. Res., Vol. 79, 1519 - 1526 (1974).

A general neutral exosphere model, which includes density and temperature variations at the exobase, is applied to the moon to obtain surface and radial density distributions for H, H_2, and He. We assume that the source for these constituents derives from accretion of solar wind ions. The surface distributions are determined by requiring that the sum of the neutral and solar wind ion fluxes for a given constituent vanish at all points on the surface. On this basis, we obtain maximum day side surface densities for H (23 cm^{-3}) H_2 (4.2×10^3 cm^{-3}), and He (1.2×10^3 cm^{-3}) and maximum night side surface densities for H (13 cm^{-3}), H_2 (1.3×10^4 cm^{-3}) and He (2.2×10^4 cm^{-3}) that are consistent with either measured values or upper limits.

094.153 **Studier av månens kjemiske sammensetning.**
E. Steinnes.
Astron. Tidssk., Årg. 7, p. 69 - 74 (1974).

094.154 **The origin of the lunar surface.** B. Steverding.
Phys. Earth Planet. Interiors, Vol. 8, 287 - 291 (1974).

A mechanism is described how the lunar landscape could have originated by a chain reaction of volcanic eruptions. The mechanism consists of a sequence of three steps: surface solidification and fragmentation, gas formation and crack-wave interactions. A quantitative analysis is presented and it is shown that the volcanic chain reaction was most likely to occur around 10 million years after surface solidification.

094.155 **Electrical properties of a synthetic lunar pyroxenite, and the internal temperature of the moon.**
H. G. Tolland.
Phys. Earth Planet. Interiors, Vol. 8, 292 - 294 (1974). – Research note.

094.156 **Heterogeneous accretion and the lunar crust.**
A. E. Ringwood.
Geochim. Cosmochim. Acta, Vol. 38, 983 - 984 (1974).

The case for heterogeneous accretion of the moon is considered in the light of recent studies of lunar internal differentiation. The evidence which was formerly held to favour heterogeneous accretion is no longer compelling.

094.157 **Lunar surface topography: computer generation and display of high-resolution radar astronomy maps.**
S. H. Zisk.
Astron. Astrophys., Suppl. Ser., Vol. 15, 355 - 361 (1974). Conference paper (see 012.011).

094.158 **Operation and performance of a lunar laser ranging station.** E. C. Silverberg.
Applied Optics, Vol. 13, 565 - 574 (1974).

The Lunar Laser Ranging Experiment has been in continuous operation for about 4 years using the McDonald Observatory 2.7-m telescope. This article tabulates the observed signal strengths, the success ratios, and the major operating restrictions that have characterized the daily performance of the experiment. These empirical data can be used to optimize the design of such installations in the future.

094.159 **The role of lava erosion in the formation of lunar rilles and Martian channels.** M. H. Carr.
Icarus, Vol. 22, 1 - 23 (1974). – Paper presented at the International Colloquium on Mars, 28 November – 1 December 1973, Pasadena, California.

Lava tubes and channels develop around active sources of low viscosity lava. The channels normally form without erosion; however, sustained flow can result in the incision of a lava channel and simulation of fluvial erosion features. Lunar sinuous rilles are examined in light of the proposed lava erosion. The mechanism explains many features of lunar rilles that were heretofore puzzling and implies erosion rates comparable to terrestrial rates. Many Mars channels also appear to form by the action of lava; however, the larger, more spectacular Mars channels do not appear to have been formed by the same process.

094.160 **On the problem of accuracy of representation of the gravitational potential of the moon.**
A. P. Yuzefovich.
Astron. vestn., Vol. 8, 77 - 81 (1974). In Russian.

Calculations of anomalous acceleration of low-orbit artificial moon satellites have been accomplished with the help of anomalous point masses. Two models have been used: a model with 83 depth masses with parameters selected from the map of radial accelerations of Lunar Orbiter, and an American model with 580 surface point masses. Calculations have been accomplished for lower orbits of Apollo 14 and Apollo 15 and compared with the results of Doppler tracking of spaceships. The advantage of the first model permitting to reflect general features of the moon's gravitational field has been noted.

094.161 **Meteoritic matter in the surface layers of the moon.**
A. V. Ivanov, K. P. Florenskij, Yu. I. Stakheev.
Meteoritika, vyp. (No.) 33, p. 73 - 78 (1974). In Russian.

094.162 **On the moon's gravity field.** W. Köhnlein.
Veröff. Astron. Inst. Bonn, No. 86, 5 + 197 pp. (1973).

The geometric and dynamic structure of the moon's external gravity field is investigated. In a first part the zonal and tesseral field is studied at elevations 0 km, 250 km, 1000 km and 10000 km above the moon's surface. From 10000 km upwards, the combined influence of the gravitational field of earth and sun is added and the investigation is extended to the moon-earth surrounding. A new set of nonzonal harmonic coefficients is deduced — up to 8,8 in degree and order — by a combination method. The accuracy of the gravity field is estimated from the discrepancy of the data sets provided by orbital analysis and by direct determination of acceleration variations of lunar space probes.

094.163 Briefly about the moon. A. P. Vinogradov.
Vestn. Mosk. un-ta. Geologiya, 1973, No. 6, p. 3 - 9. In Russian.

094.164 On the second harmonic of phase dependence of lunar radio emission. L. I. Fedoseev.
Izv. vyssh. ucheb. zavedenij. Radiofizika, Vol. 16, 1398 - 1400 (1973). In Russian. — Abstr. in Referativ. Zhurn. 51. Astron., 5.51.309 (1974).

094.165 Simultaneous Explorer 35 and Apollo 15 orbital magnetometer observations: implications for lunar electrical conductivity inversions.
G. Schubert, B. R. Lichtenstein, P. J. Coleman, Jr. C. T. Russell.
Journ. Geophys. Res., Vol. 79, 2007 - 2013 (1974).

Comparisons of simultaneous Explorer 35 and Apollo 15 subsatellite magnetometer measurements indicate that the inductive response of the moon to solar wind transients is not detectable at subsatellite altitude (100 km) on the day side of the moon. Thus, within the limitations of the present study, it appears that day side confinement of the induced fields by a thin ($\lesssim$ 100 km) current layer just above the lunar surface adequately describes global electromagnetic induction in the moon.

094.166 Techniques in Doppler gravity inversion.
R. J. Phillips.
Journ. Geophys. Res., Vol. 79, 2027 - 2036 (1974).

This paper addresses the question of to what extent the Doppler-derived gravity data can be utilized to make quantitative investigations into the nature of localized mass anomalies. In particular, the author investigates the use of a discrete surface layer representation of the anomalous lunar gravity field to derive information regarding the size, shape, and depth of the so-called lunar mascons.

094.167 On the interaction of the solar wind with the lunar limb. H. Pérez-de-Tejada.
Journ. Geophys. Res., Vol. 79, 1813 - 1818 (1974).

The viscothermal interaction between a collision-dominated magneto-gas-dynamic flow and a flat solid surface in the presence of a magnetic field aligned to the flow direction is examined under conditions representing those of the solar wind at the lunar limb. An analytical expression for the normalized magnitude of the magnetic field intensity at the surface is derived and computed for different boundary conditions imposed on the surface. A discussion of the effects that the presence of a shock front placed ahead of the lunar terminator produces on the viscothermal interaction at the lunar limb is also included.

094.168 A two-parametric division into regions of the lunar surface. N. N. Evsyukov.
Astron. Zhurn. Akad. Nauk SSSR, Vol. 51, 611 - 616 (1974). In Russian. English translation in Soviet Astron. AJ, Vol. 18, No. 3.

On the base of the albedo-colour diagram composed by the author a division into regions of the lunar surface is carried out. A simplified scheme of a two-parametric chart of the lunar disk is given.

094.169 Surveyor observations of lunar horizon-glow.
J. J. Rennilson, D. R. Criswell.
The Moon, Vol. 10, 121 - 142 (1974).

Each of the Surveyor 7, 6, and 5 spacecraft observed a line of light along its western lunar horizon following local sunset. It has been suggested that this horizon-glow (HG) is sunlight, which is forward-scattered by dust grains present in a tenuous cloud formed temporarily just above sharp sunlight/shadow boundaries in the terminator zone. Detailed analysis of the HG absolute luminance, temporal decay, and morphology confirm the cloud model. The levitation mechanism must eject 10^7 more particles per unit time into the cloud than could micrometeorites.

094.170 A new theory for the formation of the maria and Cayley type lunar regions. J. A. Bastin.
The Moon, Vol. 10, 143 - 162 (1974).

A new 'liquefaction' theory for the origin of the flat marial and Cayley areas on the lunar surface is described. It is supposed that the flat terrain in these areas resulted from periods in the development of the moon when these regions, although not liquid, had a sufficiently low viscosity for the surfaces to relax more or less completely to a level form. To account for this low viscosity a model is developed in which preferentially high temperatures were maintained close to the lunar surface. The paper examines in some detail the possibility that these high temperatures may have resulted from instabilities in the lunar heat flow pattern caused by the presence of a surface layer of very low thermal conductivity produced by the debris of early meteorite impacts. A comparison is made between current models for the formation of the lunar surface and the theory here proposed: the advantages of the latter are enumerated and discussed.

094.171 Observations and analysis of lunar radio emission at 3.09 mm wavelength.
B. L. Ulich, J. R. Cogdell, J. H. Davis, T. A. Calvert.
The Moon, Vol. 10, 163 - 174 (1974).

Observations of lunar radio emission were made at 3.09 mm wavelength (97.1 GHz) from April 18 to May 20, 1971. Absolute brightness temperatures were measured for five distinct areas—Copernicus, Sea of Serenity, Sea of Tranquility, Ocean of Storms, and a highland region near the mean center—and lunation curves were determined. Theoretical brightness temperatures throughout a lunation were calculated for Copernicus, Sea of Serenity, and the Highlands region using a thermophysical model employing variable properties of the lunar soil. Calculations were also made for comparison with earlier observations at the same frequency of the total lunar eclipse on February 10, 1971. Variations in the observed eclipse and lunation curves between regions are sufficiently great to prevent matching all the observations with a single model.

094.172 Lunar gravity: Apollo 15 Doppler radio tracking.
P. M. Muller, W. L. Sjogren, W. R. Wollenhaupt.
The Moon, Vol. 10, 195 - 205 (1974).

New detailed gravity measurements were obtained over a 10- to 70-km surface strip from $-70°$ to $+70°$ long. during low-altitude orbits. The trajectory path went over the centers of both Maria Serenitatis and Crisium, providing a complete center gravity profile of two large mascons.

094.173 Topographic survey on the lunar surface by means of Soviet automatic space probes.
B. N. Rodionov, B. V. Nepoklonov, V. V. Kiselev, A. S. Selivanov, V. M. Govorov, V. V. Zasetskij, O. G. Ivanov, F. I. Babkov, R. M. Mikhajlov, E. A. Reshetov, M. V. Abramova, A. A. Bol'shoj, K. K. Davidovskij, V. G. Somal'.
Geod. i kartografiya, 1973, No. 10, p. 29 - 41. In Russian. Abstr. in Referativ. Zhurn. 51. Astron., 6.51.294 (1974).

094.174 Isotopie-Effekte und "particle tracks" in lunarem Material und ihr Bezug zur Bestrahlungsgeschichte des Mondes. U. Herpers, W. Herr, W. A. Kaiser, H. Kulus, R. Michel, K. Thiel.

Analyse extraterrestrischen Materials, (see 012.020), p. 85 - 107 (1974).

094.175 Elementkorrelationen und die chemische Zusammensetzung von Mond und Erde. H. Wänke.
Analyse extraterrestrischen Materials, (see 012.020), p. 183 - 201 (1974).

094.176 Die Mondforschung heute in astronomischer Sicht. J. Meurers.
Analyse extraterrestrischen Materials, (see 012.020), p. 229 - 237 (1974).

094.177 Spinellchemismus als Beweis für einen weitgehend differenzierten Mond. A. El Goresy.
Analyse extraterrestrischen Materials, (see 012.020), p. 249 - 263 (1974).

094.178 The existence of horizontal stress fields and orthogonal fracture systems in the moon's crust. N. Hast.
Modern Geol., (GB), Vol. 4, 73 - 84 (1973).
  On the basis of the extensive measurements of absolute stress in the earth's crust a study of the orthogonal fracture systems in the moon's crust provides information on the stress conditions in the moon's crust that could otherwise be obtained only through such measurements on the moon itself.

094.179 Turbulent lava flow and the formation of lunar sinuous rilles. G. Hulme.
Modern Geol., (GB), Vol. 4, 107 - 117 (1973).
  It is argued that turbulent lava flow may have occurred on the moon and that, because this heats the ground much more than laminar flow, it may have melted the ground and caused the formation of sinuous rilles.

094.180 Origin of the surface features of the moon. E. Orowan.
Proc. Roy. Soc. London, Ser. A, Vol. 336, 141 - 163 (1973).
  Mathematical models for the various impact mechanisms causing lunar craters are discussed. The evidence of high-resolution photographs is shown to support the meteorite indentation theory for small craters, and the comet explosion theory for large ones. The behaviour of the moon as a brittle body under indentation is shown.

094.181 The challenge of Apollo 17. W. R. Muehlberger, E. W. Wolfe.
American Scient., Vol. 61, 660 - 669 (1973).
  The discoveries of the last manned lunar landing add much to knowledge of the moon's geological history. These discoveries are reviewed.

094.182 Apparent resistivity of the moon and its interpretation. L. L. Vanyan, I. V. Yegorov, B. A. Okulessky, M. N. Berdichevsky, M. S. Krass, V. E. Fadeyev.
Ann. Géophys., Vol. 29, 367 - 374 (1973).
  Using a model of the lunar cavity (hemisphere and ellipsoid) and of the lunar interior two methods are given to determine the apparent resistivity from measurements made simultaneously on the lunar surface and aboard a satellite. A multi-layered model is shown to agree with the experimental data, the second layer of which has reduced resistance and is interpreted as a spinel peridotites layer.

094.183 The physics of lunar and planetary surfaces. J. E. Geake.
Contemporary Phys., (GB), Vol. 15, 121 - 158 (1974).
  A survey is made of our present knowledge of those surfaces in the solar system which are accessible to inspection. The Apollo lunar surface instruments and the implications of the data they give, and also the physical properties of returned lunar samples, are discussed in some detail. Some results are given from the Venus probes, especially Venera 7, and the Mars probes, especially Mariner 9. The minor atmosphereless objects in the solar system, such as asteroids and planetary satellites, are discussed briefly.

094.184 Chemistry of the moon's surface. B. Mason.
Chem. in Brit., (GB), Vol. 9, 456 - 461 (1973).

094.185 On the characters and origin of lunar and Martian craters. P. Leonardi.
Atti Mus. Civ. Stud. Nat. Trieste, (Italy), Vol. 28, 397 - 430 (1973).

Why is the moon so dark?
Sky Telescope, Vol. 47, 380 (1974).

Pictorial guide to the moon.
See Abstr. 003.017.

Mapping of the moon — past and present.
See Abstr. 003.072.

Perturbation of the solar wind by the lunar atmosphere. See Abstr. 074.032.

The interaction between an impact-produced neutral gas cloud and the solar wind at the lunar surface.
See Abstr. 074.079.

On the thermal evolution of the terrestrial planets.
See Abstr. 091.003.

Absolute brightness temperature measurements at 2.1-mm wavelength. See Abstr. 091.011.

Moon, Local Properties

094.301 The chemical composition of the lunar surface in the region of Lunokhod 2 operation.
G. E. Kocharov, S. V. Viktorov.
Dokl. Akad. Nauk SSSR. Ser. Mat. Fiz., Vol. 214, 71 - 74 (1974). In Russian.

094.302 Geological-morphological analysis of the region of Lunokhod 2 operation.
K. P. Florenskij, A. T. Bazilevskij, A. A. Gurshtejn, V. V. Zasetskij, A. A. Pronin, V. P. Polosukhin.
Dokl. Akad. Nauk SSSR. Ser. Mat. Fiz., Vol. 214, 75 - 78 (1974). In Russian.

094.303 In-situ measurement of the rate of ^{235}U fission induced by lunar neutrons.
D. S. Woolum, D. S. Burnett.
Earth Planet. Sci. Letters, Vol. 21, 153 - 163 (1974).

The depth profile of the neutron-induced fission rate of ^{235}U was directly measured to a depth of 350 g/cm^2 by the Apollo 17 Lunar Neutron Probe Experiment. The shape of theoretical depth profile of Lingenfelter et al. fits the measured capture rates well at all depths. The excellent agreement between theory and experiment implies that conclusions drawn previously by interpreting lunar sample data with the theoretical capture rates will not require revision. In particular lunar surface processes, rather than uncertainties in the capture rates, are required to explain the relatively low neutron fluences observed for surface soil samples compared to the fluences expected for a uniformly mixed regolith.

094.304 Lunar monazite: a late-stage (mesostasis) phase in mare basalt.
J. F. Lovering, D. A. Wark, A. J. W. Gleadow, R. Britten.
Earth Planet. Sci. Letters, Vol. 21, 164 - 168 (1974).

Thorium-poor monazite occurs as an inclusion in a ferrohedenbergite grain within a mesostasis area of the relatively coarse-grained Apollo 11 basalt 10047,68. Electron probe analyses indicate chondrite-normalized REE fractionation patterns for both lunar and terrestrial monazites in which the light REE's (La to Sm) are highly enriched relative to the heavy REE's (Gd to Lu). Lunar monazite shows a distinct Eu anomaly which is absent from the terrestrial sample.

094.305 Cation distribution in low-calcium pyroxenes: dependence on temperature and calcium content and the thermal history of lunar and terrestrial pigeonites.
S. K. Saxena, S. Ghose, A. C. Turnock.
Earth Planet. Sci. Letters, Vol. 21, 194 - 200 (1974).

In low-calcium pyroxenes, namely, pigeonites and subcalcic augites, the preference of Fe^{2+} for M2 site increases with the Ca-content, but decreases with temperature. From the experimental data, approximate cation equilibrium temperatures in pigeonites can be determined. The cation distribution in lunar pigeonites usually indicates an apparent equilibration temperature of 700–860°C, while that in the Mull pigeonite indicates an equilibration temperature of 550°C.

094.306 Plagioclase saturation in lunar high-titanium basalt.
M. J. O'Hara, G. M. Biggar, P. G. Hill, B. Jefferies, D. J. Humphries.
Earth Planet. Sci. Letters, Vol. 21, 253 - 268 (1974).

High-titanium basalt collected at the Apollo 11 site was sparsely porphyritic on eruption. Phenocrysts of anorthite are reported for the first time, and were present with a few phenocrysts of olivine, pigeonite, augite, ilmenite and spinel in the liquid. Attention is drawn to the consequences of the nearly eutectic character of the crystallization of the high-titanium basalts.

094.307 The ellipsoidal and dumbbell-shaped inclusions within particulate lunar globules.
J. A. Bastin, A. Volborth.
Icarus, Vol. 21, 112 - 120 (1974).

The centrifugal motion of liquid, solid, and gaseous particles within the symmetrical lunar dumbbell and prolate droplets is studied. A method for determining the temperatures of formation of the globules is presented which depends on the comparison between the prediction of simple models, and the observed position of the particles within a globule.

094.308 Inert gas patterns in the regolith at the Apollo 15 landing site.
J. L. Jordan, D. Heymann, S. Lakatos.
Geochim. Cosmochim. Acta, Vol. 38, 65 - 78 (1974).

The authors report the results of inert gas analyses made on eight Apollo 15 fines and one 2–4 mm particle from 15603. As geochemical tracers, the inert gases can provide information on the composition of the solar wind, on the interaction of galactic radiation with materials on the lunar surface, as well as information on the evolution of the lunar regolith. Four components of these gases have been generally recognized in lunar materials: solar-wind implanted gases, inert gases produced by galactic cosmic rays and possible solar cosmic rays, those produced as a result of radioactive decay, and inert gases from the lunar atmosphere re-implanted into the regolith.

094.309 A statistical-chemical and thermodynamic approach to the study of lunar mineralogy.
S. K. Saxena, L. S. Walter.
Geochim. Cosmochim. Acta, Vol. 38, 79 - 95 (1974).

Principal components analysis is used to study the chemical compositions of pyroxenes of five Apollo 12 specimens. Important correlations recognized in the variation of oxide weight per cent are given. These correlations indicating substitutional relationships can be interpreted as representative of stable and metastable trends of crystallization by using crystal-chemical and thermodynamic information. The per cent variance of pyroxene groups with characteristic trends in each specimen can be evaluated and interpreted in terms of history of crystallization.

094.310 Noble gas investigations of lunar rocks 10017 and 10071. P. Eberhardt, J. Geiss, H. Graf, N. Grögler, U. Krähenbühl, H. Schwaller, A. Stettler.
Geochim. Cosmochim. Acta, Vol. 38, 97 - 120 (1974).

The authors report in detail the results of noble gas measurements, Kr^{81}/Kr exposure ages and K/Ar age determinations on two Apollo 11 lunar rocks (10017.43 and 10071.30). A summary of the age results was given at the Apollo 11 Lunar Science Conference and has been published in Science, Vol. 167, 558 - 560 (Eberhardt et al., 1970). All the data have been re-evaluated and additional measurements have been made. Some insignificant differences between ages reported here and those in Science have resulted.

094.311 Surface history of lunar soil and soil columns.
R. L. Fleischer, H. R. Hart, Jr., W. R. Giard.
Geochim. Cosmochim. Acta, Vol. 38, 365 - 380 (1974).

Measurements of cosmic ray track densities are presented for soil samples from Apollo 15, 16 and 17. Median track densities are used to infer total effective exposure times within ~15 cm of the lunar surface. Minimum track densities are used to derive the time of the last impact-produced rearrangement of soil grains. For samples from near various craters ages are derived of 40 m.y. for St. George, 6 (±3) m.y. for S. Ray, 25 - 90 m.y. for Plum, and 20 - 35 m.y. for Shorty.

094.312 Studies of K-Ar dating and xenon from extinct radioactivities in breccia 14318; implications for

early lunar history.
J. H. Reynolds, E. C. Alexander, Jr., P. K. Davis, B. Srinivasan.
Geochim. Cosmochim. Acta, Vol. 38, 401 - 417 (1974).

The rare gases argon and xenon were studied intensively in lunar breccia 14318, one of a family of three Apollo 14 breccias exhibiting similarities, including substantial amounts of 'parentless' xenon from the spontaneous fission of extinct ^{244}Pu. Implications of the parentless xenon from extinct sources, as seen in these different rocks, depend upon the model adopted for its evolution and storage. We present four different models, all of which are unsatisfactory in some respects, so that we are presently unable to narrow the question.

094.313 Detection of radon emission at the edges of lunar maria with the Apollo alpha-particle spectrometer.
P. Gorenstein, L. Golub, P. Bjorkholm.
Science, Vol. 183, 411 - 413 (1974).

The distribution of radioactive polonium-210 shows enhanced concentrations at the edges of lunar maria. Enhancements are seen at the edges of Mare Fecunditatis, Mare Crisium, Mare Smythii, Mare Tranquillitatis, Mare Nubium, Mare Cognitum, and Oceanus Procellarum.

094.314 On the correlation between the normal albedo of sections of lunar maria and the crater density.
V. I. Ezerskij, V. A. Ezerskaya, I. I. Latynina, V. I. Lats'ko.
Vestn. Khar'kov. Univ. No. 99 (Ser. Astron. No. 8) p. 9 - 18 (1973). In Russian.

094.315 Theory of the magnetic viscosity of lunar and terrestrial rocks.   D. J. Dunlop.
Rev. Geophys. Space Phys., Vol. 11, 855 - 901 (1973).

Lunar materials exhibit two distinct types of viscous or time-dependent magnetic behavior: igneous rocks and largely recrystallized breccias, lunar soils and low metamorphic grade breccias. Four theories of magnetic viscosity are reviewed in this paper in an attempt to interpret the very different viscous properties of these two types of rocks (Richter (1937), Néel (1949), Néel (1950), Stacey (1963)).

094.316 Lunar red spots: possible pre-mare materials.
M. C. Malin.
Earth Planet. Sci. Letters, Vol. 21, 331 - 341 (1974).

Some anomalous red features observed on the moon by Whitaker are examined in detail. Crater counts and regional stratigraphy suggest that ages for these features are comparable to that of lunar highland material. Initial gamma-ray spectrometry indicates higher than average native radioactivity is associated with certain red areas. A possible explanation for these observations is that the red objects are the surface manifestation of highly radioactive pre-mare basalts (Apollo 14/KREEP/norite material).

094.317 $^{18}O/^{16}O$ ratios in lunar fines and rocks.
M. Javoy, S. Fourcade.
Earth Planet. Sci. Letters, Vol. 21, 377 - 382 (1974).

$^{18}O/^{16}O$ ratios have been measured for Luna 20 and Apollo 15 fines and Apollo 15 rocks. Isotopic composition and fractionation between minerals are compared with previous results. Partial fluorination experiments on Luna 20 soil and Apollo 15021 extreme fines show large ^{18}O enrichments in grain surfaces. These results are discussed.

094.318 Seismic scattering and shallow structure of the moon in Oceanus Procellarum.
A. M. Dainty, M. N. Toksöz, K. R. Anderson, P. J. Pines, Y. Nakamura, G. Latham.
The Moon, Vol. 9, 11 - 29 (1974). – Communication presented at the Lunar Science Institute Conference (see 012.002).

Long, reverberating trains of seismic waves produced by impacts and moonquakes may be interpreted in terms of scattering in a surface layer overlying a non-scattering elastic medium. Model seismic experiments are used to qualitatively demonstrate the correctness of the interpretation. Three types of seismograms are found, near impact, far impact and moonquake. Only near impact and moonquake seismograms contain independent information. Details are given in the paper of the modelling of the scattering processes by the theory of diffusion.

094.319 Preliminary results of an experimental study of the magnetic effects of shocking lunar soil.
M. Fuller, F. Rose, P. J. Wasilewski.
The Moon, Vol. 9, 57 - 61 (1974). – Communication presented at the Lunar Science Institute Conference (see 012.002).

094.320 Some characteristic magnetic properties of lunar materials.
T. Nagata, R. M. Fisher, F. C. Schwerer.
The Moon, Vol. 9, 63 - 77 (1974). – Communication presented at the Lunar Science Institute Conference (see 012.002).

094.321 Mare Serenitatis: a preliminary definition of surface units by remote observations.
T. W. Thompson, R. W. Shorthill, E. A. Whitaker, S. H. Zisk.
The Moon, Vol. 9, 89 - 96 (1974). – Communication presented at the Lunar Science Institute Conference (see 012.002).

094.322 Rock types present in lunar highland soils.
A. M. Reid.
The Moon, Vol. 9, 141 - 146 (1974). – Communication presented at the Lunar Science Institute Conference (see 012.002).

094.323 A new technique of thermal conductivity measurement of lunar core samples.
K. Horai, M. Langseth, Jr., A. Wechsler, J. Winkler, D. Colvin, S. Keihm.
The Moon, Vol. 9, 243 (1974). – Communication presented at the Lunar Science Institute Conference (see 012.002). – Abstract.

094.324 Instrument and methods of investigation of engineering-physical characteristics of the lunar ground and its terrestrial analogues.
A. S. Buyalo, A. I. Vedenin, V. V. Markachev.
Vopr. standartiz., metrol. i tekhn. tochnykh izmerenij. Moskva, Izd-vo standartov, 1973, p. 249 - 255. In Russian. – Abstr. in Referativ. Zhurn. 62. Issled. kosmich. prostranstva, 3.62.468 (1974).

094.325 Isotopic evidence for a terminal lunar cataclysm.
F. Tera, D. A. Papanastassiou, G. J. Wasserburg.
Earth Planet. Sci. Letters, Vol. 22, 1 - 21 (1974).

The authors present evidence for wide-spread shock metamorphism and element redistribution in the neighborhood of approximately 3.9 AE which resulted from large-scale impacts on an ancient lunar crust. This conclusion is based on ages of highland samples from all lunar missions determined by the U–Th–Pb, Rb–Sr and K–Ar methods.

094.326 Antiferromagnetic inclusions in lunar glass.
A. N. Thorpe, F. E. Senftle, C. Briggs, C. Alexander.
Earth Planet. Sci. Letters, Vol. 22, 85 - 90 (1974).

The magnetic susceptibility of 11 glass spherules from the Apollo 15, 16, and 17 fines and two specimens of a relatively large glass spherical shell were studied as a function of temperature from room temperature to liquid helium temperatures. All but one specimen showed the presence of antiferromagnetic inclusions. Closely spaced temperature measurements of the magnetic susceptibility below 77 K on five of the specimens showed antiferromagnetic temperature transitions (Néel transitions).

094.327 Comparison of calculated and measured release patterns of trapped gases. H. Baur, U. Frick, H. Funk, D. Phinney, L. Schultz, P. Signer.
Meteoritics, Vol. 8, 325 - 326 (1973). – Abstr. Meteoritical Soc.

094.328 Age and history of the lunar highlands based on $^{87}Rb/^{87}Sr$ studies. J. L. Birck, C. J. Allégre.
Meteoritics, Vol. 8, 326 (1973). – Abstr. Meteoritical Soc.

094.329 Dimensional analysis of impact structures. M. R. Dence.
Meteoritics, Vol. 8, 343 - 344 (1973). – Abstr. Meteoritical Soc.

094.330 Rare gas diffusion studies in individual lunar soil particles and in artificially implanted glasses.
H. Ducati, S. Kalbitzer, J. Kiko, T. Kirsten, H. W. Müller.
Meteoritics, Vol. 8, 355 - 356 (1973). – Abstr. Meteoritical Soc.

094.331 Charged-particle track studies of lunar and analogous materials. S. A. Durrani.
Meteoritics, Vol. 8, 357 - 358 (1973). – Abstr. Meteoritical Soc.

094.332 ^{39}Ar-^{40}Ar ages of lunar material.
P. Eberhardt, J. Geiss, N. Grögler, P. Maurer, A. Stettler.
Meteoritics, Vol. 8, 360 - 361 (1973). – Abstr. Meteoritical Soc.

094.333 Noble gas investigations on lunar regolith fragments.
P. Eberhardt, J. Geiss, N. Grögler, L. Weber.
Meteoritics, Vol. 8, 362 (1973). – Abstr. Meteoritical Soc.

094.334 Zirconium and hafnium abundances in lunar materials and meteorites and implications of their ratios. W. D. Ehmann, L. L. Chyi.
Meteoritics, Vol. 8, 363 (1973). – Abstr. Meteoritical Soc.

094.335 Mineralogy and geochemistry of the opaques in lithic fragments from the Apollo 17 landing site.
A. El Goresy, P. Ramdohr, O. Medenbach, H.-J. Bernhardt.
Meteoritics, Vol. 8, 363 - 364 (1973). – Abstr. Meteoritical Soc.

094.336 Trapped solar wind and cosmic ray produced noble gases in Apollo 16 and 17 soil samples.
O. Eugster, P. Eberhardt, J. Geiss, N. Grögler, S. Guggisberg, M. Mörgeli.
Meteoritics, Vol. 8, 364 - 365 (1973). – Abstr. Meteoritical Soc.

094.337 Diffusional gas release patterns of trapped gases from lunar soil separates. U. Frick, H. Baur, H. Funk, D. Phinney, L. Schultz, P. Signer.
Meteoritics, Vol. 8, 367 - 369 (1973). – Abstr. Metcoritical Soc.

094.338 Release patterns of natural and low energy implanted argon from separates of lunar fines.
H. Funk, H. Baur, U. Frick, D. Phinney, L. Schultz, P. Signer.
Meteoritics, Vol. 8, 369 - 371 (1973). – Abstr. Meteoritical Soc.

094.339 High Al_2O_3 (~25%) bottle-green lunar glasses. B. P. Glass.
Meteoritics, Vol. 8, 371 - 372 (1973). – Abstr. Meteoritical Soc.

094.340 Investigation and simulation of metallic spherules from lunar soils. J. I. Goldstein, P. J. Blau.
Meteoritics, Vol. 8, 373 - 374 (1973). – Abstr. Meteoritical Soc.

094.341 Origin of inert gases in "Rusty Rock", 66095. D. Heymann, W. Hübner.
Meteoritics, Vol. 8, 378 - 380 (1973). – Abstr. Meteoritical Soc.

094.342 Rare gases in Apollo 17 soils. W. Hübner, T. Kirsten.
Meteoritics, Vol. 8, 386 - 388 (1973). – Abstr. Meteoritical Soc.

094.343 Diffuse X-ray scattering by lunar feldspars. H. Jagodzinski, M. Korekawa.
Meteoritics, Vol. 8, 389 (1973). – Abstr. Meteoritical Soc.

094.344 Chemical composition of Apollo 16 and 17 samples. J. C. Laul, R. A. Schmitt.
Meteoritics, Vol. 8, 405 - 407 (1973). – Abstr. Meteoritical Soc.

094.345 Apollo 17 stratigraphy: clues from a boulder. U. B. Marvin, J. A. Wood, J. Bower.
Meteoritics, Vol. 8, 412 - 413 (1973). – Abstr. Meteoritical Soc.

094.346 Apollo 15, 16 and 17: chemical compositions of some rocks and soils and a comparison with materials from other lunar sites. D. F. Nava.
Meteoritics, Vol. 8, 415 - 416 (1973). – Abstr. Meteoritical Soc.

094.347 On the differentiation time of KREEP and VHA basalts. L. E. Nyquist, N. J. Hubbard, B. M. Bansal, H. Wiesmann, B. M. Jahn.
Meteoritics, Vol. 8, 416 - 417 (1973). – Abstr. Meteoritical Soc.

094.348 Trapped gases in mineral separates of lunar fines. D. Phinney, H. Baur, U. Frick, H. Funk, L. Schultz, P. Signer.
Meteoritics, Vol. 8, 423 - 424 (1973). – Abstr. Meteoritical Soc.

094.349 The formation of lunar carbide from lunar iron silicates and solar wind implanted carbon.
C. T. Pillinger, B. D. Batts, G. Eglinton, A. P. Gowar, A. J. T. Jull, J. R. Maxwell.
Meteoritics. Vol. 8, 425 (1973). – Abstr. Meteoritical Soc.

094.350 Lunar rock types and weighted least-squares mixing models. E. Schonfeld.
Meteoritics, Vol. 8, 432 - 435 (1973). – Abstr. Meteoritical Soc.

094.351 Shock-induced melting in lunar anorthositic rock 60015. C. B. Sclar.
Meteoritics, Vol. 8, 437 - 438 (1973). – Abstr. Meteoritical Soc.

094.352 Questions about the ^{40}Ar excess in lunar fines.
P. Signer, H. Baur, U. Frick, H. Funk, D. Phinney, L. Schultz.
Meteoritics, Vol. 8, 441 - 443 (1973). — Abstr. Meteoritical Soc.

094.353 Fission track dating of lunar glass spherules.
D. Storzer, G. Poupeau.
Meteoritics, Vol. 8, 444 - 445 (1973). — Abstr. Meteoritical Soc.

094.354 Determination of major elements in lunar samples with 14 MeV neutrons. F. Teschke, H. Wänke.
Meteoritics, Vol. 8, 446 - 447 (1973). — Abstr. Meteoritical Soc.

094.355 Magnetic hysteresis classification of lunar samples.
P. J. Wasilewski.
Meteoritics, Vol. 8, 456 - 460 (1973). — Abstr. Meteoritical Soc.

094.356 Particle shape and magnetization of lunar samples, chondrules, chondrite meteorites, and naturally reduced terrestrial basalts. P. J. Wasilewski.
Meteoritics, Vol. 8, 461 - 463 (1973). — Abstr. Meteoritical Soc.

094.357 Shock remagnetization mechanisms and processes in lunar samples and meteorites.
P. J. Wasilewski.
Meteoritics, Vol. 8, 466 - 467 (1973). — Abstr. Meteoritical Soc.

094.358 Selten sichtbare Mondlandschaften. Cordilleren und Rook-Berge. C. Albrecht.
SuW, Vol. 13, 60 - 61 (1974).

094.359 Report from the Sea of Serenity.
A. Vinogradov, S. Sokolov.
Spaceflight, Vol. 16, 175 - 178 (1974).

This article summarises the results obtained with the self-propelled roving vehicle Lunokhod 2 during exploration of the eastern boundary of the Sea of Serenity early last year.

094.360 Unit cell parameters, compositions and $2V$ measurements of selected lunar plagioclases.
J. C. Butler, M. F. Carman, Jr., E. A. King, Jr.
The Moon, Vol. 9, 327 - 334 (1974).

Unit cell parameters of 12 selected lunar plagioclase specimens have been obtained using least-squares refinement of X-ray powder diffraction data. Compositions of splits of 11 of these plagioclase specimens were determined using the electron microprobe. Optical axial angles were measured on analyzed spots for nine of the probed specimens. Analyses cluster about An 96.5 ± 1.0 and unit cell parameters and $2V_x$ fit the values for this composition obtained in other studies.

094.361 Magnetic remanence mechanisms in iron and iron-nickel alloys, metallographic recognition criteria and implications for lunar sample research. P. Wasilewski.
The Moon, Vol. 9, 335 - 354 (1974).

This paper considers the problem of remanent magnetization acquisition in lunar samples. Nine remanent magnetization mechanisms are proposed along with the characteristic metal structures associated with each of the mechanisms. The magnetic-metallographic approach described in this paper may provide the basis for understanding magnetic remanence properties of lunar samples and meteorites.

094.362 Morphology and structure of the Taurus-Littrow highlands (Apollo 17): Evidence for their origin and evolution. J. W. Head.
The Moon, Vol. 9, 355 - 395 (1974).

The purpose of this paper is to present the results of an analysis of the morphology and major structural trends in the Taurus-Littrow region and to outline the bearing of these features on the origin and subsequent history of this area of the highlands. Data were obtained by mapping structures in the Taurus-Littrow area (valleys, rilles, lineaments, crater chains, scarps, linear crater walls, etc.) and measuring the azimuths of features or regional groups of these features. Additional data were obtained by compiling hypsographic maps and profiles from existing topographic maps.

094.363 Laboratory simulation of the herringbone pattern associated with lunar secondary crater chains.
V. R. Oberbeck, R. H. Morrison.
The Moon, Vol. 9, 415 - 455 (1974).

V-shaped ridge components of the herringbone pattern associated with lunar secondary crater chains have been simulated by simultaneous and nearly simultaneous impact of two projectiles near one another. The impact velocities and angles of the projectiles were similar to those of the fragments that produced secondary craters found at various ranges from large lunar craters. Comparison of the magnitudes of the ridge angles of both laboratory crater pairs and secondary crater chains of the crater Copernicus implies that material was ejected from Copernicus at angles in excess of 60°, measured from the normal, to form many of Copernicus' satellitic craters. Moreover, other independent calculations presented indicate that many of the fragments that produced secondary craters also ricocheted to produce tertiary craters. Application of the study to identification of isolated secondary craters and to the determination of the origin of large lunar craters is discussed.

094.364 Particle track record of Apollo 16 rocks from Plum crater. R. L. Fleischer, H. R. Hart, Jr.
Journ. Geophys. Res., Vol. 79, 766 - 768 (1974).

Particle track measurements on rocks 61016 and 61175 from Plum crater imply that they were intensely shocked 30 (± 10) m.y. ago, very likely at the time of formation of the crater.

094.365 Evidence for a ~ 4.5 aeon age of plagioclase clasts in a lunar highland breccia.
E. K. Jessberger, J. C. Huneke, G. J. Wasserburg.
Nature, Vol. 248, 199 - 202 (1974).

Argon from neutron-irradiated mineral separates and whole rock samples of a metamorphosed breccia (65015) from Apollo 16 has been analysed. The results on plagioclase show a ^{40}Ar–^{39}Ar plateau age of 3.98×10^9 yr, attributed to the time of metamorphism, and an age of $\sim 4.5 \times 10^9$ yr in the high temperature fraction.

094.366 The base surge in atomic and volcanic explosions and its significance under lunar conditions.
R. L. Daw.
Journ. British Astron. Ass., Vol. 84, 176 - 183 (1974).

094.367 The mapping of lunar radar scattering characteristics.
G. H. Pettengill, S. H. Zisk, T. W. Thompson.
The Moon, Vol. 10, 3 - 16 (1974).

This is the first of four articles describing a comprehensive series of radar maps of the entire visible lunar hemisphere carried out at wavelengths of 3.8 and 70 cm and analyzing the echoes in both orthogonal senses of circular polarization. In this paper, the basic techniques of delay-Doppler mapping by radar are developed, and the particular steps employed in mapping the moon are outlined. Succeeding articles present the results obtained and discuss the way in which these results relate to other, nonradar measurements as well as to the actual lunar surface properties.

094.368 High-resolution radar maps of the lunar surface at 3.8-cm wavelength.
S. H. Zisk, G. H. Pettengill, G. W. Catuna.
The Moon, Vol. 10, 17 - 50 (1974).

The entire earth-facing lunar surface has been mapped at a resolution of 2 km using the 3.8-cm radar of Haystack Observatory. The observations yield the distribution of relative radar backscattering efficiency with an accuracy of about 10% for both the polarized (primarily quasispecular or coherent) and depolarized (diffuse or incoherent) scattered components. The results show a variety of discrete radar features, many of which are correlated with craters or other features of optical photographs. Particular interest, however, attaches to those features with substantially different radio and optical contrasts.

094.369 Atlas of lunar radar maps at 70-cm wavelength.
T. W. Thompson.
The Moon, Vol. 10, 51 - 85 (1974).

This article is the second of a series of four articles describing the radar mapping of the moon. The intensity distribution of lunar radar echoes has been mapped for two-thirds of the earth-visible lunar surface at a wavelength of 70 cm. The surface resolution is $25-100$ km^2. Mappings with this resolution confirmed that the young craters have enhanced returns. A few craters were found to have enhanced echoes only from their rims. Backscattering differences were also observed.

094.370 A comparison of infrared, radar, and geologic mapping of lunar craters. T. W. Thompson, H. Masursky, R. W. Shorthill, G. L. Tyler, S. H. Zisk.
The Moon, Vol. 10, 87 - 117 = Lunar Sci. Inst., *Houston, Texas*, Contr. No. 16 (1974).

Between 1000 and 2000 infrared (eclipse) and radar anomalies have been mapped on the nearside hemisphere of the moon. A study of 52 of these anomalies indicates that most are related to impact craters and that the nature of the infrared and radar responses is compatible with a previously developed geologic model of crater aging processes. A few craters, however, have infrared and radar behaviors not predicted by the aging model.

094.371 Lunar surface: Identification of the dark mantling material in the Apollo 17 soil samples.
C. Pieters, T. B. McCord, M. P. Charette, J. B. Adams.
Science, Vol. 183, 1191 - 1194 (1974).

Evidence indicates that Apollo 17 sample 74001, a soil consisting of very dark spheres, is composed almost entirely of the dark mantling material that covers a large region of the southeastern boundary of Mare Serenitatis. Other Apollo 17 samples contain only a component of this material. The results are derived from telescopic and laboratory measurements of the optical properties of lunar soil.

094.372 Polarized magnetic field fluctuations at the Apollo 15 site: Possible regional influence on lunar induction. G. Schubert, B. F. Smith, C. P. Sonett, D. S. Colburn, K. Schwartz.
Science, Vol. 183, 1194 - 1197 (1974).

High-frequency (5 to 40 millihertz) induced lunar magnetic fields, observed at the Apollo 15 site near the southeastern boundary of Mare Imbrium and the southwestern boundary of Mare Serenitatis, show a strong tendency toward linear polarization in a direction radial to the Imbrium basin and circumferential to the Serenitatis basin, a property that could be indicative of a possible regional influence on the induction.

094.373 Terrestrial and lunar anorthosites – a tentative comparison. O. A. Bogatikov.
Geokhimiya. Mineralogiya. Petrografiya. Tom 7. Itogi nauki i tekhniki VINITI AN SSSR. Moskva, 1973, p. 79 - 99. In Russian. – Abstr. in Referativ. Zhurn. 51. Astron., 4.51.320; 62. Issled. kosmich. prostranstva, 4.62.162 (1974).

094.374 Apollo 16 and Apollo 17 – preliminary results.
C. Cristescu.
Progr. şti., Vol. 9, 385 - 392 (1973). In Romanian.

094.375 Distribution of gold and rhenium between nickel-iron and silicate melts: implications for the abundance of siderophile elements on the earth and moon.
K. Kimura, R. S. Lewis, E. Anders.
Geochim. Cosmochim. Acta, Vol. 38, 683 - 701 (1974).

094.376 Meteoritic metal in Apollo 16 samples.
S. J. B. Reed, S. R. Taylor.
Meteoritics, Vol. 9, 23 - 34 (1974).

One hundred metallic particles from Apollo 16 soils (61181, 65701) and rocks (60018, 60315, 66055) have been investigated microscopically and by electron microprobe analysis.

094.377 Morphology of lunar craters: A test of lunar erosional models. G. E. McGill.
Icarus, Vol. 21, 437 - 447 (1974).

This paper is a test of published theoretical and experimental studies of crater erosion by micrometeorite bombardment which predict systematic variations in the morphology of lunar craters as a function of crater diameter and crater age. Numerical, ranking-type degradation classifications indicate that the craters on Mare Imbrium and Mare Tranquillitatis confirm these predictions.

094.378 A detailed structural analysis of the Deslandres area of the moon. J. L. Whitford-Stark.
Icarus, Vol. 21, 457 - 465 (1974).

The components of the lunar grid of the Deslandres area of the moon are described and their azimuths plotted. The grid is shown to be the product of endogenic processes which may also have been important in the early geological evolution of the earth.

094.379 Effects of iron oxidation state on viscosity, lunar composition 15555.
M. Cukierman, D. R. Uhlmann.
Journ. Geophys. Res., Vol. 79, 1594 - 1598 (1974).

The viscous flow behavior of lunar composition 15555, which contains about 22.5 wt % FeO, has been determined over a range of temperature from 620° to 700°C and from 1215° to 1400°C. The material was synthesized under mildly reducing conditions to simulate the Fe^{2+}/total Fe ratio of the lunar environment.

094.380 Electrical properties of lunar soil sample 15301,38.
G. R. Olhoeft, A. L. Frisillo, D. W. Strangway.
Journ. Geophys. Res., Vol. 79, 1599 - 1604 (1974).

Electrical property measurements have been made on an Apollo 15 lunar soil sample in ultrahigh vacuum from room temperature to 827°C for the frequency spectrum from 100 Hz through 1 MHz. The dielectric constant, the total ac loss tangent, and the dc conductivity were measured. The dc conductivity is fitted by an exponential temperature distribution and becomes the dominant loss above 700°C.

094.381 Application of remote spectral reflectance measurements to lunar geology classification and determination of titanium content of lunar soils.
M. P. Charette, T. B. McCord, C. Pieters, J. B. Adams.
Journ. Geophys. Res., Vol. 79, 1605 - 1613 (1974).

Plots of reflectance slope between 0.402 and 0.564 μm

versus the intensity ratio between 0.564 and 0.948 μm are used to quantitatively define the mare, mare crater, upland, and bright upland crater spectral types previously presented by McCord et al. (1972). An additional spectral type, dark mantling material, has also been found. Quantification of lunar spectral types allows direct comparison of the spectral units with geologic units established by the U.S. Geological Survey, including the dark mantling material unit. An empirical relationship is derived relating TiO_2 content of the bulk lunar soils to the slope of the spectral curve between 0.402 and 0.564 μm.

094.382 **Carbon, nitrogen and sulfur in lunar fines 15012 and 15013: abundances, distributions and isotopic compositions.** S. Chang, J. Lawless, M. Romiez, I. R. Kaplan, C. Petrowski, H. Sakai, J. W. Smith.
Geochim. Cosmochim. Acta, Vol. 38, 853 - 872 (1974).

Lunar fines 15012,16 and 15013,3 were analyzed by stepwise pyrolysis and acid hydrolysis as well as complete combustion in oxygen to determine carbon, nitrogen and sulfur. In addition, hydrogen was analysed during pyrolysis as well as during hydrolysis. By comparison of the distribution frequencies of C, N, S, H_2 and Fe with ^{4}He, considered to have arisen from solar wind contribution, it is concluded that nitrogen and hydrogen have largely a solar origin. Carbon has a significant solar contribution, and metallic iron may have resulted from solar wind interaction with ferrous minerals on the lunar surface. Sulfur probably has a predominantly lunar origin. There is no direct evidence for meteoritic contribution to these samples. Solar wind interaction also has a marked effect on the stable isotope distribution of $^{13}C/^{12}C$, $^{15}N/^{14}N$, and $^{34}S/^{32}S$.

094.383 **$^{40}Ar-^{39}Ar$ ages and trace element contents of Apollo 14 breccias; an interlaboratory cross-calibration of $^{40}Ar-^{39}Ar$ standards.**
E. C. Alexander, Jr., P. K. Davis.
Geochim. Cosmochim. Acta, Vol. 38, 911 - 928 (1974).

Rare gas data are presented from step-wise heatings of lunar breccias 14066 and 14318 and from an interlaboratory cross-calibration of five standards used in $^{40}Ar-^{39}Ar$ dating. Concentrations of K, Ca, Ba, Br, U and I are given for 14318 and 14066. The authors also present an updating of all of the $^{40}Ar-^{39}Ar$ ages and trace element concentrations previously published. $^{40}Ar-^{39}Ar$ dating standards from Menlo Park, Pasadena, Stony Brook, Toronto and Berkeley are calibrated against each other and the internal homogeneity of their $^{40*}Ar/K$ ratios is tested. The Berkeley standard (from the St. Severin meteorite) has an age of 4.504 ± 0.020 b.y. from this intercalibration. ^{80}Kr from capture of lunar neutrons is detected in 14318.

094.384 **Lunar notes. Luna incognita for 1974; The ALPO lunar dome survey: a progress report (II); Additions to the A.L.P.O. lunar photograph library.**
J. E. Westfall, H. D. Jamieson.
Strolling Astronomer, Vol. 24, 184 - 195, 196, 197 (1974).

094.385 **On significant indices for differentiation of cratered lunar areas.**
E. V. Zabalueva, G. A. Lejkin, A. I. Kurochkina.
Astron. vestn., Vol. 8, 70 - 76 (1974). In Russian.

On a sample of three lunar areas it is shown that the interpretation of cratered lunar surface photos by crater size distribution and morphological classes is informative enough to distinguish cratered areas using the methods of mathematical statistics. The methods to evaluate the informativeness of the material are presented and physical meaning and significance of indices are discussed.

094.386 **Degassing of the moon. III. Lunar maria and entrails.** J. Classen.
Astron. vestn., Vol. 8, 82 - 86 (1974). In Russian.

094.387 **The chemical composition of regolith returned by Luna 20 from the continental region of the moon.**
Yu. A. Surkov, G. M. Kolesov, F. F. Kirnozov.
Geokhimiya, 1974, No. 1, p. 3 - 9. In Russian. – Abstr. in Referativ. Zhurn. 51. Astron., 5.51.320; 62. Issled. kosmich. prostranstva, 5.62.172 (1974).

094.388 **On the connection of structural metamorphism of a metal fragment from Mare Fecunditatis with parameters of outer influence.** R. I. Mints, T. M. Petukhova, V. M. Segal', A. V. Ivanov.
Khimiya, 1971, No. 1, p. 10 - 17. In Russian. – Abstr. in Referativ. Zhurn. 51. Astron., 5.51.321; 62. Issled. kosmich. prostranstva, 5.62.171 (1974).

094.389 **Rümker Hills: a lunar volcanic dome complex.**
E. I. Smith.
The Moon, Vol. 10, 175 - 181 (1974).

The Rümker Hills is characterized by overlapping plains-forming units with lobate scarps, volcanic domes, a 60 km ring, and a scarp which separates the plateau from surrounding mare materials. Plains-forming units are interpreted as fluid volcanic flows, and domes as viscous extrusions. One dome may be a strato-volcano. The ring system is discordant with regional structural trends and probably has a local origin.

094.390 **Apollo drill core depth relationships.**
W. D. Carrier III.
The Moon, Vol. 10, 183 - 194 (1974).

Preliminary depth relationships are presented for the Apollo 15, 16 and 17 drill core samples. For a given depth in any of these drill stems, the in situ lunar surface depth can be estimated. Ranges of uncertainty are also established, based on percent core recovery and degree of sample disturbance.

094.391 **$^{40}Ar-^{39}Ar$ chronology and cosmic ray exposure ages of the Apollo 15 samples.** L. Husain.
Journ. Geophys. Res., Vol. 79, 2588 - 2606 (1974).

Crystallization ages have been determined for a suite of lunar samples from seven stations at the Apollo 15 landing site. The samples include five large crystalline rocks, seven rake basalts, and thirteen 2- to 4-mm igneous and two breccia fragments. Rocks 15058, 15499, and 15555 have ages of 3.358 ± 0.025, 3.34 ± 0.08, and 3.34 ± 0.04 Gy. Six of the seven rake basalts have ages between 3.26 and 3.39 Gy. Similarly, seven of the thirteen 2- to 4-mm igneous fragments give ages of 3.26–3.35 Gy. The younger ages of 3.13 ± 0.06, 3.16 ± 0.05, and 3.20 ± 0.07 Gy for three igneous samples were also determined. The cosmic ray exposure ages of all the samples were measured by using correlation plots of $^{38}Ar^*/^{37}Ar$ and $^{40}Ar^*/^{39}Ar^*$. Large rocks (15058, 15415, 15499, and 15555) have exposure ages of between 80 and 135 m.y. Most rake samples and 2- to 4-mm fragments have exposure ages of between 200 and 500 m.y.

094.392 **Edelgasuntersuchungen an Mondproben.**
H. Wänke.
Analyse extraterrestrischen Materials, (see 012.020), p. 1 - 26 (1974).

094.393 **Solar wind nitrogen in lunar material.**
O. Müller.
Analyse extraterrestrischen Materials, (see 012.020), p. 53 - 64 (1974).

094.394 **Interpretation des ^{14}C-Tiefenprofils in einem Mondstein.** W. Born.

Analyse extraterrestrischen Materials, (see 012.020), p. 65 - 74 (1974).

094.395 **Zerstörungsfreie Bestimmung einiger Spurenelemente in Mond- und Meteoritenproben mit 14 MeV-Neutronen.** H. Palme.
Analyse extraterrestrischen Materials, (see 012.020), p. 147 - 161 (1974).

094.396 **The measure of the moon.** H. H. Schmitt.
Photogramm. Engineering, (*USA*), Vol. 39, 815 - 820 (1973).

094.397 **Piezo-remanent magnetization of lunar rocks.** T. Nagata.
Pure and Applied Geophys., (*Switzerland*), Vol. 110, 2022 - 2030 (1973).

094.398 **Determination of 12 rare earth elements in Luna-16 lunar sample by neutron activation analysis.**
A. Elek, G. Perneczki, E. Szabo, N. N. Dogadkin.
Radiochem. Radioanal. Letters, (*Switzerland*), Vol. 15, 123 - 133 (1973).

12 rare earth elements have been determined in a Luna-16 lunar sample by neutron activation analysis. Concentration data on La, Ce, Pr, Nd, Sm, Eu, Gd, Tb, Ho, Er, Yb and Lu in the lunar sample and in the W-1 standard rock are compared with reported values.

094.399 **Lunar notes.** J. E. Westfall.
Strolling Astronomer, Vol. 24, 247 - 253 (1974).

094.400 **Investigation of minerals and of lunar samples (14163, 14258) by simultaneous thermal and X-ray analysis.** H. G. Wiedemann, G. Bayer.
Zeitschr. Analyt. Chem., Vol. 266, No. 2, p. 97 - 109 (1973).
Abstr. in Phys. Ber., Vol. 53, No. 1-5032 (1974).

094.401 **On the "orphan-argon" from lunar fines.**
H. Baur, U. Frick, H. Funk, D. Phinney, L. Schultz, P. Signer.
Fortschr. Mineralog., [Schweizerbart'sche Verlagsbuchhandlung, Stuttgart], Vol. 50, No. 3, p. 46 - 48 (1973).

094.402 **Electron microprobe analyses of lithic fragments, glasses, chondrules, and minerals in Apollo 14 lunar samples.**
M. Prinz, C. E. Nehru, G. Kurat, K. Keil, G. H. Conrad.
Special Publ. Univ. New Mexico, Inst. Meteorit., No. 6, p. 1 - 38 (1973).

094.403 **Catalogue of Apollo 15 rake samples from stations 2 (St. George), 7 (Spur Crater), and 9a (Hadley Rille).**
E. Dowty, G. H. Conrad, J. A. Green, P. F. Hlava, K. Keil, R. B. Moore, C. E. Nehru, M. Prinz.
Special Publ. Univ. New Mexico, Inst. Meteorit., No. 8, p. 1 - 75 (1973).

094.404 **Electron microprobe analyses of spinel group minerals and ilmenite in Apollo 15 rake samples of igneous origin.**
C. E. Nehru, M. Prinz, E. Dowty, K. Keil.
Special Publ. Univ. New Mexico, Inst. Meteorit., No. 10, p. 1 - 91 (1973).

094.405 **Stratigraphy of ejecta from the lunar crater Aristarchus.** J. E. Guest.
Bull. Geol. Soc. America, Vol. 84, 2873 - 2893 (1973).

094.406 **Low orthopyroxene from a lunar deep crustal rock: a new pyroxene polymorph of space group P 21ca.**
J. R. Smyth.
Geophys. Res. Letters, Vol. 1, No. 1, p. 27 - 30 (1974).

Spectral reflectance systematics for mixtures of powdered hypersthene, labradorite, and ilmenite.
See Abstr. 093.029.

Isotopie-Effekte und "particle tracks" in lunarem Material und ihr Bezug zur Bestrahlungsgeschichte des Mondes.
See Abstr. 094.174.

Halogens in meteoritic and lunar samples.
See Abstr. 105.037.

Determination of the noble metals in geological materials by multielement neutron activation analysis.
See Abstr. 105.061.

Petrographic and grain size characteristics of suevite and lunar impact breccias. See Abstr. 105.076.

Proposed magnetic remanence mechanisms for meteorites and lunar samples with metallographic recognition criteria. See Abstr. 105.083.

On the irradiation history and origin of gas-rich meteorites. See Abstr. 105.096.

Indications of shock metamorphism in minerals of meteorites and lunar rocks. See Abstr. 105.123.

095 Lunar Eclipses

095.001 **Report on December's lunar eclipse.**
Sky Telescope, Vol. 47, 130 - 131 (1974).

095.002 **Die partielle Mondfinsternis vom 10. 12. 1973.**
B. De Bona.
Orion Schaffhausen, 32. Jahrgang, p. 26 (1974).

095.003 **Partielle Mondfinsternis am 10. 12. 1973.**
F. Dorst, W. Hänig, M. Schmuchel, G. Küveler.
SuW, Vol. 13, 137 (1974).

095.004 **Drie keer de maaneclips van 10 december.**
H. Betlem.
Zenit, Vol. 1, No. 2, p. 19 (1974).

095.005 **The partial lunar eclipse of 1973 December 10.**
J. Miller, P. Moore.
Journ. British Astron. Ass., Vol. 84, 196 (1974).

095.006 **Partial lunar eclipse on June 4, 1974.**
M. M. Dagaev.
Priroda, No. 5.74, p. 116 (1974). In Russian.

095.007 **Djelomična pomrčina Mjeseca 10. Prosinca (Decembra) 1973.g. (Partial eclipse of the moon on December 10, 1973).** G. Kren.
Vasiona, Vol. 22, 45 - 46 (1974).

095.008 **Contacts of lunar features with the shadow during the lunar eclipse of 10 December 1973.**
M. Dujnič.
Říše hvězd, Vol. 55, 56 - 57 (1974). In Slovak.

095.009 **Partial eclipse of the moon on June 4/5, 1974.**
A. Udalski, M. Zawilski.
Urania Kraków, Vol. 45, 150 - 154 (1974). In Polish.

Finsternisse 1974.   See Abstr. 079.001.

096 Lunar Occultations

096.001 **The January occultation of Saturn.**
D. W. Dunham.
Sky Telescope, Vol. 47, 62 - 64 (1974).

096.002 **Interesting winter occultations of Mars and Saturn.**
D. W. Dunham.
Sky Telescope, Vol. 47, 137 - 139 (1974).

096.003 **Observations of occultations of stars by the moon.**
G. Hilaire.
Astron. Astrophys., Suppl. Ser., Vol. 13, 395 - 403 (1974).
In French.
   This paper presents the data of observations of occultations made at the Besançon Observatory between January 1969 and September 1972.

096.004 **Saturnbedeckung durch den Mond am 11. Dezember 1973.** T. Düring.
SuW, Vol. 13, 40 (1974).

096.005 **Saturnbedeckung durch den Mond am 11. 12. 73.**
F. Dorst, A. Keil.
SuW, Vol. 13, 103 (1974).

096.006 **Occultations of Mars and Saturn.**
Sky Telescope, Vol. 47, 346 - 347 (1974).

096.007 **Occultation resolution of σ Scorpii.**
R. E. Nather, J. Churms, P. A. T. Wild.
Publ. Astron. Soc. Pacific, Vol. 86, 116 - 124 (1974).
   An occultation of the β CMa variable σ Sco by the moon was observed photoelectrically from the South African Astronomical Observatory at Cape Town and simultaneously from the S.A.A.O. outstation at Sutherland. Both observations show that the star has a companion 2.2 magnitudes fainter than the primary; the two observations combined show that the fainter star is $0\rlap{.}''49$ distant from the primary in position angle 268°. Visual observations of occultation events extending back to 1860 are combined with these observations to derive some information concerning the relative orbit of the fainter star.

096.008 **Lunar occultation light curves perturbed by random limb irregularities.** J. L. Sowers.
Astron. Journ., Vol. 79, 321 - 323 (1974).
   The effects of an irregular limb on lunar occultation diffraction patterns are presented. The profile of the lunar limb is approximated by a sequence of uniformly random perturbations from the mean edge. Conclusions are drawn as to the effects of a randomly irregular lunar profile on occultation measurements.

096.009 **Dubbelportret van een bedekking.**
B. A. Gerritsen, J. Gijsbers.
Zenit, Vol. 1, No. 2, p. 32 - 33 (1974).

096.010 **Grazing lunar occultations.** G. N. Walker.
Monthly Notes Astron. Soc. Southern Africa, Vol. 33, 39 - 41 (1974).

096.011 **Grazing occultation of ZC 2797, 1973 October 4, Villiers–South Africa.** M. D. Overbeek.
Monthly Notes Astron. Soc. Southern Africa, Vol. 33, 42 - 44 (1974).

096.012 **Grazing occultation of ZC 2902, 1973 October 5, Sphinx, near Pretoria, South Africa.** J. Hers.
Monthly Notes Astron. Soc. Southern Africa, Vol. 33, 44 - 46 (1974).

096.013 Occultations of stars by the moon in Abastumani in 1967, 1968, 1969, 1970, 1971, 1972.
O. R. Bolkvadze.
Byull. Abastumansk. Astrofiz. Obs., No. 45, p. 77 - 84 (1974). In Russian.

096.014 Observations of occultations of stars by the moon in 1972 in Tashkent.
M. R. Ehshmatov, Eh. Rakhmatov.
Tsirk. Astron. Inst., *Tashkent,* No. 43 (390), p. 28 - 29 (1973). In Russian.

096.015 Les prochaines belles occultations en France.
J. Meeus, H. Debehogne.
L'Astronomie, 88ᵉ année, p. 147 - 148 (1974).

096.016 The July occultation of Venus. D. W. Dunham.
Sky Telescope, Vol. 47, 420 - 423 (1974).

096.017 Les prochaines belles occultations en Belgique.
H. Debehogne, J. Meeus.
Ciel et Terre, Vol. 90, 148 - 149 (1974).

096.018 Saturnbedeckung vom 11. Dezember 1973.
K. Friedrich.
Blick in das Weltall, Archenhold-Sternw., Berlin-Treptow, 22. Jahrgang, p. 22 - 23 (1974).

096.019 Saturnbedeckung durch den Mond am 2./3. März 1974. A. Poirier, Ramond.
Orion Schaffhausen, 32. Jahrgang, p. 111 (1974).

096.020 Observations of star occultations by the moon.
M. N. Pyshnenko.
Dal'novost. fiz. sb., Vol. 4, 79 - 85 (1973). In Russian. – Abstr. in Referativ. Zhurn. 51. Astron., 5.51.194 (1974).

096.021 Observation of a lunar occultation of δ Geminorum.
S. D. Tremaine, E. J. Groth, M. R. Nelson.
Astron. Journ., Vol. 79, 649 - 650 (1974).

096.022 Observation de l'occultation partielle de Saturne par la lune du 11 décembre 1973.
J. Bourgeois, J. Neirinckx.
Ciel et Terre, Vol. 90, 196 - 200 (1974).

096.023 Results of lunar occultations of the galactic center region in HI, OH and H_2CO lines and in the nearby continua. A. Sandqvist.
Stars and the Milky Way system, Proc. 1972, (see 012.018), p. 157 - 163 (1974).

096.024 Occultation of Saturn by the moon on February 3rd 1974. A. Udalski, M. Zawilski.
Urania Kraków, Vol. 45, 24 - 27 (1974). In Polish.

096.025 Occultation of Saturn by the moon on March 3, 1974. A. Udalski, M. Zawilski.
Urania Kraków, Vol. 45, 55 - 58 (1974). In Polish.

096.026 Occultations of stars by the moon. L. Zajdler.
Urania Kraków, Vol. 45, 58 - 59 (1974). In Polish.

096.027 Observations of the occultation of Saturn by the moon. W. Sędzielowski.
Urania Kraków, Vol. 45, 92 (1974). In Polish.

096.028 Observations of the eclipse of Saturn by the moon on Dec. 10/11, 1973. R. Fangor.
Urania Kraków, Vol. 45, 120 - 123 (1974). In Polish.

096.029 Occultation of Saturn VI by the moon on 1974 February 3. D. W. Dunham, K. Aksnes.
IAU Circ., No. 2627 (1974).

096.030 Occultation of Saturn VI by the moon on 1974 March 2–3. D. W. Dunham.
IAU Circ., No. 2635 (1974).

096.031 Occultations of Saturn VI by the moon.
P. Muller, D. W. Dunham.
IAU Circ., No. 2646 (1974).

096.032 Occultation of 139 Juewa by the moon on 1974 May 3. D. W. Dunham.
IAU Circ., No. 2655 (1974).

096.033 Occultations of VX Sagittarii by the moon.
G. Wallerstein.
IAU Circ., No. 2669 (1974).

096.034 The occultation of Mars by the moon on May 16, 1971. C. Capen, M. Otis.
Strolling Astronomer, Vol. 24, 244 - 247 (1974).

096.035 Observations des occultations à Belgrade en 1971.
M. B. Protitch.
Bull. Obs. Astron. Belgrade, Vol. 29, F. 1, No. 125, p. 29 (1971/72).

096.036 Réductions des occultations observées à Belgrade.
M. B. Protitch, M. Simić.
Bull. Obs. Astron. Belgrade, Vol. 29, F. 1, No. 125, p. 30 (1971/72).

096.037 Lunar occultations visible at Belgrade during the period 1972–1974.
Prepared by H. M. Nautical Office, London.
Bull. Obs. Astron. Belgrade, Vol. 29, F. 1, No. 125, p. 47 - 53 (1971/72).

096.038 Occultation observations in 1972. T. Mori, Y. Ganeko, Y. Harada, M. Sasaki, M. Yamaguti.
Data Rep. Hydrographic Observations, Ser. Astron. Geod., *Tokyo,* No. 8, p. 1 - 28 (1974).

096.039 Osservatorio Astronomico "Guido Horn d'Arturo": Osservazioni di occultazioni lunari. L. Baldinelli.
Giornale Ass. Astrofili Bolognesi, *Bologna,* No. 33, p. 7 (1974)

Extended glow preceding reappearance of Saturn during a lunar occultation. See Abstr. 100.005.

The angular diameter of 87 Leonis. See Abstr. 115.008.

An angular diameter and effective temperature for V Cancri from an occultation observation. See Abstr. 115.009.

Lunar occultation stellar diameter measurements. See Abstr. 115.014.

Measurement of double stars by occultation with small telescopes. See Abstr. 118.018.

Ooty occultations of 76 radio sources. See Abstr. 141.087.

Problemi di identificazione ottica delle sorgenti X galattiche osservabili in occultazione lunare dal satellite EXOSAT (Helos). See Abstr. 142.098.

097 Mars, Mars Satellites

Mars

097.001 Evaluation of intensity of resonance line emission of inert gases in planetary atmospheres.
M. S. Biryulina, V. N. Konashenok.
Astron. vestn., Vol. 8, 20 - 25 (1974). In Russian.

Evaluations of intensity of the emission lines He 584 Å, Ne 736 Å and Ar 1048 Å in the atmospheres of Mars, Venus and Jupiter have been performed. Three possible models of atmospheric composition, depending on the supposed mechanism of its development, have been considered for Mars and Venus. A two-straight-lines model has been used for Jupiter. Calculations for the earth atmosphere have been made for the sake of comparison. The obtained results are presented in the form of high-altitude profiles of rising and descending radiations in the mentioned lines.

097.002 Abyssal sources of the megamorphology of Mars.
V. B. Nejman.
Astron. vestn., Vol. 8, 26 - 28 (1974). In Russian.

It is shown that the bright areas of Mars (continents) are composed of acid and the dark ones (maria) of basic rocks. The depth of the Martian crust is evaluated to about 40 km.

097.003 Mars in the opposition of 1965.
N. K. Andrianov.
Astron. vestn., Vol. 8, 56 - 59 (1974). In Russian.

Visual observations of Mars were performed in the period from February 16 to April 17, 1965 using the 165-mm reflector with various colored filters. On this basis a map of Mars was compiled.

097.004 Martian volcanism: Additional observations and evidence for pyroclastic activity. M. West.
Icarus, Vol. 21, 1 - 11 (1974).

Inspection of the Mariner 9 B-camera and A-camera photographs of Mars reveals numerous analogs of terrestrial and lunar volcanic features. In addition to the exceptionally large constructional features in the Tharsis region, many other large and small landforms present probably are related to endogenic processes.

097.005 Wind-blown streaks, splotches, and associated craters on Mars: Statistical analysis of Mariner 9 photographs. R. E. Arvidson.
Icarus, Vol. 21, 12 - 27 (1974).

The purpose of this paper is to look at detailed statistical patterns of aeolian deposits and associated craters to provide further insight into the nature of Martian wind-blown sediment distribution and the processes of interaction between craters and wind-sediment systems. Parameters that are dealt with here were tabulated from Mariner 9 A-frame photographs and are: (1) crater diameters, locations, and morphologies for craters with streaks or dark splotches located on one side of crater floors; (2) locations and trends of dark splotches; and (3) location, trends, and lengths of bright and dark streaks.

097.006 Mars: Clearing of the 1971 dust storm.
W. K. Hartmann, M. J. Price.
Icarus, Vol. 21, 28 - 34 (1974).

A semiquantitative analysis of clearing in the 1971 great dust storm on Mars is presented as a function of time and altitude, using Mariner 9 orange- and visual-light photos. Theoretical models of optical depth versus time are developed for various distributions of particles in the atmosphere. By interpreting settling in terms of Stokes' law, estimates of the maximum radii of dust particles throughout the atmosphere have been obtained.

097.007 Observation of OI 1300 Å radiation in the Martian atmosphere. V. G. Kurt, A. S. Smirnov, L. G. Titarchuk, S. D. Chuvahin.
Icarus, Vol. 21, 35 - 41 (1974).

Measurements from the 1225 to 1340 Å region by the ultraviolet detectors on Mars-3 are presented. Model calculations of the intensity of the OI triplet lines at 1304 Å are compared with the measurements made on December 27, 1971, and February 17, 1972. Agreement is found between experimental data and a model in which the neutral oxygen density at 100 km is $2-8 \times 10^9 \, cm^{-3}$.

097.008 Recent photographs of Mars and Jupiter.
S. Larson, J. Fountain.
Sky Telescope, Vol. 47, 17 - 18 (1974).

097.009 Color pictures of planets from black-and-white images. S. E. Jones, N. O. Cook.
Sky Telescope, Vol. 47, 57 - 59 (1974).

097.010 Photographs of a recent Martian dust storm.
A. S. Murrell, C. F. Knuckles.
Sky Telescope, Vol. 47, 168 - 169 (1974).

097.011 Helium in the Martian atmosphere: thermal loss considerations.
J. S. Levine, G. M. Keating, E. J. Prior.
Planet. Space Sci., Vol. 22, 500 - 503 (1974).

Helium concentrations in the Martian atmosphere are estimated assuming that the helium production on Mars, comparable to its production on earth, via the radioactive decay of uranium and thorium, is in steady state equilibrium with its thermal escape. The computed helium concentration at the Martian exobase (200 km) is 8×10^6 atoms cm^{-3}.

097.012 Update on Mars: Clues about the early solar system.
W. D. Metz.
Science, Vol. 183, 187 - 189 (1974). – Research news.

097.013 Application of the instrumental method to the reduction of space photographs. V. I. Chikmachev.
Kosmich. Issled., Vol. 12, 147 - 150 (1974). In Russian.
Brief information.

097.014 Phase dependence of the brightness of Mars during the opposition of 1971.
Yu. V. Aleksandrov, D. F. Lupishko, T. A. Lupishko.
Vestn. Khar'kov. Univ. No. 99 (Ser. Astron. No. 8), p. 3 - 9 (1973). In Russian.

097.015 On the application of the TV method to planetary observations.
A. N. Abramenko, V. V. Prokof'eva, N. A. Ushakova.
Izv. Krymskoj Astrofiz. Obs., Vol. 49, 98 - 104 (1974). In Russian.

Observations of planets using the TV technique are carried out at the Crimean Astrophysical Observatory since 1969. The method of observations is described. Photometry of TV photographs of Mars taken in 1969 shows that the accuracy of the brightness and contrast measurements equals to 8 and 6% respectively. The threshold contrast is 5%.

097.016 Carbon dioxide hydrate and floods on Mars.
D. J. Milton.

Science, Vol. 183, 654 - 656 (1974).

Ground ice on Mars probably consists largely of carbon dioxide hydrate. Catastrophic dissociation of carbon dioxide hydrate during some past epoch when the near surface temperature was in the range 0° to 10° would have produced chaotic terrain and flood channels.

097.017 **Photodissociation of carbon dioxide in the Mars upper atmosphere.** C. A. Barth.
Zeitschr. Naturforschung, Vol. 29a, 185 - 188 (1974).

Photodissociation of carbon dioxide produces $O(^1S)$ atoms and $CO(a^3\Pi)$ molecules in the Mars upper atmosphere. Calculations of the emission rate of the atomic oxygen 2972 Å line and the carbon monoxide Cameron bands produced by the photodissociation mechanism are factors of 3 and 10, respectively, smaller than the emission rates observed by Mariner ultraviolet spectrometers.

097.018 **Wind tunnel simulations of light and dark streaks on Mars.** R. Greeley, J. D. Iversen, J. B. Pollack, N. Udovich, B. White.
Science, Vol. 183, 847 - 849 (1974).

Wind tunnel experiments have revealed a characteristic flow field pattern over raised-rim craters which causes distinctive zones of aeolian erosion and deposition. Comparisons of the results with Mariner 9 images of Mars show that some crater-associated dark zones result from wind erosion and that some crater-associated light streaks are depositional.

097.019 **The optical oblateness of Mars.** E. J. Öpik.
Irish Astron. Journ., Vol. 11, 1 - 23 (1973).

The apparent shape of the Martian disk has been determined from precision measurements of isophotes on near-opposition (1958) photometric exposures with the 60-inch Mount Wilson reflector. Differential corrections for gibbousness at the small phase angle (5°) were applied as based on theory and previous Martian photometry. From three exposures in yellow and three in blue, a remarkably concordant pattern of the latitude deviations of optical scattering level from dynamical shape was obtained, implying a lift-up of the scattering medium with a rising motion, and a depression in altitude with descending climatic circulation.

097.020 **Magnetic fields of Mars and Venus: solar wind interactions.** N. F. Ness.
The Moon, Vol. 9, 43 - 47 (1974). — Communication presented at the Lunar Science Institute Conference (see 012.002).

Mars 2 and 3 magnetometer experiments report the existence of an intrinsic planetary magnetic field, sufficiently strong to form a magnetopause, deflecting the solar wind around the planet and its ionosphere. This is in contrast to the case for Venus where it is assumed to be the ionosphere and processes therein which are responsible for the solar wind deflection. An empirical relationship appears to exist between planetary dipole magnetic moments and their angular momentum for moon, Mars, Venus, earth and Jupiter. Implications for the magnetic fields of Mercury and Saturn are discussed.

097.021 **Craters and associated aeolian features on Mariner 9 photographs: an automated data gathering and handling system and some preliminary results.**
R. E. Arvidson, T. A. Mutch, K. L. Jones.
The Moon, Vol. 9, 105 - 114 (1974). — Communication presented at the Lunar Science Institute Conference (see 012.002).

Craters and associated aeolian features visible on Mariner 9 photographs have been examined. A brief description of the method used to accumulate data is presented, along with some preliminary results for A-frames.

097.022 **Mars: crustal structure inferred from Bouguer gravity anomalies.**
R. J. Phillips, R. S. Saunders, J. E. Conel.
The Moon, Vol. 9, 240 - 242 (1974). — Communication presented at the Lunar Science Institute Conference (see 012.002). — Abstract.

097.023 **Mars and Venus.** R. M. Goody.
The planets today. Symposium 1973, (see 012.003), p. 35 - 61 (1974).

097.024 **Ground-based television patrol of Martian cloud formations in 1969 and 1971–1972.**
A. N. Abramenko, V. V. Prokofieva.
Icarus, Vol. 21, 191 - 195 (1974).

The results of the T.V. patrol of the Martian clouds during the opposition of 1969 and 1971 are given. About 50000 T.V. pictures have been obtained in ten spectral regions (370–760 nm) during the period of August 1971 to January 1972. The telescope used for the T.V. planetary observations is described.

097.025 **Mars vor der Opposition 1973.** H. Treutner.
Orion Schaffhausen, 32. Jahrgang, p. 24 - 25 (1974).

097.026 **Martian climate: an empirical test of possible gross variations.** T. Owen.
Science, Vol. 183, 763 - 764 (1974).

A number of authors have speculated about the possibility that the average climate on Mars may once have been more clement by terrestrial standards than it is at present. There appears to be evidence for cyclic behavior in which the surface pressure periodically reaches values compatible with the flow of water in equatorial regions on the planet. It is the purpose of this note to call attention to a relatively simple test of such hypotheses—a test that may be carried out as early as February 1974.

097.027 **News of plasma envelopes of Mars and Venus.**
O. L. Vajsberg.
Zemlya i Vselennaya, 1974, No. 2, p. 19 - 22. In Russian.

097.028 **Determination of the dielectric losses of Mars surface material from its emission at millimeter wavelengths.**
Yu. N. Vetukhnovskaya, A. D. Kuzmin, B. Ya. Losovskij.
Astron. Zhurn. Akad. Nauk SSSR, Vol. 51, 413 - 416 (1974). In Russian. English translation in Soviet Astron. AJ, Vol. 18, No. 2.

8.22-mm observations of Martian radio emission near the 1971 opposition are presented and discussed.

097.029 **Martian albedo features and topography.** D. Popper.
Mercury, (Journ. Astron. Soc. Pacific), Vol. 2, No. 6, p. 10 - 11, 14 - 15, 18 (1973).

097.030 **Television observations of Mars during the period from July 14 - September 11, 1973.**
A. N. Abramenko, M. N. Naugolnaya, V. V. Prokof'eva.
Astron. Tsirk., No. 800, p. 6 - 7 (1973). In Russian.

097.031 **Dust cloud formations on Mars at perihelion in 1973.**
A. N. Abramenko, M. N. Naugolnaya, V. V. Prokof'eva.
Astron. Tsirk., No. 801, p. 1 - 2 (1973). In Russian.

097.032 **On the relationship between "blue clearing" and dust clouds on Mars.**
V. V. Prokof'eva, T. A. Chuprakova.
Astron. Tsirk., No. 801, p. 3 - 5 (1973). In Russian.

097.033 **Streams of charged particles in the vicinity of Mars.**

S. N. Vernov, B. A. Tverskoj, V. A. Yakovlev, E. V. Gorchakov, P. P. Ignat'ev, G. P. Lyubimov, O. A. Marchenko, T. E. Shvidkovskaya, N. N. Kontor, T. I. Morozova, A. G. Nikolaev, Yu. A. Rozental', E. A. Chuchkov, I. V. Getselev, L. V. Onishchenko, V. I. Tkachenko.
Kosmich. Issled., Vol. 12, 252 - 263 (1974). In Russian.

097.034 **Belgische amateurs tekenen kaart van Mars.**
A. van der Jeugt, A. Gabriël, A. Verschraegen.
Zenit, Vol. 1, No. 2, p. 29 - 31 (1974).

097.035 **Photogrammetric measurements of Olympus Mons on Mars.** M. E. Davies.
Icarus, Vol. 21, 230 - 236 (1974).

Mariner 9 took many pictures of the giant Olympus Mons during its year in orbit around Mars. Control points have been identified on the top of Olympus Mons, on the volcanic shield, and on the surrounding plains, and their locations have been measured on the television pictures. These measurements were used to compute the areographic coordinates and the planetary radii of the points. The radii at some of the points were derived from radar elevation measurements and from radio occultation measurements. The mountain rises about 21 km above its base.

097.036 **Mariner 9 ultraviolet spectrometer experiment: pressure-altitude measurements on Mars.**
C. W. Hord, K. E. Simmons, L. K. McLaughlin.
Icarus, Vol. 21, 292 - 302 (1974).

Ultraviolet spectrometer measurements of the reflectance at 3050 Å are modeled to give pressure-altitudes for Mars assuming a quiescent atmosphere. Ultraviolet light that is Rayleigh-scattered by the Mars molecular atmosphere, with allowance for uniform turbidity, is proportional to surface pressure independent of atmospheric temperature structure.

097.037 **Variable features on Mars III: comparison of Mariner 1969 and Mariner 1971 photography.**
J. Veverka, C. Sagan, L. Quam, R. Tucker, B. Eross.
Icarus, Vol. 21, 317 - 368 (1974).

Mariner 9 and Mariner 6 and 7 photography of common regions of Mars are compared, with appropriate attention to the photometric properties of the camera systems. The comparison provides a 2.5yr time baseline for study of variable albedo features. The authors find the development of bright streaks and patches; the evolution of dark crater splotches into dark streaks; and a planetwide increase in splotchiness.

097.038 **Martian lee waves revisited.** R. L. Wildey.
Nature, Vol. 249, 132 - 133 (1974).

The Martian lee waves have been interpreted as high level condensation clouds in which the brightness variations are due to a combination of true lateral variations in the local concentration of condensed material and the shadows cast on lower levels by such scattering material. The author demonstrates that the variation in slope of the mean scattering layer with respect to the local horizontal produces a photoclinometric brightness variation strong enough to explain the data with the invocation of quite modest relief in the cloud deck.

097.039 **Accurate positions of Mars and its satellite Deimos from photographic observations in Tashkent.**
L. I. Bashtova, A. G. Rakhimov.
Tsirk. Astron. Inst., *Tashkent,* No. 42 (389), p. 35 - 38 (1973). In Russian.

097.040 **The Martian structure from new data and comparison with the moon.** Yu. A. Khodak.
Izv. AN SSSR. Ser. geol., 1973, No. 10, p. 3 - 13. In Russian. Abstr. in Referativ. Zhurn. 51. Astron., 4.51.263; 62. Issled. kosmich. prostranstva, 4.62. 177 (1974).

097.041 **Photo-induced free radicals on a simulated Martian surface.** S.-S. Tseng, S. Chang.
Nature, Vol. 248, 575 - 577 (1974).

The authors describe the electron spin resonance study of free radicals in the ultraviolet irradiation of a simulated Martian surface. This study confirms the reports of Hubbard et al. that HCO_2H and possibly HCHO and other organic compounds were formed during the photolysis of atmospheric gases adsorbed on a simulated Martian surface.

097.042 **Photolysis of $CO-NH_3$ mixtures and the Martian atmosphere.** J. P. Ferris, E. A. Williams, D. E. Nicodem, J. S. Hubbard, G. E. Voecks.
Nature, Vol. 249, 437 - 439 (1974).

The findings discussed here indicate an additional possible mechanism of nitrogen conservation. If NH_3 is outgassed from the Martian crust it would be photolysed and would react with CO, yielding cyanic acid. The cyanic acid could well precipitate as ammonium cyanate which in turn would be converted to urea.

097.043 **Martian cratering and central peak statistics: Mariner 9 results.**
B. M. Cordell, R. E. Lingenfelter, G. Schubert.
Icarus, Vol. 21, 448 - 456 (1974).

Mariner 9 imagery shows that central peaked craters occur much more frequently in the Martian south polar region than in typical equatorial areas, and that both regions have crater size frequency distributions characteristic of saturation. Several arguments indicate that a preferential production mechanism, e.g., pingo formation made possible by subsurface permafrost confined to Martian polar regions, may account for the central peak excess in the south polar region.

097.044 **Differential transmission of sunlight on Mars: Biological implications.**
C. Sagan, J. B. Pollack.
Icarus, Vol. 21, 490 - 495 (1974).

A euphotic zone seems to exist at about 1 cm subsurface in the Martian epilith. At this depth visible light is still intense enough to be utilized by conceivable photosynthetic organisms; but the germicidal ultraviolet intensities at the Martian surface have been reduced to values manageable by terrestrial life.

097.045 **The effect of radiative transfer on shear-flow instability in the atmospheres on Mars and Venus.**
J. J. Dudis, S. C. Traugott.
Icarus, Vol. 21, 496 - 505 (1974).

The results of two theoretical investigations concerning the destabilizing effects of radiative transfer on stably stratified shear flows are applied to the CO_2 atmospheres of Mars and Venus.

097.046 **The atmosphere of Mars.** C. A. Barth.
Annual Rev. Earth Planet. Sci., Vol. 2, (see 003.006), 333 - 367 (1974).

097.047 **The gravity field of Mars as determined by the Mariner 9 radio-tracking data.** J. Lorell.
The Moon, Vol. 9, 245 (1974). − Communication presented at the Lunar Science Institute Conference (see 012.002). Abstract.

097.048 **Convection in a Martian magnetosphere.**
M. E. Rassbach, R. A. Wolf, R. E. Daniell, Jr.
Journ. Geophys. Res., Vol. 79, 1125 - 1127 (1974).

In this letter the authors present an estimate of the rate of convection to be expected in the Martian ionosphere-magnetosphere. Two ways of estimation are given and discussed.

097.049 **Mariner 6, 7, and 9 ultraviolet spectrometer experiment: Analysis of hydrogen Lyman alpha data.**
D. E. Anderson, Jr.
Journ. Geophys. Res., Vol. 79, 1513 - 1518 (1974).

Four Lyman α airglow measurements of the limb and disc of Mars, made by ultraviolet spectrometers on Mariner 6 and 7 in 1969 and Mariner 9 in 1971, are analyzed to determine the amount and distribution of atomic hydrogen above 80 km. The variation of atomic hydrogen with altitude is calculated by using time-independent chemical diffusion models from 80 to 250 km, and an exospheric model is used above 250 km.

097.050 **Evidence about hydrate and solid water in the Martian surface from the 1969 Mariner infrared spectrometer.** G. C. Pimentel, P. B. Forney, K. C. Herr.
Journ. Geophys. Res., Vol. 79, 1623 - 1634 (1974).

Because of the spectral range scanned ($2-14\,\mu$) and the excellent geographic resolution (130 km at closest approach), the Mariner 6 and 7 infrared spectra offer unique evidence about both gaseous and condensed phase water and their distributions. The $3\text{-}\mu$ absorption of condensed phase water was immediately obvious in the Mariner Mars (MM) 6 and 7 spectra at the first data return, and it was reported in a first publication (Herr and Pimentel, 1969). However, both quantitative deductions and spectral differentiation between mineral hydrate water and ice have required extensive laboratory simulation studies and comparative computer analyses of the spectra. We report here the results of these analyses relevant to the presence, distribution, and form of condensed phase water.

097.051 **Tegenslag voor Russen bij verkenning van Mars.**
A. van den Berg.
Zenit, Vol. 1, No. 5, p. 13 (1974).

097.052 **Mariner 6, 7 and 9 ultraviolet spectrometer experiment: analysis of hydrogen Lyman-α data.**
D. E. Anderson, Jr.
EOS Trans. American Geophys. Union, Vol. 54, 1153 (1973).

Four Lyman-α airglow measurements of the limb and disc of Mars, made by ultraviolet spectrometers on Mariner 6 and 7 in 1969 and Mariner 9 in 1971 are analyzed.

097.053 **Mars 1969 opposition period – ALPO report IV.**
C. F. Capen.
Strolling Astronomer, Vol. 24, 169 - 177 (1974).

097.054 **Mariner-9 photographie Mars.** D. Verguèse.
Sci. Progrès Découverte, No. 3444, p. 22 - 27 (1972).

097.055 **Variable features on Mars. IV. Pavonis Mons.**
C. Sagan, J. Veverka, R. Steinbacher, L. Quam, R. Tucker, B. Eross.
Icarus, Vol. 22, 24 - 47 (1974). – Paper presented at the International Colloquium on Mars, 28 November – 1 December 1973, Pasadena, California.

A remarkable set of albedo changes has been uncovered by Mariner 9 photography of the upper slopes of the shield volcano Pavonis Mons, near its summit caldera. The most likely explanation of the event is aeolian transport of fine-grained particles. Since the atmospheric pressure in this locality is ~ 1.5 mb, minimum wind velocities above the surface boundary layer of about 110 m/s are necessary, corresponding to 0.51 of the speed of sound. Slope winds in this velocity range are expected near the upper flanks of major Martian volcanic constructs.

097.056 **The ionospheres of Mars and Venus.**
R. C. Whitten, L. Colin.
Rev. Geophys. Space Phys., Vol. 12, 155 - 192 (1974).

Observations made by spacecraft in the Mariner and Venera series have provided considerable knowledge of the structure of the ionospheres and atmospheres of Mars and Venus. This paper begins with brief but complete discussions of these measurements and their interpretations. Specifically, the authors summarize the characteristics and use of UV radiometry, magnetometers, and ion probes for determining solar wind properties near planets, as well as the application of the occultation experiment. The paper concludes with a summary of current knowledge of the upper ionospheres of Mars and Venus and the information needed to understand fully their structures.

097.057 **Seasonal variations of the parameters of the Martian ionosphere and conditions for the radio wave propagation in it, solar activity phases and distances of planets from the sun.** A. N. Kazantsev.
Izv. vyssh. ucheb. zavedenij. Radiofizika, Vol. 16, 1401 - 1404 (1973). In Russian. – Abstr. in Referativ. Zhurn. 51. Astron., 5.51.273; 62. Issled. kosmich. prostranstva, 5. 62.181 (1974).

097.058 **Winds and the occultation experiment.**
S. H. Gross.
Planet. Space Sci., Vol. 22, 789 - 791 (1974).

The occultation experiment may be used to obtain dynamic meteorological information when performed from an orbiting spacecraft. It is shown that interpretation of refractivity data in this fashion does not require a composition, as normally used to obtain pressure, density and temperature profiles.

097.059 **Comment on: "The photolytic stability of the Martian atmosphere" by R. C. Whitten and J. S. Sims.**
D. M. Hunten.
Planet. Space Sci., Vol. 22, 878 - 879, with a reply by R. C. Whitten, J. S. Sims, p. 879 (1974).

097.060 **What is what on Mars. [List of designations].**
D. Ya. Martynov.
Zemlya i Vselennaya, 1974, No. 3, p. 21 - 27. In Russian.

097.061 **When does the opposition of Mars begin?**
V. A. Bronshtehn.
Zemlya i Vselennaya, 1974, No. 3, p. 70 - 73. In Russian.

097.062 **Possible origin and probable discharges of meandering channels on the planet Mars.** J. G. Weihaupt.
Journ. Geophys. Res., Vol. 79, 2073 - 2076 (1974).

Application of empirical equations derived from studies of the meandering rivers on earth shows that the Martian meanders are consistent with earth meanders in terms of wavelength/radius of curvature relationships but inconsistent in terms of wavelength/channel width and radius of curvature/channel width relationships. The mean annual discharge of an earth river comparable to the Martian channel studied is calculated to be at least 2700 m^3/s and perhaps considerably greater.

097.063 **Comments on 'Sandstorms and Eolian erosion on Mars' by Carl Sagan.** W. J. Maegley.
Journ. Geophys. Res., Vol. 79, 2145 - 2146 (1974).

097.064 **Frictional and stream velocities in sandstorms.**
C. Sagan.
Journ. Geophys. Res., Vol. 79, 2147 (1974).

097.065 **Mars from Mariner 9 (New data on the nature and structure of the planetary surface).**
A. V. Zhivago.
Izv. AN SSSR, ser. geogr., 1973, No. 6, p. 119 - 125. In Rus-

sian. — Abstr. in Referativ. Zhurn. 51. Astron., 6.51.273 (1974).

097.066 Infrared heterodyne spectroscopy of CO_2 on Mars.
D. W. Peterson, M. A. Johnson, A. L. Betz.
Nature, Vol. 250, 128 - 130 (1974).

The authors have used an infrared heterodyne spectrometer operating near 11 μm with a resolving power of 1.5×10^6 to obtain spectral line profiles of carbon dioxide absorption in the Martian atmosphere. The multichannel system determined shapes of three lines in the 10° 0–00° 1 vibration-rotation band of $^{13}C^{16}O_2$, and found equivalent widths about 50 MHz. A lower limit of 670 MHz on the equivalent width of the P(20) line of $^{12}C^{16}O_2$ was also obtained.

097.067 Heterodyne spectroscopy of CO_2 on Mars.
D. W. Peterson, M. A. Johnson.
Bull. American Astron. Soc., Vol. 6, 220 (1974). — Abstr. AAS.

097.068 What else is present in the Martian atmosphere?
T. Owen.
Comments Astrophys. Space Phys., Vol. 5, 175 - 180 (1973).

097.069 The major Martian yellow storm of 1971.
L. J. Martin.
Icarus, Vol. 22, 175 - 188 (1974).

Extensive earth-based photography produced by the International Planetary Patrol has been used to map the positions of brightened areas (clouds) during the 1971 storm on Mars. Summaries of these maps are presented to illustrate the changes that take place during the course of a Martian day, as well as the changes from one day to the next. The possible influence of Martian topography on the progress of the storm is examined. Comparisons between red- and blue-filter photographs of the storm are presented cartographically and are discussed.

097.070 The elevation of Olympus Mons from limb photography. A. B. Whitehead.
Icarus, Vol. 22, 189 - 196 (1974).

Using a novel photogrammetric technique, the relative elevations of a set of control points on the Martian volcano Olympus Mons have been measured from Mariner 9 limb pictures. The summit of Olympus Mons is found to be 22 ± 1 km above the mean level of the surrounding plain.

097.071 Pyramidal structures on Mars.
M. Gipson, Jr., V. K. Ablordeppey.
Icarus, Vol. 22, 197 - 204 (1974).

Triangular and polygonal pyramid like structures have been observed on the Martian surface. The mean diameter of the triangular pyramidal structures at the base is approximately 3.0 km, and the mean diameter of the polygonal structures is approximately 6.0 km.

097.072 Ozone and the polar hood of Mars.
C. A. Barth, M. L. Dick.
Icarus, Vol. 22, 205 - 211 (1974).

Ozone co-appears with the clouds of the polar hood in the winter hemisphere of Mars, but each is variable from day to day and location to location. Both the appearance of ozone and the polar hood clouds correlate with the temperature of the atmosphere which varies from day to day and location to location.

097.073 NERC/LPU Mars pilot study.
Published by Lunar and Planetary Unit, Dep. Environmental Sci., Univ. Lancaster, England. 25 + 26 pp. (1974).

Following the acquisition by LPU of early NASA releases of Mariner photographic material and limited support data, a pilot study of Mars was initiated by LPU with the dual aim of (1) exploring avenues of attack on the intercomparison of terrestrial, lunar and Martian surface features and processes and (2) investigating the Martian dust cycle. This report, prepared for NERC, summarises the findings of five Working Groups of LPU.

097.074 Mariner 9 ultraviolet spectrometer experiment: morning terminator observations of Mars.
J. M. Ajello, C. W. Hord.
Journ. Atmosph. Sci., Vol. 30, 1495 - 1501 (1973).

Mariner 9 ultraviolet morning terminator data indicate the presence of a high, optically thin scattering layer in the equatorial region of Mars extending to 50S. The terminator observations poleward of 50S were modeled by a homogeneous atmosphere of molecular and particulate scatterers. The shape of the morning terminator reflectivity measured as a function of shadow height indicates larger apparent scale heights than were observed in the afternoon. The difference is attributed to an increase in thickness of the scattering layer during the night.

097.075 McLaughlin and Mars. J. Veverka, C. Sagan.
American Scient., Vol. 62, 44 - 53 (1974).

The predictions of this pioneering astronomer-geologist about the albedo markings of Mars are reviewed in the light of Mariner 9 photographs of the planet's surface.

097.076 Mars 1971 apparition — ALPO report I.
C. F. Capen, T. R. Cave.
Strolling Astronomer, Vol. 24, 220 - 228 (1974).

097.077 Atmospheric pressure variation and the climate of Mars. P. J. Gierasch, O. B. Toon.
Journ. Atmosph. Sci., Vol. 30, 1502 - 1508 (1973).

If Mars has permanent CO_2 polar caps, atmospheric heat transport may cause the atmospheric pressure to be extremely sensitive to variations of solar heating at the poles. A simple climatological model is used to study the question.

097.078 Meteorite impact: a suggestion for the origin of some stream channels on Mars.
T. A. Maxwell, E. P. Otto, M. D. Picard, R. C. Wilson.
Geology, (USA), Vol. 1, No. 1, p. 9 - 10 (1973).

097.079 Martian regolith X-ray analyzer: tests results of geochemical performance.
B. C. Clark, A. K. Baird.
Geology, (USA), Vol. 1, No. 1, p. 15 - 18 (1973).

097.080 Mariner 9 photographs of small volcanic structures on Mars. R. Greeley.
Geology, (USA), Vol. 1, No. 4, p. 175 - 180 (1973).

097.081 Neues über die Geologie des Mars.
P. N. Kropotkin.
Zeitschr. Angew. Geol., Vol. 19, 377 - 378 (1973).

Planetary ephemerides based on radar, transit circle, and Mariner 9 data. See Abstr. 041.005.

The use of radar-ranging in planetary ephemeris development. See Abstr. 041.006.

Premières mesures d'albédo du nivéo-éolien sur terre. See Abstr. 081.016.

Deconvolution in planetary photometry. See Abstr. 091.023.

Radiative transfer within the mesospheres of Venus and Mars. See Abstr. 093.012.

Comparison of the far ultraviolet spectra of Venus and Mars. See Abstr. 093.013.

Atmospheric ion wakes of Venus and Mars in the solar wind. See Abstr. 093.037.

Problems associated with estimating the relative impact rates on Mars and the moon. See Abstr. 094.118.

The role of lava erosion in the formation of lunar rilles and Martian channels. See Abstr. 094.159.

On the characters and origin of lunar and Martian craters. See Abstr. 094.185.

The occultation of Mars by the moon on May 16, 1971. See Abstr. 096.034.

Mars, Satellites

097.201 **Relative positions of Phobos and Deimos.**
A. Sh. Khatisov, G. N. Salukvadze.
Byull. Abastumansk. Astrofiz. Obs., No. 45, p. 71 - 74 (1974). In Russian. – July - August 1971.

097.202 **Phobos og Deimos.** T. S. Ringnes.
Naturen 1973, Årg. 97, No. 6, p. 273 - 286 = Inst. Teor. Astrofys., Blindern – Oslo, Småtrykk No. 81 (1973).

097.203 **Infrared observations of Phobos from Mariner 9.**
I. Gatley, H. Kieffer, E. Miner, G. Neugebauer.
Astrophys. Journ., Vol. 190, 497 - 503 (1974).
Radiometric measurements of Phobos at 10 and 20 μ show that the surface is covered by a material whose conductivity is extremely low, around 10^{-6} cal cm^{-1} s^{-1} (deg-K)$^{-1}$. It is concluded Phobos is covered with a layer of dust at least 1 mm thick.

Physical properties of the natural satellites. See Abstr. 091.004.

Accurate positions of Mars and its satellite Deimos from photographic observations in Tashkent. See Abstr. 097.039.

098 Minor Planets

098.001 **The masses of the first two asteroids.** J. Schubart.
Astron. Astrophys., Vol. 30, 289 - 292 (1974).
The original observational records were used in a new reduction of early meridian observations of Ceres and Pallas. Later observations underwent systematic corrections. Normal places represent the single observations of the interval from 1802 to 1970. A new value for the mass of Ceres equal to $(5.9 \pm 0.3) \times 10^{-10}$ solar mass resulted from a differential correction of the orbit of Pallas. The reversion of the problem allowed the derivation of a preliminary value for the mass of Pallas from the observations of Ceres. This value has a comparatively large relative uncertainty, because the mass of Pallas equals only about one fourth or one fifth of the mass of Ceres.

098.002 **On the origin of the asteroids.**
W. McD. Napier, R. J. Dodd.
Monthly Notices Roy. Astron. Soc., Vol. 166, 469 - 489 (1974).
Three hypotheses concerning the origin of asteroids were examined, viz. that the asteroids are collisional fragments deriving from a few primaeval planetoids; that they are the remnants of a planet which disrupted; and that they are in process of forming a single body by random aggregation. Monte Carlo simulations and theoretical arguments were used to predict mass distributions and rotation periods. It was concluded that, most probably, the asteroids are debris arising from the collisional shattering of a few primal bodies, some of which may still survive intact.

098.003 **Asteroidal collisions.** G. W. Wetherill.
Icarus, Vol. 21, 94 - 95 (1974).
Taff (1973) has concluded that asteroidal collision rates are much lower than those found by previous authors. His calculations are found to be in error as a consequence of inclusion of an extraneous and incorrect factor of $\sim 10^{-5}$. The assumption of molecular chaos in the asteroid belt, while not strictly correct, is not an important source of error in calculations of asteroidal collision rates.

098.004 **Application of the Brown-Shook model to the Kirkwood gaps and resonant asteroids.** W. H. Ip.
Celestial Mechanics, Vol. 9, 73 - 79 (1974).
The mathematical model developed by Brown and Shook is used to explain the absence of asteroids in the Kirkwood gaps as well as the existence of some resonant asteroids with orbital periods commensurable to that of Jupiter.

098.005 **The minor planets: sizes and mineralogy.**
C. R. Chapman, D. Morrison.
Sky Telescope, Vol. 47, 92 - 95 (1974).

098.006 **Particle concentration in the asteroid belt from Pioneer 10.**
R. K. Soberman, S. L. Neste, K. Lichtenfeld.
Science, Vol. 183, 320 - 321 (1974).
The spatial concentration and size distribution for particles measured by the asteroid/meteoroid detector on Pioneer 10 between 2 and 3.5 astronomical units are presented.

098.007 **Stability of asteroid orbits passing close to Jupiter.**
J. V. Breakwell.
Bull. American Astron. Soc., Vol. 6, 206 - 207 (1974). Abstr. AAS.

098.008 **Asteroid evolution: spectrophotometric evidence.**
C. R. Chapman.
The Moon, Vol. 9, 239 - 240 (1974). – Communication presented at the Lunar Science Institute Conference (see 012.002). – Abstract.

098.009 **Minor planet motions.** P. Herget.
Celestial Mechanics, Vol. 9, 315 - 319 (1974).
Presented at a colloquium on 'Copernicus ... and modern dynamical astronomy', Washington, 1972 December.

098.010 **Minor planets and related objects. XV. Asteroid (1620) Geographos.** J. L. Dunlap.
Astron. Journ., Vol. 79, 324 - 332 (1974).

Observations were made in 1969 of lightcurves, brightness, colors, and polarization of asteroid (1620) Geographos. The absolute magnitude of the lightcurve maximum is $B_0(1,0) = 16.08$, and the colors, which do not change appreciably with phase, are $B-V = +0.87$ and $U-B = +0.50$. The lightcurves are relatively smooth with two distinct maxima and minima; there is, however, an unexplained variation in the arrival times of the minima. The largest amplitude is 2.03 mag, which is the largest observed on asteroids to date. Precession of the axis of rotation was not required to obtain this solution.

098.011 **Vlucht naar Eros in 1977/78. Planetoïden vormen interessant doelwit voor ruimteonderzoek.**
T. de Vries.
Zenit, Vol. 1, No. 3, p. 26 - 30 (1974).

098.012 **Photoelectric lightcurves of the minor planet 43 Ariadne during the 1972 opposition.**
R. Burchi, L. Milano.
Astron. Astrophys., Suppl. Ser., Vol. 15, 173 - 180 (1974).

The minor planet 43 Ariadne was observed during the 1972 opposition with the photoelectric photometer attached to the 40 cm refractor of the Teramo Observatory. The lightcurve appears fairly regular with a double maximum; the synodic period seems to be $5^h 45^m 02^s$. The standard magnitude $V(1,0)$, the phase coefficient, the albedo and the mean radius have been derived.

098.013 **L'appulse du 14 octobre 1974.** J. Meeus.
L'Astronomie, 88e année, p. 114 (1974).

098.014 **Toro: the imprisoned bull?**
R. L. Duncombe, P. M. Janiczek, P. K. Seidelmann.
Sky Telescope, Vol. 47, 381 - 383 (1974).

098.015 **Observations of minor planets at the Bucharest Observatory during 1964 and 1965.**
C. Cristescu, V. I. Vlăsceanu, B. Milet.
Stud. Cerc. Astron., Vol. 19, 109 - 118 (1974).

098.016 **Fragmentation of asteroids – a synopsis of analytical methods.** J. Dorschner.
Astron. Nachr., Vol. 295, 141 - 146 = Mitt. Univ. Sternw. Jena No. 119 (1974).

Based on the author's work a review of the possibilities as well as the limits of treating the problem of the collisional history of the asteroids by analytical methods is given. Using empirical data on rock fragmentation and general principles like symmetry and mass conservation the distribution function of the fragments arising from a single collision is analytically formulated. Quasi-stationary solutions of the equation of fragmentation are discussed for particular cases. The problem of the steady state is reduced to the solution of a transcendental equation. The results obtained show that analytical methods already offer a good theoretical understanding of the observed size distribution of the asteroids.

098.017 **Photometrische Beobachtung des Kleinen Planeten (89) Julia.** H. J. Schober, G. Lustig.
Mitt. Astron. Ges., No. 34, p. 122 - 124 (1973/74).
Presented at the "Wissenschaftliche Tagung der Astron. Ges., Oberkochen, 1973 April 24 - 27".

098.018 **Asteroiden Toros banresonans med jorden och Venus.** L. Danielsson.
Astron. Tidssk., Årg. 7, p. 24 - 30 (1974).

098.019 **Commensurabilities with Jupiter in the relation $\sigma_i(a)$ for minor planets.** S. Gąska.
Stud. Soc. Sci. Torunensis, Sectio F (Astron.), Vol. 5, 89 - 92 = Bull. Astron. Obs. Toruń, No. 51/I (1973).

098.020 **Dispersion of the inclinations of minor planets with respect to Jovian plane.**
S. Kasperczuk, L. Dybkowski.
Stud. Soc. Sci. Torunensis, Sectio F (Astron.), Vol. 5, 93 - 98 = Bull. Astron. Obs. Toruń, No. 51/II (1973).

Gąska (1970) has pointed out, that it is possible to investigate the problem of the origin of minor planets by studying the statistical parameters of their orbits. In the present work this dependence has been reexamined in a coordinate system associated with the plane of Jupiter, which enables further theoretical research of the problem of asteroids.

098.021 **Statistical investigation of orbital elements of minor planets. Part II.** L. Dybkowski, S. Kasperczuk.
Stud. Soc. Sci. Torunensis, Sectio F (Astron.), Vol. 5, 99 - 103 = Bull. Astron. Obs. Toruń, No. 51/III (1973).

Resolving proper systems of equations we get elements Ω_o and i_o of a mother-planet, nearly invariable with respect to the grouping of planetoids into groups of four. This confirms the hypothesis, that minor planets originated as a result of a disintegration of the mother-planet.

098.022 **Minor planet service.** G. A. Chebotarev.
Vestn. AN SSSR, 1973, No. 12, p. 62 - 68. In Russian. – Abstr. in Referativ. Zhurn. 51. Astron., 6.51.31 (1974).

098.023 **On the possibility of a resonance capture of the asteroid Toro by the earth.**
L. Danielsson, R. Mehra.
Stars and the Milky Way system, Proc. 1972, (see 012.018), p. 283 - 289 (1974).

098.024 **Les earth-grazers (ou EGA), des petits astres qui frolent la terre.** J. Meeus, M.-A. Combes.
L'Astronomie, 88e année, p. 194 - 220 (1974).

098.025 **Origin of the asteroids: collision of a small planet with a large comet.** G. Horedt.
Icarus, Vol. 22, 230 - 232 (1974).

It is not likely that the asteroids originated from the collision of a small planet with a large comet, because such a catastrophic event has a probability much too low.

098.026 **On the secular perturbations of asteroids.**
M. Yuasa.
Proc. symposium "Celestial mechanics", Kyoto 1973, (see 012.022), p. 9 - 19 (1974). In Japanese.

098.027 **1973 EC.** R. E. McCrosky, C. Y. Shao, G. Schwartz, M. Mattei.
IAU Circ., No. 2616 (1974).

098.028 **887 Alinda.**
R. Petrovičová, A. Mrkos, A. R. Klemola.
IAU Circ., No. 2633 (1974).

098.029 **887 Alinda.** R. E. McCrosky, G. Schwartz, J. Bulger, C. Y. Shao, M. Antal, A. Mrkos.
IAU Circ., No. 2648 (1974).

098.030 **887 Alinda.** L. Lorenzi, V. Zappalà, R. Pannunzio, G. de Sanctis.
IAU Circ., No. 2655 (1974).

098.031 **887 Alinda.**
G. Schwartz, C. Y. Shao, A. Mrkos.
IAU Circ., No. 2659 (1974).

098.032 **1971 UA.** B. G. Marsden.
IAU Circ., No. 2664 (1974).

098.033 **Observations of minor planets.**
G. Schwartz, R. E. McCrosky, C. Y. Shao, M. Mattei.
IAU Circ., No. 2679 (1974). − Concerning precise positions of 1932 HA, 1971 FA, 1972 XA.

098.034 **Minor planets at highly favorable opposition in 1974.** F. Pilcher.
Minor Planet Bull., Vol. 1, 15 - 17 (1974).

098.035 **The impossibility of observing asteroid surfaces.**
C. R. Chapman.
Minor Planet Bull., Vol. 1, 17 (1974).

098.036 **Observations of planet 487 Venetia.** F. Pilcher.
Minor Planet Bull., Vol. 1, 22 (1974).

098.037 **Observations of planet 952 Caia.** R. G. Hodgson.
Minor Planet Bull., Vol. 1, 23 (1974).

098.038 **Implications of recent diameter and mass determinations of Ceres.** R. G. Hodgson.
Minor Planet Bull., Vol. 1, 24 - 28 (1974).

098.039 **Minor planet work for smaller observatories.**
R. G. Hodgson.
Minor Planet Bull., Vol. 1, 30 - 34 (1974).

098.040 **Observations of 887 Alinda.** R. G. Hodgson.
Minor Planet Bull., Vol. 1, 36 - 37 (1974).

098.041 **Selected ephemerides.**
Minor Planet Bull., Vol. 1, 20 - 21, 29, 38 - 39 (1974).

098.042 **Lichtelektrische Beobachtungen des Planetoiden 2 Pallas in den Jahren 1968 und 1969.**
H. Haupt, A. Schroll.
Mitt. Univ.-Sternw. Graz, No. 16, 10 pp. (1974).

098.043 **Ephemerides of minor planets for 1975.**
Editor: Institut Teoreticheskoj Astronomii Akademii Nauk SSSR, under the editorship of G. A. Chebotarev.
Izdatel'stvo "Nauka", Leningradskoe Otdelenie, Leningrad. 184 pp. Price 2 Rbl. 11 Kop. (1974). In Russian and English.
Contents: Introduction, p. 3 - 8; Information on new elements, p. 9 - 10; New elements, p. 10 - 11; Elements, p. 12 - 44; Opposition dates, p. 45 - 55; Ephemerides, p. 56 - 169; Ephemerides of bright planets, p. 170 - 179; Ephemerides of some unusual planets, p. 180 - 182; Critical list, p. 183.

098.044 **A contribution to the problem of minor planets.**
V. Mišković.
Srpska akad. nauka i umetn., Beograd − Glas Srpske akad. nauka i umetn. 283, 1972, p. 79 - 164, Beograd. − Abstr. Bull. Sci. Yougoslavie, Section A, Vol. 19, 101 (1974).

098.045 **Planetoïde Apollo opnieuw ontdekt.**
G. W. E. Beekman.
Zenit, Vol. 1, No. 6, p. 5 (1974).

098.046 **Minor Planet Circulars, (MPC), Nos 3603 - 3684** (1974).
Edited by Cincinnati Observatory under the supervision of P. Herget.
A repository of nearly all new data for numbered and unnumbered minor planets: Observations, elements and ephemerides, identifications, newly assigned numbers and names, occultations.

Thoughts on the future of photographic catalogues.
See Abstr. 041.053.

Approximate axial alignment times for spinning bodies. See Abstr. 042.007.

Periodic Trojan-type orbits in the earth−sun system.
See Abstr. 042.043.

12-micron emission features of the Galilean satellites and Ceres. See Abstr. 099.207.

Observations of comets and minor planets.
See Abstr. 103.007.

Errata

098.901 **Errata: 'Theory of secular perturbations of asteroids including terms of higher orders and higher degrees"**
[Publ. Astron. Soc. Japan, Vol. 25, 399 - 445 (1973)].
M. Yuasa.
Publ. Astron. Soc. Japan, Vol. 26, 307 - 308 (1974).

099 Jupiter, Jupiter Satellites

Jupiter

099.001 **Ultraviolet photometry from the Orbiting Astronomical Observatory. XIII. The albedos of Jupiter, Uranus, and Neptune.** B. D. Savage, J. J. Caldwell.
Astrophys. Journ., Vol. 187, 197 - 208 (1974).

This paper presents OAO-2 broad-band ultraviolet photometry data for Jupiter, Uranus, and Neptune. These data have been combined with observations of 23 late-type stars to derive planetary albedos over the region 2000–4300 Å. The results for Uranus and Neptune have been compared with theoretical calculations for semi-infinite and finite pure H_2 Rayleigh-Raman scattering atmospheres. Except for the shortest wavelengths, the measurements lie below the semi-infinite theoretical calculations. The calculations for the finite atmospheres suggest that an additional absorbing constituent will be needed to explain the observed albedos.

099.002 **Jupiter: Identification of ethane and acetylene.** S. T. Ridgway.
Astrophys. Journ., (*Letters*), Vol. 187, L41 - L43 (1974).

Spectra of 1.3 cm^{-1} resolution of Jupiter have been obtained in the 10-micron window. Ethane and acetylene have been identified near 800 cm^{-1}. The molecular lines appear in emission.

099.003 **Raman scattering from H_2 in Jupiter.** H. Fast, R. Poeckert, J. R. Auman.
Astrophys. Journ., Vol. 187, 403 - 405 (1974).

The detection, by means of a correlation technique, of pure rotational Raman scattering from H_2 in Jupiter in the 3500–3900 Å region is described. The combined intensity of the $S(0)$ and $S(1)$ pure rotational Raman lines is found to be 3.5 ± 0.9 (p.e.) percent of the total observed intensity in this spectral region.

099.004 **Internal constitution and the figures of the giant planets.** V. N. Zharkov, V. P. Trubitsyn.
Phys. Earth Planet. Interiors, Vol. 8, 105 - 107 (1974).

The equations of state of the matter contained in the giant planets are investigated. The thermal state of Jupiter, Saturn, Uranus and Neptune is considered, and the results of the calculations of their figures, gravitational potentials and models of the internal structures are given.

099.005 **Investigation of Jupiter and Saturn with the 2-m reflector of the Shemakha Astrophysical Observatory.** N. B. Ibragimov.
Astron. Zhurn. Akad. Nauk SSSR, Vol. 51, 178 - 186 (1974). In Russian. English translation in Soviet Astron. AJ, Vol. 18, No. 1.

Fine structure of the absorption bands CH_4 6190 Å, NH_3 6450 Å in the spectrum of Jupiter and CH_4 6190 Å, CH_4 6800 Å in that of Saturn has been studied using spectrograms obtained with the 2-m reflector of the Shemakha Observatory. On the basis of halfwidths of lines in these bands the pressure in the atmospheres of these two planets was evaluated. Intensity variations have been studied for the methane bands at 5430, 6190, 7020, 7250, 7980, 7820 Å in the spectrum of Saturn and for the bands CH_4 7020, 7250, CH_4 7980 + NH_3 7920 + CH_4 7820 in the spectrum of Jupiter.

099.006 **Spatial variations in the Jovian 20-micrometer flux.** R. E. Murphy, R. A. Fesen.
Icarus, Vol. 21, 42 - 46 (1974).

The 17–28μm brightness temperature of the center of the disk of Jupiter is 136 ± 4K. Model calculations yield an effective temperature of 142 ± 4K at the center of the disk for a helium to hydrogen ratio He/H_2 of 0. This corresponds to an effective temperature of the entire disk of 136 ± 5K. The NEB, SEB, and STeB are shown to emit an excess flux at 20μm when compared to the neighboring zones. The relationships between 5-μm and 20-μm flux excesses and the cloud structures are discussed.

099.007 **Variation of Jupiter's CH_4 and NH_3 bands with position on the planetary disk.** R. W. Avery, J. J. Michalsky, Jr., R. A. Stokes.
Icarus, Vol. 21, 47 - 54 (1974).

The authors report the results of recent spectrophotometric scans of the 6190 Å and 7290 Å methane bands and the 6450 Å ammonia band on Jupiter as a function of position on the planetary disk. The equivalent widths are found to decrease toward the equatorial limb and to increase toward the poles relative to the center of the disk.

099.008 **Molecular band variations as a probe of the vertical structure of a Jovian atmosphere.** J. J. Michalsky, Jr., R. A. Stokes, R. W. Avery, W. C. DeMarcus.
Icarus, Vol. 21, 55 - 65 (1974).

A variational technique is used to compute synthetic spectra for models of cloudy Jovian planetary atmospheres which incorporate abrupt changes in their vertical structure. The dependence of the center-to-limb variations in equivalent widths of molecular bands upon the properties of the various scattering layers in the model is examined. A range of theoretical models are delineated on the basis of their ability to reproduce observational results for the specific case of Jupiter.

099.009 **The inner structure of giant planets.** V. P. Trubitsyn.
Zemlya i Vselennaya, 1974, No. 1, p. 50 - 55. In Russian.

099.010 **The flyby of Jupiter.** T. Gehrels.
Sky Telescope, Vol. 47, 76 - 78 (1974).

099.011 **Pioneer observes Jupiter.** R. N. Watts, Jr.
Sky Telescope, Vol. 47, 79 - 83 (1974).

099.012 **Cross-section for the dissociative photoionization of hydrogen by 584 Å radiation: the formation of protons in the Jovian ionosphere.** K. M. Monahan, W. T. Huntress, Jr., A. L. Lane, J. M. Ajello, T. E. Burke, P. Le Breton, A. Williamson.
Planet. Space Sci., Vol. 22, 143 - 149 (1974).

The cross-section for dissociative photoionization of hydrogen by 584 Å radiation has been measured, yielding a value of 5×10^{-20} cm^2. The process can be explained as a transition from the $X^1\Sigma_g^+$ ground state to a continuum level of the $X^2\Sigma_g^+$ ionized state of H_2. The branching ratio for proton (H^+) vs molecular ion (H_2^+) production at this energy is 8×10^{-3}. This process is quite likely an important source of protons in the Jovian ionosphere near altitudes where peak ionization rates are found.

099.013 **Pioneer 10 mission: Summary of scientific results from the encounter with Jupiter.** A. G. Opp.
Science, Vol. 183, 302 - 303 (1974). – Report.

099.014 **Preliminary Pioneer 10 encounter results from the Ames Research Center plasma analyzer experiment.** J. H. Wolfe, H. R. Collard, J. D. Mihalov, D. S. Intriligator.

Science, Vol. 183, 303 - 305 (1974).

Preliminary results from the Ames Research Center plasma analyzer experiment for the Pioneer 10 Jupiter encounter indicate that Jupiter has a detached bow shock and magnetopause similar to the case at earth but much larger in spatial extent. In contrast to earth, Jupiter's outer magnetosphere appears to be highly inflated by thermal plasma and therefore highly responsive in size to changes in solar wind dynamic pressure.

099.015 **Magnetic field of Jupiter and its interaction with the solar wind.** E. J. Smith, L. Davis, Jr., D. E. Jones, D. S. Colburn, P. J. Coleman, Jr., P. Dyal, C. P. Sonett.
Science, Vol. 183, 305 - 306 (1974).

Preliminary results of the Pioneer 10 vector helium magnetometer experiment.

099.016 **Protons and electrons in Jupiter's magnetic field: Results from the University of Chicago experiment on Pioneer 10.** J. A. Simpson, D. Hamilton, G. Lentz, R. B. McKibben, A. Mogro-Campero, M. Perkins, K. R. Pyle, A. J. Tuzzolino, J. J. O'Gallagher.
Science, Vol. 183, 306 - 309 (1974).

The University of Chicago's experiment on Pioneer 10 measured not only the relativistic electron flux but also the high energy proton flux about which nothing was previously known. In this report, the authors present some of the most immediate results and conclusions from their first examination of the data.

099.017 **Energetic electrons in the magnetosphere of Jupiter.** J. A. Van Allen, D. N. Baker, B. A. Randall, M. F. Thomsen, D. D. Sentman, H. R. Flindt.
Science, Vol. 183, 309 - 311 (1974).

Preliminary report of in situ observations of energetic electrons in the magnetosphere of Jupiter during the Pioneer 10 encounter 26 November to 11 December 1973.

099.018 **Energetic particle population in the Jovian magnetosphere: A preliminary note.**
J. H. Trainor, B. J. Teegarden, D. E. Stilwell, F. B. McDonald, E. C. Roelof, W. R. Webber.
Science, Vol. 183, 311 - 313 (1974).

Preliminary account of the Jovian encounter as viewed by the Pioneer 10 particle detector systems of Goddard Space Flight Center and the University of New Hampshire.

099.019 **Radiation belts of Jupiter.**
R. W. Fillius, C. E. McIlwain.
Science, Vol. 183, 314 - 315 (1974).

Pioneer 10 counted relativistic electrons throughout the magnetosphere of Jupiter, with the greatest fluxes being inside 20 Jupiter radii. The peak flux of electrons with energy greater than 50 million electron volts was 1.3×10^7 per square centimeter per second at the innermost penetration of the radiation belts.

099.020 **Pioneer 10 infrared radiometer experiment: Preliminary results.** S. C. Chase, R. D. Ruiz, G. Münch, G. Neugebauer, M. Schroeder, L. M. Trafton.
Science, Vol. 183, 315 - 317 (1974).

Thermal maps of Jupiter at 20 and 40 micrometers show structure closely related to the visual appearance of the planet. A preliminary discussion of the data, in terms of simple radiative equilibrium models, is presented.

099.021 **Pioneer 10 observations of the ultraviolet glow in the vicinity of Jupiter.** D. L. Judge, R. W. Carlson.
Science, Vol. 183, 317 - 318 (1974).

A two-channel ultraviolet photometer aboard Pioneer 10 has made several observations of the ultraviolet glow in the wavelength range from 170 to 1400 angstroms in the vicinity of Jupiter. Preliminary results are presented.

099.022 **The imaging photopolarimeter experiment on Pioneer 10.** T. Gehrels, D. Coffeen, M. Tomasko, L. Doose, W. Swindell, N. Castillo, J. Kendall, A. Clements, J. Hämeen-Anttila, C. K. Knight, C. Blenman, R. Baker, G. Best, L. Baker.
Science, Vol. 183, 318 - 320 (1974).

During the recent flyby of Jupiter, a large quantity of imaging and polarimetric data was obtained on Jupiter and the Galilean satellites over a wide range of phase angles.

099.023 **Photometry of Jupiter with interference filters.**
N. G. Litkevich, M. F. Khodyachikh.
Vestn. Khar'kov. Univ. No. 99 (Ser. Astron. No. 8), p. 18 - 26 (1973). In Russian.

099.024 **Erste Resultate der Pionier-10-Sonde: Zu heißer Sonnenwind jenseits der Marsbahn – Gigantischer Magnetschirm um den Jupiter.** H. J. Fahr.
Phys. Blätter, 30. Jahrgang, p. 66 - 68 (1974).

099.025 **Pionier 10 am Planeten Jupiter. Übersicht über die Messungen und erste wissenschaftliche Ergebnisse.** H. W. Köhler.
Phys. Blätter, 30. Jahrgang, p. 161 - 167 (1974).

099.026 **Hyperfine structure in the radio spectra of Jupiter.**
G. R. A. Ellis.
Proc. Astron. Soc. Australia, Vol. 2, 191 - 193 (1973). – Presented at the annual general meeting of the Astron. Soc. Australia – see 012.001.

099.027 **Jupiter and Saturn.** R. Hide.
The planets today. Symposium 1973, (see 012.003), p. 63 - 84 (1974).

099.028 **Cloud base levels for Jupiter and Venus and the heteromolecular nucleation theory.**
D. Stauffer, C. S. Kiang.
Icarus, Vol. 21, 129 - 146 (1974).

Recent interest on the formation of clouds centered on the question if they consist of binary mixture droplets. The authors calculate here from classical homogeneous heteromolecular nucleation theory the threshold partial pressures necessary to achieve droplet nucleation for the gas mixtures NH_3-H_2O (Jupiter), $HCl-H_2O$ and $H_2SO_4-H_2O$ (Venus), and $C_2H_5OH-H_2O$ (laboratory). In the last case, theory and experiment agree satisfactorily.

099.029 **Significance of gravitational moments for interior structure of Jupiter and Saturn.**
W. B. Hubbard, V. P. Trubitsyn, V. N. Zharkov.
Icarus, Vol. 21, 147 - 151 (1974).

The authors present the case for liquid models of Jupiter and Saturn. They then discuss the information which can be obtained about their interior structure from a knowledge of their gravitational moments. They also discuss the nature of presently available data and data which will be necessary to fully exploit future results.

099.030 **Determination of the equation of state of the molecular envelopes of Jupiter and Saturn from their gravitational moments.** V. N. Zharkov, V. P. Trubitsyn.
Icarus, Vol. 21, 152 - 156 (1974).

The authors develop a method for the determination of the density and equation of state of the matter in the outer layers of the planets Jupiter and Saturn using data for their external gravitational field.

099.031 Organic synthesis in a simulated Jovian atmosphere. III. Synthesis of aminonitriles.
P. M. Molton, C. Ponnamperuma.
Icarus, Vol. 21, 166 - 174 (1974).

The products from spark and semi-corona discharges through mixtures simulating the Jovian atmosphere were analyzed by gas chromatography combined with mass spectrometry. Although the spectra were not identical, there were notable similarities between these and the mass spectra of some compounds present in the Murray and Orgeuil meteorites. Aminonitriles may occur as minor constituents of the Jovian atmosphere and perhaps by cyclization may produce pyrimidines.

099.032 Vorläufige Ergebnisse der Jupiter-Erforschung durch Pioneer 10. E. Wiedemann.
Orion Schaffhausen, 32. Jahrgang, p. 15 - 16 (1974).

099.033 Jupiter, présentation 1972. S. Cortesi.
Orion Schaffhausen, 32. Jahrgang, p. 17 - 20 (1974).
Rapport No. 25 du «Groupement planétaire SAS».

099.034 Pioneer 10 am Jupiter vorbeigeflogen. Erste Ergebnisse. H. W. Köhler.
SuW, Vol. 13, 83 - 85 (1974).

099.035 Ammonia absorption relevant to the albedo of Jupiter. II. Interpretation. M. G. Tomasko.
Astrophys. Journ., Vol. 187, 641 - 650 (1974).

The ultraviolet albedo of Jupiter is computed by a modification of the "doubling" technique for some simple models of NH_3 inhomogeneously mixed with H_2 above an NH_3 cloud. The author discusses his results in the light of existing ultraviolet data. The Appendix contains some details of his layer adding method of computing the geometric albedo of an inhomogeneous atmosphere.

099.036 The electron diffusion coefficient in Jupiter's magnetosphere. T. Birmingham, W. Hess, T. Northrop, R. Baxter, M. Lojko.
Journ. Geophys. Res., Vol. 79, 87 - 97 (1974).

A steady state model of Jupiter's electron radiation belt is developed. The model includes injection from the solar wind, radial diffusion, energy degradation by synchrotron radiation, and absorption at Jupiter's surface.

099.037 The occultation of β Scorpii by Jupiter I. The structure of the Jovian upper atmosphere.
J. Veverka, L. H. Wasserman, J. Elliot, C. Sagan, W. Liller.
Astron. Journ., Vol. 79, 73 - 84 (1974).

Simultaneous records of the 13 May 1971 Jupiter occultation of Beta Scorpii AB and C were obtained at three wavelengths with a time resolution of 0.01 sec. An outstanding feature of these events was the occurrence of numerous light flashes, or "spikes" (Veverka et al. 1972). In this paper the authors show that the spikes can be understood as small fluctuations in the refractivity profile, and from the bright star emersion data they derive the structure of the upper atmosphere of Jupiter.

099.038 Jupiter, Saturn, and Uranus disk temperature measurements at 2.07 and 3.56 cm. B. L. Gary.
Astron. Journ., Vol. 79, 318 - 320 (1974).

Observations at 2.07 cm with Goldstone 64-m (210-ft) antenna yield the following disk temperatures: Jupiter = $173.4 \pm 5.2°K$, Saturn = $162.3 \pm 3.8°K$, and Uranus = $178.7 \pm 12.9°K$. An observation of Saturn at 3.56 cm with the same antenna yields a disk temperature of $169.7 \pm 2.3°K$.

099.039 Leeft er iets op Jupiter? T. de Vries.
Zenit, Vol. 1, No. 1, p. 10 - 12 (1974).

099.040 Jupiter's ionosphere: prospects for Pioneer 10.
S. K. Atreya, T. M. Donahue, M. B. McElroy.
Science, Vol. 184, 154 - 156 (1974).

Model Jovian ionospheres are constructed for comparison with Pioneer 10 results. Electron density maxima are predicted at a level approximately 220 kilometers above an assumed reference height where the hydrogen density is 10^{16} molecules per cubic centimeter. It may be possible to use observations of the electron density to locate the turbopause. Attention is drawn to a possible strong source of ionized sodium from Io which might lead to large electron densities at low altitudes.

099.041 Radiative-dynamical equilibrium states for Jupiter.
L. M. Trafton, P. H. Stone.
Astrophys. Journ., Vol. 188, 649 - 655 (1974).

In order to obtain accurate estimates of the radiative heating that drives motions in Jupiter's atmosphere, previous radiative equilibrium calculations are improved by including the NH_3 opacities and updated results for the pressure-induced opacities. The radiative-convective equilibrium temperature structure consistent with these changes is calculated. The radiative equilibrium calculations are used to calculate whether equilibrium states can occur on Jupiter which are similar to the baroclinic instability regimes on the earth and Mars.

099.042 Five-micron pictures of Jupiter.
J. A. Westphal, K. Matthews, R. J. Terrile.
Astrophys. Journ., (*Letters*), Vol. 188, L111 - L112 (1974).

More than 440 five-micron "video" pictures of Jupiter with $1''$ resolution were made during 1973 September, October, and December. Comparisons of these pictures with color photographs show direct, detailed correlations with the darker "purple" features. Forty-four of these pictures were made just before Pioneer 10 encounter.

099.043 On the formation of Jupiter and Saturn.
S. S. Kumar.
Astrophys. Space Sci., Vol. 28, 173 - 176 (1974).

It is proposed that Jupiter and Saturn were initially formed as small rocks which grew into their present sizes as a result of accretion of matter from the gas-dust cloud surrounding the sun. The energy released by the accretion of Jupiter and Saturn is computed. It is concluded that the 'excess' radiation from these planets is due to simple cooling and that the gravitational contraction from initially extended states most probably never occurred.

099.044 Stably trapped proton fluxes in the Jovian magnetosphere. F. V. Coroniti, C. F. Kennel, R. M. Thorne.
Astrophys. Journ., Vol. 189, 383 - 388 (1974).

A model of the energetic proton fluxes in the Jovian magnetosphere is constructed based on the inward radial diffusion of protons from the solar wind and the plasma turbulent precipitation loss of protons from the radiation belts.

099.045 The atmospheres of Jupiter and Saturn.
G. E. Hunt.
Endeavour, No. 118, Vol. 33, 23 - 28 (1974).

The giant planets Jupiter and Saturn, which lie beyond the asteroid belt, are very different from the inner terrestrial planets. They are huge, low density, rapidly rotating bodies with a permanently banded appearance of changing colours. Their atmospheres are composed mainly of hydrogen and helium in amounts similar to those found in the sun. Traces of ammonia and methane have also been detected.

099.046 Dekameterstrålning från Jupiter.
L. Lindegren.
Astron. Tidssk., Årg. 7, p. 75 - 85 (1974).

099.047 Oscillations trimestrielles de la Tache Rouge de

Jupiter. F. Link.
Comptes Rendus Acad. Sci. Paris, Sér. B, Vol. 278, 993 - 994 (1974).

Les oscillations trimestrielles de la Tache Rouge sur Jupiter paraissent être synchronisées par les conjonctions inférieures de Mercure.

099.048 **Structure of the visible Jovian clouds.**
G. E. Hunt, J. T. Bergstralh.
Nature, Vol. 249, 635 - 637 (1974).

During the 1972 apparition of Jupiter the authors studied hydrogen quadrupole line intensities at the centre of the planetary disk, by observing the S (1) lines of the H_2 (3-0) and (4-0) quadrupole bands which are respectively centred at 12268.8 and 15704.01 cm^{-1}. These observations show that the abundance of hydrogen detected at these frequencies is extremely variable because of rapid changes in the structure of the visible clouds, which are confirmed by photographs of the planet.

099.049 **The occultation of Beta Scorpii by Jupiter. III. Simultaneous high time-resolution records at three wavelengths.** W. Liller, J. L. Elliot, J. Veverka, L. H. Wasserman, C. Sagan.
Icarus, Vol. 22, 82 - 104 (1974).

Simultaneous high time-resolution records at three wavelengths of the May 13, 1971 Jupiter occultation of Beta Scorpii are presented. These observations are unique and contain important information about the structure and composition of the Jovian atmosphere.

099.050 **The occultation of β Scorpii by Jupiter. IV. Diurnal temperature variations and the methane mixing ratio in the Jovian upper atmosphere.** L. H. Wasserman.
Icarus, Vol. 22, 105 - 110 (1974).

The nighttime cooling of the Jovian atmosphere near the occultation level of 10^{14} cm^{-3} is calculated using the models of Strobel (1973) and Strobel and Smith (1973). It is consistent with the commonly accepted CH_4 mixing ratio of 7×10^{-4}.

099.051 **Jovian atmosphere: structure and composition between the turbopause and the mesopause.**
C. Sagan, J. Veverka, L. Wasserman, J. Elliot, W. Liller.
Science, Vol. 184, 901 - 903 (1974).

The occultation of the star Beta Scorpii by Jupiter was observed at high time resolution in three wavelength channels. The results suggest that the vertical eddy diffusion coefficient near the turbopause has a lower limit of 7×10^5 square centimeters per second, and that the turbopause lies above the altitude where the density is 5×10^{13} molecules per cubic centimeter. Below the turbopause, the ratio of hydrogen to helium is consistent with cosmic abundances.

099.052 **Energetic electrons in Jupiter's magnetosphere.**
F. V. Coroniti.
Astrophys. Journ., Suppl. Ser., No. 244, Vol. 27, 261 - 281 (1974).

A theoretical model for the energetic electron fluxes in the Jovian magnetosphere is developed. Electrons are transported inward from the solar wind or Jovian magnetospheric tail by radial diffusion. The radial diffusion is driven by fluctuating ionospheric dynamo electric fields associated with a neutral-wind tidal eigenmode at ionospheric altitudes. The tidal mode is excited by the electromagnetic coupling of the solar wind to the polar ionosphere. Two injection models are considered: (1) electron penetration through the dayside magnetopause–low-energy model; and (2) injection of electrons from an assumed magnetospheric tail–high-energy model. Both thermal solar-wind electrons and energetic solar-flare electrons are considered.

099.053 **Jupiter: Présentation 1973.** F. Jetzer.
Orion, 32. Jahrgang, p. 106 - 110 (1974). – Rapport No. 27 du Groupement planétaire SAS.

099.054 **Origin of Jovian decameter wave emissions – conversion from the electron cyclotron plasma wave to the ordinary mode electromagnetic wave.** H. Oya.
Planet. Space Sci., Vol. 22, 687 - 708 (1974).

Energy conversion rates from the extraordinary mode to the ordinary mode of the electromagnetic waves in the Jovian plasmasphere have been calculated for a model of the sharp boundary that is given in the vicinity of the position where $\omega = \omega_p$, for an angular frequency ω and the angular plasma frequency ω_p. The results give conversion rates of 1–50 per cent, at the most, when a wave normal direction of an incident wave is nearly parallel to the boundary normal direction and when the Jovian magnetic field vector is close to the boundary normal direction within an angle range from 10° to 15°.

099.055 **Was Jupiter the protosun's core?**
E. M. Drobyshevski.
Nature, Vol. 250, 35 - 36 (1974). – Letter.

099.056 **Computer simulation of the Pioneer 10 microwave occultation experiment.** S. G. Ungar.
Journ. Geophys. Res., Vol. 79, 1969 - 1973 (1974). – Brief report.

099.057 **The Jovian radiation belt.**
D. B. Beard, J. L. Luthey.
Astrophys. Space Sci. Library, Vol. 42, (see 012.016), 641 - 647 (1974).

099.058 **Jupiter's radiation belts.**
K. G. Stansberry, R. S. White.
Journ. Geophys. Res., Vol. 79, 2331 - 2342 (1974).

Fluxes of electrons and protons in Jupiter's radiation belts are calculated with the source (radial diffusion inward from the solar wind) and the loss (synchrotron radiation). The calculations are tested against the measured radio wave wavelength distribution, the radio wave distribution with distance from Jupiter, and the degree of polarization of the radio waves. The Fokker-Planck equation is solved by using the method of Farley and Walt with the fixed flux at the outer boundary and the zero flux at Jupiter's surface.

099.059 **Implications of the Pioneer 10 measurements of the Jovian magnetic field for theories of Io-modulated decametric radiation.** R. A. Smith, C. S. Wu.
Astrophys. Journal, (Letters), Vol. 190, L91 - L95 (1974).

The configuration of the magnetic field of Jupiter imposes constraints upon theoretical models of decametric emission triggered by Io, with respect to such features as the frequency of the emission, the density of the inner plasmasphere, and the beaming pattern.

099.060 **An interferometric study of Jupiter at 3.7 and 11.1 cm.** E. T. Olsen.
Bull. American Astron. Soc., Vol. 6, 266 (1974). – Abstr. AAS.

099.061 **Jupiter's clouds: equatorial plumes and other cloud forms in the Pioneer 10 images.**
J. W. Fountain, D. L. Coffeen, L. R. Doose, T. Gehrels, W. Swindell, M. G. Tomasko.
Science, Vol. 184, 1279 - 1281 (1974).

Pioneer 10 images of Jupiter show bright nuclei in the equatorial zone that appear to be thermally driven sources of cloud plume formations.

099.062 **Models of the giant planets.**
M. Podolak, A. G. W. Cameron.

Icarus, Vol. 22, 123 - 148 (1974).

Models of the giant planets were constructed based on the assumption that the hydrogen to helium ratio is solar in these planets. This assumption, together with arguments about the condensation sequence in the primitive solar nebula, yields models with a central core of rock and possibly ice surrounded by an envelope of hydrogen, helium, methane, ammonia, and water. Jupiter was found to have a core of about 40 earth masses and a water enhancement in the atmosphere of about 7.5 times the solar value. Saturn was found to have a core of 20 earth masses and a water enhancement in the atmosphere of about 25 times the solar value. Rock plus ice constitute 75–85% of the mass of Uranus and Neptune. Temperatures in the interiors of these planets are probably above the melting points. Some aspects of the sensitivities of these results to uncertainties in rotational flattening are discussed.

099.063 **The significance of pressure shifts for the interpretation of H_2 quadrupole lines in planetary spectra.** A. R. W. McKellar.
Icarus, Vol. 22, 212 - 219 (1974).

The effect of pressure shifts on the formation of H_2 quadrupole lines in planetary atmospheres is considered and shown to be negligible for Jupiter, small for Saturn, and rather large for Uranus.

099.064 **Possible H_2^+ ultraviolet spectrum of Jupiter.**
F.-M. Wu, C. L. Beckel, M. Shafi, C. L. Hyder.
Icarus, Vol. 22, 220 - 223 (1974).

An ultraviolet spectral probe for a hydrogen-rich planetary atmosphere, such as that of Jupiter, is suggested, utilizing discrete lines in the $H_2^+ 2p\pi_u - 1s\sigma_g$ electronic transition. For the Jovian atmosphere, the dominant mechanism for exciting H_2^+ to its $2p\pi_u$ state appears to be photoexcitation, principally through absorption of the solar Lyman-α line.

099.065 **A preliminary dynamic view of the circulation of Jupiter's atmosphere.** V. P. Starr.
Pure and Applied Geophys., (Switzerland), Vol. 110, 2108 - 2129 (1973).

The problem of the gross nature of the Jovian atmospheric circulation is examined. Since there is evidence that Jovian dark spots have statistical maxima of occurrence along the tropical shear lines flanking the equator, these are assumed to be vertical convective systems forming, in effect, convective vortex sheets which generate the high angular momentum of the equatorial zone. Various additional concepts are discussed, and many comparisons with conditions in the sun and in the earth's atmosphere are made.

099.066 **The photochemistry of NH_3 in the Jovian atmosphere.** D. F. Strobel.
Journ. Atmosph. Sci., Vol. 30, 1205 - 1209 (1973).

A quantitative study of the photochemistry of NH_3 above the Jovian tropopause is given. The NH_3 density distribution is described by two relevant vertical scales: the scale height of the background atmosphere, and the 'photomechanical' scale height.

099.067 **Abundance of NH_3 on Jupiter inferred from microwave radiometry data.** R. J. Richardson.
Journ. Spacecraft and Rockets, (USA), Vol. 10, 647 - 651 (1973).

Measurements of the thermal disk temperature of Jupiter have been made at several microwave frequencies. NH_3 abundance in a number of possible models of the Jupiter atmosphere were determined by computing disk temperature vs frequency curves for each model with varying NH_3 abundance and comparing these results with the measured disk temperatures. Microwave absorption losses were also calculated for these model atmospheres.

099.068 **Results of the Jupiter investigation.** H. W. Kohler.
Techn. Rundschau, (Switzerland), Vol. 66, No. 10, p. 27 (1974). In German.

099.069 **Jovian colors in 1973.** P. K. Mackal.
Strolling Astronomer, Vol. 24, 213 - 217 (1974).

099.070 **Jupiter in 1968–1969: rotation periods.**
P. W. Budine.
Strolling Astronomer, Vol. 24, 228 - 235 (1974).

099.071 **Jupiter in 1969–70: rotation periods.**
P. W. Budine.
Strolling Astronomer, Vol. 24, 236 - 239 (1974).

099.072 **Addendum to Mr. Budine's transit report for 1969–70.** P. K. Mackal.
Strolling Astronomer, Vol. 24, 239 - 240 (1974).

099.073 **Jovian red spots in 1973–74.** C. F. Capen.
Strolling Astronomer, Vol. 24, 240 - 241 (1974).

099.074 **The occultation of Beta Scorpii by Jupiter. II. The hydrogen-helium abundance in the Jovian atmosphere.** J. L. Elliot, L. H. Wasserman, J. Veverka, C. Sagan, W. Liller.
Astrophys. Journ., Vol. 190, 719 - 729 (1974).

The helium abundance in the Jovian atmosphere has been determined by a new method from the authors' observations of the occultation of β Scorpii by Jupiter. A new method for determining the composition of a hydrogen-helium atmosphere from the data is presented, and the principal sources of error are discussed and evaluated. The derived helium fraction is $f(He) = 0.16 (+0.19, -0.16)$ by number.

099.075 **Models of Jupiter and Saturn.**
V. N. Zharkov, A. B. Makalkin, V. P. Trubitsyn.
Astron. Tsirk., No. 812, p. 1 - 2 (1974). In Russian.

099.076 **On the composition of Jupiter and Saturn.**
V. N. Zharkov, A. B. Makalkin, V. P. Trubitsyn.
Astron. Tsirk., No. 812, p. 3 - 4 (1974). In Russian.

099.077 **Configuration of the Jovian magnetosphere.**
T. W. Hill, A. J. Dessler, F. C. Michel.
Geophys. Res. Letters, Vol. 1, No. 1, p. 3 - 6 (1974).

We present a model in which the Jovian magnetosphere is severely inflated by the centrifugal stress of partially-corotating plasma streaming out along field lines from the ionosphere. The model is consistent with observations reported from the Pioneer 10 encounter, including the disk-like field configuration, the diurnal modulation of trapped-particle fluxes, and the inferred departure from rigid corotation in the outer magnetosphere.

Jupiter, the largest planet, may also be most complicated.
IEEE Spectrum, Vol. 11, No. 3, p. 26 (1974).

Jupiter: the largest planet.
See Abstr. 003.021.

Report from Jupiter. Part 1. See Abstr. 053.011.

Synchrotron loss rates of energetic magnetospheric electrons. See Abstr. 084.297.

Inversion of gravity data for giant planets.
See Abstr. 091.007.

Evaluation of intensity of resonance line emission

of inert gases in planetary atmospheres. See Abstr. 097.001.

Recent photographs of Mars and Jupiter. See Abstr. 097.008.

Color pictures of planets from black-and-white images. See Abstr. 097.009.

Commensurabilities with Jupiter in the relation σ_i (a) for minor planets. See Abstr. 098.019.

Dispersion of the inclinations of minor planets with respect to Jovian plane. See Abstr. 098.020.

The chemistry of the solar system. See Abstr. 107.003.

Jupiter, Satellites

099.201 **Mutual phenomena of Jupiter's Galilean satellites, 1973–74.** K. Aksnes.
Icarus, Vol. 21, 100 - 111 (1974).

Two series of predictions have been published for the 1973–74 mutual phenomena of Jupiter's satellites, one (June–October, 1973) by Milbourn and Carey, and the other (February, 1973–May, 1974) by Brinkmann and Millis. The main purpose of this paper is to investigate some significant discrepancies between these two sets of predictions. New predictions are calculated for the period June 1973–May, 1974. The paper concludes with a brief discussion of the problems involved in extracting information about the positions, radii, and albedos of the satellites from observed light curves.

099.202 **On the principle of least interaction action and the Laplacean satellites of Jupiter and Uranus.**
M. W. Ovenden, T. Feagin, O. Graf.
Celestial Mechanics, Vol. 8, 455 - 471 (1974).

The principle of least interaction action, which explains the observed preference in the solar system for two-satellite resonant configurations, is shown to apply also to the Laplacean satellites of Jupiter and Uranus, in the sense that these triplet resonant structures lie close to configurations for which the time-mean of the action associated with the mutual interaction of the satellites is an overall minimum.

099.203 **Moons of Jupiter: Io seems to play an important role.** W. D. Metz.
Science, Vol. 183, 293 (1974). – Research news.

099.204 **Gravitational parameters of the Jupiter system from the Doppler tracking of Pioneer 10.**
J. D. Anderson, G. W. Null, S. K. Wong.
Science, Vol. 183, 322 - 323 (1974).

Preliminary analyses of Doppler data from Pioneer 10 during its encounter with Jupiter indicate that the mass of Io is about 20 percent greater than previously thought and that Io's mean density is about 3.5 grams per cubic centimeter. A determination of the dynamical flattening of Jupiter is found to lie in the neighborhood of 0.065, which agrees with the value determined from satellite perturbations.

099.205 **Preliminary results on the atmospheres of Io and Jupiter from the Pioneer 10 S-band occultation experiment.**
A. Kliore, D. L. Cain, G. Fjeldbo, B. L. Seidel, S. I. Rasool.
Science, Vol. 183, 323 - 324 (1974). – Report.

099.206 **A method of revitalizing Sampson's theory of the Galilean satellites.** J. H. Lieske.
Astron. Astrophys., Vol. 31, 137 - 150 (1974).

A method is developed by which Sampson's theory of motion of Jupiter's Galilean satellites can be revitalized by use of algebraic manipulation software on a digital computer. The technique seeks (a) to remove algebraic errors existing in the current Sampson theory, (b) to introduce some neglected effects due to solar interactions and the 3 : 7 commensurability between the outer two satellites, (c) to allow for non-zero amplitude and phase of the libration, (d) to allow future revision of the arbitrary constants of integration, (e) to express the final results as analytic functions of variations in the numerous arbitrary constants of integration and arbitrary parameters, and (f) to provide analytic partial derivatives by means of which one can adjust the numerical values of coefficients in the expressions for the coordinates. The level of precision desired is one arc second (jovicentric) for the coordinates (2 km, 3.3 km, 5.2 km and 9.1 km, respectively, for satellites I through IV).

099.207 **12-micron emission features of the Galilean satellites and Ceres.** O. L. Hansen.
Astrophys. Journ., (Letters), Vol. 188, L31 - L33 (1974).

Broad-band spectra between 8 and 20 μ have been obtained of the Galilean satellites Io, Ganymede, and Callisto, as well as of four asteroids: 1 Ceres, 10 Hygiea, 39 Laetitia, and 40 Harmonia. Analysis by spectral ratios indicates that the Galilean satellites and Ceres exhibit emission features near 12 μ.

099.208 **Computerized reconstruction of Sampson's theory.**
J. H. Lieske.
Bull. American Astron. Soc., Vol. 6, 207 (1974). – Abstr. AAS.

099.209 **High-resolution spectra of sodium emission from Io.**
R. A. Brown, F. H. Chaffee, Jr.
Astrophys. Journ., (Letters), Vol. 187, L125 - L126 (1974).

Four high-resolution spectra of sodium D-line emission from Io are analyzed for Doppler shift, line width, and equivalent width.

099.210 **Sodium emission from Io: Implications.**
M. B. McElroy, Y. L. Yung, R. A. Brown.
Astrophys. Journ., (Letters), Vol. 187, L127 - L130 (1974).

The surface of Io may be covered with a layer of ammonia ice containing trace amounts of sodium, potassium, and calcium; and atmospheric nitrogen could be formed as a photochemical product of ammonia photolysis.

099.211 **Eclipse of Ganymede by Callisto.**
A. R. Walker.
Monthly Notes Astron. Soc. Southern Africa, Vol. 33, 27 - 30 (1974).

099.212 **Implications of Jupiter's early contraction history for the composition of the Galilean satellites.**
J. B. Pollack, R. T. Reynolds.
Icarus, Vol. 21, 248 - 253 (1974).

Graboske et al. (1973) have shown that Jupiter's luminosity was orders of magnitude larger during its initial contraction phase than it is today. As a result, during Jupiter's earliest contraction history, ices would have preferentially been prevented from condensing within the region containing the orbits of the inner satellites. The observed variation of the mean density of the Galilean satellites with distance from Jupiter implies that the satellite formation process was operative on a time scale of about five million years. Another consequence of the high luminosity phase is that water should be the only ice present in significant proportions in any of the Galilean satellites.

099.213 **Positional observations of Ganymede during the occultation of the star SAO 186658.**
O. R. Bolkvadze, R. I. Kiladze.
Byull. Abastumansk. Astrofiz. Obs., No. 45, p. 75 - 76 (1974). In Russian.

Photographic observations of Ganymede were carried out on 19 June 1972. The differential positions of Ganymede referred to the occulted star are determined at 9 moments.

099.214 **Frost spectra: Comparison with Jupiter's satellites.**
H. H. Kieffer, W. D. Smythe.
Icarus, Vol. 21, 506 - 512 (1974).

The near infrared spectral reflectance of pure CH_4, CO_2, H_2O, H_2S, NH_3 and NH_4SH frosts has been measured. Comparison with recent Galilean satellite spectra indicates that H_2O at approximately 150°K and with about 0.01 cm grain size is the major component on the surface of JII (Europa) and JIII (Ganymede). Materials other than these frosts must be present on the surface of JI (Io) and JIV (Callisto).

099.215 **Périodicités des phénomènes des satellites de Jupiter.** J. Meeus.
Ciel et Terre, Vol. 90, 123 - 135 (1974).

099.216 **Io as an emitter of 100-keV electrons.**
R. F. Hubbard, S. D. Shawhan, G. Joyce.
Journ. Geophys. Res., Vol. 79, 920 - 928 (1974).

If the surface conductivity of Jupiter's satellite Io is greater than $5 \times 10^{-11} \Omega^{-1} cm^{-1}$ and if the height-integrated Pederson conductivity of the Jovian ionosphere is greater than 0.6 mho, then it is likely that significant Debye and photoelectron sheaths form around Io. The Io sheath model, originally proposed by Gurnett (1972) and further developed by Shawhan et al. (1972) and Hubbard (1972), is used to produce a quantitative estimate of the emitted particle energy spectrum. This spectrum has energies up to several hundred keV because photoelectrons emitted from the surface of Io are accelerated across a sheath through a potential characteristic of the motional emf developed across Io (~670 kV).

099.217 **Observations of mutual phenomena of Galilean satellites. (1).** T. Nakamura.
Tokyo Astron. Bull., Second Ser., No. 231, p. 2655 - 2674 (1974).

Mutual phenomena of Galilean satellites in 1973 were observed, and six light curves for four phenomena could be obtained at Nara besides at Tokyo Astronomical Observatory. A computer program to provide an ephemeris of the satellites based on Sampson's original theory was also constructed in order to compare the observations with the theory, especially with respect to the central times of the phenomena.

099.218 **The spatial extent of sodium emission around Io.**
L. Trafton, T. Parkinson, W. Macy, Jr.
Astrophys. Journal, (*Letters*), Vol. 190, L85 - L89 (1974).

The sodium D emissions associated with Io's spectrum also originate in a large volume of space surrounding the satellite, extending more than 10 arc sec in radius. Although this region extends well beyond the equilibrium point between Jupiter and Io, no emission torus has been detected around Jupiter. The emission, however, appears to be stronger close to Io's orbital plane and especially on the Jovian side of Io.

099.219 **Mutual eclipses of Jupiter's satellites.**
C. Blanco, S. Catalano.
Astron. Astrophys., Vol. 33, 303 - 305 (1974).

Photoelectric *UBV* light curves of three mutual eclipses of Jupiter's satellites in 1973 are given. The times of occurrence and the durations generally agree with the predictions, while the observed light losses show appreciable differences from the predicted ones. The discrepancies in the eclipse depths are interpreted in terms of bright polar caps.

099.220 **The Galilean satellites.** R. L. Millis.
Mercury, (Journ. Astron. Soc. Pacific), Vol. 3, No. 1, p. 3 - 7 (1974).

099.221 **Mouvement des satellites galiléens de Jupiter: mise sur ordinateur de la théorie de Sampson.**
D. T. Vu, J. L. Sagnier.
Bull. Groupe Recherches Géod. Spatiale, No. 11, 3+67 pp. (1974).

A transformation of Sampson's theory leading to a straightforward computation of better prediction of positions appears with some urgency as an efficient goal. The authors hope to obtain satisfactory results by programming all the steps of Sampson's theory in the original way. They present the present state of the programs with a description of necessary modifications in the original theory.

099.222 **Jupiter I (Io).** R. A. Brown.
IAU Circ., No. 2682 (1974).

099.223 **Die Galilei-Monde des Jupiter.** J. Chodak.
Ideen Exakten Wiss., [Deutsche Verlags-Anstalt, Stuttgart], 1973, No. 10, p. 613 - 619.

Jupiter mit seinen vier größten Monden. Ein neues Zusatzgerät aus Jena für Planetarien. See Abstr. 009.009.

Physical properties of the natural satellites. See Abstr. 091.004.

Satellites and magnetospheres of the outer planets. See Abstr. 091.013.

Implications of the Pioneer 10 measurements of the Jovian magnetic field for theories of Io-modulated decametric radiation. See Abstr. 099.059.

On the photometric variations of the Saturn and Jupiter satellites. See Abstr. 100.210.

Errata

099.901 **Erratum: 'Spectral data for the v_2 bands of ammonia with applications to radiative transfer in the atmosphere of Jupiter'** [Journ. Quant. Spectrosc. Radiat. Transfer, Vol. 13, 1181 - 1217 (1973)]. F. W. Taylor.
Journ. Quant. Spectrosc. Radiat. Transfer, Vol. 14, 457 (1974).

100 Saturn, Saturn Satellites

Saturn

100.001 The 7.5- to 13.5-micron spectrum of Saturn.
F. C. Gillett, W. J. Forrest.
Astrophys. Journ., (Letters), Vol. 187, L37 - L39 (1974).
A medium-resolution ($\Delta\lambda/\lambda \approx 0.015$) spectrum of Saturn in the 7.5–13.5 μ range is presented. The observations are briefly discussed.

100.002 The atmosphere of Saturn from measurements in the near ultraviolet. V. G. Tejfel'.
Astron. vestn., Vol. 8, 3 - 12 (1974). In Russian.
The reflective properties of some regions of Saturn's disk in ultraviolet are discussed. The optical thickness of the overcloud atmosphere is estimated from analysis of photographic and spectrometric observations at $\lambda\lambda$ 0.33–0.50 microns. The estimates have shown the presence of aerosol matter in Saturn's atmosphere over the theoretical upper boundary of the convective zone independent of the assumed H_2/He abundances. The latitudinal variations of the aerosol altitude are in good agreement with the results obtained from the measurements of methane absorption bands.

100.003 On the study of the rarefied outer ring of Saturn.
M. S. Bobrov.
Astron. vestn., Vol. 8, 13 - 16 (1974). In Russian.
Attention is called to the fact that the discovery of the rarefied outer ring of Saturn by Feibelman (1967) has been confirmed by the observations of Kuiper (1972). It is proposed to designate this object as E-ring in order to avoid confusion with the innermost, also rarefied, D-ring observed by Guérin (1970) and earlier by Barabashov and Semejkin (1933). The effects of the interaction of the E-ring with the inner Saturn satellites are briefly discussed. It is concluded that in cosmogonic time scale these effects are small. It is also shown that the optical thickness of the E-ring is lower than 5×10^{-5}.

100.004 Optical properties and structure of Saturn's atmosphere. III. The vertical structure of the aerosol layer derived from data of photographic and photoelectric spectrophotometry. V. G. Tejfel', G. A. Kharitonova.
Astron. Zhurn. Akad. Nauk SSSR, Vol. 51, 167 - 177 (1974). In Russian. English translation in Soviet Astron. AJ, Vol. 18, No. 1.

100.005 Extended glow preceding reappearance of Saturn during a lunar occultation.
G. Reed, F. J. Howell, T. A. Clark.
Nature, Vol. 247, 447 - 448 (1974). – Letter.

100.006 East-west asymmetry of Saturn's ring.
N. A. Kozyrev.
Astrophys. Space Sci., Vol. 27, 111 - 116 (1974).
A spectrophotometric comparison of the eastern and western parts of Saturn's ring was made using the photographs taken with the 50-in. reflector of the Crimean Astrophysical Observatory during 5 oppositions from 1968 to 1972. The excess of brightness, observable sometimes at the east, sometimes at the west, appears to be due to the following effects: the eastern excess of brightness, which does not depend on wavelength, changes with time and decreases with opening the ring and the western excess of brightness, which depends on wavelength, is best of all noticeable in the shortwave part of the spectrum but does not depend on the epoch of observations.

100.007 Photoelectric photometry of the disk center of Saturn. A. M. Gretskij.
Vestn. Khar'kov. Univ. No. 99 (Ser. Astron. No. 8), p. 64 - 67 (1973). In Russian.

100.008 The dynamical evolution of planetary rings.
A. W. Harris.
Bull. American Astron. Soc., Vol. 6, 207 (1974). – Abstr. AAS.

100.009 Comment on radar scattering from Saturn's rings.
G. H. Pettengill, T. Hagfors.
Icarus, Vol. 21, 188 - 190 (1974).
Transparent particle scattering is proposed to explain the unexpectedly large radar cross section recently observed for Saturn's rings. According to this theory, only 10% of the optically observed material in the A, B, and C rings need consist of smooth ice fragments larger than 8 cm in radius to yield the radar results.

100.010 The microwave properties of Saturn's rings.
F. H. Briggs.
Astrophys. Journ., Vol. 189, 367 - 377 (1974).
Radio interferometer observations at 1420, 2695, and 8085 MHz show that the ring particles are nearly lossless at microwave frequencies. The microwave brightness temperature of the rings is less than 20°K. The ability of the rings to obscure part of Saturn's disk indicates that most of the particles must be larger than 5 cm in diameter. A model assuming particles of water ice is consistent with the radio interferometer, radar, and millimeter wave observations.

100.011 Thermal emission of Saturn's rings and disk at 34 μm.
I. G. Nolt, J. V. Radostitz, R. J. Donnelly, R. E. Murphy, H. C. Ford.
Nature, Vol. 248, 659 - 660 (1974).
Using new filter techniques for ground-based observations, the authors have measured the relative thermal emission from the rings and disk of Saturn in the band from 29 to 43 μm. This measurement provides an important constraint for separating the thermal properties of Saturn from those of its rings.

100.012 Saturn central meridian ephemeris: 1974.
J. E. Westfall.
Strolling Astronomer, Vol. 24, 182 - 184 (1974).

100.013 Infrared radiometry of the rings of Saturn.
D. Morrison.
Icarus, Vol. 22, 57 - 65 (1974).
Broad-band radiometry with a spatial resolution of 5 arc sec is presented of Saturn and its rings. The brightness temperature of the B ring is 96 ± 3°K at 20 μm and 91 ± 3°K at 11 μm. From differences in the thermal emission of the ansae, the author suggests that the leading side of the particles has higher albedo than the trailing side. A measured drop in temperature of the B ring following eclipse of 2.0 ± 0.5°K is consistent with radii for the ring particles of 2 cm or larger.

100.014 Microwave brightness of Saturn's rings.
J. N. Cuzzi, D. van Blerkom.
Icarus, Vol. 22, 149 - 158 (1974).
It is shown that a lower limit exists on the microwave brightness of the rings of Saturn, if they are assumed to be composed of Mie scatterers of geological composition. Implications for the multiple-scattering hypothesis of the radar cross section of the rings are noted.

100.015 Saturn's rings. I. Optical thickness of rings A, B, D and structure of ring B. I. R. Ferrín.

Icarus, Vol. 22, 159 - 174 (1974).

A photometric study of high-resolution ($\sim 0\rlap{.}''3$) plates of Saturn taken at the Lowell Observatory in 1943 and 1945 is presented. N–S scans were taken over both the planet and rings. The excess brightness due to the planet seen through the rings is found by taking the difference between the central meridian (CM) scans and scans displaced by $5\rlap{.}''7$. Adopting a value for the albedo of the planet, it is possible to obtain the optical thickness $\tau_{CM}(r)$. Ring B exhibits a pronounced (7–10%) decrease in brightness from the extremity of the major axis to the CM. The author concludes that the ring particles are nonspherical and are in synchronous rotation around the planet with their long axis toward it.

100.016 **Radar contact with Saturn: clusters of rock fragments in the rings.**
Techn. Rundschau, (*Switzerland*), Vol. 66, No. 6, p. 31 (1974). In German.

Saturn and his anses. See Abstr. 004.033.

Rotación de la tierra año 1972. Resultados obtenidos en San Fernando con el Astrolabio Impersonal Danjon OPL No. 37. I. – Tiempo y latitud. II. – Observaciones de Saturno. See Abstr. 044.011.

Inversion of gravity data for giant planets. See Abstr. 091.007.

The optical properties of Venus and the Jovian planets. II. Methods and results of calculations of the intensity of radiation diffusely reflected from semi-infinite homogeneous atmospheres. See Abstr. 091.015.

Brightness temperatures of Venus, Saturn and Mercury at 3.87 mm wavelength. See Abstr. 093.001.

Saturnbedeckung vom 11. Dezember 1973. See Abstr. 096.018.

Internal constitution and the figures of the giant planets. See Abstr. 099.004.

Investigation of Jupiter and Saturn with the 2-m reflector of the Shemakha Astrophysical Observatory. See Abstr. 099.005.

The inner structure of giant planets. See Abstr. 099.009.

Jupiter and Saturn. See Abstr. 099.027.

Significance of gravitational moments for interior structure of Jupiter and Saturn. See Abstr. 099.029.

Determination of the equation of state of the molecular envelopes of Jupiter and Saturn from their gravitational moments. See Abstr. 099.030.

Jupiter, Saturn, and Uranus disk temperature measurements at 2.07 and 3.56 cm. See Abstr. 099.038.

On the formation of Jupiter and Saturn. See Abstr. 099.043.

The atmospheres of Jupiter and Saturn. See Abstr. 099.045.

Models of the giant planets. See Abstr. 099.062.

The significance of pressure shifts for the interpretation of H_2 quadrupole lines in planetary spectra. See Abstr. 099.063.

Models of Jupiter and Saturn. See Abstr. 099.075.

On the composition of Jupiter and Saturn. See Abstr. 099.076.

Saturn, Satellites

100.201 **On the origin of the commensurabilities amongst the satellites of Saturn–II.** A. T. Sinclair.
Monthly Notices Roy. Astron. Soc., Vol. 166, 165 - 179 (1974).

The possible types of resonance motion involving the inclinations of a pair of commensurable satellites are discussed, and the tidal evolution of the system when close to the commensurability is described. An explanation is given of how the satellites Enceladus and Dione could have avoided capture into one of the inclination-type resonances, a necessary condition if the tidal hypothesis of the origin of the eccentricity-type resonance that actually exists between these satellites is to be accepted.

100.202 **Titan: unidentified strong absorptions in the photometric infrared.** L. Trafton.
Icarus, Vol. 21, 175 - 187 (1974).

Titan's bands exhibit a progressive washing out with increasing wavelength relative to the corresponding bands in Saturn's spectrum. Unidentified absorption features of apparently gaseous origin appear in the long-wavelength wing of Titan's 1-μm CH_4 band complex with considerably greater strength. Greater absorption is also observed in the 1.1 μm region. Not all of the phenomena can be explained by enhanced CH_4 absorption, so Titan's spectrum implies the existence of another spectroscopically active gas. Several candidates for this gas are briefly considered.

100.203 **On the formation of the orbit–orbit resonance of Titan and Hyperion.**
G. Colombo, F. A. Franklin, I. I. Shapiro.
Astron. Journ., Vol. 79, 61 - 72 (1974).

The authors consider two possibilities for the formation of the orbit–orbit resonance of Titan and Hyperion. The first involves an approximate determination of the region in phase space corresponding to a stable libration at the 4:3 resonance. The second possibility concerns the evolution of the system from an initially circulatory state to a libratory state through tidal interaction.

100.204 **Phenomena of Saturn's satellites.** J. G. Porter.
Journ. British Astron. Ass., Vol. 84, 209 - 216 (1974). – Section report.

100.205 **The albedo of Titan.** R. L. Younkin.
Icarus, Vol. 21, 219 - 229 (1974).

The irradiance of Titan has been measured from 0.50 to 1.08 μ in 30 Å bandpasses spaced 0.01–0.02 μ apart. These narrow band photoelectric measurements of Titan were undertaken to obtain a more accurate representation of the methane band contours and strengths, values of the narrow band geometric albedo, a value for the bolometric albedo and hence the effective temperature, and to search for other possible broad shallow absorptions.

100.206 The photochemistry of hydrocarbons in the atmosphere of Titan. D. F. Strobel.
Icarus, Vol. 21, 466 - 470 (1974).

Detailed photochemical models are constructed for two model atmospheres: (1) 100% CH_4 and (2) 50% H_2, 50% CH_4. Both models predict large column densities of C_2H_2 and C_2H_6 (~1 cm atm) for eddy mixing rates ~10^5 cm^2 sec^{-1}, which are comparable to rates appropriate for Jupiter. The models confirm the interpretation by Danielson et al. (1973) of the 12μ feature in the spectra of Gillett et al. (1973) as emission by C_2H_6 in a thermal inversion region.

100.207 The radio brightness of Titan. F. H. Briggs.
Icarus, Vol. 22, 48 - 50 (1974).

Observations of Titan with the NRAO interferometer yield a brightness temperature of 115 ± 40°K at 8085 MHz, giving an estimate for the mean surface temperature of 135 ± 45°K.

100.208 Albedos and densities of the inner satellites of Saturn. D. Morrison.
Icarus, Vol. 22, 51 - 56 (1974).

Broad-band radiometry at 20 μm is presented of Rhea and Dione; the measured flux densities, together with visual photometry, indicate that both satellites have geometric albedos near 0.6 and that their radii are, respectively, 800 ± 125 and 575 ± 100 km. The density of Dione is 1.4 ± 0.6 g cm^{-3}; for Tethys, Enceladus, and Mimas the author shows that their densities are probably near unity. These satellites must therefore all be composed primarily of ices.

100.209 The atmosphere of Titan. D. M. Hunten.
Icarus, Vol. 22, 111 - 116 (1974).

A summary is given of our current knowledge of Titan's atmosphere, based on the report of the 1973 Titan Atmosphere Workshop.

100.210 On the photometric variations of the Saturn and Jupiter satellites. C. Blanco, S. Catalano.
Astron. Astrophys., Vol. 33, 105 - 111 (1974).

The results of UBV photoelectric observations of Jupiter's satellites I, II, III, and IV are given. The rotational phase angles of maximum and minimum light of Saturn's and Jupiter's satellites are found to be correlated with the distance from the relative primaries. The relation might be explained assuming that dark meteoroidal material, trapped by the planet, accumulates on a given portion of the satellite's surface determining a decrease in reflectivity of regions which were once covered with ice or snow.

100.211 The UBV orbital phase curves of Rhea, Dione, and Tethys. G. N. Blair, F. N. Owen.
Icarus, Vol. 22, 224 - 229 (1974).

Orbital phase curves are reported for Rhea (SV), Dione (SIV), and Tethys (SIII). Variations of 0.2 mag. and 0.8 mag. were found for Rhea and Dione respectively, while no variation of Tethys was detected. Some color variations are also reported.

100.212 An interaction effect between oblateness perturbation and mutual perturbation in the Enceladus-Dione system. T. Nakamura.
Proc. symposium "Celestial mechanics", Kyoto 1973, (see 012.022), p. 2 - 8 (1974). In Japanese.

100.213 Saturn VI. F. A. Franklin.
IAU Circ., No. 2628 (1974).

100.214 Saturn VI. C. Blanco, S. Catalano.
IAU Circ., No. 2679 (1974).

100.215 Infrared photometry of Titan. F. J. Low, G. H. Rieke.
Astrophys. Journal, (*Letters*), Vol. 190, L143 - L145 (1974).

Infrared photometry from 1.6 to 34 μ and over a 10-year period constrains the atmospheric models of Titan.

The atmosphere of Titan. See Abstr. 003.065.

Titan as a gravitational brake. See Abstr. 053.005.

Physical properties of the natural satellites. See Abstr. 091.004.

Satellites and magnetospheres of the outer planets. See Abstr. 091.013.

Occultation of Saturn VI by the moon on 1974 February 3. See Abstr. 096.029.

Occultation of Saturn VI by the moon on 1974 March 2–3. See Abstr. 096.030.

Occultations of Saturn VI by the moon. See Abstr. 096.031.

101 Uranus, Neptune, Pluto, Transplutonian Planets

101.001 On the long-term motion of Pluto.
P. E. Nacozy, R. E. Diehl.
Celestial Mechanics, Vol. 8, 445 - 454 (1974).

A modified periodic orbit of the third kind is introduced that is closely related to periodic orbits of the third kind as defined by Poincaré. It is shown that Pluto librates about the periodic orbit with apparent stability. This further explains the librational motion of the resonant argument of Pluto and the avoidance of a Pluto-Neptune close approach as found by Cohen and Hubbard and the long-term motion of Pluto and the librational motion of the perihelion as found by Williams and Benson.

101.002 A semi-analytical theory of the secular perturbations of Pluto. P. E. Nacozy, R. E. Diehl.
Bull. American Astron. Soc., Vol. 6, 205 (1974).– Abstr. AAS.

101.003 Miranda: her eccentricity and inclination.
R. J. Greenberg, E. A. Whitaker.
Bull. American Astron. Soc., Vol. 6, 207 (1974).– Abstr. AAS.

101.004 A possible atmosphere for Pluto. M. H. Hart.
Icarus, Vol. 21, 242 - 247 (1974).

At the temperature of Pluto ($\sim 43°K$) the only gas which would neither condense nor escape is neon. Since neon is cosmically abundant it is suggested that Pluto may have a fairly extensive atmosphere consisting of almost pure neon. The possibility that such an atmosphere exists is analyzed, along with the possibility that oceans of liquid neon may exist at the surface.

101.005 An investigation of the rotational period of the planet Pluto.
J. S. Neff, W. A. Lane, J. D. Fix.
Publ. Astron. Soc. Pacific, Vol. 86, 225 - 230 (1974).

New photometry of Pluto was obtained to resolve the longstanding ambiguity in the synodic period of Pluto. Measurements obtained at Kitt Peak National Observatory on four consecutive nights in April 1973 show that the shorter period of $1^d.1819$ is spurious and confirm the $6^d.3867 \pm 0.0003$ period found by Hardie (1964). The times of mean light crossing on the rising branch yield an improved synodic period of $6^d.3874 \pm 0.0002$ based on 20 years of observations.

101.006 On the identification of the 6420 Å absorption feature in the spectra of Uranus and Neptune.
T. Owen, B. L. Lutz, C. C. Porco, J. H. Woodman.
Astrophys. Journ., Vol. 189, 379 - 381 (1974).

The authors have found a methane band at λ6420 that appears to account satisfactorily for an absorption observed at this wavelength in spectra of Uranus and Neptune. There thus appears to be no compelling reason to continue to attribute this feature to pressure-induced absorption by hydrogen.

101.007 Orbit of Nereid and the mass of Neptune.
L. E. Rose.
Astron. Journ., Vol. 79, 489 - 490 (1974).

A new orbit for Neptune's outer satellite, Nereid, is derived using 44 observations 1949 - 69. The value for the mass of the Neptune system is found to be $m^{-1} = 19438 \pm 116$ (m.e.).

101.008 Pluto. K. Messell.
Astron. Tidssk., Årg. 7, p. 31 - 32 (1974).

101.009 On the upper atmosphere of Neptune.
J. Veverka, L. Wasserman, C. Sagan.
Astrophys. Journ., Vol. 189, 569 - 575 (1974).

The authors have reanalyzed the Mount Stromlo observations of the occultation of BD $-17°4388$ by Neptune and find that the upper atmosphere is not isothermal as suggested by Freeman and Lyngå. For a pure hydrogen atmosphere, their results give a temperature of $135°K$ at a number density of 10^{15} cm^{-3}.

101.010 Neptune II (Nereid). K. Aksnes.
IAU Circ., No. 2665 (1974).

101.011 The planet X: putting things in perspective.
R. H. Mendez.
Rev. Astron., Vol. 45, Nos. 185 - 186, p. 14 - 16 (1973).
In Spanish.

The author maintains that the trans-Plutonian planet was never more than a mathematical hypothesis. The author concludes that it is impossible that there is any planet of the mass, magnitude, average distance and orbital inclination predicted by Brady.

101.012 Pluto's zichtbare pool kan zuid en noord zijn.
G. W. E. Beekman.
Zenit, Vol. 1, No. 6, p. 23 (1974).

101.013 The search for HD in the spectrum of Uranus: an upper limit to [D/H]. B. L. Lutz, T. Owen.
Astrophys. Journ., Vol. 190, 731 - 734 (1974).

The $P(1)$ line of the 4–0 rotation-vibration band of HD was sought in the spectrum of Uranus. No feature which could be identified with any certainty was found, yielding an upper limit on the column density of HD in the $J = 1$ rotational level of $\eta N_1 (HD) < 331$ meter-amagat (m-Am). At $85°$ K this estimate corresponds to an upper limit of the total HD column density of $\eta N (HD) < 860$ m-Am and a deuterium to hydrogen ratio of $[D/H] < 4 \times 10^{-4}$.

Inversion of gravity data for giant planets.
See Abstr. 091.007.

Satellites and magnetospheres of the outer planets.
See Abstr. 091.013.

The optical properties of Venus and the Jovian planets. II. Methods and results of calculations of the intensity of radiation diffusely reflected from semi-infinite homogeneous atmospheres. See Abstr. 091.015.

Ultraviolet photometry from the Orbiting Astronomical Observatory. XIII. The albedos of Jupiter, Uranus, and Neptune. See Abstr. 099.001.

Internal constitution and the figures of the giant planets. See Abstr. 099.004.

The inner structure of giant planets.
See Abstr. 099.009.

Jupiter, Saturn, and Uranus disk temperature measurements at 2.07 and 3.56 cm. See Abstr. 099.038.

Models of the giant planets. See Abstr. 099.062.

The significance of pressure shifts for the interpretation of H_2 quadrupole lines in planetary spectra.
See Abstr. 099.063.

On the principle of least interaction action and the Laplacean satellites of Jupiter and Uranus.
See Abstr. 099.202.

102 Comets (Origin, Structure, Atmospheres, Dynamics)

102.001 **Hydrodynamics of the H_2O comet.** M. K. Wallis.
Monthly Notices Roy. Astron. Soc., Vol. 166, 181 - 189 (1974).

The heating accompanying the photodissociation of H_2O can dominate the hydrodynamics of a cometary coma. The inner collisional coma is modelled as a heated radial flow from a central source, for which the asymptotic solution has finite Mach number with expansion velocity increasing as $r^{1/3}$ and temperature as $r^{2/3}$. The heating is important in large comets with H_2O source strengths exceeding 10^{28} s^{-1} $ster^{-1}$, as comet Bennett 1970 II. Its relevance for expansion velocities, for Shul'man's halo hypothesis, for chemical processes and for interpretation of the Lyman-α profiles is briefly discussed.

102.002 **Neutral atmospheres of comets: a distributed source model.** W.-H. Ip, D. A. Mendis.
Astrophys. Space Sci., Vol. 26, 153 - 166 (1974).

It is argued that the typical nuclear region of a comet, under a variety of circumstances, consists of a central icy nucleus surrounded by an extended icy halo of grains forming a supplementary source for the observed atmospheric constituents. A complete hydrodynamic description of a neutral atmosphere corresponding to such a distributed source model when the predominant 'parent-molecule' is H_2O is given, and numerical results corresponding to two different velocity distributions of the icy halo consistent with smaller and larger grains are computed. The applicability and relevance of these models are discussed.

102.003 **The brightness of comets.** L. G. Jacchia.
Sky Telescope, Vol. 47, 216 - 220 (1974).

102.004 **Dust emission structure in comets with type II tails.** P. Benvenuti.
Astrophys. Space Sci., Vol. 27, 203 - 209 (1974).

Short exposure plates of comet Arend-Roland (1956h) are examined and compared with similar photographs of comet Bennett (1969i), comet Halley (1910 II) and comet Mrkos (1957d). It is found that the emission structure of the dust near the nucleus in the first comet is different from that in the others. Close resemblances with configurations described by Wurm and Mammano (1972) have been found for comet Halley and comet Mrkos.

102.005 **The substructure in the heads of comets with type I and type II tails.** K. Wurm.
Astrophys. Space Sci., Vol. 27, 211 - 216 (1974).

On a number of comet Halley 1910 photographs of short exposures (Helwan material) a substructure in dust emission in the inner coma becomes detectable. Narrow and diffuse dust streamers leave the nucleus area within a smaller or larger cone opened in the direction to the sun. On plates with longer exposures these structures merge together in a parabolic envelope. Structures of the type I tail are not visible for lower heliocentric distances ($r < 1$) in the inner coma; however, occasionally CN, C_2 eject circular emission areas.

102.006 **Physical properties of comets.** J. Rahe.
Naturwissenschaften, 61. Jahrgang, p. 45 - 50 (1974).

Comets can be considered as relics of the early composition of the solar nebula that furnish information about the past history of the solar system and the present conditions in interplanetary space. Current theories concerning their physical and chemical properties are summarized.

102.007 **The nature of comets.** F. L. Whipple.
Sci. American, Vol. 230, No. 2, p. 48 - 57 (1974).

Although comet Kohoutek was a disappointment to visual observers, it provided a good opportunity for the study of one of the objects that may be relics of the cloud from which the sun and the planets formed.

102.008 **On the real lengths of cometary tails. A vista in experimental astrophysics.** R. Döpel.
Vistas in astronomy, Vol. 15, (see 003.001), 131 - 144 (1973).

102.009 **Neuere Entwicklungen unserer Kenntnis der Kometen.** L. Biermann.
SuW, Vol. 13, 9 - 15 (1974).

102.010 **Cometary motions.** B. G. Marsden.
Celestial Mechanics, Vol. 9, 303 - 314 (1974). Presented at a colloquium on 'Copernicus ... and modern dynamical astronomy', Washington, 1972 December.

The history of the study of cometary motions is described, from the pre-Copernican era until the present time, with emphasis on the determination of orbits, the calculation of special perturbations, and the analysis of nongravitational effects. A brief survey is made of the statistics of cometary orbits and of the implications concerning cometary origin and evolution.

102.011 **CH_4 and CH_2 in comets.** J. H. Black.
Nature, Vol. 248, 319 (1974). – Letter.

102.012 **Kometen worden zichtbaar door fluorescerend gas.** T. de Groot.
Zenit, Vol. 1, No. 1, p. 22 - 26 (1974).

102.013 **The aging and the brightness decrease of comets.** L. Kresák.
Bull. Astron. Inst. Czechoslovakia, Vol. 25, 87 - 112 (1974).

The present concepts of the process of aging of short-period comets, which assume a rapid secular decrease of their absolute brightness, are shown to be untenable. The rates of fading of several magnitudes per century are contrary to many points of observational evidence, especially if a statistical approach is adopted and if a steady state within reasonable time-spans is required. All discrepancies are removed, if instrumental effects are properly taken into account. These are shown to be much stronger than generally assumed, producing a mean scale shift of 3 magnitudes since the end of the nineteenth century, and extreme systematic differences as large as 7 to 8 magnitudes. It is concluded that the progressive changes are entirely masked by random fluctuations, being generally less than 1 magnitude per century as long as the comet moves in a relatively stable orbit.

102.014 **Magnetisation of comets.**
A. Mendis, H. Alfvén.
Nature, Vol. 248, 36 - 37 (1974).

It seems that a comet, when sufficiently close to the sun, may be able to generate an appreciable internal magnetic field and that the magnetic fields observed in the tail may then result from processes analogous to those producing the earth's magnetotail.

102.015 **Comets, solar wind and the D/H ratio.**
H. Reeves.
Nature, Vol. 248, 398 (1974). – Letter.

102.016 **Possible maser emission in H_2O microwave lines from comets.** V. S. Strel'nitskij.
Astron. Tsirk., No. 805, 3 pp. (1973). In Russian.

102.017 **Cosmic gamma-ray emission and comet collisions with compact stars.**
O. Kh. Guseinov, V. Vanýsek.
Astrophys. Space Sci., Vol. 28, L11 - L15 (1974).

The probability that γ-ray bursts may be generated by the infall of comet-like objects on the neutron stars, as recently proposed by Harwit and Salpeter (1973), is reexamined.

102.018 **The temperature of the rotating nucleus of a comet.**
P. Egibekov.
Komety i Meteory, *Dushanbe*, No. 21, p. 3 - 17 (1972). In Russian.

The thermal regime of rotating comet nuclei is analysed in details. Particularly the temperature distribution and sublimation rate are given as functions of longitude and latitude of points on the nuclear surface. The influence of the orientation of rotation axes on the degree of symmetry of cometary brightness curves is considered.

102.019 **Peculiarities of the distribution of surface brightness in cometary heads.** M. Z. Markovich.
Komety i Meteory, *Dushanbe*, No. 22, p. 13 - 19 (1973). In Russian.

Curves of surface brightness distribution presented in a dimensionless form can facilitate the detection of fine effects in the distribution of matter in cometary heads.

102.020 **Neutral oxygen in comets.**
P. L. Byard, G. H. Newsom.
Nature, Vol. 249, 433 - 434 (1974). – Letter.

102.021 **The prediction of anomalous tails of comets.**
Z. Sekanina.
Sky Telescope, Vol. 47, 374 - 377 (1974).

102.022 **Emissionsbanden- und Kontinuums-Photometrie von Kometen.** J. Rahe.
Mitt. Astron. Ges., No. 34, p. 166 - 169 (1973/74).
Presented at the "Wissenschaftliche Tagung der Astron. Ges., Oberkochen, 1973 April 24 - 27".

102.023 **On the problem of the age of long-periodic comets.**
V. P. Tomanov.
Astron. vestn., Vol. 8, 87 - 91 (1974). In Russian.

It is shown that average statistic values of absolute magnitude of comets, of angular distance of orbital planes from the solar apex, and the availability of a cometary tail can be used to characterize the age of long-periodic and almost parabolic comets. Arguments to support the hypothesis of a comparatively recent simultaneous capture of comets are presented.

102.024 **Comets give evidence of immense eruptive processes in the solar system.** S. Vsekhsvyatskij.
Kometn. Tsirk., *Kiev*, No. 166 (1974). In Russian.

102.025 **On the influence of nongravitational forces on the motion of comets.**
K. Šteins, L. Divina, I. Revina.
Uch. Zap. Latv. Gos. Univ., Vol. 175, Astron., vyp. (No.) 8, (see 003.010), p. 38 - 61 (1973). In Russian.

102.026 **Urey-Hypothese: Kometen als Epochemacher.**
H. C. Urey.
Bild Wiss., [Deutsche Verlags-Anstalt, Stuttgart], Vol. 10, 1339 - 1354 (1973).

Comets, meteorites and men. See Abstr. 003.076.

Comets. See Abstr. 003.138.

Interplanetary material. See Abstr. 106.008.

103 Comets: Listed Objects

103.001 **Kometen.** E. Wiedemann.
Orion Schaffhausen, 32. Jahrgang, p. 76 - 77 (1974). Report on a lecture by P. Wild.

103.002 **Observations at the Gissar Observatory of the Academy of Sciences of the Tadzhik SSR.**
Kometn. Tsirk., *Kiev*, No. 157 (1974). In Russian. − Concerning comet Kojima, 1972j; comet Schwassmann-Wachmann 1.

103.003 **Definitive designations of comets of 1972.**
Kometn. Tsirk., *Kiev*, No. 158 (1974). In Russian.

103.004 **Photographic positional observations of comets.** A. Sh. Khatisov.
Byull. Abastumansk. Astrofiz. Obs., No. 45, p. 85 - 86 (1974). In Russian.
Comets Faye (1969a), Kohoutek (1969b), Tago-Sato-Kosaka (1969g), Kojima (1972j).

103.005 **Die endgültige Bezeichnung der Kometen 1972.** R. Lukas.
SuW, Vol. 13, 175 (1974).

103.006 **Kometer 1973.** H. Q. Rasmusen.
Astron. Tidssk., Årg. 7, 86 - 87 (1974).

103.007 **Observations of comets and minor planets.**
R. E. McCrosky, C. Y. Shao, G. Schwartz, M. Mattei.
IAU Circ., No. 2650 (1974). − Concerning the comets 1969 II, 1972 VIII, 1972 IX, 1972 X and the minor planets 1932 HA and 1971 FA.

103.008 **Observations of comets.** C. Torres, J. Petit.
IAU Circ., No. 2655 (1974). − Concerning comets 1971 I, 1971 IX, 1971 X, 1972 III, 1972 IV, 1972 XII.

103.009 **Observations of comets.** R. E. McCrosky, C. Y. Shao, G. Schwartz, M. Mattei, J. Bulger.
IAU Circ., No. 2673 (1974). − Concerning comets 1925 II, 1969 II, 1971 I, 1972 VIII, 1973e.

103.010 **Observations of comets.**
C. Torres, J. Petit, S. Barros, M. Wischniewsky, H. Wroblewski.
IAU Circ., No. 2681 (1974). − Concerning comets 1971 VI, 1972 II, 1973f.

103.011 **Observations of comets.** V. Barocas, M. J. Hendrie, I. M. Purcell, G. H. Rutter, R. H. South, R. L. Waterfield, N. Wood.
British Astron. Ass. Circ. No. 551 (1974). − Concerning Sandage 1972h, Kojima 1972j, P/Tuttle-Giacobini-Kresak 1973b, Kohoutek 1973e, Kohoutek 1973f, Huchra 1973h, P/Gehrels 1973n.

103.012 **Cometas − Viento solar. Atlas.**
Published by Observatorio Astronómico Universidad Nacional de Córdoba, Comisión Nacional de Estudios Geo-Heliofísicos, Argentina, 3 + 13 pp. (1973). − Concerning the comets Pereyra (1963 V), Ikeya-Seki (1965f), Mitchell-Jones-Gerber (1967f), Honda (1968c), Honda (1968e), Tago-Sato-Kosaka (1969g).
The Atlas contains observations of comet tails and their results, as part of the "Comets-Solar Wind" program, sponsored by special agreement between Observatorio Astronómico de la Universidad Nacional de Córdoba, and Comisión Nacional de Estudios Geo-Heliofísicos (Argentina).

103.100 **Comet 1970 II Bennett**

Particle sizes in comet Bennett (1970 II).
C. R. O'Dell.
Icarus, Vol. 21, 96 - 99 (1974).
The particle size distribution in the coma and tail of comet Bennett has been determined by several methods, each sensitive to a particular size range.

The cyanogen bands of comet Bennett 1970 II.
G. C. L. Aikman, W. J. Balfour, J. B. Tatum.
Icarus, Vol. 21, 303 - 316 = Contr. Dominion Astrophys. Obs., Victoria, No. 212 (1974).
The CN band spectrum of comet Bennett 1970 II was photographed on April 14, 1970. In addition to the (0, 0) band, some other faint lines were observed, which arise either from the (1, 1) band of the normal isotopic species or from the (0, 0) band of ^{13}CN. The Swings effect was investigated theoretically. Absolute intensities of the CN lines were found by comparison with the lunar spectrum. From these measurements the distribution, density and total mass of cyanogen in the gaseous coma of the comet are estimated.

Detailed polarimetry of comet Bennett 1970 II.
N. N. Kiselev.
Komety i Meteory, *Dushanbe*, No. 22, p. 8 - 12 (1973).
The results of a photographic polarimetry of the comet 1970 II are given. The difference in the polarization of the different parts of the head and the difference in the polarization of the head and of the tail are noted.

Wechselwirkungen zwischen Sonnenwind und Komet Bennett.
L. F. Burlaga, J. Rahe, B. D. Donn, M. Neugebauer.
Mitt. Astron. Ges., No. 34, p. 162 - 165 (1973/74).
Presented at the "Wissenschaftliche Tagung der Astron. Ges., Oberkochen, 1973 April 24 - 27".

Photographic observations of comet Bennett 1969i in Tashkent. A. G. Rakhimov, Eh. Rakhmatov.
Tsirk. Astron. Inst., *Tashkent*, No. 40 (387), p. 10 - 17 (1973). In Russian.

Electropolarimetry of the comets Tago-Sato-Kosaka 1969 IX and Bennett 1970 II. See Abstr. 103.121.

103.101 **Comet 1973f Kohoutek**

Spectroscopic observations of comet Kohoutek (1973f). P. Benvenuti, K. Wurm.
Astron. Astrophys., Vol. 31, 121 - 122 (1974).
Spectra of comet Kohoutek (1973f), at heliocentric distance of 1.53 A.U., are described. Besides the usual bands of CN, C_3, C_2, the appearance of two unidentified doublets at $\lambda\lambda 6148-6200$ Å and $\lambda\lambda 6977-7038$ Å is pointed out.

Tentative identification of the H_2O^+ ion in comet Kohoutek. G. Herzberg, H. Lew.
Astron. Astrophys., Vol. 31, 123 - 124 (1974).
Unidentified features in the spectrum of comet Kohoutek observed by Herbig and Benvenuti and Wurm fit with the low-temperature lines in the spectrum of the H_2O^+ ion. The presence of this ion in the tail of comet Kohoutek is thus strongly indicated.

Spectrum of comet Kohoutek 1973f.
E. K. Denisyuk, Z. V. Karyagina.
Astron. Tsirk., No. 802, p. 1 - 2 (1973). In Russian.

Photoelectrical registration of the spectrum of comet Kohoutek. A. V. Kharitonov, V. M. Tereshchenko.
Astron. Tsirk., No. 811, p. 5 - 6 (1974). In Russian.

Detection of OH at 18-centimeter wavelength in comet Kohoutek (1973f). B. E. Turner.
Astrophys. Journ., (*Letters*), Vol. 189, L137 - L139 (1974).

The 1665- and 1667-MHz lines of OH have been observed in absorption toward comet Kohoutek and are apparently optically thin. It is argued that the OH absorption arises in an extended halo about the comet, and that the OH is being pumped by solar ultraviolet radiation.

Infrared observations of comet Kohoutek near perihelion. E. P. Ney.
Astrophys. Journ., (*Letters*), Vol. 189, L141 - L143 (1974).

Observations between 0.55 and 18 μ show that after perihelion passage comet Kohoutek possessed an infrared tail and antitail.

Identification of H_2O^+ in the tail of comet Kohoutek (1973f). P. A. Wehinger, S. Wyckoff, G. H. Herbig, G. Herzberg, H. Lew.
Astrophys. Journ., (*Letters*), Vol. 190, L43 - L46 (1974).

Spectra of the tail of comet Kohoutek have been obtained in the region 5000-7500 Å. They show clearly and fairly strongly the main progression of bands of the H_2O^+ ion with the required alternation of band structure. Only lines expected to be strong at low rotational temperature (< 100° K) are observed in the cometary bands. The identification of H_2O^+ in the tail of comet Kohoutek provides the first conclusive evidence for the presence of H_2O in comets.

Comet Kohoutek 1973f.
British Astron. Ass. Circ. No. 550 (1974).

Comet Kohoutek 1973f. B. G. Marsden.
British Astron. Ass. Circ. No. 551 (1974).

Near infrared Fourier spectroscopy of comet Kohoutek (1973f). T. Hilgeman, L. L. Smith.
Bull. American Astron. Soc., Vol. 6, 220 (1974). – Abstr. AAS.

Cometa Kohoutek.
Coelum, Vol. 42, 41 - 42 (1974).

Observations de la comète Kohoutek 1973 f à l'Observatoire de Haute Provence.
C. Fehrenbach, Y. Andrillat.
Comptes Rendus Acad. Sci. Paris, Sér. B, Vol. 278, 359 - 360 (1974).

La comète de Kohoutek (1973f) observée en lumière de résonance du radical oxhydrile OH.
J. Blamont, M. Festou.
Comptes Rendus Acad. Sci. Paris, Sér. B, Vol. 278, 479 - 484 (1974).

Le spectre de la comète Kohoutek 1973f.
C. Fehrenbach, Y. Andrillat.
Comptes Rendus Acad. Sci. Paris, Sér. B, Vol. 278, 607 - 610 (1974).

Comet Kohoutek (1973f).
E. P. Ney, H. Hefele, R. G. Roosen, C. Sherrod, L. Lorenzi, H. L. Giclas, V. Zappalà, G. De Sanctis, R. Petrovičová, M. Antal, A. Mrkos, W. Ferreri.
IAU Circ., No. 2614 (1974).

Comet Kohoutek (1973f). E. P. Ney, W. F. Ney, M. Zeilik, J. Bouška, L. Jacchia, D. P. Elias, S. P. Maran, P. Feldman, L. Snyder, D. Buhl, W. Huebner, T. Dickinson, D. Milon, J. Bortle, R. S. Harrington, M. Dubin, C. Sherrod, K. Henize, P. Maley.
IAU Circ., No. 2616, with corrections in No. 2623 (1974).

Comet Kohoutek (1973f).
G. Herzberg, H. Lew, P. Benvenuti, K. Wurm, O. E. H. Rydbeck, J. Elldér, W. M. Irvine, C. Opal, G. Carruthers, W. Elliot, P. K. Smith, J. Mikolas, P. S. McIntosh, M. J. Mayo, J. Truxton.
IAU Circ., No. 2618 (1974).

Comet Kohoutek (1973f).
K. Suzuki, T. Urata, I. R. Bejtrishvili, N. S. Chernykh, L. I. Chernykh, H. Debehogne, G. Roland, C. B. Stephenson, W. Ameling, L. Krumenaker, D. Scholle, R. L. Waterfield, I. M. Purcell, T. Seki, N. Kojima, J. M. Codina, A. R. Klemola, E. P. Ney, J. Stoddart, A. Wardrop, N. Torras.
IAU Circ., No. 2619 (1974).

Comet Kohoutek (1973f). R. S. Harrington, M. Miranian, B. G. Marsden, J. H. Black, E. J. Chaisson, J. A. Ball, H. Penfield, A. E. Lilley, S. P. Maran, L. Haughney, C. B. Cosmovici, K. W. Michel, A. E. Roche, W. C. Wells, M. A. Seeds, J. L. Michael, G. H. Newsom, M. J. Mayo, J. Truxton, J. E. Bortle.
IAU Circ., No. 2621, with a correction in No. 2628 (1974).

Comet Kohoutek (1973f). J. M. Codina, N. Torras, T. Seki, H. L. Giclas, G. Klare, R. S. Harrington, S. P. Maran, A. L. Broadfoot, S. Kumar, P. Lee, D. Skoza, A. Feinstein, P. Maley, C. S. Morris.
IAU Circ., No. 2623 (1974).

Comet Kohoutek (1973f). V. Zappalà, W. Ferreri, G. De Sanctis, H. L. Giclas, P. A. Wehinger, S. Wyckoff, D. D. Meisel, R. A. Berg, R. W. Hobbs, S. P. Maran, W. J. Webster, Jr., R. S. Harrington, J. W. Christy, K. Simmons, J. Bortle, C. S. Morris, P. Maley.
IAU Circ., No. 2626, with corrections in Nos. 2628, 2636 (1974).

Comet Kohoutek (1973f). P. Benvenuti.
IAU Circ., No. 2628 (1974).

Comet Kohoutek (1973f). B. Milet, A. A. Schoenmaker, H. L. Giclas, M. L. Kantz, G. Soulié, R. S. Harrington, B. A. Morrison, G. H. Herbig, G. B. Baratta, R. Buonanno, F. Smriglio, R. Nesci, R. Viotti.
IAU Circ., No. 2629 (1974).

Comet Kohoutek (1973f).
N. Vogt, M. J. Mayo, J. Truxton, R. A. Keen, W. Elliot, J. Mikolas, P. Maley, C. S. Morris.
IAU Circ., No. 2631 (1974).

Comet Kohoutek (1973f).
H. L. Shipman, J. E. Bortle, P. Maley.
IAU Circ., No. 2632 (1974).

Comet Kohoutek (1973f). P. A. Wehinger, S. Wyckoff, R. L. Waterfield, G. E. D. Alcock, P. Maley, R. Petrovičová, A. Mrkos, H. Debehogne, G. Roland, J. M. Codina, N. Torras, W. Ferreri, L. Lorenzi, E. Anderlucci, H. L. Giclas, M. L. Kantz, R. G. Roosen, R. S. Harrington.
IAU Circ., No. 2634, with a correction in No. 2657 (1974).

Comet Kohoutek (1973f).

R. Davis, Z. Sekanina, C. E. Koppeschaar, G. E. D. Alcock, M. Beyer, M. V. Jones, V. L. Matchett, K. Simmons.
IAU Circ., No. 2637 (1974).

Comet Kohoutek (1973f).
N. Kojima, T. Seki, V. Zappalà, L. Lorenzi.
IAU Circ., No. 2640 (1974).

Comet Kohoutek (1973f). J. E. Bortle.
IAU Circ., No. 2641 (1974).

Comet Kohoutek (1973f). R. L. Waterfield, R. H. S. South, P. Monk, C. S. Morris, W. Liller.
IAU Circ., No. 2643 (1974).

Comet Kohoutek (1973f). I. R. Bejtrishvili, N. S. Chernykh, L. I. Chernykh, H. K. Raudsaar, M. Antal, J. M. Codina, N. Torras, T. J. Armijo, R. S. Harrington, G. E. D. Alcock, G. D. Thompson, V. L. Matchett, P. Maley, K. Simmons.
IAU Circ., No. 2645 (1974).

Comet Kohoutek (1973f). O. Demircan, R. Petrovičová, A. Mrkos, T. Urata, W. Ferreri.
IAU Circ., No. 2648 (1974).

Komet Kohoutek (1973f).
J. M. Codina, N. Torras, Mundet.
IAU Circ., No. 2653 (1974).

Comet Kohoutek (1973f).
T. Seki, H. L. Giclas, Mundet, J. M. Codina, N. Torras, L. Lorenzi, T. C. Armijo, R. S. Harrington, J. E. Bortle.
IAU Circ., No. 2657 (1974).

Comet Kohoutek (1973f).
C. Torres, H. Debehogne, G. Roland.
IAU Circ., No. 2658 (1974).

Comet Kohoutek (1973f).
H. Rantaseppä-Helenius, T. Seki, A. Mrkos.
IAU Circ., No. 2661 (1974).

Comet Kohoutek (1973f).
C. Rogati, H. Debehogne, E. Roemer.
IAU Circ., No. 2671 (1974).

An attempt to observe comet Kohoutek through perihelion. G. N. Sprott.
Journ. Astron. Soc. Victoria, Vol. 27, 14 - 15 (1974).

Comet Kohoutek concluded. J. B. Trainor.
Journ. Astron. Soc. Victoria, Vol. 27, 29 - 30 (1974).

Observations of comet Kohoutek. J. B. Newton.
Journ. Roy. Astron. Soc. Canada, Vol. 68, 152 - 153 (1974).

Comet Kohoutek, 1973f.
Kometn. Tsirk., Kiev, No. 157 (1974). In Russian. – Positions from observations at the Crimean Astrophysical Observatory; Observations at the Kleť Observatory; Corrections to the ephemeris of Marsden (*V. P. Konopleva*).

Comet Kohoutek, 1973f.
Kometn. Tsirk., Kiev, No. 158 (1974). In Russian. – Observations at the Skalnaté Pleso Observatory.

Comet Kohoutek, 1973f.
Kometn. Tsirk., Kiev, No. 159 (1974). In Russian. – Observations at the Crimean Astrophysical Observatory (*N. S. Chernykh, G. R. Kastel'*); Observations at Uzhgorod with the SBG-camera (*S. I. Ignatovich, V. P. Epishev*); Observations at Pulkovo (*N. M. Bronnikova*).

Comet Kohoutek, 1973f.
Kometn. Tsirk., Kiev, No. 160 (1974). In Russian. – Observations in Tartu (*H. K. Raudsaar, G. R. Kastel'*); Observations in Kitab with the AFA-camera (*G. G. Borzov, M. M. Zakirov, R. A. Mambetov, V. S. Samonov*); Observations at the Main Astronomical Observatory of the Academy of Sciences, Pulkovo (*N. M. Bronnikova, N. N. Klyuchnik*); Observations in Moscow (*B. S. Vozdvizhenskij*); Observations at the Kleť Observatory (*A. Mrkos*).

Comet Kohoutek, 1973f.
Kometn. Tsirk., Kiev, No. 161 (1974). In Russian.

Comet Kohoutek, 1973f.
Yu. E. Migach, G. R. Kastel'.
Kometn. Tsirk., Kiev, No. 162 (1974). In Russian – Observations of the Alpine expedition of the Sternberg Institute (*V. M. Kovalenko, O. N. Kovalenko*); Observations at the Kleť Observatory (*A. Mrkos*).

Comet Kohoutek, 1973f.
Kometn. Tsirk., Kiev, No. 163 (1974). In Russian.

Comet Kohoutek, 1973f.
Kometn. Tsirk., Kiev, No. 164 (1974). In Russian. – Observations in Nikolaev, Department of the Main Astronomical Observatory of the USSR Academy of Sciences (*F. F. Kalikhevich*); Observations at the Saratov University (*M. B. Bogdanov, Yu. V. Mikhajlov*); Observations in Uzhgorod (*V. P. Epishev, S. I. Ignatovich, I. I. Romanov*); Observations in České Budějovice; Electrophotometry of comet Kohoutek, 1973f (*N. N. Kiselev, G. P. Chernova*); Measurements of the brightness of comet Kohoutek, 1973f (*D. Andrienko*).

Comet Kohoutek, 1973f.
Kometn. Tsirk., Kiev, No. 165 (1974). In Russian. – Observations at the Alpine expedition of the Sternberg Institute (*V. M. Kovalenko, O. N. Kovalenko*); Spectral observations of comet 1973f.(*I. R. Bejtrishvili*).

Comet Kohoutek, 1973f.
Kometn. Tsirk., Kiev, No. 166 (1974). In Russian.

Comet Kohoutek, 1973f.
Kometn. Tsirk., Kiev, No. 167 (1974). In Russian. – Observations at the Alpine expedition of the Sternberg Institute (*V. M. Kovalenko, O. N. Kovalenko*); Observations at the Ussurij Solar Patrol Station of the USSR Academy of Sciences (*V. A. Golubev*); Polarization observations of comet Kohoutek in 8 parts of the spectrum, 423–798 mμ (*L. A. Bugaenko, O. I. Bugaenko, V. P. Konopleva, A. V. Morozhenko, M. G. Rodriges, A. Eh. Rozenbush*).

Observations photographiques de la comète Kohoutek. S. Koutchmy.
L'Astronomie, 88e année, p. 172 - 177 (1974).

Further observations of comet Kohoutek 1973f.
J. C. Bennett.
Monthly Notes Astron. Soc. Southern Africa, Vol. 33, 13 1974).

Kohoutek makes a good night out. J. Hall.
Nature, Vol. 247, 124 (1974).

Infrared observations of comet Kohoutek before perihelion passage. M. Zeilik II.
Nature, Vol. 248, 120 (1974).

The author reports broadband observations at 10 μm and

20 µm of the nucleus of comet Kohoutek. These measurements were taken about 1 week before perihelion, while the comet was about 0.35 AU from the sun.

Detection of methyl cyanide in comet Kohoutek.
B. L. Ulich, E. K. Conklin.
Nature, Vol. 248, 121 - 122 (1974).

The detection of methyl cyanide in the nucleus of comet Kohoutek at a reasonable abundance has provided the first direct confirmation of the parent-daughter molecule hypothesis.

Unpredictability of comets.
Nature, Vol. 248, 732 (1974).– Letter.

Photometry of comet Kohoutek (1973f).
G. H. Rieke, T. A. Lee.
Nature, Vol. 248, 737 - 740 (1974).

Extensive photoelectric and infrared photometry of comet Kohoutek (1973f) shows that its nucleus is 10 to 15 km in diameter and that the comet contains relatively little dust, and provides new information about the nature of cometary dust grains.

Dust in the head of comet Kohoutek.
G. R. Hopkinson, Y. Elsworth, J. F. James.
Nature, Vol. 249, 233 - 234 (1974). – Letter.

Behaviour of comet Kohoutek (1973f).
D. A. Mendis, W-H. Ip.
Nature, Vol. 249, 536 - 537 (1974).

Observations of comet Kohoutek have revealed the presence of CH_3CN and HCN in the coma and of H_2O^+ in the tail; although predicted, H_2O was not observed. The behaviour of comet Kohoutek can be understood in terms of a nuclear model in which the H_2O exceeds the more volatile species by a factor less than 6:1 so that a fraction of these are not trapped in the clathrate lattice.

Komet Kohoutek (1973f). F. Seiler.
Orion Schaffhausen, 32. Jahrgang, p. 27 (1974).

Komet Kohoutek (1973f) eine Enttäuschung?
E. Wiedemann.
Orion Schaffhausen, 32. Jahrgang, p. 27 (1974).

Comet Kohoutek 1973f in the year 1973.
J. Bouška.
Říše hvězd, Vol. 55, 25 - 29 (1974). In Czech.

Comet 1973f again. V. Vanýsek.
Říše hvězd, Vol. 55, 81 - 84 (1974). In Czech.

More news of comet Kohoutek.
Sky Telescope, Vol. 47, 24 - 25 (1974).

Kohoutek rounds the sun.
Sky Telescope, Vol. 47, 91, 95 (1974).

A scientists' comet.
Sky Telescope, Vol. 47, 153 - 158 (1974).

An album of comet Kohoutek portraits.
Sky Telescope, Vol. 47, 267 - 273 (1974).

Der Komet 1973f Kohoutek: Von wo kam er und was wird aus ihm? J. Hopmann.
Sternenbote, 17. Jahrgang, p. 68 - 71 (1974).

News of comet Kohoutek. D. Milon.
Strolling Astronomer, Vol. 24, 177 - 181 (1974).

Note on comet Kohoutek 1973f. C. S. Morris.
Strolling Astronomer, Vol. 24, 208 - 209 (1974).

Comet Kohoutek observations with an image intensifier. D. A. Sutherland.
Strolling Astronomer, Vol. 24, 218 - 219 (1974).

Komet Kohoutek (1973f).
SuW, Vol. 13, 26, 28, 33 (1974).

Die ersten Beobachtungsergebnisse an dem Kometen Kohoutek 1973f. T. Kleine.
SuW, Vol. 13, 28 (1974).

Komet Kohoutek (1973f).
F. Seiler, R. Mehrmann, W. Sorgenfrey.
SuW, Vol. 13, 98 (1974).

Observations of comet Kohoutek. A. Pilski.
Urania Kraków, Vol. 45, 90 - 92 (1974). In Polish.

Observations of comet Kohoutek in Warsaw division of the PTMA. R. Fangor.
Urania Kraków, Vol. 45, 154 - 155 (1974). In Polish.

Kometa Kohoutek i njeno posmatranje (Comet Kohoutek and its observation). M. Stupar.
Vasiona, Vol. 22, 43 - 45 (1974).

Zwakte van Kohoutek was te verwachten. J. H. Oort.
Zenit, Vol. 1, No. 2, p. 13 - 14 (1974).

Vanuit vliegtuig waargenomen. G. Comello.
Zenit, Vol. 1, No. 2, p. 14 - 15 (1974).

Helderheidschattingen van de komeet Kohoutek 1973 f. H. Feyth.
Zenit, Vol. 1, No. 4, p. 11 (1974).

Theoretical radiants of the hypothetical meteor shower of comet Kohoutek 1973f. See Abstr. 104.028.

103.102 Comet 1971 II Encke

Comets and nongravitational forces. VI. Periodic comet Encke 1786–1971. B. G. Marsden, Z. Sekanina.
Astron. Journ., Vol. 79, 413 - 419 (1974).

Full perturbations by all nine planets are included, and the nongravitational secular acceleration is allowed for by means of the procedure discussed in Paper V of this series. It is confirmed that the secular acceleration has been decreasing since the early part of the nineteenth century. The secular acceleration evidently attained a maximum around the year 1820. The physical implication of this is briefly considered. Dates of perihelion passage are calculated back to the year 1570, and predicted elements are provided for the returns 1974–1984.

An investigation of the motion of comet Encke–Backlund in 1901–1970.
N. A. Bokhan, Yu. A. Chernetenko.
Astron. Zhurn. Akad. Nauk SSSR, Vol. 51, 617 - 626 (1974). In Russian. English translation in Soviet Astron. AJ, Vol. 18, No. 3.

The motion of comet Encke–Backlund has been investigated in 22 revolutions (1901–1970). The Makover model has been adopted for taking into account non-gravitational forces. Observations for the whole period of research are presented both before the perihelion and after it with satisfactory accuracy. The decrease of the secular acceleration with some variations in the course of time has been confirmed. It is demonstrated that the non-gravitational forces throughout the whole 20. century continuously affect the eccentricity.

Periodic comet Encke. E. Roemer, G. Reskin.
British Astron. Ass. Circ. No. 551 (1974).

Prochain retour de la comète périodique Encke 1786 I. A. Heck.
Ciel et Terre, Vol. 90, 143 - 147 (1974).

Periodic comet Encke.
B. Donn, D. K. Yeomans.
IAU Circ., No. 2648 (1974).

Periodic comet Encke.
J. Bortle, E. P. Ney, J. Stoddart.
IAU Circ., No. 2664 (1974).

Periodic comet Encke. E. P. Ney.
IAU Circ., No. 2666 (1974).

Periodic comet Encke. J. Bortle.
IAU Circ., No. 2670 (1974).

Periodic comet Encke. J. C. Bennett.
IAU Circ., No. 2675 (1974).

Periodic comet Encke.
Ikawa, D. Seargent, A. F. Jones.
IAU Circ., No. 2680 (1974).

Comet Encke.
Kometn. Tsirk., *Kiev*, No. 165 (1974). In Russian.

Comet Encke.
Kometn. Tsirk., *Kiev*, No. 166 (1974). In Russian.

Prochain retour de la comète périodique Encke 1786 I. A. Heck.
Orion Schaffhausen, 32. Jahrgang, p. 78 - 79 (1974).

Comet Encke makes another appearance.
Sky Telescope, Vol. 47, 279 (1974).

103.103 Comet 1917 I Mellish

Meteors of periodic comet Mellish and the Geminids. See Abstr. 104.012.

103.104 Comet 1925 II Schwassmann-Wachmann 1

Observations of comets.
S. I. Gerasimenko, N. P. Marchenko.
IAU Circ., No. 2647 (1974).

Periodic comet Schwassmann-Wachmann 1.
P. Herget.
IAU Circ., No. 2652 (1974).

Ephemeris of comet Schwassmann-Wachmann 1, 1925 II.
Kometn. Tsirk., *Kiev*, No. 158 (1974). In Russian.

The flaring periodicity of comet Schwassmann-Wachmann 1. F. D. Miller.
Publ. Astron. Soc. Pacific, Vol. 85, 767 - 768 (1973).

The recurrence of brightness flares of comet Schwassmann-Wachmann 1 reported by Vsekhsvyatskij and attributed by him to persistent sources of solar corpuscular radiation, may reflect, at least in part, the influence of lunar phase on comet observations.

Observations of comets. See Abstr. 103.128.

103.105 Comet 1974b Bradfield

New comet Bradfield 1974b.
W. A. Bradfield, D. Gans, M. P. Candy.
British Astron. Ass. Circ. No. 552 (1974).

Comet Bradfield 1974b.
A. Jones, G. D. Thompson, M. P. Candy, B. G. Marsden.
British Astron. Ass. Circ. No. 553 (1974).

Comet Bradfield 1974b. M. P. Candy.
British Astron. Ass. Circ. No. 554 (1974).

Comet Bradfield (1974b).
W. A. Bradfield, G. Thompson, D. Gans, M. P. Candy.
IAU Circ., No. 2633 (1974).

Comet Bradfield (1974b).
M. P. Candy, T. B. Tregaskis.
IAU Circ., No. 2636 (1974).

Comet Bradfield (1974b).
M. P. Candy, D. Gans, W. Liller, D. Herald.
IAU Circ., No. 2640 (1974).

Comet Bradfield (1974b).
W. Liller, N. Walborn, R. Salmon, N. Irvine, C. Jekabsons, G. Lowe, M. P. Candy, D. J. Gans.
IAU Circ., No. 2642 (1974).

Comet Bradfield (1974b).
D. Herald, A. Jones, G. D. Thompson.
IAU Circ., No. 2647 (1974).

Comet Bradfield (1974b). K. Simmons, M. Rogers, P. Maley, J. Bortle, H. Povenmire, R. G. Roosen, T. C. Armijo.
IAU Circ., No. 2650 (1974).

Comet Bradfield (1974b).
C. Jekabsons, D. J. Gans, G. Lowe, M. P. Candy, T. Seki, T. C. Armijo, R. S. Harrington, R. G. Roosen, D. W. Dunham, R. R. D. Austin, E. P. Ney, J. Stoddart, J. E. Bortle, K. Simmons, V. L. Matchett, P. Maley.
IAU Circ., No. 2651 (1974).

Comet Bradfield (1974b).
H. L. Giclas, T. Seki, H. R. Calvird, J. C. Mikolas, P. Maley, B. Comsa, C. Sherrod.
IAU Circ., No. 2654 (1974).

Comet Bradfield (1974b). T. Urata, H. L. Giclas, C. R. Chambliss, J. E. Bortle, G. E. D. Alcock, Z. Sekanina.
IAU Circ., No. 2656 (1974).

Comet Bradfield (1974b).
W. Gieren, M. Hoffmann, E. Pendl, J. B. Tatum, P. A. Wehinger, S. Wyckoff, G. E. D. Alcock, D. Seargent, C. S. Morris, C. Sherrod, P. Maley.
IAU Circ., No. 2658 (1974).

Comet Bradfield (1974b).
D. Herald, Mundet, J. M. Codina, N. Torras, A. Mrkos, R. Petrovičová, T. Urata, T. Seki, E. P. Ney, J. Gallagher, J. Stoddart, R. G. Roosen, K. G. Andersson, J. Muirden, C. Sherrod, P. Maley, P. Rapavy.
IAU Circ., No. 2661 (1974).

Comet Bradfield (1974b).
J. E. Bortle, G. E. D. Alcock.
IAU Circ., No. 2663 (1974).

Comet Bradfield (1974b).
E. P. Ney, J. Stoddart, G. W. E. Beekman.
IAU Circ., No. 2665 (1974).

Comet Bradfield (1974b).
Maiwald, B. Wedel, Hotop, R. L. Waterfield, L. Lorenzi, W. Ferreri, K. Tsuchiya, T. Urata, H. L. Giclas, M. L. Kantz, J. B. Tatum, M. de Pascual-Martinez, B. Milet, C. Scovil, A. J. Pinau, C. R. Chambliss, G. A. Becker, A. Kiasat, J. E. Bortle, D. K. Yeomans, B. G. Marsden.
IAU Circ., No. 2667 (1974).

Comet Bradfield (1974b).
C. S. Morris, K. Simmons, C. Sherrod, P. Maley.
IAU Circ., No. 2669 (1974).

Comet Bradfield (1974b).
R. L. Waterfield, P. Monk.
IAU Circ., No. 2670 (1974).

Comet Bradfield (1974b).
R. Petrovičová, A. Mrkos, C. B. Stephenson, L. E. Krumenaker, W. Gieren, E. Pendl, Mundet, N. Torras, T. Seki, C. Sherrod, P. Maley.
IAU Circ., No. 2671 (1974).

Comet Bradfield (1974b).
W. M. Jackson, T. Clark, B. Donn, R. B. Minton.
IAU Circ., No. 2674, with a correction in No. 2675 (1974).

Comet Bradfield (1974b).
C. Torres, M. Wischniewsky, H. Wroblewski, H. Debehogne, M. Winiarski, J. M. Kreiner, M. Kurpińska, N. P. Wieth-Knudsen, J. E. Bortle, P. Maley, C. Sherrod.
IAU Circ., No. 2676 (1974).

Comet Bradfield (1974b).
K. Tsuchiya, T. Urata, J. M. Codina, N. Torras, T. Seki.
IAU Circ., No. 2681 (1974).

Comet Bradfield 1974b: an Australian amateur's second discovery. J. B. Trainor.
Journ. Astron. Soc. Victoria, Vol. 27, 28 (1974).

New comet Bradfield, 1974b.
Kometn. Tsirk., *Kiev,* No. 161 (1974). In Russian.

Comet Bradfield, 1974b.
Kometn. Tsirk., *Kiev,* No. 162 (1974). In Russian.

Comet Bradfield, 1974b.
Kometn. Tsirk., *Kiev,* No. 163 (1974). In Russian.

Comet Bradfield, 1974b.
Kometn. Tsirk., *Kiev,* No. 164 (1974). In Russian. − Observations in Uzhgorod with the SBG 42.5/78 cm-camera (*V. P. Epishev, S. I. Ignatovich, I. I. Romanov*); Observations in České Budějovice (*A. Mrkos*); Observations at the Ussurij Solar Patrol Station (*V. A. Golubev*); Photoelectric observations in Alma-Ata (*D. A. Rozhkovskij, D. I. Gorodetskij*).

Comet Bradfield, 1974b.
Kometn. Tsirk., *Kiev,* No. 165 (1974). In Russian. − Observations in České Budějovice, Kleť Observatory (*A. Mrkos*); Observations at the Ussurij Solar Patrol Station (*V. A. Golubev*); New elements from 58 observations II 14 − IV 19, by *B. G. Marsden.*

Comet Bradfield, 1974b.
Kometn. Tsirk., *Kiev,* No. 166 (1974). In Russian.

Comet Bradfield, 1974b.
Kometn. Tsirk., *Kiev,* No. 167 (1974). In Russian. − Observations at the Odessa Astronomical Observatory; Observations at the Ussurij Station (*V. A. Golubev*); Observations at the Kleť Observatory (*A. Mrkos*).

Neuer Komet Bradfield (1974b).
Orion Schaffhausen, 32. Jahrgang, p. 77 (1974).

Komet Bradfield (1974b). K. Kaila.
Orion Schaffhausen, 32. Jahrgang, p. 111 (1974).

News of the spring comet 1974b.
Sky Telescope, Vol. 47, 288 - 289 (1974).

Amateurs follow comet Bradfield.
Sky Telescope, Vol. 47, 414 - 417 (1974).

Der Komet Bradfield (1974d).
Sternenbote, 17. Jahrgang, p. 62 - 64 (1974).

103.106 **Comet 1944 III Du Toit 1**

Periodic comet Du Toit 1 (1944 III).
IAU Circ., No. 2649 (1974).

Elements and ephemeris of the periodic comet Du Toit 1, 1944 III.
Kometn. Tsirk., *Kiev,* No. 157 (1974). In Russian.

103.107 **Comet 1972h Sandage**

Ephemeris of comet Sandage, 1972h.
Kometn. Tsirk., *Kiev,* No. 157 (1974). In Russian.

Continuation of the ephemeris of comet Sandage, 1972h.
Kometn. Tsirk., *Kiev,* No. 162 (1974). In Russian.

103.108 **Comet 1973n Gehrels 2**

New periodic comet Gehrels (2) 1973n.
T. Gehrels, B. G. Marsden.
British Astron. Ass. Circ. No. 551 (1974).

Periodic comet Gehrels 2 (1973n).
R. L. Waterfield, R. H. S. South, B. G. Marsden.
IAU Circ., No. 2622 (1974).

Periodic comet Gehrels 2 (1973n).
T. Seki, T. Furuta, R. E. McCrosky, J. Bulger.
IAU Circ., No. 2659 (1974).

Periodic comet Gehrels 2, 1973n.
Kometn. Tsirk., *Kiev,* No. 158 (1974). In Russian.

103.109 **Comet 1973j Brooks 2**

Continuation of the ephemeris of comet Brooks 2, 1973j.
Kometn. Tsirk., *Kiev*, No. 158 (1974). In Russian.

103.110 **Comet 1974a Forbes**

Periodic comet Forbes 1974a.
E. Roemer, L. M. Vaughn.
British Astron. Ass. Circ. No. 551 (1974).

Periodic comet Forbes (1974a).
E. Roemer, L. M. Vaughn.
IAU Circ., No. 2625 (1974).

Periodic comet Forbes (1974a).   T. Seki.
IAU Circ., No. 2670 (1974).

Ephemeris of comet Forbes, 1929 II.
Kometn. Tsirk., *Kiev*, No. 158 (1974). In Russian.

Rediscovery of the short-period comet Forbes, 1974a.
Kometn. Tsirk., *Kiev*, No. 160 (1974). In Russian.

Short-period comet Forbes, 1974a.
Kometn. Tsirk., *Kiev*, No. 166 (1974). In Russian.

103.111 **Comet 1973o Gibson**

New comet Gibson 1973o.
J. Gibson, B. G. Marsden.
British Astron. Ass. Circ. No. 551 (1974).

Comet Gibson (1973o).
J. Gibson, M. R. Cesco, C. U. Cesco, B. G. Marsden.
IAU Circ., No. 2615 (1974).

Comet Gibson (1973o).
J. Gibson, M. R. Cesco, B. G. Marsden.
IAU Circ., No. 2620 (1974).

Comet Gibson (1973o).
J. Gibson, A. C. Gilmore, P. M. Kilmartin.
IAU Circ., No. 2635, with a correction in No. 2666 (1974).

Comet Gibson (1973o).   B. G. Marsden.
IAU Circ., No. 2664 (1974).

Comet Gibson, 1973o.
Kometn. Tsirk., *Kiev*, No. 159 (1974). In Russian.

103.112 **Comet 1973l Schwassmann-Wachmann 2**

Periodic comet Schwassmann-Wachmann 2 (1973l).
M. Mattei, G. Schwartz, B. A. Romanowicz, T. Furuta, T. Seki.
IAU Circ., No. 2658 (1974).

Periodic comet Schwassmann-Wachmann 2 (1973l).   C. B. Stephenson, L. E. Krumenaker.
IAU Circ., No. 2679 (1974).

Comet Schwassmann-Wachmann 2, 1929 I.
Kometn. Tsirk., *Kiev*, No. 159 (1974). In Russian.

103.113 **Comet 1973g Reinmuth 2**

Periodic comet Reinmuth 2 (1973g).
B. G. Marsden.
IAU Circ., No. 2644 (1974).

Ephemeris of comet Reinmuth 2, 1973g.
B. Marsden.
Kometn. Tsirk., *Kiev*, No. 163 (1974). In Russian.

103.114 **Comet 1967 IX Finlay**

Periodic comet Finlay.   D. K. Yeomans.
IAU Circ., No. 2637 (1974).

Preliminary ephemeris of comet Finlay.
Kometn. Tsirk., *Kiev*, No. 161 (1974). In Russian.

103.115 **Comet 1942 IX Stephan-Oterma**

Linking of apparitions of comet Stephan-Oterma 1867 I and 1942 IX.
E. I. Kazimirchak-Polonskaya, L. M. Belous.
Kometn. Tsirk., *Kiev*, No. 161 (1974). In Russian.

103.116 **Comet 1974c Lovas**

Comet Lovas.   M. Lovas.
IAU Circ., No. 2653 (1974).

Comet Lovas (1974c).   C. Y. Shao.
IAU Circ., No. 2654 (1974).

Comet Lovas (1974c).
H. L. Giclas, M. L. Kantz, C. Y. Shao.
IAU Circ., No. 2657 (1974).

Comet Lovas (1974c).   T. Seki, C. Y. Shao, R. E. McCrosky, G. Schwartz, B. G. Marsden.
IAU Circ., No. 2660 (1974).

Comet Lovas (1974c).   M. Lovas, A. C. Gilmore, P. M. Kilmartin, H. L. Giclas, M. L. Kantz.
IAU Circ., No. 2663 (1974).

Comet Lovas (1974c).   P. Monk, R. L. Waterfield, H. L. Giclas, M. L. Kantz, B. G. Marsden.
IAU Circ., No. 2668 (1974).

Comet Lovas (1974c).
A. Mrkos, R. Petrovičová, C. B. Stephenson, L. E. Krumenaker, R. E. McCrosky, G. Schwartz, C. Y. Shao, K. Tomita.
IAU Circ., No. 2672 (1974).

Comet Lovas (1974c).   C. Y. Shao, M. Mattei, R. E. McCrosky, K. Tomita, H. L. Giclas, M. L. Kantz.
IAU Circ., No. 2677 (1974).

Comet Lovas (1974c).
R. H. S. South, R. L. Waterfield, H. L. Giclas, M. L. Kantz.
IAU Circ., No. 2682 (1974).

Comet Lovas, 1974c.   G. R. Kastel'.
Kometn. Tsirk., *Kiev*, No. 163 (1974). In Russian.

Comet Lovas, 1974c.
Kometn. Tsirk., *Kiev*, No. 164 (1974). In Russian.

Comet Lovas, 1974c.
Kometn. Tsirk., *Kiev*, No. 165 (1974). In Russian. — Observations at the Kleť Observatory, České Budějovice (A. Mrkos); New elements and ephemeris by B. G. Marsden (IAU Circ. 2668).

Comet Lovas, 1974c.
Kometn. Tsirk., *Kiev*, No. 166 (1974). In Russian.

103.117 **Comet 1919 III Brorsen-Metcalf**

Comet Brorsen-Metcalf 1847 V = 1919 III.
L. M. Belous.
Kometn. Tsirk., *Kiev*, No. 161 (1974). In Russian.

103.118 **Comet 1972d Giacobini-Zinner**

Periodic comet Giacobini-Zinner 1972d.
C. S. Morris.
Strolling Astronomer, Vol. 24, 198 - 203 (1974).

Der periodische Komet Giacobini-Zinner (1972 d).
T. Kleine.
SuW, Vol. 13, 56 - 58 (1974).

103.119 **Comet 1906 IV Kopff**

The nongravitational motion of comet Kopff.
D. K. Yeomans.
Publ. Astron. Soc. Pacific, Vol. 86, 125 - 127 (1974).
Nongravitational terms are required in the equations of motion if the observations of comet Kopff are to be successfully represented over the 1906–70 interval. A secular decrease in the deceleration of the comet's motion is followed by a secular increase in the acceleration of the motion.

103.120 **Comet 1968 VI Honda**

Absolute spectrophotometry of the comet Honda 1968 VI. O. V. Dobrovol'skij, O. Mamadov.
Byull. Inst. Astrofiz., *Dushanbe*, No. 62, p. 3 - 9 (1974). In Russian.
Four objective prism spectra of comet Honda are studied. Isophotes are constructed for the whole spectra and for continuum and emissions separately. The integral magnitude of the comet is determined in individual emissions and continuum. Absolute contents of CN, C_3 and C_2 in the head of the comet are estimated.

Slit spectra of comet Honda (1968c) 1968 VI.
O. V. Dobrovol'skij, O. Mamadov.
Komety i Meteory, *Dushanbe*, No. 21, p. 18 - 24 (1972). In Russian.
Slit spectra of comet Honda (1968c) are analysed in the range 3800–7300 Å. The following molecular emissions are identified: CN (0,0) and (0,1) of the violet system, the Swan system of C_2, the group 4050 C_3, the α bands of ammonia (NH_2), bands of CH. Some coincidences are noted with emissions of H_2. The continuum is found to differ significantly from that of the sun. Assuming a Boltzmann distribution of the vibrational energy of C_2 the vibrational temperature of C_2 molecules is estimated to $\sim 4000°K$.

103.121 **Comet 1969 IX Tago-Sato-Kosaka**

Photoelectric observations of magnitude and colour of comet 1969 IX. N. N. Kiselev.
Komety i Meteory, *Dushanbe*, No. 22, p. 3 - 4 (1973). In Russian.

Electropolarimetry of the comets Tago-Sato-Kosaka 1969 IX and Bennett 1970 II. N. N. Kiselev.
Komety i Meteory, *Dushanbe*, No. 22, p. 5 - 7 (1973). In Russian.

103.122 **Comet 1910 II Halley**

On Halley's comet 1909c or 1910 II with particular emphasis on the observations made in Japan. (Collective review).
K. Saito, S. Shinozawa.
Tokyo Astron. Obs., Report No. 63, Vol. 16, 760 - 906 (1974). In Japanese.

103.123 **Comet 1916 I Taylor**

Short-period comet Taylor, 1916 I.
V. Emel'yanenko.
Kometn. Tsirk., *Kiev*, No. 167 (1974). In Russian.

103.124 **Comet 1972 VIII Heck-Sause**

Positions and orbital elements of comet Heck-Sause (1973a). H. Debehogne, S. Vaghi, V. Zappalà.
Acta Astron., Vol. 24, 243 - 244 (1974).
22 positions of comet Heck-Sause (1973a) are given as obtained from photographic observations made at the Observatory of Torino, and its orbital elements are derived.

Comet Heck-Sause (1972 VIII).
C. Y. Shao, R. E. McCrosky, G. Schwartz, T. Furuta, T. Seki.
IAU Circ., No. 2638 (1974).

Comet Heck-Sause (1972 VIII).
A. Heck, J. Manfroid, G. Sause, A. M. Rousseau, M. Degueldre.
IAU Circ., No. 2642 (1974).

103.125 **Comet 1844 I De Vico-Swift**

Definitive orbit of comet de Vico-Swift, 1844 I.
S. D. Shaporev.
Kometn. Tsirk., *Kiev*, No. 164 (1974). In Russian.

103.126 **Comet 1973b Tuttle-Giacobini-Kresák**

Observations of comet Tuttle-Giacobini-Kresák, 1973b at the Alpine expedition of the Sternberg Institute.
V. M. Kovalenko, O. N. Kovalenko.
Kometn. Tsirk., *Kiev*, No. 165 (1974). In Russian.

103.127 Comet 1973k Sandage

Comet Sandage (1973k). T. Seki.
IAU Circ., No. 2622 (1974).

Comet Sandage (1973k). T. Seki.
IAU Circ., No. 2639 (1974).

Comet Sandage (1973k).
J. Bulger, C. Y. Shao, R. E. McCrosky.
IAU Circ., No. 2659 (1974).

Comet Sandage (1973k). C. Y. Shao, M. Mattei, C. B. Stephenson, L. E. Krumenaker, K. Tomita.
IAU Circ., No. 2676 (1974).

103.128 Comet 1972j Kojima

Observations of comets.
C. B. Stephenson, W. Ameling, L. Krumenaker, D. Scholle.
IAU Circ., No. 2627 (1974).

Comet Kojima (1972j). B. G. Marsden.
IAU Circ., No. 2649 (1974).

Observations of comets. See Abstr. 103.104.

103.129 Comet 1972 XII Araya

Comet Araya (1972 XII). B. G. Marsden.
IAU Circ., No. 2630 (1974).

103.130 Comet 1972 IV Neujmin 3

Periodic comet Neujmin 3 (1972 IV).
C. Y. Shao, M. Mattei, J. Bulger, R. E. McCrosky.
IAU Circ., No. 2639 (1974).

103.131 Comet 1967 XII Wolf 1

Periodic comet Wolf (1967 XII). K. Tomita.
IAU Circ., No. 2639 (1974).

103.132 Comet 1973e Kohoutek

Comet Kohoutek (1973e). B. G. Marsden.
IAU Circ., No. 2646 (1974).

103.133 Comet 1967 VI Arend

Periodic comet Arend. B. G. Marsden.
IAU Circ., No. 2662 (1974).

103.134 Comet 1930 VI Schwassmann-Wachmann 3

Periodic comet Schwassmann-Wachmann 3 (1930 VI).
K. Kono.
IAU Circ., No. 2672 (1974).

Periodic comet Schwassmann-Wachmann 3 (1930 VI). H. Povenmire.
IAU Circ., No. 2676 (1974).

103.135 Comet 1973i Clark

Periodic comet Clark (1973i).
P. M. Kilmartin, R. R. D. Austin.
IAU Circ., No. 2679 (1974).

103.136 Comet 1965 I Tsuchinshan 1

Evolution of the orbits of comets Tsuchinshan 1 and Tsuchinshan 2. Y. C. Chang, J. Zhang, M. Dong.
Acta Astron. Sinica, Vol. 15, 3 - 10 (1974). In Chinese.

Comets Tsuchinshan 1 and Tsuchinshan 2 were first observed at Purple Mountain Observatory in the beginning of 1965 within a short interval of ten days. They have very similar orbits. By the end of 1971, these comets were recovered on their return to perihelion. Starting from these orbital elements, the authors tried to find out what changes have happened to these orbits during a time interval of several centuries; whether this supposed "twin comets" have any genealogical relationship between them.

103.137 Comet 1965 II Tsuchinshan 2

Evolution of the orbits of comets Tsuchinshan 1 and Tsuchinshan 2. See Abstr. 103.136.

103.138 Comet 1968 I Ikeya-Seki

Comet magnitude analysis: Ikeya-Seki 1968 I (1967n). C. S. Morris.
Strolling Astronomer, Vol. 24, 219 - 220 (1974).

104 Meteors, Meteor Streams

104.001 **Heating of meteoric bodies.**
V. G. Kruchinenko, A. N. Shajdo.
Astron. vestn., Vol. 8, 35 - 41 (1974). In Russian.

Solutions of the heat conduction equation for cylindrical and spherical meteoric bodies heated in the earth's atmosphere of limited thickness have been obtained. The results have been compared with the existing solutions for an infinitely spreading atmosphere. It follows from comparison that the solutions for a limited isothermic atmosphere asymptotically tend to solutions for an infinitely spreading atmosphere.

104.002 **On some features of meteoric radiation.**
V. A. Smirnov.
Astron. vestn., Vol. 8, 42 - 48 (1974). In Russian.

Comparison of radiation of long-wave and short-wave regions of meteoric spectra has been accomplished on the basis of data on photometric statistics of spectral lines behaviour on meteor spectra. The intensity of short-wave radiation during flight increases as compared to long-wave radiation. Diffusion features of various elements composing meteoric gas mixture, reasons for instability of irradiation in the process of meteor entry inside the atmosphere, dual flares of shower meteors, in particular, have been considered.

104.003 **Increase in the effectiveness of the TV method for observation of meteors.** M. Begkhanov, Kh. Gul'medov, S. Mukhamednazarov, V. Gun'ko.
Astron. vestn., Vol. 8, 49 - 51 (1974). In Russian.

104.004 **Comparison of theoretical and experimental data on the drag of radio meteors.**
B. L. Kashcheev, N. V. Novoselova.
Astron. vestn., Vol. 8, 52 - 55 (1974). In Russian.

Data relating to the drag of meteors up to $+12^m$ in Kharkov, up to $+10^m$ in the USA and up to $+7^m$ in Ashkhabad are compared. Considerable distinction of the Kharkov results is pointed out. The drag of meteors up to $+12^m$ is less than the drag of meteors up to $+10^m$ and up to $+7^m$ almost in the whole range of the observed velocities, which is evidently connected with drag determination errors. Theoretical data on the drag of meteors agree with the results obtained in Kharkov.

104.005 **Lyrids in 1972.** I. G. Demidova.
Astron. vestn., Vol. 8, 59 - 61 (1974). In Russian.

An hourly rate ranging from 0.63 to 2.96 km^{-2} × hour^{-1} during the maximum of the shower has been determined on the basis of visual observations performed by the Moscow branch of All-Union Astronomical-Geodetic Society. The distance between particles in the stream ranges from 1400 to 850 km.

104.006 **Observations of the Draconids in October 1972.**
R. L. Khotinok.
Astron. vestn., Vol. 8, 61 - 63 (1974). In Russian.

The summary of visual observations performed on the territory of the USSR on October 5–9, 1972 in the period of the expected meteor shower of Draconids is given.

104.007 **The influx of visual sporadic meteors.**
D. W. Hughes.
Monthly Notices Roy. Astron. Soc., Vol. 166, 339 - 343 (1974).

By combining visual and radar data the mean rate of visual meteors for an individual observer is found to be 9.7 ± 0.7 hr^{-1}. The cumulative flux of visual meteors having magnitudes brighter than M ($0 < M < -5$) is also found.

104.008 **Notes on three autumn meteor showers.**
Sky Telescope, Vol. 47, 134 - 135 (1974).

104.009 **Meteor of August 10, 1972.** R. D. Rawcliffe, C. D. Bartky, F. Li, E. Gordon, D. Carta.
Nature, Vol. 247, 449 - 450 (1974).

Calculation of the trajectory.

104.010 **Interplanetary and near-Jupiter meteoroid environments: Preliminary results from the meteoroid detection experiment.**
W. H. Kinard, R. L. O'Neal, J. M. Alvarez, D. H. Humes.
Science, Vol. 183, 321 - 322 (1974). – Report.

104.011 **The mass distribution of radio meteors and the full-wave scattering theory.**
J. Jones, J. G. Collins.
Monthly Notices Roy. Astron. Soc., Vol. 166, 529 - 542 (1974).

The underdense train approximation can result in large errors in the meteoroid mass distribution indices as determined by meteor radar systems. This paper shows how the transition from the underdense to the overdense reflection regime may be taken into account. When the method is applied to several sets of sporadic meteor data good agreement is obtained between both the corrected indices themselves and also with the theoretical value predicted by Dohnanyi.

104.012 **Meteors of periodic comet Mellish and the Geminids.** M. Kresáková.
Bull. Astron. Inst. Czechoslovakia, Vol. 25, 20 - 33 (1974).

Showers of faint telescopic meteors, occurring shortly before the annual apparition of the Geminids, have been observed on three different occasions at least. A search in all available catalogues of photographic meteor orbits suggests an association with a minor stream consisting of two components. The orbit of the long-period component excellently agrees with that of periodic comet Mellish. The short-period component revolves approximately in the same plane but with the aphelia in the asteroid belt; its orbital elements coincide within their internal dispersion with the Geminid stream, except for a reversal of the nodes.

104.013 **On the relation between the size, brightness, velocity and the spectral reddening of meteors.**
J. Rajchl.
Bull. Astron. Inst. Czechoslovakia, Vol. 25, 34 - 46 (1974).

A comprehensive diagram, demonstrating the relation between the magnitudes, radii, velocities and spectral types of meteors, has been constructed. This diagram, called the box diagram, together with the observations made hitherto, have yielded a new view of the problem of the reddening of fainter meteors.

104.014 **Radio observation of the Giacobinids 1972.**
M. Šimek.
Bull. Astron. Inst. Czechoslovakia, Vol. 25, 66 - 67 (1974).

The expected meteor shower associated with the comet Giacobini-Zinner was not confirmed. Only slight activity was observed between 15^h and 17^h UT., October 8, 1972.

104.015 **The parameters of orbits of meteor bodies from radio location observations.**
V. V. Fedynskij, Yu. I. Voloshchuk, B. L. Kashcheev, N. V. Novoselova, A. A. Tkachuk.
Doklady Akad. Nauk SSSR, Ser. Mat. Fiz., Vol. 215, 72 - 75 (1974). In Russian.

104.016 Effective sections of diffusion of meteor atoms and ions in the atmosphere for the broadened region of interaction energies. Yu. I. Portnyagin, V. S. Tokhtas'ev.
Geomagn. Aeronom., Vol. 14, 83 - 89 (1974). In Russian.

104.017 About the influence of diffusion on the precision of determination of the drift velocity of meteor traces.
B. L. Kashcheev, I. A. Delov.
Geomagn. Aeronom., Vol. 14, 172 - 174 (1974). In Russian. Brief information.

104.018 Feuerkugel am 4. Oktober 1973.
V. Kasten, M. Schnuchel, C. Göttig.
SuW, Vol. 13, 103 (1974).

104.019 Radiantenbestimmung. Hinweise auf die Existenz neuer Meteorströme. H. J. Becker.
SuW, Vol. 13, 119 - 122 (1974).

104.020 Space density of meteor bodies in the Draconid meteor stream 1972.
A. K. Terent'eva, S. S. Tryashin.
Astron. Tsirk., No. 797, p. 7 - 8 (1973). In Russian.

104.021 Light curves of very faint meteors.
J. Jones, R. L. Hawkes.
Nature, Vol. 248, 211 (1974).

On the night of September 31/October 1, 1973 in London, Ontario, 179 meteors were observed with a sensitive television equipment. The train lengths of the television meteors are considerably shorter than predicted by the compact meteoroid model and also significantly shorter than those of the photographic meteors. On the basis of these results the authors conclude that the compact meteoroid model is no longer tenable.

104.022 Strongly differentiated material in high-inclination and retrograde orbits. G. A. Harvey.
Astron. Journ., Vol. 79, 333 - 336, 345 - 346 (1974).

Much of our present understanding of the origin and evolution of the solar system has come from the study of meteorites which represent the oldest material available for analysis. The currently proposed sources for the meteoroids, especially the highly differentiated ones (main belt asteroids, Apollo asteroids, defunct short-period comet nuclei), are all in short-period, direct orbits and give earth-atmosphere entry velocities of about 40 km/sec or less. We now have direct evidence of the existence of strongly differentiated material in comet-type orbits. This evidence is a set of meteor spectrograms of essentially iron-free material entering the earth's atmosphere at high velocities ($V_e > 40$ km/sec) from high-inclination and retrograde orbits.

104.023 Werkgroep meteoren kan nog best waarnemers gebruiken! Overzicht van de campagne 1974.
B. Apeldoorn.
Zenit, Vol. 1, No. 2, p. 25 - 27 (1974).

104.024 Vorig jaar 2231 meteoren met het oog waargenomen.
B. Apeldoorn.
Zenit, Vol. 1, No. 3, p. 2 - 5 (1974).

104.025 Heldere boliden. Aan de nachthemem van april kan van alles gebeuren. P. A. Koning.
Zenit, Vol. 1, No. 3, p. 18 - 19 (1974).

104.026 Lyriden: Een zwerm met vele heldere meteoren.
B. Apeldoorn.
Zenit, Vol. 1, No. 4, p. 28 - 29 (1974).

104.027 Determination of the Quadrantid incident flux density. Part I.
O. I. Bel'kovich, V. S. Tokhtas'ev.
Bull. Astron. Inst. Czechoslovakia, Vol. 25, 112 - 115 (1974).

A modified method of computing the incident flux from radar meteor shower observations is presented. The method was applied for the 1966 Quadrantid shower.

104.028 Theoretical radiants of the hypothetical meteor shower of comet Kohoutek 1973 f.
A. K. Terent'eva.
Astron. Tsirk., No. 807, p. 6 - 8 (1974). In Russian.

104.029 Meteor ionization.
V. N. Lebedinets, V. B. Shushkova.
Geomagn. Aeronom., Vol. 14, 271 - 276 (1974). In Russian.

104.030 Observations of telescopic meteors of the Perseid stream in Sofia (1968).
S. B. Vladimirov, N. S. Nikolov.
Komety i Meteory, *Dushanbe*, No. 20, p. 14 - 16 (1971).
In Russian.

112 meteors were observed during 11.4 hours. Two radiants have been determined. The distributions of telemeteors according to their brightness and to the length of trail are given.

104.031 Results of photographic observations of meteor flares in Dushanbe.
P. B. Babadzhanov, I. M. Khaimov.
Komety i Meteory, *Dushanbe*, No. 20, p. 17 - 25 (1971).
In Russian.

On the basis of statistical studies of meteor flares it is established that flares of 1 - 2 km length and 0.02 - 0.06 sec duration are predominant. Increasing meteor velocities involve increase of the altitude of flares and decrease of their duration. Integral brightness of flares is proportional to that of meteors.

104.032 On the correlation of E_s with the Perseid meteor stream. D. Latipov.
Komety i Meteory, *Dushanbe*, No. 20, p. 26 - 28 (1971).
In Russian.

The behaviour of sporadic E layers is examined in the action period of the Perseid meteor stream of 1964. It is concluded that meteor streams cause an increase of maximum frequency and other parameters of E_s only in active periods.

104.033 The dependence of the fragmentation parameter on the inclination of meteor trains.
I. M. Khaimov.
Komety i Meteory, *Dushanbe*, No. 20, p. 29 - 32 (1971).
In Russian.

The fragmentation parameter is determined as a function of zenith distance of the radiant on the basis of 284 meteors of Dushanbe and 413 of Harvard.

104.034 Theoretical and experimental study of the rate of change of the critical radius of a meteor trail.
G. I. Kolomiets, E. I. Fialko.
Komety i Meteory, *Dushanbe*, No. 21, p. 25 - 31 (1972).
In Russian.

Results are given of a theoretical and experimental study of the rate of change of the critical radius of meteor trails. A discussion is given of measurement errors of drift velocity of meteor trails.

104.035 The influence of atmospheric dust and neutral particles of meteor matter on the formation rate of negative ions in a trail. R. Sh. Bibarsov.
Komety i Meteory, *Dushanbe*, No. 21, p. 32 - 43 (1972).
In Russian.

It is shown that the average value of the electron attachment rate in a meteor trail at a height of 95 km is 50 times that of aeronomical data. The resulting attachment rate for reactions of electron + atmospheric oxygen and electron + dust equals 5 percent that observed in a meteor trail. The meteor neutral particles affect essentially the free electron concentration in the trail.

104.036 **Comparison of the results of photographic and photoelectric observations of meteors.**
V. A. Smirnov, M. Toktogulov.
Komety i Meteory, *Dushanbe*, No. 21, p. 44 - 48 (1972). In Russian.

104.037 **Classification of ionized meteor trails.**
G. I. Kolomiets, E. I. Fialko.
Komety i Meteory, *Dushanbe*, No. 22, p. 30 - 37 (1973). In Russian.

104.038 **Distribution of meteor bodies according to kinetic energies (The case of inverse power law distribution of velocities of meteor bodies).**
E. I. Fialko, V. F. Romanyuk.
Komety i Meteory, *Dushanbe*, No. 22, p. 38 - 42 (1973). In Russian.

104.039 **On the method of synchronous measurement of meteor drift velocities.** G. I. Kolomiets.
Komety i Meteory, *Dushanbe*, No. 22, p. 43 - 45 (1973). In Russian.

104.040 **Spectrophotometric study of five bright meteors.**
K. Kh. Saidov, O. F. Zolova.
Byull. Inst. Astrofiz., *Dushanbe*, No. 59, p. 3 - 27 (1971). In Russian.
The results of a spectrophotometric study of five bright meteors and identification of spectral lines are given. Distribution of energy for different points of the meteor paths is found. Masses of meteor particles are determined on the basis of integrated curves of brightness.

104.041 **The spectra of two sporadic meteors.**
K. Kh. Saidov, O. F. Zolova.
Byull. Inst. Astrofiz., *Dushanbe*, No. 61, p. 12 - 26 (1972). In Russian.
The spectra of two sporadic meteors were studied. Lines of Na I, Mg I, Mg II, Ni I, Cr I, Cr II, Mn I, Mn II, Fe I, Fe II, Ca I, Ca II and Co I in the spectra were identified. Spectral energy distribution and masses of meteor particles were determined.

104.042 **Abundance of sodium in meteor spectra.**
K. Kh. Saidov.
Byull. Inst. Astrofiz., *Dushanbe*, No. 61, p. 27 - 30 (1972). In Russian.
Photometry was made of the sodium line in three meteor spectra. The abundance of sodium in meteor bodies is found to be $\approx 6\%$.

104.043 **Some results of studies of meteor phenomena and the sporadic E-layer.**
D. Latipov, Sh. O. Isamutdinov.
Byull. Inst. Astrofiz., *Dushanbe*, No. 62, p. 10 - 15 (1974). In Russian.

104.044 **On effective meteor temperatures from sodium lines.**
K. Kh. Saidov.
Byull. Inst. Astrofiz., *Dushanbe*, No. 62, p. 19 - 21 (1974). In Russian.
An attempt is made to determine the effective meteor coma temperature on the basis of intensity measurements of sodium lines in the meteor spectra. The effective temperature is shown to be $1500-2600°K$ in the velocity range 40–60 km/sec.

104.045 **On colour indices of bright meteors.**
K. Kh. Saidov, O. F. Zolova.
Byull. Inst. Astrofiz., *Dushanbe*, No. 62, p. 22 - 25 (1974). In Russian.
Color indices of bright meteors are investigated on the basis of their spectra. A certain dependence of color index on absolute meteor magnitude is found.

104.046 **On observational conditions of meteor streams at individual observation places.**
Sh. O. Isamutdinov, D. Latipov.
Dokl. AN TadzhSSR, Vol. 16, No. 6, p. 19 - 22 (1973). In Russian. – Abstr. in Referativ. Zhurn. 51. Astron., 4.51.352 (1974).

104.047 **Preliminary results of studying the mass distribution of Geminid meteors in 1959 - 1966.** M. Shimek.
Issled. verkhn. atmosf. Zemli. Moskva, Gidrometeoizdat, 1973, p. 222 - 225. In Russian. – Abstr. in Referativ. Zhurn. 51. Astron., 4.51.353 (1974).

104.048 **Four years of meteor spectra patrol.**
G. A. Harvey.
Sky Telescope, Vol. 47, 378 - 380 (1974).

104.049 **Radar determination of the mass index of the Geminid meteor shower.**
B. A. McIntosh, M. Šimek.
Bull. Astron. Inst. Czechoslovakia, Vol. 25, 180 - 182 (1974).
Geminid meteor echo counts from 5 years observing with the Ondřejov radar are analysed using new instrumental corrections for short duration echoes and a modified theory which allows a variable mass index s. It is found that $s = 1.4$ for meteors producing echoes of 1 sec duration, and 1.8 at 10 sec duration.

104.050 **Comment on the determination of the true duration of persistent meteor echoes.**
M. Šimek, V. Novotný.
Bull. Astron. Inst. Czechoslovakia, Vol. 25, 183 - 184 (1974).
The duration of recorded meteor echoes is affected by instrumental error which arises mainly from the slow motion of the recording film. A precise calibration has been carried out and the corrections for the Ondřejov records are presented along with a revised formula for the mass index of the Geminid meteors (Šimek, 1973).

104.051 **On the possible origin of the Taurids.**
N. V. Kulikova.
Astron. vestn., Vol. 8, 92 - 95 (1974). In Russian.
A supposition on a possible origin of Taurids is made on the basis of the results of statistic test modelling of isotropic ejection of meteoric matter from the nucleus of comet Encke-Backlund at any point of its orbit, and the calculation of the deviation of the elements of the ejected meteoric particle orbits from the elements of parent cometary orbit.

104.052 **On the mineralogical density of meteoric bodies in some streams.** V. V. Benyukh.
Astron. vestn., Vol. 8, 96 - 101 (1974). In Russian.
Densities of artificial meteoric bodies of predetermined composition, structure and form were derived by a method based on heat conduction equations. Mineralogical densities of 755 meteoric bodies mainly belonging to basic meteoric streams were determined by the same method. All density values are in the range from 1.0 to 8.0 g/cm^3.

104.053 **Results of laboratory studies of a TV system for**

observation of meteors.
S. S. Georgienko, V. A. Smirnov.
Astron. vestn., Vol. 8, 102 - 105 (1974). In Russian.

104.054 **Minor meteor showers in March 1973.**
N. V. Smirnov, T. L. Korovkina.
Astron. vestn., Vol. 8, 114 - 118 (1974). In Russian.

104.055 **Helle Meteorspuren am 18. April 1974.** R. A. Naef.
Orion, 32. Jahrgang, p. 112 (1974).

104.056 **Micrometeoric matter near the earth.**
T. N. Nazarova.
Pyl' v atmosf. i okolozemn. kosmich. prostranstve. Moskva, Nauka, 1973, p. 8 - 13. In Russian. − Abstr. in Referativ. Zhurn. 51. Astron., 5.51.342; 62. Issled. kosmich. prostranstva, 5.62.190 (1974).

104.057 **Micrometeoroids in cosmic space.** E. P. Mazets.
Pyl' v atmosf. i okolozemn. kosmich. prostranstve. Moskva, Nauka, 1973, p. 13 - 23. In Russian. − Abstr. in Referativ. Zhurn. 51. Astron., 5.51.343 (1974).

104.058 **On registering of micrometeoroids in meteor showers.** R. L. Aptekar', M. M. Bredov, S. V. Golenetskij, Yu. A. Gur'yan, V. N. Il'inskij, E. P. Mazets, V. N. Panov.
Pyl' v atmosf. i okolozemn. kosmich. prostranstve. Moskva, Nauka, 1973, p. 23 - 30. In Russian. − Abstr. in Referativ. Zhurn. 51. Astron., 5.51.344; 62. Issled. kosmich. prostranstva, 5.62.191 (1974).

104.059 **Application of a statistical test method to the theory of meteor stream formation.**
L. A. Katasev, N. V. Kulikova.
Pyl' v atmosf. i okolozemn. kosmich. prostranstve. Moskva, Nauka, 1973, p. 56 - 59. In Russian. − Abstr. in Referativ. Zhurn. 51. Astron., 5.51.347 (1974).

104.060 **Motion of smaller meteor bodies in the upper atmosphere and circumterrestrial space.**
V. N. Lebedinets, V. B. Shushkova.
Pyl' v atmosf. i okolozemn. kosmich. prostranstve. Moskva, Nauka, 1973, p. 59 - 62. In Russian. − Abstr. in Referativ. Zhurn. 51. Astron., 5.51.348 (1974).

104.061 **Telemeteor matter on the earth's orbit.**
A. M. Bakharev.
Pyl' v atmosf. i okolozemn. kosmich. prostranstve. Moskva, Nauka, 1973, p. 62 - 65. In Russian. − Abstr. in Referativ. Zhurn. 51. Astron., 5.51.349 (1974).

104.062 **Statistical investigation of orbital elements of small meteor streams.** F. Karmiński.
Stud. Soc. Sci. Torunensis, Sectio F (Astron.), Vol. 5, 105 - 109 = Bull. Astron. Obs. Toruń, No. 51/IV (1973).
The general method of this study consists of statistical investigation of orbits of these objects in the planetary system, which might have a common origin with the asteroids.

104.063 **Observations of the Giacobinids in 1972.**
K. Saito, K. Tomita.
Tokyo Astron. Obs., Report No. 63, Vol. 16, 699 - 713 (1974). In Japanese.

104.064 **A meteoroid model.** D. W. Hughes.
Journ. British Astron. Ass., Vol. 84, 272 - 274 (1974).

104.065 **A meteorite that missed the earth.** L. G. Jacchia.
Sky Telescope, Vol. 48, 4 - 9 (1974).

104.066 **Determination of density and mass of a meteor body by the radio method.**
R. Sh. Bibarsov, R. P. Chebotarev.
Dokl. AN TadzhSSR, Vol. 16, No. 8, p. 18 - 22 (1973). In Russian. − Abstr. in Referativ. Zhurn. 51. Astron., 6.51.336 (1974).

104.067 **Determination of the energy of electron affinity of particles of meteoric matter.** R. Sh. Bibarsov.
Dokl. AN TadzhSSR, Vol. 16, No. 11, p. 13 - 15 (1973). In Russian. − Abstr. in Referativ. Zhurn. 51. Astron., 6.51.337 (1974).

104.068 **Meteor astronomy.** N. I. Ivanov.
Astronomical Institute of the Academy of Sciences of the Uzbek SSR − 100 years, (see 003.011), p. 130 - 134 (1974). In Russian.

104.069 **Observations of a fire ball on July 8, 1973.**
M. Szulc.
Urania Kraków, Vol. 45, 93 (1974). In Polish.

104.070 **The Lyncids: a new meteor radiant.**
M. Savill, J. Mason.
Astron. and Space, (GB), Vol. 3, 280 - 285 (1973).

104.071 **Distribution of the minor meteor shower radiants observed during October−February 1960−72.**
Srirama Rao, M. Srirama Rao, S. Rajaratnam, A. Gopalakrishnamurthy.
Current Sci., (India), Vol. 42, 785 - 786 (1973).
Srirama Rao et al. (1969) have identified 10 groups of established minor meteor showers over Waltair during November, December and January, 1961−67. Further 17 and 12 groups of established minor meteor showers were established during February 1962−72 and October 1962−71 respectively. The distribution of these established minor meteor showers on the celestial sphere in relation to the plane of the ecliptic was studied.

104.072 **'Pisciden' een nieuwe zwerm meteoren?**
T. Oosterloo, K.-W. Treurniet, P. Roelfsema.
Zenit, Vol. 1, No. 6, p. 25, 31 (1974).

104.073 **Bijzonder heldere bolide gezien op 4 januari.**
B. Apeldoorn.
Zenit, Vol. 1, No. 6, p. 36 - 38 (1974).

104.074 **Correlation of maximum and average daily meteor echo rates per hour.** E. I. Fialko, V. N. Donij.
Astron. Tsirk., No. 818, p. 7 - 8 (1974). In Russian.

A self-operating meteor camera. See Abstr. 034.056.

The micrometeoroid detector aboard the satellite Prospero. See Abstr. 054.015.

Superrotation of upper atmosphere by global deposition of meteoroids. See Abstr. 082.009.

Vertical propagation of tides at meteor heights. See Abstr. 082.022.

The atmospheric circulation at the altitudes of registration of drifts of meteor trails and ionospheric irregularities. See Abstr. 083.031.

Periodic comet Schwassmann-Wachmann 3 (1930 VI). See Abstr. 103.134.

Interplanetary material. See Abstr. 106.008.

105 Meteorites, Meteorite Craters

105.001 Zirconium and hafnium in meteorites.
W. D. Ehmann, L. L. Chyi.
Earth Planet. Sci. Letters, Vol. 21, 230 - 234 (1974).

The abundances of zirconium and hafnium have been determined in nine stony meteorites by a new, precise neutron-activation technique. The Zr/Hf abundance ratios for the chondrites vary in a rather narrow range, consistent with previously published observations from our group. Replicate analyses of new, carefully selected clean interior samples of the C1 chondrite Orgueil yield mean zirconium and hafnium abundances of 5.2 and 0.10 ppm, respectively.

105.002 Chemical and petrographic correlations among carbonaceous chondrites.
W. R. Van Schmus, J. M. Hayes.
Geochim. Cosmochim. Acta, Vol. 38, 47 - 64 (1974).

Detailed study of the petrographic and chemical properties of carbonaceous chondrites shows that the four distinct petrographic subtypes may be related to one of two distinct chemical subdivisions. These subdivisions are recognized primarily by the relative abundances of the nonvolatile elements Si, Ca, Al, Ti, Cu and Fe. Cl, C2 and C3 (O) chondrites form one subdivision. Vigarano subtype chondrites form the other subdivision and include chondrites previously referred to as C2, C3 and C4.

105.003 Mesosiderites–I. Compositions of their metallic portions and possible relationship to other metal-rich meteorite groups.
J. T. Wasson, R. Schaudy, R. W. Bild, C.-L. Chou.
Geochim. Cosmochim. Acta, Vol. 38, 135 - 149 (1974).

The metal from 17 mesosiderites has been analyzed for Ni, Ga, Ge and Ir by the techniques of atomic-absorption spectrometry and neutron activation. Most mesosiderite metal samples fall in a narrow compositional range: Ni, 7.0–9.0 per cent; Ga, 13–16 ppm; Ge, 47–58 ppm; and Ir, 2.4–4.4 ppm. Most of those falling outside these ranges belong to Powell's (1971) least-metamorphosed type.

105.004 Origin of the high-temperature fraction of C2 chondrites. L. Grossman, E. Olsen.
Geochim. Cosmochim. Acta, Vol. 38, 173 - 187 (1974).

The coarse-grained fraction of C2 chondrites is composed mostly of single crystals and aggregates of crystals of Mg-rich olivine and pyroxene. They do not possess compelling textural evidence of being the solidification products of rapidly-quenched molten droplets. It is certain they have never been involved in planetary processes that significantly modified their original states. It is the purpose of this paper to treat quantitatively the condensation of metal in terms of several models, to apply these models to the observed compositions of C2 metal grains and to determine details of the early stages of the condensation of the solar system.

105.005 Oxygen isotope cosmothermometer revisited.
N. Onuma, R. N. Clayton, T. K. Mayeda.
Geochim. Cosmochim. Acta, Vol. 38, 189 - 191 (1974).

A previously published estimate of the oxygen isotopic composition of the gas of the early solar nebula must be revised in light of the discovery of non-chemical isotope effects in carbonaceous chondrites. The solids which accreted to form the earth, moon and ordinary chondrites probably did not equilibrate isotopically with the gas below 1000 K.

105.006 Radionuclides produits par le rayonnement cosmique dan la météorite Saint-Séverin.
R. Bibron, C. Leger, J. Tobailem, Y. Yokoyama, H. Mabuchi, N. Baillard.
Geochim. Cosmochim. Acta, Vol. 38, 197 - 205 (1974).

The activities of 18 radionuclides in 4 samples from the Saint-Séverin meteorite have been measured after chemical separation. The results are reported in a table.

105.007 A comparative study on ^{26}Al and ^{53}Mn in eighteen chondrites.
M. Heimann, P. P. Parekh, W. Herr.
Geochim. Cosmochim. Acta, Vol. 38, 217 - 234 (1974).

An investigation on the low energy cosmic ray products ^{26}Al and ^{53}Mn in 18 chondrites encompassed several side aspects in meteoritics. The ratio of observed to calculated specific activity of ^{26}Al in 13 chondrites with recovered masses ranging from $\sim$ 3 to 2000 kg remains constant within $\pm$ 15 per cent of a mean of 0.93. A plot of the ^{53}Mn specific production rate (dpm/kg Fe) versus recovered meteorite mass revealed that the compensation by the secondaries for the absorption of the primaries is also effective in the case of ^{53}Mn. Thus, the ^{53}Mn production rate remains rather constant within $\pm$ 15 per cent of a mean of 520 dpm/kg Fe for most of the stones. Applying this information the observed production rate of a cosmogenic species is a function of target element composition, of the absolute exposure time, intensity and energy spectrum of the cosmic ray flux, and the position of the specimen studied relative to the preatmospheric size and shape of the meteorite.

105.008 Surface properties of the Orgueil meteorite: implications for the early history of solar system volatiles. F. P. Fanale, W. A. Cannon.
Geochim. Cosmochim. Acta, Vol. 38, 453 - 470 (1974).

This study deals with the importance of adsorption in sheet structures for the early history of solar system volatiles. We studied Orgueil because, among specimens in hand, there is reason to believe that CI meteorites represent the most primitive, least volatile-depleted solid material in the solar system (Urey, 1953; Mason, 1960; Kerridge, 1972); however, it appears unlikely that most other types of meteorites or planetary objects were directly produced from them by metamorphism in a closed system (Urey, 1961).

105.009 Equilibrium temperatures in enstatite chondrites.
J. W. Larimer, P. R. Buseck.
Geochim. Cosmochim. Acta, Vol. 38, 471 - 477 (1974).

105.010 Tungus event revisited.
G. L. Wick, J. D. Isaacs.
Nature, Vol. 247, 139 - 140 (1974).

All the data presented support the hypotheses that the Tunguska event was caused by a comet with a head composed of iron and silicate fragments cemented together by frozen volatiles. On contact with the earth's atmosphere, the comet exploded, scattering dust and fragments over the landscape and into the atmosphere, leaving no significant crater.

105.011 Gas-rich meteorites: Possible evidence for origin on a regolith.
D. Macdougall, R. S. Rajan, P. B. Price.
Science, Vol. 183, 73 - 74 (1974).

The authors report several new features of track-rich grains observed during a detailed investigation of two well-known gas-rich meteorites, Kapoeta and Fayetteville. They outline briefly the new observations and their implications.

105.012 Thermomagnetic analysis of meteorites, 1. C1 chondrites. E. E. Larson, D. E. Watson, J. M.

Herndon, M. W. Rowe.
Earth Planet. Sci. Letters, Vol. 21, 345 - 350 (1974).

The authors have begun a systematic study of the thermomagnetic properties of all carbonaceous chondrites. In this paper they present the results of their thermomagnetic analysis of the five C1 chondrites.

105.013 **Meteoritic magnetite as a cosmothermometer.**
J. M. Herndon, M. W. Rowe.
Nature, Vol. 247, 531 - 532 (1974). – Letter.

105.014 **The metal phase in unequilibrated ordinary chondrites and its implications for calculated accretion temperatures.** R. T. Dodd.
Geochim. Cosmochim. Acta, Vol. 38, 485 - 494 (1974).

The author assumes that the metal grains in low type 3 ordinary chondrites are not significantly different in size from those available to the volatile elements during their condensation. He presents new data on the size distribution of metal particles in type 3 chondrites, reviews data pertinent to their abundance and composition. and considers the implications of the results for the calculation of accretion temperatures.

105.015 **The chemical composition of metallic spheroids and metallic particles within impactite from Barringer meteorite crater, Arizona.**
W. R. Kelly, E. Holdsworth, C. B. Moore.
Geochim. Cosmochim. Acta, Vol. 38, 533 - 543 (1974).

Atomic absorption analyses of 25 metallic spheroids from Barringer meteorite crater have been carried out for Fe, Ni, Co and Cu. In addition, electron microprobe analyses of 58 impactite metallic particles have been carried out for Fe, Ni and Co from four different impactite samples.

105.016 **Characteristics of fission tracks in zircon: applications to geochronology and cosmology.**
S. Krishnaswami, D. Lal, N. Prabhu, D. Macdougall.
Earth Planet. Sci. Letters, Vol. 22, 51 - 59 (1974).

The authors describe a new etchant for revelation of tracks in zircon and explore the practical aspects of using this ubiquitous uranium-rich accessory mineral for fission track dating. They also report the track recording and retention characteristics of zircon using the new etch. A simple chemical technique has been developed to quantitatively recover zircon crystals larger than $\sim 50~\mu m$ from 10 - 100 g of whole rock sample.

105.017 **Particles around Boxhole meteorite crater.**
P. W. Hodge, F. W. Wright.
Meteoritics, Vol. 8, 315 - 320 (1973).

A sampling program of soil surrounding the Boxhole meteorite crater reveals very much smaller amounts of meteoritic material in the soil than found for other craters, such as Henbury and Canyon Diablo.

105.018 **$^{87}Rb/^{87}Sr$ age of Juvinas basaltic achondrite and early igneous activity in the solar system.**
C. J. Allégre, J. L. Birck, S. Fourcade.
Meteoritics, Vol. 8, 323 (1973). – Abstr. Meteoritical Soc.

105.019 **Elemental fractionations among enstatite chondrites.**
P. A. Baedecker, J. T. Wasson.
Meteoritics, Vol. 8, 323 - 324 (1973). – Abstr. Meteoritical Soc.

105.020 **The meteoritic shower of Parambú, Ceará State, Brazil: mineralogy and petrology.**
A. Barreto, Z. Fonseca de Mello, G. R. Levi-Donati, G. P. Sighinolfi.
Meteoritics, Vol. 8, 324 (1973). – Abstr. Meteoritical Soc.

105.021 **Low temperature meteorite formation: some simulation experiments.** M. R. Bloch, O. Müller.
Meteoritics, Vol. 8, 327 - 328 (1973). – Abstr. Meteoritical Soc.

105.022 **Carbon-14 in stone meteorites.** W. Born.
Meteoritics, Vol. 8, 328 - 329 (1973). – Abstr. Meteoritical Soc.

105.023 **Reconsideration of the l'Aigle hypersthene chondrite.** M. Bourot, M. Christophe-Michel-Lévy.
Meteoritics, Vol. 8, 329 (1973). – Abstr. Meteoritical Soc.

105.024 **Origin of enstatite achondrites.**
W. V. Boynton, R. A. Schmitt.
Meteoritics, Vol. 8, 330 (1973). – Abstr. Meteoritical Soc.

105.025 **The electrical conductivity of carbonaceous meteorites.** A. Brecher.
Meteoritics, Vol. 8, 330 - 331 (1973). – Abstr. Meteoritical Soc.

105.026 **The pallasite Imilac, Chile.** V. F. Buchwald.
Meteoritics, Vol. 8, 333 - 334 (1973). – Abstr. Meteoritical Soc.

105.027 **Data on the structure of crater 10, Campo del Cielo, Argentina.** W. A. Cassidy, M. L. Renard.
Meteoritics, Vol. 8, 335 (1973). – Abstr. Meteoritical Soc.

105.028 **Phosphide growth in coarse structured iron meteorites.** R. S. Clarke, Jr., J. I. Goldstein.
Meteoritics, Vol. 8, 335 - 336 (1973). – Abstr. Meteoritical Soc.

105.029 **A component of primitive nuclear composition in carbonaceous chondrites.**
R. N. Clayton, L. Grossman, T. K. Mayeda.
Meteoritics, Vol. 8, 336 - 337 (1973). – Abstr. Meteoritical Soc.

105.030 **Depression of meteoritic ^{26}Al production rates by low shielding.** P. J. Cressy, Jr., G. F. Herzog.
Meteoritics, Vol. 8, 337 - 338 (1973). – Abstr. Meteoritical Soc.

105.031 **The unique cosmic-ray history of the Malakal chondrite.** P. J. Cressy, Jr., L. A. Rancitelli.
Meteoritics, Vol. 8, 338 (1973). – Abstr. Meteoritical Soc.

105.032 **A chemical investigation of minerals from L-6 chondritic meteorites.**
D. B. Curtis, R. A. Schmitt.
Meteoritics, Vol. 8, 339 - 340 (1973). – Abstr. Meteoritical Soc.

105.033 **The role of adiabatic compression in the formation of tektites.** E. J. H. David.
Meteoritics, Vol. 8, 340 - 343 (1973). – Abstr. Meteoritical Soc.

105.034 **Olivine compositions in howardites and other achondritic meteorites.**
C. Desnoyers, D. Y. Jérome.
Meteoritics, Vol. 8, 344 - 345 (1973). – Abstr. Meteoritical Soc.

105.035 **Araguainha Dome and Serra da Cangalha, Brazil: probable astroblemes.** R. S. Dietz, B. French.
Meteoritics, Vol. 8, 345 - 347 (1973). – Abstr. Meteoritical Soc.

105.036 "Aging" of fossil nuclear particle tracks in extraterrestrial material.
J. C. Dran, M. Maurette, M. Dumail, J. P. Duraud.
Meteoritics, Vol. 8, 347 - 354 (1973). — Abstr. Meteoritical Soc.

105.037 Halogens in meteoritic and lunar samples.
G. Dreibus, E. Jagoutz, H. Wänke.
Meteoritics, Vol. 8, 354 - 355 (1973). — Abstr. Meteoritical Soc.

105.038 Cosmic-ray record in the San Juan Capistrano meteorite.   R. C. Finkel, D. Lal, K. Marti.
Meteoritics, Vol. 8, 365 (1973). — Abstr. Meteoritical Soc.

105.039 Composition and origin of lithic fragments in L- and H-group chondrites.
R. V. Fodor, K. Keil.
Meteoritics, Vol. 8, 366 - 367 (1973). — Abstr. Meteoritical Soc.

105.040 Partitioning of potassium between silicates and sulfide melts.   K. A. Goettel.
Meteoritics, Vol. 8, 373 (1973). — Abstr. Meteoritical Soc.

105.041 Noble gases in three Indian meteorites.
K. Gopalan, M. N. Rao.
Meteoritics, Vol. 8, 374 (1973). — Abstr. Meteoritical Soc.

105.042 High-temperature condensates in C2 chondrites.
L. Grossman, E. Olsen.
Meteoritics, Vol. 8, 374 - 375 (1973). — Abstr. Meteoritical Soc.

105.043 Some applications of ^{26}Al and ^{53}Mn in meteoritics.
M. Heimann, W. Herr.
Meteoritics, Vol. 8, 377 (1973). — Abstr. Meteoritical Soc.

105.044 Bulk, rare earth and other abundances in the highgrade metamorphosed anorthosite complex, Fiskenaesset, West Greenland.   P. Henderson, S. J. Fishlock, J. C. Laul, T. D. Cooper, R. A. Schmitt.
Meteoritics, Vol. 8, 377 - 378 (1973). — Abstr. Meteoritical Soc.

105.045 W, Re, Os, Ir, Pt, Au, Hg, Tl, Pb, Bi, Th and U in the four recently found Antarctic meteorites Yamato a, b, c, and d, and in 5 other stony meteorites of different classes.
H. Hintenberger, K. P. Jochum, M. Seufert.
Meteoritics, Vol. 8, 380 - 382 (1973). — Abstr. Meteoritical Soc.

105.046 Spinel-bearing lithic fragments and chondrules from the Mezö-Madaras chondrite.
G. Hoinkes, G. Kurat.
Meteoritics, Vol. 8, 383 - 384 (1973). — Abstr. Meteoritical Soc.

105.047 Semestrial distribution of falls of meteorites belonging to different classes.
J. Jedwab, S. Huyberechts.
Meteoritics, Vol. 8, 389 - 391 (1973). — Abstr. Meteoritical Soc.

105.048 Specific area of carbonaceous meteorites.
J. Jedwab, R. Wollast, C. de Donder.
Meteoritics, Vol. 8, 391 - 392 (1973). — Abstr. Meteoritical Soc.

105.049 Pre- or post-accretional distribution of trace elements?   S. Jovanovic, G. W. Reed, Jr.
Meteoritics, Vol. 8, 392 - 393 (1973). — Abstr. Meteoritical Soc.

105.050 ^{26}Al in iron meteorites.   H. Kammer.
Meteoritics, Vol. 8, 393 - 394 (1973). — Abstr. Meteoritical Soc.

105.051 Composition and origin of lithic fragments in LL-group chondrites.   K. Keil, R. V. Fodor.
Meteoritics, Vol. 8, 394 - 396 (1973). — Abstr. Meteoritical Soc.

105.052 Sulfur concentrations of metallic spheroids and impactite bombs from Barringer meteorite crater, Arizona.   W. R. Kelly, J. Cripe, C. B. Moore.
Meteoritics, Vol. 8, 396 - 397 (1973). — Abstr. Meteoritical Soc.

105.053 Crystallographic studies of cloudy taenite in iron-nickel meteorites.   M. R. Kimball.
Meteoritics, Vol. 8, 397 - 399 (1973). — Abstr. Meteoritical Soc.

105.054 Siderophile elements on the earth, moon, and the eucrite parent body.
K. Kimura, R. S. Lewis, E. Anders.
Meteoritics, Vol. 8, 399 - 400 (1973). — Abstr. Meteoritical Soc.

105.055 Isotope studies in the Mundrabilla iron meteorite.
T. Kirsten.
Meteoritics, Vol. 8, 400 - 403 (1973). — Abstr. Meteoritical Soc.

105.056 Experimental determination of the melting relations of the Allende meteorite.
I. Kushiro, M. G. Seitz.
Meteoritics, Vol. 8, 403 - 405 (1973). — Abstr. Meteoritical Soc.

105.057 The pyroxene equilibrium temperatures in chondrites of various petrological types according to the Mössbauer spectroscopy data.
A. K. Lavrukhina, T. V. Malysheva.
Meteoritics, Vol. 8, 407 - 408 (1973). — Abstr. Meteoritical Soc.

105.058 Experimental evidence of low-temperature thermal metamorphism in stony meteorites.   F. A. Levi.
Meteoritics, Vol. 8, 408 - 409 (1973). — Abstr. Meteoritical Soc.

105.059 The Mills, New Mexico, chondrite — a new find.
G. R. Levi-Donati, E. Jarosewich.
Meteoritics, Vol. 8, 409 (1973). — Abstr. Meteoritical Soc.

105.060 Tracks studies in Keyes and Saint Séverin chondrites.   J. C. Lorin, G. Poupeau.
Meteoritics, Vol. 8, 410 - 411 (1973). — Abstr. Meteoritical Soc.

105.061 Determination of the noble metals in geological materials by multielement neutron activation analysis.   G. H. Morrison, R. A. Nadkarni.
Meteoritics, Vol. 8, 413 - 414 (1973). — Abstr. Meteoritical Soc.

105.062 Enrichment of volatile elements in Muong Nong-type tektites: clues for their formation history?
O. Müller, W. Gentner.

Meteoritics, Vol. 8, 414 - 415 (1973). – Abstr. Meteoritical Soc.

105.063 **The identification of early solar condensates in the Allende meteorite.**
D. A. Papanastassiou, C. M. Gray, G. J. Wasserburg.
Meteoritics, Vol. 8, 417 - 418 (1973). – Abstr. Meteoritical Soc.

105.064 **Search for pre-irradiation effects in Allegan chondrite.** P. Pellas, A. Ducatel, J. L. Berdot.
Meteoritics, Vol. 8, 418 - 419 (1973). – Abstr. Meteoritical Soc.

105.065 **Xe isotopes in carbonaceous chondrites.**
D. Phinney.
Meteoritics, Vol. 8, 420 - 422 (1973). – Abstr. Meteoritical Soc.

105.066 **Discovery of micrometeorite craters in chondrule-like objects from Kapoeta.**
R. S. Rajan, P. B. Price, D. E. Brownlee.
Meteoritics, Vol. 8, 425 - 427 (1973). – Abstr. Meteoritical Soc.

105.067 **Trace element content of chondritic metals.**
E. Rambaldi.
Meteoritics, Vol. 8, 427 - 428 (1973). – Abstr. Meteoritical Soc.

105.068 **A Swiss review about ancient timepieces.**
J. H. Reynolds.
Meteoritics, Vol. 8, 428 - 430 (1973). – Abstr. Meteoritical Soc.

105.069 **The Campo del Cielo, Argentina meteorite crater field.** A. Romaña, W. A. Cassidy.
Meteoritics, Vol. 8, 430 - 431 (1973). – Abstr. Meteoritical Soc.

105.070 **Thermomagnetic analysis of carbonaceous chondrites and ureilites.**
M. W. Rowe, J. M. Herndon, D. E. Watson, E. E. Larson.
Meteoritics, Vol. 8, 432 (1973). – Abstr. Meteoritical Soc.

105.071 **Depth dependence of spallogenic noble gases in the St. Séverin chondrite.**
L. Schultz, D. Phinney, P. Signer.
Meteoritics, Vol. 8, 435 - 436 (1973). – Abstr. Meteoritical Soc.

105.072 **Factor analysis applications to enstatite chondrite geochemistry.**
D. M. Shaw, M. E. Lipschutz, C. M. Binz, R. K. Kurimoto.
Meteoritics, Vol. 8, 438 (1973). – Abstr. Meteoritical Soc.

105.073 **Mineralogical and chemical composition of new Antarctica meteorites.** Mak. Shima, Mas. Shima.
Meteoritics, Vol. 8, 439 - 440 (1973). – Abstr. Meteoritical Soc.

105.074 **The Parambú meteorite: chemical compositions determined by wet chemical methods and by spark source mass spectrometry.**
Mas. Shima, G. P. Sighinolfi, K. P. Jochum, H. Hintenberger.
Meteoritics, Vol. 8, 440 - 441 (1973). – Abstr. Meteoritical Soc.

105.075 **Cordierite glass formed by shock in solid state in a cordierite-garnet-gneiss from the Ries crater, Germany.** V. Stähle.
Meteoritics, Vol. 8, 443 (1973). – Abstr. Meteoritical. Soc.

105.076 **Petrographic and grain size characteristics of suevite and lunar impact breccias.** D. Stöffler.
Meteoritics. Vol. 8, 443 - 444 (1973). – Abstr. Meteoritical Soc.

105.077 **The U-Th-Pb systematics of the Allende carbonaceous chondrite.**
M. Tatsumoto, D. M. Unruh, G. A. Desborough.
Meteoritics, Vol. 8, 446 (1973). – Abstr. Meteoritical Soc.

105.078 **^{40}Ar-^{39}Ar chronology of chondrites.**
G. Turner, P. H. Cadogan.
Meteoritics, Vol. 8, 447 - 448 (1973). – Abstr. Meteoritical Soc.

105.079 **The distribution of platinoids and gold between various phases of chondrites.**
A. P. Vinogradov, A. K. Lavrukhina.
Meteoritics, Vol. 8, 448 - 449 (1973). – Abstr. Meteoritical Soc.

105.080 **Sampling model of chondrule compositional variations.**
L. Walter, R. Dodd, P. Smidinger.
Meteoritics, Vol. 8, 449 - 450 (1973). – Abstr. Meteoritical Soc.

105.081 **The chemistry of the Ca, Al-rich Allende inclusions.**
H. Wänke, H. Baddenhausen, H. Palme, B. Spettel.
Meteoritics, Vol. 8, 450 - 451 (1973). – Abstr. Meteoritical Soc.

105.082 **Magnetic classification and magnetic remanence properties of individual chondrules from Allende (C3), Chainpur (LL3), Bjurbole (L4), and Allegan (H5) chondrites.** P. J. Wasilewski.
Meteoritics. Vol. 8, 451 - 456 (1973). – Abstr. Meteoritical Soc.

105.083 **Proposed magnetic remanence mechanisms for meteorites and lunar samples with metallographic recognition criteria.** P. J. Wasilewski.
Meteoritics, Vol. 8, 464 - 466 (1973). – Abstr. Meteoritical Soc.

105.084 **Fractionation of mesovolatile elements in ordinary chondrites.** J. T. Wasson, C.-L. Chou.
Meteoritics, Vol. 8, 467 (1973). – Abstr. Meteoritical Soc.

105.085 **Meteorites from the 2:1 Kirkwood gap.**
P. D. Zimmerman, G. W. Wetherill.
Meteoritics, Vol. 8, 468 - 469 (1973). – Abstr. Meteoritical Soc.

105.086 **Radiation burn of trees in the area of the Tunguska meteorite fall.** N. V. Vasil'ev, Yu. A. L'vov.
Priroda, No. 3.74, p. 99 - 101 (1974). In Russian.

105.087 **Melting relations of the Allende meteorite.**
M. G. Seitz, I. Kushiro.
Science, Vol. 183, 954 - 957 (1974).
It is suggested that the earth's core contains significant amounts of both nickel and sulfur and that a 3:2 mixture of Allende bulk sample and calcium- and aluminum-rich aggregates is closer in major element abundances than either of these components to the average composition of the moon.

105.088 **A new type of natural microspherule.**
R. Valach, J. Bauer, K. Žebera.

Journ. Geophys. Res., Vol. 79, 417 - 421 (1974).

From the synopsis of abiogenic microspherules it follows that many types of them occur in nature and that the finding of new types could be expected. In south Bohemia colorless microspherules of about 0.1 mm in diameter dissolving quickly in diluted hydrochloric acid have been found. Their properties are the same as those of the microspherules of dissolution-precipitation origin. Cosmic origin is possible but uncertain.

105.089 **Coexisting bronzite and clinobronzite and the thermal evolution of the Steinbach meteorite.**
A. M. Reid, R. J. Williams, H. Takeda.
Earth Planet. Sci. Letters, Vol. 22, 67 - 74 (1974).

Steinbach is a stony-iron meteorite with approximately equal amounts of silicate and metal that shows Widmanstätten structure. The silicate portion contains tridymite, orthobronzite, and clinobronzite that formed by inversion from high-temperature protobronzite. The assemblage orthobronzite − protobronzite − tridymite − metallic iron indicates an equilibrium temperature of 1200°C and an f_{O_2} of 10^{-12} under a total pressure of less than 2 kbar. Preservation of the high-temperature phase relations implies much more rapid cooling in the 1200 − 700°C range than the rates that have been deduced for the development of Widmanstätten structure in the 700 − 500°C range.

105.090 **Multiple fall of Příbram meteorites photographed. 11. Preatmospheric size and radiation age of the Příbram chondrite.**
A. K. Lavrukhina, A. V. Fisenko, E. M. Kolesnikov.
Bull. Astron. Inst. Czechoslovakia, Vol. 25, 122 - 126 (1974).

Measurements have been carried out of the Příbram ^{39}Ar meteorite in bulk sample and in separated metallic phase. On the basis of these data and the results of the theoretical calculation of the ^{39}Ar production rate in the Fe−Ni phase of this meteorite and the depth of the sample bedding the preatmospheric radius of the Příbram meteorite was determined.

105.091 **Experiment to measure the antimatter content of the Tunguska meteor.** H. Crannell.
Nature, Vol. 248, 396 - 398 (1974).

The author examines the hypothesis of an antimatter meteor, to ascertain whether other experimental tests can be carried out. An experiment to detect the antimatter component of the Tunguska meteor is conceptually simple. The ^{26}Al content of rocks or soil is measured as a function of the distance from the centre of the explosion. The highest concentration of ^{26}Al should be found near the centre.

105.092 **Existence of two groups in thermoluminescence of meteorites.** D. W. Sears, A. A. Mills.
Nature, Vol. 249, 234 - 235 (1974).

The thermoluminescence (TL) properties of meteorites may be applicable to the determination of exposure age, terrestrial age, thermal gradients, radiation history and preatmospheric shape. In an attempt to apply TL to the determination of terrestrial age the authors have discovered that meteorites can be divided into two groups, the existence of which is relevant to several of the applications.

105.093 **Major and trace elements in the Allende meteorite.**
P. M. Martin, B. Mason.
Nature, Vol. 249, 333 - 334 (1974).

The authors have determined major and trace element abundances for seven Ca,Al-rich aggregates, ten melilite chondrules, and an olivine chondrule and two olivine-rich aggregates from the Allende meteorite.

105.094 **Nedagolla, a remelted iron meteorite.**
G. T. Miyake, J. I. Goldstein.
Geochim. Cosmochim. Acta, Vol. 38, 747 - 755 (1974).

The Nedagolla meteorite was recognized by Axon to be a rare example of an iron which has been preterrestrially reheated to the point of melting. In this paper the authors have attempted to more accurately define the shock and thermal history of Nedagolla. This was accomplished by detailed metallographic and electron microprobe analyses and by the use of solidification and phase growth calculations.

105.095 **Determination of REE, Ba, Fe, Mg, Na and K in carbonaceous and ordinary chondrites.**
N. Nakamura.
Geochim. Cosmochim. Acta, Vol. 38, 757 - 775 (1974).

Precise determination of REE and Ba abundances in three carbonaceous (Orgueil C1, Murchison C2 and Allende C3) and seven olivine-bronzite chondrites were carried out by mass spectrometric isotope dilution technique.

105.096 **On the irradiation history and origin of gas-rich meteorites.** R. S. Rajan.
Geochim. Cosmochim. Acta, Vol. 38, 777 - 788 (1974).

With the transmission electron microscope, we have made detailed studies of the track density gradients and irradiation geometries of track-rich grains and chondrules in sections of Fayetteville and Kapoeta. We have made the same type of studies in sections of lunar breccias and grains from lunar soil for comparison. The results are reported and discussed.

105.097 **The meteoritic shower of Parambu, Ceara State, Brazil: −mineralogy and petrology.**
G. R. Levi-Donati, G. P. Sighinolfi.
Meteoritics, Vol. 9, 1 - 9 (1974).

The fall of a meteorite shower in Parambú, Ceará State, Brazil, is described. Parambú is an L-group chondrite. The mineralogical composition is olivine Fa_{28}, ortho- and clino-pyroxenes, plagioclase, maskelynite, whitlockite, nickel-iron, troilite, chromite, and ilmenite. The structure of Parambú is characteristically polymict and shows an advanced brecciation. Interesting features of metamorphism are notable, with good evidence of shock.

105.098 **Grossular in the Allende (type III carbonaceous) meteorite.** L. H. Fuchs.
Meteoritics, Vol. 9, 11 - 18 (1974).

Grossular garnet has been observed in several white inclusions in the Allende meteorite. Compositions range from $Gro_{95}Py_5$ to $Gro_{88}Py_{12}$ in five inclusions. These grossular-bearing inclusions either condensed directly as metastable liquids from the solar nebula or if initial solid condensates were liquefied by some subsequent heating process.

105.099 **The Toulon meteorite: a new chondrite from Illinois.** E. Olsen, G. I Huss.
Meteoritics, Vol. 9, 19 - 22 (1974).

Toulon is an olivine-bronzite chondrite found near Toulon, Illinois in 1962. It contains abundant, well preserved chondrules, as well as glasses that are not well devitrified. Most of the metal has been weathered out. Olivine and pyroxene are well equilibrated. It is classified as an H5 chondrite.

105.100 **A refractory glass chondrule in the Vigarano chondrite.**
A. M. Reid, R. J. Williams, E. K. Gibson, Jr., K. Fredriksson.
Meteoritics, Vol. 9, 35 - 45 (1974).

Vigarano, a type 3 carbonaceous chondrite, contains a chondrule composed of highly refractory Ca and Al rich glass with minor spinel. Experiments with synthetic analogues and a comparison with studies in the system CaO-MgO-Al_2O_3-SiO_2 indicate a temperature for formation of the chondrule at or above 1700°C followed by very rapid cooling.

105.101 Thermoluminescence and the terrestrial age of meteorites. D. W. Sears, A. A. Mills.
Meteoritics, Vol. 9, 47 - 67 (1974).

The use of thermoluminescence (TL) to determine the terrestrial age of meteorites is investigated. It is found that meteorites can be divided into two groups. One group, in which members lose their low temperature TL rather rapidly (the "low retentivity" group), may be dated up to about 100 years after fall, although with little accuracy. The other (the "high" group) is more retentive, and may still be dated several hundred years after fall. A meteorite of unknown date of fall may be assigned to the high or low group by laboratory determination of the rate of decay of the low temperature TL.

105.102 Fractionation of moderately volatile elements in ordinary chondrites. J. T. Wasson, C.-L. Chou.
Meteoritics, Vol. 9, 69 - 84 (1974).

Observed differences in the abundance ratios of moderately volatile elements found in ordinary chondrites relative to CI chondrites may have resulted from a continuous loss of nebular gas from the ordinary-chondrite formation region during condensation. Such a nebular gas loss can occur as a result of momentum exchange between solids and gases, as a result of interactions between the nebular gas and solar photons or particles at the surface of the nebula, or as a result of the settling of previously condensed solids to the median plane of the nebula.

105.103 Catalog of the collection of meteorites at the University of California, Los Angeles (the Leonard collection). J. T. Wasson, E. R. D. Scott, K. L. Robinson.
Meteoritics, Vol. 9, 85 - 98 (1974).

105.104 Astronomisch-geodätische Lotabweichungen als Beitrag zu den geophysikalischen Untersuchungen im Nördlinger Ries. G. Soltau.
Mitt. Astron. Ges., No. 34, p. 170 (1973/74). – Abstr. AG.

105.105 Solar system sources of meteorites and large meteoroids. G. W. Wetherill.
Annual Rev. Earth Planet. Sci., Vol. 2, (see 003.006), 303 - 331 (1974).

105.106 The Meteoritical Bulletin, No. 52. R. S. Clarke, Jr. (Editor).
Meteoritics, Vol. 9, 101 - 121 (1974). – New falls and discoveries of meteorites are given in this bulletin.

105.107 A new interpretation of the mechanical properties of the Gibeon meteorite.
A. A. Johnson, J. L. Remo.
Journ. Geophys. Res., Vol. 79, 1142 - 1146 (1974).

A set of results on the temperature and strain rate dependence of the tensile properties of the Gibeon meteorite reported by Gordon in 1970 is reinterpreted. It is shown that these results indicate that the meteorite material undergoes a ductile-brittle transition in low strain rate tensile tests at about 50°K. It is estimated that under asteroidal impact conditions the transition temperature would be about 200°K.

105.108 "Le Clot" bij Cabrerolles meteoorkrater in het zuiden van Frankrijk. W. Buijze.
Zenit, Vol. 1, No. 5, p. 30 - 31 (1974).

105.109 Measurement of Al^{26} in stone meteorites and its use in the derivation of orbital elements.
I. R. Cameron, Z. Top.
Geochim. Cosmochim. Acta, Vol. 38, 899 - 909 (1974).

The Al^{26} activity has been measured by gamma-ray coincidence spectrometry in a total of 30 stony meteorites. The measured Al^{26} content has been compared with calculated values based on the method developed by Lavrukhina and Ustinova (1972). Comparison of the measured value with that predicted in the absence of solar modulation permits the estimation of the aphelion of the orbit of the meteorite. The great majority of the derived aphelia lie within the range 2.05 – 2.45 AU.

105.110 Foreign inclusions in stony meteorites–II. Rare gases and oxygen isotopes in a carbonaceous chondritic xenolith in the Plainview gas-rich chondrite.
L. L. Wilkening, R. N. Clayton.
Geochim. Cosmochim. Acta, Vol. 38, 937 - 945 (1974).

Measurements of oxygen and rare gas isotopes in a carbonaceous xenolithic inclusion in the Plainview H5 chondrite indicate that the xenolith and Plainview host are of two distinct meteorite types, and that no isotopic exchange has taken place between the two materials since their juxtaposition. The radiation ages of the xenolith and the host are identical. Analyses of the rare gases in a sample of the host material adjacent to the xenolith show that Plainview is gas-rich, i.e., contains large amounts of solar-type trapped gases. The authors speculate that carbonaceous chondritic material may be more prevalent in the asteroid belt than previously suspected.

105.111 On the origin of the crater Iriston.
V. A. Romejko, N. F. San'ko.
Astron. vestn., Vol. 8, 112 - 113 (1974). In Russian.

The formation of this crater is not connected with the fall of a meteorite.

105.112 On the origin of isotopes of primordial gases.
L. K. Levskij.
Meteoritika, vyp. (No.) 33, p. 23 - 29 (1974). In Russian.

105.113 On the process of recrystallization of chondrites on the basis of data on the distribution of elements among mineral fractions.
A. K. Lavrukhina, N. K. Sazhina, A. V. Chubarova, V. G. Kashkarova.
Meteoritika, vyp. (No.) 33, p. 30 - 33 (1974). In Russian.

105.114 Fe^{2+}, Mg cation ordering in pyroxenes from olivine-hypersthene chondrites according to data of Mössbauer spectroscopy. T. V. Malysheva, A. K. Lavrukhina, S. A. Stakheeva, L. M. Satarova.
Meteoritika, vyp. (No.) 33, p. 34 - 35 (1974). In Russian.

105.115 On regularities of distribution of nickel and cobalt in meteorites and products of nickel production.
A. A. Yavnel', V. M. Grigor'eva, I. E. Gorbunova.
Meteoritika, vyp. (No.) 33, p. 36 - 41 (1974). In Russian.

105.116 The content of rare-earth elements in five normal chondrites. G. M. Kolesov.
Meteoritika, vyp. (No.) 33, p. 42 - 46 (1974). In Russian.

105.117 C^{14} in iron meteorites and determination of terrestrial ages of finds. V. A. Alekseev, A. I. Ivliev.
Meteoritika, vyp. (No.) 33, p. 47 - 48 (1974). In Russian.

105.118 Al^{26} in chondrites with small radiation age.
V. A. Alekseev, G. K. Ustinova.
Meteoritika, vyp. (No.) 33, p. 49 - 50 (1974). In Russian.

105.119 Petrological and chemical peculiarities of meteorites with different orbits.
A. K. Lavrukhina, G. K. Ustinova, G. V. Baryshnikova.
Meteoritika, vyp. (No.) 33, p. 51 - 57 (1974). In Russian.

105.120 Cosmogenic Ar^{39} in the metallic phase of chondrites and their pre-atmospheric dimensions.
A. K. Lavrukhina, A. V. Fisenko, E. M. Kolesnikov.
Meteoritika, vyp. (No.) 33, p. 58 - 60 (1974). In Russian.

105.121 Possibilities of X-ray diffraction methods for investigation of the deformed structure of iron meteorites. A. P. Pushel', Eh. S. Gorshkov.
Meteoritika, vyp. (No.) 33, p. 61 - 63 (1974). In Russian.

105.122 Some old meteorite craters in the USSR.
V. L. Masajtis.
Meteoritika, vyp. (No.) 33, p. 64 - 68 (1974). In Russian.

105.123 Indications of shock metamorphism in minerals of meteorites and lunar rocks.
E. K. Lazarenko, A. A. Yasinskaya.
Meteoritika, vyp. (No.) 33, p. 69 - 72 (1974). In Russian.

105.124 On shock waves during flight and explosion of meteorites.
V. P. Korobejnikov, P. I. Chushkin, L. V. Shurshalov.
Meteoritika, vyp. (No.) 33, p. 79 - 80 (1974). In Russian.

105.125 Reflection of air shock waves from the earth's surface during the fall of large meteorite bodies.
A. K. Stanyukovich.
Meteoritika, vyp. (No.) 33, p. 81 - 82 (1974). In Russian.

105.126 Results of an investigation of the Malakal chondrite (Sudan). S. M. El Rabaa, A. M. Daminova, M. I. D'yakonova, L. G. Kvasha, L. K. Levskij, A. V. Fisenko.
Meteoritika, vyp. (No.) 33, p. 83 - 89 (1974). In Russian.

105.127 New data on a chemical analysis of the stony meteorites-chondrites Bushkhof, Karakol, Ausson.
M. I. D'yakonova, V. Ya. Kharitonova.
Meteoritika, vyp. (No.) 33, p. 90 - 93 (1974). In Russian.

105.128 Crystallomorphology of olivine in pallasites.
V. D. Kolomenskij, T. V. Pakhomova, I. I. Shavranovskij.
Meteoritika, vyp. (No.) 33, p. 94 - 99 (1974). In Russian.

105.129 Determination of the mechanical and thermal properties of the meteorites Kunashak and Elenovka.
R. V. Medvedev.
Meteoritika, vyp. (No.) 33, p. 100 - 104 (1974). In Russian.

105.130 On the cosmic history of the Sikhote-Alin meteorite.
E. M. Kolesnikov, A. K. Lavrukhina, L. K. Levskij, A. V. Fisenko.
Meteoritika, vyp. (No.) 33, p. 105 - 108 (1974). In Russian.

105.131 On the homogeneity of the chemical composition of the Sikhote-Alin meteorite.
E. M. Kolesnikov, A. K. Lavrukhina.
Meteoritika, vyp. (No.) 33, p. 109 - 111 (1974). In Russian.

105.132 Distribution of elements between the metallic and sulfide phases of the Sikhote-Alin iron meteorite.
A. K. Lavrukhina, L. D. Revina, N. K. Sazhina, L. V. Yukina, Yu. A. Sil'vanovich, Kh. R. Rakhimov, U. Kh. Khudajbergenov.
Meteoritika, vyp. (No.) 33, p. 112 (1974). In Russian.

105.133 Investigation of the Sikhote-Alin meteorite with the methods of Mössbauer and micro-X-ray spectroscopy.
T. V. Malysheva, I. D. Shevaleevskij, V. D. Shvagerev.
Meteoritika, vyp. (No.) 33, p. 113 - 116 (1974). In Russian.

105.134 On silicates and chromite of polymineral inclusions of the Sikhote-Alin octahedrite.
N. I. Zaslavskaya, L. G. Kvasha.
Meteoritika, vyp. (No.) 33, p. 117 - 121 (1974). In Russian.

105.135 Investigation of the composition of gases from inclusions in tektites and cosmic dust.
Yu. A. Dolgov.
Meteoritika, vyp. (No.) 33, p. 122 - 129 (1974). In Russian.

105.136 Distribution of some elements in mineral fractions of the Saratov chondrite. A. K. Lavrukhina, A. V. Chubarova, N. K. Sazhina, V. G. Kashkarova.
Meteoritika, vyp. (No.) 33, p. 130 - 133 (1974). In Russian.

105.137 Investigation of metallic particles of chondrites.
A. K. Lavrukhina, N. K. Sazhina, S. A. Stakheeva.
Meteoritika, vyp. (No.) 33, p. 134 - 139 (1974). In Russian.

105.138 The iron meteorite Aprelsky.
O. A. Kirova, M. I. D'yakonova.
Meteoritika, vyp. (No.) 33, p. 140 - 142 (1974). In Russian.

105.139 On silicate inclusions and indications of shock metamorphism in the octahedrite Elga.
L. G. Kvasha, Yu. G. Lavrent'ev, N. V. Sobolev.
Meteoritika, vyp. (No.) 33, p. 143 - 147 (1974). In Russian.

105.140 Density and magnetic properties of meteorites.
E. G. Gus'kova.
Meteoritika, vyp. (No.) 33, p. 148 - 156 (1974). In Russian.

105.141 Radiants and orbits of meteorites.
A. N. Simonenko, B. Yu. Levin.
Meteoritika, vyp. (No.) 33, p. 157 - 173 (1974). In Russian.

105.142 New fragment of the Kashin meteorite.
A. K. Stanyukovich.
Meteoritika, vyp. (No.) 33, p. 174 - 175 (1974). In Russian.

105.143 New fragments of the Ochansk meteorite.
A. K. Stanyukovich.
Meteoritika, vyp. (No.) 33, p. 176 - 177 (1974). In Russian.

105.144 Partial a. f. demagnetization studies of 40 meteorites. E. E. Larson, D. E. Watson, J. M. Herndon, M. W. Rowe.
Journ. Geomagn. Geoelectr., (Japan), Vol. 25, 331 - 338 (1973).

105.145 Meteorites, remains of outside worlds.
R. Buhler.
Techn. Rundschau, (Switzerland), Vol. 66, No. 5, p. 29 - 31 (1974). In German.

A brief outline is given on the history and development of research on meteorites.

105.146 Theory of shock-wave ionization upon high-velocity impact of micrometeorites.
S. Drapatz, K. W. Michel.
Zeitschr. Naturforschung, Vol. 29a, 870 - 879 (1974).

The relevant processes in shock wave ionization of a solid Fe micrometeorite impinging on a W target are analyzed. The internal energy behind the shock wave is shown to depend on impact velocity, target and meteorite density in a simple analytical form.

105.147 The paleomagnetic record in carbonaceous chondrites: natural remanence and magnetic properties.
A. Brecher, G. Arrhenius.
Journ. Geophys. Res., Vol. 79, 2081 - 2106 (1974).

Recent results of an intensive study of the natural remanence and the magnetic properties of carbonaceous chondrites are summarized. It is convincingly demonstrated that the record of ancient magnetic fields has been preserved in these least-altered old samples of solar system material known.

105.148 **Genesis of the melt rocks at Tenoumer crater, Mauritania.** R. F. Fudali.
Journ. Geophys. Res., Vol. 79, 2115 - 2121 (1974).

A common problem encountered at probable meteorite impact craters is the occurrence of associated melt rocks, presumably produced by the impact event, that cannot be matched chemically to the target rocks. A similar situation is found at Tenoumer crater, Mauritania. Chemical analyses of 11 melt rocks range from felsic country rock values to much more mafic compositions in two well-defined chemical sequences. Neither meteoritic contamination nor vapor fractionation can explain the chemical sequences exhibited by these melt rocks. They most likely result from the combination of felsic country rocks with various amounts of an amphibolite vein that appears to strike through the center of the crater. The hypothesis is advanced that the amphibolite was preferentially incorporated in the impact-generated melt because of favorable kinetic and viscosity characteristics. The clear-cut evidence from Tenoumer makes it difficult to deny an impact origin for certain craters solely on the basis of associated melt rocks that are not isochemical with the country rocks.

105.149 **Helium, Neon und Argon in einigen Steinmeteoriten.** L. Schultz, P. Signer.
Analyse extraterrestrischen Materials, (see 012.020), p. 27 - 36 (1974).

105.150 **Massenspektrometrische Bestimmung der Spurenelemente in Steinmeteoriten.**
H. Hintenberger, K. P. Jochum, M. Seufert.
Analyse extraterrestrischen Materials, (see 012.020), p. 125 - 145 (1974).

105.151 **Neutronenaktivierungsanalytische Bestimmung von Spurenelementen in Meteoriten der Vatikanischen Sammlung.** F. Hermann, M. Wichtl.
Analyse extraterrestrischen Materials, (see 012.020), p. 163 - 172 (1974).

105.152 **Trennverfahren und radiochemische Bestimmungsmethoden für geringe Mengen seltener Erden in Chondriten.** R. R. Becker, K. Buchtela, F. Grass, R. Kittl, G. Müller.
Analyse extraterrestrischen Materials, (see 012.020), p. 173 - 181 (1974).

105.153 **Spezielle Möglichkeiten mikrochemischer Methoden als Beitrag zur Meteoritenforschung.**
H. Ballczo, A. Mitter.
Analyse extraterrestrischen Materials, (see 012.020), p. 239 - 247 (1974).

105.154 **Chemismus von Spinellen aus dem Mezö-Madaras-Chondrit.** G. Hoinkes, G. Kurat.
Analyse extraterrestrischen Materials, (see 012.020), p. 265 - 288 (1974).

105.155 **Programm zur analytischen Auswertung röntgenographischer Daten – II: Bearbeitung meteoritischer Proben.** H. H. Weinke, A. Kracher.
Analyse extraterrestrischen Materials, (see 012.020), p. 307 - 314 (1974).

105.156 **Untersuchungen am Landes-Meteorit.**
A. Kracher.
Analyse extraterrestrischen Materials, (see 012.020), p. 315 - 326 (1974).

105.157 **Craters, meteorites and black holes.**
A. Dupas, J. Vuilleumier.
Recherche, (*France*), Vol. 5, No. 41, p. 69 - 73 (1974). In French.

It is estimated that during the earth's history it would have been hit by 350000 meteorites, whereas only 160 craters have been discovered which can be presumed to be due to the impact of meteorites. The authors discuss reasons for the disproportion between these two figures. They go on to discuss two large meteorite craters recently discovered in Brazil, and then consider the hypothesis put forward by Jackson and Ryan who maintain that the Tunguska meteorite (1908) was, in fact, a black hole.

105.158 **Analysis of Dongtai meteorite.**
X. Wang, X. Li, S. Wang.
Acta Astron. Sinica, Vol. 15, 93 - 97 (1974). In Chinese.

The structure of various chondrules of this meteorite, the resulting data of mineralogical and chemical analysis, and its potassium-argon age are presented.

105.159 **Bij Utrecht sloegen in 1843 twee meteorieten in.**
G. Koppert.
Zenit, Vol. 1, No. 6, p. 38 (1974).

105.160 **Possible meteorite crater in Mexico.**
L. Maupomé.
Rev. Mexicana Astron. Astrofis., Vol. 1, No. 1, p. 81 - 86 (1974).

A circular crater with present depth of about 92 meters from the rim to the floor and with a diameter of 1180 m has been found in the state of Puebla, México. Several arguments are discussed that point to the meteoritic origin of this crater. Topographical and geological results are presented.

105.161 **The chemical compositions of the stone meteorites Yamata (a), (b), (c) and (d), and Numakai.**
Mas. Shima.
Meteoritics, Vol. 9, 123 - 135 (1974).

105.162 **The Patora meteorite, an H6 fall.**
A. L. Graham, V. K. Nayak.
Meteoritics, Vol. 9, 137 - 139 (174).

105.163 **Présentation d'une tectite de Côte-d'Ivoire avec quelques considérations sur l'origine des tectites.**
R. Woodtli.
Bull. Soc. Vaudoise, Sci. Nat., (*Switzerland*), Vol. 71, 405 - 417 (1973).

105.164 **Evidence for late formation and young metamorphism in the achondrite Nakhla.**
D. A. Papanastassiou, G. J. Wasserburg.
Geophys. Res. Letters, Vol. 1, No. 1, p. 23 - 26 (1974).

The authors have determined an age T = 1.37 AE for Nakhla and an initial $^{87}Sr/^{86}Sr$, I = 0.7023. The most reasonable model is that Nakhla formed as an igneous rock at ~ 3 AE and was metamorphosed at 1.37 AE. Several similarities exist between the Nakhla parent planet and the earth and indicate the existence of other objects with close terrestrial chemical affinities.

105.165 **The Committee on Meteorites of the USSR Academy of Sciences.** E. L. Krinov.
Meteoritics, Vol. 9, 141 - 144 (1974).

105.166 **The Mills, New Mexico, chondrite – a new find.**
G. R. Levi-Donati, E. Jarosewich.

Meteoritics, Vol. 9, 145 - 155 (1974).

105.167 **Meteorite material in the Mineralogisk-Geologisk Museum, Oslo.** W. L. Griffin.
Meteoritics, Vol. 9, 157 - 171 (1974).

105.168 **Kaaba stone: not a meteorite, probably an agate.** R. S. Dietz, J. McHone.
Meteoritics, Vol. 9, 173 - 179 (1974).

Comets, meteorites and men. See Abstr. 003.076.

Die Bedeutung der Spurenelemente für die Kosmochemie. See Abstr. 061.072.

Thermoluminescence effects in lunar materials and meteorites. See Abstr. 094.131.

Elementkorrelationen und die chemische Zusammensetzung von Mond und Erde. See Abstr. 094.175.

Zirconium and hafnium abundances in lunar materials and meteorites and implications of their ratios. See Abstr. 094.334.

Particle shape and magnetization of lunar samples, chondrules, chondrite meteorites, and naturally reduced terrestrial basalts. See Abstr. 094.356.

Shock remagnetization mechanisms and processes in lunar samples and meteorites. See Abstr. 094.357.

Zerstörungsfreie Bestimmung einiger Spurenelemente in Mond- und Meteoritenproben mit 14 MeV-Neutronen. See Abstr. 094.395.

Organic synthesis in a simulated Jovian atmosphere. III. Synthesis of aminonitriles. See Abstr. 099.031.

Planetary accretion and meteorites. See Abstr. 107.008.

A new limit on the interstellar abundance of boron. See Abstr. 131.094.

Errata

105.901 **Erratum: 'The Norton County meteoroid: a case for statistical study of meteorite fragments'** [Meteoritics, Vol. 8, 277 - 286 (1973)].
B. Lang, K. Liszewska.
Meteoritics, Vol. 9, 99 (1974).

106 Interplanetary Matter, Interplanetary Magnetic Field, Zodiacal Light

106.001 **OGO-5 measurements of the Lyman-alpha sky background in 1970 and 1971.**
G. E. Thomas, R. F. Krassa.
Astron. Astrophys., Vol. 30, 223 - 232 (1974).

The Lyman-alpha sky background emission at 1216 Å was mapped three times in 1969 and 1970, and three additional times in 1970 and 1971 by the OGO-5 ultraviolet photometer experiment. The authors report the results from the last three "spin-up" maneuvers.

106.002 **Extra-terrestrial ^{53}Mn in Antarctic ice.**
R. Bibron, R. Chesselet, G. Crozaz, G. Leger, J. P. Mennessier, E. Picciotto.
Earth Planet. Sci. Letters, Vol. 21, 109 - 116 (1974).

The reasons why ^{53}Mn (a cosmogenic radionuclide with a half-life of 3.7×10^6 y) appears as one of the best indicators of the presence of interplanetary dust are summarized. This paper reports the detection of ^{53}Mn in pre-1952 snow samples collected on the Eastern Antarctic Plateau in the vicinity of Plateau Station.

106.003 **An investigation of the motion of zodiacal dust particles – I. Radial velocity measurements on Fraunhofer line profiles.** T. R. Hicks, B. H. May, N. K. Reay.
Monthly Notices Roy. Astron. Soc., Vol. 166, 439 - 448 (1974).

An experiment to record the spectrum of the zodiacal light in the neighbourhood of the Mg I absorption line (5183.6 Å) is described. Measurements were made of the Doppler shift imposed on the absorption line by the motion of the interplanetary dust particles. Observations were concentrated on the ecliptic plane, spectra being obtained at lower elongation angles from the sun than previously achieved, and also over the entire range of high elongations including the gegenschein. The reduction methods applied to the data are described and compared, and the new results are presented.

106.004 **Interplanetary shock waves generated by solar flares.** M. Dryer.
Space Sci. Rev., Vol. 15, 403 - 468 (1974). – This paper is a revised and updated version of an invited review originally presented at the IUGG XV General Assembly, Moscow, U.S.S.R., 2 – 14 August 1971.

Recent observational and theoretical studies of interplanetary shock waves associated with solar flares are reviewed. An attempt is made to outline the framework for the genesis, life and demise of these shocks. Thus, suggestions are made regarding their birth within the flare generation process, MHD wave propagation through the chromosphere and inner corona, and maturity to fully-developed coronal shock waves. Their subsequent propagation into the ambient interplanetary medium and disturbing effects within the solar wind are discussed within the context of theoretical and phenomenological models.

106.005 **The extraterrestrial UV-background and the nearby interstellar medium.** H. J. Fahr.
Space Sci. Rev., Vol. 15, 483 - 540 (1974).

Recent measurements of the extraterrestrial UV- and EUV-radiation, and the various theoretical approaches used in explaining the measured features of these radiations are reviewed. The effects of solar radiation pressure, and of temporal variations and spatial asymmetries in the solar radiations, on the structure of the extraterrestrial $L\alpha$ sky are investigated in detail, and the various attempts to derive interstellar parameters from the interpretation of the measured $L\alpha$ intensities are discussed. From these discussions the local interstellar medium is established as a tenuous hot intercloud H I-medium. Further activities concerning the measurement of extraterrestrial UV- and EUV-radiation features are suggested that may be highly valuable in clarifying the outstanding problems.

106.006 **Rocket photometry of the inner zodiacal light.**
C. Leinert, H. Link, E. Pitz.
Astron. Astrophys., Vol. 30, 411 - 422 (1974).

Observations of the zodiacal light on circles around the sun at elongation angles between 15° and 30° have been obtained with five rocket-borne telescopes, launched from the ESRO-range at Sardinia on July 2, 1971. The brightness was measured at two broad wavelength bands centered at 4755 Å and 5915 Å. The results provide coverage of the observational gap between the earth-bound measurements of the inner zodiacal light and outer corona data. Polarization measurements at $\epsilon = 15°$ and $21°$ in the blue band give rather high values near the ecliptic and a slight decrease outside. Apparent deviations of the plane of symmetry from the ecliptic are discussed.

106.007 **Indicatrices de diffusion et distribution héliocentrique des particules zodiacales.** R. Dumont.
Comptes Rendus Acad. Sci. Paris, Sér. B, Vol. 278, 537 - 540 (1974).

Un nuage ellipsoïdal où les particules interplanétaires seraient distribuées en r^{-1} ou $r^{-1,2}$ dans le plan moyen des planètes, avec un degré d'aplatissement $a/b \cong 8$, reproduirait bien les caractéristiques photopolarimétriques de la lumière zodiacale.

106.008 **Interplanetary material.** P. M. Millman.
Journ. Roy. Astron. Soc. Canada, Vol. 68, 13 - 22 (1974). — Presented at a symposium on "Chemical evolution in the universe", organized by the Royal Society of Canada, Queen's University, Kingston, Ontario, June 6, 1973.

106.009 **Local hydrogen gas and the background Lyman-alpha pattern.** M. K. Wallis.
Monthly Notices Roy. Astron. Soc., Vol. 167, 103 - 119 (1974).

The trapping of resonant solar photons in the local hydrogen cloud contributes substantially to the Lyman-α background. The contribution, commonly ignored as 'multiply-scattered' photons, is described here as 'cavity radiation'. Two types of cavity are described and the backscattered radiation intensity calculated — a physical cavity produced by destructive processes, appropriate to hot gas; and a 'Doppler cavity' due to gravitational acceleration of nearby hydrogen atoms and relevant to cool gas. The OGO 5 Lyman-α intensities are re-interpreted as the sum of near-isotropic cavity radiation plus the direct backscatter from asymmetrically distributed gas inside the cavity.

106.010 **Variation de la luminance de la lumière zodiacale dans la direction anti-solaire.**
R. Robley, P. Prêtre.
Comptes Rendus Acad. Sci. Paris, Sér. B, Vol. 278, 671 - 674 (1974).

On met en évidence une variation saisonnière de la luminance de la lumière zodiacale dans la direction anti-solaire. Cet effet pourrait être dû aux grains de poussière localisés dans la ceinture d'astéroïdes.

106.011 **Diagnostics of the interplanetary magnetic field from the planetary K_p-index for 1932 - 1972.**
V. I. Afanas'eva, L. V. Evdokimova, N. V. Mikerina, K. G. Ivanov.
Geomagn. Aeronom., Vol. 14, 13 - 17 (1974). In Russian.

106.012 **The possibility of diagnostics of the normal component of the interplanetary magnetic field from the planetary K_p-index.** N. V. Mikerina, K. G. Ivanov.
Geomagn. Aeronom., Vol. 14, 149 - 150 (1974). In Russian. Brief information.

106.013 **Interplanetary magnetic field, solar flares and geomagnetic disturbances.** V. I. Afanas'eva.
Geomagn. Aeronom., Vol. 14, 175 - 176 (1974). In Russian. Brief information.

106.014 **Coherent propagation of charged-particle bunches in random magnetic fields.** J. A. Earl.
Astrophys. Journ., Vol. 188, 379 - 397 (1974).

When the distribution function for energetic particles moving in interplanetary space is expressed in terms of eigenfunctions of an operator which describes pitch-angle scattering by random magnetic fields, the familiar phenomenon of diffusion is the dominant solution of the transport equations provided that the spectrum of magnetic fluctuations is not too steep. However, if the power-versus-wavenumber spectrum of the random fields has a spectral index $q > 2$, the dominant solution of the same equations is a qualitatively different mode of transport in which density inhomogeneities propagate coherently along the guiding field. This paper, which extends classical transport theory to give a detailed treatment of the new mode, presents explicit formulae which describe the spatial and temporal structure of the coherent disturbance. The relationship between coherent and diffusive modes is examined with particular emphasis upon the situation in which both modes are simultaneously present. In applying these ideas to the propagation of solar cosmic rays, it is found that the prompt arrival and marked anisotropies seen in the early phases of solar proton events can be understood as a consequence of coherent propagation. However, the most striking implication of the theory is that the observed "scatter-free" propagation of solar flare electrons below 1 MeV is a manifestation of the coherent mode in which an impulsive burst of electrons travels from the sun to earth with negligible dispersion.

106.015 **Piston shock waves in the interplanetary medium.**
G. F. Krymskij, I. A. Transkij, E. K. Elshin.
Geomagn. Aeronom., Vol. 14, 196 - 200 (1974). In Russian.

106.016 **Investigation of the propagation of radio waves in the solar system by means of Soviet space vehicles.**
M. A. Kolosov, O. I. Yakovlev.
Uspekhi fiz. nauk, Vol. 111, 372 - 373 (1973). In Russian.
Abstr. in Referativ. Zhurn. 62. Issled. kosmich. prostranstva, 4.62.173 (1974).

106.017 Hydromagnetic radiation of the interplanetary plasma. A. V. Gul'el'mi, B. V. Dovbnya.
Pis'ma v ZhEhTF, Vol. 18, 601 - 604 (1973). In Russian. Abstr. in Referativ. Zhurn. 62. Issled. kosmich. prostranstva, 4.62.189 (1974).

106.018 Observations of interplanetary scintillations at 327 MHz.
A. Pramesh Rao, S. M. Bhandari, S. Ananthakrishnan.
Australian Journ. Phys., Vol. 27, 105 - 120 (1974).
The properties of the interplanetary medium and the structures of 9 compact radio sources have been derived from the observed interplanetary scintillations at 327 MHz using the Ooty radio telescope.

106.019 Stray light suppression in optical space experiments. C. Leinert, D. Klüppelberg.
Applied Optics, Vol. 13, 556 - 564 (1974).
Measurements of stray light suppression with mirrors, lenses, and baffle systems from four zodiacal light and one noctilucent cloud space experiments are reported. The method used to derive the total stray light suppression from these measurements is given.

106.020 Is the zodiacal light intensity steady?
G. B. Burnett, J. G. Sparrow, E. P. Ney.
Nature, Vol. 249, 639 - 640 (1974).
The authors believe the zodiacal cloud to be a smooth, regular feature of the solar system as one might expect from a steady influx of particles from the continual breakup taking place in the asteroid belt. There is no evidence for a cloud, with intensity fluctuations of a factor of two, composed of an interwoven maze of comet trails. No variations in intensity or polarization greater than the limit of accuracy of the experiment—usually 10% arising from comets, solar activity, lunar phase or other events in the solar system—were observed.

106.021 The zodiacal component at the pole of the ecliptic. S. N. Krylova.
Pyl' v atmosf. i okolozemn. kosmich. prostranstve. Moskva, Nauka, 1973, p. 46 - 53. In Russian. − Abstr. in Referativ. Zhurn. 51. Astron., 5.51.340 (1974).

106.022 On the circumterrestrial cosmic cloud.
V. G. Fesenkov.
Pyl' v atmosf. i okolozemn. kosmich. prostranstve. Moskva, Nauka, 1973, p. 5 - 8. In Russian. − Abstr. in Referativ. Zhurn. 51. Astron., 5.51.341 (1974).

106.023 On the influence of the earth on the motion of dust particles in the interplanetary space.
N. I. Komarnitskaya.
Pyl' v atmosf. i okolozemn. kosmich. prostranstve. Moskva, Nauka, 1973, p. 38 - 46. In Russian. − Abstr. in Referativ. Zhurn. 51. Astron., 5.51.345 (1974).

106.024 Calculation of the temperature of dust particles in the interplanetary space. L. S. Trofimova.
Pyl' v atmosf. i okolozemn. kosmich. prostranstve. Moskva, Nauka, 1973, p. 54 - 55. In Russian. − Abstr. in Referativ. Zhurn. 51. Astron., 5.51.346 (1974).

106.025 Sectorial structure of interplanetary plasma inhomogeneities. N. A. Lotova, N. V. Vereshchagina.
Izv. vyssh. ucheb. zavedenij. Radiofizika, Vol. 16, 1645 - 1651 (1973). In Russian. − Abstr. in Referativ. Zhurn. 51. Astron., 5.51.424 (1974).

106.026 Analisi di fotogrammi della luce zodiacale.
F. Ficarrotta.
Coelum, Vol. 42, 129 - 132 (1974).

106.027 Investigation of solar plasma near Mars and on the track earth − Mars by means of traps of charged particles aboard Soviet space vehicles from 1971 to 1973. I. Methods and instruments. K. I. Gringauz, V. V. Bezrukikh, G. I. Volkov, M. I. Verigin, L. N. Davitaev, V. F. Kopylov, L. S. Musatov, G. F. Sluchenkov.
Kosmich Issled., Vol. 12, 430 - 439 (1974). In Russian.

106.028 Piston shock waves in the interplanetary medium and Forbush effects.
G. F. Krymskij, I. A. Transkij, E. K. Elshin.
Geomagn. Aeronom., Vol. 14, 407 - 410 (1974). In Russian.

106.029 The interplanetary magnetic field and seismic activity. Yu. D. Kalinin.
Geomagn. Aeronom., Vol. 14, 502 - 505 (1974). In Russian.

106.030 Anomalies in the composition of interplanetary heavy ions with $0.01 < E < 40$ MeV per amu.
J. H. Chan, P. B. Price.
Astrophys. Journ., (Letters), Vol. 190, L39 - L41 (1974).
Analysis of tracks in plastic detectors on Apollo 17 shows that at energies from ~ 2 to ~ 10 MeV per amu, the composition of interplanetary heavy ions is quite different from that of either solar particles or galactic cosmic rays. C + N, Mg + Si, and $Z \geq 20$ are depleted with respect to O, and Li + Be + B are missing. The existence of a peculiar hump in the oxygen energy spectrum is confirmed.

106.031 A correlative study of ssc's, interplanetary shocks, and solar activity. J. K. Chao, R. P. Lepping.
Journ. Geophys. Res., Vol. 79, 1799 - 1807 (1974).
The authors present an observational study of 38 flare-associated shocks. The shock speeds and normals have been computed accurately for 22 of these cases through the use of multiple spacecraft (S/C) observations; the speeds and normals for the remaining 16 shocks are also obtained.

106.032 Modern ideas on the circumterrestrial space.
Ts. Ralchovski.
Priroda (NRB), Vol. 22, No. 3, p. 50 - 52 (1973). In Bulgarian.

106.033 Interplanetary control of magnetospheric dynamics.
G. K. Parks, E. H. Gardner, C. S. Lin, R. E. Newell, P. J. Wehrenberg.
Astrophys. Space Sci. Library, Vol. 42, (see 012.016), 49 - 60 (1974).

106.034 The relation between the azimuthal component of the interplanetary magnetic field and the geomagnetic field in the polar caps. L. Svalgaard.
Astrophys. Space Sci. Library, Vol. 42, (see 012.016), 61 - 84 (1974).

106.035 Interplanetary magnetic field direction and auroral zone disturbances.
L. Rossberg, W. Riedler, G. Skovli, P. Stauning, G. R. Thomas.
Astrophys. Space Sci. Library, Vol. 42, (see 012.016), 85 - 93 (1974).

106.036 Observation of interplanetary shocks on August 4, 1972. M. B. Cattaneo, P. Cerulli-Irelli, L. Diodato, A. Egidi, G. Moreno, P. C. Hedgecock.
Astrophys. Space Sci. Library, Vol. 42, (see 012.016), 555 - 562 (1974).

106.037 The dynamics of circum-solar dust grains.
P. L. Lamy.
Astron. Astrophys., Vol. 33, 191 - 194 (1974).
The inward spiraling of interplanetary dust grains under Poynting-Robertson and corpuscular pressure drags is shown

to be either counterbalanced or reduced by the effect of the net increase of the radiation pressure force caused by the decrease of the grains' radii when sublimating. Precise trajectories show that silicate grains remain in the vicinity of the sun where they describe an impressive number of orbits. Dynamical dust-free zones are clearly established and the location of regions of probable concentration are predicted.

106.038 **Correlation of interplanetary scintillation and spacecraft plasma density measurements.**
Z. Houminer, A. Hewish.
Planet. Space Sci., Vol. 22, 1041 - 1042 (1974).

Interplanetary scintillation and spacecraft measurements of proton density are observed to be strongly correlated. This result confirms the association between corotating sectors of enhanced scintillation and compression regions located at the interface between fast and slow solar wind streams.

106.039 **Possible acceleration of charged particles through the reconnection of magnetic fields in interplanetary space.** G. Gloeckler, F. M. Ipavich, E. H. Levy.
Bull. American Astron. Soc., Vol. 6, 271 (1974). – Abstr. AAS.

Protonenlawine vor interplanetaren Schockwellen.
Umschau, 74. Jahrgang, p. 193 (1974).

Dust in the atmosphere and in circumterrestrial space. See Abstr. 003.137.

Large-amplitude hydromagnetic waves. See Abstr. 062.064.

Solar sector boundary configuration from comparison of synoptic charts of the photospheric magnetic field with the observed interplanetary field. See Abstr. 071.065.

Solar proton flares and the sectorial structure of the interplanetary magnetic field. See Abstr. 073.024.

Wave-trains in the solar wind. II: Comments on the propagation of Alfvén waves in the quiet interplanetary medium. See Abstr. 074.007.

First order latitude effects in the solar wind. See Abstr. 074.017.

Further remarks on plasma instabilities produced by ions born in the solar wind. See Abstr. 074.033.

Acceleration of the solar wind by the interplanetary magnetic field. See Abstr. 074.037.

The adiabatic cooling of the protons in the solar wind: the case where the interplanetary magnetic-field is of spiral form. See Abstr. 074.038.

Pioneer solar plasma and magnetic field measurements in interplanetary space during August 2–17, 1972. See Abstr. 074.077.

Motion of the sources for type II and type IV radio bursts and flare-associated interplanetary disturbances. See Abstr. 077.019.

Implications of the reported low energy electron gradients. See Abstr. 078.007.

Interplanetary acceleration of solar cosmic rays to relativistic energy. See Abstr. 078.014.

Non-relativistic solar electrons. See Abstr. 078.021.

Relativistic electron events in interplanetary space. See Abstr. 078.022.

Observations of He I 584 Å nighttime radiation; evidence for an interstellar source of neutral helium. See Abstr. 082.029.

Berechnung des atmosphärischen Streulichts für Milchstraße und Zodiakallicht als extraterrestrische Quellen. See Abstr. 082.101.

Wavelength dependence of radio scintillation: ionosphere and interplanetary irregularities. See Abstr. 083.038.

Geomagnetic effects of interplanetary sector structure. See Abstr. 084.202.

Association of DP and DR fields with the interplanetary magnetic field variation. See Abstr. 084.225.

High-latitude electric fields and the three-dimensional interaction between the interplanetary and terrestrial magnetic fields. See Abstr. 084.226.

Relationship between interplanetary plasma parameters and geomagnetic Dst. See Abstr. 084.227.

Electric field and currents connected with Y-component of interplanetary magnetic field. See Abstr. 084.264.

Investigation of plasma in the earth's magnetosphere and in the interplanetary space on the satellites of the Prognoz series. See Abstr. 084.266.

Magnetotail variations associated with the southward interplanetary magnetic field. See Abstr. 084.271.

On the heliographic latitude dependence of the interplanetary magnetic field as deduced from the 22-year cycle of geomagnetic activity. See Abstr. 084.300.

Magnetic field observations near Mercury: preliminary results from Mariner 10. See Abstr. 092.036.

Search by Mariner 10 for electrons and protons accelerated in association with Venus. See Abstr. 093.026.

Investigation of the composition of gases from inclusions in tektites and cosmic dust. See Abstr. 105.135.

Orientation of interstellar and interplanetary grains. See Abstr. 131.014.

Structure of 194 southern declination radio sources from interplanetary scintillations. See Abstr. 141.064.

Power spectra of scintillations on irregularities of the interplanetary plasma. See Abstr. 141.127.

Spatial dependence of the pitch-angle and associated spatial diffusion coefficients for cosmic rays in interplanetary space. See Abstr. 143.003.

Cosmic ray scintillations. 2. General theory of interplanetary scintillations. See Abstr. 143.021.

Cosmic ray scintillations. 3. The low-frequency limit and observations of interplanetary scintillations. See Abstr. 143.022.

Scattering of cosmic rays on inhomogeneities of the interplanetary magnetic field. See Abstr. 143.048.

On the investigation of streams of cosmic radiation in the interplanetary medium. See Abstr. 143.050.

107 Cosmogony of the Planetary System

107.001 **Stochastic coalescence model for terrestrial planetary accretion.** M. Hallam, A. H. Marcus.
Icarus, Vol. 21, 66 - 85 (1974).
A nonequilibrium stochastic coalescence model for terrestrial planetary accretion is developed by using an approximation to the Safronov-Golovin solution for the scalar transport equation with linear kernel. According to this model, formation of comparatively massive objects occurs quite rapidly during the early stages of accretive evolution in a given terrestrial planetesimal population, while during late growth stages, an increasingly substantial fraction of total population mass becomes incorporated into progressively fewer, relatively very large bodies.

107.002 **Early chemical history of the solar system.** L. Grossman, J. W. Larimer.
Rev. Geophys. Space Phys., Vol. 12, 71 - 101 (1974).
The purpose of the present paper is to review the recent literature on chemical fractionations during the condensation of the solar system and on their consequences to the establishment of chemical differences between the different classes of chondrites and between the planets.

107.003 **The chemistry of the solar system.** J. S. Lewis.
Sci. American, Vol. 230, No. 3, p. 50 - 60, 65 (1974).
The sun, the planets and other bodies in the system formed out of a cloud of dust and gas. What processes in the cloud could account for the present composition of these objects?

107.004 **Some numerical results on the dynamical evolution of the solar system.** P. E. Nacozy.
Bull. American Astron. Soc., Vol. 6, 205 (1974).— Abstr. AAS.

107.005 **The internal evolution of planetary-sized objects.** D. C. Tozer.
The Moon, Vol. 9, 167 - 182 (1974). — Communication presented at the Lunar Science Institute Conference (see 012.002).
The importance of 'creep' in controlling the internal thermal state of large objects with physical properties corresponding to a roughly homogeneous meteoritic composition is reviewed. Some results of this study are used to justify a picture of evolution as a quasistatic process.

107.006 **Asteroid and comet experiments as follow-on to lunar exploration.** G. Arrhenius, H. Alfvén.
The Moon, Vol. 9, 243 - 244 (1974). — Communication presented at the Lunar Science Institute Conference (see 012.002). — Abstract.

107.007 **Equilibrium condensation in a solar nebula.** G. Arrhenius, B. R. De.
Meteoritics, Vol. 8, 297 - 313 (1973).
In attempts to reconstruct the environment of condensation of solar system materials, particularly exemplified by certain meteorite components, the relative temperatures of the gas and the solid are of critical importance. The relationships that determine the heat balance in a circumsolar grain-gas system are examined. Fundamental considerations show that regardless of opacity or gas density, the gas will always be at a higher temperature than the solid in such regions of the system where condensation is possible. Implications of the characteristic temperature differential between the gas and the condensing solid are discussed.

107.008 **Planetary accretion and meteorites.** R. Hutchison.
Meteoritics, Vol. 8, 388 (1973). — Abstr. Meteoritical Soc.

107.009 **Abundances of the most abundant substances in the protoplanetary cloud.** A. B. Makalkin.
Astron. Zhurn. Akad. Nauk SSSR, Vol. 51, 417 - 424 (1974). In Russian. English translation in Soviet Astron. AJ, Vol. 18, No. 2.
A brief review is presented of modern data on the abundances of 15 most abundant elements in the protoplanetary cloud, on $P - T$ conditions, temperatures of condensations and chemical reactions in the cloud. Some remarks are made on the nature of the condensation process. These data are used for the tabulation of the abundances of abundant substances (element hydrides and oxides) in the cloud outside the high temperature inner zone. Implications for the construction of models of planets and satellites are discussed.

107.010 **On the origin of the solar system, I.** G. P. Kuiper.
Celestial Mechanics, Vol. 9, 321 - 348 (1974).
Presented at a colloquium on 'Copernicus ... and modern dynamical astronomy', Washington, 1972 December.

107.011 **Quand les planètes n'étaient que poussières...** J. Lequeux, H. Reeves.
Sci. Progrès Découverte, No. 3445, p. 12 - 20 (1972).

107.012 **Heutige astronomische Vorstellungen über die Entstehung des Planetensystems.**   J. Dorschner.
Sterne, Vol. 50, 91 - 100 (1974).

107.013 **Direct simulation of collision processes. II: The growth of planetesimals.**
R. J. Dodd, W. McD. Napier.
Astrophys. Space Sci., Vol. 29, 51 - 59 = Commun. Roy. Obs. Edinburgh, No. 163 (1974).

The mass distribution of protoplanets is studied by a Monte Carlo technique. It is assumed that the bodies coalesce on collision and that their self-gravitation is important. The authors consider whether the observed planetary axial inclinations and rotation periods are consistent with the planetesimal hypothesis.

107.014 **Structure and evolutionary history of the solar system, IV.**   H. Alfvén, G. Arrhenius.
Astrophys. Space Sci., Vol. 29, 63 - 159 (1974).

Chemical compositions in the solar system; Meteorites and their precursor states; Mass distribution and the critical velocity; The structure of the groups [of bodies]; Summary.

107.015 **On the future of the two planets earth – moon.**
V. G. Demin.
Vasiona, Vol. 22, 7 - 11 (1974). In Serbo-Croatian.

107.016 **Applied ancient astronomy.**   R. R. Newton.
Applied Phys. Lab. Johns Hopkins Univ., Techn. Digest, (USA), Vol. 12, No. 1, p. 11 - 20 (1973).

Data from ancient astronomy can be applied to a study of the effects that govern the long-term evolution of the solar system.

107.017 **The solar system and its origin.**   J. R. Dormand.
Phys. Education, (GB), Vol. 8, 475 - 481 (1973).

A description is provided of the solar system and earlier theories to explain it are outlined. 8 conditions which any satisfactory theory must explain are given, and there is a brief consideration of some modern theories.

Condizioni fisiche e possibilità di vita nel sistema solare. Conversazione del ciclo "Ipotesi su civiltà extraterrestri". See Abstr. 015.017.

The dynamics of the planets and their satellites. See Abstr. 091.006.

Update on Mars: Clues about the early solar system.   See Abstr. 097.012.

On the formation of Jupiter and Saturn. See Abstr. 099.043.

Strongly differentiated material in high-inclination and retrograde orbits.   See Abstr. 104.022.

Oxygen isotope cosmothermometer revisited. See Abstr. 105.005.

Surface properties of the Orgueil meteorite: implications for the early history of solar system volatiles. See Abstr. 105.008.

Meteoritic magnetite as a cosmothermometer. See Abstr. 105.013.

Stars

111 Stellar Parallaxes

111.001 **Parallaxes of faint dwarf members of the Hyades cluster.** A. R. Upgren.
Astron. Journ., Vol. 79, 651 - 652 (1974).
Parallaxes and proper motions have been obtained for faint red dwarfs in seven fields in the Hyades cluster region which have previously been found to have a high probability of membership, but have no other parallax determination. The weighted mean yields an absolute parallax of 0.0219 ± 0.0018 and a distance modulus of 3.29 ± 0.18.

111.002 **How far away are the stars?** J. Ashbrook.
Sky Telescope, Vol. 47, 165 (1974).

111.003 **An astrometric study of Lalande 21185.** G. Gatewood.
Astron. Journ., Vol. 79, 52 - 53 (1974).
The parallax, proper motion, and acceleration in position of the sun's fourth nearest stellar neighbor, Lalande 21185, have been determined from 143 exposures obtained with the Allegheny Observatory 30-in. Thaw photographic refractor. Several tests fail to confirm the existence of the unseen companion proposed by Lippincott (1960).

111.004 **The U.S. Naval Observatory parallax program.** K. A. Strand, R. S. Harrington, C. C. Dahn.
IAU Symposium No. 61, (see 012.014), p. 159 - 166 (1974).

111.005 **Report on the Herstmonceux I.N.T. parallax programme.** B. F. Jones.
IAU Symposium No. 61, (see 012.014), p. 167 (1974). – Abstract.

111.006 **Comparison of secular parallaxes determined from proper motions of stars relative to galaxies at the Pulkovo and Lick Observatories.** N. V. Fatchikhin.
IAU Symposium No. 61, (see 012.014), p. 205 - 206 (1974).

111.007 **The Yerkes Observatory photoelectric parallax scanner.** W. F. Van Altena.
IAU Symposium No. 61, (see 012.014), p. 311 - 315 (1974).

111.008 **On inconsistencies found in long-term parallax series.** W. D. Heintz.
IAU Symposium No. 61, (see 012.014), p. 317 (1974). – Abstract.

111.009 **The trigonometric parallax calibration of the Wilson-Bappu effect.** T. E. Lutz, D. H. Kelker.
Bull. American Astron. Soc., Vol. 6, 227 (1974). – Abstr. AAS.

111.010 **Parallaxes and proper motions. IX.** A. R. Upgren, S. E. Burt, W. S. Mesrobian.
Astron. Journ., Vol. 79, 708 - 709 (1974).
Relative parallaxes and proper motions are given for 20 stars or stellar systems, ten of which have no previous parallax determination. Most of the stars are nearby dwarfs of spectral class K2 and later. The results for nine of the stars, including four physical binary stars, are discussed individually.

Retour sur Copernic, Kepler, Bessel et les parallaxes.
See Abstr. 004.028.

Proper motion and parallax programmes for large telescopes. See Abstr. 112.018.

The space distribution and radial velocities of some early-type stars in the Perseus spiral arm.
See Abstr. 155.045.

112 Proper Motions, Radial Velocities, Space Motions

112.001 **Radial velocities of nineteen carbon stars.**
Y. Yamashita.
Publ. Astron. Soc. Japan, Vol. 26, 159 - 161 = Tokyo Astron. Obs. Repr. No. 449 (1974).

112.002 **Catalog of individual radial velocities, 12^h–24^h, measured by astronomers of the Mount Wilson Observatory.** H. A. Abt.
Astrophys. Journ., Suppl. Ser., No. 234, Vol. 26, 365 - 489 (1973).

This is the second part of a compilation of the times of observation and individual radial velocities that formed the basis for the mean velocities obtained at the Mount Wilson Observatory and published in Wilson's General Catalogue of Stellar Radial Velocities. This part contains nearly 12000 velocity measures made prior to 1952 of 3700 stars.

112.003 **CPD −72°1184, a high-velocity blue star.**
D. Kilkenny.
Observatory, Vol. 94, 4 - 6 (1974).

In a survey of early-type stars at intermediate galactic latitudes, CPD −72°1184 was found to have a radial velocity in excess of 200 km/s. The evidence suggests that it is a B0 III runaway star rather than a hot subdwarf.

112.004 **On the absence of wide moving pairs among K and M dwarfs.** D. Branch.
Observatory, Vol. 94, 17 - 19 (1974).

112.005 **Stellar proper motions with reference to galaxies.**
S. Vasilevskis.
Vistas in astronomy, Vol. 15, (see 003.001), 145 - 160 (1973).

112.006 **Absolute proper motions of 14600 stars in 85 areas of the northern sky as obtained from galaxies with the Pulkovo normal astrograph.** N. V. Fatchikhin.
Trudy Glav. Astron. Obs. Pulkovo, Ser. 2, Vol. 81, 4 - 211 (1974). In Russian.

A catalogue of absolute proper motions of 14600 stars with respect to 281 galaxies in 85 areas of the northern sky is given. Altogether 1372 stars were taken from the AGK3, 779 of them being used in comparing the two catalogues. The difference of epochs is equal to 22.4 years on the average. The probable error of one absolute proper motion for the reference stars is $\pm 0''.0064$ and for the AGK3 stars $\pm 0''.0078$, and for galaxies $\pm 0''.0069$.

112.007 **Catalogue of relative proper motions of 1474 stars in the α Cyg region.** N. M. Bronnikova.
Trudy Glav. Astron. Obs. Pulkovo, Ser. 2, Vol. 81, 212 - 247 (1974). In Russian.

Relative proper motions of 1474 stars in the α Cyg region are given. They were derived from three plate pairs, taken with the normal astrograph of the Pulkovo Observatory, the difference of epochs being equal to 26, 41 and 45 years. The probable error of one determination of proper motions according to both coordinates is $\pm 0''.0014$ on the average.

112.008 **Catalogue of absolute proper motions of stars with respect to galaxies in Kapteyn's Selected Area 32.**
A. V. Bolbochanu.
Trudy Glav. Astron. Obs. Pulkovo, Ser. 2, Vol. 81, 248 - 260 (1974). In Russian.

The catalogue contains absolute proper motions of 201 stars in SA 32 and outside its limits in a circle of 60' radius 12 bright AGK3 stars, 13 stars with large proper motions and 41 bright stars up to the 12th magnitude. These absolute proper motions have been derived as mean values on two plate pairs directly from 92 reference galaxies with average magnitude $15^m.0$. The probable errors of one proper motion of a star are equal to $\pm 0''.0029$, $\pm 0''.0027$ respectively.

112.009 **Proper motions of stars in the regions of the open clusters NGC 1039 (M 34) and NGC 2682 (M 67).**
A. A. Latypov.
Tsirk. Astron. Inst., *Tashkent,* No. 41 (388), 30 pp. (1973). In Russian.

112.010 **Proper motions of stars in the region of the open cluster NGC 6633.** A. A. Latypov.
Tsirk. Astron. Inst., *Tashkent,* No. 42 (389), p. 11 - 34 (1973). In Russian.

112.011 **Proper motions of stars in the region of the open cluster NGC 6755.** A. A. Latypov.
Tsirk. Astron. Inst., *Tashkent,* No. 43 (390), p. 10 - 27 (1973). In Russian.

112.012 **Radial velocities of stars with moderately broad-lined spectra.** D. P. Hube.
Journ. Roy. Astron. Soc. Canada, Vol. 68, 143 - 145 (1974).

No systematic differences in the radial velocities of stars with broad-lined spectra have been found between measurements made at the Dominion Astrophysical, David Dunlap and Radcliffe Observatories and data listed in the General Catalogue of Stellar Radial Velocities.

112.013 **Proper Motion Survey with the forty-eight inch Schmidt telescope. XXXVI. Proper motions for 6955 faint stars.** W. J. Luyten.
Separate print Univ. Minnesota, Minneapolis, Minnesota. 64 pp. (1974).

In continuation of the lists given in Nos XXX - XXXIV data are herewith given for another 6955 stars, all of which were obtained with the automated-computerized plate scanner and measuring machine funded by NASA and built by Control Data Corporation. The author has included mainly data for those stars with motions larger than $0''.18$ annually for which no earlier determination of proper motion is available as well as for some double stars with smaller motions. Of the 6955 stars listed, the motions for 6883 are believed to be new.

112.014 **Proper motions of stars in wide surroundings of the α Persei cluster.** N. M. Artyukhina, E. P. Kalinina.
Trudy Gos. Astron. Inst. Shternberga, Vol. 44, 3 - 102 (1973). In Russian.

Photographic proper motions and magnitudes for 5468 stars brighter than $13^m.0$ pg are derived in the region with radius $4°.5$ around the center of the α Per cluster. The proper motions were derived using Astrographic Catalogues and six plates taken with the wide-angle astrograph of the Sternberg Astronomical Institute in Moscow. The mean square error for one component of proper motion is $\pm (0''.006 - 0''.007)$. The catalogue is complete to $12^m.0$ pg. Probable members of the cluster are indicated in a table.

112.015 **Proper motions of stars in 10 selected areas of the sky near extragalactic nebulae.**
L. P. Panteleeva.
Trudy Gos. Astron. Inst. Shternberga, Vol. 44, 103 - 146 (1973). In Russian.

Proper motions and magnitudes of 1540 stars and 32 nebulae are obtained from 49 plates of the 15" astrograph of the Moscow Observatory. The probable errors of the proper motions are ± (0."0020 – 0."0032).

112.016 **The radial velocity of δ Delphini.** K.-C. Leung.
Astron. Journ., Vol. 79, 626 (1974).
δ Delphini was observed on seven nights with the 72-in. telescope at a dispersion of 15 Å/mm. The amplitude of the radial velocity curve is about 4 km/sec, and none of the spectra show double lines.

112.017 **Radial velocities from objective-prism plates in the direction of the Large Magellanic Cloud. List of 398 stars, LMC members. List of 1434 galactic stars in the LMC direction.** C. Fehrenbach, M. Duflot.
Astron. Astrophys., Suppl. Ser., Vol. 13, 173 - 267 (1974). In French.
Radial velocities of 1832 stars from the Fehrenbach and Duflot catalogues (1970, 1973).

112.018 **Proper motion and parallax programmes for large telescopes.** C. A. Murray.
IAU Symposium No. 61, (see 012.014), p. 151 - 158 (1974). Invited paper.

112.019 **The Palomar Proper Motion Survey.** W. J. Luyten.
IAU Symposium No. 61, (see 012.014), p. 169 - 173 (1974).

112.020 **Remarks on the definition and determination of proper motions.** W. Dieckvoss.
IAU Symposium No. 61, (see 012.014), p. 187 - 191 (1974). Invited paper.

112.021 **Proper motions with respect to galaxies.** A. N. Deutsch, A. R. Klemola.
IAU Symposium No. 61, (see 012.014), p. 193 - 200 (1974). Invited paper.

112.022 **Report on the Yale-Columbia southern proper motion program.** A. J. Wesselink.
IAU Symposium No. 61, (see 012.014), p. 201 - 203 (1974).

112.023 **The errors of absolute photographic proper motions of stars relative to galaxies.** A. A. Kiselev.
IAU Symposium No. 61, (see 012.014), p. 207 - 208 (1974).

112.024 **Progress report on the Herstmonceux programme of measurement of proper motions in the northern Selected Areas.** C. A. Murray.
IAU Symposium No. 61, (see 012.014), p. 221 - 225 (1974).

112.025 **Proper motions of OB stars in both hemispheres.** H. Yasuda.
IAU Symposium No. 61, (see 012.014), p. 227 (1974). – Abstract.

112.026 **Estimation de la précision des mouvements propres donnés dans un catalogue d'étoiles à l'aide des observations visuelles de couples stellaires optiques.** J. Dommanget.
IAU Symposium No. 61, (see 012.014), p. 229 - 232 (1974).

112.027 **Effect of convection on radial velocities.** V. L. Trimble.
Bull. American Astron. Soc., Vol. 6, 227 (1974). – Abstr. AAS.

112.028 **Observations of proper motions near $l = 140°$.** R. J. Dodd.
Publ. Roy. Obs. Edinburgh, Vol. 9, (No. 2), 63 - 84 (1974).
A list of Cartesian coordinates measured by the GALAXY measuring machine, galactic coordinates, proper motion components in galactic longitude and latitude, and U, B, V photometry is presented for 553 stars in a region near the galactic equator at longitude 140°. The proper motions were derived by direct mapping of stars from the first epoch astrograph plate to the second epoch Schmidt plate and by use of the AGK3 catalogue as an intermediate stage. The methods of analysis are described and the results of the two reductions compared.

An astrometric study of Lalande 21185.
See Abstr. 111.003.

Parallaxes and proper motions. IX.
See Abstr. 111.010.

Spectral observations of HD 187399.
See Abstr. 114.035.

The spectrum of CH Cygni from 1961 to 1973.
See Abstr. 114.165.

Luminosities and motions of peculiar B-type stars.
See Abstr. 115.016.

Do OB runaways have collapsed companions?
See Abstr. 117.038.

Radial velocities of three visual binaries.
See Abstr. 118.004.

Space velocities of nearby classical cepheids.
See Abstr. 122.017.

A spectroscopic survey of southern hemisphere white dwarfs – IV. Radial velocities and space motions.
See Abstr. 126.001.

Internal motions in H II regions. I. The radial velocity field of NGC 6164-6165. See Abstr. 131.554.

Detection of pulsar proper motion.
See Abstr. 141.330.

Proper motions, cluster membership and reddening in NGC 6611. See Abstr. 153.019.

Evolved stars in open clusters. See Abstr. 153.028.

The Hyades convergent point. See Abstr. 153.034.

The space distribution and radial velocities of some early-type stars in the Perseus spiral arm. See Abstr. 155.045.

Stellar kinematics and galactic research involving proper motions. See Abstr. 155.057.

113 Stellar Magnitudes, Colors, Photometry

113.001 **Ultraviolet photometry from the Orbiting Astronomical Observatory. XIV. An extension of the survey of Lyman-α absorption from interstellar hydrogen.**
E. B. Jenkins, B. D. Savage.
Astrophys. Journ., Vol. 187, 243 - 255 (1974).

Data from 26 new stars scanned by the Wisconsin far-ultraviolet spectrometer aboard OAO-2 have been combined with an earlier survey of interstellar Lα absorption reported by Savage and Jenkins, thereby giving the H I density to a total of 95 stars. The authors have again studied the relation between H I column densities and $B - V$ color excesses.

113.002 **On the nature of BD$-10°4662$.**
R. R. Zappala.
Astrophys. Journ., Vol. 187, 257 - 259 (1974).

Infrared observations show that BD$-10°4662$ has excesses in the $H - K$ and $K - L$ colors similar to ordinary TTauri stars. Emission at Ca II H and K confirms the close relationship to that group of objects.

113.003 **The Vilnius photometric system for three-dimensional classification of stars and determination of interstellar reddening.** V. Straižys.
Bull. Vilnius Astron. Obs., No. 36, p. 3 - 16 (1973).

113.004 **Photoelectric photometry of stars in the system UPXYZVS. VII.** A. Bartkevičius, A. Gurklytė, G. Kavaliauskaitė, A. Kazlauskas, R. Kalytis, Z. Sviderskienė, J. Sperauskas, J. Sūdžius, V. Jasevičius.
Bull. Vilnius Astron. Obs., No. 36, p. 17 - 24 (1973).
In Russian.

The catalogue is a direct continuation of the previous one (1972) and contains the results of photoelectric photometry of 175 stars in the Vilnius photometric system, as well as their two-dimensional classification.

113.005 **Photometry of three new population II giants in the Vilnius system.** A. Bartkevičius.
Bull. Vilnius Astron. Obs., No. 36, p. 25 - 32 (1973).
In Russian.

It is shown that the three stars HD 88609, HD 195636 and BD + 41°3735 have enlarged color indices U−P (λ_0 3450−λ_0 3740) in comparison with normal giants of the same T_e. As was shown earlier (1970), this is a property of population II giants with great metal deficiency. The observational data of the three stars mentioned as well as of three metal-deficient giants earlier known are presented in tables. The diagrams U−P, V−S; Q_{UPY}, Q_{XYV} and Q_{PYZ}, Q_{XYZ}, suitable for one-to-one identification of population II giants, are shown in figures.

113.006 **Ultra-violet photometry of early type stars from TD 1satellite observations.**
K. Nandy, G. I. Thompson, C. M. Humphries.
Monthly Notices Roy. Astron. Soc., Vol. 166, 297 - 304 (1974).

Observations of over 200 stars brighter than $V = 7\overset{m}{.}0$ in the spectral type range from O to A5 have been used to verify an ultra-violet photometric system proposed in an earlier paper. The present observations contain a number of giants and supergiants and the effects of luminosity on the ultra-violet colours $(U1 - V)$, $(U3 - U2)$ and the parameter ϕ have been studied. The scatter in the relation between the classification parameter ϕ and quoted MK spectral types is also discussed.

113.007 **Infrared photometry of red giant stars in the four globular clusters M 13, M 71, M 92 and M 5. I.**
R. M. Rusev.
Astron. Zhurn. Akad. Nauk SSSR, Vol. 51, 122 - 131 (1974). In Russian. English translation in Soviet Astron. AJ, Vol. 18, No. 1.

In summer 1971 some infrared photographs of M 13, M 71, M 92 and M 5 were obtained. Eggen showed that the $(B - V)$ colour of the Johnson-Morgan system is not an effective criterion for the temperature classification of red giant stars with $(B - V) \gtrsim + 1\overset{m}{.}5$. With the aid of the $I - (B - V)$ and $I - (B - I)$ diagrams of the program clusters it is shown that the I magnitudes and the $(B - I)$ colours are effective criterions of the luminosity and the temperature classifications of red giant stars in globular clusters. Moreover, it is shown that in globular clusters very red giant stars exist.

113.008 **Photometry in the Cygnus X-1 field.** B. Margon.
Astron. Astrophys., Vol. 30, 467 - 469 (1974).

UBV photometry and extinction corrections are presented for 50 stars in the immediate field of the X-ray source Cygnus X-1. An observation to search for rapid photometric variability in HDE 226868, the optical component of the system, is also reported.

113.009 **Studies of blue objects at high galactic latitudes. III. Faint blue objects in the field of BD + 15°2469.**
C. Barbieri, P. Benvenuti.
Astron. Astrophys., Suppl. Ser., Vol. 13, 269 - 291 (1974).

This paper gives a list of 306 stellar or semistellar objects with negative $U−B$ colour index found on the third field (part of the Virgo cluster of galaxies) of the Asiago Survey of Blue Objects at High Galactic Latitudes. Comparing this field with the previous two, the authors find a remarkably higher number of small ultraviolet galaxies, therefore presumably associated with the Virgo cluster. In a second table several ultraviolet condensations found on bright galaxies of the field are listed. Individual objects of particular interest are also described.

113.010 **UV excess stars and radio sources in the Aquila ring.** J. Isserstedt.
Astron. Astrophys., Vol. 31, 207 - 210 (1974).

With MK spectral classification and photoelectric UBV photometry the UV-excesses, defined by $\Delta(U-B)=(U-B)-E_{U-B}-(U-B)_0$, were determined for each star of the Aquila ring. The UV-excesses are similar to the excesses of young open clusters. The scatter of radial velocities for stars with UV-excesses is smaller than that for all stars. There are some radio sources on the ring. In addition, the ring seems to be surrounded by a shell of radio sources with a diameter of 37 pc, which might be a very old supernova remnant.

113.011 **Photoelectric and spectroscopic observations of WRA 795.**
N. V. Vidal, D. T. Wickramasinghe, B. A. Peterson, M. S. Bessell.
Astrophys. Journ., Vol. 188, 163 - 165 (1974).

Photoelectric and spectroscopic observations of WRA 795 are presented. Light variations with a period of about 4 hours are revealed. The spectra also show variations in emission at Hα, Hβ, and Hγ and possibly in the strength of some absorption lines in time scales of half an hour.

113.012 **Eine photographische UBV-Photometrie in Centaurus.** C. McGruder III.
Diss. Naturwiss. Gesamtfakultät, Ruprecht-Karl-Univ. Heidelberg. 5 + 68 pp. (1974).

The author presents a UBV photometry of a region in

Centaurus near $l = 307°$, $b = 0°$ covering an area of 5 square degree. The spatial distribution of OB stars is derived and the spiral structure of our Galaxy is discussed.

113.013 Five-color photometry of nearby red dwarfs.
A. R. Upgren, S. J. Kerridge.
Publ. Astron. Soc. Pacific, Vol. 85, 721 - 723 (1973).

Photoelectric photometry is presented for 47 stars in the $(R-I)$ color index on the Kron system and, for most of them, in standard UBV color indices as well. Almost all of the stars are nearby K and early M dwarfs selected from the McCormick lists, which have, or soon will have, a trigonometric parallax determined in the course of the astrometric program of the Van Vleck Observatory and which, with few exceptions, do not have previous photometry.

113.014 A new UBV comparison of M67 and Hyades stars.
C. Sturch.
Publ. Astron. Soc. Pacific, Vol. 85, 724 - 729 (1973).

New UBV observations reduce the discrepancies in the author's M67 and Hyades photometry reported earlier. Differences from the original Hyades photometry are negligible. Color excesses for three RR Lyrae stars near M67 are discussed and $E(B-V) = 0\overset{m}{.}07 \pm 0\overset{m}{.}03$ is adopted for the cluster.

113.015 Colour measurements in R-I and the use of the extended–S20 photocathode.
R. G. Bingham, A. W. J. Cousins.
Monthly Notes Astron. Soc. Southern Africa, Vol. 33, 15 - 20 (1974).

113.016 Blue objects near M31. II. Frequency distribution and statistics. G. A. Richter.
Astron. Nachr., Vol. 295, 27 - 31 (1974). In German.

This paper presents a statistical investigation of 999 blue objects [$(U-B) \leq 0.00$, limiting magnitude $U = 20$, $B = 21$], the majority of which is newly discovered on M31-plates of the large Schmidt telescope of the Karl Schwarzschild Observatory Tautenburg.

113.017 The calibration of uby photometry.
E. C. Olson.
Publ. Astron. Soc. Pacific, Vol. 86, 80 - 89 (1974).

Intermediate-band Strömgren color indices $(b-y)$ and $(u-b)$ are calculated using the known photometric sensitivity functions and normalizing with spectrum scanner observations. Colors are calculated for three grids of stellar model atmospheres, and compared to photoelectric observations spanning the B to G spectral range.

113.018 UBV photometry of four magnetic stars.
J. E. Winzer.
Astron. Journ., Vol. 79, 124 - 127 (1974).

Differential UBV observations of four magnetic Ap stars are presented. No significant variations are detected for HD 2453 over a 14-month interval. A possible period of $10\overset{d}{.}61$ is suggested for 9 Tau = HD 22374, although this period gives a poorly defined light curve. Well-defined light curves have been obtained for the other two stars, HD 171586 and HD 192913, which have periods of $2\overset{d}{.}1436$ and $16\overset{d}{.}478$, respectively.

113.019 Statistical analysis of the $uvby\beta$ catalogue.
E. Lindemann.
Astron. Astrophys., Suppl. Ser., Vol. 15, 209 - 214 (1974).

From the 7603 stars with data in the $uvby\beta$ catalogue 740 were measured by two or more authors in $b-y$, m_1, c_1 and 613 in Hβ. A statistical study of the deviations from their mean is presented. Standard deviations by spectral types and by publication reference are briefly discussed. Individual discussion of the stars having largely discordant values is presented.

113.020 On the relationship between the apparent magnitudes given in several catalogues and the UBV system. F. Ochsenbein.
Astron. Astrophys., Suppl. Ser., Vol. 15, 215 - 252 (1974).

Relationships between (BV) measurements and apparent magnitudes given in the Henry Draper catalogue, B. Boss's General Catalogue, AGK 2-3 catalogue, Cape Zone Catalogue $-40°$ to $-52°$ and Cape Photographic Catalogue for the equinox 1950 are derived from a comparison to the Photoelectric Catalogue. The apparent magnitudes in the catalogue of stellar identifications are now given in a coherent system for about 360000 stars brighter than 11th magnitude. A total of more than 700 errors were detected in the Photoelectric Catalogue; they are listed in tables. A new version of this catalogue, with new identifications and corrected measurements, is now available at the Strasbourg Data Center.

113.021 Photoelectric photometry of shell stars.
H. F. Haupt, A. Schroll.
Astron. Astrophys., Suppl. Ser., Vol. 15, 311 - 319 (1974). In German.

Stars showing shell phenomena were suspected of indicating active phases by light variations that are more easily measured than their spectra. Therefore 83 shell stars and carefully selected comparison stars have been observed during 1968/69 at the Vienna University Observatory and at Observatoire de Haute-Provence. A table contains the final results.

113.022 Infrared photometry of southern emission-line stars.
D. A. Allen, I. S. Glass.
Monthly Notices Roy. Astron. Soc., Vol. 167, 337 - 350 (1974).

A search for 2.2-μm continua in southern compact planetary nebulae and emission-line stars has proved about 50 per cent successful. The objects detected are probably not conventional planetary nebulae and appear to form two groups, one of which is identified with symbiotic stars and the other with early-type forbidden-line stars.

113.023 The Coalsack. II. Photometry of suspected flare stars. W. B. Weaver.
Astrophys. Journ., Vol. 189, 81 - 83 (1974).

$UBVI_{ph}$ photographic photometry was obtained for most of the rapid variables discovered in the Coalsack by Andrews. No significant number of these stars are nebular flare stars, UV Ceti stars or T Tauri stars, but their colors and objective-prism spectra are consistent with ordinary reddened early-type stars.

113.024 A search for faint violet stars in southern galactic latitudes. S. Jaidee, G. Lyngå.
Ark. Astron., Vol. 5, 345 - 379 (1974).

A search for faint violet stars on two-image Schmidt plates has yielded a catalogue of 296 stars over 540 square degrees. Some of these have been observed photometrically in UBV as well as in red.

113.025 Photoelectric photometry of early-type stars in a field in Ophiuchus.
A. Ardeberg, S. Wramdemark.
Ark. Astron., Vol. 5, 387 - 390 (1974).

Photoelectric measurements of magnitudes and colours on the standard UBV system are presented for 39 early-type stars in an area approximately centred on the open cluster NGC 6633.

113.026 Photoelectric photometry of luminous B-type stars in a Milky Way field in Cepheus.

A. Ardeberg, K. Särg.
Ark. Astron., Vol. 5, 391 - 395 (1974).
   Photoelectric measurements of magnitudes and colours on the standard UBV system are presented for 40 stars of spectral type B, situated in a Milky Way field in Cepheus. The observed stars are mainly of OB type.

113.027 **Four-color photometry of early-type stars.**
A. G. D. Philip.
Dudley Obs., *Albany, New York*, Rep. No. 8, 42 pp. (1973).
   Review paper given to Commission 45 of the International Astronomical Union on August 28, 1973 in Sydney, Australia.

113.028 **Photographic photometry of carbon stars in the field centered at $l = 204°.4$, $b = -1°.3$.**   Z. Alksne.
Investigation of the sun and red stars, (see 003.008), p. 7 - 22 (1974). In Russian.
   In the low galactic latitude field of 5° diameter centered at $l = 204°.4$, $b = -1°.3$ all available (17) carbon stars have been monitored for light variations in three pass-bands. R magnitudes have been measured on 35 plates, V magnitudes on 25 films and B magnitudes on 8 plates. 13 carbon stars were found to be variable, one among them of Mira type, three semiregular (SR) and nine irregular (Lb).

113.029 **Photometric investigations of carbon stars in the field centered at $l = 173°.5$, $b = -1°.5$.**
L. Duncāns.
Investigation of the sun and red stars, (see 003.008), p. 23 - 43 (1974). In Russian.
   Stellar magnitudes for 25 carbon stars are determined by photographic photometry in the region of the Milky Way round the cluster NGC 1893. Light variations based on 45 observations in B, V, R for 11 stars are defined. Light curves are given for these stars and for DY Aur and S Aur in V and B in the period October 1969 to April 1972.

113.030 **Variability of infrared carbon stars.**
A. Alksnis.
Investigation of the sun and red stars, (see 003.008), p. 44 - 62 (1974). In Russian.
   Based on the results of photographic photometry performed with the Schmidt camera of the Radioastrophysical Observatory and on photometric data, mainly infrared, published by several authors, the light variations of infrared carbon stars CIT 5, CIT 6, CIT 13 and IRC+10216 are presented.

113.031 **The detection of Ap stars in the Geneva photometric system.**   B. Hauck.
Astron. Astrophys., Vol. 32, 447 - 448 (1974).
   Using the photometric diagrams of the Geneva system, the author shows that it is possible to detect the Ap stars by photometry. Two stars (HD 47756 and HD 72524) considered as normal stars in the literature appear as "photometrical" Ap stars.

113.032 **Photovisual photometry of semiregular and irregular carbon stars.**   J. Krempeć.
Stud. Soc. Sci. Torunensis, Sectio F (Astron.), Vol. 5, 51 - 67 = Bull. Astron. Obs. Toruń, No. 50/III (1973).
   The paper contains photovisual photometry of twenty semiregular and irregular carbon stars. The observations were made at the Toruń Observatory. The obtained photovisual magnitudes are given for each star in tables. Two semiregular carbon stars, ST And and S Cam behaved rather regularly in the examined period of time. The photovisual light curves of these stars are shown in figures.

113.033 **Report on the Celescope ultraviolet observations from the OAO-2 satellite and associated research at the Smithsonian Astrophysical Observatory.**
E. Avrett, R. Davis, W. Deutschman, K. Haramundanis, R. Kurucz, C. Payne-Gaposchkin, E. Peytremann, R. Schild.
Astron. Inst. Utrecht, Internal Rep. ROF 72, (see 012.012), p. C2.1 - C2.19 (1974).

113.034 **Ultraviolet photometry of early type stars from TD-1 satellite observations.**
C. M. Humphries, K. Nandy, G. I. Thompson.
Astron. Inst. Utrecht, Internal Rep. ROF 72, (see 012.012), p. C14.1 (1974). – Abstract.

113.035 **JHKL photometry of 145 southern stars.**
I. S. Glass.
Monthly Notes Astron. Soc. Southern Africa, Vol. 33, 53 - 58 (1974).

113.036 **Infrared photometry and polarimetry of cool stars. I.**
G. V. Khozov, T. N. Khudyakova.
Trudy Astron. Obs., *Leningrad*, Vol. 30 (= Uchenye Zapiski Leningr. Un-ta, No. 373 = Seriya Matem. Nauk, vyp. (No.) 50), p. 48 - 69 (1974). In Russian.
   The results of infra-red photometric and polarimetric observations of 20 cool stars are presented. The observations were carried out in 1969–1971 in the standard I ($\lambda_{eff} \approx 0.9\ \mu$) and K ($\lambda_{eff} \approx 2.2\ \mu$) bands.

113.037 **Stellar flux measurements of early type stars at 950 Å.**   B. E. Troy, Jr., C. Y. Johnson, J. M. Young, J. C. Holmes.
Bull. American Astron. Soc., Vol. 6, 215 (1974). – Abstr. AAS.

113.038 **An intrinsic color and absolute magnitude calibration for F-type stars.**   D. L. Crawford.
Bull. American Astron. Soc., Vol. 6, 224 (1974). – Abstr. AAS.

113.039 **Four-color observations of early-type stars. IV. South Galactic Pole.**   A. G. D. Philip.
Astrophys. Journ., Vol. 190, 573 - 578 (1974).
   Forty-eight B- and A-type stars with visual apparent magnitudes from 4.3 to 14.5 have been measured photoelectrically in the Strömgren four-color system. In a 34 square degree region surrounding the South Galactic Pole, all stars classified as spectral type A5 and earlier were measured. Of the 26 early-type stars in this area, 15 were classified as members of population I, four were population II, four were an intermediate population, two were B-type, and one was peculiar.

113.040 **Untersuchung der interstellaren Verfärbung an O- und B-Sternen.**   W. Wiemer.
Thesis Univ. Bonn (1973).

113.041 **List of spectroscopic and photometric catalogues lately published or to be published. IV.**   B. Hauck.
Centre de Données Stellaires, Inform. Bull. No. 6, p. 14 - 21 (1974).

113.042 **Improved version of the photoelectric catalogue (Blanco et al., 1968, Publ. U.S. Naval Obs..2nd Series, Vol. 21).**   F. Ochsenbein.
Centre de Données Stellaires, Inform. Bull. No. 6, p. 22 - 23 (1974).

113.043 **Multicolor photometry of metallic-line stars. I. ν^1 Draconis and ν^2 Draconis.**
E. E. Mendoza V, S. F. González B.
Rev. Mexicana Astron. Astrofis., Vol. 1, No. 1, p. 67 - 74 (1974).
   This paper describes UBVRI-photometric observations of the double stars BS6554 (ν^1 Dra) and BS6555 (ν^2 Dra). The photometry indicates that probably both are variable stars.

The variability amplitude is around 0.1 mag. At the first glance, the variables are irregular.

113.044 Five-color photometry of nearby red dwarfs. II.
A. R. Upgren.
Publ. Astron. Soc. Pacific, Vol. 86, 294 - 295 (1974).

Photoelectric photometry is given for 46 stars in the $(R - I)$ color index on the Kron system and in the standard UBV color indices. Most stars are nearby K and early M dwarfs selected from the McCormick lists for which a trigonometric parallax is determined by the astrometric program of the Van Vleck Observatory.

113.045 Photographische Photometrie einiger Hα-Sterne im Gebiet des Nordamerikanebels.
R. Hudec, K. Juza, S. Rössiger.
MVS, *Sonneberg*, Vol. 6, 131 - 132 (1974).

For 10 Hα-stars from Welin's list, V- and B-magnitudes were determined by photographical photometry. The distances derived from them suggest the conclusion that at least some of these stars lie within the space occupied by the aggregate around the nebula NGC 7000. For 4 variable stars listed by Welin, the light variations could not be confirmed.

High-speed photometry.   See Abstr. 031.011.

Television photometry: the Mariner 9 experience. See Abstr. 031.024.

Applications of Schmidt telescopes – with emphasis on direct and spectral surveys.   See Abstr. 032.048.

The southern sky surveys – a review of the ESO Sky Survey Project.   See Abstr. 041.023.

CPD −72°1184, a high-velocity blue star. See Abstr. 112.003.

The Balmer discontinuity of the Ap stars. See Abstr. 114.006.

Absolute flux measurements in the rocket ultraviolet.   See Abstr. 114.007.

Pre-main-sequence stars. III. Herbig Be/Ae stars and other selected objects.   See Abstr. 114.029.

The nature of the light variations of HD 188041. See Abstr. 114.043.

Photometric variations of the helium-weak star HR1063.   See Abstr. 114.060.

The emission-line variable HBV 475. See Abstr. 114.061.

On the period and peculiarity of the cool Ap star HD 203006 (ϑ^1 Mic).   See Abstr. 114.067.

A survey for Hα emission objects in the Milky Way. I. Vulpecula−Cygnus.   See Abstr. 114.087.

Photoelectric measurements of the Swan band of C_2 in some G and K stars.   See Abstr. 114.088.

On the relation between carbon star spectral types and colors.   See Abstr. 114.098.

On the C_2, CN, and CO indices of carbon stars. See Abstr. 114.159.

Strong-line K stars. I. Photometry. See Abstr. 114.173.

A luminosity index for bright K0−K5 stars. See Abstr. 115.003.

Multicolor photometry of eclipsing binaries with "undersize" subgiant secondaries.   See Abstr. 121.049.

An infrared survey of RW Aurigae stars. See Abstr. 122.062.

Three-color photometry of the flare star EV Lacertae.   See Abstr. 122.097.

Ultraviolet photometry from the Orbiting Astronomical Observatory. XV. The strongly magnetic variable HD 215441.   See Abstr. 122.101.

Photographic surface photometry of the nebulae surrounding V380 Ori and R Mon.   See Abstr. 132.002.

On the ratio of total to selective absorption in the open cluster IC 2581.   See Abstr. 153.011.

Proper motions, cluster membership and reddening in NGC 6611.   See Abstr. 153.019.

Four-color and Hβ photometry for open clusters. IX. NGC 6871.   See Abstr. 153.022.

The old open cluster NGC 2420. See Abstr. 153.023.

Intermediate-band photometry of M67. See Abstr. 153.024.

Integrated colors of star clusters in the Large Magellanic Cloud.   See Abstr. 153.025.

Four-color and Hβ photometry for open clusters. X. The α Persei cluster.   See Abstr. 153.029.

On the variability of dwarf K- and M-type stars in the Pleiades and Hyades.   See Abstr. 153.030.

The open cluster NGC 6281.   See Abstr. 153.033.

Analysis of composite spectra – the ultra-violet excess of stars in globular clusters.   See Abstr. 154.004.

Photometry of southern globular clusters – III. Bright stars in 47 Tucanae, NGC 6397 and NGC 288. See Abstr. 154.017.

Infrared observations of stars in the field of NGC 7099.   See Abstr. 154.020.

Photometry to the main-sequence of 47 Tucanae. See Abstr. 154.022.

The oldest disk stars.   See Abstr. 155.022.

Photometry of stars in the nuclear bulge of the Galaxy through a low-absorption window at $l = 0°, b = -8°$. See Abstr. 155.046.

Catalogue of dB and gK stars in the Palomar – Groningen Variable-Star Field No. 3.   See Abstr. 155.078.

OB stars detection in the Large Magellanic Cloud. See Abstr. 159.003.

114 Stellar Spectra, Temperatures, Spectroscopy

114.001 Infrared spectra of γ^2 Velorum and ζ Puppis.
T. G. Barnes, D. L. Lambert, A. E. Potter.
Astrophys. Journ., Vol. 187, 73 - 82 (1974).

Observations of γ^2 Vel and ζ Pup are presented for the spectral region 5800–11,000 cm^{-1} (1.7–0.9 μ), at resolutions of 2 and 4 cm^{-1}, respectively. This constitutes the first reported spectroscopy of a Wolf-Rayet star beyond 1.1 μ.

114.002 Stellar molecular abundances. II. The violet depression in carbon stars.
L. W. Hartmann, J. F. Dolan.
Astrophys. Journ., Vol. 187, 151 - 153 (1974).

The behavior of the theoretical abundance of the AlH molecule as a function of the C/O abundance ratio is not consistent with the suggestion that AlH is the source of the violet depression found in the continuous spectra of carbon stars. The results are consistent with the suggestion that a molecular species such as SiC is responsible.

114.003 Emission lines in the spectrum of Zeta Ophiuchi.
V. S. Niemelä, R. H. Méndez.
Astrophys. Journ., (Letters), Vol. 187, L23 - L26 (1974).

Hα and He I λ5875 emissions and asymmetric profiles in the absorption lines have been discovered on recent spectra of ζ Oph. A description of the observations and a discussion of the possible interpretations are made.

114.004 Two new physical processes in the far-ultraviolet spectrum of Zeta Tauri.
S. R. Heap, T. P. Stecher.
Astrophys. Journ., (Letters), Vol. 187, L27 - L29 (1974).

The authors have detected absorption features in the far-ultraviolet spectrum of ζ Tau (B3pe), one of which corresponds to a new transition involving quasi-H$^-$, and the other to a transition involving quasi-H$_2^+$.

114.005 Short-term spectral variability of γ^2 Velorum. Photometric observations.
A. Sanyal, W. Weller, S. Jeffers.
Astrophys. Journ., (Letters), Vol. 187, L31 - L33 (1974).

Photometric observations of γ^2 Vel using a narrow-band interference filter centered on He II λ4686 have been obtained. Analysis of these data shows short-period variability with a period of 154 ± 35 s.

114.006 The Balmer discontinuity of the Ap stars.
M. Gerbaldi, B. Hauck, N. Morguleff.
Astron. Astrophys., Vol. 30, 105 - 109 (1974).

The Ap stars of the Osawa catalogue measured in the Geneva and Barbier-Morguleff system show that the Balmer discontinuity is affected by the particular character of the star.

114.007 Absolute flux measurements in the rocket ultraviolet. R. C. Bohlin, D. Frimout, C. F. Lillie.
Astron. Astrophys., Vol. 30, 127 - 134 (1974).

A two-channel spectrometer was calibrated in the wavelength region 1200 - 3400 Å and flown on an Aerobee rocket to observe the stars α Lyr, η UMa, and ζ Oph. Standard tungsten lamps provided the absolute calibration down to 2250 Å and a photodiode calibrated by the National Bureau of Standards was the reference at shorter wavelengths. The flux from η UMa agrees with the prediction of a hydrogen line blanketed model atmosphere within 10% between 1700 and 3400 Å and within 4% over most of this wavelength region.

114.008 Infrared observations of HD 65750, a red giant in a reflection nebula. R. M. Humphreys, E. P. Ney.
Astron. Astrophys., Vol. 30, 159 - 160 (1974).

Photometric observations from 0.4 μ to 18 μ of the red giant HD 65750 in the reflection nebula IC 2220 reveal the presence of the silicate emission feature at 10 and 20 μ. The energy distribution of HD 65750 very closely resembles the Mira variables in the infrared as opposed to pre-main sequence objects. This similarity supports the suggestion that it is an evolved star losing mass perhaps in the process of forming a planetary nebula.

114.009 Astronomical application of information compression. A. Bijaoui.
Astron. Astrophys., Vol. 30, 199 - 202 (1974). In French.

In the case of gaussian processes, methods based upon the correlation matrix allow an analysis in terms of a few statistically independent parameters and can lead to a reduction in the observational error. Such a method is described and expressions are obtained for the probable error of measured quantities. The method is applied to the analysis of the equivalent widths of Fe I lines in the spectra of F, G and K type stars in terms of a parameter which is related to the Fe/H abundance ratio for a star of a given type.

114.010 The iron-rich stars HR 511 and HR 7670.
J. B. Hearnshaw.
Astron. Astrophys., Vol. 30, 203 - 209 (1974).

An iron abundance of [Fe/H] = + 0.36 has been obtained from a differential line blanketed model atmosphere analysis of the Fe I spectrum of the low velocity K 0 V star HR 511. Similarly, an abundance of [Fe/H] = + 0.26 has been obtained for HR 7670, a high velocity G 6 subgiant. The first star has a large parallax, and it may lie slightly below the main sequence defined by the majority of nearby stars. The second star has an estimated age of about 8 × 10^9 yr, a result contrary to the hypothesis that the oldest stars in the Galaxy should be metal poor.

114.011 On a possible three-year cycle of η Carinae.
A. Feinstein, H. G. Marraco.
Astron. Astrophys., Vol. 30, 271 - 273 (1974).

Recent UBV observations of η Carinae combined with slightly older measures show that it has been increasing in brightness since 1940. This new data suggest also a long-period variation with an amplitude of 0^m.20, which is confirmed by the observations published by the southern visual observers. The combined data give a period of P = 1110 ± 8 days.

114.012 A note on the use of the strength of the Si II doublet $\lambda\lambda$6347, 6371 as a luminosity indicator in B9–A2 supergiants. J. D. Rosendhal.
Astrophys. Journ., Vol. 187, 261 - 263 (1974).

Measurements of the strength of the red Si II lines $\lambda\lambda$6347, 6371 indicate that among the B9–A2 supergiants there is a linear relation between absolute visual magnitude and line strength for stars more luminous than M_V = −5.

114.013 Lines of neutral barium and the abundance of barium in two K supergiants.
A. R. Hyland, J. R. Mould.
Astrophys. Journ., Vol. 187, 277 - 282 (1974).

The abundance of barium in the late supergiants ϵ Peg and HR 4050 relative to the classical barium star HR 5058 is derived from the strengths of the weak Ba I lines, obtained from high-resolution echelle plates. Both stars are found to have essentially solar values of Ba/H and Ba/Fe.

114.014 Supergiant binary stars.

R. M. Humphreys, E. P. Ney.
Astrophys. Journ., (Letters), Vol. 187, L75 - L79 (1974).

The energy radiated at infrared wavelengths by HD 101584 (F2e Iap) is difficult to reconcile with thermal reradiation by atmospheric grains. It is suggested that the F supergiant has an M supergiant companion. Other possible binaries containing an infrared star are discussed.

114.015 **A broad absorption region in the ultraviolet spectra of early-type stars.**
G. I. Thompson, C. M. Humphries, K. Nandy.
Astrophys. Journ., (Letters), Vol. 187, L81 - L84 (1974).

Spectrophotometric measurements at 35 Å resolution with the S2/68 Ultraviolet Sky-Survey Telescope of satellite TD1 have shown the existence of a broad region of absorption centered at 1920 Å in the spectra of many early B-type stars. The absorption, which extends between approximately 1800 and 2150 Å, becomes pronounced for stars of high luminosity.

114.016 **On the strength of Hα in the O-stars.** W. Osborn.
Monthly Notices Roy. Astron. Soc., Vol. 166, 463 - 468 (1974).

Observations of the strength of the Hα line in O-stars have been compared with the predictions of both LTE and non-LTE model atmospheres computed by Auer & Mihalas. It is found that non-LTE effects are definitely important for O-stars. For main sequence stars the equivalent widths predicted by the non-LTE models are in excellent agreement with the observations, but for high luminosity stars the predicted widths appear to be too large.

114.017 **The coolest Wolf-Rayet stars.**
B. L. Webster, I. S. Glass.
Monthly Notices Roy. Astron. Soc., Vol. 166, 491 - 497 (1974).

Four emission-line stars, M4$-$18, He 2-113, CPD $-56°$ 8032 and V348 Sgr, can be considered as a class, with low excitation surrounding nebulae, infra-red radiation from dust grains and characteristic spectra dominated by C II emission lines. They appear to form a cool extension to the carbon sequence of the population II Wolf-Rayet stars. V348 Sgr is of particular interest as it has features in common with both the R Coronae Borealis stars and the planetary nebulae.

114.018 **Detection of NML Cygnus at 2.8 cm.**
W. M. Goss, A. Winnberg, H. J. Habing.
Astron. Astrophys., Vol. 30, 349 - 350 (1974).

The authors have detected a radio source of 12 ± 2 mfu at 2.8 cm apparently associated with the late M-type supergiant NML Cygnus. They discuss the detection and sketch briefly some explanations.

114.019 **A new helium variable.** L. Balona, W. L. Martin.
Monthly Notices Roy. Astron. Soc., Vol. 166, 35P - 38P (1974).

The B-type star HD 175362 has been found to be a spectrum variable with a period of 3.7 days. The strongest variation occurs for the helium lines, but silicon and other lines are also suspected to vary, both in strength and width.

114.020 **Equivalent width of interstellar molecular lines. IV: Analyses of the line spectrum of H$_2$ in the interstellar spectrum of δ Scorpii.**
K. S. Krishna Swamy, S. P. Tarafdar.
Astrophys. Space Sci., Vol. 26, 179 - 182 (1974).

The authors have made an analysis of the observed equivalent widths of the lines of Lyman and Werner bands by Smith. The H$_2$ column densities of 3×10^{19} to 10^{20} and the excitation temperature of 80 to 150 K satisfy the observations. The authors have also discussed the importance of getting the excitation temperature from lines of H$_2$ and other heteronuclear diatomic molecules for the same star and from different regions of space.

114.021 **On the spectra of Si II and Si III in ζ Draconis.**
B. E. J. Pagel.
Astron. Astrophys., Vol. 30, 471 - 473 (1974).

On the basis of equivalent widths given by Underhill (1973) for the B 6 III star ζ Dra, it is shown that the Si II/Si III ionisation equilibrium is in accordance with simple-minded LTE estimates but that two lines attributed to the doubly excited multiplet 13 of Si III are anomalously strong and apparently misidentified.

114.022 **A spectroscopic study of the blue halo silicon star HD 97859.** R. Stalio.
Astron. Astrophys., Vol. 31, 89 - 95 (1974).

A study of the atmosphere of the high latitude blue star HD 97859 was carried out at 12.4 Å/mm. The results obtained are presented.

114.023 **Heavy metals in M-giants.** P. J. Huggins.
Astron. Astrophys., Vol. 31, 103 - 107 (1974).

A reanalysis is made of data presented for the M 2-giant star β Peg by Yamashita (1965). The large overabundance of heavy metals found by Yamashita is not confirmed, and the implications of this result for other cool stars are discussed.

114.024 **Wolf-Rayet stars in the galactic window around $l = 307°$.** G. Lyngå, B. Stenholm.
Astron. Astrophys., Vol. 31, 111 - 112 (1974).

Around $l = 307°$ there are several WR stars at large distances with unexpectedly low colour excesses. This is suggested to be due to a gap between spiral arms, a galactic window.

114.025 **The unusual strength of the lines of Cr I and Fe I in the A 2p star HD 110066.** P. L. Selvelli.
Astron. Astrophys., Vol. 31, 113 - 116 (1974).

A direct comparison between the equivalent widths of the Ap star HD 110066 and the sun has been made in order to show the high intensity of these lines. The Cr I lines are systematically stronger than those in the solar spectrum while the Fe I lines are, on the average, slightly weaker. These facts suggest a great overabundance of these elements.

114.026 **Variable K line profile of 73 Draconis.**
K. Sadakane.
Publ. Astron. Soc. Japan, Vol. 26, 93 - 102 (1974).

Observations were made to analyze quantitatively the variable intensity and profile of the Ca II K line of the A2p star 73 Draconis. The measurements of profiles and equivalent widths of this line and of Hγ are given. Equivalent widths of Ca I 4227 and of other strong metallic lines are also measured. A simple model to interpret the observed K line profile variation is proposed.

114.027 **Analysis of the energy distribution of the Be star π Aqr.** H. L. Nordh, S. G. Olofsson.
Astron. Astrophys., Vol. 31, 343 - 345 (1974).

The energy distribution of the Be star π Aqr and its variability are analyzed in the spectral region 3500–10400 Å.

114.028 **Spectral energy distributions of T Tauri stars.**
L. V. Kuhi.
Astron. Astrophys., Suppl. Ser., Vol. 15, 47 - 89 (1974).

The spectral energy distributions of T Tauri stars and related objects have been measured with a photoelectric spectrum scanner from 0.33 to 1.11 μ. Reddening corrections derived from the spectral types show an increase towards earlier type presumably due to the presence of circumstellar material. The corrected energy distributions indicate clearly the peculiar nature of the ultraviolet and blue continuous emission from T Tauri stars. Emission-line fluxes are also measured for the brighter hydrogen, calcium and helium lines. A strong correlation is found between the ultraviolet flux and the Ca II K emission and indicates a chromospheric origin for the various emission phenomena.

114.029 **Pre-main-sequence stars. III. Herbig Be/Ae stars and other selected objects.** M. Breger.
Astrophys. Journ., Vol. 188, 53 - 58 (1974).

The majority of the Herbig Be/Ae stars show considerable linear polarization in contrast to other hot young stars. A dependence on evolutionary phase exists in that the unpolarized stars are on the zero-age main sequence. Only little variability in apparent magnitude on the time scale of days exists, but long-term variability is common. In the Taurus region and in NGC 2264 two stars are especially unusual: RY Tau shows strong, time-variable polarization increasing strongly toward the ultraviolet; and W90, situated below the main sequence in NGC 2264, shows variability in apparent magnitude as well as amount and wavelength dependence of polarization. The new data are interpreted as indicating variations in the optical depth of the circumstellar shell.

114.030 **The energy distribution of the very red star in NGC 6231.** R. Schild, J. B. Oke, L. Searle.
Astrophys. Journ., Vol. 188, 71 - 74 (1974).

From multichannel spectrometer data and infrared observations, the energy distribution of NGC 6231 - 92 is compared to model stellar atmospheres. The data are consistent with models hotter than T_{eff} = 6000°K when the standard reddening law for Perseus is applied. Nine new UBV photometric observations and infrared observations made in 1969, 1970, and 1972 contain no evidence of variability. The new results support Herbig's suggestion that the star is a heavily reddened F supergiant behind NGC 6231.

114.031 **Veiling and the presence of circumstellar gas and dust in some infrared stars.** R. M. Humphreys.
Astrophys. Journ., Vol. 188, 75 - 85 (1974).

Four high-luminosity M supergiants, VY CMa, VX Sgr, S Per, and HD 97671, are discussed to see whether gaseous emission might be the cause of their spectral and infrared peculiarities. Both spectroscopic data and photometric observations from 0.4 to 18 μ are combined for a study of the observed properties of these peculiar M supergiants.

114.032 **Free-free and free-bound emission in low-surface-gravity stars.** R. C. Gilman.
Astrophys. Journ., Vol. 188, 87 - 94 (1974).

The energy distributions of four cool supergiants, HD 97671, S Per, VX Sgr, and VY CMa, have been fitted using the emission spectra of the H^- ion, of ionic free-free scattering, and of pure magnesium silicate grains, as well as an underlying stellar blackbody. Densities and temperatures in the emission region were derived on the basis of these fits.

114.033 **The oxygen abundance in the metal-deficient star HD 122563.**
D. L. Lambert, C. Sneden, L. M. Ries.
Astrophys. Journ., Vol. 188, 97 - 100 (1974).

The forbidden neutral oxygen line at 6300.3 Å has been detected in the spectrum of the metal-deficient star HD 122563. The [O I] equivalent width and a model-atmosphere analysis provide an oxygen abundance log $[N(O)/N(H)]$ = -5.38 ± 0.15 or a logarithmic overabundance of oxygen relative to the metals of [O/Fe] = +0.6.

114.034 **The recent shell event of Zeta Ophiuchi.**
N. J. Irvine.
Astrophys. Journ., (Letters), Vol. 188, L19 (1974).

Between July 1971 and July 1973, ζ Oph produced a shell optically thick in the hydrogen lines. Mass outflow observed in the rocket-ultraviolet prior to 1971 is probably related to the formation of the shell.

114.035 **Spectral observations of HD 187399.**
G. V. Akhundova, N. A. Ivanova, Kh. I. Novruzova.
Astrofizika, Vol. 9, 605 - 608 (1973). In Russian. English translation in Astrophysics, Vol. 9, No. 4.

Five spectra of the shell star HD 187399 with dispersions 30 Å/mm and 8 Å/mm in August 1970 and June-August 1971 were obtained. The hydrogen line profiles and radial velocities are determined. The expansion velocity of the extended envelope of the star is of the order of 75–100 km/sec.

114.036 **On the variability of hydrogen lines in the spectrum of the peculiar star HD 184905.**
R. N. Kumajgorodskaya, N. M. Chunakova.
Astrofizika, Vol. 9, 608 - 611 (1973). In Russian. English translation in Astrophysics, Vol. 9, No. 4.

Variability (double wave) of different parameters of hydrogen lines in the spectrum of the peculiar star HD 184905 during the period of 2^d17 is found. The largest variations occur in the central parts of these lines.

114.037 **Remarques sur les spectres continus théoriques d'étoiles O et B.** R. Krikorian.
Comptes Rendus Acad. Sci. Paris, Sér. B, Vol. 278, 131 - 133 (1974).

En utilisant comme critères de comparaison la discontinuité de Balmer D et le paramètre $\Delta \Phi_0$, nous confrontons les résultats obtenus à partir des derniers modèles non ETL d'étoiles O et B aux observations Chalonge-Divan.

114.038 **Deux étoiles de dixième magnitude, riches en hélium ionisé et très déficientes en hydrogène, à haute latitude galactique.**
J. Berger, A.-M. Fringant, E. Rebeirot.
Comptes Rendus Acad. Sci. Paris, Sér. B, Vol. 278, 227 - 230 (1974).

BD 48° 1777 et BD 37° 1977 ont un spectre caractérisé par les raies d'absorption intenses de He II, l'hydrogène semble absent; la première est une sous-naine O du type de BD 75° 325, la deuxième ne serait pas sous-lumineuse et s'apparente étroitement à BD 37° 442.

114.039 **Microturbulent velocities of southern F-type supergiants.** J. C. Castley.
Proc. Astron. Soc. Australia, Vol. 2, 200 - 202 (1973). – Presented at the annual general meeting of the Astron. Soc. Australia – see 012.001.

114.040 **The 10-micron excess of α Herculis.**
N. J. Woolf.
Publ. Astron. Soc. Pacific, Vol. 85, 730 - 731 (1973).

The star α Her has been spectrophotometrically compared with α Boo in the 8–14 micron band. It shows a small (30% of continuum at peak) excess emission whose spectral shape is similar to that of cooler M giants and M supergiants, the Orion nebula, and comet Bennett 1969.

114.041 **HD 148688: an oxygen-rich supergiant.**
M. Jaschek, E. Brandi.
Publ. Astron. Soc. Pacific, Vol. 85, 736 - 738 (1973).

The discovery of an oxygen-rich supergiant is reported. Variations in the intensity of the lines belonging to light elements, helium, nitrogen, and carbon, were found. The time scale is of the order of days.

114.042 **An unidentified absorption in the spectra of C-S stars.** P. M. Rybski.
Publ. Astron. Soc. Pacific, Vol. 85, 751 - 755 (1973).

An unidentified, broad absorption feature has been noticed in low-dispersion spectra of some carbon Mira variable stars and some C-S stars. Resembling a red-degraded molecular absorption visible from 6372 Å to 6394 Å and being strongest at 6379 Å, the feature exhibits little phase-dependent behavior among the C-S stars. Evidence is presented that, though similar in appearance, the feature in the C-S stars may not be the same as that in the carbon Mira variables where calcium iodide is identified tentatively as the species responsible for the feature.

114.043 **The nature of the light variations of HD 188041.**
T. J. Jones, S. C. Wolff.
Publ. Astron. Soc. Pacific, Vol. 85, 760 - 763 (1973).

Four-color ($uvby$) observations of the peculiar A star HD 188041 have been obtained at the Mauna Kea Observatory. A combination of flux redistribution from the ultraviolet and local line-blocking effects can explain the light variations of this star.

114.044 **On the optical stability of X Persei.**
A. Frohlich, I. Nevo.
Monthly Notices Roy. Astron. Soc., Vol. 167, 221 - 226 (1974).

The optical candidate for the X-ray source 2U 352+30, X Per was observed photoelectrically four times during 1972 December. The authors present measurements of the broadband colours of X Per and of its optical behaviour on time scales from one to a few hundred seconds.

114.045 **Observations of the linear polarization in the Hβ emission feature of γCas.**
D. Clarke, I. S. McLean.
Monthly Notices Roy. Astron. Soc., Vol. 167, 27P - 30P (1974).

Linear polarization measurements have been made across the Hβ emission line of γ Cas. The results show a significant reduction in the degree of polarization at the emission peak relative to the values in the wings and the continuum. This depolarization effect in the emission feature exhibits variability.

114.046 **Spectrophotometry of stars of the Pleiades cluster. II. Absorption lines.**
T. P. Vykhrestyuk, V. G. Karetnikov.
Astrometriya i Astrofizika, *Kiev*, vyp. (No.) 20, (see 003.002), p. 13 - 18 (1973). In Russian.

Equivalent widths of the spectral lines of hydrogen and K Ca II were determined on the basis of spectrograms of 57 stars in the Pleiades cluster. Electronic densities in stellar atmospheres were calculated by the methods of Inglis-Teller and Unsöld. A comparison of the obtained values with the characteristics of «normal» stars was made. There is no good agreement in some cases.

114.047 **The peculiar star He 2-177: a slow nova and a possible X-ray source.** E. D. Carlson, K. G. Henize.
Astrophys. Journ., (*Letters*), Vol. 188, L47 - L50 (1974).

The spectrum of the peculiar emission-line star He 2-177 is found to resemble that of the slow nova RR Tel in remarkable detail. The evolution in the spectrum since 1965 closely resembles the evolution of RR Tel between 1951 and 1954. A newly determined position of the X-ray source 2U 1639–62 makes its identification with He 2-177 less probable than before.

114.048 **Time variation of the H$_2$O maser and infrared continuum in late-type stars.**
P. R. Schwartz, P. M. Harvey, A. H. Barrett.
Astrophys. Journ., Vol. 187, 491 - 496 (1974).

Time variations in the microwave H$_2$O emission from late-type stars are shown to be periodic and are correlated with optical and infrared variations. The microwave variability may result from changes in the maser pump rate induced by changes in the infrared flux.

114.049 **OAO-2 observations of the helium spectrum variable a Centauri.** M. R. Molnar.
Astrophys. Journ., Vol. 187, 531 - 537 (1974).

Ultraviolet photometry and spectrometer data are presented for a Cen, which exhibits large variations in the strength of its He I lines. A model atmosphere analysis indicates that the photometric variations are not due to flux redistribution from back-warming effects, but can be interpreted as due to the rotation of a distorted stellar disk in the shape of a prolate spheroid.

114.050 **Spectroscopic studies of O-type stars. IV. Lines in the red region.** P. S. Conti.
Astrophys. Journ., Vol. 187, 539 - 549 (1974).

Equivalent widths of several lines in the red spectral region have been obtained for 77 O and Of type stars. Comparison with the non-LTE plane-parallel models of Auer and Mihalas has been made.

114.051 **On the abundance of europium.**
M. R. Hartoog, C. R. Cowley, S. J. Adelman.
Astrophys. Journ., Vol. 187, 551 - 554 (1974).

The inclusion of the effects of hyperfine splitting can significantly lower the abundance estimate of Eu from singly ionized lines which lie on the flat portion of the curve of growth. In the 21 cool Ap stars studied by Adelman and the five Am stars studied by Smith the Eu abundance was reduced by 0.4 dex on the average. This makes the Eu abundance comparable to that of its neighboring rare earths Sm and Gd in the Ap stars and less than Sm and Gd in the Am stars, but still substantially overabundant with respect to solar values.

114.052 **On the problem of V1016 Cygni and the evolutionary stage of the symbiotic stars.**
G. B. Baratta, A. Cassatella, R. Viotti.
Astrophys. Journ., Vol. 187, 651 - 659 (1974).

A spectrophotometric study of V1016 Cygni has been carried out using several objective-prism spectra taken during 1971–1972, covering the spectral range 3425–9000 Å. No spectral variations have been detected during this period, which indicates that this object has reached a stationary phase. It is shown that the recent history of V1016 Cyg can be best represented by a model of an evolved star surrounded by a large gas and dust envelope of few hundredths of a solar mass. It is concluded that the phenomenon of the symbiotic stars could be interpreted as a manifestation of advanced stages of a very late star.

114.053 **Observación y estudio teórico del espectro de la estrella peculiar HD 18474.**
M. J. Fernández-Figueroa.
Urania Barcelona, Año 58, Nos. 277 - 278, p. 3 - 53 (1973).

The spectrum of the peculiar G star HD 18474 is investigated. Accurate fine analyses for HD 18474 yield metal abundance ratios (relative to hydrogen) that do not differ significantly from solar values. These results do not agree with preceding studies. Carbon abundance is strong deficient in comparison with the sun. The lines of CH were used to obtain an estimate of excitation temperature and abundance which was found to be nearly the same as the mean relative carbon abundance. These results suggest a deficient carbon content in the pre-stellar medium.

114.054 **Lines of H I and He II in the spectra of Wolf-Rayet stars.** M. Ilmas, T. Nugis.

Tartu Astr. Obs., Teated No. 44, p. 19 - 52 (1973). In Russian.

The equations of stationarity for the level populations of H I and He II are solved on the predicted conditions in the envelopes of the WR stars HD 192163 and HD 191765 (both WN 6), HD 192103 (WC 7) and HD 192641 (WC 6-7).

114.055 Analysis of the He I lines in the spectra of some Wolf-Rayet stars. T. Nugis.
Tartu Astr. Obs., Teated No. 44, p. 53 - 81 (1973). In Russian.

The system of the equations of stationarity for the level populations of He I is solved on the predicted conditions in the He II zone. The results are presented in tables.

114.056 Solution of equations of stationarity for N V, C IV, triplets of N IV and doublets of N III. T. Nugis.
Tartu Astr. Obs., Teated No. 44, p. 83 - 143 (1973). In Russian.

The system of the equations of stationarity for the level populations of N V, C IV, the triplets of N IV and the doublets of N III is solved on the predicted conditions in the envelopes of some WR Stars.

114.057 Solution of equations of stationarity for singlets of C III. T. Nugis, T. Feklistova.
Tartu Astr. Obs., Teated No. 44, p. 145 - 165 (1973). In Russian.

The system of the equations of stationarity is solved for the singlets of C III. The results are discussed in connection with the stars HD 192103 (WC 7) and HD 192641 (WC 6-7).

114.058 Obtaining normalized line profiles with an electronic digital computer. G. I. Abbasov.
Astron. Tsirk., No. 797, p. 6 - 7 (1973). In Russian.

114.059 Blue CN-absorption measurements of close binary stars. R. H. Koch.
Astron. Journ., Vol. 79, 34 - 41 (1974).

The blue CN-absorption has been observed photoelectrically for numerous single stars and visual binary components covering wide ranges of spectral type and luminosity class. These objects are defined as standards for the present program. For 97 close binaries in assorted evolutionary stages, similar observations have been obtained and compared to those of the standards. Anomalous indices are noted for a few objects which have UV excesses, three supergiant pairs, and several strongly interacting short-period binaries.

114.060 Photometric variations of the helium-weak star HR1063. J. E. Winzer.
Astron. Journ., Vol. 79, 45 - 46 (1974).

UBV observations of the helium-weak star HR1063 = HD21699 are presented. It is found to be variable in light with a period of either $2^d.4761$ or $1^d.6780$ and amplitudes of 0.03, 0.04, and 0.05 mag in V, B, and U, respectively. This is the only helium-weak star in which periodic variations have so far been established.

114.061 The emission-line variable HBV 475. F. M. Stienon, M. R. Chartrand III, C. Y. Shao.
Astron. Journ., Vol. 79, 47 - 51, 105 (1974).

Photometric and spectroscopic studies of the peculiar emission star HBV 475 have been made. A photographic light curve for the past 80 y and objective-prism spectral data before and after the 1966 outburst are presented and discussed. The characteristics of this object resemble those of a symbiotic variable undergoing outbursts.

114.062 On the reality of the high lithium abundances in carbon stars. J. G. Cohen.
Publ. Astron. Soc. Pacific, Vol. 86, 31 - 32 (1974).

Because of recent advances in model atmospheres for cool stars, there may be some doubt as to the validity of previous estimates of the Li abundance in the super-Li stars. Through the behavior of the resonance lines of K I, the author shows that these analyses are correct, and that the super-Li stars represent a real abundance peculiarity of Li.

114.063 MK spectral types for some bright F stars. A. Cowley, D. Fraquelli.
Publ. Astron. Soc. Pacific, Vol. 86, 70 - 73 (1974).

MK spectral types are presented for 100 stars previously classified F0. This comprises a portion of a larger project to assign accurate spectral types to the stars in the Bright Star Catalogue.

114.064 A search for technetium in the Ba II star ζ Capricorni. A. M. Boesgaard, R. A. Fesen.
Publ. Astron. Soc. Pacific, Vol. 86, 76 - 77 (1974).

A search for singly ionized technetium (Tc II) in the barium star ζ Cap (G5 III) at a dispersion of 3.3 Å mm^{-1} is reported. η Her (G7 III-IV) and the sun were used for spectral comparison. No evidence for the presence of Tc II in ζ Cap is found and an upper limit of log N(Tc) ≤ 1.6 is derived.

114.065 On the equivalent width of CH line in Eta Aquilae. G. C. Joshi, K. R. Bondal, M. C. Pande.
Bull. Astron. Inst. Czechoslovakia, Vol. 25, 120 - 121 (1974).

Equivalent widths of a line belonging to the 0–0 bands of the electronic transition $^2\Delta - ^2\Pi$ of the CH molecule have been calculated for various phases of light variation in η Aql.

114.066 Uranium in HR8911. P. Galeotti, E. Lovera.
Nature, Vol. 249, 130 - 131 (1974).

The authors report preliminary results concerning some anomalies in the spectrum, which were obtained during analysis of the Cr-Eu-Sr peculiar star HR8911 (κ Piscium) and which show evidence for the presence of r-process elements in this object.

114.067 On the period and peculiarity of the cool Ap star HD 203006 (ϑ^1 Mic).
H. M. Maitzen, J. Breysacher, R. Garnier, C. Sterken, N. Vogt.
Astron. Astrophys., Vol. 32, 21 - 28 (1974). In German.

Measurements carried out in UBV and with a multicolor medium band filter system (Maitzen, 1973) combined with $uvby$ measurements of Morrison and Wolff (1971) yield a rotational period of 2.1219 days. The intensity distribution with wavelength relative to the comparison star HD 203585 shows the flux to be reduced in the v band at the primary maximum of the Eu variation. A flux depression occurs at λ 5240 Å. Both phenomena are compared with similar features in the intensity distribution of the longer period Eu star HD 125248. On the base of the magnetic accretion theory (Havnes and Conti, 1971) it is discussed whether a difference in the product of magnetic field strength and age is responsible for the different photometric behaviour of both stars.

114.068 The spectrum of the Ap star HD 111133. I. S. Engin.
Astron. Astrophys., Vol. 32, 93 - 101 (1974).

The spectrum of the Cr-Eu-Sr Ap star HD 111133 has been studied. A differential analysis by means of the curve of growth has been made using o Peg as standard star. Moreover an atmospheric model has been computed using the program ATLAS 5. The results of the rough and fine analyses are generally in good agreement.

114.069 High-resolution spectra of cool stars in the 10- and 20-micron regions. R. Treffers, M. Cohen.
Astrophys. Journ., Vol. 188, 545 - 552 (1974).

The spectra of two carbon stars, IRC +10216 and IRC +30219 (CIT 6), and of three oxygen-rich stars, VY CMa, α Ori, and W Hya, are presented with 2 cm^{-1} resolution. The carbon stars show an emission band attributable to circumstellar particles of silicon carbide. VY CMa and α Ori have excess emission at both 10 and 20 μ, characteristic of circumstellar silicate particles. No atomic or molecular features are seen.

114.070 **Are 2-micron absorptions and 11-micron emissions of M stars related?** T. D. Faÿ, Jr.
Astrophys. Journ., Vol. 188, 553 - 557 = Contr. Louisiana State Univ. Obs., *Baton Rouge*, No. 91 (1974).

Three properties of M stars are compared: (1) [3.5] − [11] colors, (2) two measures of 2.3-μ band absorption, and (3) two measures of 1.9-μ band absorption. It is found that the 2.3-μ bands of 21 M giants and supergiants correlate with [3.5] − [11] colors. A comparable relation exists between the mean 1.9-μ bands and [3.5] − [11] colors of 23 M giants and Miras. The sum of the 1.9-μ + 2.3-μ band intensities correlates with [3.5] − [11] color for 31 M stars. Binaries and Ia supergiants appear to depart from these relations.

114.071 **Rotational distortion of stellar absorption lines. I. Parameters from photographic spectra.**
T. R. Stoeckley, C. S. Morris.
Astrophys. Journ., Vol. 188, 579 - 594 (1974).

A differential corrections procedure is described for deriving the half-width, shape, central depth, and equivalent width of rotationally broadened absorption-line profiles from high-dispersion photographic spectra. A method is described for combining several profiles into a single, improved profile which can be reanalyzed. The capabilities of the procedure are illustrated with exhaustive tests on actual profiles and on synthetically generated noisy profiles.

114.072 **High-spectral-resolution measurements of the H I λ1216 and Mg II λ2800 emissions from Arcturus.**
H. W. Moos, J. L. Linsky, R. C. Henry, W. McClintock.
Astrophys. Journ., (*Letters*), Vol. 188, L93 - L95 (1974).

High-spectral-resolution scans of H I λ1216 and Mg II λλ2796, 2803 obtained using the ultraviolet spectrometer aboard the Copernicus satellite show broad and very asymmetrical emission profiles. The ratio of the line widths to the solar values is consistent with a law similar to the Wilson-Bappu relation for the calcium K reversal. A fit of the interstellar absorption profile indicates that the average H density toward this nearby star is low, 0.02−0.1 cm^{-3}.

114.073 **Radio emission from pre-main-sequence stars.**
J. H. Spencer, P. R. Schwartz.
Astrophys. Journ., (*Letters*), Vol. 188, L105 - L106 (1974).

Thermal radio emission has been detected from circumstellar nebulae of the pre-main-sequence stars T Tauri and LkHα-101. The radio sources have angular size of less than 1.''0.

114.074 **Equivalent widths and depths of line formation in the spectra of stars with peculiar chemical composition. I. Lines of magnesium, silicon, calcium and strontium.**
V. V. Leushin.
Soobshch. Spets. Astrofiz. Obs. AN SSSR, *Zelenchukskaya*, vyp. (No.) 5, p. 18 - 45 (1972). In Russian.

For homogeneous model atmospheres with Θ_{eff} = 0.4, 0.5, 0.6 and lg g = 4.0 calculations are performed of the equivalent widths and formation depths of some spectral lines for various abundances of elements. The calculated equivalent widths are compared with observed ones. The abundances of the investigated elements obtained by different methods are compared in the atmospheres of a number of normal and peculiar stars. A relation is given between the optical depth and the geometrical one in the models considered.

114.075 **Infrared fluxes, spectral types, and temperatures for very cool stars.**
H. M. Dyck, G. W. Lockwood, R. W. Capps.
Astrophys. Journ., Vol. 189, 89 - 100 (1974).

From narrow- and broad-band photometry between 0.55 and 10.2 μ, complete energy distributions, photometric spectral types, and total fluxes are derived for a sample of cool giants, Mira variables, and IRC stars. The effective temperature scale for cool stars is reexamined and its sources of error are discussed.

114.076 **Metallicism in border regions of the Am domain. III. Analysis of the hot stars Alpha Geminorum A and B and Theta Leonis.** M. A. Smith.
Astrophys. Journ., Vol. 189, 101 - 111 (1974).

The fine curve-of-growth analysis previously applied to a number of Am and Fm stars is extended to some of the hottest known Am stars. The author compared the abundances of α Gem A and B and θ Leo with α Lyr. The results suggest that not only are α Gem A and B metallic lined but so is the heretofore A2 V spectral standard θ Leo. In addition, with a T_{eff} of 10, 200°K, α Gem becomes the hottest known Am star. The most important result to emerge herein is the discovery of a moderate deficiency of helium in α Gem A (and probably also in θ Leo and Sirius). This represents the confirmation of an important prediction of the subsurface diffusion model.

114.077 **Photoelectric profile measurements of Hα and Hβ in Be stars.** D. F. Gray, J. M. Marlborough.
Astrophys. Journ., Suppl. Ser., No. 240, Vol. 27, 121 - 130 (1974).

The authors present photoelectric measurements of Hα and Hβ in 14 Be stars. They find from a shape analysis of the wings of Hα that a Gaussian broadening mechanism is reasonable whereas a damping profile is not. It is suggested on this basis that the width and shape of Hα can be reasonably interpreted as a combination of shell rotation and electron scattering.

114.078 **Selected line identifications in the ultraviolet spectrum of Gamma Equulei.** S. J. Adelman.
Astrophys. Journ., Suppl. Ser., No. 242, Vol. 27, 183 - 201 (1974).

Selected line identifications for the cool Ap star γ Equ in the region λλ 3093−3806 are given. Atomic species of importance include Co II, Ni II, Nb II, Mo II, Rh I, Pd I, Ce III, Pr III, Nd III, Sm III, Tb II, Dy II, Ho II, Er II, Tm II, and Yb II. Some implications of their identifications are discussed.

114.079 **Some morphological properties of WN spectra.**
N. R. Walborn.
Astrophys. Journ., Vol. 189, 269 - 271 (1974).

An extension of the Hiltner-Schild line-width classification for WN spectra is introduced, in order to describe more fully the range of line widths among these stars. A simplified excitation classification, based solely upon the relative strengths of N III, IV, and V emission features in the classification region, is used. The mutual independence of excitation, emission-line width, and emission-line strength in WN spectra is emphasized, with reference to an illustration of wide, 62 Å mm^{-1} spectrograms of six southern WN stars.

114.080 **Short-period radial-velocity variations in π Aquarii.**
A. E. Ringuelet, M. E. Machado.
Astrophys. Journ., Vol. 189, 285 - 288 (1974).

The central star in π Aquarii, B0 Ve, undergoes short-period radial-velocity variations with a period of 0.087 days.

114.081 **Observations of Hα in HDE 226868.**
R. J. Brucato, R. R. Zappala.

Astrophys. Journ., (Letters), Vol. 189, L71 - L74 (1974).

High-dispersion spectrograms in the red have been obtained for HDE 226868, the optical candidate for Cyg X-1. Preliminary results are presented.

114.082 The peculiar A star HD 215441. S. J. Adelman.
Astrophys. Journ., Suppl. Ser., No. 242, Vol. 27, 203 - 245 (1974).

Line identifications are given for the sharp-lined strong magnetic Ap star HD 215441 in the spectral region $\lambda\lambda$ 3638–4644 near magnetic maximum. Lines of Cl II and Ti III were found in addition to those of atomic species previously identified. Comparison of the spectrum with those of normal B and other hot magnetic Ap stars shows that this star shares the anomalies of other silicon stars.

114.083 Spontaneous fission on the Ap star HR 465.
T. Ohnishi.
Nature, Vol. 249, 532 - 533 (1974).

The author examines the possibility that the rare earth anomaly on HR 465 originates from asymmetric spontaneous fission.

114.084 On the spectral variations of P Cygni in 1973.
R. Viotti, R. Nesci.
Inform. Bull. Variable Stars, (I.A.U. Commission 27), Konkoly Obs., Budapest, No. 878, 2 pp. (1974).

114.085 CD −44° 3318: an emission-line star involved in a dark nebula. N. Sanduleak.
Inform. Bull. Variable Stars, (I.A.U. Commission 27), Konkoly Obs., Budapest, No. 880 (1974).

114.086 A general catalogue of cool carbon stars.
C. B. Stephenson.
Publ. Warner and Swasey Obs., Cleveland, Ohio, Vol. 1, No. 4, 79 pp. (1973).

This catalogue is intended to list all cool carbon stars having known positions of at least roughly the precision of the Henry Draper Catalogue. Cool carbon stars will be defined as stars whose spectra at low dispersion (say a resolution no better than 1–2 Å) are known to show bands of the Swan system of the C_2 molecule; or, if the spectral region of the Swan system is inadequately observed, they show the red or infrared bands of CN in strength adequate to infer that the Swan bands almost certainly would be seen if their presence could be tested. The closing date for the author's literature search was 1973.0. The catalogue contains 3219 stars.

114.087 A survey for Hα emission objects in the Milky Way. I. Vulpecula–Cygnus.
G. V. Coyne, T. A. Lee, E. De Graeve.
Vatican Obs. Publ., Vol. 1, (No. 5), 181 - 196 (1974).

A survey for stars which have emission at Hα has been conducted with a 12 degree objective prism on the Vatican Schmidt telescope. The survey reaches 2 to 3 magnitudes fainter than existing surveys. More than 100 new emission stars have been discovered and photoelectric UBV photometry has been obtained for them. From the two-color plot it is possible to separate out those Hα emission stars which are more probably young objects.

114.088 Photoelectric measurements of the Swan band of C_2 in some G and K stars.
J. B. Alexander, D. Branch.
Monthly Notices Roy. Astron. Soc., Vol. 167, 539 - 549 (1974).

Photoelectric measurements with interference filters are used to examine the behaviour of C_2 in a variety of G and K stars. Observations indicate that C_2 is enhanced in metal-rich stars and giants. It appears probable that α Ser and μ Leo are appreciably more metal rich than the sun. In some barium stars, there is no evidence for any abnormalities in the carbon abundance.

114.089 The carbon isotope ratio in Arcturus.
R. Griffin.
Monthly Notices Roy. Astron. Soc., Vol. 167, 645 - 662 (1974).

The C^{12}/C^{13} ratio in Arcturus is shown to be $5\,1/2$ to 6 by analysis of photographic spectrograms of the red CN (2, 0) band.

114.090 Photoelectric spectrum scans and coudé spectroscopy of HD 153919 (= 2 U 1700–37).
B. W. Bopp, G. Grupsmith.
Monthly Notices Roy. Astron. Soc., Vol. 167, 65P - 68P (1974).

Photoelectric scanner observations and coudé spectroscopy of HD 153919 are reported. None of the scans obtained on four nights show any evidence of broad emission wings near 4100 or 4350 Å, as has been previously reported by Walker and Hensberge et al. The Hβ emission is variable in intensity from night to night and perhaps on time scales as short as a few minutes. The coudé spectra show the C III–N III and He II emission features to have broad (± 40 Å) wings.

114.091 Resultado de las observaciones de α Peg efectuadas desde el satélite europeo TD1.
M. Rego Fernández, M. J. Fernández-Figueroa.
Revista Real Acad. Ci. Exactas, Fís. Nat., Madrid, Vol. 67, 629 - 658 = Univ. Complutense Madrid, Fac. Ci., Seminario Astron. Geod., Publ. No. 75 (1973).

114.092 Temperature determination of B7–B8 type stars by means of low dispersion ultraviolet observations.
L. Houziaux.
Centre Univ. État Mons, Fac. Sci., Dép. Astrophys., Commun. No. 39, 9 pp. (1974).

Temperatures obtained for 29 B7–B8 objects are given.

114.093 Light variations in seven Wolf-Rayet stars.
A. F. J. Moffat, W. Haupt.
Astron. Astrophys., Vol. 32, 435 - 439 (1974).

Extensive photometric observations in the continuum and emission lines of seven Wolf-Rayet stars yield no significant short period (2 min $\lesssim P \lesssim$ 1 h) oscillations as one might expect if they are massive He- or C-burning stars at the upper limit of pulsational stability. However, four of the stars, three of them "single", show variations over the three hour measuring interval, which tend to be larger in the emission lines than in the continuum. This is interpreted in terms of the binary nature of the Wolf-Rayet phenomenon.

114.094 Spectrum variability of σ Ori E. K. Hunger.
Astron. Astrophys., Vol. 32, 449 - 451 (1974).

The hydrogen-deficient star σ Ori E is found to be spectrum variable. The time scale of the variations seems to be less than a day or probably one hour. The variability is similar to that found in α Centauri.

114.095 The spectroscopic analysis of bright carbon stars, Y CVn and U Hya in the one-micron region.
M. Hirai.
Proc. Japan Acad., Vol. 49, 534 - 539 = Contr. Dep. Astron., Univ. Tokyo, No. 165 (1973).

In this paper the spectra of the bright carbon stars, U Hya (N2; $C7_3$) and Y CVn (N3; $C5_4$) were studied in the region between 1.00 and 1.08 μm.

114.096 Method for measurement of the polarization in the infrared spectral region.

Yu. N. Lipskij, V. V. Novikov, A. P. Popov.
Soobshch. Gos. Astron. Inst. Shternberga, No. 184, p. 3 - 16 (1973). In Russian.

The method of measuring the linear polarization in the infrared spectral region is described. The calculations of analyzing systems of two types and results of an investigation of working characteristics of the spectropolarimeter are given.

114.097 **New catalogue of A stars with peculiar spectra (Ap) and with metallic lines (Am).**
C. Bertaud, M. Floquet.
Astron. Astrophys., Suppl. Ser., Vol. 16, 71 - 153 (1974). In French.

A first table contains 1049 stars that seem to be well classified. A second table gives information on 189 stars with more doubtful classification. Some remarks following these two tables indicate other important features, such as nature of visual or spectroscopic binary, optical or spectral variability and presence of a magnetic field.

114.098 **On the relation between carbon star spectral types and colors.**
R. K. Honeycutt, T. D. Faÿ, Jr., W. H. Warren, Jr.
Astron. Journ., Vol. 79, 631 - 633 = Publ. Goethe Link Obs., Indiana Univ., *Bloomington, Indiana*, No. 158 (1974).

The authors report observations of the $[0.57]-[0.68]$-μ (RED) colors of 32 carbon stars and compare carbon star RED colors and Yamashita spectral classes. The goal is to describe the different relations between carbon star spectral class and color and to consider whether or not these relations are consistent with a temperature gradient sequence for carbon stars.

114.099 **Scanner observations of carbon stars from 5000 to 7000 Å.**
T. D. Faÿ, Jr., W. H. Warren, Jr., H. R. Johnson, R. K. Honeycutt.
Astron. Journ., Vol. 79, 634 - 641 = Publ. Goethe Link Obs., Indiana Univ., *Bloomington, Indiana*, No. 157 (1974).

Photoelectric spectrum scans at 20-Å resolution of 26 N-type and 11 R-type carbon stars from 5000 to 7000 Å are reduced to flux and presented in graphic form. A monochromatic color index $[0.57]-[0.68]$ for this spectral region is defined and correlated with $(V-R)$ and $[3.5]-[11.0]$ color indices. The authors devise new photometric indices to measure CN and C_2 band strengths. The use of their CN indices provides a new photometric estimate of the $^{12}C/^{13}C$ ratio. Their 0.56-μ C_2 index correlates positively with the 1-μ CN index and anticorrelates with the 2.3-μ CO index for a wide variety of carbon stars. Comparison of these correlations with predictions of molecular column densities from model atmospheres indicates that these interrelations can be understood as a variation of C/O among carbon stars.

114.100 **The nature of infrared excesses in extreme Be stars.**
R. Schild, F. Chaffee, J. A. Frogel, S. E. Persson.
Astrophys. Journ., Vol. 190, 73 - 83 (1974).

The purpose of this paper is to specify further the properties of the extreme Be stars, particularly with regard to their continuous energy distributions, and to strengthen the previously noted connection with the pole-on stars and the core-contraction stage of post-main-sequence evolution.

114.101 **High-resolution Fourier spectra of stars and planets.**
P. Connes, G. Michel.
Astrophys. Journ., (Letters), Vol. 190, L29 - L32 (1974).

High-resolution near-infrared spectra (1-2.5 μ) have been recorded with a fast-stepping large-maximum-path-difference interferometer coupled to the Palomar 500-cm telescope. Samples of stellar and planetary spectra are given.

114.102 **Cyanogen-band strengths of giant stars in 47 Tucanae.**
R. D. McClure, W. Osborn.
Astrophys. Journ., Vol. 189, 405 - 407 (1974).

Photoelectric photometry on the DDO system has been obtained for stars on the giant branch of the globular cluster 47 Tucanae in order to obtain a quantitative measure of the cyanogen-band strength. A value of metal abundance inferred from the CN strengths is in very good agreement with that inferred from the ultraviolet excesses. This value is approximately $1/4$–$1/5$ Hyades abundance.

114.103 **Spectral classification from infrared spectra of moderate dispersion.** H. Albers.
Astrophys. Journ., Vol. 189, 463 - 467 (1974).

The various kinds of spectra that can be recognized on widened infrared objective-prism spectra having a dispersion of 700 Å mm^{-1} are discussed. A temperature sequence is shown for the giant M stars as well as the luminosity effects which can be used to identify dwarf and supergiant M stars. The criteria which can be used to select high-luminosity stars in other parts of the spectral sequence are described. Other groups of stars which can be identified from these spectra are illustrated.

114.104 **Profiles of emission lines in Be stars. III. Further study of the long-period V/R variation.**
E. Albert, S.-S. Huang.
Astrophys. Journ., Vol. 189, 479 - 483 (1974).

The long-period V/R emission variations in three Be stars, HD 20336, 25 Ori, and β^1 Mon have been studied according to the mathematical formulation of the eccentric rotating model presented in a previous paper. The predictions of the theory fit satisfactorily the general trend of observed data. Periods, eccentricities, and sizes of the emitting ring with respect to the stellar radius are thereby derived and found to be consistent with results of the previous investigation of 105 Tauri.

114.105 **Carbon and nitrogen abundances in metal-poor stars.**
C. Sneden.
Astrophys. Journ., Vol. 189, 493 - 507 (1974).

Carbon and nitrogen abundances have been derived for 10 metal-poor stars. For each star, a combination of coarse and fine analyses have been used to determine a model atmosphere. Then, using the model and atomic- and molecular-line parameters in a spectrum-synthesis program, artificial CH, CN, and NH spectra have been computed. By comparing the synthetic and observed spectra, carbon and nitrogen abundances have been obtained. The consequences for galactic nucleosynthesis theories using the present results are discussed.

114.106 **FG Sagittae: the s-process episode.**
G. E. Langer, R. P. Kraft, K. S. Anderson.
Astrophys. Journ., Vol. 189, 509 - 521 = Lick Obs. Bull., No. 653 (1974).

The spectrum of the supergiant FG Sge has been studied from a series of high dispersion 120-inch (3-m) coudé spectrograms obtained during the interval 1969–1972, thus continuing the work of Herbig and Boyarchuk. Abnormally strong absorption lines of Y II, Zr II, Ce II, La II, and other s-process species began to appear in the spectrum of the central star some time after 1967 and have progressively strengthened. Present abundances per gram of these elements are about 25 times the solar value. The present evolutionary state of FG Sge is discussed qualitatively on the basis of two model scenarios. Observational and theoretical difficulties with each of these scenarios are discussed.

114.107 **Ultraviolet stellar spectra obtained with the Utrecht orbiting stellar spectrophotometer S 59 aboard the ESRO TD-1 A satellite.** R. Hoekstra, K. A. van der Hucht, C. de Jager, T. Kamperman, H. J. Lamers.

Space Research XIII, p. 871 - 877 = Utrechtse Sterrekundige Overdrukken No. 244 (1973).

114.108 Ultraviolet stellar astronomy. A. D. Code.
Astron. Inst. Utrecht, Internal Rep. ROF 72, (see 012.012), p. C1.1 (1974). − Abstract.

114.109 Observations of ultraviolet stellar spectra by the Utrecht Orbiting Stellar Spectrophotometer S59.
K. A. van der Hucht, H. J. Lamers.
Astron. Inst. Utrecht, Internal Rep. ROF 72, (see 012.012), p. C3.1 - C3.9 (1974).

114.110 The near ultraviolet spectrum of early type stars obtained with S59. H. J. Lamers, K. A. van der Hucht, R. Hoekstra, R. Faraggiana, M. Hack.
Astron. Inst. Utrecht, Internal Rep. ROF 72, (see 012.012), p. C4.1 - C4.13 (1974).

114.111 Line identifications in the near ultraviolet spectrum of the peculiar A star ϵ Ursae Majoris.
M. Burger, K. A. van der Hucht.
Astron. Inst. Utrecht, Internal Rep. ROF 72, (see 012.012), p. C6.1 - C6.14 (1974).

114.112 Low resolution ultraviolet spectroscopy of several stars from Apollo 17. R. C. Henry, P. D. Feldman, W. G. Fastie, H. W. Moos, A. Weinstein.
Astron. Inst. Utrecht, Internal Rep. ROF 72, (see 012.012), p. C7.1 - C7.24 (1974).

114.113 The continuum of A type stars between 1350 and 2800 Å. D. Malaise, M. Gros, D. Macau.
Astron. Inst. Utrecht, Internal Rep. ROF 72, (see 012.012), p. C9.1 - C9.7 (1974).

114.114 Line features in the UV spectra of A-stars.
F. Praderie, D. Macau, C. Jamar.
Astron. Inst. Utrecht, Internal Rep. ROF 72, (see 012.012), p. C10.1 - C10.15 (1974).

114.115 S 2/68 observations of B-stars continua shortward of 2500 Å. J. M. Vreux, D. Malaise, J. P. Swings.
Astron. Inst. Utrecht, Internal Rep. ROF 72, (see 012.012), p. C11.1 (1974). − Abstract.

114.116 Absorption features in the UV spectra of B-stars.
J. P. Swings, C. Jamar, J. M. Vreux.
Astron. Inst. Utrecht, Internal Rep. ROF 72, (see 012.012), p. C12.1 (1974). − Abstract.

114.117 Temperature determination of B7−B8 type stars by means of low dispersion ultraviolet observations.
L. Houziaux.
Astron. Inst. Utrecht, Internal Rep. ROF 72, (see 012.012), p. C15.1 - C15.9 (1974).

114.118 Spectral observations of stars and interstellar gas in the balloon-ultraviolet. A. Boksenberg, B. Kirkham, W. A. Towlson, T. E. Venis, B. Bates, P. P. D. Carson, G. R. Courts.
Astron. Inst. Utrecht, Internal Rep. ROF 72, (see 012.012), p. C22.1 - C22.20 (1974).

114.119 FG Sagittae: Rosetta Stone for nucleosynthesis?
R. P. Kraft.
Sky Telescope, Vol. 48, 18 - 22 (1974).

114.120 The spectral energy distributions of BD + 34° 3815 (Cyg X−1) and X Per.
I. N. Glushneva, V. T. Doroshenko, T. S. Fetisova.
Astron. Zhurn. Akad. Nauk SSSR, Vol. 51, 526 - 531 (1974).
In Russian. English translation in Soviet Astron. AJ, Vol. 18, No. 3.

The spectral energy distributions of BD + 34° 3815 and X Per, identified with X-ray sources, are obtained. The roles of a gaseous envelope with T_{eff} = 20000° K and a hot star in the continuum formation are estimated in the spectral region 3200 − 7600 Å. A conclusion on the existence of additional absorption in the nearest neighbourhood of X Per is made.

114.121 On the nature of the emission object M 1-2 = VV 8.
V. P. Arkhipova, R. I. Noskova, A. Ganbarov.
Astron. Zhurn. Akad. Nauk SSSR, Vol. 51, 532 - 535 (1974).
In Russian. English translation in Soviet Astron. AJ, Vol. 18, No. 3.

UBV observations of the peculiar emission object VV 8 = M 1-2, obtained in 1971 - 72 at the Crimean station of the Sternberg Astronomical Institute, show a rather red star with strong ultraviolet excess. In the photometric history investigated, no brightness variations have been found. The absolute energy distribution in the continuum from 4000 Å to 8000 Å was measured and compared with the results of O'Dell. The photometric data were interpreted by means of a model of a symbiotic star proposed by Boyarchuk. The absolute visual magnitude of the cool component is found to be $-1^{m}.5$. The distance of VV 8, about 7 kpc, is determined.

114.122 On the problem of spectral classification of late-type stars. N. S. Komarov, V. E. Panchuk.
Astron. Zhurn. Akad. Nauk SSSR, Vol. 51, 593 - 596 (1974).
In Russian. English translation in Soviet Astron. AJ, Vol. 18, No. 3.

Some criteria of three-dimensional spectral classification of cool stars are considered. It is shown that an analysis of intensities in molecular bands and atomic lines is urgent.

114.123 Absolute ultraviolet spectrophotometry from the TD-1 satellite. III. The continuum of A type stars between 1350 and 2500 Å.
D. Malaise, M. Gros, D. Macau.
Astron. Astrophys., Vol. 33, 79 - 86 (1974).

For 52 stars for which well calibrated spectra exist from the TD-1 satellite, the authors perform a statistical study of their ultraviolet spectrum. The analysis of the values of two spectral indices, built to describe two different parts of the upper stellar atmosphere, leads to the following conclusions: 1) the spectral classification breaks down at 1400 Å, where large variations are present from star to star within the same spectral type; 2) an extra opacity is needed to reconcile the theory with the observations around 2100 Å.

114.124 Absolute ultraviolet spectrophotometry from the TD-1 satellite. IV. Line features in UV-spectra of A stars. C. Jamar, D. Macau, F. Praderie.
Astron. Astrophys., Vol. 33, 87 - 91 (1974).

About 20 absorption features are detected in S 2/68 spectra of A type stars between 1350 and 2550 Å. Tentative identifications are given as well as a discussion of their behaviour from star to star. Ap stars and supergiants are specially considered.

114.125 Absolute ultraviolet spectrophotometry from the TD-1 satellite. V. Bolometric corrections for early type stars of luminosity class V.
F. Beeckmans, D. Macau, D. Malaise.
Astron. Astrophys., Vol. 33, 93 - 98 (1974).

A few main sequence stars for which the authors have accurate absolute ultraviolet spectrophotometric data and visible energy distributions are used for the determination of semi-empirical bolometric corrections. For the spectral range where no observational data are available, they use fluxes predicted by LTE models of Kurucz and Carbon and Gingerich. Effective temperatures are deduced from the bolometric fluxes for stars with known diameter.

114.126 Weakening of absorption lines in the near infrared spectrum of χ Cyg.   T. E. Derviz.
Vestn. Leningr. un-ta, 1973, No. 19, p. 130 - 133. In Russian. Abstr. in Referativ. Zhurn. 51. Astron., 6.51.464 (1974).

114.127 He I λ4471 profiles in B stars: calculations with an improved line-broadening theory.
D. Mihalas, A. J. Barnard, J. Cooper, E. W. Smith.
Astrophys. Journ., Vol. 190, 315 - 318 (1974).

Theoretical profiles for the He I λ4471 line in B-star spectra have been computed using an improved broadening theory of Barnard, Cooper, and Smith, together with level populations determined by a self-consistent solution by Auer and Mihalas of the coupled transfer and statistical-equilibrium equations. The results presented show that this revision of the broadening theory leads to computed stellar profiles which are in much better agreement with observed profiles than any previously obtained.

114.128 Ultraviolet spectra of Capella.
G. A. Gurzadyan.
Nature, Vol. 250, 204 - 205 (1974).

In December 1973, during observations from the space observatory Orion 2 on Soyuz 13, three spectrograms of Capella were obtained in the wavelength region 2000–3000 Å. An analysis of the spectrograms is presented.

114.129 Narrow-band observations of super-metal-rich stars.
P. M. Williams.
Stars and the Milky Way system, Proc. 1972, (see 012.018), p. 174 - 177 (1974).

114.130 High resolution ultraviolet stellar spectra obtained with the Orbiting Stellar Spectrophotometer S 59.
K. A. van der Hucht, H. J. Lamers.
Stars and the Milky Way system, Proc. 1972, (see 012.018), p. 182 - 187 (1974).

114.131 Atmosphère de 22 Bootis, étoile à raies métalliques géante.   C. Burkhart, C. Van't Veer.
Comptes Rendus Acad. Sci. Paris, Sér. B, Vol. 278, 1103 - 1106 (1974).

Les premiers résultats de l'analyse de l'atmosphère de 22 Bootis montrent qu'une étoile évoluée peut encore présenter, de façon nette, le caractère «étoile à raies métalliques» (Am).

114.132 Time variations in the OH microwave and infrared emission from late-type stars.
P. M. Harvey, K. P. Bechis, W. J. Wilson, J. A. Ball.
Astrophys. Journ., Suppl. Ser., No. 248, Vol. 27, 331 - 357 (1974).

The OH microwave and infrared variability of 14 OH/IR stars were studied. Many of the sources show periodic flux variations between 1 and 10 μ and in the 1612-MHz OH lines with no appreciable phase difference between the infrared and microwave. The best explanation for this correlation seems to be that the OH masers are radiatively pumped. A correlation does exist between the period of variation, the amplitude of variation at 2.2 μ, the [3.5 μ] − [10 μ] color index, the separation in velocity of the two OH emission features, and the existence of 1.35-cm H_2O emission in the Mira type stars observed.

114.133 The ultraviolet spectrum of Eta Canis Majoris, B5 Ia.
A. B. Underhill.
Astrophys. Journ., Suppl. Ser., No. 249, Vol. 27, 359 - 389 (1974).

A list of lines visible in OAO-3 spectrum scans of the spectrum of η CMa, B5 Ia, is presented for the regions 1008–1425 Å and 1890–3105 Å. Line-blocking factors for each 10 Å of spectrum from 1010 to 1420 Å and blocking factors for each 20 Å of spectrum between 1900 and 3080 Å are given in tables.

114.134 Absolute flux measurements in the rocket ultraviolet.
R. C. Bohlin, D. Frimout, C. F. Lillie.
Bull. American Astron. Soc., Vol. 6, 215 (1974). − Abstr. AAS.

114.135 Stellar ultraviolet energy distributions.
C. F. Lillie, R. C. Bohlin.
Bull. American Astron. Soc., Vol. 6, 215 (1974). − Abstr. AAS.

114.136 Mariner 9 UV observations of strong stellar lines in OB stars.   M. R. Molnar.
Bull. American Astron. Soc., Vol. 6, 215 (1974). − Abstr. AAS.

114.137 A new objective prism for classification of faint stars.   V. M. Blanco.
Bull. American Astron. Soc., Vol. 6, 216 (1974). − Abstr. AAS.

114.138 Magnetic tape of Bidelman's spectral data file.
S. B. Parsons, J. D. Wray, G. F. Benedict.
Bull. American Astron. Soc., Vol. 6, 217 (1974). − Abstr. AAS.

114.139 The astronomical data file for the Celescope Catalog.   R. J. Davis.
Bull. American Astron. Soc., Vol. 6, 218 (1974). − Abstr. AAS.

114.140 Preliminary results of a stellar search for the unidentified molecular maser line near 3.48 mm.
L. E. Snyder, N. Kaifu, D. Buhl.
Bull. American Astron. Soc., Vol. 6, 220 (1974). − Abstr. AAS.

114.141 Is thermal dust emission responsible for the infrared continuum of Eta Carinae?   J. P. Apruzese.
Bull. American Astron. Soc., Vol. 6, 222 (1974). − Abstr. AAS.

114.142 Rapid spectral line variations in Gamma-Two Velorum.   F. B. Wood, W. H. Schneider, R. R. Austin.
Bull. American Astron. Soc., Vol. 6, 222 (1974). − Abstr. AAS.

114.143 Observations of the barium star binary system HD 196673.   R. B. Culver, P. A. Ianna.
Bull. American Astron. Soc., Vol. 6, 223 (1974). − Abstr. AAS.

114.144 Formation of the sodium D lines in Arcturus.
W. L. Kelch.
Bull. American Astron. Soc., Vol. 6, 226 (1974). − Abstr. AAS.

114.145 Line profile of the O I infrared triplet in Vega.
L. R. Doherty, L. W. Hartmann.
Bull. American Astron. Soc., Vol. 6, 226 (1974). − Abstr. AAS.

114.146 Two-dimensional spectral classification of the southern HD stars.   N. Houk.
Bull. American Astron. Soc., Vol. 6, 226 - 227 (1974). Abstr. AAS.

114.147 On interstellar depletion.   G. B. Field.
Bull. American Astron. Soc., Vol. 6, 262 (1974). Abstr. AAS.

114.148 Radio emission from forbidden-line stars with circumstellar dust.
P. A. Feldman, C. R. Purton, K. A. Marsh, D. A. Allen.
Bull. American Astron. Soc., Vol. 6, 263 (1974). − Abstr. AAS.

114.149 Non-LTE silicon spectra of B stars.   L. W. Kamp.

Bull. American Astron. Soc., Vol. 6, 266 (1974). Abstr. AAS.

114.150 Rapid spectroscopic variations in HZ Herculis. B. W. Bopp, G. Grupsmith, R. S. McMillan, P. A. Vanden Bout, H. A. Wootten.
Bull. American Astron. Soc., Vol. 6, 277 (1974). – Abstr. AAS.

114.151 Osservazioni ultraviolette. M. Hack.
Mem. Soc. Astron. Italiana, Nuova Ser., Vol. 44, 625 - 627 (1973/74).

114.152 MK classifications for F- and G-type stars. III. E. A. Harlan.
Astron. Journ., Vol. 79, 682 - 686 = Lick Obs. Bull., No. 655 (1974).

MK spectral classifications are given for 461 stars of HD types F2–G5, and having m_v = 7.5 or brighter. The classifications were made on slit spectrograms of dispersion 75 Å mm at $H\gamma$.

114.153 Spectroscopic observation of the carbon stars Y Canum Venaticorum and U Hydrae in the one-micron region. M. Hirai.
Publ. Astron. Soc. Japan, Vol. 26, 163 - 188 (1974).

Line identification in the spectra of the carbon stars Y CVn and U Hya has been carried out in the region between 1.00 and 1.08 µm. About 600 lines were measured for both stars; about two thirds of the lines are identified as CN lines. Most of the remaining lines are identified as $C^{13}N$, C_2, and atomic lines. Evidence for the presence of C_2H_2 is presented for U Hya. The results of line identification are given in a table. A comparison of the spectra of Y CVn, U Hya, 19 Psc, and WZ Cas is made.

114.154 New spectral classifications for variables wrongly suspected of being carbon stars. C. B. Stephenson.
Inform. Bull. Variable Stars, (I. A. U. Commission 27), Konkoly Obs., Budapest, No. 861 (1974).

114.155 Interstellar lines in the ultraviolet spectrum of Delta Scorpii. A. M. Smith.
Astrophys. Journ., Vol. 190, 565 - 571 (1974).

Data are presented for interstellar lines arising in C^0, C^+, N^0, O^0, Al^+, Si^+, and Fe^+ which have been obtained in an ultraviolet spectrogram of δ Sco obtained by rocket techniques. Excepting the iron transition, these lines are analyzed to yield column densities from which abundances relative to atomic hydrogen are determined. Using the measured column densities of neutral and singly ionized carbon an average ratio of the electron density to hydrogen atom density in the clouds equal to 2.1×10^{-4} cm^{-3} is derived.

114.156 The holmium Ap star HD 51418. T. J. Jones, S. C. Wolff, W. K. Bonsack.
Astrophys. Journ., Vol. 190, 579 - 583 (1974).

The peculiar A star HD 51418 differs from other Ap-type stars in that its spectrum is dominated by lines of the heavy rare earths, including particularly holmium and dysprosium. The magnetic field of HD 51418 varies from + 750 to − 200 gauss, and magnetic minimum coincides in phase with rare-earth minimum and with minimum light. The observed radial velocity variations give strong evidence that the spectrum variations can be accounted for by the oblique rotator model.

114.157 Observations of the profile of the Ca II infrared triplet line λ8498 in late-type stars. C. M. Anderson.
Astrophys. Journ., Vol. 190, 585 - 590 (1974).

The profiles of the Ca II infrared triplet line λ8498, observed at a resolution of 0.28 Å for 28 stars of spectral type F8 and later, are presented. Although there does not appear to be any correlation of line parameters with the strength of the emission observed in the H- and K-lines, the profile is clearly sensitive to the luminosity of the star.

114.158 The carbon monoxide band strength and $^{12}C/^{13}C$ ratio in K giants. S. T. Ridgway.
Astrophys. Journ., Vol. 190, 591 - 596 (1974).

The first overtone vibration-rotation bands of CO have been studied in the spectra of 15 giants, G8-K5. CN-strong stars are found to be characterized by CO band-strength excess, and it is argued that a true abundance difference must be responsible. Spectral synthesis is employed to deduce $^{12}C/^{13}C$ ratio of about 10 for four normal K giants and α Boo. The normal giants appear to have a carbon abundance near the solar value.

114.159 On the C_2, CN, and CO indices of carbon stars. T. D. Faÿ, Jr.
Astrophys. Journ., Vol. 190, 597 - 599 (1974).

The purpose of this paper is to describe the positions of two types of carbon stars on the CN/C_2 versus CO/C_2 diagram: (1) the J (^{13}C rich) stars and (2) the Mira carbon stars compared to the other carbon stars.

114.160 Resolution of the praseodymium abundance anomaly in the Ba II stars. M. S. Allen, C. R. Cowley.
Astrophys. Journ., Vol. 190, 601 - 604 (1974).

Previous abundance analyses of Ba II stars have usually yielded large excesses of praseodymium over its rare-earth neighbors. Several of these analyses have been revised to include the effect of hyperfine structure in the spectrum of Pr II. The revised abundances show excellent agreement with predictions of the s-process.

114.161 The V1057 Cygni OH source: time variation, polarization properties, and accurate position. K. Y. Lo, K. P. Bechis.
Astrophys. Journal, (Letters), Vol. 190, L125 - L128 (1974).

The position of the 1720-MHz OH source associated with the peculiar star V1057 Cygni was determined to be in agreement with the star position. Polarization studies indicate that the line profile is consistent with one right- and one left-circularly polarized component, and an unpolarized component at an intermediate frequency, suggesting a Zeeman pattern that has undergone Faraday depolarization. The intensity of the OH source has decreased by a factor of 5 in five months.

114.162 M supergiants in the Large Magellanic Cloud. R. M. Humphreys.
Astrophys. Journal, (Letters), Vol. 190, L133 - L135 (1974).

Eight red stars in the Large Magellanic Cloud have been classified as high-luminosity M supergiants. Their absolute visual magnitudes range from −7.7 to −6.6 mag. These are the first extragalactic M supergiants to be recognized by MK spectral classification.

114.163 Variations of the emission line profiles in the O6ef star Lambda Cephei. P. S. Conti, S. A. Frost.
Astrophys. Journal, (Letters), Vol. 190, L137 - L140 (1974).

Dramatic variations in the emission line profiles of $H\alpha$ and λ4686 He II from eight consecutive nights of observations are reported. Other absorption lines appear to be constant during this time. Also described is an innovative computer method for obtaining line profiles from spectrograms.

114.164 Corrections for the type and/or punched card version of the Kennedy MK classification Supplement. W. H. Warren, Jr.
Centre de Données Stellaires, Inform. Bull. No. 7, p. 38 - 39 (1974).

114.165 **The spectrum of CH Cygni from 1961 to 1973.**
A. J. Deutsch, L. Lowen, S. C. Morris, G. Wallerstein.
Publ. Astron. Soc. Pacific, Vol. 86, 233 - 236 (1974).

Radial velocities of absorption lines from 30 spectrograms taken between 1964 and 1973 yield no evidence of orbital motion. Evidence is presented that a circumstellar envelope is always present and that an "outburst" occurs when the envelope is excited by a disturbed region on the surface of the star. Some comparisons are made of CH Cyg with the VV Cephei stars, symbiotic objects, and the symbiotic long-period variable R Aqr.

114.166 **On the continuous spectrum of HD 184927.**
N. A. Higginbotham, F. H. Schiffer III, R. J. Dufour.
Publ. Astron. Soc. Pacific, Vol. 86, 272 - 275 = Contr. Louisiana State Univ. Obs., *Baton Rouge*, No. 92 (1974).

New spectrophotometric scans from $\lambda 3390$ to $\lambda 5840$ are presented for the helium-rich star HD 184927. An apparent effective temperature is determined by comparing the unreddened continuum with continua predicted by model atmospheres. Both the scans and the effective temperature are found to be in discordance with previously published results.

114.167 **Aluminum oxide in stellar spectra: an infrared electronic transition.** R. E. Luck, D. L. Lambert.
Publ. Astron. Soc. Pacific, Vol. 86, 276 - 278 (1974).

The possibility of detecting the infrared electronic transition $A^2\Pi_i - X^2\Sigma^+$ of AlO in spectra of cool stars is reviewed. It is shown that the transition may be of importance in Mira-type variables.

114.168 **A stellar continuum feature near 8800 Å.**
V. Oinas.
Publ. Astron. Soc. Pacific, Vol. 86, 321 - 323 (1974).

Detailed spectrophotometric scans of stars covering a wide range in spectral type reveal a wide absorption feature near 8800 Å, appearing in stars of type F and increasing in strength with decreasing temperature.

114.169 **Variations in the spectra of A-type supergiants.**
W. Buscombe.
Observatory, Vol. 94, 120 - 122 (1974).

114.170 **A catalogue of Hγ measures by R. M. Petrie.**
D. Crampton, A. Leir, F. Younger.
Publ. Dominion Astrophys. Obs., Victoria, Vol. 14, (No. 8), 151 - 184 (1973).

A catalogue of equivalent widths of Hγ measured by R. M. Petrie in the spectra of 1171 stars is presented. The catalogue represents a compilation of all the published and unpublished values available to us. It is shown that no systematic differences exist between the equivalent widths measured by Petrie and those currently being derived at the Dominion Astrophysical Observatory.

114.171 **Line list for $\lambda\lambda 4143$-4712 Å of an A-type peculiar star HD 221568.** K. Kodaira.
Ann. Tokyo Astron. Obs., Second Ser., Vol. 14, (No. 2), 85 - 119 (1974).

About 1800 absorption features are measured in 5.3 Å/mm spectrograms ($\lambda\lambda 4143$-4712 Å) of HD 221568 at the "red" phase ($P = 0.89$). About half of these features are identified, the other half remains only unsatisfactorily identified. Discussions are given concerning the nature of the strong unidentified lines, and the possible identifications of PmII lines and the lines of actinoids.

114.172 **On the absorption spectra of titanium oxide molecules in the atmospheres of M-stars.**
V. E. Panchuk.
Astron. Tsirk., No. 820, p. 5 - 7 (1974). In Russian.

114.173 **Strong-line K stars. I. Photometry.** V. Oinas.
Astrophys. Journ., Suppl. Ser., No. 250, Vol. 27, 391 - 404 (1974).

Absolute spectrophotometric photometry has been performed for 26 super-metal-rich (SMR) and normal giants and dwarfs of spectral type K. Emergent fluxes from model atmospheres in conjunction with blanketing corrections derived from high-dispersion spectra are used to obtain effective temperatures. Hα wing strengths are also discussed in relation to temperatures and metal abundances, and the material is used to construct an H-R diagram for the program stars.

114.174 **Strong-line K stars. II. Chemical abundances.**
V. Oinas.
Astrophys. Journ., Suppl. Ser., No. 250, Vol. 27, 405 - 414 (1974).

The observations presented in Paper I (Oinas 1973) are used to derive chemical abundances and atmospheric parameters of 26 "super-metal-rich" and normal comparison stars. The model atmospheres are also described.

Astrospectroscopy – the language of the universe.
See Abstr. 003.085.

Frühe Spektralanalyse von Fraunhofer bis Kirchhoff.
See Abstr. 004.057.

Electronic transitions of the LaO molecule.
See Abstr. 022.057.

Measurements of velocity dispersions and Doppler shifts from digitized optical spectra. See Abstr. 031.005.

Applications of Schmidt telescopes – with emphasis on direct and spectral surveys. See Abstr. 032.048.

The composition and evolutionary status of the helium-rich stars. See Abstr. 064.002.

Formation of the luminosity-sensitive O I multiplet at 7774 Å. See Abstr. 064.004.

Detectability and nature of CNO anomalies in early-type main-sequence stars. See Abstr. 064.007.

Stellar spectral synthesis in the ultraviolet.
See Abstr. 064.013.

Relaxation oscillations in red-giant envelopes and the symbiotic stars. See Abstr. 064.017.

Determination of titanium oxide and atomic oxygen abundances in the atmospheres of M-type stars.
See Abstr. 064.021.

Fe I line formation in solar-type giants and dwarfs.
See Abstr. 064.027.

The temperature and velocity distribution in Wolf-Rayet stars. See Abstr. 064.030.

A fine analysis of the helium-rich star HD 184927.
See Abstr. 064.051.

An interpretation of the puzzling observations of FG Sagittae. See Abstr. 065.094.

Nucleosynthesis of lithium in low-energy flares.
See Abstr. 065.095.

Telluric lines as radial velocity standards. See Abstr. 071.011.

The Vilnius photometric system for three-dimensional classification of stars and determination of interstellar reddening. See Abstr. 113.003.

Photoelectric and spectroscopic observations of WRA 795. See Abstr. 113.011.

List of spectroscopic and photometric catalogues lately published or to be published. IV. See Abstr. 113.041.

Multicolor photometry of metallic-line stars. I. ν^1 Draconis and ν^2 Draconis. See Abstr. 113.043.

Photographische Photometrie einiger Hα-Sterne im Gebiet des Nordamerikanebels. See Abstr. 113.045.

Photoelectric equivalent widths of the O I λ 7774 line and M_v's of selected F supergiants. See Abstr. 115.005.

An independent calibration of a spectrographic absolute magnitude system. See Abstr. 115.007.

Computed luminosities for T Tauri and related objects. See Abstr. 115.015.

The variations of the magnetic Ap star 49 Camelopardalis. See Abstr. 116.001.

On the interpretation of the magnetic curves of the Ap stars as determined by the photographic technique. See Abstr. 116.005.

The magnetic field of HR 6870. See Abstr. 116.007.

Measurements of magnetic fields in young main-sequence stars. See Abstr. 116.008.

Detection of crossover effect in the Ap star HD 98088. See Abstr. 116.012.

Elementhäufigkeit in magnetischen Sternen. See Abstr. 116.014.

On the stars HD 55549, HD 77105, HD 5373 and HD 56495. See Abstr. 117.028.

Reinvestigation of certain long-period A-type binaries. See Abstr. 119.005.

Complex infrared emission features in the spectrum of Beta Lyrae. See Abstr. 121.076.

Technetium in UX Draconis. See Abstr. 122.044.

Variable and peculiar stars in the Magellanic Clouds. See Abstr. 122.117.

Detection of lithium in the dMle star Gliese 182. See Abstr. 122.131.

The structure of four 1612 MHz OH emission sources. See Abstr. 131.023.

Optical interstellar line studies of a nearby cold cloud. See Abstr. 131.054.

A study of interstellar polarization at the $\lambda\lambda$4430 and 5780 features in HD 183143. See Abstr. 131.055.

The Coalsack. III. A search for T Tauri stars. See Abstr. 131.066.

Magnesium and iron abundances in the Gum nebula region determined from interstellar absorption lines in the spectrum of γ^2 Vel. See Abstr. 132.027.

New southern planetary nebulae previously classified as emission-line stars. See Abstr. 133.012.

Coordinated observations of OJ 287 at radio and optical wavelengths. See Abstr. 141.050.

Spectroscopy of objects near Texas radio-source positions. See Abstr. 141.097.

VY Canis Majoris. IV. The emission bands of ScO. See Abstr. 141.604.

A dust-shell model of the infrared object HD 45677. See Abstr. 141.605.

Optical spectra of HD 153919 (3U1700−37). See Abstr. 142.076.

Intermediate-band photometry of M67. See Abstr. 153.024.

Evolved stars in open clusters. See Abstr. 153.028.

The UV-bright stars in ω Centauri. See Abstr. 154.021.

The space distribution of M giants in the Warner and Swasey luminosity function field LF 15. See Abstr. 155.018.

The galactic structure in Cepheus near NGC 7235. See Abstr. 155.039.

Catalogue of dB and gK stars in the Palomar − Groningen Variable-Star Field No. 3. See Abstr. 155.078.

OB stars detection in the Large Magellanic Cloud. See Abstr. 159.003.

Errata

114.901 Erratum: 'Calculation of hyperfine structure of scandium and vanadium for stellar spectral analysis' [Astron. Astrophys., Vol. 23, 311 - 316 (1973)]. M. G. Edmunds. Astron. Astrophys., Vol. 30, 177 (1974).

114.902 Erratum: 'A study of B 6 stars' [Astron. Astrophys., Vol. 25, 161 - 174 (1973)]. A. B. Underhill. Astron. Astrophys., Vol. 30, 481 (1974).

114.903 Erratum: 'New emission-line stars in OH clouds' [Publ. Astron. Soc. Pacific, Vol. 85, 193 - 199 (1973)]. G. L. Grasdalen, L. V. Kuhi, E. A. Harlan. Publ. Astron. Soc. Pacific, Vol. 85, 817 (1973).

115 Stellar Luminosities, Masses, Diameters, HR-Diagrams and Others

115.001 Four stellar-diameter measurements by a new technique: amplitude interferometry.
D. G. Currie, S. L. Knapp, K. M. Liewer.
Astrophys. Journ., Vol. 187, 131 - 134 (1974).

Diameters are reported for four late-type giant stars, α Boo, α Ori, α Tau, and β Peg. The diameters were obtained with a new kind of interferometer designed expressly to operate in the presence of atmospheric fluctuations. The new technique, called amplitude interferometry, is briefly described. The results include measurements of α Ori at several wavelengths.

115.002 The Hγ-absolute magnitude calibration.
L. Balona, D. Crampton.
Monthly Notices Roy. Astron. Soc., Vol. 166, 203 - 217 (1974).

A new calibration is given relating the equivalent width of Hγ to the absolute magnitude of the star. With the exception of the O stars, the earlier calibration given by Petrie is essentially confirmed. The calibration of the MK system based on the new calibration is significantly different in some regions from that tabulated by Blaauw, but in excellent agreement with that recently published by Walborn.

115.003 A luminosity index for bright K0–K5 stars.
J. P. Chaturvedi.
Bull. Astron. Inst. Czechoslovakia, Vol. 25, 15 - 19 (1974).

An index for luminosity classification of K0–K5 stars based upon broad band *UBRI* observations has been obtained and discussed. This index has been employed to obtain the MK luminosity classifications for 21 bright stars for which only the spectral classification is so far available.

115.004 The angular diameters of 32 stars.
R. Hanbury Brown, J. Davis, L. R. Allen.
Monthly Notices Roy. Astron. Soc., Vol. 167, 121 - 136 (1974).

The complete results of the observational programme on single stars carried out with the stellar intensity interferometer at Narrabri are presented. The measurements are analysed to yield the angular diameters of 32 stars in the spectral range O5f to F8. Information is also presented on nine multiple stars.

115.005 Photoelectric equivalent widths of the O I λ 7774 line and M_v's of selected F supergiants.
P. W. Baker.
Publ. Astron. Soc. Pacific, Vol. 86, 33 - 37, with a correction p. 232 = Lick Obs. Bull., No. 639 (1974).

Photoelectric scanner measurements of the equivalent width of the O I λ 7774 triplet, for 52 F supergiants brighter than m_v = 9.0, are presented. The linear relation between the equivalent width and absolute visual magnitudes found by Osmer (1972) for 10 stars with known distances is confirmed and this calibration is used to find the absolute visual magnitudes of 51 F supergiants. A comparison of the equivalent widths for 17 stars in common with Osmer shows a good agreement, on the average.

115.006 Luminosity and velocity distribution of high-luminosity red stars. IV. The G-type giants.
O. J. Eggen.
Publ. Astron. Soc. Pacific, Vol. 86, 129 - 161 (1974).

The G-type giants and subgiants brighter than $V_0 = 5\overset{m}{.}0$ are discussed together with the Hyades group giants, the very young-disk-population cluster M11, the Wolf 630 group and M67 in the old-disk population, and the available narrow-band indices which measure surface gravity and CN strength in red-giant stars. Also included are discussions of the luminosity function of young disk-population giants, the bright members of the Pleiades group, a few possible giant members of the solar group, and color-color relations as a function of luminosity.

115.007 An independent calibration of a spectrographic absolute magnitude system.
K. M. Yoss, T. E. Lutz.
Astron. Journ., Vol. 79, 480 - 482 (1974).

An independent calibration of the zero point and scale of 562 spectrographic absolute magnitudes is made through mean-secular parallaxes. The spectrographic values were obtained from spectrograms taken at David Dunlap Observatory, and have mean errors of about ± 0.4 mag. No significant systematic errors in either zero point or scale are found in the spectrographic absolute magnitudes.

115.008 The angular diameter of 87 Leonis.
D. W. Dunham, D. S. Evans, W. H. Sandmann.
Astron. Journ., Vol. 79, 483 - 484 (1974).

The angular diameter of 87 Leo, (HR 4432) a K4III star, was determined on 12 May 1973 at occultation using the McDonald 36-in. telescope. The trace has been analyzed and yields a diameter of 3.7 arcms for a uniform disk and 4.1 arcms for a fully darkened disk with an uncertainty of 10% leading to an effective temperature near 3590 K.

115.009 An angular diameter and effective temperature for V Cancri from an occultation observation.
J. T. McGraw, J. R. P. Angel.
Astron. Journ., Vol. 79, 485 - 488 (1974).

An occultation observation in the near infrared of the S-type Mira variable V Cancri made on 11 April 1973 at McDonald Observatory yields an angular diameter near 2.8 milliseconds of arc for a fully darkened disk at phase 0.74 on the visual light curve. An estimated value of the bolometric correction leads to an effective temperature near 2055 K. The linear radius is found to be $400 \pm 115\ R_\odot$.

115.010 The effects of limb darkening on measurements of angular size with an intensity interferometer.
R. Hanbury Brown, J. Davis, R. J. W. Lake, R. J. Thompson.
Monthly Notices Roy. Astron. Soc., Vol. 167, 475 - 483 (1974).

The necessary correction to obtain the true limb-darkened angular diameter may be made by means of an approximate formula and a limb-darkening coefficient. More precise corrections, used in analyzing the results from Narrabri Observatory, are given for a grid of model stellar atmospheres and these are compared with those given by the approximate formula. Observations of the secondary maximum of the correlation curve for Sirius are reported and compared with predictions based on a model atmosphere.

115.011 The "Hyades anomaly" and the *uvby* luminosity calibration. D. C. Barry.
Astron. Journ., Vol. 79, 616 - 622 (1974).

This paper presents evidence that the Hyades anomaly may be removed through the empirical determination of a metallicity dependent luminosity calibration from large parallax field stars and its subsequent application to the late F through early G dwarfs in five clusters.

115.012 Mean absolute magnitudes of carbon stars and related objects.   J. H. Baumert.
Astrophys. Journ., Vol. 190, 85 - 90 (1974).

Statistical methods which use proper motions and radial velocities have been employed in conjunction with narrow-band photometry to determine the mean absolute magnitudes of carbon stars and related objects at 1.04 μ.

115.013 Bolometric corrections for B-stars.
F. Beeckmans, D. Macau, D. Malaise.
Astron. Inst. Utrecht, Internal Rep. ROF 72, (see 012.012), p. C13.1 - C13.9 (1974).

115.014 Lunar occultation stellar diameter measurements.
W. I. Beavers, J. J. Eitter.
Bull. American Astron. Soc., Vol. 6, 216 (1974). – Abstr. AAS.

115.015 Computed luminosities for T Tauri and related objects.   C. L. Imhoff, E. E. Mendoza V.
Rev. Mexicana Astron. Astrofis., Vol. 1, No. 1, p. 25 - 33 (1974).

Improved and additional data for the T Tauri-like objects, with more realistic assumptions concerning extinction, allow to calculate new values for the luminosities of eleven stars. Both the spectroscopic and photometric luminosities are considered; in several cases, they are found to disagree. These discrepancies may be resolved if the extinction affecting these stars is peculiar, with a large value of the ratio of total to selective absorption.

115.016 Luminosities and motions of peculiar B-type stars.
O. J. Eggen.
Publ. Astron. Soc. Pacific, Vol. 86, 241 - 262 (1974).

The luminosities of peculiar A- and B-type stars with well-determined apparent motions have been obtained from a calibration of intermediate-band photometry. This calibration is derived from members of the Pleiades group and extended to all stars brighter than visual magnitude 6.5. Some 15% of the B-type stars in the Pleiades group are peculiar. The incidence of spectroscopic binaries is also discussed.

On the origin of the blue halo stars.
See Abstr. 065.002.

The theoretical low mass main sequence.
See Abstr. 065.004.

The mass-angular momentum-diagram and the black hole limit.   See Abstr. 066.090.

An intrinsic color and absolute magnitude calibration for F-type stars.   See Abstr. 113.038.

A note on the use of the strength of the Si II doublet $\lambda\lambda 6347, 6371$ as a luminosity indicator in B9–A2 supergiants.   See Abstr. 114.012.

Absolute ultraviolet spectrophotometry from the TD-1 satellite. V. Bolometric corrections for early type stars of luminosity class V.   See Abstr. 114.125.

Observations of the profile of the Ca II infrared triplet line $\lambda 8498$ in late-type stars.   See Abstr. 114.157.

M supergiants in the Large Magellanic Cloud.
See Abstr. 114.162.

BDS 1269: a possible physical pair with a metal-deficient primary.   See Abstr. 118.032.

The instability strip of population II cepheids.
See Abstr. 122.095.

Analysis of composite spectra – the ultra-violet excess of stars in globular clusters.   See Abstr. 154.004.

Die räumliche Verteilung, Kinematik und Alter der sonnennahen Sterne.   See Abstr. 155.053.

116 Stellar Magnetic Field, Figure, Rotation

116.001 **The variations of the magnetic Ap star 49 Camelopardalis.**   W. K. Bonsack, C. A. Pilachowski, S. C. Wolff.
Astrophys. Journ., Vol. 187, 265 - 270 (1974).

The magnetic field of 49 Cam is found to vary cyclically between the limits of ± 1.7 kilogauss with a period of $4^d.285$. Central depths of lines of Fe, Cr, and several rare earths are found to vary synchronously, with maximum strength corresponding to magnetic maximum. The photometric variation patterns are complex. The results are discussed in terms of the rigid-rotator model.

116.002 **The orientation of magnetic axes in Ap stars: An alternative interpretation of the component with small obliquity.**   E. F. Borra.
Astrophys. Journ., Vol. 187, 271 - 274 (1974).

Zeeman-analyzed line profiles have been computed for oblique-rotator models and decentered-dipole configurations of the magnetic-field geometry. It is argued on the basis of these profiles that periodic magnetic variables with nonreversing magnetic fields can be interpreted as having the magnetic axis nearly orthogonal to the rotation axis.

116.003 **The intrinsic linear polarization of 53 Camelopardalis and α^2 Canum Venaticorum.**
J. C. Kemp, R. D. Wolstencroft.
Monthly Notices Roy. Astron. Soc., Vol. 166, 1 - 18 (1974).

The intrinsic phase-dependent linear polarization of the magnetic Ap stars 53 Cam and α^2 CVn has been studied. Mechanisms for the polarization are discussed, with emphasis on qualitative features predicted by the general model of a symmetric, oblique rotator.

116.004 **On the distortion of magnetic stars and their inferred magnetic fields.**   J. J. Monaghan.
Monthly Notices Roy. Astron. Soc., Vol. 167, 163 - 168 (1974).

The magnetic field of 53 Cam is analysed by using a general expansion of an axisymmetric field and by allowing the star to be either spherical or non-spherical. The nonspherical model gives a marginally better fit. The results are consistent with an internal toroidal field much larger than the external poloidal field, and they suggest that future attempts at inferring the fields of magnetic stars should allow for a non-spherical surface.

116.005 **On the interpretation of the magnetic curves of the Ap stars as determined by the photographic technique.**   E. F. Borra.
Astrophys. Journ., Vol. 188, 287 - 290 (1974).

Theoretical Zeeman-analyzed line profiles in magnetic stars have been computed for decentered dipole configurations of the magnetic field. It is found that if the core of the lines are emphasized during the reduction of Zeeman-analyzed photographic plates, the longitudinal field obtained deviates from its true value. The deviations predicted are successful in producing the anharmonic variations observed in several Ap stars.

116.006 **Theoretical aspects of light variations of magnetic stars.**   K. D. Rakosch, R. Sexl, W. W. Weiss.
Astron. Astrophys., Vol. 31, 441 - 446 (1974).

It is shown by the example of α^2 CVn that it is possible to calculate the amplitude of light variations of magnetic stars over a large spectral range (1500 Å –6000 Å), if one assumes different integral effective values of gravity and temperature for both magnetic poles and the magnetic equator. The influence of the magnetic field on effective gravity is shown to be large. A van Allen belt-type extension of the stellar photosphere along the magnetic equator is suggested to be the cause of the observed light variations.

116.007 **The magnetic field of HR 6870.**
T. J. Jones, S. C. Wolff.
Publ. Astron. Soc. Pacific, Vol. 86, 67 - 69 (1974).

Zeeman spectroscopic observations demonstrate that the peculiar B star HR 6870 has a magnetic field. The measured field strengths range from −260 gauss to −1080 gauss with no reversal of sign. This star is probably a spectroscopic binary, since it appears to show low-amplitude (~ 1 km sec^{-1}) velocity variations.

116.008 **Measurements of magnetic fields in young main-sequence stars.**   A. M. Boesgaard.
Astrophys. Journ., Vol. 188, 567 - 569 (1974).

The Zeeman analyzer at the coudé spectrograph of the 224-cm (2.2-m) telescope on Mauna Kea has been used to search for magnetic fields in young main-sequence stars at a dispersion of 3.3 Å mm^{-1}. Eight stars of spectral types F0 V–K0 V were selected as young stars by one or more criteria: intense Ca II emission cores, high Li content, high axial rotation.

116.009 **The linear polarization of 53 Camelopardalis.**
G. D. Finn, J. C. Kemp.
Monthly Notices Roy. Astron. Soc., Vol. 167, 375 - 391 (1974).

Equations are derived for the Stokes parameters of light scattered by a layer of elementary electric dipoles lying on the surface of a magnetic star based on the classical Hanle effect. Using appropriate geometric parameters, components of the Stokes vector are integrated analytically over the frequencies of a normal Zeeman triplet and numerically over the visible disk of the star. The results of this calculation are compared with the recent broadband observations and Kemp & Wolstencroft for the linear polarization of 53 Cam.

116.010 **Magnetic accretion and atmospheric peculiarities in early type stars.**   O. Havnes.
Astron. Astrophys., Vol. 32, 161 - 176 (1974).

The magnetic accretion theory (Havnes and Conti, 1971) for magnetic peculiar A stars is here tested quantitatively. Magnetic stars accrete atoms and ions from the interstellar gas. The accretion rate of different elements is dependent on their ionization state as the gas encounters the stellar magnetosphere. The ionization state of the gas around stars whose effective temperatures range from 8000 to 14000°K is calculated as are also the grain temperatures and evaporation times of grains. The calculated abundance peculiarities formed by magnetic accretion are generally found to be in good agreement with the observed values.

116.011 **Magnētiskās zvaigznes.**   E. Grasbergs.
Zvaigžņotā debess, 1974. gada pavasaris, p. 18 - 21.

116.012 **Detection of crossover effect in the Ap star HD 98088.**   G. Hensberge.
Astron. Astrophys., Vol. 32, 457 - 459 (1974).

On a high-dispersion Zeeman plate of the Eu-Cr-Sr star HD 98088 several lines of Fe, Cr and Ti show the crossover effect. The mean effective field strength in the magnetic regions where the lines arise is −8400±450 and +7600±400 Gauss, respectively. Different ions show slightly different field strengths.

116.013 The extraordinarily slow magnetic variation of Gamma Equulei.
W. K. Bonsack, C. A. Pilachowski.
Astrophys. Journ., Vol. 190, 327 - 330 (1974).

Sixty-one observations of the Zeeman effect in the spectrum of γ Equ are reported, covering the years 1963 to 1973. These are discussed together with Babcock's results for 1946–56. The magnetic field has apparently been declining monotonically from a broad and ill-defined maximum of +500 gauss near 1952 to a value of −150 gauss in 1973. If the star's magnetic cycle is symmetrical about a zero field, the period may be as long as 72 years. No evidence for short-period fluctuations about the mean curve exists in the newer data.

116.014 Elementhäufigkeit in magnetischen Sternen.
K. D. Rakos.
Analyse extraterrestrischen Materials, (see 012.020), p. 213 - 227 (1974).

An iterative technique for the rectification of observed distributions. See Abstr. 031.061.

Rotating magnetosphere: far-field solutions. See Abstr. 062.026.

Convection in A stars: convection suppressed in magnetic A stars? See Abstr. 065.028.

Differential rotation in degenerate stars. See Abstr. 065.043.

Static criteria for stability of arbitrarily rotating stars. See Abstr. 065.051.

Generation of a stellar magnetic field by convection in the outer regions. See Abstr. 065.080.

The decay of magnetic fields in stars. See Abstr. 065.082.

Dynamical phases of rotating supermassive stars. See Abstr. 065.096.

On the stability of axisymmetric systems to axisymmetric perturbations in general relativity. V. Differentially rotating configurations. See Abstr. 066.104.

UBV photometry of four magnetic stars. See Abstr. 113.018.

On the period and peculiarity of the cool Ap star HD 203006 (ϑ^1 Mic). See Abstr. 114.067.

Rotational distortion of stellar absorption lines. I. Parameters from photographic spectra. See Abstr. 114.071.

The peculiar A star HD 215441. See Abstr. 114.082.

The holmium Ap star HD 51418. See Abstr. 114.156.

The V1057 Cygni OH source: time variation, polarization properties, and accurate position. See Abstr. 114.161.

The evolution of magnetic binaries with mass loss. II. Calculations with magnetic field. See Abstr. 117.032.

On the inclination of rotation axes in visual binaries. See Abstr. 118.014.

Ultraviolet photometry from the Orbiting Astronomical Observatory. XV. The strongly magnetic variable HD 215441. See Abstr. 122.101.

G240–72: a new magnetic white dwarf with unusual polarization. See Abstr. 126.021.

117 Binary and Multiple Stars, Theory

117.001 The formation of gas clouds around close binary systems.
M. Bielicki, S. Piotrowski, K. Ziołkowski.
Astrophys. Space Sci., Vol. 26, 173 - 177 (1974).

The problem is considered of the slow outflow of gas from a close binary system with subsequent formation of a shell or cloud of matter around the whole system. It appears that with a small change of velocity introduced to the most external parts of the gaseous ring around one of the components the gas particles can flow out from the binary system leaving it through the external Lagrangian point. This process can lead to the formation of a shell around the binary star.

117.002 Tabular mass solutions for close binaries.
P. Giannone, M. A. Giannuzzi.
Astrophys. Space Sci., Vol. 27, 245 - 251 (1974).

The solutions for the mass of single-line close binaries with main-sequence primaries are given in a tabular form as a result of the application of the mass-luminosity relation. Linear interpolation in a compact bi-variate table suffices to give the values of mass of the two components for a wide range of masses and to a very good approximation. The indeterminacy of the solutions depending on the formulation of the mass-luminosity relation for the primaries is discussed, as well as the condition of filling up their Roche lobes by the secondaries.

117.003 Mass effects in triple star stability.
R. S. Harrington.
Bull. American Astron. Soc., Vol. 6, 205 (1974). – Abstr. AAS.

117.004 On the mass determination of collapsed objects in close binaries: applications to black holes and neutron stars. Y. Kondo.
Astrophys. Space Sci., Vol. 27, 293 - 301 (1974).

Observational information required for determination of the masses of collapsed objects in close binaries is examined. Assumptions commonly used to evaluate the masses or their lower limits, in the absence of the required observational data, are critically discussed.

117.005 On secular stability of tidally distorted stars of arbitrary structure. Z. Kopal.
Astrophys. Space Sci., Vol. 27, 389 - 418 (1974).

The aim of the present paper is to develop a theory which should make it possible to investigate secular stability of close binary systems, consisting of tidally-distorted components of arbitrary internal structure, by a minimization of the potential energy of the system as a whole. In the second section appropriate expressions for the total potential energy of a close binary are formulated. Section 3 is concerned mainly with the nature of the tide-generating potential, and its effects on the shape of each star. In Section 4, the amplitudes of partial tides raised by this potential are specified, for stars of arbitrary structure, correctly to terms of second order in superficial distortion; and in Section 5 the author investigates the effects of interaction between rotation and tides to the same degree of approximation. Section 6 contains an explicit formulation of different constituents adding up to the total potential energy of the system.

117.006 Mass loss from convective envelopes of giant components in close binary systems.
M. Plavec, R. K. Ulrich, R. S. Polidan.
Publ. Astron. Soc. Pacific, Vol. 85, 769 - 805 (1973).

In many spectroscopic binaries, the more massive component will reach the Roche limit when it is a giant star with a deep convective envelope. Mass loss ensuing in this case is studied for a red giant initially of 7 solar masses. The Jendrzejec mode of a steady laminar outflow of mass from the convective envelope leads to very high rates of mass transfer, up to 10^{-1} solar masses per year. Two additional modes of mass outflow have been investigated in order to see whether these rates could actually be lower. Four sequences of evolutionary models have been computed.

117.007 Evolution of magnetic binaries with mass loss. I. Computations without a magnetic field.
E. M. Drobyshevski, B. I. Reznikov.
Acta Astron., Vol. 24, 29 - 43 (1974).

Computations have been performed for the evolution of a series of systems in which the primary is a red giant with hydrogen burning shell source. The computed mass losses and the momentum loss agree well with estimates based on observations. Simple expressions have been derived for computing the component masses and the system period in the course of the evolution.

117.008 On the observability of extrasolar planetary systems.
E. Argyle.
Icarus, Vol. 21, 199 - 201 (1974).

An unresolved companion to a variable star can in principle be detected by taking the apparent autocovariance function of a suitable photometric record of the system.

117.009 A simple model of fluctuations in accretion discs.
G. T. Bath, W. D. Evans, J. Papaloizou.
Monthly Notices Roy. Astron. Soc., Vol. 167, 7P - 9P (1974).

A relation between the period and lifetime of fluctuations in accretion discs is derived and compared with the observations of transient periodicities in X-ray sources and dwarf novae.

117.010 Evolution of primary components of massive close binary systems.
A. V. Tutukov, L. R. Yungelson, A. Ya. Klajman.
Nauchn. Informatsii, vyp. (No.) 27, p. 3 - 57 (1973). In Russian.

The evolution of initially more massive components of four massive close binaries of $10 M_\odot + 9.4 M_\odot$, $16 M_\odot + 15 M_\odot$, $32 M_\odot + 30 M_\odot$, $64 M_\odot + 60 M_\odot$ has been computed. The Ledoux criterion for convective instability was used for the handling of semiconvection. The influence of neutrino emission on the advanced evolution of primaries was investigated.

117.011 Evolution of close binaries and Wolf-Rayet stars.
A. V. Tutukov, L. R. Yungelson.
Nauchn. Informatsii, vyp. (No.) 27, p. 58 - 69 (1973). In Russian.

The evolutionary stage of Wolf-Rayet stars in close binary systems is discussed. The lower mass limit of the progenitor stars for WR members of such systems is found to be about $15 M_\odot$.

117.012 Evolution of massive close binaries.
A. V. Tutukov, L. R. Yungelson.
Nauchn. Informatsii, vyp. (No.) 27, p. 70 - 85 (1973). In Russian.

The evolution of a massive close binary is discussed, starting from main sequence up to the formation of two relativistic objects and disruption of the system.

117.013 Evolution of a close binary with a relativistic component. A. V. Tutukov, L. R. Yungelson.

Nauchn. Informatsii, vyp. (No.) 27, p. 86 - 92 (1973). In Russian.

The evolution of semi-detached close binary systems containing a neutron or collapsed star has been investigated. It is shown that all systems with masses of component exceeding $\sim 10\,M_\odot$ may become X-ray sources.

117.014 Evolution of the secondary component of a close binary system. L. R. Yungelson.
Nauchn. Informatsii, vyp. (No.) 27, p. 93 - 98 (1973). In Russian.

The evolution of the secondary component of a close binary system has been investigated. The initial mass of components was $1.5\,M_\odot$ and $1\,M_\odot$, the initial semiaxis of the orbit $4.25\,R_\odot$.

117.015 Tidal flow in components of binaries with the rotation axis inclined to the orbital plane and magnetic field generation. A. Z. Dolginov.
Astron. Zhurn. Akad. Nauk SSSR, Vol. 51, 388 - 394 (1974). In Russian. English translation in Soviet Astron. AJ, Vol. 18, No. 2.

Explicit analytical expressions for the tidal flow velocities into the component of a binary stellar system are obtained. It is shown that the inclination of the rotation axis to the orbital plane leads to essential change in the behaviour of the tidal flow. Such flow can be effective in the process of magnetic field generation.

117.016 On the reflection effect in spectral lines. L. Toomasson, I. Pustylnik.
Peremennye Zvezdy, Vol. 19, 279 - 294 (1973). In Russian.

The influence of the Lyman continuum radiation of a primary star on the physical conditions in the atmosphere of a cool companion in a close binary system is investigated. Equations of statistical equilibrium are formulated and solved for an idealized hydrogen atom consisting of three discrete energy levels and continuum. The values of the Menzel's dimensionless factors b_1, b_2, b_3 are evaluated for some combinations of the effective temperatures of the component stars, the dilution factor of infalling radiation of a hot star, electron density and electron temperature.

117.017 Anomalous redshifts in binary stars. L. V. Kuhi, J.-C. Pecker, J. P. Vigier.
Astron. Astrophys., Vol. 32, 111 - 114 (1974).

The problem of the anomalous redshifts of the γ-velocities of Wolf-Rayet binaries as deduced from the emission lines is reviewed. The earlier explanations (Wilson, 1940) are rediscussed and it is shown that, while the absorption hypothesis is very attractive, similar redshifts can also be produced by inelastic photon boson scattering.

117.018 The intrinsic polarization of YY Eri. V. A. Oshchepkov.
Byull. Abastumansk. Astrofiz. Obs., No. 45, p. 51 - 62 (1974). In Russian.

The polarization features of YY Eri were observed during 6 nights in October and November of 1971. 222 measurements were obtained. The polarization turned out to be variable. Qualitatively the polarization maximum at phase 0.25 may be explained by scattering of light on a gaseous stream emerging from the primary component.

117.019 Ellipsoid-ellipsoid model for the interpretation of the light curves of close binary systems (VII). Recurrent formulae for the functions $D_k^l(x)$. V. Ureche.
Studia Univ. Babeş–Bolyai, Ser. Math.-Mech., Anul 19, Fasc. 1, p. 89 - 94 (1974).

117.020 The age of α Centauri.
A. M. Boesgaard, W. Hagen.
Astrophys. Journ., Vol. 189, 85 - 87 (1974).

The rotational velocity, Li content, and Ca II emission intensity are used to determine the age of the α Cen triple system by comparison with Skumanich's decay curves. The authors conclude that α Cen A is at least 3 billion years old, which is inconsistent with the expected age of 4×10^8 years for α Cen C, a dM4e flare star.

117.021 Numerical simulation of the gas flow in close binary systems.
K. H. Prendergast, R. E. Taam.
Astrophys. Journ., Vol. 189, 125 - 136 (1974).

A hydrodynamical calculation of the gas flow in a semi-detached close binary system is presented. The steady-state solutions for three separate cases are considered, viz., synchronous rotation of both components, nonsynchronous rotation of the secondary component, and nonsynchronous rotation of the primary component. Finally, a theoretical velocity curve is constructed, and an attempt is made to compare it with an observational velocity curve.

117.022 On the sphere-ellipsoid model. Application for the close binary system XZ Andromedae.
V. Ureche, H. Minţi.
Stud. Cerc. Astron., Vol. 19, 67 - 77 (1974). In Romanian.

117.023 Die Eindeutigkeit von Doppelsternentwicklung.
J. Hazlehurst.
Mitt. Astron. Ges., No. 34, p. 81 (1973/74). – Abstr. AG.

117.024 Über Veränderungen der Position von Gasströmen in halbgetrennten Systemen. K. Walter.
Mitt. Astron. Ges., No. 34, p. 125 (1973/74). – Abstr. AG.

117.025 On a possible influence of magnetic fields on the structure of a disk formed during accretion of plasma in binary systems. L. A. Pustilnik, V. F. Shvartsman.
IAU Symposium No. 64, (see 012.008), p. 213 (1974). – Abstract.

117.026 The fission theory of binary stars. II. Stability to third-harmonics disturbances. N. R. Lebovitz.
Astrophys. Journ., Vol. 190, 121 - 130 (1974).

The stability of the compressible Riemann ellipsoids to modes of disturbance associated with the third ellipsoidal harmonics is worked out, and the implications of the stability theory for the evolution of a contracting, ellipsoidal mass are discussed.

117.027 Calculations of proximity effects in close binary systems. E. Budding.
Astrophys. Space Sci., Vol. 29, 17 - 39 (1974).

The proximity effects in the light curves of close binary systems are investigated. The treatment follows, to a large extent, that summarized by Kopal (1959) and makes extensive use of the alpha-functions and related integrals provided by that author. The 'ellipticity' and 'reflection' effects are studied individually and different expressions are checked and compared. Test data are drawn continually from the well known system SZ Cam for which a 'rectified' light curve is ultimately produced.

117.028 On the stars HD 55549, HD 77105, HD 5373 and HD 56495. M. A. Khamdi.
Soobshch. AN GruzSSR, Vol. 72, 321 - 323 (1973). In Russian. – Abstr. in Referativ. Zhurn. 51. Astron., 6.51.602 (1974).

117.029 On the interpretation of the He II $\lambda 4686$ emission line in HDE 226868 (Cygnus X-1).

G. F. Bisiacchi, D. Dultzin, C. Firmani, S. Hacyan.
Astrophys. Journal, (Letters), Vol. 190, L59 - L62 (1974).

Two models are studied to explain the origin and behavior of the observed He II emission line λ4686 in the binary system 226868 (Cyg X-1). It is shown that observational data can be quantitatively explained if the line is emitted by a jet of matter flowing from the primary to the secondary. This, in turn, permits to estimate the ratio μ (mass of the unseen secondary/total mass), which turns out to be ~ 0.3.

117.030 **Nonperiodic optical flickering in HZ Herculis.**
T. J. Moffett, R. E. Nather, P. A. Vanden Bout.
Astrophys. Journal, (Letters), Vol. 190, L63 - L66 (1974).

High-speed simultaneous dual-channel photometry of HZ Her and a nearby comparison star reveal nonperiodic optical flickering in the HZ Her system on a time scale of 15–300 s. The amplitude of the flickering appears to be correlated with orbital phase. Optical emission from a hot spot in a disk of material around the X-ray source cannot account for the flickering.

117.031 **The W UMa-type systems as contact binaries. II. A- and W-type systems.** S. M. Ruciński.
Acta Astron., Vol. 24, 119 - 151 (1974).

Properties of W UMa-type systems are discussed in terms of their division into two types. Study of the difference in depth of minima Δ required a new set of theoretical computations to be made with the reflection effect included. Comparison of the observational and theoretical Δ permitted determination of $X = \Delta T/T$ for 25 systems (ΔT measures the temperature deviations of secondary components relative to the conventional contact model); the distinct separation of both types in the parameter X is clearly established.

117.032 **The evolution of magnetic binaries with mass loss. II. Calculations with magnetic field.**
E. M. Drobyshevski, B. I. Reznikov.
Acta Astron., Vol. 24, 189 - 205 (1974).

The tensions of the magnetic field in the stream which connects the primary component with a gas disc rotating around the secondary produces a force moment decelerating the disc. Two situations are considered: (1) the limiting case of a "strong" magnetic field (when magnetic pressure in the stream equals gas pressure), and (2) the more realistic case of a turbulent field (where magnetic flux in the stream is determined by the flux of the turbulent field in the convective envelope of the primary).

117.033 **The undersize subgiants.** D. S. Hall.
Acta Astron., Vol. 24, 215 - 234 (1974).

Twenty five binaries suspected of having undersize subgiants are examined. The undersize subgiant problem is solved in the sense that there is no evidence to support the notion that a subgroup of post-main-sequence mass-exchange remnants exists in which the subgiant is smaller than its Roche lobe. Those binaries which clearly are detached are not mass-exchange remnants; they are RS CVn-type systems, which seem to be pre-main-sequence objects. Those binaries which really are mass-exchange remnants are not detached; they only appear detached if one believes solutions of radial velocity curves distorted by gas-streaming effects and if one fails to appreciate how sensitive computed absolute dimensions are to uncertainties in the relative radius of the subgiant.

117.034 **Determination of upper mass limits for massive X-ray binaries.** G. Hensberge, E. P. J. van den Heuvel.
Astron. Astrophys., Vol. 33, 311 - 313 (1974).

Upper mass limits for the visible companions of binary X-ray sources are determined from the observed system parameters. This shows the OB-supergiant components of Cyg X-1 and SMC X-1 to be less massive than about $25\,M_\odot$ and $21\,M_\odot$, respectively.

117.035 **The influence of an axial rotation and a tidal action on the non-radial oscillations of a star, with reference to the β Canis Majoris stars.** J. Denis, P. Smeyers.
Stars and the Milky Way system, Proc. 1972, (see 012.018), p. 36 - 38 (1974).

117.036 **Apsidal motion in close binaries with variable orbital period.** I. Todoran.
Stars and the Milky Way system, Proc. 1972, (see 012.018), p. 42 - 52 (1974).

117.037 **Die W-UMa-Sterne, III.** M. Fernandes.
BAV Rundbrief, 23. Jahrgang, p. 1 - 4 (1974).

117.038 **Do OB runaways have collapsed companions?**
J. D. Bekenstein, R. L. Bowers.
Astrophys. Journ., Vol. 190, 653 - 659 (1974).

The authors present a list of 55 OB runaway candidates. Assuming the conventional explanation of a runaway in terms of a supernova explosion in a close binary, they use considerations of mass exchange to place an upper bound on the mass of an hypothetical collapsed companion for some of the candidates. In 11 cases the available data are not inconsistent with the presence of a companion massive enough to be a black hole.

117.039 **Evolución en binarias cerradas.** J. Sahade.
Revista Real Acad. Ci. Exactas, Fís. Nat., Madrid, Vol. 67, (No. 1), 27 - 49 = Inst. Astron. Fís. Espacio, (IAFE), Buenos Aires, Argentina, Tirada Aparte No. 7 (1973).

The dynamical decay of unstable 4-body systems.
See Abstr. 042.058.

Relaxation oscillations in red-giant envelopes and the symbiotic stars. See Abstr. 064.017.

Gravity-darkening in stars for general rotation laws.
See Abstr. 064.020.

On a nebular model for Wolf-Rayet stars.
See Abstr. 065.048.

Is the sun an astrometric binary?
See Abstr. 080.017.

The occultation of β Scorpii by Jupiter I. The structure of the Jovian upper atmosphere.
See Abstr. 099.037.

The occultation of Beta Scorpii by Jupiter. III. Simultaneous high time-resolution records at three wavelengths. See Abstr. 099.049.

The occultation of β Scorpii by Jupiter. IV. Diurnal temperature variations and the methane mixing ratio in the Jovian upper atmosphere. See Abstr. 099.050.

An astrometric study of Lalande 21185.
See Abstr. 111.003.

Supergiant binary stars. See Abstr. 114.014.

DL Virginis — an eclipsing binary as part of a triple system. See Abstr. 121.087.

The hot spot in the close binary system SS Cyg.
See Abstr. 122.069.

Supernova: the result of the death spiral of a white

dwarf into a red giant.   See Abstr. 125.008.

Infrared stars in binary systems.
See Abstr. 141.611.

Multiple star systems and X-ray sources.
See Abstr. 142.049.

A model for the formation of binary stars and binary galaxies.   See Abstr. 151.055.

Die Realität der Sterntrupps.   See Abstr. 155.076.

118   Visual Double and Multiple Stars

118.001   Orbital elements of twelve visual double stars.
P. Baize.
Astron. Astrophys., Suppl. Ser., Vol. 13, 65 - 80 (1974). In French.
The orbital elements of twelve visual double stars (ADS 281, 896, 1371, 3991, 7158, 8419, 8680, 9182, 11300, 15447, 16467 and Kuiper 39) are given, with, for each pair, the observations and their discussion, the residuals $O-C$, the dynamical parallaxes and the ephemeris.

118.002   Orbit of the binary Mlr 4.   P. Muller.
Astron. Astrophys., Suppl. Ser., Vol. 13, 99 - 100 (1974). In French.
Orbital elements, $O-C$ and ephemeris 1972-82 of the binary Mlr 4.

118.003   Numerical experiments concerning the determination of the orbits of close visual doubles considering second order terms in the observing errors.
H. Eichhorn, W. G. Clary.
Monthly Notices Roy. Astron. Soc., Vol. 166, 433 - 438 (1974).
The authors assumed several sets of orbital elements, computed ephemerides from them, added observing errors of known dispersion to the computed locations and applied a least squares routine to this material to recover the elements from which the ephemerides were computed. This was first done in the conventional way with condition equations that are linearized in the observing errors, and next with condition equations that consider second order terms in the observing errors following an algorithm worked out by the authors. The results show that the consideration of second order terms in the observing errors always lead to recovered elements that are considerably close to the true ones than those computed by the conventional, first order approach.

118.004   Radial velocities of three visual binaries.
J. Andersen.
Astron. Astrophys., Suppl. Ser., Vol. 13, 355 - 357 (1974).
Radial-velocity observations of the primary components of ADS 7967 and ADS 8327 and of both components of IDS 19316N5957 are reported. They confirm that the last of these is an optical pair.

118.005   Métodos de cálculo de órbitas de sistemas dobles visuales; Programa para el cálculo de efemérides de estrellas dobles visuales.   C. Morales Piñeiro.
Urania Barcelona, Año 58, Nos. 277 - 278, p. 54 - 62, 63 - 71 (1973).

118.006   ε Lyrae: de dubbele dubbele.   M. Drummen.
Zenit, Vol. 1, No. 4, p. 8 - 11 (1974).

118.007   New double stars (11th series) discovered at Nice.
P. Couteau.
Astron. Astrophys., Suppl. Ser., Vol. 15, 253 - 260 (1974). In French.
The author gives a list of 100 double stars discovered at the 50 cm and 74 cm refractors.

118.008   Revised elements of eight visual binaries.
W. D. Heintz.
Astron. Journ., Vol. 79, 397 - 400 (1974).
Orbital elements, dynamical parallaxes, and masses have been computed for 8 double stars. σ Ori (ADS 4241) is the most massive visual binary known.

118.009   Systemelemente von Boss 4496 = BV 420 = V 539 Ara für den visuellen Spektralbereich.
E. Schöffel.
Mitt. Astron. Ges., No. 34, p. 126 - 131 (1973/74).
Presented at the "Wissenschaftliche Tagung der Astron. Ges., Oberkochen, 1973 April 24 - 27".

118.010   Parallax and orbital motion of Epsilon Eridani.
P. van de Kamp.
Astron. Journ., Vol. 79, 491 - 492 (1974).
Measurement of 900 plates on Epsilon Eridani taken with the Sproul 24-in. refractor on 238 nights over the interval 1938 - 1972 give a value of $0.''302 \pm 0.''002$ (p.e.) for the relative parallax. The star appears to have perturbation with a period of 25 yr and a semiaxis major of $0.''0191 \pm 0.''0018$; the mass of the unseen companion has a minimum value of $0.006\ M_{\odot}$.

118.011   Fourth list of stars observed at the Lamont-Hussey Observatory of the University of Michigan.
F. Holden.
Publ. Obs. Univ. Michigan, Ann Arbor, Vol. 12, No. 1, 18 pp. (1972).

118.012   Une solution graphique du problème géométrique des étoiles doubles.   P. Rossier.
Orion, 32. Jahrgang, p. 104 - 106 (1974).

118.013   Long period double stars.

A. Bücher, P. Laques.
Astron. Astrophys., Vol. 33, 15 - 20 (1974).

Stock and Wroblemski have used an objective prism to carry out a spectroscopic analysis of large separation double stars; they have concluded that a number of these are physically related. The authors have used this information, together with other observational data, to compute the periods of these probable binaries. This work confirms Stock and Wroblemski's conclusions about the physical nature of the majority of these main sequence stars. The authors have also made a classification in terms of the period.

118.014 On the inclination of rotation axes in visual binaries. E. W. Weis.
Astrophys. Journ., Vol. 190, 331 - 337 (1974).

Spectral types and projected rotational velocities are determined for a number of visual binary systems with known orbits. These data, along with data for other visual binaries taken from the literature, are used to demonstrate a tendency for the rotation axes to be aligned perpendicular to the orbital planes in such systems. This tendency appears strong in F stars while the evidence for A stars is rather weak. The question of coplanarity of orbits in triple systems is briefly discussed.

118.015 Kinematic properties of selected visual binaries. G. A. Bakos.
Stars and the Milky Way system, Proc. 1972, (see 012.018), p. 60 - 64 (1974).

118.016 The visual double star catalogues. C. E. Worley.
Bull. American Astron. Soc., Vol. 6, 217 (1974). – Abstr. AAS.

118.017 Sulla ripartizione delle masse fra le componenti dei sistemi binari. M. G. Fracastoro.
Mem Soc. Astron. Italiana. Nuova Ser., Vol. 44, 643 - 647 (1973/74).

The frequency of the mass-ratios occurring among 367 visual binaries and 243 spectroscopic binaries has been examined, and some comparative considerations are made preliminarily.

118.018 Measurement of double stars by occultation with small telescopes. A. M. Sinzi, Y. Harada.
Rep. Hydrograph. Res., (Japan), No. 9, p. 33 - 38 (1974).

118.019 Orbite nouvelle. P. Couteau.
Circ. d'Inform. (U.A.I. Commission des Étoiles Doubles), Obs. Meudon, No. 62 (1974).

118.020 Étoiles doubles nouvelles découvertes à Béograd (Lunette de 65 cm). G. M. Popovic.
Circ. d'Inform. (U.A.I. Commission des Étoiles Doubles), Obs. Meudon, No. 62 (1974).

118.021 Étoiles doubles découvertes à Nice (Lunettes de 50 et 74 cm). P. Couteau, P. Muller.
Circ. d'Inform. (U.A.I. Commission des Étoiles Doubles), Obs. Meudon, No. 62 (1974).

118.022 Orbites nouvelles.
W. D. Heintz, P. Couteau, v.d. Wiele.
Circ. d'Inform. (U.A.I. Commission des Étoiles Doubles), Obs. Meudon, No. 63 (1974).

118.023 Trajectoires rectilignes. J. Dommanget.
Circ. d'Inform. (U.A.I. Commission des Étoiles Doubles), Obs. Meudon, No. 63 (1974).

118.024 Étoiles doubles nouvelles (Lunette de 50 cm de Nice, Lunette de 65 cm de Belgrade).
P. Couteau, G. M. Popovic.
Circ. d'Inform. (U.A.I. Commission des Étoiles Doubles), Obs. Meudon, No. 63 (1974).

118.025 Untersuchungen an 12 visuellen Doppelsternen. IV. J. Hopmann.
Sitzungsber. Österreich. Akad. Wiss., Math.-nat. Kl., Abt. II, Vol. 181, 301 - 328 = Astron. Mitt. Wien No. 13 (1973).

The aim of this paper is the determination of parallaxes, masses, absolute magnitudes and so on of long-period binaries. If possible also orbits are computed.

118.026 Drei bemerkenswerte visuelle Mehrfachsterne. J. Hopmann.
Sitzungsber. Österreich. Akad. Wiss., Math.-nat. Kl., Abt. II, Vol. 181, 9 - 27 = Astron. Mitt. Wien No. 14 (1973).

Discussed are three remarkable multiple systems: The triple system ADS 8695 = 35 Comae, the triple system ADS 9436 = Aitken 14, and the quadruple system ADS 7114 = ιUMa.

118.027 The orbits of four visual double stars. (ADS 1359, 2377, 9126 and 16873). G. M. Popović.
Bull. Obs. Astron. Belgrade, Vol. 29, F. 1, No. 125, p. 31 - 36 (1971/72).

118.028 Orbite du système ADS 2531 = A 829. D. M. Olević.
Bull. Obs. Astron. Belgrade, Vol. 29, F. 1, No. 125, p. 36 (1971/72).

118.029 Mesures micrométriques des étoiles doubles faites en 1971 à l'Observatoire de Belgrade. (Série 21).
G. M. Popović, D. J. Zulević, D. M. Olević.
Bull. Obs. Astron. Belgrade, Vol. 29, F. 1, No. 125, p. 37 - 42 (1971/72).

118.030 The trajectory of the rectilinear relative motion of a component of the visual double star ADS 9047 = β 614 = IDS 1349N1038, mag. = 8.1–11.8, sp. F0.
D. M. Olević.
Bull. Obs. Astron. Belgrade, Vol. 29, F. 1, No. 125, p. 43 (1971/72).

118.031 Orbite de l'étoile double ADS 8050 = A 1591. V. Erceg.
Bull. Obs. Astron. Belgrade, Vol. 29, F. 1, No. 125, p. 43 - 44 (1971/72).

118.032 The new double stars discovered in Belgrade with the Zeiss refractor 65/1055 cm. Supplement III.
G. M. Popović.
Bull. Obs. Astron. Belgrade, Vol. 29, F. 1, No. 125, p. 44 - 47 (1971/72).

118.033 The orbits of the visual binaries ADS 1315, ADS 8901, λ179, ADS 9264, ADS 12033, ADS 15176.
G. A. Starikova.
Astron. Tsirk., No. 819, p. 7 - 8 (1974). In Russian.

118.034 BDS 1269: a possible physical pair with a metal-deficient primary. G. E. Mechler.
Publ. Astron. Soc. Pacific, Vol. 86, 279 - 280 (1974).

The brighter component of BDS 1269 has four-color and Hβ indices characteristic of a metal-deficient star, while the indices of the secondary are those of a normal early F-type star. The primary may be (1) a horizontal-branch star, and, if so, the brightest known, (2) the only such star physically associated with a normal population I object, and (3) slightly variable.

Estimation de la précision des mouvements propres donnés dans un catalogue d'étoiles à l'aide des observations visuelles de couples stellaires optiques. See Abstr. 112.026.

Blue CN-absorption measurements of close binary stars. See Abstr. 114.059.

119 Spectroscopic Binaries

119.001 **Investigation of the Wolf-Rayet spectroscopic binary HD 152270; line identifications and radial velocities in the spectral region 3500–6000 Å.**
W. Seggewiss.
Astron. Astrophys., Vol. 31, 211 - 221 (1974).

The spectrum of the Wolf-Rayet star HD 152270 has been investigated on plates taken with the ESO coudé-spectrograph and the new Cassegrain-spectrograph of Boyden Observatory. Both the central absorptions and the emission lines exhibit radial velocity curves with a period of 8.893 days. They differ in phase by exactly 0.5, demonstrating the binary nature of the star. According to its radial velocity HD 152270 is a member of the young open cluster NGC 6231.

119.002 **Investigation of radial velocities of the stars 71 Tau, 49 Ori, 8π Vir and ξ Aql.**
G. V. Akhundova, O. Kh. Gusejnov.
Astrofizika, Vol. 9, 479 - 486 (1973). In Russian. English translation in Astrophysics, Vol. 9, No. 4.

The invisible components of the spectroscopic binaries 71 Tau, 49 Ori, 8π Vir and ξ Aql may be relativistic objects (black holes or neutron stars). These conclusions were made on the basis of uncertain information on orbital elements of these systems. To obtain more accurate elements the authors have investigated radial velocities of these stars for three years on spectra taken with the 2-m telescope of the Shemakha Observatory.

119.003 **BD + 47 781, a spectroscopic binary with H and K emission.** M. P. FitzGerald.
Journ. Roy. Astron. Soc. Canada, Vol. 68, 23 - 30 = Contr. Univ. Waterloo Obs., No. 24 (1974).

Orbital elements are given for the eighth magnitude, $8^d.038$, sub-giant spectroscopic-eclipsing binary BD + 47 781 (=BV 307). One component has strong Ca II emission with the same velocities as its absorption lines. Both components are of class G5IV, with an apparent magnitude difference of about $0^m.6$. BD + 47 781 is a member of a class of about 26 spectroscopic-eclipsing systems known to have Ca II emission.

119.004 **On the light curve of HD 77581.**
D. T. Wickramasinghe, J. Whelan.
Monthly Notices Roy. Astron. Soc., Vol. 167, 21P - 25P (1974).

An analysis of the 'ellipsoidal variable' type light curve of HD 77581, on the assumption of co-rotation, indicates that the primary must be nearly in contact with its Roche lobe. An orbital inclination $i \sim 90°$ and a mass ratio $q \sim 0.09$ give reasonable agreement with the light curve, mass function and the X-ray eclipse duration. There is considerable evidence for gas streaming within the system and some evidence for a transient disc around the entire system.

119.005 **Reinvestigation of certain long-period A-type binaries.** H. A. Abt, S. G. Levy.
Astrophys. Journ., Vol. 188, 291 - 294 (1974).

The published orbital elements of 16 normal late A-type stars have been reinvestigated with new radial velocities. New or confirmed elements are given for several systems. There is no longer any convincing evidence for secondary stars (or black holes) having masses greater than those of their primaries.

119.006 **Parallax and orbit for the suspected subluminous binary 2 Andromedae.** K. W. Kamper, Jr.
Astron. Journ., Vol. 79, 54 - 56 (1974).

The author obtains a relative parallax of $0\rlap{.}''009 \pm 0\rlap{.}''007$ from Allegheny observations which indicates that the primary is most likely not subluminous. In addition, examination of the available radial velocities indicates that the evidence for velocity variation is weak. However, there remains fair evidence for light variations in the B component.

119.007 **Spectroscopic observations of the Hyades spectroscopic binary BD + 24°692.** A. Young.
Publ. Astron. Soc. Pacific, Vol. 86, 59 - 62 (1974).

The star BD + 24°692 is confirmed to be a double-lined spectroscopic binary consisting of two lower main-sequence K stars. Another candidate for a low-mass binary is also given.

119.008 **Orbital elements of the spectroscopic binary HD 92168.** N. Ginestet, A. Pédoussaut, J.-M. Carquillat, R. Nadal.
Astron. Astrophys., Suppl. Ser., Vol. 15, 133 - 139 (1974). In French.

Two series of recent observations (Observatoire de Haute-Provence), associated with first observations (Wilson and Victoria Observatories), gave the orbital elements of this binary for the first time. The secondary component would be a dM star; the separation of components would be near 20 $R_\odot$.

119.009 **Orbit of the manganese star HR 8704.**
R. J. Wolff.
Publ. Astron. Soc. Pacific, Vol. 86, 173 - 175 (1974).

Radial velocity measurements from 11 coudé spectrograms are used to derive the orbital elements of this double-line spectroscopic binary.

119.010 **Spectroscopic orbit of the radio star b Persei.**
S. C. Wolff, R. J. Wolff.
Publ. Astron. Soc. Pacific, Vol. 86, 176 - 178 (1974).

Radial velocities from nine coudé spectrograms, obtained in September 1973, are used to derive orbital elements for the radio star b Per.

119.011 Apsidal motion in the Ap star HD 98088.
S. C. Wolff.
Publ. Astron. Soc. Pacific, Vol. 86, 179 - 180 (1974).

New observations of the radial velocity variation of the Ap star HD 98088 are used to show that the period of apsidal motion is at least 700 years. The new observations, combined with earlier ones, are used to derive revised orbital elements.

119.012 The orbit of HR 7694. M. M. Dworetsky.
Publ. Astron. Soc. Pacific, Vol. 86, 183 - 186 (1974).

HR 7694 (HD 191110) was discovered to be a double-lined peculiar star of the Hg type during a search for peculiar objects. The orbit is circular with a period of 9.346 days. The chemical compositions of the two components appear to be different.

119.013 Vorläufige Ergebnisse einer Analyse interstellarer Absorptionslinien im Spektrum von γ^2 Velorum.
M. Grewing, C. M. Walmsley, C. Wulf-Mathies.
Mitt. Astron. Ges., No. 34, p. 160 - 161 (1973/74).
Presented at the "Wissenschaftliche Tagung der Astron. Ges., Oberkochen, 1973 April 24 - 27".

119.014 A possible eclipse in HD 82191. J. D. Fernie.
Inform. Bull. Variable Stars, (I.A.U. Commission 27), Konkoly Obs., Budapest, No. 872 (1974).

119.015 Orbital elements and dimensions of two double-lines eclipsing binaries: W Gru and UX Men.
M. Imbert.
Astron. Astrophys., Vol. 32, 429 - 434 (1974). In French.

The orbital elements have been determined for the eclipsing binary W Gru from 20 double-lines spectra. The analysis of the light curve deduced from observations by Gaposchkin, combined with the spectroscopic elements, enables to estimate the absolute dimensions of the system. The system appears to be detached. For the double-lines eclipsing binary UX Men, the orbital elements have been determined from 10 radial velocity plates.

119.016 Properties and nature of shell stars. 4. 88 Herculis: the 87-day period confirmed.
P. Harmanec, P. Koubský, J. Krpata.
Astron. Astrophys., Vol. 33, 117 - 121 (1974).

Radial-velocity variations of 88° Her as measured on the Haute Provence coudé spectrograms of 1963–1972 by Doazan (1973) confirm the periodic character of the changes reported by the authors previously. Combining these measurements with others they were able to determine a much more precise value of the period: 86.59 ± 0.06 days. The star is considered to be a spectroscopic binary and an improved set of orbital elements is given.

119.017 The double star Krüger 60. Z. Paprotny.
Urania Kraków, Vol. 45, 44 - 47 (1974). In Polish.

A brief note on searching for periodicities in data obtained at irregular intervals. See Abstr. 031.084.

Observation of a lunar occultation of δ Geminorum. See Abstr. 096.021.

Luminosities and motions of peculiar B-type stars. See Abstr. 115.016.

Detection of crossover effect in the Ap star HD 98088. See Abstr. 116.012.

Systemelemente von Boss 4496 = BV 420 = V 539 Ara für den visuellen Spektralbereich. See Abstr. 118.009.

Sulla ripartizione delle masse fra le componenti dei sistemi binari. See Abstr. 118.017.

Orbital elements and absolute dimensions of the eclipsing system LY Aurigae. See Abstr. 121.003.

Spectrographic observations of the 1971 eclipse of 32 Cygni. See Abstr. 121.008.

A spectroscopic study of LY Aurigae. See Abstr. 121.014.

Absolute dimensions of 140 close binary systems. See Abstr. 121.015.

The region of the K line in the spectrum of β Lyrae. See Abstr. 121.086.

The distribution of neutral hydrogen around α Virginis. See Abstr. 131.526.

Optical observations of HD 77581 and a model for the system HD 77581 – 2U 0900 – 40. See Abstr. 142.115.

120 Variable Stars: Catalogues, Ephemerides, Miscellanea

120.001 **Modelling the light curves and determinations of periods of variable stars.** N. E. Kurochkin.
Peremennye Zvezdy, Vol. 19, 117 - 127 (1973). In Russian.

A computer program for the determinations of periods of variable stars is presented. The program is based on modelling the light curve and on methods of the pattern recognition.

120.002 **Application of computers to determination of epochs by the Hertzsprung method.**
V. M. Grigorevskij, V. D. Motrich, N. S. Zgonyajko, R. I. Butenko, S. S. Vykhrestyuk.
Peremennye Zvezdy, Vol. 19, 145 - 154 (1973). In Russian.

An algorithm and peculiarities of calculations by the Hertzsprung method with electronic computers are given. As an example automatic and by hand reductions of observational data for the variable stars V532 Cyg and CSV 503 are cited.

120.003 **Use of electronic computers in studying periods of light variations of variable stars.**
V. M. Grigorevskij.
Peremennye Zvezdy, Vol. 19, 155 - 163 (1973). In Russian.

The paper deals with the use of electronic computers at the Odessa Astronomical Observatory for the solution of a number of problems related to the study of variable stars, such as: the determination of the period, the estimate of the precision of a photometric curve, the Hertzsprung method, the study of Blazhko effect and others.

120.004 **Progressi nell'interpretazione e studio delle stelle variabili.** L. Rosino.
Coelum, Vol. 42, 102 - 120 (1974).

120.005 **About the accuracy of visual observations of variable stars.** Z. Pokorný, J. Šilhán.
Říše hvězd, Vol. 55, 7 - 9 (1974). In Czech.

121 Eclipsing Variables

121.001 Spectroscopic observations of HD 153919 (2U 1700−37). S. C. Wolff, N. D. Morrison. Astrophys. Journ., Vol. 187, 69 - 72 (1974).

High-dispersion spectrograms are used to derive radial velocities and magnetic-field intensities for HD 153919, the optical counterpart of 2U 1700−37, and to determine its spectroscopic properties.

121.002 Interpretation of Epsilon Aurigae. II. Infrared excess, secondary light variations, and plausible formation of a planetary system. S.-S. Huang. Astrophys. Journ., Vol. 187, 87 - 92 (1974).

Infrared excess based on the disk model proposed in a previous paper has been computed. It has been found that the disk alone will emit infrared radiation below the margin of detection. However, if individual condensations are present, the combined result of the disk proper and the condensations yields results of infrared excesses that are consistent with observations. The presence of condensations also makes the secondary light variation understandable. An elementary theory has been developed that analyzes such light variations. The result of the analysis yields the size of the orbit of the condensation around the secondary component.

121.003 Orbital elements and absolute dimensions of the eclipsing system LY Aurigae. G. E. McCluskey, Jr., Y. Kondo. Astrophys. Journ., Vol. 187, 93 - 97 (1974).

Orbital solutions have been obtained for the early-type eclipsing binary LY Aurigae from the light curves obtained with the OAO-2 by Heap and from the V light curve obtained from ground-based observations by Mayer and Horak. The solutions take into account the existence of a nearby (0.''5 separation) companion not accounted for by previous investigators. The spectroscopic observations by Mayer and Batten were used to compute absolute dimensions for the binary orbit and for each component.

121.004 Possible detection of very soft X-rays from SS Cygni. S. Rappaport, W. Cash, R. Doxsey, J. McClintock, G. Moore. Astrophys. Journ. (Letters), Vol. 187, L5 - L7 (1974).

A new source of ultrasoft X-rays ($E < 1/4$ keV) has been detected in the Cygnus region. The case is made for its identification with the close-binary variable SS Cygni.

121.005 The orbit evolution of the eclipsing binary system BD + 16°516 and the rotation period of its white dwarf. J. G. Hills, T. M. Dale. Astron. Astrophys., Vol. 30, 135 - 139 (1974).

The young hot white dwarf BD + 16°516 B is part of an eclipsing binary system having an orbital period of 0.52 days. Due to the short orbital period one can anticipate the white dwarf having a very short rotational period. The result is that one limb of the fast rotating star is brighter than the other, and this brightness difference can be measured from eclipse data to allow a determination of the rotation velocity of the star and consequently of its rotational period. A comparison between the observed and theoretical rotational period can be used to place narrow limits on the initial orbital period and consequently on the initial orbital separation and the initial mass of the companion of the white dwarf prior to any mass exchange or mass loss from the binary system.

121.006 BD − 18° 3437, a new short period eclipsing binary. N. B. Sanwal, M. B. K. Sarma, M. Parthasarathy, K. D. Abhyankar. Astron. Astrophys., Suppl. Ser., Vol. 13, 81 - 90 (1974).

BD − 18° 3437 has been found to be a short period variable and from the nature of its light and colour variation the authors conclude that it is an eclipsing binary of W UMa type with depths of minima of $0.^m20$ and $0.^m18$ and a period of 0.3166386 days. After rectification the depths of the minima were found to be too shallow indicating that the light variation of this system is mainly due to the ellipticity and reflection effects.

121.007 UBV photometry of RV CrV. K. D. Abhyankar, M. Parthasarathy, N. B. Sanwal, M. B. K. Sarma. Astron. Astrophys., Suppl. Ser., Vol. 13, 101 - 108 (1974).

UBV observations of RV CrV made on 11 nights in 1971 are reported. An improved period of 0.7472521 day is found. After rectifying the V light curve by the Russell-Merrill method a tentative photometric solution is obtained by the second method of Kopal. On combining the results with the spectroscopic elements found by Struve and Gratton it appears that the system might be a contact binary.

121.008 Spectrographic observations of the 1971 eclipse of 32 Cygni. G. Bisiacchi, U. Flora, M. Hack. Astron. Astrophys., Suppl. Ser., Vol. 13, 109 - 118 (1974).

Equivalent widths and contours of the chromospheric K-line, together with radial velocities of the metallic lines, hydrogen lines and K-line, plus the ratio between the continuous flux of B-type and K-type stars have been determined using 12.4 Å/mm spectrograms taken during the atmospherical and totality phases of the 1971 eclipse. Cyclic oscillations of the atmosphere with periods of a few days are indicated by the deviations of the radial velocity of the Fe I lines and Balmer lines with respect to the computed orbital velocity.

121.009 Photometric orbit of VZ CVn and the variation of its light curve. C. Ibanoğlu. Astron. Astrophys., Suppl. Ser., Vol. 13, 119 - 126 (1974).

A total of 928 photoelectric observations of VZ CVn were obtained in yellow and in blue light during the 1971 and 1972 seasons. Some variations in its light curve were detected and are discussed. Nine epochs of minima were obtained and from these improved light elements were derived. The light curves were analyzed and consistent geometric elements for the system were derived from each light curve.

121.010 Differential UBV photometry of SW Lac. H. Muthsam, K. D. Rakos. Astron. Astrophys., Suppl. Ser., Vol. 13, 127 - 132 (1974).

Differential UBV observations of the W UMa-type variable SW Lac obtained in autumn 1960 with the 42″ reflector at Lowell Observatory are presented. Times of minimum light are calculated from the observations.

121.011 Period variations of the eclipsing binary RZ Cassiopeiae. T. Herczeg, H. Frieboes-Conde. Astron. Astrophys., Vol. 30, 259 - 263 (1974).

Three new photoelectric minima of RZ Cassiopeiae are given together with a discussion of the period changes based upon the entire available photoelectric material.

121.012 RU Monocerotis. New results. Variations in the system. D. Ya. Martynov. Astron. Zhurn. Akad. Nauk SSSR, Vol. 51, 107 - 115 (1974). In Russian. English translation in Soviet Astron. AJ, Vol. 18, No. 1.

From new photoelectric observations in 1968−71 new photometric elements of the system were deduced. A convincing proof of the value of the period of apsidal rotation — 285 years approximately — has been obtained.

121.013 HD 52942, a new early-type eclipsing binary with a double-line spectrum.
A. F. J. Moffat, N. Vogt.
Astron. Astrophys., Vol. 30, 381 - 384 (1974).

HD 52942 is a close detached binary located in a reflection nebula in the CMa OB 1 association. The components are nearly equal (B 2.5 IV–V) and undergo partial eclipses with a period of $1^d 27306$.

121.014 A spectroscopic study of LY Aurigae.
J. Andersen, A. H. Batten, R. W. Hilditch.
Astron. Astrophys., Vol. 31, 1 - 4 (1974).

Radial velocities derived from 44 spectrograms are analyzed to derive new elements for LY Aurigae. Using the light-curve solution of McCluskey and Kondo (1973) the authors derive physical parameters for the system.

121.015 Absolute dimensions of 140 close binary systems.
P. Giannone, M. A. Giannuzzi.
Astrophys. Space Sci., Vol. 26, 289 - 304 (1974).

The absolute dimensions of 140 close binaries have been computed. These systems have been extracted from recent catalogues of photometric and spectroscopic elements.

121.016 Optimization analysis of the light curves of eccentric eclipsing binary systems. E. Budding.
Astrophys. Space Sci., Vol. 26, 371 - 385 (1974).

The iterative optimization procedures described in a previous paper by the author have been extended to cover unrectified light curves of eccentric eclipsing binary systems. Two stars have been studied in detail: α Coronae Borealis and V 477 Cygni. In both cases improved element sets have been obtained. Moreover, in the case of V477 Cygni the obtained values of the longitude of periastron for three different light curves are found to provide a useful basis for discussion of the apsidal motion of this system.

121.017 Binary systems in star clusters. I. V701 Sco.
K.-C. Leung.
Astron. Astrophys., Suppl. Ser., Vol. 13, 315 - 324 (1974);

The eclipsing binary V701 Sco in the galactic cluster NGC 6383 was observed photoelectrically for eight nights with B, V filters. Complete coverage of the light curves was obtained. The standard Russell–Merrill method of rectification was applied to the light curves. It was found that the rectified depths of the minima were too shallow to allow a definitive solution of the orbital elements.

121.018 Photoelectric observations of β Lyr in 1971.
B. Cester, M. Pucillo.
Astron. Astrophys., Suppl. Ser., Vol. 13, 405 - 414 (1974).

The results of photoelectric observations in UBV and with six interferential filters of β Lyr and of a set of standard stars made during the international campaign in July-August 1971 are given and commented.

121.019 Variations of line profiles of eclipsing binary systems during the eclipse. K. Sato.
Publ. Astron. Soc. Japan, Vol. 26, 65 - 91 (1974).

Variations of line profiles of eclipsing binary systems during the eclipse are discussed. The formula by Kopal (1959) is extended to include the effect of the center-limb variation of line profile. The effects of tidal distortion and differential rotation upon the line profile are estimated for two eclipsing binaries outside the eclipse. The observed rotational effects are compared with the calculated ones for the two eclipsing binaries R CMa and YZ Cas. The variation of residual intensity at the deepest point in the line profile of R CMa is compared with the observation by Kitamura (1969).

121.020 Photoelectric observations of CH Cyg in 1969.
T. S. Belyakina, E. S. Brodskaya.
Izv. Krymskoj Astrofiz. Obs., Vol. 49, 58 - 64 (1974). In Russian.

During August 25 to August 31 1969 simultaneous photoelectic observations of CH Cyg were carried out using the 70-cm and 33-cm reflectors. Continuous records of CH Cyg in the regions λλ 5120 - 5320Å, 4155 - 4280Å and 3350 - 3650 Å were made with a photoelectric spectrocolorimeter mounted at the 70-cm reflector. The results of the observations are presented. Assuming that CH Cyg is a binary star, the luminosity of its components has been estimated.

121.021 The close binary Beta Lyrae. I. Some spectroscopic results from the second international campaign.
S. Kříž, F. Ždárský.
Bull. Astron. Inst. Czechoslovakia, Vol. 25, 1 - 5 (1974).

The profiles of emission lines Hα, Hγ, He I 4472 and He I 6678 are given and discussed. An attempt was made to find lines of the secondary component in the spectrum, however, the result was negative.

121.022 The close binary Beta Lyrae. II. Absolute dimensions and properties of components. S. Kříž.
Bull. Astron. Inst. Czechoslovakia, Vol. 25, 6 - 15 (1974).

The masses, absolute dimensions and luminosities of the components of Beta Lyrae were determined. It is shown that the secondary component is probably a star of the main sequence of the spectral type B0–B2.

121.023 The spectrum of the secondary component of U Cephei. B. W. Baldwin.
Publ. Astron. Soc. Pacific, Vol. 85, 714 - 717 (1973).

An attempt to detect abundance anomalies in the secondary component of U Cep is reported. A spectral comparison of the secondary of U Cep with single stars of similar spectral type and luminosity class reveals no substantial spectral differences, with the exception that the calcium H and K lines in the spectrum of the secondary of U Cep are weak and appear to be filled in by emission.

121.024 Outside-eclipse spectral changes in β Lyrae.
G. W. Wolf, S. Sobieski, Y. Kondo.
Publ. Astron. Soc. Pacific, Vol. 85, 718 - 720 (1973).

Ten spectra of β Lyr at first quadrature were obtained in 1971 at a dispersion of 39 Å mm^{-1}. Variations of the He I λ 4471 line and a possibly related spectral feature are discussed.

121.025 New minima and light elements for the eclipsing binary SW Lyncis. A. U. Landolt.
Publ. Astron. Soc. Pacific, Vol. 85, 742 - 745 = Contr. Louisiana State Univ. Obs., Baton Rouge, No. 89 (1973).

New photoelectrically determined times of minima are given for SW Lyn. These are combined with minima in the literature to provide new light elements.

121.026 The O9+Of eclipsing binary V729 Cygni in Cygnus OB2. D. S. Hall.
Acta Astron., Vol. 24, 69 - 78 (1974).

This variable was observed photoelectrically; 143 UBV observations have been obtained and are discussed here.

121.027 UBV photometry of the eclipsing binary ZZ Cygni. D. S. Hall, R. O. Cannon III.
Acta Astron., Vol. 24, 79 - 88 (1974).

174 UBV observations are presented and discussed.

121.028 The light-curves of 13 eclipsing variables.

R. Szafraniec.
Acta Astron., Vol. 24, 89 - 117 (1974).

Visual observations were reduced and light-curves determined for 13 eclipsing variables: V806 Oph; CP, ET, FK, FL, FR, GG, OS, V343 Ori; UX, AQ, BG, BN Peg.

121.029 The nature of long-period variations in the close binary system β Lyr. V. Ya. Alduseva.
Astron. Zhurn. Akad. Nauk SSSR, Vol. 51, 403 - 412 (1974). In Russian. English translation in Soviet Astron. AJ, Vol. 18, No. 2.

The nature of the additional emission during integrated brightness β Lyrae-system maxima is investigated. The source of this emission is shown to be optically thin in continuum with electron temperature of $T_e \simeq 40000°$. It is suggested that such a gaseous layer might be the consequence of interactions of gaseous layers of the disk-shaped envelope surrounding the second component of β Lyrae.

121.030 A search for Lyman-alpha emission in Beta Lyrae from Copernicus. Y. Kondo, G. E. McCluskey, Jr.
Astrophys. Journ., (Letters), Vol. 188, L63 - L65 (1974).

High-resolution (0.2 Å) spectrophotometric observations of the complex eclipsing binary β Lyrae were obtained with the Princeton Telescope Spectrometer on the Copernicus satellite. The authors discuss the search for Lα emission in β Lyrae and compare the Copernicus results with the OAO-2 observations of the same binary system. The possible Lα emission features observed from OAO-2 are identified as blends of the emission lines of other elements in the vicinity of Lα.

121.031 Variations of the shape of primary minima for Algol-type stars.
S. L. Piotrowski, M. Różyczka.
Peremennye Zvezdy, Vol. 19, 107 - 116 (1973). In Russian.

From about 200 eclipsing binaries 17 were chosen for investigations of possible cycle-to-cycle variations of the primary minima shape. Only detached and semi-detached systems were considered. In some cases of the 17 binaries, observed changes between individual minima seem to have a physical significance. These are the stars R CMa, TX UMa and, possibly, V346 Aql, BR Cyg and RZ Cas.

121.032 On studies of eclipsing binaries in the USSR within the years 1969 – 1972.
V. P. Tsesevich, A. M. Shulberg.
Peremennye Zvezdy, Vol. 19, 199 - 204 (1973). In Russian.

121.033 A direct method of light curve solution of an eclipsing binary system with extended atmosphere. Calculation and application of weights. Computer programmes.
A. M. Cherepashchuk.
Peremennye Zvezdy, Vol. 19, 227 - 252 (1973). In Russian.

A direct method for light curve solution of an eclipsing binary system with an extended atmosphere is described. All unknown values: the functions describing the structure of the disk of the peculiar component, and the satellite radius and inclination of the orbital plane are obtained using the minimum of $|O-C|$. The value of $|O-C|$ is the weighted sum of square deviations of the theoretical light curve from the observed one. It allows to obtain a reliable solution and its error. Computer programmes in Fortran for light curve solution are given.

121.034 Ultraviolet photometry from the Orbiting Astronomical Observatory. XI. The 1971 eclipse of 32 Cygni. L. R. Doherty, J. F. McNall, A. V. Holm.
Astrophys. Journ., Vol. 187, 521 - 530 (1974).

The authors observed the 1971 eclipse of 32 Cyg at regular intervals from November 1 to 15 with the wide-band filter photometers aboard OAO-2. Data have been reduced for seven filter bands with effective wavelengths between 4250 and 1430 Å. During totality, the observed 2460 Å radiation was 2 mag brighter than expected for a normal K supergiant. Excess radiation was also observed in the 1550 and 1430 Å bands. A possible origin of this excess is emission in the extended K atmosphere due to recombination of Mg and Si ions.

121.035 Photoelectric minima of the eclipsing variable T Leo Minoris. A. Okazaki, M. Kitamura.
Inform. Bull. Variable Stars, (I.A.U. Commission 27), Konkoly Obs., Budapest, No. 862 (1974).

121.036 The spectral types of the members of VV Cephei systems from photometric analysis.
A. S. Wawrukiewicz, T. A. Lee.
Publ. Astron. Soc. Pacific, Vol. 86, 51 - 55 (1974).

The photometric data for ten binary systems that have a highly luminous red primary with a less luminous blue companion have been analyzed and the spectral types of the member stars determined. Eight of these systems are VV Cephei-type systems, one, α Sco, has a more distant blue companion, while XX Per has a blue companion that is probably too distant to induce emission. The authors have determined that the blue companion of WY Gem is an intrinsic variable.

121.037 LX Persei, an eclipsing binary with H and K emission. E. J. Weiler.
Publ. Astron. Soc. Pacific, Vol. 86, 56 - 58 (1974).

Masses and MK classes are determined for the eclipsing spectroscopic binary LX Per (BD + 47°0781, $8^m.55 - 9^m.55$). The system has a period of 8.0 days. LX Per is likely a member of the group of eclipsing binaries with strong H and K emission whose prototype is RS CVn.

121.038 A spectroscopic orbit for the Am eclipsing system AN Andromedae. A. Young.
Publ. Astron. Soc. Pacific, Vol. 86, 63 - 66 (1974).

A new spectroscopic orbit for the eclipsing binary AN And is presented based upon 23 intermediate-dispersion spectrograms. A change in the γ velocity from an earlier orbit is noted and discussed.

121.039 V453 Cyg, an early type Algol binary.
H. L. Cohen.
Astron. Astrophys., Suppl. Ser., Vol. 15, 181 - 208 = Contr. Rosemary Hill Obs., Univ. Florida, Gainesville, Florida, No. 42 (1974).

539 photoelectric UBV observations of the massive, double-line eclipsing binary V453 Cyg were obtained at Lowell Observatory in 1966. The observations show that (1) the system is Algol-type, (2) perturbations of the light curve are small, (3) the eclipses are complete with nearly equal depths and nearly flat bottoms, (4) orbital eccentricity effects are obvious only for e cos ω, and (5) $(B-V)$ and $(U-B)$ are constant. A consistent set of geometrical elements fits both eclipses in all colors. Available spectroscopic data and membership in NGC 6871 can be used to estimate absolute dimensions and magnitudes. The system appears to be composed of two almost spherical stars of nearly identical surface temperatures. V453 Cyg is a distinctly detached system with components near the main sequence turn-off of NGC 6871.

121.040 A time of minimum for HD 217312.
C. D. Scarfe, D. J. Barlow.
Journ. Roy. Astron. Soc. Canada, Vol. 68, 96 - 98 (1974).

121.041 Rediscussion of eclipsing binaries. X. The B stars AG Persei and CW Cephei. D. M. Popper.
Astrophys. Journ., Vol. 188, 559 - 565 (1974).

New spectrographic material is presented for the double-lined B-type eclipsing binaries AG Per and CW Cep. Small but significant revisions of the masses and radii are found to be required. A collection of the most precisely known properties of B-type binary components is tabulated.

121.042 On the duration of totality in the eclipse of Zeta Aurigae. M. Kitamura.
Astrophys. Space Sci., Vol. 28, L17 - L23 (1974).

From a discussion of narrow-band observations of Zeta Aurigae in the 1963–64 eclipse, the duration of totality has been precisely found to be $\leq$ 36.901 days, which is appreciably smaller than the value 37.48 days currently used. The present result reveals that a hypothesis of gradual expansion of the K-type supergiant component could not be tenable.

121.043 Electrophotometric observations of the 1971 eclipse of the eclipsing variable 32 Cyg.
N. L. Magalashvili, Ya. I. Kumsishvili.
Byull. Abastumansk. Astrofiz. Obs., No. 45, p. 37 - 44 (1974). In Russian.

In October–November of 1971 photoelectric observations of 32 Cyg were performed with narrow-band filters at 5000 Å, 4240 Å, 3500 Å. The results are given in tables and figures.

121.044 Three-colour photoelectric observations of the eclipsing variable CU Tauri.
N. L. Magalashvili, Ya. I. Kumsishvili.
Byull. Abastumansk. Astrofiz. Obs., No. 45, p. 45 - 50 (1974). In Russian.

Electrophotometric UBV observations of CU Tauri have been carried out. The data of individual observations appear in a table. From three minima improved elements have been derived.

121.045 Search for "black holes" in close binary systems.
L. I. Snezhko.
Soobshch. Spets. Astrofiz. Obs. AN SSSR, *Zelenchukskaya*, vyp. (No.) 6, p. 3 - 18 (1972). In Russian.

The calculations available for the evolution of close binary systems (CBS) are considered in connection with the problem of detection of "black holes" in binary systems. It is shown that the evolution theory of CBS does not contradict the possibility of formation of components – the "black holes".

121.046 Radial velocity curves of HD 101799.
R. F. Sisteró, M. E. Castore de Sisteró.
Astron. Journ., Vol. 79, 391 - 396, 425 (1974).

Radial velocities of the W UMa star HD 101799 were measured on spectrograms at 39 Å/mm. Spectroscopic elements are given from a weighted mean of five radial velocity curves derived from measurements of individual absorption lines ($\lambda\lambda$ 4005, 4045, 4325, H_δ, and H_γ). The masses of the components and absolute dimensions were found by combining these elements with the photometric orbital parameters. The primary component is of spectral type $F8$, while the secondary is a $F5$ star; the system nearly fills the critical equipotential Roche surface, probably in a noncontact configuration.

121.047 On the possible apsidal advance in HS Herculis.
C. D. Scarfe, D. J. Barlow.
Publ. Astron. Soc. Pacific, Vol. 86, 181 - 182 (1974).

Times of both minima of HS Her were obtained in the summer of 1973. The results support the proposal of Hall and Hubbard that the orbit is slightly eccentric, but fail to confirm their apsidal period.

121.048 The eclipsing binary system KO Aquilae.
C. Blanco, S. Cristaldi.
Publ. Astron. Soc. Pacific, Vol. 86, 187 - 194 (1974).

From observations the authors found the period of the KO Aql system to be increasing by 3×10^{-8} days per period. V light observations carried out during 1965 show a significant secondary minimum which becomes shallower in the 1967 and 1969 observations. The solutions of the light curves obtained were possible only by assuming the primary minimum to be an occultation.

121.049 Multicolor photometry of eclipsing binaries with "undersize" subgiant secondaries. R. C. Barnes.
Publ. Astron. Soc. Pacific, Vol. 86, 195 - 202 (1974).

Strömgren $uvby$ photometric observations of 10 eclipsing binaries thought to contain undersize subgiant secondaries are presented and analyzed. The data suggest that the actual number of undersize systems may be less than previously thought.

121.050 Photometric observations of the eclipsing binary LT Herculis.
E. G. Ebbighausen, G. Penegor.
Publ. Astron. Soc. Pacific, Vol. 86, 203 - 207 (1974).

Photoelectric observations in the B and V spectral regions are given for the interesting eclipsing binary LT Her whose orbital period is close to 26 hours. Remarks are made regarding peculiar features of the light curves and spectrum of the secondary component.

121.051 The long period, high latitude, eclipsing systems V 748 Cen (= Cen X-4?) and BL Tel.
A. M. van Genderen, I. S. Glass, M. W. Feast.
Monthly Notices Roy. Astron. Soc., Vol. 167, 283 - 298 (1974).

Extensive photoelectric photometry (Walraven system) as well as infra-red photometry and spectroscopy of V 748 Cen (= S 5003 Cen) are discussed. This object is found to be a binary showing total eclipses. The secondary is an M-type star. Infra-red photometry of BL Tel shows the presence of a cool secondary. The considerable distance of these systems from the galactic plane and their relation to such binaries as AR Pav and (nova) T CrB are briefly discussed. The possiblity that the primary of V 748 Cen like that of some related systems is highly evolved and that the heating effect of the gas stream was unusually strong near the epoch of the appearance of the transient (nova-like) X-ray source Cen X-4 makes the identification with this source at least plausible.

121.052 Spectroscopy of the eclipsing variable AR Pavonis.
A. D. Thackeray, J. B. Hutchings.
Monthly Notices Roy. Astron. Soc., Vol. 167, 319 - 336 (1974).

Spectra of the remarkable symbiotic object AR Pavonis covering 20 years (12 eclipse periods) are analysed for radial velocities and line-intensities. The eclipsed body shows an emission O-type spectrum. The secondary is believed to be M3 III (possibly variable). A model is proposed in which the M star is filling its Roche lobe and losing mass to the secondary through a stream which is mainly responsible for the cF absorption. Masses of order 2.5 and $1.2_\odot$ for the primary and secondary are regarded as most likely.

121.053 The light curve and orbital elements of XZ Sgr.
G. F. G. Knipe.
Monthly Notices Roy. Astron. Soc., Vol. 167, 369 - 374 (1974).

The Algol-type eclipsing binary XZ Sgr was observed in Johannesburg during 1971. Approximately 600 observations each were obtained in B and V, and 78 in U, mostly near primary minimum. XZ Sgr consists of an A3 primary and a G5 secondary component. Using Smak's spectroscopic data, the masses and radii of the stars are found to be $1.8 M_\odot$, $1.59 R_\odot$ and $0.26 M_\odot$, $2.44 R_\odot$.

121.054 **The secondary component of Beta Lyrae.**
R. E. Wilson.
Astrophys. Journ., Vol. 189, 319 - 329 (1974).

Quantitative analysis of β Lyr light curves shows that thin-disk and "no-disk" models for the secondary can be ruled out, leaving (presumably) only a thick-disk model. There is no evidence that the secondary component is underluminous for its mass, provided the observations are interpreted in terms of a thick secondary disk, most of whose luminosity emerges at the poles. The polar effective temperature of the disk is several thousand degrees K higher than that of the primary component, and is therefore the obvious candidate for the elusive source of excitation for the B2-B5 shell spectrum. Because the secondary is not underluminous, there is, at present, no reason to postulate the presence of a collapsed star at its center. Reasons are given to explain why the secondary spectrum has not been seen.

121.055 **Field-scanning photoelectric observations of the eclipsing variable YY Gem.**
E. Budding, M. Kitamura.
Inform. Bull. Variable Stars, (I.A.U. Commission 27), Konkoly Obs., Budapest, No. 900, 2 pp. (1974).

121.056 **HZ Her, une variable énigmatique.**
A. Oberstatter.
L'Astronomie, 88ᵉ année, p. 99 - 102 (1974).

121.057 **Eclipsing binary RT Andromedae (II).**
A. Dumitrescu.
Stud. Cerc. Astron., Vol. 19, 89 - 97 (1974). In Romanian.

RT And is found to be an eclipsing system of two dwarf stars belonging to the spectral class F8V and K0V. The corresponding radius of Roche lobe for primary component takes the value $r_g^* = 0.390$. Some characteristics of the light of the system can be assigned to random mass loss from the primary star. A hot spot on the secondary star in the side of impact of the stream coming from primary, should explain the anomalies of the light curve.

121.058 **Photoelectric photometry of TV Cassiopeiae.**
J. Papoušek.
Bull. Astron. Inst. Czechoslovakia, Vol. 25, 152 - 160 (1974).

1583 photoelectric observations in B and V and 1469 in U were carried out at the Brno Observatory in 1972. Nine new times of primary minimum light are presented. The photoelectric orbital solution is ambiguous, two different solutions exist. The limb darkening coefficient for the primary component was derived from the light curve.

121.059 **Copernicus spectra of β Lyrae.**
M. Hack, J. B. Hutchings, Y. Kondo, G. E. McCluskey, M. Plavec, R. S. Polidan.
Nature, Vol. 249, 534 - 536 (1974).

The authors have observed the far ultraviolet spectrum of β Lyrae using the Princeton spectrometer, on the Copernicus satellite. The results are reported and discussed.

121.060 **V445 Cas is an eclipsing variable.** L. Szabados.
Inform. Bull. Variable Stars, (I.A.U. Commission 27), Konkoly Obs., Budapest, No. 867, 2 pp. (1974).

121.061 **Photoelectric photometry of the eclipsing binaries RZ Cas and AR Lac.** C. R. Chambliss.
Inform. Bull. Variable Stars, (I.A.U. Commission 27), Konkoly Obs., Budapest, No. 883, 3 pp. (1974).

121.062 **Polarimetric observations of AW UMa.**
V. A. Oshchepkov.
Inform. Bull. Variable Stars, (I.A.U. Commission 27), Konkoly Obs., Budapest, No. 884, 4 pp. (1974).

121.063 **AY Mus: a triple system?** S. Söderhjelm.
Inform. Bull. Variable Stars, (I.A.U. Commission 27), Konkoly Obs., Budapest, No. 885, 2 pp. (1974).

121.064 **Eclipse timings of cataclysmic variables.**
G. S. Mumford.
Inform. Bull. Variable Stars, (I.A.U. Commission 27), Konkoly Obs., Budapest, No. 889, 2 pp. (1974).

121.065 **HD 271227, a new β Lyrae type star just outside the error box of LMC X-2.** A. M. van Genderen.
Inform. Bull. Variable Stars, (I.A.U. Commission 27), Konkoly Obs., Budapest, No. 891, 2 pp. (1974).

121.066 **UBV observations of 31 Cygni during the 1972 eclipse.** R. M. Williamon.
Inform. Bull. Variable Stars, (I.A.U. Commission 27), Konkoly Obs., Budapest, No. 892 (1974).

121.067 **Ultra-violet observations of the eclipsing binary star δ Pictoris from the TD-1A satellite.**
R. G. Evans.
Monthly Notices Roy. Astron. Soc., Vol. 167, 517 - 525 (1974).

The B1 star δ Pictoris has been observed 18 times by the sky survey telescope on the ESRO TD-1A satellite. The spectrophotometric observations in the wavelength range 1360-2740 Å are described and a model of this eclipsing binary system is presented.

121.068 **Photometry of the eclipsing system AR Pavonis.**
P. J. Andrews.
Monthly Notices Roy. Astron. Soc., Vol. 167, 635 - 644 (1974).

UBV measures of the eclipsing symbiotic variable AR Pav are presented covering the years 1967 to 1973. By combining these results with those of Mayall an improved period of 604.6 days is derived. A comparison of the eclipse curves when the star is bright and faint leads to a model similar to that of Thackeray & Hutchings with a secondary, filling its Roche lobe, losing mass to the primary.

121.069 **On the evolutionary scheme W UMa → CV → type I SN.** B. Warner.
Monthly Notices Roy. Astron. Soc., Vol. 167, 61P - 64P (1974).

Considerations of space densities and lifetimes indicate that W UMa stars evolve into cataclysmic variable (CV) stars, and that these later turn into one of the two kinds of type I supernovae. It is suggested that the mechanism for the latter evolutionary stage is the transfer of helium from the remnant of the secondary to the primary, resulting in pushing the white dwarf primary over the Chandrasekhar limit.

121.070 **Photoelectric observations of the eclipsing binaries HS Herculis and RU Monocerotis.**
D. Ya. Martynov.
Soobshch. Gos. Astron. Inst. Shternberga, No. 185, 44 pp. (1973). In Russian.

121.071 **Narrow-band hydrogen line photometry of three early-type eclipsing binary systems.**
E. C. Olson, E. W. Weis.
Astron. Journ., Vol. 79, 642 - 648 (1974).

The authors use narrow-band Hγ photoelectric photometry to explore the radiative interaction as a function of phase in Z Vul, and to detail anomalies due to circumstellar contamination in AI Dra and RZ Cas.

121.072 **Interpretation of Epsilon Aurigae. III. Study of the light curve based on disk models.** S.-S. Huang.

Astrophys. Journ., Vol. 189, 485 - 491 (1974).

The light curve expected from the eclipse of a star by a companion possessing a semitransparent disk has been investigated. Two computational procedures for deriving the light curve based on the geometrically thin-disk model are given here. Using these procedures, the author has calculated the light curves of ϵ Aur based on different sets of parameters including those given previously by Wilson. He has also studied the light curve of ϵ Aur based on a geometrically thick disk model, and has advanced two proposals in order to understand the change in the shape of the light curve from one eclipse to the next.

121.073 **A new method of light curves solution for eclipsing binary systems with extended atmospheres.**
A. M. Cherepashchuk.
Astron. Zhurn. Akad. Nauk SSSR, Vol. 51, 542 - 551 (1974). In Russian. English translation in Soviet Astron. AJ, Vol. 18, No. 3.

A new method for light curve solution for an eclipsing binary system with an extended atmosphere is described. The method is based on the combined solution of both minima of the light curve.

121.074 **Modern aspects on the eclipsing variables.**
S. Svolopoulos.
In honorem S. Placidis, (see 003.009), p. 355 - 372 (1974). In Greek.

121.075 **A remarkable star in a highly reddened group of OB$^+$-stars in Norma.** G. Klare, T. Neckel.
Astron. Astrophys., Vol. 33, 143 - 145 (1974).

An interesting star (its spectrum suggests an object of the VV Cephei type) was found in a group of highly reddened OB$^+$-stars in Norma. The polarization behaviour in the surroundings of this group is discussed.

121.076 **Complex infrared emission features in the spectrum of Beta Lyrae.**
T. H. Morgan, A. E. Potter, Y. Kondo.
Astrophys. Journ., Vol. 190, 349 - 352 (1974).

Spectra of β Lyrae over the spectral region 5800–11000 cm^{-1} (1.76–0.9 μ) at two different phases have been obtained. They show a remarkable emission-absorption complex at 9231 cm^{-1}, a highly structured emission at Pβ, and several additional broad weak emissions.

121.077 **The radius-luminosity relation in eclipsing binaries.**
C. Popovici, A. Dumitrescu.
Stars and the Milky Way system, Proc. 1972, (see 012.018), p. 53 - 59 (1974).

121.078 **Comparison of spectrophotometric observations of the eclipsing binary, W W Auriga, with predictions of the theory of non-grey stellar atmospheres.**
J. K. Davidson, J. S. Neff.
Bull. American Astron. Soc., Vol. 6, 222 (1974). – Abstr. AAS.

121.079 **Spectrophotometry of VV Cephei-type systems.**
J. Piccirillo.
Bull. American Astron. Soc., Vol. 6, 222 (1974). – Abstr. AAS.

121.080 **uvbyβ observations of four short-period binaries.**
G. G. Spear.
Bull. American Astron. Soc., Vol. 6, 222 (1974). – Abstr. AAS.

121.081 **AR Cassiopeiae: results of the 1968–69 IAU campaign.** T. Herczeg.
Bull. American Astron. Soc., Vol. 6, 223 (1974). – Abstr. AAS.

121.082 **Photographic photometry of two unusual eclipsing binaries.** T. Herczeg, J. Kerrigan.
Bull. American Astron. Soc., Vol. 6, 223 (1974). – Abstr. AAS.

121.083 **An evolutionary model of β Lyr.** J. Ziolkowski.
Bull. American Astron. Soc., Vol. 6, 263 - 264 (1974). – Abstr. AAS.

121.084 **Photoelectric minima of eclipsing binaries.**
S. Mancuso, L. Milano.
Inform. Bull. Variable Stars, (I. A. U. Commission 27), Konkoly Obs., Budapest, No. 904 (1974).

121.085 **Coherent oscillations in UX Ursae Majoris.**
R. E. Nather, E. L. Robinson.
Astrophys. Journ., Vol. 190, 637 - 651 (1974).

Rapid oscillations appeared in the light curve of UX UMa and were studied from 1973 March 3 through 8, during which time they varied smoothly in period from 30.03 to 28.54s, and exhibited a rapid phase shift of $-360°$ associated with each eclipse. The phase shift can be understood in terms of nonradial pulsations with $l = 2$ and $m = -2$ or $+2$. The eclipse curve is shown to be composite, with the oscillations associated with the primary. It is proposed that the blue star in UX UMa is a pulsating white dwarf.

121.086 **The region of the K line in the spectrum of β Lyrae.**
A. H. Batten, G. L. Diamond, W. A. Fisher.
Publ. Astron. Soc. Pacific, Vol. 86, 237 - 240 (1974).

High-dispersion spectrograms of β Lyr obtained at the Hale and Dominion Astrophysical Observatories have been examined in a search for the secondary component of the K line. No such component could be certainly detected.

121.087 **DL Virginis – an eclipsing binary as part of a triple system.** E. Schöffel, D. M. Popper.
Publ. Astron. Soc. Pacific, Vol. 86, 267 - 271 (1974).

The A-type eclipsing binary, DL Vir, period 1^d3, shares a common center of mass with a G8 giant, their period being in the vicinity of 2500^d, with a large uncertainty. The eclipsing companion is not visible in the spectrum, although the disappearance of the A-type lines at primary minimum shows that the eclipse is total. The short-period, single-lined binary orbit is fairly well determined.

121.088 **Photoelectric observations of HD 77581.**
N. V. Vidal.
Publ. Astron. Soc. Pacific, Vol. 86, 317 - 320 (1974).

The light curve of HD 77581 (3U0900–40) in V is presented.

121.089 **Periodenänderungen und dritte Komponente im System RT Persei.** P. Ahnert.
MVS, Sonneberg, Vol. 6, 143 - 153 (1974).

The processing of 316 minima of this eclipsing variable, which were observed from 1905 to 1973, shows regular period changes with a period of 37.2 years, evidently caused by a third component. These regular period variations are superposed by irregularities internally produced by mass transport. New elements of the light change are deduced; orbit and mass of the third component are estimated.

121.090 **Lists of minima of eclipsing binaries.**
Compiled by R. Diethelm, R. Germann, K. Locher, H. Peter, H. Wittwer, S. Forster, P. Morger.
BBSAG Bull., No. 13, p. 1 - 3; No. 14, p. 1 - 2; No. 15, p. 1 - 3 (1974). – 46th - 48th list of Swiss Astronomical Society's Eclipsing Variable Observers.

121.091 **Visual lightcurve of V 839 Ophiuchi.**
R. Diethelm.
BBSAG Bull., No. 13, p. 3 (1974).

121.092 **Duration and magnitude of the minimum of VW Hydrae.** K. Locher.
BBSAG Bull., No. 13, p. 3 (1974).

121.093 **The period of Z Vulpeculae.** R. Diethelm.
BBSAG Bull., No. 13, p. 4 (1974).

121.094 **Note on the O-C of VW Cygni.** K. Locher.
BBSAG Bull., No. 14, p. 2 (1974).

121.095 **New elements for RW Comae.** R. Diethelm.
BBSAG Bull., No. 15, p. 3 - 4 (1974).

121.096 **Note on the O-C of RV Lyrae.** K. Locher.
BBSAG Bull., No. 15, p. 4 (1974).

121.097 **Narrow-band photometry and spectroscopic observations of β Lyrae.**
M. Kiyokawa, S. Kikuchi.
Tokyo Astron. Bull., Second Ser., No. 233, p. 2687 - 2704 (1974).
This report concerns the Japanese contribution to the international campaign of photoelectric and spectroscopic observations of β Lyrae proposed by Commission 42 of IAU (Batten 1970) and carried out mainly in April, July and August of 1971.

121.098 **Algol-type variable CSV 61.**
A. S. Sharov, P. N. Kholopov.
Peremennye Zvezdy, Prilozhenie, Vol. 1, 403 - 406 (1973). In Russian.

121.099 **Photographic observations of the eclipsing binary stars CL Aur, IK Aur, AQ Tau.** Yu. A. Fadeev.
Peremennye Zvezdy, Prilozhenie, Vol. 1, 433 - 438 (1973). In Russian.

121.100 **Observations of the eclipsing variable star Wr 104 – CSV 7265.**
V. P. Tsesevich, V. G. Karetnikov.
Peremennye Zvezdy, Prilozhenie, Vol. 1, 459 - 475 (1973). In Russian.

121.101 **Three-colour photoelectric observations of the binary systems HZ Her and V1357 Cyg connected with X-ray sources Her X-1 and Cyg X-1.**
A. M. Cherepashchuk, V. M. Kovalenko, O. N. Kovalenko, A. V. Mironov.
Astron. Tsirk., No. 812, p. 5 - 7 (1974). In Russian.

121.102 **Elements of ten eclipsing stars.** V. P. Tsesevich.
Astron. Tsirk., No. 816, p. 8 (1974). In Russian.

On the problem of matter outflow from Wolf-Rayet stars. See Abstr. 064.023.

X-ray heating in HZ Her. See Abstr. 064.029.

Occultation resolution of σ Scorpii. See Abstr. 096.007.

Blue CN-absorption measurements of close binary stars. See Abstr. 114.059.

Photoelectric spectrum scans and coudé spectroscopy of HD 153919 (= 2 U 1700–37). See Abstr. 114.090.

Observations of the barium star binary system HD 196673. See Abstr. 114.143.

On the mass determination of collapsed objects in close binaries: applications to black holes and neutron stars. See Abstr. 117.004.

The intrinsic polarization of YY Eri. See Abstr. 117.018.

Orbital elements and dimensions of two double-lines eclipsing binaries: W Gru and UX Men. See Abstr. 119.015.

The effects of the 35-day X-ray cycle on the light curve of HZ Herculis. See Abstr. 122.102.

Supernova explosions in close binary systems, with an application to Centaurus X-3. See Abstr. 125.007.

Are binary systems involved in supernova explosions always disrupted? See Abstr. 125.014.

An investigation of accretion of matter onto white dwarfs as a possible X-ray mechanism. See Abstr. 126.018.

Pulsars and close binary systems. See Abstr. 141.313.

Black holes in binary systems: Instability of disk accretion. See Abstr. 142.003.

Extended observations of > 7-keV X-rays from Centaurus X-3 by the OSO-7 satellite. See Abstr. 142.006.

Spectroscopic observations of HZ Herculis. See Abstr. 142.007.

Do cosmic rays heat HZ Herculis? See Abstr. 142.008.

A determination of the cooling time and the speed of the surface currents of HZ Herculis. See Abstr. 142.009.

2U 1700–37: Another black hole? See Abstr. 142.010.

Spectroscopic variations in Cyg X-2. See Abstr. 142.013.

Comments on magnetic accretion and Her X-1. See Abstr. 142.016.

Observations of a variable emission core in the Hβ line of HZ Herculis. See Abstr. 142.017.

Reflection and reprocessing of X-ray source radiation by the atmosphere of the normal star in a binary system. See Abstr. 142.018.

A ten-day observation of Hercules X-1 from the OSO-7 satellite. See Abstr. 142.019.

Optical spectra and the mass of SMC X-1. See Abstr. 142.022.

Collapsed objects and galactic X-ray sources. See Abstr. 142.023.

Black holes and X-ray binaries. See Abstr. 142.025.

The synthesis of close-binary light curves. VI. X-ray and collapsar binaries. See Abstr. 142.028.

Upper limit on 2.5-second pulsations from Hercules X-1. See Abstr. 142.029.

Soft X-ray variability of binary X-ray stars. See Abstr. 142.030.

A slaved disk model for Hercules X-1. See Abstr. 142.034.

The nature of Cygnus X-3: a prototype for old population binary X-ray sources. See Abstr. 142.050.

Search for high frequency optical variations in Vela XR-1. See Abstr. 142.052.

Binary X-ray sources. See Abstr. 142.054.

Coordinated campaign for observations of X-ray binaries, I.A.U. Commission 42 in coordination with Commission 44. See Abstr. 142.057.

Model for the binary X-ray source 2U1700-37. See Abstr. 142.059.

A preliminary report on the light variation of Sk. 160 = SMC X-1. See Abstr. 142.063.

Spectroscopic observations of the optical companion to Centaurus X-3. See Abstr. 142.069.

Variations of the optical emission-line spectrum of the eclipsing binary X-ray star HD 153919 (= 2 U 1700-37). See Abstr. 142.071.

Possible periodic variations in the secondary minimum of HD 153919 ≡ 2U 1700-37. See Abstr. 142.072.

Skylab S019 ultraviolet observations of HD153919 = 3U1700-37. See Abstr. 142.074.

Optical observations of the spectrum of the binary X-ray source 2U0900-40. See Abstr. 142.077.

Observations of the eclipsing binary Cen X-3 by OSO-7. See Abstr. 142.087.

Photoelectric spectrum scans of HD 153919 = 2U1700-37. See Abstr. 142.088.

On the distance to Cygnus X-1 (HDE 226868). See Abstr. 142.112.

Optical observations of HD 77581 and a model for the system HD 77581 - 2U 0900 - 40. See Abstr. 142.115.

A new measurement of the Hercules X-1 X-ray pulse profile. See Abstr. 142.116.

Limits on rapid X-ray pulsing in X-ray binaries. See Abstr. 142.117.

122 Physical Variables, Flare Stars, Pulsation Theory

122.001 Relativistic terms in nonlinear pulsation theory.
C. G. Davis.
Astrophys. Journ., Vol. 187, 175 - 177 (1974).

The relativistic terms of radiation-transfer hydrodynamics have been included in the nonlinear equations of pulsation theory. Included are the velocity-dependent terms in the equations to account for the interaction of the material flow with the radiation field. The effect of these relativistic terms is studied in a model of W Virginis.

122.002 *UBV* photometry of EU Tau.
N. B. Sanwal, M. Parthasarathy.
Astron. Astrophys., Suppl. Ser., Vol. 13, 91 - 97 (1974).

UBV observations of EU Tau are presented, and it is confirmed that EU Tau is a classical cepheid variable with a period of 2^d10250.

122.003 *UBV* observations of Mira variables near maximum light. D. Koester.
Astron. Astrophys., Suppl. Ser., Vol. 13, 133 - 141 (1974).

UBV observations are presented for 14 Mira variables near maximum light and further observations of about 15 variables made at the European Southern Observatory in Chile.

122.004 Emission-line variability in quiescent flare stars.
B. W. Bopp.
Monthly Notices Roy. Astron. Soc., Vol. 166, 79 - 87 (1974).

Spectroscopic monitoring of five flare stars (UV Ceti, YZ CMi, BY Dra, YY Gem and BD +48°1958A) reveals that the stars exhibit marked decreases in the Balmer and Ca II emission line strengths over intervals of one day or less. The variations can be explained by a non-uniform, patchy distribution of emitting regions on a rotating star.

122.005 Metal abundance and the luminosities of cepheids.
S. C. B. Gascoigne.
Monthly Notices Roy. Astron. Soc., Vol. 166, 25P - 27P (1974).

Semi-empirical arguments are advanced to show that, for a fixed colour and period, a cepheid with $Z = 0.005$ will be less luminous than one with $Z = 0.02$ by about 0.4 mag. Considered as distance indicators cepheids are therefore rather sensitive to metal abundance. Some implications of this are discussed.

122.006 On the spatial distribution of flare stars in the Pleiades cluster. P. N. Kholopov.
Astron. Zhurn. Akad. Nauk SSSR, Vol. 51, 116 - 121 (1974). In Russian. English translation in Soviet Astron. AJ, Vol. 18, No. 1.

The density distribution of 230 flare variables brighter than 18^m5 pg and 81 variables fainter than 18^m5 pg in minimum of brightness in the Pleiades cluster is studied.

122.007 Intrinsic colours of cepheid variables. IV.
N. S. Nikolov, G. R. Ivanov.
Astron. Zhurn. Akad. Nauk SSSR, Vol. 51, 132 - 138 (1974). In Russian. English translation in Soviet Astron. AJ, Vol. 18, No. 1.

122.008 On the radius and temperature of Mira.
D. Koester.
Astron. Astrophys., Vol. 30, 391 - 394 (1974).

The combination of radial velocity and near-infrared photoelectric observations of Mira supports the explanation of the variations in terms of volume pulsation and leads to the derivation of a mean radius and a tentative temperature scale.

122.009 The statistics of the β Cephei stars. J. R. Percy.
Astron. Astrophys., Vol. 30, 465 - 466 (1974).

Information on the variability of B 1 – B 2 stars has been gathered from the literature. Of the 42 B 1 – B 2 stars within 200 pc of the sun, at least 8, or 19%, are β Cephei stars.

122.010 On the three-colour behaviour of V 1057 Cyg after its outburst. F. Gieseking.
Astron. Astrophys., Vol. 31, 117 - 119 (1974).

Three-colour measurements of the peculiar variable V 1057 Cyg after its outburst, derived by means of photographic photometry, are discussed. The resultant three-colour behaviour of V 1057 Cyg is compared with the corresponding observed changes of spectral type.

122.011 Emission lines in the spectra of southern long-period variables. W. Buscombe, L. W. Simon.
Astrophys. Space Sci., Vol. 27, 157 - 165 (1974).

Quantitative measures are presented for the intensity of emission features in the spectra of 32 southern red variables.

122.012 The luminosities of the Beta Canis Majoris variables.
D. H. P. Jones, R. R. Shobbrook.
Monthly Notices Roy. Astron. Soc., Vol. 166, 649 - 661 (1974).

The luminosities of six β CMa variables are deduced from their membership of Scorpio–Centaurus; moreover the distance of α Vir has been measured interferometrically. The luminosities of the remaining variables are estimated from Hβ and Hγ after careful elimination of the effects of duplicity and systematic error. Intrinsic luminosity and colour are closely correlated. The present discussion confirms the identification of the β CMa instability strip with the 'S bend' phase of stellar evolution.

122.013 A spectroscopic study of the R CrB type star XX Camelopardalis in quiet state.
M. Ya. Orlov, M. H. Rodríguez.
Astron. Astrophys., Vol. 31, 203 - 206 (1974).

A quantitative analysis of the R CrB-type star XX Cam has been performed by the curve-of-growth method using coudé-spectrograms (dispersion 8 Å/mm, λλ 4000–5000 Å) obtained at maximum brightness.

122.014 Spectral and photoelectric observations of the flare star EQ Peg.
A. N. Kulapova, N. I. Shakhovskaya.
Izv. Krymskoj Astrofiz. Obs., Vol. 49, 65 - 72 (1974). In Russian.

According to a photoelectric patrol of the combined light of EQ Peg (Gl 896 AB) components, it has been found that the flare activity of this star is 5–10 times less than that of the most active flare stars UV Cet, YZ CMi, EV Lac, AD Leo. The variability of Ca II emission in the spectra of the brightest component (Gl 896 A) gave some indications of the possible flare activity of this component. The existent opinion that only faintest components of UV Cet type binaries are flare stars must be abandoned.

122.015 On the nature of the variability of BY Dra.
R. E. Gershberg, N. I. Shakhovskaya.
Izv. Krymskoj Astrofiz. Obs., Vol. 49, 73 - 79 (1974). In Russian.

All observational features of BY Dra can be described

within the framework of a spectral binary model with strong UV Cet-type activity on the main component surface.

122.016 On the spectral duplicity of δ Cep C.
E. A. Vitrichenko, V. A. Marsakov, P. N. Kholopov, G. S. Tsarevskij.
Izv. Krymskoj Astrofiz. Obs., Vol. 49, 80 - 85 (1974). In Russian.

Sixty-five radial velocity measurements of δ Cep C (blue visual companion of cepheid δ Cep A) have been obtained. Evidence is given that δ Cep C is a spectral binary. The most probable periods of the system are $0\overset{d}{.}89986$ and $0\overset{d}{.}46458$. The secondary component of δ Cep C may have very low mass (near 0.1 solar mass or less).

122.017 Space velocities of nearby classical cepheids.
R. Wielen.
Astron. Astrophys., Suppl. Ser., Vol. 15, 1 - 33 (1974).

The space velocities of the 45 classical cepheids with distances $r \leq 1$ kpc are derived and discussed.

122.018 Photometric observations of δ Scuti stars.
I. HR 8006, HR 9039. J. M. Le Contel, J. C. Valtier, J. P. Sareyan, A. Baglin, G. Zribi.
Astron. Astrophys., Suppl. Ser., Vol. 15, 115 - 132 (1974).

The first results of an observational program on short period variable stars are presented. The main characteristics of the equipment, especially optimized for precise differential photoelectric photometry are described. Light curves of the δ Scuti star HR 8006 are presented. Detection has been performed successfully on HR 9039 (82 Peg) which is a new δ Scuti type star of very small amplitude in light variations. An estimated period of 0.06 day is derived from the observations. The complexity of the light curve in most δ Scuti stars is emphasized. It is proposed to attribute this complexity to non-linear atmospheric effects, and not to a mixture of harmonic modes of oscillations.

122.019 Infrared variability of V1016 Cygni. P. M. Harvey.
Astrophys. Journ., Vol. 188, 95 - 96 (1974).

The infrared flux variations of V1016 Cyg are similar to those of Mira variables and possibly suggest the presence of a late M component in V1016 Cyg.

122.020 Flare stars in the Pleiades. IV.
V. A. Ambartsumyan, L. V. Mirzoyan, E. S. Parsamyan, O. S. Chavushyan (*H. S. Chavushian*), L. K. Erastova, E. S. Kazaryan, G. B. Oganyan (*Ohanian*), I. I. Yankovich (*Jankovich*).
Astrofizika, Vol. 9, 461 - 478 (1973). In Russian. English translation in Astrophysics, Vol. 9, No. 4.

The results of photographic observations of stellar flares in the Pleiades region carried out at Byurakan mainly during the autumn 1972 and the winter 1972–1973 are given. 40 new flare stars and 66 repeated flares of known flare stars have been found. The general numeration of flare stars in the Pleiades, including also the flare stars found at other observatories during the same period, is continued. The total number of flare stars in this region including the number of stars which can be found in the future is estimated; this number is of the order of 1000.

122.021 Radio emission from R Aquarii.
P. C. Gregory, E. R. Seaquist.
Nature, Vol. 247, 532 - 534 (1974).

R Aquarii has been known as a variable star since its discovery by Harding in 1811. Here the authors report the detection of variable radio emission from this variable.

122.022 Photoelectric observations of the flare star UV Cet during the 1973, September 20 - October 5 international patrol. S. Cristaldi, M. Rodonò.
Inform. Bull. Variable Stars, (I.A.U. Commission 27), Konkoly Obs., Budapest, No. 857, 14 pp. (1974).

122.023 Photoelectric observations of four δ Scuti-type stars. R. A. Fesen.
Publ. Astron. Soc. Pacific, Vol. 85, 732 - 735 (1973).

Photoelectric photometry of four previously observed δ Scuti-type variables, with standard Strömgren filters ($uvby$), is presented.

122.024 Simultaneous optical and radio observations of the flare star EV Lacertae. J. C. Webber, K. M. Yoss, D. Deming, K. S. Yang, R. F. Green.
Publ. Astron. Soc. Pacific, Vol. 85, 739 - 741 (1973).

Simultaneous optical and radio observations of the flare star EV Lac have been carried out. The only optical flare detected was followed by a burst of radio emission which lagged about 4 minutes behind the optical event.

122.025 A distant cepheid variable in Centaurus.
K. E. Ebisch.
Publ. Astron. Soc. Pacific, Vol. 85, 746 - 750 (1973).

Observations of variable stars near $l = 295°$, $b = 0°$ are described, including a new cepheid for which the distance is estimated to be 11.8 kpc.

122.026 R Coronae Borealis. R. Germann.
Orion Schaffhausen, 32. Jahrgang, p. 80 (1974).

122.027 A possible theoretical explanation of the erratic period changes of the RR Lyrae variables in globular clusters. P. G. Laskarides.
Astrophys. Space Sci., Vol. 27, 485 - 488 (1974).

122.028 On a new characteristic of intrinsic stellar variability. F. I. Lukatskaya.
Astrometriya i Astrofizika, *Kiev*, vyp. (No.) 20, (see 003.002), p. 3 - 9 (1973). In Russian.

V, B and U, B relations for simultaneous UBV values of individual stellar flares reveal differences in the composition of radiation on the ascending and descending branches of the flares. Radiation on the descending branches is considered as a sum of ascending and additional radiation. The latter is called «residual». $U-B$ and $B-V$ colour indices of residual radiations are determined.

122.029 Period variation of the cepheid Zeta Geminorum.
H. A. Abt, S. G. Levy.
Astrophys. Journ., (*Letters*), Vol. 188, L75 - L76 (1974).

In anticipation of the coming lunar occultations of ζ Gem on 1974 March 4 and 31, the authors derive a new ephemeris based on 70 years of radial-velocity measurements. The period decreases at a steady rate of $3\overset{s}{.}1$ per year, which is 4 times the evolutionary rate for a $7 M_\odot$ cepheid in the fundamental mode. However, it is also possible that the period decrease is due to a light-time effect in a long-period (> 225 years) binary having a low-mass secondary.

122.030 On period variations of some cepheids.
V. M. Grigorevskij, V. D. Motrich.
Peremennye Zvezdy, Vol. 19, 165 - 179 (1973). In Russian.

The period variations of four I_s-type cepheids (in the sense of Efremov's (1968) classification) are discussed. It is shown that for these stars the O–C-diagram representation by a parabola is at least as likely as its representation by a broken line.

122.031 Variable stars in the globular cluster M13.
R. M. Rusev.
Peremennye Zvezdy, Vol. 19, 181 - 185 (1973). In Russian.

The results of investigations of eight variable stars in the globular cluster M13 are given. More accurate data for the elements of the light curves in the photographic B system are obtained. It is shown that V15 is a long-period variable with $P = 140^d.3$.

122.032 Determination of individual luminosities of galactic cepheids.   N. N. Yakimova.
Peremennye Zvezdy, Vol. 19, 187 - 198 (1973). In Russian.

The author suggests two indirect rules for determination of the relative individual luminosities of galactic cepheids: 1) through relative light amplitudes (A_V, A_B) and colour amplitudes (A_{B-V}); 2) through light amplitudes and maximum colour $(B-V)_{0\,max}$. Both methods have been calibrated by means of the luminosities in the system of average radii determined by Kurochkin.

122.033 Intrinsic colours of cepheid variables. V.
N. S. Nikolov, G. R. Ivanov.
Peremennye Zvezdy, Vol. 19, 207 - 225 (1973). In Russian.

On the basis of the calibrated "spectrum–$(U-B)_0$" relation (Nikolov, Ivanov, 1973) the intrinsic U–B colours and excesses E_{U-B} of about 140 cepheid variables were derived.

122.034 On the variable component of Z Andromedae.
F. I. Lukatskaya.
Peremennye Zvezdy, Vol. 19, 253 - 257 (1973). In Russian.

A large series of visual observations of Z And was examined by a statistical analysis as well as photoelectric UBV observations. Considering the type of the autocorrelative function of brightness changes, the slopes of the linear regressive V, B- and U, B-dependences and the type of displacement of V, B-dependence for rise and descent of the flare with maximum near J.D. 2438850, it is deduced that the main characteristics of light variability of Z And are caused by the later component of the system.

122.035 Statistical analysis of the light curves of fourteen Is- and In T-type variables.   F. I. Lukatskaya.
Peremennye Zvezdy, Vol. 19, 258 - 271 (1973). In Russian.

The distribution and autocorrelation functions of visual brightness are given for T Cha, R CrA, T CrA, TY CrA and XY Per. Results of such analyses for other nine variables of Is and In T types are also given. Some conclusions about statistical descriptions of light variation processes for variables of these types are made.

122.036 Analysis of period changes of two cepheids of spherical component.   M. B. Bogdanov.
Peremennye Zvezdy, Vol. 19, 295 - 301 (1973). In Russian.

Histograms, autocorrelation functions and power spectral density functions were computed for the period changes of two cepheids: κ Per and AP Her. Analyses show that the period changes of these stars contain some periodic components.

122.037 The space density of cataclysmic variable stars.
B. Warner.
Monthly Notes Astron. Soc. Southern Africa, Vol. 33, 21 - 26 (1974).

122.038 Is the variable V39 in the dwarf galaxy IC 1613 a normal cepheid with period $27^d.965$?
P. N. Kholopov.
Astron. Tsirk., No. 797, p. 5 - 6 (1973). In Russian.

122.039 Determination of cepheid intrinsic colours from the periods and amplitudes.   G. R. Ivanov.
Astron. Tsirk., No. 798, p. 3 - 6 (1973). In Russian.

122.040 New flare star SVS 1989.
A. K. Alksnis, A. S. Sharov.
Astron. Tsirk., No. 800, p. 8 (1973). In Russian.

122.041 Variation and extinction of the young star RR Tauri.   S. Rössiger, W. Wenzel.
Astron. Nachr., Vol. 295, 47 - 52 (1974). In German.

Photoelectric UBV observations of RR Tauri show three components of its brightness variations: permanent short-term fluctuations, slow variations of the "maximum brightness" and nonperiodic minima. The hypothesis that these minima are produced by clouds of circumstellar condensation products is suggested. From photometry and spectral-type determination of a number of surrounding stars the authors conclude, that in "maximum brightness" no circumstellar extinction can be detected, provided the spectral type is B9V.

122.042 The light variations of RX Andromedae.
P. Szkody.
Publ. Astron. Soc. Pacific, Vol. 86, 38 - 42 (1974).

Short time-resolution photometry of RX And was done over a complete orbital cycle to determine the light variations. No eclipses or periodicity on the order of seconds were detected. Variations of the order of 0.1 to 0.6 magnitude occurred on the order of 3 minutes. Under the assumption that free-free emission from a hot spot is responsible for most of the light in the binary system, the possible changes in the relevant parameters of temperature and electron density are computed to account for the observed variations.

122.043 Photometry of UW Arietis and V600 Herculis.
M. Jerzykiewicz.
Publ. Astron. Soc. Pacific, Vol. 86, 43 - 50 (1974).

Photoelectric observations of UW Ari and V600 Her are presented. It is shown that, contrary to what is reported in the literature, neither UW Ari nor V600 Her is a β Cephei variable. UW Ari is found to be constant in light. The light variation of V600 Her is either erratic or is due to photometric effects of binary motion in an eccentric orbit.

122.044 Technetium in UX Draconis.   W. Hackos, Jr.
Publ. Astron. Soc. Pacific, Vol. 86, 78 - 79 (1974).

Absorption lines of the unstable element technetium are present in the spectrum of the N-type semiregular variable UX Dra.

122.045 Investigations into suggestions of a long period for S Apodis; S Apodis, primary minima 1890 to 1973 August 31.   F. M. Bateson.
Southern Stars, Vol. 25, 98 - 99 (1973). – Conference abstract.

122.046 On the RR Lyrae-type variable HV Vulpeculae with anomalous Blazhko effect.   V. P. Tsesevich.
Astron. Tsirk., No. 803, p. 7 - 8 (1973). In Russian.

122.047 On the possible connection of UV Ceti-type stars with observed γ-flares.   E. A. Karitskaya.
Astron. Tsirk., No. 810, p. 3 - 4 (1974). In Russian.

122.048 Supplement to the Catalogue of RR Lyrae Stars.
V. P. Tsesevich.
Astron. Tsirk., No. 811, p. 1 - 2 (1974). In Russian.

122.049 The luminosities of population II cepheids.
E. Böhm-Vitense.
Astrophys. Journ., Vol. 188, 571 - 577 (1974).

A method is described for the distance determination of pulsating stars from measured colors and radial velocities. A period-luminosity relation for population II stars with periods between 17.3 and 28.5 days is derived from the stars W Vir,

TW Cap, and star No. 42 in M5.

122.050 On the metal abundance spectral index of RR Lyrae-type stars. I. F. Aaniya.
Byull. Abastumansk. Astrofiz. Obs., No. 45, p. 3 - 24 (1974). In Russian.

The classification of RR Lyrae type variables performed at Abastumani Astrophysical Observatory is summarized. The values of the metal abundance spectral index ΔS in the extreme phases are given for 46 variables.

122.051 Two-colour electrophotometric observations of BK Dra. I. F. Alaniya.
Byull. Abastumansk. Astrofiz. Obs., No. 45, p. 25 - 36 (1974). In Russian.

260 BV observations of BK Dra have been obtained during 9 nights. The results are discussed and presented in tables and a plot.

122.052 Investigation of the light curves of cepheids with periods from 0.9–3.0 days. O. P. Vasil'yanovskaya.
Byull. Inst. Astrofiz., *Dushanbe*, No. 60, p. 3 - 20 (1972). In Russian.

The form of the light curves of 117 cepheids belonging to different star systems were studied. The light curves of 7 cepheids from the Galaxy were expanded in the component oscillations.

122.053 Investigations of classical cepheids with periods longer than 20 days. T. G. Nikulina.
Byull. Inst. Astrofiz., *Dushanbe*, No. 60, p. 21 - 25 (1972). In Russian.

Standard curves B, B–V, U–B of cepheids in the flat component of the Galaxy for periods 21^d85-37^d62 and 38^d76-50^d16 are given. The changes of periods for 9 cepheids are examined.

122.054 On the type of variability of UY CMa. G. E. Erleksova.
Byull. Inst. Astrofiz., *Dushanbe*, No. 60, p. 31 - 33 (1972). In Russian.

The light variation of UY CMa was studied on the basis of the author's photographic observations and those published by other observers. It is shown that UY CMa is a yellow semi-regular variable (SRd). Period changes have been detected. The summary of light minima is given.

122.055 Investigation of classical cepheids with periods of 9–11 days. T. G. Nikulina.
Byull. Inst. Astrofiz., *Dushanbe*, No. 60, p. 36 - 41 (1972). In Russian.

Standard light curves, B–V, U–B of cepheids in the flat component of the Galaxy with periods 9–11 days are given. The changes of periods for 3 cepheids are examined. A comparison with cepheids in the spherical component is presented.

122.056 On four classical cepheids. S. N. Zhurkina.
Byull. Inst. Astrofiz., *Dushanbe*, No. 60, p. 42 - 43 (1972). In Russian.

The results of a photometric study of X, RS, VZ and AQ Pup are given.

122.057 On five classical cepheids. T. G. Nikulina.
Byull. Inst. Astrofiz., *Dushanbe*, No. 60, p. 44 - 45 (1972). In Russian.

The results of a photometric study of VX Cyg, RY Sco, WZ Sgr, T Mon, SV Vul are given.

122.058 Cepheid SU Cygni. Yu. V. Borisov.
Byull. Inst. Astrofiz., *Dushanbe*, No. 61, p. 7 - 11 (1972). In Russian.

The results of visual observations of SU Cyg 1965–1969 are given.

122.059 Investigation of classical cepheids with periods of 16–21 days. T. G. Nikulina.
Byull. Inst. Astrofiz., *Dushanbe*, No. 62, p. 40 - 47 (1974). In Russian.

Three standard light curves for cepheids of the flat component of the Galaxy with periods 16–21 days are constructed. Possibility is pointed out of variation of light curves in different cycles. Graphs are given of change of period for two cepheids.

122.060 The RR Lyrae stars in the Large Magellanic Cloud cluster NGC 1835. J. A. Graham, M. T. Ruiz.
Astron. Journ., Vol. 79, 363 - 367, 421 (1974).

The globular cluster NGC 1835 in the Large Magellanic Cloud is shown to be very rich in variable stars of the RR Lyrae type. Periods and mean blue magnitudes are given for 18 variables, 11 of type ab, and seven of type c. The period distribution is characteristic of an Oosterhoff Type I cluster.

122.061 The ratio of total selective absorption from RS Puppis. J. D. Fernie.
Publ. Astron. Soc. Pacific, Vol. 86, 231 (1974).

By means of a geometrically determined distance for the cepheid RS Pup, the ratio of total to selective interstellar absorption is found to be $R = A_V/E_{B-V} = 2.85 \pm 0.5$.

122.062 An infrared survey of RW Aurigae stars. I. S. Glass, M. V. Penston.
Monthly Notices Roy. Astron. Soc., Vol. 167, 237 - 249 (1974).

An infrared photometric survey of 89 RW Aur type variables in both hemispheres has been made. JHKL magnitudes and colours are listed and discussed.

122.063 X-ray emission from cataclysmic variable stars. B. Warner.
Monthly Notices Roy. Astron. Soc., Vol. 167, 47P - 50P (1974).

An estimate of the X-ray flux from the hot spot model of cataclysmic variables is in agreement with the recent detection of soft X-rays from SS Cyg. Rates of mass transfer are $\sim 10^{-8} M_\odot \text{yr}^{-1}$ in dwarf novae and $\sim 10^{-7} M_\odot \text{yr}^{-1}$ in classical novae. Some implications of these rates are discussed. Possible X-ray candidates for RW Tri and U Gem are proposed.

122.064 High-frequency stellar oscillations. X. The rapid blue variable CD –42°14462. J. E. Hesser, B. M. Lasker, P. S. Osmer.
Astrophys. Journ., Vol. 189, 315 - 318 (1974).

The spectrally peculiar degenerate blue variable CD –42°14462 is both a coherent high-frequency variable, with periods of 29.08 and 30.15 s and mean amplitudes 0.00028 mag, and an erratic low-frequency variable. Variable Hα emission and Hβ and Hγ absorption lines are observed; continuum colors place CD –42°14462 near the 12000°K blackbody line. The observations can be interpreted as g-mode oscillations on one member of a white-dwarf binary system.

122.065 Duration of dMe flares. W. E. Kunkel.
Nature, Vol. 248, 571 - 573 (1974). – Letter.

122.066 A preliminary study of DO Aquarii. I. Todoran.
Stud. Cerc. Astron., Vol. 19, 25 - 30 (1974).

From the 468 photographic observations two light curves are drawn and the new photometric elements are based on 24 observed maxima.

122.067 Quasiperiodischer Lichtwechsel einiger T Tauri-Sterne.   H. Mauder.
Mitt. Astron. Ges., No. 34, p. 125 (1973/74). — Abstr. AG.

122.068 Short-lived flare activity of the Hyades flare star H II 2411.   M. Rodonò.
Astron. Astrophys., Vol. 32, 337 - 341 (1974).

High-speed photometry of the Hyades flare star H II 2411 has shown that its flare frequency is even greater than those of many UV Cet-type stars. Moreover, high-frequency oscillations with flare-like light curve have been detected during a rather intense flare. These observations could validate Gurzadian's (1970) suggestion about the evolution of flare activity with star age from a permanent phenomenon to a sporadic one.

122.069 The hot spot in the close binary system SS Cyg.   N. F. Vojkhanskaya.
Astron. Tsirk., No. 801, p. 5 - 7 (1973). In Russian.

122.070 A search for flare stars in the Orion nebula region.   C. Roslund.
Ark. Astron., Vol. 5, 381 - 385 (1974).

A systematic search for flare type variables in the region of the Orion nebula has been carried out with the 20/26-inch Schmidt telescope of the Uppsala Southern Station at Mount Stromlo. Fifteen stars showing rapid flare-ups have been detected on plates taken in blue light during a total of nearly forty hours of effective observation time. Characteristics of the flares are given together with the positions of the stars.

122.071 The correlation between the infrared and photographic spectra of RV Tauri stars.   T. Lloyd Evans.
Monthly Notices Roy. Astron. Soc., Vol. 167, 17P - 19P (1974).

Spectral classification of four southern RV Tauri stars does not support a reported correlation between the infrared and photographic spectra.

122.072 Periodical component in the light curve of R Coronae Austrinae.   I. M. Ishchenko.
Inform. Bull. Variable Stars, (I.A.U. Commission 27), Konkoly Obs., Budapest, No. 865, 3 pp. (1974).

122.073 Photoelectric observations of EV Lac during the 1973 international patrol.
J. Arsenijević, A. Kubičela, I. Vince.
Inform. Bull. Variable Stars, (I.A.U. Commission 27), Konkoly Obs., Budapest, No. 866 (1974).

122.074 Continuous photoelectric observations of EV Lac during the 1973 international patrol.
B. N. Andersen, B. R. Pettersen.
Inform. Bull. Variable Stars, (I.A.U. Commission 27), Konkoly Obs., Budapest, No. 874, 4 pp. (1974).

122.075 UBV photoelectric photometry for flare stars in the Pleiades cluster.   B. Iriarte Erro.
Inform. Bull. Variable Stars, (I.A.U. Commission 27), Konkoly Obs., Budapest, No. 875, 2 pp. (1974).

122.076 Photoelectric observations of the flare star YZ CMi.   K. Osawa, K. Ichimura, T. Okada, K. Okida, M. Yutani, H. Koyano.
Inform. Bull. Variable Stars, (I.A.U. Commission 27), Konkoly Obs., Budapest, No. 876, 2 pp. (1974).

122.077 A probable periodicity in the light variation of the LMC supergiant HD 33579.   A. M. van Genderen.
Inform. Bull. Variable Stars, (I.A.U. Commission 27), Konkoly Obs., Budapest, No. 877, 4 pp. (1974).

Photometry of the brightest supergiant of the LMC HD 33579 in 1971, 1972, 1973 and 1974 shows evidence for a periodicity of around 90 days. The nature is probably pulsation.

122.078 Flares of UV Ceti, 1970.
W. E. Kunkel, N. Zárate.
Inform. Bull. Variable Stars, (I.A.U. Commission 27), Konkoly Obs., Budapest, No. 879, 7 pp. (1974).

122.079 Eleven red variable stars in the Cygnus cloud, VV 480–490.   W. J. Miller.
Ric. Astron., Specola Vaticana, Castel Gandolfo, Vol. 8, (No. 22), 445 - 467 (1973).

Six newly-discovered and five previously-discovered red variable stars have been studied using 13,917 observations on plates from the Vatican, Hamburg, Heidelberg, Harvard and Hale Observatories. Two of the stars are irregular red variables, and the other nine are long period variables, including five known variables whose periods have hitherto not been published. The results are summarized in three tables of data and in ten pages of fragmentary light curves in the form of ten-day means.

122.080 The light variations of HD 34626.   J. R. Percy.
Inform. Bull. Variable Stars, (I.A.U. Commission 27), Konkoly Obs., Budapest, No. 894, 3 pp. (1974).

122.081 Possible beats in two Delta Scuti stars.
S. J. Horan, J. L. Michael, M. A. Seeds.
Inform. Bull. Variable Stars, (I.A.U. Commission 27), Konkoly Obs., Budapest, No. 896 (1974).

122.082 Photoelectric observations of the flare activity of red dwarfs.   N. I. Shakhovskaya.
Inform. Bull. Variable Stars, (I.A.U. Commission 27), Konkoly Obs., Budapest, No. 897, 3 pp. (1974).

122.083 Some new flares in the Pleiades region.
G. Szécsényi-Nagy.
Inform. Bull. Variable Stars, (I.A.U. Commission 27), Konkoly Obs., Budapest, No. 898, 3 pp. (1974).

122.084 Light variations in CO Aurigae.
D. L. DuPuy, R. C. Brooks.
Observatory, Vol. 94, 71 - 73 (1974).

Analysis of UBVR photoelectric observations of CO Aur show no significant indications of periodic features. A previous classification as an irregular variable is supported.

122.085 A short-period cepheid variable in the globular cluster NGC 6752.   S. W. Lee.
Observatory, Vol. 94, 74 - 79 (1974).

To determine the period of light variation of V1, 43 photographic plates were taken in 1972 with the 26-inch Yale-Columbia refractor at Mount Stromlo Observatory, and UBV photoelectric observations were obtained during 1972 and 1973 with the 24-inch and 40-inch reflectors at the Siding Spring Observatory.

122.086 UZ Librae: a possible spotted flare star?
D. S. Evans, B. W. Bopp.
Observatory, Vol. 94, 80 (1974). — Note.

122.087 Studies in the colours and magnitudes of short period cepheids.   S. C. Joshi.
Agra Univ. Journ. Res. (Sci.), Vol. 20, (Part 3), 97 - 102 = Uttar Pradesh State Obs., Naini Tal, Repr. No. 65 (1971).

122.088 Variations in the spectrum of β Cephei.
B. A. Goldberg, G. A. H. Walker, G. J. Odgers.

Astron. Astrophys., Vol. 32, 355 - 362 (1974).

Spectrophotometric observations of the pulsating star β Cephei were made with a detection system employing Image Isocon and Silicon Diode Vidicon television cameras as detectors. The Isocon was used to monitor the lines of Si III $\lambda\lambda$ 4553 and 4568 for $\simeq$ 0.8 cycle of the pulsation (P = 4.6 hours) and the Vidicon to monitor Hα for $\simeq$ 1.0 cycle. The profiles of the Si III lines were found to vary systematically throughout the cycle, their asymmetries changing with relation to the radial velocities. The result provides the most conclusive evidence so far that the line profile variations are caused by radial pulsations of the stellar atmosphere. Non-central filling-in of the Hα absorption line by emission occurs near the phase of maximum radius. A general picture of the line profile variation is given based on both past and present observations.

122.089 **The short-term light variations of S Dor and HD 37836.** I. Appenzeller.
Astron. Astrophys., Vol. 32, 469 - 470 (1974).

New observations confirm that the two peculiar high luminosity stars S Dor and HD 37836 in the Large Magellanic Cloud show detectable light variations on time scales of a few hours. The observed variations seem to be of irregular nature. They do not support the assumption that the two stars are vibrationally unstable massive main-sequence stars.

122.090 **Flare stars in clusters and associations.**
V. A. Ambartsumyan.
Zemlya i Vselennaya, 1974, No. 3, p. 14 - 20. In Russian.

122.091 **Photometric and spectrophotometric investigations of RU Camelopardalis.**
S. Krawczyk, J. Krełowski.
Stud. Soc. Sci. Torunensis, Sectio F (Astron.), Vol. 5, 69 - 88 = Bull. Astron. Obs. Toruń, No. 50/IV (1973).

Photographic and photovisual photometry of RU Cam has been made by means of intercomparisons with the NPS. Photographic and photovisual light curves with amplitudes of about $0^{m}_{.}2$ have been obtained. The brightness values m_{pg} and m_{pv} of about 50 stars in the vicinity of RU Cam have been determined and transformed to the B, V system. A two-parameter spectral classification has been made. These data were used to estimate the interstellar extinction. Central depths of characteristic bands and of some lines have been measured. A discussion of results and a comparison with other carbon stars have been made.

122.092 **Photoelectric monitoring observations of flare stars. IX. AD Leo and UV Cet (1973).**
K. Ichimura, Y. Shimizu, K. Okida.
Tokyo Astron. Bull., Second Ser., No. 230, p. 2649 - 2653 (1974).

Photoelectric monitoring observations of flare stars AD Leo and UV Cet were carried out at the Okayama Station of the Tokyo Astronomical Observatory during the periods of international cooperative observations in 1973.

122.093 **New observations of some RR Lyrae variables in the globular cluster M3 (NGC 5272).** C. Cacciari.
Astron. Astrophys., Suppl. Ser., Vol. 16, 63 - 69 (1974).

Periods and light curves of 13 RR Lyrae variables in M3 are examined using all the available observations and a set of new observations taken at the Asiago Observatory in the years 1955 and 1956. For 3 variables alternative periods are given.

122.094 **Photometric study of the pulsating variable star EH Lib.** A. Terzan, B. Rutily.
Astron. Astrophys., Suppl. Ser., Vol. 16, 155 - 161 (1974). In French.

New photoelectric observations in U, B, V of the pulsating variable star EH Lib were carried out with the 1 m (f/15) photometric telescope of ESO. They enabled the authors 1) to derive eight new maxima, 2) to calculate new values of epoch and period, 3) to set up a new $U-B/B-V$ diagram.

122.095 **The instability strip of population II cepheids.**
S. Demers, W. E. Harris.
Astron. Journ., Vol. 79, 627 - 630 (1974).

The position of the instability strip of population II cepheids in the H-R diagram is established. The results are based on the absolute magnitudes and intrinsic colors of 25 population II cepheids in globular clusters and 11 field variables. Compared with theoretical models, the position of the blue edge of the strip indicates that the helium content of population II cepheids is $Y \simeq 0.3$.

122.096 **Photoelectric observations of SS Cygni.**
M. Bretz, L. V. Mirzoyan, V. S. Oskanyan.
Astrofizika, Vol. 10, 39 - 52 (1974). In Russian. – English translation in Astrophysics, Vol. 10, No. 1.

UBV photometry was obtained on the decreasing branch of nova-like outburst of SS Cyg near minimum. The colours were measured at different parts of the light curve and during four rapid "flares". It is shown that these rapid flares are superimposed on flares of comparatively large amplitude and duration, which are of mean scale between the cyclical outbursts and the rapid flares. Radial velocities of SS Cyg measured on two spectrograms agreed with the radial velocity curve when the phase-function of Walker and Chincarini was used.

122.097 **Three-color photometry of the flare star EV Lacertae.** T. R. Flesch, J. P. Oliver.
Astrophys. Journ., (Letters), Vol. 189, L127 - L129 (1974).

Four flares have been observed during simultaneous monitoring in the u, b and r wavelength regions. One event shows substantial dips in the r below the quiescent level of the star; this same event is markedly deficient in u light. It is possible, but not likely, that these dips result from inverse Compton scattering.

122.098 **Polarimetric and photometric investigation of variable stars. II.**
V. A. Dombrovskij, T. A. Polyakova.
Trudy Astron. Obs., Leningrad, Vol. 30 (= Uchenye Zapiski Leningr. Un-ta, No. 373 = Seriya Matem. Nauk, vyp. (No.) 50), p. 89 - 103 (1974). In Russian.

The results of polarimetric and photometric observations of μ Cep, V CVn, RY Tau during 1971 and of AE Cap, R Gem, VY CMa during several years are given. There are also presented the results of test observations of the high-luminosity red variables: R Aqr, RX Boo, RS Cnc, T Cnc, T Cep, o Cet, T Dra, RY Dra, X Her, U Her, T Lyr, α Ori, Z Psc. The variations of polarization with change of brightness and wavelength are discussed briefly.

122.099 **The variable polarization of radiation of HR Del, Z And and Z Cam.**
E. T. Belokon', O. S. Shulov.
Trudy Astron. Obs., Leningrad, Vol. 30 (= Uchenye Zapiski Leningr. Un-ta, No. 373 = Seriya Matem. Nauk, vyp. (No.) 50), p. 103 - 110 (1974). In Russian.

Polarimetric and photometric observations of HR Del, Z And and Z Cam are reported. The three stars demonstrate variable polarization.

122.100 **A note on the determination of the absolute magnitude of RR Lyrae variables.**
S. V. M. Clube, D. H. P. Jones.
Astron. Astrophys., Vol. 33, 153 - 155 (1974).

The recent determination of the absolute magnitude of RR Lyrae variables by Heck is likely to be seriously influenced by an error in the Rigal-Jung maximum likelihood technique for deriving a statistical parallax. The nature of this error is discussed especially in relation to the procedure used by Clube and Jones.

122.101 **Ultraviolet photometry from the Orbiting Astronomical Observatory. XV. The strongly magnetic variable HD 215441.** D. S. Leckrone.
Astrophys. Journ., Vol. 190, 319 - 326 (1974).

The Wisconsin Experiment Package on OAO-2 was used to obtain light curves of the strongly magnetic, variable Ap star HD 215441 over the wavelength range 1550–4250 Å. The observations of HD 215441 demonstrate that an Ap star need not be a striking spectrum variable at blue-visible wavelengths for its photometric variations to be controlled by opacity variations in the ultraviolet.

122.102 **The effects of the 35-day X-ray cycle on the light curve of HZ Herculis.** S. A. Grandi, P. M. N. O. Hintzen, E. B. Jensen, A. E. Rydgren, J. S. Scott, P. M. Stickney, J. A. J. Whelan, S. P. Worden.
Astrophys. Journ., Vol. 190, 365 - 373 (1974).

The authors report extensive photoelectric UBV photometry of the variable star HZ Her, originally undertaken to discover possible correlations between variations in its 1.7-day light curve and the 35-day cycle of the X-ray source Her X-1. Correlations recently reported by other groups are confirmed. These, as well as other features observed, are provisionally analyzed using a model consisting of a primary star filling its Roche lobe and being illuminated by X-rays.

122.103 **Are BL Lac-type objects nearby black holes?** S. L. Shapiro, J. L. Elliot.
Nature, Vol. 250, 111 - 113 (1974).

Isolated black holes accreting interstellar gas can account for the salient properties of the BL Lac-type objects. The observed frequency spectra in the visible and infrared, the rapid variations in intensity and polarisation, and the absence of discrete features in the optical spectra are all consequences of the adopted black hole accretion model.

122.104 **Large flare on the red dwarf star UV Ceti.** B. Lovell, L. N. Mavridis, M. E. Contadakis.
Nature, Vol. 250, 124 - 125 (1974).

On October 11, 1972 the authors were able to obtain simultaneous radio and optical recordings of an unusually large flare on the star UV Ceti (L726–8AB), during which the apparent magnitude of the star increased by more than 4.55 mag.

122.105 **Model for a flare on the star UV Ceti.** F. D. Kahn.
Nature, Vol. 250, 125 - 127 (1974).

On October 11, 1972 a large flare occurred on the star UV Ceti. It was observed simultaneously at photoelectric frequencies and at the radio frequency of 408 MHz. These observations provide enough information to describe a reasonably good model of the outer layers of the star, and of the overall mechanics of the flare.

122.106 **Variable stars and stellar evolution.** R. Kippenhahn.
Stars and the Milky Way system, Proc. 1972, (see 012.018), p. 1 - 16 (1974). – Invited lecture.

122.107 **On the stability of the light curves of galactic cepheids.** G. Asteriadis, L. N. Mavridis, A. Tsioumis.
Stars and the Milky Way system, Proc. 1972, (see 012.018), p. 17 - 30 (1974).

122.108 **On the pulsation amplitude of cepheid variables.** N. Nikolov, T. Tsvetkov.
Stars and the Milky Way system, Proc. 1972, (see 012.018), p. 31 - 35 (1974).

122.109 **Non-stationary stars.** V. S. Shevchenko.
Astronomical Institute of the Academy of Sciences of the Uzbek SSR – 100 years, (see 003.011), p. 113 - 121 (1974). In Russian.

122.110 **Some new data on the rate of stellar flaring in the Coalsack.** D. J. MacConnell.
Bull. American Astron. Soc., Vol. 6, 211 (1974). – Abstr. AAS.

122.111 **Theoretical aspects of mixed-mode variable stars.** R. F. Stellingwerf.
Bull. American Astron. Soc., Vol. 6, 212 (1974). – Abstr. AAS.

122.112 **Ultraviolet behavior of the cepheid β Doradus: Evidence for shock waves.** J. L. Hutchinson.
Bull. American Astron. Soc., Vol. 6, 221- 222 (1974). Abstr. AAS.

122.113 **The reality of a half-cycle phase shift in 1 Mon.** K. C. Leung.
Bull. American Astron. Soc., Vol. 6, 222 - 223 (1974). Abstr. AAS.

122.114 **Search of OSO-3 data for X-ray emission from stellar flares.** V. Tsikoudi, H. Hudson.
Bull. American Astron. Soc., Vol. 6, 264 (1974). – Abstr. AAS.

122.115 **Old disk flare stars.** G. J. Veeder.
Astron. Journ., Vol. 79, 702 - 704 (1974).

At least ten of the known UV Ceti-type flare stars in the solar neighborhood have old disk space motions. Five others have large space motions and are probably also members of the old disk population. The existence of flare stars with old disk space motions and also the existence of some flare stars below the main sequence complicates the picture of these stars as a pre-main-sequence evolutionary stage.

122.116 **Some new data on the rate of stellar flaring in the Coalsack.** D. J. MacConnell, M. S. Dixon.
Astron. Journ., Vol. 79, 705 - 707, 765 (1974).

The authors report the discovery of three new flare stars in the Coalsack region during 23.3 h of monitoring using the ultraviolet multiple-image search method. These observations do not support the high rate of flaring reported by Sanduleak and by Andrews.

122.117 **Variable and peculiar stars in the Magellanic Clouds.** M. W. Feast.
Proc. ESO/SRC/CERN conference, (see 012.021), p. 169 - 176 (1974).

122.118 **The variable cepheid stars and the latest results of observations of R Muscae.** J. R. Garcia, G. Galassi.
Rev. Astron., Vol. 45, Nos. 185 - 186, p. 35 - 46 (1973). In Spanish.

122.119 **Atmospheric properties of short-period cepheids and the nature of the beat cepheids.** E. G. Schmidt.
Monthly Notices Roy. Astron. Soc., Vol. 167, 613 - 620 (1974).

Atmospheric parameters are obtained for three short-period cepheids (TU Cas, SZ Tau and α UMi) from high dispersion spectra. Additionally, for these three stars and three long-period cepheids (RX Aur, X Cyg and T Mon) radial velocities are measured for metal lines and for Hα. It is found that the beat cepheid TU Cas does not resemble the other stars as closely as expected indicating that it is not a classical cepheid as had been thought. This resolves the problem of the masses of beat cepheids. It is also concluded that differential velocities in the atmospheres of cepheids do not contribute greatly to the observed microturbulence.

122.120 **The spectrum of BC Draconis.** N. J. Irvine.
Inform. Bull. Variable Stars, (I.A. U. Commission 27), Konkoly Obs., Budapest, No. 858 (1974).

122.121 **On the light variations of CI Cyg in 1973.**
T. S. Belyakina.
Inform. Bull. Variable Stars, (I. A. U. Commission 27), Konkoly Obs., Budapest, No. 863 (1974).

122.122 **Flares on AD Leonis and YZ Canis Minoris.**
R. C. Kapoor, S. D. Sinvhal.
Inform. Bull. Variable Stars, (I. A. U. Commission 27), Konkoly Obs., Budapest, No. 901, 2 pp. (1974).

122.123 **New periods for ST Pic and XZ Cet.** J. F. Dean.
Inform. Bull. Variable Stars, (I. A. U. Commission 27), Konkoly Obs., Budapest, No. 902 (1974).

122.124 **The secondary period of AE UMa.** B. Szeidl.
Inform. Bull. Variable Stars, (I. A. U. Commission 27), Konkoly Obs., Budapest, No. 903, 3 pp. (1974).

122.125 **Pulsating stars.** J. P. Cox.
Rep. Progr. Phys., Vol. 37, 356 - 698 (1974).

In this review the main emphasis is on purely radial pulsations of spherically symmetric, non-rotating, non-magnetic stars. After a short review of the relevant stellar time scales, the main observational data on pulsating stars are summarized, and some elements of the basic theory of pulsating stars presented. Considerable emphasis is given to the linear theory of radial pulsations. Nonlinear pulsation calculations are reviewed. An attempt is made to assess present understanding of the causes and nature of pulsations in the various known types of pulsating stars and some important unsolved problems are summarized and discussed.

122.126 **On the beat phenomenon in the Beta Cephei stars.**
R. G. Deupree.
Astrophys. Journ., Vol. 190, 631 - 636 (1974).

The hypothesis of the degeneracy between a radial and a nonradial mode as an explanation of the beat phenomenon in the β Cephei stars is examined by means of a linear, adiabatic pulsation analysis using realistic stellar models. The results are compared with observational equivalents, and the agreement for a number of the β Cephei stars showing this effect gives support to this hypothesis. On the assumption that this is the correct explanation of the beats, some conclusions may be drawn regarding the properties of the β Cephei stars.

122.127 **Short-duration radio flares of UV Ceti stars.**
S. R. Spangler, S. D. Shawhan, J. M. Rankin.
Astrophys. Journal, (Letters), Vol. 190, L129 - L131 (1974).

The UV Ceti-type stars YZ Canis Minoris, AD Leonis, Wolf 424, and V371 Orionis were monitored with the Arecibo radio telescope at meter wavelengths. A total of 26 flares were recorded on YZ CMi, AD Leo, and Wolf 424. The observed flares had shorter durations and smaller amplitudes than those previously reported in the literature.

122.128 **Photoelectric observations of some flare stars.**
J. Arsenijević, A. Kubičela, T. Angelov.
Bull. Obs. Astron. Belgrade, Vol. 29, F. 1, No. 125, p. 53 - 58 (1971/72).

122.129 **Photoelectric flare stars observations in 1971.**
J. Arsenijević, A. Kubičela.
Bull. Obs. Astron. Belgrade, Vol. 29, F. 1, No. 125, p. 58 - 60 (1971/72).

122.130 **RR Lyrae variables in the Magellanic Clouds.**
A. D. Thackeray.
Monthly Notes Astron. Soc. Southern Africa, Vol. 33, 66 - 70 (1974).

122.131 **Detection of lithium in the dMle star Gliese 182.**
B. W. Bopp.
Publ. Astron. Soc. Pacific, Vol. 86, 281 - 283 (1974).

The results of a search for the $\lambda 6707$ lithium line in several flare and dMe stars are reported. Of 14 stars surveyed, 13 have no detectable lithium. However one star, Gliese 182, has the line present with EW ~ 200 mÅ.

122.132 **LR Sagittarii.** N. V. Vidal.
Publ. Astron. Soc. Pacific, Vol. 86, 308 - 310 (1974).

The identification and recent brightening of LR Sgr are discussed. A spectrum taken in its bright phase shows a faint Hα line in emission.

122.133 **VY Scl and the Z Cam phenomenon.**
B. Warner, G. W. van Citters.
Observatory, Vol. 94, 116 - 120 (1974).

122.134 **Mira variables in four metal-rich globular clusters.**
P. J. Andrews, M. W. Feast, T. Lloyd Evans, A. D. Thackeray, J. W. Menzies.
Observatory, Vol. 94, 133 - 135 (1974).

Periods are given for one Mira variable in each of NGC 5927 (310^d), NGC 6553 (265^d) and NGC 6637 (195^d). Radial velocities are given for one star in each of NGC 5927 and NGC 6388. The results appear to confirm a dependence of period on metal content for Miras.

122.135 **Photoelectric observations of δ Del in 1971 - 1972.**
B. A. Murnikov.
Peremennye Zvezdy, Prilozhenie, Vol. 1, 407 - 410 (1973). In Russian.

122.136 **V437 Cassiopeiae.** G. F. Tokareva.
Peremennye Zvezdy, Prilozhenie, Vol. 1, 453 - 454 (1973). In Russian.

122.137 **Y Vulpeculae.** E. V. Kazarovets, S. Yu. Shugarov.
Peremennye Zvezdy, Prilozhenie, Vol. 1, 455 - 457 (1973). In Russian.

122.138 **BK Cassiopeiae.**
V. P. Goranskij, E. V. Kazarovets, S. Yu. Shugarov.
Peremennye Zvezdy, Prilozhenie, Vol. 1, 477 - 480 (1973). In Russian.

122.139 **Observations of three variable stars.**
O. A. Chekanikhina.
Peremennye Zvezdy, Prilozhenie, Vol. 1, 481 - 486 (1973). In Russian.

122.140 **Is DZ And an R CrB type variable?**
M. Ya. Orlov, M. H. Rodriguez.
Astron. Tsirk., No. 813, p. 1 - 2 (1974). In Russian.

122.141 **Mira variables in globular clusters.**
B. V. Kukarkin.
Astron. Tsirk., No. 813, p. 4 - 5 (1974). In Russian.

122.142 **On the RR Lyrae-type star DP Sagittae.**
V. P. Tsesevich.
Astron. Tsirk., No. 813, p. 5 - 6 (1974). In Russian.

122.143 On the fading of UV Cassiopeiae. I. Daube.
Astron. Tsirk., No. 813, p. 6 - 7 (1974). In Russian.

122.144 On the variable star SVS 864 = CSV 1487.
V. P. Tsesevich.
Astron. Tsirk., No. 820, p. 7 - 8 (1974). In Russian.

Cosmic gamma-ray bursts from directed stellar flares.   See Abstr. 061.035.

The CNO abundances of the CH stars in Omega Centauri.   See Abstr. 064.005.

Emission Balmer decrement and electron density of stellar chromospheres.   See Abstr. 064.047.

Carbon and eruptive stars: surface enrichment of lithium, carbon, nitrogen, and ^{13}C by deep mixing.
See Abstr. 065.047.

A comparison of variable and nonvariable stars in the cepheid strip.   See Abstr. 065.061.

The Doppler splitting of spectral lines in pulsating stars.   See Abstr. 065.086.

An excitation mechanism for pulsations in Beta Cephei stars.   See Abstr. 065.093.

Models of asymptotic-giant-branch stars.
See Abstr. 065.127.

The radial velocity of δ Delphini.
See Abstr. 112.016.

The Coalsack. II. Photometry of suspected flare stars.   See Abstr. 113.023.

Photographic photometry of carbon stars in the field centered at $l = 204°.4$, $b = -1°.3$.   See Abstr. 113.028.

Photometric investigations of carbon stars in the field centered at $l = 173°.5$, $b = -1°.5$.   See Abstr. 113.029.

Variability of infrared carbon stars.
See Abstr. 113.030.

Photovisual photometry of semiregular and irregular carbon stars.   See Abstr. 113.032.

Multicolor photometry of metallic-line stars.I. ν^1 Draconis and ν^2 Draconis.   See Abstr. 113.043.

On a possible three-year cycle of η Carinae.
See Abstr. 114.011.

Infrared fluxes, spectral types, and temperatures for very cool stars.   See Abstr. 114.075.

Light variations in seven Wolf-Rayet stars.
See Abstr. 114.093.

Ultraviolet stellar astronomy.
See Abstr. 114.108.

Time variations in the OH microwave and infrared emission from late-type stars.   See Abstr. 114.132.

An angular diameter and effective temperature for V Cancri from an occultation observation.
See Abstr. 115.009.

Possible detection of very soft X-rays from SS Cygni.   See Abstr. 121.004.

The intrinsic polarization in the light of μ Cephei.
See Abstr. 131.085.

On the light variability of the planetary nebula Hb $6 = 7 + 1°1$, identified with the irregular variable AS Sagittarii.   See Abstr. 133.009.

Untersuchungen über die Sternassoziation Cyg T 1.
See Abstr. 152.003.

On the variability of dwarf K- and M-type stars in the Pleiades and Hyades.   See Abstr. 153.030.

Structure of the subsystem of long-period cepheids in the galactovertical direction.   See Abstr. 155.015.

Ages of δ Cephei stars and the density of gravitating masses near the sun.   See Abstr. 155.016.

Errata

122.901 Errata: 'Étude photométrique de BL Lacertae. II. Variations d'éclat de mars 1969 à janvier 1971'
[Astron. Astrophys., Vol. 24, 357 - 368 (1973)].
C. Bertaud, G. Wlérick, P. Véron, B. Dumortier, J. Bigay, G. Paturel, M. Duruy, P. de Saevsky, A. Durand.
Astron. Astrophys., Vol. 31, 475 (1974).

123 Variable Stars: Lists of Observations, Individual Observations

123.001 **Variable stars in the Large Magellanic Cloud–II.** E. M. Lindsay.
Monthly Notices Roy. Astron. Soc., Vol. 166, 703 - 710 (1974).
Data are given for nine new variables in the Large Magellanic Cloud and for some known Harvard variables.

123.002 **R Coronae Borealis.** R. Lukas.
SuW, Vol. 13, 105 (1974).

123.003 **1000 Tage Z UMa.** E. Heiser.
SuW, Vol. 13, 138 - 139 (1974).

123.004 **Light curve of HR 6773.** G. F. G. Knipe.
Monthly Notes Astron. Soc. Southern Africa, Vol. 33, 14 (1974).

123.005 **R Coronae Borealis plotseling zwakker.** G. Comello, H. Betlem.
Zenit, Vol. 1, No. 4, p. 19 (1974).

123.006 **Chi Cygni: een der meest waargenomen variabelen.** H. Feijth.
Zenit, Vol. 1, No. 4, p. 32 - 34 (1974).

123.007 **R Trianguli: gedurende gehele lichtwisseling te volgen.** H. Feijth.
Zenit, Vol. 1, No. 5, p. 6 - 8 (1974).

123.008 **The peculiar variable star V361 Per.** L. Szabados.
Inform. Bull. Variable Stars, (I.A.U. Commission 27), Konkoly Obs., Budapest, No. 868, 4 pp. (1974).

123.009 **On the period of U Cephei.** G. A. Bakos, J. Tremko.
Inform. Bull. Variable Stars, (I.A.U. Commission 27), Konkoly Obs., Budapest, No. 869 (1974).

123.010 **1973 photometry of V725 Sgr.** S. Demers, B. F. Madore.
Inform. Bull. Variable Stars, (I.A.U. Commission 27), Konkoly Obs., Budapest, No. 870, 2 pp. (1974).

123.011 **Observations and preliminary periods for variable stars in NGC 1261.** A. Wehlau, L. Davis, S. Demers.
Inform. Bull. Variable Stars, (I.A.U. Commission 27), Konkoly Obs., Budapest, No. 871, 4 pp. (1974).

123.012 **DK Sge.** K. Häussler.
Inform. Bull. Variable Stars, (I.A.U. Commission 27), Konkoly Obs., Budapest, No. 882 (1974).

123.013 **GL Tauri.** G. Romano.
Inform. Bull. Variable Stars, (I.A.U. Commission 27), Konkoly Obs., Budapest, No. 886 (1974).

123.014 **Observations of 2 CSV-stars on sky patrol plates.** H. Busch.
Inform. Bull. Variable Stars, (I.A.U. Commission 27), Konkoly Obs., Budapest, No. 887 (1974).

123.015 **Observations of 6 CSV-stars on sky patrol plates.** K. Häussler.
Inform. Bull. Variable Stars, (I.A.U. Commission 27), Konkoly Obs., Budapest, No. 887, 3 pp. (1974).

123.016 **UZ Octantis.** M. E. Castore de Sistero, R. F. Sistero.
Inform. Bull. Variable Stars, (I.A.U. Commission 27), Konkoly Obs., Budapest, No. 888, 2 pp. (1974).

123.017 **Notes on the Be-star HR 1508 and the eclipsing binaries AA Ceti and BT Eri.** B. Grønbech.
Inform. Bull. Variable Stars, (I.A.U. Commission 27), Konkoly Obs., Budapest, No. 890, 2 pp. (1974).

123.018 **Osservazione visuale di stelle variabili.** C. Romoli, C. Pezzarossa.
Coelum, Vol. 42, 84 - 88 (1974).

123.019 **Observations of variable stars July - December 1973. Report No. 25.** L. Plaut, H. Feijth.
Nederlandse Vereniging voor Weer- en Sterrenkunde. Kapteyn Astron. Lab. Groningen – Netherlands. 20 pp. (1974).
This report gives 3043 visual observations of 153 Mira-type stars, semi-regular type stars, U Gem and Z Cam type stars, and miscellaneous type stars. A few light curves of Mira variables are added.

123.020 **Long series of observations on GL, KP, and V343 Cyg.** D. Hoffleit.
Inform. Bull. Variable Stars, (I.A.U. Commission 27), Konkoly Obs., Budapest, No. 893, 2 pp. (1974).

123.021 **Period and lightcurve of V1 in the globular cluster NGC 6752.** A. J. Wesselink.
Inform. Bull. Variable Stars, (I.A.U. Commission 27), Konkoly Obs., Budapest, No. 895, 3 pp. (1974).

123.022 **Two semi-regular variables: T Psc, 1926–60 and V UMa, 1934–72.** J. E. Isles.
Journ. British Astron. Ass., Vol. 84, 287 - 296 (1974). – Report of Variable Star Section of BAA.

123.023 **Posmatranje promenljive zvezde δ Cephei (Observations of the variable star δ Cephei).** B. Đorđević.
Vasiona, Vol. 22, 20 - 22 (1974).

123.024 **On the elements of some variable stars in Cygnus cloud.** G. Pinto, G. Romano.
Mem. Soc. Astron. Italiana, Nuova Ser., Vol. 44, 649 - 664 (1973/74).
Photographic observations of 14 variable stars in the Cygnus cloud for which no reliable elements were available have been re-elaborated. The results of the analysis are presented.

123.025 **R Coronae Borealis.** P. W. Hornby, Shigehisa, Hosokawa, Hirasawa, S. Sakuma.
IAU Circ., No. 2617 (1974).

123.026 **R Coronae Borealis.** T. Kuwabara, S. Sakuma, K. Gomi, U. Hopp.
IAU Circ., No. 2624 (1974).

123.027 **R Coronae Borealis.** U. Hopp, J. Bortle, R. Lukas.
IAU Circ., No. 2639 (1974).

123.028 **R Coronae Borealis.** M. Seeds, J. Michael.
IAU Circ., No. 2642 (1974).

123.029 **R Coronae Borealis.**
M. Seeds, S. Horan, J. Michael.
IAU Circ., No. 2652 (1974).

123.030 **R Coronae Borealis.**   U. Hopp.
IAU Circ., No. 2654 (1974).

123.031 **R Coronae Borealis.**   J. E. Bortle, U. Hopp, R. Lukas, E. Mayer.
IAU Circ., No. 2660 (1974).

123.032 **AL Comae Berenices.**
C. B. Ford, C. Scovil, E. Mayer, T. A. Cragg.
IAU Circ., No. 2666 (1974).

123.033 **R Coronae Borealis.**   R. Lukas.
IAU Circ., No. 2666 (1974).

123.034 **R Coronae Borealis.**   U. Hopp, R. Lukas.
IAU Circ., No. 2672 (1974).

123.035 **R Coronae Borealis.**   D. W. Dawson, E. F. Tedesco.
IAU Circ., No. 2678 (1974).

123.036 **Observations of variable stars.**
E. Mayer, C. Ford, M. Marcario.
IAU Circ., No. 2680 (1974).

123.037 **R Coronae Borealis.**
R. G. Hodgson, Hosokawa, T. Urata.
British Astron. Ass. Circ. No. 550 (1974).

123.038 **Lichtwechsel von T UMa.**   H. Schubert.
BAV Rundbrief, 23. Jahrgang, p. 5 - 6 (1974).

123.039 **Five more variable stars in NGC 6402.**
A. Wehlau, L. Davis, N. Potts.
Inform. Bull. Variable Stars, (I. A. U. Commission 27), Konkoly Obs., Budapest, No. 859 (1974).

123.040 **Photoelectric observations of 44 Tau.**
A. Terzan, V. Oskanian.
Inform. Bull. Variable Stars, (I. A. U. Commission 27), Konkoly Obs., Budapest, No. 899, 4 pp. (1974).

123.041 **Two variable stars in globular cluster M79.**
Y. Chu.
Acta Astron. Sinica, Vol. 15, 98 - 100 (1974). In Chinese.
  During the period of 1957–1958, 9 plates of the globular cluster M79 (NGC 1904) were taken with the 40-cm refractor of the Zô-Sè Section of Shanghai Observatory, and two new variable stars were discovered in this cluster.

123.042 **Neue Typenbezeichnungen von 9 irregulären Veränderlichen.**   L. Meinunger.
MVS, *Sonneberg*, Vol. 6, 133 - 134 (1974).
  A revision is given for 9 variable stars which are not correctly classified in the GCVS (Moscow 1971). KT Cygni is a very interesting hot R CrB-like star.

123.043 **Bearbeitung von 50 Veränderlichen am Südhimmel (Feld η Arae, Teil III).**   H. Geßner.
MVS, *Sonneberg*, Vol. 6, 138 - 139 (1974).

123.044 **Instantane Elemente des Mira-Sterns W Lyrae.**
M. Heß.
MVS, *Sonneberg*, Vol. 6, 139 (1974).

123.045 **Visuelle Beobachtungen von Mirasternen.**
P. Ahnert.
MVS, *Sonneberg*, Vol. 6, 140 (1974).

123.046 **Instantane Elemente des Mira-Sterns RS Herculis.**
M. Werner.
MVS, *Sonneberg*, Vol. 6, 140 (1974).

123.047 **DN Aquarii.**   H. Geßner.
MVS, *Sonneberg*, Vol. 6, 141 - 142 (1974).

123.048 **Verbesserte Elemente des Mira-Sterns T Herculis.**
H.-J. Jerominek.
MVS, *Sonneberg*, Vol. 6, 154 (1974).

123.049 **Investigation of seven variable stars in Taurus and Auriga. I.**   Yu. A. Fadeev.
Peremennye Zvezdy, Prilozhenie, Vol. 1, 411 - 423 (1973). In Russian.

123.050 **Investigation of variable stars in Taurus and Auriga. II.**   Yu. A. Fadeev.
Peremennye Zvezdy, Prilozhenie, Vol. 1, 424 - 432 (1973). In Russian.

123.051 **Investigation of 17 variable stars.**   N. E. Kurochkin.
Peremennye Zvezdy, Prilozhenie, Vol. 1, 439 - 451 (1973). In Russian.

123.052 **List of variable stars.**
Peremennye Zvezdy, Prilozhenie, Vol. 1, 487 - 490 (1973). In Russian.

123.053 **Variable star notes.**   J. A. Mattei.
Journ. Roy. Astron. Soc. Canada, Vol. 68, 48 - 52, 112 - 116, 169 - 172 (1974).

Fr. Walter Miller's variable star researches.
See Abstr. 005.019.

Photographic photometry in the field of NGC 2264.
See Abstr. 153.013.

Errata

123.901 **Erratum: 'Interesting variable star'** [Astron. Tsirk., No. 793, p. 1 - 2 (1973). In Russian].
A. S. Sharov.
Astron. Tsirk., No. 799, p. 8 (1973). In Russian.

124 Novae

124.001 Contracting envelopes of novae after outbursts.
K. Nariai.
Publ. Astron. Soc. Japan, Vol. 26, 57 - 64 = Tokyo Astron. Obs. Repr. No. 447 (1974).

The evolution of a star with an envelope of a very small fraction of mass and with an insulated hard core is studied. The models explain the behavior of light curves of novae not only in the visual-photographic region but also those in the ultraviolet region recently published by Code (1972).

124.002 Modèle d'éjection de matière des novae.
I. Malakpur.
Astrophys. Space Sci., Vol. 27, 467 - 484 (1974).

The author has found that the gas of the premaximum, diffuse-enhanced and Orion systems is produced by the nova itself and is ejected by shock waves in the form of discrete layers. The principal envelope originates in the gas responsible for the premaximum system, as a result of considerable, sudden increase of the radiation pressure of $L\alpha$, a certain time after the beginning of the nova explosion. He discusses the cause of the nova explosion and concludes that there is no satisfactory theory of novae.

124.003 Infra-red photometry of some old novae.
M. W. Feast, I. S. Glass.
Monthly Notices Roy. Astron. Soc., Vol. 167, 81 - 85 (1974).

J, H, K, L photometry is given for some old novae (T CrB, RS Oph, RT Ser). Some observations are given of V 1017 Sgr, RR Pic and V603 Aql.

124.004 Gamma-ray lines from novae.
D. D. Clayton, F. Hoyle.
Astrophys. Journ., (Letters), Vol. 187, L101 - L103 (1974).

Nova explosions may be accompanied by detectable levels of radioactivity. Positron annihilation and nuclear de-excitation following decays of ^{13}N, ^{14}O, ^{15}O, and ^{22}Na lead to gamma-ray line fluxes whose measurement will clarify the model.

124.005 Novae in the Andromeda nebula in 1972.
A. S. Sharov, A. K. Alksnis.
Astron. Tsirk., No. 799, p. 3 - 4 (1973). In Russian.

124.006 Two recurrent novae: T CrB, 1948–71 and RS Oph, 1963–71. J. E. Isles.
Journ. British Astron. Ass., Vol. 84, 203 - 208 (1974). – Section report.

124.007 Novae 1967–1971. R. Lukas.
SuW, Vol. 13, 177 (1974).

124.008 The role of gravitational radiation in the evolution of dwarf novae. J. Faulkner.
IAU Symposium No. 64, (see 012.008), p. 97 (1974). – Abstract.

124.009 On the continuous outflow of matter during the post-maximum stage of a nova.
V. G. Gorbatskij, L. N. Ivanov.
Astrofizika, Vol. 10, 73 - 83 (1974). In Russian. – English translation in Astrophysics, Vol. 10, No. 1.

The effect of outburst of a nova on a satellite – red dwarf – is considered as the cause of powerful and durable outflow of matter that forms an extended envelope. It is found that the luminosity of nova Aquilae 1918 was about 10^{40} erg/sec for several months after the outburst. The considerable part of the energy was emitted within the X-ray region. The quantity of matter outflowing from the surface of the red dwarf due to heating by X-ray radiation was estimated. As it has been revealed such heating is the main cause of the observed outflow.

124.010 Ultraviolet detection of the nova variables V603 Aquilae and RR Pictoris.
J. S. Gallagher, A. V. Holm.
Astrophys. Journ., (Letters), Vol. 189, L123 - L126 (1974).

During the lifetime of the OAO-2, ultraviolet photometric observations of 6 old novae were attempted. Only V603 Aql and RR Pic were positively measured; and by comparing ultraviolet colors of the novae with those of normal stars the authors are able to deduce color temperatures of 25000° and in excess of 35000°K, respectively. They cannot find a unique luminosity from the data but are able to estimate the minimum bolometric luminosity to be about $10 L_\odot$ for both stars. The two observations of RR Pic indicated that the star varied by 0.5 mag during a time interval of 1.7 hours while the relative continuum energy distribution remained unchanged; V603 Aql was not observed to be variable.

124.011 Nova in Large Magellanic Cloud. F. M. Bateson.
Inform. Bull. Variable Stars, (I. A. U. Commission 27), Konkoly Obs., Budapest, No. 860 (1974).

The peculiar star He 2-177: a slow nova and a possible X-ray source. See Abstr. 114.047.

A simple model of fluctuations in accretion discs. See Abstr. 117.009.

The light variations of RX Andromedae. See Abstr. 122.042.

124.100 Nova DQ Herculis

DQ Herculis.
R. E. Nather, J. I. Smak, J. T. McGraw.
IAU Circ., No. 2677 (1974).

DQ Herculis. J. C. Kemp, J. B. Swedlund.
IAU Circ., No. 2679 (1974).

An oblique rotator model for DQ Herculis.
G. T. Bath, W. D. Evans, J. E. Pringle.
Monthly Notices Roy. Astron. Soc., Vol. 166, 113 - 121 (1974).

It is suggested that the old nova DQ Herculis and the known pulsing X-ray sources belong to a common class of magnetic, accreting, compact components in close semi-detached binary systems. The 71-s periodicity observed in the white dwarf component of DQ Herculis is interpreted as due to accretion hot spots on an oblique magnetic rotator.

124.101 Nova VW Hydri

Further observation of rapid oscillations in VW Hyi.
B. Warner, G. W. Van Citters, A. J. Brickhill, P. R. Hurly, A. R. Walker.
Inform. Bull. Variable Stars, (I. A. U. Commission 27), Konkoly Obs., Budapest, No. 860 (1974).

An orbital period for VW Hydri.
B. F. Marino, W. S. G. Walker.
Inform. Bull. Variable Stars, (I.A.U. Commission 27), Konkoly Obs., Budapest, No. 864, 3pp. (1974).

Observations of rapid blue variables—XIII: Rapid pulsations in VW Hyi during outburst.
B. Warner, A. J. Brickhill.
Monthly Notices Roy. Astron. Soc., Vol. 166, 673 - 681 (1974).

Photometric observations on 1972 December 25 — following an outburst of dwarf nova VW Hyi — show the presence of sinusoidal oscillations. These increased in period from 28 to 34s within 4 hr. Good agreement is found from a comparison of these pulsations (and ones observed in other dwarf novae) with recent computations of non-radial quadrupole oscillations of white dwarf stars.

124.102 Nova Herculis 1963

On the spectrum of the nitrogen explosion of nova Herculis 1963.
K. Khan, T. Mo, Z. Liu.
Acta Astron. Sinica, Vol. 15, 43 - 50 (1974). In Chinese.

Evolution of the spectrum of nova Herculis 1963 in the nebular phase.
Y. Andrillat, S. Collin-Souffrin.
Astron. Astrophys., Vol. 31, 347 - 351 (1974). In French.

The authors determine equivalent widths and relative intensities of the emission lines in the spectrum of nova Her 63 in the range 3700–8500 Å from April to September 1963.

124.103 Nova Delphini 1967

Light curve (m_{pg}) of HR Del (nova Delphini 1967).
A. Terzan, M. Bally, A. Durand.
Astron. Astrophys., Suppl. Ser., Vol. 15, 107 - 113 (1974). In French.

The light curve (m_{pg}) of HR Del is drawn for the period going from July 10th 1967 to November 5th 1972.

On the consequences of having become a nova — The case of HR Del 1967.
J. S. Gallagher III.
Bull. American Astron. Soc., Vol. 6, 211 (1974). – Abstr. AAS.

On the nova Delphini 1967.
N. N. Kiselev.
Byull. Inst. Astrofiz., *Dushanbe*, No. 60, p. 26 - 30 (1972). In Russian.

On the basis of the author's photoelectric observations and observations published in the literature the light curves V, B–V, U–B are presented. The photometric behaviour of the nova in the different stages of its evolution has been considered. The distance of the nova has been determined. The author's photoelectric observations are given.

An atlas of nova Delphini 1967.
W. Seitter.
Stars and the Milky Way system, Proc. 1972, (see 012.018), p. 39 - 41 (1974).

The variable polarization of radiation of HR Del, Z And and Z Cam.
See Abstr. 122.099.

124.104 Nova IV Cephei 1971

Observations of nova IV Cephei 1971.
G. C. L. Aikman, R. W. Hilditch, F. Younger.
Publ. Astron. Soc. Pacific, Vol. 85, 756 - 759 (1973).

Spectrographic and three-color photometric observations of nova IV Cephei 1971 are presented. A stationary circumstellar helium shell visible spectroscopically during the nova's early decline may be evidence of an earlier outburst from this star.

124.105 Nova RR Telescopii

Possible evidence for continued ejection of RR Telescopii.
M. Friedjung.
Astrophys. Space Sci., Vol. 29, L5 - L6 (1974).

Fe II and [Fe II] intensity measures enable the size of the region containing singly ionized iron to be estimated, calculations for a phase near maximum light support a continued ejection model.

Energy levels of Ca VII from observations of RR Telescopii.
A. D. Thackeray.
Monthly Notices Roy. Astron. Soc., Vol. 167, 87 - 93 (1974).

Three lines attributed to [Ca VII] have been found present in RR Tel since 1967. Their measured wavelengths lead to slightly revised energy levels of Ca VII.

124.106 Nova V1017 Sagittarii

Observations of the recent outburst of nova V1017 Sagittarii.
N. V. Vidal, A. W. Rodgers.
Publ. Astron. Soc. Pacific, Vol. 86, 26 - 30 (1974).

Photometric and spectrographic observations of the recent outburst of V1017 Sgr are reported. The B magnitude varied from the 11th to 14th magnitude during the period February to June 1973. The shape and amplitude of the light curve is similar to the light curve observed at the 1901 outburst.

124.107 Nova FH Serpentis

Ultraviolet photometry from the Orbiting Astronomical Observatory. X. Nova FH Serpentis 1970.
J. S. Gallagher III, A. D. Code.
Astrophys. Journ., Vol. 189, 303 - 314 (1974).

Results of ultraviolet filter photometry between 1430 and 4250 Å and low-resolution spectral scans in the wavelength region 2500–3600 Å obtained with the Wisconsin package on the OAO-A2 are reported for nova FH Serpentis 1970. The observations were made over an interval of 53 days commencing at about the time of maximum visible light.

125 Supernovae, Supernova Remnants

125.001 **The formation of deuterium and the light elements by spallation in supernova shocks.** S. A. Colgate. Astrophys. Journ., Vol. 187, 321 - 332 (1974).

The author first discusses the hydrodynamic properties of supernova shock waves independent of supernova mechanisms and shows how such a shock wave will become strong enough to spall helium nuclei. He then discusses the structure of the envelopes of various expected presupernova models and emphasizes those models where the density of the envelope is low. The structure of the shock transition of an energetic shock wave in a low-density envelope then becomes the central problem of the paper. The spallation occurring in such a shock front is related to various supernova models, and the resulting light-element nucleosynthesis is compared with abundances.

125.002 **Early gamma rays from supernovae.** S. A. Colgate. Astrophys. Journ., Vol. 187, 333 - 335 (1974).

The recent γ-ray pulses observed from the Vela satellites are interpreted as γ-ray pulses from type II hydrogen-spectrum supernovae in distant galaxies. The origin of the γ-rays is presumed to be the bremsstrahlung and inverse-Compton-heated photons originating from the high-ion-temperature shock precursor as it breaks through the surface of the pre-supernova star.

125.003 **Type I supernovae.** J. C. Wheeler. Astrophys. Journ., Vol. 187, 337 - 340 (1974).

Dynamical calculations of collapsing degenerate carbon cores give results which are quite similar to conditions found in Tycho's (1572) supernova, the prototypical type I event. The models and their implications are discussed.

125.004 **What kind of supernovae is responsible for observed remnants?** I. S. Shklovsky. Astron. Zhurn. Akad. Nauk SSSR, Vol. 51, 3 - 7 (1974). In Russian. English translation in Soviet Astron. AJ, Vol. 18, No. 1.

It is shown that the mass of the supernova responsible for the Cygnus Loop is less than $1.5 - 2\,M_\odot$. The same conclusion holds for the majority of old remnants of supernovae which must belong to type I. The peculiarity of the Crab nebula can be explained by abnormally small expansion velocity.

125.005 **The interaction between supernova remnants and the interstellar medium.** K. V. Bychkov. Astron. Zhurn. Akad. Nauk SSSR, Vol. 51, 21 - 25 (1974). In Russian. English translation in Soviet Astron. AJ, Vol. 18, No. 1.

Some dense, cold knots in supernova remnants, like "knots" of Cas A and filaments of the Crab nebula can be separated from their remnants. They move then independently through the interstellar medium and can cover a large distance. The stability of such formations and the possibility of their observation in radio, optical and UV rays is considered.

125.006 **Neutron stars in supernova remnants.** P. R. Amnuel, O. Kh. Gusejnov (*O. H. Guseinov*). Astron. Astrophys., Vol. 31, 37 - 40 (1974).

Data on supernova remnants is analysed in view of a search for black holes. The energetics of the remnants, the increase of magnetic flux, and data about the directions of the magnetic fields are discussed with the assumption that, in each remnant, there is a neutron star (pulsar).

125.007 **Supernova explosions in close binary systems, with an application to Centaurus X-3.** W. Sutantyo. Astron. Astrophys., Vol. 31, 339 - 342 (1974).

The effect of a spherically symmetric supernova explosion on the orbital period of a close binary system is considered. The upper and lower limit of the mass ratio before the explosion can be determined from current theories of close binary evolution. It is shown that the final period (after the explosion) as estimated from observational data, yields an order of magnitude estimate for the upper limit of the ejection velocities of the supernova shell. Application to the neutron star in the X-ray binary Cen X-3 indicates that this star is likely to have originated in a type II supernova with an ejection velocity of the order of 10^9 cm/s.

125.008 **Supernova: the result of the death spiral of a white dwarf into a red giant.** W. M. Sparks, T. P. Stecher. Astrophys. Journ., Vol. 188, 149 - 153 (1974).

The proposed model is a binary consisting of a white dwarf and a star evolving toward the red-giant branch. Conditions are given under which the revolution period of the binary and the rotation period of the red giant will reach a synchronous state and under which no stable synchronous orbit is possible. For the case of a nonstable synchronous orbit, the evolution of the decay of the orbit is given. The authors suggest that a supernova explosion will result from a collision with the core and leave a neutron stellar remnant. The relationship to binary X-ray stars is discussed.

125.009 **Line ^{57}Co gamma rays: new diagnostic of supernova structure.** D. D. Clayton. Astrophys. Journ., Vol. 188, 155 - 157 (1974).

The nuclear gamma-ray lines emitted when ^{57}Co decays to ^{57}Fe should be detectable for several years following galactic supernovae, and the time structure of their intensity relative to those emitted following the ^{56}Ni and ^{56}Co decays can reveal the structure of the exploding star. An analysis of a simplified analytic model is presented as an example.

125.010 **The Vela supernova remnant in ultraviolet light.** E. W. Miller. Publ. Astron. Soc. Pacific, Vol. 85, 764 - 766 (1973).

Photographs of the Vela supernova remnant in ultraviolet light have been combined to form a mosaic which shows for the first time the complete circular outline of the nebula. The center of the X-ray shell is found to coincide well with the optical center of the nebula; however, the position of the Vela pulsar, PSR 0833–45, is approximately $1°.5$ west of the center.

125.011 **Supernovae and neutron stars.** S. A. Colgate. IAU Symposium No. 53, (see 012.004), p. 183 - 187 (1974).

125.012 **The optical emission of the filaments in Tycho.** K. V. Bychkov. Astron. Zhurn. Akad. Nauk SSSR, Vol. 51, 317 - 320 (1974). In Russian. English translation in Soviet Astron. AJ, Vol. 18, No. 2.

The shock wave that moves in the density fluctuations of the interstellar gas originates the vision of the filaments in 3C10 (Tycho). The radial velocities are due to the velocity of the shock. The hot gas that moves on the fluctuations and compresses it explains the 'proper motions' of the filaments. The shape and dimensions of the filaments in optic- and radio waves are discussed. The question of distribution of the radio-emission on the disk of the nebula is touched. The anomalous weak emission from the center of the nebula is explained.

125.013 **X-ray emission from an inward-propagating shock in young supernova remnants.** C. F. McKee.
Astrophys. Journ., Vol. 188, 335 - 339 (1974).

An inward-propagating shock wave is shown to accompany the deceleration of an ejected supernova shell by the ambient medium. The gas behind this shock can be a strong source of soft thermal X-rays. This model is applied to the young supernova remnants that are known to be X-ray sources and is shown to account for most of the observed features of the X-ray spectrum of Cas A.

125.014 **Are binary systems involved in supernova explosions always disrupted?** W. Sutantyo.
Nature, Vol. 248, 208 - 209 (1974). — Letter.

125.015 **Polarization of supernova remnants KE 69, G 357.7-0.1, HC 26, and γ Cygni at 6- and 2.8-cm wavelengths.** M. R. Kundu, T. Velusamy, P. E. Hardee.
Astron. Journ., Vol. 79, 132 - 138 (1974).

Distributions of brightness and linear polarization over the supernova remnants KE 69, G 357.7−0.1, HC 26, and γ Cygni were observed at 2.8 and 6 cm with beams of 3 and 6 arcmin, respectively. γ Cygni was observed only at 2.8 cm. KE 69 shows an incomplete shell structure and a high degree of polarization (~20%). The rotation measures in these sources were computed from the position angles at 6 and 2.8 cm. The distributions of the projected magnetic fields were derived from the observed position angles at 2.8 cm. The field patterns in KE 69 and G 357.7−0.1 are roughly tangential and uniform, respectively. HC 26 and γ Cygni do not show any definite field pattern.

125.016 **The evolution of supernova remnants. I. Spherically symmetric models.** R. A. Chevalier.
Astrophys. Journ., Vol. 188, 501 - 516 (1974).

The evolution of supernova remnants is studied using a spherically symmetric hydrodynamic code with a magnetic field approximately included. Properties of the emitted radiation are investigated. Particular attention is paid to the late phases when a dense neutral shell forms which accretes matter from both the interstellar medium and the hot interior.

125.017 **On the expansion of supernova shells.** C. Maceroni, M. Salvati, F. Pacini.
Astrophys. Space Sci., Vol. 28, 205 - 212 (1974).

The authors have investigated a simple model for the effects of a central pulsar on the expansion of supernova shells. Some numerical results relevant to the Crab nebula are reported.

125.018 **Ap-Sterne und Supernovae.** B. Baschek, W. H. Kegel.
Mitt. Astron. Ges., No. 34, p. 84 (1973/74). — Abstr. AG.

125.019 **High resolution 21-cm observations of Tycho's remnant.** R. G. Strom, R. M. Duin.
Galaxies and relativistic astrophysics, Proc. 1972, (see 012.010), p. 214 - 217 (1974).

125.020 **Quand une étoile explose.** J. Lequeux.
Sci. Progrès Découverte, No. 3449, p. 36 - 43 (1972).

125.021 **Supernova anno 1667 is door niemand gezien.** C. de Jager.
Zenit, Vol. 1, No. 5, p. 2 - 4 (1974).

125.022 **Location of supernovae in parent galaxies.** M. F. McCarthy.
Ric. Astron., Specola Vaticana, Castel Gandolfo, Vol. 8, (No. 19), 411 - 428 (1973).

A study has been made of the location of supernovae in external galaxies. Using the Warsaw Observatory Catalogue of Supernovae together with additional lists by Sargent and coworkers plus the I.A.U. Circulars published before March 1971, 197 parent galaxies have been examined. For parent galaxies with best available radial velocity data (87 galaxies, each with individual redshifts >500 km/sec) the frequency distribution of the supernovae shows one concentration near 3 kiloparsecs from the center and another near 7 kiloparsecs.

125.023 **Indicative distances of 148 supernovae from the nuclei of their parent galaxies.** M. F. McCarthy.
Ric. Astron., Specola Vaticana, Castel Gandolfo, Vol. 8, (No. 20), 429 - 437 (1973).

Zwicky's method of indicative distances has been used to derive the distances from the nucleus of their parent galaxies for 148 supernovae. Three groups of supernovae are considered: supernovae whose parent galaxies have velocities greater than 500 km/sec; those with velocities less than 500 km/sec; and supernovae whose parent galaxies lack individual measured velocities, but which belong to clusters of galaxies for which a mean radial velocity is available.

125.024 **Observations of intensity and linear polarization in supernova remnants at 11 cm wavelength.** T. Velusamy, M. R. Kundu.
Astron. Astrophys., Vol. 32, 375 - 390 (1974).

Distributions of intensity and linear polarization at 11 cm in twenty-four galactic sources in the northern sky are presented. The observations were made with angular resolution of ~5' using the 300-ft. telescope of the National Radio Astronomy Observatory. Twenty sources were mapped in both total and polarized intensities and four were mapped in total intensity only. The results indicate that the galactic sources G 53.9+0.3, G 54.4−0.3, G 69.0+2.7 and G 93.4+1.8 are possible supernova remnants which until now are not listed as such in any of the catalogues. Observations of linear polarization at 6 cm in two sources, 3 C 434.1 and G 93.4+1.8, are also presented. The characteristics of the observed sources are discussed.

125.025 **Determination of the types of galactic supernova remnants (SNR).** G. V. Akhundova, O. Kh. Gusejnov, Sh. Yu. Rakhamimov.
Astrofizika, Vol. 10, 85 - 91 (1974). In Russian. — English translation in Astrophysics, Vol. 10, No. 1.

A method of estimating the type of SNR in their connection with open clusters and OB-associations is suggested. In a 2.5 kpc volume the space distribution of 32 SNR is considered. It gives an opportunity to determine the type of SNR.

125.026 **Numerical models of the evolution of supernova remnants: the shell-formation stage.** W. C. Straka.
Astrophys. Journ., Vol. 190, 59 - 66 (1974).

Numerical calculations of the behavior of supernova remnants have been carried out at several initial energies, with emphasis on the thermally unstable shell-formation phase. An approximate formula to fit the Cox cooling curves has been used. It is found that the time of onset of rapid shell formation is related to the time at which the temperature just behind the shock front in the Sedov phase decreases to the point that recombination processes exceed thermal bremsstrahlung ($\sim 2 \times 10^6$ °K).

125.027 **Can supernovae produce deuterium?** R. I. Epstein, W. D. Arnett, D. N. Schramm.
Astrophys. Journ., (Letters), Vol. 190, L13 - L16 (1974).

An investigation has been made of the nucleosynthesis of the light elements in the shock waves of type II supernovae. The preliminary results basically confirm Colgate's estimates of the energy that is required to produce deuterium. It has also been found that there may be a very large production of ^7Li, ^9Be, and ^{11}B, in which case Colgate's model is more likely to be a source of these isotopes than of deuterium. Some possible consequences of supernova shock waves on the heavy-element abundances, the cosmic-ray composition, and the diffuse γ-ray background are briefly discussed.

125.028 **Large-scale effects of supernova remnants on the Galaxy: generation and maintenance of a hot network of tunnels.** D. P. Cox, B. W. Smith.
Astrophys. Journ., (Letters), Vol. 189, L105 - L108 (1974).

It is found that a supernova rate on the order of 1 per 50 years in the gaseous disk of our Galaxy is sufficient to generate and maintain throughout the interstellar medium a mesh of interconnected tunnels containing very low-density gas. This tunnel system would have $n \lesssim 10^{-2} \text{ cm}^{-3}$, $T \sim 10^6 \text{ °K}$, very low magnetic field strength, tunnel radii ~ 10 pc, and would occupy roughly half the interstellar volume.

125.029 **Review of the research on supernovae.** F. Zwicky.
Astrophys. Space Sci. Library, Vol. 45, (see 012.015), 1 - 16 (1974).

125.030 **The Asiago supernova search.** L. Rosino, G. di Tullio.
Astrophys. Space Sci. Library, Vol. 45, (see 012.015), 19 - 27 (1974).

125.031 **The supernova search at Zimmerwald.** P. Wild.
Astrophys. Space Sci. Library, Vol. 45, (see 012.015), 29 - 32 (1974).

125.032 **The Palomar supernova search.** W. L. W. Sargent, L. Searle, C. T. Kowal.
Astrophys. Space Sci. Library, Vol. 45, (see 012.015), 33 - 49 (1974).

125.033 **Supernova survey at the Konkoly Observatory.** L. Detre.
Astrophys. Space Sci. Library, Vol. 45, (see 012.015), 51 - 53 (1974).

125.034 **A search for radio emission from young extragalactic supernova remnants.** A. G. de Bruyn.
Astrophys. Space Sci. Library, Vol. 45, (see 012.015), 55 - 59 (1974).

125.035 **An experiment to search for prompt emissions from supernovae at microwave frequencies.** J. V. Jelley, W. P. S. Meikle, R. W. P. Drever, G. G. C. Palumbo, F. Bonoli, H. Smith, T. Delaney.
Astrophys. Space Sci. Library, Vol. 45, (see 012.015), 61 - 71 (1974).

125.036 **Historical observations of supernovae.** F. R. Stephenson.
Astrophys. Space Sci. Library, Vol. 45, (see 012.015), 75 - 85 (1974).

125.037 **Photometry of supernovae.** H. Arp.
Astrophys. Space Sci. Library, Vol. 45, (see 012.015), 89 - 97 (1974).

125.038 **On the light curve of type 1 supernovae.** R. Barbon, F. Ciatti, L. Rosino.
Astrophys. Space Sci. Library, Vol. 45, (see 012.015), 99 - 102 (1974).

125.039 **Recent observations of supernovae at Asiago.** R. Barbon, F. Ciatti, L. Rosino.
Astrophys. Space Sci. Library, Vol. 45, (see 012.015), 115 - 118 (1974).

125.040 **The interpretation of the spectra of supernovae.** L. Searle.
Astrophys. Space Sci. Library, Vol. 45, (see 012.015), 125 - 130 (1974).

125.041 **The spectrum of SN 1972e and the fluorescence theory of supernova light.** L. Sartori, B. C. Chiu, P. Morrison.
Astrophys. Space Sci. Library, Vol. 45, (see 012.015), 131 - 133 (1974).

125.042 **Statistics of supernovae.** G. A. Tammann.
Astrophys. Space Sci. Library, Vol. 45, (see 012.015), 155 - 185 (1974).

125.043 **On the frequency of type I and type II supernovae.** N. Dallaporta.
Astrophys. Space Sci. Library, Vol. 45, (see 012.015), 187 - 194 (1974).

125.044 **Space distribution of supernovae in parent galaxies.** M. F. McCarthy.
Astrophys. Space Sci. Library, Vol. 45, (see 012.015), 195 - 202 (1974).

125.045 **Notes on type II supernovae.** G. de Vaucouleurs.
Astrophys. Space Sci. Library, Vol. 45, (see 012.015), 203 - 208 (1974).

125.046 **Supernovae and extragalactic distances.** D. Branch.
Astrophys. Space Sci. Library, Vol. 45, (see 012.015), 209 - 214 (1974).

125.047 **Remarks on outlying supernovae and the structure of their parent galaxies.** G. A. Tammann.
Astrophys. Space Sci. Library, Vol. 45, (see 012.015), 215 - 220 (1974).

125.048 **Observation of X-ray emission from supernova remnants.** P. Gorenstein.
Astrophys. Space Sci. Library, Vol. 45, (see 012.015), 223 - 242 (1974).

125.049 **X-ray observations of supernova remnants by Copernicus.** D. J. Adams, J. C. Zarnecki.
Astrophys. Space Sci. Library, Vol. 45, (see 012.015), 243 - 250 (1974).

125.050 **Preliminary results on the evolution of supernova remnants and their X-ray spectrum.** V. N. Mansfield, E. E. Salpeter.
Astrophys. Space Sci. Library, Vol. 45, (see 012.015), 251 - 260 (1974).

125.051 **Physical conditions in the filaments of SNR from their optical spectra: IC 443.** S. D'Odorico.
Astrophys. Space Sci. Library, Vol. 45, (see 012.015), 283 - 293 (1974).

125.052 **High resolution 21 cm continuum observations of some galactic supernova remnants.** R. M. Duin, R. G. Strom, H. van der Laan.
Astrophys. Space Sci. Library, Vol. 45, (see 012.015), 295 - 301 (1974).

125.053 **Observations of the structures of Tycho's and Kepler's supernova remnants at 2.7 and 5.0 GHz.**

S. N. Henbest.
Astrophys. Space Sci. Library, Vol. 45, (see 012.015), 303 - 305 (1974).

125.054 Search for gamma ray lines from supernovae and supernova remnants.
E. L. Chupp, D. J. Forrest, A. N. Suri, R. Adams, C. Tsai.
Astrophys. Space Sci. Library, Vol. 45, (see 012.015), 311 - 316 (1974).

125.055 Optical observations of supernova remnants Simeiz 59 and W 28. T. A. Lozinskaya.
Astrophys. Space Sci. Library, Vol. 45, (see 012.015), 317 - 320 (1974).

125.056 Supernovae and their remnants. L. Woltjer.
Astrophys. Space Sci. Library, Vol. 45, (see 012.015), 323 - 327 (1974).

125.057 Carbon detonation supernovae. J.-R. Buchler.
Astrophys. Space Sci. Library, Vol. 45, (see 012.015), 329 - 332 (1974).

125.058 The continuum radio spectra of supernova remnants.
J. R. Dickel.
Astrophys. Space Sci. Library, Vol. 45, (see 012.015), 333 - 336 (1974).

125.059 The evolution of young supernova remnants.
S. F. Gull.
Astrophys. Space Sci. Library, Vol. 45, (see 012.015), 337 - 341 (1974).

125.060 The magnetic field distribution in supernova remnants. J. R. Dickel, D. K. Milne.
Astrophys. Space Sci. Library, Vol. 45, (see 012.015), 343 - 354 (1974).

125.061 Dust envelope in young supernova remnants.
S. Aiello, A. Bonetti, F. Mencaraglia, E. Massaro.
Astrophys. Space Sci. Library, Vol. 45, (see 012.015), 365 - 372 (1974).

125.062 Pre-detonation lifetime and mass of supernovae from galaxy density wave theory. E. P. Moore.
Astrophys. Space Sci. Library, Vol. 45, (see 012.015), 373 (1974). – Abstract.

125.063 Deuterium and boron from supernovae?
H. Reeves.
Astrophys. Space Sci. Library, Vol. 45, (see 012.015), 381 - 382 (1974).

125.064 Numerical models for supernova remnants.
V. N. Mansfield, E. E. Salpeter.
Astrophys. Journ., Vol. 190, 305 - 313 (1974).
Numerical calculations are presented for supernova ejecta with initially constant total energy E_0 expanding into an interstellar medium of uniform density n_0, assuming perfect spherical symmetry. Scaling laws are presented for changes in E_0/n_0 with constant $E_0 n_0^2$. Results are given for the time development of various quantities, including the emitted X-ray spectrum, for three different values of $E_0 n_0^2$.

125.065 A two dimensional "carbon detonation" supernova model. J. H. Mahaffy, C. J. Hansen.
Bull. American Astron. Soc., Vol. 6, 212 (1974). – Abstr. AAS.

125.066 Models for the light curves of type II supernovae.
S. W. Falk, W. D. Arnett.
Bull. American Astron. Soc., Vol. 6, 269 (1974). – Abstr. AAS.

125.067 Gamma-ray lines from radioactive ^{57}Co in young supernova. D. D. Clayton.
Bull. American Astron. Soc., Vol. 6, 273 (1974). – Abstr. AAS.

125.068 The X-ray spectrum of the Vela X supernova remnant. F. R. Harnden, Jr., P. Gorenstein.
Bull. American Astron. Soc., Vol. 6, 273 (1974). – Abstr. AAS.

125.069 The structure of supernova shocks.
T. Weaver, G. Chapline, L. Wood, J. Silk.
Bull. American Astron. Soc., Vol. 6, 274 (1974). – Abstr. AAS.

125.070 Comparative X-ray spectra of three supernova remnants as observed by OSO-7.
P. F. Winkler, G. W. Clark, T. H. Markert, G. F. Sprott.
Bull. American Astron. Soc., Vol. 6, 274 (1974). – Abstr. AAS.

125.071 Do pulsars make supernovae? Calculations of light curves for type II events.
J. P. Ostriker, P. Bodenheimer.
Bull. American Astron. Soc., Vol. 6, 274 (1974). – Abstr. AAS.

125.072 Distribution of the galactic supernova remnants.
K. Kodaira.
Publ. Astron. Soc. Japan, Vol. 26, 255 - 261 (1974).
The derivation of the distribution of the galactic supernova remnants by Ilovaisky and Lequeux (1972) is critically examined. The author proposes an empirical method for the correction of observational effects and presents a revised surface-density curve of the galactic supernova remnants which is similar to the surface-density curve of the supernovae themselves reported by Johnson and MacLeod (1963) for spiral galaxies. Some consequences of the revision are briefly discussed.

125.073 Supernova in anonymous galaxy. C. T. Kowal.
IAU Circ., No. 2624 (1974).

Nucleosynthesis in supernova outbursts and the chemical composition of the envelopes of neutron stars.
See Abstr. 061.004.

Electron capture in carbon dwarf supernovae.
See Abstr. 065.029.

On the evolutionary scheme W UMa → CV → type I SN. See Abstr. 121.069.

Statistical time-dependent model for the interstellar gas. See Abstr. 131.057.

Detection of the [Fe XIV] coronal line at 5303 Å in the Cygnus Loop. See Abstr. 132.011.

The evolution of the radio spectrum of Cassiopeia A.
See Abstr. 141.030.

The distance of NGC 7319 in Stephan's Quintet and its supernova. See Abstr. 158.099.

125.100 Supernova in NGC 5457

Variable radio emission from the extragalactic supernova-1970g in M 101.
W. M. Goss, R. J. Allen, R. D. Ekers, A. G. de Bruyn.
Astrophys. Space Sci. Library, Vol. 45, (see 012.015), 145 - 152 (1974).

UBV observations of supernova 1970g in NGC 5457.

J. E. Winzer.
Journ. Roy. Astron. Soc. Canada, Vol. 68, 36 - 39 (1974).

UBV observations of the 1970 supernova in NGC5457 (M101) during the month following its discovery are presented. The supernova declined from V = 11.65, B–V = +0.29, U–B = –0.26 to V = 12.18, B–V = +0.87, U–B = +0.08 during this interval.

125.101 Supernova in NGC 5236

Observations of the 1968 type II supernova in NGC 5236. R. Wood.
Astrophys. Space Sci. Library, Vol. 45, (see 012.015), 119 - 122 (1974).

The 1968 type II supernova in NGC 5236. R. Wood, P. J. Andrews.
Monthly Notices Roy. Astron. Soc., Vol. 167, 13 - 29 (1974).

The 1968 supernova in NGC 5236 was observed spectroscopically and photometrically with the Radcliffe reflector following its discovery. *UBV* magnitudes were obtained on two dates one month apart. They allow modification of Wamsteker's results by making a more accurate correction for the background. The 15 spectrograms cover a period of 57 days and show the supernova to be typical of type II. The spectral evolution and the redward drift of some features are discussed.

125.102 Supernova in NGC 5253

Photometry and spectroscopy of the 1972 supernova in NGC 5253. A. Ardeberg, M. de Groot.
Astrophys. Space Sci. Library, Vol. 45, (see 012.015), 103 - 113 (1974).

Objective prism spectra of SN 1972e near maximum phase. M. F. McCarthy.
Astrophys. Space Sci. Library, Vol. 45, (see 012.015), 135 - 141 (1974).

Detection of X-ray emission from supernova 1972e. C. R. Canizares, J. E. Neighbors.
Bull. American Astron. Soc., Vol. 6, 269 (1974). – Abstr. AAS.

The fluorescent origin of the light of SN 1972e. B. Chiu, P. Morrison, W. Shields.
Bull. American Astron. Soc., Vol. 6, 269 (1974). – Abstr. AAS.

On the possible detection of an energetic photon pulse from the recent supernova in NGC 5253. H. Ögelman, M. E. Özel.
Galaxies and relativistic astrophysics, Proc. 1972 (see 012.010), p. 228 - 232 (1974).

Ultraviolet photometry of the supernova (1972) in NGC 5253. A. V. Holm, C.-C. Wu, J. J. Caldwell.
Publ. Astron. Soc. Pacific, Vol. 86, 296 - 303 (1974).

The supernova (1972) in NGC 5253 was observed with the wide-band filter photometers of OAO-2 between 1972 May 20 and 1972 July 22. The authors measured the energy radiated between 4250 Å and 2460 Å to be 2.9×10^{48} ergs. Only 30% of this energy was radiated shortward of 3320 Å. From the OAO-2 observations and published ground-based data the authors estimate the minimum total radiated energy of the supernova to be 7×10^{49} ergs.

The near infrared spectrum of supernova 1972e in NGC 5253. M. F. McCarthy, G. Araya.
Ric. Astron., Specola Vaticana, *Castel Gandolfo*, Vol. 8, (No. 21), 439 - 444 (1973).

Observations made on two nights with the Curtis Schmidt and objective prisms of the bright supernova in NGC 5253 show the presence of a strong emission band near 8650 Å, a deep depression near 8250 Å and a less intense emission band which extends from 7150 to 8100 Å. The coincidence of the 8650 Å feature with the strong Ca II, N I and O I emission lines in the same region in novae is noted.

125.103 Supernova in NGC 1058

A check on the size of the region producing Fe II emission lines of the 1961 supernova in NGC 1058. M. Friedjung.
Astrophys. Space Sci. Library, Vol. 45, (see 012.015), 143 - 144 (1974).

125.104 Supernova in NGC 493

Upper limits on the X-ray emission from the supernova in NGC 493 during the period of maximum light. G. F. Sprott, H. V. Bradt, G. W. Clark, W. H. G. Lewin, H. W. Schnopper, L. Pigatto, L. Rosino.
Bull. American Astron. Soc., Vol. 6, 269 (1974). – Abstr. AAS.

125.105 Supernova in NGC 3627

Supernova in NGC 3627. L. Rosino.
IAU Circ., No. 2615 (1974).

Supernova in NGC 3627. R. Barbon, F. Ciatti, L. Rosino.
IAU Circ., No. 2624 (1974).

125.106 Supernova in IC 43

Supernova in IC 43. C. T. Kowal.
IAU Circ., No. 2620 (1974).

125.107 Supernova in NGC 4156

Supernova in NGC 4156. H. C. Arp.
IAU Circ., No. 2632 (1974).

125.108 Supernova in NGC 5161

Supernova in NGC 5161. C. T. Kowal.
IAU Circ., No. 2640 (1974).

125.109 Supernova in NGC 3310

Supernovae. P. van der Kruit, H. C. Arp.
IAU Circ., No. 2641 (1974).

125.110 Supernova in NGC 3916

Supernovae. M. Lovas.
IAU Circ., No. 2653 (1974).

125.111 Supernova in NGC 4038 - 39

Supernova in NGC 4038–39. K. Locher.
IAU Circ., No. 2663, with an addendum in No. 2664 (1974).

Supernovae. See Abstr. 125.110.

125.112 Supernova in NGC 4414

Supernovae. W. Burgat.
IAU Circ., No. 2664 (1974).

Supernova in NGC 4414.
P. Wild, R. Wood, B. Patchett, C. Scovil, E. Mayer, L. Rosino.
IAU Circ., Nos. 2666 and 2674 (1974).

Supernova in NGC 4414. J. E. Bortle, E. Mayer.
IAU Circ., No. 2668 (1974).

Supernova in NGC 4414. U. Hopp.
IAU Circ., No. 2671 (1974).

Supernova in NGC 4414.
S. Wyckoff, J. E. Bortle.
IAU Circ., No. 2678 (1974).

Errata

125.901 **Erratum: 'Polarimetric aperture synthesis observations of four small-diameter supernova remnants'**
[Astron. Journ., Vol. 78, 879 - 884, 983 - 986 (1973)].
B. R. Hermann, J. R. Dickel.
Astron. Journ., Vol. 79, 420 (1974).

126 Low-luminosity Stars, Subdwarfs, White Dwarfs

126.001 **A spectroscopic survey of southern hemisphere white dwarfs–IV. Radial velocities and space motions.** G. Wegner.
Monthly Notices Roy. Astron. Soc., Vol. 166, 271 - 280 (1974).

New radial velocity measurements are reported for 31 southern white dwarfs. The K-term which is interpreted as the mean gravitational redshift, was estimated on the assumption of a mean solar motion for the white dwarfs. The K-term, space motions, and velocity ellipsoids were also derived for the white dwarfs from their radial velocities after combining the present data with Greenstein & Trimble's results for northern white dwarfs. Some of the stellar evolutionary interpretations of these observations are briefly discussed.

126.002 **The cooling of liquefying hot white dwarfs.**
U. De Angelis, L. De Cesare, A. Forlani, G. Platania.
Astrophys. Space Sci., Vol. 27, 117 - 135 (1974).

The consequences of gas-liquid phase transitions in the core of hot white dwarf stars are discussed. Expressions for the latent heat and the liquefaction curve $T_l = T_l(\rho)$ are obtained. Then a model for a hot white dwarf is introduced and the corresponding liquefaction sequences are built on the H-R diagram; relations luminosity-central temperature and effective temperature-central temperature are also given for liquefying white dwarfs. Finally the cooling curves are obtained for such stars taking into account the effect of latent heat emission.

126.003 **On the interpretation of white dwarf spectra.**
F. D'Antona, I. Mazzitelli.
Astrophys. Space Sci., Vol. 27, 137 - 146 (1974).

Following the line of research outlined by Strittmatter and Wickramasinghe (1971) and recently by Shipman (1972), attention was drawn to the study of convective zones in white dwarfs for various masses and chemical compositions, in order to understand some observational features of their spectra. According to the results, convection strongly depends on the mass of the star, for $M \lesssim 0.5\,M_\odot$, and low hydrogen abundances ($X \simeq 0.2$) are sufficient to decrease the extension of the He-convection zones with respect to pure-He models. It is shown that the existence of DB white dwarfs seems not in agreement with the accretion rates predicted by Bondi (1952), and that DB's must evolve into H-lacking white dwarfs.

126.004 **Chemical compositions of cool helium and carbon white dwarfs.** T. C. Grenfell.
Astron. Astrophys., Vol. 31, 303 - 315 (1974).

An atmospheric abundance analysis is carried out for the white dwarfs vMa 2 (DG), G 47–18 (λ 4670), and G 99–37 (λ 4670). Spectroscopic observations of Oke and Shipman to calibrate the spectral data are combined with non-gray model atmospheres including convection. A spectrum synthesis is performed over the entire wavelength range where spectral features are observed in order to avoid an arbitrary choice of the continuum level and to account for overlapping atomic lines and molecular bands.

126.005 **Absolute spectral energy distributions for white dwarfs.** J. B. Oke.
Astrophys. Journ., Suppl. Ser., No. 236, Vol. 27, 21 - 35 (1974).

Absolute spectral energy distributions are presented for 38 white dwarfs of different spectral types. The wavelength range covered is approximately from λ3200 or λ3300 to somewhere between λ9500 and λ10500.

126.006 Continuum polarization in magnetic white dwarfs.
F. K. Lamb, P. G. Sutherland.
IAU Symposium No. 53, (see 012.004), p. 265 - 285. (1974).

126.007 Polarized radiation from white dwarfs and atoms in strong magnetic fields. R. F. O'Connell.
IAU Symposium No. 53, (see 012.004), p. 287 - 300 (1974).

126.008 Discovery of time-varying circular and linear polarization in the white-dwarf suspect GD 229.
J. B. Swedlund, R. D. Wolstencroft, J. J. Michalsky, Jr., J. C. Kemp.
Astrophys. Journ., (Letters), Vol. 187, L121 - L123 (1974).

Both circular and linear polarization have been found in the white-dwarf suspect GD 229. Both appear to change with time over periods of a few minutes and of about one day. No definite periodicities have been identified. In broadband measurements, sign changes of the circular polarization are seen in the red but not in the blue.

126.009 G29-38 — a variable white dwarf.
O. S. Shulov, E. N. Kopatskaya.
Astron. Tsirk., No. 810, p. 1 - 2 (1974). In Russian.

126.010 On the linear polarization of GD 229.
J. C. Kemp, G. V. Coyne, J. B. Swedlund, R. D. Wolstencroft.
Astrophys. Journ., (Letters), Vol. 189, L79 - L80 (1974).

The linear polarization of the peculiar white-dwarf suspect GD 229 has been defined more carefully as $p = (3.2 \pm 0.8)$ percent, $\theta = 80° \pm 5°$, in the blue to visible, with little apparent variability.

126.011 Statistical population indices of white dwarfs.
W. Iwanowska.
Stud. Soc. Sci. Torunensis, Sectio F (Astron.), Vol. 5, 43 - 49 = Bull. Astron. Obs. Toruń, No. 50/II (1973).

Statistical population indices were determined for 142 white dwarfs from their tangential velocities as well as for 67 of them from their full galactic velocity components. Population II white dwarfs are absent in the $B-V<0$ interval, they are more numerous in later and DC classes. Besides they are more numerous among white dwarfs than among other kinds of stars. Hydrogen lines are stronger in population I white dwarfs.

126.012 The evolution of rotating white dwarfs with outflow of matter. A. V. Ovsepyan (Hovsepian).
Astrofizika, Vol. 10, 99 - 107 (1974). In Russian. — English translation in Astrophysics, Vol. 10, No. 1.

The outflow of matter due to rotation during the cooling of white dwarfs of masses 1 $M_\odot$ to 3 $M_\odot$ is considered by the energetic method. It is shown that as a result of the outflow during the cooling of a star the mass could be lost up to 10% $M_\odot$. For masses more than 2 $M_\odot$ the outflow of matter stops due to a great loss of angular momentum of rotation. The time dependence of the amount of ejected matter for the mass $M = 1.1 M_\odot$ is given.

126.013 Discovery of light variability in the white dwarf G 29-38. O. S. Shulov, E. N. Kopatskaya.
Astrofizika, Vol. 10, 117 - 120 (1974). In Russian. — English translation in Astrophysics, Vol. 10, No. 1.

Rapid variations in the light of the white dwarf G 29-38 = EG No. 159, have been observed with two periods 13.6 ± 0.2 min. and 10.20 ± 0.05 min.

126.014 Observations for broad-band circular polarization in white dwarfs and nuclei of planetary nebulae.
A. Rich, W. L. Williams.
Astrophys. Journ., Vol. 190, 117 - 120 (1974).

Observations have been made to search for broad-band optical circular polarization in the radiation from eight white dwarfs, four planetary nuclei, and the peculiar blue variable CoD $-42°14462$. All results are negative with the exception of a possible effect in the central star of He 2-131.

126.015 The wavelength dependence of circular polarization in GD 229.
J. D. Landstreet, J. R. P. Angel.
Astrophys. Journ., (Letters), Vol. 190, L25 - L26 (1974).

Circular polarization measurements of GD 229 in narrow wavelength bands show structure down to the instrumental resolution of 160 Å. The data do not indicate any significant variability on a night-to-night basis in any region of the spectrum.

126.016 The spectrum of the polarized white dwarf GD 229.
J. L. Greenstein, M. Schmidt, L. Searle.
Astrophys. Journ., (Letters), Vol. 190, L27 - L28 (1974).

The spectrum of the polarized white dwarf GD 229 shows broad absorptions at λλ3890, 4150, 5300, 6410 and 7120. Variability of the absorption at λ4150 is suspected. Similarities and differences with the polarized white dwarf Gr +70°8247 are discussed.

126.017 The effects of rapid, differential rotation on the spectra of white dwarfs. R. L. Milton.
Astrophys. Journ., Vol. 189, 543 - 554 = Lick Obs. Bull., No. 649 (1974).

A grid of hydrogen-rich model atmospheres, including the effects of both line blanketing and convection has been calculated for a range of temperatures and gravities appropriate for DA white dwarfs. These atmospheres have been fitted to a series of interior models for differentially rotating white dwarfs of large mass ($1.58 M_\odot \leqslant M \leqslant 4.06 M_\odot$) to determine the observable spectra for such objects.

126.018 An investigation of accretion of matter onto white dwarfs as a possible X-ray mechanism.
A. J. DeGregoria.
Astrophys. Journ., Vol. 189, 555 - 561 (1974).

The accretion flow onto white dwarfs is investigated by numerically integrating the full nonlinear hydrodynamic equations under the assumption of spherical symmetry to determine the short-term time-dependent behavior, to determine the effects on the flow when radial modes of oscillation of the star are excited, to see if X-rays are produced, and to determine the radiation spectrum.

126.019 A new list of 52 degenerate stars. VII.
J. L. Greenstein.
Astrophys. Journ., (Letters), Vol. 189, L131 - L133 (1974).

Conclusion of observations with the prime-focus spectrograph of the Hale reflector leads to a list of 20 white dwarfs confirmed spectroscopically. An additional 32 white and yellow degenerates are included in another table; these stars were observed with the multichannel spectrophotometer. Degenerate stars are recognized by narrow-band colors, and by hydrogen and helium lines. Some 10 stars are classified as free of lines, DC, or are known to be circularly polarized.

126.020 On the temperature and density distribution in the radiative layer of white dwarfs.
P. I. Seidova, T. A. Ehminzade.
Astron. Zhurn. Akad. Nauk SSSR, Vol. 51, 536 - 541 (1974). In Russian. English translation in Soviet Astron. AJ, Vol. 18, No. 3.

Results of a calculation of radiative envelopes of white dwarfs with masses 1, 0.8, 0.6 $M_\odot$ and luminosities 1, 10^{-1},

10^{-2}, 10^{-3} $L_\odot$ are given. The opacity coefficients are taken according to Cox and Stewart. The relation $L \sim T^{2.8}$ between luminosity and central temperature has been found.

126.021 G240−72: a new magnetic white dwarf with unusual polarization.
J. R. P. Angel, P. Hintzen, P. A. Strittmatter, P. G. Martin. Astrophys. Journal, (*Letters*), Vol. 190, L71 - L72 (1974).

G240−72 (= LP 44-113), a DC white dwarf, is found to show elliptical polarization. There is no evidence of variability. The circular component, typically 0.5 percent, changes sign with wavelength, being negative in blue light and positive in red. There is a relatively strong component of linear polarization, 1.4 percent in blue light.

126.022 Spectrophotometric observations of cool white dwarfs.   J. Liebert.
Bull. American Astron. Soc., Vol. 6, 265 (1974). − Abstr. AAS.

126.023 Bremsstrahlung model of polarized radiation from magnetic white dwarfs.   R. F. O'Connell.
Phys. Letters A, (*Netherlands*), Vol. 46A, 249 - 250 (1973).

Motivated by the desire to provide a model for the polarized radiation observed from some white dwarfs, the author calculates the intensities of right and left circularly polarized bremsstrahlung radiation emitted by electron in the atmospheres of magnetic white dwarfs.

The transfer of circularly polarized radiation.
See Abstr. 063.002.

Rotation in late stages of stellar evolution.
See Abstr. 065.064.

Are the UV stars nuclear-powered?
See Abstr. 065.089.

The orbit evolution of the eclipsing binary system BD + 16°516 and the rotation period of its white dwarf.
See Abstr. 121.005.

The effect of UV stars on the interstellar medium. III. Computer simulations of an interstellar medium with UV stars.   See Abstr. 131.018.

The oldest disk stars.   See Abstr. 155.022.

Interstellar Matter, Gaseous Nebulae, Planetary Nebulae

131 Interstellar Matter, Polarization of Starlight, H I, H II Regions

Interstellar Matter, Polarization of Starlight

131.001 Accurate positions of OH sources.
C. G. Wynn-Williams, M. W. Werner, W. J. Wilson.
Astrophys. Journ., Vol. 187, 41 - 44 (1974).

The Owens Valley interferometer has been used at 1665, 1667, and 1720 MHz to measure the positions of 11 OH emission sources.

131.002 Detection of interstellar lithium in the direction of 55 Cygni. P. A. Vanden Bout, G. Grupsmith.
Astrophys. Journ., (Letters), Vol. 187, L9 - L11 (1974).

The $\lambda 6708$ line of Li I has been detected in the interstellar spectrum of 55 Cyg with a strength of 5.2 ± 0.8 mÅ equivalent width. Determination of the ionization equilibrium from observations of Ca I $\lambda 4226$ and Ca II H and K yields a column density $N(\text{Li}) = 3.4 \times 10^{12}$ cm^{-2}.

131.003 Interstellar circular polarization: A report of eight new positive results. J. J. Michalsky, Jr., J. B. Swedlund, R. A. Stokes, R. W. Avery.
Astrophys. Journ., (Letters), Vol. 187, L13 - L14 (1974).

A survey for interstellar polarization effects carried out during 1973 March–September at Battelle Observatory has resulted in the discovery of circular polarization in the light of eight stars. The authors report the first detection of interstellar circular polarization in stars later than spectral class A and at least one exception to the usual wavelength dependence of the effect.

131.004 Microwave spectrum of ^{13}C methanol.
S. S. Haque, R. M. Lees, J. M. S. Clair, Y. Beers, D. R. Johnson.
Astrophys. Journ., (Letters), Vol. 187, L15 - L17 (1974).

Laboratory measurements of the frequencies of some astronomically interesting transitions of the ^{13}C isotopic species of methanol are reported. Most lines of the $J = 2 \leftarrow 1$, $3 \leftarrow 2$, $4 \leftarrow 3$ μ_a-type $\Delta k = 0$ transitions in the 94-, 142-, and 189-GHz regions have been measured, as well as a number of μ_b-type transitions in the 14–50 GHz region.

131.005 Extinction by interstellar radicals.
C. D. Andriesse, J. de Vries.
Astron. Astrophys., Vol. 30, 51 - 58 (1974).

The authors calculate the scattering by interstellar free radicals (Platt's particles). The results are applied to interpret recent data on the far-ultraviolet extinction, which apparently is due to scattering rather than absorption processes. The calculation is based on two one-dimensional models of free electrons confined to a small box with size R, the first with an infinite- and the second with a finite binding potential U_0. An exact analytic treatment is given for the first model and an approximative treatment for the second model based on a perturbation calculation. The authors derive transition frequencies and probabilities between the ground and the first excited state as a function of the model parameters R and U_0. Using two plausible distribution functions for the size of interstellar radicals the authors calculate continuous extinction spectra, made up of distinct quantum transitions within individual radicals. A possible fit is shown to extinction data for ζ Per below 1600 Å. They conclude that scattering by free electrons in rather small radicals can account for the observed far-ultraviolet extinction and the high albedo.

131.006 Thermal and ionisation equilibrium of an X-ray heated intercloud medium.
M. Grewing, C. M. Walmsley.
Astron. Astrophys., Vol. 30, 281 - 287 (1974).

New X-ray measurements are used to compute the ionisation equilibrium of the heavy elements in the intercloud gas. The results are compared with ultra-violet observations from the Copernicus satellite. It is found that a hot intercloud medium of density 0.2 cm^{-3} is compatible with the X-ray results if depleted intercloud abundances derived from the ultra-violet observations are used. The ionisation equilibrium predicted on this basis is in contradiction with observations only in the case of nitrogen.

131.007 The chemistry of sulfur in interstellar clouds.
M. Oppenheimer, A. Dalgarno.
Astrophys. Journ., Vol. 187, 231 - 235 (1974).

A chemical scheme is proposed for the creation and destruction of interstellar molecules that contain sulfur. The predicted abundances of H$_2$S, OCS, CS, H$_2$CS, and SO are consistent with those observed. It is suggested that SN, SN$^+$, and SO$^+$ may be present in detectable amounts in dense clouds.

131.008 The interstellar abundance of titanium.
G. Wallerstein, D. Goldsmith.
Astrophys. Journ., Vol. 187, 237 - 242 (1974).

From spectrograms of 15 stars combined with data for eight further stars reported by others, the authors have determined the column densities of Ti II for 11 stars and upper limits for 12 stars. Observations and theory can be used to show that little or no Ti I or Ti III is expected to be present in interstellar gas. Data and arguments are presented to show that much of the interstellar titanium and calcium is probably locked up in the grains.

131.009 Microwave detection of interstellar CH.
B. E. Turner, B. Zuckerman.
Astrophys. Journ., (Letters), Vol. 187, L59 - L62 (1974).

The $^2\Pi_{1/2}$, $J = 1/2$ Λ-doublet transition of CH has been detected in emission in the direction of Cas A, at a frequency of 3349.185 MHz. For an assumed 10 percent population inversion the CH/H abundance ratios in the Orion and Perseus arm clouds are 3×10^{-9} and 10^{-9}, respectively, if the observed line is the $F = 1 \to 1$ hyperfine component.

131.010 Radiative transfer, excitation, and cooling of molecular emission lines (CO and CS).
N. Z. Scoville, P. M. Solomon.
Astrophys. Journ., (Letters), Vol. 187, L67 - L71 (1974).

The authors consider the radiative transfer of molecular lines in interstellar clouds having flow velocities large compared with random motions. The equilibrium level populations of

CO and CS are calculated including the effects of both self-radiation and collisions with hydrogen molecules. Analytic expressions are also developed for the excitation of a two-level molecule. Because of the velocity gradients in the cloud, the observed emission will originate not only from the near boundary but also from interior regions where excitation is greatly enhanced by scattered photons.

131.011   **The dipole moment of isocyanic acid, HNCO, and its astrophysical consequences.**
W. H. Hocking, M. C. L. Gerry, G. Winnewisser.
Astrophys. Journ., (Letters), Vol. 187, L89 - L91 (1974).

The component of the dipole moment of $D^{14}N^{12}C^{16}O$ along its b-inertial axis (μ_b) has been found from measurements of the Stark effect to be 1.35 ± 0.1 debye. This value is an order of magnitude larger than previously estimated, and revises the estimates of the minimum hydrogen density required to collisionally excite HNCO molecules in the direction of Sgr B2.

131.012   **The ionization rate in interstellar clouds.**
M. V. Penston.
Monthly Notices Roy. Astron. Soc., Vol. 166, 21P - 23P (1974).

Recent ultra-violet observations suggest that the heating rate in interstellar clouds greatly exceeds that estimated hitherto.

131.013   **5 GHz observations of six OH objects.**
S. Harris.
Monthly Notices Roy. Astron. Soc., Vol. 166, 29P - 34P (1974).

Six OH objects have been observed at 5 GHz with the Cambridge 5-km telescope. Those of type I (W75N(OH), ON1 and ON2) were all detected, whereas those of type IIb (NML Cyg, IRC + 50137 and ON4) were not, which is in general agreement with previous results.

131.014   **Orientation of interstellar and interplanetary grains.**   A. Z. Dolginov.
Astron. Zhurn. Akad. Nauk SSSR, Vol. 51, 56 - 65 (1974). In Russian. English translation in Soviet Astron. AJ, Vol. 18, No. 1.

Explicit expressions for the orientation distribution function of interstellar and interplanetary dust grains in anisotropic corpuscular or radiation fluxes with consideration for the magnetic field influence are obtained. It is noted that small inhomogeneous dust grains should possess a specific rotation of polarization. The orientation of these dust grains is considered. The time required for the polarization is estimated. A possibility of explaining interstellar polarization and polarization of cometary radiation is discussed.

131.015   **A synthesis of our present knowledge of interstellar dust.**   P. S. Wesson.
Space Sci. Rev., Vol. 15, 469 - 482 (1974).

The present state of knowledge as regards interstellar dust is reviewed in Section 1 (Introduction); Section 2 (Composition of dust grains); Section 3 (Size of grains); Section 4 (Charge and temperature of grains); Section 5 (Distribution and origin of grains); Section 6 (Cosmogonical and cosmological aspects of interstellar grains).

131.016   **The 2200 Å extinction feature and the shape-distribution of graphite grains.**
N. C. Wickramasinghe, K. Nandy.
Astrophys. Space Sci., Vol. 26, 123 - 129 (1974).

The observed profile of the 2200 Å interstellar extinction feature is consistent with extinction by polydispersions in shape of small graphite spheroids.

131.017   **Composition of grain mantles in interstellar clouds.**
W. W. Duley.
Astrophys. Space Sci., Vol. 26, 199 - 205 (1974).

The question of grain mantle composition in dense clouds is examined in the light of new observational and theoretical results on atomic and molecular concentrations in the gas phase. It is shown that if grain temperatures are less than about 20 K these mantles will be primarily solid CO. Methods of identifying CO coated grains are discussed.

131.018   **The effect of UV stars on the interstellar medium. III. Computer simulations of an interstellar medium with UV stars.**   K. G. Castle, J. G. Hills.
Astron. Astrophys., Vol. 30, 455 - 464 (1974).

The effect of hot prewhite dwarfs (UV stars) on the interstellar medium (IM) has been investigated through computer simulations. In the first series of simulations the UV stars were introduced at random into an IM of uniform density. The structure of the IM resulting from the overlap of the Strömgren spheres produced by the UV stars was studied by means of computer-generated plots showing two-dimensional slices of the IM. The second series of computer simulations allowed for the decrease in the space density of the UV stars and in the density of the gas in the IM as a function of height above the galactic plane.

131.019   **OH 26.5 + 0.6 – a strong OH source at 1612 MHz.**
C. Andersson, L. E. B. Johansson, W. M. Goss, A. Winnberg, Nguyen-Quang-Rieu.
Astron. Astrophys., Vol. 30, 475 - 477 (1974).

A strong OH source (OH 26.5 + 0.6) has been found at 1612 MHz. The peak flux density is greater than 400 f.u. and is comparable to that of NML Cygnus; in terms of peak flux OH 26.5 + 0.6 is one of the five strongest OH emission sources detected to date.

131.020   **The OH absorption against the Carina nebula.**
H. R. Dickel, J. V. Wall.
Astron. Astrophys., Vol. 31, 5 - 10 (1974).

A map of the 1667 MHz OH absorption against the Carina nebula has been obtained with the Parkes 64-m telescope; the OH molecules are in two concentrations, neither of which is centred on the continuum peak of the nebula. Comparisons of these observations both with 4830 MHz H_2CO absorption and with optical absorption indicate that the OH and H_2CO molecules are closely associated with the dust obscuring the nebula.

131.021   **Interstellar lines in the southern hemisphere.**
J. J. Rickard.
Astron. Astrophys., Vol. 31, 47 - 61 (1974).

Observations of more than 300 interstellar K-line components seen in 168 southern stars are examined for information about galactic structure and turbulent motions in the interstellar gas. The data of several observers are analyzed together with new observations of 67 stars by the author. A catalogue of interstellar lines, including data relevant to the stars, is presented in a table. A figure shows the radial velocities of all components vs. galactic longitude.

131.022   **Pulse broadening due to multiple scattering in the interstellar medium—III.**   I. P. Williamson.
Monthly Notices Roy. Astron. Soc., Vol. 166, 499 - 512 (1974).

Two previous papers have discussed the pulse shapes expected from various model configurations of the interstellar medium and these theoretical pulse shapes are here compared with the observations of broadened pulses from five pulsars. It is found that the pulse shape generated by a model in which scattering occurs uniformly throughout the medium does not fit the observations, nor is the initial rise of the

131.023 The structure of four 1612 MHz OH emission sources.
M. R. W. Masheder, R. S. Booth, R. D. Davies.
Monthly Notices Roy. Astron. Soc., Vol. 166, 561 - 583 (1974).

Three M-type supergiant infra-red/OH objects, NML Cyg, VY CMa and VX Sag, and the W43A source have been studied with the Mark II–Mark III interferometer in the 1612 MHz OH satellite line. Detailed structure has been determined for the NML Cyg and VY CMa sources. A model of these objects has been constructed from the observations which indicates an efflux of 5×10^{-3} solar masses per year from the vicinity of the central star. The implications of the results for theories of circumstellar clouds is discussed.

131.024 Detection of interstellar lines of para-formaldehyde and OCS and searches for other molecular lines at 73 GHz.
K. Akabane, M. Morimoto, K. Nagane, K. Miyazawa, T. Miyaji, H. Tabara, H. Hirabayashi, N. Kaifu, Y. Chikada.
Publ. Astron. Soc. Japan, Vol. 26, 1 - 7 = Tokyo Astron. Obs. Repr. No. 445 (1974).

The $1_{01}-0_{00}$ line of para-formaldehyde and $J = 6$ to 5 line of OCS were detected in Sgr B2. The formaldehyde line was also detected in Ori A (OH source). Searches were made without success for these lines in W51, Heiles cloud 4, and IRC+10216 and also for other lines of methylamine, HCCCN, and SO in all sources mentioned above.

131.025 A search for OD in the galactic center.
M. Allen, D. A. Cesarsky, R. M. Crutcher.
Astrophys. Journ., Vol. 188, 33 - 34 (1974).

The authors have unsuccessfully searched for the $\Delta F = 0$ Λ-doublet transitions of the $^2\Pi_{3/2}, J = 3/2$ state of OD. The resulting upper limit of the OD/OH column density ratio is 1/400.

131.026 Ion-molecule reactions, molecule formation, and hydrogen-isotope exchange in dense interstellar clouds.
W. D. Watson.
Astrophys. Journ., Vol. 188, 35 - 42 (1974).

The influence of (positive) ion-molecule reactions on the formation and dissociation of molecules in dense (particle density $\gtrsim 10^4$ cm^{-3}) interstellar clouds is investigated. Unlike reactions between neutral molecules, no activation energy barrier inhibits these ion-molecule reactions at low interstellar gas temperatures. Based on an ionization rate due only to the observed high-energy cosmic rays, certain chemical processes are shown to occur on a time scale comparable with or faster than other relevant time scales.

131.027 On the association of C$^+$ and COH$^+$ with H$_2$.
F. C. Fehsenfeld, D. B. Dunkin, E. E. Ferguson.
Astrophys. Journ., Vol. 188, 43 - 44 (1974).

Three-body association-rate constants of C$^+$ and COH$^+$ with H$_2$ have been measured at 90°K. These results give some information relative to the same associations by radiative processes. These radiative processes have been proposed as possible precursor steps in the formation of CH and formaldehyde in interstellar clouds. Possible implications for interstellar molecule production are discussed.

131.028 Optical transmission in a circumstellar dust cloud.
T. W. Jones.
Publ. Astron. Soc. Pacific, Vol. 85, 811 - 814 (1973).

It is shown from the equation of radiative transfer that the visual attenuation of an optically thick, spherically symmetric circumstellar cloud of dust can be small compared to that produced by an equal optical depth of interstellar dust. The reasons for this are the moderately large albedos of the grains and their tendency to emphasize forward scattering.

131.029 Incremental polarization of starlight at different locations in the Galaxy.
L. A. Fowler, M. Harwit.
Monthly Notices Roy. Astron. Soc., Vol. 167, 227 - 236 (1974).

The authors have used existing data, compiled by other investigators, to obtain the incremental polarization of starlight produced at different distances from the sun and at different galactic latitudes and longitudes. Their results are shown as a series of plots. They find that, within spiral arms, the orientation of the incremental polarization vector is correlated with the direction of the arms.

131.030 Interstellar molecular oxygen.
G. M. Rudnitskij.
Astron. Zhurn. Akad. Nauk SSSR, Vol. 51, 329 - 334 (1974). In Russian. English translation in Soviet Astron. AJ, Vol. 18, No. 2.

The mechanisms of formation and destruction for molecular oxygen in the interstellar medium are considered. Concentrations of O_2 under various conditions and possible intensities of O_2 emission lines in the millimeter wavelength range are estimated.

131.031 Investigation of the variability of 18-cm line maser sources.
E. E. Lekht.
Astron. Zhurn. Akad. Nauk SSSR, Vol. 51, 341 - 345 (1974). In Russian. English translation in Soviet Astron. AJ, Vol. 18, No. 2.

In July 1972 observations of galactic sources of maser emission W49 A, NGC 6334 A, B at 1665 and W75 S at 1665 and 1667 MHz have been obtained with the large radio telescope at Nançay. Temporal variations of the flux of radiation have been searched for. No variability with small time-scale has been found. Profiles of the sources investigated with a 600-Hz spectral resolution are obtained. The present work confirms the previously detected variability of NGC 6334 A, B, and W75 S at 1665 MHz.

131.032 A possible mechanism of the formation of maser OH and H$_2$O sources associated with compact H II zones.
V. V. Burdyuzha, T. V. Ruzmajkina.
Astron. Zhurn. Akad. Nauk SSSR, Vol. 51, 346 - 353 (1974). In Russian. English translation in Soviet Astron. AJ, Vol. 18, No. 2.

It is shown that maser OH and H$_2$O clouds near compact H II zones could be created by thermal instability.

131.033 The sizes of cosmic masers.
E. L. Klochan, V. S. Strel'nitskij, V. M. Tomozov.
Astron. Zhurn. Akad. Nauk SSSR, Vol. 51, 354 - 362 (1974). In Russian. English translation in Soviet Astron. AJ, Vol. 18, No. 2.

The relation between the observed properties of OH and H$_2$O maser sources and the actual size of the amplifying medium is investigated.

131.034 Is X-ogen HCO$^+$?
E. Herbst, W. Klemperer.
Astrophys. Journ., Vol. 188, 255 - 256 (1974).

A detailed calculation of the rotational spectrum of the ion H^{13}CO$^+$ is performed. The first rotational transition is estimated to occur at $86,720 \pm 10$ MHz. It is suggested that the H^{13}CO$^+$/H^{12}CO$^+$ line-intensity ratio may be close to the ^{13}C/^{12}C cosmic-abundance ratio.

131.035 Interferometric scans of interstellar K I lines.
L. M. Hobbs.

Astrophys. Journ., (Letters), Vol. 188, L67 - L69 (1974).

Interferometric scans of the interstellar λ7699 line of neutral potassium are reported toward 26 stars. The advantages of the K I line, as compared with other optical interstellar lines, are discussed. A comparison of the interstellar λ7699 and Lα lines toward 10 stars shows that the K/H ratio in the interstellar gas is normal.

131.036 **Further evidence for an interstellar source of nighttime He I 584 Å radiation.**
F. Paresce, S. Bowyer, S. Kumar.
Astrophys. Journ., (Letters), Vol. 188, L71 - L73 (1974).

The intensity of He I, 584 Å radiation in the night sky was measured from a sounding rocket at altitudes between 160 and 217 km. The results obtained confirm previous measurements of this radiation. The observed intensity and spatial variations can be entirely accounted for by resonance scattering of solar 584 Å radiation from interstellar helium.

131.037 **Depletion of interstellar sodium and calcium.**
R. E. White.
Astrophys. Journ., Vol. 187, 449 - 452 (1974).

New measurements of the Ca II/Ca I ratios in interstellar clouds lead to the conclusion that not only Ca, but also Na, is underabundant in these regions. The extreme underabundance of Ca, and some part of the abundance range, probably reflects the composition of the grain cores.

131.038 **Interstellar abundances: gas and dust.**
G. B. Field.
Astrophys. Journ., Vol. 187, 453 - 459 (1974).

Data on abundances of interstellar atoms, ions, and molecules in front of ζ Oph are assembled and analyzed.

131.039 **Interstellar polarization from a medium with changing grain alignment.** P. G. Martin.
Astrophys. Journ., Vol. 187, 461 - 472 (1974).

If the direction of grain alignment in the interstellar medium changes along the line of sight to a star, circular polarization can be produced. Models are developed for the polarization from such a medium, as the basis for interpreting existing observations. Emphasis is placed on the numerous relationships between circular polarization, linear polarization, and extinction from which information about the interstellar grains can be derived. Detailed applications are made to both stars and clusters of stars. Some new measurements of circular polarization are reported and observational prospects for the future are evaluated.

131.040 **On the opacity of the interstellar medium to ultrasoft X-rays and extreme-ultraviolet radiation.**
R. Cruddace, F. Paresce, S. Bowyer, M. Lampton.
Astrophys. Journ., Vol. 187, 497 - 504 (1974).

The opacity of the interstellar medium at soft X-ray and extreme-ultraviolet wavelengths has been reexamined in the light of recent evidence that the density of interstellar hydrogen in the solar neighborhood may be quite low. It is shown that the distance at which significant attenuation occurs is not negligible, especially in the 50–300 Å region.

131.041 **Ionization of carbon and nitrogen in the intercloud medium.**
M. Kafatos, H. Gerola, S. Hatchett, R. McCray.
Astrophys. Journ., (Letters), Vol. 187, L113 - L116 (1974).

If the intercloud medium in the solar neighborhood was suddenly ionized less than 10^6 years ago, the subsequent relaxation of the gas can account for the very low observed column density of C III and N III yet still leave a significant fraction (~0.05) of ionized hydrogen and of C I in a cold ($T \approx 20°$ K) intercloud medium. The OAO-C Copernicus observations in the direction of λ Sco are consistent with cosmic abundances and a uniform-density medium, except for C III whose predicted column density from the C III ionization zone surrounding the star is a factor ~10 greater than the observed result.

131.042 **Observations of structure in the interstellar polarization curve: Preliminary results.** G. E. Mavko, D. S. Hayes, J. M. Greenberg, W. A. Hiltner.
Astrophys. Journ., (Letters), Vol. 187, L117 - L119 (1974).

The authors have measured the wavelength dependence of the interstellar polarization curve with higher wavelength resolution than has generally been used before. Their data shows structure which has been previously unknown, and which may have important implications with respect to the composition of the interstellar grains.

131.043 **A deviation-defect method for the detection of optically thick neutral hydrogen.** P. L. Baker.
Astrophys. Journ., Vol. 187, 223 - 229 (1974).

A detection technique called the deviation-defect method has been developed and tested against interferometric measurements of absorption spectra. All optically thick gas features detected by the deviation-defect method are confirmed by the interferometer results. The new method has been applied to a large sample of the sky, and a very extensive optically thick hydrogen layer in the galactic anticenter has been mapped.

131.044 **Der kosmische Staub und seine Rolle bei der Sternentstehung.** J. Dorschner.
Jenaer Rundschau, (Jena Review), 18. Jahrgang, p. 323 - 330 (1973).

131.045 **Interstellar reddening and globular clusters.**
B. V. Kukarkin, N. N. Kireeva.
Astron. Tsirk., No. 797, p. 3 - 5 (1973). In Russian.

131.046 **Two new OH emission sources in the W43 region.**
V. I. Slysh.
Astron. Tsirk., No. 798, p. 1 (1973). In Russian.

131.047 **On Pottasch's interpretation of the interstellar Na I/Ca II ratio.** R. E. White.
Astron. Astrophys., Vol. 31, 459 - 460 (1974).

The interpretation of the interstellar Na I/Ca II ratio given by Pottasch (1972) leads to the prediction of Ca I absorption in conflict with observation.

131.048 **Optical data for selected Barnard objects.**
B. J. Bok, C. C. McCarthy.
Astron. Journ., Vol. 79, 42 - 44, 101 - 103 (1974).

Total photographic absorptions, radii, and minimum cosmic dust masses have been obtained for eight Barnard objects.

131.049 **Interstellar molecules – a review of recent developments.** B. E. Turner.
Journ. Roy. Astron. Soc. Canada, Vol. 68, 55 - 88 (1974).
Presented at a symposium on 'Chemical evolution in the universe' organized by the Royal Society of Canada, Queen's University, Kingston, Ontario, June 6, 1973.

131.050 **Absorption and emission by insterstellar CH at 9 cm.**
B. J. Robinson, F. F. Gardner, M. W. Sinclair, J. B. Whiteoak.
Nature, Vol. 248, 31 - 32, with a correction, p. 626 (1974).

The authors have observed the 3,335.475 MHz transition of CH in several southern galactic sources. In RCW38 this line is seen in absorption, while the two satellite lines are seen in emission. In several sources the distribution of CH is found to be extended.

131.051 **Observations of the interstellar Na 3302 doublet and**

the interstellar sodium abundance.
K. S. de Boer, S. R. Pottasch.
Astron. Astrophys., Vol. 32, 1 - 5 (1974).

Observations are presented of the interstellar neutral sodium UV doublet at 3302 Å (to be called Na U doublet). Together with the scarce U line data in the literature, these form the basis for deriving accurate Na° column densities. After allowance for ionization balance the authors find that in these clouds with strong D lines, the ratio of total sodium to hydrogen is close to the solar abundance value. Using both doublets (U and D) they find that the internal doppler velocities are of the order of $1-3$ km s^{-1} in most clouds.

131.052 **Polarization observations in the Coma cluster Mel 111.** T. Markkanen.
Astron. Astrophys., Vol. 32, 89 - 91 (1974).

The polarization of light for sixteen Coma cluster stars has been measured. The amount of polarization ranges from $0.01-0.10\%$ with a mean error of ± 0.02. No correlation between polarization and intrinsic stellar parameters has been found. Assuming the polarization to be of interstellar origin, the corresponding lower limits of extinction and colour excesses are estimated.

131.053 **The He$^+$/H$^+$ ratio in dark clouds.**
R. L. Brown, J. Gómez-González.
Astrophys. Journ., Vol. 188, 475 - 480 (1974).

The relative abundance of the ions He$^+$ and H$^+$ in dark clouds is determined by interaction with many atomic and molecular species so that the ratio $n(\text{He}^+)/n(\text{H}^+)$ may be quite different from what is expected in tenuous H I regions. The authors conclude that use of the observational $n(\text{He}^+)/n(\text{H}^+)$ ratio to discriminate between cosmic-ray and X-ray ionization of the interstellar medium is highly unreliable unless all the physical parameters of the recombination line region are precisely known.

131.054 **Optical interstellar line studies of a nearby cold cloud.** R. M. Crutcher, K. W. Riegel.
Astrophys. Journ., Vol. 188, 481 - 488 (1974).

Interstellar Ca II lines in the spectra of OB stars in a large region toward the galactic center are shown to arise in the cold hydrogen cloud seen in self-absorption in the 21-cm line. The distances of two of these stars are 150 and 180 pc, which are thus upper limits to the cloud distance. The authors have derived H I and Na I densities for the line of sight to μ Sgr and can thus compute the electron density. The temperature at this position is also derived. Finally, the existence of interstellar Na I over a wider velocity range than that covered by the cold hydrogen is demonstrated.

131.055 **A study of interstellar polarization at the $\lambda\lambda$ 4430 and 5780 features in HD 183143.**
P. G. Martin, J. R. P. Angel.
Astrophys. Journ., Vol. 188, 517 - 521 (1974).

The authors report measurements of HD 183143 which confirm the absence of any polarization change at $\lambda 4430$. Because of the possibility that the narrower interstellar diffuse lines have a different origin from the $\lambda 4430$ band, they have also observed the polarization through the line $\lambda 5780$, obtaining a similar null result.

131.056 **Von der Phänomenologie zur Physik des kosmischen Staubes.** J. Dorschner.
Sterne, 50. Jahrgang, p. 20 - 29 (1974).

131.057 **Statistical time-dependent model for the interstellar gas.**
H. Gerola, M. Kafatos, R. McCray.
Astrophys. Journ., Vol. 189, 55 - 66 (1974).

The authors present models for temperature and ionization structure of low, uniform-density interstellar gas in a galactic disk which is exposed to soft X-rays from supernova outbursts occurring randomly in space and time. The calculation yields probability distribution functions for ionized fraction x, temperature T, and their various observable moments. These time-dependent models predict a bimodal temperature distribution of the gas with structure in x, T that agrees with various observations.

131.058 **The H$_2$O source in Sagittarius B2.**
J. A. Waak, C. H. Mayer.
Astrophys. Journ., Vol. 189, 67 - 68 (1974).

Observations of water molecule radiation from Sgr B2 show a broad, complex spectrum, and a source region which is extended in declination.

131.059 **On the recombination-line observations toward supernova 3C391.** E. J. Chaisson.
Astrophys. Journ., Vol. 189, 69 - 72 (1974).

Radio recombination-line observations toward the supernova 3C391 are interpreted in terms of emission from a line-of-sight H II region. Measurements of continuum radiation at high microwave frequencies lend some support to this alternative model.

131.060 **Detection of possible maser emission near 3.48 millimeters from an unidentified molecular species in Orion.** L. E. Snyder, D. Buhl.
Astrophys. Journ., (Letters), Vol. 189, L31 - L33 (1974).

The authors have detected a group of unidentified molecular emission lines near 86,245 MHz (~ 3.48 mm) in the direction of the Orion nebula molecular cloud. The narrow line widths and the compactness of the Orion emission region suggest that they have observed the first millimeter-wave maser.

131.061 **Discovery and CO observations of a new molecular source near M17.**
C. Lada, D. F. Dickinson, H. Penfield.
Astrophys. Journ., (Letters), Vol. 189, L35 - L37 (1974).

A rich new source of molecular line radiation has been found near M17. Extended emission from HCN, SO, and millimeter H$_2$CO has been detected. Emission lines of ^{12}CO and ^{13}C^{16}O have been mapped throughout this region, and ^{12}C^{18}O has been detected in two locations.

131.062 **^{12}C/^{13}C abundance ratios from observations of interstellar H$_2$^{13}C^{16}O.**
B. Zuckerman, D. Buhl, P. Palmer, L. E. Snyder.
Astrophys. Journ., Vol. 189, 217 - 220 (1974).

A study of the interstellar H$_2$ ^{12}C^{16}O/H$_2$^{13}C^{16}O abundance ratio in the direction of seven H II regions suggests that the galactic ^{12}C/^{13}C ratio is approximately terrestrial except, possibly, in massive clouds near the galactic center.

131.063 **Molecular CH, CH$^+$, and H$_2$ in the interstellar gas.**
W. D. Watson.
Astrophys. Journ., Vol. 189, 221 - 225 (1974).

Reactions of molecular H$_2$ with CH$^+$ and CH$_2^+$ are likely to lead to the formation of interstellar CH at a rate that is consistent with observation. Rapid destruction of CH$^+$ by the reaction CH$^+$ + H$_2$ $\to$ CH$_2^+$ + H indicates that CH$^+$ and H$_2$ observed in the same direction occur at different locations. CH is expected to occur with the H$_2$.

131.064 **Detection of radio recombination-line emission from the Rho Ophiuchi dark cloud.**
R. L. Brown, G. R. Knapp.
Astrophys. Journ., Vol. 189, 253 - 258 (1974).

Recombination-line observations in the direction of the ρ Oph dark cloud at 21 cm are described in which the authors have detected the carbon 166α and 167α lines. The excellent agreement between the velocity of these lines and the velocity of molecular lines from the ρ Oph cloud lead to an unequivocal identification of the regions of recombination-line emission with the dark cloud itself.

131.065 On the velocity structure of the interstellar clouds near Rho Ophiuchi.
J. G. Cohen, G. Wallerstein.
Astrophys. Journ., Vol. 189, 259 - 261 (1974).

Accurate radial velocities of the optical interstellar features for some of the B stars embedded in the ρ Oph complex are obtained by averaging results of several photographic high-dispersion spectra. In at least one case, there is a real difference between the radial velocity of the atomic features (Na I and Ca II) and the molecular ones (CH and CN).

131.066 The Coalsack. III. A search for T Tauri stars.
W. B. Weaver.
Astrophys. Journ., Vol. 189, 263 - 267 (1974).

Stars found to have Hα emission in the Coalsack on objective-prism plates were examined with photographic $UBVI_{ph}$ photometry and blue objective-prism plates in order to confirm the presence of a T Tauri association suggested by Pik-Sin The. This association was not confirmed.

131.067 The interstellar depletion mystery, or where have all those atoms gone? J. M. Greenberg.
Astrophys. Journ., (Letters), Vol. 189, L81 - L85 (1974).

The observed depletion of intermediate-weight elements O, C, and N from the interstellar medium is shown to be significantly greater than can be accounted for by accretion on interstellar dust. A number of possible explanations are presented, ranging from the existence in interstellar space of many "snowballs" intermediate in size between dust grains and comets to the existence of many far more complicated interstellar molecules than have been detected.

131.068 Dust and gas in the Orion molecular cloud: observations of 1-millimeter continuum and 2-centimeter H_2CO emission.
P. M. Harvey, I. Gatley, M. W. Werner, J. H. Elias, N. J. Evans II, B. Zuckerman, G. Morris, T. Sato, M. M. Litvak.
Astrophys. Journ., (Letters), Vol. 189, L87 - L91 (1974).

The Orion nebula region has been mapped over a $3' \times 6'$ region in the continuum at 1 mm and in the $2_{11} \to 2_{12}$ transition of H_2CO at 2 cm. The similarity of the two maps is strong evidence that the 1-mm radiation originates in the molecular cloud. The column density of hydrogen molecules through the cloud, derived from the H_2CO results, is $\sim 2 \times 10^{23}$ cm^{-2}. The mass of the region observed is estimated to be $\sim 500\, M_\odot$. The radial velocities and linewidths of the H_2CO emission are discussed with respect to collapse, rotation, and turbulence.

131.069 Study of potential interstellar molecules: nuclear quadrupole coupling constants of the nitrogen atom in pyrrole. K. Bolton, R. D. Brown.
Australian Journ. Phys., Vol. 27, 143 - 146 (1974).

The nuclear quadrupole coupling constants of the nitrogen atom in pyrrole have been redetermined by analysis of the hyperfine structure for several low J rotational transitions. Small modifications to the previously reported rotational constants are given.

131.070 Can OH-sources be chemical masers?
W. H. Kegel.
Astron. Astrophys., Vol. 32, 227 - 228 (1974).

It is discussed to which extent the population densities N_m and N_n of the maser levels in OH-sources are determined by the chemical reactions by which the molecules are formed. Since the 18 cm OH emission is due to transitions in the ground state Λ-doublet, it is concluded – assuming stationarity – that the processes that depopulate the lower maser level are more important than those by which the molecules are formed.

131.071 Thermal equlibrium in a molecular hydrogen cloud.
A. Kudo.
Sci. Rep. Tôhoku Univ., First Ser., Vol. 56, 152 - 169 (1973).

The thermal balance in molecular hydrogen clouds is considered. Ionization rate by low-energy cosmic rays and X-rays is calculated, taking ionization, dissociation and excitation due to secondary energetic electrons into consideration. Heating rates by these processes are calculated, including the energy due to molecular reactions and the effect of energetic electrons. The cooling rates of atomic and molecular species which are efficient in dense clouds are estimated. The equilibrium temperature at the center of the cloud and the hydrogen abundance as functions of hydrogen density are computed for several ionization rates. The results are compared with observations.

131.072 The extinction feature at λ 2200 Å and the diffuse interstellar lines. J. Dorschner.
Astron. Nachr., Vol. 295, 147 - 149 = Mitt. Univ. Sternw. Jena No. 120 (1974).

Correlations of the diffuse interstellar lines λλ 5780, 5797, and 6284 Å with the λ 2200 Å feature as well as the interstellar extinction in the visual and in the UV have been found.

131.073 Spiralarme und Wolkenstruktur. K. Rohlfs.
Mitt. Astron. Ges., No. 34, p. 86 - 89 (1973/74).
Presented at the "Wissenschaftliche Tagung der Astron. Ges., Oberkochen, 1973 April 24 - 27".

131.074 Das Ionisationsgleichgewicht eines durch weiche Röntgenstrahlung ionisierten interstellaren Mediums.
M. Grewing, C. M. Walmsley.
Mitt. Astron. Ges., No. 34, p. 158 - 160 (1973/74).
Presented at the "Wissenschaftliche Tagung der Astron. Ges., Oberkochen, 1973 April 24 - 27".

131.075 A comment of the paper "The shapes of neutral globules associated with diffuse nebulae".
J. E. Dyson.
Astron. Astrophys., Vol. 32, 349 - 350 (1974).

The radiation field boundary condition used by Dyson (1973) is re-considered. A more correct boundary condition is set up and the consequences are examined.

131.076 Electric charge and acceleration of suprathermal grains. N. C. Wickramasinghe.
Astrophys. Space Sci., Vol. 28, L25 - L29 (1974).

Dust grains expelled from cool stars into the interstellar medium at speeds $\gtrsim 10^8$ cm s^{-1} are charged by secondary electron emission to potentials in the range 8–40 V. The conditions for the acceleration of such charged grains by the Fermi process are discussed.

131.077 Un radical, son rayonnement et le maser.
Nguyen-Quang-Rieu.
Sci. Progrès Découverte, No. 3442, p. 24 - 32 (1972).

131.078 Study of probability density of a signal from interstellar hydroxyl radio lines.
M. I. Pashchenko, G. M. Rudnitskij, V. I. Slysh.
Izv. vyssh. ucheb. zavedenij. Radiofizika, Vol. 16, 1344 - 1349 (1973). In Russian. – Abstr. in Referativ. Zhurn. 51. Astron., 5.51.639 (1974).

131.079 Hydroxyl and formaldehyde production in the

interstellar clouds observed by Davies and Matthews. D. A. Williams.
Observatory, Vol. 94, 66 - 70 (1974).

This study shows that, whatever process seems suitable for molecule formation in the clouds described by Davies and Matthews, the author expects H_2 to be present. Grain-surface reactions, if fully efficient, will produce the required molecular densities if the total density is high enough. At this total density, appreciable amounts of H_2 will be present.

131.080 A remarkable "bright dark nebula" in Chamaeleon. M. P. FitzGerald.
Astron. Astrophys., Vol. 32, 465 - 468 = Contr. Univ. Waterloo Obs., No. 32 (1974).

A remarkable "bright dark nebula" is reported at $l = 303°74$, $b = -15°95$. The following parameters are derived for it: distance 500 pc; $z = 140$ pc; A_{pg} (centre) $\geq 3\overset{m}{.}7$; mass $\sim 30 M_\odot$.

131.081 A correlation study of carbon ions and hydroxyl molecules toward galactic nebulae. E. J. Chaisson.
Astron. Journ., Vol. 79, 555 - 564 (1974).

New observations of the C 92α recombination line and of the $^2\Pi_{(3/2)}$, $J = 3/2$ rotational transition of OH are reported in the direction of the five strongest carbon-line sources: W3A, W10, W12, W22, and W51A. For every carbon line, there is an OH absorbing feature with a very similar radial velocity; a linear-regression analysis yields a correlation coefficient of 0.999. The widths of the velocity correlated features are also related.

131.082 Wavelength dependence of polarization. XXV. Rotation of the position angle by the interstellar medium. G. V. Coyne.
Astron. Journ., Vol. 79, 565 - 580 (1974).

Multicolor polarimetric data are given for 105 early-type stars. These data plus those previously published in this series of papers are analyzed with respect to the rotation of the position angle of the interstellar polarization with wavelength. Altogether, 24 stars (13 % of those measured in the northern hemisphere) for which the polarization is predominantly interstellar show this effect. The observations support the explanation previously proposed of a change both in the direction of grain alignment and in grain size along the line of sight to those stars.

131.083 Wavelength dependence of polarization. XXVI. The wavelength of maximum polarization as a characteristic parameter of interstellar grains. G. V. Coyne, T. Gehrels, K. Serkowski.
Astron. Journ., Vol. 79, 581 - 589 (1974).

A unique representation of the wavelength dependence of the interstellar polarization is obtained by using the parameter λ_{max}, which is the wavelength at which the maximum polarization p_{max} occurs, to normalize the wavelengths at which the observations are made. Values of λ_{max} and p_{max} are given for all 202 stars for which measures of the interstellar polarization have been made at the University of Arizona. The observed normalized wavelength dependence of interstellar polarization is compared to that predicted by some classical models of the interstellar grains.

131.084 Wavelength dependence of polarization. XXVII. Interstellar polarization from 0.22 to 2.2 μm. T. Gehrels.
Astron. Journ., Vol. 79, 590 - 593 (1974).

The linear polarization of ζ Oph was observed at 0.225 and 0.286 μm with a balloon telescope, between 0.3 and 1.0 μm with standard equipment, and at 1.6 and 2.2 μm with a new infrared polarimeter. A measurement of κ Cas at 0.225 μm is also reported.

131.085 The intrinsic polarization in the light of μ Cephei. T. A. Polyakova.
Astrofizika, Vol. 10, 53 - 63 (1974). In Russian. – English translation in Astrophysics, Vol. 10, No. 1.

The interstellar polarization of μ Cep with $p = 1.5\%$, $\theta_E = 47°$ is redetermined. The intrinsic polarization for 1957–1973 and curves p (λ) (where possible) are derived. A connection between variation of light and intrinsic polarization has been revealed. A qualitative explanation of the results obtained is given.

131.086 Dust grains in a hot gas. I. Basic physics. J. R. Burke, J. Silk.
Astrophys. Journ., Vol. 190, 1 - 10 (1974).

The interaction of graphite grains with a hot gas is investigated. Detailed computations, based on experimental data and simple theoretical models, are presented of the energy transfer by gas particle collisions and of the sputtering rates and grain lifetimes, as functions of gas temperature and grain radius. The electric charge on the grains is calculated, and the effect of electric forces on mechanical stability is discussed. The rate at which the gas cools by this mechanism is evaluated.

131.087 Dust grains in a hot gas. II. Astrophysical applications. J. Silk, J. R. Burke.
Astrophys. Journ., Vol. 190, 11 - 17 (1974).

The results of the work of Burke and Silk on gas-grain cooling and sputtering rates are applied to various astrophysical environments where dust and hot gas may coexist. The effect is studied of swept-up interstellar grains on the evolution of young supernova remnants, and the infrared luminosity is computed as a function of the age of the remnant. The authors propose an interpretation of far-infrared sources embedded in compact H II regions or dense clouds, in terms of the supernova phenomenon, with specific application to η Carinae. They also reexamine the question of the existence of dust in intergalactic matter in galaxy clusters.

131.088 Astrophysical masers. IV. Line widths. P. Goldreich, J. Kwan.
Astrophys. Journ., Vol. 190, 27 - 34 (1974).

The standard theory of line formation in astrophysical masers is outlined. The results of a comparison between the predictions of this theory and the observed line widths are taken as evidence that the theory is incomplete. The standard theory is extended to include effects arising from infrared line radiation trapped between the maser levels and other molecular rotation levels. It is shown that when these effects are included, the discrepancies between theory and observations are resolved.

131.089 Chlorine-bearing molecules in interstellar clouds. M. Jura.
Astrophys. Journ., (Letters), Vol. 190, L33 - L34 (1974).

The reaction $Cl^+ + H_2 \rightarrow HCl^+ + H$ is rapid, and it leads to the formation of the molecules HCl^+, H_2Cl^+, and HCl. It may be possible to detect HCl^+ near 3600 Å, and HCl may be observable at 1291 Å. Also, photodissociation of HCl^+ may be a significant source of protons within clouds, and chemical heating resulting from the destruction of chlorine-bearing molecules may account for perhaps 1 - 10 percent of the expected cooling from diffuse interstellar clouds.

131.090 Line spectra in interstellar clouds. I. The Perseus 2 cloud. F. H. Chaffee, Jr.
Astrophys. Journ., Vol. 189, 427 - 440 (1974).

An observational study of interstellar lines in the spectra of ξ Per, ζ Per, and o Per in the region 3100 Å $\leq \lambda \leq$ 4300 Å is presented. Spectral lines arising from Ca II, Na I, Ti II, Fe I,

CN, CH, and CH$^+$ have been detected and studied by means of a Cassegrain echelle spectrograph. From several arguments a diameter of 16 pc and a mean hydrogen density of 30 cm^{-3} can be assigned to the cloud. An ionization model of the cloud has been constructed based on White's measured electron densities. The inferred abundances of the atomic constituents suggest that calcium, titanium, and iron are underabundant by an average of 1200, 800, and 5, respectively; sodium is found to be nearly solar. The observed molecular column densities seem inconsistent with the predictions of Solomon and Klemperer, and this suggests that physical processes other than the simple gas-phase reactions considered by these authors may be responsible for determining interstellar diatomic molecular equilibria. For both atomic and molecular constituents, the Perseus 2 cloud is found to be rather heterogeneous—variations up to factors of 10 in some trace-element abundances exist from point to point in the cloud.

131.091 **Molecular clouds.** P. Goldreich, J. Kwan.
Astrophys. Journ., Vol. 189, 441 - 453 (1974).

It is proposed that molecular clouds are in a state of gravitational collapse. The coupled equations of statistical equilibrium and radiative transfer from diatomic molecules in a collapsing cloud are solved for arbitrary optical depths in the rotational lines. It is shown that most of the observed CS and SiO lines and the stronger CO lines are optically thick. The rate at which energy is radiated in the CO lines is found to exceed the rate at which work is done by the adiabatic compression of the collapsing gas. This result implies the existence of an energy source which maintains the temperature of the gas against the cooling due to radiative energy losses. It is suggested that collisions between gas molecules and warm dust grains transfer energy to the gas.

131.092 **OH observations near the reflection nebulae NGC 2068 and NGC 2071.** L. E. B. Johansson, B. Höglund, A. Winnberg, Nguyen-Q-Rieu, W. M. Goss.
Astrophys. Journ., Vol. 189, 455 - 458 (1974).

Narrow OH emission lines at 1667 MHz, apparently from a class 1 source, have been observed near the reflection nebula NGC 2071. OH emission corresponding to the dust cloud north and east of NGC 2024 is also seen. At 1720 MHz, the dust-cloud component appears in absorption.

131.093 **Hartree-Fock bound states for molecule-ions HeC^{2+} and HeC$^+$.**
S. W. Harrison, G. A. Henderson, L. J. Massa, P. Solomon.
Astrophys. Journ., Vol. 189, 605 - 607 (1974).

Roothaan-Hartree-Fock calculations on the molecule-ions HeC^{2+} and HeC$^+$ suggest that they are stable, but weakly bound, systems. Spectroscopic constants for the $^1\Sigma^+$ ground state of HeC^{2+} and the presumed $^2\Pi$ ground state of HeC$^+$ are given. It is suggested that these molecule-ions may exist in interstellar space.

131.094 **A new limit on the interstellar abundance of boron.**
D. C. Morton, A. M. Smith, T. P. Stecher.
Astrophys. Journ., (Letters), Vol. 189, L109 - L111 (1974).

The authors report a new upper limit on interstellar boron in the direction of ζ Oph, obtained by making 25 scans across the B II resonance line at 1362.461 Å with the Copernicus satellite. They find B/H $\leq 7.6 \times 10^{-11}$ by number, implying that boron in the interstellar gas is depleted by a factor of about 60 relative to boron in carbonaceous chondrites. These results are related to theories of the formation of grains, and the flux of cosmic rays.

131.095 **A new source of intense molecular emission in the Rho Ophiuchi complex.** P. J. Encrenaz.
Astrophys. Journ., (Letters), Vol. 189, L135 - L136 (1974).

A new source of intense molecular line emission has been found in a dark cloud in the ρ Ophiuchi complex. The densities implied by the observations are higher than 10^5 cm^{-3}.

131.096 **Review of ultraviolet interstellar extinction and interstellar lines.** L. F. Smith.
Astron. Inst. Utrecht, Internal Rep. ROF 72, (see 012.012), p. C17.1 - C17.36 (1974).

131.097 **Observations of interstellar atoms by the Copernicus satellite.** J. B. Rogerson, Jr.
Astron. Inst. Utrecht, Internal Rep. ROF 72, (see 012.012), p. C19.1 (1974). – Abstract.

131.098 **Observations of interstellar molecules by the Copernicus satellite.** E. B. Jenkins.
Astron. Inst. Utrecht, Internal Rep. ROF 72, (see 012.012), p. C20.1 (1974). – Abstract.

131.099 **Interstellar absorption lines observed with the orbiting spectrophotometer S-59.** K. S. de Boer.
Astron. Inst. Utrecht, Internal Rep. ROF 72, (see 012.012), p. C23.1 - C23.8 (1974).

131.100 **OH absorption in the direction of W51. I. Observations.** V. I. Slysh.
Astron. Zhurn. Akad. Nauk SSSR, Vol. 51, 470 - 478 (1974). In Russian. English translation in Soviet Astron. AJ, Vol. 18, No. 3.

The large radio telescope in Nançay was used for mapping the source complex W51 in the continuum and OH absorption lines 1667 and 1665 MHz. Continuum and line isophotes are presented as well as right ascension – radial velocity diagrams. 1667 MHz maser emission was discovered coinciding with the 1665 MHz maser emission source.

131.101 **The formation of special type structures behind a shock wave front in the interstellar gas.**
V. S. Avedisova.
Astron. Zhurn. Akad. Nauk SSSR, Vol. 51, 479 - 488 (1974). In Russian. English translation in Soviet Astron. AJ, Vol. 18, No. 3.

The development of thermal and convective instabilities behind a shock wave front generated by stellar wind of WR stars is considered including magnetic field effects. The condensation mode of thermal instability in an infinite layer is derived. The structures of nebulae NGC 6888, NGC 2359 and S308 are considered.

131.102 **Density of the interstellar gas and strength of the magnetic field of the Small Magellanic Cloud.**
A. V. Zasov.
Astron. Zhurn. Akad. Nauk SSSR, Vol. 51, 502 - 507 (1974). In Russian. English translation in Soviet Astron. AJ, Vol. 18, No. 3.

A possibility of theoretical determination of gas density and magnetic field strength is considered for a disk of galaxies if their internal mass distribution, surface density of gas, synchrotron brightness and/or velocity dispersion along the z-axis are known. As an example the results of the calculation for the Small Magellanic Cloud are given.

131.103 **On chemical and collisional pumping of interstellar OH maser.** V. V. Burdyuzha, S. Ya. Umanskij.
Astron. Zhurn. Akad. Nauk SSSR, Vol. 51, 669 - 671 (1974). In Russian. English translation in Soviet Astron. AJ, Vol. 18, No. 3.

On the basis of the calculated velocity constant for transitions between rotational sublevels of the OH molecule it is shown that the demands on the density in chemical and collisional mechanisms of pumping are conflicting.

131.104 **A nonlinear model for the intensity, line width, and coherence of astrophysical masers.**
R. A. Rosen.
Astrophys. Journal, (Letters), Vol. 190, L73 - L76 (1974).

The effect of pumping on the steady-state coherent lossless light wave trains which can propagate through a two-level resonant medium is studied. The results are applied to the physical conditions present in astrophysical clouds from which masing is observed, and it is shown that present observations are consistent with the form of coherent radiation predicted by the model described.

131.105 **Observations of interstellar CH microwave line ratios.** O. E. H. Rydbeck, J. Elldér, W. M. Irvine, A. Sume, Å. Hjalmarson.
Astron. Astrophys., Vol. 33, 315 - 319 (1974).

Emission line ratios for the three transitions of the ground state Λ-doublet of interstellar CH indicate significant departures from equilibrium (LTE) population ratios in H I regions, as well as in the direction of H II regions.

131.106 **Molecules as probes of the interstellar matter.**
P. G. Mezger.
Stars and the Milky Way system, Proc. 1972, (see 012.018), p. 88 - 114 (1974). – Invited lecture.

131.107 **Molecules in the widely distributed interstellar clouds.** R. D. Davies, R. J. Cohen, A. J. Wilson.
Stars and the Milky Way system, Proc. 1972, (see 012.018), p. 115 - 119 (1974).

131.108 **Very long baseline interferometer measurements of H_2O masers in Orion A, W49N, W3OH and VY CMa.**
G. D. Papadopoulos.
Stars and the Milky Way system, Proc. 1972, (see 012.018), p. 129 - 130 (1974). – Abstract.

131.109 **High resolution search for interstellar lithium.**
H. E. Utiger.
Stars and the Milky Way system, Proc. 1972, (see 012.018), p. 131 - 139 (1974).

131.110 **Diffusion of elements in the Galaxy.**
E. Basinska, W. Iwanowska.
Stars and the Milky Way system, Proc. 1972, (see 012.018), p. 165 - 173 (1974).

131.111 **The V1057 Cygni OH sources: accurate position and polarization properties.**
K. Y. Lo, K. P. Bechis.
Bull. American Astron. Soc., Vol. 6, 220 - 221 (1974). Abstr. AAS.

131.112 **H_2CO and HI observations of dark clouds in NGC 2264.** Y. K. Minn, J. M. Greenberg.
Bull. American Astron. Soc., Vol. 6, 221 (1974). – Abstr. AAS.

131.113 **CO in the Cygnus X region.** H. R. Dickel, J. R. Dickel, W. J. Wilson, E. E. Epstein.
Bull. American Astron. Soc., Vol. 6, 221 (1974). – Abstr. AAS.

131.114 **Extended studies of microwave emission from interstellar CH.**
O. E. H. Rydbeck, J. Ellder, W. M. Irvine.
Bull. American Astron. Soc., Vol. 6, 221 (1974). – Abstr. AAS.

131.115 **On the origin of polyatomic molecules in dense clouds.** B. Donn, L. J. Stief, J. Ormes.
Bull. American Astron. Soc., Vol. 6, 221 (1974). – Abstr. AAS.

131.116 **The interstellar medium in the line-of-sight to α Virgo.** D. G. York.
Bull. American Astron. Soc., Vol. 6, 225 (1974). – Abstr. AAS.

131.117 **Radio detection of the three ground state lines of interstellar CH.**
O. E. H. Rydbeck, J. Ellder, W. M. Irvine.
Bull. American Astron. Soc., Vol. 6, 261 (1974). – Abstr. AAS.

131.118 **Stable equilibria of the interstellar gas and magnetic field in the galactic gravitational field.**
T. C. Mouschovias.
Bull. American Astron. Soc., Vol. 6, 262 - 263 (1974). Abstr. AAS.

131.119 **The interpretation of radio molecular spatial maps – simple linear molecules.** F. O. Clark.
Bull. American Astron. Soc., Vol. 6, 263 (1974). – Abstr. AAS.

131.120 **Study of correlations among OH, H_2CO, HI and dust toward W 44.** P. C. Myers.
Bull. American Astron. Soc., Vol. 6, 263 (1974). – Abstr. AAS.

131.121 **Interstellar circular polarization: results from a survey of 84 stars.** R. A. Stokes, J. B. Swedlund, R. W. Avery, J. J. Michalsky.
Astron. Journ., Vol. 79, 678 - 681 (1974).

Circular polarization observations of 84 bright stars were carried out on the Battelle 79-cm telescope during the one-year period ending 1973 August. The survey was aimed at detecting circular polarization at the 0.01% level due to interstellar dust. Thirteen stars were found to be circularly polarized.

131.122 **Organische Moleküle im interstellaren Raum.**
E. W. Salpeter.
Analyse extraterrestrischen Materials, (see 012.020), p. 203 - 212 (1974).

131.123 **Molecules in interstellar space.** V. Vanýsek.
Říše hvězd, Vol. 55, 121 - 123 (1974). In Czech.

131.124 **Cooling of dense interstellar clouds by carbon monoxide and carbon atoms.** T. Nakano.
Publ. Astron. Soc. Japan, Vol. 26, 189 - 196 (1974).

The cooling of dense clouds by the rotational transition lines of CO and the magnetic dipole transition lines of C atoms is investigated by examining the excitation states of CO and C and the radiation field in the clouds. It is found that the low brightness temperature near 6 K observed for the 1–0 rotational line of CO in many clouds can be explained by the balance between cooling by CO and C and heating by the star light in the mean interstellar space.

131.125 **The use of large southern telescopes for studying interstellar matter.** J. Lequeux.
Proc. ESO/SRC/CERN conference, (see 012.021), p. 127 - 129 (1974).

131.126 **Interstellar molecules and the origin of life.**
M. Shimizu.
Viva Origino, Vol. 2, No. 1, p. 1 - 8 (1973). In Japanese.

Recent theories on the formation mechanism of interstellar molecules are surveyed. The similarity of the composition of the cometary coma and tail to that of the interstellar molecules is pointed out and the cometary nuclei are suggested to be correlated with the solar nebula. Some theories on the formation of the primitive earth in the solar nebula are briefly described. Thermal origin of prebiological molecules in the primitive ocean is discussed.

131.127 **U Orionis.** L. Pataki, J. Kolena.

IAU Circ., No. 2680 (1974).

131.128 Interstellar hydroxyl and water masers and formaldehyde masers and dasars.
D. ter Haar, M. A. Pelling.
Rep. Progr. Phys., Vol. 37, 481 - 561 (1974).

A review of the observational data on anomalous emission by OH, H_2O and H_2CO molecules and on anomalous absorption by H_2CO molecules. Various theories advanced to account for these data and some models of a few particular sources are discussed.

131.129 On the origin of interstellar molecules.
A. Dauvillier.
Bull. Soc. Roy. Sci. Liège, (*Belgium*), Vol. 42, 413 - 416 (1973).

The author, after classifying the interstellar molecules so far discovered, shows that they could not be formed in a high vacuum and at low temperatures. He considers that they are the result of paleovolcanic processes in the absence of oxygen and at elevated temperatures and that these products were evaporated from planetary surfaces during the course of stellar evolution.

131.130 Galactic clouds of organic molecules. D. Buhl.
Origin of Life, (*Netherlands*), Vol. 5, 29 - 40 (1974).

The discovery of immense organic clouds of gas embedded in the dusty regions of our Galaxy is of tremendous importance to the origin of life. It is within these clouds that the formation of stars and planetary systems is believed to take place.

131.131 A partial-coalescence model for colliding interstellar clouds. J. S. Goldstein, A. J. Mazzella, Jr.
Nuovo Cimento B, Ser. 11, Vol. 21 B, 142 - 150 (1974).

Previous statistical models for the mass distribution of interstellar clouds have assumed that two colliding clouds will coalesce to form one cloud. In contrast to the complete-coalescence model, the partial-coalescence model considered here gives a much lower number of large-mass clouds.

131.132 Interstellar dust. J. Lequeux.
Recherche, (*France*), Vol. 5, No. 41, p. 65 - 69 (1974). In French.

The author describes interstellar DWT, and its detection. It is found throughout interstellar regions, closely mixed with gas. The author discusses the origin of the dust, and the role of giants and supergiants.

131.133 Cosmic background radiation at 1.32 millimeters.
D. J. Hegyi, W. A. Traub, N. P. Carleton.
Astrophys. Journ., Vol. 190, 543 - 544 (1974).

The authors have remeasured the $R(2)$ line of the 3874 Å band of interstellar CN toward ζ Oph and find an equivalent width of 0.120 ± 0.074 mÅ, which implies an $R(2)-R(1)$ rotational temperature of $3.1°$ ($+0.5°$, $-0.7°$) K. Summing these observations with their earlier spectrum and refitting the summation, they find the $R(2)$ equivalent width to be 0.089 ± 0.051 mÅ, which implies a temperature of $2.9°$ ($+0.4°$, $-0.5°$) K for the cosmic background radiation at 1.32 mm.

131.134 Observational evidence for the excitation of HCN and H_2O in protostellar molecular clouds.
F. O. Clark, D. Buhl, L. E. Snyder.
Astrophys. Journ., Vol. 190, 545 - 556 (1974).

Detailed modeling of the molecular cloud and excitation of HCN in the Orion nebula have been compared with HCN observations. This results in information about the density distribution of the cloud, collisional interaction of H_2 with HCN, number densities, kinetic temperature, cloud mass, upper age limit, and molecular isotopic ratios. The results are combined with *VLB* observations of H_2O masers in Orion and detailed H_2O excitation calculations of others to produce a consistent interpretation of the Orion molecular cloud.

131.135 On detecting cold, low-density interstellar gas.
J. C. Weisheit.
Astrophys. Journal, (*Letters*), Vol. 190, L121 - L123 (1974).

The author points out that Hobb's very high resolution scans of interstellar Na I D lines in the spectra of hot stars are consistent with the existence of cold intercloud gas in the direction of the Orion association, and he describes additional Na I observations which could detect other regions of cold intercloud gas.

131.136 Dark nebulae in the Small Magellanic Cloud.
P. W. Hodge.
Publ. Astron. Soc. Pacific, Vol. 86, 263 - 266 (1974).

A search for dark nebulae on a set of Schmidt telescope plates of the Small Magellanic Cloud has resulted in a catalog of 45 objects, ranging from $1'$ to $6'$ in major dimension. The average size is 42 pc, and the average extinction, a few tenths of a magnitude. They are generally about half the size of those found in a similar survey of the Large Magellanic Cloud.

131.137 Interaction of stars and interstellar matter.
L. A. Pavlova.
Astron. Tsirk., No. 814, p. 1 - 2 (1974). In Russian.

131.138 On the different methods of determination of anisotropy of star light absorption.
Eh. Ya. Vil'koviskij, L. A. Pavlova.
Astron. Tsirk., No. 814, p. 2 - 4 (1974). In Russian.

131.139 Formation of interstellar molecules on the surface of grains.
T. Iguchi, A. Sakata, H. Murakami, N. Nakagawa.
Chem. Letters, (*Japan*), No. 5, p. 505 - 508 (1973).

Hot hydrogen atoms: initiators of reactions of interest in interstellar chemistry and evolution.
See Abstr. 022.049.

A reflector type beam switching for millimeter wave telescope, and its application to a search for interstellar molecule of CH_3NH_2. See Abstr. 033.049.

X-ray transition radiation formed in molecular clouds. See Abstr. 061.057.

Polytropic sheets, cylinders and spheres with negative index. See Abstr. 061.060.

Die Bedeutung der Spurenelemente für die Kosmochemie. See Abstr. 061.072.

A plasma hypothesis for anomalous OH emission. See Abstr. 062.018.

Sonnenfleckentätigkeit und interstellare Materie. See Abstr. 072.013.

Further remarks on plasma instabilities produced by ions born in the solar wind. See Abstr. 074.033.

Photopolarimetry of planets and stars.
See Abstr. 091.005.

Results of lunar occultations of the galactic center region in HI, OH and H_2CO lines and in the nearby continua. See Abstr. 096.023.

Local hydrogen gas and the background Lyman-alpha pattern. See Abstr. 106.009.

Untersuchung der interstellaren Verfärbung an O- und B-Sternen. See Abstr. 113.040.

Detection of NML Cygnus at 2.8 cm. See Abstr. 114.018.

Equivalent width of interstellar molecular lines. IV: Analyses of the line spectrum of H_2 in the interstellar spectrum of δ Scorpii. See Abstr. 114.020.

Spectral energy distributions of T Tauri stars. See Abstr. 114.028.

Pre-main-sequence stars. III. Herbig Be/Ae stars and other selected objects. See Abstr. 114.029.

Time variation of the H_2O maser and infrared continuum in late-type stars. See Abstr. 114.048.

Spectral observations of stars and interstellar gas in the balloon-ultraviolet. See Abstr. 114.118.

High resolution ultraviolet stellar spectra obtained with the Orbiting Stellar Spectrophotometer S59. See Abstr. 114.130.

Preliminary results of a stellar search for the unidentified molecular maser line near 3.48 mm. See Abstr. 114.140.

Interstellar lines in the ultraviolet spectrum of Delta Scorpii. See Abstr. 114.155.

The V1057 Cygni OH source: time variation, polarization properties, and accurate position. See Abstr. 114.161.

The ratio of total to selective absorption from RS Puppis. See Abstr. 122.061.

Some new data on the rate of stellar flaring in the Coalsack. See Abstr. 122.110.

Polarization of supernova remnants KE 69, G 357.7−0.1, HC 26, and γ Cygni at 6- and 2.8-cm wavelengths. See Abstr. 125.015.

Further observations of interstellar gas in the Gum nebula. See Abstr. 132.021.

Isotopic abundances and line formation in the Orion nebula. See Abstr. 132.029.

Pulsar hydrogen line absorption and the electron density in the interstellar medium. See Abstr. 141.325.

Interstellar scattering of the Vela pulsar. See Abstr. 141.343.

A search for soft X-ray sources in the galactic anticenter. Absorption of X-rays from the Crab nebula. See Abstr. 142.033.

Observations of the circular polarization in HDE 226868 = Cyg X-1. See Abstr. 142.066.

Cosmic gamma-ray bursts from relativistic dust grains. See Abstr. 142.081.

Molecular clouds in the galactic nucleus. See Abstr. 155.004.

Sensitivity of the star formation rate to the interstellar gas abundance of heavy elements. See Abstr. 155.025.

The local complex of O and B stars. I. Distribution of stars and interstellar dust. See Abstr. 155.033.

Isotopic abundances and their variations within the Galaxy. See Abstr. 155.035.

The linear polarization of part of the northern sky at 49-cm wavelength. See Abstr. 155.041.

Gas kinematics near the galactic center from CO observations. See Abstr. 155.054.

The Parker instability in differentially-rotating disks. See Abstr. 155.056.

The outer spiral structure of the Galaxy and the high-velocity clouds. See Abstr. 155.062.

The galactic background radiation and thermal electrons in the Galaxy. See Abstr. 155.071.

Models of three-dimensional variation of galactic absorption. See Abstr. 155.077.

An almost complete survey of 21 cm line radiation for $|b| \geqslant 10°$. I. Atlas of contour maps. See Abstr. 157.005.

An almost complete survey of 21 cm line radiation for $|b| \geqslant 10°$. II. The accurate data on machine readable magnetic tape. See Abstr. 157.006.

H I, H II Regions

131.501 Carbon recombination lines and interstellar hydrogen clouds. A. K. Dupree.
Astrophys. Journ., Vol. 187, 25 - 39 (1974).

Observations of radiofrequency recombination lines of carbon and hydrogen are used in conjunction with theoretical predictions of excitation and emission processes, to obtain the temperatures, densities, sizes, and energy requirements of three H I regions. Such parameters are required for comparison with various theories of heating, ionization, and cooling in the interstellar medium. Because the carbon-emitting clouds are associated with regions producing molecular lines, the conclusions reached here can also better define relationships between H I clouds and molecular sources.

131.502 Interaction of hot stars and of the interstellar medium. IV. Two bright Hα-knots associated with the H II region Sharpless 88. M. C. Lortet-Zuckermann.
Astron. Astrophys., Vol. 30, 67 - 72 (1974).

Two bright Hα-knots near the diffuse nebula Sharpless 88 are described. They both contain one or two central stars, one of which has been found from UBV photometry to be a reddened blue star. Attention is drawn to the relative positions of these knots, a closeby OH source and two radio continuum sources, all located in an obscured region definitely distinct from the more extended optical nebula Sharpless 88.

131.503 Ionization of cool H I regions.
G. Steigman, B. Z. Kozlovsky, M. J. Rees.
Astron. Astrophys., Vol. 30, 87 - 93 (1974).

The diffuse, soft ($E \gtrsim 0.3$ keV) X-ray flux contributes significantly to the ionization of atoms and ions in the interstellar medium. In cool H I regions ions such as C IV and N III should be sufficiently abundant to be detectable. The authors have restricted their attention to the ionization of carbon and nitrogen in typical interstellar clouds. They also compute the ionization due to a large flux of low energy cosmic rays ($E \approx 1 - 10$ MeV).

131.504 Infrared observation of DR 21 at 2.2 μ.
F. Sibille, J. Bergeat, M. Lunel.
Astron. Astrophys., Vol. 30, 181 - 183 (1974).

The radio source DR 21 has been measured at 2.2 μ. The flux density in a 40" beam is 0.1 fu. The proposed mechanism is pure "free-free + free-bound" emission of the ionized gas. Derived absorption at 2.2 μ is 4 magnitudes. A close spectral similarity between DR 21 and IRS 3/W 3 B is suggested.

131.505 Interaction of hot stars and of the interstellar medium. V. The compact optical regions Sharpless 255, 257 and 269.
M. Chopinet, L. Deharveng-Baudel, M. C. Lortet-Zuckermann.
Astron. Astrophys., Vol. 30, 233 - 240 (1974).

Hβ photographs obtained using a Lallemand electronic camera, low dispersion spectra in the range 4300–7300 Å and high dispersion spectra near Hα have been obtained for three galactic nebulae. The Hα radial velocity field of Sh 2-255 and the identification of the exciting star of Sh 2-269 are discussed. All three nebulae, as inferred from their relatively high density and from their association with OH and H$_2$O sources might represent an early stage of evolution of H II regions.

131.506 A statistical model for the intercloud H I gas. I. Theory.
U. Mebold, O. Hachenberg, C. A. Laury-Micoulaut.
Astron. Astrophys., Vol. 30, 329 - 338 (1974).

The small scale fluctuations observed in the emission of the intercloud HI gas have been explained in the literature by a gas model in which this gas is composed of a large number of turbulence cells which are randomly distributed in space and velocity. In the present paper analytical formulae are derived from which the most interesting parameters of the proposed model can be derived. An application of these formulae to published data shows that these data are not accurate enough to give significant results for the parameters – especially the kinetic temperature – of the observed gas.

131.507 H II regions [observational history].
H. J. Habing.
Proc. NATO Advanced Study Institute on the 'Interstellar medium', Schliersee, Germany, 1973 April 2 - 13 [D. Reidel Publishing Company, Dordrecht-Holland – Boston-U.S.A.], p. 91 - 125 (1974).

Gives a broad picture of recent developments in the study of H II regions.

131.508 Studies of neutral-hydrogen cloud structure.
G. L. Verschuur.
Astrophys. Journ., Suppl. Ser., No. 238, Vol. 27, 65 - 112 (1974).

The properties of 200 neutral hydrogen clouds are discussed and the data suggest that clouds are not moving as randomly as has heretofore been suggested. Clouds are often filamentary or parts of filaments. Typical cloud masses are 10 $M_\odot$ for a 3 pc diameter cloud and the mass of the H I in the filaments is of this order per length of 3 pc in the filaments. The observations for two regions surveyed are given here in toto to show the nature of the small-scale structure in the interstellar H I.

131.509 On the nature of maser OH sources connected with H II regions. V. V. Burdyuzha.
Astron. Zhurn. Akad. Nauk SSSR, Vol. 51, 26 - 29 (1974). In Russian. English translation in Soviet Astron. AJ, Vol. 18, No. 1.

On the basis of an analysis of observations and the pumping mechanism by far-infrared radiation of galactic maser OH a model of maser sources is built connected with H II regions.

131.510 The presence of dust in H II regions.
G. Tenorio-Tagle.
Astrophys. Space Sci., Vol. 26, 111 - 114 (1974).

The gas to dust ratio distribution in M 16 is determined. The gas to dust ratio reaches (as in M 8 and in Orion) a constant value throughout the nebula with the exception of the inner part. In all the available cases (even in Orion) the amount of dust seems to increase in the inner part of the nebula (say, 4' from the centre).

131.511 On the nature of some non radio emitting Sharpless H II regions. M. Felli, M. Perinotto.
Astrophys. Space Sci., Vol. 26, 115 - 122 (1974).

Thirty four objects appearing in the Sharpless Catalogue of H II Regions with radio flux densities below 0.1 fu at 1400 MHz are studied. It is shown that two of them are planetary nebulae, two are reflection nebulae and two weak emission-reflection nebulae. From statistical arguments it is concluded that six may be evolved planetaries while the remaining are faint nearby H II regions excited by stars of relatively advanced spectral type.

131.512 A model of compact H II regions emitting in the infrared. S. R. Pottasch.
Astron. Astrophys., Vol. 30, 371 - 379 (1974).

The compact H II regions, which are strong emitting sources in the infrared, are studied. The conclusions drawn are presented.

131.513 OH and H$_2$CO absorption line measurements toward sources in W 31. T. L. Wilson.

Astron. Astrophys., Vol. 31, 83 - 87 (1974).

The H_2CO absorption has been mapped in the direction of the H II regions in W 31, and new OH profiles have been taken toward the three continuum peaks. These results confirm the discrepancy between kinematic distances obtained from the H 109α emission and from OH and H_2CO absorption velocities. The velocities of the H II regions deviate from the Schmidt rotation model by as much as 36 km s^{-1}, and kinematic distances derived from recombination lines are unreliable for these sources.

131.514 Helium abundances in four galactic H II regions from radio recombination line observations.
W. K. Huchtmeier, R. A. Batchelor.
Australian Journ. Phys., Vol. 26, 901 - 902 (1973).

Helium abundances in four galactic H II regions have been derived from radio recombination line observations.

131.515 A comparison catalogue of H II-regions.
P. Maršálková.
Astrophys. Space Sci., Vol. 27, 3 - 110 (1974).

For complete information about the apparent distribution of H II-regions along the whole galactic equator a catalogue of 698 known emission nebulae is given on the basis of 13 lists and catalogues of these objects. For H II-regions, only optical data were taken into account. Where several authors describe the complex of nebulosities a more detailed specification of objects is preferred. Coincidences with SNR and planetaries were sought for independently of identifications given in the source-catalogues. The list of objects from the source-catalogues of H II-regions excluded as SNR, planetaries and galaxies is also given.

131.516 An interferometer study of H I absorption in Centaurus A. M. C. H. Wright.
Astron. Astrophys., Vol. 31, 283 - 287 (1974).

Interferometer observations of the narrow H I absorption features around 560 km s^{-1} in the radio source Centaurus A (NGC 5128) have failed to reveal fine scale structure. Accordingly the absorption must be distributed over the radio emitting region in a phenomenon which may be similar to the high-latitude clouds in our Galaxy. The existence of a neutral hydrogen emission feature at 760 km s^{-1} is also confirmed.

131.517 A very high-velocity cloud. M. C. H. Wright.
Astron. Astrophys., Vol. 31, 317 - 322 (1974).

λ 21 cm observations have revealed a H I cloud some $5° \times 3°$ in angular extent in a direction towards $l = 128°$, $b = -33°$ and at an LSR velocity of ~ -400 km s^{-1}. The cloud has a peak column density 4.2×10^{19} atoms cm^{-2} and a H I mass in excess of $200 \times (D[kpc])^2 M_\odot$. The appearance of the cloud and the velocity dispersions within a concentration are similar to those found in other high-velocity clouds and this cloud is discussed in terms of theories relating to the high-velocity cloud phenomenon.

131.518 Étude statistique sur les vitesses radiales des raies Hα et [N II] dans deux régions H II. R. Louise.
Comptes Rendus Acad. Sci. Paris, Sér. B, Vol. 278, 177 - 179 (1974).

A partir des anneaux d'interférence en Hα et [N II] obtenus sur IC 1318 et IC 405 on cherche une relation entre les vitesses radiales de ces deux raies en faisant une statistique sur plus de 200 points pour chaque nébuleuse. Cette statistique montre que ces deux raies ont des mouvements turbulents identiques et justifie a posteriori la méthode de détermination de la température dans les régions H II à partir des profils des raies Hα et [N II].

131.519 The spectrum of the compact H II region RCW 117.
M. W. Feast.
Observatory, Vol. 94, 13 (1974).

This compact H II region is found to have a total visual absorption of about 11^m and an electron density of about 4.5×10^3 cm^{-3}.

131.520 Heat and ionization sources in the interstellar medium. J. Silk.
Publ. Astron. Soc. Pacific, Vol. 85, 704 - 713 (1973).

This is a review of the astrophysical sources of heat and ionization in interstellar clouds and in the intercloud medium. Observational data on density, ionization, and temperatures of H I regions are summarized, and the heat and ionization requirements are specified in terms of heat input H and hydrogen ionization rate ζ_H. A wide variety of sources and mechanisms are reviewed and contrasted, and corresponding estimates of H and ζ_H are presented in tabular form.

131.521 The nature of small H II regions deduced from their radio emission.
Y. Terzian, B. Dennison, B. Balick.
Publ. Astron. Soc. Pacific, Vol. 85, 806 - 810 (1973).

The authors report radio observations of several apparently small thermal sources at $\lambda 11$ cm and discuss their physical nature. Most of these objects are either small H II regions or planetary nebulae.

131.522 Observation of Ne II in the compact H II region G333.6−0.2. D. K. Aitken, B. Jones.
Monthly Notices Roy. Astron. Soc., Vol. 167, 11P - 15P (1974).

Observations of the spectrum of G333.6−0.2 between 8μ and 13μ show emission from the Ne II line at 12.80μ. A neon abundance in the range 7.63−7.93 is found, in agreement with recent estimates of the cosmic abundance of neon. The continuum radiation is considered to be thermal emission.

131.523 Infrared studies of H II regions and OH sources.
C. G. Wynn-Williams, E. E. Becklin, G. Neugebauer.
Astrophys. Journ., Vol. 187, 473 - 485 (1974).

This paper presents the observational results of a high spatial-resolution mapping and photometric study in the wavelength range $1.65 \leq \lambda \leq 20\mu$ of four H II regions and seven OH sources.

131.524 Infrared emission from the southern H II region H2-3. E. E. Becklin, J. A. Frogel, D. E. Kleinmann, G. Neugebauer, S. E. Persson, C. G. Wynn-Williams.
Astrophys. Journ., Vol. 187, 487 - 489 (1974).

The H II region H2-3 has been found to have a large infrared flux with a luminosity of $1-2 \times 10^5 L_\odot$ between 1.65 and 25μ. Most of this flux comes from a single component with a diameter of $110''$ (2 pc). At 2.2μ there is an unresolved source which is identified as the exciting star of the nebula. The 3- to 25-μ radiation is shown to be consistent with dust heated by Lα radiation within the nebula, but much of the 40- to 350-μ radiation probably originates from outside the H II region.

131.525 A survey of fine structure in galactic H II regions.
M. Felli, G. Tofani, L. R. D'Addario.
Astron. Astrophys., Vol. 31, 431 - 439 (1974).

A sample of 140 galactic H II regions has been observed with an east-west equivalent half-power beam-width of 16.1 s of arc at a wavelength of 2.8 cm. Angular structure smaller than one minute of arc and stronger than 1 fu was found in 26 regions, 19 of which had not previously been observed with such a high resolution. Taking into account the sensitivity limit of the survey, it is estimated that roughly 28% of all H II regions show compact angular structure.

131.526 The distribution of neutral hydrogen around

α Virginis.   I. Fejes.
Astron. Journ., Vol. 79, 25 - 30 (1974).

21-cm line observations of interstellar neutral hydrogen in the region $13^h04^m \leq RA \leq 13^h44^m$, $-16° \leq D \leq -6°$ are presented. A peculiar 5°–8° diameter deficiency is found in the HI column density distribution around α Virginis, which may be interpreted as the result of ionization effects of the star.

131.527 Infrared emission from H II regions.
C. G. Wynn-Williams, E. E. Becklin.
Publ. Astron. Soc. Pacific, Vol. 86, 5 - 25 (1974). – Invited review article.

131.528 Helium abundance at the galactic center.
M. Jura, E. L. Wright.
Astrophys. Journ., Vol. 188, 473 - 474 (1974).

The authors suggest that the low abundance of ionized helium in H II regions at the galactic center can be explained by selective absorption of helium-ionizing photons by dust within the ionized gas.

131.529 Spectrophotometric observations of the compact H II region K3-50 and of NGC 6857.
S. E. Persson, J. A. Frogel.
Astrophys. Journ., Vol. 188, 523 - 527 (1974).

Spectrophotometric observations of K3-50 from 0.5 to 1.1 μ demonstrate that there is a discrepancy between the values of the interstellar extinction determined from radio, infrared, and optical measurements. Similar observations of the nearby nebula NGC 6857 show that it differs in several respects from K3-50. In particular, the extinction toward NGC 6857 is considerably less than that observed toward K3-50.

131.530 On ionization in H I regions.   L. M. Hobbs.
Astrophys. Journ., (Letters), Vol. 188, L107 - L109 (1974).

A comparison is made of the K I, Na I, Ca II, and H I interstellar absorption lines formed in relatively low-density interstellar clouds in H I regions. The data indicate that the fractional ionization n_e/n_H is nearly constant for a range in densities n_H covering a factor of 15. If ionization balance exists in the clouds, then ionization of hydrogen by either low-energy cosmic rays or soft X-rays cannot be the principal source of the observed free electrons, and the hydrogen ionization rate ζ is less than 10^{-16} s^{-1} for these processes.

131.531 The structure of the dense cores of H II regions.
M. A. Dopita, J. E. Dyson, J. Meaburn.
Astrophys. Space Sci., Vol. 28, 61 - 90 (1974).

The latest observational and theoretical evidence which suggests the existence of a large number of partially ionized globules in the cores of H II regions is presented. The consequences of this proposition to the structure of the Orion nebula are considered. It is suggested that the dense ionized core of M 42 is caused by two major ionization fronts eating into a large mass of neutral material which contains many dense neutral globules. The 'hourglass' in M 8 is shown to be very similar.

131.532 21-cm observations in the directions of 43 hot stars and 34 X-ray sources.
E. J. Grayzeck, F. J. Kerr.
Astron. Journ., Vol. 79, 368 - 375 (1974).

An atlas of 21-cm emission profiles is presented for the positions of 43 hot luminous stars and 34 X-ray sources. The observations were obtained using the NRAO 140-ft radiotelescope, which has a half-power beamwidth at λ = 21 cm of 21 arcmin, and the 413-channel autocorrelation receiver. A table of hydrogen column densities, N_H, under the assumption that the gas can be considered to be optically thin, is also given for the 77 observed directions.

131.533 1-millimeter observations of the galactic H II regions M42 and DR 21.
P. A. R. Ade, P. E. Clegg, J. D. G. Rather.
Astrophys. Journ., (Letters), Vol. 189, L23 - L26 (1974).

The authors have made measurements of the 1.4-mm fluxes from the galactic H II regions DR 21 and the Orion nebula. Measurements at 350 μ have been used in the interpretation of these data.

131.534 A second survey of H II regions in galaxies.
P. W. Hodge.
Astrophys. Journ., Suppl. Ser., No. 239, Vol. 27, 113 - 119 (1974).

An Hα survey of 27 galaxies used interference filters to detect H II regions in galaxies of various types, especially Sb and SB galaxies, not well covered in previous surveys. Plates, charts, and tables give positions of the H II regions in those with the best Hα discrimination.

131.535 Recombination lines from H I gas toward Orion A.
E. J. Chaisson, C. J. Lada.
Astrophys. Journ., Vol. 189, 227 - 237 (1974).

Long-integration, high spectral-resolution observations of the 158α and 166α recombination-line spectra are reported toward Ori A. Recombination lines of ^{1}H, of ^{12}C, and possibly of a superposition of heavier atoms are observed to originate in a predominantly neutral region that exists somewhere along the line of sight to the nebula. The recombination-line parameters that characterize the H I gas are derived.

131.536 A 2700 MHz map of IC 434 and the surrounding Orion region.   J. L. Caswell, W. M. Goss.
Astron. Astrophys., Vol. 32, 209 - 216 (1974).

A 2700 MHz map of a region extending from Orion A (NGC 1976) to Orion B (NGC 2024) is presented; it shows extensive weak emission, predominantly from the faint diffuse nebula IC 434. The sensitivity is comparable with that of existing optical photographs and the radio contours show good agreement with the extent of optical emission.

131.537 Messungen an HII-Regionen im fernen Infrarot.
D. Lemke.
Mitt. Astron. Ges., No. 34, p. 95 (1973/74). – Abstr. AG.

131.538 IR-excess radiation and the absorption characteristics of dust in galactic H II regions.
P. G. Mezger, L. F. Smith, E. Churchwell.
Astron. Astrophys., Vol. 32, 269 - 282 (1974).

The IR-excess of H II regions is derived as a function of the absorption cross sections of dust grains and the radiation spectrum of the associated star cluster in the wavelength ranges $228 < \lambda < 504$ Å, $504 < \lambda < 912$ Å and $\lambda > 912$ Å. The relative absorption efficiencies in the three wavelength ranges are estimated. The consequences of the newly established absorption cross sections to the physical state and observable properties of the H II regions and the surrounding interstellar matter are discussed.

131.539 Helium abundance in galactic H II regions.
E. Churchwell, P. G. Mezger, W. Huchtmeier.
Astron. Astrophys., Vol. 32, 283 - 308 (1974).

The authors present observations of He and H radio recombination lines of 39 galactic H II regions, most of which require ionization by stars of spectral type O9 or earlier. They obtain values for the abundance of ionized He in the range $0.06 \lesssim \langle N(He^+)/N(H^+)\rangle \lesssim 0.10$ for spiral arm H II regions and

$\langle N(He^+)/N(H^+)\rangle \lesssim 0.02$ for three giant H II regions in the galactic center, while the abundance of doubly ionized He $\langle N(He^{++})/N(H^+)\rangle < 0.008$.

131.540 H 90 α radio recombination line observations of RCW 38. W. K. Huchtmeier.
Astron. Astrophys., Vol. 32, 335 - 336 (1974).

The H 90 α radio recombination line has been observed at a grid of nine points over the H II region RCW 38. A gradient in radial velocity over the source is found, which probably indicates expansional motions.

131.541 A new search for H I in the Southern Coalsack. F. J. Kerr, P. F. Bowers, R. H. Harten.
Australian Journ. Phys., Vol. 27, 285 - 287 (1974).

Since neutral hydrogen in dust clouds can often be detected from narrow self-absorption dips, two 21 cm profiles with a velocity resolution of 0.2 km s^{-1} have been obtained in the direction of the Southern Coalsack, in the hope of finding such a self-absorption dip. No neutral hydrogen could be detected. It is likely that most of the hydrogen in the Coalsack is molecular.

131.542 Observations of H I in dense interstellar dust clouds: I. A survey of 88 clouds. G. R. Knapp.
Astron. Journ., Vol. 79, 527 - 540 (1974).

A large number of interstellar dust clouds, with various values of visual absorption, was surveyed in the 21-cm line of neutral hydrogen using high-velocity resolution, in a search for narrow absorption lines in the H I emission profiles which could be attributed to the presence of cold H I in the dust clouds. It was found that somewhat less than half of the dust clouds observed were associated with such self-absorption dips. The radial velocities of this group of dust clouds were used to study the kinematics of the local dust cloud population. It was found that the radial velocities of the dust clouds are very close to the H I peak emission velocities in the same directions, showing that the dust clouds are part of the general local system of interstellar matter. The observations were analyzed using simple models for the density and temperature distribution of the atomic hydrogen in the dust clouds.

131.543 Observations of H I in dense interstellar dust clouds: II. The cloud Khavtassi 3. G. R. Knapp.
Astron. Journ., Vol. 79, 541 - 554, 661 (1974).

High-frequency-resolution H I observations of the dark cloud Khavtassi 3 show that the H I in the cloud, seen as a narrow self-absorption component at +13 km/sec, correlates well in position with the darkest parts of the cloud. Simple models of the H I distribution in the cloud, combined with the absorption profile for the continuum source W44, which is behind Kh 3, show that the neutral atomic hydrogen in this cloud is less than 1 % of the total hydrogen content of the cloud. Several other self-absorption features were found in the observations, corresponding to H I clouds at different velocities and distances along the line of sight. An H I cloud at +30 km/sec was mapped.

131.544 On microwave recombination lines from H I regions. B. Zuckerman, J. A. Ball.
Astrophys. Journ., Vol. 190, 35 - 51 (1974).

The authors observed recombination lines attributed to carbon at 21-, 30-, 36-, and 43-cm wavelengths toward IC 1795 (W3), Orion A and NGC 2024; and a narrow hydrogen line toward NGC 2024. A comparison of a considerable body of carbon recombination-line and radio molecular-line data indicates that these two types of lines are probably formed in dense contiguous regions—the carbon lines in a thin layer facing a hot star and the molecular lines in the rest of the cloud shielded from the stellar radiation by the carbon-emitting slab. The high-frequency carbon lines toward Orion A are probably due to spontaneous rather than stimulated emission. In NGC 2024 the H I regions responsible for the observed hydrogen and carbon lines are probably dense ($n_H > 10^3$ cm^{-3}).

131.545 Radio detection of a compact H II-region associated with far-infrared and molecular line emission in Cygnus. H. J. Wendker, J. W. M. Baars.
Astron. Astrophys., Vol. 33, 157 - 159 (1974).

A recently detected source of far-infrared and molecular line radiation in Cygnus has been observed with the Westerbork Synthesis Radio Telescope at 6 and 21 cm wavelength, yielding flux densities of 85 and 67 × 10^{-29} W m^{-2} Hz^{-1}, respectively, and an angular size of about 3″. The authors find the radio source to be a compact H II-region, which could be excited by an early B-type star. No optical object can be seen at its position.

131.546 On the homogeneity and the kinetic temperature of the intercloud HI-gas.
O. Hachenberg, U. Mebold.
Stars and the Milky Way system, Proc. 1972, (see 012.018), p. 120 - 123 (1974).

131.547 Infrared photometry of galactic H II regions.
H. Olthof, J. J. Wijnbergen, T. J. Helmerhorst, R. J. van Duinen.
Stars and the Milky Way system, Proc. 1972, (see 012.018), p. 243 - 247 (1974).

131.548 Chemical abundances of H II regions in the LMC.
M. Peimbert, S. Torres-Peimbert.
Bull. American Astron. Soc., Vol. 6, 224 (1974). – Abstr. AAS.

131.549 Observations of a peculiar neutral hydrogen phenomenon associated with the North Polar Spur.
G. L. Verschuur.
Bull. American Astron. Soc., Vol. 6, 225 (1974). – Abstr. AAS.

131.550 Predicted infrared spectra of H II regions.
P. A. Aannestad.
Bull. American Astron. Soc., Vol. 6, 263 (1974). – Abstr. AAS.

131.551 Negative velocity clouds near IC 443 and the Cygnus Loop. L. K. DeNoyer.
Bull. American Astron. Soc., Vol. 6, 263 (1974). – Abstr. AAS.

131.552 Survey of faint H II regions in the nearest galaxies by means of the large reducteur focal of the 193 cm telescope of the Haute Provence Observatory.
J. Boulesteix, G. Courtès, A. Laval, G. Monnet, with the technical collaboration of H. Petit.
Proc. ESO/SRC/CERN conference, (see 012.021), p. 221 - 230 (1974).

131.553 Steps toward the Hubble constant. I. Calibration of the linear sizes of extragalactic H II regions.
A. Sandage, G. A. Tammann.
Astrophys. Journ., Vol. 190, 525 - 538 (1974).

Measurements of the angular sizes of the core and halo parts of H II regions are given for 11 galaxies in the local and M81-NGC 2403 groups whose distances are known from cepheid variables. The authors have calibrated the linear diameters versus the luminosity class of the parent galaxy. The mean size of the largest H II region (core plus halo diameters) is 550 pc for Sc I galaxies, decreasing to 110 pc for luminosity class V galaxies.

131.554 Internal motions in H II regions. I. The radial velocity field of NGC 6164-6165. P. Pişmiş.
Rev. Mexicana Astron. Astrofis., Vol. 1, No. 1, p. 45 - 54

(1974).
On two Fabry-Perot interferograms, radial velocities from the Hα line at about 100 points in the HII region complex, NGC 6164-6165, are obtained. The velocity field shows considerable complexity, yet, axial symmetry around the exciting star HD 148937, O6f, is clearly evident.

131.555 Equal-velocity contour diagrams of neutral hydrogen 21-cm line radiation for the region of $48.5° < l < 51.2°$, $-1° < b < 1°$. F. Sato, K. Akabane.
Ann. Tokyo Astron. Obs., Second Ser., Vol. 14, (No. 2), 120-139 (1974).

Thirty-five equal-velocity contour diagrams derived from the second edition of the Maryland–Green Bank Galactic 21-cm Line Survey are presented. Radial velocities from -5 to 80 km s^{-1} are covered with intervals of 2 or 3 km s^{-1}. At some radial velocities, clear absorption features are seen due to continuum sources, including the complex source W51.

The extraterrestrial UV-background and the nearby interstellar medium. See Abstr. 106.005.

Ultraviolet photometry from the Orbiting Astronomical Observatory. XIV. An extension of the survey of Lyman-α absorption from interstellar hydrogen. See Abstr. 113.001.

The Vela supernova remnant in ultraviolet light. See Abstr. 125.010.

On the recombination-line observations toward supernova 3C391. See Abstr. 131.059.

Observations of interstellar CH microwave line ratios. See Abstr. 131.105.

Carina nebula: a possible interpretation of the molecular observations. See Abstr. 132.003.

On the comet-like structure in NGC 7635. See Abstr. 132.010.

Infrared map of the Orion nebula. See Abstr. 132.017.

Observations of major ionization fronts in M42 and M8. See Abstr. 132.020.

The heating of young planetary nebulae by trapped Lyman-α radiation. See Abstr. 133.015.

On three galactic H$_2$O sources. See Abstr. 141.603.

Interpretation of far infrared observations. See Abstr. 141.614.

On the formation of interstellar cloud complexes, OB associations and giant H II regions. See Abstr. 151.048.

Détection de nouvelles formations d'hydrogène ionisé, de grandes dimensions angulaires, dans le bras local et dans le bras du Sagittaire de notre Galaxie. See Abstr. 155.010.

Rising and falling motion of gas above the spiral arm. See Abstr. 155.043.

A new look at the interstellar hydrogen through a very-wide-field photographic Hα survey of the whole Milky Way. See Abstr. 155.044.

Studies of neutral hydrogen cloud structure in the vicinity of the North Polar Spur. See Abstr. 155.048.

Ionized gas in the direction of the galactic center I: kinematics and physical conditions in the nuclear disk. See Abstr. 155.063.

Galaxy counts and dust in the galactic disk. See Abstr. 155.067.

Comparison of CO and HI observations on a small scale. See Abstr. 155.068.

The HI distribution in nearby galaxies. See Abstr. 158.070.

Radio observations of H II regions in external galaxies. I. H II regions in M 33. See Abstr. 158.082.

Physical studies of emission regions in M 33. See Abstr. 158.104.

Upper limit to the neutral hydrogen content of the elliptical galaxy NGC 4472. See Abstr. 158.113.

The Magellanic Stream. See Abstr. 159.004.

Errata

131.901 Errata: 'Observations of recombination-line emission in the direction of two supernova remnants'. [Astrophys. Journ., (Letters), Vol. 183, L143 (1973)]. D. A. Cesarsky, C. J. Cesarsky. Astrophys. Journ., (Letters), Vol. 187, L47 (1974).

131.902 Errata: 'Interstellar sulfur monoxide'. [Astrophys. Journ., (Letters), Vol. 184, L59 - L64 (1973)]. C. A. Gottlieb, J. A. Ball. Astrophys. Journ., (Letters), Vol. 187, L47 (1974).

131.903 Errata: To 'Interaction between the interstellar medium and the solar wind' [Astrophys. Space Sci., Vol. 20, 3 - 18 (1973)]. M. K. Wallis. Astrophys. Space Sci., Vol. 27, 516 (1974).

131.904 Erratum: 'Implications of the Copernicus observations of unreddened stars' [Astrophys. Journ., (Letters), Vol. 186, L33 - L36 (1973)]. J. C. Weisheit, C. B. Tarter. Astrophys. Journ., (Letters), Vol. 188, L77 (1974).

132 Emission Nebulae, Reflection Nebulae

132.001 The methanol source in Orion at 1.2 centimeters.
M. F. Chui, A. C. Cheung, D. Matsakis, C. H. Townes, A. G. Cardiasmenos.
Astrophys. Journ., (Letters), Vol. 187, L19 - L21 (1974).

Methanol line radiation from Orion near 25 GHz has been reexamined and found to be much stronger than expected from earlier reports. The line widths are also narrower than other nonmasing molecular radiation. These results indicate either that there is more than one small source of methanol radiation in Orion, or that these transitions are inverted and varying with time.

132.002 Photographic surface photometry of the nebulae surrounding V380 Ori and R Mon. M. T. Brück.
Monthly Notices Roy. Astron. Soc., Vol. 166, 123 - 154 (1974).

The present observations of the nebulosities surrounding V380 Ori and R Mon in U B V Hα and I show that both nebulae are bluer than the stars and that NGC 2261 surrounding R Mon has a distinct colour pattern.

132.003 Carina nebula: a possible interpretation of the molecular observations. H. R. Dickel.
Astron. Astrophys., Vol. 31, 11 - 16 (1974).

A model is put forward to interpret the observations of the molecules in the direction of the Carina nebula. The visual absorption is mapped and it is shown that the concentrations of dust, molecules, and neutral hydrogen are probably located at the H II region. The observed pattern of molecular radial velocities can be interpreted by an outward expansion from the H II region at a rate of about 7 km s^{-1}. Values for the thickness, density, and the mass of the various components of the Carina nebula are estimated from neutral hydrogen observations.

132.004 The structure of the Orion nebula. I. Observations of the C 85α recombination line.
B. Balick, R. H. Gammon, L. H. Doherty.
Astrophys. Journ., Vol. 188, 45 - 52 (1974).

The C 85α recombination line of carbon has been mapped in Orion A with 3' resolution. The observed C II distribution is extended and asymmetric about a peak position near the radio continuum peak. Carbon emission is observed throughout the radio nebula with extensions toward the dark bay and dust lane east and northeast of the trapezium. The narrow carbon line width ($\Delta v \simeq 4$ km s^{-1}) and constant velocity ($v \simeq 9$ km s^{-1}) indicate a dynamically distinct and quiescent C II emitting region compared with the H II region. A model is proposed in which the H II region is partially embedded in the near side of a large neutral gas cloud.

132.005 Absorption of radio waves in the Gum nebula.
H. V. Cane.
Proc. Astron. Soc. Australia, Vol. 2, 197 - 198 (1973). — Presented at the annual general meeting of the Astron. Soc. Australia — see 012.001.

132.006 Optical thickness in the He I singlet spectrum of nebulae. R. R. Robbins, A. P. Bernat.
Astrophys. Journ., Vol. 188, 309 - 314 (1974).

An integral equation expressing the transfer of resonance line radiation in the He I singlets has been solved for nebular models consisting of differentially expanding uniform spheres. Solutions have been obtained numerically for a variety of temperatures, optical depths, expansion velocities, and scattering functions appropriate to nebulae. Comparison of the transfer solutions to observed helium-singlet line intensities gives optical depths for various nebulae in the resonance lines in the ground-state bound-free continuum. Further discussion of the results demonstrates that transfer solutions of this type provide a sensitive observational test of computed ionization-stratification models for nebulae.

132.007 Der Rosettennebel. F. Börngen, W. Högner, P. Lochno, N. Richter.
Jenaer Rundschau, (Jena Review), 18. Jahrgang, p. 331 - 332 (1973).

132.008 Orionnevel: van foto tot equidensiet.
J.-C. Lameer.
Zenit, Vol. 1, No. 3, p. 37 - 39 (1974).

132.009 Search for the exciting star for the nebula near γ Cyg. V. P. Arkhipova, N. A. Gorynya, V. F. Esipov, T. A. Lozinskaya, A. S. Sharov.
Astron. Tsirk., No. 810, p. 4 - 6 (1974). In Russian.

132.010 On the comet-like structure in NGC 7635.
H. M. Johnson.
Astron. Astrophys., Vol. 32, 17 - 19 (1974).

A plate of the galactic nebula NGC 7635 clearly shows a comet-like structure inside the ring structure. Its unusually strong, continuous spectrum is illustrated.

132.011 Detection of the [Fe XIV] coronal line at 5303 Å in the Cygnus Loop. B. E. Woodgate, H. S. Stockman, Jr., J. R. P. Angel, R. P. Kirshner.
Astrophys. Journ., (Letters), Vol. 188, L79 - L82 (1974).

Emission in the [Fe XIV] coronal line at 5303 Å has been detected in strong X-ray emitting regions of the Cygnus Loop. The fluxes agree well with calculations based on a thermal model for X-ray emission with normal iron abundance. Upper limits to fluxes in the line were obtained for the center of the Cygnus Loop and for Puppis A and IC 443.

132.012 Electronographic measurements of the ionised helium abundance in the core of the Orion nebula.
M. A. Dopita, K. H. Elliott, J. Meaburn.
Astrophys. Space Sci., Vol. 28, 163 - 171 (1974).

Measurements have been made with an electronographic image tube spectrograph of two helium and three hydrogen lines at twenty nine positions in the dense core of the Orion nebula. Values for the helium abundance derived show that a large amount of the helium is neutral. The total abundance of helium is shown to be in the range $0.104 < N(\text{He})/N(\text{H}) < 0.132$.

132.013 Emission lines in the Orion nebula near 8500 Å.
I. J. Danziger, M. Aaronson.
Publ. Astron. Soc. Pacific, Vol. 86, 208 - 210 (1974).

An identification of forbidden lines of Cl II at 8579.5 Å and Fe II at 8617 Å is reported for the Orion nebula. The unusual behavior of the O I allowed transition at 8556.4 Å is also discussed.

132.014 A measurement of electron temperature in the Orion nebula from the intensity ratio of forbidden transitions in S III. P. Foukal.
Publ. Astron. Soc. Pacific, Vol. 86, 211 - 214 (1974).

The temperature dependence of the intensity ratio R of three forbidden transitions within the $3p^2$ ground configuration of S III was calculated, and the ratio has been observed in the Orion nebula. The observed value of $R = 56 \pm 25\%$ leads to an electron temperature of $9700°$K $\pm 1000°$K, after a reasonable reddening correction is made.

132.015 Photoelectric spectrophotometry of the Cygnus Loop.   J. S. Miller.
Astrophys. Journ., Vol. 189, 239 - 248 = Lick Obs. Bull., No. 644 (1974).

This paper presents a discussion of scanner measurements of line and continuum intensities made at three places in the Cygnus Loop. It gives a presentation of the observations, and contains a discussion of physical parameters derived from the observational data and a comparison with published theoretical calculations.

132.016 Photoelectric photometry of gaseous nebulae. II. The temperature and ionization structure of M8 and M20.   M. A. Dopita.
Astron. Astrophys., Vol. 32, 121 - 126 (1974).

An analysis has been made of the photoelectrically measured intensities of the $H\alpha$, $H\beta$, [N II] λ 6584 Å and [O III] λ 5007 Å nebular lines at 96 points in M8 and 31 points in M20. This has yielded the variation in temperature and ionization conditions in these nebulae. In the case of M8 very low oxygen and nitrogen abundances in the ionized gas were inferred when compared to solar abundance, and it is concluded that in both M8 and M42 (observed earlier) a large fraction of the oxygen is locked up in the dust. There is strong evidence of large optical depth variations apparently caused by neutral inclusions of material in the nebulae.

132.017 Infrared map of the Orion nebula.
D. Lemke, F. J. Low, C. Thum.
Astron. Astrophys., Vol. 32, 231 - 233 (1974).

The Orion nebula complex has been mapped at 21 μ with a 1 arc min beam. The flux from the entire region was found to be 1.5×10^{-21} W m^{-2} Hz^{-1}. Therefore the Orion nebula is the brightest 21 μ object outside the solar system studied so far.

132.018 Fine structures in galactic and extragalactic nebulae.   J. Meurers.
Astron. Nachr., Vol. 295, 151 - 157 (1974).

The equidensity-method represents not only new possibilities for analyzing photographic images, but this method is also suitable in the way of an equidensity of first order for marking images on the photographic plate. In the following investigation this possibility is used to measure fine structures in the border of the North America nebula and in the arms of some galaxies and to discuss the results from a statistical point of view. Artificial galaxies produced by random in an experiment are compared with real galaxies with respect to fine structures. Possible influences of the equidensity-method upon the marking of objects are investigated. The usefulness of the method for the marking is shown.

132.019 The variation of the He nα : H nα radio recombination line ratios as a function of the relative sizes of the He$^+$ and H$^+$ regions deduced from a model of Orion A.   A. S. J. Batchelor.
Astron. Astrophys., Vol. 32, 343 - 348 (1974).

Calculations have been made with the hydrogen model of Orion A by Brocklehurst and Seaton (1971–1972), as modified to include helium by Batchelor and Brocklehurst (1972), to derive the He nα : H nα radio recombination line ratios for both LTE and non-LTE cases, where the beamsize of the observing telescope and the relative size of the He Strömgren sphere are independently varied. The calculations were performed for telescope beams encompassing the entire model, decreasing in stages to very small pencil beams. The size of the He Strömgren sphere relative to the H Strömgren sphere was varied from coincidence to an almost zero He$^+$ region.

132.020 Observations of major ionization fronts in M42 and M8.   K. H. Elliott, J. Meaburn.
Astrophys. Space Sci., Vol. 28, 351 - 364 (1974).

Image tube filter photographs taken in the light of the low ionization energy [N II] and high ionization energy [O III] lines of the core of M42 are presented. The whole core of M42 appears to be a complex of ionization fronts. An [N II] photograph of M8 also suggests that the whole of this H II region is composed of a mass of fronts penetrating into a huge neutral region containing dust, gas and molecules.

132.021 Further observations of interstellar gas in the Gum nebula.   A. D. Thackeray.
Observatory, Vol. 94, 55 - 60 (1974).

Following the announcement by Wallerstein and Silk of the presence of high-velocity Ca II components in HD 74455 and HD 75821, some 3° from the Vela pulsar, Thackeray and Warren pointed out the presence of additional high-velocity components in the same two stars. More early-type stars nearer to the Vela pulsar have now been observed by the writer with the Radcliffe coudé spectrograph, mostly at 6.8 Å/mm. No further cases of interstellar velocities as large as those in HD 74455 and HD 75821 have been found, but some new instances of doubling with separations of order 30 km/s are reported.

132.022 Absolute spectrophotometric study of the continuous spectrum of the Orion nebula.   C. T. Hua.
Astron. Astrophys., Vol. 32, 423 - 428 (1974).

Absolute data for the spectral region $\lambda\lambda$ 5000–3100 are deduced from the recent direct comparison between the energy distribution in the continuous spectrum of the central part of the Orion nebula and the standard source previously compared to blackbody radiation. These results confirm the author's former spectrophotometric study of the nebula, i.e. values of the absolute gradients and Balmer discontinuity. Using models by Brown and Mathews, and the energy distribution in the Balmer continuum, the author derived the contribution of scattered light and the value of the electron temperature.

132.023 Spectral observations of some cometary nebulae.   M. Mendez, Eh. S. Parsamyan.
Astrofizika, Vol. 10, 65 - 71 (1974). In Russian. – English translation in Astrophysics, Vol. 10, No. 1.

On the 40"-telescope of the Tonantzintla Observatory spectral observations of four cometary nebulae have been made and are reported here in detail. A spectral classification of cometary nebulae shows that the spectra of the nebulae are not defined by the spectral type of the illuminating star. It is supposed that the cometary nebulae state is one of the possible post-fuor stages.

132.024 Monte Carlo model of reflection nebulae: intensity gradients.   T. Roark, B. Roark, G. W. Collins II.
Astrophys. Journ., Vol. 190, 67 - 72 (1974).

Theoretical photometric surface-intensity gradients are computed for three-dimensional model reflection nebula. A nonbiased Monte Carlo analog to the multiple-scattering problem is used. Comparison of the results with those obtained by earlier, semianalytic approaches indicates the latter do not adequately take account of multiple scattering within such model nebulae. Observed intensity gradients are briefly discussed with respect to theoretical interpretations.

132.025 Possible evidence for a large magnetic field in the Orion infrared nebula.
C. A. Beichman, E. J. Chaisson.
Astrophys. Journ., (Letters), Vol. 190, L21 - L24 (1974).

Two independent observational results suggest that a magnetic field of milligauss strength permeates the Orion infrared nebula/molecular cloud. First, the large degree of infrared polarization is best explained in terms of preferential absorption by elongated silicate grains aligned by a ~ 10-milligauss field. Second, the microwave OH emission features appear to show a classical Zeeman pattern, suggesting a longitudinal field strength of about 3 milligauss.

132.026 On the use of mean escape probabilities to solve transfer problems in nebulae.
A. P. Bernat, R. R. Robbins.
Astrophys. Journ., Vol. 189, 459 - 462 (1974).

Results from an accurate numerical solution are compared with the solutions predicted by one particular formulation of mean escape probabilities. It is found that the latter do not predict numerical solutions well and that it is difficult to determine a priori how to improve the agreement. Mean escape probabilities can be extremely useful, however, in solving additional "nearby" cases when one solution is known to high accuracy.

132.027 Magnesium and iron abundances in the Gum nebula region determined from interstellar absorption lines in the spectrum of γ^2 Vel.
M. Grewing, C. Wulf-Mathies, H. J. Lamers, C. M. Walmsley.
Astron. Inst. Utrecht, Internal Rep. ROF 72, (see 012.012), p. C21.1 - C21.14 (1974).

132.028 Technique for the reduction of nebular line profiles observed with photoelectric Fabry-Perot monochromators. M. A. Dopita.
Astron. Astrophys., Vol. 33, 147 - 150 (1974).

A method is given for deriving temperatures of nebulae by Fourier analysis of an observed hydrogen line profile and that of a forbidden line of a heavier element. This is a factor of two hundred times faster than the convolution and fitting technique used hitherto.

132.029 Isotopic abundances and line formation in the Orion nebula.
P. G. Wannier, P. J. Encrenaz, R. W. Wilson, A. A. Penzias.
Astrophys. Journal, (Letters), Vol. 190, L77 - L79 (1974).

Measurements of the spatial variation of $H^{12}C^{14}N$ and $H^{13}C^{14}N$ line emission from the molecular cloud associated with the Orion nebula indicate that $H^{12}C^{14}N$ has a high central opacity. This result seems to contradict a recent suggestion that the observed hyperfine structure of the line indicates a low opacity, which in turn would yield a $^{13}C/^{12}C$ abundance significantly different from other determinations.

132.030 Search for deuterium in Orion and detection of high-velocity features.
W. A. Traub, N. P. Carleton, D. J. Hegyi.
Astrophys. Journal, (Letters), Vol. 190, L81 - L84 (1974).

The authors have searched for $D\alpha$ emission near the bright $H\alpha$ line in Orion. The observations revealed a high-velocity $H\alpha$ feature near -60 km s^{-1}, whose intensity varied across the nebula, ranging from 0.0005 to 0.005 relative to $H\alpha$; another much weaker feature, possibly also $H\alpha$, was observed at about -100 km s^{-1}. The authors are at present only able to derive an upper limit $D/H < 1.3 \times 10^{-4}$ for the number ratio in Orion.

132.031 Ionization equilibrium of Herbig-Haro objects and transition radiation. G. A. Gurzadyan.
Astron. Astrophys., Vol. 33, 307 - 310 (1974).

If one assumes that fast electrons with an energy of the order of 1.5 MeV exist in Herbig-Haro objects, and that these electrons generate transition radiation through their electrodynamic interaction with dust particles, one can, in principle, account for the observed luminosities and spectral structures of these anomalous objects. The theory has been applied to Herbig-Haro object No. 1 and its energy and optical characteristics have been determined.

132.032 Detection of the Fe XIV coronal line at 5303 Å in the Cygnus Loop.
B. E. Woodgate, H. S. Stockman, Jr., J. R. P. Angel, R. P. Kirshner.
Bull. American Astron. Soc., Vol. 6, 273 (1974). – Abstr. AAS.

132.033 Measurement of the spatial structure of the X-ray emission of the Cygnus Loop.
H. Helava, M. C. Weisskopf, R. S. Wolff.
Bull. American Astron. Soc., Vol. 6, 273 (1974). – Abstr. AAS.

132.034 CO and CS in the Orion nebula. H. S. Liszt, R. W. Wilson, A. A. Penzias, K. B. Jefferts, P. G. Wannier, P. M. Solomon.
Astrophys. Journ., Vol. 190, 557 - 564 (1974).

The authors present maps, in $^{12}C^{16}O$, $^{13}C^{16}O$, and $^{12}C^{32}S$ line emission, of the central (~ 20') portion of the Orion A molecular cloud. The dominant feature in the maps is a dense $4' \times 9'$ emission "ridge" which appears to be rotating about an east-west axis running through the position of the Kleinmann-Low infrared source. The authors argue that their line profiles and maps are most readily understood in terms of simple, systematic, radial velocity structure in the cloud.

132.035 Vibrationally excited silicon monoxide in the Orion nebula. J. H. Davis, G. N. Blair, H. Van Till, P. Thaddeus.
Astrophys. Journal, (Letters), Vol. 190, L117 - L119 (1974).

A 1° K line has been detected in the center of the molecular cloud in the Orion nebula which is probably the 129,363.1-MHz, $J = 3 \rightarrow 2$ rotational transition of SiO in its first excited ($v = 1$) vibrational state.

Stellar winds and related phenomena in surrounding nebulae.   See Abstr. 064.058.

The Coalsack. II. Photometry of suspected flare stars.   See Abstr. 113.023.

FG Sagittae: Rosetta Stone for nucleosynthesis? See Abstr. 114.119.

A search for flare stars in the Orion nebula region. See Abstr. 122.070.

The OH absorption against the Carina nebula. See Abstr. 131.020.

Detection of possible maser emission near 3.48 millimeters from an unidentified molecular species in Orion. See Abstr. 131.060.

Discovery and CO observations of a new molecular source near M17.   See Abstr. 131.061.

The Coalsack. III. A search for T Tauri stars. See Abstr. 131.066.

Dust and gas in the Orion molecular cloud: observations of 1-millimeter continuum and 2-centimeter H_2CO emission.   See Abstr. 131.068.

A comment of the paper "The shapes of neutral globules associated with diffuse nebulae". See Abstr. 131.075.

A correlation study of carbon ions and hydroxyl molecules toward galactic nebulae.   See Abstr. 131.081.

Molecular clouds.   See Abstr. 131.091.

OH observations near the reflection nebulae NGC 2068 and NGC 2071.   See Abstr. 131.092.

The formation of special type structures behind a shock wave front in the interstellar gas.
See Abstr. 131.101.

Observational evidence for the excitation of HCN and H_2O in protostellar molecular clouds.
See Abstr. 131.134.

Interaction of hot stars and of the interstellar medium. V. The compact optical regions Sharpless 255, 257 and 269.   See Abstr. 131.505.

The presence of dust in H II regions.
See Abstr. 131.510.

Spectrophotometric observations of the compact H II region K3-50 and of NGC 6857.   See Abstr. 131.529.

The structure of the dense cores of H II regions.
See Abstr. 131.531.

1-millimeter observations of the galactic H II regions M42 and DR 21.   See Abstr. 131.533.

Recombination lines from H I gas toward Orion A.
See Abstr. 131.535.

A 2700 MHz map of IC 434 and the surrounding Orion region.   See Abstr. 131.536.

Negative velocity clouds near IC 443 and the Cygnus Loop.   See Abstr. 131.551.

On the interpretation of continuum flux observations from thermal radio sources – I. Continuum spectra and brightness contours.   See Abstr. 141.078.

On the interpretation of continuum flux observations from thermal radio sources – II. Three-dimensional models.   See Abstr. 141.079.

X-ray pulsar in the Cygnus Loop?
See Abstr. 142.082.

133   Planetary Nebulae

133.001   Observations of the [O III] λ 5007 Å emission line from the Helix nebula.   K. Taylor.
Astron. Astrophys., Vol. 30, 45 - 49 (1974).

Observations of the [O III] λ 5007 Å emission line were made of the Helix planetary nebula (NGC 7293) using a pressure-scanned double-etalon, PEPSIOS-type monochromator combined with the 106-cm reflector at the Pic-du-Midi Observatory. 32 positions on the object were observed with a minute of arc beam on the sky. Each position displayed multi-peaked profiles with up to 4 components appearing in the deconvolution of the scans. However, systematic behaviour of the separate features from position to position were not observed and no evidence for a simple expanding shell model could be found.

133.002   Monochromatic observations of the ring nebula in Lyra (NGC 6720).   R. Louise.
Astron. Astrophys., Vol. 30, 189 - 197 (1974). In French.

The ring nebula in Lyra (NGC 6720) was observed through narrow-band interference filters centered respectively on Hα, Hβ, [O III] (λ = 5007 Å), [N II] (λ = 6584 Å), [O I] (λ = 6300 Å) and [S II] (λ 6715 Å + λ 6732 Å). For each monochromatic image, photometric reductions were made for two directions, one along the major axis of the nebula and one along its minor axis. Some geometrical parameters were defined from the photometric profile. These parameters were compared to each other for the various observed lines. An attempt to study the variation of [O III]/[O I] was made. The shell model proposed by Curtis (1918) and Gurzadyan (1970) was compared with the observations. Two other models were suggested: – a quasi-flat ring as proposed by Minkowski and Osterbrock (1960) and more recently by Hua and Louise (1970), – an oblate spheroid, denser at the equator than at the poles. A programme of high spectral resolution observations is needed to differentiate between these two models.

133.003   Filamentary structure in planetary nebulae.
G. O. Boeshaar.
Astrophys. Journ., Vol. 187, 283 - 298 (1974).

Small-scale inhomogeneities in planetary nebulae were studied via photographic spectrophotometry in order to determine the characteristics of these features with respect to the ambient nebular gas. Untrailed spectrograms of eight planetaries yielded two types of data. First, relative intensity distributions along the projections of the spectrographic slit on the nebulae were produced to show the spatial variations in the emission from different atoms and ions. Then, relative emission-line fluxes were derived and used to find the electron temperatures and densities of filamentary regions and their surroundings.

133.004   Evolution of central stars of planetary nebulae towards the crystallizing white dwarf stage.
U. De Angelis, L. De Cesare, A. Forlani, G. Platania.
Astrophys. Space Sci., Vol. 27, 217 - 225 (1974).

The evolution of the central stars of planetary nebulae, interpreted as hot white dwarfs with liquefying cores, towards the cold white dwarf stage is discussed and theoretical evolutionary tracks are built for such central stars as they cool towards the crystallizing region. The conclusions seem to hint a picture in which crystalline white dwarfs can be looked at as final stages of the central stars of planetary nebulae.

133.005   High helium abundances in two planetary nebulae.
J. B. Kaler.
Astrophys. Journ., (Letters), Vol. 188, L15 - L17 (1974).

Two planetary nebulae, Me 2-2 and Hu 1-2, are found to have helium to hydrogen ratios of 0.14 and 0.15, respectively, well in excess of the ratios for H II regions. Absolute Hβ fluxes are measured for both nebulae.

133.006 Photographies de quelques nébuleuses planétaires dans la région du proche infrarouge.
Y. Andrillat, M. Duchesne.
Comptes Rendus Acad. Sci. Paris, Sér. B, Vol. 278, 223 - 225 (1974).

Lors des essais de la caméra électronique dans le proche infrarouge, nous avons obtenu quelques photographies de nébuleuses planétaires NGC 7026, 6826, 6543 à l'aide de filtres interférentiels isolant [S III] (9069 Å), P8 (9546 Å) + [S III] (9532 Å), et avec un filtre correspondant à $\lambda > 7400$ Å. Pour une nébuleuse donnée, des différences très nettes de structure apparaissent suivant le filtre utilisé.

133.007 Infrared emission from planetary nebulae. II. Refined and supplemented data on the emission from planetary nebulae in the 1.0 - 2.5 micron region.
G. S. Khromov.
Astron. Zhurn. Akad. Nauk SSSR, Vol. 51, 335 - 340 (1974). In Russian. English translation in Soviet Astron. AJ, Vol. 18, No. 2.

Supplemented and refined data on the continuous infrared emission in the spectral region $1.0 - 2.5 \mu$ from 19 planetary nebulae are presented.

133.008 Radio-synthesis observations of planetary nebulae.
Y. Terzian, B. Balick, C. Bignell.
Astrophys. Journ., Vol. 188, 257 - 277 (1974).

Continuum radio observations of planetary nebulae are presented with angular resolutions as small as ~ 2″. These observations were made at NRAO at 2.7 and 8.1 GHz with aperture-synthesis techniques. Detailed radio maps are presented for 14 nebulae and results of observations are also given for many others.

133.009 On the light variability of the planetary nebula Hb 6 = 7 + 1°1, identified with the irregular variable AS Sagittarii. V. P. Arkhipova.
Peremennye Zvezdy, Vol. 19, 273 - 277 (1973). In Russian.

The stellar planetary nebula VV 128 = Hb 6, identified with the irregular variable AS Sgr, did not show any spectral variations during 1961–1970. The integrated magnitude of the nebula $m_{pg} = 13.2$ is in good agreement with the mean brightness of AS Sgr, quoted by Leavitt and considerably higher than that of Swope. As the radiation of AS Sgr is concentrated mainly in strong emission lines, its brightness estimates can be influenced by slight variations of the photometric system.

133.010 Nebular photometry with an echelle spectrometer: [O III] line ratios in NGC 1976 and NGC 6853.
T. J. Bohuski, R. J. Dufour, D. E. Osterbrock.
Astrophys. Journ., Vol. 188, 529 - 532 (1974).

Results are presented for measurements of the [O III] $\lambda 5007/\lambda 4363$ intensity ratio in NGC 1976, NGC 6853, NGC 650-1, and several small planetary nebulae.

133.011 Interferometric survey of planetary nebulae.
J. H. Cahn, R. H. Rubin.
Astron. Journ., Vol. 79, 128 - 131 (1974).

A survey of radio fluxes and positions of planetary nebulae is reported using the NRAO three-element interferometer. The survey represents an attempt to quickly obtain fluxes and positions of weak sources in the range 0.010–0.100 f.u. Successful observations are reported on 69 planetary nebulae.

133.012 New southern planetary nebulae previously classified as emission-line stars. N. Sanduleak.
Publ. Astron. Soc. Pacific, Vol. 86, 215 - 216 (1974).

Eight objects, originally classified as emission-line stars in earlier Hα-emission-line objective-prism surveys, have been reclassified as planetary nebulae as a result of two recently conducted objective-prism surveys of the southern Milky Way which utilized blue-green sensitive plates.

133.013 Shell flashes in planetary nuclei.
H. M. Van Horn, C. J. Hansen.
Nature, Vol. 249, 429 - 431 (1974).

A satisfactory explanation for the postulated high internal fields in the planetary nuclei has not yet been suggested. The authors point out here that a natural mechanism for the production of strong magnetic fields may in fact be present in at least some planetary nuclei. A prominent feature of the evolution of such stars is the occurrence of one or more thermal runaways, 'flashes', in an unstable shell source which produces nuclear energy. Near the peak of a flash a transient convection zone is invariably developed, and if the star has even a modest amount of differential rotation, the generation of magnetic field in a stellar dynamo can occur.

133.014 The remarkable object M 2–9.
S. van den Bergh.
Astron. Astrophys., Vol. 32, 351 - 352 (1974).

The peculiar nebula M 2–9 is found to exhibit brightness changes on a timescale $< 1 \times 10^8$ s.

133.015 The heating of young planetary nebulae by trapped Lyman-α radiation. D. George.
Monthly Notices Roy. Astron. Soc., Vol. 167, 485 - 491 (1974).

Using a simple model, the effect of trapped Lyman-α radiation on the energy balance of a young planetary nebula with an H I envelope is calculated. It is shown that the Lyman-α heats the gas and that the effect in the H II region is small. However, the heating is so strong in neutral gas that a small neutral globule inside the H II region might not be able to exist in thermal equilibrium at low temperature. In any case, this source of heating could well be important in determining the structure of neutral globules in young planetary nebulae.

133.016 Statistical population indices of planetary nebulae.
W. Iwanowska.
Stud. Soc. Sci. Torunensis, Sectio F (Astron.), Vol. 5, 33 - 41 = Bull. Astron. Obs. Toruń, No. 50/I (1973).

Statistical population indices for planetary nebulae have been redetermined using new distance values published by Cahn and Kaler (1971). The coefficients of correlation between statistical population indices and the abundance ratios He/H and O/H compiled by Kaler (1970) were calculated. No significant correlation has been found for He/H ratio, a positive correlation for O/H ratio is established, i.e. O/H is higher for population I nebulae.

133.017 Planetary nebulae with multiple shells.
J. B. Kaler.
Astron. Journ., Vol. 79, 594 - 596 (1974).

It is shown that the central stars of planetary nebulae with giant halos (ratio of outer to inner radius >5) are much cooler than are those of nebulae which consist of double shells with a radius ratio ~2. For double shell nebulae, the change of ratio as a function of radius is consistent with the ejection of shells at different times. It appears that all of Hromov and Kohoutek's (1968) type I nebulae, and perhaps 1/3 of all planetaries, possess double shells. Large halos appear around NGC 7009 and 7662, which make these examples of triple shell planetaries.

133.018 **Electron densities, radii and masses of planetary nebulae.** J. H. Lutz.
Bull. American Astron. Soc., Vol. 6, 213 - 214 (1974).
Abstr. AAS.

133.019 **Variabler planetarischer Nebel FG Sagittae 1973.** W. Wenzel.
MVS, *Sonneberg*, Vol. 6, 135 - 137 (1974).
Photoelectrically determined light and colour curves of this variable central part of a planetary nebula for 1973 are discussed and are compared with the star's behaviour since 1962.

133.020 **On evolutionary differences between planetary nebulae of two groups.** E. B. Kostyakova.
Astron. Tsirk., No. 817, p. 1 - 3 (1974). In Russian.

Stellar winds and related phenomena in surrounding nebulae. See Abstr. 064.058.

An interpretation of the puzzling observations of FG Sagittae. See Abstr. 065.094.

Models for nuclei of planetary nebulae and ultraviolet stars. See Abstr. 065.101.

Infrared photometry of southern emission-line stars. See Abstr. 113.022.

FG Sagittae: the s-process episode. See Abstr. 114.106.

Observations for broad-band circular polarization in white dwarfs and nuclei of planetary nebulae. See Abstr. 126.014.

A confusing radio source near the planetary nebula IC 2149. See Abstr. 141.018.

Infrared emission by dust in NGC 1068 and three planetary nebulae. See Abstr. 141.612.

134 Crab Nebula

134.001 **The structure of the Crab nebula—IV. The variation of polarization with wavelength.** A. S. Wilson.
Monthly Notices Roy. Astron. Soc., Vol. 166, 617 - 631 (1974).
This paper attempts to interpret recent data on the distribution of polarization in the Crab nebula in terms of the internal structure of the nebula. After a brief summary of the relevant observations, the author considers the depolarization by the bright filaments in more detail. The depolarization near the optical bays and that of regions well away from both bright filaments and bays is considered. Some consequences of the results for current ideas on the nebular magnetic field are discussed briefly.

134.002 **The origin of the magnetic field and relativistic particles in the Crab nebula.**
M. J. Rees, J. E. Gunn.
Monthly Notices Roy. Astron. Soc., Vol. 167, 1 - 12 (1974).
No fully satisfactory detailed model for the electrodynamic link between pulsar and nebula has yet been given, and the authors address themselves to this problem in the present paper. Their proposal involves, in essence, a modified and extended version of the general scheme first suggested by Piddington (1957) and subsequently discussed by Kardashev (1965), according to which the magnetic field has been 'wound up' by a central spinning object. It is now clear that this object should be identified with the pulsar NP 0531, and the authors therefore consider how the 'wound up field' concept can be reconciled with existing ideas on pulsars, and with the view that the pulsar also continuously replenishes the supply of relativistic electrons responsible for the non-thermal optical and X-ray output from the nebula.

134.003 **Effect of Faraday rotation on the circular polarization of the Crab nebula.** M. J. Gerver.
Astrophys. Journ., Vol. 189, 249 - 252 (1974).
The effect of Faraday rotation on the circular polarization of an electromagnetic wave propagating through a magnetized plasma is calculated for various limits of the plasma and wave parameters appropriate to a 30-Hz wave in the Crab nebula.

134.004 **Radio filaments in the Crab nebula from high resolution maps.** A. S. Wilson.
Galaxies and relativistic astrophysics, Proc. 1972,(see 012.010), p. 218 - 226 (1974).

134.005 **A model of the Crab nebula derived from dual-frequency radio measurements.**
K. W. Weiler, G. A. Seielstad.
Galaxies and relativistic astrophysics, Proc. 1972, (see 012.010), p. 227 (1974). — Abstract.

134.006 **Gamma ray observations from the Crab nebula and NP-0532.**
B. Parlier, B. Agrinier, M. Forichon, J. P. Leray, G. Boella, L. Maraschi, R. Buccheri, N. R. Robba, L. Scarsi.
Astrophys. Space Sci. Library, Vol. 45, (see 012.015), 267 - 281 (1974).

134.007 **Distribution and motion of the relativistic electrons in the Crab nebula.** A. S. Wilson.
Astrophys. Space Sci. Library, Vol. 45, (see 012.015), 355 - 364 (1974).

134.008 **On the formation of wisps in the Crab nebula.** A. Ferrari.
Astrophys. Space Sci. Library, Vol. 45, (see 012.015), 375 - 379 (1974).

134.009 **TV photometry of the Crab nebula.** K. Davidson, P. Crane.
Bull. American Astron. Soc., Vol. 6, 263 (1974). – Abstr. AAS.

134.010 **Parametric instabilities in the Crab nebula.** C. E. Max, J. Arons.
Bull. American Astron. Soc., Vol. 6, 273 - 274 (1974). Abstr. AAS.

Non-linear Compton magneto-bremsstrahlung. See Abstr. 061.039.

Energy conversion between longitudinal and transverse waves by mode-mode coupling in a relativistic plasma. See Abstr. 062.077.

Radiation reaction in non-linear Compton effect. See Abstr. 063.014.

On the expansion of supernova shells. See Abstr. 125.017.

X-ray spectrum of NP0532. See Abstr. 141.326.

Optical polarization of the Crab nebula pulsar. III. New observations, predictions, and the possibility of variability. See Abstr. 141.333.

Radio properties of the Crab pulsar: Where is the emission region located? See Abstr. 141.334.

Optical radiation from the Crab pulsar. See Abstr. 141.336.

A search for soft X-ray sources in the galactic anticenter. Absorption of X-rays from the Crab nebula. See Abstr. 142.033.

Radio Sources, Quasars, Pulsars, Infrared, X-Ray, Gamma-Ray Sources, Cosmic Radiation

141 Radio Sources, Quasars, Pulsars, Infrared Sources

Radio Sources, Quasars

141.001 **Acceleration of QSO clouds by radiation pressure.** R. Opher.
Astrophys. Journ., Vol. 187, 5 - 9 (1974).

The acceleration of a cloud by resonance-line scattering is discussed in detail. It is proven, under fairly general assumptions, that resonance-line scattering cannot produce clouds of the requisite final velocities $v_f \sim c$ for QSOs with $z_{abs} \ll z_{em}$. As an alternative to resonance-line scattering, the possibility that matter can be accelerated to velocities $v_f \sim c$ by free electron scattering in the outer region of the QSO core is discussed.

141.002 **A sustained pulse from ON 231.**
J. T. Pollock, D. L. Hall, C. Ambruster, P. D. Usher.
Astron. Astrophys., Vol. 30, 41 - 43 (1974).

Blue photographic plates from 1931 to 1952 reveal the existence of a 1 magnitude pulse of half-width of about 3 years emitted by the quasi-stellar object ON 231 (W Com). Rapid variability of amplitude about 1 magnitude is superposed upon the slower variations. The optical variability and radio spectrum of ON 231 are discussed, and the possibility of the existence of a second optical pulse is explored.

141.003 **Redshift magnitude bands for quasi-stellar sources.** P. Véron, M. P. Véron.
Astron. Astrophys., Vol. 30, 155 - 157 (1974).

Tifft (1973) has shown that in a redshift-magnitude diagram, quasi-stellar sources form a system of parallel bands. If one removes from the sample used by Tifft the objects which have no photoelectric magnitude and those which are strongly variable, and if one corrects the magnitudes for galactic absorption, one finds that the statistical weight of this band system is very low.

141.004 **The spectra of 373 radio sources.**
M. P. Véron, P. Véron, A. Witzel.
Astron. Astrophys., Suppl. Ser., Vol. 13, 1 - 53 (1974).

The authors have studied the spectra of 373 radio sources, using most published flux densities between 10 and 10 000 MHz. Taking into account only the straight spectra, they have determined a scaling factor for all the used catalogues of flux densities, in order to get a relative but consistent set of scales. These scaling factors are in good agreement with the published scaling factors for frequencies smaller than 3000 MHz. Using their calibration method, however, the authors find that the published scales at frequencies larger than 5000 MHz are generally too low and that there is then no trend for the spectra to become steeper at higher frequencies contrary to the finding of other authors.

141.005 **The radio structure of quasars. II.**
J. F. C. Wardle, G. K. Miley.
Astron. Astrophys., Vol. 30, 305 - 315 (1974).

The structures of thirty-nine quasars with known redshifts have been studied with NRAO interferometer at wavelengths of 11.1 and 3.7 cm. Twelve sources were unresolved, five sources were slightly resolved, and structures have been derived at both frequencies for the remainder. Nearly half of the well resolved sources were found to have more than two components. The "largest angular size – redshift" diagram is extended to 166 quasars, and is interpreted in terms of the "ram pressure" model for the containment of extended radio sources. It is shown that for the largest quasars at large redshifts there is a serious conflict between their radiative and dynamic lifetimes.

141.006 **Southern radio sources possibly identified with X-ray sources.** N. Sanduleak, J. F. Dolan.
Astrophys. Journ., (Letters), Vol. 187, L73 - L74 (1974).

Three radio sources in the southern Milky Way have published coordinates which nearly coincide with well-determined positions of three X-ray sources in the 3U catalog. They may represent the radio counterparts of these X-ray sources.

141.007 **Fine structure in radio sources at 81.5 MHz–III. The survey.** A. C. S. Readhead, A. Hewish.
Mem. Roy. Astron. Soc., Vol. 78, 1 - 49 (1974).

A survey of radio sources which exhibit interplanetary scintillations has been carried out with the Cambridge 18000 m² array at 81.5 MHz. The methods of observation, and of determination of angular structure on a scale $0''.2$ to $2''.0$, are discussed and a catalogue lists the structure of ~1500 4C sources between declinations $-12°$ and $+90°$, roughly 60 per cent of which scintillate.

141.008 **Observations of ten extragalactic radio sources with very steep spectra.** A. Slingo.
Monthly Notices Roy. Astron. Soc., Vol. 166, 101 - 111 (1974).

The Half-Mile telescope at the Mullard Radio Astronomy Observatory has been used at 1421 MHz to map ten sources which have very steep radio spectra. Of the seven sources which are coincident with clusters of galaxies, only one shows evidence for the steep spectrum being associated with a halo. The curvature noted in the spectra of two of these sources is interpreted as being due to synchrotron losses, following the withdrawal of the energy input in the form of relativistic electrons.

141.009 **A search for flux density variations in the central components of the extended extragalactic radio sources Virgo A and 3C 111.** B. L. Fanaroff.
Monthly Notices Roy. Astron. Soc., Vol. 166, 1P - 7P (1974).

Measurements of the λ 6 cm and λ 11 cm flux densities of the central component and jet of Virgo A and of the central component of 3C 111 have been made over a period of 2 yr. Evidence has been found of variations in the central component of 3C 111 and an upper limit set on variations in the nucleus and jet of Virgo A.

141.010 **Optical identifications for 45 radio sources based** on accurate positions. D. G. Hoskins, H. S. Murdoch, R. L. Adgie, J. H. Crowther, H. Gent. Monthly Notices Roy. Astron. Soc., Vol. 166, 235 - 247 (1974).

Optical identifications are given for 32 of 45 sources based on accurate radio positions measured with the Royal Radar Establishment interferometer. Both the optical and the radio positions used are sufficiently accurate (~ 0.5 arcsec) that reliable identifications can be made on the basis of position alone. A high proportion of the sources have 'abnormal' radio spectra. For these the identification content is high (80 per cent) including four identifications with stellar objects of neutral colour.

141.011 **Variation of radio source counts with direction, for the 3CR and 4C surveys.** T. J. Pearson. Monthly Notices Roy. Astron. Soc., Vol. 166, 249 - 257 (1974).

In an investigation of variations with direction of the radio source counts $N(S)$ for the 3CR and 4C surveys, no anisotropy has been found which is not explicable in terms of random statistical differences between limited samples of sources; in particular, there is no significant difference between the slopes of the source counts in the northern and southern galactic hemispheres.

141.012 **Radio source structure, clusters of galaxies and the complex sources.** T. Hooley. Monthly Notices Roy. Astron. Soc., Vol. 166, 259 - 270 (1974).

Correlations between the properties of extragalactic radio sources and membership of clusters of galaxies are examined for a complete sample of radio galaxies.

141.013 **The analysis of quasar samples.** J. C. Jackson. Monthly Notices Roy. Astron. Soc., Vol. 166, 281 - 295 (1974).

This paper is largely about technique; the method devised by Lynden-Bell to allow for observational selection effects in complete quasar samples is extended to deal simultaneously with several samples, each with its own selection criteria; expressions giving the uncertainty in the answers obtained are also derived. The method is illustrated using 3CR and 4C quasar samples, and resulting density evolution, colour and optical luminosity functions are presented and discussed.

141.014 **Observations of Cygnus A with the 5-km radio telescope.** P. J. Hargrave, M. Ryle. Monthly Notices Roy. Astron. Soc., Vol. 166, 305 - 327 (1974).

Maps of the distribution of radio brightness and polarization in Cygnus A at 5 GHz have been made with an angular resolution of $2\rlap.{''}0 \times 3\rlap.{''}1$ arc. On the assumption of equipartition between relativistic electrons and magnetic field, and from a study of the spectral distribution, it is concluded that there must be a continuous replenishment of energetic electrons within each of the two main compact components of the source. The evidence now available is inconsistent with several of the physical models which have been proposed. The models most likely to provide a satisfactory explanation of the details observed seem to be those in which energy is continually transported to the components from the nucleus in the form of energetic particles or low frequency waves.

141.015 **Isotropy of radio source populations from comparison of number–flux density curves.** L. M. Golden. Monthly Notices Roy. Astron. Soc., Vol. 166, 383 - 408 (1974).

To study the distribution of radio source populations with position over the sky, the 4C sky is divided into regions and the slope of the source count curve of the sources in each is computed. The isotropic nature of this distribution is shown in several ways. A recent report of an anisotropic distribution of radio sources is discussed in light of these results. In addition, the dispersion among the slopes of the source count curves is examined in terms of a hypothetical clustering of radio sources.

141.016 **Non-aligned components in extragalactic radio sources.** A. Harris. Monthly Notices Roy. Astron. Soc., Vol. 166, 449 - 461 (1974).

The author uses more recent high resolution data to re-examine the complex structure within the components of double sources and to suggest ways in which it can be accounted for. Evidence of non-alignment is found in source components on a variety of scales. The analysis of the observations and the implications of these results for models of radio sources are described.

141.017 **Initial results from a radio continuum survey of the area around NGC 7331 and Stephan's Quintet.** A. R. Gillespie. Monthly Notices Roy. Astron. Soc., Vol. 166, 11P - 15P (1974).

Observations of a $2° \times 2°$ region centred on $\alpha = 22^h 34^m 40^s$, $\delta = 36° 10'$ at 1421 MHz have revealed an excess of radio sources in a small area north of Stephan's Quintet. It is shown that the density of sources observed is not likely to be due to chance, and some implications of this result are discussed.

141.018 **A confusing radio source near the planetary nebula IC 2149.** G. Sistla, M. A. Kaftan-Kassim. Monthly Notices Roy. Astron. Soc., Vol. 166, 17P - 19P (1974).

High-resolution dual-frequency observations of the planetary nebula IC 2149 show that the nebula is confused by a nearby source, thus accounting for the apparent non-thermalicity. The position and spectrum of the confusing source are given.

141.019 **Polarization of radio emission of Vir A at 4 cm.** N. S. Soboleva. Astron. Zhurn. Akad. Nauk SSSR, Vol. 51, 71 - 74 (1974). In Russian. English translation in Soviet Astron. AJ, Vol. 18, No. 1.

With the large Pulkovo radiotelescope observations of polarization of radio emission of Vir A at 4 cm have been carried out. It is found that the region of polarized radiation is of less size than the region of nonpolarized radiation; and the maximum of polarized radiation is shifted to the east relative to the maximum of nonpolarized radiation. The electron density inside the polarized region is estimated.

141.020 **Statistics of apparent distances between QSO's and bright galaxies.** R. Kippenhahn, H. L. de Vries. Astrophys. Space Sci., Vol. 26, 131 - 135 (1974).

Two experiments with points randomly distributed over the sky were carried out in order to check the significance of the apparent associations between bright galaxies and quasi-stellar objects found by Burbidge et al. (1971). The experiments confirm the statistical relevance of these associations.

141.021 **On the redshifts of quasars.** B. Abramenko. Astron. Astrophys., Vol. 31, 109 - 110 (1974).

The notion that the large redshifts of quasars are entirely cosmological is criticised. The author shows that the Hubble diagram of Bahcall and Hills is consistent with a part of the redshift being intrinsic to quasars.

141.022 **Spectra for 107 radio sources selected at 408 MHz.**

H. S. Murdoch, D. G. Hoskins.
Australian Journ. Phys., Vol. 26, 867 - 880 (1973).

Radio spectra are presented for 107 sources which have previously been examined for optical identifications based on 408 MHz positions obtained with the Molonglo cross. Flux density measurements are given at five frequencies from 318 to 2695 MHz. The relationship between spectrum and optical identification is discussed.

141.023 **The structures of some extragalactic sources at 5000 MHz.** J. V. Wall, D. J. Cole.
Australian Journ. Phys., Vol. 26, 881 - 896 (1973).

The large scale structures of 30 extragalactic sources have been investigated using the Parkes 64 m telescope at 5000 MHz. Contour maps are presented for the resolved sources and comparisons are made with observations at lower frequencies.

141.024 **Close pairs of QSOs.**
J. N. Bahcall, L. Woltjer.
Nature, Vol. 247, 22 - 23 (1974).

The authors show that the pairs of QSOs observed so far may be comfortably explained as random coincidences and that the small probabilities that are often quoted depend on the use of a posteriori statistics.

141.025 **Radio spectra for three QSOs.** H. S. Murdoch.
Nature, Vol. 247, 443 - 444 (1974).

Radio spectra are presented for three sources of current interest: 1548 + 114 (4C 1150), 1331 + 170, and 1442 + 101 (OQ172).

141.026 **Redshift-magnitude relations for quasars.**
C. J. Krieger.
Astrophys. Space Sci., Vol. 27, 241 - 243 (1974).

Two cosmological models represent the observed magnitudes and redshifts of 150 quasars well with $\langle M_v \rangle = -22.75$ and -22.67, respectively. The luminosity function is derived.

141.027 **Models of extragalactic radio sources with a continuous energy supply from a central object.**
P. A. G. Scheuer.
Monthly Notices Roy. Astron. Soc., Vol. 166, 513 - 528 (1974).

This paper explores the dynamics of radio source models in which energy is carried from a nucleus to the radio components by a relativistic beam. Only a small fraction of the energy supply can be radiated away from the tip of the beam, and the rest lingers on in a cavity surrounding the beam.

141.028 **The central radio sources in three rich clusters of galaxies.** W. Jaffe, G. C. Perola.
Astron. Astrophys., Vol. 31, 223 - 234 (1974).

The authors present maps at 1415 MHz and 5 GHz obtained with the Westerbork Synthesis Radio Telescope of 5C4.81 and 5C4.85 (in the Coma cluster), 3C338 (in Abell 2199) and 4C17.66 (in the Hercules cluster). They discuss the relationships between the observed properties of the radio sources and the physical conditions inside the clusters.

141.029 **A theory of galactic nuclei and quasi-stellar objects.**
R. Opher.
Astrophys. Journ., Vol. 188, 201 - 205 (1974).

A theory is presented based on the assumption that interactions are cosmologically induced, resulting in depressed coupling constants in very compact objects which are identified with QSOs and galactic nuclei. The resulting model is that of a nuclear-burning shell surrounding an inert core. Predictions of the theory are discussed and compared with existing astrophysical data.

141.030 **The evolution of the radio spectrum of Cassiopeia A.**
W. A. Dent, H. D. Aller, E. T. Olsen.
Astrophys. Journ., (*Letters*), Vol. 188, L11 - L13 (1974).

Measurements of the decay of Cas A at 8.0 GHz yield a rate of -0.70 ± 0.07 percent per year. A comparison with determinations at lower frequencies establishes that the decay rate is frequency dependent and that the radio spectrum of Cas A is becoming flatter with time.

141.031 **Radio structure of the 'double quasar' 4C11.50.**
D. Stannard, P. N. Wilkinson, R. S. Warwick, R. J. Davis.
Nature, Vol. 247, 527 - 528 (1974).

Wampler et al. (1973) have recently announced the discovery of a close pair of optical quasars associated with the radio source 4C11.50. Here, the authors investigate further the alignment of the radio structure, seeking in particular any evidence of a radio connection between the two optical quasars.

141.032 **Reassessment of suggested frequency dependent positions for 4C sources.**
D. G. Hoskins, H. S. Murdoch.
Proc. Astron. Soc. Australia, Vol. 2, 194 - 195 (1973). – Presented at the annual general meeting of the Astron. Soc. Australia – see 012.001.

141.033 **The Parkes 2700 MHz survey – on-line reduction.**
J. V. Wall.
Proc. Astron. Soc. Australia, Vol. 2, 195 - 197 (1973). – Presented at the annual general meeting of the Astron. Soc. Australia – see 012.001.

141.034 **The magnitude distribution of QSOs.**
J. Katgert.
Observatory, Vol. 94, 20 - 21 (1974). – Letter.

141.035 **The redshift–angular size relation for double radio sources.** A. Kruszewski, I. Semeniuk.
Acta Astron., Vol. 24, 17 - 28 (1974).

Basing on available material, a list of angular separations for double radio sources with known redshifts is prepared. No reasonable cosmological model can be fitted to the observed redshift–angular size diagram for such objects. Much better fit can be obtained when the evolutionary effects due to adiabatic expansion and ram pressure are taken into account.

141.036 **Quasar counts and the lagging core model ('white holes').** Y. Ne'eman.
IAU Symposium No. 53, (see 012.004), p. 319 - 320 (1974).

141.037 **Spectra of radio sources with inhomogeneous magnetic field.** Yu. P. Ochelkov, O. F. Prilutskij.
Astron. Zhurn. Akad. Nauk SSSR, Vol. 51, 321 - 328 (1974). In Russian. English translation in Soviet Astron. AJ, Vol. 18, No. 2.

The spectra of radio sources with inhomogeneous magnetic field are considered. The injection spectrum of the electrons is assumed to be a power law spectrum. It is shown that the space inhomogeneity of the magnetic field can lead to essential changes of the form of the spectrum of the synchrotron radiation.

141.038 **Confusion and flux-density error distributions.**
J. J. Condon.
Astrophys. Journ., Vol. 188, 279 - 286 (1974).

The amplitude distribution of apparent flux densities due to background or "confusion" sources observed by a pencil-beam telescope in a universe randomly populated by unresolved sources obeying a power-law number-flux-density relation is derived. The confusion probability distributions are used to compute error distributions for confusion-limited flux-density measurements. A method for observing moving or variable sources weaker than the confusion limit is analyzed.

141.039 Early evolution of radio outbursts and a possible transient emission of high-energy photons.
F. Pacini, M. Salvati.
Astrophys. Journ., (Letters), Vol. 188, L55 - L58 (1974).

The authors have investigated a model for radio outbursts assuming constant injection of fast particles and magnetic field. They discuss the expected spectral evolution, and predict the existence of a related transient emission of high-energy photons.

141.040 On the theory of radio variability of galactic and quasar nuclei. L. M. Ozernoj, L. E. Ulanovsky.
Astron. Zhurn. Akad. Nauk SSSR, Vol. 51, 8 - 20 (1974).
In Russian. English translation in Soviet Astron. AJ, Vol. 18, No. 1.

Three basically different types of models explaining the extragalactic radio sources variability are possible, the dominating energy density component being thermal plasma (type A), magnetic field (type B) or relativistic particles (type C). The authors present a theory of radio variability for B- and C-type models. The B-type model is found to be in good agreement with observations, while the C-type model as well as the A-type model fail to explain some quantitative properties of variable radio sources.

141.041 Flux densities of some radio sources at 14400 MHz.
V. R. Amirkhanyan, A. G. Gorshkov, V. K. Konnikova, M. G. Larionov, V. G. Mirovskij, I. A. Strukov, V. S. Ehtkin.
Astron. Tsirk., No. 802, p. 7 - 8 (1973). In Russian.

141.042 Search for optical identifications in the 5C3-radio survey. I. Observations.
G. A. Richter, G. M. Richter, L. Richter, N. B. Richter.
Astron. Nachr., Vol. 295, 19 - 26 (1974).

In this paper the search for optical identifications in the 5C-radio surveys is continued. Near the positions of 135 radio sources of the 5C3 catalogue all present optical objects up to the plate limit (about $B \approx 21$) have been measured photometrically in UBVr and astrometrically on plates of the 134/200 cm Schmidt telescope of the Karl Schwarzschild Observatory Tautenburg. A total of 111 candidates for possible optical identifications were found. The number of real identifications will be, of course, much smaller.

141.043 Search for radio emission from faint variable starlike objects. A. H. Bridle, E. B. Fomalont.
Astron. Journ., Vol. 79, 6 - 7 (1974).

A high sensitivity search for 2695- and 8085-MHz radio emission from optically-variable quasar candidates has had negative results.

141.044 The NGC 520 chain of quasars. M. F. Barnothy.
Astron. Journ., Vol. 79, 8 - 10 (1974).

Statistical analysis reveals that in areas of the sky where the object density equals that found around NGC 520, there is a 70% chance to find a configuration similar to the NGC 520 quasar chain. The configuration does not support, therefore, the assumption that the four quasars were ejected from NGC 520 galaxy, or that the redshift of the four quasars were not cosmological.

141.045 The problem of satellite radio sources in the vicinity of spiral galaxies. M. L. De Jong.
Publ. Astron. Soc. Pacific, Vol. 86, 90 - 91 (1974).

The question of whether bright spiral galaxies have radio sources lying outside of their optical dimensions is examined. Evidence against such a general conclusion is given.

141.046 Flux-density measurements of selected radio sources relative to Cassiopeia A at 21.84 GHz.
M. J. Klein.
Astron. Journ., Vol. 79, 139 - 143 (1974).

New measurements of the flux-density ratios of DR 21 and six other sources relative to Cassiopeia A at 21.84 GHz are reported. Based on the assumption that the flux density of Cas A was 266.6 flux units at the mean epoch 1971.1, the flux densities of the program sources are 18.6 ± 1.4 f.u. for DR 21, 420 ± 28 f.u. for Taurus A, 300 ± 20 f.u. for Orion A, 448 ± 50 f.u. for M 17, 56.5 ± 4.0 f.u. for Cygnus A, 21.7 ± 1.6 f.u. for Virgo A, and 41.2 ± 2.5 f.u. for 3C273.

141.047 The Ohio Survey: VI. J. R. Ehman, R. S. Dixon, C. M. Ramakrishna, J. D. Kraus.
Astron. Journ., Vol. 79, 144 - 317 (1974).

A 1415 MHz continuum survey with the Ohio State University 110 by 21-m radio telescope has been made between declinations of 0° and 40° north and between 25° and 31° south covering 8190 deg^2 of sky. This is the sixth installment of the Ohio Survey. It fills large gaps in some of the declination zones of the five previously published installments of the survey. There are 81 maps of the regions surveyed and a list of 5955 sources at or above 0.18 flux units.

141.048 Observations of Sgr A at 160 MHz with 1.9 arc min resolution. G. A. Dulk, O. B. Slee.
Nature, Vol. 248, 33 - 34 (1974).

The authors extend the high resolution observations of Sgr A down to 160 MHz. The results are presented and discussed in detail.

141.049 Super-relativistic phase velocities of radio source components. R. H. Sanders.
Nature, Vol. 248, 390 - 392 (1974).

A sudden injection of relativistic electrons at the centre of an extended dipolar magnetic field could lead to an apparent faster-than-light expansion of two synchrotron emitting regions. Subsequent injections of electrons would produce components which move outwards along the same axis.

141.050 Coordinated observations of OJ 287 at radio and optical wavelengths.
T. D. Kinman, J. F. C. Wardle, E. K. Conklin, B. H. Andrew, G. A. Harvey, J. M. Macleod, W. J. Medd.
Astron. Journ., Vol. 79, 349 - 357 (1974).

Observations of the total flux of OJ 287 were made at 11.1, 4.5, 3.7, 2.8, 0.95, and 0.35 cm and at 0.44 μ during early 1972 when the object had its greatest recorded optical brightness. The linear polarization was measured at 3.7 and 11.1 cm and at 0.4 μ. Significant changes in the total flux in times of the order of a day were found at 4.5 cm and shorter wavelengths. These rapid fluctuations at short centimeter wavelengths were in phase to within a small fraction of a day. No correlation was seen between the optical and radio fluctuations. Correlograms of the optical total flux and Stokes parameters give evidence for an eight day periodicity. The consequences of these observations are briefly discussed.

141.051 Structure of 3C139.2, 3C154, 3C172 and 3C215 at 327 MHz.
V. K. Kapahi, Gopal-Krishna, M. N. Joshi.
Monthly Notices Roy. Astron. Soc., Vol. 167, 299 - 309 (1974).

Accurate radio positions and high-resolution brightness distributions derived from lunar occultations observed at 327 MHz are presented for 3C139.2, 154, 172 and 215. Possible identifications with faint objects are suggested for 3C154 and

3C172. The QSS 3C215 is a symmetric double with no structure finer than 1 second of arc, and has the lowest known ratio of component separation to component size among the 3C quasars.

141.052 **The morphology of extragalactic radio sources of high and low luminosity.**
B. L. Fanaroff, J. M. Riley.
Monthly Notices Roy. Astron. Soc., Vol. 167, 31P - 35P (1974).

The relative positions of the high and low brightness regions in the extragalactic sources in the 3CR complete sample are found to be correlated with the luminosity of these sources.

141.053 **Radio observations of QSOs in the directions of Abell clusters of galaxies.**
D. Stannard, I. W. A. Browne.
Monthly Notices Roy. Astron. Soc., Vol. 167, 37P - 39P (1974).

Radio observations are reported of nine optically selected quasars in the directions of Abell clusters of galaxies. Only one is found to be a radio source stronger than 40 mfu at 962 MHz, and this has a redshift much greater than that of the associated cluster.

141.054 **High radio polarization of 4C 47.08.**
R. W. Porcas, A. M. Treverton, A. Wilkinson.
Monthly Notices Roy. Astron. Soc., Vol. 167, 41P - 42P (1974).

Measurements are presented of the radio position and linear polarization of the unusual radio source 4C 47.08. A neutral stellar object identified with the source has been shown to have a featureless optical spectrum.

141.055 **Measurement of angular dimensions of discrete radio sources by means of an independent reception interferometer at 408 MHz with a resolving power of 0.1 arc sec.**
V. A. Alekseev, M. A. Antonets, I. F. Belov, A. A. Varypaev, E. D. Gatehlyuk, V. A. Efanov, B. N. Lipatov, I. G. Moiseev, S. M. Mkrtchyan, V. A. Sanamyan, A. S. Sizov, V. S. Troitskij, B. V. Shchekotov, A. I. Chikin.
Izv. vyssh. ucheb. zavedenij. Radiofizika, Vol. 16, 1314 - 1317 (1973). In Russian. − Abstr. in Referativ. Zhurn. 51. Astron., 4.51.113 (1973).

141.056 **Measurement of angular dimensions of discrete sources by means of an independent reception interferometer at 25 MHz with a radio Grakovo-Zimenki base.**
S. A. Volokhov, E. D. Gatehlyuk, P. S. Zhivora, B. S. Ivanov, V. A. Alekseev, M. A. Antonets, S. Ya. Braude, E. N. Vinyajkin, S. A. Kamenskaya, A. E. Kryukov, P. A. Mel'yanovskij, A. V. Men', A. S. Sizov, A. I. Teplykh, V. S. Troitskij, A. I. Chikin, M. V. Yankavtsev.
Izv. vyssh. ucheb. zavedenij. Radiofizika, Vol. 16, 1318 - 1319 (1973). In Russian. − Abstr. in Referativ. Zhurn. 51. Astron., 4.51.114 (1974).

141.057 **Are quasars dusty?** C. F. McKee, V. Petrosian.
Astrophys. Journ., Vol. 189, 17 - 21 (1974).

Two separate techniques have been applied to estimate the amount of dust in the line of sight toward a number of quasars. No evidence for dust was found. The possible relationship between quasars and galaxies is discussed. A new upper limit on the dust content of the intergalactic medium is obtained.

141.058 **Radiative acceleration of gas clouds near quasistellar objects and Seyfert galaxy nuclei.**
W. G. Mathews.
Astrophys. Journ., Vol. 189, 23 - 31 (1974) = Lick Obs. Bull. No. 645

Hydrogen photoionization dominates the momentum transfer by ultraviolet radiation to gas clouds in the emission-line region of quasi-stellar objects and Seyfert nuclei. Constraints on the nature of the intercloud medium and the effects of differential radiative forces are discussed. The radiative acceleration of fully ionized clouds is proportional to the gas density, which may explain why permitted spectral lines are often broader than forbidden lines.

141.059 **The distance of BL Lacertae.**
J. B. Oke, J. E. Gunn.
Astrophys. Journ., (Letters), Vol. 189, L5 - L8 (1974).

The extended object around BL Lacertae has been observed spectrophotometrically, with the use of both an annulus and round apertures. The annulus spectrum shows the absorption lines normally seen in spectra of ordinary giant galaxies; these lines yield a redshift $z = 0.07$. If $H_0 = 60$ km s^{-1} Mpc^{-1} is used, the absolute visual magnitude of the galaxy is -22.9 and the distance is 350 Mpc. On the assumption that the redshift $z = 0.07$, the central source is similar in spectral index and luminosity to quasars such as 3C 48, 3C 279, and 3C 345, provided a cosmological distance is assumed for these latter quasars.

141.060 **Further observations of apparent changes in the structure of 3C 273 and 3C 279.**
K. I. Kellermann, B. G. Clark, D. B. Shaffer, M. H. Cohen, D. L. Jauncey, J. J. Broderick, A. E. Niell.
Astrophys. Journ., (Letters), Vol. 189, L19 - L22 (1974).

During 1972 the authors continued to use the Goldstack interferometer to observe the variable radio sources 3C 273 and 3C 279. It is possible to fit the data with models consisting of three fixed collinear components which individually vary in intensity and size, or with expanding double models.

141.061 **On gravitational-lens quasars.** L. N. K. de Silva.
Astrophys. Journ., Vol. 189, 177 - 179 (1974).

It is shown that the intensification due to a transparent gravitational lens is inadequate for quasars to be gravitational-lens images of Seyfert nuclei.

141.062 **QSO envelopes: optically thin, low density, and normal abundances?**
J. D. Scargle, L. J. Caroff, C. B. Tarter.
Astrophys. Journ., Vol. 189, 181 - 186 = Lick Obs. Bull., No. 654 (1974).

The authors consider the family of photoionization models which are optically thin to the ionizing radiation, have low enough density that collisional de-excitation is unimportant, and have normal (cosmic) chemical abundances. The resulting model is in reasonably good agreement with the typical or composite QSO spectrum and with line intensities for two recently observed individual QSOs.

141.063 **Photoelectric spectrophotometry of OQ 172 and OH 471.** J. B. Oke.
Astrophys. Journ., (Letters), Vol. 189, L47 - L49 (1974).

Absolute spectral energy distributions for the large-redshift quasars OQ 172 and OH 471 are discussed along with similar data for two other quasars: 4C 05.34 and PHL 957.

141.064 **Structure of 194 southern declination radio sources from interplanetary scintillations.**
S. M. Bhandari, S. Ananthakrishnan, A. Pramesh Rao.
Australian Journ. Phys., Vol. 27, 121 - 128 (1974).

Interplanetary scintillation observations of 194 southern declination radio sources have been made at 327 MHz. The angular size and the fraction of the flux present in the scintillating component have been estimated. More than half of the observed sources scintillate and contain ⩾10% of their flux in angular size smaller than 0."5.

141.065 Optical identification of southern radio sources. II.
B. M. Lasker, M. G. Smith.
Australian Journ. Phys., Vol. 27, 135 - 138 (1974).

From a study of the fields of 11 southern radio sources, six identifications are suggested with faint galaxies and three with compact objects, while two fields appear blank to the limits of the photographs. Additional material for one previously identified field is also given.

141.066 Radio source identified with a neutral stellar object near an unusual galaxy.
H. S. Murdoch, W. B. McAdam, R. W. Hunstead.
Nature, Vol. 248, 491 - 492 (1974).

The authors report here the identification of a neutral stellar object of 19 mag close to an unusual blue galaxy, using an accurate radio position obtained at 408 MHz with the Mills Cross of the Molongo Observatory.

141.067 Statistical analysis of close pairs of QSOs.
E. M. Burbidge, G. R. Burbidge, S. L. O'Dell.
Nature, Vol. 248, 568 - 569 (1974).

The authors reconsider the pairs already discovered and then make an a priori calculation based on realistic parameters such that a test involving the discovery of further close pairs of QSOs may enable a significant statement to be made.

141.068 Upper mass limit for quasar. B. Paczynski.
Nature, Vol. 249, 329 - 330 (1974).

The radio source 4C11.50 is reported to be a pair of quasars only 5 arcs apart. The author points out that if the redshifts are cosmological the mass of 4C11.50a must be less than about $4 \times 10^{12} M_\odot$.

141.069 Optical identifications and radiospectra of the deep NRAO 5 GHz survey sources in the B 2 survey area.
R. Fanti, A. Ficarra, L. Formiggini, I. Gioia, L. Padrielli.
Astron. Astrophys., Vol. 32, 155 - 160 (1974).

Optical identifications are given with galaxies and blue objects of a survey of faint sources made with the NRAO 300-ft. telescope at 5 GHz. The comparison between the spectral indices distribution of the sources of this sample and the distribution of strong sources at the same frequency shows a dependence of the mean spectral index on the flux density.

141.070 Optical identifications from the Parkes 2700 MHz survey: the 03^h, 19^h, and 23^h zones, declinations $-33°$ to $-75°$. J. V. Wall, R. D. Cannon.
Australian Journ. Phys., Astrophys. Suppl., No. 31, 40 pp. = Separate print Division Radiophys., C.S.I.R.O., Sydney (1973).

Identifications are suggested for 105 radio sources in the third part of the Parkes 2700 MHz survey. The identifications, 63 with galaxies and 42 with quasi-stellar objects, were obtained from examination of V, B plate pairs taken with the Uppsala Schmidt telescope at Mount Stromlo Observatory. Additional position measurements with the Parkes 64 m telescope were made on 199 of the 459 catalogue sources covered by these plate pairs. Identification statistics for the area are compared with those of the first two parts of the Parkes 2700 MHz survey.

141.071 Intensity dependent radio spectral distributions in the Parkes 2700-MHz survey.
J. J. Condon, D. L. Jauncey.
Astron. Journ., Vol. 79, 437 - 452 (1974).

On the basis of 318-MHz flux densities measured at Arecibo, two-point spectral indices $\alpha(318,2700)$ were determined for all 382 sources north of $\delta = -2°$ in the Parkes 2700-MHz survey. Supplementary 606-MHz observations were made to reduce confusion errors and provide better definition to weak-source curved spectra. The detailed spectra of selected individual sources are treated. Statistical investigations of the distributions of the spectral indices $\alpha(318,2700)$ are reported and their implications are discussed.

141.072 Identification of southern hemisphere radio sources. III. Optical objects in the radio fields. P. K. Lü.
Astron. Journ., Vol. 79, 453 - 455, 503 - 509 (1974).

Further identifications and measurements of optical objects in the fields of southern radio sources are presented. Identifications are suggested for nine QSO's or possible QSO's, twenty-three galaxies and four n-type galaxies.

141.073 Narrow-band observations of galactic and extragalactic sources at 1 mm.
P. E. Clegg, P. A. R. Ade, M. Rowan-Robinson.
Nature, Vol. 249, 530 - 532 (1974).

The authors report the result of a search for discrete extragalactic sources of continuum radiation near 1.2 mm. They attempted observations of a total of 90 known extragalactic objects: three were probably detected (3C84, 3C120, 3C273) and two were possibly detected (3C345, 0727−11). These five objects comprise two quasars, a Seyfert galaxy, an N galaxy and an unidentified radio source. They also observed a number of well known galactic regions and found millimetre emission in several of them.

141.074 Astrophysical implications of extragalactic radio sources. M. J. Rees.
Galaxies and relativistic astrophysics, Proc. 1972, (see 012.010), p. 190 - 209 (1974). − Invited lecture.

141.075 Digital dispersion spectroscopy: a technique for the study of very rapid variability of celestial radio sources. J. G. Ables.
Astron. Astrophys., Suppl. Ser., Vol. 15, 371 (1974). − Conference paper (see 012.011).

141.076 Radio spectra of OH471 and OQ172.
M. R. Gearhart, J. D. Kraus, B. H. Andrew, G. Blake, P. Scott, M. Ryle, S. Ya. Braude, N. K. Sharykin, I. N. Zhouck, A. H. Bridle, E. K. Conklin, J. N. Douglas, O. Hachenberg, M. Thiel. P. Kaufmann, C. R. Purton, P. A. Feldman, K. A. Marsh, M. A. Stull, K. Price, J. W. Warner, G. Assousa, B. Balick.
Nature, Vol. 249, 743 - 746 (1974).

Radio spectra for the highly redshifted quasars OH471 and OQ172 are presented for frequencies between 16.7 MHz and 85.3 GHz. The spectra are complex. Both have a peak at or near 1.4 GHz and a second peak near or above 85 GHz. Both have a low frequency cutoff, the one for OH471 being especially sharp. OH471 sometimes seems to be variable at 31 GHz on a time scale of days but the source appears to be stable at 10.7 and 22 GHz.

141.077 Measurement of angular dimensions of the radio source Cas A at 9 MHz frequency with an independent reception interferometer on 1.5 and 7 km bases.
V. A. Alekseev, M. A. Antonets, E. N. Vinyajkin, S. A. Volokhov, Eh. D. Gatehlyuk, P. S. Zhivora, B. S. Ivanov, S. A. Kamenskaya, A. E. Kryukov, G. R. Pashkova, A. S. Sizov, A. I. Teplykh, V. S. Troitskij, A. I. Chikin, M. V. Yankavtsev.
Izv. vyssh. ucheb. zavedenij. Radiofizika, Vol. 16, 1307 - 1313 (1973). In Russian. − Abstr. in Referativ. Zhurn. 51. Astron., 5.51.620 (1974).

141.078 On the interpretation of continuum flux observations from thermal radio sources − I. Continuum

spectra and brightness contours.   M. Salem, M. J. Seaton.
Monthly Notices Roy. Astron. Soc., Vol. 167, 493 - 510 (1974).

The main result is to show that the distribution of surface brightness as a function of solid angle can be deduced from the spectrum of the total flux density on evaluating an inverse Laplace transform. A value of T_e must be assumed. This value can be adjusted so as to obtain a best fit to the distribution of surface brightness obtained directly from high resolution observations. The theory is illustrated by considering results for the Orion nebula and for the compact H^+ region DR 21.

141.079   **On the interpretation of continuum flux observations from thermal radio sources – II. Three-dimensional models.**   M. Salem.
Monthly Notices Roy. Astron. Soc., Vol. 167, 511 - 515 (1974).

Given a function $\Omega(E)$, which is the total angular area containing matter with emission measure greater than or equal to E, a unique model of a thermal radio source may be constructed for a given geometry. The construction of a spherical model with filling factor unity is described. Such a model is constructed for the Orion nebula, and compared with a model computed by a different method. The significance of models of this type is briefly discussed.

141.080   **Radio observations of the 'double QSO' 4C 11.50.**   D. Wills, W. D. Cotton.
Monthly Notices Roy. Astron. Soc., Vol. 167, 75P - 77P (1974).

The radio source 4C 11.50 is double at 365 MHz and at 2695 MHz; at the latter frequency one component agrees in position with the brighter of the close pair of QSOs studied by Wampler et al. At 8085 MHz, only one component was observed, and its position again agrees with that of the brighter QSO. An upper limit of 0.05×10^{-26} W m^{-2} Hz^{-1} at 2695 and 8085 MHz can be set to emission from the fainter QSO or from the red galaxy about 10″ arc away from the brighter QSO.

141.081   **Spectroscopic observations of eight QSO candidates among Markarian objects.**
D. Wills, B. J. Wills.
Monthly Notices Roy. Astron. Soc., Vol. 167, 79P - 81P (1974).

Eight of the ten objects suggested as possible QSOs in Markarian's fourth list of objects with ultraviolet continua have been observed spectroscopically; all eight are galactic stars, predominantly DA type white dwarfs.

141.082   **Absolute magnitudes for quasars with cosmological redshifts.**   A. Evans, D. Falla.
Observatory, Vol. 94, 45 - 50 (1974).

Assuming quasar redshifts are cosmological in origin, the authors calculate values of absolute magnitude M for 164 quasars, in different intervals of redshift z, for $q_0 = +1$. The distribution of M values obtained is found to be dependent upon z. The results are in agreement with a model in which the M distribution is related to the light curve for the individual quasar, provided that two factors are included: these are the observational faint limit to apparent magnitude ($m \leq 19.5$), and the cosmological redshift dependence of the light travel time from the source.

141.083   **The radio magnitude-redshift relationship for QSOs.**   A. Evans.
Observatory, Vol. 94, 50 - 55 (1974).

The radio-frequency Hubble diagram for QSOs is reexamined in the light of a recent suggestion by McCrea, and is found to be consistent with McCrea's hypothesis and the cosmological interpretation of QSO redshifts.

141.084   **Colours and redshifts of QSOs.**   D. Basu.
Observatory, Vol. 94, 61 - 66 (1974).

The correlation between colours and redshifts of QSOs has been studied by several authors. The influence of the emission lines on the observed colours can certainly not be ruled out and any physical interpretation of the relationship between colour and redshift must obviously be based on continuum colours and not on observed colours. "Pure continuum" colours have recently been computed and published for 105 QSOs by Evans. The author was therefore prompted to examine the relationship between colour and redshift using these data.

141.085   **Kosmosa objekti ar straujām optiskā un radiostarojuma maiņām.**   A. Alksnis.
Zvaigžņotā debess, 1973./74. gada ziema, p. 4 - 13.

141.086   **A 408 MHz fan-beam survey at low galactic latitudes. I. The observations and a catalogue of small diameter sources.**   C. Fanti, M. Felli, A. Ficarra, C. J. Salter, G. Tofani, P. Tomasi.
Astron. Astrophys., Suppl. Ser., Vol. 16, 43 - 61 (1974).

Using the east-west arm of the Bologna Northern Cross radio telescope, fully sampled observations have been made of an extensive region at low galactic latitudes. The galactic longitude range is $15° \leq l \leq 245°$, with a latitude coverage of at least 7° at each longitude. The observations, their analysis, and the calibrations and errors of the survey, are fully discussed. A catalogue is given of 586 small-diameter radio sources with measured flux densities above 1.2×10^{-26} W m^{-2} Hz^{-1}.

141.087   **Ooty occultations of 76 radio sources.**
V. K. Kapahi, M. N. Joshi, N. V. G. Sarma.
Astron. Journ., Vol. 79, 515 - 526, 659 (1974).

Positional and structural information derived from lunar occultations observed at 327 MHz is presented for 76 radio sources, most of them of flux density less than 2×10^{-26} Wm^{-2} Hz^{-1}. Only 16 sources are optically identified.

141.088   **The photometric history of the object identified with PKS 0537–441.**   W. Liller.
Astrophys. Journ., (Letters), Vol. 189, L101 - L102 (1974).

Harvard photographs reveal that since 1892 the B-magnitude of the object identified with PKS 0537–441 has fluctuated erratically between 12.6 and 16.5 with peak brightnesses coming in 1917 and in 1944–1945. If any long-term periodicity exists, a maximum brightness should have occurred in late 1972. There may also be a 60-day periodicity.

141.089   **On the time evolution of quasar luminosity and redshift distribution.**
R. Meiseles, B.-Z. Kozlovsky, G. Shaviv.
Astrophys. Space Sci., Vol. 29, 221 - 231 (1974).

The possibility to predict the redshift distribution on the basis of certain assumptions on the time-dependence of quasar luminosity is discussed. It is found that for a certain general class of models that resembles the spinar model for quasars only specific combinations of parameters give rise to acceptable results. Further consequences are discussed.

141.090   **Componenti compatte nelle radiosorgenti extragalattiche.**   A. Braccesi.
Mem. Soc. Astron. Italiana, Suppl. Vol. 44, p. S 251 - S 269 (1974).

141.091   **Quasars.**   G. Banos.
In honorem S. Placidis, (see 003.009), p. 251 - 276 (1974). In Greek.

141.092   **Positions of extragalactic radio sources from very long baseline interferometry.**

C. C. Counselman III.
IAU Symposium No. 61, (see 012.014), p. 119 - 124 (1974). Invited paper.

141.093 Position solution of compact radio sources using long coherence VLBI.
J. S. Gubbay, A. J. Legg, D. S. Robertson.
IAU Symposium No. 61, (see 012.014), p. 125 - 129 (1974).

141.094 The Texas radio astrometric survey.
J. N. Douglas.
IAU Symposium No. 61, (see 012.014), p. 141 (1974). – Abstract.

141.095 Absolute positions of radio sources obtained with the NRAO interferometer. P. Brosche.
IAU Symposium No. 61, (see 012.014), p. 143 (1974). – Abstract.

141.096 Determination of an extragalactic radio source position catalogue suitable for use in monitoring geophysical phenomena. J. L. Fanselow, J. G. Williams.
IAU Symposium No. 61, (see 012.014), p. 145 (1974). – Abstract.

141.097 Spectroscopy of objects near Texas radio-source positions. D. Wills, B. J. Wills.
Astrophys. Journ., Vol. 190, 271 - 277 (1974).

Spectroscopic observations of 62 objects near accurate Texas radio-source positions are reported. Among a sample of 17 blue stellar objects within 3″ of the Texas radio positions all 17 are found to be QSOs. A further seven QSOs, and two stars, are found among nine blue stellar objects lying between 3″ and 5″ of the radio positions. Observations of 12 red and neutral stellar objects within 5″ of Texas, and other, accurate radio positions show two of them to be QSOs and three to be rather luminous galaxies, comparable with low-luminosity QSOs. New redshifts are given for 30 QSOs and galaxies.

141.098 The absence of radio emission from HZ Herculis.
L. Hartmann, A. S. Lapedes.
Astrophys. Journal, (*Letters*), Vol. 190, L67 - L68 (1974).

The absence of pulsed emission in HZ Herculis at 430 MHz found in this study is consistent with current theoretical models. However, the absence of a radio halo or supernova remnant emitting continuum radiation indicates either that the pulsar is more than 10^5 years old, or that there was no supernova remnant associated with the pulsar's birth.

141.099 Redshifts of 1548 + 115a and 1548 + 115b in the theory of the generalised gravitational potential.
N. H. Cherry.
Nature, Vol. 250, 127 - 128 (1974). – Letter.

141.100 The Arecibo 611 MHz multi-beam sky survey.
D. J. Pleticha, J. M. Durdin, D. L. Jauncey, M. J. Yerbury, J. J. Condon, C. Hazard.
Bull. American Astron. Soc., Vol. 6, 212 (1974). – Abstr. AAS.

141.101 No redshifts in quasi-stellar objects.
Y. P. Varshni.
Bull. American Astron. Soc., Vol. 6, 213 (1974). – Abstr. AAS.

141.102 A model for Cassiopeia A spectral evolution.
F. W. Peterson.
Bull. American Astron. Soc., Vol. 6, 213 (1974). – Abstr. AAS.

141.103 Arecibo 611 MHz multi-beam sky survey tapes.
J. M. Durdin, D. J. Pleticha, D. L. Jauncey, M. J. Yerbury, J. J. Condon, C. Hazard.
Bull. American Astron. Soc., Vol. 6, 217 (1974). – Abstr. AAS.

141.104 Broad emission lines in high-redshift quasars.
R. Ptak, R. Stoner.
Bull. American Astron. Soc., Vol. 6, 261 (1974). – Abstr. AAS.

141.105 Relativistic and cold electron densities in compact radio sources. A. G. Pacholczyk, T. L. Swihart.
Bull. American Astron. Soc., Vol. 6, 262 (1974). – Abstr. AAS.

141.106 Is there dust in quasars? V. Petrosian, C. McKee.
Bull. American Astron. Soc., Vol. 6, 278 - 279 (1974). – Abstr. AAS.

141.107 Analysis of QSO variability.
J. D. Scargle, L. J. Caroff.
Bull. American Astron. Soc., Vol. 6, 279 (1974). – Abstr. AAS.

141.108 A simple model for a recent radio outburst of Cygnus X-3.
K. A. Marsh, C. R. Purton, P. A. Feldman.
Bull. American Astron. Soc., Vol. 6, 280 (1974). – Abstr. AAS.

141.109 The Parkes 2700 MHz survey (sixth part). Catalogue for the declination zone $-30°$ to $-35°$.
A. J. Shimmins, J. G. Bolton.
Australian Journ. Phys., Astrophys. Suppl. No. 32, 55 pp. = Separate print Division Radiophys., C. S. I. R. O., Sydney (1974).

This paper presents a catalogue of 939 extragalactic radio sources obtained from a sky survey at 2700 MHz. The area surveyed is between declinations $-30°$ and $-35°$, and covers all right ascensions except for two regions within approximately $10°$ of the galactic plane. For an area of 0.390 sr, the catalogue is complete to a limiting flux density of 0.18 f.u. at 2700 MHz ($\sim$ 1100 sources per steradian) and is thought to be at least 90% complete at a flux density of 0.15 f.u. ($\sim$ 1400 sources per steradian). Flux densities at 5009 MHz have been measured for 291 of the strong sources. Identifications with galaxies have been made for 80 of the sources, 26 sources have been confirmed as quasi-stellar objects and a further 39 are suggested as possible quasi-stellar objects.

141.110 Quasars. Z. Sima.
Říše hvězd, Vol. 55, 1 - 3 (1974). In Czech.

141.111 Quasi-stellar object research. P. A. Strittmatter.
Proc. ESO/SRC/CERN conference, (see 012.021), p. 109 - 113 (1974).

141.112 Complete samples of 3CR and 4C quasars.
M. Schmidt.
Proc. ESO/SRC/CERN conference, (see 012.021), p. 253 - 257 (1974).

141.113 Westerbork Synthesis Radio Telescope programmes in the perspective of observational cosmology.
H. van der Laan, R. S. Le Poole.
Proc. ESO/SRC/CERN conference, (see 012.021), p. 259 - 264 (1974).

141.114 Results from the Parkes 2700 MHz survey.
J. V. Wall.
Proc. ESO/SRC/CERN conference, (see 012.021), p. 265 - 274 (1974).

141.115 Statistical studies of QSOs. D. Wills.
Proc. ESO/SRC/CERN conference, (see 012.021), p. 275 - 290 (1974).

141.116 PKS 0735+17. C. Pollas, B. Dumortier.
IAU Circ., No. 2678 (1974).

141.117 Westerbork ontdekt objecten: groter dan ooit waargenomen. 3C 236 en DA 240: Radiostelsels groter dan 5 miljoen lichtjaren. R. G. Strom.
Zenit, Vol. 1, No. 6, p. 2 - 3 (1974).

141.118 Radioastronomie auf Kurzwellen. S. Braude, A. Menj (Men').
Ideen Exakten Wiss., [Deutsche Verlags-Anstalt, Stuttgart], 1973 No. 9, p. 581 - 589.

141.119 Spectroscopic observations of objects identified with radio sources. P. A. Strittmatter, R. F. Carswell, G. Gilbert, E. M. Burbidge.
Astrophys. Journ., Vol. 190, 509 - 514 (1974).

Results of a spectroscopic survey of radio source identifications are reported. Redshifts have been determined for 20 quasars and seven radio galaxies. A large proportion of identifications with neutral-colored stellar objects appear to have continuous spectra of the BL Lacertae type.

141.120 On the significance of the Cerenkov process in quasi-stellar objects. J. D. Colvin.
Astrophys. Journ., Vol. 190, 515 - 519 (1974).

A two-stage model for the optical flux from quasi-stellar objects is considered. It is shown that a very high energy density of photons near the electron plasma frequency is produced by the Cerenkov process when an assumed distribution of ultra-relativistic electrons produced in an explosive event passes through a small ($R < 10^{14}$ cm), moderately dense ($N_e \simeq 10^8$ cm^{-3}), magnetized source region. These photons serve as the ambient radiation field for the inverse Compton process to work.

141.121 3C 66A: a bright new quasi-stellar object. B. J. Wills, D. Wills.
Astrophys. Journal, (Letters), Vol. 190, L97 - L98 (1974).

The strong unresolved radio source 3C 66A, lying 6.5 from 3C 66B (identified with a 13-mag galaxy), has been identified as a 15.2-mag blue stellar object having QSO colors and a continuous optical spectrum.

141.122 Molecular-hydrogen absorption features in the spectrum of quasi-stellar object 4C 05.34.
R. W. Carlson.
Astrophys. Journal, (Letters), Vol. 190, L99 - L100 (1974).

It is suggested that short-wavelength absorption features in the spectrum of 4C 05.34 correspond to the $R(0)$ lines of the Lyman and Werner band systems of H$_2$. The derived redshift of these features is $z = 2.6354$.

141.123 Optical observations of the radio source 0735 + 178. R. F. Carswell, P. A. Strittmatter, R. E. Williams, T. D. Kinman, K. Serkowski.
Astrophys. Journal, (Letters), Vol. 190, L101 - L104 (1974).

It is shown that the variable radio source 0735 + 178 varies substantially in both optical flux and polarization on time scales less than a month. It is therefore assigned to the class of objects containing BL Lac and OJ 287. The spectrum of 0735 + 178 does, however, contain two sharp absorption features at 3981 and 3991 Å which are identified with the Mg II λ2798 doublet at a redshift $z = 0.424$. This confirms that 0735 + 178 is extragalactic. The presence of highly redshifted absorption lines in 0735 + 178 also strengthens the hypothesis that these objects are similar to quasars.

141.124 The double quasar 1548 + 115a, b as a gravitational lens. J. R. Gott III, J. E. Gunn.
Astrophys. Journal, (Letters), Vol. 190, L105 - L108 (1974).

The two quasars 1548 + 115a and 1548 + 115b are separated by only 5" of arc. The authors consider the possibility that the light from the distant quasar is enhanced by the gravitational-lens effect of the nearby quasar and any galaxy in which it may reside. The observed separation allows one immediately to set a strict upper limit of $7 \times 10^{12} M_\odot$ on the mass of the system containing 1548a. It is likely that the lens forms a fainter but still observable secondary image of 1548b on the other side of 1548a.

141.125 Observational selection in the identification of quasars and claims for anisotropy. L. M. Golden.
Observatory, Vol. 94, 122 - 126 (1974).

It has been claimed that quasars are distributed anisotropically with position over the sky. Although the effects of observational selection in these deductions are important, they are often either ignored or dealt with inadequately. By discussing several claims of anisotropy, it is shown that two such selection effects are largely responsible.

141.126 New QSO. M. A. Kazaryan, R. F. Carswell, Eh. E. Khachikyan.
Astron. Tsirk., No. 813, p. 2 - 4 (1974). In Russian.

141.127 Power spectra of scintillations on irregularities of the interplanetary plasma. T. D. Shishova.
Astron. Tsirk., No. 819, p. 1 - 3 (1974). In Russian.

141.128 Quasars and the redshift. K. P. Tritton.
Sci. Progr., (GB), No. 242, Vol. 61, 275 - 288 (1973).

Very long baseline interferometry in the southern hemisphere. See Abstr. 031.009.

Total power statistical confusion surveys. See Abstr. 033.042.

Radio and optical astrometry. See Abstr. 041.054.

Physics of compact nonthermal sources. I. Theory of radiation processes. See Abstr. 062.025.

Omnidirectional induced Compton scattering by relativistic electrons. See Abstr. 063.009.

Radiation reaction in non-linear Compton effect. See Abstr. 063.014.

Transfer of resonance-line radiation in differentially expanding atmospheres. III. Formation of P Cygni-type lines by a doublet line or two partially "blended" lines. See Abstr. 064.025.

A measurement of solar gravitational microwave deflection with the Westerbork Synthesis Telescope. See Abstr. 066.002.

Action d'un champ de gravitation statique sur un milieu dispersif isotrope. See Abstr. 066.010.

Anisotropic spheres in general relativity. See Abstr. 066.028.

Observations of interplanetary scintillations at 327 MHz. See Abstr. 106.018.

UV excess stars and radio sources in the Aquila ring. See Abstr. 113.010.

Blue objects near M31. II. Frequency distribution and statistics. See Abstr. 113.016.

Detection of NML Cygnus at 2.8 cm.
See Abstr. 114.018.

Radio emission from pre-main-sequence stars.
See Abstr. 114.073.

Radio emission from forbidden-line stars with circumstellar dust.   See Abstr. 114.148.

Spectroscopic orbit of the radio star b Persei.
See Abstr. 119.010.

Radio emission from R Aquarii.
See Abstr. 122.021.

Short-duration radio flares of UV Ceti stars.
See Abstr. 122.127.

On the recombination-line observations toward supernova 3C391.   See Abstr. 131.059.

Infrared observation of DR 21 at 2.2 μ.
See Abstr. 131.504.

Scorpius X-1: origin of the radio and hard X-ray emissions.   See Abstr. 142.002.

Sixth radio outburst of Cygnus X-3.
See Abstr. 142.036.

Osservazioni radio di sorgenti X.
See Abstr. 142.100.

The gravitational slingshot and the structure of extragalactic radio sources.   See Abstr. 151.050.

Investigation of the brightness of two extragalactic objects.   See Abstr. 158.032.

A blue galactic nucleus with a featureless spectrum.
See Abstr. 158.089.

Photographic photometry of compact extragalactic objects. II.   See Abstr. 158.091.

Radio observations of Abell 1314.
See Abstr. 160.012.

The Hubble relation for nonstandard candles and the origin of the redshift of quasars.   See Abstr. 162.018.

Pulsars

141.301 "Lorentz force-free" pulsar rotating fields.
V. G. Endean.
Astrophys. Journ., Vol. 187, 359 - 360 (1974).

It is shown that the "Lorentz force-free" rotating electromagnetic field problem may be reduced to an equation for the magnetic field of the form $(V \times B^*) \times B = 0$, where B^* is related in a simple manner to B. Some implications of this new equation are discussed in the light of two special cases, details of which have already been published.

141.302 Search for gravitational radiation from pulsars.
T. S. Mast, J. E. Nelson, J. A. Saarloos.
Astrophys. Journ., (Letters), Vol. 187, L49 - L51 (1974).

A search for gravitational radiation from pulsars and other periodic radiators was performed using a seismometer and the earth as a detector. The amplified output of a vertical seismometer was sampled, recorded on magnetic tape, and Fourier transformed. Data were collected at two quiet sites in California and searched over the range of 0.1–125 Hz both for any signals of narrow bandwidth and for signals at the periods and half-periods of 81 known pulsars. No signals were found.

141.303 Hot pulsar magnetospheres.
R. N. Henriksen, D. R. Rayburn.
Monthly Notices Roy. Astron. Soc., Vol. 166, 409 - 424 (1974).

The equilibrium of the co-rotating zone of the pulsar magnetosphere is investigated. It is suggested that the particle density is much higher than the charge density and that the required heat input to maintain hydrostatic equilibrium is supplied by plasma turbulence in the ion-sound mode. The radiation properties of the resulting hot plasma are investigated. Finally a criterion is given for determining the extent of the co-rotating zone.

141.304 The suspected periodicity in the radio frequency spectrum of PSR 0329 + 54.
D. A. Jones, A. G. Lyne.
Monthly Notices Roy. Astron. Soc., Vol. 166, 39P - 42P (1974).

A periodicity in the radio frequency spectrum of PSR 0329 + 54 has been reported by Sturrock et al. Approximately 12 times the data available to them were analysed, with no sign of any obvious, consistent periodicity. It is concluded that the apparent periodicity is due to random scintillation in the interstellar medium.

141.305 A search for OH absorption in the spectrum of PSR 0329 + 54.   J. A. Galt.
Astron. Astrophys., Vol. 31, 235 (1974).

The pulsar's spectrum has been measured at the 1.667 GHz OH line over the region in which absorption lines would be spread by galactic rotation. At a resolution of 1.8 km s^{-1} no absorption was detected to a limiting rms optical depth of 0.1 (2.5 times the rms deviation).

141.306 TV investigations of ultra-short period optical radiation sources.
A. N. Abramenko, O. P. Gollandskij, V. V. Prokof'eva.
Izv. Krymskoj Astrofiz. Obs., Vol. 49, 89 - 97 (1974). In Russian.

A TV method for search and study of optical radiation sources having ultra-short period light variations (with periods from 0.001 to 1 sec) is suggested. The method was checked by study of the pulsar NP 0532 as well as by study of a laboratory variable light source. Preliminary results on search for an ultra-short periodic light source in the vicinity of the X-ray source Cyg X-1 are given.

141.307 Computer hardware and software for on-line pulsar hunting. P. T. Rayner, P. S. Whitham.
Proc. Astron. Soc. Australia, Vol. 2, 202 - 203 (1973). — Presented at the annual general meeting of the Astron. Soc. Australia — see 012.001.

141.308 Simultaneous observations of pulsar intensity variations at Parkes and Ootacamund.
O. B. Slee, J. G. Ables, R. A. Batchelor, S. Krishna-Mohan, V. R. Venugopal, G. Swarup.
Monthly Notices Roy. Astron. Soc., Vol. 167, 31 - 47 = Radiophys. Publ., C. S. I. R. O., Sydney, RPP 1706 (1974).

Simultaneous observations at 326.5 MHz of the intensity variations of seven pulsars have been conducted over a 8000-km baseline between Australia and India. The theoretical interpretation of the drifting diffraction pattern is developed in some detail and a practical method is determined for measuring the velocity component along the baseline in the presence of intrinsic pulsar variability and random evolution of the pattern. For the majority of the pulsars observed, the intensity fluctuations at the two stations were significantly decorrelated for all time shifts. Three of the seven pulsars provided correlation functions which could be processed by the proposed method to give the velocity of pattern drift along the baseline.

141.309 Matter in superstrong magnetic fields.
M. Ruderman.
IAU Symposium No. 53, (see 012.004), p. 117 - 131 (1973).

141.310 Pulsar observations and neutron star models.
G. Börner, J. M. Cohen.
IAU Symposium No. 53, (see 012.004), p. 227 - 236 (1974).

141.311 The pulsar phenomenon.
G. S. Bisnovatyj-Kogan.
Zemlya i Vselennaya, 1974, No. 2, p. 23 - 28. In Russian.

141.312 Pulsar magnetosphere with anomalous low conductivity of relativistic turbulence plasma.
S. A. Kaplan, V. N. Tsytovich, V. Ya. Ehjdman.
Astron. Zhurn. Akad. Nauk SSSR, Vol. 51, 363 - 372 (1974). In Russian. English translation in Soviet Astron. AJ, Vol. 18, No. 2.

Turbulence of relativistic circumpulsar plasma drastically reduces its conductivity especially in the vicinity of the pulsar surface where the gyrofrequency is larger than the plasma frequency. The consequences of this effect are discussed.

141.313 Pulsars and close binary systems.
G. S. Bisnovatyj-Kogan, B. V. Komberg.
Astron. Zhurn. Akad. Nauk SSSR, Vol. 51, 373 - 381 (1974). In Russian. English translation in Soviet Astron. AJ, Vol. 18, No. 2.

It is supposed that the absence of radio pulsars in double systems results from decreasing of the magnetic field of a neutron star during accretion of matter on it from the central component. The evolution of close binary systems with approximately equal values of masses is considered qualitatively and the possibility of first supernova explosion of the larger mass companion and destruction of the pair is shown. An analysis of the spatial distribution of pulsars is made and some pairs of close pulsars are indicated at high galactic latitudes.

141.314 Pulse intensity histograms of pulsars.
K. H. Hesse, R. Wielebinski.
Astron. Astrophys., Vol. 31, 409 - 413 (1974).

Pulse intensity histograms have been calculated for 11 pulsars observed at 2695 MHz. These and previously published histograms have been grouped into three categories according to position and number of histogram maxima. The relationship between the histogram group and some of the observed properties of pulsars is discussed.

141.315 Individual pulse polarization properties of three pulsars.
J. M. Rankin, D. B. Campbell, D. C. Backer.
Astrophys. Journ., Vol. 188, 609 - 613 (1974).

Both average and individual pulse polarization measurements at 430 MHz are presented for three pulsars: PSR 0823 + 26, PSR 0834 + 06, and PSR 2303 + 30. In each case the average Stokes parameter wave forms are found to depend on the individual pulse polarization properties in a complicated manner.

141.316 On the passage of radiation through inhomogeneous, moving media. III. The steady-state fields of inertial charge distributions. I. Lerche.
Astrophys. Journ., Vol. 188, 627 - 635 (1974).

Motivated by the conjectures that the differentially shearing magnetospheres of pulsars are responsible for the observed pulsed-radiation production, the author has investigated as simply as possible the basic physics and topology of electric field lines emanating from an inertial charge distribution embedded in a differentially shearing medium. For the two simple cases of a uniformly charged rod and a uniformly charged plate he computes and sketches the basic field-line patterns.

141.317 Mechanical properties of the neutron lattice in PSR 0833–45. P. B. Jones.
Astrophys. Space Sci., Vol. 28, 213 - 223 (1974).

The mechanical properties of a possible solid core in the pulsar PSR 0833–45 are examined in relation to the observed changes in period and their interpretation by Pines, Shaham and Ruderman (1972). Qualitative arguments based on known characteristics of solid He^3 and assuming a many-body Hamiltonian with non-directional two-particle potentials indicate that the neutron lattice should have a high value of Poisson's ratio and allow dislocation glide at low applied stresses.

141.318 Counter-rotation of polarization vectors in pulsars.
F. G. Smith.
Monthly Notices Roy. Astron. Soc., Vol. 167, 43P - 46P (1974).

The plane of polarization rotates within individual subpulses of PSR 0809 + 74 in the opposite sense to the rotation in the integrated profile. This observation favours the identification of the subpulses with a basic beaming process, while the integrated profile represents a distribution of sources in longitude.

141.319 Pulsar flux-density spectra.
D. C. Backer, J. R. Fisher.
Astrophys. Journ., Vol. 189, 137 - 145 (1974).

The authors have obtained time-averaged flux-density spectra of 10 pulsars with simultaneous observations at six frequencies between 250 and 8085 MHz. The effects of interstellar scintillation on the accuracy of the measurements are discussed. All of the pulsars exhibited a power-law spectrum at high frequencies with spectral indices between −1.6 and −3.0. Several of the pulsars show a more positive spectral index at the low-frequency end of the data.

141.320 Free precession of neutron stars.
D. Pines, J. Shaham.
Nature, Vol. 248, 483 - 486 (1974).

Wobble in pulsars can provide a clue to the structure and properties of the associated neutron stars.

141.321 The magnetic fields of pulsars. D. M. Sedrakian.

IAU Symposium No. 64, (see 012.008), p. 187 (1974). – Abstract.

141.322 Pulsars. F. Pacini.
Galaxies and relativistic astrophysics, Proc. 1972, (see 012.010), p. 210 - 213 (1974). – Invited lecture.

141.323 Removal of dispersion distortion from pulsar radio signals. T. H. Hankins.
Astron. Astrophys., Suppl. Ser., Vol. 15, 363 - 365 (1974). Conference paper (see 012.011).

141.324 Force-free pulsar magnetosphere – I. The steady, axisymmetric theory for the charge-separated plasma. I. Okamoto.
Monthly Notices Roy. Astron. Soc., Vol. 167, 457 - 474 (1974).

In the steady, axisymmetric theory of the pulsar magnetosphere with the charge-separated plasma the critical field line separating the region of negative charge from the region of positive charge must be a straight line parallel to the equator owing to the rigidity of the pulsar surface. Thus, the theory with the complete charge separation yields an unphysical field structure for the dipole condition or the split monopole at the surface.

141.325 Pulsar hydrogen line absorption and the electron density in the interstellar medium.
J. Gómez-Gonzàlez, M. Guélin.
Astron. Astrophys., Vol. 32, 441 - 446 (1974).

Nine new pulsars have been observed for 21-cm line absorption. Distance limits have been derived for six of them. The new distances, together with previous pulsar data, are used for an analysis of the electron distribution in the interstellar medium. The electron density, averaged over a few kiloparsec path close to the galactic plane, is found to be uniform and is 0.03 cm^{-3}.

141.326 X-ray spectrum of NP0532.
R. M. Thomas, R. Rothenflug.
Nature, Vol. 249, 812 - 814 (1974).

There is a change in the X-ray spectral index of the Crab pulsar NP0532 near 10 keV. The authors here examine this in a more quantitative fashion than before, using most of the available published data, together with some unpublished data.

141.327 Pulsar magnetic axis alignment and counter-alignment. W. W. Macy, Jr.
Astrophys. Journ., Vol. 190, 153 - 163 (1974).

A model is proposed for the alignment or counteralignment of the magnetic axis of a pulsar with its symmetry axis by the motion of the magnetic distortion through the star due to the same kind of secular instability that has been proposed to cause polar wandering on the earth. The evolution of various observable characteristics is determined by simultaneously integrating the equations for the rate of change of the magnetic-axis inclination and for the angular momentum. The damping time of the precession about the average rotation axis is considered for two different crust materials.

141.328 Detection of soft X-ray emission from PSR 0833–45.
J. L. Culhane, A. M. Cruise, C. G. Rapley, F. J. Hawkins.
Astrophys. Journ., (*Letters*), Vol. 190, L9 - L12 (1974).

An observation of the Vela pulsar in the energy range 0.7–1.5 keV is presented. A comparison of the flux in this band with the spectrum of the source 3U 0833–45 for the 2–10 keV band suggests a low-energy turnover in the spectrum. The measurement is compared with other observations and possible explanations of the turnover are discussed.

141.329 An upper limit on soft X-ray pulsations from the pulsar PSR 0833–45.
W. E. Moore, P. C. Agrawal, G. Garmire.
Astrophys. Journ., (*Letters*), Vol. 189, L117 - L118 (1974).

A soft X-ray observation of the pulsar PSR 0833–45 places a 2 σ upper limit of 0.3 percent on the fraction, relative to the flux from the entire Vela X nebula, of pulsed flux in the 0.5- to 1.0-keV energy band. This limit is approximately a factor of 7 below the value observed by Harnden and Gorenstein.

141.330 Detection of pulsar proper motion.
R. N. Manchester, J. H. Taylor, Y. Y. Van.
Astrophys. Journ., (*Letters*), Vol. 189, L119 - L122 (1974).

Proper motion has been detected for PSR 1133+16 by means of pulse arrival-time observations made over a 4-year period. The observed value of $\mu = 0\overset{''}{.}58 \pm 0\overset{''}{.}22$ yr^{-1} implies a transverse pulsar velocity of 380 km s^{-1} for an assumed distance of 130 pc.

141.331 Electromagnetic processes in the atmospheres of pulsars. A. K. Yukhimuk.
Kosmich. issledovaniya na Ukraine. Resp. mezhved. sb., 1973, vyp. (No.) 3, p. 30 - 36. In Russian. – Abstr. in Referativ. Zhurn. 51. Astron., 6.51.516 (1974).

141.332 On the structure of the pulse of NP 0532 at γ-ray energies. L. Maraschi, A. Treves.
Astrophys. Space Sci. Library, Vol. 45, (see 012.015), 307 - 309 (1974).

141.333 Optical polarization of the Crab nebula pulsar. III. New observations, predictions, and the possibility of variability. D. C. Ferguson, W. J. Cocke, T. Gehrels.
Astrophys. Journ., Vol. 190, 375 - 380 (1974).

New and very precise measurements of the optical polarization of the Crab nebula pulsar are presented. Better fits by the relativistic vector model than were previously reported are possible due to an improved method of plotting the data. A prediction of the true polarization arising outside the pulsar is made. The possibility of secular changes in the pulsar's polarization is discussed.

141.334 Radio properties of the Crab pulsar: Where is the emission region located?
R. M. Price, J. M. Sutton.
Bull. American Astron. Soc., Vol. 6, 213 (1974). – Abstr. AAS.

141.335 Low energy X-ray observations of PSR 0833–45.
A. M. Cruise, J. L. Culhane, C. G. Rapley, F. Hawkins.
Bull. American Astron. Soc., Vol. 6, 271 - 272 (1974). Abstr. AAS.

141.336 Optical radiation from the Crab pulsar.
P. A. Sturrock, V. Petrosian, J. S. Turk.
Bull. American Astron. Soc., Vol. 6, 272 (1974). – Abstr. AAS.

141.337 Break-up of pulsar low-frequency radiation into filaments. D. G. Wentzel.
Bull. American Astron. Soc., Vol. 6, 272 (1974). – Abstr. AAS.

141.338 Comments on inner superfluid and information from pulsar glitches. T. Takatsuka.
Progr. Theor. Phys. Japan, Vol. 50, 1755 - 1757 (1973).

Pulsar glitches are shown to provide information not only on the masses of neutron stars but also on their internal temperatures. Among the models proposed to explain the glitches, the two-component model simplifies the neutron stars as composed of the outer crust and inner superfluid and attributes the origin of the glitches to starquakes.

141.339 **The propagation and broadening of pulses in weakly irregular media [pulsar radiation].**
B. J. Uscinski.
Proc. Roy. Soc. London, Ser. A, Vol. 336, 379 - 392 (1974).

The radio signals received from pulsars are pulses which are frequently broadened as a result of scattering in the interstellar medium. This broadening is examined theoretically using a diffraction theory of wave propagation in weakly irregular media. Typical pulse shapes are calculated and the results are compared with the theory of pulse broadening based on geometrical optics.

141.340 **Results of search for 10^{12} eV gamma rays from pulsars.** D. M. Jennings, G. White, N. A. Porter, E. O'Mongain, D. J. Fegan, J. White.
Nuovo Cimento B, Ser. 11, Vol. 20B, 71 - 82 (1974).

A fast night-sky Cerenkov detector has been used in a search for γ-rays of energy 10^{12} eV from the ten pulsars. No strong evidence for continuous emission was found, and the upper flux limits are given. Analysis for γ-rays pulsed at the pulsar period shows an anomalous effect from the Crab nebula which may correspond to a true flux.

141.341 **Pulsars.** F. G. Smith.
Rep. Progr. Phys., Vol. 35, 399 - 461 (1972).

This review is concerned with the observational data on pulsars, and the interpretations which can be made directly from them. Introductory sections describe the discovery of the first pulsars, the search techniques now in use, the theory of neutron stars, and the relation between the Crab nebula pulsar and the nebula itself. A general description of the pulse observations, with a discussion of the accurate timing, leads to a detailed account of the radio observations, and particularly the pulse shapes and polarizations. Scintillation and Faraday rotation measurements are discussed, leading to a description of the ionized interstellar medium and the galactic magnetic field, and to the suggestion that pulsars have a high velocity relative to the interstellar gas. Finally, the population of the pulsars within the Galaxy is discussed in relation to the probable association between pulsars and the supernova remnants.

141.342 **Self-consistent solution for an axisymmetric pulsar model.** L. G. Kuo-Petravic, M. Petravic, K. V. Roberts.
Phys. Rev. Letters, Vol. 32, 1019 - 1022 (1974).

The self-consistent solution of a closed set of relativistic fluid equations and Maxwell's equations describing the motion of a neutron star in the magnetosphere is obtained for the case when both the rotation and magnetic axes are in alignment. Flow patterns for polar and equatorial particles are presented as projections of their respective velocity fields.

141.343 **Interstellar scattering of the Vela pulsar.**
D. C. Backer.
Astrophys. Journ., Vol. 190, 667 - 671 (1974).

The frequency dependence of the parameters of interstellar scattering between 837 and 8085 MHz for the Vela pulsar are consistent with thin-screen models of strong scattering. The magnitudes of the parameters indicate an anomalous turbulence along the path when they are compared with results for other pulsars with comparable column densities of free electrons in the line of sight. This anomaly is due presumably to the Gum nebula.

141.344 **On pulsar emission diagrams.**
O. Kh. Gusejnov, I. M. Yusifov.
Astron. Tsirk., No. 819, p. 5 - 7 (1974). In Russian.

Cosmic gamma-ray bursts from relativistic dust grains. See Abstr. 061.034.

On the passage of radiation through inhomogeneous, moving media. I. The plane, differentially sheared medium. See Abstr. 062.027.

On the passage of radiation through inhomogenous, moving media. II. The rotating, differentially sheared medium. See Abstr. 062.028.

Index of refraction of plasma in motion. See Abstr. 062.076.

Neutron star structure from pulsar observations. See Abstr. 065.040.

Compounds in neutron-star crusts. See Abstr. 065.054.

Absorption of high energy heavy nuclei and γ rays at the surface of hot neutron stars. See Abstr. 065.063.

The Vela supernova remnant in ultraviolet light. See Abstr. 125.010.

Do pulsars make supernovae? Calculations of light curves for type II events. See Abstr. 125.071.

Polarized radiation from white dwarfs and atoms in strong magnetic fields. See Abstr. 126.007.

Gamma ray observations from the Crab nebula and NP-0532. See Abstr. 134.006.

Origin of energetic cosmic rays. II. The possibility of a contribution from pulsars. See Abstr. 143.066.

Structure of the local galactic magnetic field. See Abstr. 156.002.

Infrared Sources

141.601 An infrared source associated with a Herbig-Haro object. K. M. Strom, S. E. Strom, G. L. Grasdalen.
Astrophys. Journ., Vol. 187, 83 - 86 (1974).

An infrared source of angular extent $\leq 6''$ has been found associated with a Herbig-Haro object in the Corona Austrina dark cloud. The infrared spectrum of this object, between 5 and 12.6 μ, is very similar to that characterizing three irregular variables of the Orion population located within the same dark cloud. It is proposed that the optical manifestation of this Herbig-Haro object represents the light scattered from a T Tauri-like star embedded deep within the CrA dark cloud.

141.602 Detection of ^{17}O in IRC + 10216.
D. M. Rank, T. R. Geballe, E. R. Wollman.
Astrophys. Journ., (Letters), Vol. 187, L111 - L112 (1974).

Lines of the fundamental vibration-rotation band of $^{12}C^{17}O$ have been observed near 4.7 μ in IRC + 10216. The presence of $^{12}C^{17}O$ is also indicated in α Her and α Sco.

141.603 On three galactic H_2O sources.
A. Baudry, W. J. Welch.
Astron. Astrophys., Vol. 31, 471 - 473 (1974).

A search for 22 GHz water-vapour emission in OH infrared stars, and in complex H II regions, has been conducted at the Hat Creek Observatory of the University of California. Three new H_2O sources displaying 1.35 cm time variation are briefly discussed. Continuum observations (22 GHz) of stars are also reported.

141.604 VY Canis Majoris. IV. The emission bands of ScO.
G. H. Herbig.
Astrophys. Journ., Vol. 188, 533 - 538 (1974).

The recent laboratory analyses of the blue-green system of ScO by Adams, Klemperer, and Dunn, and of the orange system by Stringat, Athénour, and Féménias, have been used to compute theoretical profiles for the unresolved ScO 0–0 band of the $A\ ^2\Pi - X\ ^2\Sigma$ (orange) system, under assumptions of LTE and optical thinness. The sharply peaked, almost line-like structure in VY CMa is shown to be due to the crowding of rotational structure near the band heads.

141.605 A dust-shell model of the infrared object HD 45677.
J. P. Apruzese.
Astrophys. Journ., Vol. 188, 539 - 543 (1974).

A dust-shell model of the infrared star HD 45677 was constructed by means of a generalization of the mathematical techniques of Huang. The model shows that an isotropically scattering, uniform-density spherical circumstellar dust shell is able to account for the observed emergent energy distribution. This model is compared with that of Dyck and Milkey, who propose free-free emission from a circumstellar gas shell to account for the observed infrared excess.

141.606 Structure of the OH/infrared object NML Cygnus. I. Analysis of the near-infrared image.
G. H. Herbig, J. Lorre.
Astrophys. Journ., Vol. 189, 73 - 74 (1974) = Lick Obs. Bull. No. 641.

Direct photographs of the image of NML Cyg obtained at an effective wavelength of about 8000 Å at the prime focus of the Lick 120-inch (3-m) reflector have been analyzed photometrically in an attempt to determine if the image of the star is extended or nebulous. It was concluded that no nebulosity comparable with that at VY CMa is present, and that the image of NML Cyg is entirely stellar.

141.607 Structure of the OH/infrared object NML Cygnus. II. Analysis of the OH interferometry.
G. H. Herbig.
Astrophys. Journ., Vol. 189, 75 - 79 (1974) = Lick Obs. Bull. No. 642.

The 1612-MHz OH interferometry of NML Cyg published by Davies, Masheder, and Booth has been analyzed in terms of pure radial expansion. It is found that the positions and radial velocities of most of the very complex OH structure, which covers an area of about $2''.5$ by $3''.5$, can be explained by simultaneous ejection about 100 ± 20 years ago (if the distance is 500 pc) from a center lying near one edge of the source. The expansion velocity is 46 ± 2 km s^{-1}.

141.608 Spectrophotometric observations of a highly absorbed object in Cygnus.
K. M. Merrill, B. T. Soifer.
Astrophys. Journ., (Letters), Vol. 189, L27 - L30 (1974).

Spectrophotometric observations from 2.8 to 14 μ with resolution $\Delta\lambda/\lambda \sim 0.015$ are reported for an object (AFCRL No. 809-2992) found in the AFCRL infrared sky survey. Significant absorption is found at 3.1 μ and in a very broad feature at $\sim 10\ \mu$. The 3.1-μ feature is attributed to absorption by ice, while the 10-μ feature is fitted well with an absorption profile defined by the emission spectrum of the Trapezium.

141.609 The $^2\Pi_{1/2}, J = 1/2$ state of OH, and the infrared stars associated with H_2O/OH microwave emission.
A. Baudry.
Astron. Astrophys., Vol. 32, 191 - 195 (1974).

Significant upper limits to the 6 cm $^2\Pi_{1/2}, J = 1/2$ transitions of OH have been obtained in infrared stars displaying H_2O microwave emission and in some unusual OH sources (1612 MHz/OH star-like sources). The absence of strong excited-state OH lines in stars is used to derive some circumstellar physical parameters, and to show that the OH gas component must be far from the star. In connection with the lack of strong 6 cm emission, arguments are given to suggest that low gas density and collisional pumping models may account for basic properties in type I OH stars.

141.610 Relevance of electron pair production in the interpretation of IR and X extragalactic sources.
S. Bonometto, F. Lucchin.
Galaxies and relativistic astrophysics, Proc. 1972, (see 012.010), p. 233 - 239 (1974).

141.611 Infrared stars in binary systems.
R. M. Humphreys, E. P. Ney.
Astrophys. Journ., Vol. 190, 339 - 347 (1974).

The purpose of this paper is to give a more complete discussion of HD 101584 and the properties of the proposed binary system. The peculiar F supergiants 89 Her, v Sgr, and R CrB are also discussed as possible binary systems containing an infrared star. All four of these stars have infrared excesses with characteristics similar to the infrared radiation from several cool objects which are known single stars.

141.612 Infrared emission by dust in NGC 1068 and three planetary nebulae. R. F. Jameson, A. J. Longmore, J. A. McLinn, N. J. Woolf.
Astrophys. Journ., Vol. 190, 353 - 357 (1974).

Photometric observations of NGC 1068, NGC 7027, IC 418, and BD +30°3639 between 5 and 27 μ are reported. The observations are explained as thermal radiation from a form of dust that shows spectral features, but which has not yet been identified, and resembles that in the carbon-rich RV Tauri star AC Her. Arguments are deduced for its being a condensate from a carbon- and nitrogen-rich gas in an excited environment.

141.613 34-micron observations of Eta Carinae, G333.6–0.2, and the galactic center.
E. Sutton, E. E. Becklin, G. Neugebauer.
Astrophys. Journal, (Letters), Vol. 190, L69 - L70 (1974).

Preliminary 34-μ observations show that (a) the majority of the infrared flux from η Carinae can be attributed to a 250° K circumstellar dust shell, (b) the infrared energy distribution of G333.6−0.2 exhibits a smooth continuum rising to 34 μ, and (c) the central 30" area of the galactic center is a strong 34-μ emitter with an energy distribution similar to that of G333.6−0.2.

141.614 **Interpretation of far infrared observations.**
S. R. Pottasch.
Stars and the Milky Way system, Proc. 1972, (see 012.018), p. 209 - 231 (1974). − Invited lecture.

141.615 **Possibilità di misure infrarosse di sorgenti X.**
P. Salinari.
Mem. Soc. Astron. Italiana, Nuova Ser., Vol. 44, 615 - 616 (1973/74).

141.616 **BM Scorpii and a possible cluster of infrared sources.** E. P. Ney, R. M. Humphreys.
Publ. Astron. Soc. Pacific, Vol. 86, 304 - 307 (1974).

BM Sco and six other infrared sources near the cluster M6 were observed at infrared wavelengths. The presence of these infrared sources near M6 is not statistically significant and is not sufficient to associate them with the cluster.

On circumstellar shells of protostellar origin.
See Abstr. 065.013.

The evolution of a massive protostar.
See Abstr. 065.014.

Veiling and the presence of circumstellar gas and dust in some infrared stars. See Abstr. 114.031.

Free-free and free-bound emission in low-surface-gravity stars. See Abstr. 114.032.

High-resolution spectra of cool stars in the 10- and 20-micron regions. See Abstr. 114.069.

Infrared fluxes, spectral types, and temperatures for very cool stars. See Abstr. 114.075.

Time variations in the OH microwave and infrared emission from late-type stars. See Abstr. 114.132.

Interpretation of Epsilon Aurigae. II. Infrared excess, secondary light variations, and plausible formation of a planetary system. See Abstr. 121.002.

Accurate positions of OH sources.
See Abstr. 131.001.

Infrared studies of H II regions and OH sources.
See Abstr. 131.523.

Infrared emission from the southern H II region H2-3. See Abstr. 131.524.

Spectrophotometric observations of the compact H II region K3-50 and of NGC 6857.
See Abstr. 131.529.

1-millimeter observations of the galactic H II regions M42 and DR 21. See Abstr. 131.533.

CO and CS in the Orion nebula.
See Abstr. 132.034.

Infrared spectrum of NGC 1068.
See Abstr. 158.034.

Errata

141.901 Erratum: "Stimulated linear acceleration radiation: a pulsar radio emission mechanism" [Astrophys. Journ., Vol. 184, 291 - 300 (1973)]. W. J. Cocke.
Astrophys. Journ., Vol. 187, 211 (1974).

141.902 Erratum: 'Spectra of some Ohio radio sources: List IV' [Astrophys. Journ., Vol. 185, 137 - 144 (1973)]. B. H. Andrew, J. R. Ehman, M. R. Gearhart, J. D. Kraus.
Astrophys. Journ., Vol. 189, 165 - 166 (1974).

142 X-Ray, Gamma-Ray Sources

142.001 **Interstellar absorption of X-rays.** E. L. Fireman.
Astrophys. Journ., Vol. 187, 57 - 60 (1974).

The photoabsorption cross-sections of some medium-weight elements that had been ignored in calculations of interstellar X-ray absorption, as well as the effect of interstellar dust grains on X-ray absorption, are computed.

142.002 **Scorpius X-1: origin of the radio and hard X-ray emissions.**
R. Ramaty, C. C. Cheng, S. Tsuruta.
Astrophys. Journ., Vol. 187, 61 - 68 (1974).

The consequences of models for the central radio source and the hard X-ray (> 30 keV) emitting region in Sco X-1 are examined. The radio emission could result from noncoherent synchrotron radiation, and the X-rays may be produced by bremsstrahlung. The authors show that both these mechanisms require a mass outflow from Sco X-1.

142.003 **Black holes in binary systems: Instability of disk accretion.** A. P. Lightman, D. M. Eardley.
Astrophys. Journ., (Letters), Vol. 187, L1 - L3 (1974).

The authors have tested the stability of a thin, orbiting accretion disk near a black hole. Under conditions appropriate for a binary X-ray source, with the usual assumptions about viscosity, the disk is always secularly unstable on time scales of a few seconds or less. The authors mention possibilities for alternative models; perhaps the secular instability explains chaotic time variations in Cygnus X-1.

142.004 **A search for hard X-rays from the transient source 2U 1543−47.**
G. Pizzichini, A. Spizzichino, G. R. Vespignani.
Astron. Astrophys., Vol. 30, 161 - 165 (1974).

The location of the transient source 2U 1543−47 discovered by Uhuru in August 17, 1971 was scanned by a hard X-ray detector aboard OSO-6 in May and September 1971. The time of appearance of the source in the 27−189 keV range is limited to later than May 22 - 28, in agreement with a similar result at lower energies. For September 7 - 11 the observation gives a 3σ upper limit of $\sim 2 \times 10^{-3}$ and 4×10^{-4} photon $(cm^2\ s\ keV)^{-1}$ at 35 keV and 60 keV, respectively, to a possible hard spectral tail above 20 keV.

142.005 **Search for gamma-ray emission from the galactic anti-center region.**
M. Forichon, B. Agrinier, J. P. Leray, B. Parlier, G. Boella, L. Maraschi, R. Buccheri, B. Sacco, L. Scarsi.
Astron. Astrophys., Vol. 30, 323 - 327 (1974).

The analysis procedure and the results of a search for discrete gamma-ray sources ($E > 20$ MeV) in the galactic anti-center region, carried out in the years 1969−70−71 with a balloon-flown spark-chamber telescope, are reported. Except for the Crab nebula no other gamma-ray source was observed. Upper limits are given for the gamma-ray emission above 20 MeV of the objects in the observed sky region which have been reported as gamma-ray sources (Pe − γ 1, 3C 120, galactic disc).

142.006 **Extended observations of >7-keV X-rays from Centaurus X-3 by the OSO-7 satellite.**
W. A. Baity, M. P. Ulmer, W. A. Wheaton, L. E. Peterson.
Astrophys. Journ., Vol. 187, 341 - 344 (1974).

The UCSD X-ray telescope on board OSO-7 provided 43 days of continuous coverage of the variable X-ray source Cen X-3 at energies above 7 keV during 1971 December and 1972 January. Cen X-3 is usually described as a close binary system, with one of the components being a rapidly rotating neutron star accreting gas from the primary. Information on the spectra and time variations provides constraints on this model.

142.007 **Spectroscopic observations of HZ Herculis.**
D. Crampton.
Astrophys. Journ., Vol. 187, 345 - 348 (1974).

Spectroscopic observations of HZ Her obtained over a period of 1 year are presented. The space motion and distance of HZ Her are briefly discussed.

142.008 **Do cosmic rays heat HZ Herculis?**
K. Brecher, P. Morrison.
Astrophys. Journ., Vol. 187, 349 - 350 (1974).

It is suggested that the optical variability of HZ Her is due to surface heating produced by cosmic rays arising in the X-ray source Her X-1.

142.009 **A determination of the cooling time and the speed of the surface currents of HZ Herculis.**
R. E. Dahab.
Astrophys. Journ., Vol. 187, 351 - 353 (1974).

A simple model has been used to compute the time for an element of the surface of HZ Herculis to cool once it is removed from the illumination of the X-ray source. This time turned out to be on the order of 10 seconds, i. e., extremely short compared with the orbital period. Additionally, estimates of the velocity and extent of currents due to the uneven heating of the stellar surface have been made. The results in orders of magnitude are 4 km s^{-1} for the velocity and 40 km for the lateral extent. These numbers are too small to cause any likely observable effect.

142.010 **2U 1700−37: Another black hole?**
M. S. Bessell, B. A. Peterson, D. T. Wickramasinghe, N. V. Vidal.
Astrophys. Journ., Vol. 187, 355 - 358 (1974).

Analyses of the reddening and interstellar features indicate that HD 153919, the primary star of the binary system containing 2U 1700−37, is a bright distant Of star. Comparison with the Of stars in Sco OB1 suggests a mass of 65 $M_\odot$ for HD 153919, which implies a mass of about 2.7 $M_\odot$ for 2U 1700−37 from the solution of the radial-velocity curve. As this mass is greater than the probable upper limit for neutron stars, the X-ray source 2U 1700−37 could be another black hole.

142.011 **Cosmic gamma-ray burst detected with an instrument on board the OGO-5 satellite.**
J. L'Heureux.
Astrophys. Journ., (Letters), Vol. 187, L53 - L56 (1974).

Gamma-ray bursts of cosmic origin have recently been detected by instruments on the Vela satellites. The author confirms the detection of the 1971 June 30 event with an instrument on board the OGO-5 satellite.

142.012 **Model for Cygnus X-3.** J. E. Pringle.
Nature, Vol. 247, 21 - 22 (1974).

The author shows that Cyg X-3 can be interpreted by means of the 'standard model' for galactic X-ray sources in terms of mass loss from a large companion star and accretion onto a compact object.

142.013 **Spectroscopic variations in Cyg X-2.**
B. W. Bopp.
Nature, Vol. 247, 139 (1974). − Letter.

142.014 **Is Cyg X-1 a neutron star?**
A. C. Fabian, J. E. Pringle, J. A. J. Whelan.
Nature, Vol. 247, 351 - 352 (1974).

The authors suggest that the source of X-ray emission Cygnus X-1 may be a neutron star in orbit about an early-type relatively unevolved star, both of which are in orbit about the massive 09.7Iab supergiant star HD 226868. They conclude that there is, as yet, no need to invoke the existence of black holes to explain the properties of X-ray sources.

142.015 **Near infrared observations of Cygnus X-3.**
J. N. Bahcall, N. A. Bahcall.
Nature, Vol. 247, 446 - 447 (1974).

The authors present the results of a series of observations made at near infrared wavelengths, $\lambda_{eff} \cong 0.85$ μm on ten nights in July and August 1973.

142.016 **Comments on magnetic accretion and Her X-1.**
R. N. Henriksen, M. Reinhardt.
Astron. Astrophys., Vol. 31, 195 - 201 (1974).

This paper is a sequel to the investigation by Henriksen et al. (1973) on magnetic accretion with special emphasis on Her X-1. The present work was necessitated by new observations of Her X-1, which support both the authors' theory of magnetic accretion and their model of the X-ray binary.

142.017 **Observations of a variable emission core in the Hβ line of HZ Herculis.** R. Barbon, P. Benvenuti, P. L. Bernacca, F. Ciatti, A. Treves.
Astron. Astrophys., Vol. 31, 237 - 238 (1974).

Spectroscopic observations of HZ Her indicate the existence of variable emission in Hβ, which is suggested to be associated with the formation of the disk surrounding the neutron star. The spectral behaviour of Hβ is not correlated with any of the periodic phenomena of the X-ray binary.

142.018 **Reflection and reprocessing of X-ray source radiation by the atmosphere of the normal star in a binary system.** M. M. Basko, R. A. Sunyaev, L. G. Titarchuk.
Astron. Astrophys., Vol. 31, 249 - 263 (1974).

The authors consider the case of a semi-infinite plane atmosphere in a stream of hard X-rays, and solve numerically the X-ray transfer equation. Photoabsorption and Compton scattering are taken into account. The albedo of the reflected X-rays is found; their spectral, polarizational, spatial and temporal characteristics have been obtained. The results obtained are discussed with specific reference to Her X 1, Cyg X3 and 2U 1700−37.

142.019 **A ten-day observation of Hercules X-1 from the OSO-7 satellite.** J. E. McClintock, G. W. Clark, W. H. G. Lewin, H. W. Schnopper, C. R. Canizares, G. F. Sprott.
Astrophys. Journ., Vol. 188, 159 - 162 (1974).

The MIT X-ray detectors aboard the OSO-7 spacecraft viewed Hercules X-1 from 1972 November 14 to November 24. The observations are reported in detail.

142.020 **2U 0900−40: a black hole?**
D. T. Wickramasinghe, N. V. Vidal, M. S. Bessell, B. A. Peterson, M. E. Perry.
Astrophys. Journ., Vol. 188, 167 - 172 (1974).

An analysis of the radial-velocity curve, spectrum, and reddening of HD 77581 indicates that its X-ray companion 2U 0900−40 has a mass $\gtrsim 3 M_\odot$. As this mass is greater than that currently accepted for neutron stars, the companion may be a black hole.

142.021 **A preliminary catalog of transient cosmic gamma-ray sources observed by the Vela satellites.**
I. B. Strong, R. W. Klebesadel, R. A. Olson.
Astrophys. Journ., (Letters), Vol. 188, L1 - L3 (1974).

The times of occurrence are given for all cosmic γ-ray bursts detected thus far by at least two satellites of the Vela satellite system. Preliminary information on source directions, energy fluxes, and other characteristics is also given.

142.022 **Optical spectra and the mass of SMC X-1.**
P. S. Osmer, W. A. Hiltner.
Astrophys. Journ., (Letters), Vol. 188, L5 - L7 (1974).

Photographic spectra of Sanduleak 160 show 4686 emission with a large amplitude velocity variation (~260 km s^{-1}) that is approximately in phase with the X-ray object and evidence for an orbital motion up to 40 km s^{-1} of the primary itself. Analysis of the results gives a mass of near $2 M_\odot$ for the X-ray object. The new data also provide additional constraints for models of the system.

142.023 **Collapsed objects and galactic X-ray sources.**
B. A. Peterson.
Proc. Astron. Soc. Australia, Vol. 2, 178 - 183 (1973). — Invited paper — see 012.001.

142.024 **The search for Cen X-3.** M. E. Perry, B. A. Peterson, D. T. Wickramasinghe, N. V. Vidal.
Proc. Astron. Soc. Australia, Vol. 2, 188 - 190 (1973). — Presented at the annual general meeting of the Astron. Soc. Australia — see 012.001.

142.025 **Black holes and X-ray binaries.** N. V. Vidal, D. T. Wickramasinghe, M. S. Bessell, B. A. Peterson.
Proc. Astron. Soc. Australia, Vol. 2, 190 - 191 (1973). — Presented at the annual general meeting of the Astron. Soc. Australia — see 012.001.

142.026 **Strange X-ray sources.** P. R. Amnuehl'.
Zemlya i Vselennaya, 1974, No. 2, p. 29 - 35. In Russian.

142.027 **On the value of the magnetic field at the surface of a neutron star — X-ray pulsar.**
G. S. Bisnovatyj-Kogan.
Astron. Zhurn. Akad. Nauk SSSR, Vol. 51, 443 - 444 (1974). In Russian. English translation in Soviet Astron. AJ, Vol. 18, No. 2.

Evidence is given which indicates that the magnetic field magnitude on the surface of an X-ray pulsar is $5 \times 10^8 - 10^{10}$ gauss.

142.028 **The synthesis of close-binary light curves. VI. X-ray and collapsar binaries.** J. B. Hutchings.
Astrophys. Journ., Vol. 188, 341 - 348 (1974).

General relations are presented for the interpretation of single distorted star light curves. These are then applied to five X-ray binaries (Cyg X-1, Her X-1, Vela X-1, SMC X-1, and 2U 1700−37). Together with considerations of X-ray and spectroscopic data, models and masses are derived for the systems. Similar considerations are applied to two non-X-ray binaries (HD 72754 and HD 187399) and suggest that massive collapsed stars are present in them.

142.029 **Upper limit on 2.5-second pulsations from Hercules X-1.** Y. Avni, J. N. Bahcall, P. C. Joss, D. Q. Lamb, E. Schreier, H. Tananbaum.
Astrophys. Journ., (Letters), Vol. 188, L35 - L36 (1974).

No pulsed power is detected with a period of 2.5 seconds in the 1972 January Uhuru observations of Her X-1. An upper limit of 7 percent is derived for the amount of 2.5-s pulsed modulation relative to the observed amount of 1.24-s pulsed modulation.

142.030 **Soft X-ray variability of binary X-ray stars.**
J. Buff, R. McCray.

Astrophys. Journ., (Letters), Vol. 188, L37 - L40 (1974).

If X-ray stars in binary systems are neutron stars or black holes whose luminosity results from accretion from a stellar wind, they should have soft X-ray cutoffs in the observable range (0.25 keV $\lesssim \epsilon \lesssim 5$ keV) which vary in a predictable way according to the radius and inclination of the binary orbit.

142.031 **OSO-7 observations of a high-latitude X-ray source associated with Abell cluster A2052.** C. J. Heinz, G. W. Clark, W. H. G. Lewin, H. W. Schnopper, G. F. Sprott.
Astrophys. Journ., (Letters), Vol. 188, L41 - L43 (1974).

An X-ray source with a flux of 1.6×10^{-10} ergs s^{-1} cm^{-2} (2–10 keV) has been observed at high galactic latitude by the MIT X-ray detectors on the OSO-7 satellite. Designated GX 8 + 50, its position is $\alpha = 15^h 14^m 2$, $\delta = 6°51'$. A likely identification is the Abell cluster A2052, whose center coincides with the radio source 3C 317.

142.032 **The bright stars associated with galactic X-ray sources.** S. Sofia.
Astrophys. Journ., (Letters), Vol. 188, L45 - L46 (1974).

A comparison of 44 galactic compact X-ray sources in the third Uhuru X-ray catalog having error boxes $\lesssim 0.02$ deg^2 with the stars listed in the Smithsonian Astrophysical Observatory catalog shows eight positional coincidences. Already three of those stars have been positively identified with the corresponding X-ray sources, and a fourth identification has been proposed. This Letter identifies the other bright stars which are good candidates to be X-ray counterparts.

142.033 **A search for soft X-ray sources in the galactic anticenter. Absorption of X-rays from the Crab nebula.**
R. W. Hill, G. A. Burginyon, F. D. Seward, J. P. Stoering, A. Toor.
Astrophys. Journ., Vol. 187, 505 - 510 (1974).

Results are presented from the third part of a three-part survey of the galactic plane for cosmic X-ray sources in the energy range 0.2–18 keV. This part of the survey covers the region from Cassiopeia to Puppis, galactic longitudes 110°–270°. The sources detected by the experiment were Cas A, the Crab nebula, Vel X, and Pup A. Spectral parameters were derived for the Crab using a power-law photon spectrum. Upper limits are given for the intensities of several other supernova remnants scanned during the experiment.

142.034 **A slaved disk model for Hercules X-1.** W. J. Roberts.
Astrophys. Journ., Vol. 187, 575 - 584 (1974).

A markedly inclined rotational axis for HZ Herculis could have a dramatic effect on the observed behavior of the binary system. The author presents a qualitative discussion of possible connections between certain observations and a precession of the axis of HZ Her. An accreting disk is oriented by the axial direction of HZ Her. The changing orientation of this disk can economically account for the periodic appearance and disappearance of the pulsar every 35.5 days.

142.035 **The third Uhuru catalog of X-ray sources.**
R. Giacconi, S. Murray, H. Gursky, E. Kellogg, E. Schreier, T. Matilsky, D. Koch, H. Tananbaum.
Astrophys. Journ., Suppl. Ser., No. 237, Vol. 27, 37 - 64 (1974).

A new edition of the catalog of X-ray sources observed with Uhuru is presented. About 125 days of data have been analyzed for the 3U catalog, yielding a total of 161 X-ray sources. The distribution of sources is similar to that obtained for earlier editions of this catalog. Location error regions for many of the sources previously listed in the 2U catalog have been significantly reduced in size.

142.036 **Sixth radio outburst of Cygnus X-3.**
K. A. Marsh, C. R. Purton, P. A. Feldman.
Nature, Vol. 248, 319 - 320 (1974).

The authors present observations of a sixth radio outburst of Cygnus X-3, which occurred in June and July 1973.

142.037 **An estimate of the distance to X-ray source 2U1700−37 from measurements of the absorption in the region.** M. E. Perry, B. A. Peterson.
Astron. Journ., Vol. 79, 1 - 5, 99 - 100 (1974).

Four color (ubvy) and Hβ photometry was carried out on several B stars in the vicinity of the X-ray source 2U1700−37. Absolute magnitudes were derived from a $[c_1]$, β relationship. A distance of ~ 3.8 kpc and an absolute magnitude of ~ -7.8 were derived for the primary object in the HD 153919 system.

142.038 **Observations of galactic X-ray sources.**
H. Gursky.
Black holes. Les Houches 1972, (see 012.007), p. 291 - 341 (1973).

142.039 **Zit er een zwart gat in de Zwaan?**
T. de Vries, E. P. van den Heuvel.
Zenit, Vol. 1, No. 2, p. 16 - 18 (1974).

142.040 **Upper limit to the flux of soft X rays from λ-Sco.**
K. T. Strong, M. W. Colley, J. L. Culhane.
Nature, Vol. 248, 34 - 35, 90 (1974).

A 3 σ upper limit was obtained to the source luminosity of $4.0 (\pm 1.0) \times 10^{32}$ erg s^{-1} for the energy range 0.5 to 1.5 keV.

142.041 **Flashes of X-ray novae.**
P. R. Amnuehl', O. Kh. Gusejnov, Sh. Yu. Rakhamimov.
Astron. Tsirk., No. 803, p. 1 - 3 (1973). In Russian.

142.042 **Correlation analysis of X-ray emission from Cygnus X-1.** A. C. Brinkman, D. R. Parsignault, E. Schreier, H. Gursky, E. M. Kellogg, H. Tananbaum, R. Giacconi.
Astrophys. Journ., Vol. 188, 603 - 608 (1974).

Normalized autocorrelation and cross-correlation coefficients have been calculated for the X-ray emission of Cygnus X-1 in two energy bins, 2.1–5.1 keV and 5.1–12 keV. The analysis shows a strong correlation between the pulsations in the two energy bins, with the low-energy pulsations lagging behind the high-energy ones, or alternatively a shorter duration of the pulse trains in the high-energy bin.

142.043 **On the nature of the unidentified high-latitude Uhuru sources.** S. S. Holt, E. A. Boldt, P. J. Serlemitsos, S. S. Murray, R. Giacconi, E. M. Kellogg, T. A. Matilsky.
Astrophys. Journ., (Letters), Vol. 188, L97 - L101 (1974).

It is found that the unidentified high-latitude Uhuru sources can have either of two very different explanations. They must either reside at great distances with luminosity $\gtrsim 10^{46}$ ergs s^{-1}, or be contained in the Galaxy with luminosity $\lesssim 10^{34}$ ergs s^{-1}. The two possibilities are indistinguishable with the available data.

142.044 **Observation of soft X-ray emission from a region near Per X-1.**
P. C. Agrawal, K. Long, G. P. Garmire.
Astrophys. Space Sci., Vol. 28, 185 - 191 (1974).

Soft X-ray emission from the X-ray source Per X-1 was observed in the 0.4–2 keV energy interval from a rocket borne X-ray detector. Spectral analysis of the data indicates that in the 0.4–2 keV band the X-ray emission from Per X-1 can be fitted either with a power law of slope $-(4.8 \pm 1.2)$ or a thermal bremsstrahlung spectrum with a kT value of $(0.26^{+0.13}_{-0.08})$ keV. Such a steep spectrum is inconsistent with the spectrum measured above 2 keV. The measured flux in

0.4–2keV band corresponds to X-ray luminosity of 3×10^{45} ergs s^{-1} for Per X-1.

142.045 Fluctuations in the integrated X-ray background.
M. Rowan-Robinson, A. Fabian.
Monthly Notices Roy. Astron. Soc., Vol. 167, 419 - 435 (1974).

The theory of fluctuations in the integrated background radiation from discrete sources is developed, including the effects of telescope beam profile, extended sources, dispersion in luminosity of sources, and cosmological evolution in the source population. The present observational limit on the fluctuations probably rules out the possibility that all the background comes from normal galaxies distributed in superclusters containing ~ 10^4 galaxies, although about 60 per cent could do so.

142.046 Accretion flows in galactic X-ray sources. I. Optically thin spherically symmetric model.
J. Buff, R. McCray.
Astrophys. Journ., Vol. 189, 147 - 155 (1974).

Spherically symmetric accretion flows in the presence of a strong X-ray source are discussed. Gas temperature is determined locally by a balance between X-ray heating and radiative cooling. The rate of mass accretion is regulated by the location of the sonic surface, which depends on X-ray luminosity and spectrum. No steady flow is possible unless the source has a soft X-ray cutoff with energy $\gtrsim 2$ keV. Radiation pressure on the gas due to photoionization may exceed that due to Compton scattering by a large factor.

142.047 Observation of a celestial hard X-ray burst in coincidence with a gamma-ray burst.
G. G. C. Palumbo, G. Pizzichini, G. R. Vespignani.
Astrophys. Journ., (Letters), Vol. 189, L9 - L11 (1974).

One of the celestial γ-ray bursts recently discovered by the Vela satellites has been observed in the 27–189 keV energy range by a directional X-ray detector aboard OSO-6, on 1969 October 7. Possible spectral shapes are discussed.

142.048 Millisecond temporal structure in Cygnus X-1.
R. E. Rothschild, E. A. Boldt, S. S. Holt, P. J. Serlemitsos.
Astrophys. Journ., (Letters), Vol. 189, L13 - L16 (1974).

Evidence is presented for the X-ray variability of Cyg X-1 on time scales down to 1 ms. Several "bursts" of millisecond duration are observed. Such bursts may be associated with turbulence in disk accretion at the innermost orbits for a black hole.

142.049 Multiple star systems and X-ray sources.
J. N. Bahcall, F. J. Dyson, J. I. Katz, B. Paczyński.
Astrophys. Journ., (Letters), Vol. 189, L17 - L18 (1974).

Two classes of triple-star models are proposed which allow X-ray sources of the type of Cygnus X-1 to be interpreted without invoking a black hole. The large mass which is usually assigned to the X-ray source Cyg X-1 belongs in the models to a normal star.

142.050 The nature of Cygnus X-3: a prototype for old population binary X-ray sources.
A. Davidsen, J. P. Ostriker.
Astrophys. Journ., Vol. 189, 331 - 338 (1974).

The authors present a model for Cyg X-3 based on a short-lived phase in the evolution of a U Geminorum-type binary where rapid mass transfer and mass loss occurs. X-rays are generated by accretion on the white dwarf component, while both stars are enveloped in a thick stellar wind emanating from the red dwarf. Electron scattering in the cloud is responsible for the modulation of the X-rays with the orbital period of 4.8 hours. The model also provides a framework in which certain aspects of the associated nonthermal radio source may be understood. Finally, the authors suggest that the model may be applicable to Sco X-1 as well.

142.051 An unusual X-ray source in Scutum.
R. Hill, G. Burginyon, R. Grader, A. Toor, J. Stoering, F. Seward.
Astrophys. Journ., (Letters), Vol. 189, L69 - L70 (1974).

Two previously unreported X-ray sources have been observed. The spectrum of one of them, Sct X-1 ($l^{II} = 24°41$), shows the strongest absorption of any source that the authors have ever observed. Both sources are probably variable.

142.052 Search for high frequency optical variations in Vela XR-1. J. E. Hesser, B. M. Lasker, R. H. Miller.
Nature, Vol. 248, 492 (1974). – Letter.

142.053 Galactic contributions to the isotropy of the cosmic X-ray background. A. C. Fabian.
Astron. Astrophys., Vol. 32, 127 - 131 (1974).

Observations of the cosmic X-ray background in the energy ranges of 1–3 keV and 3–9 keV have been analyzed for fluctuations. A positive fluctuation in the 1–3 keV samples is interpreted in terms of a galactic contribution. Although such a contribution is small in intensity terms, it will produce a relatively large measure of fluctuations. The galactic distribution of these sources is discussed, and it is deduced that they are unlikely to give rise to many of the weak high latitude X-ray sources unless they have a scale height $\langle z \rangle$, greater than 600 pc.

142.054 Binary X-ray sources. R. Giacconi.
IAU Symposium No. 64, (see 012.008), p. 147 - 180 (1974).

142.055 Long-term observations of Cygnus X-2 from OSO-7.
M. P. Ulmer, A. Sammuli, W. A. Baity, W. A. Wheaton, L. E. Peterson.
Astrophys. Journ., Vol. 189, 339 - 341 (1974).

This paper reports the results of observations of Cyg X-2 made with the UCSD OSO-7 X-ray telescope during the intervals 1972 January 9–16, 1972 January 21–26, and 1972 July 8–11. Random factor-of-two variations were found in the data.

142.056 Comment on accretion and compact X-ray source models. P. Boynton, J. Deeter, D. Gerend.
IAU Symposium No. 64, (see 012.008), p. 216 (1974) – Abstract.

142.057 Coordinated campaign for observations of X-ray binaries, I.A.U. Commission 42 in coordination with Commission 44. K. Gyldenkerne, Y. Kondo.
Inform. Bull. Variable Stars, (I.A.U. Commission 27), Konkoly Obs., Budapest, No. 873, 2 pp. (1974).

142.058 Les sources X, ces inconnues. S. A. Ilovaisky.
Sci. Progrès Découverte, No. 3450, p. 12 - 18 (1972).

142.059 Model for the binary X-ray source 2U1700-37.
J. A. de Freitas Pacheco.
Nature, Vol. 249, 637 - 638 (1974).

The author suggests that in the case of the source 2U1700-37, the primary star, an O7f type star (HD 153919), is not filling its Roche lobe and that the matter which is accreted by the neutron star is due to a strong stellar wind from the primary star.

142.060 Millisecond temporal structure in Cyg X-1.
R. E. Rothschild, E. A. Boldt, S. S. Holt, P. J. Serlemitsos.
Goddard Space Flight Center, Greenbelt, Maryland, GSFC

Document X-661-73-393, Preprint, 1 + 13 pp. (1973).

Evidence is presented for the X-ray variability of Cyg X-1 on time scales down to a millisecond. Several "bursts" of millisecond duration are observed. The duty cycle for bursting is estimated to be $\gtrsim 2 \times 10^{-4}$ averaged over the entire 49 second exposure, although the maximum burst activity is associated with a region of enhanced emission lasting about 1/3 second. Such bursts may be associated with turbulence in disk accretion at the innermost orbits for a black hole.

142.061 **A new measurement of the Her X-1 X-ray pulse profile.**
S. S. Holt, E. A. Boldt, R. E. Rothschild, P. J. Serlemitsos.
Goddard Space Flight Center, Greenbelt, Maryland, GSFC Document X-661-74-11, Preprint, 11 + 2 pp. (1974).

The authors have measured a triple-peaked 1.24 sec pulse profile in a 1-minute rocket borne exposure to Her X-1, in contrast to the double-peaked profiles expected from models which maximize the X-ray emission at the magnetic equator of an accreting neutron star. The profile exhibits statistically significant energy dependence, with the emission $\gtrsim 12$ keV having narrower peaks which lag (by approximately 5 % of the pulse period) the corresponding peaks at lower energies. Approximately one-third of the total emission from the source is non-pulsed.

142.062 **New position of Cen X-3 from Copernicus.**
J. H. Parkinson, K. O. Mason, P. W. Sanford.
Nature, Vol. 249, 746 - 747 (1974). – Letter.

142.063 **A preliminary report on the light variation of Sk. 160 = SMC X-1.** A. M. van Genderen.
Monthly Notices Roy. Astron. Soc., Vol. 167, 57P - 59P (1974).

Photometric observations of the X-ray binary Sk. 160 = SMC X-1 are presented. By matching the present observations with those of Butler & Byrne the author finds a period of 3.89267 ± 0.00010 (e.e.) days. With this period the X-ray minimum coincides perfectly with one of the minima.

142.064 **Energy spectrum and evidence for extragalactic origin of diffuse gamma-radiation in the MeV-range.**
V. Schönfelder, G. Lichti.
Max-Planck-Inst. Phys. Astrophys., Inst. Extraterr. Phys., München, MPI-PAE/Extraterr. 95, 16 pp. (1974).

Results on the diffuse gamma-ray flux between 1.5 and 10 MeV from a balloon borne double Compton telescope are presented. The diffuse flux exceeds a straight line extrapolation from the X-ray range by a factor of 4 at 2.5 MeV. From the constancy of the gamma counting rate from regions of the sky with galactic latitudes between $b^{II} = 0°$ and $b^{II} = -30°$ an extragalactic origin of more than 85% of the measured diffuse flux can be deduced.

142.065 **Further simultaneous hard X-ray and optical observations of Sco X-1**
M. Matsuoka, S. Miyamoto, J. Nishimura, M. Oda, Y. Ogawara, Y. Ohtsuka, S. Hayakawa, I. Kasahara, F. Makino, Y. Tanaka, P. C. Agrawal, B. V. Sreekantan, R. Manabe, Y. Mikami, B. Lokanadham, C. Raghavender Rao, S. Sreedhar Rao, R. Swaminathan.
Nature, Vol. 250, 38 - 40 (1974).

Properties of Sco X-1 obtained by the simultaneous X-ray and optical observations are summarised as follows: (1) The X-ray intensity in the energy range 20 to 40 keV shows a positive correlation with optical luminosity in the bright phase of Sco X-1. (2) The apparent temperature of Sco X-1 derived from the spectrum in the energy range 20 to 40 keV does not appreciably change in comparison with considerable spectral variations in the energy range 1 to 10 keV. (3) Flare activity is associated with an increase of the total mass of a hot plasma.

142.066 **Observations of the circular polarization in HDE 226868 = Cyg X-1.**
O. S. Shulov, E. N. Kopatskaya.
Astrofizika, Vol. 10, 120 - 123 (1974). In Russian. – English translation in Astrophysics, Vol. 10, No. 1.

Observations suggest that a small circular polarization increasing with wavelength may exist in HDE 226868, the optical counterpart of Cyg X-1.

142.067 **Optical studies of Uhuru sources. VIII. Observations of 92 possible counterparts of X-ray sources.**
C. A. Jones, T. Chetin, W. Liller.
Astrophys. Journ., (Letters), Vol. 190, L1 - L3 (1974).

A total of 92 stars located at or near Uhuru X-ray source positions were observed photometrically from the Cerro Tololo Inter-American Observatory. The authors summarize their measurements in a $(U - B, B - V)$-diagram and tabulate the data for the more interesting stars.

142.068 **Observation of enhanced soft X-ray emission from the vicinity of the North Polar Spur.**
P. A. J. de Korte, J. A. M. Bleeker, A. J. M. Deerenberg, Y. Tanaka, K. Yamashita.
Astrophys. Journ., (Letters), Vol. 190, L5 - L8 (1974).

Enhanced soft X-ray emission from the vicinity of the North Polar Spur has been found by rocket-borne soft X-ray detectors in the energy range 0.15–1.2 keV. The location of the regions of enhanced emission and the observed photon fluxes are in fair agreement with a recent experiment of Bunner et al.

142.069 **Spectroscopic observations of the optical companion to Centaurus X-3.** J. J. Rickard.
Astrophys. Journ., (Letters), Vol. 189, L113 - L115 (1974).

Image-tube spectra of the optical companion of Cen X-3 have been used to estimate its spectral type to be O9.5–B0.5 Ib. Its distance is calculated to be 10 kpc, which is consistent with the observed $E(B-V)$, the strengths of interstellar features, and the 21-cm H I column density in the direction toward the star. The column density is of the order of 5×10^{21} atoms cm^{-2}. The soft X-ray flux observed from Cen X-3 and the model proposed by Blecker et al. are probably inconsistent with such a large amount of interstellar matter. Two possible solutions to this problem are suggested.

142.070 **Simultaneous hard X-ray and optical observations of Sco X-1 and a model of semi-opaque X-ray sources.** I. Kasahara
Astrophys. Space Sci., Vol. 29, 191 - 210 (1974).

Simultaneous observations of hard X-ray and optical emission from Sco X-1 were carried out at Hyderabad, India, on April 16 and 19, 1972. The author describes instrumentation, the performance of flights, the procedure of data reduction and the experimental results. The last section is devoted to a brief discussion on a hot plasma model of Sco X-1, where the physical parameters that characterize Sco X-1 are derived taking into account the effect of radiative transfer.

142.071 **Variations of the optical emission-line spectrum of the eclipsing binary X-ray star HD 153919 (= 2 U 1700–37).** J. Dachs, H. J. Schober.
Astron. Astrophys., Vol. 33, 49 - 54 (1974).

The strengths of several emission lines in the spectrum of the Of-type star HD 153919 have been studied for periodic variations. No regular variations with phase exceeding the mean error of the measurements are found for the strength of the emission lines of N III, C III and He II λ 4686 Å. From ten measurements of the equivalent width of the Hα emission line, a periodic variation is indicated with a maximum at or shortly after the centre of the occultation of the X-ray source by the

Of star. These results confirm earlier conclusions of Hutchings (1968) that the Hα emission in Of-type stars is produced in an expanding envelope which is much more extended than the layers emitting the N III, C III and He II λ 4686 Å lines. The most likely explanation of the strong variation of the Hα emission appears to be a synchronous rotation of the optical Of-star and its asymmetric envelope with the orbital revolution of the companion X-ray source.

142.072 **Possible periodic variations in the secondary minimum of HD 153919 ≡ 2U 1700−37.**
J. A. de Freitas Pacheco, J. E. Steiner, G. R. Quast.
Astron. Astrophys., Vol. 33, 131 - 133 (1974).

UBV photoelectric photometry of HD 153919 has been performed during southern winter 1973 at ITA Observatory. The comparison of these data with other published data suggests that the secondary minimum of HD 153919 varies with a period of about 32 days.

142.073 **Copernicus observations of variations in the X-ray flux from Cygnus X-1.**
P. W. Sanford, K. O. Mason, F. J. Hawkins, P. Murdin, A. Savage.
Astrophys. Journal, (*Letters*), Vol. 190, L55 - L58 (1974).

The Mullard Space Science Laboratory X-ray instrumentation on board Copernicus has observed Cyg X-1 for a continuous period of 6.5 days from 1973 November 5 to November 11. The authors report observations of a maximum in the X-ray intensity corresponding to the time of closest approach of the secondary component of the optical candidate to the earth.

142.074 **Skylab S019 ultraviolet observations of HD 153919 = 3U1700−37.** Y. Kondo, K. G. Henize, J. D. Wray, S. B. Parsons, G. F. Benedict, P. M. Rybski.
Bull. American Astron. Soc., Vol. 6, 223 (1974). − Abstr. AAS.

142.075 **Observation of an X-ray transient event on Apollo 16.**
J. I. Trombka, R. L. Schmadebeck, I. Adler, E. Eller, A. E. Metzger, P. Gorenstein, D. Gilman.
Bull. American Astron. Soc., Vol. 6, 262 (1974). − Abstr. AAS.

142.076 **Optical spectra of HD 153919 (3U1700−37).**
F. Boley, D. E. Mook.
Bull. American Astron. Soc., Vol. 6, 264 (1974). − Abstr. AAS.

142.077 **Optical observations of the spectrum of the binary X-ray source 2U0900−40.**
D. E. Mook, F. I. Boley.
Bull. American Astron. Soc., Vol. 6, 264 (1974). − Abstr. AAS.

142.078 **Extended observations of > 7 keV X-rays from Cen X-3 by the OSO-7 satellite.**
W. A. Baity, M. P. Ulmer, W. A. Wheaton, L. E. Peterson.
Bull. American Astron. Soc., Vol. 6, 264 (1974). − Abstr. AAS.

142.079 **OGO-5 observations of cosmic X-ray bursts.**
S. R. Kane, W. A. Mahoney, K. A. Anderson.
Bull. American Astron. Soc., Vol. 6, 270 (1974). − Abstr. AAS.

142.080 **Evidence for a galactic origin of transient gamma-ray bursts.** I. B. Strong, R. W. Klebesadel.
Bull. American Astron. Soc., Vol. 6, 270 (1974). − Abstr. AAS.

142.081 **Cosmic gamma-ray bursts from relativistic dust grains.** J. E. Grindlay, G. G. Fazio.
Bull. American Astron. Soc., Vol. 6, 270 (1974). -- Abstr. AAS.

142.082 **X-ray pulsar in the Cygnus Loop?**
R. D. Bleach, R. C. Henry, G. Fritz, S. D. Shulman,
J. F. Meekins, H. Friedman.
Bull. American Astron. Soc., Vol. 6, 271 (1974). − Abstr. AAS.

142.083 **X-ray survey of Scutum − Aquila region.**
R. W. Hill, G. Burginyon, R. Grader, J. P. Stoering, F. D. Seward, A. Toor.
Bull. American Astron. Soc., Vol. 6, 272 (1974). − Abstr. AAS.

142.084 **Deuterium formation hypothesis for the diffuse gamma-ray excess at 1 MeV.** M. Leventhal.
Bull. American Astron. Soc., Vol. 6, 275 (1974). − Abstr. AAS.

142.085 **Observations of the 4.8 hour pulsations of Cyg X-3 above 7 keV.**
M. P. Ulmer, W. A. Baity, W. A. Wheaton, L. E. Peterson.
Bull. American Astron. Soc., Vol. 6, 276 (1974). − Abstr. AAS.

142.086 **Long term variability in the X-ray intensity of 3U 1908+0.** J. E. McClintock, T. H. Markert.
Bull. American Astron. Soc., Vol. 6, 276 (1974). − Abstr. AAS.

142.087 **Observations of the eclipsing binary Cen X-3 by OSO-7.** G. W. Clark, M. Ansour, F. Li.
Bull. American Astron. Soc., Vol. 6, 276 (1974). − Abstr. AAS.

142.088 **Photoelectric spectrum scans of HD 153919 = 2U1700−37.**
B. W. Bopp, G. Grupsmith, P. A. Vanden Bout.
Bull. American Astron. Soc., Vol. 6, 276 - 277 (1974). Abstr. AAS.

142.089 **Temporal and spectral studies of Her X-1, Cyg X-1, and Cyg X-3.**
R. Rothschild, E. Boldt, S. Holt, P. Serlemitsos.
Bull. American Astron. Soc., Vol. 6, 277 (1974). − Abstr. AAS.

142.090 **A search for hard X-ray emission from Hercules X-1.**
W. N. Johnson III, R. D. Bleach, J. D. Kurfess.
Bull. American Astron. Soc., Vol. 6, 277 (1974). − Abstr. AAS.

142.091 **A hard X-ray observation of Cygnus X-1.**
J. D. Kurfess, R. D. Bleach, W. N. Johnson III.
Bull. American Astron. Soc., Vol. 6, 277 (1974). − Abstr. AAS.

142.092 **Optical spectra and the mass of SMC X-1.**
P. S. Osmer, W. A. Hiltner.
Bull. American Astron. Soc., Vol. 6, 277 (1974). − Abstr. AAS.

142.093 **π° − decay gamma rays from neutron stars.**
J. M. Cohen, R. Ramaty, S. A. Stephens.
Bull. American Astron. Soc., Vol. 6, 279 (1974). − Abstr. AAS.

142.094 **Hydrodynamic atmospheres for companions of binary X-ray sources.** L. Anderson, J. Arons.
Bull. American Astron. Soc., Vol. 6, 280 (1974). − Abstr. AAS.

142.095 **Natura delle sorgenti X galattiche.**
G. Cavallo.
Mem. Soc. Astron. Italiana, Nuova Ser., Vol. 44, 553 - 569 (1973/74).

142.096 **Fisica delle sorgenti X extragalattiche.**
A. Cavaliere.
Mem. Soc. Astron. Italiana, Nuova Ser., Vol. 44, 571 - 597 (1973/74).

142.097 **Identificazioni ottiche di sorgenti X extragalattiche.**
R. Fanti.
Mem. Soc. Astron. Italiana, Nuova Ser., Vol. 44, 599 - 601 (1973/74).

142.098 **Problemi di identificazione ottica delle sorgenti X galattiche osservabili in occultazione lunare dal satellite EXOSAT (Helos).** C. Reina.
Mem. Soc. Astron. Italiana, Nuova Ser., Vol. 44, 603 - 607 (1973/74).

142.099 **Prospettive di ricerca sulle sorgenti di raggi X nella regione ottica. Variabilità ultrarapida e spettroscopia.** P. L. Bernacca.
Mem. Soc. Astron. Italiana, Nuova Ser., Vol. 44, 609 - 610 (1973/74).

142.100 **Osservazioni radio di sorgenti X.** G. Grueff.
Mem. Soc. Astron. Italiana, Nuova Ser., Vol. 44, 611 - 613 (1973/74).

142.101 **Millisecond X-ray pulses from Cyg X-1.**
M. Oda, K. Takagishi, M. Matsuoka, S. Miyamoto, Y. Ogawara.
Pùbl. Astron. Soc. Japan, Vol. 26, 303 - 306 (1974).

Rocket data of Cyg X-1 obtained by Rappaport et al. (1971) on May 1, 1971 have been analyzed. The result indicates that on several occasions during the observation millisecond pulses appeared in clusters with considerable statistical significance.

142.102 **IAU coordinated observing program for X-ray binaries.** Y. Kondo.
IAU Circ., No. 2617 (1974).

142.103 **Centaurus X-3.**
D. T. Wickramasinghe, M. S. Bessell, N. V. Vidal.
IAU Circ., No. 2631 (1974).

142.104 **Cygnus X-3.** M. Ryle.
IAU Circ., No. 2670 (1974).

142.105 **Optical identifications of X-ray sources.** H. Mauder.
IAU Circ., No. 2673 (1974).

142.106 **Optical identifications of X-ray sources.**
D. T. Wickramasinghe, N. V. Vidal, A. W. Rodgers, M. S. Bessell, V. M. Lyutyj, Yu. N. Efremov.
IAU Circ., No. 2675 (1974).

142.107 **Scorpius X-1.**
A. M. Cherepashchuk, V. M. Lyutyj, N. I. Shakura.
IAU Circ., No. 2679 (1974).

142.108 **Circinus X-1.** I. Glass.
IAU Circ., No. 2680 (1974).

142.109 **Scorpius X-1.** P. W. Sanford.
IAU Circ., No. 2682 (1974).

142.110 **Plasma theory of Scorpius X-1 – magnetohydrodynamic and optical analyses.**
S. Ichimaru, T. Nakano.
Progr. Theor. Phys., (*Japan*), Vol. 50, 1867 - 1878 (1973).

A theoretical model of Sco X-1 phenomena is developed, based on the idea of a magnetic white dwarf undergoing accretion. An optical analysis of the accreting plasma is carried out with emphasis on its luminosity and optical depth. The origin of the superthermal hard X-rays is accounted for in terms of the stochastic acceleration of electrons in the turbulent plasma. A possible mechanism of radio flaring is also suggested.

142.111 **A model of pulsating X-ray sources. Accretion by pulsating white dwarfs.** H. Yokoo, R. Hoshi.
Progr. Theor. Phys. (*Japan*), Vol. 51, 418 - 427 (1974).

Accretion of matter by a pulsating white dwarf is studied as a model of pulsating X-ray sources.

142.112 **On the distance to Cygnus X-1 (HDE 226868).**
J. Bregman, D. Butler, E. Kemper, A. Koski, R. P. Kraft, R. P. S. Stone.
IAU Symposium No. 64, (see 012.008), p. 181 = Contr. Lick Obs., No. 385 (1974). – Abstract.

142.113 **X and gamma rays in the sky background.**
T. Montmerle, C. Ryter.
Bull. Inform. Sci. and Techn. (*France*), No. 187, p. 13 - 27 (1973). In French.

Available data on the sky background radiation have been compiled taking into account stringent selection criteria. Those data to which the various corrections made by the authors are not explicitly mentioned are rejected. The remaining data are believed to form a consistent sample. They are displayed on a graph. It is concluded that the sky background is easily accounted for by a very general mechanism like coupled synchrotron and Compton effect of distant objects.

142.114 **The extended X-ray source in Virgo and its relation to M87.** R. C. Catura, L. W. Acton, H. M. Johnson, W. T. Zaumen.
Astrophys. Journ., Vol. 190, 521 - 523 (1974).

Spectral data in the 0.2–1.6 keV range are presented for the extended X-ray source in the Virgo cluster. When combined with *Uhuru* data, the simplest function which fits the composite spectrum from 0.2 to 10 keV is one describing bremsstrahlung from an isothermal plasma at 36×10^6 °K. Production of the X-rays by inverse Compton scattering is also considered and models relating the X-ray source to M87 are discussed.

142.115 **Optical observations of HD 77581 and a model for the system HD 77581 – 2U 0900–40.**
L. D. Petro, W. A. Hiltner.
Astrophys. Journ., Vol. 190, 661 - 666 (1974).

From an analysis of photometric observations a more precise value of the period of HD 77581 – 2U 0900–40 has been determined, namely, 8.972 days. From an analysis of the photometric, spectroscopic, and X-ray eclipse observations the mass ratio is approximately 0.08 and $\sin i = 1.0$. Evidence for an elliptical orbit and further evidence for mass loss are presented.

142.116 **A new measurement of the Hercules X-1 X-ray pulse profile.** S. S. Holt, E. A. Boldt, R. E. Rothschild, J. L. R. Saba, P. J. Serlemitsos.
Astrophys. Journal, (*Letters*), Vol. 190, L109 - L111 (1974).

The authors have measured a triple-peaked 1.24-s pulse profile in a 1-minute rocket-borne exposure to Her X-1, in contrast to the double-peaked profiles expected from models which maximize the X-ray emission at the magnetic equator of an accreting neutron star. The profile exhibits statistically significant energy dependence, with the emission $\gtrsim 12$ keV having narrower peaks which appear to lag the corresponding peaks at lower energies. Approximately one-third of the total emission from the source is nonpulsed.

142.117 **Limits on rapid X-ray pulsing in X-ray binaries.**
G. Spada, H. Bradt, R. Doxsey, A. Levine, S. Rappaport.
Astrophys. Journal, (*Letters*), Vol. 190, L113 - L115 (1974).

The X-ray variability of Cir X-1 (3U 1516–56) and Vela X-1 (3U 0900–40) has been investigated down to 1 ms time resolution in a recent sounding rocket experiment. Sporadic flaring activity on time scales from 1 to several seconds was observed, corresponding to intensity changes of about 20 percent in Cir X-1 and up to 100 percent in Vela X-1. No periodic, aperiodic, or quasi-periodic pulsations were found with time constants less than 1 s. This and earlier work place upper limits for rapid periodic pulsing on five of the eight known X-ray binary sources (Cir X-1, Cyg X-1, Vela X-1, Her X-1 and 3U 1700−37).

142.118 The statistical analysis of stellar X-ray sources.
S. C. Saunders.
Technometrics, (*USA*), Vol. 15, 341 - 351 (1973).

The Poynting-Robertson effect and Eddington limit for electrons scattering with hard photons.
See Abstr. 022.011.

Programmi sperimentali di astronomia in raggi X.
See Abstr. 034.073.

Fluctuations in the X-ray background.
See Abstr. 061.003.

Molecular hydrogen in X-ray astronomy.
See Abstr. 061.048.

Multiplex methods and advantages in X-ray astronomy. See Abstr. 061.049.

Interpretation of double structure in the celestial γ-ray bursts. See Abstr. 061.050.

Panoramica dello stato attuale dell'astronomia X.
See Abstr. 061.069.

X-ray heating in HZ Her. See Abstr. 064.029.

X-rays from hot, dense coronas.
See Abstr. 064.053.

γ-ray lines from accreting neutron stars.
See Abstr. 065.075.

Photometry in the Cygnus X-1 field.
See Abstr. 113.008.

On the optical stability of X Persei.
See Abstr. 114.044.

The peculiar star He 2-177: a slow nova and a possible X-ray source. See Abstr. 114.047.

Observations of Hα in HDE 226868.
See Abstr. 114.081.

Photoelectric spectrum scans and coudé spectroscopy of HD 153919 (= 2 U 1700−37).
See Abstr. 114.090.

The spectral energy distributions of BD +34°3815 (Cyg X-1) and X Per. See Abstr. 114.120.

On the mass determination of collapsed objects in close binaries: applications to black holes and neutron stars.
See Abstr. 117.004.

A simple model of fluctuations in accretion discs.
See Abstr. 117.009.

On the interpretation of the He II λ4686 emission line in HDE 226868 (Cygnus X-1). See Abstr. 117.029.

Nonperiodic optical flickering in HZ Herculis.
See Abstr. 117.030.

Determination of upper mass limits for massive X-ray binaries. See Abstr. 117.034.

On the light curve of HD 77581.
See Abstr. 119.004.

Spectroscopic observations of HD 153919 (2U 1700−37). See Abstr. 121.001.

Possible detection of very soft X-rays from SS Cygni. See Abstr. 121.004.

HZ Her, une variable énigmatique.
See Abstr. 121.056.

Photoelectric observations of HD 77581.
See Abstr. 121.088.

Three-colour photoelectric observations of the binary systems HZ Her and V1357 Cyg connected with X-ray sources Her X-1 and Cyg X-1. See Abstr. 121.101.

X-ray emission from cataclysmic variable stars.
See Abstr. 122.063.

The effects of the 35-day X-ray cycle on the light curve of HZ Herculis. See Abstr. 122.102.

Search of OSO-3 data for X-ray emission from stellar flares. See Abstr. 122.114.

Supernova explosions in close binary systems, with an application to Centaurus X-3. See Abstr. 125.007.

The Vela supernova remnant in ultraviolet light.
See Abstr. 125.010.

X-ray emission from an inward-propagating shock in young supernova remnants. See Abstr. 125.013.

Are binary systems involved in supernova explosions always disrupted? See Abstr. 125.014.

Observation of X-ray emission from supernova remnants. See Abstr. 125.048.

X-ray observations of supernova remnants by Copernicus. See Abstr. 125.049.

Preliminary results on the evolution of supernova remnants and their X-ray spectrum. See Abstr. 125.050.

The X-ray spectrum of the Vela X supernova remnant. See Abstr. 125.068.

An investigation of accretion of matter onto white dwarfs as a possible X-ray mechanism.
See Abstr. 126.018.

21-cm observations in the directions of 43 hot stars and 34 X-ray sources. See Abstr. 131.532.

Detection of the [Fe XIV] coronal line at 5303 Å in the Cygnus Loop. See Abstr. 132.011.

Detection of the Fe XIV coronal line at 5303 Å in the Cygnus Loop. See Abstr. 132.032.

Measurement of the spatial structure of the X-ray emission of the Cygnus Loop. See Abstr. 132.033.

Gamma ray observations from the Crab nebula and NP-0532. See Abstr. 134.006.

Southern radio sources possibly identified with X-ray sources. See Abstr. 141.006.

Early evolution of radio outbursts and a possible transient emission of high-energy photons. See Abstr. 141.039.

The absence of radio emission from HZ Herculis. See Abstr. 141.098.

A simple model for a recent radio outburst of Cygnus X-3. See Abstr. 141.108.

TV investigations of ultra-short period optical radiation sources. See Abstr. 141.306.

X-ray spectrum of NP0532. See Abstr. 141.326.

Detection of soft X-ray emission from PSR 0833−45. See Abstr. 141.328.

An upper limit on soft X-ray pulsations from the pulsar PSR 0833−45. See Abstr. 141.329.

On the structure of the pulse of NP 0532 at γ-ray energies. See Abstr. 141.332.

Results of a search for 10^{12} eV gamma rays from pulsars. See Abstr. 141.340.

Relevance of electron pair production in the interpretation of IR and X extragalactic sources. See Abstr. 141.610.

Possibilità di misure infrarosse di sorgenti X. See Abstr. 141.615.

Galactic positrons and γ-rays, and the cosmic-ray mean path-length in the Galaxy. See Abstr. 143.031.

The spectrum of diffuse cosmic X-rays observed by OSO-3 between 7 and 100 keV. See Abstr. 143.034.

The open cluster NGC 6281. See Abstr. 153.033.

SAS-2 observations of the high-energy gamma radiation from the Vela region. See Abstr. 155.058.

Emissione X da nuclei galattici compatti. See Abstr. 158.096.

Observation of X-ray emission from M31. See Abstr. 158.102.

Inverse Compton radiation and the magnetic field in clusters of galaxies. See Abstr. 160.005.

The correlation of radio emission and optical type with X-ray emission from clusters of galaxies. See Abstr. 160.016.

Copernicus X-ray observations of NGC 1275 and the core of the Perseus cluster. See Abstr. 160.017.

Inverse Compton radiation and the magnetic field in clusters of galaxies. See Abstr. 160.027.

143 Cosmic Radiation

143.001 Interpretation of the chemical composition of very high energy cosmic rays.
C. J. Cesarsky, J. Audouze.
Astron. Astrophys., Vol. 30, 119 - 126 (1974).

The presently available data on the chemical composition of the high energy cosmic rays have been analyzed in the frame of the most widely used theory of cosmic ray propagation. It appears from this analysis that 1) the mean path length of the cosmic rays in the Galaxy decreases when the energy increases, 2) the sources of high energy cosmic rays are richer in Fe relatively to the M(C + O) elements than the sources of lower energy cosmic rays. Other hypotheses on the sources of cosmic rays are also discussed. In Appendix A the dependence of the mean path length on the helium content of the interstellar gas is studied, while the effects of a possible variation of the nuclear cross sections with the energy are considered in Appendix B.

143.002 An energy-dependent confinement model for galactic cosmic rays. J. A. Holmes.
Monthly Notices Roy. Astron. Soc., Vol. 166, 155 - 163 (1974).

Recent observations of cosmic ray composition suggest that the average path length decreases with energy. A confinement model is proposed to explain this, based on the scattering of cosmic rays by hydromagnetic waves in regions outside the plane of the Galaxy where the density of neutral matter is low enough to allow the waves to form. The effect of this confinement model on the spectrum of cosmic ray electrons is investigated.

143.003 Spatial dependence of the pitch-angle and associated spatial diffusion coefficients for cosmic rays in interplanetary space. H. J. Völk, G. Morfill, W. Alpers, M. A. Lee.
Astrophys. Space Sci., Vol. 26, 403 - 430 (1974).

The spatial dependence of the pitch-angle and associated spatial diffusion coefficients for cosmic ray particles in interplanetary space is calculated in the WKB approximation. The model considers only Alfvén waves of solar origin to be responsible for scattering of moderate energy particles. After developing the general theory results are presented for the asymptotic case corresponding to radial distances r greater than about 1 to 2 AU.

143.004 Investigation of cosmic radiation at small heights.
B. M. Makhmudov.
UzSSR Fanlar Akad. dokl., Dokl. AN UzSSR, 1973, No. 9, p. 16 - 18. In Russian. — Abstr. in Referativ. Zhurn. 62. Issled. kosmich. prostranstva, 3.62.246 (1974).

143.005 The leakage of cosmic rays from the Galaxy.
I. McIvor, J. Skilling.
Monthly Notices Roy. Astron. Soc., Vol. 167, 49 - 54 (1974).

A model is constructed for the confinement of cosmic ray particles within the Galaxy by hydromagnetic waves, themselves generated by a slow leakage of such particles. An 'equation of state' is derived for the cosmic rays, as they escape, and, among other consequences of the model, the leakage rate is found to increase with particle energy.

143.006 Time-dependent distribution of the neutron density recorded on a neutron supermonitor.
N. P. Chirkov, V. L. Yanchukovskij.
Geomagn. Aeronom., Vol. 14, 152 - 154 (1974). In Russian. Brief information.

143.007 Possible evidence for structured acceleration of cosmic rays on a galactic scale from recent γ-ray observations.
F. W. Stecker, J. L. Puget, A. W. Strong, J. H. Bredekamp.
Astrophys. Journ., (Letters), Vol. 188, L59 - L61 (1974).

Recent data from SAS-2 on the galactic γ-ray line flux as a function of longitude reveal a broad maximum in the region $|l| \lesssim 30°$. These data imply that the low energy (1–10 GeV) galactic cosmic-ray flux varies with the radial distance, ϖ, from the galactic center and is about an order of magnitude higher than the local value in a toroidal region for ϖ between 4 and 5 kpc. The authors show that this enhancement can be plausibly accounted for by Fermi acceleration and compression caused by a hydrodynamic shock driven by the expanding gas in the "3-kpc" arm and invoked in some versions of galactic structure theory.

143.008 The anomalous abundance of cosmic-ray nitrogen and oxygen nuclei at low energies.
F. B. McDonald, B. J. Teegarden, J. H. Trainor, W. R. Webber.
Astrophys. Journ., (Letters), Vol. 187, L105 - L108 (1974).

Recent measurements using the Goddard-University of New Hampshire cosmic-ray telescope on the Pioneer 10 spacecraft have revealed an anomalous spectrum of nitrogen and oxygen nuclei relative to other nuclei such as He and C, in the energy range 3–30 MeV per nucleon. The intensity of nitrogen and oxygen nuclei is enhanced by a factor of up to 20 relative to their abundance in galactic or solar cosmic rays. It is argued that this is most likely a new extrasolar component of cosmic rays.

143.009 Super high-energy cosmic radiation.
K. M. V. Apparao.
Sky Telescope, Vol. 47, 306 - 307 (1974).

143.010 The change in the long-term modulation function of cosmic rays in 1969.
N. Iucci, M. Parisi, G. Villoresi.
Journ. Geophys. Res., Vol. 79, 659 - 660 (1974).

The abrupt change in the long-term modulation function of cosmic rays that occurred in 1969 is interpreted as being due to a simultaneous sharp decrease of the solar activity at middle heliolatitudes (~30°).

143.011 Cosmic-ray energy spectrum from 10^{15} to 10^{18} eV.
T. K. Gaisser.
Nature, Vol. 248, 122 - 124 (1974). — Letter.

143.012 On a multidimensional statistical analysis of the accuracy of continuous recording data on cosmic radiation. L. I. Dorman, I. A. Pimenov, A. B. Rodionov, B. A. Shakov.
Ionosphere and solar-terrestrial relations, (see 003.003), p. 137 - 146 (1972). In Russian.

143.013 Upper limit on the low energy interstellar cosmic-ray flux. P. Mészáros.
Nature, Vol. 248, 35 - 36 (1974).

The author points out that there exists an indirect experimental upper limit on the interstellar low energy flux, based on determinations of the state of ionization of the trace elements in the interstellar gas by the OAO-3 satellite Copernicus.

143.014 Empirical models of cosmic ray propagation in the Galaxy. V. S. Ptuskin.
Astrophys. Space Sci., Vol. 28, 3 - 16, 17 - 30 (1974). In

Russian and English.

Homogeneous and flat diffusion models of the propagation of cosmic rays through the Galaxy are considered. Both models are equivalent from the point of view of the description of the elemental composition of cosmic rays. This conclusion is justifiable both for galactic and metagalactic theories of the occurrence of cosmic rays. A connection is established between the effective path length of the matter which nuclei encounter in the Galaxy and the parameters of the diffusion model with an inhomogeneous distribution of an interstellar gas. Initial results allow the interpretation of a relationship of cosmic ray element composition to their energies. Restrictions which must be placed on such a relationship are shown from the different data on cosmic rays at the earth.

143.015 **The chemical composition of cosmic ray nuclei above 1.3 GeV n^{-1} and 23 GeV n^{-1}.**
G. D. Badhwar, R. W. Osborn.
Astrophys. Space Sci., Vol. 28, 101 - 109 (1974).

Measurements made with a balloon-borne counter telescope having an exposure factor of 774 m² sr s are reported. The data analysis indicates that the integral flux ratios above 1.3 GeV n^{-1} and 23 GeV n^{-1} are consistent with energy independence.

143.016 **Cosmic rays.** B. Peters.
Cosmical geophysics, (see 003.004), p. 327 - 338 (1973).

143.017 **Kinetics of cosmic particles in front of a moving "magnetic mirror". II.**
L. I. Dorman, V. Kh. Shogenov.
Geomagn. Aeronom., Vol. 14, 355 - 357 (1974). In Russian. Brief information.

143.018 **The spectrum of cosmic-ray Forbush decreases in the region of high energies.**
A. A. Bishara, L. I. Dorman.
Geomagn. Aeronom., Vol. 14, 357 - 360 (1974). In Russian. Brief information.

143.019 **Possibilité de mesurer l'intensité du rayonnement cosmique passé par le rapport (samarium-146/samarium-147).** J.-C. Le Roulley, Y. Yokoyama, G. Lambert.
Comptes Rendus Acad. Sci. Paris, Sér. B, Vol. 278, 869 - 872 (1974).

La mesure directe par spectrométrie α de l'abondance relative des isotopes 146 et 147 du samarium à la surface de la Lune ou des météorites doit permettre d'estimer l'intensité moyenne du rayonnement cosmique au cours des derniers 10^8 ans.

143.020 **Origin of cosmic rays, atomic nuclei and pulsars in explosions of massive stars.** W. D. Arnett.
IAU Symposium No. 64, (see 012.008), p. 182 (1974). – Abstract.

143.021 **Cosmic ray scintillations. 2. General theory of interplanetary scintillations.** A. J. Owens.
Journ. Geophys. Res., Vol. 79, 895 - 906 (1974).

The motion of charged particles in a stochastic magnetic field with nonzero mean is considered via a generalized quasilinear expansion of Liouville's equation. The general result is an equation relating cosmic ray scintillations to magnetic fluctuations and to cosmic ray gradients. The resonant interaction between particles and the random magnetic field is considered in detail, and the effect of nonlinear terms in the equations is considered. The nonlinear terms are important in damping out initial conditions and in determining conditions near cyclotron resonances. The application of the theory to the propagation of cosmic rays during quiet times in interplanetary space is considered. It is concluded that cosmic ray scintillations in interplanetary space may provide useful information about interplanetary particles and fields and also about nonlinear plasma interactions.

143.022 **Cosmic ray scintillations. 3. The low-frequency limit and observations of interplanetary scintillations.**
A. J. Owens, J. R. Jokipii.
Journ. Geophys. Res., Vol. 79, 907 - 912 (1974).

Statistically significant broad-band fluctuations, or 'scintillations,' in the high-energy (~1 GeV) cosmic ray intensity observed by neutron monitors are interpreted. The scintillations are caused by fluctuations in the interplanetary magnetic field. The theory of the scintillations is presented for the low-frequency limit, $f \lesssim 10^{-4}$ Hz, including the effects of the earth's rotation on the fluxes observed by the neutron monitors. The observations and the theory are in good agreement.

143.023 **The 1972 cosmic ray electron spectrum above 0.5 GeV.** J. J. Burger, B. N. Swanenburg.
Journ. Geophys. Res., Vol. 79, 1533 - 1534 (1974).

The cosmic ray electron spectrum above 0.5 GeV has been measured during a special operation of the Ogo 5 satellite from June 30 to July 13, 1972. The results are presented and compared with earlier data provided by the same instrument over the period March 1968 to August 1971.

143.024 **An interpretation of the observed oxygen and nitrogen enhancements in low energy cosmic rays.**
L. A. Fisk, B. Kozlovsky, R. Ramaty.
Goddard Space Flight Center, Greenbelt, Maryland, GSFC Document X-660-73-383, Preprint, 1 + 8 pp. (1973).

It is proposed that the enhancements of cosmic-ray oxygen and nitrogen observed at ~10 MeV/nucleon could result from neutral interstellar particles which are swept into the solar cavity by the motion of the sun through the interstellar medium, and are subsequently ionized and accelerated.

143.025 **The anomalous abundance of cosmic ray nitrogen and oxygen nuclei at low energies.**
F. B. McDonald, B. J. Teegarden, J. H. Trainor, W. R. Webber.
Goddard Space Flight Center, Greenbelt, Maryland, GSFC Document X-660-73-392, Preprint, 2 + 12 pp. (1973).

Recent measurements using the Goddard-University of New Hampshire cosmic ray telescope on the Pioneer 10 spacecraft have revealed an anomalous spectrum of nitrogen and oxygen nuclei relative to other nuclei such as He and C, in the energy range 3–30 MeV/nuc. The intensity of nitrogen and oxygen nuclei is enhanced by a factor of up to 20 relative to their abundance in galactic or solar cosmic rays. It is argued that this is most likely a new extra-solar component of cosmic rays.

143.026 **On cosmic ray diffusion and anisotropy.**
J. Skilling, I. McIvor, J. A. Holmes.
Monthly Notices Roy. Astron. Soc., Vol. 167, 87P - 91P (1974).

The authors investigate the effects of a turbulent magnetic field on the 'leaky box' model of galactic cosmic ray confinement, finding that within the galactic disk cosmic rays diffuse in three dimensions with a mean free path of about 30 pc. If cosmic rays are produced by supernovae, then we observe at earth rays from about 2000 separate sources. The local value of their anisotropy is determined by conditions local to those magnetic field lines which pass near the sun.

143.027 **The cosmic ray albedo in the circumterrestrial space.**
A. N. Charakhch'yan, G. A. Bazilevskaya, Yu. I. Stozhkov, T. N. Charakhch'yan.
Geomagn. Aeronom., Vol. 14, 411 - 416 (1974). In Russian.

143.028 **Charge composition of cosmic rays between 4 and 100 GV.** R. L. Golden, J. H. Adams, G. D. Badhwar, C. L. Deney, P. J. Lindstrom, H. H. Heckman.
Nature, Vol. 249, 814 - 816 (1974).

The authors report the measurement of the ratio of L nuclei to M nuclei and of VH nuclei to M nuclei using a magnetic spectrometer. Their results are based on two balloon flights, the first from Palestine, Texas, on August 23, 1969, and the second on November 13, 1970, from Parana, Argentina.

143.029 **Isotropy of the arrival directions of ultra high energy cosmic rays.** J. Linsley, A. A. Watson.
Nature, Vol. 249, 816 - 817 (1974). − Letter.

143.030 **An interpretation of the observed oxygen and nitrogen enhancements in low-energy cosmic rays.**
L. A. Fisk, B. Kozlovsky, R. Ramaty.
Astrophys. Journ., (Letters), Vol. 190, L35 - L37 (1974).

It is proposed that the enhancements of cosmic-ray oxygen and nitrogen observed at $\sim$ 10 MeV per nucleon could result from neutral interstellar particles which are swept into the solar cavity by the motion of the sun through the interstellar medium, and which are subsequently ionized and accelerated.

143.031 **Galactic positrons and γ-rays, and the cosmic-ray mean path-length in the Galaxy.**
C. Dilworth, L. Maraschi, G. C. Perola.
Astron. Astrophys., Vol. 33, 43 - 48 (1974).

The interstellar equilibrium spectrum of cosmic-ray positrons arising from meson decay is calculated with two propagation models [(a) slow leakage from the galactic disk, and (b) isotropic diffusion] and using the information on cosmic ray distribution in the Galaxy obtained from an analysis of the γ-ray data. Comparison is made with the experimental information on the interstellar positron flux. The results are discussed in relation with those obtained from the data on the fragmentation of the nucleonic component.

143.032 **Burst of cosmic γ-emission from observations on Cosmos 461.**
E. P. Mazets, S. V. Goleneckij, V. N. Il'inskij.
Pis'ma v ZhurnEhTF, Vol. 19, 126 - 128 (1974). In Russian.
Abstr. in Referativ. Zhurn. 51. Astron., 6.51.529 (1974).

143.033 **High-energy cosmic ray intensity increase of non-solar origin and the unusual Forbush decrease of August 1972.** S. P. Agrawal, A. G. Ananth, M. M. Bemalkhedkar, L. V. Kargathra, U. R. Rao, H. Razdan.
Journ. Geophys. Res., Vol. 79, 2269 - 2280 (1974).

A series of spectacular cosmic ray events, which included two relativistic solar particle enhancements and three major Forbush decreases, was registered by ground-based cosmic ray monitors beginning on August 4, 1972. Among these, the Forbush decrease that occurred on August 4–5 exhibited extremely interesting and complex behavior. This paper describes the detailed observational features and presents a unified model to explain these in terms of a transient modulating region associated with the passage of a shock front.

143.034 **The spectrum of diffuse cosmic X-rays observed by OSO-3 between 7 and 100 keV.**
D. A. Schwartz, L. E. Peterson.
Astrophys. Journ., Vol. 190, 297 - 303 (1974).

The authors consider the effects of long-lived induced radioactivity, energy dependence of the effective aperture, and fluorescence radiation from the shield on their previous estimate of the diffuse X-ray counting rates observed from OSO-3. They confirm their previous conclusion that the spectral data cannot be represented by a single power-law function.

143.035 **The Fokker-Planck coefficient for pitch-angle scattering of cosmic rays.**
L. A. Fisk, M. L. Goldstein, A. J. Klimas, G. Sandri.
Astrophys. Journ., Vol. 190, 417 - 428 (1974).

For the case of homogeneous, isotropic magnetic field fluctuations, it is shown that most theories which are based on the quasi-linear and adiabatic approximations yield the same integral for the Fokker-Planck coefficient for the pitch-angle scattering of cosmic rays. It is also shown, however, that this integral in most cases has been evaluated incorrectly in the past. The implications of these corrections on our ability to relate cosmic-ray diffusion coefficients to observed properties of the interplanetary magnetic field are discussed.

143.036 **Apparatus for investigation of cosmic radiation aboard the scientific station Proton 4.**
N. L. Grigorov, I. D. Rapoport, I. A. Savenko, L. F. Kalinkin, G. P. Kakhidze.
Izuchenie kosmich. luchej na iskusstven. sputnikakh Zemli. Moskva, Nauka, 1973, p. 49 - 94. In Russian. − Abstr. in Referativ. Zhurn. 62. Issled. kosmich. prostranstva, 6.62.61 (1974).

143.037 **Investigations of cosmic radiation particles aboard the artificial earth satellite Intercosmos 6.**
L. A. Vedeshin, R. A. Nymmik, I. D. Rapoport, A. F. Titenkov.
Vestn. AN SSSR, 1973, No. 11, p. 59 - 66. In Russian.
Abstr. in Referativ. Zhurn. 62. Issled. kosmich. prostranstva, 6.62.114 (1974).

143.038 **Investigation of cosmic radiation on artificial earth satellites.**
In-t kosmich. issled. AN SSSR. Nauka, Moskva. 172 pp. Price 79 Kop. (1973). In Russian. − Review in Referativ. Zhurn. 62. Issled. kosmich. prostranstva, 6.62.158 (1974).

143.039 **Perspectives of investigation of particles of high-energy cosmic radiation aboard heavy artificial earth satellites.** N. L. Grigorov.
Izuchenie kosmich. luchej na iskusstven. sputnikakh Zemli. Moskva, Nauka, 1973, p. 3 - 9. In Russian. − Abstr. in Referativ. Zhurn. 62. Issled. kosmich. prostranstva, 6.62.159 (1974).

143.040 **Investigation of the energy distribution of primary particles of cosmic radiation in the energy region of $10^{10} - 10^{14}$ eV aboard the space stations Proton 1, 2, 3.**
N. L. Grigorov, V. E. Nesterov, I. D. Rapoport, I. A. Savenko, G. A. Skuridin.
Izuchenie kosmich. luchej na iskusstven. sputnikakh Zemli. Moskva, Nauka, 1973, p. 34 - 48. In Russian. − Abstr. in Referativ. Zhurn. 62. Issled. kosmich. prostranstva, 6.62.160 (1974).

143.041 **Investigation of the chemical composition and of energy spectra of galactic cosmic radiation aboard artificial earth satellites.** N. L. Grigorov, N. N. Volodichev, I. A. Savenko, A. A. Suslov.
Izuchenie kosmich. luchej na iskusstven. sputnikakh Zemli. Moskva, Nauka, 1973, p. 126 - 137. In Russian. − Abstr. in Referativ. Zhurn. 62. Issled. kosmich. prostranstva, 6.62.161 (1974).

143.042 **Distribution of galactic cosmic radiation and dynamics of structural formations in the solar wind.**
AN SSSR. Yakutsk. fil. Sib. otd. In-t kosmofiz. issled. i aehron. Yakutsk. 295 pp. Price 1 Rbl. 30 Kop. (1973). In Russian.

143.043 **The first spherical harmonic in the distribution of**

cosmic radiation. G. V. Skripin.
Raspredelenie galakt. kosmich. luchej i dinamika struktur. obrazovanij v solnechn. vetre. Yakutsk, 1973, p. 16 - 42.
In Russian. — Abstr. in Referativ. Zhurn. 62. Issled. kosmich. prostranstva, 6.62.164 (1974).

143.044 **The second spherical harmonic in the distribution of cosmic radiation.** P. A. Krivoshapkin, G. F. Krymskij, A. I. Kuz'min, G. V. Skripin.
Raspredelenie galakt. kosmich. luchej i dinamika struktur. obrazovanij v solnechn. vetre. Yakutsk, 1973, p. 43 - 104.
In Russian. — Abstr. in Referativ. Zhurn. 62. Issled. kosmich. prostranstva, 6.62.165 (1974).

143.045 **The second spherical harmonic in the distribution of cosmic radiation (model ideas).**
G. F. Krymskij, P. A. Krivoshapkin, A. I. Kuz'min, G. V. Skripin.
Raspredelenie galakt. kosmich. luchej i dinamika struktur. obrazovanij v solnechn. vetre. Yakutsk, 1973, p. 105 - 117.
In Russian. — Abstr. in Referativ. Zhurn. 62. Issled. kosmich. prostranstva, 6.62.166 (1974).

143.046 **On the third harmonic in the daily variation of cosmic radiation.**
G. F. Krymskij, G. V. Skripin, V. G. Grigor'ev.
Raspredelenie galakt. kosmich. luchej i dinamika struktur. obrazovanij v solnechn. vetre. Yakutsk, 1973, p. 118 - 125.
In Russian. — Abstr. in Referativ. Zhurn. 62. Issled. kosmich. prostranstva, 6.62.167 (1974).

143.047 **Equations of modulation of cosmic radiation and approximations.**
G. F. Krymskij, I. A. Transkij.
Raspredelenie galakt. kosmich. luchej i dinamika struktur. obrazovanij v solnechn. vetre. Yakutsk, 1973, p. 126 - 145.
In Russian. — Abstr. in Referativ. Zhurn. 62. Issled. kosmich. prostranstva, 6.62.168 (1974).

143.048 **Scattering of cosmic rays on inhomogeneities of the interplanetary magnetic field.**
G. F. Krymskij, I. A. Transkij.
Raspredelenie galakt. kosmich. luchej i dinamika struktur. obrazovanij v solnechn. vetre. Yakutsk, 1973, p. 146 - 153.
In Russian. — Abstr. in Referativ. Zhurn. 62. Issled. kosmich. prostranstva, 6.62.169 (1974).

143.049 **Dynamical effects in cosmic radiation.**
A. M. Altukhov, G. F. Krymskij, A. I. Kuz'min.
Raspredelenie galakt. kosmich. luchej i dinamika struktur. obrazovanij v solnechn. vetre. Yakutsk, 1973, p. 198 - 248.
In Russian. — Abstr. in Referativ. Zhurn. 62. Issled. kosmich. prostranstva, 6.62.170 (1974).

143.050 **On the investigation of streams of cosmic radiation in the interplanetary medium.** A. M. Altukhov, G. F. Krymskij, A. I. Kuz'min, I. A. Transkij.
Raspredelenie galakt. kosmich. luchej i dinamika struktur. obrazovanij v solnechn. vetre. Yakutsk, 1973, p. 249 - 266.
In Russian. — Abstr. in Referativ. Zhurn. 62. Issled. kosmich. prostranstva, 6.62.171 (1974).

143.051 **A standing shock wave and stationary modulation of cosmic radiation.** G. F. Krymskij.
Raspredelenie galakt. kosmich. luchej i dinamika struktur. obrazovanij v solnechn. vetre. Yakutsk, 1973, p. 267 - 277.
In Russian. — Abstr. in Referativ. Zhurn. 62. Issled. kosmich. prostranstva, 6.62.172 (1974).

143.052 **Short-periodic variations of cosmic radiation.**
G. F. Krymskij.
Raspredelenie galakt. kosmich. luchej i dinamika struktur. obrazovanij v solnechn. vetre. Yakutsk, 1973, p. 278 - 288.
In Russian. — Abstr. in Referativ. Zhurn. 62. Issled. kosmich. prostranstva, 6.62.173 (1974).

143.053 **Track structure.** R. Katz.
Bull. American Astron. Soc., Vol. 6, 214 - 215 (1974). — Abstr. AAS.

143.054 **The anomalous abundance of oxygen and nitrogen in the low energy galactic cosmic rays.**
F. B. McDonald, B. J. Teegarden, J. H. Trainor, W. R. Webber.
Bull. American Astron. Soc., Vol. 6, 274 - 275 (1974). Abstr. AAS.

143.055 **Changing composition of cosmic ray primaries in the TeV energy range.** P. Kotzer, S. N. Anderson, J. R. Florian, L. D. Kirkpatrick, J. J. Lord, J. W. Martin.
Bull. American Astron. Soc., Vol. 6, 275 (1974). — Abstr. AAS.

143.056 **The effects of non-linear terms in cosmic-ray diffusion theory.** A. J. Owens.
Bull. American Astron. Soc., Vol. 6, 278 (1974). — Abstr. AAS.

143.057 **Sector synchronous modulations of galactic cosmic rays.** J. T. A. Ely.
Bull. American Astron. Soc., Vol. 6, 278 (1974). — Abstr. AAS.

143.058 **A strong-coupling theory of cosmic ray propagation.** A. J. Klimas.
Bull. American Astron. Soc., Vol. 6, 278 (1974). — Abstr. AAS.

143.059 **A photometric determination of the charge spectrum of low energy primary cosmic rays in the region $17 \leqslant Z \leqslant 28$.** C. E. Long.
Thesis Univ. Minnesota, Minneapolis. [Available from Univ. Microfilms, Ann Arbor, Mich., USA. Order No. 73-10602], 111 pp. (1972).

A measurement is reported of the charge spectrum of primary cosmic ray particles of low energy detected over Ft. Churchill, Canada in a stack of nuclear photographic emulsions. After application of various corrections, the resulting density distribution shows two peaks which are interpreted as being those of particles of charge 20 and 26, calcium and iron. Estimates are given for the number of particles found of each integral charge in the range considered.

143.060 **The possible source of cosmic rays.** L. F. Vegas.
Bol. Acad. Ciencias Fis. Mat. Nat. Republica Venezuela, No. 97, Vol. 32, 45 - 48 (1972). In Spanish and English.

143.061 **Charge composition of high energy heavy primary cosmic ray nuclei.** R. D. Price.
Goddard Space Flight Center, Greenbelt, Maryland, GSFC Document X-661-74-71, 12 + 163 pp. (1974).

A detailed study of the charge composition of primary cosmic radiation for about 5000 charged nuclei from neon to iron with energies greater than 1.16 GeV/nucleon is presented.

143.062 **The cosmic ray spectrum at energies above 10^{17} eV.**
D. M. Edge, A. C. Evans, H. J. Garmston, R. J. O. Reid, A. A. Watson, J. G. Wilson, A. M. Wray.
Journ. Phys. A, General Phys., Vol. 6, 1612 - 1634 (1973).

143.063 **A comment on the dust grain model for primary cosmic rays.** K. Terasaki-Okada.
Res. Inst. Theor. Phys. Hiroshima Univ., (Japan), Vol. 73, No. 4, p. 1 - 5 (1973).

143.064 **Mass dependence of the energy spectrum of primary**

cosmic rays. B. J. Daniel, C. J. Hume, L. K. Ng, M. G. Thompson, M. R. Whalley, J. Wdowczyk, A. W. Wolfendale.
Journ. Phys. A, General Phys., Vol. 7, L20 - L24 (1974).

143.065 **Origin of energetic cosmic rays. I. Galactic diffusion in the energy range 10^{14}-10^{17} eV.**
M. C. Bell, J. Kota, A. W. Wolfendale.
Journ. Phys. A, General Phys., Vol. 7, 420 - 436 (1974).

143.066 **Origin of energetic cosmic rays. II. The possibility of a contribution from pulsars.**
S. Karakula, J. L. Osborne, J. Wdowczyk.
Journ. Phys. A, General Phys., Vol. 7, 437 - 443 (1974).

143.067 **Multiple scattering of charged particles in irregular magnetic fields.** K. O. Thielheim.
Journ. Phys. A, General Phys., Vol. 7, 444 - 448 (1974).

Cosmic ray physics: nuclear and astrophysical aspects. See Abstr. 003.056.

Cosmic rays in the interplanetary space. See Abstr. 003.087.

High energy astronomy. See Abstr. 061.075.

Implications of the reported low energy electron gradients. See Abstr. 078.007.

A new limit on the interstellar abundance of boron. See Abstr. 131.094.

Turbulence-enhanced synchrotron radiation in the Galaxy. See Abstr. 155.023.

Stellar Systems

151 Kinematics and Dynamics of Stellar Systems

151.001 **Stellar stability for distribution function "holes" in phase space.** H. W. Bloomberg.
Astron. Astrophys., Vol. 30, 59 - 66 (1974).

The stability of a one-dimensional stellar system, where the interactions are collective, is considered. Eigenmode equations for a double water bag configuration are solved for the case that the distribution function inside the inner (lower energy) bag is less than it is outside the bag. An overstability is found to exist for equilibria where the period of particles on the outer contour is nearly equal to the period of particles on the inner contour. The instability occurs at integer values of the particle excursion frequencies. Expressions are derived for the dependence of the growth rate on the equilibrium configuration.

151.002 **Motions of satellites associated with large spiral systems.** A. Lauberts.
Astron. Astrophys., Vol. 30, 73 - 80 (1974).

This paper describes a study of the orbital kinematics of satellites moving in the gravitational field of a prominent spiral galaxy. As a typical model of a large spiral system, our own Galaxy has been chosen. The satellites are assumed to possess small masses with negligible reactions on the central galaxy. Different hypotheses have been considered in order to explain a pronounced polar oriented distribution of apparent position angles. Special attention has been devoted to the hypothesis of mass ejections in roughly the polar directions from the nucleus of the parent system. In the course of time, however, the radial-type orbits degrade to lower latitudes, the rate depending on the orbital periods of the ejected bodies. Because this behaviour throws severe doubts on the ejection idea, a more convential orbital geometry has also been considered, in which the satellites are moving in somewhat less elongated, but stable, elliptical orbits.

151.003 **The stability of self-gravitating systems with quadratic potential. III. The stability of bi- and three-axial collisionless stellar ellipsoids.**
A. G. Morozov, V. L. Polyachenko, I. G. Shukhman.
Astron. Zhurn. Akad. Nauk SSSR, Vol. 51, 75 - 82 (1974). In Russian. English translation in Soviet Astron. AJ, Vol. 18, No. 1.

Freeman's model of barred spiral galaxies as three-axial ellipsoids is considered. The boundary between the regions of stable and unstable solutions is obtained. The stability of models for observed ratios of semi-axes is proved. The growth rates of unstable solutions are calculated. Exact eigenfunctions and eigenvalues of a collisionless biaxial stellar ellipsoid of arbitrary degree of flattening are obtained also.

151.004 **Star encounters in spherical stellar systems.** V. M. Danilov.
Astron. Zhurn. Akad. Nauk SSSR, Vol. 51, 83 - 91 (1974). In Russian. English translation in Soviet Astron. AJ, Vol. 18, No. 1.

A formula for the probability of star encounters with known variation of stellar velocity in a system with a potential of regular forces depending only on the distance from the centre of mass in the system was deduced in a general form. Using this probability the integral equation for the distribution function for the module of stellar velocities in all parts of the system is given.

151.005 **On an approach to the problem of the third integral for stationary stellar systems.** L. P. Osipkov.
Astron. Zhurn. Akad. Nauk SSSR, Vol. 51, 92 - 101 (1974). In Russian. English translation in Soviet Astron. AJ, Vol. 18, No. 1.

For the investigation of the third integral $K(P, Z, \rho, z)$ polar coordinates (V, α): $P = V \cos \alpha$, $Z = V \sin \alpha$ are introduced instead of usual projections of the velocity vector (P, Z). It is investigated in what way the isolating properties of the integral K are displayed in these coordinates.

151.006 **Density perturbations set up by a rotating galactic dipole.** P. S. Wesson.
Astrophys. Space Sci., Vol. 26, 189 - 197 (1974).

The response of material to a rotating magnetic dipole, considered as primeaval, the axis of which lies in the galactic plane of a model galaxy, is examined. The flow pattern is found to be always characterised by two streamers of high-velocity matter emerging in the plane of the galaxy. The accompanying density distribution suggests a ready analogy with spiral galaxies, especially of SBc and SBb type; the main implication of the hypothesis, however, is that galactic dipoles will inevitably set up density perturbations of a form suitable for the generation of spiral arms via the mechanism of density waves.

151.007 **A class of disk-like models for self-gravitating stellar systems.** M. Miyamoto.
Astron. Astrophys., Vol. 30, 441 - 454 (1974).

A construction scheme is proposed for disk-like models of self-gravitating steady-state stellar systems by using Fricke's (1952) type of the expansion for the velocity distribution function $f(E, J)$ with respect to the energy integral E and the angular momentum integral J.

151.008 **Potential energy of gravitationally interacting spherical galaxies.** A. Potdar, G. M. Ballabh.
Astrophys. Space Sci., Vol. 26, 353 - 363 (1974).

A theory has been developed for obtaining the potential energy of two interpenetrating spherically symmetric galaxies of unequal dimensions due to their mutual gravitational interaction. The mass distribution in both the galaxies is assumed to be that of a polytrope of integral index. A basic function that occurs in the theory has been tabulated for the cases of polytropes of indices $n = 0$ and 4 for four ratios of the radii.

151.009 **Dynamical models for the formation and evolution of spherical galaxies.** R. B. Larson.
Monthly Notices Roy. Astron. Soc., Vol. 166, 585 - 616 (1974).

Models for the collapse of spherical protogalaxies have been computed using a two-fluid hydrodynamical approach to treat both the gas and the stars. The effects of gaseous energy losses, star formation, stellar mass loss and heavy element (Z) production are included and are followed over the lifetime of the galaxy. With reasonable choices for the parameters, the models reproduce well the observed structure of

NGC 3379 and other nearly spherical galaxies. Various characteristics of the models suggest that the quasar phenomenon may be identifiable with the formation of the nucleus of a giant elliptical galaxy.

151.010 The structure of barred stellar systems.
C. Hunter.
Monthly Notices Roy. Astron. Soc., Vol. 166, 633 - 648 (1974).

Some self-consistent stellar dynamic models of rotating barred systems due to Freeman are analysed. The complete range of models is delineated and their important non-dimensional properties are plotted graphically.

151.011 Formation and evolution of inhomogeneities in non-linear theory of gravitational instability.
A. G. Doroshkevich, S. F. Shandarin.
Astrofizika, Vol. 9, 549 - 565 (1973). In Russian. English translation in Astrophysics, Vol. 9, No. 4.

The process of the formation and evolution of disc-like condensations of matter (protoclusters of galaxies) appearing in the non-linear stage of growth of adiabatic perturbations is considered on the basis of numerical one-dimensional calculations. The influence of various parameters (such as the initial spectrum of perturbations, the parameters of the cosmological model, the time of the origin of the disc) on the properties and parameters of the contracted gas is investigated.

151.012 Tidal effects of the Galaxy on the orbits of stars in spherical stellar systems.
D. W. Keenan, K. A. Innanen.
Bull. American Astron. Soc., Vol. 6, 207 (1974). – Abstr. AAS.

151.013 Resonant stellar orbits in spiral galaxies.
P. O. Vandervoort.
Bull. American Astron. Soc., Vol. 6, 207 - 208 (1974). Abstr. AAS.

151.014 Is there a third integral of motion? P. Pişmiş.
Bull. American Astron. Soc., Vol. 6, 208 (1974). Abstr. AAS.

151.015 Correlation energy in stellar systems.
R. H. Miller.
Bull. American Astron. Soc., Vol. 6, 208 (1974). – Abstr. AAS.

151.016 Some barred collisionless stellar systems.
C. Hunter.
Bull. American Astron. Soc., Vol. 6, 208 (1974). – Abstr. AAS.

151.017 Galactic spiral arms and evolution.
J. H. Piddington.
Proc. Astron. Soc. Australia, Vol. 2, 170 - 173 (1973). – Invited paper – see 012.001.

151.018 Spiral structure viewed as a density wave.
A. J. Kalnajs.
Proc. Astron. Soc. Australia, Vol. 2, 174 - 177 (1973). – Invited paper – see 012.001.

151.019 Uniform density equilibrium model of self-gravitating stellar systems. A. Ahmad.
Astrophys. Space Sci., Vol. 27, 343 - 350 (1974).

A class of equilibrium solutions of the Vlasov equation for self-gravitating systems is discussed. The density and the potential are derived in form of Jacobi polynomials, which in a special case give rise to a model with uniform density.

151.020 Computer simulations of star cluster dynamics.
S. J. Aarseth.
Vistas in astronomy, Vol. 15, (see 003.001), 13 - 37 (1973).

151.021 Numerical difficulties with the gravitational N-body problem. R. H. Miller.
Proc. conference on numerical solution of ordinary differential equations, 1972, (see 012.005), p. 260 - 275 (1974).

151.022 On the numerical integration of the N-body problem for star clusters. R. Wielen.
Proc. conference on numerical solution of ordinary differential equations, 1972, (see 012.005), p. 276 - 290 (1974).

151.023 A variable order method for the numerical integration of the gravitational N-body problem. G. Janin.
Proc. conference on numerical solution of ordinary differential equations, 1972, (see 012.005), p. 291 - 303 (1974).

151.024 The method of the doubly individual step for N-body computations. A. Hayli.
Proc. conference on numerical solution of ordinary differential equations, 1972, (see 012.005), p. 304 - 312 (1974).

151.025 Integration of the N-body gravitational problem by separation of the force into a near and a far component. A. Ahmad, L. Cohen.
Proc. conference on numerical solution of ordinary differential equations, 1972, (see 012.005), p. 313 - 336 (1974).

151.026 Numerical experiments on the statistics of the gravitational field. L. Cohen, A. Ahmad.
Proc. conference on numerical solution of ordinary differential equations, 1972, (see 012.005), p. 337 - 359 (1974).

151.027 Integration errors and their effects on macroscopic properties of calculated N-body systems.
H. Smith, Jr.
Proc. conference on numerical solution of ordinary differential equations, 1972, (see 012.005), p. 360 - 373 (1974).

151.028 Numerical integration of gravitational N-body systems with the use of explicit Taylor series.
M. Lecar, R. Loeser, J. R. Cherniack.
Proc. conference on numerical solution of ordinary differential equations, 1972, (see 012.005), p. 451 - 470 (1974).

151.029 Dynamical friction and the motion of stars in spherical clusters. A. S. Baranov, Yu. V. Batrakov.
Astron. Zhurn. Akad. Nauk SSSR, Vol. 51, 310 - 316 (1974). In Russian. English translation in Soviet Astron. AJ, Vol. 18, No. 2.

The effects of dynamical friction on the motion of stars in spherical clusters are considered. An approximate expression for the force of the dynamical friction has been deduced. Differential equations determining the evolution of the orbit of a star in the cluster under the mutual influence of a regular force field and of dynamical friction have been derived. An approximate solution of these equations has been obtained for two cases of the density distribution – a uniform and an exponential one.

151.030 Die Dichtewellentheorie der Spiralstruktur.
K. Rohlfs.
SuW, Vol. 13, 114 - 118 (1974).

151.031 Stellar dynamics. I. R. King.
Celestial Mechanics, Vol. 9, 349 - 357 (1974).
Presented at a colloquium on 'Copernicus ... and modern dynamical astronomy', Washington, 1972 December.

This review attempts to place stellar dynamics in relation to other dynamical fields and to describe some of its important techniques and present-day problems.

151.032 Linear stability of single phase space holes.

J. P. Dorémus, G. Baumann.
Astron. Astrophys., Vol. 32, 47 - 49 (1974).

The authors study the linear stability of one-dimensional stellar systems with a double water-bag distribution function. They compute the eigenfrequencies of the system and compare their results with the numerical experiments of Cuperman and Tzur. Further, they consider only the asymptotic eigenfrequencies and give an expression for the instability growth rate.

151.033 **Dynamical friction in gravitational systems.**
A. Ahmad, L. Cohen.
Astrophys. Journ., Vol. 188, 469 - 471 (1974).

Results of numerical experiments are presented showing the existence of dynamical friction in gravitational systems. Comparisons are made with the theory of Chandrasekhar and von Neumann.

151.034 **The effectiveness of different relaxation mechanisms of stellar systems.** S. G. Pomagaev.
Byull. Inst. Astrofiz., *Dushanbe*, No. 61, p. 3 - 6 (1972). In Russian.

A comparative estimation is given of the effectiveness of two different mechanisms of relaxation: relaxation caused by encounters of stars with massive objects and relaxation caused by cooperative phenomena.

151.035 **Effect of population II stars and three-dimensional motion on spiral structure.**
R. W. Hockney, D. R. K. Brownrigg.
Monthly Notices Roy. Astron. Soc., Vol. 167, 351 - 357 (1974).

A three-dimensional computer model of a fully self-consistent isolated disk galaxy of finite thickness is described using approximately 50000 simulated population I stars. Population II stars are represented by a fixed imposed gravitational field. This model is used to study the stability of a thin-disk galaxy for varying fractions of population II stars. The authors compare the behaviour of the three-dimensional model, in which motion of the population I stars out of the plane is allowed, and the behaviour of the simpler two-dimensional model, in which the motion of the stars is confined to the plane of the disk. The lifetime of the spirals is significantly longer in the more realistic three-dimensional model.

151.036 **On gaseous flows in disk galaxies.** A. F. Saaf.
Astrophys. Journ., Vol. 189, 33 - 38 (1974).

The equations of motion and continuity which determine the structure of the gaseous component of a thin disk galaxy have been studied for the case in which the spiral structure is stationary in a coordinate system rotating with the spiral pattern. Approximate solutions are presented for both the 2θ and 4θ modes. The solutions for the 2θ mode coincide with the hydrodynamical results of the purely linearized theory of Lin and Shu.

151.037 **The appearance of relativistic stellar systems.**
K. G. Suffern.
Australian Journ. Phys., Vol. 27, 93 - 104 (1974).

A formula is derived for the projected density distribution on a photographic plate which would arise from a given static spherically-symmetric relativistic stellar system. For weak gravitational fields, a corresponding post-Newtonian expression is derived which is relatively simple to use once a particular proper stellar number density has been specified for the relativistic system under consideration.

151.038 **Equilibrium structures of relativistic star clusters.**
F. Occhionero, A. San Martini.
Astron. Astrophys., Vol. 32, 203 - 208 (1974).

We investigate the spherically symmetric equilibrium structures available to a monoenergetic collection of point masses under the influence of their own gravity in extreme relativistic conditions. The family of solutions, parametrized by R_s/R, the ratio between the Schwarzschild and the coordinate radius (or equivalently by z_c, the central redshift), attains a 3.6% maximum of the fractional binding energy for $R_s/R \cong 0.3$ ($z_c \cong 0.5$). Beyond this limit the relativistic cluster is probably dynamically unstable for radial perturbations and therefore uninteresting from an astrophysical point of view. On the contrary, the equilibrium sequence for $0 \lesssim R_s/R \lesssim 0.3$ represents a possible evolutionary track for quasi-static contraction.

151.039 **Vortex flow of matter around a singularity and a galactic hypothesis of Jeans.** P. S. Wesson.
Astrophys. Space Sci., Vol. 28, 289 - 302 (1974).

The object of this work is to show that the dynamics of galaxies, including the permanence of galactic structure, can be explained as resulting from the nucleation of hydrodynamical matter-flow by compact or collapsed objects, which have associated with themselves singular sources of vorticity. The motivation for this comes from Jeans' remark that the centres of spiral galaxies look like the places where matter is being injected into our part of the universe from some unknown region of space (Jeans, 1928).

151.040 **The theory of spiral structure. Resonances.**
G. Contopoulos.
Galaxies and relativistic astrophysics, Proc. 1972, (see 012.010), p. 104 - 113 (1974). – Invited lecture.

151.041 **On spiral generating.** D. Lynden-Bell.
Galaxies and relativistic astrophysics, Proc. 1972, (see 012.010), p. 114 - 119 (1974). – Invited lecture.

151.042 **On the possible role of the stellar drift motions for the dynamics and structure of differentially rotating stellar systems.** M. N. Maksumov.
Galaxies and relativistic astrophysics, Proc. 1972, (see 012.010), p. 120 - 127 (1974).

151.043 **On galaxy parameters as derived from primeval turbulence.** N. Dallaporta, F. Lucchin.
Galaxies and relativistic astrophysics, Proc. 1972, (see 012.010), p. 140 - 141 (1974). – Abstract.

151.044 **Periodic and conditionally-periodic trajectories in conservative systems under minimal smoothness conditions. I.** V. A. Antonov.
Trudy Astron. Obs., *Leningrad*, Vol. 30 (= Uchenye Zapiski Leningr. Un-ta, No. 373 = Seriya Matem. Nauk, vyp. (No.) 50), p. 111 - 132 (1974). In Russian.

The area-preserving mapping of a ring is discussed. A special class of "regular" periodic cycles is introduced. The necessary conditions of their existence are found.

151.045 **Mono-trajectory flat systems of gravitating bodies.**
M. A. Belozerova.
Trudy Astron. Obs., *Leningrad*, Vol. 30 (= Uchenye Zapiski Leningr. Un-ta, No. 373 = Seriya Matem. Nauk, vyp. (No.) 50), p. 132 - 140 (1974). In Russian.

The equation of motion for a multi-trajectory flat self-gravitating system is derived. Mono-trajectory self-gravitating flat systems are considered in detail. For selected values of a parameter several characteristics of mono-trajectory systems are calculated, namely the distance of pericluster, the distance of apocluster, the angle between the directions from the centre to pericluster and nearest apocluster. Selected orbits are plotted. Asymptotic solutions are obtained for the cases of radial and almost circular motions.

151.046 **Formal integrals of Hamiltonian systems in resonance and near-resonance cases.**
G. Contopoulos.
In honorem S. Placidis, (see 003.009), p. 139 - 150 (1974).

151.047 **Epicyclic motions and their applications.**
B. Barbanis.
In honorem S. Placidis, (see 003.009), p. 285 - 300 (1974). In Greek.

151.048 **On the formation of interstellar cloud complexes, OB associations and giant H II regions.**
T. C. Mouschovias, F. H. Shu, P. R. Woodward.
Astron. Astrophys., Vol. 33, 73 - 77 (1974).

The authors propose that large cloud complexes, OB associations, and giant H II regions form as a result of the initiation of the magnetic Rayleigh-Taylor instability in the interstellar medium by the passage of a galactic shock. Total masses of about $10^6 \, M_\odot$, formation of unbound systems, and alignment along spiral arms with typical separation of about 1 kpc are natural consequences of this point of view.

151.049 **The problem of the spiral structure of galaxies.**
L. S. Marochnik, A. A. Suchkov.
Uspekhi fiz. nauk, Vol. 112, 275 - 308 (1974). In Russian. Abstr. in Referativ. Zhurn. 51. Astron., 6.51.645 (1974).

151.050 **The gravitational slingshot and the structure of extragalactic radio sources.**
W. C. Saslaw, M. J. Valtonen, S. J. Aarseth.
Astrophys. Journ., Vol. 190, 253 - 270 (1974).

The authors propose a simple dynamical mechanism to eject massive objects from the nuclei of galaxies. When three or more massive objects interact strongly near the center, the configuration becomes gravitationally unstable and splits up. In the case of three objects, two of them will usually form a compact binary and fling out the third. To study the interaction for a large range of initial conditions, the authors have numerically computed the orbits of 25000 triple systems and of 250 two-binary systems. Their general properties are analyzed and compared with the observed structures of extragalactic radio sources. The authors also discuss observational constraints on radio emission from the massive objects, and further observations of special interest.

151.051 **On the third integral of motion in stellar dynamics. II.** J. S. Stodółkiewicz.
Acta Astron., Vol. 24, 153 - 164 (1974).

It is shown that the class of potentials which are axially symmetric and symmetric with respect to a certain plane perpendicular to the axis of symmetry and which have "local third integrals" is identical with the class of potentials obtained in Paper I (Stodółkiewicz 1972).

151.052 **Encounters between galaxies of equal size.**
A. Lauberts.
Astron. Astrophys., Vol. 33, 231 - 240 (1974).

This paper deals with gravitational interactions between two stellar systems of equal size. Each system is composed of a compact central part and a surrounding spherical halo, both of the same mass. Most of the attention has been devoted to slow, close encounters, which give rise to conspicuous deformations due to strong tidal interactions. Comparison is also made with the "impulsive approximation", and it is concluded that the relative motions must be very rapid for this approximation to work well.

151.053 **The gravitational N-body problem for star clusters.**
R. Wielen.
Stars and the Milky Way system, Proc. 1972, (see 012.018), p. 326 - 354 (1974). — Invited lecture.

151.054 **Galaxy-quasar associations and globular clusters.**
J. Barnothy.
Bull. American Astron. Soc., Vol. 6, 212 (1974). – Abstr. AAS.

151.055 **A model for the formation of binary stars and binary galaxies.** G. Byrd.
Bull. American Astron. Soc., Vol. 6, 223 - 224 (1974). Abstr. AAS.

151.056 **Stabilization of spiral structure.** F. Hohl.
Bull. American Astron. Soc., Vol. 6, 225 (1974). Abstr. AAS.

151.057 **Overstability of density waves due to mass exchange between stars and gas.** S. Kato.
Publ. Astron. Soc. Japan, Vol. 26, 207 - 220 (1974).

It was shown in a previous paper (Kato 1972) that small amplitude tightly-wound density waves in a rotating star-gas disk can become overstable, if the stars and gas interact during the oscillations by processes of star formation and mass loss from stars. Star formation at the condensed phase of waves works in favor of the overstability of the waves. In this paper that result is confirmed and errors made in the previous paper are corrected. The mechanism of this overstability is clarified and some numerical results are presented.

151.058 **On the homologous finite deformation of stellar systems.** S. Kikuchi.
Proc. symposium "Celestial mechanics", Kyoto 1973, (see 012.022), p. 43 - 53 (1974). In Japanese.

151.059 **Velocity distribution function for a stellar disk.**
S. Manabe.
Proc. symposium "Celestial mechanics", Kyoto 1973, (see 012.022), p. 54 - 66 (1974). In Japanese.

151.060 **Spiral structure, dust clouds, and star formation.**
F. H. Shu.
American Scient., Vol. 61, 524 - 536 (1973).

The author discusses the density-wave theory of spiral structure, distributions of light and mass, galactic shocks, large-scale aspects of the flow, compression on a small scale, dust lanes and dust clouds.

151.061 **On the stability of a disk galaxy.** R. H. Miller.
Astrophys. Journ., Vol. 190, 539 - 542 (1974).

Numerical experiments on strictly axisymmetric disk galaxies are reported which confirm the stability threshold predicted by Toomre. This provides strong evidence that the difference between the Toomre criterion and the experimental behavior of disk-galaxy models in a computer is attributable to non-axisymmetric disturbances. The implications of this result for disk-galaxy models in general are discussed, along with applications to models of the Galaxy.

The problem of three bodies.
See Abstr. 042.038.

Galactic models and stellar orbits.
See Abstr. 158.107.

The kinematics and dynamics of M51. I. The observations. See Abstr. 158.131.

The kinematics and dynamics of M51. II. Axisymmetric properties. See Abstr. 158.132.

The kinematics and dynamics of M51. III. The spiral structure. See Abstr. 158.133.

Errata

151.901 Corrigenda: "Hydromagnetic stability of thin self-gravitating disks and spiral structure" [Australian Journ. Phys., Vol. 22, 505 - 519 (1969)]. R. J. Hosking. Australian Journ. Phys., Vol. 26, 903 (1973).

151.902 Erratum: 'Stability of a self gravitating system with phase space density function of energy and angular momentum' [Astron. Astrophys., Vol. 29, 401 - 407 (1973)]. J. P. Dorémus, G. Baumann, M. R. Feix. Astron. Astrophys., Vol. 32, 479 (1974).

152 Stellar Associations

152.001 **The B-stars in the UMa-stream.** I. N. Latyshev. Astron. Tsirk., No. 798, p. 6 - 7 (1973). In Russian.

152.002 **Stars of the lower part of the main sequence in the UMa-stream.** I. N. Latyshev. Astron. Tsirk., No. 798, p. 7 - 8 (1973). In Russian.

152.003 **Untersuchungen über die Sternassoziation Cyg T 1.** F. Gieseking. Veröff. Astron. Inst. Bonn, No. 87, 40 pp. (1973).

After a detailed discussion of the qualities of the astrograph 300/1500 of the Observatorium Hoher List, the results of photometric investigations of the T-association Cyg T 1 ($\alpha_{1900} = 20^h 47^m .4$; $\delta_{1900} = 44°.0$) are presented. The results of an extensive blink-survey in the field of Cyg T 1, including some remarks on the completeness of the variable-search, are given and the light curves of 20 named and newly discovered young irregular variables are discussed. The two variables DS Cyg and Nr. 10 are found to be especially remarkable because of their extreme colour indices. After the discussion of the distribution of the emission-line objects, variables, suspected variables and stars with early spectral-types in the region of Cyg T 1, the results of a three colour photometry of a large number of these stars are given.

UV excess stars and radio sources in the Aquila ring. See Abstr. 113.010.

A remarkable star in a highly reddened group of OB$^+$-stars in Norma. See Abstr. 121.075.

On the formation of interstellar cloud complexes, OB associations and giant H II regions. See Abstr. 151.048.

The chemical composition of stars as derived from photoelectric observations of narrow-band indices of equivalent widths. See Abstr. 153.031.

Die Realität der Sterntrupps. See Abstr. 155.076.

153 Galactic Clusters

153.001 **Spectroscopic comparison of open clusters. I. The reddening, blanketing, and metallicity of M67.**
D. C. Barry, R. H. Cromwell.
Astrophys. Journ., Vol. 187, 107 - 115 (1974).

Equivalent widths are determined for 13 hydrogen and metallic features in the spectra of M67 dwarfs and subgiants using 1-mm widened image-tube spectrograms having 48 Å mm^{-1} dispersion. These are directly compared with equivalent widths of the same features in F and G dwarfs in the Hyades and Coma clusters and nearby field dwarfs and subgiants. Assuming that equal hydrogen-line strengths indicate equal effective temperatures for dwarfs in the various clusters, relative metallicities, differential blanketing, and interstellar reddening values are determined. Comparisons with values available in the literature empirically establish the validity of the method.

153.002 **Photographic photometry of the galactic cluster NGC 6819.** G. Auner.
Astron. Astrophys., Suppl. Ser., Vol. 13, 143 - 171 (1974).

Photographic photometry was done on 891 stars in the area of the open cluster NGC 6819 down to a limiting magnitude of $V = 16^m.6$ in the UBV system. Colour excess and distance modulus are determined to have values of $E(B-V) = 0^m.28$ and $(m-M)_0 = 11^m.76$ respectively. Strong arguments for the existence of a main sequence gap at $V = 15^m.35$ are found. A projected density diagram is set up by star counting in concentric rings.

153.003 **Hydrogen content of young stellar clusters. I. Methods of observation and reduction.**
H. M. Tovmassian.
Australian Journ. Phys., Vol. 26, 829 - 835 (1973).

Observations of 16 open galactic clusters in their continuum emission and at the neutral hydrogen line have been made with the Parkes 64 m radio telescope in an attempt to determine the total amount of hydrogen gas associated with them. In this, the first of a series of five papers, the observing procedure and the method of data reduction are described.

153.004 **Hydrogen content of young stellar clusters. II. Clusters NGC 2175, 2264, 2353, and 2362.**
H. M. Tovmassian, E. T. Shahbazian.
Australian Journ. Phys., Vol. 26, 837 - 842 (1973).

Measurements of the total amount of gaseous hydrogen associated with the O-type stellar clusters NGC 2175, 2264, 2353, and 2362 are reported.

153.005 **Hydrogen content of young stellar clusters. III. Clusters NGC 3293, 6167, 6193, 6200, and 6204.**
H. M. Tovmassian, E. T. Shahbazian, S. E. Nersessian.
Australian Journ. Phys., Vol. 26, 843 - 851 (1973).

The results of 21 cm line and continuum observations of the O-type stellar clusters NGC 3293, 6193, and 6204 and of the doubtful clusters NGC 6167 and 6200 are presented.

153.006 **Hydrogen content of young stellar clusters. IV. Clusters NGC 6231, 6383, 6514, and 6531.**
H. M. Tovmassian, S. E. Nersessian, E. T. Shahbazian.
Australian Journ. Phys., Vol. 26, 853 - 860 (1973).

The results of 21 cm line and continuum observations of the O-type stellar clusters NGC 6231, 6383, 6514, and 6531 are presented.

153.007 **Hydrogen content of young stellar clusters. V. Clusters NGC 6604, 6611, and 6823.**
H. M. Tovmassian, S. E. Nersessian.
Australian Journ. Phys., Vol. 26, 861 - 866 (1973).

Measurements of the total amount of hydrogen gas associated with the O-type clusters NGC 6604, 6611, and 6823 are reported.

153.008 **RGU photometry of the open clusters NGC 581 (=M 103), Tr 1 and NGC 659.** H. Steppe.
Astron. Astrophys., Suppl. Ser., Vol. 15, 91 - 105 (1974).

By means of RGU three-colour photographic photometry the open clusters NGC 581 (=M 103), Tr 1 and NGC 659 have been investigated. Distance, absorption and age of the clusters have been redetermined.

153.009 **NGC 2287 and the Pleiades group.** O. J. Eggen.
Astrophys. Journ., Vol. 188, 59 - 70 (1974).

Intermediate-band photometry in NGC 2516 and intermediate and broad-band photometry in NGC 2287 are presented. A preliminary luminosity calibration for early-type stars based on $(\beta, [u - b])$ is discussed and applied to both the brighter members of cluster stars in the Pleiades group and the field group members isolated by a survey of the kinematics of all bright ($V_0 \leqslant 5.0$) B-type stars. A composite (M_{bol}, log T_e)-diagram is constructed for the group members and the distributions of the stars in the (X, Z) and (X, Y) planes are briefly discussed.

153.010 **Probable OB star members in eleven Berkeley clusters.** N. Sanduleak.
Publ. Astron. Soc. Pacific, Vol. 86, 74 - 75, with a correction p. 340 (1974).

The Setteducati-Weaver catalog of new open star clusters discovered on the Palomar Sky Survey plates was compared recently with the catalogs of OB stars published by the Hamburg and Warner and Swasey Observatories. This resulted in the identification of eleven clusters which appear to contain OB stars and which may thus serve as important additional spiral arm tracers in the galactic longitude range of 75° to 135°.

153.011 **On the ratio of total to selective absorption in the open cluster IC 2581.** A. F. J. Moffat.
Astron. Astrophys., Vol. 32, 103 - 106 (1974).

Based on the ideas of Becker (1966), a quantitative method is presented which tests the reality of $R = A_V/E_{B-V}$ derived by the variable extinction method in open clusters. Applied to IC 2581 it is shown that Turner's (1973) value of $R = 5.5 \pm 0.3$ is spuriously too large. This is caused mainly by (normal) errors in the photometry.

153.012 **Hγ equivalent widths in the open cluster NGC 7243.**
G. Hill, W. Fisher, A. Allison.
Astron. Journ., Vol. 79, 376 - 378 (1974).

Measurements of the equivalent width of Hγ are presented for 23 stars in the young open cluster NGC 7243. By using the $M_v[W, (B - V)_0]$ calibration of Balona and Crampton (1974) a cluster distance modulus of 9.4 ± 0.1 (m.e.) has been determined. The $M_v(\beta)$ calibration of Crawford (1972) and the Hγ calibration of Balona and Crampton are compared for NGC 7243 and the Pleiades.

153.013 **Photographic photometry in the field of NGC 2264.**
R. H. Koch, P. M. Perry.
Astron. Journ., Vol. 79, 379 - 386 (1974).

A new photographic investigation of NGC 2264 is described. For the historical variables and numerous new variables, the amplitudes of variation or limiting magnitudes are given. A small selection of the variables may possibly be eclipsing pairs, but it is suggested that shell activity may be the dominant cause of the light variations for these objects even if they are binaries.

153.014 The distance to the Hyades cluster.
W. F. van Altena.
Publ. Astron. Soc. Pacific, Vol. 86, 217 - 222 (1974). – A review paper prepared for IAU Commission No. 33, August 1973.

The results of numerous investigations of the distance to the Hyades cluster are reviewed. The principal result is that all "secondary" distance indicators yield distance moduli greater than those determined from proper motions. A weighted value of $(m-M) = +3.21 \pm 0.03$ (s.e.) is adopted as the best present distance modulus for the Hyades cluster.

153.015 CN strengths in M67 and NGC 188.
B. E. J. Pagel.
Monthly Notices Roy. Astron. Soc., Vol. 167, 413 - 417 (1974).

Using $(R-I)$, $(B-V)$ and Spinrad and Taylor's T-index as temperature criteria, it turns out that the CN blocking fractions measured by them in M67 and NGC 188 giants are consistent with metal abundances intermediate between those of the sun and the Hyades except in the case of stars in the 'clump' at $(B-V)_0 \simeq 1.05$, $M_v \simeq 1.0$.

153.016 Age determination for some open clusters.
R. Dinescu, V. Muzylev.
Stud. Cerc. Astron., Vol. 19, 79 - 87 (1974). In Romanian.

The paper presents estimations of the age for ten open clusters, by comparison of the star positions in the $H-R$ diagram with the theoretical isochrones constructed on the basis of the evolutionary tracks for $1.5\,M_\odot - 15\,M_\odot$.

153.017 Stellar evolution near the main sequence: on some systematic differences between cluster sequences and model calculations. A. Maeder.
Astron. Astrophys., Vol. 32, 177 - 190 (1974).

Detailed comparisons in the colour-magnitude diagram are made between computed isochrones and observed sequences of some well known old open clusters, namely NGC 188, M 67, NGC 3680, NGC 752, NGC 2360 and Praesepe. As basic data, we have considered different evolutionary models, calibrations of T_{eff} confirmed by recent models and the numerous results on the abundances of the clusters; the effects of the spread due to binaries are also examined in an appendix.

153.018 Statistische Untersuchungen zur Entwicklung junger Sternhaufen und ihrer Mitglieder. W. Götz.
Veröff. Sternw. Sonneberg, Vol. 8, (No. 3), 131 - 171 (1973).

The results of the present paper are obtained by summarizing and comparing a number of investigations on the structure of young open clusters and the phases of formation and evolution of their members. It is shown that there is a continuous star formation and evolution in the investigated clusters, NGC 6611, NGC 6530 I, NGC 2264 and M 45. Certain groups of stars, which are defined by their positions in the (U-B)–(B-V)-diagram, were found in each of these clusters. The behaviour of the stars is represented in evolution diagrams. Some spectroscopical statements are made.

153.019 Proper motions, cluster membership and reddening in NGC 6611. L. W. Kamp.
Astron. Astrophys., Suppl. Ser., Vol. 16, 1 - 23 (1974).

Proper motions are derived with a mean error of 0.07 arcsec/100 years for 142 stars brighter than $V = 13\overset{m}{.}2$ in the field of NGC 6611, using one first-epoch and two second-epoch plates taken on the Yerkes 40" refracting telescope. The probability of cluster membership of each star is computed from the distribution of the proper motions, yielding 35 members. Using the normal coefficient of selective absorption (0.72 for O stars) a systematic difference is observed of about $0\overset{m}{.}06$ between derived intrinsic colors $(B-V)$ and those appropriate to the spectral types. This is corrected by deriving a law of selective extinction for this cluster which yields normal intrinsic colors and applying this to the cluster members. A ratio of total to selective extinction is derived of 3.9 ± 0.2 and a cluster distance of 1.6 ± 0.2 kpc. The cluster age is estimated to be on the order of 5×10^6 years.

153.020 NGC 3105, a young very distant cluster in Vela.
A. F. J. Moffat, M. P. FitzGerald.
Astron. Astrophys., Suppl. Ser., Vol. 16, 25 - 32 = Contr. Univ. Waterloo Obs. No. 30 (1974).

The following parameters are derived for NGC 3105: distance 8.0 ± 1.5 kpc, reddening $E_{B-V} = 1.09 \pm 0.03$, diameter 5.2 pc, earliest spectral type b1. NGC 3105 contains 2 blue and 2 red supergiants, 1 (possibly 2) strong emission B star, and a possible high luminosity cepheid.

153.021 NGC 2345, a moderately young open cluster in Canis Major. A. F. J. Moffat.
Astron. Astrophys., Suppl. Ser., Vol. 16, 33 - 42 (1974).

Photoelectric UBV and spectroscopic observations yield for NGC 2345 a distance of 1.75 kpc and earliest spectral type B4 after corrections for variable reddening. The cluster contains seven bright giants, two of rare A-type and five of K-type. One of the red bright giants has a blue variable companion.

153.022 Four-color and Hβ photometry for open clusters. IX. NGC 6871.
D. L. Crawford, J. V. Barnes, W. H. Warren, Jr.
Astron. Journ., Vol. 79, 623 - 625 (1974).

Data for 24 stars that are members of the cluster yield a distance modulus of $V_0 - M_v = 11.5$. The average color excess is $E(b-y) = 0.38$. The X-ray source Cyg X-1, which is near NGC 6871 in the sky, is probably well beyond the cluster in distance from the sun.

153.023 The old open cluster NGC 2420.
R. D. McClure, W. T. Forrester, J. Gibson.
Astrophys. Journ., Vol. 189, 409 - 421 (1974).

New photographic and photoelectric photometry on the UBV system is presented for stars in the old open cluster NGC 2420. This photometry confirms the ultraviolet excess of $\delta(U-B) = 0.09$ mag and almost negligible reddening reported by West. Intermediate band photometry on the DDO system indicates that the giants in the cluster have weak cyanogen bands ($\delta CN = -0.06$). Both types of photometry imply a low heavy element abundance for the cluster of $[Fe/H]_{Hyd} = -0.5$. A distance modulus of $(m - M)_0 = 11.4 \pm 0.2$ mag and an age of $3.3 \pm 0.5 \times 10^9$ years are derived. The radial velocity of the cluster from measurements of three giant stars is $> + 100$ km s^{-1}. An extreme barium II star was found in the cluster. Its derived absolute magnitude is $M_v = 0.0$ mag.

153.024 Intermediate-band photometry of M67.
K. A. Janes.
Astrophys. Journ., Vol. 189, 423 - 426 (1974).

Observations on the DDO photometric system have been obtained of 24 stars on the giant and subgiant branches of the old cluster M67. The following values are obtained: $E(B-V) = 0.051 \pm 0.004$ mag, $(m-M) = 9.26 \pm 0.09$ mag, and $\delta CN = 0.032 \pm 0.005$ mag. The results of Eggen and Sandage are confirmed and M67 has a normal composition. Furthermore, there is no evidence that the "clump" stars are mixed or have lost mass.

153.025 Integrated colors of star clusters in the Large Magellanic Cloud. A. Bernard, J. H. Bigay.
Astron. Astrophys., Vol. 33, 123 - 130 (1974). In French.

Photometry of the integrated light of 95 star clusters

in the UBV is reported. The results of the integrated $uvby$ observations for 29 bright clusters previously measured in the UBV-photometry (van den Bergh and Hagen, 1968) are given. The $U - B$ versus $B - V$ diagram shows that the clusters form two groups: young clusters with $B - V < 0.5$, and a few old red clusters with $B - V > 0.6$, thus confirming previous results obtained by different investigators.

153.026 A possible extension of NGC 5822/3.
B. Hidajat, A. Kuncoro.
Bull. American Astron. Soc., Vol. 6, 224 (1974). — Abstr. AAS.

153.027 An enriched M, R–I main sequence and distance modulus for the Hyades cluster. A. R. Upgren.
Bull. American Astron. Soc., Vol. 6, 224 (1974). — Abstr. AAS.

153.028 Evolved stars in open clusters. G. L. Hagen.
Bull. American Astron. Soc., Vol. 6, 266 (1974). — Abstr. AAS.

153.029 Four-color and Hβ photometry for open clusters. X. The α Persei cluster.
D. L. Crawford, J. V. Barnes.
Astron. Journ., Vol. 79, 687 - 697 (1974).
Data are given for 89 stars, mostly B-, A-, and F-type, in the vicinity of the F5Ib supergiant α Persei; most are cluster members. The authors derive an average interstellar reddening of $E(b-y) = 0.07$; the reddening is somewhat variable over the field. The corrected distance modulus is $V_0 - M_v = 6.1$, corresponding to 166 pc. The heavy element abundance is solar-like, and the Balmer discontinuity is found to be sensitive to rotational velocity effects.

153.030 On the variability of dwarf K- and M-type stars in the Pleiades and Hyades.
E. L. Robinson, R. P. Kraft.
Astron. Journ., Vol. 79, 698 - 701 = Lick Obs. Bull., No. 661 (1974).
Small variations in brightness with time scales of the order of days are found in several Pleiades dwarfs in the spectral-type range K3V–M0V; Hyades stars of similar types are constant. The brightness variations are accompanied by small or nonexistent color variations. The material at hand suggests that brightness variations are fairly common in late K- and early M-type dwarfs of age 3×10^7 yr, but essentially disappear at an age near 5×10^8 yr.

153.031 The chemical composition of stars as derived from photoelectric observations of narrow-band indices of equivalent widths. P. E. Nissen.
Proc. ESO/SRC/CERN conference, (see 012.021), p. 201 - 205 (1974).

153.032 UBV-Photometrie der offenen Sternenhaufen NGC 7419, K 10 und NGC 6192. G. Handschel.
Thesis Univ. München, 64 pp. (1973).

153.033 The open cluster NGC 6281.
A. Feinstein, J. C. Forte.
Publ. Astron. Soc. Pacific, Vol. 86, 284 - 288 (1974).
Stars in the field of NGC 6281, including the bright star HD 153919, have been observed photoelectrically in the UBV system. A distance of 560 ± 30 parsecs, a mean reddening of $E_{B-V} = 0.^m15 \pm 0.02$, and an estimated age of 10^8 years have been derived for cluster members. Possible light variations of the star HD 153919, recently identified as an X-ray source are noted.

153.034 The Hyades convergent point. S. V. M. Clube.
Observatory, Vol. 94, 126 - 133 (1974).

Parallaxes of faint dwarf members of the Hyades cluster. See Abstr. 111.001.

A new UBV comparison of M67 and Hyades stars.
See Abstr. 113.014.

The energy distribution of the very red star in NGC 6231. See Abstr. 114.030.

Spectrophotometry of stars of the Pleiades cluster. II. Absorption lines. See Abstr. 114.046.

Luminosity and velocity distribution of high-luminosity red stars. IV. The G-type giants.
See Abstr. 115.006.

The "Hyades anomaly" and the $uvby$ luminosity calibration. See Abstr. 115.011.

Spectroscopic observations of the Hyades spectroscopic binary BD + 24°692. See Abstr. 119.007.

Investigation of the Wolf-Rayet spectroscopic binary HD 152270; line identifications and radial velocities in the spectral region 3500–6000 Å.
See Abstr. 119.001.

Binary systems in star clusters. I. V701 Sco.
See Abstr. 121.017.

On the spatial distribution of flare stars in the Pleiades cluster. See Abstr. 122.006.

Flare stars in the Pleiades. IV.
See Abstr. 122.020.

UBV photoelectric photometry for flare stars in the Pleiades cluster. See Abstr. 122.075.

Interstellar polarization from a medium with changing grain alignment. See Abstr. 131.039.

Polarization observations in the Coma cluster Mel 111. See Abstr. 131.052.

The galactic orbit of the old open cluster NGC 2420.
See Abstr. 155.024.

Die Realität der Sterntrupps. See Abstr. 155.076.

Color-magnitude diagrams for eleven young clusters in the Magellanic Clouds. See Abstr. 159.001.

Errata

153.901 Erratum: 'A study of the open cluster NGC 1778' [Astron. Astrophys., Suppl. Ser., Vol. 10, 1 - 10 (1973)]. R. Barbon, S. M. Hassan.
Astron. Astrophys., Suppl. Ser., Vol. 13, 293 (1974).

154 Globular Clusters

154.001 The globular cluster NGC 2298. G. Alcaino. Astron. Astrophys., Suppl. Ser., Vol. 13, 55 - 63 (1974).

A UBV photometric investigation of the southern globular cluster NGC 2298 was carried out at the Cerro Tololo Inter-American Observatory using the 1.5-m reflector for the photoelectric and the photographic work. Fourteen stars were observed photoelectrically to apparent magnitude $V = 16.22$. With this sequence, 60 stars were calibrated photographically. The apparent distance modulus is $(m-M)_{app} = 15.70$. The reddening is $E(B-V) = 0.08$. The true distance modulus is $(m-M)_0 = 15.46$. The giant branch rises 3 magnitudes above the horizontal branch at $(B-V)_0 = 1.4$, characteristic of a metal poor object.

154.002 The globular cluster NGC 4372. G. Alcaino. Astron. Astrophys., Suppl. Ser., Vol. 13, 345 - 353 (1974).

A UBV photometric investigation of the southern globular cluster NGC 4372 was carried out using the 1-m telescope of Cerro La Silla (ESO) for the photoelectric work, and the 1.5-m telescope of Cerro Tololo (AURA) together with the 1-m telescope of Cerro Las Campanas (CARSO) for the photographic work. Fourteen stars were observed photoelectrically with a limiting apparent magnitude $V = 15^m.88$. Using this sequence 186 stars were measured photographically. The derived apparent distance modulus is $(m-M)_{app} = 15^m.1$; the reddening $E(B-V) = 0^m.37$. The true distance modulus is $(m-M)_0 = 13^m.99$. The giant branch rises $3^m.1$ above the horizontal branch at $(B-V)_0 = 1^m.4$, characteristic of a metal poor object.

154.003 The horizontal branch in globular clusters of the Galaxy and Magellanic Clouds and in Sculptor type dwarf galaxies. A. M. Ehjgenson (*Eigenson*). Astrofizika, Vol. 9, 589 - 593 (1973). In Russian. English translation in Astrophysics, Vol. 9, No. 4.

A preliminary comparison of the horizontal branch in globular clusters of the Galaxy with that of Magellanic Cloud clusters and Sculptor type dwarf galaxies is made. There is a higher percentage of younger globular clusters in the Clouds than in the Galaxy. A conclusion about the similarity of the stage of stellar evolution in globular clusters and in Sculptor type dwarfs is derived.

154.004 Analysis of composite spectra - the ultra-violet excess of stars in globular clusters. R. G. Bingham, W. L. Martin. Monthly Notices Roy. Astron. Soc., Vol. 167, 137 - 161 (1974).

Ultra-violet excesses which are nearly independent of interstellar reddening are estimated for 38 globular clusters. A model-fitting process is used. The data are details of the observed upper part of each individual $B - V$, V array and the integrated $U - B$ colours. By noting the effects of changes in the data, the authors assess errors in a way which may be applicable in other work on the integrated colours of clusters and galaxies. Inconsistencies in the available data are found.

154.005 UBV photometry of 26 globular clusters. G. V. Zajtseva, V. M. Lyutyj, B. V. Kukarkin. Astron. Zhurn. Akad. Nauk SSSR, Vol. 51, 438 - 440 (1974). In Russian. English translation in Soviet Astron. AJ, Vol. 18, No. 2.

The results of measurements of 26 globular clusters in the UBV system are published. The clusters NGC 6366, 6535 and 6539 have relatively rich metal abundances.

154.006 On the metal abundance in ω Centauri. K.-H. Schmidt, S. van den Bergh. Astron. Nachr., Vol. 295, 101 - 103 (1974). In German.

Radiation pressure during the high-luminosity collapse phase of cluster evolution may, under certain circumstances, lead to the ejection of interstellar grains. For the most luminous clusters such ejection might produce significant depletion of heavy elements. It is suggested that the metal abundance dispersion that has recently been detected among the giant stars in ω Centauri might be accounted for by such radiation pressure induced heavy element depletion.

154.007 The integrated colors of globular clusters. W. E. Harris, S. van den Bergh. Astron. Journ., Vol. 79, 31 - 33 (1974).

This paper presents new observations of the integrated UBV colors of 29 globular clusters. These data are combined with previous observations to yield an up-to-date catalog of all published UBV colors of galactic globular clusters. Observations of integrated magnitudes through various aperture sizes are also given for 11 clusters.

154.008 The colour-magnitude diagram of the globular cluster M4. V. G. Moshkalev. Astron. Tsirk., No. 806, p. 3 - 4 (1974). In Russian.

154.009 Red variable stars as indicators of distances of globular clusters. B. V. Kukarkin, N. N. Kireeva. Astron. Tsirk., No. 808, p. 7 - 8 (1974). In Russian.

154.010 UBVI colours of globular clusters and metal abundance. B. V. Kukarkin, N. N. Kireeva. Astron. Tsirk., No. 811, p. 4 - 5 (1974). In Russian.

154.011 The mass of the globular cluster NGC 6388. G. Illingworth, K. C. Freeman. Astrophys. Journ., (*Letters*), Vol. 188, L83 - L86 (1974).

The central velocity dispersion has been derived for the globular cluster NGC 6388 from high-dispersion coudé spectra of the integrated light. A new method using Fourier techniques for accurately determining the velocity dispersion from integrated light spectra is described. The mass was obtained from King's models, using photoelectric surface photometry and star counts to define the surface density distribution. The resulting mass is $(1.3 \pm 0.3) \times 10^6 M_\odot$, and the mass-luminosity ratio $(M/L_V) = 1.7 \pm 0.4$ solar units.

154.012 The color-magnitude diagram of M14. C. Smith Kogon, A. Wehlau, S. Demers. Astron. Journ., Vol. 79, 387 - 390, 423 (1974).

B and V magnitudes of about 1000 stars in the globular cluster M14 have been measured. The data have been used to obtain a color-magnitude diagram which extends to slightly fainter than the RR Lyrae stars. Attention is drawn to the lack of horizontal branch stars to the red of the variable star gap such as would be expected for this cluster. On the basis of colors observed for RR Lyrae stars the reddening is estimated at 0.35, considerably lower than that given by other authors. Such a value for the reddening leads to an increased distance modulus of $(m - M)_0 = 15.7 \pm 0.6$.

154.013 Near-infrared photometry of globular clusters - II. The metal-rich cluster 47 Tucanae (NGC 104). T. Lloyd Evans. Monthly Notices Roy. Astron. Soc., Vol. 167, 393 - 411 (1974).

Photographic photometry on the VI system is presented for a complete sample of 185 stars brighter than $I = 13.0$ in an annulus between $3'$ and $8'$ from the centre of 47 Tucanae. Photometry is also given for 47 red giant stars nearer or further from the centre and for most of the 23 red variables now known. New spectral types are given for 11 stars near the red giant tip.

154.014 **The turn-off point of Omega Centauri.**
R. D. Cannon, M. Kontizas.
Monthly Notices Roy. Astron. Soc., Vol. 167, 51P - 56P (1974).

Photoelectric and photographic photometry to below $V = 19$ in the globular cluster ω Cen indicate that the main sequence turn-off point occurs at about $V = 18.3$, nearly four magnitudes below the horizontal branch. ω Cen thus differs from most other clusters which have been studied, since this magnitude difference is usually near 3.4 mag. Some implications are discussed briefly.

154.015 **A C-M diagram for M80 obtained with a new photographic calibration technique.**
W. E. Harris, R. Racine.
Astron. Journ., Vol. 79, 472 - 479, 511 - 513 (1974).

New photographic photometry in B and V to a limiting magnitude $V \simeq 17.8$ is presented for the southern globular cluster Messier 80 (NGC 6093), the cluster which was the seat of the nova T Sco of 1860. The color-magnitude diagram is given and discussed. The cluster distance modulus is used to redetermine the luminosity of the nova T Sco. In an Appendix the authors describe a new photographic calibration technique used for this work.

154.016 **M13: bolhoop in Hercules.** M. Drummen.
Zenit, Vol. 1, No. 5, p. 10 - 12 (1974).

154.017 **Photometry of southern globular clusters – III. Bright stars in 47 Tucanae, NGC 6397 and NGC 288.**
R. D. Cannon.
Monthly Notices Roy. Astron. Soc., Vol. 167, 551 - 580 (1974).

New photoelectric UBV data are presented for about 60 stars down to $V = 16$ in 47 Tuc, 80 stars to $V = 16.5$ in NGC 6397 and 50 stars to $V = 17.5$ in NGC 288. A comparison is made between these three clusters and two studied previously, ω Cen and NGC 6752, and four northern clusters studied extensively by Sandage. The interstellar reddenings, distances, ultraviolet excesses and other parameters are derived.

154.018 **Distribution of orbital eccentricities of the globular clusters.** C. J. Peterson.
Astrophys. Journ., (Letters), Vol. 190, L17 - L20 (1974).

Under the assumption that the size of a globular cluster is limited by the tidal force field of the Galaxy, the author has computed the perigalacticon distances for 41 clusters. These new data have been combined with the observed radial velocities and positions in a simple model to estimate the distribution of orbital eccentricities.

154.019 **Interstellar reddening and globular clusters.**
B. V. Kukarkin, N. N. Kireeva.
Astron. Zhurn. Akad. Nauk SSSR, Vol. 51, 588 - 592 (1974). In Russian. English translation in Soviet Astron. AJ, Vol. 18, No. 3.

A method of determination of intrinsic colours and interstellar reddening of globular clusters from photometrical observations in $UBVI$ bands is presented.

154.020 **Infrared observations of stars in the field of NGC 7099.** V. Castellani, R. De Amicis, F. Smriglio.
Acta Astron., Vol. 24, 235 - 242 (1974).

Iris measurements of infrared plates are reported for 66 stars in the field of the globular cluster NGC 7099. By combining the results with the UBV photometry done by Dickens, it is suggested that I-photometry can give useful information about the membership of the cluster. The general behaviour of I-luminosity is also discussed for cluster stars with V-magnitudes lower than $V = 15.5$.

154.021 **The UV-bright stars in ω Centauri.** J. Norris.
Bull. American Astron. Soc., Vol. 6, 216 (1974). Abstr. AAS.

154.022 **Photometry to the main-sequence of 47 Tucanae.**
F. D. A. Hartwick, J. E. Hesser.
Bull. American Astron. Soc., Vol. 6, 216 (1974). – Abstr. AAS.

154.023 **Southern globular clusters.** R. J. Dickens.
Proc. ESO/SRC/CERN conference, (see 012.021), p. 71 - 78 (1974).

154.024 **Young globular clusters in the Magellanic Clouds.**
K. C. Freeman.
Proc. ESO/SRC/CERN conference, (see 012.021), p. 177 - 187 (1974).

154.025 **The main sequence turn-off points of southern globular clusters.** R. D. Cannon.
Proc. ESO/SRC/CERN conference, (see 012.021), p. 207 - 208 (1974).

154.026 **A survey of star clusters in the Magellanic Clouds.**
G. Alcaino.
Proc. ESO/SRC/CERN conference, (see 012.021), p. 209 - 210 (1974).

154.027 **M 13: bolhoop in Hercules, II.** M. Drummen.
Zenit, Vol. 1, No. 6, p. 10 - 12 (1974).

154.028 **Photometry of the globular clusters of NGC 185.**
P. W. Hodge.
Publ. Astron. Soc. Pacific, Vol. 86, 289 - 293 (1974).

Photometry in the UBV system for five recognized globular clusters in NGC 185 shows that their colors are normal and the luminosities are low. A brief rediscussion of the use of globular clusters as extragalactic distance indicators is based on the recent photometry of globular clusters in the M31 group.

154.029 **BV photometry of globular clusters in M87.**
H. D. Ables, E. B. Newell, E. J. O'Neil, Jr.
Publ. Astron. Soc. Pacific, Vol. 86, 311 - 316 (1974).

The authors present the first results of an electrographic study of extragalactic globular clusters. B and V magnitudes have been measured for several faint objects (78 in V and 37 in B) around M87 to test the reliability of the photographic magnitudes and colors measured by Racine and Hanes.

154.030 **The colour-magnitude diagram of the globular cluster NGC 6934.** A. V. Mironov.
Astron. Tsirk., No. 817, p. 5 - 7 (1974). In Russian.

The CNO abundances of the CH stars in Omega Centauri. See Abstr. 064.005.

Infrared photometry of red giant stars in the four globular clusters M 13, M 71, M 92 and M 5. I. See Abstr. 113.007.

A new UBV comparison of M67 and Hyades stars. See Abstr. 113.014.

Cyanogen-band strengths of giant stars in 47 Tucanae.
See Abstr. 114.102.

A possible theoretical explanation of the erratic period changes of the RR Lyrae variables in globular clusters.
See Abstr. 122.027.

Variable stars in the globular cluster M13.
See Abstr. 122.031.

The RR Lyrae stars in the Large Magellanic Cloud cluster NGC 1835.   See Abstr. 122.060.

A short-period cepheid variable in the globular cluster NGC 6752.   See Abstr. 122.085.

New observations of some RR Lyrae variables in the globular cluster M3 (NGC 5272).   See Abstr. 122.093.

Mira variables in four metal-rich globular clusters.
See Abstr. 122.134.

Mira variables in globular clusters.
See Abstr. 122.141.

Interstellar reddening and globular clusters.
See Abstr. 131.045.

Galaxy-quasar associations and globular clusters.
See Abstr. 151.054.

Integrated colors of star clusters in the Large Magellanic Cloud.   See Abstr. 153.025.

The mass of M31 as determined from the motions of its globular clusters.   See Abstr. 158.101.

155 Structure and Evolution of the Galaxy

155.001 Observation of gamma-radiation from the galactic center region.
G. H. Share, R. L. Kinzer, N. Seeman.
Astrophys. Journ., Vol. 187, 45 - 56 (1974).

The authors report results of an investigation of the galactic center region with a detector sensitive to γ-rays above about 10 MeV and having an angular resolution of about $1^1/_2°$ above 15 MeV. These measurements confirm the emission of γ-radiation from along the galactic equator and, furthermore, indicate that the radiation is concentrated in a narrow band, about 3° wide, in the vicinity of the galactic center. A composite energy spectrum of this emission is derived from this low-energy observation and from the observations above 100 MeV. The authors also discuss their investigation of the various suspected point sources reported in the vicinity of the galactic center.

155.002 Evolution of the nearby stellar population and its kinematics. P. Biermann, B. M. Tinsley.
Astron. Astrophys., Vol. 30, 1-12 (1974).

Following the discovery of a great number of low-velocity M dwarfs (Weistrop, 1972 and others), Biermann (1973) formulated the idea that the velocity dispersions of stars are a property largely given to the stars at formation and modified over galactic time-scales. Using these ideas and the formalism of Tinsley (1972), a quantitative model of the stellar population in the solar neighborhood is investigated which includes the evolution of stellar kinematics. The model is able to account for all major properties of the stellar disk population (halo stars excluded). The tentative conclusion is drawn that the spiral arm inclination and/or shock strength was considerably greater in the past history of the Galaxy.

155.003 On the nature of the nonthermal radio emission from the galactic center. T. W. Jones.
Astron. Astrophys., Vol. 30, 37 - 40 (1974).

It has been shown, when the most recent observations at low and high frequencies are included, and the multiple nature of Sgr A are accounted for, the principal source has a definitely nonthermal shape with spectral index $\alpha \simeq 0.4$-0.5. It is further argued that, on the basis of its general appearance, energetics and recent observations which no longer put Sgr A exactly at the galactic center, Sgr A could be a supernova remnant near the galactic center rather than a miniature of energetic extragalactic sources.

155.004 Molecular clouds in the galactic nucleus.
N. Z. Scoville, P. M. Solomon, K. B. Jefferts.
Astrophys. Journ., (Letters), Vol. 187, L63 - L66 (1974).

Observations of the 2.6-mm CO line have been obtained along the galactic equator covering longitudes $|l| \leq 3°$ and velocities $|V| \leq 300$ km s^{-1}. The similarities in distribution and brightness temperature between the CO and 100-μ emissions suggest that the grains and the gas are colocated and nearly in thermal equilibrium. Kinematic models for the gas in the nucleus are briefly discussed.

155.005 A search for possible stellar members of the α-spiral arm. R. J. Dodd.
Astrophys. Space Sci., Vol. 26, 513 - 519 (1974).

The proper motions of groups of stars, selected according to magnitude and colour, near $(l, b) = (140°, 0°)$, were analysed in an attempt to identify possible members of the α-spiral arm defined by Verschuur from HI observations. Other than late type giants the only objects whose distance corresponded to that of this arm were a small group of reddened early type stars.

155.006 Das Produkt AR_0 und weitere galaktische Parameter.
W. Lohmann.
Astrophys. Space Sci., Vol. 27, 227 - 231 (1974) = Mitt. Astron. Rechen-Inst. Heidelberg, Ser. A.

According to the tangential method the product AR_0 is determined with 145.7 km s^{-1} from measurements of the line profiles of the 21-cm line of the neutral hydrogen by Weaver and Williams (1973). The recent individual measurements of Oort's constant A and of the distance R_0 of the sun from the galactic centre yields 138.5 km s^{-1}. The mean value 142.1 km s^{-1} leads to $A = 14.56$ km s^{-1}kpc^{-1} and $R_0 = 9.76$ kpc. At the galactocentric distance R near R_0 the angular velocity is represented by $\omega(R) = 25.84 - 2.98 (R - 9.76) + 0.075 (R - 9.76)^2$. The mass of the Galaxy amounts to $1.92 \times 10^{11} M_\odot$.

155.007 Gaseous structure in the galactic center region.
N. Kaifu, T. Iguchi, T. Kato.
Publ. Astron. Soc. Japan, Vol. 26, 117 - 128 (1974).

The existence of an almost complete ring of neutral hydrogen similar to the 270-pc expanding ring determined from OH and H$_2$CO data is indicated. The total mass of the expanding ring is estimated to be $1 \times 10^7 M_\odot$ and the kinetic energy of expansion to be 4×10^{54} erg. The interpretation based on the shock wave model of the expanding ring is briefly discussed. Discussions of the infrared and nonthermal radio emissions suggest that these emissions may be radiated from the dense region of the expanding ring.

155.008 A high-resolution map of the galactic-center region.
J. E. Kapitzky, W. A. Dent.
Astrophys. Journ., Vol. 188, 27 - 32 = Contr. Five College Obs., Univ. Mass., Amherst, No. 166 (1974).

The galactic-center region has been mapped at 15.5 GHz with an effective resolution of 2$''$25. The greater sensitivity and resolution of this map over previous ones reveals that the region is extremely complex consisting of a large number of densely clustered discrete sources.

155.009 Tentative identification of main-sequence stars in the nuclear bulge of the Galaxy.
S. van den Bergh.
Astrophys. Journ., (Letters), Vol. 188, L9 - L10 (1974).

The main-sequence population in the nuclear bulge of the Galaxy has been tentatively identified at $V > 19.5 \pm 0.3$ in a field at $l = 0°$, $b = -8°$. From these observations it is concluded that the maximum contribution to the star counts in this direction is provided by stars at a true distance modulus $(m-M)_0 = 14.8 \pm 0.5$ $(R = 9.0 [+2.1, -1.8]$ kpc).

155.010 Détection de nouvelles formations d'hydrogène ionisé, de grandes dimensions angulaires, dans le bras local et dans le bras du Sagittaire de notre Galaxie.
J.-P. Sivan.
Comptes Rendus Acad. Sci. Paris, Sér. B, Vol. 278, 127 - 129 (1974).

Le bilan aussi complet que possible de l'émission Hα galactique a été fait au moyen de clichés de 60° de champ. Cette note présente les plus spectaculaires et les plus riches en informations nouvelles: émission diffuse générale dans le bras du Sagittaire, nouvelles régions H II très étendues dans ce bras et dans le bras local, extension du complexe d'Orion entre les latitudes galactiques 0 et $-55°$.

155.011 The corrugation of the galactic layer.
R. J. Quiroga.
Astrophys. Space Sci., Vol. 27, 323 - 342 (1974).

The 21-cm line intensities in a (Z, R) distribution is

studied at the locus of tangential points of the inner parts of the Galaxy using both northern and southern data. A corrugation effect is observed in the galactic neutral hydrogen layer with an average wave length of 2 kpc and a wave amplitude of 70 pc. The patterns obtained for the I and IV quadrant indicate that the inner and the outer parts of the spiral arms are located, respectively, below and above the galactic plane. Also, with high angular resolution the corrugation pattern suggests the existence of 'faults' in a geological sense in the inner arm zones.

155.012 **The kinematics of galactic spiral structure.**
W. B. Burton.
Publ. Astron. Soc. Pacific, Vol. 85, 679 - 703 (1973). – Invited paper read at the 139th meeting, American Astronomical Society, Las Cruces, New Mexico, 10 January 1973.

The emphasis in the first part of the present article is on the observable kinematic characteristics of the neutral hydrogen in the Milky Way. In the second part some results obtained for other galaxies are discussed.

155.013 **Statistical principles of galactic optical astronomy. Part II.** H. Eelsalu.
Akademiya Nauk Ehstonskoj SSR. Tartuskaya Astrofizicheskaya Observatoriya im. W. Struve, 88 pp. = Tartu Astron. Obs., Teated No. 45 (1974). In Russian.

155.014 **On the space density of faint M-stars in the vicinity of the sun.** M. Jôeveer.
Tartu Astr. Obs., Teated No. 46, p. 3 - 17 (1974). In Russian.

The mean value and the dispersion of the distribution of the absolute magnitudes for the faint M-stars discovered by Sanduleak are estimated. Contrary to Murray and Sanduleak it is concluded that the space density of the faint M-stars is too low for explaining the phenomenon of missing mass near the sun.

155.015 **Structure of the subsystem of long-period cepheids in the galactovertical direction.** M. Jôeveer.
Tartu Astr. Obs., Teated No. 46, p. 18 - 34 (1974). In Russian.

The space distribution and motions in the z-direction of the classical population I cepheids are studied. For the local galactic dynamical parameter C an estimate C = 80 km/s/kpc is obtained, the corresponding value of the galactic mass density near the sun is $0.115\,M_\odot/\mathrm{pc}^3$.

155.016 **Ages of δ Cephei stars and the density of gravitating masses near the sun.** M. Jôeveer.
Tartu Astr. Obs., Teated No. 46, p. 35 - 49 (1974). In Russian.

The dependence of the dispersion σ_z of the galactovertical distances of δ Cephei stars on the ages of the stars is discussed. The value $0.05 - 0.075\,M_\odot/\mathrm{pc}^3$ for the galactic mass density in the vicinity of the sun has been obtained.

155.017 **The space density of A stars in a region in Cassiopeia.** S. W. McCuskey.
Astron. Journ., Vol. 79, 107 - 115, 339 - 343 (1974).

New objective prism spectral types and UBV photometric data for B8–A3 stars, and a few others, with $V < 12$ mag are given for an area of 21.3 sq deg centered at R. A. $2^h 14^m$; Dec. $+60°$ (1950). McCuskey and Houk (1971) found in this region a high space density of these A stars extending from $r = 250$ to 650 pc. An analysis of this new data indicates that the high space density previously found was due to uncertainty in the variation of the interstellar absorption with distance, which had been derived from photometric data distributed over considerable ranges in galactic longitude and latitude.

155.018 **The space distribution of M giants in the Warner and Swasey luminosity function field LF 15.**
P. S. Thé, R. F. A. Staller, E. J. A. Meurs.
Astron. Astrophys., Suppl. Ser., Vol. 15, 141 - 171 (1974).

A rectangular field of $3° \times 2°$ centered on the open cluster NGC 6067 in the Warner and Swasey luminosity function field LF 15 has been surveyed for M-type stars. The survey goes to a limiting infrared magnitude of about 12.5. The mean infrared absorption in the field was studied using data of M2, M3 and M4 stars and those published by Drilling (1968). Space densities have been computed for the M2–M4, M5–M6.5 and M7–M9 groups using Schalen's method. The first group seems to be more connected to the Sagittarius arm, as delineated by optical spiral tracers, then the other two groups.

155.019 **The possible relation of the 3-kiloparsec arm to explosions in the galactic nucleus.**
R. H. Sanders, K. H. Prendergast.
Astrophys. Journ., Vol. 188, 489 - 500 (1974).

One- and two-dimensional hydrodynamical calculations have been carried out in order to simulate the effects of an energetic explosion in the center of the Galaxy. In both cases the galactic gravitational field, differential rotation, and radiative cooling were approximately simulated. The essential result is that a radially oscillating ring of cold gas is produced in galactic plane and that this oscillation persists for some time longer than the initial expansion time of the hot gas. The results are discussed with regard to their possible relevance to the observed 3-kpc arm.

155.020 **Infrared polarization of the galactic nucleus.**
H. M. Dyck, R. W. Capps, C. A. Beichman.
Astrophys. Journ., (Letters), Vol. 188, L103 - L104 (1974).

The authors have detected 2.4 ± 0.4 percent linear polarization at $11\,\mu$ in the direction of the galactic center. The electric vector lies approximately perpendicular to the plane of the Galaxy.

155.021 **An A-star concentration in Cassiopeia.**
P. Pesch, S. W. McCuskey.
Astron. Journ., Vol. 79, 116 - 123 (1974).

A study by $uvby$, Hβ photometry, MK spectral types, and radial-velocity data for 99 A stars at $l = 133°, b = -1°$ in Cassiopeia shows that: (a) the interstellar absorption $A_v \sim 1.3$ mag over $r = 200-500$ pc; (b) the frequency distribution with distance of 86 A stars peaks at $r = 250$ pc and then declines to $r = 500$ pc; (c) the A stars have a group motion relative to the LSR similar to the motion of the sun in the (U, V) plane, but depart considerably from the solar motion in the W component; and (d) the space velocity distribution for A stars not members of the cluster Stock 2 differs considerably from that found by Schlesinger (1966) for B9–A5 stars.

155.022 **The oldest disk stars.** O. J. Eggen.
Publ. Astron. Soc. Pacific, Vol. 86, 162 - 172 (1974).

A summary of some photometric and kinematic data for the oldest disk stars is presented together with a detailed discussion of one of the oldest disk-population groups; the Arcturus group. The slope of the lower main sequence of the old-disk population appears to be a function of the eccentricity of the star's galactic orbit and this phenomenon, together with the apparent lack of stars intermediate between the Hyades and subdwarf sequences, requires understanding before moduli of many old clusters can be regarded as reliable.

155.023 **Turbulence-enhanced synchrotron radiation in the Galaxy.** R. Cowsik, J. Mitteldorf.
Astrophys. Journ., Vol. 189, 51 - 53 (1974).

The intensity of synchrotron radiation from the galactic disk depends both on the magnetic field strength and on the density of the cosmic-ray electrons. Evidence based on Faraday rotation of pulsar signals indicates that the average galactic magnetic field B_0 is ~3 microgauss with fluctuations ΔB of the same order of magnitude. In this field cosmic-ray electrons with the same density as that observed near the solar system produce insufficient synchrotron radiation to account for the galactic radio background. If, however, the magnetic field fluctuations arise because of compressions of the average fields by motions of the interstellar gas, then betatron processes induce a correlated increase of the relativistic electrons in the high field regions. The average synchrotron emissivity nonlinearly increases to make the estimated emissivities compatible with radio-astronomical data.

155.024 **The galactic orbit of the old open cluster NGC 2420.** D. W. Keenan, K. A. Innanen.
Astrophys. Journ., Vol. 189, 205 - 207 (1974).

The galactic orbit of NGC 2420 has been computed in three different mass models of the Galaxy. Among other possible explanations, its eccentric orbit and age may possibly be reconciled as the result of a gravitational perturbation by an encounter with one of the Magellanic clouds.

155.025 **Sensitivity of the star formation rate to the interstellar gas abundance of heavy elements.**
R. J. Talbot, Jr.
Astrophys. Journ., Vol. 189, 209 - 215 (1974).

The critical pressure required to produce the thermal instability in the interstellar gas is approximately inversely proportional to the heavy-element abundance. A model for the rate of star formation is presented which is based upon this relationship plus the assumption that star formation accompanies the formation of interstellar gas clouds of high density. The model is shown to agree with the following observations: the mass fraction remaining in the form of interstellar gas; the U, Th, Re, Os, Pu, and I radioactive chronologies; the metal abundance ot disk stars formed throughout the Galaxy's history, including spatial variations at a given epoch; the efficiency of star formation; and both a mass-to-light ratio and a supernova rate which are compatible with those of the Galaxy.

155.026 **Galactic arm structure and gamma-ray astronomy.**
G. F. Bignami, C. E. Fichtel.
Astrophys. Journ., (Letters), Vol. 189, L65 - L67 (1974).

In an attempt to explain the observed unexpectedly high-energy gamma-radiation over a broad region of the galactic plane in the general direction of the galactic center, a model is proposed wherein the galactic cosmic rays are preferentially located in the high-matter-density regions of galactic arm segments, as a result of the weight of the matter in these arms tying the magnetic fields and hence the cosmic rays to these regions.

155.027 **The center of the Galaxy.**
R. H. Sanders, G. T. Wrixon.
Sci. American, Vol. 230, No. 4, p. 66 - 74, 76 - 77 (1974).

Coded in the radio, infrared and X-ray emissions from the invisible nucleus of our galaxy is mounting evidence that it is periodically the scene of titanic explosions.

155.028 **Kinematical parameters for main sequence and giant type III stars.** A. E. Gómez.
Astron. Nachr., Vol. 295, 133 - 139 (1974).

The parameters of solar motion and velocity ellipsoid are derived from radial velocities of spectroscopically well defined samples of main sequence and of giant type III stars. A short discussion of the statistical method applied is given. The results indicate that the parameters of the velocity ellipsoid chosen arbitrarily for a group of stars do not always represent correctly the sample observed.

155.029 **Flächenpolarimetrie der südlichen Milchstraße.**
J. Staude, K. Wolf, T. Schmidt.
Mitt. Astron. Ges., No. 34, p. 95 - 98 (1973/74). – Presented at the "Wissenschaftliche Tagung der Astron. Ges., Oberkochen, 1973 April 24 - 27".

155.030 **Die Häufigkeit roter Zwergsterne in der Umgebung des galaktischen Südpols.** W. Gliese.
Mitt. Astron. Ges., No. 34, p. 100 - 101 (1973/74).
Presented at the "Wissenschaftliche Tagung der Astron. Ges., Oberkochen, 1973 April 24 - 27".

155.031 **Kinematics and age of stars.** M. Mayor.
Astron. Astrophys., Vol. 32, 321 - 327 (1974).

From a sample of 1010 A and F stars with $uvby$ photometry, the author has computed the space velocity, age and metallicity, and has studied the relationship between kinematics and age. The kinematic-age relations obtained are used to estimate the variation of calcium re-emission with age.

155.032 **Heteroclinic stellar orbits and "wild" behaviour in our Galaxy.** L. Martinet.
Astron. Astrophys., Vol. 32, 329 - 333 (1974).

The eigencurves are constructed through some hyperbolic fixed points of the phase plane $(\varpi, \dot{\varpi})$ for different values of the energy E in the axisymmetric models of the Schmidt galactic potentials. The existence of heteroclinic stellar orbits is proved for small values of the angular momentum J and the author is able to explain the appearance of the "wild" (semi-ergodic) region which arises in the $(\varpi, \dot{\varpi})$ plane, by observing how the network of eigencurves related to the unstable 3-, 5- and 7-periodic orbits evolves as the energy E increases from 130000 km^2 s^{-2} to 115000 km^2 s^{-2}. Implications for the orbit computation in various models of the galactic potential are indicated.

155.033 **The local complex of O and B stars. I. Distribution of stars and interstellar dust.**
R. Stothers, J. A. Frogel.
Astron. Journ., Vol. 79, 456 - 471 (1974).

The O-B5 stars, supergiants, young clusters, and associations within 1 kpc of the sun populate two flat systems inclined to each other by 19°–22°. The present paper discusses the historical background, statistical significance, composition, spatial arrangement of the contents, and interstellar extinction in the two belts.

155.034 **An upper limit to the mass loss from the centre of the Galaxy.** A. Poveda, C. Allen.
IAU Symposium No. 64, (see 012.008), p. 36 (1974). – Abstract.

155.035 **Isotopic abundances and their variations within the Galaxy.** M. Bertojo, M. F. Chui, C. H. Townes.
Science, Vol. 184, 619 - 623 (1974).

Information on isotopic abundances in interstellar material from microwave molecular spectra, along with a small amount of information from other spectral regions, is assembled and analyzed to obtain as clear a view of these abundances as is presently practical.

155.036 **A galactic model with a pulsating active nucleus. II. Chemical evolution.** T. Ohnishi.
Astrophys. Space Sci., Vol. 28, 325 - 350 (1974).

The chemical evolution of the Galaxy with a pulsating active nucleus is investigated. The surface densities of gas, stellar remnants, stars and chemical species such as helium and heavy elements ($Z \geq 6$) are calculated as functions of the position in the Galaxy and of the evolutional time of the Galaxy. Using this model one can account for the observed phenomena such as the smooth dependence of the elemental abundance in the halo population on the distance from the galactic center, the high abundance of heavy elements in quasar spectra, etc.

155.037 **Ces rayons gamma qui arrivent du centre de la Galaxie.** G. Vedrenne.
Sci. Progrès Découverte, No. 3443, p. 33 - 41 (1972).

155.038 **Distribution of giant M and carbon stars in a region in Puppis ($l = 246°$, $b = -0.°6$).**
K. A. Kirton, M. P. FitzGerald.
Journ. Roy. Astron. Soc. Canada, Vol. 68, 154 - 162 = Contr. Univ. Waterloo Obs., No. 34 (1974).

A 12.19 square degree region in Puppis has been surveyed for giant M stars and carbon stars using objective prism spectra obtained in the blue and infrared. For 333 M stars and 37 carbon stars, and an additional 24 stars in an adjacent region, spectra and V and I photographic magnitudes were obtained to a limiting magnitude of I = 11.7. The stars (classified on the Case system) decrease in number with distance from the sun.

155.039 **The galactic structure in Cepheus near NGC 7235.**
W. R. Kubinec.
Publ. Warner and Swasey Obs., Cleveland, Ohio, Vol. 1, No. 3, 41 + A13 + B7 + C7 + D12 + E3 pp. (1973).

The Cepheus region of the Milky Way is very complex. A region centered at $\alpha = 22^h 16^m$, $\delta = +56°30'$ (1950); $l = 103°$, $b = 0°$, and covering the 8 sq deg region bounded by $102° \leq l \leq 104°$, $b = \pm 2°$ has been selected for detailed study. This region contains the young galactic cluster NGC 7235 and the H II region S132 (Sharpless 1959). Both of these objects are supposedly associated with the Perseus spiral arm. A catalogue and identification charts of 1500 stars in Cepheus near NGC 7235; tabulation of pertinent data for the OB stars in Cepheus; carbon- and M-stars; an analysis of the spectral groups B2–B3, B4–B6, G5–G6, and G7–K2 are given in appendices.

155.040 **Some results of measurements of the galactic centre region at mm- and m-wavelengths.**
V. I. Ariskin, I. I. Berulis, V. N. Brezgunov, R. D. Dagkesamanskij, R. L. Sorochenko, V. A. Udal'tsov.
Izv. vyssh. ucheb. zavedenij. Radiofizika, Vol. 16, 1334 - 1341 (1973). In Russian. – Abstr. in Referativ. Zhurn. 51. Astron., 5.51.696 (1974).

155.041 **The linear polarization of part of the northern sky at 49-cm wavelength.**
J. R. Baker, A. Wilkinson.
Monthly Notices Roy. Astron. Soc., Vol. 167, 581 - 592 (1974).

A survey of the linearly polarized component of galactic background radiation at 49-cm wavelength is presented for the area $23^h < \alpha(1950.0) < 08^h$, $20° < \delta(1950.0) < 90°$. The results have been convolved to a resolution of $1.°3$ for comparison with the Leiden 49-cm polarization survey, and in this form the Jodrell Bank survey is illustrated as a vector map and as contours of polarized brightness temperature.

155.042 **A new determination from OB stars of the galactic rotation constants and the distance to the galactic centre.** L. A. Balona, M. W. Feast.
Monthly Notices Roy. Astron. Soc., Vol. 167, 621 - 634 (1974).

The overall kinematics of the OB stars has been rediscussed using a new calibration of Hγ equivalent widths and MK luminosity classes by Balona & Crampton. As well as published material the authors have used data from new Hγ equivalent width measurements in 590 southern OB stars. The principal results are $A = 16.8$ km s^{-1} kpc^{-1}, $u_0 = +7.4$ km s^{-1} and $v_0 = +13.0$ km s^{-1}. A distance, $R_\odot$, to the galactic centre of 9.0 kpc with a range of about 7.7–10.9 kpc is found.

155.043 **Rising and falling motion of gas above the spiral arm.** M. Tosa, Y. Sofue.
Astron. Astrophys., Vol. 32, 461 - 464 (1974).

Evidence is given for the existence of rising and falling motions of the neutral hydrogen gas in the space above the spiral arm. Systematic variations of the radial velocity of gas with galactic latitudes are examined in detail with special regard to the Perseus arm. They are interpreted in terms of vertical motion of gas ejected out of the spiral arm in the z-direction and falling freely toward the galactic plane. The ejection velocity is estimated to about 90 km/s.

155.044 **A new look at the interstellar hydrogen through a very-wide-field photographic Hα survey of the whole Milky Way.** J. P. Sivan.
Astron. Astrophys., Suppl. Ser., Vol. 16, 163 - 172 (1974).

In order to extend the limit of detection of the general Hα emission in the galactic plane, a photographic survey was carried out with a new, 60°-field, high-luminosity camera using a very narrow interference filter centered on Hα. A mosaic of fourteen wide-field Hα plates is presented. This atlas sums up, as completely as possible, the Hα emission features for the entire Milky Way. The survey reveals a number of new large-scale regions of faint emission. They are catalogued and described. The survey also reveals unsuspected filamentary extensions of the Orion complex.

155.045 **The space distribution and radial velocities of some early-type stars in the Perseus spiral arm.**
S. W. McCuskey, P. Pesch, G. A. Snyder.
Astron. Journ., Vol. 79, 597 - 602 (1974).

Distances for 73 early-type stars and radial velocities for 43 of them have been determined in the region LF5 ($l = 129°$, $b = -2°$). Most of these stars are of luminosity classes III and V, and are shown to be at distances such that they fall within the stellar grouping commonly called the Perseus spiral arm. The results of an analysis of these data are presented.

155.046 **Photometry of stars in the nuclear bulge of the Galaxy through a low-absorption window at $l = 0°$, $b = -8°$.** S. van den Bergh, E. Herbst.
Astron. Journ., Vol. 79, 603 - 615, 663 - 665 (1974).

A color-magnitude diagram down to $V \sim 20$ has been obtained in a low-absorption window at $l = 0°$, $b = -8°$. This diagram is discussed in detail. Photoelectric observations of distant RR Lyrae variables yields $E_{B-V} = 0.25 \pm 0.05$. With this reddening value, the distance to the nuclear bulge becomes 9.2 ± 2.2 kpc.

155.047 **A redetermination of the galactic H I half-thickness and a discussion of some dynamical consequences.**
P. D. Jackson, S. A. Kellman.
Astrophys. Journ., Vol. 190, 53 - 58 (1974).

The authors have redetermined the half-thickness, $z_{1/2}$, of the neutral hydrogen layer of our Galaxy over a significant portion of the disk (1.4 kpc $\leq R \leq$ 16.8 kpc). Using this thickness in conjunction with a recent galactic mass model, they are able to determine the quantity Q over a large portion of the galactic disk, where Q^2 is the ratio at the galactic plane of the sum of the gas, magnetic, and cosmic-ray pressures to the gas density.

155.048 **Studies of neutral hydrogen cloud structure in the vicinity of the North Polar Spur.** G. L. Verschuur.
Astrophys. Journ., Suppl. Ser., No. 245, Vol. 27, 283 - 306 (1974).

The H I cloud structure in two directions on or near the top of the North Polar Spur have been studied. It is impossible to recognize discrete clouds and derive their parameters in the way done in a previous paper. Instead the clouds appear to show elongations parallel to the magnetic field lines thought to be associated with the spur.

155.049 The far ultraviolet background radiation from +70° to −70° galactic latitude.
C. S. Weller, G. R. Carruthers, R. C. Henry.
Astron. Inst. Utrecht, Internal Rep. ROF 72, (see 012.012), p. C24.1 (1974). − Abstract.

155.050 A galactic model with a pulsating active nucleus. III: Evolution of rare-light elements.
T. Ohnishi.
Astrophys. Space Sci., Vol. 29, 3 - 16 (1974).

The accumulation and distribution of rare-light elements in the Galaxy is investigated according to a model of the Galaxy at the center of which exists a pulsating active nucleus with decreasing activity with time.

155.051 Determination of the kinematical parameters for stars in the vicinity of the sun ($r < 25$ pc).
D. K. Karimova, E. D. Pavlovskaya.
Astron. Zhurn. Akad. Nauk SSSR, Vol. 51, 597 - 606 (1974). In Russian. English translation in Soviet Astron. AJ, Vol. 18, No. 3.

The effect of kinematical selection for dK and dM stars is found and the procedure of its calculation is developed. It is demonstrated that the neglect of the mentioned effect can overestimate the velocity dispersion to about 5 % for dK stars and even to 10 % for dM stars. The distribution function of the u-, v-, w-components is examined; its asymmetry for the v-components of dK and dM stars is confirmed. The parameters of the velocity ellipsoid are determined.

155.052 Evidenza circa fenomeni esplosivi nucleari ricavata dalle popolazioni stellari. V. Castellani.
Mem. Soc. Astron. Italiana, Suppl. Vol. 44, p. S 301 - S 316 (1974).

155.053 Die räumliche Verteilung, Kinematik und Alter der sonnennahen Sterne. H. Jahreiß.
Diss. Naturwiss. Gesamtfakultät Ruprecht-Karl Univ. Heidelberg. 3 + 96 pp. (1974).

155.054 Gas kinematics near the galactic center from CO observations. R. H. Sanders, G. T. Wrixon.
Astron. Astrophys., Vol. 33, 9 - 14 (1974).

A fine grid 2.6 mm CO line survey of a small region near the galactic center has been carried out with the 36' antenna of the NRAO. The survey at $b = 00°00'$ covers the longitude range $359°50' < l < 00°10'$ and the velocity range -238 km s$^{-1} < V < 238$ km s^{-1}. The results are displayed in the form of a longitude velocity contour map of antenna temperature. A model of the spatial distribution of CO emitting regions which explains the bright emission components is presented.

155.055 Continuum radio structure of the galactic disk.
R. M. Price.
Astron. Astrophys., Vol. 33, 33 - 38 (1974).

An analysis of the 150 MHz continuum profile along the galactic equator suggests that the nonthermal radio emission of the disk has two major components, a base disk and a spiral component. Each contributes approximately one-half of the observed brightness at 150 MHz. The volume emissivity of the base disk decreases with increasing distance from the galactic center. The spiral features have volume emissivities that are at least four times the value of the base disk in the same region. The results of the model suggest that the sun is near the inner edge of a nonthermal spiral feature.

155.056 The Parker instability in differentially-rotating disks.
F. H. Shu.
Astron. Astrophys., Vol. 33, 55 - 72 (1974).

The author investigates the Parker instability for a differentially rotating system comprised of thermal gas, magnetic field, and cosmic-ray particles. The rotation axis coincides with the direction of the vertical gravity, and the rotation is modelled to occur with linear shear. The general initial-value problem is formulated, and the condition for normal modes is obtained. A dispersion relation is obtained for the limiting case when the growth rate (or frequency) of the wave is large in comparison with the kinematic shear rate. This dispersion relation suffices to show, in the absence of dissipation, that no finite amount of shear and rotation can ever completely stabilize Parker's mode. Eigenvalues and eigenfunctions are computed analytically for a number of limiting cases of interest, and a few numerical examples are given.

155.057 Stellar kinematics and galactic research involving proper motions. S. V. M. Clube.
IAU Symposium No. 61, (see 012.014), p. 217 - 220 (1974). Invited paper.

155.058 SAS-2 observations of the high-energy gamma radiation from the Vela region.
D. J. Thompson, G. F. Bignami, C. E. Fichtel, D. A. Kniffen.
Astrophys. Journal, (Letters), Vol. 190, L51 - L53 (1974).

Data from a scan of the galactic plane by the SAS-2 high-energy ($E > 30$ MeV) γ-ray experiment in the region $250° < l^{II} < 290°$ and the observed distribution is discussed in terms of possible origins.

155.059 Gould's Belt. P. O. Lindblad.
Stars and the Milky Way system, Proc. 1972, (see 012.018), p. 65 - 75 (1974). − Invited lecture.

155.060 On the distribution of stars in the Carina − Centaurus region. A. Sundman.
Stars and the Milky Way system, Proc. 1972, (see 012.018), p. 76 - 82 (1974).

155.061 Conclusions on galactic kinematics from proper motions with respect to galaxies. W. Fricke.
Stars and the Milky Way system, Proc. 1972, (see 012.018), p. 83 - 87 (1974).

155.062 The outer spiral structure of the Galaxy and the high-velocity clouds. R. D. Davies.
Stars and the Milky Way system, Proc. 1972, (see 012.018), p. 124 - 128 (1974).

155.063 Ionized gas in the direction of the galactic center I: kinematics and physical conditions in the nuclear disk. P. G. Mezger, E. B. Churchwell, T. A. Pauls.
Stars and the Milky Way system, Proc. 1972, (see 012.018), p. 140 - 156 (1974).

155.064 A model for the galactic center.
V. de Sabbata, P. Fortini, C. Gualdi.
Stars and the Milky Way system, Proc. 1972, (see 012.018), p. 164 (1974). − Abstract.

155.065 8−13 μm spectrum of the galactic centre.
D. Aitken, B. Jones.
Stars and the Milky Way system, Proc. 1972, (see 012.018), p. 248 (1974). − Abstract.

155.066 A comparative study of periodic orbits ($m < 4$) in various three-dimensional models of our Galaxy.
L. Martinet, F. Mayer.
Stars and the Milky Way system, Proc. 1972, (see 012.018), p. 355 - 357 (1974).

155.067 Galaxy counts and dust in the galactic disk.
F. J. Kerr, G. R. Knapp.
Bull. American Astron. Soc., Vol. 6, 212 (1974) − Abstr. AAS.

155.068 Comparison of CO and HI observations on a small scale.   S. C. Simonson, W. J. Wilson.
Bull. American Astron. Soc., Vol. 6, 224 (1974). − Abstr. AAS.

155.069 Tentative identification of main sequence stars in the nuclear bulge of the Galaxy.   S. van den Bergh.
Bull. American Astron. Soc., Vol. 6, 224 - 225 (1974). Abstr. AAS.

155.070 Comparison of s- and r-process abundances as a probe of galactic evolution.
B. M. Tinsley, D. N. Schramm.
Bull. American Astron. Soc., Vol. 6, 281 (1974). − Abstr. AAS.

155.071 The galactic background radiation and thermal electrons in the Galaxy.   H. Hirabayashi.
Publ. Astron. Soc. Japan, Vol. 26, 263 - 287 (1974).
Galactic background radiation was observed at 4.2 GHz and 15.5 GHz for 9 points on the galactic equator. Brightness temperature for the "ridge" of the background radiation is typically 0.07 K at 15.5 GHz. Results of the new observations are compared with those at lower frequencies to determine the spectrum of the galactic background radiation between 1.4 and 15.5 GHz. By combining the amount of thermal emission with the strength of the recombination lines the kinetic temperature and the emission measure of thermal electrons in the diffuse interstellar gas are determined.

155.072 The center of the Galaxy.   P. G. Mezger.
Proc. ESO/SRC/CERN conference, (see 012.021), p. 79 - 107 (1974).

155.073 Kinematic model for stellar subsystems of the Galaxy.   J. O. Peralta.
Doctoral thesis, Instituto de Física, Universidad Nacional de Tucuman, Argentina (1973). In Spanish.
A kinematic model of the Galaxy has been developed that includes the effects of non-circular orbits and orbits inclined to the galactic plane. The model allows one to calculate the variation with galactic longitude and latitude of the mean velocity and velocity dispersion of a given stellar subsystem as a function of the radius, thickness, and distribution of orbital eccentricities of the system. Therefore, these system parameters for a stellar population may be derived from observations of radial velocities in a number of given regions, the model indicating the areas most appropriate. Application of the model to observed radial velocity measures of planetary nebulae shows that the model represents well the observed distribution of velocities with longitude, including that it predicts several observed features of the distribution that cannot be obtained by models based on only circular orbits.

155.074 A note on the density wave model of the Galaxy.
B. Basu, R. Bandyopadhaya, S. N. Paul, A. K. Roy, G. Saha.
Indian Journ. Phys., Vol. 47, 614 - 626 (1973).
The general gas dynamical equations have been solved in the wave form and the general dispersion relation has been deduced. This dispersion relation has been used with simplifying assumptions plausible for special regions of the Galaxy, and results obtained have been shown to be able to interpret some observed dynamical behaviours as well as the distributional property of the gas in those special regions.

155.075 Unterscheidbarkeit der galaktischen Rotation von der Präzession. Ein "Copernicanisches" Problem des 20. Jahrhunderts.   K. Ferrari d'Occhieppo.
Beiträge zur Kopernikusforschung, Katalog Oberösterreich. Landesmuseum No. 86, p. 86 - 92 (1973).

155.076 Die Realität der Sterntrupps.   J. Hopmann.
Sitzungsber. Österreich. Akad. Wiss., Math.-nat. Kl., Abt. II, Vol. 181, 111 - 149 = Astron. Mitt. Wien No. 11 (1973).
The author discusses a formula for the probability of stargroups etc. Most, but not all, of these troops seem to be real. A discussion of the numerous moving groups in the Catalog of D. Hoffleit is given. The surrounding of our sun (up to distances of some 300 pc) seems to be predominantly a mixture of numerous such stargroups, in continuation of the ideas of Klinkerfues (1870) and Kapteyn (1920). Finally the author discusses the numerous cases of coexistence in single or double troops of main sequence stars mixed with red giants.

155.077 Models of three-dimensional variation of galactic absorption.
K. Ferrari d'Occhieppo, H. Jenkner.
Sitzungsber. Österreich. Akad. Wiss., Math.-nat. Kl., Abt. II, Vol. 181, 157 - 173 = Astron. Mitt. Wien No. 12 (1973). In German.
This merely theoretical investigation is devoted to the question, as to what extent distances derived photometrically from adopted distance-moduli may, under realistic conditions, be influenced by the non-random distribution of interstellar matter near the galactic plane.

155.078 Catalogue of dB and gK stars in the Palomar − Groningen Variable-Star Field No. 3.
I. Radiman, B. Hidajat.
Publ. Bosscha Obs., *Lembang*, No. 7, 12 + 6 pp. (1973).
In order to determine the distribution of intermediate-age stars in the direction of the galactic center, the authors have searched for main sequence B8−A2 and giant G8−K2 stars in the direction of the Palomar−Groningen Variable-Star Field no. 3 (R.A. 18^h25^m; Dec. $-33°$; $l = 0°.8$; $b = -9°.9$). Magnitudes and colors for 662 stars are presented.

155.079 The dB and gK stars near the direction of the galactic center.   B. Hidajat, I. Radiman.
Proc. ITB, Vol. 7, 77 - 86 = Bandung Inst. Technol., Dep. Sci., Contr. Bosscha Obs., *Lembang*, No. 46 (1973).
An objective prism survey of the Palomar−Groningen Variable Stars Field No. 3 has been undertaken. A concentration of dB stars has been found at the distances between 150 to 250 pcs. The color excess in this direction is found to remain very low.

Fine analytic abundance determination of Magellanic Cloud A-supergiants and its importance for the discrimination of theories for the chemical evolution of the Galaxy.
See Abstr. 064.052.

Differences between the evolutionary tracks of young stars in the Galaxy and in the Magellanic Clouds.
See Abstr. 065.097.

Berechnung des atmosphärischen Streulichts für Milchstraße und Zodiakallicht als extraterrestrische Quellen.
See Abstr. 082.101.

Results of lunar occultations of the galactic center region in HI, OH and H_2CO lines and in the nearby continua.
See Abstr. 096.023.

Stellar proper motions with reference to galaxies.
See Abstr. 112.005.

Eine photographische UBV-Photometrie in Centaurus.   See Abstr. 113.012.

Four-color observations of early-type stars. IV. South Galactic Pole.   See Abstr. 113.039.

Wolf-Rayet stars in the galactic window around $l = 307°$.   See Abstr. 114.024.

Luminosity and velocity distribution of high-luminosity red stars. IV. The G-type giants. See Abstr. 115.006.

A remarkable star in a highly reddened group of OB$^+$-stars in Norma.   See Abstr. 121.075.

Space velocities of nearby classical cepheids. See Abstr. 122.017.

The space density of cataclysmic variable stars. See Abstr. 122.037.

Large-scale effects of supernova remnants on the Galaxy: generation and maintenance of a hot network of tunnels.   See Abstr. 125.028.

Interstellar lines in the southern hemisphere. See Abstr. 131.021.

A search for OD in the galactic center. See Abstr. 131.025.

Incremental polarization of starlight at different locations in the Galaxy.   See Abstr. 131.029.

Spiralarme und Wolkenstruktur. See Abstr. 131.073.

Molecules as probes of the interstellar matter. See Abstr. 131.106.

Diffusion of elements in the Galaxy. See Abstr. 131.110.

Helium abundance at the galactic center. See Abstr. 131.528.

A new search for H I in the Southern Coalsack. See Abstr. 131.541.

34-micron observations of Eta Carinae, G333.6−0.2, and the galactic center.   See Abstr. 141.613.

Search for gamma-ray emission from the galactic anti-center region.   See Abstr. 142.005.

Evidence for a galactic origin of transient gamma-ray bursts.   See Abstr. 142.080.

Possible evidence for structured acceleration of cosmic rays on a galactic scale from recent γ-ray observations. See Abstr. 143.007.

Tidal effects of the Galaxy on the orbits of stars in spherical stellar systems.   See Abstr. 151.012.

On the formation of interstellar cloud complexes, OB associations and giant H II regions. See Abstr. 151.048.

Stabilization of spiral structure. See Abstr. 151.056.

On the stability of a disk galaxy. See Abstr. 151.061.

An almost complete survey of 21 cm line radiation for $|b| \geqslant 10°$. I. Atlas of contour maps. See Abstr. 157.005.

156 Galactic Magnetic Field

156.001 Polarisationsmessungen in galaktischen Breiten $b < -45°$. R. Schröder.
Mitt. Astron. Ges., No. 34, p. 98 - 99 (1973/74). — Presented at the "Wissenschaftliche Tagung der Astron. Ges., Oberkochen, 1973 April 24 - 27".

156.002 **Structure of the local galactic magnetic field.** R. N. Manchester.
Astrophys. Journ., Vol. 188, 637 - 644 (1974).

Rotation measures have been determined for 18 pulsars, which brings the total number known to 38. It is concluded from these results that the magnetic field in the local region consists of a longitudinal component of strength 2.2 ± 0.4 microgauss directed toward $l = 94° \pm 11°$, together with superposed irregularities of comparable field strength. The conclusions are compared with those from observations of rotation measures of extragalactic sources and optical polarization, and their implications on proposed models for the origin of the galactic field are discussed.

156.003 **Characteristics of the local galactic magnetic field determined from background polarization surveys.**
A. Wilkinson, F. G. Smith.
Monthly Notices Roy. Astron. Soc., Vol. 167, 593 - 611 (1974).

The range of resolutions and frequencies at which background polarization studies are now available makes it possible for the first time to investigate the frequency-dependent behaviour of both the percentage and angle of polarization over a large area. This confirms that a considerable random magnetic field component exists, along with a system of magnetic fields showing large-scale order. The high latitude distribution of polarized radiation is associated with the galactic Loops, the polarization characteristics of which may be qualitatively explained by a shell source model.

Stable equilibria of the interstellar gas and magnetic field in the galactic gravitational field. See Abstr. 131.118.

Turbulence-enhanced synchrotron radiation in the Galaxy. See Abstr. 155.023.

The Parker instability in differentially-rotating disks. See Abstr. 155.056.

Comment on "Galactic magnetic fields: cellular or filamentary structure?". See Abstr. 158.002.

Abstracts 11.157.001 - 11.157.007, 157 Cross References

157 Galactic Radio Radiation

157.001 **The spectrum of the galactic non-thermal background radiation – I. Observations at 151.5 and 408 MHz.** G. Sironi.
Monthly Notices Roy. Astron. Soc., Vol. 166, 345 - 353 (1974).

Observations of the galactic radio background radiation have been carried out at 151.5 and 408 MHz with dipole arrays having a beamwidth of 17° × 12°. The results indicate that between 17.5 and 408 MHz the galactic spectral index is frequency dependent and its value changes over the sky. Evidence for the existence of two or more components of the galactic radio-emission is suggested.

157.002 **The spectrum of the galactic non-thermal background radiation – II. Observations at 408, 610 and 1407 MHz.** A. S. Webster.
Monthly Notices Roy. Astron. Soc., Vol. 166, 355 - 371 (1974).

Between 408, 610 and 1407 MHz the mean differential temperature spectral index of the galactic non-thermal radiation is found to be close to 2.80, although the exact value varies with direction. Several regions with anomalous spectra are associated with previously-known features of the continuum background.

157.003 **Correlation search for dispersed radio emission from the galactic centre.**
P. J. Edwards, R. B. Hurst, M. P. C. McQueen.
Nature, Vol. 247, 444 - 446 (1974).

The authors report the results of a search for fluctuations in the intensity of radio emission from the vicinity of the galactic centre using a two-channel cross-correlation technique.

157.004 **A further 408 MHz survey of the northern sky.**
C. G. T. Haslam, W. E. Wilson, D. A. Graham, G. C. Hunt.
Astron. Astrophys., Suppl. Ser., Vol. 13, 359 - 394 (1974).

The 100-m radiotelescope of the Max-Planck-Institut für Radioastronomie has been used to make a fully sampled survey at 408 MHz of part of the northern sky. The region covered was chosen to complement an earlier survey made at Jodrell Bank (Haslam et al. 1970). The area covered is 12h< R.A.<04h, −8°<Dec.<+48°. The angular resolution of the survey is 37 arc minutes. Two independent surveys have been made of the area and the rms temperature difference in the colder regions is 1.7 K. The temperature scale and zero level have been determined from comparison with other surveys and have errors of 10% and ±3 K respectively.

157.005 **An almost complete survey of 21 cm line radiation for $|b|\geqslant 10°$. I. Atlas of contour maps.**
C. Heiles, H. J. Habing.
Astron. Astrophys., Suppl. Ser., Vol. 14, 1 - 555 (1974).

The 85-foot telescope at Hat Creek, California, has been used to survey the 21 cm line radiation outside of the strip bounded by $|b|\geqslant 10°$ and north of declination −30˚ over the velocity range from −92 to +75 km s^{-1}. The profiles, separated by 0.6° in galactic latitude and 0.3° in longitude, number approximately 140000 and are presented in the form of contour maps of antenna temperature versus velocity and longitude for constant latitude.

157.006 **An almost complete survey of 21 cm line radiation for $|b|\geqslant 10°$. II. The accurate data on machine readable magnetic tape.** C. Heiles.
Astron. Astrophys., Suppl. Ser., Vol. 14, 557 - 564 (1974).

The almost fully sampled Hat Creek λ 21 cm line survey of the northern sky for $|b|\geqslant 10°$ has been subjected to a correction procedure which minimizes internal errors. Comparison with previously existing determinations of zero level define the absolute survey baseline to within 0.1°K. These corrected data are available on a set of 3 magnetic tapes.

157.007 **On the linear polarization of the galactic radio emission at 210 MHz.**
P. A. Kapustin, A. A. Petrovskij, L. V. Pupysheva, V. A. Razin.
Izv. vyssh. ucheb. zavedenij. Radiofizika, Vol. 16, 1325 - 1333 (1973). In Russian. – Abstr. in Referativ. Zhurn. 51. Astron., 5.51.641 (1974).

A high-resolution map of the galactic-center region. See Abstr. 155.008.

Continuum radio structure of the galactic disk. See Abstr. 155.055.

The galactic background radiation and thermal electrons in the Galaxy. See Abstr. 155.071.

158 Single and Multiple Galaxies, Peculiar Objects

Single and Multiple Galaxies

158.001 **H I emission from Stephan's Quintet.**
G. S. Shostak.
Astrophys. Journ., Vol. 187, 19 - 23 (1974).

Neutral hydrogen emission profiles of two members of Stephan's Quintet, NGC 7320 and NGC 7319 (with redshifts of 744 and 6620 km s^{-1}, respectively) are presented. By requiring that the ratios M_{HI}/M_T and M_T/L_{pg} for these galaxies be normal for their classification, distances can be derived which are independent of the redshifts. Detection of a low-luminosity anonymous galaxy about 68' west of the Quintet, having a redshift of 886 km s^{-1}, further strengthens the contention that NGC 7320 is a member of a group which includes the large spiral NGC 7331 and which is at a distance of 10–20 Mpc.

158.002 **Comment on "Galactic magnetic fields: cellular or filamentary structure?"** [Astrophys. Journ., Vol. 179, 771 - 780 (1973)]. E. N. Parker.
Astrophys. Journ., Vol. 187, 191, with a reply by F. C. Michel, p. 193 (1973).

158.003 **Optical and 21-cm line spectroscopic studies of some compact galaxies.**
N. Carozzi, P. Chamaraux, R. Duflot-Augarde.
Astron. Astrophys., Vol. 30, 21 - 29 (1974).

Spectra of two blue compact galaxies (VII Zw 403 and VII Zw 301) and of two blue galaxies with compact parts (NGC 4861 and I Zw 41) were obtained; all of them show emission lines. NGC 4861 appears as a normal rotating late-type galaxy. In the double galaxy I Zw 41 evidence is obtained of rotation in the southern component while no significant redshift difference is observed between the two members. VII Zw 301 has the mean characteristics of the class of the compact galaxies with emission lines described by Sargent. Lastly, VII Zw 403 is found to have a slight blueshift of -110 km s^{-1}. The characteristics of its spectrum, particularly the evidence for absorption lines and the strength of the continuum seem to rule it out as a galactic object. It should perhaps be attributed to the class of dwarf compacts surveyed by Arp and Sargent.

158.004 **On the compactness and redshifts of companion galaxies members of groups of galaxies.**
S. Collin-Souffrin, J.-C.Pecker, H. M. Tovmassian.
Astron. Astrophys., Vol. 30, 351 - 352 (1974).

In groups of galaxies, although the brightest galaxy of the group has often a compact nucleus, the companion galaxies are more clearly redshifted when they are a compact nucleus themselves. Although the sample of galaxies studied in this paper is rather limited, it seems that compactness, altogether, favors redshift.

158.005 **Observations of the infrared radiation from the nuclei of NGC 1068 and NGC 4151.**
W. A. Stein, F. C. Gillett, K. M. Merrill.
Astrophys. Journ., Vol. 187, 213 - 217 (1974).

The purpose of the present discussion is to describe the results of monitoring NGC 1068 and NGC 4151 at $\lambda_0 = 11\mu$ with $\Delta\lambda \simeq 2\mu$ over the last 3 years and to discuss the spectral energy distribution of NGC 1068 between 8 and 13 μ. The data presented here on NGC 4151 suggests variations out to wavelengths as long as 11μ. Limits can be set on the magnitude and time scale of changes of flux from NGC 1068. No definite changes of 11-μ flux have been observed from this source.

158.006 **Spectroscopic observations of NGC 4676.**
A. Stockton.
Astrophys. Journ., Vol. 187, 219 - 221 (1974).

New spectroscopic observations of the peculiar double galaxy NGC 4676 are presented. These observations reverse the sense of the velocity difference between the two galaxies previously determined by Burbidge and Burbidge and thus remove a major objection to the tidal model for this system proposed by Toomre and Toomre. The observations are consistent with a rapid, brief period of star formation during the tidal encounter.

158.007 **The formation of galaxies in the non-linear theory of gravitational instability.**
A. G. Doroshkevich, S. F. Shandarin.
Astron. Zhurn. Akad. Nauk SSSR, Vol. 51, 41 - 49 (1974).
In Russian. English translation in Soviet Astron. AJ, Vol. 18, No. 1.

The problem of the formation of galaxies in the frame of non-linear theory of gravitational instability is discussed. It is supposed that a gas contracted in a shock wave becomes turbulent owing to hydrodynamic instability and fragmentates into clouds of globular cluster type and then into galaxies owing to the gravitational instability of the thin disc. The parameters of the developing objects are estimated.

158.008 **On the dependence of the mass-to-luminosity ratio on colour index for rotating galaxies.**
A. V. Zasov, L. V. Mossakovskaya.
Astron. Zhurn. Akad. Nauk SSSR, Vol. 51, 66 - 70 (1974).
In Russian. English translation in Soviet Astron. AJ, Vol. 18, No. 1.

The dependence is found between the ratio of mass to luminosity and the colour-index C_0 for the inner parts of non-elliptical galaxies within the radius R_m, which corresponds to the maximal velocity of galactic rotation. It is concluded that the mass distributions of stars in the considered regions are the same for most of the galaxies or change monotonously along the colour sequence.

158.009 **Neutral hydrogen, spiral structure and density waves in M 31.** J. Guibert.
Astron. Astrophys., Vol. 30, 353 - 370 (1974).

The 21 cm line observations of M 31 made at Nançay indicate that the neutral hydrogen is confined within the Holmberg dimensions of the galaxy. The large scale hydrogen distribution presents a ring-shaped structure. A well defined spiral structure is superimposed on this large scale distribution, most of the H I arms being in good agreement with the optical arms. The density wave theory has been applied to the region of the N.F.semi-major axis situated 10 kpc from the centre. Using the dispersion relation proposed by Lin and Shu, and the observed arm spacing of the spiral structure, the angular velocity of the spiral pattern is found.

158.010 **Aperture synthesis study of H I in the galaxy IC 10.**
G. S. Shostak.
Astron. Astrophys., Vol. 31, 97 - 101 (1974).

Synthesis observations of the neutral hydrogen in the partially obscured, irregular galaxy IC 10 are presented. The data are consistent with a flattened gas distribution undergoing general rotation. Morphology of the H I distribution is similar to that of the Small Magellanic Cloud. IC 10 is approximately 3 times as massive as the SMC. Evidence is presented which suggests that the obscuration in the direction of IC 10

158.011 **Interstellar hydrogen in galaxies.** M. S. Roberts.
Science, Vol. 183, 371 - 378 (1974).
Radio observations of neutral hydrogen yield valuable information on the properties of galaxies.

158.012 **Galaxy magnitudes. I: The dependence of photoelectric measures on aperture.**
G. E. Kron, C. D. Shane.
Astrophys. Space Sci., Vol. 27, 233 - 240 (1974).
In this, the first of a short series of papers on the magnitudes of galaxies, the dependence of magnitude on the aperture used in photoelectric measures is discussed. Mean results from an empirical study are presented.

158.013 **Studies of Ir II galaxies. II. NGC 3955 and NGC 4433.** F. R. Chromey.
Astron. Astrophys., Vol. 31, 165 - 177 (1974).
Long-slit image-tube spectrograms of the Ir II galaxies NGC 4433 and NGC 3955 show that the ionized gas in both objects is of low density and excitation and must be executing large-scale non-circular motions. Dynamical models are presented which permit mass estimates for the inner portions of each galaxy (10.7 and 8.1 × $10^{10} M_\odot$ for NGC 4433 and NGC 3955, respectively), as well as some estimate of the magnitude of the non-circular velocities. Eleven-color intermediate band photometry is combined with the spectroscopic data to derive emission and absorption line intensities, as well as estimates of the reddening internal to each galaxy. These measurements permit construction of synthetic stellar populations for each galaxy, which show that both objects contain appreciable numbers of young stars, but that NGC 3955 may be deficient in stars hotter than B 3.

158.014 **A high resolution map of the distribution of neutral hydrogen in the spiral galaxy M81.**
A. H. Rots, W. W. Shane.
Astron. Astrophys., Vol. 31, 245 - 248 (1974).
Preliminary results of HI line observations of M81 with the Westerbork Synthesis Radio Telescope at 50″ resolution are presented.

158.015 **UBV photometry on thirty-eight southern galaxies.**
G. Alcaino.
Astron. Astrophys., Suppl. Ser., Vol. 13, 305 - 313 (1974).
The integrated magnitudes and colours of thirty-eight southern galaxies brighter than $V \approx 14$ have been observed in the UBV system with the 1-m photometric telescope of the European Southern Observatory. All observations have been done through a 21 second of arc diaphragm. The galaxies have been selected from Sersic's "Atlas" (1968). Plots of the uncorrected U–B vs B–V colours for all galaxy types define a mean galaxy sequence in agreement with earlier data.

158.016 **Reddening effect in the central regions of normal galaxies.** L. P. Metik, I. I. Pronik.
Izv. Krymskoj Astrofiz. Obs., Vol. 49, 86 - 88 (1974). In Russian.
The reddening of the central regions for 81 normal galaxies has been estimated by comparing central regions colour indices to those of the bluest galaxies (Tifft 1961, 1963, 1969). If the reddening effect in the central regions is due to interstellar dust, the upper limit of dust content in the central regions equals to some $10^5 M_\odot$.

158.017 **The starlike nucleus of NGC 6207.**
D. W. Weedman, R. F. Carswell.
Astrophys. Journ., Vol. 188, 1 - 2 (1974).
Photometric and spectroscopic observations confirm that the apparent stellar nucleus of the spiral galaxy NGC 6207 is in fact a foreground star.

158.018 **The motions in the central region of NGC 4736: Evidence for an expanding ring.**
P. C. van der Kruit.
Astrophys. Journ., Vol. 188, 3 - 17 (1974).
Extensive spectroscopic observations of the central part of NGC 4736 are presented. The analysis shows that there are noncircular motions and that a Lindblad resonance dispersion ring as the interpretation must be rejected on geometrical grounds. The most likely model is one in which the bright central ring of H II regions is expanding with a velocity of about 30 km s^{-1}.

158.019 **The structure of the Fornax dwarf galaxy.**
P. W. Hodge, D. W. Smith.
Astrophys. Journ., Vol. 188, 19 - 25 (1974).
New photoelectric surface photometry and new star counts are combined with previously published data to re-examine the structure of the Fornax dwarf galaxy.

158.020 **Galaxies with ultraviolet continuum. VI.**
B. E. Markarian, V. A. Lipovetskij.
Astrofizika, Vol. 9, 487 - 494 (1973). In Russian. English translation in Astrophysics, Vol. 9, No. 4.
The sixth list of galaxies having intensive ultraviolet continuum is presented. This list contains data for 97 objects. The presence of emission lines is either established or suspected among 63 of them. The presence of Seyfert characteristics can be predicted for three objects. Four other objects are candidates of QSOs.

158.021 **The correlation absolute magnitude-color index for the nuclei of normal spirals.**
S. G. Iskudaryan.
Astrofizika, Vol. 9, 503 - 508 (1973). In Russian. English translation in Astrophysics, Vol. 9, No. 4.
Two groups of nuclei of normal spirals exist on the absolute magnitude-color index diagram at $M_{pg} = -16^m0$, regardless of the form of the nuclei. Apparently, such a division is connected with a definite stage of development of spiral galaxies, when the nuclei are in a subsequent stage of higher activity. The starlike nuclei on the same diagram form a common sequence. However, the higher the luminosity of nuclei of Sa, Sc type galaxies, the bluer the nuclei and vice versa. The starlike nuclei of Sb galaxies have intermediate values.

158.022 **Markarian 474 (NGC 5683) – a Seyfert galaxy of type one (Sy 1).** H. C. Arp, Eh. E. Khachikyan.
Astrofizika, Vol. 9, 509 - 514 (1973). In Russian. English translation in Astrophysics, Vol. 9, No. 4.
The results of spectral observations of Markarian 474 and NGC 5682 are presented. The redshifts and the absolute luminosity of the galaxies have been obtained. It is assumed that Markarian 474 is a Seyfert type galaxy. The problem of physical connection of Markarian 474 and NGC 5682 is discussed.

158.023 **Some observational aspects of the prestellar matter theory.** Yu. K. Melik-Alaverdyan.
Astrofizika, Vol. 9, 595 - 603 (1973). In Russian. English translation in Astrophysics, Vol. 9, No. 4.
It is suggested that the decay of prestellar matter, which according to Ambartsumyan is responsible for the activity of galactic nuclei, may be accompanied by the conversion of elementary particles, particulary, by the decay of mesons to γ-quanta. The observational aspects of this hypothesis are discussed. It is shown that the originated γ-quanta may cause the infrared emission observed in the nuclei of Seyfert galaxies. The expected flux of γ-quanta is estimated.

158.024 **Detection of formaldehyde in external galaxies.**
F. F. Gardner, J. B. Whiteoak.
Nature, Vol. 247, 526 - 527 (1974).

Further to their observations of OH in the edge-on spiral galaxies NGC253 and NGC4945, the authors report the detection of H_2CO absorption in both galaxies—the first evidence of this molecule in external galaxies.

158.025 **Étude spectroscopique de quelques galaxies compactes.** N. Carozzi, P. Chamaraux, R. Duflot.
Comptes Rendus Acad. Sci. Paris, Sér. B, Vol. 278, 429 - 431 (1974).

Parmi les quatorze galaxies compactes bleues étudiées, sept possèdent un spectre d'absorption et cinq présentent des raies en émission. Sur le diagramme de Hubble ces galaxies se situent, aux erreurs de mesures près, entre les galaxies normales et les QSS.

158.026 **On the structure and rotation of NGC 1313.**
G. J. Carranza, E. L. Agüero.
Observatory, Vol. 94, 7 - 9 (1974).

The purpose of this paper is to discuss some morphological and kinematical features of the late-type spiral galaxy NGC 1313, observed with the 154-cm reflector at Bosque Alegre. The size of this interesting galaxy (nearly 7') allowed the use of high-luminosity optics attached to the light-collector.

158.027 **Roberts' redshift effect.** B. M. Lewis.
Observatory, Vol. 94, 9 - 13 (1974).

158.028 **NGC 5195: a symmetric barred spiral galaxy?**
H. Spinrad, E. Harlan.
Publ. Astron. Soc. Pacific, Vol. 85, 815 - 816 (1973).

The purpose of this note is to point out briefly that NGC 5195, the companion system to NGC 5194 in the M51 pair, is morphologically similar to an SBa(r) galaxy, and its previous classification as Irr II (Sandage 1961) or I(p)-E(p) (Morgan 1958) is most likely a result of asymmetric dust absorption over the face of NGC 5195.

158.029 **Optical variability of three Seyfert galaxies.**
H. Netzer.
Monthly Notices Roy. Astron. Soc., Vol. 167, 1P - 5P (1974).

A photographic investigation of the three Seyfert galaxies NGC 4051, NGC 3227 and MARK 79 has been carried out. A new reduction process is described, which enables one to detect variations of $0.^m07$. The galaxies NGC 4051 and Markarian 79 are found to be variable on a time scale of one month, with an amplitude larger than $0.^m2$.

158.030 **The degree of investigation of galaxies with Seyfert and Seyfert-like spectra.**
G. Ivanišević, B. Vorontsov-Velyaminov.
Astron. Zhurn. Akad. Nauk SSSR, Vol. 51, 300 - 309 (1974). In Russian. English translation in Soviet Astron. AJ, Vol. 18, No. 2.

The list of objects presented aims to facilitate observations that should be made in order to complete the data. Galaxies probably adjoining Seyfert galaxies are added to show the links and the probable stages of development.

158.031 **On the steady flow of gas from the nuclei of Seyfert galaxies.** A. M. Wolfe.
Astrophys. Journ., Vol. 188, 243 - 254, with an addendum, p. 441 (1974).

Heating of gas by energy released during explosive events produces hot ($T \sim 5 \times 10^6$ °K) winds in Seyfert nuclei. A study of stability of the thermal equilibrium state leads to a two-component model: a hot stable phase (the wind), in pressure equilibrium with a cooler stable phase (clouds). Effects of the wind upon the central galactic region, such as the sweeping out of interstellar H^0, the radial motions of spiral arms, and the production of large-scale turbulent clouds, are also considered.

158.032 **Investigation of the brightness of two extragalactic objects.** N. E. Kurochkin.
Peremennye Zvezdy, Vol. 19, 128 - 143 (1973). In Russian.

An investigation of the brightness of the nucleus of the Seyfert galaxy NGC 5548 and the quasar (2134 + 004) was carried out on Sternberg Institute plates. There are no essential light variations in the case of NGC 5548 during 1911–1970. Possible brightening of NGC 5548 by $\sim 0.^m3$ occurred in 1970 (J. D. 2440649–715). If this brightening is real, it may be due to a supernova flare in the nucleus of NGC 5548. The light of the QSO (2134 + 004) was constant during 1960–1970.

158.033 **On the nuclear motions in NGC 4151.**
K. S. Anderson.
Astrophys. Journ., Vol. 187, 445 - 447 (1974).

The emission lines arising from the nuclear regions of NGC 4151 indicate nuclear gas motions. It is suggested that these motions are rotational rather than radial in nature and that the emitting gas is essentially confined to the galactic plane. A rough estimate is made of the nuclear mass.

158.034 **Infrared spectrum of NGC 1068.** R. F. Jameson, A. J. Longmore, J. A. McLinn, N. J. Woolf.
Astrophys. Journ., (Letters), Vol. 187, L109 (1974).

The authors have observed the long-wavelength ($\sim 18\mu$) turnover in the spectrum of the Seyfert galaxy NGC 1068.

158.035 **Rotation and masses of the galaxies NGC 145, 275 and 949.** E. K. Denisyuk, N. N. Pavlova.
Astron. Tsirk., No. 797, p. 1 - 3 (1973). In Russian.

158.036 **New objects with broad emission lines from the fifth list of Markarian galaxies.**
E. K. Denisyuk, V. A. Lipovetskij.
Astron. Tsirk., No. 798, p. 2 - 3 (1973). In Russian.

158.037 **On the possible variability of nuclei of the galaxies Mr 42, Mr 69, Mr 205.** L. P. Metik, I. I. Pronik.
Astron. Tsirk., No. 800, p. 3 - 4 (1973). In Russian.

158.038 **The bar-like objects in the centres of galaxies as a possible generator of spiral density waves.**
V. I. Korchagin, L. S. Marochnik.
Astron. Tsirk., No. 800, p. 4 - 6 (1973). In Russian.

158.039 **Photographic photometry of compact galaxies near the north galactic pole.** P. Notni.
Astron. Nachr., Vol. 295, 33 - 45 (1974).

Photographic photometry in U, B, r of 265 compact galaxies from a sample selected by Richter, Richter, and Schneller (1972) is presented. The colours of most of them resemble the colour of E-galaxies. A few objects show a blue or ultraviolet excess if compared with E-galaxies. Some of these may be galaxies with an active nucleus (Seyfert or N-type).

158.040 **Radio observations of normal galaxies.**
A. M. Le Squéren, J. Crovisier.
Astron. Astrophys., Vol. 31, 447 - 450 (1974).

Observations of 15 spiral and irregular galaxies at 4850 MHz are described. The radio spectra between 21 cm and 6 cm of 18 disk components and 9 nuclear components are compared and discussed. The spectral index of the radio disk component appears to be steeper than that of the radio nucleus.

Implications for the origin of relativistic electrons responsible of the synchrotron radio radiation are discussed.

158.041 **Eighteen possible galaxies in Puppis.**
M. P. FitzGerald.
Astron. Astrophys., Vol. 31, 467 - 470 = Contr. Univ. Waterloo Obs., No. 31 (1974).

Eighteen possible galaxies are identified within $2°.5$ of the galactic plane in a region in Puppis at $l = 245°$ and $b = 0°$. These lie in the galactic plane and slightly to the north of it. A rough estimate of the maximum photographic absorption in the direction of the galaxies is $4^m.5$.

158.042 **A digital analysis of the color structure within Messier 51.** S. P. Worden.
Publ. Astron. Soc. Pacific, Vol. 86, 92 - 98 (1974).

Blue and visual photographic plates of Messier 51 have been digitally analyzed to give computer produced (B−V) maps of M51. The technique used to accomplish this is described and this material is compared with the preliminary Westerbork 21-cm neutral hydrogen data. The indication of a ring of blue stars in the nucleus of M51 is reported.

158.043 **Spiral and irregular galaxies associated with rich regular clusters.** H. J. Rood.
Publ. Astron. Soc. Pacific, Vol. 86, 99 - 103 (1974).

Thirty galaxies brighter than $m_p = 15^m.7$ in a 23 square-degree region centered on the Coma cluster are classified as spirals and irregulars on 48-inch Schmidt plates. The homogeneous radial distribution of spirals and irregulars in a 2-square-degree region centered on Abell 2199 suggests that all or nearly all of these galaxies are members of the field. A discussion of uncertainties in the analyses is given.

158.044 **Er lijkt massa zoek te zijn in de lokale groep van melkwegstelsels.** F. P. Israel.
Zenit, Vol. 1, No. 2, p. 9 - 12, 40 (1974).

158.045 **M35 nevelig eilandje met linten aan weerskanten.** D. Drummen.
Zenit, Vol. 1, No. 2, p. 36 - 39 (1974).

158.046 **NGC 2403: Melkwegstelsel op tien miljoen lichtjaren afstand.** M. Drummen.
Zenit, Vol. 1, No. 3, p. 20 - 23 (1974).

158.047 **Directional dependence of the red shift of galaxies.**
V. Bahýl'.
Bull. Astron. Inst. Czechoslovakia, Vol. 25, 115 - 120 (1974).

The dependence of Hubble's constant on direction and distance is statistically studied. Different values of Hubble's constant were obtained for different directions, however, their scatter does not exceed the limits of accuracy of determining the constant itself. A slight decrease in the value of Hubble's constant with distance is shown, which might be explained under the assumption of intergalactic absorption. The accuracy of determining the distance modulus by means of van den Bergh's method of luminosity classes and by means of the method of H II region sizes is discussed.

158.048 **Evolution of dense galactic nuclei through dwarf star collisions.** M. M. Shara, G. Shaviv.
Nature, Vol. 248, 398 - 400 (1974).

The authors draw attention to the fact that a very important type of collision in dense galactic nuclei is one in which a white dwarf is involved.

158.049 **Constraints on a model of a supermassive black hole as a source of activity of nuclei of galaxies.**
L. M. Ozernoj.
Astron. Tsirk., No. 804, p. 1 - 3 (1973). In Russian.

158.050 **On the nature of "tailed" radio galaxies.**
V. N. Kurilchik.
Astron. Tsirk., No. 806, p. 5 - 7 (1974). In Russian.

158.051 **Spectral observations of Markarian galaxies. V.**
E. K. Denisyuk.
Astron. Tsirk., No. 809, p. 1 - 2 (1974). In Russian.

158.052 **Spectral observations of Markarian galaxies. VI.**
E. K. Denisyuk.
Astron. Tsirk., No. 809, p. 2 - 3 (1974). In Russian.

158.053 **Luminosities and dimensions of some double and interacting galaxies.**
B. A. Vorontsov-Velyaminov, L. Baranova, L. Ruzan.
Astron. Tsirk., No. 809, p. 7 - 8 (1974). In Russian.

158.054 **Dynamical evidence for the presence of hidden mass in galaxies.**
J. Einasto, E. Saar, A. Kaasik, P. Traat.
Astron. Tsirk., No. 811, p. 3 - 4 (1974). In Russian.

158.055 **Two faint companions to M 81.**
F. Bertola, P. Maffei.
Astron. Astrophys., Vol. 32, 117 - 119 (1974).

A description of two faint galaxies in the field of M 81 is given. The brighter is of spheroidal type while the other one is of Magellanic type. They have counterparts in IC 1613 and in the Fornax system in the Local Group.

158.056 **Variable N galaxies as composite systems.**
T. F. Adams.
Astrophys. Journ., Vol. 188, 463 - 468 (1974).

Procedures are developed on the basis of Sandage's composite N galaxy model for interpreting the trajectories traced out by variable N galaxies in the color-color diagram. The same procedures are then applied to published data for the peculiar variable BL Lac. The author argues that BL Lac traces a similar trajectory in the color-color diagram, and considers the implications of assuming that it is a variable N galaxy at low redshift.

158.057 **Catalogue of isolated pairs of galaxies in the northern hemisphere.** I. D. Karachentsev.
Soobshch. Spets. Astrofiz. Obs. AN SSSR, Zelenchukskaya, vyp. (No.) 7, 92 pp. (1972). In Russian.

The catalogue contains 603 isolated pairs of galaxies with apparent magnitudes of components $\leq 15^m.7$ and with declination $> -3°$; 332 of them show the signs of interaction.

158.058 **Spiralnebelbilder als Zufallsgebilde.**
J. Meurers, G. Eder.
Naturwissenschaften, 61. Jahrgang, p. 214 (1974).

158.059 **A radio survey of interacting galaxies.**
A. E. Wright.
Monthly Notices Roy. Astron. Soc., Vol. 167, 251 - 272 (1974).

A radio survey of 44 interacting systems of galaxies drawn from the Atlas and Catalogue of Interacting Galaxies by Vorontsov-Velyaminov has been made at 2700 and 5000 MHz using the Parkes 64-m telescope. The results have been compared with a larger sample of non-interacting, bright galaxies observed in a similar way at 2700 MHz. It is concluded that the interacting systems investigated show no anomalous radio properties. The author interprets this conclusion as evidence against a nuclear ejection origin for the observed deformations in interacting systems.

158.060 **Flux densities of bright galaxies at 2700 and 5000 MHz.** A. E. Wright.

Monthly Notices Roy. Astron. Soc., Vol. 167, 273 - 282 (1974).

Radio flux density measurements made at Parkes are presented for 240 bright galaxies lying south of declination +20°; 193 were observed at 2700 MHz and 47 at 5000 MHz. The measurements furnish a comparison sample for a recent survey of interacting galaxies.

158.061 Observations of Markarian galaxies at 75 cm. P. Tomasson, V. G. Malumyan.
Izv. vyssh. ucheb. zavedenij. Radiofizika, Vol. 16, 1342 - 1343 (1973). In Russian. − Abstr. in Referativ. Zhurn. 51. Astron., 4.51.892 (1974).

158.062 The southern Seyfert galaxies NGC 1566 and NGC 3783. P. S. Osmer, M. G. Smith, D. W. Weedman.
Astrophys. Journ., Vol. 189, 187 - 194 (1974).

High-resolution spectrophotometry, photometry, and photographs are presented for the bright southern hemisphere Seyfert galaxies NGC 1566 and NGC 3783.

158.063 The absorption-line spectrum of NGC 4151. K. S. Anderson.
Astrophys. Journ., Vol. 189, 195 - 203 (1974).

Spectroscopic observations of the blueshifted absorption-line spectrum of NGC 4151 are presented.

158.064 Le quintette de Stephan. P. Chamaraux.
L'Astronomie, 88ᵉ année, p. 115 - 140 (1974).

158.065 The nature of the distribution of galaxies. P. J. E. Peebles.
Astron. Astrophys., Vol. 32, 197 - 202 (1974).

A striking result has emerged from estimates of the covariance function for the distribution of galaxies. Within the limits of random and systematic errors, there is no evidence of anti-correlation of galaxy positions on any observed scale − there are not holes around clusters. Rather, the covariance function varies as a simple power law over a large range of the argument. To the accuracy we can estimate it, this statistic gives no evidence of a natural division between groups and clusters of galaxies, or between clusters and superclusters.

158.066 A study of physical groups of galaxies. E. Holmberg.
Ark. Astron., Vol. 5, 305 - 343 (1974).

This paper presents the results of an investigation of 174 physical groups of galaxies, most of them presumably comparable to the Milky Way and M31 groups, and the M81 group. The groups selected are centered on prominent spiral galaxies, for which the distance moduli can be estimated. The survey work has been based on the Palomar Sky Atlas, the prints being evaluated down to the practical limit, as regards galaxies; the limiting diameter of the group members is 0.6 kpc, and the limiting absolute pg magnitude about −10.6.

158.067 A list of peculiar galaxies, interacting pairs, groups and clusters south of declination −43°. J. L. Sérsic.
Astrophys. Space Sci., Vol. 28, 365 - 373 (1974).

The coordinates, dimensions and short descriptions of 186 objects are given after a search of the Maksutov plate collection in the Observatorio Astronómico, Universidad de Chile, Santiago.

158.068 Recent radio work in nearby galaxies. J. H. Oort.
Galaxies and relativistic astrophysics, Proc. 1972, (see 012.010), p. 1 - 14 (1974). − Invited lecture.

158.069 Neutral hydrogen observations of external galaxies with the Mark IA radio telescope. R. D. Davies, R. J. Stephenson.
Galaxies and relativistic astrophysics, Proc. 1972, (see 012.010), p. 15 - 18 (1974).

158.070 The HI distribution in nearby galaxies. D. T. Emerson.
Galaxies and relativistic astrophysics, Proc. 1972, (see 012.010), p. 19 - 20 (1974). − Abstract.

158.071 Compact structures associated with the radio galaxies 3 C 66, 264 and 315. K. J. E. Northover.
Galaxies and relativistic astrophysics, Proc. 1972, (see 012.010), p. 21 (1974). − Abstract.

158.072 Recent optical studies on nearby galaxies. G. Monnet.
Galaxies and relativistic astrophysics, Proc. 1972, (see 012.010), p. 22 - 33 (1974). − Invited lecture.

158.073 Gas motions in the nuclei of Seyfert galaxies. M.-H. Ulrich.
Galaxies and relativistic astrophysics, Proc. 1972, (see 012.010), p. 34 - 36 (1974).

158.074 Electronographic observations of the optical structure of radio galaxies. C. D. Mackay.
Galaxies and relativistic astrophysics, Proc. 1972, (see 012.010), p. 37 - 38 (1974). − Abstract.

158.075 Galactic nuclei. G. R. Burbidge.
Galaxies and relativistic astrophysics, Proc. 1972, (see 012.010), p. 62 - 64 (1974). − Invited lecture.

158.076 Galactic nuclei. L. M. Ozernoy.
Galaxies and relativistic astrophysics, Proc. 1972, (see 012.010), p. 65 - 83 (1974). − Invited lecture.

158.077 The chemical evolution of the galaxies. A. Unsöld.
Galaxies and relativistic astrophysics, Proc. 1972, (see 012.010), p. 84 - 103 (1974). − Invited lecture.

158.078 The dynamics of nearby galaxies. P. J. Warner.
Galaxies and relativistic astrophysics, Proc. 1972, (see 012.010), p. 128 (1974). − Abstract.

158.079 Structural changes in globular galaxies due to collisions. S. M. Alladin, K. S. Sastry, G. M. Ballabh.
Galaxies and relativistic astrophysics, Proc. 1972, (see 012.010), p. 129 - 132 (1974).

158.080 Waves in rotation curves of galaxies as population effects. P. Pişmiş.
Galaxies and relativistic astrophysics, Proc. 1972, (see 012.010), p. 133 - 139 (1974).

158.081 Morphological properties of some bright southern galaxies. R. F. Garrison, N. R. Walborn.
Journ. Roy. Astron. Soc. Canada, Vol. 68, 117 - 142 (1974).

158.082 Radio observations of H II regions in external galaxies. I. H II regions in M 33.
F. P. Israel, P. C. van der Kruit.
Astron. Astrophys., Vol. 32, 363 - 374 (1974).

Observations at 1415 MHz with the Westerbork Synthesis Radio Telescope of 67 H II regions or groups of H II regions in M 33 are presented and analysed.

158.083 Remarkable pairs of double galaxies. H. C. Arp, J. Heidmann, Eh. E. Khachikyan.

Astrofizika, Vol. 10, 7 - 12 (1974). In Russian. — English translation in Astrophysics, Vol. 10, No. 1.

The results of a spectroscopic observation of a double object 2' north of Markarian galaxies 261 and 262 are presented. Their component spectra are quite the same and show high excitation. The measurement of the redshift shows that the distance of the double object and Markarian's pair is the same. It is concluded that all these objects form one physical group as a result of an explosive action.

158.084 **On the form of gaseous ejections in active galaxies.** A. A. Rumyantsev.
Astrofizika, Vol. 10, 109 - 116 (1974). In Russian. — English translation in Astrophysics, Vol. 10, No. 1.

The front of a strong shock wave propagating in a gaseous medium of galactic nuclei, deformed by their rotation, takes a form extended in the direction of poles. The matter is thrown off by the shock wave mainly in the same direction. In the presence of a magnetic field equatorial ejections and belts could arise.

158.085 **On a possible mechanism of formation of deuterium in the nucleus of a galaxy.**
Yu. K. Melik-Alaverdyan.
Astrofizika, Vol. 10, 123 - 126 (1974). In Russian. — English translation in Astrophysics, Vol. 10, No. 1.

Some observational consequences of the possible γ-activity of galactic nuclei are considered. The supposed γ-activity of the nucleus of our Galaxy permits to explain the high deuterium content in the direction of the galactic centre.

158.086 **Polarization of optical radiation of extragalactic objects.** V. A. Hagen-Thorn.
Astrofizika, Vol. 10, 127 - 157 (1974). In Russian. — English translation in Astrophysics, Vol. 10, No. 1.

This article presents an almost comprehensive review of the papers which deal with the polarimetric study of extragalactic objects in optical wavelengths.

158.087 **On the stellar content and reddening in the nucleus of NGC 5195.** J. W. Warner.
Astrophys. Journ., Vol. 190, 19 - 26 (1974).

Spectrophotometric observations of the nucleus of NGC 5195 were made using photographic methods. Absorption-line equivalent widths and continuum measures were compared with models of the stellar content composed of 11 stellar groups. Significant differences were noted for the calcium, magnesium, and sodium lines implying possible overabundances in these elements and interstellar contributions to the calcium and sodium lines. All models were found to be bluer than the observed continuum suggesting that some interstellar absorption is present in front of or in the nucleus. The [N II]/Hα ratio is estimated for two regions centered on the nucleus and is found to be greater than unity. Similarities and differences with M82 are noted.

158.088 **The polarization of normal galaxies at radio wavelengths.** J. F. C. Wardle, R. A. Sramek.
Astrophys. Journ., Vol. 189, 399 - 404 (1974).

The integrated linear polarization of 13 elliptical and four spiral galaxies has been measured at 2695 and 8085 MHz. The spiral galaxies are all weakly polarized. Among the elliptical galaxies, those with compact "active" radio sources are weakly polarized, while those with extended "passive" radio sources tend to be quite strongly polarized. The observed depolarization of NGC 6047 implies a density of interstellar ionized hydrogen of about 2×10^{-4} cm^{-3}.

158.089 **A blue galactic nucleus with a featureless spectrum.** E. Ye. Khachikian, D. W. Weedman.
Astrophys. Journ., (Letters), Vol. 189, L99 - L100 (1974).

The galaxy Markarian 501 is found to have a bright blue nucleus with colors very similar to those of the Seyfert galaxy NGC 1275. The spectrum of the nucleus of Markarian 501 appears featureless, however, so it may be related to objects such as BL Lacertae which have featureless, nonthermal spectra. Markarian 501 is within 3' of the listed position of the radio source 4C 39.49.

158.090 **Die Masse von M33.** W. Lohmann.
Astrophys. Space Sci., Vol. 29, 61 - 62 = Astron. Rechen-Institut. Heidelberg, Mitt. Ser. A (1974).

According to Bottlinger's law the mass of M33 is determined to $(1.29 \pm 0.23) \times 10^{10} M_\odot$ from measurements of the velocities of the neutral hydrogen by Warner et al. (1973).

158.091 **Photographic photometry of compact extragalactic objects. II.** M. K. Babadzhanyants, S. K. Vinokurov, V. A. Hagen-Thorn, E. V. Semenova.
Trudy Astron. Obs., *Leningrad*, Vol. 30 (= Uchenye Zapiski Leningr. Un-ta, No. 373 = Seriya Matem. Nauk, vyp. (No.) 50), p. 69 - 89 (1974). In Russian.

The results of photographic observations of six compact extragalactic objects made in 1971 are given. For three of them (3C 345, 3C 371 and 3C 390.3) detailed UBV light-curves are obtained. The unusual brightness of the quasar 3C 345 is noted. There are indications that in several cases the brightness has measurable changes within one night. The colours change only slightly.

158.092 **Variability of the emission line spectrum of the nucleus of the Seyfert galaxy NGC 1275.**
I. I. Pronik.
Astron. Zhurn. Akad. Nauk SSSR, Vol. 51, 457 - 463 (1974). In Russian. English translation in Soviet Astron. AJ, Vol. 18, No. 3.

It is shown that during 1930–1972 an essential variation of the relative emission line intensities in the gaseous part of the nucleus of NGC 1275 took place, indicating a variability of its physical conditions.

158.093 **Spectrophotometric investigation of the nucleus of the Seyfert galaxy Markarian 79.**
I. M. Yankulova, Eh. A. Dibaj, V. F. Esipov.
Astron. Zhurn. Akad. Nauk SSSR, Vol. 51, 464 - 469 (1974). In Russian. English translation in Soviet Astron. AJ, Vol. 18, No. 3.

The results of a spectrophotometric investigation of the Seyfert galaxy Markarian 79 nucleus are given using data obtained with the 125-cm reflector. The profiles of emission lines are typical for Seyfert galaxies. The widths of hydrogen lines correspond to velocities $\approx \pm 4000$ km/sec. Three emission zones of different elements are present in the nucleus: zone H II; zone O III, Ne III; zone O II, N II, S II. Using the forbidden line intensities the values of electron density, temperature and mass of the emitting gas are calculated.

158.094 **Orientation of angular momenta of galaxies.**
S. F. Shandarin.
Astron. Zhurn. Akad. Nauk SSSR, Vol. 51, 667 - 669 (1974). In Russian. English translation in Soviet Astron. AJ, Vol. 18, No. 3.

It is shown that it is possible in principle to choose between the adiabatic and turbulent theory of formation of galaxies with the help of observations of position angles and axial ratios of galaxies.

158.095 **I nuclei galattici.** L. Gratton.
Mem. Soc. Astron. Italiana, Suppl. Vol. 44, p. S 221 - S 250 (1974).

158.096 **Emissione X da nuclei galattici compatti.**

F. Occhionero.
Mem. Soc. Astron. Italiana, Suppl. Vol. 44, p. S 271 - S 285 (1974).

158.097 **Aspetti teorici dell'emissione γ dei nuclei galattici.** S. Bonometto.
Mem. Soc. Astron. Italiana, Suppl. Vol. 44, p. S 287 - S 299 (1974).

158.098 **Spectroscopic study of some compact galaxies.**
N. Carozzi, P. Chamaraux, R. Duflot.
Astron. Astrophys., Vol. 33, 113 - 116 (1974). In French.

The spectra of fourteen compact blue galaxies were recorded with a RCA C 33011 image tube attached to the E spectrograph at the Observatory of Haute Provence. Seven of them have absorption lines and five show emission lines. As far as possible radial velocities and intensities of some lines have been measured. The Hubble diagram is drawn for these galaxies and those of Barbon.

158.099 **The distance of NGC 7319 in Stephan's Quintet and its supernova.** J. Heidmann.
Astrophys. Space Sci. Library, Vol. 45, (see 012.015), 73 - 74 (1974).

158.100 **Late stages of stellar evolution in the light of elliptical galaxies.**
W. K. Rose, B. M. Tinsley.
Astrophys. Journ., Vol. 190, 243 - 251 (1974).

The predicted spectral energy distributions, broad-band colors, and strengths of spectral features of the integrated light of giant elliptical galaxies are studied using a variety of assumptions concerning evolution of solar-mass stars after core helium ignition.

158.101 **The mass of M31 as determined from the motions of its globular clusters.**
F. D. A. Hartwick, W. L. W. Sargent.
Astrophys. Journ., Vol. 190, 283 - 284 (1974).

An application of the virial theorem to van den Bergh's velocity measurements for the globular clusters results in a mass for M31 which is consistent with that derived by Rubin and Ford from the rotation curve. It is shown that there is a correlation between the clusters' velocity dispersion and their metallicity in the sense that metal-poor clusters have a larger velocity dispersion than metal-rich ones.

158.102 **Observation of X-ray emission from M31.**
S. Bowyer, B. Margon, M. Lampton, R. Cruddace.
Astrophys. Journ., Vol. 190, 285 - 289 (1974).

The authors have observed X-ray emission from M31 in the 0.5–5.0 keV band, using rocket-borne proportional counters. The integrated flux corresponds to $(2.2 \pm 0.6) \times 10^{39}$ ergs s^{-1} at M31, and is confined chiefly to a region coinciding with the optical image of the galaxy. The spectrum has a slope similar to the bright galactic X-ray sources.

158.103 **On the tidal origin of the bridge of Arp 295.**
A. Stockton.
Astrophys. Journal, (Letters), Vol. 190, L47 - L49 (1974).

Rotation curves are presented for the two principal galaxies of the interacting system Arp 295. The relative radial velocities and the ratio of masses of the two components are in good agreement with the tidal model for the system proposed by Toomre and Toomre.

158.104 **Physical studies of emission regions in M33.**
G. Comte, G. Monnet.
Astron. Astrophys., Vol. 33, 161 - 176 (1974).

The authors present photographic and spectrophotometric data for bright H II regions, diffuse arm emission regions and interarm general emission in the spiral galaxy M33 (NGC 598). Spectrophotometry provides an Hα/[N II] 6584 Å line intensity ratio for 18 bright H II regions, as well as Hα/[S II] 6717 Å + 6731 Å and [S II] 6717 Å/[S II] 6731 Å line intensity ratios for 12 of them. The diffuse arm extended emission studied at three points in the southern arm gives Hα/[N II] line intensity ratios close to the values observed in the nearby bright regions, reflecting similar physical conditions. The authors discuss the possible causes of the interarm ionization. Stars with UV excess are evaluated based on a 30000°K blackbody model in the UV range of the spectrum: the corresponding predictions for the disk energy distribution are ruled out by observation.

158.105 **Study of the central region of M 31. II. Synthetic models of stellar populations.** M. Joly.
Astron. Astrophys., Vol. 33, 177 - 186 (1974).

The author tries to obtain synthetic models of stellar populations which allow to interpret the equivalent widths of absorption lines and the galactic flux observed in the central part of M 31 (inside 1.5 kpc). The galactic equivalent widths are compared with the equivalent widths of the same lines from a sample of stars which were classified as to their spectral type, luminosity and metallic abundance. Stellar populations are obtained from a numerical analysis which derives synthetic models.

158.106 **Evolution of galaxies and related parameters.**
L. Bottinelli, L. Gouguenheim.
Astron. Astrophys., Vol. 33, 269 - 276 (1974).

The integral properties of 131 galaxies for which hydrogen and total masses are available, are investigated in terms of evolution properties. It is definitely shown that the various morphological types cannot be accounted for by a decrease of the initial gas density in the protogalaxy when the morphological type goes from lenticular to irregular.

158.107 **Galactic models and stellar orbits.** J. Einasto.
Stars and the Milky Way system, Proc. 1972, (see 012.018), p. 291 - 325 (1974). – Invited lecture.

158.108 **Observations of the kinematics in the nuclear region of M81.** J. W. Goad.
Bull. American Astron. Soc., Vol. 6, 212 - 213 (1974). Abstr. AAS.

158.109 **Detection of giant dust complexes near NGC 7538 in the Perseus arm.** Y. K. Minn, J. M. Greenberg.
Bull. American Astron. Soc., Vol. 6, 225 (1974). – Abstr. AAS.

158.110 **Further spectrophotometric observations of faint radio galaxies.** H. Spinrad, H. E. Smith.
Bull. American Astron. Soc., Vol. 6, 261 (1974). – Abstr. AAS.

158.111 **The space density of Markarian galaxies including a region of the south galactic hemisphere.**
J. P. Huchra, W. L. W. Sargent.
Bull. American Astron. Soc., Vol. 6, 262 (1974). – Abstr. AAS.

158.112 **The mass of galaxies and the density of the universe.**
J. P. Ostriker, A. Yahil.
Bull. American Astron. Soc., Vol. 6, 262 (1974). – Abstr. AAS.

158.113 **Upper limit to the neutral hydrogen content of the elliptical galaxy NGC 4472.**
G. R. Knapp, F. J. Kerr.
Astron. Journ., Vol. 79, 667 - 670 (1974).

The large elliptical galaxy NGC 4472 was observed in the 21-cm line of neutral hydrogen in a search for HI emission. No HI was found in the galaxy. The HI content of the galaxy is then less then 0.006% by mass, much lower than the amount of interstellar matter which is expected to exist in the galaxy due to mass loss by stars. Several possible explanations are considered.

158.114 Faint envelopes of galaxies.
J. Kormendy, J. N. Bahcall.
Astron. Journ., Vol. 79, 671 - 677, 755 - 764 (1974).

Photographic surface photometry is presented for 12 fields containing small and moderately rich groups of galaxies, as well as isolated galaxies of various morphological types. Surface brightnesses of 26–28 green mag per square arcsec are reached using isodensity tracings with small density steps. Observations confirm the existence of faint envelopes $\gtrsim 100$ kpc in diameter. When measured at the same surface brightness, spiral galaxies are not substantially smaller than ellipticals of the same luminosity. Some groups of galaxies dominated by ellipticals are imbedded in common luminous envelopes up to 1.3 Mpc in diameter.

158.115 Zur Entwicklung der Metagalaxis.   H. Oleak.
Astron. in der Schule, 11. Jahrgang, p. 50 - 53 (1974).

158.116 Hot spots in the central region of NGC 2903.
S. Oka, K. Wakamatsu, K. Sakka, M. Nishida, J. Jugaku.
Publ. Astron. Soc. Japan, Vol. 26, 289 - 298 (1974).

Photographic and spectroscopic observations have been made of the knots (hot spots) in the central region of NGC 2903. The blue color and Balmer absorption lines of these knots show that they contain many early-type stars. The knots also contain late-type stars and are regarded as young giant clusters of stars accompanied by gas excited by hot stars. A velocity difference of about 150 km s^{-1} is found between two of the knots.

158.117 Future research on nearby galaxies.
S. van den Bergh.
Proc. ESO/SRC/CERN conference, (see 012.021), p. 115 - 121 (1974).

158.118 Desiderata from theory of chemical evolution of galaxies.   B. E. J. Pagel.
Proc. ESO/SRC/CERN conference, (see 012.021), p. 131 - 142 (1974).

158.119 Peut-on prolonger la courbe (magnitude V, log z)?
G. Wlérick, G. Lelièvre.
Proc. ESO/SRC/CERN conference, (see 012.021), p. 211 - 219 (1974).

158.120 Galaxies.   M. Różyczka.
Urania Kraków, Vol. 45, 130 - 136, 162 - 169 (1974). In Polish.

158.121 Some recent results from galactic and stellar evolution theory.   R. J. Talbot, Jr, W. D. Arnett.
Astrophys. Journ., Vol. 190, 605 - 608 (1974).

Recent progress in stellar and galactic evolution theory give excellent agreement between observational abundances and the theoretical yields for elements heavier than helium. This includes the secondary production of ^{14}N and the variation of N/O over the face of M101 which was observed by Searle.

158.122 A catalogue of galaxies having radial velocities.
L. W. Fredrick, W. A. Gutsch.
Publ. Leander McCormick Obs., Univ. Virginia, *Charlottesville*, Vol. 15, (Part 7), 39 - 56 (1974).

This catalogue was generated originally to study the distribution of radial velocities. It is intended as a search aid bringing together material from many different sources.

158.123 Results of a study of interacting galaxies.
B. A. Vorontsov-Vel'yaminov.
Astron. Tsirk., No. 814, p. 5 - 6 (1974). In Russian.

158.124 Photoelectric observations of some compact Zwicky galaxies.   V. P. Arkhipova, M. V. Savel'eva.
Astron. Tsirk., No. 814, p. 6 - 8 (1974). In Russian.

158.125 Galaxies of the M51 type.
B. A. Vorontsov-Vel'yaminov.
Astron. Tsirk., No. 815, p. 1 - 2 (1974). In Russian.

158.126 Comet-like galaxies in the optical and radio range.
B. A. Vorontsov-Vel'yaminov.
Astron. Tsirk., No. 816, p. 1 - 2 (1974). In Russian.

158.127 Formation of small satellites of galaxies.
B. A. Vorontsov-Vel'yaminov.
Astron. Tsirk., No. 816, p. 2 - 3 (1974). In Russian.

158.128 A possible explanation of intensity variations of the hydrogen lines in the spectra of nuclei of Seyfert galaxies.   I. M. Yankulova.
Astron. Tsirk., No. 816, p. 4 - 5 (1974). In Russian.

158.129 Nests of galaxies, their fragmentation.
B. A. Vorontsov-Vel'yaminov.
Astron. Tsirk., No. 817, p. 3 - 5 (1974). In Russian.

158.130 The periodicity in light variations of the N-galaxy 3C 371.   M. K. Babadzhanyants, E. T. Belokon'.
Astron. Tsirk., No. 819, p. 3 - 5 (1974). In Russian.

158.131 The kinematics and dynamics of M51. I. The observations.   R. B. Tully.
Astrophys. Journ., Suppl. Ser., No. 251, Vol. 27, 415 - 435 (1974).

A photographic Fabry-Perot interferometer has been used to obtain a detailed velocity map of the spiral galaxy M51. The instrument, observations, and reduction procedures are described and the Hα intensity distribution and line-of-sight velocities are reported wherever the intensity exceeds 1×10^{-14} ergs s^{-1} cm^{-2}. Intensities as low as 3×10^{-15} ergs s^{-1} cm^{-2} have been observed.

158.132 The kinematics and dynamics of M51. II. Axisymmetric properties.   R. B. Tully.
Astrophys. Journ., Suppl. Ser., No. 251, Vol. 27, 437 - 448 (1974).

A simple model in which the distribution of matter is described by two Gaussian components is consistent with the rotational velocities observed in the spiral galaxy M51. The two components can be associated with the disk of the galaxy and a less-flattened nuclear lens. A third Gaussian component, associated with the optical semistellar nucleus, is required to explain the wide emission lines observed at the center of M51. Together, the three components lead to a surface density of matter which is inversely dependent on the radius over the inner disk. This configuration has been theoretically anticipated because it is radially stable. There is a discussion of the distribution of angular momentum, rotational kinetic energy, and ionized hydrogen and a discussion of the thickness of a stable gaseous disk in this galaxy.

158.133 The kinematics and dynamics of M51. III. The spiral structure.   R. B. Tully.
Astrophys. Journ., Suppl. Ser., No. 251, Vol. 27, 449 - 471

(1974).

A detailed model is presented which explains the origin and nature of the spiral structure of M51 and the peculiar velocities observed in the nuclear region of the galaxy. In this model, the outer spiral arms are material arms, the tidal consequences of the close passage of NGC 5195. A density wave has been generated by the innermost material clumping arising from the encounter and this wave has propagated inward until it reached the inner Lindblad resonance at a radius of 600 pc. There, an elongated dispersion ring has formed, accounting for the apparent "outflow" of gas in the nuclear region of M51. The time scale for the propagation of the density wave is less than 10^8 years and is comparable to the time scale for the close approach of the companion. Over most of its existence, M51 was not such a spectacular spiral galaxy as it is today.

Measurements of velocity dispersions and Doppler shifts from digitized optical spectra. See Abstr. 031.005.

Analysis of electronographic images. See Abstr. 031.058.

Toroidal figures of equilibrium. See Abstr. 042.087.

Omnidirectional induced Compton scattering by relativistic electrons. See Abstr. 063.009.

Novae in the Andromeda nebula in 1972. See Abstr. 124.005.

Location of supernovae in parent galaxies. See Abstr. 125.022.

Indicative distances of 148 supernovae from the nuclei of their parent galaxies. See Abstr. 125.023.

Remarks on outlying supernovae and the structure of their parent galaxies. See Abstr. 125.047.

An interferometer study of H I absorption in Centaurus A. See Abstr. 131.516.

A second survey of H II regions in galaxies. See Abstr. 131.534.

Survey of faint H II regions in the nearest galaxies by means of the large reducteur focal of the 193 cm telescope of the Haute Provence Observatory. See Abstr. 131.552.

Steps toward the Hubble constant. I. Calibration of the linear sizes of extragalactic H II regions. See Abstr. 131.553.

Fine structures in galactic and extragalactic nebulae. See Abstr. 132.018.

Radio source structure, clusters of galaxies and the complex sources. See Abstr. 141.012.

Initial results from a radio continuum survey of the area around NGC 7331 and Stephan's Quintet. See Abstr. 141.017.

Statistics of apparent distances between QSO's and bright galaxies. See Abstr. 141.020.

A theory of galactic nuclei and quasi-stellar objects. See Abstr. 141.029.

The NGC 520 chain of quasars. See Abstr. 141.044.

The problem of satellite radio sources in the vicinity of spiral galaxies. See Abstr. 141.045.

Radiative acceleration of gas clouds near quasi-stellar objects and Seyfert galaxy nuclei. See Abstr. 141.058.

On gravitational-lens quasars. See Abstr. 141.061.

Structure of 194 southern declination radio sources from interplanetary scintillations. See Abstr. 141.064.

Narrow-band observations of galactic and extra-galactic sources at 1 mm. See Abstr. 141.073.

Spectroscopic observations of eight QSO candidates among Markarian objects. See Abstr. 141.081.

Spectroscopy of objects near Texas radio-source positions. See Abstr. 141.097.

Spectroscopic observations of objects identified with radio sources. See Abstr. 141.119.

3C 66A: a bright new quasi-stellar object. See Abstr. 141.121.

Infrared emission by dust in NGC 1068 and three planetary nebulae. See Abstr. 141.612.

On the nature of the unidentified high-latitude Uhuru sources. See Abstr. 142.043.

The extended X-ray source in Virgo and its relation to M87. See Abstr. 142.114.

The problem of the spiral structure of galaxies. See Abstr. 151.049.

The gravitational slingshot and the structure of extragalactic radio sources. See Abstr. 151.050.

Encounters between galaxies of equal size. See Abstr. 151.052.

The horizontal branch in globular clusters of the Galaxy and Magellanic Clouds and in Sculptor type dwarf galaxies. See Abstr. 154.003.

Photometry of the globular clusters of NGC 185. See Abstr. 154.028.

BV photometry of globular clusters in M87. See Abstr. 154.029.

Simultaneous optical and radio observations of BL Lac and 3C 120. See Abstr. 158.303.

Formation of galaxies and clusters of galaxies by self-similar gravitational condensation. See Abstr. 162.012.

Peculiar Objects

158.301 **Optical and radio properties of a new BL Lacertae class object.**
J. Crovisier, A. M. Le Squéren, J. T. Pollock, P. D. Usher.
Astron. Astrophys., Vol. 30, 175 - 176 (1974).

It is shown that the source B2 0912 + 29, which has a featureless optical spectrum, a flat radio spectrum, and exhibits optical variations, is a member of the BL Lacertae class of QSO's.

158.302 **Étude de BL Lacertae. Estimation de la distance de l'astre.** G. Wlérick, D. Michet, G. Lelièvre.
Comptes Rendus Acad. Sci. Paris, Sér. B, Vol. 278, 245 - 248 (1974).

A partir de clichés électronographiques, nous avons pu séparer BL Lac en deux composantes: un noyau ponctuel et une nébulosité assimilable à une galaxie. Cette assimilation permet d'estimer la distance de l'astre: 300 Mpc ± 80 Mpc, valeur 300 fois plus grande que les estimations précédentes. On en déduit que les propriétés de BL Lac sont voisines de celles de la radiosource 3 C 120 et que les dimensions linéaires de la source radioélectrique sont de l'ordre du parsec. En outre, BL Lacertae doit appartenir à un petit amas de galaxies.

158.303 **Simultaneous optical and radio observations of BL Lac and 3C 120.** H. Tabara, S. Kikuchi, Y. Mikami, N. Kawano, N. Kawajiri, T. Ojima, M. Inoue, M. Konno, K. Tomino, T. Daishido.
Tokyo Astron. Bull., Second Ser., No. 228, p. 2633 - 2637 (1974).

The peculiar object BL Lac (VRO 42.22.01) and the Seyfert galaxy 3C 120 are known as rapid variables in optical, infrared and radio regions. In order to study the correlation between the optical and the radio variations of these objects, the authors have made simultaneous observations in the photographic B and V colors and of the flux density at 7.2 cm. The observations were made from November 30 to December 10, 1972, which covered the period of the international coordinated observations of BL Lac and 3C 120.

158.304 **The region around Eta Carinae.** A. Feinstein.
Rev. Astron., Vol. 45, Nos. 185 - 186, p. 29 - 34 (1973). In Spanish.

The author gives a description of Eta Carinae and presents its light curve. Eta Carinae is surrounded by a large group of stars of spectral types O and B, that is, with high surface temperatures, forming stellar cumuli.

Blue objects near M31. II. Frequency distribution and statistics. See Abstr. 113.016.

Are BL Lac-type objects nearby black holes? See Abstr. 122.103.

The distance of BL Lacertae. See Abstr. 141.059.

Spectroscopic observations of objects identified with radio sources. See Abstr. 141.119.

A blue galactic nucleus with a featureless spectrum. See Abstr. 158.089.

Peut-on prolonger la courbe (magnitude V, log z)? See Abstr. 158.119.

Errata

158.901 **Erratum: 'Further high-resolution radio continuum observations of bright spiral galaxies at 1415 MHz'** [Astron. Astrophys., Vol. 29, 249 - 262 (1973)].
P. C. van der Kruit.
Astron. Astrophys., Vol. 31, 477 (1974).

158.902 **Errata: 'The distances to members of Stephan's Quintet determined from 21-cm measures'** [Bull. American Astron. Soc., Vol. 5, 430 (1973)]. G. S. Shostak.
Bull. American Astron. Soc., Vol. 6, 268 (1974). – Abstr. AAS.

159 Magellanic Clouds

159.001 **Color-magnitude diagrams for eleven young clusters in the Magellanic Clouds.** J. W. Robertson.
Astron. Astrophys., Suppl. Ser., Vol. 15, 261 - 309 (1974).

Photographic B and V observations are presented for 1 SMC and 10 LMC clusters. The color-magnitude diagrams for these clusters show fairly tight groups of red and blue giants in the mass range 5 to 25 $M_\odot$.

159.002 **The Magellanic Clouds.** B. E. Westerlund.
Galaxies and relativistic astrophysics, Proc. 1972, (see 012.010), p. 39 - 61 (1974). – Invited lecture.

159.003 **OB stars detection in the Large Magellanic Cloud.** N. Martin, J. Rousseau.
Astron. Astrophys., Vol. 33, 135 - 138 (1974). In French.

The authors used a previously described method and continued the very hot stars detection in LMC. They have detected more than 250 new O–B2 stars and classified again 400 Sanduleak stars. Among them. about 200 stars are O–B2 stars. A color-color diagram, and the star positions on the LMC chart show the main points of the results.

159.004 **The Magellanic Stream.**
D. S. Mathewson, M. N. Cleary, J. D. Murray.
Astrophys. Journ., Vol. 190, 291 - 296 (1974).

A southern sky survey of H I in the velocity range -340 to $+380$ km s^{-1} has shown that a long filament of H I extends from the region between the Magellanic Clouds down to the south galactic pole and connects with the long H I filament discovered recently by Wannier and Wrixon and van Kuilenburg. There is also some evidence that this continues on the other side of the Magellanic Clouds and crosses the galactic plane at $l^{II} = 306°$. This filament, which follows very closely a great circle over its entire 180° arc across the sky, is given the name "the Magellanic Stream".

159.005 **Ionized hydrogen mass in the Clouds of Magellan.** N. W. Broten.
Bull. American Astron. Soc., Vol. 6, 262 (1974). – Abstr. AAS.

159.006 **The future of Magellanic Cloud research.** J. A. Graham.
Proc. ESO/SRC/CERN conference, (see 012.021), p. 159 - 167 (1974).

159.007 **The Magellanic Clouds – some initial research programmes for the new large telescopes.**
D. S. Mathewson.
Proc. ESO/SRC/CERN conference, (see 012.021), p. 189 - 195 (1974).

Fine analytic abundance determination of Magellanic Cloud A-supergiants and its importance for the discrimination of theories for the chemical evolution of the Galaxy. See Abstr. 064.052.

Differences between the evolutionary tracks of young stars in the Galaxy and in the Magellanic Clouds. See Abstr. 065.097.

Radial velocities from objective-prism plates in the direction of the Large Magellanic Cloud. List of 398 stars, LMC members. List of 1434 galactic stars in the LMC direction. See Abstr. 112.017.

M supergiants in the Large Magellanic Cloud. See Abstr. 114.162.

The RR Lyrae stars in the Large Magellanic Cloud cluster NGC 1835. See Abstr. 122.060.

A probable periodicity in the light variation of the LMC supergiant HD 33579. See Abstr. 122.077.

The short-term light variations of S Dor and HD 37836. See Abstr. 122.089.

Variable and peculiar stars in the Magellanic Clouds. See Abstr. 122.117.

RR Lyrae variables in the Magellanic Clouds. See Abstr. 122.130.

Variable stars in the Large Magellanic Cloud–II. See Abstr. 123.001.

Density of the interstellar gas and strength of the magnetic field of the Small Magellanic Cloud. See Abstr. 131.102.

Dark nebulae in the Small Magellanic Cloud. See Abstr. 131.136.

Chemical abundances of HII regions in the LMC. See Abstr. 131.548.

Integrated colors of star clusters in the Large Magellanic Cloud. See Abstr. 153.025.

Young globular clusters in the Magellanic Clouds. See Abstr. 154.024.

A survey of star clusters in the Magellanic Clouds. See Abstr. 154.026.

160 Clusters of Galaxies

160.001 **Radio observations of two clusters of galaxies.**
F. F. Donivan, Jr., T. D. Carr, G. C. Omer, Jr.
Astrophys. Journ., Vol. 187, 11 - 17 (1974).

Maps including the Coma and Hercules clusters of galaxies were made at 195, 318, and 611 MHz, and were compared with optical counts of galaxies in the clusters. An extended region of radio emission approximately 20' in diameter was observed at the center of the Coma cluster. A distributed source region was also observed which appears to include the Hercules cluster.

160.002 **Limits on variation of G from clusters of galaxies.**
D. S. Dearborn, D. N. Schramm.
Nature, Vol. 247, 441 - 443 (1974).

In order to set a limit on the rate of change of the gravitational constant ($\dot{G}/G$), the dynamical effects on an isolated cluster of galaxies caused by non zero values of $\dot{G}/G$ were studied numerically. For a range of initial conditions, the maximum rate of change in G was determined which would still allow the existence of clusters of galaxies at the current epoch. A similar study was made for globular clusters and the results were found to be comparable.

160.003 **Compact groups of compact galaxies.**
R. K. Shakhbazyan.
Astrofizika, Vol. 9, 495 - 501 (1973). In Russian. English translation in Astrophysics, Vol. 9, No. 4.

The first trial list of 30 compact groups of compact galaxies discovered on 25 charts of Palomar Survey is presented. Photographs of 30 compact groups of compact galaxies are given.

160.004 **X-ray and radio emission by clusters of galaxies.**
B. N. G. Guthrie.
Astrophys. Space Sci., Vol. 27, 489 - 496 (1974).

Analysis of new X-ray and optical data confirms that the X-ray luminosity of a cluster of galaxies is strongly dependent on its richness. The radio power of clusters at 1445 MHz is independent of richness, but is greater on the average for clusters with dominant cD galaxies than for those without. The radio emission depends on the activity of one of the brightest galaxies; dominant cD galaxies are often responsible for radio emission, especially if they have double or multiple nuclei.

160.005 **Inverse Compton radiation and the magnetic field in clusters of galaxies.**
D. E. Harris, W. Romanishin.
Astrophys. Journ., Vol. 188, 209 - 216 (1974).

Radio observations below 200 MHz are reported for the cluster Abell 401. These data together with *Uhuru* X-ray measurements and relevant data for other clusters of galaxies are interpreted in terms of an ensemble of relativistic electrons producing radio emission by the synchrotron process and X-rays by the inverse Compton process. Formulae are presented which relate the observations to the relativistic electron spectra. The authors believe that the critical test of the inverse Compton model will come from X-ray observations between 10 and 100 keV and have calculated X-ray intensities predicted for these energies.

160.006 **High-energy X-rays from the Perseus cluster.**
A. Bui-Van, K. Hurley, G. Vedrenne.
Astrophys. Journ., Vol. 188, 217 - 219 (1974).

X-radiation has been observed from the Perseus cluster at energies between 53 and 93 keV. The spectrum is consistent with a power law, and a model for inverse Compton X-ray production is presented.

160.007 **The definition, visibility, and significance of redshift-magnitude bands.** W. G. Tifft.
Astrophys. Journ., Vol. 188, 221 - 232 (1974).

Redshift-magnitude bands as they occur in the Coma cluster are formally defined, and the original bands observed in 1972 are shown to have a likelihood of random occurrence of only 0.005 independent of their direction. The properties of the Coma bands are used to predict band properties for the A2199 cluster. The resultant power-spectrum test of a preliminary A2199 sample shows agreement which has a random likelihood of occurrence of only 0.001. The A2199 cluster also shows a band-related morphological separation as in Coma.

160.008 **The Perseus cluster: galaxy distribution, anisotropy, and the mass/luminosity ratio.** N. A. Bahcall.
Astrophys. Journ., Vol. 187, 439 - 444 (1974).

Counts of galaxies in the Perseus cluster are used to determine the center, the distribution of galaxies to different limiting magnitudes, the core radius, and the apparent anisotropy. The mass and luminosity of the cluster core are calculated, and the ratio of central mass to visual light is found to be about $700(H_0/100)$ solar units with conventional assumptions.

160.009 **On the distribution of galaxies in clusters.**
B. I. Gorbachev.
Astron. Tsirk., No. 803, p. 3 - 5 (1973). In Russian.

160.010 **Empirical properties of the mass discrepancy in groups and clusters of galaxies. II.** H. J. Rood.
Astrophys. Journ., Vol. 188, 451 - 461 (1974).

Data compiled by de Vaucouleurs, Karachentsev, Holmberg, Burbidge and Sargent, and others show that the mass ratio (ratio of virial-theorem mass to luminous mass) of groups and clusters is generally correlated with radius and velocity dispersion. "Solar-system–type" groups have mass ratios which increase with projected velocity and distance of a secondary galaxy relative to the primary galaxy. The non-Doppler redshift hypothesis to resolve the mass discrepancy problem is examined.

160.011 **Counts of galaxies in seven distant clusters.**
T. W. Noonan.
Astron. Journ., Vol. 79, 358 - 362 (1974).

This paper presents counts of galaxies in each of seven distant clusters with redshifts between 0.17 and 0.46. Luminosity segregation is found in two of the clusters. The data is inconclusive as regards color segregation.

160.012 **Radio observations of Abell 1314.**
J. C. Webber.
Publ. Astron. Soc. Pacific, Vol. 86, 223 - 224 (1974).

The cluster of galaxies Abell 1314 has been observed at three radio frequencies. Tentative optical identifications for the two radio sources detected are made.

160.013 **The stability of galaxy clusters: neutral hydrogen observations.**
D. S. De Young, M. S. Roberts.
Astrophys. Journ., Vol. 189, 1 - 9 (1974).

New upper limits are obtained for the 21-cm emission from the Cancer, Coma, Pegasus I, and Perseus clusters of galaxies. In all cases these limits correspond to a galactic mass ($\sim 10^{11} M_\odot$) of optically thin neutral hydrogen. These observations rule out the possibility that the Coma cluster is gravitationally bound by rotationally supported, optically thick clouds of neutral hydrogen.

160.014 Remarks on the magnitude-redshift bands in the Coma cluster.
J. M. Barnothy, M. F. Barnothy.
Astrophys. Journ., Vol. 189, 11 - 16 (1974).

The band pattern found by Tifft in the magnitude-redshift diagram of the galaxies in the Coma cluster has no statistical significance, and could be caused by a random variation in the velocity and apparent magnitude of the objects.

160.015 A distance limit for NGC 7318B in Stephan's Quintet. G. S. Shostak.
Astrophys. Journ., (Letters), Vol. 189, L1 - L3 (1974).

Observations at 21 cm of NGC 7318B in Stephan's Quintet show that the H I flux from this galaxy is no more than 0.008 f.u. The resulting limit on M_{HI} is incompatible with a distance as small as 15 Mpc, which is the accepted distance of the low-redshift member of the Quintet, NGC 7320.

160.016 The correlation of radio emission and optical type with X-ray emission from clusters of galaxies.
F. N. Owen.
Astrophys. Journ., (Letters), Vol. 189, L55 - L58 (1974).

A correlation of X-ray emission from clusters of galaxies with radio emission and with Rood and Sastry optical cluster type is reported. In light of this correlation the two most likely X-ray emission mechanisms − inverse Compton scattering and thermal bremsstrahlung − are discussed.

160.017 Copernicus X-ray observations of NGC 1275 and the core of the Perseus cluster.
A. C. Fabian, J. C. Zarnecki, J. L. Culhane, F. J. Hawkins, A. Peacock, K. A. Pounds, J. H. Parkinson.
Astrophys. Journ., (Letters), Vol. 189, L59 - L63 (1974).

The Perseus cluster of galaxies has been studied with 0.5−1.5 keV and 1.0−3.1 keV grazing-incidence X-ray telescope systems on the Orbiting Astronomical Observatory, Copernicus. The observations shed further light on the detailed distribution of X-ray emission in the central regions of the Perseus cluster. A brief discussion is added which considers the new results together with the Uhuru and other published X-ray data.

160.018 Amas de galaxies en expansion? J.-C. Pecker.
L'Astronomie, 88e année, p. 41 (1974).

160.019 Observations of Abell clusters of galaxies at 1400 MHz. F. N. Owen.
Astron. Journ., Vol. 79, 427 - 436 (1974).

Observations are reported of 503 Abell clusters of galaxies at 1400 MHz with a limiting flux density of 0.10 f.u. A total of 259 sources were detected within one Abell cluster radius. Contour maps are presented for 20 clusters. Three of the most interesting are discussed in more detail.

160.020 Sjachbazian I: unieke cluster in Grote Beer.
G. W. Beekman.
Zenit, Vol. 1, No. 5, p. 24 (1974).

160.021 Evidence for the existence of second-order clusters of galaxies. M. Kalinkov.
Galaxies and relativistic astrophysics, Proc. 1972, (see 012.010), p. 142 - 161 (1974).

160.022 Studies of rich clusters of galaxies − I. Galaxy counts for nine clusters of intermediate redshift and the angular diameter−redshift relation.
T. B. Austin, J. V. Peach.
Monthly Notices Roy. Astron. Soc., Vol. 167, 437 - 456 (1974).

Counts of galaxies in the fields of nine rich clusters with z ranging from 0.13 to 0.20 have been made on V plates taken with the Palomar 48-in. Schmidt telescope. The surface density of galaxies as a function of distance from the cluster centre is discussed in the context of three previously suggested approaches to a definition of angular size. The mean distance of galaxies from the line of sight through the cluster centre, $\bar{r}$, is a measure characterizing the size of the cluster as a whole. The intrinsic cluster and core radii in Mpc, are found to be related through an empirical equation. This equation is used to define a new measure of angular radius $\bar{r}_c$ by reducing the observed value of $\bar{r}$ to that which would be observed for a cluster of standard radius. A plot of $\bar{r}_c$ against z is given which shows a standard deviation of the residuals of $\bar{r}_c$ from a mean line with $q_0 = +1$ of only 18 per cent.

160.023 Compact groups of compact galaxies. II.
R. K. Shakhbazyan, M. B. Petrosyan.
Astrofizika, Vol. 10, 13 - 20 (1974). In Russian. − English translation in Astrophysics, Vol. 10, No. 1.

A second list of compact groups of compact galaxies is presented. The list contains data on 54 new objects of this class. Identification charts for all 54 groups of the list are given in red colour.

160.024 Four-colour photometry of compact groups of compact galaxies. F. Börngen, A. T. Kalloglyan.
Astrofizika, Vol. 10, 21 - 27 (1974). In Russian. − English translation in Astrophysics, Vol. 10, No. 1.

Integral brightnesses and B−V, V−r, r−i colours for objects in four compact groups of compact galaxies discovered at the Byurakan Observatory are given. Measurements have been made with iris photometer on plates obtained with the Tautenburg two-meter telescope. Most of the compact galaxies in the investigated groups are very red with B−V > $1^m.0$.

160.025 On the radio emission of clusters of galaxies.
G. M. Tovmasyan, M. S. Shirbakyan.
Astrofizika, Vol. 10, 29 - 38 (1974). In Russian. − English translation in Astrophysics, Vol. 10, No. 1.

It is shown that radio emission occurs from 6 to 7 times more often in clusters of galaxies which contain bright galaxies of cD-type or compact, peculiar or close double galaxies in comparison with clusters which do not contain outstanding bright galaxies or when bright galaxies are usually ellipticals. It is also shown that the percentage of radio emitting clusters of galaxies increases with the number of cluster members. Radio luminosities of detected radio sources are of the order of weak radio galaxies.

160.026 On the distribution of clusters of galaxies.
G. Reaves.
Astron. Zhurn. Akad. Nauk SSSR, Vol. 51, 520 - 525 (1974). In Russian. English translation in Soviet Astron. AJ, Vol. 18, No. 3.

The author demonstrates the inhomogeneity and the incompleteness of existing catalogues of clusters as well as the uncertainty of their recognition. He claims that for reliable conclusions on the structure of the metagalaxy it is necessary to avoid the defects of the catalogues mentioned above or at least to study them and to take them into account.

160.027 Inverse Compton radiation and the magnetic field in clusters of galaxies.
D. E. Harris, W. Romanishin.
Bull. American Astron. Soc., Vol. 6, 275 (1974). − Abstr. AAS.

160.028 Shachbazian I − an unusual cluster of galaxies.
Z. Klimek.

Urania Kraków, Vol. 45, 34 - 37 (1974). In Polish.

160.029 **Current constraints on hidden mass in the Coma cluster.** J. Tarter, J. Silk.
Quarterly Journ. Roy. Astron. Soc., Vol. 15, 122 - 140 (1974).

Various possible means of accounting for the dynamical mass of the Coma cluster are reviewed. The authors intentionally choose to emphasize conventional forms of hidden mass, specifically gas and low-luminosity stars, while adopting conventional mass-to-luminosity ratios for the observed luminous material in galaxies. They have attempted to combine all previous work relating to assumed gas parameters in order to examine precisely what constraints current observations, together with theoretical and observational upper limits, place on any gas in the Coma cluster.

Studies of blue objects at high galactic latitudes. III. Faint blue objects in the field of BD + 15°2469. See Abstr. 113.009.

Radio source structure, clusters of galaxies and the complex sources. See Abstr. 141.012.

The central radio sources in three rich clusters of galaxies. See Abstr. 141.028.

Radio observations of QSOs in the directions of Abell clusters of galaxies. See Abstr. 141.053.

OSO-7 observations of a high-latitude X-ray source associated with Abell cluster A2052. See Abstr. 142.031.

The extended X-ray source in Virgo and its relation to M87. See Abstr. 142.114.

Spiral and irregular galaxies associated with rich regular clusters. See Abstr. 158.043.

The nature of the distribution of galaxies. See Abstr. 158.065.

Intergalactic matter in clusters of galaxies. See Abstr. 161.002.

The small-scale anisotropy of the cosmic light. See Abstr. 162.011.

Formation of galaxies and clusters of galaxies by self-similar gravitational condensation. See Abstr. 162.012.

Errata

160.901 **Addendum: 'Apparent distribution and velocities of galaxies in the Virgo cluster and its surroundings'** [Astron. Astrophys., Vol. 28, 109 - 118 (1973)]. G. de Vaucouleurs, A. de Vaucouleurs.
Astron. Astrophys., Vol. 30, 479 (1974).

161 Intergalactic Matter

161.001 **On the existence and the density of intergalactic dust.** K.-H. Schmidt.
Astron. Nachr., Vol. 295, 163 - 168 (1974).

A relation between the redshift z of QSO's and their colour index $(B-V)'$ corrected for line emission and galactic absorption is interpreted in terms of an intergalactic selective extinction. The observed amount of extinction corresponds to a density of intergalactic dust grains of about 10^{-34} to 10^{-33} g cm^{-3} at $z = 0$. The life-time of these particles is estimated to be longer than the Hubble-time at present. But at large z the life-time of the grains considerably depends on the flux density of cosmic rays. Some implications of the existence of intergalactic dust are discussed.

161.002 **Intergalactic matter in clusters of galaxies.** G. B. Field.
Bull. American Astron. Soc., Vol. 6, 275 (1974). — Abstr. AAS.

Cosmic far ultraviolet background. See Abstr. 061.015.

The stability of galaxy clusters: neutral hydrogen observations. See Abstr. 160.013.

162 Structure and Evolution of the Universe, Cosmology

162.001 A time-symmetric, matter, antimatter, tachyon cosmology. J. R. Gott III.
Astrophys. Journ., Vol. 187, 1 - 3 (1974).

A generalized, time-symmetric, big-bang cosmology is proposed. In big-bang cosmologies the creation of the universe is associated with a singularity or singular region of spacetime. Solutions to the Einstein field equations are given which show that there are three distinct universes naturally associated with such a singularity.

162.002 Lepton creation and the Dirac relationship between fundamental constants. J. V. Narlikar.
Nature, Vol. 247, 99 - 100 (1974).

Cavallo (1973) has proposed an interpretation of the large number ratios between some of the fundamental constants occurring in physics and cosmology, in terms of rotating cosmological models. The purpose of this communication is to suggest another interpretation of these large dimensionless numbers in terms of lepton creation.

162.003 Generalisation of Hubble's law.
D. H. Weinstein, J. Keeney.
Nature, Vol. 247, 140 (1974). – Letter.

162.004 Primeval turbulence from matter-antimatter annihilation. N. Dallaporta, L. Danese, F. Lucchin.
Astrophys. Space Sci., Vol. 27, 497 - 512 (1974).

The present work is intended to test whether matter-antimatter annihilation theory at the origin of the universe, as it is developed by Omnès and his collaborators, may be considered to yield the required energy source of the turbulence. To this aim, the main results of the work of Stecker and Puget concerning the behaviour of the most important quantities of the matter-antimatter annihilation theory are used, in order to describe the properties of the energy source for the turbulence down to recombination time, and extended to take a somewhat more detailed account of their z and mean matter density dependences. The energy feeding source thus obtained is then introduced into the general framework of the galaxy formation theory from cosmic turbulence taking account of energy dissipation as developed in the authors' previous works.

162.005 Nicht-statische Weltmodelle. D. Wiedemann.
Orion Schaffhausen, 32. Jahrgang, p. 71 - 75 (1974).

162.006 Why the sky is dark at night. E. R. Harrison.
Phys. Today, Vol. 27, No. 2, p. 30 - 33, 35 - 36 (1974).

In a universe uniformly filled with stars we would expect the sky to be ablaze with light from all directions, according to a 250-year-old paradox we are just beginning to understand.

162.007 Evolution of inhomogeneous cosmological models.
W. B. Bonnor.
Monthly Notices Roy. Astron. Soc., Vol. 167, 55 - 61 (1974).

It is shown that an open Robertson-Walker (R-W) model can evolve from a variety of initial states. In fact there exist spherically symmetric inhomogeneous models which approach R-W ones as $t \to \infty$, but whose density on an initial space-like surface is an arbitrary function of the radial variable.

162.008 Matter-antimatter separation and the antimatter problem in cosmology. R. Omnès.
IAU Symposium No. 53, (see 012.004), p. 301 - 317 (1974).

162.009 Evolution of cosmological turbulence. I. Inertial transformation of the eddy velocity spectrum.
A. A. Kurskov, L. M. Ozernoj.
Astron. Zhurn. Akad. Nauk SSSR, Vol. 51, 270 - 280 (1974). In Russian. English translation in Soviet Astron. AJ., Vol. 18, No. 2.

Evolution of cosmological turbulence up to the moment of plasma recombination is studied.

162.010 Some peculiarities of the evolution of homogeneous anisotropic cosmological models. V. N. Lukash.
Astron. Zhurn. Akad. Nauk SSSR, Vol. 51, 281 - 292 (1974). In Russian. English translation in Soviet Astron. AJ, Vol. 18, No. 2.

The problem of isotropization of homogeneous cosmological models with hydrodynamic stress-energy tensor is considered. Analytical solutions for all characteristic cases of isotropization are given. Numerical calculations of the isotropization moment are given on the basis of up-to-date data on the degree of a large-scale anisotropy of the relict emission.

162.011 The small-scale anisotropy of the cosmic light.
S. A. Shectman.
Astrophys. Journ., Vol. 188, 233 - 242 (1974).

The author has measured the spatial power spectrum of the faint fluctuations of the nightsky brightness due to distant clusters of galaxies. This spectrum depends strongly on the luminosity density of the universe and on the covariance structure of the galaxy distribution. Based on the luminosity function for nearby galaxies, he obtains a density contrast $\langle \rho^2 \rangle / \langle \rho \rangle^2 = 29$ for galaxy clustering at a correlation length $l = 1$ Mpc.

162.012 Formation of galaxies and clusters of galaxies by self-similar gravitational condensation.
W. H. Press, P. Schechter.
Astrophys. Journ., Vol. 187, 425 - 438 (1974).

The authors consider an expanding Friedmann cosmology containing a "gas" of self-gravitating masses. The masses condense into aggregates which they identify as single particles of a larger mass. They propose that after this process has proceeded through several scales, the mass spectrum of condensations becomes "self-similar" and independent of the spectrum initially assumed. The results of numerical experiments on 1000 bodies are presented; these appear to show new nonlinear effects: condensations can "bootstrap" their way up in size faster than the linear theory predicts. The self-similar model predicts relations between the masses and radii of galaxies and clusters of galaxies, as well as their mass spectra. The authors compare the predictions with available data, and find some rather striking agreements.

162.013 The principle of correspondence in cosmology. I. The isotropic cosmos as Newtonian limit of relativistic theories of gravitation. H.-J. Treder.
Astron. Nachr., Vol. 295, 1 - 8 (1974). In German.

According to the principle of correspondence (in Heisenberg's formulation) each general relativistic theory of gravitation must give a Newtonian representation for an isotropic cosmos with the Robertson-Walker-metric. Indeed, the Friedmann equations can be interpreted as the expression for the Hamiltonian H of a closed Newtonian system of the cosmic fundamental particles, written in the rest-system of the center of gravity. In this Hamiltonian H only the relative-coordinates and the relative-velocities of the particles are present and one can write H without absolute quantities but only with Milne's relative-quantities.

162.014 **The final states of continuously expanding universes.** H.-J. Treder.
Astron. Nachr., Vol. 295, 9 - 10 (1974).

The relativistic expanding universes with negative curvatures $k < 0$ and $\lambda = 0$ or with positive cosmological constant $\lambda > 0$ become for $t \to \infty$, $R \to \infty$ empty space-times, asymptotically. This empty space-time is for $k < 0$, $\lambda = 0$ the Minkowski-world (in Milne's coordinates) and for $\lambda > 0$ a de-Sitter world.

162.015 **The principle of correspondence in cosmology. II. Classical and relativistic representations of world models.** H.-J. Treder.
Astron. Nachr., Vol. 295, 55 - 71 (1974). In German.

According to the equivalence between the Friedmann equation of relativistic cosmology and the condition for the time-independence $\dot{H} = 0$ of the Hamiltonian H of an isotropic particle system in the Newtonian mechanics the author constructs the corresponding classical Hamiltonians to the relativistic world-models. Each cosmological model which is resulting from a physically meaningful gravitation theory must give a Friedmann equation as the cosmological formulation of the time-independence condition of the energy H for the corresponding Newtonian N-particle system.

162.016 **Wie entwickelte sich die Welt?** W. Kundt.
Umschau, 74. Jahrgang, p. 273 - 277 (1974).

The most important measuring methods are reviewed and their consequences discussed.

162.017 **The influence of turbulence on the cosmological expansion.**
G. M. Vereshkov, L. S. Marochnik, N. V. Pelikhov.
Astron. Tsirk., No. 803, p. 5 - 7 (1973). In Russian.

162.018 **The Hubble relation for nonstandard candles and the origin of the redshift of quasars.** V. Petrosian.
Astrophys. Journ., Vol. 188, 443 - 449 (1974).

It is shown that the magnitude-log (redshift) relation for brightest quasars can have a slope different from the value expected for standard candles. The value of this slope depends on the luminosity function and its evolution. Therefore the difference of this slope from the expected value cannot be used as evidence against the cosmological origin of the redshift of the quasars.

162.019 **Warum ist das Weltall isotrop?**
C. B. Collins, S. W. Hawking, translated with a comment by O. Heckmann.
SuW, Vol. 13, 152, 154 - 156 (1974).

162.020 **Observations in locally inhomogeneous cosmological models.** C. C. Dyer, R. C. Roeder.
Astrophys. Journ., Vol. 189, 167 - 175 (1974).

Locally inhomogeneous "Swiss cheese" cosmological models have been extended to the cases in which the cosmological constant is nonzero. Application of the optical scalar equations in such models has yielded estimates of the distortions of distant objects and revised distance-redshift relations. Results of numerical integrations also show that in the Friedmann models the "zero shear" models previously discussed by the authors yield a good approximation as far as observational quantities are concerned.

162.021 **The gravitational-instability picture and the nature of the distribution of galaxies.** P. J. E. Peebles.
Astrophys. Journ., (Letters), Vol. 189, L51 - L53 (1974).

The covariance function for the distribution of galaxies is close to a power law, exhibiting no characteristic scales or features, over a broad range of the argument. This has a simple interpretation in the gravitational-instability picture for the evolution of irregularities in an expanding universe. A prediction for a new test of the picture is described.

162.022 **The effect of a linear rotational perturbation on the isotropy of the cosmic background radiation.**
A. M. Anile, S. Motta.
Astron. Astrophys., Vol. 32, 137 - 139 (1974).

The effect of a linear rotational perturbation of an Einstein-de Sitter universe on the anisotropy of the cosmic background radiation is investigated. The predicted temperature variations are compared with the observed upper limits (Conklin and Bracewell, 1968).

162.023 **The riddle of the redshifts.** G. Burbidge.
Mitt. Astron. Ges., No. 34, p. 19 - 42 (1973/74).
Invited paper presented at the "Wissenschaftliche Tagung der Astron. Ges., Wien, 1972 September 18 - 23".

162.024 **Magnetization, matter-antimatter symmetry and the baryon-photon ratio in the universe.** M. A. Melvin.
IAU Symposium No. 64, (see 012.008), p. 101 (1974). – Abstract.

162.025 **Modelli cosmologici.** G. Romano.
Coelum, Vol. 42, 13 - 22, with a correction p. 183 (1974).

162.026 **Interacting matter and radiation in homogeneous isotropic world models.**
W. R. Knight, O. Bergmann.
International Journ. Theor. Phys., (GB), Vol. 9, 47 - 54 (1974).

162.027 **A dust ball in empty space.**
M. P. Korkina, V. G. Martynenko.
Ukr. fiz. zhurn., Vol. 18, 1948 - 1954 (1973). In Russian.
Abstr. in Referativ. Zhurn. 51. Astron., 5.51.761 (1974).

162.028 **The effect of a lumpy matter distribution on the growth of irregularities in an expanding universe.**
P. J. E. Peebles.
Astron. Astrophys., Vol. 32, 391 - 397 (1974).

Given that matter is distributed in a lumpy fashion, in galaxies and clusters of galaxies in the present universe, perhaps in small lumps at an early epoch, one would suppose that the grainy distribution has some effect on the growth of irregularities. It has been suggested that there is a substantial minimum rate of growth of structure once grains form. However, it is shown here that the effect has been overestimated, so that it seems questionable whether it could have played an interesting role in the origin of galaxies.

162.029 **Upper limit on photon-photon interaction from physical cosmology.**
J. L. Puget, E. Schatzman.
Astron. Astrophys., Vol. 32, 477 - 478 (1974).

It is shown that if the microwave background has been emitted as black-body radiation, enough is known about its spectrum to put upper limits on photon-photon interaction relevant to the problem of the origin of redshifts.

162.030 **On the evolution of cosmological turbulence. II. Dissipation of subsonic motions and alternative ways for post-recombination evolution of turbulence.**
A. A. Kurskov, L. M. Ozernoj.
Astron. Zhurn. Akad. Nauk SSSR, Vol. 51, 508 - 519 (1974). In Russian. English translation in Soviet Astron. AJ, Vol. 18, No. 3.

The dissipation of cosmological turbulence up to the onset of recombination ($z = z_{rec} \approx 1.4 \times 10^3$) is considered and the distortions produced in the spectrum of relic radiation are calculated.

162.031 **An estimate of the size of the universe from the topological view-point.** D. D. Sokolov, V. F. Shvartsman.
Zhurn. ehksperim. i teor. fiz., Vol. 66, 412 - 420 (1974). In Russian. – Abstr. in Referativ. Zhurn. 51. Astron., 6.51.692 (1974).

162.032 **Is the cosmological constant really constant?** A. D. Linde.
Pis'ma v ZhurnEhTF, Vol. 19, 320 - 322 (1974). In Russian. Abstr. in Referativ. Zhurn. 51. Astron., 6.51.693 (1974).

162.033 **Charged cosmology.** G. W. Barry.
Astrophys. Journ., Vol. 190, 279 - 282 (1974).
It is proposed that the universe has a net charge consisting of a uniform intergalactic gas of protons or electrons. No fields are induced if the universe has Robertson-Walker symmetry. The principal effect is to give the photon a mass equal to the plasma frequency. Application of these ideas to the weak interaction leads to a weak coupling constant which increases with time.

162.034 **Structure (topology) of the universe.** Ya. B. Zeldovich.
Comments Astrophys. Space Phys., Vol. 5, 169 - 173 (1973).

162.035 **The integrated background radiation from galaxies in general relativistic world models.** K. Hara.
Publ. Astron. Soc. Japan, Vol. 26, 299 - 301 (1974).
In general relativistic world models the integrated background intensity from galaxies is computed numerically under the conditions that absorption effects due to galaxies and intergalactic matter, and the evolution effect of galaxies are neglected. The results are illustrated by contours on the (q_0, σ_0) plane.

162.036 **Observational cosmology – radio astronomy and the most distant galaxies.** M. S. Longair.
Proc. ESO/SRC/CERN conference, (see 012.021), p. 243 - 252 (1974).

162.037 **Cosmic evolution; stars, galaxies, and the universe.**
Mercury, (Journ. Astron. Soc. Pacific), Vol. 3, No. 1, p. 18 - 19 (1974). – An abstract of remarks by A. Sandage in his summer 1973 lecture in the series, 'The next billion years', sponsored by NASA/Ames Research Center, the A.S.P., and others. The abstract was prepared for the series sponsors by E. Burgess.

162.038 **Recent developments in cosmology.** M. J. Rees.
Phys. Bull., (GB), Vol. 24, 651 - 653 (1973).

162.039 **Limit on the rest masses from big bang cosmology.** A. S. Szalay. G. Marx.
Acta Phys. Acad. Sci. Hungaricae, Vol. 35, 113 - 129 (1974).
The age of the universe and the Hubble deceleration of galaxies depends upon the average mass density. The temperature of the electromagnetic background radiation determines also the neutrino particle density. These empirical informations put an upper limit on the rest masses of the neutrinos which are more restrictive than the laboratory values.

162.040 **Effective homogeneity of some inhomogeneous cosmologies.** F. Lund.
Phys. Rev. D, Particles and Fields, Vol. 8, 4229 - 4230 (1973).
Previously developed methods for the quantisation of spherically symmetric and dust filled geometries are applied to the inhomogeneous generalisations of Kantowski-Sachs and Bianchi type I models. The models turn out to have only a finite number of degrees of freedom and are thus effectively homogeneous.

162.041 **Big-bang model without singularities.** G. L. Murphy.
Phys. Rev. D, Particles and Fields, Vol. 8, 4231 - 4233 (1973).
A simplified universe containing fluid having pressure and viscosity is considered for which Einstein's equations are integrated exactly to yield a family of solutions characterized by a constant C. $C = 0$ is the steady state solution, and the solutions with $C > 0$ are that of physical interest.

162.042 **Electromagnetic waves in an expanding universe.** B. Mashhoon.
Phys. Rev. D, Particles and Fields, Vol. 8, 4297 - 4302 (1973).
A quantum theory of electromagnetic radiation in a Robertson-Walker universe is presented and it is shown that, whilst the annihilation and creation operators for the photon field depend upon time, the total number operator does not. Other properties of the theory are given.

162.043 **Notes on the cosmological principle.** J. Fujimura.
Sci. Rep. Yokohama National Univ., (Japan), Ser. I, No. 20, p. 15 - 20 (1973).
Roles of the cosmological principle are discussed, with the consequences in the framework of simple Newtonian theory. The fact that the result obtained coincides with Friedman's equation seems to indicate that the success of Friedman's model originates mainly in the cosmological principle, and not necessarily in the theory of general relativity.

162.044 **The universe is the unity of infinitude and finiteness.** S. Bian.
Acta Phys. Sinica, Vol. 23, No. 2, p. 83 - 94 (1974).

162.045 **Isotropic universe models and de-Sitter geometry.** G. Caviglia, G. Luzzatto, E. Massa.
Atti Accad. Sci. Torino, Ser. I, Vol. 107, 681 - 698 (1973). In Italian.

162.046 **Cosmological models and the large numbers hypothesis.** P. A. M. Dirac.
Proc. Roy. Soc. London, Ser. A, Vol. 338, 439 - 446 (1974).
The large numbers hypothesis asserts that all the large dimensionless numbers occurring in nature are connected with the present epoch, expressed in atomic units, and thus vary with time. It requires that the gravitational constant G shall vary, and also that there shall be continuous creation of matter. The consistent following out of the hypothesis leads to the possibility of only two cosmological models.

162.047 **Two sets of exact solution cosmological models.** R. F. Sisteró.
Univ. Nacional Tucumán, Rev. Ser. A, Vol. 22, 93 - 100 = Obs. Astron. Univ. Nacional Córdoba, Argentina, Tirada Aparte, No. 185 (1972).

162.048 **On the dissipation of vortex velocities in a low-density universe.** A. A. Kurskov.
Astron. Tsirk., No. 818, p. 1 - 2 (1974). In Russian.

162.049 **Contributions à l'étude de la symétrie matière-antimatière en cosmologie.** J.-L. Puget.
Thesis Sci. Phys., Paris–Meudon, AO-CNRS-8800, 129 pp. (1973).

162.050 **La transition de phase dans le modèle d'univers symétrique.** S. Caser.
Thesis Sci. Phys., Paris. AO-CNRS-9132, 31 pp. (1973).

162.051 **La coalescence dans le modèle d'univers symétrique.** R. Aldrovandi.
Thesis Sci. Phys., Paris. AO-CNRS-9133, 23 pp. (1973).

From the black hole to the infinite universe. See Abstr. 003.050.

Cosmology now. See Abstr. 003.068.

Kants Kosmologie und der physische Teil des naturwissenschaftlichen Weltbildes. See Abstr. 004.054.

Die Kosmogonie Immanuel Kants. I. See Abstr. 004.055.

Neutrinos in the universe. See Abstr. 061.019.

Die diffuse kosmische Gamma-Strahlung. See Abstr. 061.054.

Tetrad field equations and a generalized Friedmann equation. See Abstr. 066.031.

Observational background of Treder's gravitation theory in cosmology and the role of viscosity. See Abstr. 066.041.

A class of solutions of Einstein-Maxwell equations with the cosmological constant. See Abstr. 066.071.

f gravity and gravitational singularities. See Abstr. 066.080.

Electromagnetic fields in general relativity: a constructive procedure. See Abstr. 066.130.

General relativity, unitarian Einstein theory, expansion and age of the universe. See Abstr. 066.143.

The radio magnitude-redshift relationship for QSOs. See Abstr. 141.083.

Westerbork Synthesis Radio Telescope programmes in the perspective of observational cosmology. See Abstr. 141.113.

Roberts' redshift effect. See Abstr. 158.027.

Errata

162.901   Erratum: 'New limit on small-scale irregularities of 'blackbody' radiation' [Astrophys. Journ., (*Letters*), Vol. 180, L47 - L48 (1973)].   Y. N. Parijskij. Astrophys. Journ., (*Letters*), Vol. 188, L113 (1974).

162.902   Erratum: 'The role of the electron-neutrino interaction in the primordial gas' [Astron. Astrophys., Vol. 26, 123 - 125 (1973)].   H. F. Hecht. Astron. Astrophys., Vol. 32, 235 (1974).

Author Index

The authors are listed in alphabetical order
according to the initial letter following the first names.

Aannestad, P. A.
131.550
Aaronson, M.
132.013
Aarseth, S. J.
151.020 .050
Abalakin, V. K.
047.013
Abbasov, G. I.
114.058
Abdel-Wahab, S.
084.205
Abdusamatov, Kh. I.
072.011 .040
Abell, G.
003.014
Abell, G. O.
008.074
Abelson, P. H.
053.002
Abeyasekere, D.
033.091
Abhyankar, K. D.
121.006 .007
Ables, H. D.
154.029
Ables, J. G.
008.124
021.017
033.029
141.075 .308
Ablordeppey, V. K.
097.071
Abraham, H. J. M.
045.001
Abramenko, A. N.
097.015 .024 .030 .031
141.306
Abramenko, B.
141.021
Abramova, M. V.
094.127 .173
Abramowicz, M. A.
065.033
Abt, H. A.
112.002

Abt, H. A.
119.005
122.029
Abubakirov, N.
004.013
Acquista, C.
065.017
Acton, L. W.
073.053
076.026 .038
142.114
Adam, A.
004.078
Adam, N. V.
084.245
Adamczewski, J.
003.015
Adams, D. J.
125.049
Adams, J. B.
094.371 .381
Adams, J. H.
143.028
Adams, R.
125.054
Adams, T. F.
158.056
Adati, K.
031.083
Adcock, B. S.
010.008
Ade, P. A. R.
080.005
131.533
141.073
Adelman, S. J.
064.041
114.051 .078 .082
Adgie, R. L.
141.010
Adler, I.
142.075
Adler, J. E. M.
002.029
Afanas'eva, P. M.
041.030

Afanas'eva, V. I.
106.011 .013
Agapov, E. S.
055.001
Agekyan, T. A.
042.003
Agrawal, P. C.
141.329
142.044 .065
Agrawal, S. P.
143.033
Agreen, R. W.
046.028
Agrinier, B.
134.006
142.005
Agueero, E. L.
158.026
Ahern, F. J.
034.023
Ahmad, A.
151.019 .025 .026 .033
Ahmad, K.
042.072
Ahnert, P.
121.089
123.045
Aiello, S.
125.061
Aikman, G. C. L.
103.100
124.104
Aitchison, C. S.
033.062
Aitken, D.
155.065
Aitken, D. K.
131.522
Aizenman, M. L.
065.105
Ajello, J. M.
097.074
099.012
Akabane, K.
033.049 .094
131.024 .555

Akasofu, S.-I.
 084.001 .025 .036 .256
Akhmedov, Sh. B.
 077.048
Akhundova, G. V.
 114.035
 119.002
 125.025
Akimov, V. V.
 051.016
Aksnes, K.
 096.029
 099.201
 101.010
Alaev, L. B.
 004.058
Alaniya, I. P.
 122.050 .051
Alarcon, C. E.
 034.087
Albers, H.
 114.103
Albert, E.
 114.104
Albrecht, C.
 094.358
Alcaino, G.
 154.001 .002 .026
 158.015
Alcock, G. E. D.
 103.101 .105
Aldrovandi, R.
 162.051
Alduseva, V. Ya.
 121.029
Alekhin, Yu. V.
 091.017
Aleksandrov, V. S.
 054.018
Aleksandrov, Yu. V.
 097.014
Aleksandrovskij, N. M.
 033.016
Alekseev, N. V.
 034.018
Alekseev, V. A.
 105.117 .118
 141.055 .056 .077
Aleshin, V. I.
 077.010
 080.013
Alexander, C.
 094.326
Alexander, D. R.
 064.032
Alexander, J. B.
 114.088
Alexander Jr., E. C.
 094.312 .383
Alexandropoulos, N. G.
 034.078
Alferov, A. M.
 082.057
Alferov, G. V.
 053.017
Alfven, H.
 094.004
 102.014
 107.006 .014
Alimov, O.
 083.032 .033 .034

Alissandrakis, C.
 077.070
Alksne, A.
 008.107
Alksne, Z.
 010.017.
 113.028
Alksnis, A.
 011.027
 113.030
 141.085
Alksnis, A. K.
 122.040
 124.005
Alladin, S. M.
 158.079
Allan, R. R.
 084.208
Allegre, C. J.
 094.328
 105.018
Allen, B. J.
 065.026
Allen, C.
 155.034
Allen, D. A.
 113.022
 114.148
Allen, L. R.
 115.004
Allen, M.
 131.025
Allen, M. S.
 114.160
Allen, R. J.
 021.018
 034.002
 125.100
Aller, H. D.
 141.030
Aller, L. H.
 007.000
 071.025 .029 .030
Allison, A.
 153.012
Alodzhants, G. P.
 022.014
Alojants, G. P.
 See Alodzhants, G. P.
Alpers, W.
 143.003
Alpert, N. L.
 003.016
Altenhoff, W. J.
 031.076
Alter, D.
 003.017 .018
Altrock, R. C.
 071.052 .055
 080.008
Altschuler, M. D.
 074.087
Altukhov, A. M.
 143.049 .050
Alvarez, H.
 077.016 .071 .901
Alvarez, J. M.
 104.010
Alvarez, M.
 076.025

Ambartsumyan, V. A.
 003.019
 122.020 .090
Ambruster, C.
 141.002
Ameling, W.
 103.101 .128
Amirkhanyan, V. R.
 034.046
 141.041
Amnuehl', P. R.
 125.006
 142.026 .041
Amte, P. G.
 065.119
Ananth, A. G.
 143.033
Ananthakrishnan, S.
 106.018
 141.064
Anastassiadis, M. A.
 079.100
Anderlucci, E.
 103.101
Anders, E.
 094.375
 105.054
Andersen, B. N.
 122.074
Andersen, F.
 084.304
Andersen, J.
 118.004
 121.014
Anderson, B.
 031.033
Anderson, C. M.
 114.157
Anderson, D.
 063.041
Anderson, D. L.
 094.129
Anderson, J. C.
 066.140
Anderson, J. D.
 092.040
 093.022
 099.204
Anderson, K. A.
 076.011 .022
 084.033 .231 .284
 142.079
Anderson, K. R.
 094.111 .318
Anderson, K. S.
 114.106
 158.033 .063
Anderson, L.
 142.094
Anderson, R. C.
 022.029
Anderson, S. N.
 143.055
Anderson Jr., D. E.
 097.049 .052
Andersson, C.
 131.019
Andersson, K. G.
 103.105
Andreev, B. N.
 093.034

Andrejchikov, B. M.
093.002
Andrejko, A. V.
071.039
Andresen, R. D.
051.023
Andrew, B. H.
141.050 .076 .902
Andrews, P. J.
008.102
121.068
122.134
125.101
Andrianov, N. K.
097.003
Andrianov, S. A.
077.049
Andrienko, D.
103.101
Andriesse, C. D.
131.005
Andrillat, Y.
103.101
124.102
133.006
Andzans, A.
005.008
Angel, J. R. P.
115.009
126.015 .021
131.055
132.011 .032
Angelov, T.
122.128
Anger, C. D.
084.038
Angione, R. J.
082.041
Angstroem, A.
082.068
Anguita, C.
041.038 .044
Anile, A. M.
063.026
162.022
Anisimov, V. F.
055.001
Ansour, M.
142.087
Antal, M.
098.029
103.101
Antonets, M. A.
141.055 .056 .077
Antonov, A. V.
033.013
Antonov, V. A.
151.044
Antonova, L. A.
082.053
Antonucci, E.
074.009 .030 .061 .092
Aoki, S.
041.026
Apeldoorn, B.
104.023 .024 .026 .073
Apparao, K. M. V.
143.009
Appenzeller, I.
065.014 .067
073.046

Appenzeller, I.
122.089
Apruzese, J. P.
114.141
141.605
Aptekar', R. L.
104.058
Apushkinskij, G. P.
033.051
077.047
Aranoff, S.
066.137
Araya, G.
125.102
Ardeberg, A.
113.025 .026
125.102
Arena, P.
077.037 .038
Argence, E.
066.010
Argyle, E.
117.008
Argyrakos, I.
003.009
041.032
Ariskin, V. I.
155.040
Arkani-Hamed, J.
094.002
Arkhangel'skij, Yu. B.
031.010
Arkhipova, V. P.
114.121
132.009
133.009
158.124
Armaly, B. F.
063.024
Armijo, T. C.
103.101 .105
Armstrong, K. R.
034.042
Arnett, W. D.
061.068
065.087 .110 .111 .901
066.108
125.027 .066
143.020
158.121
Arnold, D. A.
046.026
Arnold, J. R.
094.124
Arnold, K.
081.032
Arnoldy, R. L.
084.026
Arnould, M.
065.049 .104
Arnush, D.
083.044
Arny, T.
008.005
Arons, J.
134.010
142.094
Arora, K. K.
077.902
Arp, H.
007.000

Arp, H.
012.015
125.037
Arp, H. C.
125.107 .109
158.022 .083
Arrhenius, G.
094.004
105.147
107.006 .007 .014
Arsen'eva, O. A.
033.015
Arsenijevic, J.
122.073 .128 .129
Artem'ev, A. V.
014.011 .014
080.034
Artyukh, V. S.
033.008
Artyukhina, N. M.
112.014
Arvidson, R. E.
097.005 .021
Asare, A.
014.029
Asbridge, J. R.
074.041 .054 .073 .080
092.035
093.021
Ash, R.
003.138
Ashbrook, J.
004.020 .065
005.001
007.000
092.002 .003
094.105
111.002
Ashby, D. E. T. F.
061.009
Ashcraft, C.
032.015
Ashcraft, V.
079.100
Ashenfelter, T. E.
082.003
Asimov, I.
003.020 .021 .022 .083
Asnin, S. K.
052.043
Assousa, G.
141.076
Asteriadis, G.
122.107
Athay, R. G.
073.006
Atreya, S. K.
099.040
Atwood, B.
031.059
Aubier, M. G.
077.027
Audouze, J.
143.001
Audouze, J. M.
065.070
Auer, R.-D.
084.224
Aufgebauer, P.
005.013

Auman, J. R.
 064.057
 099.003
Auner, G.
 153.002
Austin, R. R.
 114.142
Austin, R. R. D.
 103.105 .135
Austin, T. B.
 160.022
Avduevskij, V. S.
 093.007 .038 .039
Avedisova, V. S.
 131.101
Avery, L. W.
 073.075
Avery, B. W.
 099.007 .008
 131.003 .121
Avni, Y.
 142.029
Avrett, E.
 113.033
Axford, W. I.
 082.025
 091.013
Axisa, F.
 077.026
Ayres, T. R.
 064.055
Ayubasheva, S. I.
 085.010

Baars, J. W. M.
 033.001
 131.545
Baart, E. E.
 009.006
Babadzhanov, P. B.
 083.031
 104.031
Babadzhanyants, M. K.
 158.091 .130
Babaev, A. H.
 073.007
Babayev, A. H.
 See Babaev, A. H.
Babcock, H. W.
 008.098
Babcock, R. R.
 077.036
Babcock, T. A.
 036.002
Babichenko, S. I.
 054.018
Babkov, F. I.
 094.127 .173
Bacchus, P.
 041.059
Backer, D. C.
 141.315 .319 .343
Backus, G. E.
 084.301
Badalyan, O. G.
 073.002
Baddenhausen, H.
 105.081
Badhwar, G. D.
 061.010

Badhwar, G. D.
 143.015 .028
Baedecker, P. A.
 105.019
Baglin, A.
 122.018
Bahcall, J. N.
 141.024
 142.015 .029 .049
 158.114
Bahcall, N. A.
 142.015
 160.008
Bahnsen, A.
 084.281 .286
Bahyl', V.
 158.047
Baillard, N.
 105.006
Baird, A. K.
 097.079
Baity, W. A.
 142.006 .055 .078 .085
Baize, P.
 118.001
Baker, D.
 051.003
 053.011
 054.009 .016
Baker, D. N.
 099.017
Baker, J. C.
 084.297
Baker, J. R.
 155.041
Baker, K. D.
 084.010
Baker, L.
 099.022
Baker, P. L.
 131.043
Baker, P. W.
 115.005
Baker, R.
 099.022
Baker, R. H.
 003.045
Bakharev, A. M.
 104.061
Bakhrakh, L. D.
 033.011
Bakhshiyan, B. Ts.
 053.004
Bakhshyan, G. G.
 061.057
Bakos, G. A.
 118.015
 123.009
Baldinelli, L.
 031.087
 096.039
Baldini, R.
 075.005
Baldwin, B. W.
 121.023
Balfour, W. J.
 103.100
Balick, B.
 131.521
 132.004
 133.008

Balick, B.
 141.076
Balint, E.
 042.086
Balklavs, A.
 003.008
 010.017
Ball, J. A.
 103.101
 114.132
 131.544 .902
Ballabh, G. M.
 151.008
 158.079
Ballczo, H.
 105.153
Ballif, J. R.
 084.228
Bally, M.
 124.103
Balmino, G.
 081.006 .028
Balodis, J.
 031.064
 036.008
Balona, L.
 114.019
 115.002
Balona, L. A.
 155.042
Balsley, B. B.
 083.047
Bame, S. J.
 074.041 .073 .080
 084.257
 092.035
 093.021
Banducci
 092.024
Bandyopadhaya, R.
 155.074
Banos, C.
 032.036
Banos, G.
 141.091
Bansal, B. M.
 094.347
Bar, V.
 073.068
Baranne, A.
 032.047
Baranov, A. S.
 151.029
Baranova, L.
 158.053
Baranovskij, E. A.
 072.005
Baranovskij, I. V.
 071.050
Baratta, G. B.
 103.101
 114.052
Barbanis, B.
 012.010
 151.047
Barbaro, G.
 065.100
Barbieri, C.
 008.008
 032.014
 113.009

Barbon, R.
125.038 .039 .105
142.017
153.901
Barbour, J. B.
066.036
Bardeen, J. M.
062.070
066.022 .023 .067
Bardfield, W. A.
103.105
Bareau, C.
072.024
079.100
Barisch, S. F.
015.003
Barkat, Z. K.
065.018
Barker, K. D.
066.019
Barkstrom, B. R.
091.021
Barletti, R.
082.076
Barlier, F.
081.017
Barlow, D. J.
121.040 .047
Barnard, A. J.
114.127
Barnes, A.
062.064
074.037
Barnes, J. V.
153.022 .029
Barnes, R. C.
121.049
Barnes, T. G.
114.001
Barnothy, J.
080.043
151.054
Barnothy, J. M.
091.002
160.014
Barnothy, M. F.
141.044
160.014
Barocas, V.
010.012
103.011
Barraclough, D. R.
084.216
Barreto, A.
105.020
Barrett, A. H.
114.048
Barros, S.
103.010
Barry, D. C.
115.011
153.001
Barry, G. W.
162.033
Bartenwerfer, D.
080.060
Barth, C. A.
097.017 .046 .072
Bartkevicius, A.
113.004 .005

Bartky, C. D.
104.009
Baryshnikova, G. V.
105.119
Basart, J. P.
033.047
Baschek, B.
064.007
125.018
Bash, F. N.
033.038
Bashkirtsev, V. S.
073.020
Bashkov, V. I.
066.102
Bashtova, L. I.
097.039
Basinska, E.
131.110
Basistov, G. G.
033.004
Basko, M. M.
142.018
Bastin, J. A.
094.170 .307
Basu, B.
155.074
Basu, D.
141.084
Batchelor, A. S. J.
132.019
Batchelor, R. A.
131.514
141.308
Bates, B.
114.118
Bates, D. R.
015.006
Bates, H. F.
083.049
Bates, L.
082.003
Bates, R. H. T.
031.070
033.032
Bateson, F. M.
010.024
011.022
013.004
122.045
124.011
Bath, G. T.
117.009
124.100
Batrakov, Yu. V.
151.029
Batten, A. H.
121.014 .086
Batts, B. D.
094.349
Baturin, V. V.
072.019
Baudry, A.
141.603 .609
Bauer, J.
105.088
Bauer, S. J.
093.043
Baum, P. J.
062.083
073.070

Baumann, G.
151.032 .902
Baumert, J. H.
115.012
Baumgarte, J.
042.034
Baur, H.
094.327 .337 .338 .348
.352 .401
Baxter, R.
099.036
Bay, Z.
022.023
Bayer, G.
094.400
Bazilevskaya, G. A.
143.027
Bazilevskij, A. T.
094.302
Bazykin, V. V.
055.013
Bean, B. R.
031.077
Beard, D. B.
084.206 .207 .276 .278
099.057
Beaudet, P. R.
021.010
Beavers, W. I.
008.004
115.014
Bechis, K. P.
114.132 .161
131.111
Beck, H. G.
032.010
Beck, R.
021.013
036.001
Beckel, C. L.
099.064
Becker, G.
010.017
035.007
Becker, G. A.
103.105
Becker, H. J.
104.019
Becker, K.
052.008
Becker, R. H.
073.005 .008
077.070
Becker, R. R.
105.152
Becker, R. S.
022.049
Becker, U.
022.063
Beckers, J. M.
074.011
Becklin, E. E.
032.046
131.522 .524 .527
141.613
Beckmann, P.
033.095
Bedford, D. K.
054.015
Bednyakov, A. A.
084.235

Beeckmans, F.
114.125
115.013
Beekman, G. W.
160.020
Beekman, G. W. E.
010.015
098.045
101.012
103.105
Beelen, W.
065.104
Beer, A.
003.001
Beers, Y.
131.004
Beet, E. A.
010.012
014.041
Begkhanov, M.
104.003
Begot, J.
079.100
Behannon, K. W.
092.036
093.023
Behr, A.
031.027
Behring, W. E.
062.004
071.056 .066
074.046
Beichman, C. A.
132.025
155.020
Bejtrishvili, I. R.
103.101
Bekenstein, J. D.
066.121
117.038
Bel, N.
062.081
Belcher, J. W.
062.063
Beletskij, V. V.
051.016
Belikovich, V. V.
083.006
Belinskaya, S. I.
082.020 .106
Bel'kovich, O. I.
104.027
Bell, M. C.
143.065
Bell, R. A.
064.005
Bellman, R.
063.021
Bellyustin, N. S.
062.046
Belokon', E. T.
122.099
158.130
Belon, A. E.
084.029
Belotserkovskij, O. M.
051.012
Belous, L. M.
103.115 .117
Belov, I. F.
141.055

Belozerova, M. A.
151.045
Belton, M. J. S.
092.017 .038 .039
093.024 .025
Belvedere, G.
080.052
Belyakina, T. S.
121.020
122.121
Belyashin, A. P.
084.266
Bemalkhedkar, M. M.
143.033
Benech, B.
082.061
Benedict, G. F.
114.138
142.074
Benediktov, E. A.
083.006
Benest, D.
042.041
Benhocine, M.
041.020
Benkova, N. P.
084.245
Bennett, J. C.
103.101 .102
Benning, W.
046.022
Bentley, A. N.
094.146
Benvenuti, P.
102.004
103.101
113.009
142.017
Benyukh, V. V.
104.052
Berdichevskij, M. N.
084.268
094.182
Berdot, J. L.
105.064
Berezin, Yu. E.
078.003
084.212
Berezinsky, V. S.
061.078
Berg, L.-E.
022.007
Berg, R. A.
103.101
Bergeat, J.
131.504
Berger, J.
114.038
Bergmann, O.
162.026
Bergmann, P. G.
066.060
Bergstralh, J. T.
099.048
Bergstroem, J.
062.016
Berkey, F. T.
084.013
Berlin, A. B.
033.015
077.042

Berlin, A. S.
033.015
Berlyand, N. G.
084.270
Bernacca, P. L.
142.017 .099
Bernard, A.
153.025
Bernat, A. P.
132.006 .026
Bernat, T. P.
033.021
Bernhard, H.
092.008
Bernhardt, H.-J.
094.335
Bernot, M.
075.016
Berruyer, N.
065.013
Berry, H. G.
022.044 .052
Bertaud, C.
114.097
122.901
Bertelli, G.
065.083
Bertojo, M.
155.035
Bertola, F.
158.055
Bertotti, B.
012.019
Berulis, I. I.
077.043 .046
155.040
Bessell, M. S.
113.011
142.010 .020 .025 .103
.106
Besson, R.
035.010
Best, A.
084.244
Best, G.
099.022
Bethe, H. A.
061.022
Betlem, H.
095.004
123.005
Betti, A.
010.041
Bettis, D. G.
012.005
021.001 .007
Betz, A. L.
097.066
Beyer, M.
103.101
Bezotosnyj, A. A.
034.021
077.022
Bezrukikh, V. V.
084.266
106.027
Bezus, V. A.
084.409
Bhandari, S. M.
106.018
141.064

Bhardwaj, S. N.
022.052
Bhatia, P. K.
062.009 .012
Bhonsle, R. V.
074.004
Bian, S.
162.044
Bibarsov, R. Sh.
104.035 .066 .067
Bibik, E. B.
052.022
Bibron, R.
105.006
106.002
Bidelman, W. P.
008.035
Bielicki, M.
117.001
Biermann, L.
003.111
102.009
Biermann, P.
065.003
155.002
Bigay, J.
122.901
Bigay, J.-H.
031.071
153.025
Biggar, G. M.
094.306
Bignami, G. F.
155.026 .058
Bignell, C.
133.008
Bijaoui, A.
114.009
Bild, R. W.
105.003
Billaud, G.
041.048
Billings, D. E.
076.025
Bingham, R. G.
113.015
154.004
Binz, C. M.
105.072
Biraud, F.
033.056
Biraud, Y.
012.011
Birck, J. L.
094.328
105.018
Bird, M. K.
084.278
Birkle, K.
082.043
Birmingham, T.
099.036
Biryulina, M. S.
097.001
Bishara, A. A.
143.018
Bishop, R. H.
084.010 .019
Bisiacchi, G.
121.008

Bisiacchi, G. F.
117.029
Bisnovatyj-Kogan, G. S.
061.004
065.051
066.030
141.311 .313
142.027
Bjordal, J.
084.220
Bjorkholm, P.
094.116 .313
Blaauw, A.
008.056
012.010 .018 .021
Black, D. C.
065.010
Black, J. H.
102.011
103.101
Black, W.
042.062
Blackwell, K. C.
041.008
Blaettner, W. G.
082.070
Blagonravov, A. A.
051.008
Blair, D. G.
033.021
Blair, G. N.
100.211
132.035
Blake, G.
141.076
Blake, J. B.
065.112
078.018
084.282 .402
Blamont, J.
103.101
Blamont, J. E.
082.033
084.035
Blanco, C.
010.017
064.050
099.219
100.210 .214
121.048
Blanco, V. M.
114.137
Bland, R.
033.020
Blau, P. J.
094.340
Bleach, R. D.
142.082 .090 .091
Bleeker, J. A. M.
142.068
Bleiweiss, M.
077.074
Blenman, C.
099.022
Bless, R. C.
008.076
Bleuler, K.
011.001
Blinnikov, S. I.
065.051

Blitzer, L.
042.015
052.006
Bloch, M. R.
105.021
Block, L. P.
084.240
Bloomberg, H. W.
151.001
Blum, E. J.
033.056
Blum, P.
082.107
Blum, P. W.
082.078
Blumenthal, G. R.
022.011
Bobrov, M. S.
100.003
Bochkarev, N. G.
062.049
Bodechtel, J.
003.024
Bodenheimer, P.
125.071
Boeckl, R.
061.071
Boehm, S.
075.011
Boehm-Vitense, E.
122.049
Boehme, A.
075.011
Boehme, S.
002.021
Boella, G.
134.006
142.005
Boenkova, N. M.
083.043
Boerner, G.
065.042 .076
141.310
Boerngen, F.
032.011
132.007
160.024
Boesgaard, A. M.
114.064
116.008
117.020
Boeshaar, G. O.
133.003
Bogatikov, O. A.
094.373
Bogdanov, A. V.
074.034
Bogdanov, M. B.
103.101
122.036
Bogdanov, V. V.
084.265
Bogott, F.
084.226
Bohlin, J. D.
073.065 .071
074.103
075.008
Bohlin, R. C.
114.007 .134 .135

Bohuski, T. J.
 133.010
Boischot, A.
 077.067
Bok, B. J.
 005.009
 012.014
 131.048
Bokhan, N. A.
 103.102
Boksenberg, A.
 032.034 .902
 114.118
Bolbochanu, A. V.
 112.008
Boldt, E.
 142.089
Boldt, E. A.
 142.043 .048 .060 .061
 .116
Boley, F.
 142.076 .077
Bolkvadze, O. R.
 096.013
 099.213
Boller, B. R.
 084.234
Bol'shoj, A. A.
 094.127 .173
Bolton, J. G.
 008.124
 141.109
Bolton, K.
 131.069
Bonanomi, J.
 008.090
Bonazzola, S.
 066.048
Bondal, K. R.
 114.065
Bondarenko, L. N.
 003.025
 008.084
Bondarenko, N. P.
 062.020 .021
Bondarev, A.
 003.106
Bondarev, B. V.
 044.015
Bonetti, A.
 125.061
Bonnor, W. B.
 162.007
Bonoli, F.
 125.035
Bonometto, S.
 063.014
 141.610
 158.097
Bonometto, S. A.
 061.006 .074
Bonov, A.
 092.016
Bonov, D.
 047.011
Bonsack, W. K.
 114.156
 116.001 .013
Booth, R. S.
 131.023

Bopp, B. W.
 114.090 .150
 122.004 .086 .131
 142.013 .088
Borghesi, A.
 012.015
Borgman, J.
 061.061
Boriakoff, V.
 033.041
Borisov, Eh. A.
 052.036
Borisov, Yu. V.
 034.027
 122.058
Born, G. H.
 052.003 .014
Born, R.
 073.045
Born, W.
 094.394
 105.022
Borodin, N. F.
 093.014 .038
Borovik, V. N.
 077.011
Borra, E. F.
 116.002 .005
Bortle, J.
 103.101 .102
 123.027
Bortle, J. E.
 103.101 .105
 123.031
 125.112
Borzov, G. G.
 103.101
Bosqued, J. M.
 084.009
Bottinelli, L.
 158.106
Boughn, S. P.
 033.021
Bouigue, R.
 031.054
 041.043
Boulesteix, J.
 131.552
Bourgeois, J.
 096.022
Bourot, M.
 105.023
Boury, A.
 065.020
Bouska, J.
 015.018
 103.101
Bova, B.
 003.145
Bower, J.
 094.345
Bowers, F. K.
 033.028
 034.038
Bowers, P. F.
 131.541
Bowers, R. L.
 066.028
 117.038
Bowles, J. A.
 032.054

Bowling, S. B.
 084.263
Bowman, M. R.
 082.024
Bowyer, C. S.
 082.030
Bowyer, S.
 034.044
 061.015
 082.029
 131.036 .040
 158.102
Boyd, R. L. F.
 008.073
Boynton, P.
 142.056
Boynton, P. E.
 066.008
Boynton, W. V.
 105.024
Bozis, G.
 042.073
Braccesi, A.
 036.010
 141.090
Bracewell, R. N.
 008.122
 033.006 .036
Bradbury, R.
 003.026
Bradfield, W. A.
 103.105
Bradford, H. M.
 077.020
Bradt, H.
 142.117
Bradt, H. V.
 061.063
 125.104
Braegger, H.
 034.015
Braginskij, S. I.
 084.269
Braginskij, V. B.
 033.018 .023
 066.085
Brancewicz, H.
 079.100
Branch, D.
 112.004
 114.088
 125.046
Brandi, E.
 114.041
Branley, F. M.
 003.027
Bratenahl, A.
 062.083
 073.070
Bratijchuk, M. V.
 031.017
 054.006
Braude, S.
 141.118
Braude, S. Ya.
 141.056 .076
Braunsfurth, E.
 021.014
Bravo, S.
 078.027

Bray, R. J.
003.028
Breakwell, J. V.
042.056
098.007
Brecher, A.
105.025 .147
Brecher, K.
061.035
065.109
142.008
Bredekamp, J. H.
143.007
Bredov, M. M.
104.058
Breger, M.
114.029
Bregman, J.
142.112
Breig, W. F.
063.016
Brejdo, I. I.
036.012
Brennan, M. H.
062.051
Bretagnon, P.
042.001
Brett, R.
094.130
Bretterbauer, K.
046.023
Bretz, M.
122.096
Breuer, R. A.
063.026
066.135
Breysacher, J.
114.067
Brezgunov, V. N.
155.040
Brice, N.
062.065
Brickhill, A. J.
124.101
Briden, J. C.
081.013
Bridge, H. S.
092.035
093.021
Bridle, A. H.
141.043 .076
Briggs, C.
094.326
Briggs, F. H.
100.010 .207
Brini, D.
061.011
Brinkman, A. C.
142.042
Britten, R.
094.304
Brittin, W. E.
012.004
Broadfoot, A. L.
092.038
093.025
103.101
Broderick, J.
066.035
Broderick, J. J.
141.060

Brodskaya, E. S.
121.020
Broenstad, K.
084.220
Bromley, D. A.
022.021
Bronnikova, N. M.
103.101
112.007
Bronshtehn, V. A.
010.033
014.035
097.061
Brooking, N.
033.089
Brooks, J. W.
008.124
033.029
Brooks, R. C.
122.084
Brosche, P.
066.090
141.095
Broten, N. W.
159.005
Broucke, R.
052.013
Broucke, R. A.
042.025
Brouw, W. N.
033.054
Browell, E. V.
022.029
Brown, J. C.
073.061
076.003
Brown, P. Lancaster
See Lancaster Brown, P.
Brown, R.
033.052
Brown, R. A.
099.209 .210 .222
Brown, R. D.
131.069
Brown, R. Hanbury
See Hanbury Brown, R.
Brown, R. L.
063.032 .040
131.053 .064
Brown, R. R.
084.220
Browne, I. W. A.
141.053
Brownlee, D. E.
105.066
Brownrigg, D. R. K.
151.035
Brucato, R. J.
114.081
Brueck, H. A.
008.043
Brueck, M. T.
132.002
Brueckner, G. E.
073.065 .071
074.103
075.008
Bruin, F.
075.010
Brun, A.
032.020

Brunet, J.-P.
005.015
Brunn, D. L.
092.040
093.022
Bryan, H. W.
054.015
Brzozowski, J.
022.019
Bubet, D.
031.071
Buccheri, R.
134.006
142.005
Buchanan, D. J.
061.009
Buchet, J. P.
022.045
Buchet-Poulizac, M. C.
022.045
Buchler, J.-R.
065.018 .037 .106
125.057
Buchroeder, R. A.
031.038
Buchtela, K.
105.152
Buchwald, V. F.
105.026
Budding, E.
117.027
121.016 .055
Budine, P. W.
099.070 .071
Budz'ko, V. K.
032.026
Buecher, A.
118.013
Buehler, F.
082.025
Buerger, P. F.
063.002
Buff, J.
142.030 .046
Bufton, J. L.
082.074
Bugaenko, L. A.
103.101
Bugaenko, O. I.
103.101
Bugaevskij, A. V.
074.006 .071
Buhl, D.
103.101
114.140
131.060 .062 .130 .134
Buhler, R.
105.145
Bui-Van, A.
061.032
160.006
Buijze, W.
105.108
Bujakiewicz, A.
046.024
Bulashevich, Yu. P.
084.259
085.010
Bulger, J.
098.029
103.009 .108 .127 .130

AUTHOR INDEX - VOL. 11

Buonanno, R.
103.101
Buontempo, M. E.
041.008
Buravtsev, A. K.
083.039
Burba, G. A.
011.009
094.141
Burbidge, E. M.
141.067 .119
Burbidge, G.
162.023
Burbidge, G. R.
032.050
141.067
158.075
Burchi, R.
098.012
Burdyuzha, V. V.
131.032 .103 .509
Bures, M.
062.016
Burgat, W.
125.112
Burger, J. J.
143.023
Burger, M.
114.111
Burgess, E.
162.037
Burginyon, G.
142.051 .083
Burginyon, G. A.
142.033
Burgoyne, N.
042.008
Burie, J.
022.022
Burinskij, A. Ya.
066.097
Burke, J. R.
131.086 .087
Burke, T. E.
099.012
Burke, W. J.
084.274
Burkhart, C.
114.131
Burlaga, L. F.
074.081
103.100
Burman, R. R.
062.057
Burnett, D. S.
094.303
Burnett, G. B.
106.020
Burns, J. A.
042.007
Burov, A. F.
034.008
Burrows, J. R.
078.015
Burt, S. E.
111.010
Burton, R. K.
084.255
Burton, W. B.
155.012

Busch, H.
123.014
Buschmann, E.
046.030
Buscombe, W.
114.169
122.011
Buseck, P. R.
105.009
Bushuev, E. I.
052.026
Bushuev, K. D.
051.015
Bustati, N. G.
075.010
Butcher, J. C.
021.006
Butenko, R. I.
120.002
Buti, B.
062.078
Butler, D.
142.112
Butler, D. M.
093.037
Butler, J. C.
094.360
Buurman, J.
072.008
Buyalo, A. S.
094.324
Buznikov, A. A.
082.110
Byard, P. L.
102.020
Byatt, T. P.
010.012
Bychkov, K. V.
125.005 .012
Bykov, M. F.
041.017 .065
Byrd, G.
151.055
Byrne, J. C.
062.057

Cacciari, C.
122.093
Caccin, B.
071.022
072.021
Cadez, A.
066.139
Cadogan, P. H.
105.078
Cahill, T. C.
033.086
Cahill Jr., L. J.
084.290
Cahn, J. H.
133.011
Cailleux, A.
081.016
Cain, D. L.
099.205
Cain, P. D.
010.024
Caldecott, R.
033.063

Caldwell, J. J.
099.001
125.102
Callahan, P. S.
074.001
Calvani, M.
066.001
Calvert, T. A.
094.171
Calvird, H. R.
103.105
Cameron, A. G. W.
003.101
012.004
065.029
099.062
Cameron, I. R.
105.109
Cameron, W. S.
094.139
Campbell, D. B.
093.028
141.315
Canal, R.
065.095
Candy, M. P.
103.105
Cane, H. V.
132.005
Canfield, R. C.
071.052
073.006 .031 .072
Canizares, C. R.
125.102
142.019
Cannon, C. J.
063.028 .033
064.030 .031
071.055
Cannon, R. D.
032.049
141.070
154.014 .017 .025
Cannon, W. A.
105.008
Cannon III, R. O.
121.027
Canto, J.
022.075
Canuto, V.
061.041
065.038 .092
Cao, T.
074.112
Capen, C.
096.034
Capen, C. F.
097.053 .076
099.073
Capitaine, N.
044.002
Capps, R. W.
114.075
155.020
Caputo, F.
065.077
Carapiperis, L.
003.009
085.012
Carbon, D. F.
064.003

Carder, R. W.
003.072
Cardiasmenos, A. G.
132.001
Cardoen, D.
079.100
Cardona, G.
084.009
Carignan, G. R.
082.031 .035
Carleton, N. P.
003.029
131.133
132.030
Carlson, E. D.
114.047
Carlson, H. C.
082.081
Carlson, R. W.
082.044
099.021
141.122
Carman, E. H.
085.002
Carman Jr., M. F.
094.360
Carnevale, R. F.
080.051
Caroff, L. J.
141.062 .107
Caroubalos, C.
077.034 .035
Carozzi, N.
158.003 .025 .098
Carquillat, J.-M.
119.008
Carr, B. J.
066.069
Carr, M. H.
094.159
Carr, T. D.
160.001
Carranza, G. J.
158.026
Carrasco, G.
041.038 .045
Carrier III, W. D.
094.390
Carrington, T.
022.037
Carrozzi, N.
032.047
Carruthers, G.
103.101
Carruthers, G. R.
014.045
155.049
Carson, P. P. D.
114.118
Carswell, R. F.
141.119 .123 .126
158.017
Carta, D.
104.009
Carter, B.
066.021
Cartwright, D. E.
081.014
Carver, J. H.
094.140

Casamassima, P.
075.013
Caser, S.
162.050
Cash, W.
121.004
Cassatella, A.
114.052
Casse, J. L.
033.002
Cassidy, W. A.
105.027 .069
Cassinelli, J. P.
064.012 .054
Castellani, V.
065.077
154.020
155.052
Castelli, J. P.
077.036
Castillo, N.
099.022
Castle, K. G.
131.018
Castleman Jr., A. W.
082.019
Castley, J. C.
114.039
Castor, J. I.
064.037
Castore De Sistero,
M. E.
121.046
123.016
Caswell, J. L.
008.124
131.536
Catalano, S.
064.050
099.219
100.210 .214
Catinoto, E.
075.004
Cattaneo, M. B.
106.036
Catto, P. J.
094.103
Catuna, G. W.
094.368
Catura, R. C.
076.026 .038
142.114
Caulk, H. M.
077.061
Cavaliere, A.
142.096
Cavallo, G.
142.095
Cave, T. R.
097.076
Caviglia, G.
162.045
Cayrel, R.
041.071
Cazzola, P.
065.079
Cefola, P. J.
052.012
Celaya, L. E.
034.081

Celeani, G.
075.004
Ceppatelli, G.
082.075
Cerulli-Irelli, P.
106.036
Cesarsky, C. J.
131.901
143.001
Cesarsky, D. A.
131.025 .901
Cesco, M. R.
103.111
Cess, R. D.
093.012
Cester, B.
121.018
Chaffee, F.
114.100
Chaffee Jr., F. H.
099.209
131.090
Chaisson, E. J.
103.101
131.059 .081 .535
132.025
Chakravarty, S.
065.122
Chamaraux, P.
158.003 .025 .064 .098
Chamberlain, J. W.
082.060
091.024
Chambliss, C. R.
103.105
121.061
Chan, J. H.
106.030
Chandra, S.
082.096
Chandrasekhar, S.
062.001
065.034 .078
066.052 .127
Chang, P.
063.038
Chang, S.
094.382
097.041
Chang, Y. C.
103.136
Chao, J. K.
077.019
084.204
106.031
Chapelle, J.
022.030
Chapline, G.
125.069
Chapline, G. F.
061.056
066.110
Chapman, C. R.
098.005 .008 .035
Chapman, G. A.
071.058
072.048
076.029
Chapman, R. D.
076.005

Chapman, W. B.
094.151
Chappell, C. R.
084.273 .288
Chappell, W. R.
015.019
Chapront, J.
042.020 .042
Chapront, M.
042.020
Charakhch'yan, A. N.
143.027
Charakhch'yan, T. N.
143.027
Charette, M. P.
094.371 .381
Charland, Y.
065.050
Chartrand III, M. R
114.061
Charugin, V. M.
063.009
Chase, L. M.
084.033
Chase, R.
074.091 .093 .094
Chase, S. C.
092.034
093.020
099.020
Chastel, A. A.
053.016
Chaturvedi, J. P.
115.003
Chavushian, H. S.
  See Chavushyan, O. S.
Chavushyan, O. S.
122.020
Chebotarev, G. A.
098.022 .043
Chebotarev, R. P.
083.031
104.066
Chechetkin, V. M.
061.004
Chekanikhina, O. A.
122.139
Chen, C.
004.068
064.018
Chen, H. K.
022.032
Chen, M.
004.068
Cheng, C. C.
073.088
142.002
Cheremukhina, Z. P.
093.014
Cherepashchuk, A. M.
121.033 .073 .101
142.107
Cherevko, T. N.
084.245
Cherki, G.
073.009
Chernetenko, Yu. A.
103.102
Cherniack, J. R.
151.028

Chernova, G. P.
103.101
Chernykh, L. I.
103.101
Chernykh, N. S.
055.001
103.101
Chernyshev, V. I.
033.009 .010
077.068
Cherry, N. H.
141.099
Chertoprud, V. E.
082.056
085.006 .018
Chesnokova, M. P.
014.033
Chesselet, R.
106.002
Chetin, T.
142.067
Cheung, A. C.
132.001
Chevalier, R. A.
125.016
Chevreton, M.
066.048
Chichmar, V. V.
082.038
Chikada, Y.
033.049 .094
131.024
Chikin, A. I.
141.055 .056 .077
Chikmachev, V. I.
051.009
097.013
Chin, P.
083.015
Chiosi, C.
065.083 .113
Chipman, E.
074.011
Chipman, E. G.
076.014
Chirkov, N. P.
143.006
Chitre, S. M.
065.038
Chiu, B.
125.102
Chiu, B. C.
125.041
Chiuderi, C.
061.041
062.002
074.010
077.057
Chivers, H. J. A.
082.025
Chivers, S.
010.008
Chodak, J.
099.223
Choe, J. Y.
084.206 .207 .276
Chong, S. L.
033.060
Chopinet, M.
131.505

Chou, C. K.
061.041
Chou, C.-L.
105.003 .084 .102
Chow, Y. L.
033.034
Christensen, E. J.
052.003 .014
Christophe-Michel-Levy,
M.
105.023
Christy, J. W.
103.101
Christy-Sackmann, I.-J.
065.094
Chromey, F. R.
158.013
Chrzanowski, P. L.
066.135
Chu, Y.
123.041
Chubarova, A. V.
105.113 .136
Chubaryan, E. V.
022.014
Chuchkov, E. A.
097.033
Chudnovskij, L. S.
083.004
Chuguev, G. P.
031.010
Chui, M. F.
132.001
155.035
Chulten, Ts.
073.004 .023
Chunakova, N. M.
114.036
Chung, D. H.
081.003
094.122
Chupp, E. L.
076.017
125.054
Chuprakova, T. A.
097.032
Chuprunova, O. V.
045.010
Churchwell, E.
131.538 .539
Churchwell, E. B.
155.063
Churms, J.
096.007
Chushkin, P. I.
105.124
Chuvahin, S. D.
097.007
Chyi, L. L.
094.334
105.001
Ciatti, F.
125.038 .039 .105
142.017
Cimahovica, N.
011.013
Cimino, M.
075.013 .014
Ciobanu, N.
044.007

Cirse, Z.
009.014
Cirsmaru, M.
046.008
Cissoko, M.
062.017 .039
Claflin, E. S.
084.404
Clark, A. D.
084.214
Clark, B. C.
097.079
Clark, B. G.
033.046
141.060
Clark, F. O.
131.119 .134
Clark, G. W.
061.063
125.070 .104
142.019 .031 .087
Clark, R. W.
003.030
Clark, T.
103.105
Clark, T. A.
100.005
Clarke, A. C.
003.026
Clarke, D.
014.048
031.037 .040
114.045
Clarke Jr., R. S.
105.028 .106
Clarricoats, P. J. B.
033.060 .068 .069 .089
Clary, W. G.
031.002
118.003
Classen, J.
094.101 .386
Claydon, R.
033.070
Clayton, D.
065.110
Clayton, D. D.
080.018
124.004
125.009 .067
Clayton, P. D.
032.023
Clayton, R. N.
105.005 .029 .110
Cleary, M. N.
159.004
Clegg, P. E.
080.005
131.533
141.073
Clements, A.
099.022
Cleminshaw, C. H.
003.018
Cline, T. L.
061.058
Clouet, B.
010.028
Cloutier, P. A.
093.037

Cloutman, L. D.
062.068
Clube, S. V. M.
041.051 .061
122.100
153.034
155.057
Cobos, F.
034.081
Cochran, J. E.
052.901
Cocke, W. J.
141.333 .901
Code, A. D.
114.108
124.107
Codina, J. M.
103.101 .105
Coffeen, D.
099.022
Coffeen, D. L.
099.061
Cogdell, J. R.
033.064
094.171
Cogger, L. L.
082.081
Cohen, C. J.
042.066
Cohen, G. G.
034.078
Cohen, H. L.
121.039
Cohen, J. G.
114.062
131.065
Cohen, J. M.
065.042
066.124 .130
141.310
142.093
Cohen, L.
022.013 .027
071.056
073.074
151.025 .026 .033
Cohen, M.
114.069
Cohen, M. H.
141.060
Cohen, R. J.
131.107
Colburn, D. S.
074.077
084.271
094.372
099.015
Cole, D. J.
141.023
Cole, F. W.
003.031 .032
Cole, T. W.
008.124
Coleman Jr., P. J.
074.017
094.123 .165
099.015
Colgate, S. A.
125.001 .002 .011
Colin, L.
097.056

Collard, H. R.
074.077
099.014
Collenot, M.
003.033
Colley, B. W.
142.040
Collin-Souffrin, S.
124.102
158.004
Collinder, P.
004.037
Collins, C. B.
162.019
Collins, D. G.
082.070
Collins, J. G.
104.011
Collins II, G. W.
063.002
132.024
Colombo, G.
100.203
Colvin, D.
094.323
Colvin, J. D.
141.120
Colvin, R. S.
033.039
Combes, M.-A.
098.024
Comello, G.
103.101
123.005
Comsa, B.
103.105
Comte, G.
032.047
158.104
Condon, J. J.
033.040 .042
141.038 .071 .100 .103
Conel, J. E.
093.029
097.022
Conklin, E. K.
103.101
141.050 .076
Conlon, T. W.
061.008
Connes, P.
114.101
Conrad, G. H.
094.402 .403
Contadakis, M. E.
122.104
Conti, P. S.
114.050 .163
Contopoulos, G.
003.013
010.017
042.079
151.040 .046
Cook, A. H.
003.034
Cook, G. E.
081.021
Cook, N. O.
097.009
Coombs, A. E.
010.008

Coombs, A. E.
080.053
Cooper, B. F.
008.124
Cooper, B. F. C.
033.029
Cooper, J.
022.043
114.127
Cooper, T. D.
105.044
Corben, P. M.
036.004
Corbin, T. E.
041.042
Cordell, B. M.
097.043
Cormack, W. A.
034.077
Cornwall, J. M.
084.250
Coroniti, F. V.
099.044 .052
Corrado, B.
003.035
Cortesi, S.
099.033
Cosmovici, C. B.
012.015
103.101
Cotter, C. H.
082.082
Cotton, W. D.
141.080
Couch, R. G.
065.087 .111
Coughlin, J. B.
033.088
Coulman, C. E.
031.020
Coulson, K. L.
034.043
Counselman III, C. C.
141.092
Courtes, G.
032.047
034.055
131.552
Courts, G. R.
114.118
Cousins, A. W. J.
113.015
Couteau, P.
118.007 .019 .021 .022 .024
Covington, A. E.
077.012
Cowan, R. D.
022.027
062.004 .071
073.001 .074
Cowley, A.
114.063
Cowley, A. P.
064.001
Cowley, C. R.
064.027 .056
114.051 .160
Cowsik, R.
155.023

Cox, D. P.
125.028
Cox, J. P.
065.059 .105
122.125
Cox, R. E.
032.015
Coyne, G. V.
114.087
126.010
131.082 .083
Cragg, T. A.
123.032
Cram, L. E.
063.028
Cram, T.
021.015
Crampton, D.
114.170
115.002
142.007
Crane, P.
134.009
Crannell, H.
105.091
Crawford, D. F.
033.027
Crawford, D. L.
113.038
153.022 .029
Cressy Jr., P. J.
105.030 .031
Cripe, J.
105.052
Cristaldi, S.
121.048
122.022
Cristea, M.
062.042
Cristescu, C.
094.374
098.015
Criswell, D. R.
094.169
Croce, V.
075.013
Cromwell, R. H.
153.001
Cronyn, W. M.
033.075
Crooker, N.
083.053
Crosbie, A. L.
063.016
Crovisier, J.
158.040 .301
Crowther, J. H.
141.010
Crozaz, G.
106.002
Cruddace, R.
131.040
158.102
Cruikshank, D. P.
005.002
091.004
Cruise, A. M.
141.328 .335
Crutcher, R. M.
131.025 .054

Csada, I. K.
071.026
Cukierman, M.
094.379
Culhane, J. L.
141.328 .335
142.040
160.017
Culver, R. B.
003.036
114.143
Cuperman, S.
065.036
074.021
Curkendall, D. W.
052.001
Curott, D. R.
031.059
Currie, D. G.
115.001
Curtis, D. B.
105.032
Curtis, G. W.
079.100
Cushman, G.
073.073
Cushman, R.
042.008
Cutchis, P.
085.007
Cuzzi, J. N.
092.029
100.014
Czechowsky, P.
084.006 .028
Czerny, A.
010.017

Da Rosa, A. V.
083.015
Dachs, J.
142.071
D'Addario, L. R.
033.039
131.525
Dagaev, M. M.
014.010
095.006
Dagkesamanskij, R. D.
033.008
155.040
Dagley, P.
084.217
Dahab, R. E.
142.009
Dahlbacka, G. H.
061.056
066.110
Dahn, C. C.
111.004
Dainty, A. M.
094.111 .318
Dainty, J. C.
034.050
Daishido, T.
158.303
Dale, T. M.
121.005
Dalgarno, A.
008.025

Dalgarno, A.
022.006
131.007
Dalgatov, O. I.
003.037
Dallaporta, N.
065.100
125.043
151.043
162.004
Daltabuit, E.
022.075
Daminova, A. M.
105.126
Danby, J. M. A.
042.037
Danchick, R.
021.008
Dandekar, B. S.
084.012
Danese, L.
162.004
D'Angelo, N.
084.286
Daniel, B. J.
143.064
Daniell Jr., R. E.
093.037
097.048
Daniels, D. G.
034.022
Danielson, G. E.
092.017 .039
093.024
Danielsson, L.
098.018 .023
Danilin, V. A.
091.012 .019
Danilov, V. M.
151.004
Danilov, Yu.
003.106
D'Anna, E.
012.015
066.122
D'Antona, F.
126.003
D'Antona, F. A.
065.077
Danziger, I. J.
132.013
D'Arcy, B. J.
083.050
Datlowe, D.
076.027
Datlowe, D. W.
076.008
Datta Majumdar, S.
062.043
Dattner, A.
010.015
Daube, I.
011.010
047.003
122.143
Dautcourt, G.
004.025
066.042
Dauvillier, A.
044.010
131.129

Davey, W. R.
065.059
David, E. J. H.
105.033
Davidovskij, K. K.
094.173
097.127
Davidsen, A.
061.015
142.050
Davidson, J. K.
121.078
Davidson, K.
134.009
Davidson, W.
066.037
Davies, D.
022.005
Davies, J. G.
031.030
Davies, K.
083.002 .014 .040
Davies, M. E.
092.017 .039
093.024
097.035
Davies, P. C. W.
066.088
Davies, R.
033.062
Davies, R. D.
131.023 .107
155.062
158.069
Davis, C. G.
122.001
Davis, J.
022.042
074.091 .093
082.103
115.004 .010
Davis, J. H.
033.064
094.171
132.035
Davis, L.
123.011 .039
Davis, P. K.
094.312 .383
Davis, R.
103.101
113.033
Davis, R. J.
114.139
141.031
Davis, T. N.
084.025
Davis, W. F.
031.031
Davis Jr., L.
099.015
Davitaev, L. N.
106.027
Daw, R. L.
094.366
Dawson, D. W.
123.035
Day, B. P.
081.008
Day, G. A.
008.124

Day, R. W.
071.031
Daywitt, W. C.
033.092
De, Bibhas R.
107.007
De Amicis, R.
154.020
De Angelis, U.
126.002
133.004
De Biase, G. A.
034.058
De Boer, K. S.
022.028
131.051 .099
De Bona, B.
095.002
De Bruyn, A. G.
125.034 .100
De Cesare, L.
126.002
133.004
De Donder, C.
105.048
De Feiter, L. D.
034.054
073.058
076.018
De Felice, F.
066.001
De Freitas Pacheco,
J. A.
142.059 .072
De Graaf, T.
061.019
De Graeve, E.
114.087
De Groot, M.
125.102
De Groot, T.
034.019
102.012
De Jager, C.
012.012
034.001
061.038
071.007
072.050
076.018
084.014
094.145
114.107
125.021
De Jonckheere, C. G.
082.008
094.146
De Jong, F.
010.009
De Jong, M. L.
141.045
De Korte, P. A. J.
142.068
De La Cotardiere, P.
092.021
De Loach, A. C.
076.029
De Lucia, F. C.
022.076 .077
De Medina, P.
003.084

De Pascual-Martinez, M.
  103.105
De Sabbata, V.
  066.050
  155.064
De Saevsky, P.
  122.901
De Sanctis, G.
  098.030
  103.101
De Silva, L. N. K.
  141.061
De Vaucouleurs, A.
  160.901
De Vaucouleurs, G.
  082.041
  125.045
  160.901
De Vegt, C.
  041.009 .019 .057 .062
De Vries, H. L.
  141.020
De Vries, J.
  131.005
De Vries, T.
  031.074
  053.012 .013
  098.011
  099.039
  142.039
De Young, D. S.
  160.013
Dean, J. P.
  122.123
Dearborn, D. S.
  160.002
Debarbat, S.
  041.048
Debehogne, H.
  079.100
  096.015 .017
  103.101 .105 .124
DeCou, A. B.
  031.042
Dedic, H.
  065.027
Deerenberg, A. J. M.
  142.068
Deeter, J.
  142.056
DeForest, S.
  084.001
Degregoria, A. J.
  126.018
Degtyarev, M. A.
  082.040
Degueldre, M.
  103.124
Deharveng, J. M.
  032.047
Deharveng-Baudel, L.
  131.505
Deinzer, W.
  061.047
Dejaiffe, R.
  094.009
Delaney, T.
  125.035
Delano, K. J.
  003.038

Delashmit, W. H.
  033.066
Delaurier, J. M.
  084.302
Delmas, C.
  052.019
Delov, I. A.
  082.028
  104.017
DeLuccia, M.R.
  042.022
  091.010
DeMarcus, W. C.
  099.008
Demarque, P.
  008.091
  065.002
  080.044
Demers, S.
  122.095
  123.010 .011
  154.012
Demidova, I. G.
  104.005
Demidovich, E. G.
  014.028
Demin, V. G.
  042.004
  107.015
Deming, D.
  122.024
Demircan, O.
  103.101
Demmel, G.
  033.090
Demura, A. V.
  022.034
Den, O. E.
  032.003
Dence, M. R.
  011.018
  094.329
Deney, C. L.
  142.028
Denis, A.
  022.045
Denis, J.
  117.035
Denisyuk, E. K.
  103.101
  158.035 .036 .051 .052
Dennison, B.
  066.035
  131.521
Denoyer, L. K.
  131.551
Denshchikova, L. I.
  084.266
Dent, W. A.
  141.030
  155.008
Deprit, A.
  047.012
Dere, K. P.
  076.015 .021
Deridder, G.
  071.006
Derkach, K. N.
  041.004
Derviz, T. E.
  114.126

Desai, U. D.
  061.058
Desborough, G. A.
  105.077
Desesquelles, J.
  022.045
Desnoyers, C.
  105.034
Despain, K. H.
  065.047 .094
Dessens, J.
  082.061
Dessler, A. J.
  099.077
Destombes, J.-L.
  022.022
Detre, L.
  125.033
Deubner, F.-L.
  021.012
  071.046
Deupree, R. G.
  122.126
Deutsch, A. J.
  114.165
Deutsch, A. N.
  112.021
Deutschman, W.
  113.033
Deuze, J. L.
  063.043
Devaux, C.
  063.043
DeWitt, B. S.
  012.007
  066.017
DeWitt, C.
  012.007
DeWitt-Morette, C.
  012.008
Di Biaggio, G.
  032.047
Di Cicco, D.
  082.095
Di Tullio, G.
  125.030
Diamond, G. L.
  121.086
Dibaj, Eh. A.
  158.093
Dichtl, G.
  046.021
Dick, M. L.
  097.072
Dicke, R. H.
  072.035
  080.023 .024
Dickel, H. R.
  131.020 .113
  132.003
Dickel, J. R.
  125.058 .060 .901
  131.113
Dickens, R. J.
  064.005
  154.023
Dickinson, D. F.
  131.061
Dickinson, R.
  092.040
  093.022

Dickinson, R. E.
082.010
Dickinson, T.
103.101
Dieckvoss, W.
041.019
112.020
Diehl, R. E.
101.001 .002
Dieminger, W.
084.028
Diethelm, R.
121.090 .091 .093 .095
Dietz, R. S.
105.035 .168
Dieverge, B.
003.149
Dijk, J.
033.065
Dilworth, C.
054.022
143.031
Dimitrijevic, M. S.
004.067
053.018
Dimov, N. A.
032.006
Dinescu, A.
046.009
Dinescu, R.
153.016
Dinulescu, V.
083.036
Dinwoodie, C.
010.012
Diodato, L.
074.039
106.036
Dionysiou, D.
066.091
Dirac, P. A. M.
162.046
Dirikis, M.
011.010 .027
047.001
Dirikis, M. A.
010.033
Divina, L.
102.025
Divine, N.
078.043
Dixon, M. S.
122.116
Dixon, R. S.
141.047
Djokic, M.
032.060
Djordevic, B.
123.023
Djurkovic, P. M.
010.029
014.036 .037
Djurovic, D.
044.009
Dlugach, J. M.
091.015
Dmitrenko, V. V.
084.409
Do Cao, G.
022.045

Dobaczewska, W.
011.041
Dobronravin, P. P.
055.001
Dobrovol'skij, O. V.
103.120
Dobrzycki, J.
012.024
Dobysh, G. I.
033.008
Dodd, R.
105.080
Dodd, R. J.
098.002
107.013
112.028
155.005
Dodd, R. T.
105.014
Dodge, J. C.
077.072
D'Odorico, S.
125.051
Doe, A. L.
062.079
Doe, L.
074.097
Doepel, R.
102.008
Dogadkin, N. N.
094.398
Doggett, L. E.
042.022
091.010
Doherty, L. H.
132.004
Doherty, L. R.
114.145
121.034
Dolan, J. F.
114.002
141.006
Dolginov, A. Z.
064.046
117.015
131.014
Dolginova, Yu. M.
073.014
Dolgov, Yu. A.
105.135
Dombrovskij, V. A.
122.098
Domina, G.
075.004
Domingo, V.
078.031 .032
084.280
Dommanget, J.
041.028 .058 .063
112.026
118.023
Donahue, T. M.
082.098
099.040
Donath, F. A.
003.006
Donati-Falchi, A.
071.022
072.021
Dong, M.
103.136

Donij, V. N.
104.074
Donivan Jr., F. F.
160.001
Donn, B.
103.102 .105
131.115
Donn, B. D.
103.100
Donnelly, R. F.
076.028
Donnelly, R. J.
100.011
Dooling, D.
054.008
Doose, L.
099.022
Doose, L. R.
099.061
Dopita, M. A.
131.531
132.012 .016 .028
Doremus, J. P.
151.032 .902
Dorman, J.
094.108 .120 .121
Dorman, L. I.
078.006
143.012 .017 .018
Dormand, J. R.
107.017
Dorobantu, R.
044.008
Doroshenko, V. T.
114.120
Doroshkevich, A. G.
010.017
033.023
151.011
158.007
Doroshkievich, A. G.
See Doroshkevich, A. G.
Dorschner, J.
098.016
107.012
131.044 .056 .072
Dorst, F.
079.100 .101
095.003
096.005
Dorst, H. J.
021.014
Dos Santos, A. F.
031.086
Doschek, G. A.
022.027
062.004 .071
071.056 .066
073.074
Doubochine, G. N.
See Duboshin, G. N.
Douglas, B. C.
046.011
052.046
Douglas, J. N.
033.037 .038
141.076 .094
Dovbnya, B. V.
106.017
Dowty, E.
094.403 .404

Doxsey, R.
 121.004
 142.117
Dragesco, J.
 010.028
 092.007
Drake, C. L.
 007.000
Drake, G. W. F.
 061.043
Dran, J. C.
 105.036
Drapatz, S.
 022.016
 105.146
Drawin, H. W.
 022.055
Dreibus, G.
 105.037
Drejson, S. R.
 082.085
Drever, R. W. P.
 033.020
 125.035
Dreze, A.
 079.100
Drobyshevski, E. M.
 099.055
 117.007 .032
Drobzhev, V. I.
 083.021
Drozdovskij, A. A.
 033.048
Drozyner, A.
 051.019
Drummen, D.
 158.045
Drummen, M.
 118.006
 154.016 .027
 158.046
Drummond, A. J.
 082.069
Dryer, M.
 106.004
Dubach, J.
 093.004
Dubin, M.
 103.101
Dubinin, E. M.
 084.249
Dubinsky, J.
 034.020
Duboshin, G. N.
 042.010 .057
 052.033
Dubov, E. E.
 073.059
Dubrovich, V. K.
 066.032
Ducatel, A.
 105.064
Ducati, H.
 094.330
Duchesne, M.
 133.006
Ducuroir, M.
 010.031
Dudis, J. J.
 097.045

Dudley, R. M.
 066.133
Duering, T.
 096.004
Duflot, M.
 112.017
Duflot, R.
 032.047
 158.025 .098
Duflot-Augarde, R.
 158.003
Dufour, R. J.
 114.166
 133.010
Duggal, S. P.
 078.014
Duin, R. M.
 125.019 .052
Dujnic, M.
 095.008
Duley, W. W.
 131.017
Dulk, G. A.
 074.062 .063
 077.005
 141.048
Dul'kin, L. Z.
 054.012
Dultzin, D.
 117.029
Duma, A. S.
 094.137
Dumail, M.
 105.036
Dumitrescu, A.
 121.057 .077
Dumont, R.
 106.007
Dumortier, B.
 122.901
 141.116
Duncans, L.
 113.029
Duncombe, R. L.
 098.014
Dunham, D. W.
 096.001 .002 .016 .029
  .030 .031 .032
 103.105
 115.008
Dunkin, D. B.
 131.027
Dunlap, J. L.
 098.010
Dunlop, D. J.
 094.315
Dunn, G. H.
 083.045
Dunn, P. J.
 046.028
Dunn, R. B.
 071.057
 073.078
Dunne, J. A.
 053.014
 092.033
Dupas, A.
 105.157
Dupree, A. K.
 131.501

Dupuy, D. L.
 122.084
Durand, A.
 122.901
 124.103
Durasova, M. S.
 033.053
 077.045
Duraud, J. P.
 105.036
Durdin, J. M.
 033.040
 141.100 .103
Durney, A. C.
 078.009
Durney, B. R.
 080.036
Durrani, S. A.
 094.131 .331
Durst, C.
 021.014
Duruisseau, J.-P.
 066.038
Duruy, M.
 122.901
Dvoretsky, M. M.
 119.012
D'yakonova, M. I.
 105.126 .127 .138
D'yakov, B. N.
 032.027
Dyal, P.
 099.015
Dybkowski, L.
 098.020 .021
Dybwad, J. P.
 002.029
Dyck, H. M.
 114.075
 155.020
Dyer, C. C.
 162.020
Dyer, C. S.
 061.065
Dymanus, A.
 022.002
Dyson, F. J.
 142.049
Dyson, J. E.
 131.075 .531
Dyson, P. L.
 082.007
 083.037
Dysthe, K. B.
 061.042
Dzervitis, U.
 004.022
Dzhumanaliev, N. D.
 052.011
Dzyubenko, N. I.
 074.050 .075
 097.100

Eardley, D. M.
 066.049
 142.003
Earl, J. A.
 073.069
 106.014

Ebbighausen, E. G.
121.050
Eberhardt, P.
094.310 .332 .333 .336
Ebisch, K. E.
122.025
Ebner, H.
041.009 .062
Eckert, H.-P.
014.002
Ecklund, W. L.
083.047
Eddy, J. A.
004.063
074.005
Edelson, R. E.
092.040
093.022
Eder, G.
158.058
Edge, D. M.
143.062
Edgerley, D.
022.901
Edlen, E.
022.053
074.012
Edmunds, M. G.
114.901
Edwards, P. J.
157.003
Edwards, T. W.
065.008 .009
Eelsalu, H.
013.006
155.013
Efanov, V. A.
033.011
092.019
141.055
Efimov, A. I.
093.006
Efinger, H. J.
066.037
Efremov, Yu. N.
142.106
Egeland, A.
003.004
084.030
Eggen, O. J.
115.006 .016
153.009
155.022
Egibekov, P.
102.018
Egidi, A.
074.016
106.036
Eglinton, G.
094.349
Ehjdman, V. Ya.
141.312
Ehjedel'man, E. D.
065.080
Ehjgenson, A. M.
154.003
Ehl'yasberg, P. E.
052.016
053.004
085.009

Ehman, J. R.
141.047 .902
Ehmann, W. D.
094.334
105.001
Ehminzade, T. A.
126.020
Ehrnsperger, W.
046.021
Ehshmatov, M. R.
096.014
Ehtkin, V. S.
141.041
Eiband, A. M.
032.054
Eiby, G. A.
007.000
Eichhorn, H.
003.152
031.002 .055
118.003
Eigenson, A. M.
000.000
Einasto, J.
158.054 .107
Eisenstein, J. P.
071.047
Eitter, J. J.
115.014
Ekberg, J. O.
022.018
074.012
Ekers, R. D.
021.018
066.002
125.100
Ekmann, G.
072.028
Ekonomov, A. P.
093.039
El Eid, M. F.
065.085
El Goresy, A.
094.177 .335
El Rabaa, S. M.
105.126
El-Raey, M.
077.032
Elander, N.
022.019
Elek, A.
094.398
Eleman, F.
084.239
Elias, D. P.
103.101
Elias, J. H.
131.068
Eliass, M.
011.028
Eliseev, G. F.
077.003 .050
Eliseeva, L. A.
072.007
Elitzur, M.
062.076
Ellder, J.
103.101
131.105 .114 .117
Eller, E.
142.075

Elliot, J.
099.037 .051
Elliot, J. L.
099.049 .074
122.103
Elliot, W.
103.101
Elliott, K. H.
132.012 .020
Elliott, W. A.
015.005
Ellis, G. R. A.
099.026
Ellis, P. G.
022.020
Elmabsout, B.
042.029
Elsaesser, H.
003.112
Elshin, E. K.
106.015 .028
Elsmore, B.
031.051
Elsner, E.
003.039
Elsworth, Y.
103.101
Ely, J. T. A.
143.057
Embleton, B. J. J.
045.003
Emel'yanenko, M. T.
014.023
Emel'yanenko, V.
103.123
Emerson, D. T.
158.070
Emming, J. G.
034.001
Encrenaz, P. J.
131.095
132.029
Endean, V. G.
141.301
Engin, S.
114.068
Engvold, O.
071.036
073.049
Entzian, G.
085.004 .005
Epishev, V. P.
103.101 .105
Epps, H. W.
034.045
Epstein, E. E.
131.113
Epstein, J.
015.015
Epstein, R. I.
061.068
125.027
Eraker, J. H.
092.037
093.026
Erastova, L. K.
122.020
Erceg, V.
118.031
Erdi, B.
042.065

Ereneeva, A. I.
006.000
Eremenko, R. P.
054.020
Ergma, E.
065.032
Erickson, W. C.
033.059 .067
077.041
Erleksova, G. E.
032.018
122.054
Erlichson, H.
014.046
Ermakov, V. I.
003.040
Erman, P.
022.019
Ermuratskij, P. V.
033.016
Eross, B.
097.037 .055
Esipov, V. F.
132.009
158.093
Esposito, P. B.
066.140
092.040
093.022
Essen, L.
003.041
Esteva, J. M.
022.012
Eugster, O.
094.336
Evans, A.
141.082 .083
Evans, A. C.
143.062
Evans, D. S.
084.030
115.008
122.086
Evans, J. C.
071.054
Evans, J. V.
083.015
Evans, R. G.
032.034 .902
121.067
Evans, T. L.
122.071 .134
154.013
Evans, W. D.
117.009
124.100
Evans II, N. J.
131.068
Evdokimova, L. V.
106.011
Everhart, E.
042.027
Evrard, G.
075.006
Evseev, S. V.
081.011
Evsyukov, N. N.
094.106 .107 .168
Ewing, M.
094.108 .120 .121

Eyni, M.
074.038
Ezerskaya, V. A.
094.314
Ezerskij, V. I.
006.000
094.314

Fabian, A.
142.045
Fabian, A. C.
142.014 .053
160.017
Fabian, W.
022.038
Fabiano, E. B.
084.258
Facer, R. A.
081.004
Fadeev, Yu. A.
121.099
123.049 .050
Fadeyev, V. E.
094.182
Faelthammar, C.-G.
074.040
084.241 .248
Fagot, J.
079.100
Fahr, H. J.
074.110
099.024
106.005
Fahrbach, U.
034.031 .063
Fainberg, J.
077.053
Fairbank, W. M.
033.021
Falciani, R.
071.022 .024
072.021
Falk, L.
062.037
Falk, S. W.
125.066
Palla, D.
141.082
Falworth, G.
054.026
Fanale, F. P.
105.008
Fanaroff, B. L.
141.009 .052
Fang, T.-M.
022.006
Fangor, R.
034.074
096.028
103.101
Fanselow, J. L.
141.096
Fanti, C.
141.086
Fanti, R.
141.069
142.097
Faraggiana, R.
114.110

Farley, J. D.
080.045
Farmer, A. D.
053.009
Fassio-Canuto, L.
061.041
Fast, H.
099.003
Fastie, W. G.
114.112
Fatchikhin, N. V.
111.006
112.006
Fatkullin, M. N.
083.043
Faulkner, D. J.
065.025
Faulkner, J.
124.008
Fawcett, B. C.
073.001
Fay Jr., T. D.
114.070 .098 .099 .159
Fazio, G. G.
061.034
142.081
Feagin, T.
099.202
Feast, M. W.
121.051
122.117 .134
124.003
131.519
155.042
Fedorov, E. P.
003.002
Fedoseev, L. I.
034.046
094.164
Fedynskij, V. V.
104.015
Fegan, D. J.
141.340
Fehrenbach, C.
103.101
112.017
Fehrenbach, M.
084.007
Fehsenfeld, F. C.
131.027
Feijth, H.
123.006 .007 .019
Feinstein, A.
103.101
114.011
153.033
158.304
Feissel, M.
044.001
Feix, M. R.
151.902
Fejes, I.
131.526
Fejgel'son, E. M.
093.015
Fejgin, V. M.
083.016
Feklistova, T.
114.057
Felber, H.-J.
004.056

Feldgate, D. G.
085.014
Feldman, I. A.
063.004
Feldman, P.
103.101
Feldman, P. A.
114.148
141.076 .108
142.036
Feldman, P. D.
114.112
Feldman, U.
022.027
062.004 .071
071.056 .066
073.074
074.046
Feldman, W. C.
074.041 .073 .080
093.021
Fellgett, P.
032.028
Felli, M.
131.511 .525
141.086
Fellous, J. L.
082.022
Pennell, J. P.
084.402
Fercher, A. F.
031.069
Ferdman, S.
003.042
Ferguson, D. C.
141.333
Ferguson, E. E.
131.027
Ferguson, P. M.
036.002
Fernandes, M.
117.037
Fernandez, J. B.
044.011
Fernandez-Figueroa, M. J.
114.053 .091
Fernie, J. D.
119.014
122.061
Ferrari, A.
134.008
Ferrari, A. J.
052.014
Ferrari D'Occhieppo, K.
004.080
007.000
009.020
155.075 .077
Ferraro, A. J.
083.018 .019
Ferreri, W.
103.101 .105
Ferrin, I. R.
100.015
Ferris, J. P.
097.042
Ferro, I.
072.040
Fesen, R. A.
099.006

Fesen, R. A.
114.064
122.023
Fesenkov, V. G.
106.022
Festou, M.
103.101
Fetisova, T. S.
114.120
Feyth, H.
103.101
Fiala, A. D.
041.068
Fialko, E. I.
104.034 .037 .038 .074
Ficarra, A.
141.069 .086
Ficarrotta, F.
106.026
Fichera, E.
042.063
045.012
046.014
052.029 .030
Fichtel, C. E.
155.026 .058
Field, G. B.
114.147
131.038
161.002
Fillius, R. W.
099.019
Fink, U.
092.001
Finkel, R. C.
105.038
Finn, G. D.
080.009
116.009
Finsen, W. S.
005.022
007.000
Pinzi, A.
080.022
Fiocco, G.
082.023
Fiore, C. N.
003.043
Fireman, E. L.
142.001
Firmani, C.
034.081
117.029
Firnett, P. J.
062.022
Firsov, V. V.
076.016
Firth, J. G.
076.004
Fischer, G.
008.090
Fischer, S.
034.020
Fisenko, A. V.
105.090 .120 .126 .130
Fisher, J. R.
033.059 .067
141.319
Fisher, R. M.
094.320

Fisher, R. R.
073.048
074.086
Fisher, W.
153.012
Fisher, W. A.
121.086
Fishkova, L. M.
082.117
Fishlock, S. J.
105.044
Fisk, L. A.
143.024 .030 .035
Fitzgerald, M. P.
119.003
131.080
153.020
155.038
158.041
Fix, J. D.
064.032
101.005
Fjeldbo, G.
092.040
093.022
099.205
Fjordheim, O.
084.236
Flaherty, B. J.
083.015
Flechon, J.
082.017
Fleischer, R. L.
094.311 .364
Flesch, T. R.
122.097
Fliegel, H. F.
052.018
Flindt, H. R.
099.017
Floquet, M.
114.097
Flora, U.
121.008
Floree, R.
075.006
Florenskij, K. P.
094.161 .302
Florian, J. R.
143.055
Florsch, G.
007.000
Flower, D. R.
022.010
Flowers, E.
065.102
Fodor, R. V.
105.039 .051
Fogel', A. L.
033.010
Fomalont, E. B.
141.043
Fonseca De Mello, Z.
105.020
Fontanella, J.-C.
082.015 .071
Forbes, E. G.
002.012
004.006 .011 .040
Ford, C.
123.036

Ford, C. B.
123.032
Ford, H. C.
100.011
Forichon, M.
134.006
142.005
Forlani, A.
126.002
133.004
Formiggini, L.
141.069
Formisano, V.
084.283
Forney, P. B.
097.050
Forrest, D. J.
076.017
077.036
125.054
Forrest, W. J.
100.001
Forrester, W. T.
153.023
Forster, S.
121.090
Fort, B.
074.089
Forte, J. C.
153.033
Fortini, P.
066.050
155.064
Fortini Baroni, L.
066.050
Fosbury, R. A. E.
073.028
Fossat, E.
071.027
Foukal, P.
132.014
Foukal, P. K.
076.034 .037
Foukal, P. V.
073.016
Fountain, J.
057.008
Fountain, J. W.
099.061
Fouquart, Y.
063.020
Fourcade, S.
094.317
105.018
Fourikis, N.
033.050
Fowler, J. W.
064.024 .041
Fowler, L. A.
131.029
Fowler, R. E.
003.044
Fowler, R. G.
032.034 .902
Fowler, W. A.
061.079
065.023
Fracastoro, M. G.
010.017
015.017
118.017

Franchuk, N. G.
077.043 .046
Francmane, S.
011.010
Francmanis, J.
009.014
010.017
011.028
Frank, L. A.
084.293
Franklin, F. A.
100.203 .213
Frantsman, Yu. L.
010.033
011.032
Franz, O. G.
041.060
Fraquelli, D.
114.063
Frazer, M. C.
062.006
Frazier, E. N.
083.046
Fredrick, L. W.
003.045
041.060
158.122
Fredriksson, K.
105.100
Freeman, F. F.
076.004
Freeman, J. W.
074.079
Freeman, K. C.
032.052
154.011 .024
French, B.
105.035
Frenkel', L. A.
066.011
Fresa, A.
004.060
Frick, M.
046.002 .003
Frick, U.
094.327 .337 .338 .348
.352 .401
Fricke, K.
065.069 .070
Fricke, K. J.
065.096
Fricke, W.
002.021
041.033 .035
155.061
Fricker, P. E.
091.003
Fridman, P. A.
033.007
Frieboes-Conde, H.
121.011
Fried, B. D.
083.044
Fried, J.
031.039
Friedemann, C.
003.046
Friedjung, M.
124.105
125.103

Friedman, H.
085.017
142.082
Friedman, J. L.
066.111 .120
Friedrich, K.
096.018
Frimout, D.
114.007 .134
Fringant, A.-M.
114.038
Frisillo, A. L.
094.113 .151 .380
Fritts, M. J.
061.040
Fritz, G.
142.082
Fritz, R. B.
083.015
Fritz, T. A.
084.290
Fritzova-Svestkova, L.
077.063
Froehlich, H.
046.010
Froelich, F.
093.033
Frogel, J. A.
114.100
131.524 .529
155.033
Frohlich, A.
114.044
Frolov, V. A.
033.008
Frolova, N. B.
041.031
Frost, S. A.
114.163
Fu, K. Y.
062.077
066.105
Fuchs, H.-U.
004.057
Fuchs, L. H.
105.098
Fudali, R. F.
105.148
Fuerst, E.
077.028
Fuerstenberg, F.
075.011
Fujimoto, M.-K.
065.022
Fujimura, F.
162.043
Fukatsu, M.
073.092
Fukaya, R.
031.078
Fukushima, N.
084.213
Fuligni, F.
061.011
Fuller, M.
094.109 .319
Funk, H.
094.327 .337 .338 .348
.352 .401
Furman, D. R.
076.019

Furuta, T.
103.108 .112 .124
Fusipecci, F.
036.010

Gabriel, A.
097.034
Gabriel, A. H.
061.045
076.004
Gabriel, M.
065.006 .020 .065 .074
Gagnepain, J.-J.
035.009 .011
Gail, H. P.
061.046
063.012
Gaisser, T. K.
143.011
Gaizauskas, B. R.
004.051
Gaizauskas, V.
073.075
Gajewski, R.
080.027
Galassi, G.
122.118
Galeotti, P.
114.066
Galkin, S. L.
066.082
Gall, R.
078.027
084.229
Gallagher, J.
103.105
Gallagher, J. S.
124.010
Gallagher III, J. S.
124.103 .107
Gallaher, L. J.
042.035
Gal'per, A. M.
084.409
Galt, J. A.
141.305
Gammon, R. H.
132.004
Ganbarov, A.
114.121
Ganea, I.-M.
066.040
Ganeko, Y.
096.038
Ganguli, G.
034.079
Gans, D.
103.105
Gans, D. J.
103.105
Gaposchkin, E. M.
013.016
046.027
052.045
081.030 .031
Garcia, C. J.
074.095
Garcia, J. R.
122.118

Gardner, E. H.
106.033
Gardner, F. F.
131.050
158.024
Gardner, I. S. K.
032.902
Gardner, J. L.
022.068
Gardner, S. K.
032.034
Garfinkel, B.
042.014
082.111
Garibyan, G. M.
061.057
Garmire, G.
141.329
Garmire, G. P.
082.032
142.044
Garmston, H. J.
143.062
Garnier, R.
114.067
Garrett, H. B.
084.201
Garrison, R. F.
158.081
Garwood, G. J.
071.054
Gary, B. L.
099.038
Gary, S. P.
078.001
Gascoigne, S. C. B.
122.005
Gaska, S.
098.019
Gass, I. G.
081.013
Gate, L. F.
063.031
Gatehlyuk, E. D.
141.055 .056 .077
Gatewood, G.
111.003
Gatley, I.
097.203
131.068
Gattinger, R. L.
082.004
Gault, D.
092.017
093.024
Gault, D. E.
092.039
Gaur, V. P.
072.017
Gauss, F. S.
032.038
Gavrilov, I. V.
094.102
Geake, J. E.
094.183
Gearhart, M. R.
141.076 .902
Geballe, T. R.
141.602
Gebbie, K. B.
073.026 .068

Gehrels, T.
003.047
091.005
099.010 .022 .061
103.108
131.083 .084
141.333
Geiss, J.
094.310 .332 .333 .336
Gel'frejkh, G. B.
033.012
077.011 .042
Gelsing, R. J. H.
033.088
Gendrin, R.
062.035
084.253
Gent, H.
141.010
Gentner, W.
105.062
George, D.
133.015
Georgienko, S. S.
104.053
Gerard, J.-C.
082.104
Gerashchenko, O. A.
034.062
Gerasimenko, S. I.
103.104
Gerasimov, G. I.
083.026
Gerassimenko, M.
074.090 .106
Gerbal, D.
066.029
Gerbaldi, M.
114.006
Gerend, D.
142.056
Gergely, T.
074.064
Gergely, T. E.
077.018 .069
Gerlach, U. H.
066.142
Gerlovin, I. L.
043.001 .003
Germann, R.
121.090
122.026
Gerola, H.
131.041 .057
Gerritsen, B. A.
096.009
Gerry, M. C. L.
131.011
Gershberg, R. E.
064.047
122.015
Gershengorn, G. I.
083.030
Gerstenberger, M.
047.016
Gerver, M. J.
134.003
Gessner, H.
123.043 .047
Getselev, I. V.
097.033

Ghezloun, A.
041.029
Ghobrial, S. I.
033.076
Ghose, S.
094.305
Giacaglia, G. E. O.
052.007
Giacconi, R.
142.035 .042 .043 .054
Giachetti, R.
077.057
Giannone, P.
117.002
121.015
Giannuzzi, M. A.
117.002
121.015
Giard, W. R.
094.311
Gibbins, C. J.
033.084
Gibbons, G. W.
066.070 .116
Gibbs, S. L.
004.041
Gibson, J.
103.111
153.023
Gibson Jr., E. K.
105.100
Giclas, H. L.
103.101 .105 .116
Giddings, J. W.
045.003
Gierasch, P. J.
080.035
097.077
Gieren, W.
103.105
Gierloff-Emden, H.-G.
003.024
Gieseking, F.
122.010
152.003
Gigolashvili, M. Sh.
073.035
Gijsbers, J.
096.009
Gilbert, A.
022.036
Gilbert, G.
141.119
Gill, G.
033.086
Gillespie, A. R.
141.017
Gillett, F. C.
130.001
158.005
Gillis, J.
065.019
Gillispie, C. C.
003.048
Gilman, D.
142.075
Gilman, P. A.
074.060
Gilman, R. C.
114.032

Gilmore, A. C.
103.111 .116
Ginestet, N.
119.008
Gingerich, O.
004.050
Gingold, R. A.
065.025
Gintsburg, M. A.
062.007
Ginzburg, Eh. I.
083.042
Ginzburg, V. L.
061.018
066.085
Gioia, I.
141.069
Gipson Jr., M.
097.071
Girard, A.
082.015 .071
Gisler, G. R.
061.012
Glackin, D. L.
071.033
Gladush, V. D.
066.100
Glass, B. P.
094.339
Glass, I.
142.108
Glass, I. S.
113.022 .035
114.017
121.051
122.062
124.003
Glass, M.
082.022
Gleadow, A. J. W.
094.304
Gleissberg, W.
003.149
Gliese, W.
012.014
041.036
155.030
Gloeckler, G.
106.039
Glushneva, I. N.
114.120
Gnezdilov, A. A.
077.023 .058
Goad, J. W.
158.108
Godden, G.
073.073
Godin, G.
003.049
Godoli, G.
010.017
075.004
Goebel, L. H.
022.063
Goettel, K. A.
105.040
Goettig, C.
104.018
Goetz, W.
153.018

Gokhale, M. H.
072.029
Goldberg, B. A.
122.088
Golden, L. M.
141.015 .125
Golden, R. L.
143.028
Goldenberg, H. M.
080.023
Gol'dman, V. M.
080.039
Goldreich, P.
131.088 .091
Goldsmith, D.
003.050
014.038
131.008
Goldsmith, S.
022.013
Goldstein, B. E.
094.144
Goldstein, B. R.
004.043 .048
Goldstein, J. I.
094.340
105.028 .094
Goldstein, J. S.
131.131
Goldstein, M. L.
077.056
143.035
Goldstein, R. M.
092.011
Golenetskij, S. V.
061.031 .053
104.058
143.032
Golikov, V. I.
066.103
Golitsyn, G. S.
003.051
Gollandskij, O. P.
141.306
Golub, L.
074.094 .100
076.009 .023
094.116 .313
Golub, P. A.
034.017
Golubchina, O. A.
033.012
Golubev, V. A.
103.101 .105
Golubkov, V. V.
051.016
Gomez, A. E.
155.028
Gomez-Gonzalez, J.
131.053
141.325
Gomi, K.
123.026
Goned, A.
084.205
Gontarev, O. G.
034.021
077.022
Gonzalez, W. D.
084.226

Gonzalez B., S. F.
113.043
Gonze, C.
075.006
Gooding, R. H.
081.019
Goodman, N.
046.005
Goodwin, K.
082.073 .074
Goody, R. M.
097.023
Goossens, M.
065.011
Gopal-Krishna
141.051
Gopalakrishnamurthy, A.
104.071
Gopalan, K.
105.041
Gopasyuk, S. I.
080.011
Goranskij, V. P.
122.138
Gorbachev, B. I.
160.009
Gorbatskij, V. G.
124.009
Gorbunova, I. E.
105.115
Gorchakov, E. V.
078.012
097.033
Gordienko, G. I.
083.022 .023 .024
Gordon, E.
104.009
Gordon, M. A.
012.025
Gorenstein, P.
094.116 .313
125.048 .068
142.075
Gornitz, V.
094.142
Gorodetskij, D. I.
103.105
Gorshkov, A. G.
141.041
Gorshkov, Eh. S.
105.121
Gorshkov, V. L.
041.030
Goryainova, T. V.
041.027
Gorynya, N. A.
132.009
Goscinski, O.
022.020
Gose, W. A.
012.026
Gosling, J. T.
074.005
Goss, W. M.
114.018
125.100
131.019 .092 .536
Goto, T.
082.112 .113
Gott III, J. R.
141.124

Gott III, J. R.
162.001
Gottlieb, C. A.
131.902
Gough, P. T.
031.070
Gough, R. A.
033.080
Gough, T. T.
082.018
Gouguenheim, L.
066.039
158.106
Govindarajan, M. S.
033.072
Govorov, V. M.
094.127 .173
Govar, A. P.
094.349
Goyal, P.
065.125
Graber, M. A.
045.005
Graboske Jr., H. C.
065.004
Grader, R.
142.051 .083
Graf, H.
094.310
Graf, O.
042.036
099.202
Graf, W.
072.046
Graham, A. L.
105.162
Graham, D. A.
157.004
Graham, J. A.
122.060
159.006
Gramont, L.
082.015 .071
Grandi, S. A.
122.102
Grant, I.
003.138
Grant, I. P.
063.042
Granveaud, M.
044.031
Grasbergs, E.
116.011
Grasdalen, G. L.
114.903
141.601
Grass, F.
105.152
Gratton, L.
158.095
Grau, C.
005.018
Gravitis, E.
011.011
Gray, C. M.
105.063
Gray, D. F.
114.077
Grayzeck, E. J.
131.532

Grebenikov, E. A.
042.033
Grebenkemper, C. J.
077.073
Grebinskij, A. S.
077.043
Greeley, R.
097.018 .080
Green, A. E. S.
034.079
Green, J. A.
094.403
Green, M.
034.084
Green, R. F.
122.024
Greenberg, J. M.
131.042 .067 .112
158.109
Greenberg, R. J.
101.003
Greenstein, G.
065.039
Greenstein, J. L.
032.045
126.016 .019
Greenwald, R. A.
083.047
Gregory, P. C.
122.021
Grenfell, T. C.
126.004
Gretskij, A. M.
100.007
Greve, A.
073.051
076.010
Grewing, M.
119.013
131.006 .074
132.027
Grib, A. A.
066.014
Gribbin, J.
061.059
Griem, H. R.
003.052
Griffin, R.
114.089
Griffin, W. L.
105.167
Griffith, J. S.
002.031
042.009
Griffiths, J. B.
061.076
Griffiths, J. C.
094.143
Grigor'ev, V. G.
143.046
Grigor'eva, M. I.
033.011
Grigor'eva, V. M.
105.115
Grigor'eva, V. P.
078.011
Grigorevskij, V. M.
120.002 .003
122.030
Grigorivs, M.
009.014

Grigorov, N. L.
 034.026
 078.002 .020
 084.407 .408 .409
 143.036 .039 .040 .041
Grigoryan, P. A.
 077.068
Grindlay, J. E.
 061.034
 142.081
Gringauz, K. I.
 084.266
 106.027
Grisendi, T.
 075.005
Grishchuk, L. P.
 033.023
Grobov, V. A.
 052.017
Groegler, N.
 094.310 .332 .333 .336
Groenbech, B.
 123.017
Gros, M.
 114.113 .123
Groschopf, G.
 014.004
Gross, P. G.
 065.088
 091.009
Gross, S. H.
 097.058
Grossman, A. S.
 065.004
Grossman, L.
 105.004 .029 .042
 107.002
Grossmann, W.
 004.044
Groth, E. J.
 096.021
Grueff, G.
 142.100
Gruen, M.
 051.018
Gruenwaldt, H.
 084.287
Grupsmith, G.
 114.090 .150
 131.002
 142.088
Gryazev, N. I.
 085.009
Grygar, J.
 013.011
Gualdi, C.
 066.050
 155.064
Gubarev, V. S.
 003.053
Gubbay, J. S.
 031.009
 141.093
Gubbins, D.
 061.051
Guelin, M.
 141.325
Guenther, B.
 082.098
Guentzel-Lingner, U.
 002.021

Guerin, C.
 035.013
Guest, J. E.
 094.405
Guggisberg, S.
 094.336
Guibert, J.
 158.009
Guidice, D. A.
 008.010
 077.036
Guidry, P. J.
 071.001
Guilbaut, M.
 041.074
Guillemet, B.
 082.016
Guinot, B.
 035.017
 044.001
Gulbrandsen, A.
 074.066
Gul'el'mi, A. V.
 106.017
Gull, S. F.
 125.059
Gull, T. R.
 093.017
Gul'medov, Kh.
 104.003
Gulyaeva, T. L.
 083.012
Gumen, V. F.
 034.008
Gun'ko, V.
 104.003
Gunn, J. E.
 134.002
 141.059 .124
Gunn, N.
 080.017
Gupta, J. C.
 083.054
Gurevich, L. Eh.
 065.080
Gurklyte, A.
 113.004
Gurman, J. B.
 071.008
Gurnett, D. A.
 084.293
Gurshtein, A. A.
 094.141
Gurshtejn, A. A.
 094.302
Gurski, T. R.
 034.045
Gursky, H.
 142.035 .038 .042
Gurtovenko, E. A.
 071.013
Gur'yan, Yu. A.
 104.058
Gurzadyan, G. A.
 114.128
 132.031
Gusak, A. I.
 066.146
Guseinov, O. H.
 See Gusejnov, O. Kh.

Guseinov, R. E.
 073.007
Gusejnov, M. D.
 072.004
Gusejnov, O. Kh.
 102.017
 119.002
 125.006 .025
 141.344
 142.041
Guseva, T. A.
 054.013
Gush, H. P.
 066.136
Gus'kova, E. G.
 105.140
Guslyakov, V. T.
 093.034
Gustafsson, B.
 064.044
Guthrie, B. N. G.
 160.004
Gutsch, W. A.
 158.122
Gutzwiller, M. C.
 008.148
Gvozdev, M. I.
 032.006
Gyldenkerne, K.
 142.057

Habing, H. J.
 114.018
 131.507
 157.005
Hachenberg, O.
 131.506 .546
 141.076
Hack, M.
 114.110 .151
 121.008 .059
Hackos Jr., W.
 122.044
Hacyan, S.
 021.033
 117.029
Haddad, G. N.
 022.067
Haddock, F. T.
 077.016 .071 .901
Hadjidemetriou, J. D.
 012.010
 042.074
Haemeen-Anttila, J.
 099.022
Haenig, W.
 095.003
Haeusler, B.
 084.406
Haeussler, K.
 123.012 .015
Hagan, L.
 008.142
Hagar, C. F.
 009.011
Hagen, G. L.
 065.097
 153.028
Hagen, H. M. Z.
 066.118

Hagen, W.
117.020
Hagen-Thorn, V. A.
158.086 .091
Hagfors, T.
093.028
100.009
Hainebach, K. L.
065.110
Haisch, B. M.
064.012
Hall, C. F.
053.003
Hall, D. L.
141.002
Hall, D. S.
117.033
121.026 .027
Hall, J.
103.101
Hall, J. S.
008.047
Hallam, A.
003.054
Hallam, M.
094.006
107.001
Hallissey, M.
084.027
Hamaker, J. P.
034.002
Hamill, P. J.
042.015
Hamilton, D.
099.016
Hamilton, W. O.
033.021
Hammerschlag, A.
034.001
Hamon, A.
010.028
Han, R. Y.
084.019
Hanbury Brown, R.
008.124
115.004 .010
Handschel, G.
153.032
Hankins, T. H.
141.323
Hanni, R. S.
066.128
Hansen, C. J.
008.022
012.004
061.040
125.065
133.013
Hansen, F.
074.064
Hansen, J. E.
093.031
Hansen, J. P.
061.028
Hansen, O. L.
092.041
099.207
Hansen, R.
074.027
Hansen, R. T.
073.081 .089

Hansen, R. T.
074.028 .043 .063 .095
.096 .107 .115
Hansen, S.
074.027
Hansen, S. F.
074.028 .096 .107 .115
Hansen, T. L.
082.083
Hanser, F. A.
078.030 .034
Hanson, T.
079.100
Hanson, W. B.
083.037
Hapke, B.
092.017 .039
093.024
Haque, S. S.
131.004
Hara, K.
162.035
Harada, Y.
096.038
118.018
Haramundanis, K.
113.033
Hardee, P. E.
125.015
Hardorp, J.
065.064
Hardy, D. A.
003.107
Hardy, R. N.
033.017
Hargrave, P. J.
141.014
Harlan, E.
158.028
Harlan, E. A.
114.152 .903
Harmanec, P.
119.016
Harmer, C. F. W.
034.029
Harnden Jr., F. R.
125.068
Harrington, J. P.
063.007
Harrington, R. S.
042.058
103.101 .105
111.004
117.003
Harris, A.
141.016
Harris, A. W.
100.008
Harris, D. E.
160.005 .027
Harris, I.
082.078 .107
Harris, K. K.
084.272
Harris, S.
131.013
Harris, W. E.
122.095
154.007 .015
Harrison, E. R.
061.012

Harrison, E. R.
162.006
Harrison, S. W.
131.093
Harrison, T. G.
065.008 .009
Hart, M. H.
080.006
101.004
Hart Jr., H. R.
094.311 .364
Harten, A.
074.021
Harten, R. H.
131.541
Hartle, R. E.
074.033
092.035
093.021 .043
094.152
Hartley, M.
082.018
Hartmann, L.
141.098
Hartmann, L. W.
064.054
114.002 .145
Hartmann, R.
072.025
Hartmann, W. K.
097.006
Hartoog, M. R.
114.051
Hartung, J. B.
094.132
Hartwick, F. D. A.
154.022
158.101
Harvey, G. A.
104.022 .048
141.050
Harvey, J.
032.042
034.071
071.059
074.097
Harvey, J. W.
071.008
Harvey, K.
071.059
Harvey, K. L.
073.055
Harvey, O. L.
047.005
Harvey, P. M.
114.048 .132
122.019
131.068
Harwit, M.
003.055
066.035
131.029
Hasegawa, A.
061.052
Haser, L.
022.016
Haslam, C. G. T.
033.025
157.004
Hassan, S. M.
153.901

Hast, N.
094.178
Hastie, W.
003.133
Hata, S.
074.070
Hatanaka, Y.
042.081
Hatchett, S.
131.041
Hauck, E.
021.028
113.031 .041
114.006
Hauge, O.
061.055
071.005 .036
Haughney, L.
103.101
Haupt, H.
047.015
098.042
Haupt, H. F.
113.021
Haupt, W.
114.093
Haussecker, K.
034.063
Havnes, O.
116.010
Hawkes, R. L.
104.021
Hawking, S. W.
066.020 .027 .069 .070
162.019
Hawkins, F.
141.335
Hawkins, F. J.
141.328
142.073
160.017
Hawkins, M. R. S.
031.058
Hayakawa, S.
003.056
142.065
Hayes, D. S.
131.042
Hayes, J. M.
105.002
Hayes, E. W.
073.001
Hayli, A.
151.024
Hays, D.
065.004
Hays, P. B.
084.003
Hazard, C.
033.040
141.100 .103
Hazard, D. R.
003.134
Hazlehurst, J.
117.023
He, X.
077.079
Head, J. W.
094.362
Healy, A. W.
032.023

Heap, S. R.
114.004
Heaps, M. G.
083.051
Hearn, A. G.
064.028
Hearn, D. R.
061.063
Hearnshaw, J. B.
114.010
Heasley, J. N.
073.066
Hecht, H. F.
162.902
Heck, A.
064.034
103.102 .124
Heckman, H. H.
143.028
Heckmann, O.
162.019
Hedgecock, P. C.
106.036
Hedin, A. E.
082.031 .035 .102
Heeran, M. P.
085.002
Heeschen, D. S.
008.032 .052 .133
Hefele, H.
103.101
Hegyi, D. J.
012.032
131.133
132.030
Heidmann, J.
158.083 .099
Heikkila, W. J.
084.021 .296
Heiles, C.
157.005 .006
Heimann, M.
105.007 .043
Heinemann, M. A.
074.055
Heintz, W. D.
008.123
111.008
118.008 .022
Heintzmann, H.
065.085
Heinz, C. J.
142.031
Heinz, H.
084.299
Heinzel, P.
011.021
Heiser, A. M.
008.089
Heiser, E.
123.003
Helava, H.
132.033
Heller, M.
010.017
Helliwell, R. A.
084.247
Hellsten, T.
062.015
Hellyer, B.
004.023

Helmerhorst, T. J.
131.547
Helminger, P.
022.076 .077
Helszajn, J.
033.093
Henbest, S. N.
125.053
Henderson, G. A.
131.093
Henderson, P.
105.044
Hendl, R. G.
080.047
Hendrickson, R. A.
084.292
Hendrie, M. J.
103.011
Henize, K.
103.101
Henize, K. G.
114.047
142.074
Henkel, M.
046.002
Henn, F.
002.021
Henon, M.
042.002
Henriksen, K.
084.236
Henriksen, R. N.
141.303
142.016
Henry, R. C.
114.072 .112
142.082
155.049
Hensberge, G.
116.012
117.034
Henze, W.
077.074
Herald, D.
103.105
Herbig, G. H.
103.101
141.604 .606 .607
Herbst, E.
131.034
155.046
Herczeg, T.
121.011 .081 .082
Herget, P.
098.009 .046
103.104
Herlt, E.
066.058
Herman, M.
063.043
Hermann, B. R.
125.901
Hermann, F.
105.151
Hernandez, G.
082.063
084.024
Herndon, J. M.
104.144
105.012 .013 .070

Herpers, U.
  094.174
Herr, K. C.
  097.050
Herr, W.
  094.174
  105.007 .043
Herrmann, D. B.
  006.000
  011.018
Hers, J.
  007.000
  035.004
  096.012
Herzberg, G.
  103.101
Herzo, D.
  034.066
Herzog, G. F.
  105.030
Hess, M.
  123.044
Hess, W.
  099.036
Hesse, K. H.
  141.314
Hesser, J. E.
  122.064
  142.052
  154.022
Hesstvedt, E.
  082.045
Hetherington, N. S.
  004.010
Heuseler, H.
  003.057
Hewish, A.
  106.038
  141.007
Hewitt, T. G.
  064.025
Heymann, D.
  022.901
  094.308 .341
Heyvaerts, J.
  074.003
  077.064
Heyvaerts, J. F.
  053.016
Hickey, J. R.
  031.044
Hicks, T. R.
  106.003
Hidajat, B.
  065.124
  153.026
  155.078 .079
Hide, R.
  099.027
Hieda, Y.
  022.071
Hiei, E.
  073.092
Higginbotham, N. A.
  064.051
  114.166
Higgs, L. A.
  021.022
  033.030
Hilaire, G.
  096.003

Hilditch, R. W.
  121.014
  124.104
Hildner, E.
  073.041
  074.005
Hilf, E. R.
  065.085
Hilgeman, T.
  031.032
  103.101
Hill, G.
  010.017
  153.012
Hill, H.
  075.001
  092.027
Hill, H. A.
  032.023
Hill, P. G.
  094.306
Hill, R.
  142.051
Hill, R. W.
  142.033 .083
Hill, S. J.
  032.040
Hill, T. W.
  099.077
Hillebrandt, W.
  065.085 .126
Hills, J. G.
  065.089
  121.005
  131.018
Hiltner, W. A.
  008.006
  131.042
  142.022 .092 .115
Hines, C. O.
  074.059
Hintenberger, H.
  105.045 .074 .150
Hintzen, P.
  126.021
Hintzen, P. M. N. O.
  122.102
Hirabayashi, H.
  131.024
  155.071
Hirai, M.
  022.070
  114.095 .153
Hirasawa
  123.025
Hirayama, T.
  041.064
  073.030 .033
Hirose, C.
  022.059
Hirschberg, J. G.
  034.041
Hirshberg, J.
  074.054
Hirth, W.
  077.028
Hjalmarson, A.
  131.105
Hlava, P. F.
  094.403

Hlond, M.
  031.049
  034.051
Hoag, A. A.
  013.014
Hobbs, L. M.
  131.035 .530
Hobbs, R. W.
  077.061
  103.101
Hoch, R. J.
  084.004
Hocking, W. H.
  131.011
Hockney, R. W.
  151.035
Hodge, P. W.
  003.058 .059
  105.017
  131.136 .534
  154.028
  158.019
Hodges Jr., R. R.
  094.150
Hodgson, R. G.
  003.060 .061
  098.037 .038 .039 .040
  123.037
Hoebel, P.
  031.085
Hoeg, E.
  031.079
  032.021 .037
  041.046
Hoegbom, J. A.
  033.031 .054
Hoeglund, B.
  131.092
Hoegner, W.
  036.003
  132.007
Hoekstra, R.
  034.001
  114.107 .110
Hoffleit, D.
  010.035
  123.020
Hoffleit, E. D.
  041.052
Hoffman, J. H.
  094.150
Hoffman, N. M.
  064.053
Hoffman, R. A.
  084.290
Hoffmann, M.
  103.105
Hofmann, W.
  034.031
Hogan, D.
  082.073 .074
Hohl, F.
  151.056
Hoinkes, G.
  105.046 .154
Holcomb, D. F.
  003.062
Holden, F.
  118.011
Holdsworth, E.
  105.015

Hollars, D. R.
  034.083
Hollweg, J. V.
  062.064
  074.056 .058
Holm, A. V.
  121.034
  124.010
  125.102
Holman, B. K.
  085.014
Holmberg, E.
  158.066
Holmberg, S.
  062.016
Holmes, J. A.
  143.002 .026
Holmes, J. C.
  113.037
Holt, J.
  075.008
Holt, J. N.
  080.003
Holt, S.
  142.089
Holt, S. S.
  142.043 .048 .060 .061
    .116
Holter, O.
  003.004
Holzer, R. E.
  084.254
Hones Jr., E. W.
  084.032 .233 .257 .274
Honeycutt, R. K.
  114.098 .099
Hong, J.-H.
  022.049
Hong, K.
  022.049
Honkasalo, T.
  046.019
Honorez, M.
  021.023
Hooghoudt, B. G.
  033.001
Hooke, A. J.
  053.010
Hooley, T.
  141.012
Hooper Jr., C. F.
  062.033
Hopgood, P. A.
  082.007
Hopkinson, G. R.
  103.101
Hopmann, J.
  103.101
  118.025 .026
  155.076
Hopp, U.
  123.026 .027 .030 .031
    .034
  125.112
Horai, K.
  094.323
Horak, H. G.
  082.070
Horan, D. M.
  076.015 .021

Horan, S.
  123.029
Horan, S. J.
  122.081
Hord, C. W.
  097.036 .074
Horedt, G.
  042.077
  061.060
  098.025
Hori, G.
  012.022
  042.053 .083
Hornby, P. W.
  123.025
Horstman, H.
  034.072
Horstman, H. M.
  061.011
Horstman-Moretti, E.
  061.011
Horton, B. H.
  094.140
Hoshi, R.
  142.111
Hosking, R. J.
  066.026
  151.901
Hoskins, D. G.
  141.010 .022 .032
Hosokawa
  123.025 .037
Hotop
  103.105
Hough, J.
  033.020
Houk, N.
  114.146
Houminer, Z.
  106.038
Hourani, H.
  075.010
Houston, J.
  071.056
Houziaux, L.
  014.047
  032.034 .902
  041.022
  114.092 .117
Hovestadt, D.
  078.042
  084.406
Hovsepian, A. V.
  See Ovsepyan, A. V.
Howard, A. J.
  065.023
Howard, B. E.
  021.009
Howard, E. G.
  074.098
Howard, H. T.
  092.040
  093.022
  094.125
Howard, R. A.
  074.064 .105
Howard, Ro.
  071.060 .065
  075.016
Howard, Ru.
  074.027 .096

Howard, W. M.
  061.007
Howell, F. J.
  100.005
Howlett, C.
  082.003
Howse, D.
  003.063
Hoyle, F.
  003.064
  004.015
  124.004
Hoyng, P.
  061.017
Hrachova-Rezacova, V.
  034.020
Hric, L.
  092.023
Hsi, T.
  004.068
Hu, F.
  074.112
Hua, C. T.
  132.022
Huang, S.-S.
  114.104
  121.002 .072
Huang, Y.-N.
  083.008
Hubbard, E. C.
  042.066
Hubbard, J. S.
  097.042
Hubbard, N. J.
  094.133 .347
Hubbard, R. F.
  099.216
Hubbard, W. B.
  091.007 .020
  099.029
Hube, D. P.
  112.012
Huber, M. C. E.
  022.056
  073.016
  076.034 .037
Huchra, J. P.
  158.111
Huchtmeier, W.
  131.539
Huchtmeier, W. K.
  131.514 .540
Hudec, R.
  113.045
Hudson, H.
  076.027
  122.114
Hudson, H. S.
  073.077
  076.008
  080.001 .048
Hudson, J.
  021.020
Huebner, W.
  004.042
  094.341 .342
  103.101
Huenecke, W.
  036.001
Huggins, P. J.
  114.023

Hughes, D. W.
104.007 .064
Hughes, J. A.
031.053 .080
Hughes, P. C.
052.002
Hughes, V. A.
077.020
Hughes III, H. G.
066.135
Huguenin, G. R.
066.003
Hulme, G.
094.179
Hultqvist, B.
084.016 .243
Hultqvist, L.
071.004
Hume, C. J.
143.064
Humes, D. H.
104.010
Hummer, D. G.
063.005 .006
064.036
Humphreys, R. M.
114.008 .014 .031 .162
141.611 .616
Humphries, C. M.
032.034 .902
113.006 .034
114.015
Humphries, D. J.
094.306
Hundt, E.
063.012
Huneke, J. C.
094.365
Hunger, K.
114.094
Hunstead, R. W.
041.055
141.066
Hunt, A. J.
033.029
Hunt, G. C.
157.004
Hunt, G. E.
099.045 .048
Hunt, G. R.
002.029
Hunten, D. M.
003.065
091.026
093.036
097.059
100.209
Hunter, A. N.
085.014
Hunter, C.
151.010 .016
Huntress Jr., W. T.
099.012
Hurford, G. J.
078.038 .041
Hurley, K.
061.032
160.006
Hurly, P. R.
124.101

Hurst, R. B.
157.003
Hurukawa, K.
041.064
Hurwitz, L.
084.258
Husain, L.
094.391
Huseynov, M. J.
000.000
Huss, G. I
105.099
Hut, P.
065.053
066.093 .145
Hutchings, J. B.
008.138
121.052 .059
142.028
Hutchinson, J. L.
122.112
Hutchison, R.
107.008
Huyberechts, S.
105.047
Hyder, C. L.
099.064
Hyland, A. R.
114.013
Hynds, R. J.
078.017
Hynek, J. A.
008.046 .068

Ianna, P. A.
114.143
Iatskiv, Ia. S.
See Yatskiv, Ya. S.
Ibanoglu, C.
121.009
Ibragimov, N. B.
099.005
Ichimaru, S.
003.066
142.110
Ichimura, K.
122.076 .092
Icke, V.
031.048
Idlis, G. M.
004.059
Iglesias, G. E.
084.011
Ignat'ev, P. P.
097.033
Ignatovich, S. I.
103.101 .105
Iguchi, T.
131.139
155.007
Ikawa
103.102
Il'inskij, V. N.
061.031 .053
104.058
143.032
Ilk, K. H.
046.021
Illingworth, G.
154.011

Ilmas, M.
114.054
Ilovaisky, S. A.
142.058
Imbert, M.
119.015
Imhof, W. L.
083.047
Imhoff, C. L.
115.015
Inciong, S. V.
047.008 .014
Ingalls, R. P.
093.018
Ingersoll, A. P.
093.009
Ingram, D. S.
094.005
Innanen, K. A.
151.012
155.024
Innerebner, G.
004.018
Inoue, M.
158.303
Inoue, T.
042.082
Intriligator, D. S.
074.019 .082
099.014
Ip, W.-H.
098.004
102.002
103.101
Ipavich, F. M.
106.039
Iriarte Erro, B.
122.075
Irigoyen, M.
042.061
Irvine, N.
103.105
Irvine, N. J.
114.034
122.120
Irvine, W. M.
010.017
103.101
131.105 .114 .117
Irwin, J. B.
009.001
Isaacs, J. D.
105.010
Isaka, H.
082.016
Isamutdinov, Sh. O.
104.043 .046
Ishchenko, I. M.
122.072
Ishiguro, M.
033.033
Iskandarova, V. M.
082.050 .051 .052
Iskudaryan, S. G.
158.021
Isles, J. E.
010.012
123.022
124.006
Israel, F. P.
158.044 .082

Isserstedt, J.
113.010
Istoshin, N. A.
014.026
Iucci, N.
143.010
Ivanisevic, G.
158.030
Ivannikova, M. N.
073.063
Ivanov, A. V.
094.161 .388
Ivanov, B. S.
141.056 .077
Ivanov, G. R.
122.007 .033 .039
Ivanov, K. G.
062.048
074.024
106.011 .012
Ivanov, L. N.
124.009
Ivanov, N. A.
085.010
Ivanov, N. I.
104.068
Ivanov, O. G.
094.127 .173
Ivanov, V. V.
083.004
Ivanov, Yu. M.
055.011
Ivanov-Kholodnyj, G. S.
076.007 .016
082.053
Ivanova, N. A.
114.035
Ivanovskij, A. I.
012.028
Ivanovskij, O. G.
094.001
Iversen, I. B.
084.281
Iversen, J. D.
097.018
Ivliev, A. I.
105.117
Iwanowska, W.
126.011
131.110
133.016
Izmajlov, S. V.
052.024 .035
Izvekov, B. K.
033.008

Jacchia, L.
103.101
Jacchia, L. G.
082.097
102.003
104.065
Jackisch, G.
004.038 .070
Jackson, J. C.
141.013
Jackson, J. H.
003.017
Jackson, P. D.
155.047

Jackson, W. M.
103.105
Jaeger, E. F.
084.015
Jaervi, P.
055.012
Jaffe, W.
141.028
Jagodzinski, H.
094.343
Jagoutz, E.
105.037
Jahn, B. M.
094.347
Jahreiss, H.
155.053
Jaidee, S.
113.024
Jamar, C.
032.034 .902
114.114 .116 .124
James, G. L.
033.077
James, J. F.
103.101
James, T. C.
082.036
James, T. H.
036.002
Jameson, R. F.
141.612
158.034
Jamieson, H. D.
094.384
Jancovici, B.
061.028
Janes, K. A.
153.024
Janiczek, P. M.
098.014
Janin, G.
151.023
Janin, L.
035.002
Jankovich, I. I.
See Yankovich, I. I.
Jansen Van Beek, G.
084.302
Janssens, T. J.
076.029
Jappel, A.
003.013
Jarosevich, E.
105.059 .166
Jarrett, A. H.
009.007
Jarzebowski, L.
004.074
Jaschek, M.
114.041
Jasevicius, V.
113.004
Jastrow, R.
003.067
Jauncey, D. L.
033.040 .042
066.035
141.060 .071 .100 .103
Javorka, E.
004.071

Javoy, M.
094.317
Jedwab, J.
105.047 .048
Jefferies, B.
094.306
Jefferies, J. T.
008.061
080.009
Jeffers, S.
114.005
Jefferts, K. B.
132.034
155.004
Jefferys, W. H.
042.026 .080
Jekabsons, C.
103.105
Jelley, J. V.
061.050
125.035
Jenkins, E. B.
113.001
131.098
Jenkins, E. F.
008.139
Jenkner, H.
155.077
Jennings, D. M.
141.340
Jensen, E.
007.000
080.019
Jensen, E. B.
122.102
Jensen, H. B.
065.023
Jerome, D. Y.
105.034
Jerominek, H.-J.
123.048
Jerzykiewicz, M.
122.043
Jessberger, E. K.
094.365
Jetzer, F.
099.053
Jewsbury, C. P.
065.061
Jochems, P. J. W.
033.088
Jochum, K. P.
105.045 .074 .150
Joeveer, M.
155.014 .015 .016
Johanning, D.
062.047
Johansson, L. E. B.
131.019 .092
John, L.
003.068
John, R. W.
066.061
John, T. L.
022.005
Johnson, A. A.
105.107
Johnson, C. Y.
113.037
Johnson, D. R.
131.004

Johnson, D. W.
084.038
Johnson, F. S.
094.150
Johnson, H. M.
132.010
142.114
Johnson, H. R.
064.004
114.099
Johnson, M. A.
097.066 .067
Johnson, R. G.
084.272
Johnson III, W. N.
142.090 .091
Johnston, D. H.
094.008
Johnston, M.
066.125
Jokipii, J. R.
061.067 .080
078.040
143.022
Joly, M.
158.105
Jones, A.
103.105
Jones, A. P.
103.102
Jones, A. V.
082.004
Jones, B.
131.522
155.065
Jones, B. B.
076.004
Jones, E. F.
111.005
Jones, C. A.
142.067
Jones, D. A.
141.304
Jones, D. E.
084.228
099.015
Jones, D. H. P.
122.012 .100
Jones, J.
104.011 .021
Jones, K. L.
097.021
Jones, M. V.
103.101
Jones, P. B.
141.317
Jones, R. V.
009.021
Jones, S. E.
097.009
Jones, S. S. D.
046.004
Jones, T. B.
083.040
Jones, T. J.
114.043 .156
116.007
Jones, T. W.
062.025
073.077
131.028

Jones, T. W.
155.003
Jones Jr., H. S.
033.079
Jordan, C.
074.113
076.004
Jordan, J. F.
052.014
Jordan, J. L.
094.308
Jordan, S. D.
073.076
077.061
Joshi, C. H.
046.025
Joshi, G. C.
114.065
Joshi, M. N.
141.051 .087
Joshi, S. C.
032.031
122.087
Joshi, V. J.
066.086
Joss, P. C.
065.107
142.029
Jouannic, J.
035.013
Jovanovic, M.
032.058
044.021
Jovanovic, S.
105.049
Joyce, G.
099.216
Judge, D. L.
099.021
Jugaku, J.
158.116
Jukes, T. H.
015.011
Julienne, P. S.
082.103
Jull, A. J. T.
094.349
Jung, H.
004.079
Jung, J.
002.024 .025
021.029 .030
041.071 .074
Junkes, J.
008.029
Jupp, A. H.
042.011 .012
Jura, M.
131.089 .528
Juren, C.
061.042
Juza, K.
113.045

Kaasik, A.
158.054
Kaburaki, O.
074.052
Kadyrov, A.
055.011

Kaehler, H.
065.015 .072
Kafatos, M.
131.041 .057
Kafka, P.
066.047
Kaftan-Kassim, M. A.
141.018
Kaganovskij, G. M.
041.021
Kahler, S.
073.086
076.024
Kahn, F. D.
122.105
Kai, K.
074.044
077.007 .008
Kaifu, N.
114.140
131.024
155.007
Kaila, K.
103.105
Kaiser, W. A.
094.174
Kajmakov, E. A.
022.040 .041
Kakhidze, G. P.
143.036
Kakuta, C.
082.113
Kalachev, P. D.
032.006
Kalachev, V. L.
052.027
Kalbitzer, S.
094.330
Kalenov, N. E.
054.003
Kaler, J. B.
133.005 .017
Kalikhevich, F. F.
103.101
Kalinin, Yu. D.
106.029
Kalinina, E. P.
112.014
Kalinkin, L. F.
084.408 .409
143.036
Kalinkov, M.
160.021
Kalkofen, W.
063.011 .018
Kalloglyan, A. T.
160.024
Kalman, G.
061.029
Kalmykov, A. M.
009.016
Kalnajs, A. J.
151.018
Kalytis, R.
113.004
Kalzhanov, B.
032.032
Kamaletdinova, N. F.
066.099
Kamenskaya, S. A.
141.056 .077

Kamide, Y.
084.213 .225
Kammer, H.
105.050
Kamp, L. W.
114.149
153.019
Kamper Jr., K. W.
119.006
Kamperman, T.
114.107
Kamperman, T. M.
034.001
Kan, J. R.
084.256
Kandel, R. S.
063.037
Kandpal, C. D.
032.031
Kane, R. P.
078.016
084.227
Kane, S. R.
076.030
142.079
Kanno, M.
073.050
Kantz, M. L.
103.101 .105 .116
Kao, M.
062.083
Kapahi, V. K.
141.051 .087
Kapitzky, J. E.
155.008
Kaplan, I. R.
094.382
Kaplan, S. A.
141.312
Kaplon, M. F.
061.010
Kapoor, R. C.
122.122
Kapralov, S. G.
042.076
Kaptsov, V. G.
084.266
Kapustin, P. A.
157.007
Karachentsev, I. D.
158.057
Karakula, S.
143.066
Karas, R. H.
084.220
Karetnikov, V. G.
114.046
121.100
Kargathra, L. V.
143.033
Karimov, K. A.
082.055
Karimova, D. K.
155.051
Karitskaya, E. A.
122.047
Karmanov, S. I.
051.016
054.018
Karminski, F.
104.062

Karnitskij, P. N.
014.021
Karoli, A. R.
031.044
Karpinskij, I. P.
054.018
Karpinskij, V. N.
054.012
071.039
Karpov, V. P.
078.006
Karyagina, Z. V.
103.101
Kasahara, I.
142.065 .070
Kashcheev, B. L.
104.004 .015 .017
Kashkarova, V. G.
105.113 .136
Kasinskij, V. V.
076.006
Kasper, U.
066.018
Kasperczuk, S.
098.020 .021
Kastel', G. R.
103.101 .116
Kastelein, W.
004.021
Kasten, P.
082.014
Kasten, V.
104.018
Kastner, S. O.
022.048
073.052
Katasev, L. A.
104.059
Katerfeld, G. N.
093.032
Katgert, J.
141.034
Kato, S.
151.057
Kato, T.
155.007
Kats, M.
011.016
Kattawar, G. W.
063.030
Katz, J. I.
065.101
142.049
Katz, L.
078.030 .034
Katz, R.
143.053
Kaufman, A. S.
074.038
Kaufmann, P.
077.002
141.076
Kavaliauskaite, G.
113.004
Kaverin, N. S.
073.056
Kawajiri, N.
158.303
Kawano, N.
158.303

Kawashima, N.
062.060
Kazantsev, A. N.
091.012 .019
097.057
Kazarovets, E. V.
122.137 .138
Kazaryan, E. S.
122.020
Kazaryan, M. A.
141.126
Kazimirchak-Polonskaya, E. I.
103.115
Kazlauskas, A.
113.004
Kearney, P. D.
074.041 .073
Keating, G. M.
097.011
Kechiyants, A. M.
065.058
Keen, R. A.
103.101
Keenan, D. W.
151.012
155.024
Keeney, J.
162.003
Keesey, M. S. W.
041.005
Kegel, W. H.
063.012
125.018
131.070
Kegeles, L. S.
066.124 .130
Keihm, S.
094.323
Keil, A.
096.005
Keil, K.
094.402 .403 .404
105.039 .051
Keiser, W. E.
003.016
Kelch, W. L.
114.144
Kelker, D. H.
111.009
Kelleher, R.
085.014
Keller, C. F.
074.099
Keller, H.-U.
009.004
Kellermann, K. I.
003.129
141.060
Kelley, M. C.
084.226
Kellman, S. A.
065.010
155.047
Kellogg, E.
142.035
Kellogg, E. M.
142.042 .043
Kellogg, P. J.
062.056

Kelly, W. R.
105.015 .052
Kemp, J. C.
116.003 .009
124.100
126.008 .010
Kemper, E.
142.112
Kendall, J.
099.022
Kenderdine, S.
031.035
Kennedy, J. E.
006.000
010.018
Kennel, C. F.
083.044
099.044
Kent, D. W.
078.010
Kepka, O.
077.030
Kerr, F. J.
008.037
131.532 .541
155.067
158.113
Kerr, J. L.
033.073
Kerridge, S. J.
113.013
Kerrigan, J.
121.082
Kersley, L.
083.015
Kerzhanovich, V. V.
093.011 .034 .038
Khachatryan, N. R.
077.068
Khachikian, E. Ye.
See Khachikyan, Eh. E.
Khachikyan, Eh. E.
141.126
158.022 .083 .089
Khadzhiolov, A. I.
005.014
Khaimov, I. M.
104.031 .033
Khaliullin, Kh. F.
064.023
Khamdi, M. A.
117.028
Khan, K.
124.102
Khan, M. A.
081.018
Kharchilava, D. F.
082.051 .052
Kharitonov, A. V.
103.101
Kharitonova, G. A.
100.004
Kharitonova, V. Ya.
105.127
Kharkov, A. A.
066.141
Khar'kov, A. A.
066.094
Khatisov, A. Sh.
054.013
097.201

Khatisov, A. Sh.
103.004
Khatskevich, I. G.
051.016
Khavenson, N. G.
053.004
Khetselius, V. G.
082.109
Khetsuriani, Ts. S.
074.069
Khilov, E. D.
022.060
Khlystov, A. I.
071.051
Khodak, Yu. A.
003.069
097.040
Khodyachikh, M. F.
099.023
Khokhlov, M. Z.
084.266
Kholopov, P. N.
121.098
122.006 .016 .038
Kholshevnikov, K. V.
042.068
Khomenko, L. P.
034.009
Khomenko, Yu. A.
074.075
Khotinok, R. L.
104.006
Khozov, G. V.
113.036
Khrisanov, N. P.
014.020
Khromov, G. S.
013.007
133.007
Khrutskaya, E. V.
041.014
Khudajbergenov, U. Kh.
105.132
Khudyakova, T. N.
113.036
Kiang, C. S.
099.028
Kiasat, A.
103.105
Kibler, J. F.
052.044
Kieffaber, L. M.
082.079 .080
Kieffer, H.
097.203
Kieffer, H. H.
099.214
Kienle, H.
005.005
Kiepenheuer, K. O.
073.017
082.059
Kiesl, W.
012.020
061.072
Kiewiet De Jonge, J. H.
008.100
Kifune, T.
022.071
Kiknadze, I. N.
051.016

Kiko, J.
094.330
Kikuchi, N.
082.116
Kikuchi, S.
121.097
151.058
158.303
Kiladze, R. I.
054.013
092.014
099.213
Kilkenny, D.
112.003
Kilmartin, P. M.
103.111 .116 .135
Kilmister, C. W.
003.070
Kim, C. Y.
076.040
Kim, J. S.
084.215 .401
Kim, V. F.
083.042
Kimball, M. R.
105.053
Kimura, K.
094.375
105.054
Kimura, Y.
022.071
Kinard, W. H.
104.010
King, D. A.
004.045
King, I. R.
151.031
King, J. W.
084.209
King Jr., E. A.
094.360
King-Hele, D. G.
046.007
054.005
081.001 .019 .021
Kinman, T. D.
034.084 .085
141.050 .123
Kinoshita, H.
041.026
042.053
044.019
Kinzer, R. L.
061.033
155.001
Kiosa, M. N.
042.033
Kippenhahn, R.
122.106
141.020
Kirakosyan, R.
033.014
Kirchhoff, W. H.
022.076 .077
Kireeva, N. N.
131.045
154.009 .010 .019
Kirichuk, V. V.
041.007
045.002

Kirillov-Ugryumov, V. G.
084.409
Kirk, J. G.
074.051
Kirkham, B.
114.118
Kirkpatrick, L. D.
143.055
Kirnozov, F. F.
094.387
Kirova, O. A.
105.138
Kirsch, E.
078.035
Kirsch, Y.
003.145
Kirshner, R. P.
132.011 .032
Kirsten, T.
094.330 .342
105.055
Kirton, K. A.
155.038
Kisabeth, J. L.
084.022
Kiselev, A. A.
112.023
Kiselev, F. I.
042.004
Kiselev, I. F.
052.025
Kiselev, M. I.
052.011
Kiselev, N. N.
032.019
103.100 .101 .121
124.103
Kiselev, V. V.
094.127 .173
Kishko, S. M.
022.025
Kislyakov, A. G.
033.009 .010
077.068
093.001
Kitamura, M.
121.035 .042 .055
Kittl, R.
105.152
Kivelson, M. G.
084.279
Kiyokawa, M.
121.097
Kjeldseth Moe, O.
072.031 .032
Klajman, A. Ya.
117.010
Klare, G.
103.101
121.075
Klaus, G.
031.015
073.022
Klebesadel, R. W.
142.021 .080
Klein, M. J.
141.046
Klein, O.
066.007
Kleine, T.
103.101 .118

Kleinmann, D. E.
131.524
Klejman, E. B.
062.024
Klemola, A. R.
098.028
103.101
112.021
Klemperer, W.
131.034
Klemperer, W. K.
033.035
Klepikova, L. A.
041.011
Kletnieks, Ya. M.
010.033
Klevenskij, Yu. N.
014.018
Klimas, A. J.
143.035 .058
Klimek, Z.
160.028
Klingler, R. J.
034.038
Klinkmann, W.
071.018
Kliore, A.
099.205
Kliore, A. J.
092.040
093.022
Klobuchar, J. A.
083.015
Klochan, E. L.
131.033
Klock, B. L.
032.038
Klosko, S. M.
046.011
Klotz, W.-D.
022.063
Klueppelberg, D.
106.019
Klynning, L.
022.007
Klyuchnik, N. N.
103.101
Knapp, G. R.
131.064 .542 .543
155.067
158.113
Knapp, S. L.
115.001
Knerr, R.
033.061
Kniffen, D. A.
061.064
155.058
Knight, C. K.
099.022
Knight, W. R.
162.026
Knipe, G. F. G.
121.053
123.004
Knohl, E.-D.
031.069
Knuckles, C. F.
097.010
Knudsen, W. C.
083.029

Knuth, R.
085.004 .005
Kobold, F.
007.000
Kobrin, M. M.
077.010 .045
Kobylinski, Z.
011.004 .029 .030
Koch, D.
142.035
Koch, K. R.
046.010
Koch, R. H.
114.059
153.013
Kochan, H.
084.028
Kocharov, G. E.
094.301
Kocherga, O. D.
062.037
Kochhar, R. K.
042.017
Kodaira, K.
114.171
125.072
Koeckelenbergh, A.
075.006
Koehler, H.
031.028
080.007
Koehler, H. W.
011.007
054.007
099.025 .034
Koehnlein, W.
094.162
Koester, D.
122.003 .008
Koga, R.
034.066
Kogan, L. R.
033.011
Kogan-Laskina, E. I.
084.408
Kogut, J.
077.902
Kohler, H. W.
053.020
099.068
Kojima, N.
103.101
Kokhan, E. K.
034.004
Kokin, G. A.
012.028
Kokurin, Yu. L.
032.006
Kolaczek, B.
011.041
Kolchinskij, I. G.
031.016 .081
Kolde, K.
003.071
Kolena, J.
131.127
Kolenkiewicz, R.
046.028
Kolesnik, I. G.
065.044

Kolesnikov, E. M.
 105.090 .120 .130 .131
Kolesov, A. K.
 091.022
Kolesov, G. M.
 094.387
 105.116
Kolesov, G. Ya.
 034.026
 078.020
Kolomenskij, V. D.
 105.128
Kolomiets, G. I.
 104.034 .037 .039
Kolomijtseva, G. I.
 084.245
Kolosov, M. A.
 083.003
 094.001
 106.016
Koman, G. G.
 052.034
Komarnitskaya, N. I.
 106.023
Komarov, N. S.
 114.122
Komberg, B. V.
 141.313
Komesaroff, M. M.
 008.124
Kompaniets, Eh. P.
 052.026
Konashenok, V. N.
 097.001
Kondo, Y.
 117.004
 121.003 .024 .030 .059
  .076
 142.057 .074 .102
Kondrashova, N. N.
 071.013
Kondratyev, K. Ya.
 082.110
Konecny, F.
 007.000
Kong, J. A.
 033.052
Konin, V. V.
 041.010 .012
Koning, P. A.
 104.025
Konnikova, V. K.
 141.041
Konno, M.
 158.303
Kono, K.
 103.134
Kononovich, Eh. V.
 014.019
Konopikhin, A. A.
 094.141
Konopleva, V. P.
 103.101
Kontizas, M.
 154.014
Kontor, N. N.
 097.033
Koomen, M.
 074.027 .096
Koomen, M. J.
 074.063 .105

Kopal, Z.
 002.038
 003.072
 117.005
Kopatskaya, E. N.
 126.009 .013
 142.066
Kopecka, F.
 072.016
Kopecky, M.
 064.016
 072.016 .026
Koppert, G.
 105.159
Koppeschaar, C. E.
 103.101
Kopylov, V. F.
 084.266
 106.027
Korchagin, V. I.
 158.038
Korekawa, M.
 094.343
Korkina, M. P.
 066.095 .100
 162.027
Korkotyan, G. A.
 034.017
Kormendy, J.
 158.114
Kornienko, G. I.
 072.043
 073.064
Korobchuk, O. V.
 077.043
Korobejnikov, V. P.
 105.124
Korobova, Z. B.
 034.025 .047
 072.019 .042
Korol'kov, D. V.
 033.015
Korotkov, S. V.
 031.010
 034.011
Korovkina, T. L.
 104.054
Korshunov, A. I.
 077.010
Korsun', A. A.
 044.014
Korzhavin, A. N.
 077.042
Koshelev, V. V.
 082.020 .106
 083.039
Koski, A.
 142.112
Kostenko, V. I.
 033.011
Koster, J. R.
 083.015
Kostik, R. I.
 071.044
Kostyakova, E. B.
 133.020
Kostyuk, N. D.
 073.013
Kota, J.
 143.065

Kotrc, P.
 064.016
 072.026
Kotsakis, D.
 003.009
 032.035
Kotzer, P.
 143.055
Koubsky, P.
 051.018
 119.016
Koutchmy, O.
 074.018
Koutchmy, S.
 072.024
 074.018 .050
 079.100
 103.101
Kovalenko, O. N.
 103.101 .126
 121.101
Kovalenko, V. M.
 103.101 .126
 121.101
Kovalenskaya, M. A.
 034.005
Kovalevskij, I. V.
 083.041
Kovalevsky, J.
 013.008
 042.050
 094.011
Kovals, I.
 051.010
Kovbasyuk, L. D.
 014.012
Kovrygina, L. M.
 034.018
Kovtyukh, A. S.
 084.403
Kowal, C. T.
 125.032 .073 .106 .108
Koyano, H.
 122.076
Kozai, Y.
 081.031
Kozak, P. P.
 071.050
Kozarenko, B. I.
 082.118
Kozhevnikov, N. I.
 032.032
 072.015 .036
 082.039 .092
Kozin, I. D.
 083.026
Kozina, P. E.
 083.025
Kozlov, I. S.
 042.005
Kozlov, V. I.
 031.047
Kozlov, V. P.
 082.110
Kozlov, Yu. A.
 074.035
Kozlovsky, B.
 073.015
 078.039
 143.024 .030

Kozlovsky, B.-Z.
 131.503
 141.089
Kozyrev, N. A.
 092.013
 100.006
Kracher, A.
 105.155 .156
Kraehenbuehl, U.
 094.310
Kraft, R. P.
 114.106 .119
 142.112
 153.030
Krahn, D.
 002.021
Kralj, M.
 032.057
Krall, N. A.
 003.073
Krasnenko, N. P.
 041.075
Krasnopol'skij, V. A.
 082.054 .086
Krasovskij, A. A.
 052.026
Krass, M. S.
 094.182
Krassa, R. F.
 106.001
Krat, V. A.
 043.001 .003
 051.006
 054.012
 071.038
Kraus, J. D.
 008.038
 141.047 .076 .902
Krause, F.
 062.031 .054
Kravtsov, O. V.
 062.020
Krawczyk, S.
 122.091
Kreiner, J. M.
 079.100
 103.105
Krelowski, J.
 122.091
Kremenetskij, S. D.
 077.044
Krempec, J.
 113.032
Kremser, G.
 078.035
Kren, G.
 092.030
 095.007
Kreplin, R. W.
 076.015 .021
Kresak, L.
 102.013
Kresakova, M.
 104.012
Kreysa, E.
 034.033
Krieger, A.
 071.064
 073.086
 074.090 .091 .093 .094
  .100 .104 .106

Krieger, A.
 076.023 .024
Krieger, A. S.
 076.009
Krieger, C. J.
 141.026
Krikorian, R.
 114.037
Krinberg, I. A.
 083.030
Krinov, E. L.
 105.165
Krishna Swamy, K. S.
 See Swamy, K. S. K.
Krishna-Mohan, S.
 141.308
Krishnaswami, S.
 105.016
Krivoshapkin, P. A.
 143.044 .045
Krivsky, L.
 073.018 .019 .091
 077.021
Kriz, S.
 064.040
 121.021 .022
Krogh, F. T.
 021.005
Kron, G. E.
 158.012
Kropotkin, P. N.
 097.081
Krpata, J.
 119.016
Kruchinenko, V. G.
 104.001
Kruglov, Yu. M.
 093.034
Krumenaker, L.
 103.101 .128
Krumenaker, L. E.
 103.105 .112 .116 .127
Krupene, P. N.
 022.079
Krupp, B. M.
 022.046
Kruszewski, A.
 141.035
Krylov, A. G.
 054.014
Krylova, S. N.
 106.021
Krymskij, G. F.
 043.045
 074.088
 106.015 .028
 143.044 .046 .047 .048
  .049 .050 .051 .052
Kryukov, A. E.
 141.056 .077
Kuan, T. S.
 063.038
Kubicela, A.
 122.073 .128 .129
Kubinec, W. R.
 155.039
Kucherov, V. I.
 073.034
Kuchowicz, B.
 061.077
 065.112

Kudlek, M.
 003.074
Kudo, A.
 131.071
Kudritzki, R. P.
 064.039
Kudryavtsev, M. V.
 078.011
Kuehne, C.
 031.026
Kueveler, G.
 095.003
Kugaenko, B. V.
 085.009
Kuhi, L. V.
 114.028 .903
 117.017
Kuijpers, J.
 077.039
Kuiper, G. P.
 107.010
Kukarkin, B. V.
 004.001
 122.141
 131.045
 154.005 .009 .010 .019
Kulagin, S. G.
 014.012
Kulapova, A. N.
 122.014
Kuleshov, K. F.
 072.020
Kuleshova, V. P.
 084.211
Kulikov, Yu. Yu.
 034.046
Kulikova, N. V.
 104.051 .059
Kulikovskij, P. G.
 013.003
Kulus, H.
 094.174
Kumajgorodskaya, R. N.
 114.036
Kumar, S.
 034.044
 082.029 .030
 092.038
 093.025 .036
 103.101
 131.036
Kumar, S. S.
 099.043
Kumer, J. B.
 063.015
 082.036
Kumsishvili, Ya. I.
 121.043 .044
Kunasz, P. B.
 063.005 .006
Kuncoro, A.
 153.026
Kundt, W.
 162.016
Kundu, M.
 074.064
Kundu, M. R.
 073.005 .008
 077.001 .018 .041 .069
  .070
 125.015 .024

Kunert, A.
011.017
Kunilov, M. V.
077.013
Kunitsyn, A. L.
042.059
Kunitzsch, P.
003.075
004.075
Kunkel, W. E.
122.065 .078
Kuo-Petravic, L. G.
141.342
Kuperus, M.
073.010
074.072
Kupferman, P.
073.062 .082
Kuprevich, N. F.
082.026
Kurat, G.
094.402
105.046 .154
Kurbanov, A.
033.048
Kurchenko, Yu. A.
078.006
Kurfess, J. D.
142.090 .091
Kurgan, T. B.
032.026
Kurianova, A. N.
See Kur'yanova, A. N.
Kurilchik, V. N.
158.050
Kurimoto, R. K.
105.072
Kurmaeva, A. Kh.
082.026
Kurmakaev, Z. Kh.
066.005
Kurochka, L. N.
073.018
Kurochkin, N. E.
120.001
123.051
158.032
Kurochkina, A. I.
094.385
Kurokawa, H.
073.050
Kurokawa, S.
022.071
Kurpinska, M.
103.105
Kurskov, A. A.
162.009 .030 .048
Kurt, V. G.
078.002 .011
097.007
Kurtenbach, D.
082.073 .074
Kurucz, R.
113.033
Kurucz, R. L.
022.074
064.013
071.903
Kur'yanova, A. N.
041.037

Kushiro, I.
105.056 .087
Kustaanheimo, P. E.
052.028
Kuteva, Z. N.
034.040
Kuwabara, T.
123.026
Kuwashima, M.
084.037
Kuz'menko, K. N.
006.000
032.004
041.002
Kuzmin, A. D.
093.035
097.028
Kuzmin, V. A.
083.030
Kuz'min, A. I.
143.044 .045 .049 .050
Kuz'mina, V. A.
083.017
Kuznetsov, A. I.
066.146
Kuznetsov, D. A.
072.044
Kuznetsova, R. I.
091.016
Kvasha, L. G.
105.126 .134 .139
Kvifte, G.
082.062
Kwan, J.
131.088 .091
Kwast, T.
054.024
094.012

La Bonte, B. J.
071.057
Labeyrie, A.
031.036
Labeyrie, J.
061.075
Labrum, N. R.
008.124
Labs, D.
031.039
Lacis, A. A.
093.031
Laclare, F.
082.114
Lacroute, P.
041.056 .059 .073
Lada, C.
131.061
Lada, C. J.
131.535
Lafferty, D. L.
022.001
Lai, S. T.
061.029
Lakatos, S.
094.308
Lake, R. J. W.
115.010
Lakhina, G. S.
062.078

Lal, D.
105.016 .038
Lalonde, L. M.
033.040
Lam, T. T.
063.024
Lamb, D. Q.
142.029
Lamb, F. K.
126.006
Lamb, U.
003.084
Lambeck, K.
081.017
Lambert, D. L.
114.001 .033 .167
Lambert, G.
143.019
Lambrecht, H.
004.055
Lameer, J.-C.
132.008
Lamers, H. J.
012.012
034.001
114.107 .109 .110 .130
132.027
Lammlein, D.
094.119 .120 .121
Lammlein, D. R.
094.108
Lamport, J. E.
092.037
093.026
Lampton, M.
034.044
061.015
131.040
158.102
Lamy, P. L.
106.037
Lancaster Brown, P.
003.076
Landecker, T. L.
033.028
Landini, M.
074.106
Landman, D. A.
034.067
Landmark, B.
083.028
Landolt, A. U.
008.009
121.025
Landstreet, J. D.
126.015
Lane, A. L.
099.012
Lane, W. A.
101.005
Lang, B.
105.901
Lang, K. R.
077.060
Lange, J. J.
034.039
Langel, R. A.
084.294 .295
Langer, G. E.
114.106

Langkavel, A.
039.005
Langseth Jr., M.
094.323
Langton, R. J. J.
075.020
Lanzano, P.
042.067
Lanzerotti, L. J.
078.023 .033
Lapchinskij, V. G.
003.037
Lapedes, A. S.
141.098
Lapshin, V. I.
073.034
Laques, P.
118.013
Larimer, J. W.
105.009
107.002
Larionov, M. G.
141.041
Larson, E. E.
105.012 .070 .144
Larson, H. P.
092.001
Larson, R. B.
151.009
Larson, S.
097.008
Lashkin, V. I.
052.032
Laska, L.
034.020
Laskarides, P. G.
034.057
122.027
Lasker, B. M.
122.064
141.065
142.052
Lasota, J. P.
010.017
Latham, G.
094.119 .120 .121 .318
Latham, G. V.
094.108
Latimer, J.
046.027
Latipov, D.
083.034
104.032 .043 .046
Lats'ko, V. I.
094.314
Lattimer, J.
066.112
Latynina, I. I.
094.314
Latypov, A. A.
032.007
041.066
112.009 .010 .011
Latyshev, I. N.
152.001 .002
Laub, A. J.
021.002
Lauberts, A.
151.002 .052
Laucenieks, L.
052.038 .039

Laucenieks, L.
055.010
Laul, J. C.
094.344
105.044
Laury-Micoulaut, C.
062.081
Laury-Micoulaut, C. A.
131.506
Lauter, E. A.
075.011
Lauterborn, D.
065.007 .012
Laval, A.
131.552
Lavergnat, J.
062.040 .041
Lavrent'ev, Yu. G.
105.139
Lavrinov, G. A.
033.053
077.045
Lavrova, E. V.
084.211
Lavrukhina, A. K.
091.016
105.057 .079 .090 .113
.114 .119 .120 .130
.131 .132 .136 .137
Lawless, J.
094.382
Lawley, E.
084.217
Lawrence, R. J. C.
010.008
Lawton, A. T.
015.013
033.005
051.002
Lazarenko, E. K.
105.123
Lazareva, L. P.
077.050
Lazarus, A. J.
092.035
093.021
Le Breton, P.
099.012
Le Contel, J. M.
122.018
Le Poole, R. S.
141.113
Le Roulley, J.-C.
143.019
Le Squeren, A. M.
158.040 .301
Learner, R. C. M.
022.081
Lebedev, E. I.
077.045
Lebedev, L.
003.077
Lebedinets, V. N.
104.029 .060
Lebovitz, N. R.
117.026
Lecar, M.
042.024
151.028
Leckrone, D. S.
064.041

Leckrone, D. S.
122.101
Lecolazet, R.
011.037
Lederle, T.
041.072
Lee, D. L.
066.049
Lee, H. S.
083.018 .019
Lee, J.-S.
063.001
Lee, M.
033.019
Lee, M. A.
143.003
Lee, P.
064.051
103.101
Lee, R. H.
074.008
Lee, S. W.
122.085
Lee, T. A.
103.101
114.087
121.036
Lee, T. N.
073.060
Leer, E.
080.028
Lees, R. M.
131.004
Legen'ka, A. D.
083.043
Leger, C.
105.006
Leger, G.
106.002
Legg, A. J.
141.093
Legg, A. T.
031.009
Lehmann, G.
035.016
Lehmann, H.-R.
062.047
084.244
Lehnert, B.
062.015 .016 .036
Lehr, C. G.
046.026
Leinert, C.
053.008
106.006 .019
Leir, A.
114.170
Lejkin, G. A.
055.013
094.385
Lekht, E. E.
131.031
Lelievre, G.
158.119 .302
Lemaire, J.
083.007
Lemke, D.
034.031 .063
131.537
132.017

Lena, P.
 079.100
Lenhoff, C. J.
 002.029
Lenoble, J.
 091.030
Lentz, G.
 099.016
Leonardi, P.
 094.185
Leondes, C. T.
 052.001
Leonov, A. A.
 053.007
Leontyev, S. V.
 084.264
Leovy, C. B.
 093.040
Lepping, R. P.
 092.036
 093.023
 106.031
Lequeux, J.
 107.011
 125.020
 131.125 .132
Leray, J. P.
 134.006
 142.005
Lerch, F. J.
 046.012
Lerche, I.
 061.080
 062.027 .028 .055
 141.316
Leroux, F.
 079.105
Leroy, J. L.
 071.034
Lessnoff, G. W.
 033.020
Letsch, H.
 009.009
Leung, K.-C.
 112.016
 121.017
 122.113
Leung, Y. C.
 061.025
Leushin, V. V.
 114.074
Levashev, A. E.
 066.084
Levaux, H.
 082.003
Levchenko, M. T.
 033.051
Leventhal, M.
 142.084
Levi, F. A.
 105.058
Levi-Donati, G. R.
 105.020 .059 .097 .166
Levich, B. G.
 003.078
Levin, B. Yu.
 105.141
Levine, A.
 142.117
Levine, J. S.
 097.011

Levine, R. H.
 062.067
 074.083 .087
 080.041 .061
Levitan, E. P.
 014.013 .017 .040
Levitskij, L. S.
 072.006
 078.005
Levskij, L. K.
 105.112 .126 .130
Levy, D.
 003.050
Levy, E. H.
 062.003
 065.084
 094.112
 106.039
Levy, G. S.
 092.040
 093.022
Levy, M.
 071.015
Levy, S. G.
 119.005
 122.029
Lew, H.
 103.101
Lewin, W. H. G.
 061.063
 125.104
 142.019 .031
Lewis, B. M.
 158.027
Lewis, B. R.
 022.038 .039
Lewis, J. S.
 107.003
Lewis, R. S.
 094.375
 105.054
Lezniak, J. A.
 074.078
 078.007
L'Heureux, J.
 142.011
Li, F.
 104.009
 142.087
Li, X.
 105.158
Li, Y.
 044.020
Li, Y. T.
 063.038
Li, Z.
 045.016
Liang, E. P. T.
 066.028
Liang, S.
 044.020
Lichtenfeld, K.
 098.006
Lichtenstein, B. R.
 094.123 .165
Lichti, G.
 142.064
Lidov, M. L.
 042.030
 052.015

Liebert, J.
 032.061
 046.029
 126.022
Liebscher, D.-E.
 066.018
Liemohn, H. B.
 084.251
Liese, R.
 085.015
Lieske, J. H.
 099.206 .208
Liewer, K. M.
 115.001
Lightman, A. P.
 066.049
 142.003
Liipola, E. L.
 042.060
Lijgant, M. K.
 031.010
Likin, O. B.
 034.026
 078.002 .020
Liller, W.
 099.037 .049 .051 .074
 103.101 .105
 141.088
 142.067
Lilley, A. E.
 103.101
Lillie, C. F.
 114.007 .134 .135
Lilliequist, C. G.
 074.102
Lin, B.
 045.016
Lin, C. D.
 022.004
Lin, C. S.
 106.033
Lin, R. P.
 073.077
 078.021
 084.033 .231 .284
Lincoln, J. V.
 075.021
Lindalen, H. R.
 084.030
Lindblad, P. O.
 155.059
Linde, A. D.
 162.032
Lindegren, L.
 099.046
Lindeman, R. A.
 074.079
Lindemann, E.
 113.019
Lindner, K.
 003.079
Lindsay, E. M.
 123.001
Lindsey, C. A.
 080.001
Lindsey, W. C.
 002.026
Lindstrom, P. J.
 143.028
Lingenfelter, R. E.
 097.043

Link, F.
 099.047
Link, H.
 053.008
 106.006
Linnell, A. P.
 008.042
 032.040
Linsky, J. L.
 064.055
 071.023 .045
 073.078
 114.072
Linsley, J.
 143.029
Lipatov, A. S.
 084.267
Lipatov, B. N.
 141.055
Lipovetskij, V. A.
 083.016
 158.020 .036
Lippincott, S. L.
 008.123
Lipschutz, M. E.
 105.072
Lipskij, Yu. N.
 051.009
 074.006 .071
 114.096
Lisitsa, V. S.
 022.034
Liszewska, K.
 105.901
Liszt, H. S.
 132.034
Lites, B. W.
 063.003
 064.027
Litkevich, N. G.
 099.023
Littleton, J. E.
 065.106
Litvak, M. M.
 131.068
Liu, H.-S.
 081.025
Liu, J. H. C.
 009.002
Liu, S.-Y.
 073.043 .079
Liu, X.
 077.079
Liu, Z.
 124.102
Livadas, G. C.
 009.022
 082.088 .089
Livesey, R. J.
 014.006
Livingston, D. M.
 003.080
Livingston, W.
 032.042
 034.071
 074.097
Livshits, G. Sh.
 082.105
Livshits, M. A.
 073.002
 074.074

Lloyd Evans, T.
 See Evans, T. L.
Lo, K. Y.
 114.161
 131.111
Locher, K.
 121.090 .092 .094 .096
 125.111
Lochno, P.
 132.007
Lockwood, G. W.
 114.075
Lodenquai, J.
 065.092
Loeser, R.
 151.028
Logachev, Yu. I.
 034.005
 078.002 .011
 094.235
Logan, L. M.
 002.029
Loginov, P. P.
 041.067
Lohmann, W.
 155.006
 158.090
Lojko, M.
 099.036
Lokanadham, B.
 142.065
Loncarevic, M.
 044.021
Long, A. C.
 052.012
Long, C. E.
 143.059
Long, D. A.
 002.029
Long, K.
 142.044
Longair, M. S.
 162.036
Longmore, A. J.
 141.612
 158.034
Loomer, E. I.
 084.302
Loparev, B. N.
 034.008 .009 .013
Lopez Garcia, A.
 008.137
Lord, E. A.
 066.080
Lord, J. J.
 143.055
Lorell, J.
 097.047
Lorenzi, L.
 098.030
 103.101 .105
Lorin, J. C.
 105.060
Lorre, J.
 141.606
Lortet-Zuckermann, M. C.
 131.502 .505
Losovskij, B. Ya.
 033.051
 097.028

Lot, P.
 007.000
Lotova, N. A.
 106.025
Loughhead, R. E.
 003.028
 073.039
Louise, R.
 131.518
 133.002
Louisnard, N.
 082.015 .071
Lovas, F. J.
 022.079
Lovas, M.
 103.116
 125.110
Lovelace, R. V. E.
 066.035
Lovell, B.
 008.078
 122.104
Lovelock, J. E.
 082.058
Lovera, E.
 114.066
Lovering, J. F.
 094.304
Low, B. C.
 062.069
 073.042
Low, F. J.
 034.042
 094.114
 100.215
 132.017
Lowe, G.
 103.105
Lowen, L.
 114.165
Lowes, F. J.
 084.218 .238
Loyola, P.
 041.038
Lozinskaya, T. A.
 125.055
 132.009
Lubkin, E.
 080.057
Lubomirski, H.
 013.015
Lubyshev, B. I.
 077.011
Lucaroni, L.
 065.079
Lucchin, F.
 061.006 .074
 141.610
 151.043
 162.004
Luchkov, B. I.
 084.409
Luck, M. McK.
 046.001
Luck, R. E.
 114.167
Lucy, L. B.
 031.061
Ludwig, A. C.
 033.074

Ludwig, H.
046.021
Lue, P. K.
141.072
Lui, A. T. Y.
084.257
Lukacevic, I.
062.045
Lukas, B.
103.005
123.002 .027 .031 .033
.034
124.007
Lukash, V. N.
162.010
Lukashevich, N. L.
093.015
Lukashkin, V. M.
083.013
Lukatskaya, F. I.
122.028 .034 .035
Lukoschus, D.
033.090
Lumme, K.
091.023
Lund, F.
162.040
Lundquist, C. A.
013.016
Lunel, M.
131.504
Luo, S.
044.020
Lupishko, D. F.
097.014
Lupishko, T. A.
097.014
Lupoj, K.
014.034
Lustig, G.
098.017
Luthey, J. L.
099.057
Lutomirski, R. F.
031.043
Lutsenko, V. N.
034.005
078.002
Lutz, B. L.
101.006 .013
Lutz, J. H.
133.018
Lutz, T. E.
111.009
115.007
Luyten, J. R.
021.018
Luyten, W. J.
112.013 .019
Luzzatto, G.
162.045
L'vov, Yu. A.
105.066
Lyatsky, W. B.
084.264
Lyk'yanov, B.
003.077
Lynch, D. K.
071.061
Lynden-Bell, D.
008.024

Lynden-Bell, D.
151.041
Lynds, B. T.
008.031 .065
Lyne, A. G.
141.304
Lynga, G.
113.024
114.024
Lysov, V. P.
093.034
Lyttleton, R. A.
081.009
091.029
093.003 .027
Lyu Van Lyong
073.003
Lyubarskij, A. N.
085.003
Lyubimov, G. P.
097.033
Lyubimov, Yu. K.
094.104
Lyutyj, V. M.
142.106 .107
154.005

Maanders, E. J.
033.065
Mabuchi, H.
105.006
Macau, D.
032.034 .902
114.113 .114 .123 .124
.125
115.013
Macau, J. P.
032.034 .902
Maccagni, D.
073.009
MacCallum, M. A. H.
066.059
MacConnell, D. J.
122.110 .116
MacDougall, D.
105.011 .016
Maceroni, C.
125.017
Machado, M. E.
073.080
114.080
Mackal, P. K.
099.069 .072
Mackay, C. D.
158.074
Macklin, B. L.
065.026
MacLennan, C. G.
078.033
MacLeod, J. M.
141.050
MacPhie, R. H.
033.071
MacQueen, R. M.
074.005 .008
MacRae, D. A.
008.106
Macris, C. J.
004.066

Macy Jr., W.
099.218
Macy Jr., W. W.
141.327
Maddison, R. E. W.
003.086
Madore, B. F.
123.010
Maeder, A.
153.017
Maeder, D.
033.022
Maegley, W. J.
097.063
Maffei, P.
002.028
013.010
158.055
Magalashvili, N. L.
121.043 .044
Magnan, C.
063.029
064.048
Mahaffey, C. T.
034.085
Mahaffy, J. H.
125.065
Mahn, A.
072.013
Mahoney, W. A.
076.011 .022
142.079
Mahra, H. S.
032.031
Maitzen, H. M.
114.067
Maiwald
103.105
Makalkin, A. B.
099.075 .076
107.009
Makarova, R. K.
034.010
Makeev, V. V.
091.018
Makhmudov, B. M.
143.004
Maki, A. G.
022.078
Makino, F.
142.065
Makridis, A.
003.009
Maksumov, M. N.
151.042
Malaise, D.
032.034 .902
114.113 .115 .123 .125
115.013
Malakpur, I.
124.002
Malcolme-Lawes, D. J.
084.005
Maley, P.
103.101 .105
Malin, M. C.
094.316
Malin, S. R. C.
084.214 .238
Malissa Jr., H.
012.020

Malkiehl', G. S.
034.026
Mallegni
092.024
Malone, R. C.
065.062 .101
Maltby, P.
072.028 .031 .032
Malumyan, V. G.
158.061
Malville, J. M.
073.067
Malysheva, T. V.
105.057 .114 .133
Mamadov, O.
103.120
Mambetov, R. A.
103.101
Manabe, R.
142.065
Manabe, S.
151.059
Manchester, R. N.
141.330
156.002
Mancuso, S.
121.084
Mandelman, M.
022.037
Mandel'shtam, S. L.
076.039
Manfroid, J.
064.034
103.124
Mangelsen, A. L.
051.025
Mankin, W. G.
034.075
074.008
Manno, V.
054.023
Mansfield, V. N.
125.050 .064
Mansurov, S. M.
073.024
Mansurova, L. G.
073.024
Mantz, A. W.
034.039
Manujlova, R. O.
082.085
Maran, S. P.
034.045
077.061
103.101
Maranian, M.
103.101
Maraschi, L.
134.006
141.332
142.005
143.031
Marcario, M.
123.036
Marchal, C.
042.040
Marchenko, N. P.
103.104
Marchenko, O. A.
097.033

Marchenko, T. I.
032.006
Marcolungo, P.
061.006
Marcucci, G.
075.005
Marcus, A. H.
107.001
Marduel, G.
035.010
Marek, K.-H.
055.014
Margon, B.
061.048
113.008
158.102
Margulis, L.
082.058
Marilli, E.
064.050
Marino, B. F.
124.101
Marion, J. B.
003.081
Mark, H.
051.024
Markachev, V. V.
094.324
Markarian, B. E.
158.020
Markeev, A. P.
042.070
Markellos, V. V.
042.039 .062
Markert, T. H.
125.070
142.086
Markkanen, T.
131.052
Markov, M. A.
066.066
Markova, L. G.
052.009
Markovich, M. Z.
102.019
Marks, A.
079.100
Marks, D. W.
066.106
Marlborough, J. M.
064.001
114.077
Marlenskij, A. D.
032.001
Marliere, C.
022.022
Marochnik, L. S.
151.049
158.038
162.017
Marouf, A.
041.020
Marov, M. Ya.
093.011 .015 .034 .038
.039
Marraco, H. G.
114.011
Marsakov, V. A.
122.016
Marsalkova, P.
131.515

Marsden, B.
103.113
Marsden, B. G.
098.032
102.010
103.101 .102 .105 .108
.111 .113 .116 .128
.129 .132 .133
Marsh, G. E.
066.034
Marsh, J. G.
046.011
055.002
Marsh, K. A.
114.148
141.076 .108
142.036
Marti, K.
082.025
105.038
Martin, J. W.
143.055
Martin, L. J.
097.069
Martin, N.
159.003
Martin, P. G.
126.021
131.039 .055
Martin, P. M.
105.093
Martin, S. F.
073.055
Martin, W. L.
092.040
093.022
114.019
154.004
Martina, E. F.
078.018
Martinet, L.
155.032 .066
Martirosyan, A. A.
004.012
Martres, M.-J.
010.028
074.089
075.022
Martsvaladze, N. M.
082.046 .117
Martynenko, V. G.
162.027
Martynov, D. Ya.
003.025
005.003
097.060
121.012 .070
Marussi, A.
081.023
Marvin, U.
003.082
Marvin, U. B.
094.345
Marx, G.
162.039
Masajtis, V. L.
105.122
Mascart, P.
082.016
Maseide, K.
084.030

Masevich, A. G.
010.017
Masheder, M. R. W.
131.023
Mashhoon, B.
162.042
Maslovskis, A.
009.014
Masnou-Seeuws, F.
022.047
Mason, B.
094.184
105.093
Mason, J.
104.070
Mason, K. O.
142.062 .073
Massa, E.
162.045
Massa, L. J.
131.093
Massaro, E.
125.061
Massebeuf, M.
082.022
Massey, H.
012.023
Mast, T. S.
141.302
Masursky, H.
094.370
Matas, V.
042.045 .046 .047
Matchett, V. L.
103.101 .105
Mateshvili, Yu. D.
074.036
Mathews, W. G.
008.115
141.058
Mathewson, D. S.
159.004 .007
Matilsky, T.
142.035
Matilsky, T. A.
142.043
Matsakis, D.
132.001
Matsoukas, D. A.
079.100
083.015
Matsumoto, M.
063.039
Matsuoka, M.
142.065 .101
Mattei, J. A.
123.053
Mattei, M.
098.027 .033
103.007 .009 .112 .116
.127 .130
Matthews, K.
099.042
Mattig, W.
071.014 .028
072.038
080.025
Mattoo, S. K.
074.004
Matveenko, L.
033.003

Matveenko, L. I.
033.011
Matveev, G. D.
054.013
Matzner, R. A.
066.017 .045
Mauder, H.
122.067
142.105
Maula, E.
094.047
Maupome, L.
105.160
Maurer, P.
094.332
Maurette, M.
105.036
Mavko, G. E.
131.042
Mavridis, L. N.
012.018
122.104 .107
Max, C. E.
062.080 .082
134.010
Maxia, C.
004.052 .053
Maxwell, J. R.
094.349
Maxwell, T. A.
097.078
May, B. H.
106.003
Mayaud, P.-N.
084.219
Mayeda, T. K.
105.005 .029
Mayer, B.
034.056
Mayer, C. H.
131.058
Mayer, E.
123.031 .032 .036
125.112
Mayer, F.
155.066
Mayer, W. F.
061.063
Mayfield, E. B.
074.108
076.031
Mayo, M. J.
103.101
Mayor, M.
155.031
Mayr, H. G.
082.031 .035
Mazets, E. P.
061.031 .053
104.057 .058
143.032
Mazodier, B.
035.018
044.031
Mazur, M.
079.100
Mazurek, T. J.
065.029
Mazzella Jr., A. J.
131.131

Mazzitelli, I.
126.003
Mazzoni
092.024
Mazzucconi, F.
075.005
McAdam, W. B.
141.066
McAdoo, D. C.
042.007
McAllister, H. C.
071.020
McAshan, M. S.
033.021
McCabe, M. K.
073.081
074.063
McCall, R.
003.083
McCarthy, C. C.
131.048
McCarthy, M. F.
125.022 .023 .044 .102
McClintock, J.
121.004
McClintock, J. E.
061.063
142.019 .086
McClintock, W.
114.072
McClure, J. P.
083.037
McClure, R. D.
114.102
153.023
McCluskey, G. E.
121.059
McCluskey Jr., G. E.
121.003 .030
McClymont, A. N.
076.003
McCord, T. B.
008.025
094.371 .381
McCormac, B.
008.095
McCoy, J. E.
084.033
McCray, R.
131.041 .057
142.030 .046
McCrea, W. H.
007.000
012.003
McCrosky, R. E.
098.027 .029 .033
103.007 .009 .108 .116
.124 .127 .130
McCuskey, S. W.
155.017 .021 .045
McDiarmid, I. B.
078.015
McDonald, F. B.
074.078
099.018
143.008 .025 .054
McDonough, T. R.
007.000
McDowell, M. R. C.
007.000

McElhinny, M. W.
045.003
McElroy, M. B.
092.038
093.025 .041
099.040 .210
McEntire, R. W.
084.292
McGee, J. D.
034.014
McGee, R. X.
008.124
McGehee, R.
042.049
McGill, G. E.
094.377
McGraw, J. T.
115.009
124.100
McGruder III, C.
113.012
McGuire, R. E.
084.033
McHone, J.
105.168
McIlwain, C.
084.001
McIlwain, C. E.
099.019
McInnes, B.
082.018
McIntosh, B. A.
104.049
McIntosh, P. S.
103.101
McIvor, I.
143.005 .026
McKee, C.
141.106
McKee, C. F.
125.013
141.057
McKeith, C. D.
076.010
McKeith, N. E.
076.010
McKellar, A. R. W.
099.063
McKenzie, D. L.
074.108
076.029
McKibben, R. B.
099.016
McLaughlin, L. K.
097.036
McLean, D. J.
077.009
McLean, I. S.
076.003
114.045
McLeish, C. W.
033.030
McLinn, J. A.
141.612
158.034
McMillan, R. S.
114.150
McNall, J. F.
121.034
McNally, D.
014.051

McPherron, R. L.
084.255
McQueen, M. P. C.
157.003
McVittie, G. C.
066.078
McWatters, K. D.
084.003
Meaburn, J.
131.531
132.012 .020
Mead, J.
041.069
Mebold, U.
131.506 .546
Mechler, G. E.
118.034
Meco, M.
082.076
Medd, W. J.
141.050
Medenbach, O.
094.335
Medvedev, R. V.
105.129
Medvedev, Yu. A.
092.012
Meekins, J. F.
142.082
Meerts, W. L.
022.002
Meeus, J.
054.019
096.015 .017
098.013 .024
099.215
Meffroy, J.
003.119
Megill, L. R.
082.003
084.010 .019
Meglen, R. R.
015.010
Megrelishvili, T. G.
082.047
Mehlman, G.
022.012
Mehra, R.
098.023
Mehrmann, R.
103.101
Meier, R. R.
082.065 .066
Meikle, W. P. S.
125.035
Meinunger, L.
123.042
Meisel, D. D.
103.101
Meiseles, R.
141.089
Meissner, R.
094.110
Melchior, P.
002.016
081.020 .022
Melik-Alaverdyan, Yu. K.
158.023 .085
Melioranskij, A. S.
078.011
084.409

Mel'nikov, O. A.
003.085
022.060
Melokhrino, E. I.
014.025
Melrose, D. B.
062.018 .052 .058
074.026 .047 .048
077.004 .033 .054 .055
Melville, J. G.
084.228
Melvin, M. A.
162.024
Mel'yanovskij, P. A.
141.056
Men', A.
141.118
Men', A. V.
033.013
141.056
Mencaraglia, F.
125.061
Mendell, W. W.
094.114
Mendes, G.
081.031
Mendez, M.
132.023
Mendez, R. H.
101.011
114.003
Mendillo, M.
083.015
Mendis, A.
102.014
Mendis, D. A.
091.013
102.002
103.101
Mendoza V., E. E.
021.032 .033
113.043
115.015
Meng, C.-I.
084.231 .271 .284
Mengel, J. G.
065.002
080.044
Mennessier, J. P.
106.002
Mentzer, C. A.
033.063
Menzel, K.
004.082
Menzies, J. W.
122.134
Merat, P.
066.087
080.002
Mercier, J. P.
073.009
Mergentaler, J.
075.009
Meriwether, J. W.
084.003
Merrill, K. M.
141.608
158.005
Mesnage, P.
035.008

Mesrobian, W. S.
111.010
Messell, K.
101.008
Mestehrton, A. P.
083.003
Mestel, L.
065.019
Meszaros, P.
143.013
Metik, L. P.
158.016 .037
Metz, W. D.
092.032
097.012
099.203
Metzger, A. E.
094.124
142.075
Metzner, A. W. K.
002.005
Meurers, J.
094.176
132.018
158.058
Meurs, E. J. A.
155.018
Mewaldt, R. A.
078.038 .041
Meyer, K.
021.002
Meyer, K. R.
042.064
Meyer, R. X.
076.031
Meynent, M.-J.
005.015
Mezger, P. G.
131.106 .538 .539
155.063 .072
Michael, J.
123.028 .029
Michael, J. L.
103.101
122.081
Michalitsanos, A. G.
073.038 .062 .082 .083
080.058
Michalsky, J. J.
131.121
Michalsky Jr., J. J.
099.007 .008
126.008
131.003
Michard, R.
008.096
Michaud, G.
065.050
Michel, F. C.
062.026
099.077
158.002
Michel, G.
114.101
Michel, K. W.
022.016
083.052
103.101
105.146
Michel, P. H.
003.086

Michel, R.
094.174
Michel'son, N. N.
032.012
Michet, D.
158.302
Mickey, D. L.
071.009
Mickler, E. H.
003.074
Middlehurst, B. M.
094.151
Miezis, J.
014.029
Migach, Yu. E.
103.101
Migliavacca, R.
004.049
Mihalas, D.
064.036
073.066
076.032
114.127
Mihalas, D. M.
071.053
Mihalov, J. D.
074.077
099.014
Mijatov, M.
032.059
Mikami, Y.
142.065
158.303
Mikerina, N. V.
106.011 .012
Mikesell, A. H.
066.017
Miketinac, M. J.
065.057
Mikhajlov, A. A.
013.002
Mikhajlov, R. M.
094.127 .173
Mikhajlov, Yu. V.
103.101
Mikhajlova, O. M.
036.012
Mikirtumova, G. G.
082.048 .049
Mikkelsen, D. R.
065.108
Mikolas, J.
103.101
Mikolas, J. C.
103.105
Mikulasek, Z.
006.000
080.054
Milan, J.-L.
035.018
Milano, L.
098.012
121.084
Milet, B.
098.015
103.101 .105
Miley, G. K.
013.001
141.005
Milkey, R. W.
064.004

Milkey, R. W.
071.053
076.032
Miller, B. D.
066.016
Miller, D. E.
082.008
094.146
Miller, E. K.
033.055
Miller, E. W.
125.010
Miller, F. D.
103.104
Miller, J.
095.005
Miller, J. C.
065.034
Miller, J. S.
132.015
Miller, K. L.
082.099
Miller, M. D.
065.122
Miller, M. J.
046.001
Miller, R. A.
071.035
Miller, R. H.
142.052
151.015 .021 .061
Miller, S.
093.042
Miller, V. G.
073.020
Miller, W. J.
122.079
Milligan, J. E.
076.029
Millikan, A. G.
036.009
Millis, R. L.
099.220
Millman, P. M.
106.008
Mills, A. A.
105.092 .101
Milne, D. K.
008.124
125.060
Milogradov-Turin, J.
013.009
Milon, D.
103.101
Milone, E. F.
076.033
Milone, L. A.
008.039
Milton, D. J.
097.016
Milton, R. L.
126.017
Minasyants, G. S.
072.034
Mineev, Yu. V.
034.018
Miner, E.
097.203
Miner, E. D.
092.034
093.020

Minn, Y. K.
    131.112
    158.109
Minnett, H. C.
    008.124
Minti, H.
    117.002
Minton, R. B.
    103.105
Mints, R. I.
    094.388
Mironov, A. V.
    121.101
    154.030
Miroshnichenko, L. I.
    003.087
    011.025 .026
Mirovskij, V. G.
    141.041
Mirtov, B. A.
    082.027 .087
Mirzoyan, L. V.
    122.020 .096
Misezhnikov, G. S.
    033.011
Mishin, E. V.
    074.020
Miskovic, V.
    098.044
Misner, C. W.
    066.044 .135
Mistrik, I.
    053.015
Mitalas, R.
    065.001
Mitchell, B.
    009.012
Mitra, A. P.
    083.018 .019
Mitra, V.
    082.009
Mitteldorf, J.
    155.023
Mitter, A.
    105.153
Mityakov, N. A.
    077.013
Miyaji, T.
    033.049
    131.024
Miyake, G. T.
    105.094
Miyamoto, F.
    045.013
Miyamoto, M.
    151.007
Miyamoto, S.
    142.065 .101
Miyazawa, K.
    033.094
    131.024
Mkrtchyan, S. M.
    141.055
Mo, T.
    124.102
Modestov, V. A.
    052.005
Moe, O. K.
    073.071
Moergeli, M.
    094.336

Moesgaard, K. P.
    004.036
Moffat, A. F. J.
    114.093
    121.013
    153.011 .020 .021
Moffett, T. J.
    117.030
Mogilevskij, Eh. I.
    072.033
Mogro-Campero, A.
    099.016
Moiseev, I. G.
    033.011
    092.019
    141.055
Mokrov, V. S.
    084.266
Molchanov, A. P.
    077.046
Moldovanu, A.
    084.202
Molnar, M. R.
    114.049 .136
Molsen, K.
    066.077
Molton, P. M.
    099.031
Monaghan, J. J.
    022.054
    116.004
Monahan, K. M.
    099.012
Monfils, A.
    032.034 .902
    082.104
Monin, Yu. G.
    033.011
Monk, P.
    103.101 .105 .116
Monnet, G.
    032.047
    131.552
    158.072 .104
Monsignori-Fossi, B. C.
    074.106
Montgomery, M. D.
    084.287
Montmerle, T.
    142.113
Mook, D. E.
    142.076 .077
Moore, C. B.
    105.015 .052
Moore, C. H.
    021.021
    033.045
Moore, E. L.
    066.003
Moore, E. P.
    125.062
Moore, G.
    121.004
Moore, H.
    021.011
Moore, P.
    003.088
    010.012
    041.076
    095.005

Moore, R. B.
    094.403
Moore, R. L.
    071.049 .062
Moore, W. E.
    141.329
Moorey, G. G.
    033.029
Moos, H. W.
    093.013
    114.072 .112
Morales Pineiro, C.
    118.005
Moran, P. E.
    042.062
Morando, B.
    047.010
Moranzino, C.
    044.024
Morbey, C. L.
    031.084
Moreels, G.
    082.004
Moreno, G.
    074.015 .039
    106.036
Morfill, G.
    078.017 .025
    143.003
Morfill, G. E.
    078.009
Morgan, D. J.
    022.005
Morgan, P. J.
    046.001
Morgan, T. H.
    121.076
Morgenstern Horing, N. J.
    065.017
Morger, P.
    121.090
Morguleff, N.
    114.006
Mori, T.
    096.038
Mori, T. T.
    077.078
Morimoto, M.
    131.024
Morioka, A.
    084.037
Morozhenko, A. V.
    103.101
Morozhenko, N. N.
    073.027 .029 .047
Morozov, A. G.
    151.003
Morozov, E. I.
    078.020
Morozov, S. F.
    093.014
Morozov, V. N.
    064.015
Morozova, E. I.
    034.026
Morozova, T. I.
    078.012
    097.033
Morris, C. S.
    103.101 .105 .118 .138

Morris, C. S.
114.071
Morris, G.
131.068
Morris, S. C.
114.165
Morrison, B. A.
103.101
Morrison, D.
002.032
091.004
092.034
093.020
098.005
100.013 .208
Morrison, F.
081.015
Morrison, G. H.
105.061
Morrison, N. D.
002.032
121.001
Morrison, P.
003.062
061.035
125.041 .102
142.008
Morrison, R. H.
094.363
Morrow, R.
062.051
Morton, D. C.
022.009 .028
131.094
Moser, E.
034.015
Moshkalev, V. G.
154.008
Moshkin, B. E.
093.039
Moshnikov, I. S.
031.047
Moss, D. L.
062.073
Mossakovskaya, L. V.
158.008
Mossberg, T.
022.052
Motorina, N. N.
052.021 .023
Motrich, V. D.
120.002
122.030
Motrunich, I. I.
031.017
054.006
Motta, S.
162.022
Mould, J. R.
114.013
Moulton, F. R.
003.089
Mount, G. H.
071.023 .045
Mouschovias, T. C.
131.118
151.048
Moutsoulas, M.
002.038
051.014

Moyer, H. G.
052.042
Mozer, F. S.
084.226
Mozhzherin, V. M.
055.001
Mrkos, A.
098.028 .029 .031
103.101 .105 .116
Muehlberger, W. R.
094.181
Mueller, B.
003.090
Mueller, E. A.
014.049
Mueller, G.
105.152
Mueller, H. W.
094.330
Mueller, I. I.
046.025
Mueller, O.
094.393
105.021 .062
Muench, G.
092.034
093.020
099.020
Mueuersepp, P.
003.091
Muirden, J.
103.105
Muirhead, H.
003.092
Mukhamednazarov, S.
104.003
Mukhina, M. M.
033.011
Mukins, E.
051.010 .011
Mulholland, J. D.
002.005
042.028
043.004
Mullan, D. J.
065.028
072.014 .023
Muller, C. A.
033.002
Muller, P.
096.031
118.002 .021
Muller, P. M.
094.172
Muller Zum Hagen, H.
066.144
Mulyukova, N. B.
085.006
Mumford, G. S.
121.064
Mundet
103.101 .105
Munro, R. H.
074.005
Murakami, H.
131.139
Murayama, T.
084.230
Murdin, P.
142.073

Murdoch, H. S.
141.010 .022 .025 .032
.066
Murnikov, B. A.
122.135
Murphy, G. L.
162.041
Murphy, R. E.
099.006
100.011
Murray, B.
003.026
Murray, B. C.
092.017 .039
093.024
Murray, C. A.
012.014
112.018 .024
Murray, J. D.
159.004
Murray, S.
142.035
Murray, S. S.
142.043
Murrell, A. S.
097.010
Musatov, L. S.
084.266
106.027
Muschler, W.
062.014
Musman, S.
071.048 .063
Musorina, L. M.
082.105
Mussio, P.
021.024
Mustaeva, F. G.
075.002
Mutch, T. A.
097.021
Muthsam, H.
121.010
Muzalevskij, Yu. S.
054.012
Muzylev, V.
153.016
Myasnikov, V. A.
034.008
Myers, P. C.
131.120

Nacozy, P. E.
101.001 .002
107.004
Nadal, R.
119.008
Nadezhin, D. K.
065.044
Nadkarni, R. A.
105.061
Naef, R. A.
011.023
104.055
Nagane, K.
131.024
Nagata, T.
094.320 .397
Nagel, D. J.
062.004 .071

Nagel, E.
  046.021
Nagy, A. P.
  084.003
Najita, K.
  083.008
Nakada, M. P.
  074.031 .057
Nakagawa, N.
  131.139
Nakagawa, Y.
  080.040 .041 .061 .062
Nakagomi, Y.
  073.033
Nakai, H.
  041.026
  042.053
Nakamura, N.
  105.095
Nakamura, T.
  099.217
  100.212
Nakamura, Y.
  094.108 .120 .121 .318
Nakano, T.
  131.124
  142.110
Nakayama, K.
  073.050
Namba, O.
  093.030
Nance, J. L.
  033.047
Nandi, A.
  084.302
Nandy, A.
  065.103
Nandy, K.
  032.034 .902
  113.006 .034
  114.015
  131.016
Nankivell, G. R.
  031.022
Napier, P. J.
  033.032
Napier, W. McD.
  098.002
  107.013
Narasimhan, M. S.
  033.072
Nariai, H.
  066.132
Nariai, K.
  124.001
Narlikar, J. V.
  162.002
Nash, D. B.
  093.029
Nasi, E.
  065.113
Nather, R. E.
  031.011
  096.007
  117.030
  121.085
  124.100
Naugolnaya, M. N.
  097.030 .031
Naumova, A. A.
  041.038

Nava, D. F.
  094.346
Navara, P.
  054.004
Nayak, V. K.
  105.162
Nazarenko, A. I.
  052.009
Nazarova, T. N.
  104.056
Naze Tjoetta, J.
  080.004
Neal, K. L.
  033.075
Neckel, T.
  121.075
Ne'eman, Y.
  061.027
  141.036
Nefed'eva, A. I.
  041.001
  082.100 .115
Neff, J. S.
  008.062
  031.066
  101.005
  121.078
Negele, J. W.
  061.021
Negus, C. R.
  076.004
Nehru, C. E.
  094.402 .403 .404
Neighbors, J. E.
  125.102
Neirinckx, J.
  096.022
Nejman, V. B.
  097.002
Nelson, D. F.
  084.008
Nelson, G. J.
  077.005
Nelson, J. E.
  141.302
Nelson, M. R.
  096.021
Nemecek, Z.
  034.020
Nemiro, A. A.
  041.038
Nepoklonov, B. V.
  094.127 .173
Nersessian, S. E.
  153.005 .006 .007
Nesci, R.
  103.101
  114.084
Nesis, A.
  071.028
  080.025
Nesmyanovich, A. T.
  074.050 .075
  079.100
Ness, N. F.
  092.036
  093.023
  097.020
Neste, S. L.
  098.006

Nesterov, V. E.
  051.016
  084.407
  143.040
Nesterov, V. V.
  046.015
Netzer, H.
  158.029
Neugebauer, G.
  092.034
  093.020
  097.203
  099.020
  131.523 .524
  141.613
Neugebauer, M.
  084.232 .279
  103.100
Neupert, W. M.
  034.035
  073.052
  074.101
  076.005 .020
Neuzil, L.
  054.011
Nevo, I.
  114.044
Nevskaja, N. I.
  004.077
Newby, J. S.
  074.051
Newell, E. B.
  031.067
  154.029
Newell, R. E.
  106.033
Newelski, L.
  031.068
Newkirk Jr., G. A.
  074.005
Newman, E. T.
  066.065
Newsom, G. H.
  022.081
  102.020
  103.101
Newton, B. H.
  033.080
Newton, J. B.
  103.101
Newton, R. R.
  004.024 .081
  044.006 .032
  107.016
Newton, S. J.
  033.005
Ney, E. P.
  103.101 .102 .105
  106.020
  114.008 .014
  141.611 .616
Ney, W. P.
  103.101
Ng, L. K.
  143.064
Nguyen-Quang-Rieu
  131.019 .077 .092
Ni, W.-T.
  065.090
Nicholls, R. W.
  022.015 .031

Nicholson, W.
041.051
Nicodem, D. E.
097.042
Nicolas, K. R.
073.071
Nicclini, T.
004.061
022.050
032.030
Nicolson, I.
003.150
Niell, A. E.
141.060
Nielsen, R. P.
034.037
Niemelae, V. S.
114.003
Nien Dak Sze
093.041
Nikitin, S. A.
010.016
Nikolaev, A. G.
097.033
Nikolaev, R. P.
054.012
Nikolaeva, N. S.
078.002
Nikoloff, I.
041.046
Nikolov, N.
122.108
Nikolov, N. S.
104.030
122.007 .033
Nikolow, A. S.
031.025
Nikulin, I. F.
092.010
Nikulina, T. G.
122.053 .055 .057 .059
Nimitz, K. S.
052.012
Nishida, A.
084.233
Nishida, M.
158.116
Nishikawa, T.
022.071
Nishimura, J.
142.065
Nissen, P. E.
153.031
Nita, I. D.
073.054
Nix, J. R.
061.007
Nobili, L.
066.001
Nobis, H. M.
003.093
Noels, A.
065.006 .020
Noens, J. C.
074.002 .068
Noerdlinger, P. D.
064.025
Nojkina, A. I.
093.007
Nolt, I. G.
100.011

Nomoto, K.
065.021
Noonan, T. W.
160.011
Nordh, H. L.
114.027
Nordlund, A.
064.043
Nordsieck, K. H.
034.082
Norman, R. A.
033.074
Normandin, M.
007.000
Norris, J.
154.021
North, J. D.
004.004
North, R. D.
042.009
Northover, K. J. E.
158.071
Northrop, T.
099.036
Noskova, R. I.
114.121
Nosov, A. A.
033.009
Notni, P.
158.039
Nourse, A. E.
003.094
Novak, M.
034.029
Novikov, I. D.
033.023
066.024
Novikov, S. B.
074.035
079.100
Novikov, V. V.
114.096
Novikova, G. V.
036.013
Novoselova, N. V.
104.004 .015
Novotny, V.
104.050
Novozhilov, N. I.
082.072
Novruzova, Kh. I.
114.035
Noyes, R. W.
073.016 .087
076.034 .037
Nuccio, R. M.
091.014
Nugis, T.
065.048
114.054 .055 .056 .057
Null, G. W.
099.204
Nussbaumer, H.
022.010
Nutku, Y.
066.045
Nymmik, R. A.
143.037
Nyquist, L. E.
094.347

Obashev, S. O.
032.032
Obenson, G.
046.017
Oberbeck, V. R.
094.363
Oberstatter, A.
121.056
Obridko, V. N.
072.033 .039 .041
073.024
O'Brien, J. T.
062.033
O'Brien, B. S.
094.140
Occhionero, F.
151.038
158.096
Ochelkov, Yu. P.
063.009
141.037
Ochsenbein, F.
021.031
113.020 .042
O'Connell, R. F.
022.003 .073
126.007 .023
O'Connor, G. G.
094.140
O'Connor, S.
022.081
Oda, M.
142.065 .101
O'Dell, C. R.
093.017
103.100
O'Dell, S. L.
062.025
141.067
Odgers, G. J.
122.088
Oegelman, H.
125.102
Oehman, Y.
031.004
Oepik, E.
002.030
Oepik, E. J.
097.019
Oesterwinter, C.
042.066
Oezel, M. E.
125.102
Oezisik, M. N.
063.017
Offner, A.
031.041
O'Gallagher, J. J.
099.016
Oganyan, G. B.
122.020
Ogata, A.
031.083
Ogawara, Y.
142.065 .101
Ogilvie, K. W.
074.081
092.035
093.021
Ogrins, M.
031.062 .063

O'Handley, D. A.
052.018
Ohanian, G. B.
See Oganyan, G. B.
O'Hara, M. J.
094.306
Ohle, K.-H.
085.004 .005
Ohlson, J. E.
033.087
Ohnishi, T.
065.121
114.083
155.036 .050
Ohtsuka, Y.
142.065
Ohyabu, N.
062.060
Oinas, V.
114.168 .173 .174
Ojima, T.
158.303
Ojringel, I. M.
062.024
Oka, S.
158.116
Okada, T.
122.076
Okamoto, I.
064.009
141.324
Okamura, S.
062.060
Okazaki, A.
121.035
Oke, J. B.
008.098
013.012
114.030
126.005
141.059 .063
O'Keefe III, J. A.
081.005
094.003
Okida, K.
122.076 .092
Okui, S.
033.083
Okulessky, B. A.
094.182
Olbers, D. J.
074.007
Oleak, H.
066.041
158.115
O'Leary, B.
092.017 .039
093.024
Olesen, J. O.
082.067
Oleson, J. R.
032.023
Olevic, D. M.
118.028 .029 .030
Olhoeft, G. R.
094.113 .380
Olifer, N. S.
041.002
Oliver, J. P.
122.097

Oliver, W. L.
082.042
Olivieri, G.
075.016
Olofsson, S. G.
114.027
Olsen, E.
105.004 .042 .099
Olsen, E. T.
099.060
141.030
Olson, E. C.
113.017
121.071
Olson, J. V.
084.254
Olson, R. A.
142.021
Olson, W. P.
084.252 .277
Olthof, H.
131.547
Olver, A. D.
033.060 .089
Olyanyuk, V. P.
077.046
Omelina, N. A.
044.005
Omer Jr., G. C.
160.001
Omholt, A.
003.004
084.017 .018
Omnes, R.
003.095
080.059
162.008
O'Mongain, E.
066.107
141.340
O'Neal, R. L.
104.010
O'Neil Jr., E. J.
154.029
O'Neill, T. G.
033.030
Ong, R. S. B.
083.005
Onishchenko, L. V.
093.034
097.033
Onuma, N.
105.005
Oort, A. H.
085.011
Oort, J. H.
103.101
158.068
Oosterloo, T.
104.072
Opal, C.
103.101
Opher, R.
141.001 .029
Opp, A. G.
099.013
Oppenheimer, M.
131.007
Oprescu, G.
044.007

Oran, E.
082.103
Ordanovich, A. E.
054.003
Oren, L.
022.013
Orgel, L. E.
015.012
Orhaug, T.
031.034
Orlov, E. P.
032.006
Orlov, M. Ya.
122.013 .140
Orlova, T. V.
071.044
Ormes, J.
131.115
Orowan, E.
094.180
Orozco, A.
084.229
Orrall, F. Q.
034.067
071.009
Orte, A.
044.011
Orton, G. S.
093.009
Osaki, Y.
065.093
Osawa, K.
122.076
Osborn, R. W.
143.015
Osborn, W.
114.016 .102
Osborne, J. L.
143.066
Oshchepkov, V. A.
117.018
121.062
Osipkov, L. P.
151.005
Oskanian, V.
123.040
Oskanyan, V. S.
122.096
Osmer, P. S.
064.002
122.064
142.022 .092
158.062
Ostan, F.
041.022
Oster, L.
063.036
Osterbrock, D. E.
008.115
133.010
Ostriker, J. P.
065.055
125.071
142.050
158.112
Ostroverkhaya, L. Yu.
071.050
Oszczak, S.
009.019
Otis, M.
096.034

Otten, G. R.
071.047
Ottewell, G.
047.018
Otto, E. P.
097.078
Ovenden, M. W.
099.202
Overbeek, M. D.
096.011
Ovsepyan, A. V.
065.058
126.012
Owen, F. N.
100.211
160.016 .019
Owen, T.
091.027
097.026 .068
101.006 .013
Owens, A. J.
078.040
143.021 .022 .056
Oya, H.
099.054
Ozernoj, L. M.
011.031
066.076
141.040
158.049 .076
162.009 .030
Ozolins, G.
011.028

Pacholczyk, A. G.
141.105
Pacini, F.
125.017
141.039 .322
Paciorek, J.
073.021
Paczynski, B.
010.017
065.032
141.068
142.049
Padgett, J. L.
094.139
Padrielli, L.
141.069
Page, D. E.
012.016
078.031 .032
084.280
Page, L. W.
003.096
Pagel, B. E. J.
114.021
153.015
158.118
Paik, H. J.
033.021
Pajdusakova, L.
098.120
Pakhomov, V. V.
073.056
077.045
Pakhomova, T. V.
105.128

Pakvor, I.
032.055 .056
Palamarchuk, A. V.
078.006
Palej, A. B.
036.014
Pallavicini, R.
073.086
076.024
Palme, H.
094.135 .395
105.081
Palmer, P.
131.062
Palmieri, T. M.
032.002
061.049 .066
Palmiotto, F.
074.015
Paloschi, S.
082.076
Palumbo, G. G. C.
125.035
142.047
Pamjatnikh, A.
See Pamyatnykh, A.
Pamyatnykh, A.
064.008 .022
Panagakos, N.
093.042
Panasyuk, M. I.
084.403
Panchuk, V. E.
064.026
114.122 .172
Pande, M. C.
071.016 .017
072.017
114.065
Pandharipande, V. R.
061.023
Pannunzio, R.
098.030
Panov, P. V.
077.003
Panov, V. N.
104.058
Panovkin, B. N.
015.002
051.007
Pansecchi, L.
092.024
Panteleeva, L. P.
112.015
Papadopoulos, G. D.
131.108
Papadopoulos, K.
077.056
Papagiannis, M. D.
077.902
Papaloizou, J.
117.009
Papanastassiou, D. A.
094.325
105.063 .164
Papousek, J.
121.058
Paprotny, Z.
119.017
Paquet, P.
021.023

Parekh, P. P.
105.007
Paresce, F.
034.044
082.029 .030
131.036 .040
Parijskij, Y. N.
162.901
Paris, R. B.
065.019
Parisi, M.
143.010
Parker, E. N.
008.033 .145
071.041 .042
072.037
158.002
Parker, R. A. R.
093.017
Parkinson, J. H.
142.062
160.017
Parkinson, T.
099.218
Parks, G. K.
106.033
Parlier, B.
134.006
142.005
Parra, F.
044.011
Parra, R.
054.017
Parry, W. E.
003.097
Parsamyan, Eh. S.
122.020
132.023
Parshin, A. A.
034.008
Parsian, I.
065.005 .071
Parsignault, D. R.
142.042
Parsons, S. B.
114.138
142.074
Parthasarathy, M.
121.006 .007
122.002
Pasachoff, J. M.
008.146
Paschmann, G.
084.287
Pashchenko, G. I.
009.003
Pashchenko, M. I.
131.078
Pashkova, G. R.
141.077
Pasoli, E.
003.098
Pataki, L.
131.127
Patashinskij, A. Z.
066.094 .141
Patchett, B.
125.112
Patchett, J.
010.008

Patel, L. K.
  066.138
Paterno, L.
  080.052
Patriarchi, P.
  077.037 .038
Patrick, T. J.
  032.054
Paturel, G.
  122.901
Patz, D. L.
  032.023
Paul, M. K.
  081.034
Paul, S. N.
  155.074
Paulikas, G. A.
  008.045
  078.004 .018
  084.402
Pauls, T. A.
  155.063
Pavicevic, M. K.
  022.069
Pavlenko, P. P.
  034.006
Pavlov, G. G.
  042.054
Pavlov, T.
  003.099
Pavlova, L. A.
  131.137 .138
Pavlova, N. N.
  158.035
Pavlovskaya, E. D.
  155.051
Pavlovski, K.
  011.035
Payne-Gaposchkin, C.
  005.007
  113.033
Peach, J. V.
  160.022
Peacock, A.
  160.017
Peale, S. J.
  091.001
  092.031
Pearlman, M. R.
  046.026
  082.073 .074
Pearson, T. J.
  141.011
Pechinskaya, N. I.
  034.004
Pecker, J.-C.
  003.105
  004.027 .028
  010.037
  015.014
  061.044
  062.044
  064.014
  066.087
  080.002
  117.017
  158.004
  160.018
Peckover, R. S.
  061.009

Peddie, N. W.
  084.258
Pedersen, A.
  084.289
Pedoussaut, A.
  119.008
Peebles, P. J. E.
  158.065
  162.021 .028
Pegeot, C.
  035.011
Peimbert, M.
  131.548
Peixoto, J. P.
  085.011
Pelikhov, N. V.
  162.017
Pellas, P.
  105.064
Pellat, R.
  062.040 .041
Pellet, A.
  032.047
Pelletier, G. A.
  033.034
Pelling, M. A.
  131.128
Pendl, E.
  103.105
Penegor, G.
  121.050
Penfield, H.
  103.101
  131.061
Peng, S. Y.
  084.401
Pennas, P. J.
  082.089
Penner, S. S.
  022.032 .036
Penrose, R.
  066.053
Penston, M. V.
  122.062
  131.012
Penzar, I.
  080.064
Penzias, A. A.
  008.060
  132.029 .034
Peraiah, A.
  063.042
Peralta, J. O.
  155.073
Percy, J. R.
  122.009 .080
Perdrix, J. L.
  010.008
Perez-De-Tejada, H.
  094.167
Perinotto, M.
  131.511
Perkins, M.
  099.016
Perko, L. M.
  042.056
Perlin, I. E.
  042.035
Perneczki, G.
  094.398

Pernet, J.
  004.029
Perola, G. C.
  061.069
  141.028
  143.031
Perona, G. E.
  083.020
Perotti, P.
  073.009
Perry, M. E.
  142.020 .024 .037
Perry, P. M.
  036.007
  153.013
Persides, S.
  066.057 .079 .092
Persson, S. E.
  114.100
  131.524 .529
Peruanskij, S. S.
  021.003
Pesch, P.
  155.021 .045
Pesek, L.
  003.140
Pesek, R.
  011.039
Peter, H.
  121.090
Peters, B.
  143.016
Peters Jr., L.
  033.063
Peterson, B. A.
  113.011
  142.010 .020 .023 .024
    .025 .037
Peterson, C. J.
  154.018
Peterson, D. M.
  064.002
Peterson, D. W.
  097.066 .067
Peterson, F. W.
  141.102
Peterson, L. E.
  076.008
  094.124
  142.006 .055 .078 .085
  143.034
Peterson, M.
  066.125
Peterson, R. J.
  061.040
Petit, H.
  131.552
Petit, J.
  103.008 .010
Petkovic, S.
  032.057
Petrakov, V. N.
  082.038
Petrasso, R.
  074.090 .104
Petravic, M.
  141.342
Petri, W.
  004.035
  093.016

Petro, L. D.
142.115
Petrosian, V.
141.057 .106 .336
162.018
Petrosyan, M. B.
160.023
Petrov, G. M.
041.039
Petrova, N. N.
072.003
Petrova, N. S.
072.044
Petrovicova, R.
098.028
103.101 .105 .116
Petrovskij, A. A.
157.007
Petrowski, C.
094.382
Petrukhin, N. S.
080.038
Petrzilka, V. A.
062.075
Pettengill, G. H.
093.018
094.367 .368
100.009
Pettersen, B. R.
122.074
Petukhova, T. M.
094.388
Peytremann, E.
064.049 .901
113.033
Pezzarossa, C.
123.018
Pfaffe, H.
003.100
Pfitzer, K. A.
084.277
Pfleiderer, J.
021.014
Pfotzer, G.
078.035
Pham-Van, J.
041.020
Philip, A. G. D.
032.013 .041
113.027 .039
Phillips, E.
011.019
Phillips, J. G.
003.018
022.008 .058
Phillips, K. J. H.
076.020
Phillips, R. J.
094.166
097.022
Phillips, W. G.
066.126
Philpotts, J. A.
094.134
Phinney, D.
094.327 .337 .338 .348
.352 .401
105.065 .071
Phissamay, B.
082.033

Picard, M. D.
097.078
Picciotto, E.
106.002
Piccirillo, J.
121.079
Piddington, J. H.
151.017
Pierce, K.
032.042
Pierucci, M.
066.143
Pieters, C.
094.371 .381
Pietronero, L.
066.131
Pigatto, L.
125.104
Pike, C. P.
084.023
Pikel'ner, S. B.
064.058
073.025
Pilachowski, C. A.
116.001 .013
Pilcher, P.
098.034 .036
Pilipp, W. G.
062.066
Pillinger, C. T.
094.349
Pilnik, G. P.
044.004
Pilowski, K.
032.029
Pilski, A.
103.101
Pimenov, I. A.
143.012
Pimentel, G. C.
097.050
Pinau, A. J.
103.105
Pines, D.
065.040
141.320
Pines, P. J.
094.318
Pinigin, G. I.
032.039
Pinter, S.
074.045
Pinto, G.
123.024
Pinus, V. K.
066.011 .094 .141
Piotrowski, S.
117.001
Piotrowski, S. L.
121.031
Pipes, J. G.
022.029
Piragas, K. A.
066.099
Pirani, F. A. E.
042.085
Pisarenko, N. F.
034.005 .026
078.002 .011 .020
Pismis, P.
131.554

Pismis, P.
151.014
158.080
Pitz, E.
053.008
106.006
Pivovarov, V. T.
034.011 .012
Pizzella, G.
066.122
Pizzichini, G.
021.025
076.012
142.004 .047
Pizzo, V.
074.101
Plakhov, Yu. V.
042.044 .075
Plakidis, S.
003.009
Plaskett, H. H.
005.010
Platania, G.
126.002
133.004
Plathner, D.
032.043
Platov, Yu. V.
073.056
079.100
Plaud, P.
035.013
Plaut, L.
123.019
Plavec, M.
117.006
121.059
Plebanski, J. F.
066.071
Pleticha, D. J.
141.100 .103
Pletnev, V. D.
084.265
Plotnikov, A. V.
091.012 .019
Pluzhnikov, V. Kh.
006.000
032.004
041.002
Pneuman, G. W.
074.014
Po, S.
004.068
Podgorny, I. M.
084.249
Podolak, M.
099.062
Podstrigach, T. S.
033.053
Poeckert, R.
099.003
Pogodin, I. E.
077.046
Pogrebnoj, V. N.
084.237
Poirier, A.
096.019
Pokorny, Z.
006.000
120.005

Pokras, V. M.
051.016
Pokrovsky, A. G.
082.110
Pokrzywnicki, J.
015.019
Poland, A. I.
073.065 .066
074.005
Polanuer, M. D.
094.138
Poletto, G.
071.064
Polidan, R. S.
117.006
121.059
Polishchuk, R. F.
066.083
Pollack, J. B.
097.018 .044
099.212
Pollas, C.
141.116
Pollock, J. T.
141.002
158.301
Polnarev, A. G.
066.004
Polosukhin, V. P.
094.302
Polozhentsev, D. D.
041.037 .038
Polozhentseva, T. A.
041.038
Polyachenko, V. L.
151.003
Polyakhova, E. N.
054.020
Polyakov, V. M.
091.018
Polyakova, T. A.
122.098
131.085
Pomagaev, S. G.
151.034
Pomerantz, M. A.
078.014
Pomraning, G. C.
063.025
Ponizovskij, Z. L.
004.001
Ponnamperuma, C.
003.101
099.031
Ponomarev, D. N.
041.031
Ponomarev, E. G.
014.027
Poon, P. T. Y.
063.022 .027
Pope, T. P.
034.070
Popov, A. P.
114.096
Popov, G. M.
031.008
Popov, N. A.
045.010
Popov, O. S.
074.075
079.100

Popov, O. V.
034.008 .013
Popov, V. G.
082.038
Popov, V. S.
003.085
Popova, E. I.
008.149
Popova, R. I.
032.026
Popovic, B.
042.088
Popovic, G. M.
118.020 .024 .027 .029
.032
Popovici, C.
046.008
121.077
Poppen, R. F.
092.001
Popper, D.
097.029
Popper, D. M.
121.041 .087
Poquerusse, M.
077.035
Porcas, R. W.
141.054
Porco, C. C.
101.006
Porfir'ev, L. F.
054.010
Porter, B. F.
015.003
Porter, J. G.
100.204
Porter, N. A.
004.032
141.340
Portnyagin, Yu. I.
104.016
Pospergelis, M. M.
074.071
Postal, R. B.
092.040
093.022
Potdar, A.
151.008
Pottasch, S. R.
022.028
131.051 .512
141.614
Potter, A. E.
114.001
121.076
Potter, W. H.
077.016 .071 .901
Pottinger, D.
003.150
Potts, N.
123.039
Poulain, P.
074.085
Poulton, G. T.
033.077
Pounds, K. A.
160.017
Poupeau, G.
094.353
105.060

Poveda, A.
155.034
Povenmire, H.
103.105 .134
Powell, C.
051.001
Pozdnov, U. N.
084.266
Prabhu, N.
105.016
Praderie, F.
114.114 .124
Prasad, B.
065.115
Prasad, S. S.
076.019
Prasanna, A. R.
066.119
Pratt, N. M.
034.060
Pratt, R. J.
078.028
084.260 .261 .262
Pravdyuk, L. M.
071.039
Prendergast, K. H.
117.021
155.019
Prentice, A. J. R.
080.012
Press, F.
094.111
Press, W. H.
066.055
162.012
Pretre, P.
106.010
Price, K.
141.076
Price, K. M.
077.073
Price, M. J.
097.006
Price, P. B.
105.011 .066
106.030
Price, R. D.
143.061
Price, R. M.
141.334
155.055
Priest, E. R.
074.014
077.064
Prikhod'ko, V. A.
042.051
Prilutskij, O. F.
061.030
141.037
Pringle, J. E.
124.100
142.012 .014
Prinz, D. K.
076.001
Prinz, M.
094.402 .403 .404
Prior, E. J.
097.011
Prodan, Yu. I.
046.015

Proell, H.-J.
021.013
Proelss, G. W.
083.048
Prokakis, T.
072.022
077.059
Prokhin, V. L.
051.016
Prokof'ev, V. A.
062.005
Prokof'ev, V. K.
055.001
Prokof'eva, I. A.
031.006 .007 .056
Prokof'eva, N. A.
077.045
Prokof'eva, V. V.
097.015 .024 .030 .031
.032
141.306
Prokofieva, V. V.
See Prokof'eva, V. V.
Prokudina, V. S.
072.018
Pronik, I. I.
158.016 .037 .092
Pronik, V. I.
031.008
Pronin, A. A.
094.302
Protic, V.
044.017
Protitch, M. B.
096.035 .036
Protsenko, B. A.
054.012
Proverbio, E.
004.052 .053
008.028
035.005
045.006 .007 .008 .009
.014
Prud'homme, M.
066.029
Prytkov, N. M.
033.053
Pskovskij, Yu. P.
074.006 .071
Ptacek, V.
044.018 .028
Ptak, R.
141.104
Ptuskin, V. S.
143.014
Pucillo, M.
121.018
Pudov, O. Ya.
077.046
Puget, J. L.
143.007
162.029 .049
Puglianc, A.
046.014
Pupareli, M.
034.087
Pupysheva, L. V.
157.007
Purcell, I. M.
103.011 .101

Purcell, J. D.
073.065 .071
074.103
075.008
076.036
Purohit, S. C.
064.060
Purton, C. R.
114.148
141.076 .108
142.036
Pushel', A. P.
105.121
Pustilnik, L. A.
117.025
Pustylnik, I.
117.016
Pyle, K. R.
099.016
Pyragas, K. A.
062.020 .021
Pyshnenko, M. N.
096.020
Pytte, T.
084.220

Quam, L.
097.037 .055
Quast, G. R.
142.072
Quenby, J. J.
078.009
Querci, F.
064.011
Querci, M.
064.011
Quesada, V.
035.005
045.009 .014
Quiroga, R. J.
155.011

Raadu, M. A.
073.010
Rachkovskij, D. N.
063.010
Racine, R.
154.015
Radiman, I.
155.078 .079
Radostitz, J. V.
100.011
Radziemski, L. J.
074.102
Radzievskij, V. V.
014.015
Raedler, K.-H.
062.029 .030
Raeithel-Thaler, M.
003.030
Raggi, R.
062.015
Raghavarao, R.
083.035
Raghavender Rao, C.
142.065
Rahe, J.
102.006 .022
103.100

Raimond, E.
033.044
066.002
Rajan, R. S.
105.011 .066 .096
Rajaraman, R.
065.123
Rajaratnam, S.
104.071
Rajchl, J.
104.013
Rakhamimov, Sh. Yu.
125.025
142.041
Rakhimov, A. G.
041.066
097.039
103.100
Rakhimov, Kh. R.
105.132
Rakhmatov, Eh.
096.014
103.100
Rakos, K. D.
116.014
121.010
Rakosch, K. D.
116.006
Ralchovski, Ts.
106.032
Ramakrishna, C. M.
141.047
Ramanathan, V.
093.012
Ramaty, R.
065.076
073.015
076.013
078.039
080.032
142.002 .093
143.024 .030
Rambaldi, E.
105.067
Ramdohr, P.
094.335
Ramette, J.
022.055
Ramond
096.019
Rampling, R.
078.017
Ramsey, H. E.
073.085
Ramsey, L. W.
064.004
Rancitelli, L. A.
105.031
Randall, B. A.
099.017
Rangaswamy, S.
083.015
Rank, D. M.
141.602
Rankin, J. M.
122.127
141.315
Rannou, F.
042.021
Ranson, P.
022.030

Rantaseppae-Helenius, H.
103.101
Rao, A. Pramesh
106.018
141.064
Rao, M. N.
105.041
Rao, U. R.
143.033
Rapaport, M.
042.016
Rapavy, P.
103.105
Rapley, C. G.
141.328 .335
Rapoport, I. D.
051.016
084.407
143.036 .037 .040
Rapoport, V. O.
077.013
Rapp, R. H.
081.033
Rappaport, S.
061.063
121.004
142.117
Rasmusen, H. Q.
103.006
Rasool, S. I.
099.205
Rassbach, M. E.
097.048
Rastogi, R. G.
084.221
Rather, E. D.
033.045
Rather, J. D. G.
080.005
131.533
Raudsaar, H. K.
103.101
Ravest, J. F.
010.012
Ravetz, J. R.
004.064
005.004
Raviart, A.
073.009
Rawcliffe, R. D.
104.009
Rayburn, D. R.
141.303
Rayner, P. T.
141.307
Razani, A.
063.019
Razdan, H.
143.033
Razin, V. A.
157.007
Read, P. A.
010.024
Readhead, A. C. S.
141.007
Reasenberg, R. D.
092.040
093.022
Reasoner, D. L.
084.274

Reaves, G.
160.026
Reay, N. K.
106.003
Rebeirot, E.
114.038
Reber, C. A.
082.031 .035 .102
Redburn, F. L.
031.065
Reddish, V. C.
031.057
036.004
065.116
Reece, J. S.
055.002
Reed, E. I.
084.035
Reed, G.
100.005
Reed, S. J. B.
094.376
Reed Jr., G. W.
105.049
Reedy, R. C.
094.124
Rees, M. H.
084.002
Rees, M. J.
061.012
066.075
131.503
134.002
141.074
162.038
Reeves, E. M.
051.017
073.016
076.034 .037
Reeves, H.
102.015
107.011
125.063
Refsdal, S.
065.005 .065 .066 .071
Rego Fernandez, M.
114.091
Reid, A. M.
094.322
105.089 .100
Reid, J. S.
014.044
Reid, R. J. O.
143.062
Reigber, C.
046.021
081.029
Reina, C.
065.075
142.098
Reinbold, S. J.
052.014
Reinert, W.
034.032
Reinhard, R.
078.037
Reinhardt, M.
061.036
066.043
142.016

Reitmeyer, W. L.
008.068
Reitsema, H. J.
034.083
Reiz, A.
012.021
Reme, H.
084.009
Remizov, A. P.
084.266
Remo, J. L.
105.107
Remond, A.
035.015
Renard, M. L.
105.027
Rengarajan, T. N.
065.063
Rennilson, J. J.
094.169
Rense, W. A.
073.073
Reshetov, E. A.
094.127 .173
Reskin, G.
103.102
Revelli, C.
009.018
Revina, I.
102.025
Revina, L. D.
105.132
Reynolds, J. H.
094.312
105.068
Reynolds, R. T.
091.003
099.212
Reznikov, B. I.
117.007 .032
Reznikov, I. G.
022.051
Rhoades, C. E.
065.120
Riani, I.
074.010
Ribes, E.
073.046
Ribes, J. C.
033.056
Rich, A.
126.014
Rich, F. J.
084.274
Richards, P.
031.033
Richardson, J. A.
061.063
Richardson, R. J.
099.067
Richstone, D. O.
065.046 .055
Richter, A. K.
074.007
Richter, G.
032.025
Richter, G. A.
113.016
141.042
Richter, G. M.
141.042

Richter, L.
141.042
Richter, N.
005.006 .011
132.007
Richter, N. B.
141.042
Richter, W.
032.043
Rickard, J. J.
131.021
142.069
Riddle, A. C.
073.055 .081 .089
074.013 .042 .043
077.062 .075
Ridgway, S. T.
099.002
114.158
Ridley, B. A.
082.003
Ridpath, I.
003.107
Riedler, W.
078.035
106.035
Riegel, K. W.
131.054
Riegler, G. R.
082.032
Rieke, G. H.
100.215
103.101
Ries, L. M.
114.033
Righini, A.
082.075
Righini, G.
010.027
Rigutti, M.
071.024
Rihlova, L.
045.011
Riley, J. M.
141.052
Ringeard, G.
008.092
015.014
Ringnes, T. S.
097.202
Ringuelet, A. E.
114.080
Ringwood, A. E.
094.156
Rios, M.
065.023
Ripley, G.
082.095
Risover, L. M.
077.044
Ritter, H.
065.065
Rivin, Yu. R.
084.222
Rizcv, E. F.
077.022
Rizzi, A. W.
074.111
Roark, B.
132.024

Roark, T.
132.024
Robb, W. D.
022.064
Robba, N. R.
134.006
Robbins, D. E.
074.054
Robbins, R. R.
132.006 .026
Roberti, G.
071.024
Roberts, J. A.
008.124
Roberts, K. V.
141.342
Roberts, M. S.
158.011
160.013
Roberts, W. J.
142.034
Roberts, W. O.
007.000
Robertson, D. S.
031.009
141.093
Robertson, I. W. H.
084.038
Robertson, J. W.
159.001
Robinson, B. J.
131.050
Robinson, C. L.
003.102
Robinson, D. C.
066.144
Robinson, E. L.
121.085
153.030
Robinson, G. D.
082.069
Robinson, J. H.
092.026 .028
Robinson, K. L.
105.103
Robinson, L. B.
034.059
Roble, R. G.
082.010
084.024
Robley, R.
106.010
Roche, A. E.
103.101
Roddier, F.
034.076
Rodgers, A. W.
124.106
142.106
Rodionov, A. B.
143.012
Rodionov, B. N.
094.127 .173
Rodionova, Zh. F.
051.009
Rodono, M.
064.050
122.022 .068
Rodriges, M. G.
103.101

Rodriguez, M. H.
122.013 .140
Roeder, R. C.
162.020
Roederer, J. G.
084.031 .210 .298
Roelfsema, P.
104.072
Roelof, E. C.
074.078
099.018
Roelofs, T. H.
083.008
Roemer, E.
103.101 .102 .110
Roessiger, S.
113.045
122.041
Roessler, F.
082.013
Roettger, J.
083.011
Rogati, C.
103.101
Rogers, A. E. E.
093.018
Rogers, E. H.
084.008
Rogers, H. H.
042.084
Rogers, M.
103.105
Rogerson Jr., J. B.
131.097
Rohart, F.
022.022
Rohlfs, K.
131.073
151.030
Rohr, H.
010.025
Rojkhvarger, Z. B.
078.013
Roland, G.
079.100
103.101
Roldugin, V. K.
084.007
Roman, N. G.
008.142
Romana, A.
004.062
105.069
Romanishin, W.
160.005 .027
Romano, G.
123.013 .024
162.025
Romanov, A.
003.077
Romanov, I. I.
103.101 .105
Romanowicz, B. A.
103.112
Romanyuk, V. F.
104.038
Romejko, V. A.
105.111
Romick, G. J.
083.047
084.029

Romiez, M.
094.382
Romoli, C.
123.018
Ronan, C. A.
003.103
015.001
Rood, H. J.
158.043
160.010
Roos, D. G.
052.043
Roosen, R. G.
082.002
103.101 .105
Roozeveld Van Der Ven,
J. F.
032.043
Rose, F.
094.319
Rose, L. E.
101.007
Rose, W. K.
003.104
065.084
158.100
Rosen, A.
084.291
Rosen, R. A.
131.104
Rosenbauer, H.
084.287
Rosenberg, G. V.
082.108
Rosenberg, H.
074.072
077.057
Rosenblum, A.
066.043 .062
Rosendhal, J. D.
065.061
114.012
Rosino, L.
008.008
032.014
120.004
125.030 .038 .039 .104
.105 .112
Roslund, C.
122.070
Ross, C. L.
074.005
Ross, J. E.
071.025 .029 .030
Rossa, M. I.
014.008
Rossbach, M.
032.022
080.020
Rossberg, L.
106.035
Bosset, R.
082.016
Rossier, P.
118.012
Rostoker, G.
084.022 .275
Rostovskaya, A. A.
054.003
Roth, M. L.
065.074 .081

Rothenflug, R.
141.326
Rothschild, R.
142.089
Rothschild, R. E.
142.048 .060 .061 .116
Rothwell, P. L.
078.030
Rots, A. H.
158.014
Roud, M.
092.006
Roughton, N. A.
061.040
Rousseau, A. M.
103.124
Rousseau, J.
159.003
Routly, P. M.
034.061
Rowan-Robinson, M.
141.073
142.045
Rowe, J. N.
083.018 .019
Rowe, M. W.
022.901
105.012 .013 .070 .144
Roxburgh, I. W.
062.019
065.030 .117
074.053
080.016
Roy, A. E.
091.025
Roy, A. K.
155.074
Roy, J.-R.
073.038 .083
Roze, L.
011.012
Rozelot, J. P.
003.105
074.002 .068
Rozenberg, G. V.
082.049
Rozenbush, A. Eh.
103.101
Rozental, I. L.
061.030
Rozental', Yu. A.
097.033
Rozhdestvenskij, M. K.
093.034 .038
Rozhkovskij, D. A.
103.105
Rozyczka, M.
066.113
121.031
158.120
Ruban, V. A.
066.096
Rubashev, B. M.
072.010
085.003
Rubin, R. H.
133.011
Rubincam, D. P.
094.010
Rublev, S. V.
064.033

Rubo, G. A.
079.100
Rubtsov, L. N.
083.031
Rucinski, S. M.
117.031
Rudd, M. E.
022.065
Ruderman, M.
065.092
141.309
Ruderman, M. A.
065.040 .091
082.093
Rudge, A. W.
033.085
Rudina, M. P.
003.003
083.022 .023 .024
Rudnicki, K.
007.000
Rudnitskij, G. M.
131.030 .078
Rudraiah, N.
062.072
Rueda, A.
015.021
Ruediger, G.
062.031 .054
Rueger, W.
031.019
Ruekl, A.
006.000
Rufenach, C. L.
033.075
083.038
Ruffini, R.
065.120
066.025 .125 .128
Rugge, H. R.
074.025
Ruiz, M. T.
122.060
Ruiz, R. D.
099.020
Rumsey, N. J.
031.021 .023
Rumyantsev, A. A.
158.084
Runcorn, S. K.
094.128
Rundel, R. D.
082.037
Runnells, D. D.
015.010
Rupp, Eh.
091.016
Rusev, R. M.
113.007
122.031
Rusin, V.
079.100
Russell, C. T.
074.076
084.232 .255 .279 .300
094.123 .165
Russell, J. L.
004.034
Rust, A. E.
031.055

Rust, D. M.
 072.046
 073.078 .080
Rustad, B. M.
 073.049
Rusu, I.
 032.024
Rutily, B.
 122.094
Rutter, G. H.
 103.011
Ruzan, L.
 158.053
Ruzmaikin, A. A.
 066.030
Ruzmajkina, T. V.
 131.032
Ryabov, O. L.
 093.038
Ryabov, V. P.
 054.006
Rybin, V. V.
 091.018
Rybka, E.
 003.106
 007.000
Rybka, P.
 003.106
 041.024 .025
Rybski, P. M.
 114.042
 142.074
Rydbeck, O. E. H.
 103.101
 131.105 .114 .117
Ryder, L.
 080.056
Rydgren, A. E.
 122.102
Ryle, M.
 141.014 .076
 142.104
Ryter, C.
 142.113
Ryumin, S. P.
 084.235

Saaf, A. F.
 151.036
Saar, E.
 158.054
Saari, D. G.
 042.031
 052.004
Saarloos, J. A.
 141.302
Saba, J. L. R.
 142.116
Sacco, B.
 142.005
Sacharov, V. D.
 003.108
Sackmann, I.-J.
 065.047
Sadakane, K.
 114.026
Sadikov, A.
 073.037
Sadler, D. H.
 082.082

Saerg, K.
 113.026
Safronova, U. I.
 022.033
Sagan, C.
 003.026 .109
 007.000
 012.031
 097.037 .044 .055 .064
 .075
 099.037 .049 .051 .074
 101.009
Sagdeev, R. Z.
 051.005
Sagnier, J. L.
 099.221
Saha, G.
 155.074
Sahade, J.
 117.039
Sahal-Brechot, S.
 074.049
Saidov, K. Kh.
 104.040 .041 .042 .044
 .045
Saint Clair, J. M.
 131.004
Saio, H.
 065.098
Saito, K.
 082.091
 103.122
 104.063
Saito, M.
 064.010
Saito, T.
 084.037
Sakai, H.
 094.382
Sakamoto, K.
 061.037
Sakashita, S.
 061.005
Sakata, A.
 131.139
Sakka, K.
 158.116
Sakuma, S.
 123.025 .026
Sakurai, K.
 077.019 .040
 078.019
 084.204
Salem, M.
 141.078 .079
Salema, C. E. R. C.
 033.068 .069
Salie, H.
 003.139
Salinari, P.
 141.615
Salisbury, J. W.
 002.029 .038
Salman-Zade, R. Kh.
 022.060
Salmon, R.
 103.105
Salpeter, E. E.
 065.101
 125.050 .064

Salpeter, E. W.
 131.122
Salter, C. J.
 141.086
Salukvadze, G. N.
 079.100
 097.201
Salvati, M.
 125.017
 141.039
Samir, U.
 051.021
Sammuli, A.
 142.055
Samonov, V. S.
 103.101
Samovol, V. A.
 083.003
 094.001
Samoznaev, L. N.
 083.003
 094.001
Samson, J. A. R.
 022.067 .068
San Martini, A.
 151.038
Sanamyan, V. A.
 141.055
Sanchez, M.
 044.011
Sandage, A.
 131.553
Sandakova, E. V.
 014.016
Sanders, R. H.
 141.049
 155.019 .027 .054
Sandig, H.-U.
 041.013
Sandmann, W. H.
 115.008
Sandner, W.
 092.008 .025
Sandqvist, A.
 096.023
Sandri, G.
 143.035
Sanduleak, N.
 114.085
 133.012
 141.006
 153.010
Sanford, P. W.
 142.062 .073 .109
San'ko, N. F.
 105.111
Sanwal, N. B.
 121.006 .007
 122.002
Sanyal, A.
 114.005
Sapienza, G.
 075.004
Sarangi, S.
 022.035
Sareyan, J. P.
 122.018
Sargent, W. L. W.
 125.032
 158.101 .111

Sarma, M. B. K.
121.006 .007
Sarma, N. V. G.
141.087
Sartori, L.
125.041
Sasaki, M.
096.038
Saslaw, W. C.
151.050
Sastry, G. P.
062.043
Sastry, K. S.
158.079
Satarova, L. M.
105.114
Sato, F.
131.555
Sato, H.
066.072
Sato, K.
121.019
Sato, T.
131.068
Sattarov, I.
034.024 .048
071.019
Satyvaldiev, V.
032.018
Saunders, I.
084.238
Saunders, R. S.
097.022
Saunders, S. C.
142.118
Sause, G.
103.124
Sautkin, V. A.
032.006
Savage, A.
142.073
Savage, B. D.
099.001
113.001
Savage, R. C.
084.008
Savel'eva, M. V.
158.124
Savenko, I. A.
034.005 .026
078.002 .011 .020
084.235 .407 .408 .409
141.040
143.036 .041
Savich, N. A.
083.003
094.001
Savill, M.
104.070
Savrasov, Yu. S.
052.031
Savrov, L. A.
081.010
Sawyer, C.
073.040 .084
074.115
080.021
Sawyer, R. F.
061.026
Saxena, S. K.
094.305 .309

Sayers, J.
083.050
Sazhin, M. V.
033.023
Sazhin, S. S.
062.074
Sazhina, N. K.
105.113 .132 .136 .137
Sazhina, S. A.
034.062
Scaddan, R. J.
034.050
Scalise Jr., E.
077.002
Scarf, F. L.
062.038
Scarfe, C. D.
121.040 .047
Scargle, J. D.
141.062 .107
Scarsi, L.
134.006
142.005
Schaedler, J.
031.015
Schaifers, K.
010.010
012.009
Schatten, K. H.
092.036
093.023
Schatzman, E.
003.110 .111
064.059
065.114
080.029
162.029
Schaudy, R.
105.003
Schechter, P.
162.012
Scheepmaker, A.
034.065
Scheffer, U.
002.021
Scheffler, H.
003.112
Schennum, G. H.
033.078
Scherer, M.
083.007
Scherrer, P. H.
071.065
077.032
Scherrer, V.
074.103
Scherrer, V. E.
073.065 .071
075.008
Scheuer, P. A. G.
061.003
141.027
Schick, B.
003.113
Schielicke, R.
034.052
Schiff, D.
061.028
Schiff, H. I.
082.003

Schiffer III, F. H.
114.166
Schild, R.
113.033
114.030 .100
Schiminovich, S.
002.005
Schlebbe, H.
071.014
Schlegel, K.
083.010
Schlesinger, B. M.
065.024
Schlosser, W.
071.018
Schmadebeck, R. L.
142.075
Schmadel, L. D.
031.018
079.001
Schmahl, E. J.
073.016
076.034 .037
Schmeidler, F.
005.017
Schmid, P. E.
083.015
Schmid-Burgk, J.
061.046
063.012 .013
Schmidt, D. S.
042.013
Schmidt, E.
015.004
Schmidt, E. G.
065.061
122.119
Schmidt, H. F.
046.021
Schmidt, J.
065.068
Schmidt, K.-H.
154.006
161.001
Schmidt, M.
126.016
141.112
Schmidt, T.
155.029
Schmiedekamp, A. B.
065.087 .111
Schmitt, H. H.
094.396
Schmitt, R. A.
094.344
105.024 .032 .044
Schmuchel, M.
095.003
Schneider, A.
003.141
Schneider, G.
003.113
Schneider, M.
046.021
Schneider, W. H.
114.142
Schneider, W. P.
076.033
Schnopper, H. W.
061.063
125.104

Schnopper, H. W.
142.019 .031
Schnuchel, M.
104.018
Schober, H. J.
098.017
142.071
Schoedel, J. P.
083.015
Schoeffel, E.
118.009
121.087
Schoeneich, W.
031.025
Schoenfelder, V.
061.054
142.064
Schoenmaker, A. A.
103.101
Scholer, M.
078.026
084.406
Scholle, D.
103.101 .128
Scholz, D.
075.011
Scholz, G.
074.065
Scholz, M.
064.007
Schombert, J. L.
041.041 .042
Schonfeld, E.
094.350
Schoolman, S. A.
073.085
Schoonveld, L.
022.057
Schrage, D.
032.042
Schraml, J.
033.043
Schramm, D.
066.112
Schramm, D. N.
061.016 .068
065.112 .901
066.108
125.027
155.070
160.002
Schreier, E.
142.029 .035 .042
Schroeder, D. J.
034.003
Schroeder, M.
092.034
099.020
Schroeder, R.
156.001
Schroll, A.
098.042
113.021
Schubart, J.
098.001
Schubert, G.
094.123 .165 .372
097.043
Schubert, H.
123.038

Schukowski, M.
014.039
Schulte Jr., H. F.
033.055
Schultz, G. V.
034.032 .033
Schultz, J.
041.077
Schultz, L.
094.327 .337 .338 .348
.352 .401
105.071 .149
Schultz, R. B.
072.027
Schulz, M.
084.405
Schumacher, F.
033.081
Schumacher, H.
003.114
031.085
Schurle, H.
031.069
Schutz, B. F.
066.120
Schutz, S.
084.226
Schutz Jr., B. F.
062.070
066.111
Schwaller, H.
094.310
Schwartz, D. A.
143.034
Schwartz, G.
098.027 .029 .031 .033
103.007 .009 .112 .116
.124
Schwartz, K.
094.372
Schwartz, P. R.
114.048 .073
Schwarz, C. R.
081.015
Schwentek, H.
083.011
Schwerer, F. C.
094.320
Sciuto, E.
075.004
Sciuto, V.
075.004
Sckopke, N.
084.287
Sclar, C. B.
094.351
Scott, A. F. D.
082.011
Scott, E. R. D.
105.103
Scott, J. S.
122.102
Scott, P.
141.076
Scott, R. L.
036.006
Scovil, C.
103.105
123.032
125.112

Scoville, N. Z.
131.010
155.004
Scquista, C.
065.017
Scrascia, L.
063.014
Scudder, J. D.
092.035
093.021
Scuflaire, R.
065.020
Seagraves, P. H.
073.087
Seaquist, E. R.
122.021
Seargent, D.
103.102 .105
Searle, L.
114.030
125.032 .040
126.016
Sears, D. W.
105.092 .101
Sears, R. L.
061.020
Sears, R. W.
015.003
Seaton, M. J.
141.078
Sebring, P. B.
008.143
Sedlmayr, E.
071.002
Sedmak, G.
031.001
034.058
Sedova, L. P.
082.118
Sedrakyan, D. M.
022.014
141.321
Sedzielowski, W.
096.027
Seeds, M.
123.028 .029
Seeds, M. A.
103.101
122.081
Seeman, N.
061.033
155.001
Segal', V. M.
094.388
Seggewiss, W.
004.019
119.001
Seidel, B.
092.040
093.022
Seidel, B. L.
099.205
Seidelmann, P. K.
012.006
042.022
091.010
098.014
Seidova, P. I.
126.020
Seielstad, G. A.
134.005

Seifert, H. J.
066.118 .144
Seiler, P.
103.101
Seitter, W.
124.103
Seitz, M. G.
105.056 .087
Sejnowski, T. J.
066.013 .063 .064
Sekanina, Z.
102.021
103.101 .102 .105
Sekhnal, L.
052.037
Seki, T.
103.101 .105 .108 .110
.112 .116 .124 .127
Sekiguchi, H.
077.007
Sekiguchi, N.
045.013
Selivanov, A. S.
094.127 .173
Seliverstov, V. V.
066.015
Sellers, B.
078.030 .034
Selvelli, P. L.
114.025
Semeniuk, I.
141.035
Semenov, A. I.
082.012 .108
Semenova, E. V.
158.091
Semertzidis, V. A.
082.088
Sen, R. N.
003.136
Senftle, F. E.
094.326
Sentman, D. D.
077.025
099.017
Sepehnoori, K.
021.001
Septunovs, G.
009.013
Serkowski, K.
131.083
141.123
Serlemitsos, P.
142.089
Serlemitsos, P. J.
142.043 .048 .060 .061
.116
Serrao, J. M. P.
022.062
Sersic, J. L.
158.067
Serson, P. H.
084.303
Sesplaukis, T. T.
092.040
093.022
Seufert, M.
105.045 .150
Severny, A. B.
082.094
052.019

Seversike, L. K.
052.003
Sevilla, M. J.
032.016
054.017
Sevryukov, B. N.
033.016
Seward, F.
142.051
Seward, F. D.
142.033 .083
Sexl, H.
003.115
Sexl, R.
116.006
Sexl, R. U.
003.115
Seymour, P. A. H.
014.050
Shabanov, M. F.
051.009
Shafer, Yu. G.
011.034
Shaffer, D. B.
141.060
Shafi, M.
099.064
Shagaev, M. V.
082.021
Shaham, J.
065.040
141.320
Shahbazian, E. T.
153.004 .005 .006
Shajdo, A. N.
104.001
Shakhbazyan, R. K.
160.003 .023
Shakhbazyan, Yu. L.
054.012
Shakhovskaya, N. I.
121.014
122.015 .082
Shakirov, K. S.
094.007
Shakov, B. A.
143.012
Shakura, N. I.
064.006
065.045
142.107
Shamolin, V. M.
078.011
Shandarin, S. F.
151.011
158.007 .094
Shandra, Yu. P.
033.009
Shane, C. D.
158.012
Shane, W. W.
158.014
Shankland, R. S.
004.016
Shanthakumar, M.
062.072
Shao, C. Y.
098.027 .029 .031 .033
103.007 .009 .116 .124
.127 .130
114.061

Shapere, D.
003.116
Shapiro, I. I.
092.040
093.022
100.203
Shapiro, S. L.
066.033
122.103
Shapiro, V. A.
085.010
Shaporev, S. D.
103.125
Shaposhnikov, V. E.
062.023
Shara, M. M.
158.048
Share, G. H.
061.033
155.001
Sharkov, V. I.
022.040 .041
Sharma, R. C.
062.059
064.038 .045
Sharov, A. S.
121.098
122.040
123.901
124.005
132.009
Sharp, L. R.
094.123
Sharp, R. D.
084.272
Sharp, W. E.
082.064
Sharpe, H.
094.113
Sharpe, H. A.
094.148
Sharpless, S.
008.109
Sharykin, N. K.
141.076
Shaulov, R. Ya.
080.039
Shaviv, G.
141.089
158.048
Shavranovskij, I. I.
105.128
Shavrina, A. V.
064.021
Shaw, A. W.
082.003
084.019
Shaw, D. M.
105.072
Shawhan, S. D.
077.025
099.216
122.127
Shawl, S. J.
008.069
Shcheglov, P. V.
074.035
082.117
Shcheglov, V. P.
003.011
004.017

Shcheglov, V. P.
008.126
046.016
081.026
Shchekotov, B. V.
141.055
Shchepkin, L. A.
083.039
Shcherbina-Samojlova, I. S.
002.004
Shcherbovskij, B. Ya.
034.018
Shea, M. A.
078.024 .036
Shea, W.
004.069
Sheather, P. H.
032.054
Shectman, S. A.
162.011
Sheeley, N. R.
073.065
074.103
075.008
Sheeley Jr., N. R.
073.071
Shefov, N. N.
082.108
Shejkhet, A. I.
094.001
Shelley, E. G.
084.272
Shel'ting, B. D.
072.033
Shen, B. S. P.
008.099
Shenogin, A. A.
033.051
Shenton, D. B.
076.004
Sheridan, K. V.
008.124
074.044 .062
Sherman, M. D.
066.103
Shermanzon, Eh. M.
084.409
Sherrod, C.
103.101 .105
Shestopalov, I. P.
034.005
078.002 .011
084.235
Shevaleevskij, I. D.
105.133
Shevchenko, V. S.
082.109
122.109
Shevchenko, V. V.
051.009
Shibasaki, H.
036.005
Shields, W.
125.102
Shigehisa
123.025
Shih, C.-Y.
094.133
Shima, Mak.
105.073

Shima, Mas.
105.073 .074 .161
Shimabukuro, F. I.
077.076 .078
Shimek, M.
104.047
Shimizu, M.
131.126
Shimizu, Y.
122.092
Shimmins, A. J.
141.109
Shine, R. A.
080.049
Shinozawa, S.
103.122
Shipman, H. L.
103.101
Shirbakyan, M. S.
160.025
Shirk, E. K.
078.045
Shirley, D. L.
092.049
093.022
Shirmin, G. I.
042.032
Shishkina, V. N.
041.038
Shishova, T. D.
141.127
Shklovskij, I. S.
065.099
Shklovsky, I. S.
125.004
Shkolenko, Yu. A.
010.016
Shmulevskij, V. N.
073.020
Shnejder, A. A.
082.055
Shnygin, Yu. N.
093.034
Shobbrook, R. R.
032.051
122.012
Shogenov, V. Kh.
143.017
Shorthill, R. W.
094.321 .370
Shostak, G. S.
158.001 .010 .902
160.015
Shpital'naya, A. A.
071.038
072.044
Shtajnert, K. G.
See Steinert, K. G.
Shtejngrad, Z. A.
066.101
Shtejnshlejger, V. B.
033.011
Shtern, D. Ya.
083.003
094.001
Shu, F. H.
151.048 .060
155.056
Shugarov, S. Yu.
122.137 .138

Shukhman, I. G.
151.003
Shulberg, A. M.
121.032
Shulman, S. D.
142.082
Shulov, O. S.
122.099
126.009 .013
142.066
Shurshalov, L. V.
105.124
Shushkova, V. B.
104.029 .060
Shvachkin, K. M.
093.006
Shvagerev, V. D.
105.133
Shvalagin, I. V.
031.017
054.006
Shvartsman, V. F.
011.031
066.068
117.025
162.031
Shved, G. M.
082.085
Shvetsov, A. A.
034.046
Shvidkovskaya, T. E.
097.033
Sibille, F.
131.504
Sicha, M.
034.020
Sidorenko, A. I.
083.003
094.001
Sidorenkov, N. S.
044.014
082.077
085.001
Sieber, D.
064.061
065.073
Siegal, B. S.
094.143
Siewert, C. E.
063.017
Sighinolfi, G. P.
105.020 .074 .097
Sigl, R.
046.021
Signer, P.
094.327 .337 .338 .348
.352 .401
105.071 .149
Signorini, C.
074.016
Sigunov, A. I.
054.018
Silant'ev, N. A.
064.046
Silhan, J.
120.005
Silk, J.
125.069
131.086 .087 .520
160.029

Silk, J. K.
071.064
073.086
074.094 .100 .104
076.009 .023 .024
Sil'vanovich, Yu. A.
105.132
Silverberg, E. C.
094.158
Sim, M. E.
036.004
Sima, Z.
141.110
Simek, M.
104.014 .049 .050
Simic, M.
096.036
Simkin, S. M.
031.005
Simmons, G.
033.052
Simmons, K.
103.101 .105
Simmons, K. E.
097.036
Simnett, G. M.
073.032
078.022
Simon, G. W.
071.057
073.087
080.026
Simon, J. L.
042.020 .042
Simon, L. W.
122.011
Simon, P.
079.100
Simon, R.
074.093
Simoncini, A.
035.005
Simonenko, A. N.
105.141
Simonneau, E.
064.042
Simonsen, E.
004.002
Simonson, S. C.
155.068
Simpson, J. A.
092.037
093.026
099.016
Sims, J. S.
093.004
097.059
Sinclair, A. T.
091.006 .028
100.201
Sinclair, M. W.
131.050
Singatullin, R. S.
066.098
Singer, S.
084.257
Singh, M.
062.008
Singh, R. P.
062.040

Singleton, D. G.
083.009
Singstad, I.
084.220
Sinha, A. K.
082.096
Sinha, K.
071.016 .017
Sinha, K. P.
066.080
Sinitsyn, V. M.
085.009
Sinvhal, S. D.
032.031
122.122
Sinzi, A. M.
118.018
Siquig, R. A.
065.007 .012
Siroka, M.
003.117
Siroky, J.
003.117
Sironi, G.
061.070
157.001
Siscoe, G.
083.053
Siscoe, G. L.
074.055
084.901
092.035
093.021
Sistero, R. F.
121.046
123.016
162.047
Sistla, G.
141.018
Sivan, J. P.
155.010 .044
Sivaram, C.
066.080
Sivaraman, K. R.
071.032
Sivaraman, M. R.
083.035
Sizov, A. S.
141.055 .056 .077
Sjogren, W. L.
094.115 .172
Skalafuris, A. J.
065.086
080.037
Skilling, J.
143.005 .026
Skotnikov, M. M.
093.007
Skovli, G.
106.035
Skoza, D.
103.101
Skripin, G. V.
143.043 .044 .045 .046
Skumanich, A.
073.079
Skuridin, G. A.
084.407
143.040
Slaughter, C.
032.042

Slaughter, C.
034.071
Sledzinski, J.
011.041
Slee, O. B.
008.124
141.048 .308
Slettebak, A.
008.038 .040
Slingo, A.
141.008
Slonim, Yu. M.
003.005
073.036
075.007
Slottje, C.
077.024
Slowey, J. W.
052.020
Sluchenkov, G. F.
106.027
Slysh, V. I.
131.046 .078 .100
Smagghe, N.
015.009
Smak, J. I.
124.100
Smart, D. F.
078.024 .036
Smelds, I.
011.027
034.049
Smerd, S. F.
008.124
Smeyers, P.
065.011
117.035
Smidinger, P.
105.080
Smirnov, A. S.
097.007
Smirnov, A. Yu.
061.078
Smirnov, N. V.
104.054
Smirnov, V. A.
104.002 .036 .053
Smirnov, V. V.
054.010
Smirnova, T. V.
093.035
Smith, A. G.
008.049
036.006
Smith, A. M.
114.155
131.094
Smith, B. F.
074.077
094.372
Smith, B. W.
125.028
Smith, D. E.
046.028
Smith, D. F.
077.015 .052
Smith, D. W.
158.019
Smith, E. I.
094.389

Smith, E. J.
 084.232
 099.015
Smith, E. W.
 022.043
 114.127
Smith, F. G.
 141.318 .341
 156.003
Smith, H.
 125.035
Smith, H. E.
 158.110
Smith, J. R.
 010.003
 075.020
Smith, J. W.
 094.382
Smith, L. F.
 131.096 .538
Smith, L. G.
 082.099
Smith, L. L.
 103.101
Smith, M. A.
 071.043
 114.076
Smith, M. G.
 141.065
 158.062
Smith, M. L.
 081.024
Smith, P. H.
 084.290
Smith, P. K.
 103.101
Smith, R. A.
 077.056
 099.059
Smith, R. C.
 064.020
Smith, R. L.
 065.047
Smith, S. B.
 009.015
Smith, W. H.
 022.009
Smith Jr., H.
 151.027
Smith Kogon, C.
 154.012
Smithson, R. C.
 034.068
Smolders, P. L.
 003.118
Smol'kov, G. Ya.
 073.020
 077.044
Smriglio, F.
 103.101
 154.020
Smyth, J. R.
 094.406
Smyth, M. J.
 061.062
Smythe, W. D.
 099.214
Sneath, P. H. A.
 012.030
Sneden, C.
 114.033 .105

Snezhko, L. I.
 121.045
Snider, J. L.
 066.081
 071.047
Snyder, A. L.
 084.025
Snyder, C. W.
 074.079
Snyder, G. A.
 155.045
Snyder, L.
 103.101
Snyder, L. E.
 012.025
 114.140
 131.060 .062 .134
Soberman, R. K.
 098.006
Sobieski, S.
 121.024
Sobolev, N. V.
 105.139
Sobolev, V. M.
 022.024 .061
 054.012
 071.037
 073.012
Sobolev, V. V.
 061.014
 063.008
Soboleva, N. S.
 141.019
Soederhjelm, S.
 079.100
 121.063
Soeraas, F.
 084.030 .242
Sofia, S.
 142.032
Sofue, Y.
 155.043
Soifer, B. T.
 141.608
Sokolov, D. D.
 162.031
Sokolov, S.
 094.359
Sokolov, V. S.
 073.057
Sokolovskaya, Z. K.
 011.017
Solc, M.
 011.021
Solodyna, C. V.
 062.063
Solomon, P.
 131.093
Solomon, P. M.
 131.010
 132.034
 155.004
Solomon, S. C.
 094.117
Solonitsyna, N. F.
 083.022 .023 .024
Solonskij, Yu. A.
 022.060
Solovej, B. G.
 083.031

Solov'ev, A. A.
 071.012
 080.010
Soltau, G.
 105.104
Somal', V. G.
 094.127 .173
Sonett, C. P.
 008.133
 074.077
 094.372
 099.015
Soohoo, L. B.
 034.043
Sorgenfrey, W.
 103.101
Sorkin, A.
 053.005
Sorochenko, R. L.
 155.040
Sorokin, V. P.
 093.034
Sosnovets, Eh. N.
 084.403
Soulie, G.
 103.101
South, R. H.
 103.011
South, R. H. S.
 103.101 .108 .116
Southwood, D. J.
 084.203
Sowers, B. L.
 034.039
Sowers, J. L.
 096.008
Spada, G.
 034.073
 142.117
Spangenberg, W. W.
 015.016
Spangler, S. R.
 122.127
Sparks, W. M.
 125.008
Sparrow, J. G.
 106.020
Spear, G. G.
 121.080
Specht, H.
 078.035
Speiser, T. W.
 084.015
Spencer, J. H.
 114.073
Spencer, N. W.
 082.031
Sperauskas, J.
 113.004
Spettel, B.
 105.081
Spicer, D. S.
 073.088
Spilhaus Jr., A. F.
 081.007
Spinrad, H.
 158.028 .110
Spitzer Jr., L.
 008.103
Spizzichino, A.
 076.012

Spizzichino, A.
082.022
142.004
Spreiter, J. R.
074.111
Sprott, G. F.
125.070 .104
142.019 .031
Sprott, G. N.
103.101
Spruit, H. C.
080.015
Spulgis, G.
034.049
Sramek, R. A.
158.088
Sreedhar Rao, S.
142.065
Sreekantan, B. V.
142.065
Srinivasan, B.
094.312
Srirama Rao
104.071
Srirama Rao, M.
104.071
Srivastava, J. B.
032.031
Srivastava, K. M.
062.053
Sroczynska, M.
074.067 .084
Stabell, R.
065.005 .071
Stache, P.
003.100
Stadsnes, J.
084.220
Staehle, V.
105.075
Staettmayer, P.
031.085
Stagni, R.
032.014
Stakheev, Yu. I.
094.161
Stakheeva, S. A.
105.114 .137
Stalio, R.
114.022
Stallcop, J. R.
061.001 .002
Staller, R. F. A.
155.018
Standish Jr., E. M.
041.006
Stanek, B.
003.140
Stange, L.
055.015
Stankevich, K. S.
066.006
Stanley, G. J.
008.018
Stannard, D.
141.031 .053
Stansberry, K. G.
099.058
Stanyukovich, A. K.
105.125 .142 .143

Starace, A. F.
022.066
Starikova, G. A.
118.033
Starkov, G. V.
084.007
Starkova, A. G.
082.087
Starobinsky, A. A.
066.056
Starr, V. P.
099.065
Staude, H. J.
082.101
Staude, J.
011.008
155.029
Staudte, J. H.
035.012
Stauffer, D.
099.028
Stauning, P.
083.046
106.035
Staveland, L.
080.033
Stavinschi, V.
044.008
Stebbings, R. F.
082.037
Stecher, T. P.
114.004
125.008
131.094
Stecker, F. W.
012.034
143.007
Stefanini, L.
014.030
Stefanovich, A. E.
054.018
Stehli, F. G.
003.006
Stehpa, A. P.
032.026
Steigman, G.
131.503
Stein, W. A.
062.025
158.005
Steinbacher, R.
097.055
Steinberg, J. L.
077.034 .035
Steiner, J. E.
142.072
Steinert, K. G.
032.033
Steinitz, R.
073.026 .068
Steinnes, E.
094.153
Steins, K.
036.008
052.038 .039
102.025
Stellingwerf, R. F.
122.111
Stellmacher, G.
072.024

Stelzried, C. T.
092.040
093.022
Sten, T. A.
084.030
Stenbaek-Nielsen, H. C.
084.034
Stenflo, J. O.
080.042
Stenflo, L.
061.042
062.050
Stenholm, B.
079.100
114.024
Stephani, H.
066.058
Stephens, S. A.
142.093
Stephenson, C. B.
103.101 .105 .112 .116
.127 .128
114.086 .154
Stephenson, F. R.
125.036
Stephenson, R. J.
158.069
Stepien, K.
032.008
Steppe, H.
153.008
Sterken, C.
114.067
Sternlieb, A.
065.036
Stettler, A.
094.310 .332
Stevens, G. A.
061.013 .017
Steverding, B.
094.154
Stewart, M. B.
082.001
Stewart, P.
061.039
Stewart, R. T.
074.063 .064
Stickney, P. M.
122.102
Stief, L. J.
131.115
Stiefel, E.
042.034
Stienon, F. M.
114.061
Stier, J.
014.007
Stilwell, D. E.
099.018
Stix, M.
061.047
Stockman Jr., H. S.
132.011 .032
Stockton, A.
158.006 .103
Stoddard, J.
103.105
Stoddart, J.
103.101 .102
Stodolkiewicz, J. S.
151.051

Stoeckley, T. R.
114.071
Stoeffler, D.
105.076
Stoer, J.
021.004
Stoering, J.
142.051
Stoering, J. P.
142.033 .083
Stokes, R. A.
008.105
066.008
099.007 .008
131.003 .121
Stolboushkin, S. K.
084.235
Stollman, J.
065.053
066.093 .145
Stolov, H. L.
084.234
Stone, E. C.
078.038 .041
Stone, P. H.
099.041
Stone, R. G.
077.053
Stone, R. P. S.
142.112
Stonehouse, B. E.
010.024
Stoner, R.
141.104
Stong, C. L.
034.007
Storms, J.
010.039
Storzer, D.
094.132 .353
Stothers, R.
155.033
Stozhkov, Yu. I.
143.027
Straizys, V.
113.003
Straka, R. M.
077.902
Straka, W. C.
125.026
Strand, K. A.
008.142
111.004
Strandbaek, K.
079.100
Strangway, D. W.
007.000
012.002
094.113 .148 .380
Strel'nitskij, V. S.
022.051
102.016
131.033
Strickland, D. J.
084.002
Stringer, W. J.
084.029
Stringfellow, M. F.
085.008
Strittmatter, P. A.
032.053

Strittmatter, P. A.
064.029
126.021
141.111 .119 .123
Strobel, D. F.
099.066
100.206
Strom, K. M.
141.601
Strom, R. G.
092.017 .020 .039
093.024
125.019 .052
141.117
Strom, S. E.
141.601
Strong, A. W.
143.007
Strong, I. B.
142.021 .080
Strong, K. T.
142.040
Strukov, I. A.
141.041
Studer, W.
010.025
Studnicka, J.
034.020
Stull, M. A.
141.076
Stumpff, K.
003.119
Stumpff, P.
033.043
Stupar, M.
009.017
103.101
Sturch, C.
113.014
Sturrock, P. A.
141.336
Stutzman, W. L.
033.058
Su, D.
031.072
Subtil, J. L.
022.044
Suchkov, A. A.
151.049
Sudzius, J.
113.004
Suffern, K. G.
151.037
Sugawa, C.
082.116
Sugimoto, D.
065.021
Sukharev, L. A.
032.039
Sukhodol'skij, S. A.
033.008
Sullivan, W.
003.026
Sulzmann, K. G. P.
022.036
Sume, A.
131.105
Summers, A. L.
091.003
Sumners, C.
009.010

Sundaram, S.
022.057
Sundman, A.
155.060
Sung, C.
065.016
Sung, C.-H.
062.062
Sunyaev, R. A.
066.074
142.018
Suomi, V.
092.017 .039
093.024
Suri, A. N.
076.017
125.054
Surkov, Eh. P.
072.043
073.064
Surkov, Yu. A.
051.008
093.002
094.387
Surmelian, G. L.
022.073
Sushanin, I. V.
022.025
Suslov, A. A.
034.026
078.020
143.041
Susurin, G. Eh.
066.084
Sutantyo, W.
125.007 .014
Sutherland, D. A.
103.101
Sutherland, P. G.
065.091
126.006
Sutherland, R. A.
034.079
Sutton, A.
014.043
Sutton, E.
141.613
Sutton, J. M.
141.334
Suvorov, E. V.
062.023
Suzuki, K.
082.044
103.101
Svalgaard, L.
072.047
074.009 .029 .061 .092
   .109
085.013
106.034
Svechnikov, N. Yu.
078.002
Svensson, L. A.
074.012
Svestka, Z.
077.063
Sviderskiene, Z.
113.004
Svolopoulos, S.
121.074

Swaminathan, R.
142.065
Swamy, K. S. K.
114.020
Swanenburg, B. N.
143.023
Swartz, M.
073.052
Swarup, G.
141.308
Svedlund, J. B.
124.100
126.008 .010
131.003 .121
Sweetnam, D. N.
092.040
093.022
Sweigart, A. V.
065.002 .060 .088
080.044
Swensson, J. W.
022.053
Swerdlow, N.
004.008
Swihart, T. L.
141.105
Swindell, W.
099.022 .061
Swings, J. P.
114.115 .116
Sy, W. N.-C.
077.006 .017
Syrogos, H.
033.086
Syunyaev, R. A.
064.006
Szabados, L.
121.060
123.008
Szabo, E.
094.398
Szacherska, M. K.
021.027
046.006
Szafraniec, R.
121.028
Szalay, A. S.
162.039
Szebehely, V.
042.038 .078
Szecsenyi-Nagy, G.
122.083
Szeidl, B.
122.124
Szkody, P.
122.042
Szulc, M.
104.069
Szymanski, H. A.
003.016
Szymanski, W.
072.009
Szymczak, C. J.
003.120

Taam, R. E.
117.021
Tabakova, Z. N.
054.012

Tabara, H.
131.024
158.303
Tabarroni, G.
003.098
Taber, R. C.
033.021
Tabor, J. E.
074.099
Taibo, R.
041.038
Takada, M.
022.070
Takagi, S.
045.015
Takagishi, K.
142.101
Takahashi, F.
084.037
Takakura, T.
077.066
Takatsuka, T.
065.118
141.338
Takechi, A.
082.090
Takeda, H.
105.089
Talbot Jr., R. J.
155.025
158.121
Tammann, G. A.
125.042 .047
131.553
Tamojkin, V. V.
062.046
Tanaka, H.
075.016
Tanaka, K.
080.062
Tanaka, Y.
142.065 .068
Tananbaum, H.
142.029 .035 .042
Tandberg-Hanssen, E.
073.067 .089
074.043
Tapley, B. D.
094.005
Tarafdar, S. P.
114.020
Tarenghi, M.
065.075
Tarkhov, E. N.
084.223
Tarter, C. B.
131.904
141.062
Tarter, J.
160.029
Tartois, L.
010.028
Tassoul, J.-L.
065.027
Tataru, E.
034.030
Tataru, M.
034.030
Tatsch, J. H.
003.121

Tatsumoto, M.
105.077
Tatum, J. B.
103.100 .105
Taubenheim, J.
085.004 .005
Tausworthe, R. C.
002.026
Tavastsherna, K. N.
041.047
Tavastsherna, K. S.
073.093
Tayler, R. J.
010.022
Taylor, B. G.
051.022
Taylor, F. W.
099.901
Taylor, G. E.
035.003
Taylor, J.
066.117
Taylor, J. G.
066.088
Taylor, J. H.
021.016
141.330
Taylor, K.
031.003
133.001
Taylor, S. R.
094.376
Tedesco, E. F.
123.035
Teegarden, B. J.
074.078
099.018
143.008 .025 .054
Tejfel', V. G.
100.002 .004
Teleki, G.
012.023
031.052 .075 .077 .082
045.017
082.113
Tem, Eh. L.
082.105
Temnyj, V. V.
078.003
084.212
Tengstroem, E.
011.003
Tennfors, E.
062.016
Tenorio-Tagle, G.
131.510
Teplykh, A. I.
141.056 .077
Ter Haar, D.
131.128
Tera, F.
094.325
Terasaki-Okada, K.
143.063
Terent'eva, A. K.
104.020 .028
Tereshchenko, V. M.
103.101
Terez, E. I.
034.086
082.094

Terez, G. A.
034.086
Terrile, R. J.
099.042
Terzan, A.
122.094
123.040
124.103
Terzian, Y.
131.521
133.008
Teschke, F.
094.354
Testerman, L.
080.050
Tetruashvili, Eh. I.
073.035
074.069
Teukolsky, S. A.
066.054
Thackeray, A. D.
121.052
122.130 .134
124.105
132.021
Thaddeus, P.
132.035
The, P. S.
155.018
Thekaekara, M. P.
080.030
Thewlis, J.
003.122
Thiel, K.
094.174
Thiel, M.
141.076
Thielheim, K. O.
143.067
Thierry-Mieg, J.
066.048
Thom, A.
004.007 .009 .030
Thom, A. S.
004.007
Thom, Al. S.
004.030
Thom, Ar. S.
004.030
Thomas, B. M.
008.124
Thomas, G. E.
092.015
094.152
106.001
Thomas, G. R.
078.028
084.260 .261 .262
106.035
Thomas, L.
082.024
Thomas, R. J.
076.005 .020
082.098
Thomas, R. M.
141.326
Thompson, A. R.
033.006 .036
Thompson, D. J.
155.058

Thompson, G. D.
103.101 .105
Thompson, G. I.
032.034 .902
113.006 .034
114.015
Thompson, J. E. S.
003.123
Thompson, M. G.
143.064
Thompson, M. H.
003.067
Thompson, P. A.
046.013
Thompson, R. J.
115.010
Thompson, T. W.
094.321 .367 .369 .370
Thomsen, M. F.
099.017
Thomson, M. M.
007.000
Thorne, K. S.
066.024
Thorne, R. M.
099.044
Thorp, J. M.
046.026
Thorpe, A. N.
094.326
Thuan, T. X.
065.055
Thum, C.
132.017
Tichy, M.
034.020
Tiemann, E.
022.080
Tifft, W. G.
160.007
Tikhomirova, L. N.
034.011 .012
Tilford, S. G.
076.033
Timashkova, G. M.
032.039
041.038
Timofeev, B. V.
077.045
Timofeev, G. A.
078.012
Timofeeva, G. M.
033.015
Timofeeva, T. S.
093.006
Timoshkova, E. I.
042.068
Timothy, A.
071.064
073.086
074.090 .091 .093 .094
.100 .104 .106
076.023
Timothy, A. F.
076.009
Timothy, J. G.
073.016
076.034 .037
Tinsley, B. M.
065.052
155.002 .070

Tinsley, B. M.
158.100
Tishchenko, V. M.
072.019
Tishchenkov, N. T.
033.015
Titarchuk, L. G.
097.007
142.018
Titenkov, A. F.
143.037
Titheridge, J. E.
083.001
Title, A. M.
034.028 .069 .070
Titov, V. B.
042.069
Titulaer, C.
007.000
Tjoetta, S.
080.004
Tkachenko, V. I.
097.033
Tkachuk, A. A.
104.015
Tlamicha, A.
033.024
077.021 .030 .031
Tobailem, J.
105.006
Todoran, I.
117.036
122.066
Tofani, G.
131.525
141.086
Tojo, A.
082.091
Tokareva, G. F.
122.136
Tokhtas'ev, V. S.
104.016 .027
Tokis, J. N.
042.018 .019
Toksoez, M. N.
094.008 .111 .126 .318
Toktogulov, M.
104.036
Tolendino, L. F.
002.029
Tolland, H. G.
094.155
Tolstoy, I.
003.124
Tomanov, V. P.
102.023
Tomasi, P.
141.086
Tomasko, M.
099.022
Tomasko, M. G.
099.035 .061
Tomasson, P.
158.061
Tomimatsu, A.
066.072
Tomino, K.
158.303
Tomita, K.
103.116 .127 .131
104.063

Tomozov, V. M.
 074.074
 076.006
 131.033
Toomasson, L.
 117.016
Toon, O. B.
 097.077
Toor, A.
 142.033 .051 .083
Top, Z.
 105.109
Torelli, M.
 075.013
Toroshelidze, T. I.
 074.036
Torras, N.
 103.101 .105
Torres, C.
 103.008 .010 .101 .105
Torres-Peimbert, S.
 131.548
Tosa, M.
 155.043
Toure, I.
 082.017
Tousey, R.
 073.065 .071 .087
 074.103 .105
 075.008
 076.033 .036
Tovmassian, H. M.
 153.003 .004 .005 .006
  .007
 158.004
 160.025
Tovmasyan, G. M.
 See Tovmassian, H. M.
Towlson, W. A.
 114.118
Townes, C. H.
 132.001
 155.035
Tozer, D. C.
 107.005
Traat, P.
 158.054
Trafton, L.
 099.218
 100.202
Trafton, L. M.
 099.020 .041
Trainor, J. B.
 103.101 .105
Trainor, J. H.
 074.078
 099.018
 143.008 .025 .054
Transkij, I. A.
 074.088
 106.015 .028
 143.047 .048 .050
Trask, N.
 092.017 .039
 093.024
Traub, W. A.
 131.133
 132.030
Traugott, S. C.
 097.045

Trautman, A.
 012.008
Travers, P.
 082.095
Traversi, C.
 011.005
Traving, G.
 063.012
Treanor, P. J.
 005.019
 008.029
Treder, H.-J.
 004.054
 005.012
 015.008
 066.089 .114 .115
 162.013 .014 .015
Trefall, H.
 084.220
Treffers, R.
 114.069
Tregaskis, T. B.
 010.008
 103.105
Tregear, W. H. G.
 010.008
Treguer, L.
 073.009
Trehan, S. K.
 042.017
 062.008
Tremaine, S. D.
 096.021
Tremko, J.
 123.009
Treskova, L. E.
 077.042
Treumann, R.
 062.047
 084.244
Treurniet, K.-W.
 104.072
Treutner, H.
 031.014 .046
 034.016
 097.025
Treverton, A. M.
 141.054
Treves, A.
 065.075
 141.332
 142.017
Trevese, D.
 066.122
Tricker, R. A. R.
 014.042
Trimble, V. L.
 112.027
Tritton, K. P.
 141.128
Trivelpiece, A. W.
 003.073
Troesch, B. A.
 062.022
Trofimova, L. S.
 106.024
Troitskij, B. V.
 083.021
Troitskij, V. S.
 141.055 .056 .077

Troitsky, A. V.
 093.001
Trombka, J. I.
 012.034
 094.124
 142.075
Troy Jr., B. E.
 113.037
Trubitsyn, V. P.
 099.004 .009 .029 .030
  .075 .076
Trumbo, D.
 034.071
Truran, J. W.
 065.029
Truxton, J.
 103.101
Tryashin, S. S.
 104.020
Tsai, C.
 125.054
Tsang, L.
 033.052
Tsap, T. T.
 080.011
Tsarevskij, G. S.
 122.016
Tscharnuter, W.
 065.014 .067
Tschauner, J.
 042.055
Tseng, S.-S.
 097.041
Tsesevich, V. P.
 121.032 .100 .102
 122.046 .048 .142 .144
Tsiang, C. B. H.
 046.026
Tsikoudi, V.
 122.114
Tsioumis, A.
 122.107
Tsirel', V. S.
 084.270
Tskhaj, A. A.
 072.019
Tskhovrebadze, A. S.
 073.035
Tsubaki, T.
 073.050
Tsuchiya, K.
 103.105
Tsuji, T.
 064.011
Tsunemoto, K.
 022.071
Tsuruta, S.
 065.041 .076 .092
 142.002
Tsvetkov, T.
 122.108
Tsyganov, A. N.
 033.051
 077.047
Tsytovich, V. N.
 141.312
Tucker, R.
 097.037 .055
Tucker, R. H.
 012.014
 041.050

Tulinov, V. F.
083.016
Tully, J. A.
022.062
Tully, R. B.
158.131 .132 .133
Tumanian, B. E.
004.031
Tumanyan, B. V.
014.022
Tuominen, A.
055.012
Tuominen, J.
072.002
Tupper, B. O. J.
066.031 .129
Turchin, V. I.
033.010
Turchinovich, I. E.
074.022
Turk, J. S.
141.336
Turkeeva, B. A.
083.027
Turner, B. E.
103.101
131.009 .049
Turner, G.
105.078
Turner, R. F.
076.004
Turnock, A. C.
094.305
Tursunov, O. S.
041.016
Tutukov, A. V.
011.032
117.010 .011 .012 .013
Tuzzolino, A. J.
099.016
Tverskoj, B. A.
097.033
Tyler, G. L.
092.040
093.022
094.125 .370
Tyson, J. A.
066.046
Tzirulnik, L. B.
073.007

Uberall, H.
080.055
Uchida, Y.
074.052
Udalski, A.
095.009
096.024 .025
Udal'tsov, V. A.
155.040
Udovich, N.
097.018
Ueno, S.
063.022 .027 .034
Uhlmann, D. R.
094.379
Uiblein, P.
004.080
Ukhova, O. K.
014.024 .032

Ulanovsky, L. E.
141.040
Ulich, B. L.
091.011
094.171
103.101
Ullaland, S.
084.220
Ulmer, M. P.
142.006 .055 .078 .085
Ulmschneider, P.
073.044
Ulrich, M.-H.
158.073
Ulrich, R. K.
080.014
117.006
Umanskij, S. Ya.
131.103
Underhill, A. B.
114.133 .902
Underwood, J. H.
034.035
074.108
076.029
Ungar, S. G.
099.056
Ungstrup, E.
084.281 .286
Unno, W.
065.022
073.046
Unruh, D. M.
105.077
Unsoeld, A.
158.077
Upgren, A. R.
008.080
111.001 .010
113.013 .044
153.027
Uphoff, C.
052.041
Uras, S.
045.006 .007 .008
Urata, T.
103.101 .105
123.037
Urbarz, H.
074.023
077.029
Ureche, V.
117.019 .022
Urey, H. C.
102.026
Uscinski, B. J.
141.339
Ushakova, N. A.
097.015
Usher, P. D.
141.002
158.301
Usov, V. V.
061.030
Ustinova, G. K.
105.118 .119
Ustinovshchikov, V. M.
078.020
Utiger, H. E.
131.109

Utkin, V. I.
085.010

Vaghi, S.
103.124
Vaiana, G.
071.064
073.086
074.090 .091 .093 .094
  .100 .104 .106
076.023
Vaiana, G. S.
076.009 .024
Vajsberg, O. L.
074.034
078.003
084.212
097.027
Vakhrameev, I. Ya.
054.012
Vakulov, P. V.
034.018
Valach, R.
105.088
Valbousquet, A.
041.071 .073
Valentine, D. A.
061.010
Valentine, H.
014.006
Validov, M. A.
054.012
Valley, G. C.
062.013
Valtier, J. C.
122.018
Valtonen, M. J.
151.050
Vampola, A. L.
078.029
084.282
Van, Y. Y.
141.330
Van Allen, J. A.
078.008
099.017
Van Altena, W. F.
008.033 .145
041.060
111.007
153.014
Van Beek, H. F.
034.053 .054
076.018
Van Blerkom, D.
100.014
Van Citters, G. W.
122.133
124.101
Van De Kamp, P.
118.010
Van Den Berg, A.
097.051
Van Den Bergh, S.
065.097
133.014
154.006 .007
155.009 .046 .069
158.117

Van Den Heuvel, E. P. J.
  117.034
  142.039
Van Der Borght, R.
  072.001
Van Der Hucht, K. A.
  012.012
  034.001
  114.107 .109 .110 .111
    .130
Van Der Jeugt, A.
  097.034
Van Der Kruit, P.
  125.109
Van Der Kruit, P. C.
  158.018 .082 .901
Van Der Laak, H. J. M.
  033.088
Van Der Laan, H.
  125.052
  141.113
Van Der Waerden, B. L.
  003.125
  004.046
Van Duinen, R. J.
  131.547
Van Flandern, T. C.
  043.002
Van Genderen, A. M.
  121.051 .065
  122.077
  142.063
Van Helden, A.
  004.033
Van Hemelrijck, E.
  079.100
Van Herk, G.
  041.034 .040
Van Horn, H. M.
  065.043
  133.013
Van Mieghem, J.
  003.126
Van Otterloo, P.
  054.025
Van Rensbergen, W.
  071.006
Van Schmus, W. R.
  105.002
Van Someren Greve, H. W.
  033.026
Van Till, H.
  132.035
Van Venrooij, M. A. M.
  036.011
Van Venrooy, M. A. M.
  031.018
Van Wageningen, G.
  015.007
Vandakurov, Yu. V.
  062.061
Vanden Bout, P. A.
  114.150
  117.030
  131.002
  142.088
Vandervoort, P. O.
  151.013
Van't Veer, C.
  114.131

Vanyan, L. L.
  094.182
Van'yan, L. L.
  084.267
Vanysek, V.
  061.073
  102.017
  103.101
  131.123
Varanasi, P.
  022.035
Vardavas, I. M.
  063.033
  064.031
Varin, M. P.
  041.038
Varina, V. A.
  041.038
Varshni, Y. P.
  141.101
Vartanyan, Yu. L.
  065.058
Varvarov, N. A.
  003.127
Varypaev, A. A.
  141.055
Vasil'ev, M. B.
  083.003
  094.001
Vasil'ev, N. V.
  105.086
Vasil'ev, V. P.
  073.034
Vasil'eva, A. I.
  052.026
Vasil'eva, G. Ya.
  072.044
Vasilevskis, S.
  034.059
  112.005
Vasilkova, E. A.
  084.007
Vasil'yanovskaya, O. P.
  122.052
Vasudevan, R.
  063.021
Vauclair, G.
  064.008
Vauclair, S.
  064.008 .059
  065.050
Vaughn, L. M.
  103.110
Vedeneev, Yu. B.
  033.053
Vedenin, A. I.
  094.324
Vedeshin, L. A.
  143.037
Vedrenne, G.
  155.037
  160.006
Veeder, G. J.
  122.115
Vegas, L. F.
  143.060
Vehrenberg, H.
  003.023
Veio, F.
  034.016

Veis, G.
  046.027
Velez, C. E.
  052.012
Velinov, P.
  091.008
Veltri, P.
  062.002
Velusamy, T.
  073.005 .008
  077.001
  125.015 .024
Venis, T. E.
  114.118
Venkatarangan, P.
  078.008
Venkatesan, D.
  078.008
Venugopal, V. R.
  141.308
Verdier De Genouillac,
  G.
  074.003
Vereshchagina, N. V.
  106.025
Vereshkov, G. M.
  077.014
  162.017
Vergasov, R. I.
  033.015
Verguese, D.
  097.054
Verheest, F.
  062.034
Verhoogen, J.
  045.004
Verhulst, F.
  042.048
Verigin, M. I.
  106.027
Verma, B. G.
  065.115
Vernazza, J. E.
  073.016
  076.034 .037
Vernov, S. N.
  097.033
Veron, M. P.
  141.003 .004
Veron, P.
  122.901
  141.003 .004
Verschraegen, A.
  097.034
Verschuur, G. L.
  003.128 .129
  131.508 .549
  155.048
Vesely, V.
  034.020
Vesic, D.
  044.021
Vespignani, G. R.
  076.012
  142.004 .047
Vetukhnovskaya, Yu. N.
  097.028
Veverka, J.
  093.010
  097.037 .055 .075
  099.037 .049 .051 .074

Veverka, J.
131.009
Viala, Y. P.
061.060
Vidal, N. V.
113.011
121.088
122.132
124.106
142.019 .020 .024 .025
.103 .106
Vidal-Madjar, A.
082.033
Vidyakin, V. V.
042.006
Vigier, J. P.
066.087
080.002
117.017
Viktorov, S. V.
094.301
Vilhu, O.
065.031
Vil'koviskij, Eh. Ya.
131.138
Villa, G.
073.009
Villoresi, G.
143.010
Vince, I.
122.073
Vincent, M.
035.015
Vinge, V.
053.005
Vinitskij, A. V.
083.039
Vinogradov, A.
094.359
Vinogradov, A. P.
094.163
105.079
Vinogradov, V. A.
094.001
Vinckurov, S. K.
158.091
Vinti, J. P.
042.023
Vinyajkin, E. N.
141.056 .077
Viotti, R.
103.101
114.052 .084
Visconti, G.
082.023
Vishnyakov, A. F.
042.052
Vishveshwara, C. V.
066.124
Vitini, I.
044.011
Vitinskij, Yu. I.
072.010 .012 .045
Vitkevich, V. V.
033.008
Vitrichenko, E. A.
122.016
Vlachos, D.
041.032
Vladimirov, S. B.
104.030

Vladimirskij, B. M.
072.006
Vlasceanu, V. I.
098.015
Voecks, G. E.
097.042
Voelk, H. J.
084.224
143.003
Vogt, N.
103.101
114.067
121.013
Vogt, R. E.
078.038 .041
Vojkhanskaya, N. F.
122.069
Vojskovskij, M. I.
085.009
Volborth, A.
094.307
Volkov, G. I.
084.266
106.027
Volkov, I. I.
085.009
Volodichev, N. N.
034.026
078.020
143.041
Volokhov, S. A.
141.056 .077
Voloshchuk, Yu. I.
104.015
Volsky, L. H.
012.004
Volynskij, B. A.
014.031
Volz, F. E.
082.034
Von Kusserow, H.-U.
061.047
Von Seggern, H. O.
031.085
Von Zahn, U.
083.048
Vondrak, R. R.
074.079
084.011
094.149
Vorob'ev, A. I.
034.026
Vorob'ev, V. G.
034.013
Voronov, V. N.
033.014
093.001
Vorontsov-Vel'yaminov,
B. A.
158.030 .053 .123 .125
.126 .127 .129
Vorpahl, J.
076.035
Voskobojnikov, A. Eh.
011.033
Vovelle, F.
035.014
Vozdvizhenskij, B. S.
103.101
Vrabec, D.
072.048

Vreux, J. M.
114.115 .116
Vsekhsvyatskij, S.
102.024
Vsekhsvyatskij, S. K.
074.050
097.100
Vu, D. T.
099.221
Vuilleumier, J.
105.157
Vyalshin, G. F.
073.012
Vykhrestyuk, S. S.
120.002
Vykhrestyuk, T. P.
114.046
Vyshlov, A. S.
083.003
094.001

Waak, J. A.
131.058
Wade, C.
022.048
Wade, C. M.
041.054
Waenke, H.
094.135 .175 .354 .392
105.037 .081
Wagner, C. A.
052.046
Wagner, C.-U.
075.011
Wagner, W. J.
074.028 .107
Wagoner, R. V.
065.062
066.049
Wahsner, R.
004.039
Wakamatsu, K.
158.116
Walborn, N.
103.105
Walborn, N. R.
114.079
158.081
Wald, R.
066.123
Wald, R. M.
066.124
Waldmeier, M.
008.150
031.013
074.114
075.003 .015 .016
079.100
Walker, A. R.
099.211
Walker, D. H.
010.008
Walker, D. M. C.
081.019
082.005 .006
Walker, G. A. H.
122.088
Walker, G. N.
096.010

Walker, M. F.
 013.013
Walker, W. S. G.
 124.101
Walker Jr., A. B. C.
 074.025 .108
Wall, J. V.
 131.020
 141.023 .033 .070 .114
Wallace, L.
 091.024
Wallerstein, G.
 008.118
 065.108
 096.033
 114.165
 131.008 .065
Wallis, D. D.
 084.038
Wallis, M. K.
 074.032
 102.001
 106.009
 131.903
Walls, F. L.
 083.045
Walmsley, C. M.
 119.013
 131.006 .074
 132.027
Walpole, P. H.
 092.037
 093.026
Walraven, R. L.
 034.043
Walter, H. G.
 043.005
Walter, K.
 117.024
Walter, L.
 105.080
Walter, L. S.
 094.309
Walton, J. R.
 022.901
Wang, C.
 004.068
Wang, C. G.
 061.025
Wang, C. S.
 048.215
 084.401
Wang, H. T.
 076.013
 080.032
Wang, S.
 105.158
Wang, X.
 105.158
Wang, Y.
 031.072
Wang, Z.
 074.112
Wannier, P. G.
 132.029 .034
Waranius, F. B.
 002.038
Ward, F.
 080.051
Ward, J.
 022.043

Wardle, J. F. C.
 141.005 .050
 158.088
Wardrop, A.
 103.101
Wark, D. A.
 094.304
Warman, J.
 034.080
Warner, B.
 008.026
 121.069
 122.037 .063 .133
 124.101
Warner, J. W.
 141.076
 158.087
Warner, P. J.
 158.078
Warren Jr., W. H.
 114.098 .099 .164
 153.022
Warwick, R. S.
 141.031
Wasilewski, P.
 094.361
Wasilewski, P. J.
 094.319 .355 .356 .357
 105.082 .083
Wassef, A. M.
 046.018
Wasserburg, G. J.
 094.325 .365
 105.063 .164
Wasserman, L.
 093.010
 099.051
 101.009
Wasserman, L. H.
 099.037 .049 .050 .074
Wasson, J. T.
 105.003 .019 .084 .102
 .103
Waterfield, R. L.
 103.011 .101 .105 .108
 .116
Watson, A. A.
 143.029 .062
Watson, D. E.
 105.012 .070 .144
Watson, W. D.
 131.026 .063
Wattenberg, D.
 005.020
Watts Jr., R. N.
 051.004 .013
 053.001 .006
 054.001 .002
 099.011
Waugh, A. E.
 003.130
Wawrukiewicz, A. S.
 121.036
Wdowczyk, J.
 143.064 .066
Weaver, T.
 125.069
Weaver, T. A.
 061.056
 066.110

Weaver, W. B.
 113.023
 131.066
Webb, D.
 074.100
Webb, D. J.
 081.002
Webber, J. C.
 122.024
 160.012
Webber, R. F.
 033.066
Webber, W. R.
 074.078
 078.007
 099.018
 143.008 .025 .054
Weber, J.
 033.019
Weber, L.
 094.333
Weber, R. L.
 003.131
Weber, S. E.
 079.100
Webrova, L.
 044.028
Webster, A. R.
 083.015
Webster, A. S.
 157.002
Webster, B. L.
 114.017
Webster Jr., W. J.
 077.061
 103.101
Wechsler, A.
 094.323
Wedel, B.
 014.009
 032.017
 079.100
 103.105
Weedman, D. W.
 158.017 .062 .089
Weekes, T. C.
 066.107
Wefer, F.
 077.074
Wefer, F. L.
 077.080
Wegner, G.
 126.001
Wehinger, P. A.
 103.101 .105
Wehlau, A.
 123.011 .039
 154.012
Wehlau, W. H.
 008.072
Wehrenberg, P. J.
 106.033
Weichert, D. H.
 066.147
Weigert, A.
 065.015 .072
Weigt, G. I.
 034.043
Weihaupt, J. G.
 097.062

Weiler, E. J.
   121.037
Weiler, K. W.
   066.002
   134.005
Weill, G.
   084.007
Weinke, H. H.
   105.155
Weinstein, A.
   114.112
Weinstein, D. H.
   162.003
Weis, E. W.
   118.014
   121.071
Weisheit, J. C.
   022.072
   131.135 .904
Weiss, K.
   074.025
Weiss, W. W.
   034.034
   116.006
Weisskopf, M. C.
   132.033
Weissman, P. R.
   042.043
Welch, W. J.
   141.603
Weller, C. S.
   082.065 .066
   155.049
Weller, W.
   114.005
Wellington, K. J.
   034.002
   066.002
Wells, M. B.
   082.070
Wells, W. C.
   103.101
Wendker, H. J.
   131.545
Wentzel, D. G.
   141.337
Wenzel, K.-P.
   078.031 .032
   084.280 .285
Wenzel, W.
   122.041
   133.019
Werner, M.
   123.046
Werner, M. W.
   131.001 .068
Werner, W.
   034.001
Werntz, C.
   080.055
Wesseling, K. H.
   033.047
Wesselink, A. J.
   112.022
   123.021
Wesson, P. S.
   131.015
   151.006 .039
West, M.
   097.004

West, R. M.
   008.056
   032.048
   041.023
Westerhout, G.
   031.029
Westerlund, B. E.
   082.084
   159.002
Westervelt, P. J.
   066.051
Westfall, J. E.
   094.384 .399
   100.012
Westphal, J. A.
   099.042
Wetherell, W. B.
   032.901
Wetherill, G. W.
   003.006
   042.043
   094.118
   098.003
   105.085 .105
Whalen, J. A.
   084.023
Whalley, M. R.
   143.064
Whang, Y. C.
   092.036
   093.023
Wheaton, W. A.
   142.006 .055 .078 .085
Wheeler, J. A.
   012.008
Wheeler, J. C.
   061.024
   065.018
   125.003
Whelan, J.
   119.004
Whelan, J. A. J.
   122.102
   142.014
Whipple, F. L.
   013.016
   102.007
Whitaker, E. A.
   007.000
   094.321
   101.003
White, B.
   097.018
White, G.
   141.340
White, J.
   141.340
White, J. A.
   022.023
White, O. R.
   072.027
White, R. E.
   131.037 .047
White, R. S.
   034.066
   062.083
   084.404
   099.058
White III, K. P.
   077.077

Whitehead, A. B.
   097.070
Whitehead, D. H.
   010.008
Whiteoak, J. B.
   131.050
   158.024
Whiteside, D. T.
   003.132
   004.005
Whitford-Stark, J. L.
   094.147 .378
Whitham, P. S.
   141.307
Whiting, E. E.
   022.015
Whitlock, R. R.
   062.071
Whitney, C.
   063.023
Whitrow, G. J.
   003.133
Whitten, R. C.
   093.004
   097.056 .059
Whyte, D. A.
   033.028
Wibberenz, G.
   078.037
Wichtl, M.
   105.151
Wick, G. L.
   105.010
Wickersham Jr., A. F.
   084.020
Wickramasinghe, D. T.
   113.011
   119.004
   142.010 .020 .024 .025
   .103 .106
Wickramasinghe, N. C.
   131.016 .076
Wickwar, V. B.
   082.081
Widing, K. G.
   076.036
Wiedemann, D.
   162.005
Wiedemann, E.
   010.017 .025
   011.024
   013.005
   031.012 .045
   066.012
   099.032
   103.001 .101
Wiedemann, H. G.
   094.400
Wiehr, E.
   032.022
   034.036
Wiele, V. D.
   118.022
Wielebinski, R.
   141.314
Wielen, R.
   122.017
   151.022 .053
Wiemer, W.
   113.040

Wiesmann, H.
 094.347
Wieth-Knudsen, N. P.
 103.105
Wijnbergen, J. J.
 131.547
Wilcox, J. M.
 071.065
 074.029 .109
 085.016
Wild, J. P.
 008.124
Wild, P.
 007.000
 125.031 .112
Wild, P. A. T.
 096.007
Wildey, R. L.
 097.038
Wilhelmsson, H.
 022.017
Wilkening, L. L.
 105.110
Wilkins, G. A.
 091.006 .028
Wilkinson, A.
 141.054
 155.041
 156.003
Wilkinson, P. N.
 141.031
Will, C. M.
 066.049 .104
Williamon, R. M.
 121.066
Williams, C. A.
 042.014
Williams, D. A.
 131.079
Williams, D. J.
 084.290
Williams, E. A.
 097.042
Williams, I. P.
 092.018
Williams, J. G.
 141.096
Williams, P. M.
 064.035
 114.129
Williams, R. E.
 141.123
Williams, R. J.
 105.089 .100
Williams, W. L.
 126.014
Williamson, A.
 099.012
Williamson, I. P.
 131.022
Williamson, M. R.
 081.030 .031
Williamson, R. G.
 052.046
Willis, D. M.
 078.028
 084.260 .261 .262
Wills, B. J.
 141.081 .097 .121
Wills, D.
 141.080 .081 .097 .115

Wills, D.
 141.121
Wilson, A. J.
 131.107
Wilson, A. M.
 064.019
 071.001
Wilson, A. S.
 134.001 .004 .007
Wilson, J. G.
 143.062
Wilson, J. R.
 066.134
Wilson, L.
 092.004
Wilson, L. W.
 022.026
Wilson, M. D.
 078.015
Wilson, P. R.
 062.010
 072.023
Wilson, R.
 032.034 .902
Wilson, R. C.
 097.078
Wilson, R. E.
 121.054
Wilson, R. N.
 031.060
 032.044
Wilson, R. W.
 132.029 .034
Wilson, T. L.
 131.513
Wilson, W. E.
 157.004
Wilson, W. J.
 077.078
 114.132
 131.001 .113
 155.068
Wimberly, R. N.
 094.115
Winckler, J. R.
 084.246 .292
Windsor, R. A.
 062.056
Winge Jr., C. R.
 074.017
Winiarski, M.
 103.105
Winkler, J.
 094.323
Winkler, P. F.
 125.070
Winnberg, A.
 114.018
 131.019 .092
Winnenburg, W.
 014.003
Winnewisser, G.
 131.011
Winningham, J. D.
 084.021
Winterberg, F.
 066.126
Winterbottom, A. N.
 054.005
Winzer, J. E.
 113.018

Winzer, J. E.
 114.060
 125.100
Wirth, P. B.
 004.076
Wischniewsky, M.
 103.010 .105
Withbroe, G. L.
 071.008
 073.016
 076.034 .037
Witkowski, J. M.
 007.000
Witt, A. N.
 008.129
Witten, L.
 066.073
Witten Jr., T. A.
 065.054
Wittmann, A.
 071.021
 072.030 .051
 080.031
 092.005 .009
Wittwer, H.
 121.090
Witzel, A.
 141.004
Wlerick, G.
 122.901
 158.119 .302
Woehl, H.
 080.020
 092.009
Wofsy, S. C.
 022.006
Wohlert, W.
 014.001
Wohn, J.
 046.026
Wolczek, O.
 010.014
Wolf, B.
 064.052
Wolf, E.
 003.135
Wolf, G. W.
 121.024
Wolf, K.
 155.029
Wolf, R. A.
 084.263
 097.048
Wolfe, A. M.
 158.031
Wolfe, E. W.
 094.181
Wolfe, J. H.
 074.077
 099.014
Wolfe, J. L.
 021.022
Wolfendale, A. W.
 143.064 .065
Wolff, P.
 041.018
Wolff, R. J.
 119.009 .010
Wolff, R. S.
 076.002
 132.033

Wolff, S. C.
  114.043 .156
  116.001 .007
  119.010 .011
  121.001
Wolfson, C. J.
  076.038
Wollast, R.
  105.048
Wollenhaupt, W. R.
  094.115 .172
Wollman, E. R.
  141.602
Wolstencroft, R. D.
  116.003
  126.008 .010
Wolterbeek, J.
  009.008
Woltjer, L.
  012.015
  125.056
  141.024
Wong, C.-Y.
  042.087
Wong, S. K.
  099.204
Woo, Chia-Wei
  065.122
Wood, F. B.
  008.049
  114.142
Wood, G. E.
  092.040
Wood, H.
  007.000
  041.053
Wood, J. A.
  094.136 .345
Wood, L.
  125.069
Wood, N.
  103.011
Wood, P. J.
  033.082
Wood, P. R.
  064.017
  065.127
Wood, R.
  125.101 .112
Woodgate, B. E.
  132.011 .032
Woodman, J. H.
  101.006
Woodrow, J. E. J.
  064.057
Woods, J. A.
  003.134
Woodtli, R.
  105.163
Woodward, P. R.
  151.048
Woolf, N. J.
  114.040
  141.612
  158.034
Woolum, D. S.
  094.303
Woosley, S. E.
  065.110
Wootten, H. A.
  114.150

Worden, S. P.
  122.102
  158.042
Worley, C. E.
  005.016
  118.016
Worley, R.
  064.020
Worrall, G.
  064.019
Woszczyk, A.
  006.000
  093.008
Wramdemark, S.
  113.025
Wray, A. M.
  143.062
Wray, J. D.
  114.138
  142.074
Wright, A. E.
  158.059 .060
Wright, E. L.
  131.528
Wright, F. W.
  105.017
Wright, G. A. E.
  065.082
Wright, M. C. H.
  131.516 .517
Wrixon, G. T.
  155.027 .054
Wroblewski, H.
  103.010 .105
Wroe, H.
  032.034 .902
Wu, C. S.
  074.033
  099.059
Wu, C.-C.
  125.102
Wu, F.-M.
  099.064
Wulf-Mathies, C.
  119.013
  132.027
Wurm, K.
  102.005
  103.101
Wyant, J. C.
  034.040
Wyckoff, S.
  103.101 .105
  125.112
Wyller, A. A.
  034.064
Wynn-Williams, C. G.
  131.001 .523 .524 .527
Wynne, C. G.
  032.009

Xanthakis, J.
  003.009
Xu, H.
  045.016

Yabsley, D. E.
  008.124

Yabushita, S.
  065.035
Yahil, A.
  158.112
Yakimova, N. N.
  122.032
Yakovkin, N. A.
  073.004 .023
Yakovlev, O. I.
  093.006
  106.016
Yakovlev, V. A.
  097.033
Yamaguti, M.
  096.038
Yamashita, K.
  142.068
Yamashita, Y.
  112.001
Yan, S.
  044.020
Yanchukovskij, V. L.
  143.006
Yang, C.
  061.057
Yang, K. S.
  122.024
Yaniv, A.
  022.901
Yankavtsev, M. V.
  141.056 .077
Yankovich, I. I.
  122.020
Yankulova, I. M.
  158.093 .128
Yanovitskaya, G. T.
  094.102
Yanovitskij, E. G.
  091.015
Yarskaya, M. V.
  052.015
Yasevich, B. V.
  044.016
Yashkin, S. N.
  042.071
Yasinskaya, A. A.
  105.123
Yasnov, L. V.
  077.043 .046 .049
Yasuda, H.
  031.078
  112.025
Yates, G. K.
  078.030 .034
Yatsenko, S. P.
  082.038
Yatskiv, Ya. S.
  041.037
Yavnel', A. A.
  105.115
Ye, S.
  044.020
Yeates, C. M.
  092.035
Yegorov, I. V.
  094.182
Yeh, K. C.
  083.015
Yen, J. L.
  021.019

AUTHOR INDEX - VOL. 11

Yen, T.
　004.068
Yener, Y.
　063.017
Yeomans, D. K.
　103.102 .105 .114 .119
Yerbury, M. J.
　033.040
　141.100 .103
Yeung, P. S.
　063.032 .040
Yokoo, H.
　142.111
Yokoyama, Y.
　105.006
　143.019
York, D. G.
　022.028
　131.116
Yoshimura, H.
　062.011
Yoss, K. M.
　115.007
　122.024
You, J.
　074.112
Young, A.
　119.007
　121.038
Young, A. T.
　031.024 .050
　071.011
　093.005 .008
Young, J. M.
　113.037
Young, L. G.
　093.008
Young, R. S.
　015.020
Younger, P.
　114.170
　124.104
Younkin, R. L.
　100.205
Yourgrau, W.
　066.087
Yousef, S.
　077.066
Yu, C. P.
　062.079
Yu, M. Y.
　083.005
Yuasa, M.
　012.022
　098.026 .901
Yudin, O. I.
　033.053
Yudina, I. V.
　071.010 .040
Yueh, T. Y.
　063.038
Yuen, P. C.
　083.008
Yuk Ling Yung
　093.041
Yukhimuk, A. K.
　141.331
Yukina, L. V.
　105.132
Yumi, S.
　041.064

Yung, Y. L.
　099.210
Yungelson, L. R.
　117.010 .011 .012 .013
　　.014
Yura, H. T.
　031.043
Yurovskaya, L. I.
　072.007
Yurovskij, Yu. F.
　079.102
Yusifov, I. M.
　141.344
Yutani, M.
　122.076
Yuzefovich, A. P.
　094.160

Zabalueva, E. V.
　094.385
Zagar, F.
　011.036
Zagars, J.
　052.040
　081.027
Zagars, N.
　081.027
Zaidins, C. S.
　061.040
Zajdler, L.
　007.000
　096.026
Zajtsev, V. V.
　077.013
Zajtsev, Yu. I.
　051.005
Zajtseva, G. V.
　154.005
Zakharov, V. D.
　003.136
Zakirov, I. V.
　091.017
Zakirov, M. M.
　103.101
Zandanov, V. G.
　077.042
Zane, R.
　034.067
Zanoni, C. A.
　032.023
Zappala, R. A.
　075.004
Zappala, R. R.
　113.002
　114.081
Zappala, V.
　098.030
　103.101 .124
Zarate, N.
　122.078
Zarnecki, J. C.
　125.049
　160.017
Zasetskij, V. V.
　094.127 .173 .302
Zaslavskaya, N. I.
　105.134
Zasov, A. V.
　131.102
　158.008

Zastenker, G. N.
　084.266
Zaumen, W. T.
　066.009 .109
　073.053
　076.026
　142.114
Zavrazhina, N. M.
　052.017
Zavilski, M.
　095.009
　096.024 .025
Zdarsky, F.
　121.021
Zebera, K.
　105.088
Zech, G.
　002.021
Zeilik, M.
　103.101
Zeilik II, M.
　103.101
Zel'dovich, M. A.
　034.005
Zel'dovich, Ya. B.
　033.023
　066.004
　162.034
Zelenka, A.
　080.063
Zenkert, A.
　014.005
Zentsova, A. S.
　063.035
Zepe, M.
　005.021
Zertsalov, A. A.
　078.003
　084.212
Zgonyajko, N. S.
　120.002
Zhandarov, M. E.
　034.012
Zhang, J.
　103.136
Zharikov, V. A.
　091.017
Zharkov, V. N.
　099.004 .029 .030 .075
　　.076
Zhdanov, M. S.
　084.268
Zheleznyakov, V. V.
　062.023
　077.065
Zhernokleev, N. L.
　032.006
Zhestkov, A. G.
　082.118
Zhivago, A. V.
　097.065
Zhivora, P. S.
　141.056 .077
Zhongolovich, I. D.
　047.019
Zhouck, I. N.
　141.076
Zhuchenko, Yu. M.
　083.016
Zhurkina, S. N.
　122.056

Zielenbach, J. W.
  052.018
Zieleniewski, J.
  004.073
Zielinski, J. B.
  081.012
Ziglin, S. L.
  042.030
Zigunov, V. N.
  054.003
Zikides, M.
  035.006
  042.079
Zill, B.
  009.023
Zimmer, H.
  092.022
  093.019
Zimmerman, B. A.
  065.023
Zimmerman, P. D.
  105.085
Zinnow, K. P.
  002.029
Ziolkowski, J.
  121.083
Ziolkowski, K.
  004.072
  010.017
  021.026
  117.001

Zirin, H.
  073.090
  075.008
Zirker, J. B.
  080.026
Zirm, K.
  078.035
Zisk, S. H.
  094.157 .321 .367 .368
        .370
Zisman, G. A.
  080.039
Zlotin, G. N.
  051.016
Zlotnik, E. Ya.
  062.032
  077.065
Zohar, S.
  092.011
Zolova, O. F.
  104.040 .041 .045
Zonn, W.
  010.017
Zotkin, I. T.
  006.000
Zribi, G.
  122.018

Zubieva, T. G.
  034.005
Zucht, D.
  079.100
Zuckerman, B.
  131.009 .062 .068 .544
Zuev, N. G.
  032.005
  041.003 .015
Zuev, V. S.
  032.006
Zujkov, V. N.
  073.011
Zulevic, D. J.
  118.029
Zverev, M. S.
  041.029 .037 .038 .049
Zvereva, A. M.
  082.094
Zvonarev, K. A.
  005.023
Zwicky, F.
  065.056
  125.029
Zygielbaum, A. I.
  092.040
  093.022

Subject Index

A Stars
  Kinematics
    155.031
A Stars
  MK Types
    155.021
A Stars
  Peculiar
    064.008 .041
    065.050
    113.031
    114.006 .025 .026
        .043 .051 .067
        .068 .078 .082
        .083 .097 .111
        .156 .171
    116.001 .002 .005
        .010 .012 .013
    119.009 .011
A Stars
  Photometry
    155.021
A Stars
  Space Density
    155.017
A Stars
  Spectrophotometry
    114.123 .124
A Stars
  UV Spectra
    114.114 .123 .124
Absolute Magnitudes
    115.000
Absolute Magnitudes
  Calibration
    115.002 .007
Absolute Magnitudes
  Carbon Stars
    115.012
Absolute Magnitudes
  F Supergiants
    115.005
Absolute Magnitudes
  Quasars
    141.082

Absolute Magnitudes
  RR Lyrae Stars
    122.100
Absolute Magnitudes
  Visual Binaries
    118.025
Absolute Magnitudes
  White Dwarfs
    065.089
Absorption
  Interstellar Matter
    131.513
    132.003
    141.325
    158.041
Absorption Coefficient
    061.002
Absorption Coefficient
  Solar Atmosphere
    080.063
Absorption Lines
  Emission-Line Stars
    114.003
Absorption Lines
  Peculiar A Stars
    114.171
Absorption Lines
  Solar Spectrum
    073.051
Accretion
  Neutron Stars
    065.045 .075 .099
Accretion
  White Dwarfs
    065.099
Accretion
  X-Ray Sources
    065.075
Achondrites
    105.164
Airglow
    082.000
Albedo
  Jupiter
    099.001 .035

Albedo
  Mars
    097.075
Albedo
  Mercury
    092.004
Albedo
  Moon
    094.140 .146 .314
Albedo
  Neptune
    099.001
Albedo
  Saturn
    091.015
Albedo
  Saturn Satellites
    100.205 .208
Albedo
  Uranus
    091.015
    099.001
Albedo
  Venus
    091.015
Alfven Waves
  Interplanetary Matter
    062.013
    074.007
Alfven Waves
  Plasma
    062.046 .051
Alfven Waves
  Solar Atmosphere
    080.038
Alfven Waves
  Solar Wind
    074.007 .056 .058
Algol Systems
    121.031
Almanacs
    047.000
Ammonia
  Jupiter Atmosphere
    099.007 .035 .066
        .067

SUBJECT INDEX - VOL. 11

Ammonia
  Venus Atmosphere
    093.035
Amplifiers
  033.051
Andromeda Nebula
  113.016
Andromeda Nebula
  Globular Clusters
    158.101
Andromeda Nebula
  Mass
    158.101
Andromeda Nebula
  Novae
    124.005
Andromeda Nebula
  Stellar Populations
    158.105
Andromeda Nebula
  X Rays
    158.102
Andromeda Nebula
  21 cm Radiation
    158.009
Antimatter
  065.022
  105.091
  162.004 .008
Artificial Satellites
  054.000
Artificial Satellites
  Observations
    055.000
Associations
  010.000
Associations Stellar
  152.000
Asteroid Belt
  098.006
Asteroids
  Collisions
    098.003
Asteroids
  Fragmentation
    098.016
Asteroids
  Origin
    098.002 .025
Astrodynamics
  052.000
Astrographs
  032.007 .033
Astrometry
  012.014
  041.000
Astronomical Accessories
  034.000
Astronomical Constants
  043.000
Astronomical Geodesy
  046.000
Astronomical Instruments
  032.000
Astronomical Optics
  031.000
Atlases
  041.000
Atmosphere
  Earth
    082.000

Atmosphere
  Moon
    094.149 .150
Atmospheres
  Be Stars
    064.012
Atmospheres
  Carbon Stars
    064.011
Atmospheres
  Comets
    102.002
Atmospheres
  Early Type Stars
    064.002 .016
    114.007 .050
Atmospheres
  Eclipsing Variables
    121.073
Atmospheres
  F Dwarfs
    064.044
Atmospheres
  G Giants
    064.044
Atmospheres
  K Giants
    064.044
Atmospheres
  Late Type Stars
    064.003 .011 .026
    114.032
Atmospheres
  M Stars
    064.021
    114.172
Atmospheres
  Magnetic Stars
    116.006
Atmospheres
  Massive Stars
    065.021
Atmospheres
  Metal-Poor Stars
    114.105
Atmospheres
  Metallic-Line Stars
    114.131
Atmospheres
  OB Stars
    064.007
Atmospheres
  Peculiar A Stars
    064.041
Atmospheres
  Pulsars
    141.331
Atmospheres
  Quasi-Stellar Objects
    064.025
Atmospheres
  Supergiants
    064.039
Atmospheres
  Wolf-Rayet Stars
    064.030 .037
Aurorae
  084.000
Automation
  031.000

B Stars
  Bolometric Corrections
    115.013
B Stars
  Colors
    155.078
B Stars
  Distribution
    155.033
B Stars
  Line Profiles
    114.127
B Stars
  Peculiar
    065.050
    115.016
    116.007
B Stars
  Photometry
    113.026
B Stars
  Spectra
    114.021 .037
    155.079
B Stars
  Temperatures
    114.092 .117
B Stars
  UV Spectra
    114.116
Background Radiation
  066.000
  162.022
Background Radiation
  Cosmic
    061.032
    066.003 .006 .136
    131.133
    162.011
Background Radiation
  Galactic
    155.071
Barium Stars
  065.047
  114.013 .064 .160
  115.012
Barnard Objects
  131.048
Be Stars
  Atmospheres
    064.012
Be Stars
  Envelopes
    064.010 .040
    114.027
Be Stars
  Infrared Excesses
    114.100
Be Stars
  Line Profiles
    114.077 .104
Be Stars
  Rotation
    114.077
Be Stars
  Spectra
    114.027 .080
Be Stars
  UV Spectra
    114.004

Beta Cephei Stars
　065.093
　096.007
　122.009 .088 .126
Beta Cephei Stars
　Photometry
　122.043
Beta CMa Variables
　122.012
Beta Lyrae Stars
　121.018 .021 .022
　　　.024 .029 .030
　　　.054 .065 .076
　　　.086
Beta Lyrae Stars
　Photometry
　121.097
Beta Lyrae Stars
　UV Spectra
　121.059
Bibliographical Publ
　002.000
Binaries
　117.000
Binaries
　Black Holes
　117.029
　121.045
　142.003
Binaries
　Close Binaries
　064.020
　117.001 .002 .004
　　　.005 .006 .010
　　　.011 .012 .013
　　　.014 .016 .019
　　　.021 .022 .027
　121.015
Binaries
　Evolution
　117.007 .032 .039
　121.014
Binaries
　Gaseous Streams
　117.018
Binaries
　Infrared Stars
　141.611
Binaries
　Magnetic Fields
　117.015 .025 .032
Binaries
　Mass Loss
　117.007
　121.005
Binaries
　Neutron Stars
　065.062
Binaries
　Polarization
　117.018
Binaries
　Pulsars
　141.313
Binaries
　Quasi-Stellar Objects
　141.024
Binaries
　Radio Sources
　141.016 .035

Binaries
　Rotation
　117.026
Binaries
　Supernovae
　117.038
　125.007 .014
Binaries
　White Dwarfs
　121.005
　122.064
　126.018
Binaries
　Wolf-Rayet Stars
　114.093
　117.017
　119.001
Binaries
　X-Ray Sources
　114.081
　117.034
　121.001 .101
　122.102
　125.007
　141.098
　142.003 .006 .007
　　　.009 .010 .016
　　　.017 .018 .019
　　　.022 .025 .028
　　　.030 .034 .049
　　　.050 .054 .059
　　　.063 .069 .071
　　　.072 .073 .115
　　　.116 .117
Biography
　005.000
BL Lacertae
　122.103
　141.059
　158.056 .301 .303
BL Lacertae
　Distance
　158.302
Black Holes
　012.007
　066.001 .021 .023
　　　.024 .027 .033
　　　.066 .067 .076
　　　.078 .079 .088
　　　.090 .091 .092
　　　.116 .117 .118
　　　.120 .123 .124
　　　.125 .127 .128
　　　.130 .135 .139
　　　.142 .144
　122.103
Black Holes
　Binaries
　117.029
　121.045
　142.003
Black Holes
　Close Binaries
　117.004
Black Holes
　Gamma Rays
　061.056
Black Holes
　X-Ray Sources
　113.008
　142.010 .020 .025

Black Holes
　X-Ray Sources
　142.046 .048 .049
　　　.069 .101
Blue Objects
　113.009 .016
　141.097
Bolometric Corrections
　B Stars
　115.013
Bolometric Corrections
　Early Type Stars
　114.125
Books
　003.000

C-M Diagrams
　Galactic Clusters
　153.020 .023
C-M Diagrams
　Globular Clusters
　154.008 .012 .015
　　　.030
C-M Diagrams
　Star Clusters
　159.001
C-S Stars
　Spectra
　114.042
Calendars
　047.000
Canonical
　Transformations
　042.016
Carbon Dioxide
　Mars Atmosphere
　097.016 .017 .066
Carbon Dioxide
　Planetary Atmospheres
　091.017
Carbon Dioxide
　Venus Atmosphere
　093.041
Carbon Monoxide
　Galactic Center
　155.054
Carbon Stars
　065.047
Carbon Stars
　Absolute Magnitudes
　115.012
Carbon Stars
　Atmospheres
　064.011
Carbon Stars
　Catalogues
　114.086
Carbon Stars
　Colors
　114.098
Carbon Stars
　Distribution
　155.038
Carbon Stars
　Evolution
　065.024 .100
Carbon Stars
　Infrared Photometry
　113.030

Carbon Stars
  Infrared Spectra
    114.069
Carbon Stars
  Lithium Abundance
    114.062
Carbon Stars
  Photometry
    113.028 .029 .032
    114.159
Carbon Stars
  Radial Velocities
    112.001
Carbon Stars
  Spectra
    114.002 .095 .099
        .153
Carbon Stars
  Spectral Types
    114.098
Carina Nebula
    131.020
    132.003
Cataclysmic Variables
    121.064
    122.037
Cataclysmic Variables
  X Rays
    122.063
Catalogues
  Carbon Stars
    114.086
Catalogues
  Galaxies
    158.057
Catalogues
  Metallic-Line Stars
    114.097
Catalogues
  Peculiar A Stars
    114.097
Catalogues
  Proper Motions
    041.000
    112.006 .007 .008
        .014 .015
Catalogues
  Radial Velocities
    112.002 .017
Catalogues
  Radio Sources
    141.086
Catalogues
  Star Positions
    041.000
Catalogues
  X-Ray Sources
    142.035 .043
Celestial Mechanics
    012.022
    042.000
Cepheids
    065.007
    122.001 .016 .036
        .039 .052 .053
        .055 .056 .057
        .058 .059 .061
        .087 .118 .119
        .123

Cepheids
  Ages
    155.016
Cepheids
  Colors
    122.007 .033
Cepheids
  Distances
    122.025
Cepheids
  Globular Clusters
    122.085
Cepheids
  Long Period
    122.091
Cepheids
  Luminosities
    122.005 .032 .049
Cepheids
  Lunar Occultations
    122.029
Cepheids
  Metal Abundances
    122.005
Cepheids
  Periods
    122.029 .030
Cepheids
  Photometry
    122.002
Cepheids
  Space Distribution
    155.015
Cepheids
  Space Motions
    155.015
Cepheids
  Space Velocities
    122.017
Cepheids
  Spectra
    065.061
CH
  Interstellar Matter
    131.105
CH Stars
  Models
    064.005
Chemical Composition
  Cosmic Rays
    143.001
Chemical Composition
  Earth
    081.000
Chemical Composition
  Jupiter
    099.076
Chemical Composition
  Jupiter Atmosphere
    099.045
Chemical Composition
  Moon
    094.175 .177
Chemical Composition
  Saturn
    099.076
Chemical Composition
  Saturn Atmosphere
    099.045

Chemical Composition
  Solar System
    107.014
Chemical Composition
  Stellar Spectra
    114.074
Chemical Composition
  Supergiants
    065.083
Chemical Composition
  White Dwarfs
    126.004
Chondrites
    105.002 .004 .005
        .007 .009 .012
        .014 .090 .095
        .097 .099 .100
        .102 .110 .113
        .114 .116 .118
        .120 .126 .136
        .137 .147 .152
        .154
Chromosphere
  Acoustic-Gravity Waves
    073.040
Chromosphere
  Active Regions
    073.038
Chromosphere
  Densities
    072.005
Chromosphere
  Emission Lines
    073.092
Chromosphere
  Heating
    073.002
    074.052
Chromosphere
  Line Formation
    073.026
Chromosphere
  Line Profiles
    073.028
Chromosphere
  Magnetic Fields
    072.020
Chromosphere
  Models
    073.021 .037
Chromosphere
  Mottles
    073.039
Chromosphere
  Oscillations
    077.025
Chromosphere
  Solar
    073.000
Chromosphere
  Spectra
    073.043 .050
Chromosphere
  Spectrophotometry
    073.012
Chromosphere
  Temperatures
    072.005
Chromosphere-Corona
  Transition Region
    074.010

Chromospheres
  Stellar
    064.031 .047 .048
    .050
Chronology
  004.000
Circumstellar Matter
  064.032
  065.013
  114.028 .029 .031
    .032 .052 .069
    .100 .132 .165
  122.019
  141.605 .613
Circumstellar Matter
  Radiative Transfer
    131.028
Circumstellar Shells
  Radiative Transfer
    064.061
Clocks
  035.000
Close Binaries
  117.002
  121.015
Close Binaries
  Black Holes
    117.004
Close Binaries
  Evolution
    117.010 .011 .012
    .013 .014
Close Binaries
  Gas Clouds
    117.001
Close Binaries
  Gaseous Streams
    117.021
Close Binaries
  Light Curves
    064.020
    117.019 .027
Close Binaries
  Mass Loss
    117.006
Close Binaries
  Models
    117.022
Close Binaries
  Neutron Stars
    117.004
Close Binaries
  Reflection Effects
    117.016
Close Binaries
  Secular Stability
    117.005
Clouds
  Interstellar Matter
    131.065 .073
Clouds
  Jupiter Atmosphere
    099.028 .048 .061
Clouds
  Mars Atmosphere
    097.024 .038
Clouds
  Venus Atmosphere
    093.005 .042
    099.028

Clusters
  Galactic
    153.000
Clusters
  Globular
    154.000
Clusters
  Moving Clusters
    153.000
Clusters
  Open Clusters
    153.000
Clusters of Galaxies
  160.000
Clusters of Galaxies
  Coma
    131.087
    158.043
    160.001 .007 .014
    .029
Clusters of Galaxies
  Distribution
    160.026
Clusters of Galaxies
  Formation
    162.012
Clusters of Galaxies
  Galaxy Distribution
    160.008
Clusters of Galaxies
  Intergalactic Matter
    160.013
Clusters of Galaxies
  Magnetic Fields
    160.005
Clusters of Galaxies
  Protoclusters
    151.011
Clusters of Galaxies
  Quasars
    141.053
Clusters of Galaxies
  Radial Velocities
    160.014
Clusters of Galaxies
  Radio Galaxies
    160.012
Clusters of Galaxies
  Radio Radiation
    160.001 .004 .005
    .016 .019 .025
Clusters of Galaxies
  Radio Sources
    141.012 .028
Clusters of Galaxies
  Redshifts
    160.007 .010 .014
    .015
Clusters of Galaxies
  Sizes
    160.022
Clusters of Galaxies
  Virgo
    113.009
Clusters of Galaxies
  X Rays
    160.004 .006 .016
    .017
Clusters of Galaxies
  X-Ray Sources
    142.031 .114

Clusters of Galaxies
  X-Ray Sources
    160.005
Clusters of Galaxies
  21 cm Radiation
    158.001
    160.013
Coalsack
  131.066
Coalsack
  Flare Stars
    122.116
Collapse
  Gravitation
    012.008
    066.053 .105 .117
    .119 .141
Collapse
  Supermassive Stars
    065.068
Collapse
  White Dwarfs
    125.003
Collapsing Stars
  Accretion
    066.030
Colloquia Proceedings
  012.000
Colloquia Reports
  011.000
Color Indices
  Stellar Atmospheres
    113.017
Colors
  113.000
Colors
  B Stars
    155.078
Colors
  Carbon Stars
    114.098
Colors
  Cepheids
    122.007 .033
Colors
  Galaxies
    158.042 .106
Colors
  Globular Clusters
    154.007
Colors
  K Stars
    155.078
Colors
  Minor Planets
    098.010
Colors
  Quasi-Stellar Objects
    141.084
Colors
  Seyfert Galaxies
    158.303
Coma
  Clusters of Galaxies
    131.087
    158.043
    160.001 .007 .014
    .029
Comet 1844 I
  De Vico-Swift
    103.125

Comet 1847 V
  Brorsen-Metcalf
  103.117
Comet 1867 I
  Stephan-Oterma
  103.115
Comet 1906 IV Kopff
  103.119
Comet 1910 II Halley
  103.122
Comet 1916 I Taylor
  103.123
Comet 1917 I Mellish
  103.103
Comet 1919 III
  Brorsen-Metcalf
  103.117
Comet 1925 II
  Schwassmann-Wachmann 1
  103.104
Comet 1930 VI
  Schwassmann-Wachmann 3
  103.134
Comet 1942 IX
  Stephan-Oterma
  103.115
Comet 1944 III Du Toit 1
  103.106
Comet 1965 I
  Tsuchinshan 1
  103.136
Comet 1965 II
  Tsuchinshan 2
  103.137
Comet 1967 VI Arend
  103.133
Comet 1967 IX Finlay
  103.114
Comet 1967 XII Wolf 1
  103.131
Comet 1968 I Ikeya-Seki
  103.138
Comet 1968 VI Honda
  103.120
Comet 1968c Honda
  103.120
Comet 1969i Bennett
  103.100
Comet 1969 IX
  Tago-Sato-Kosaka
  103.121
Comet 1970 II Bennett
  103.100
Comet 1971 II Encke
  103.102
Comet 1972 IV Neujmin 3
  103.130
Comet 1972 VIII
  Heck-Sause
  103.124
Comet 1972 XII Araya
  103.129
Comet 1972d
  Giacobini-Zinner
  103.118
Comet 1972h Sandage
  103.107
Comet 1972j Kojima
  103.128
Comet 1972l Araya
  103.129

Comet 1973a Heck-Sause
  103.124
Comet 1973b
  Tuttle-Giacobini-Kresak
  103.126
Comet 1973e Kohoutek
  103.132
Comet 1973f Kohoutek
  103.101
Comet 1973g Reinmuth 2
  103.113
Comet 1973i Clark
  103.135
Comet 1973j Brooks 2
  103.109
Comet 1973k Sandage
  103.127
Comet 1973l
  Schwassmann-Wachmann 2
  103.112
Comet 1973n Gehrels 2
  103.108
Comet 1973o Gibson
  103.111
Comet 1974a Forbes
  103.110
Comet 1974b Bradfield
  103.105
Comet 1974c Lovas
  103.116
Cometary Nebulae
  132.023
Comets
  102.000
Comets
  Atmospheres
  102.002
Comets
  Brightnesses
  102.013 .019
Comets
  Listed Objects
  103.000
Comets
  Long Period
  102.023
Comets
  Magnetic Fields
  102.014
Comets
  Motion
  102.010 .025
Comets
  Nuclei
  102.018
Comets
  Photometry
  102.022
Comets
  Short Period
  102.013
Comets
  Solar Wind
  102.015
Comets
  Tails
  102.004 .005 .008
Computing
  021.000
Congress Proceedings
  012.000

Congress Reports
  011.000
Convection
  Main-Sequence Stars
  065.028
Convection
  Solar Interior
  062.011
  080.015 .035 .036
  .037
Convection
  Stellar Atmospheres
  064.018 .038 .043
Convection
  Stellar Interiors
  065.030 .060 .088
Cool Stars
  Infrared Photometry
  113.036
Cool Stars
  Polarization
  113.036
Coronographs
  032.003
  073.033
Cosmic Rays
  143.000
Cosmic Rays
  Chemical Composition
  143.001
Cosmic Rays
  Electrons
  143.002 .023
Cosmic Rays
  Element Abundances
  143.008 .014 .030
Cosmic Rays
  Energy Spectra
  143.064
Cosmic Rays
  Galactic Disk
  143.026
Cosmic Rays
  Galactic Distribution
  143.007
Cosmic Rays
  High Energy
  143.033
Cosmic Rays
  Interplanetary Space
  106.014
  143.003
Cosmic Rays
  Interstellar Clouds
  131.503
Cosmic Rays
  Interstellar Matter
  143.024
Cosmic Rays
  Interstellar Space
  143.013
Cosmic Rays
  Nuclei
  143.015 .025 .028
  .061
Cosmic Rays
  Origin
  143.065 .066
Cosmic Rays
  Positrons
  143.031

SUBJECT INDEX - VOL. 11

Cosmic Rays
  Propagation
    143.001 .007 .014
    .031 .035
Cosmic Rays
  Pulsars
    143.066
Cosmic Rays
  Solar Corona
    074.026
Cosmic Rays
  Solar Flares
    078.005 .045
Cosmic Rays
  Solar Modulation
    074.019
    143.010 .033
Cosmic Rays
  X-Ray Sources
    142.008
Cosmogony
  Planetary System
    107.000
Cosmological Models
  066.041 .130
  162.001 .002 .007
  .010 .013 .014
  .015 .017 .020
  .026 .034 .035
  .040 .041 .043
  .045 .046 .047
Cosmology
  162.000
Crab Nebula
  134.000
Crab Nebula
  Filaments
    134.004
Crab Nebula
  Gamma Rays
    134.006
    142.005
Crab Nebula
  Magnetic Fields
    134.002 .003
Crab Nebula
  Polarization
    134.001 .003
Crab Nebula
  Pulsar
    134.002 .006
    141.306 .326 .333
Crab Nebula
  Relativistic Electrons
    134.007
Crab Nebula
  X Rays
    142.033
Cygnus Loop
  Iron Abundance
    132.011
Cygnus Loop
  Spectrophotometry
    132.015
Cygnus Loop
  X Rays
    132.011

Dark Clouds
  131.053 .095

Dark Clouds
  21 cm Radiation
    131.064
Dark Nebulae
  131.080
Dark Nebulae
  Magellanic Clouds
    131.136
Data Processing
  021.000
Delta Scuti Stars
  122.023 .081
Delta Scuti Stars
  Photometry
    122.018
Diameter
  Sun
    080.031
Diameters
  Late Type Stars
    115.001
Diameters
  Stars
    115.000
Diffuse Nebulae
  Electron Temperatures
    132.014
Diffuse Nebulae
  Globules
    131.075
Distance
  BL Lacertae
    158.302
Distance
  Galactic Center
    155.042
Distances
  Cepheids
    122.025
Distances
  Galactic Clusters
    153.014
Distances
  Galaxies
    158.099
Distances
  Globular Clusters
    154.009
Distances
  Pulsars
    141.325
Distances
  Sunspots
    072.043
Distances
  Supernovae
    125.023
    158.099
Distances
  X-Ray Sources
    142.037 .112
Draconids
  104.006 .020
Dust
  H II Regions
    131.510
Dust
  Intergalactic Matter
    141.057
    161.001

Dust
  Interplanetary Matter
    106.037
Dust
  Interstellar Matter
    131.015 .056 .132
    .512
    151.060
Dust
  Mars Atmosphere
    097.006 .069
Dust
  Venus Atmosphere
    093.011
Dust Clouds
  Interstellar Matter
    131.542 .543
Dust Clouds
  Mars Atmosphere
    097.031 .032
Dust Clouds
  21 cm Radiation
    131.542
Dwarf Novae
  Photometry
    122.042
Dynamical Systems
  042.021
Dynamics
  Globular Clusters
    154.018
Dynamics
  Jupiter Atmosphere
    099.030 .065
Dynamics
  Solar Wind
    074.088
Dynamics
  Star Clusters
    151.020 .029 .038
Dynamics
  Stellar Systems
    042.050
    151.000

Early Type Stars
  Atmospheres
    064.002 .016
    114.007 .050
Early Type Stars
  Bolometric Corrections
    114.125
Early Type Stars
  CNO Anomalies
    064.007
Early Type Stars
  Emission Lines
    114.003
Early Type Stars
  Luminosities
    153.009
Early Type Stars
  Photometry
    113.025 .027 .039
Early Type Stars
  Polarization
    131.082
Early Type Stars
  Radial Velocities
    155.045

Early Type Stars
  Space Distribution
    155.045
Early Type Stars
  Spectra
    114.127
Early Type Stars
  UV Photometry
    113.006
Early Type Stars
  UV Spectra
    114.015 .110 .125
    .133
Earth
  Atmosphere
    082.000
Earth
  Chemical Composition
    081.000
Earth
  Figure
    081.000
Earth
  Gravity
    081.000
Earth
  Magnetic Field
    084.000
Earth
  Rotation
    044.000
Earth Atmosphere
  Density
    082.000
Earth Atmosphere
  Turbulence
    031.050
Earth Satellites
  Motion
    052.023
    054.010
Earth Satellites
  Orbits
    052.009 .012 .014
    .015 .020 .021
    .024 .031 .035
    .036 .037 .038
    .041 .045 .046
Earth-Moon System
  Evolution
    094.004
Earth-Moon System
  Mass
    091.010
Eclipses
  Galilean Satellites
    099.219
Eclipses
  Lunar
    095.000
Eclipses
  Solar
    079.000
Eclipsing Variables
    121.000
Eclipsing Variables
  Atmospheres
    121.073
Eclipsing Variables
  B Type
    121.041

Eclipsing Variables
  Calcium Emission
    119.003
Eclipsing Variables
  Early Type
    121.003
Eclipsing Variables
  Element Abundances
    121.023
Eclipsing Variables
  Galactic Clusters
    121.017
Eclipsing Variables
  Infrared Excesses
    121.002
Eclipsing Variables
  Infrared Spectra
    121.076
Eclipsing Variables
  Light Curves
    121.009 .016 .028
    .033 .039 .048
    .050 .053 .057
    .072 .073 .088
Eclipsing Variables
  Masses
    121.041
Eclipsing Variables
  Models
    121.067 .072
Eclipsing Variables
  Orbits
    121.003 .038 .047
    .087
Eclipsing Variables
  Oscillations
    121.085
Eclipsing Variables
  Periods
    121.006 .011 .089
Eclipsing Variables
  Photometry
    121.007 .009 .025
    .027 .039 .048
    .049 .050 .051
    .058 .061 .068
    .071
Eclipsing Variables
  Spectra
    121.013 .014 .019
    .024 .037 .052
    142.071
Eclipsing Variables
  UV Photometry
    121.034
Eclipsing Variables
  UV Spectrophotometry
    121.067
Einstein Equations
    066.007 .082 .121
Element Abundances
    061.000
Element Abundances
  Cosmic Rays
    143.008 .014 .030
Element Abundances
  Eclipsing Variables
    121.023
Element Abundances
  Gaseous Nebulae
    132.016

Element Abundances
  Helium-Rich Stars
    064.002
Element Abundances
  Interplanetary Matter
    106.030
Element Abundances
  Interstellar Clouds
    131.090
Element Abundances
  Interstellar Gas
    155.025
Element Abundances
  Interstellar Matter
    131.008 .037 .038
    .041 .047 .067
Element Abundances
  Jupiter Atmosphere
    099.051
Element Abundances
  Jupiter Satellites
    099.210
Element Abundances
  K Stars
    114.174
Element Abundances
  Magnetic Stars
    116.014
Element Abundances
  Metallic-Line Stars
    114.051 .076
Element Abundances
  Meteorites
    094.395
    105.001 .045 .087
    .093 .151
Element Abundances
  Moon
    094.375
Element Abundances
  Peculiar A Stars
    114.051
Element Abundances
  Photosphere
    071.004
Element Abundances
  Solar Corona
    074.025
Element Abundances
  Solar Cosmic Rays
    078.042
Element Abundances
  Solar Spectrum
    071.036
Element Abundances
  Solar Wind
    074.110
Element Abundances
  Stellar Atmospheres
    064.042 .051 .059
Element Abundances
  Stellar Spectra
    114.066
Element Abundances
  Supergiants
    114.013
Elements
  Origin
    061.000
Emission Nebulae
    132.000

Emission-Line Stars
  Absorption Lines
    114.003
Emission-Line Stars
  Ages
    117.020
Emission-Line Stars
  H Alpha
    114.087
Emission-Line Stars
  Infrared Photometry
    113.022
    122.019
Emission-Line Stars
  Photometry
    114.121
Emission-Line Stars
  Planetary Nebulae
    133.012
Emission-Line Stars
  Polarization
    114.029 .161
Emission-Line Stars
  Rotation
    114.163
Emission-Line Stars
  Spectra
    113.011
    114.047
Emission-Line Stars
  UV Spectra
    121.030
Ephemerides
    047.000
Ephemerides
  Minor Planets
    098.043
Eta Carinae
    114.011
    131.087
    141.613
    158.304
Evolution
  Binaries
    117.007 .032 .039
    121.014
Evolution
  Carbon Stars
    065.024 .100
Evolution
  Close Binaries
    117.010 .011 .012
    .013 .014
Evolution
  Earth-Moon System
    094.004
Evolution
  Galactic Clusters
    153.018
Evolution
  Galaxies
    158.106
Evolution
  Galaxy
    155.000
Evolution
  High-Luminosity Stars
    065.002
Evolution
  Low-Mass Stars
    065.004

Evolution
  Magnetic Stars
    065.082
Evolution
  Main-Sequence Stars
    065.005 .124
Evolution
  Moon
    094.002 .008 .148
Evolution
  Planetary Nebulae
    133.020
Evolution
  Planets
    091.003
    107.005
Evolution
  Population II Stars
    065.060
Evolution
  Protostars
    065.014
Evolution
  Stellar Interiors
    065.101
Evolution
  Stellar Models
    065.015 .113
Evolution
  Subdwarf B Stars
    065.002
Evolution
  Subdwarf O Stars
    065.002
Evolution
  Supergiants
    065.094
Evolution
  Supermassive Stars
    065.096
Evolution
  Supernova Remnants
    125.016 .026 .050
    .059
Evolution
  W U Ma Stars
    121.069
Evolution
  White Dwarfs
    126.012
Evolution of Stars
    065.000
Exhibitions
    009.000
Expeditions Reports
    011.000
Extinction
    082.000
Extinction
  Galactic Clusters
    153.011
Extinction
  Interstellar Matter
    131.005 .016 .072
    .096
Extraterrestrial
  Research
    051.000
Extreme UV
  Solar Flares
    076.005

F Dwarfs
  Atmospheres
    064.044
F Dwarfs
  Luminosities
    115.011
F Stars
  Kinematics
    155.031
F Stars
  MK Types
    114.152
F Stars
  Spectral Types
    114.063
F Supergiants
  Absolute Magnitudes
    115.005
Faculae
    072.000
Faculae
  Distribution
    072.035
Faculae
  Photosphere
    072.021
Figure
  Celestial Bodies
    042.000
Figure
  Earth
    081.000
Figure
  Planets
    091.000
Figure
  Stars
    116.000
Figure
  Sun
    080.000
Filaments
    071.033
Filaments
  Crab Nebula
    134.004
Filaments
  Planetary Nebulae
    133.003
Filters
    034.025 .042 .047
    .075
Flare Stars
    064.047
    122.000 .004 .014
    .015 .022 .024
    .028 .040 .047
    .065 .074 .075
    .076 .078 .082
    .083 .092 .104
    .105 .122 .128
    .129 .131
Flare Stars
  Coalsack
    122.116
Flare Stars
  Hyades
    122.068
Flare Stars
  Orion Nebula
    122.070

Flare Stars
  Photometry
    113.023
    122.097
Flare Stars
  Pleiades
    122.006 .020
Flare Stars
  Radio Radiation
    122.127
Flare Stars
  Solar Neighborhood
    122.115
Fluid Spheres
  Magnetic Fields
    062.008
Forbidden Lines
  Planetary Nebulae
    133.010
Four-Body Problem
  042.358
Fourier Series
  042.020
Fourier Transforms
  031.005
  033.054
Fraunhofer Lines
  Solar Spectrum
    071.006 .007 .021
         .044 .050
Frequency Standards
  035.000
Fundamental Catalogues
  Systematic Errors
    041.015 .029

G Dwarfs
  Luminosities
    115.011
G Giants
  Atmospheres
    064.044
G Stars
  MK Types
    114.152
G Stars
  Peculiar
    114.053
G Stars
  Photometry
    114.088
Galactic Anti-Center
  Gamma Rays
    142.005
Galactic Center
    155.007 .072
Galactic Center
  Carbon Monoxide
    155.054
Galactic Center
  Distance
    155.042
Galactic Center
  Explosions
    155.019
Galactic Center
  Gamma Rays
    155.001 .037

Galactic Center
  Helium Abundance
    131.528
Galactic Center
  Infrared Polarization
    155.020
Galactic Center
  Infrared Radiation
    141.613
    155.027
Galactic Center
  Lunar Occultations
    096.023
Galactic Center
  Radio Radiation
    155.003 .008 .027
         .040
    157.003
Galactic Center
  Supernova Remnants
    155.003
Galactic Center
  X Rays
    155.027
Galactic Clusters
    153.000
Galactic Clusters
  Ages
    153.016
Galactic Clusters
  Bright Giants
    153.021
Galactic Clusters
  C-M Diagrams
    153.020 .023
Galactic Clusters
  Distances
    153.014
Galactic Clusters
  Eclipsing Variables
    121.017
Galactic Clusters
  Evolution
    153.018
Galactic Clusters
  Extinction
    153.011
Galactic Clusters
  Formation
    065.003
Galactic Clusters
  Hydrogen Abundance
    153.003 .004 .005
         .006 .007
Galactic Clusters
  Infrared Sources
    114.030
Galactic Clusters
  Late Type Stars
    153.024
Galactic Clusters
  Line Profiles
    153.012
Galactic Clusters
  Metal Abundances
    153.015 .023
Galactic Clusters
  OB Stars
    153.010

Galactic Clusters
  Photometry
    113.014
    153.002 .008 .009
         .013 .019 .020
         .022 .023 .024
         .029 .032 .033
Galactic Clusters
  Polarization
    131.052
Galactic Clusters
  Proper Motions
    112.009 .010 .011
    153.019
Galactic Clusters
  Spectra
    153.001
Galactic Clusters
  Stellar Evolution
    153.017
Galactic Disk
  Cosmic Rays
    143.026
Galactic Disk
  Gas Dynamics
    155.047
Galactic Disk
  Old Disk Stars
    155.022
Galactic Disk
  Radio Structure
    155.055
Galactic Disk
  Synchrotron Radiation
    155.023
Galactic Disk
  21 cm Radiation
    155.047
Galactic Magnetic Field
    156.000
Galactic Nebulae
  Exciting Stars
    131.505
    132.009
Galactic Nebulae
  Fine Structures
    132.018
Galactic Nebulae
  Line Profiles
    132.028
Galactic Nebulae
  Photometry
    113.045
    132.002
Galactic Nebulae
  Radiative Transfer
    132.006 .026
Galactic Nebulae
  Structure
    132.010
Galactic Nucleus
    155.004
Galactic Plane
  Gamma Rays
    155.026 .058
Galactic Plane
  H Alpha Surveys
    155.044
Galactic Plane
  Interstellar Matter
    155.077

Galactic Plane
  Neutral Hydrogen
    155.012
Galactic Plane
  X-Ray Sources
    142.033
Galactic Polarization
  Surveys
    155.041
Galactic Radio Radiation
    157.000
Galactic Rotation
    155.042 .075
Galactic Structure
    121.075
    131.021
    151.050 .061
    153.020
    155.004 .009 .010
        .013 .018 .019
        .024 .025 .031
        .039 .047 .048
    156.002
    158.041 .121
Galaxies
  Catalogues
    158.057
Galaxies
  Central Regions
    158.016
Galaxies
  Chemical Evolution
    158.077
Galaxies
  Classification
    158.028
Galaxies
  Colors
    158.042 .106
Galaxies
  Compact
    158.003 .025 .039
        .098
    160.003 .023 .024
Galaxies
  Compact Groups
    160.023 .024
Galaxies
  Companions
    151.002
    158.004 .055
Galaxies
  Counts
    160.011
Galaxies
  Density Waves
    158.009 .038
Galaxies
  Distances
    158.099
Galaxies
  Distribution
    158.065
    162.021
Galaxies
  Elliptical
    158.088 .100 .113
Galaxies
  Envelopes
    158.114

Galaxies
  Evolution
    158.106
Galaxies
  Fine Structures
    132.018
Galaxies
  Formation
    158.007 .094
    162.012
Galaxies
  H I Content
    158.113
Galaxies
  H I Regions
    158.010 .011 .014
Galaxies
  H II Regions
    131.534 .552 .553
    158.082 .104
Galaxies
  Infrared Radiation
    158.005
Galaxies
  Interacting
    151.008 .052
    158.059 .067 .103
Galaxies
  Internal Motions
    158.013
Galaxies
  Irregular
    158.013 .043
Galaxies
  Markarian Galaxies
    158.022 .036 .051
        .052 .061
Galaxies
  Mass-Lumin Relation
    158.008
Galaxies
  Masses
    158.035 .090
Galaxies
  Models
    158.132
Galaxies
  Nearby Galaxies
    158.068 .072 .117
Galaxies
  Nuclei
    141.029 .040
    158.005 .017 .018
        .021 .023 .037
        .048 .049 .075
        .076 .084 .085
        .087 .095 .097
        .105 .116
Galaxies
  Photometry
    158.015 .019 .039
        .042 .056 .091
        .116
Galaxies
  Physical Groups
    158.066
Galaxies
  Polarization
    158.088

Galaxies
  Radial Velocities
    158.122
Galaxies
  Radio Radiation
    158.040 .060
Galaxies
  Radio Sources
    141.045
    158.088
Galaxies
  Redshifts
    131.553
    141.097
    158.004 .006 .047
    162.023
Galaxies
  Rotation
    158.008 .026 .035
        .080
Galaxies
  Seyfert Galaxies
    141.058 .061
    158.029 .030 .031
        .032 .033 .034
        .062 .063 .073
        .089 .092 .093
        .303
Galaxies
  Spectra
    158.003
Galaxies
  Spectrophotometry
    158.104
Galaxies
  Star Formation
    065.010
    158.006
Galaxies
  Stellar Birthrate
    065.010
Galaxies
  Stellar Evolution
    158.100 .121
Galaxies
  Structure
    158.026
Galaxies
  UV Radiation
    158.020
Galaxies
  Velocity Maps
    158.131
Galaxies
  X Rays
    158.096
Galaxies
  21 cm Radiation
    158.003 .042
Galaxies Multiple
    158.000
Galaxies Single
    158.000
Galaxy
  Chemical Evolution
    155.036
Galaxy
  Differential Rotation
    155.056

Galaxy
Electron Density
155.071
Galaxy
Evolution
155.000
Galaxy
Gas Clouds
155.043
Galaxy
Interstellar Gas
155.071
Galaxy
Mass
155.006
Galaxy
Models
155.050 .073 .074
Galaxy
Nuclear Bulge
155.046
Galaxy
Stellar Orbits
155.032
Galaxy
Structure
155.000
Galilean Satellites
091.004
099.201 .206 .212
.217 .220 .221
.223
Galilean Satellites
Eclipses
099.219
Galilean Satellites
Infrared Spectra
099.207
Galilean Satellites
Spectra
099.214
Gamma Radiation
Cosmic
061.054
143.032
Gamma Radiation
Origin
142.064
Gamma Rays
Black Holes
061.056
Gamma Rays
Bursts
061.050
Gamma Rays
Crab Nebula
134.006
142.005
Gamma Rays
Galactic Anti-Center
142.005
Gamma Rays
Galactic Center
155.001 .037
Gamma Rays
Galactic Plane
155.026 .058
Gamma Rays
Novae
124.004

Gamma Rays
Pulsars
061.034
141.332 .340
Gamma Rays
Solar
076.000
Gamma Rays
Stellar Flares
061.035
Gamma Rays
Supernova Remnants
125.054
Gamma Rays
Supernovae
125.002 .009 .054
Gamma Rays
X-Ray Sources
142.021 .047
155.001
Gamma-Ray Astronomy
061.000
Gamma-Ray Sources
142.000
Gamma-Ray Sources
Bursts
142.011
Gas Clouds
Close Binaries
117.001
Gas Clouds
Galaxy
155.043
Gas Dynamics
Galactic Disk
155.047
Gaseous Nebulae
Element Abundances
132.016
Gaseous Nebulae
Photometry
132.016
Gaseous Nebulae
Temperatures
132.016
Gaseous Spheres
Oscillations
065.011
Geminids
104.012 .047 .049
.050
Geomagnetic Field
084.000
Geopotential
081.001 .005 .006
.019 .021 .028
.031
Giacobinids
104.014 .063
Globular Clusters
154.000
Globular Clusters
Andromeda Nebula
158.101
Globular Clusters
C-M Diagrams
154.008 .012 .015
.030
Globular Clusters
Cepheids
122.085

Globular Clusters
Colors
154.007
Globular Clusters
Distances
154.009
Globular Clusters
Dynamics
154.018
Globular Clusters
Horizontal Branches
154.001 .002 .003
Globular Clusters
Infrared Photometry
154.020
Globular Clusters
Late Type Stars
114.102
Globular Clusters
Magellanic Clouds
122.060
154.024
Globular Clusters
Masses
154.011
Globular Clusters
Metal Abundances
114.102
154.006
Globular Clusters
Metal-Rich
122.134
Globular Clusters
Photometry
113.014
154.002 .005 .007
.013 .014 .015
.017 .019 .028
.029
Globular Clusters
Red Giants
113.007
154.013
Globular Clusters
RR Lyrae Stars
122.027 .060 .093
Globular Clusters
Structure
154.001
Globular Clusters
UV Excess Stars
154.004
Globular Clusters
UV Stars
065.089
Globules
131.048
Globules
Diffuse Nebulae
131.075
Globules
H II Regions
131.531
Globules
Moon
094.307
Grains
Interplanetary Matter
131.014

Grains
  Interstellar Clouds
    131.017
Grains
  Interstellar Matter
    131.014 .015 .016
      .076 .083 .086
      .087
Granulation
    071.010 .014 .015
      .018 .028 .037
      .038 .039 .040
      .041 .042 .046
      .048
    073.016
Gravitation
  Collapse
    012.008
    066.053 .105 .117
      .119 .141
Gravitation
  Pulsars
    141.302
Gravitation Theory
    066.000
Gravitational Constant
    043.001 .003
    066.085
    160.002
Gravitational Deflection
    066.035
Gravitational
  Instability
    061.000
    062.059
    162.021
Gravitational Radiation
    012.008
    066.016 .044 .046
Gravitational Waves
    033.021 .023
    066.004 .013 .032
      .091 .122 .138
      .147
Gravity
  Earth
    081.000
Gravity
  Moon
    094.115 .160 .162
      .166 .172
Gravity
  Planets
    091.007
Gravity
  Venus
    093.022
Gum Nebula
    132.005 .027
Gum Nebula
  Interstellar Gas
    132.021

H Alpha
  Emission-Line Stars
    114.087
H Alpha
  O Stars
    114.016

H Alpha
  Solar Flares
    073.013
H Alpha Surveys
  Galactic Plane
    155.044
H I Clouds
    131.543
H I Clouds
  Intercloud Gas
    131.506
H I Clouds
  Structure
    131.508
H I Content
  Galaxies
    158.113
H I Regions
  Galaxies
    158.010 .011 .014
H I Regions
  Interstellar Clouds
    131.530
H I Regions
  Ionization
    131.503
H I Regions
  Radio Galaxies
    131.516
H I Regions
  Recombination Lines
    131.501 .535 .544
H I Regions
  Search
    131.541
H I Regions
  21 cm Radiation
    131.517 .526
H II Regions
    131.018 .507 .509
    132.010 .017
    151.048
    155.044
H II Regions
  Compact
    131.032 .512 .519
      .529 .545
H II Regions
  Distribution
    131.515
H II Regions
  Dust
    131.510
H II Regions
  Electron Densities
    131.505
H II Regions
  Electron Temperatures
    131.536
H II Regions
  Exciting Stars
    131.502
H II Regions
  Galactic
    131.525
    155.010
H II Regions
  Galaxies
    131.534 .552 .553
    158.082 .104

H II Regions
  Globules
    131.531
H II Regions
  Helium Abundance
    131.514 .538 .539
H II Regions
  Infrared Excesses
    131.538 .539
H II Regions
  Infrared Photometry
    131.523 .524 .547
H II Regions
  Infrared Radiation
    131.527 .533 .545
H II Regions
  Internal Motions
    131.554
H II Regions
  Line Profiles
    131.518
H II Regions
  Neon Abundance
    131.522
H II Regions
  Radio Radiation
    131.504 .511 .521
      .525 .536
    158.082
H II Regions
  Recombination Lines
    131.540
H II Regions
  Spectra
    131.519 .522
H II Regions
  Spectrophotometry
    131.529
H II Regions
  Velocities
    131.513
H II Regions
  Velocity Fields
    131.554
H II Regions
  Water-Vapor Emission
    141.603
Helium
  Cosmic Abundance
    061.012
Helium Abundance
  Galactic Center
    131.528
Helium Abundance
  H II Regions
    131.514 .538 .539
Helium Abundance
  Interstellar Matter
    131.036
Helium Abundance
  Orion Nebula
    132.012 .019
Helium Abundance
  Peculiar A Stars
    065.050
Helium Abundance
  Peculiar B Stars
    065.050
Helium Abundance
  Planetary Nebulae
    133.005

Helium Burning
  Stellar Evolution
    065.065
Helium Content
  Solar Wind
    074.073
Helium Stars
    114.094
Helium-Burning Stars
  Secular Stability
    065.065
Helium-Rich Stars
  Element Abundances
    064.002
Helium-Rich Stars
  Fine Analyses
    064.051
Helium-Rich Stars
  Spectra
    114.166
Helium-Weak Stars
  Photometry
    114.060
Herbig-Haro Objects
    132.031
    141.601
High-Luminosity Stars
  Evolution
    065.002
High-Luminosity Stars
  Luminosity Function
    115.006
High-Velocity Clouds
    131.517
High-Velocity Stars
    112.003
    117.038
History of Astronomy
    004.000
Horizontal Branches
  Globular Clusters
    154.001 .002 .003
Horizontal-Branch Stars
    065.088
HR Diagrams
    114.010
    115.000
HR Diagrams
  Population II Cepheids
    122.095
Hubble Constant
    131.553
Hubble Diagrams
  Compact Galaxies
    158.098
Hyades
    113.014
    153.014 .034
Hyades
  Flare Stars
    122.068
Hyades
  K Dwarfs
    153.030
Hyades
  M Dwarfs
    153.030
Hyades
  Red Dwarfs
    111.001

Hyades
  Spectroscopic Binaries
    119.007
Hydrodynamics
    062.000
Hydromagnetic Waves
  Plasma
    062.047
Hydromagnetic Waves
  Solar Wind
    062.064

Ideal Resonance Problem
    042.011 .012 .014
Image Tubes
    034.014 .082
Infrared Astronomy
    061.000
Infrared Excesses
  Be Stars
    114.100
Infrared Excesses
  Eclipsing Variables
    121.002
Infrared Excesses
  H II Regions
    131.538 .539
Infrared Maps
  Moon
    094.370
Infrared Photometry
    034.063
Infrared Photometry
  Carbon Stars
    113.030
Infrared Photometry
  Cool Stars
    113.036
Infrared Photometry
  Emission-Line Stars
    113.022
    122.019
Infrared Photometry
  Globular Clusters
    154.020
Infrared Photometry
  H II Regions
    131.523 .524 .547
Infrared Photometry
  Late Type Stars
    114.048
Infrared Photometry
  Novae
    124.003
Infrared Photometry
  Orion Nebula
    132.017
Infrared Photometry
  Red Giants
    113.007
    114.008
Infrared Photometry
  RW Aurigae Stars
    122.062
Infrared Polarization
  Galactic Center
    155.020
Infrared Radiation
  Galactic Center
    141.613

Infrared Radiation
  Galactic Center
    155.027
Infrared Radiation
  Galaxies
    158.005
Infrared Radiation
  H II Regions
    131.527 .533 .545
Infrared Radiation
  Mars Satellites
    097.203
Infrared Radiation
  Mercury
    092.034
Infrared Radiation
  Moon
    094.114
Infrared Radiation
  Planetary Nebulae
    133.006 .007
Infrared Radiation
  Supergiants
    114.014
Infrared Radiation
  Venus
    093.020
Infrared Radiation
  X-Ray Sources
    142.015
Infrared Sources
    065.013 .014
    114.075 .132
    121.002
    131.001 .087 .524
         .529 .533
    132.034
    141.600 .601 .610
         .616
Infrared Sources
  Galactic Clusters
    114.030
Infrared Sources
  Models
    141.605
Infrared Sources
  Photometry
    141.612
Infrared Sources
  Spectra
    114.031
    141.604
Infrared Sources
  Spectrophotometry
    114.032
    141.608
Infrared Sources
  Structure
    141.606 .607
Infrared Spectra
  Carbon Stars
    114.069
Infrared Spectra
  Eclipsing Variables
    121.076
Infrared Spectra
  Galilean Satellites
    099.207
Infrared Spectra
  Jupiter Atmosphere
    099.002

Infrared Spectra
  Late Type Stars
    114.069 .103
    141.602
Infrared Spectra
  Mercury
    092.001
Infrared Spectra
  Minor Planets
    099.207
Infrared Spectra
  Planets
    114.101
Infrared Spectra
  Saturn Atmosphere
    100.001
Infrared Spectra
  Seyfert Galaxies
    158.034
Infrared Spectra
  Wolf-Rayet Stars
    114.001
Infrared Spectroscopy
    114.101
Infrared Stars
  Binaries
    141.611
Infrared Stars
  OH Emission
    141.609
Infrared Stars
  Water-Vapor Emission
    141.603
Instability
  Gravitational
    061.000
Instability
  Interstellar Matter
    151.048
    155.056
Institutes
    008.000
Instruments
  Astronomical
    032.000
Interacting Galaxies
  Radio Surveys
    158.059
Interferometers
    034.040 .050 .076
Interferometry
  Very-Long-Baseline
    031.009
Intergalactic Matter
    161.000
    162.033
Intergalactic Matter
  Clusters of Galaxies
    160.013
Intergalactic Matter
  Dust
    141.057
    161.001
Intergalactic Matter
  Quasars
    141.057
International
  Cooperation
    013.000

Interplanetary Magnetic
  Field
    106.000
Interplanetary Matter
    106.000
Interplanetary Matter
  Alfven Waves
    062.013
    074.007
Interplanetary Matter
  Dust
    106.037
Interplanetary Matter
  Element Abundances
    106.030
Interplanetary Matter
  Grains
    131.014
Interplanetary Matter
  Lyman Alpha
    106.001 .009
Interplanetary Matter
  Nuclear Reactions
    091.016
Interplanetary Matter
  Scintillations
    106.018 .038
    143.021 .022
Interplanetary Matter
  Shock Waves
    106.031 .036
Interplanetary Matter
  Solar Wind
    074.037
    082.029
    106.037
Interplanetary Space
  Cosmic Rays
    106.014
    143.003
Interplanetary Space
  Shock Waves
    106.004
Interstellar Clouds
  Collisions
    131.131
Interstellar Clouds
  Cooling
    131.124
Interstellar Clouds
  Cosmic Rays
    131.503
Interstellar Clouds
  Element Abundances
    131.090
Interstellar Clouds
  Grains
    131.017
Interstellar Clouds
  H I Regions
    131.530
Interstellar Clouds
  Heating
    131.012 .520
Interstellar Clouds
  Ionization
    131.520 .530
Interstellar Clouds
  Masers
    131.104

Interstellar Clouds
  Molecules
    131.010 .079 .089
    .090 .091 .124
Interstellar Dust
  Distribution
    155.033
Interstellar Gas
  Densities
    131.057 .102
Interstellar Gas
  Element Abundances
    155.025
Interstellar Gas
  Galaxy
    155.071
Interstellar Gas
  Gum Nebula
    132.021
Interstellar Gas
  Ionization
    131.057
Interstellar Gas
  Shock Waves
    131.101
Interstellar Gas
  Temperatures
    131.057
Interstellar Lines
    114.020 .155
    131.021 .024 .054
    .055 .065 .072
    .090 .096 .135
Interstellar Matter
    012.018
    131.000
Interstellar Matter
  Absorption
    131.513
    132.003
    141.325
    158.041
Interstellar Matter
  Absorption Lines
    131.099
Interstellar Matter
  Boron Abundance
    131.094
Interstellar Matter
  CH
    131.105
Interstellar Matter
  Clouds
    131.065 .073
Interstellar Matter
  Cosmic Rays
    143.024
Interstellar Matter
  Dust
    131.015 .056 .132
    .512
    151.060
Interstellar Matter
  Dust Clouds
    131.542 .543
Interstellar Matter
  Element Abundances
    131.008 .037 .038
    .041 .047 .067

Interstellar Matter
  Extinction
    131.005 .016 .072
    .096
Interstellar Matter
  Galactic Plane
    155.077
Interstellar Matter
  Grains
    131.014 .015 .016
    .076 .083 .086
    .087
Interstellar Matter
  Heating
    131.006
Interstellar Matter
  Helium Abundance
    131.036
Interstellar Matter
  Instability
    151.048
    155.056
Interstellar Matter
  Intercloud Gas
    131.135
Interstellar Matter
  Isotopic Abundances
    155.035
Interstellar Matter
  Line Profiles
    131.035
Interstellar Matter
  Lithium Abundance
    131.002
Interstellar Matter
  Lyman Alpha
    113.001
Interstellar Matter
  Masers
    131.060
Interstellar Matter
  Molecules
    022.049 .059
    131.004 .007 .009
    .011 .024 .025
    .026 .027 .030
    .034 .049 .050
    .060 .061 .062
    .063 .064 .068
    .069 .071 .093
    .095 .126 .129
    .133 .134 .139
    132.001 .003 .025
    .029 .034
    155.004
Interstellar Matter
  OH
    131.078 .081 .092
Interstellar Matter
  OH Absorption
    131.020
Interstellar Matter
  Opacities
    131.040
    142.001
Interstellar Matter
  Organic Molecules
    131.122
Interstellar Matter
  Polarization
    131.003 .039 .042

Interstellar Matter
  Polarization
    131.055 .083 .084
    .121
Interstellar Matter
  Pulsars
    131.022
    141.343
Interstellar Matter
  Radio Lines
    131.034
Interstellar Matter
  Recombination Lines
    131.053 .081
Interstellar Matter
  Sodium Abundance
    131.051
Interstellar Matter
  Solar Wind
    143.030
Interstellar Matter
  Spectrophotometry
    131.002
Interstellar Matter
  Supernova Remnants
    125.005 .028
Interstellar Matter
  UV Spectra
    113.001
Interstellar Matter
  UV Stars
    131.018
Interstellar Matter
  Water
    131.058
Interstellar Matter
  X Rays
    131.040 .074
    142.001
Interstellar Matter
  21 cm Radiation
    131.043 .054 .508
    .532
Interstellar Matter
  21 cm Survey
    157.005 .006
Interstellar Radicals
    131.005
Interstellar Reddening
    113.040
    114.030
    122.061
    131.003 .039 .045
    153.001
    154.019
Interstellar Space
    131.000
Interstellar Space
  Cosmic Rays
    143.013
Ionosphere
    083.000
Ionosphere
  Jupiter
    099.012 .040
Ionosphere
  Mars
    093.037
    097.056 .057

Ionosphere
  Venus
    093.004 .022 .036
    .037 .043
    097.056
Iron Meteorites
    105.094 .121 .138

Jupiter
    012.003
    099.000
Jupiter
  Albedo
    099.001 .035
Jupiter
  Brightness Temeperature
    099.006
Jupiter
  Chemical Composition
    099.076
Jupiter
  Formation
    099.043
Jupiter
  Interior
    099.004 .029
Jupiter
  Ionosphere
    099.012 .040
Jupiter
  Magnetic Field
    099.015 .016 .059
Jupiter
  Magnetosphere
    084.297
    099.017 .018 .036
    .044 .052 .077
Jupiter
  Models
    099.075
Jupiter
  Photometry
    099.023
Jupiter
  Pictures
    099.042
Jupiter
  Pioneer 10 Encounter
    099.013 .014 .015
    .016 .017 .018
    .019 .020 .021
    .022 .204 .205
Jupiter
  Radiation Belts
    099.019 .044 .057
    .058
Jupiter
  Radio Radiation
    099.054 .059
Jupiter
  Radio Spectra
    099.026
Jupiter
  Red Spot
    099.047
Jupiter
  Spectra
    099.005

Jupiter
  Stellar Occultations
    099.037 .049 .050
    .074
Jupiter
  Temperatures
    099.038
Jupiter Atmosphere
    099.074
Jupiter Atmosphere
  Ammonia
    099.007 .035 .066
    .067
Jupiter Atmosphere
  Chemical Composition
    099.045
Jupiter Atmosphere
  Clouds
    099.028 .048 .061
Jupiter Atmosphere
  Dynamics
    099.030 .065
Jupiter Atmosphere
  Element Abundances
    099.051
Jupiter Atmosphere
  Infrared Spectra
    099.002
Jupiter Atmosphere
  Methane
    099.007
Jupiter Atmosphere
  Models
    099.008
Jupiter Atmosphere
  Organic Matter
    099.031
Jupiter Atmosphere
  Radiative Heating
    099.041
Jupiter Atmosphere
  Scattering
    099.003
Jupiter Atmosphere
  Spectra
    099.063
Jupiter Atmosphere
  Structure
    099.037 .051
Jupiter Atmosphere
  Temperatures
    099.050
Jupiter Atmosphere
  UV Spectra
    099.064
Jupiter Satellites
    099.200
Jupiter Satellites
  Early History
    099.212
Jupiter Satellites
  Element Abundances
    099.210
Jupiter Satellites
  Light Variations
    100.210
Jupiter Satellites
  Line Profiles
    099.209

Jupiter Satellites
  Spectra
    099.218

K Dwarfs
  Hyades
    153.030
K Giants
  Atmospheres
    064.044
K Stars
  Angular Diameters
    115.008
K Stars
  Colors
    155.078
K Stars
  Element Abundances
    114.174
K Stars
  Luminosities
    115.003
K Stars
  Photometry
    114.088 .173
K Stars
  Spectra
    155.079
Kinematics
  A Stars
    155.031
Kinematics
  F Stars
    155.031
Kinematics
  M Dwarfs
    065.003
Kinematics
  Main-Sequence Stars
    155.028
Kinematics
  Nearby Stars
    155.051 .053
Kinematics
  Stellar Systems
    151.000

Late Type Stars
  Atmospheres
    064.003 .011 .026
    114.032
Late Type Stars
  Diameters
    115.001
Late Type Stars
  Envelopes
    065.127
Late Type Stars
  Galactic Clusters
    153.024
Late Type Stars
  Globular Clusters
    114.102
Late Type Stars
  Infrared Photometry
    114.048
Late Type Stars
  Infrared Spectra
    114.069 .103

Late Type Stars
  Infrared Spectra
    141.602
Late Type Stars
  Line Profiles
    114.072 .157
Late Type Stars
  Luminosities
    114.157
Late Type Stars
  Magellanic Clouds
    114.162
Late Type Stars
  Masers
    114.048 .132
Late Type Stars
  Mass Loss
    064.032
Late Type Stars
  Metal Abundances
    114.105
Late Type Stars
  Models
    065.046
Late Type Stars
  Photometry
    114.070 .075
Late Type Stars
  Spectra
    114.122 .158
Late Type Stars
  Structure
    065.077
Late Type Stars
  UV Spectra
    114.072
Latitude Determination
    045.000
Light Deflection
    066.002
Limb Darkening
  Sun
    080.005
Line Broadening
  Solar Spectrum
    071.006
Line Broadening
  Stark Effect
    022.055
Line Formation
    063.003 .012 .013
Line Formation
  Stellar Models
    064.027
Line Intensities
  Prominences
    073.029
Line Profiles
  B Stars
    114.127
Line Profiles
  Be Stars
    114.077 .104
Line Profiles
  Chromosphere
    073.028
Line Profiles
  Galactic Clusters
    153.012

Line Profiles
　Galactic Nebulae
　　132.028
Line Profiles
　H II Regions
　　131.518
Line Profiles
　Interstellar Matter
　　131.035
Line Profiles
　Jupiter Satellites
　　099.209
Line Profiles
　Late Type Stars
　　114.072 .157
Line Profiles
　Magnetic Stars
　　116.005
Line Profiles
　Masers
　　131.088
Line Profiles
　Peculiar A Stars
　　114.026
Line Profiles
　Planetary Nebulae
　　133.001
Line Profiles
　Prominences
　　073.047
Line Profiles
　Pulsating Stars
　　065.086
　　122.088
Line Profiles
　Rotating Stars
　　114.071
Line Profiles
　Solar Corona
　　074.011 .046
Line Profiles
　Solar Limb
　　080.031 .033
Line Profiles
　Solar Spectrum
　　071.007 .020 .066
　　080.008 .009
Line Profiles
　Stellar Atmospheres
　　064.004 .013 .025
Line Profiles
　Stellar Spectra
　　114.065
Lithium Abundance
　Interstellar Matter
　　131.002
Low-Luminosity Stars
　　126.000
Low-Mass Stars
　Evolution
　　065.004
Luminosities
　Cepheids
　　122.005 .032 .049
Luminosities
　Early Type Stars
　　153.009
Luminosities
　F Dwarfs
　　115.011

Luminosities
　G Dwarfs
　　115.011
Luminosities
　K Stars
　　115.003
Luminosities
　Late Type Stars
　　114.157
Luminosities
　Peculiar B Stars
　　115.016
Luminosities
　Quasars
　　141.026 .089
Luminosities
　Stars
　　115.000
Luminosities
　T Tauri Stars
　　115.015
Luminosities
　W Virginis Stars
　　122.049
Luminosities
　X-Ray Sources
　　142.040
Luminosity Function
　High-Luminosity Stars
　　115.006
Lunar Eclipses
　　095.000
Lunar Occultations
　　096.000
Lunar Occultations
　Cepheids
　　122.029
Lunar Occultations
　Galactic Center
　　096.023
Lunar Occultations
　Radio Sources
　　141.051 .087
Lunar Occultations
　Visual Binaries
　　118.018
Lunar Probes
　　053.000
Lyman Alpha
　Interplanetary Matter
　　106.001 .009
Lyman Alpha
　Interstellar Matter
　　113.001
Lyman Alpha
　Mars Atmosphere
　　097.049 .052
Lyman Alpha
　Solar Flares
　　073.018
Lyman Alpha
　Spectroheliograms
　　076.001
Lyncids
　　104.070
Lyrids
　　104.005

M Dwarfs
　Hyades
　　153.030
M Dwarfs
　Kinematics
　　065.003
M Giants
　Distribution
　　155.038
M Giants
　Metal Abundances
　　114.023
M Giants
　Space Distribution
　　155.018
M Stars
　Atmospheres
　　064.021
　　114.172
M Stars
　Space Density
　　155.014
M 87
　　154.029
Maclaurin Spheroids
　　042.017
Macroturbulence
　　063.012
Magellanic Clouds
　　159.000
Magellanic Clouds
　Dark Nebulae
　　131.136
Magellanic Clouds
　Globular Clusters
　　122.060
　　154.024
Magellanic Clouds
　Late Type Stars
　　114.162
Magellanic Clouds
　Magnetic Fields
　　131.102
Magellanic Clouds
　OB Stars
　　159.003
Magellanic Clouds
　Peculiar Stars
　　122.117
Magellanic Clouds
　RR Lyrae Stars
　　122.130
Magellanic Clouds
　Star Clusters
　　153.025
　　159.001
Magellanic Clouds
　Stellar Evolution
　　065.097
Magellanic Clouds
　Supergiants
　　114.162
　　122.077
Magellanic Clouds
　Variables
　　122.089 .117
Magellanic Clouds
　21 cm Radiation
　　159.004

Magnetic Accretion
   Peculiar A Stars
      116.010
Magnetic Accretion
   X-Ray Sources
      142.016
Magnetic Field
   Earth
      084.000
Magnetic Field
   Interplanetary
      106.000
Magnetic Field
   Jupiter
      099.015 .016 .059
Magnetic Field
   Mars
      097.020
Magnetic Field
   Mercury
      092.036 .037
Magnetic Field
   Moon
      094.109 .112 .148
Magnetic Field
   Orion Nebula
      132.025
Magnetic Field
   Stars
      116.000
Magnetic Field
   Venus
      093.023
      097.020
Magnetic Fields
   Binaries
      117.015 .025 .032
Magnetic Fields
   Chromosphere
      072.020
Magnetic Fields
   Clusters of Galaxies
      160.005
Magnetic Fields
   Comets
      102.014
Magnetic Fields
   Cosmic
      061.036
      062.029
Magnetic Fields
   Crab Nebula
      134.002 .003
Magnetic Fields
   Fluid Spheres
      062.008
Magnetic Fields
   Magellanic Clouds
      131.102
Magnetic Fields
   Main-Sequence Stars
      116.008
Magnetic Fields
   Neutron Stars
      062.066
      065.084
Magnetic Fields
   Peculiar A Stars
      116.005

Magnetic Fields
   Peculiar B Stars
      116.007
Magnetic Fields
   Photosphere
      071.008 .012 .022
         .026 .041 .042
      073.038
      077.032
Magnetic Fields
   Plasma
      062.012
Magnetic Fields
   Prominences
      073.010 .020
Magnetic Fields
   Pulsars
      141.327
Magnetic Fields
   Radio Sources
      141.037
Magnetic Fields
   Solar Corona
      074.029 .044 .048
         .061 .078 .087
Magnetic Fields
   Solar Flares
      073.042 .057
      077.026
Magnetic Fields
   Solar Wind
      074.014 .017 .021
         .084
Magnetic Fields
   Stellar Envelopes
      064.015
Magnetic Fields
   Stellar Interiors
      065.080
Magnetic Fields
   Sunspots
      072.004 .007 .027
         .030 .033 .037
         .040
Magnetic Fields
   Supernova Remnants
      125.060
Magnetic Fields
   White Dwarfs
      063.002
      126.006 .007
Magnetic Fields
   X-Ray Sources
      142.027
Magnetic Stars
      065.017
      114.156
      116.001 .002 .004
         .010 .012 .013
      121.001
      126.021
Magnetic Stars
   Atmospheres
      116.006
Magnetic Stars
   Element Abundances
      116.014
Magnetic Stars
   Evolution
      065.082

Magnetic Stars
   Light Curves
      116.006
Magnetic Stars
   Line Profiles
      116.005
Magnetic Stars
   Photometry
      113.018
Magnetic Stars
   Polarization
      116.003 .009
Magnetic Stars
   Rotation
      062.026
Magnetic Stars
   Spectra
      114.082
Magnetic Variables
      122.101
Magnetohydrodynamics
      062.000
Magnetosphere
   Jupiter
      084.297
      099.017 .018 .036
         .044 .052 .077
Magnetosphere
   Mars
      097.048
Magnetosphere
   Solar Wind
      074.076
Magnetosphere
   Venus
      093.026
Magnetospheres
   Planets
      091.013
Magnetospheres
   Pulsars
      062.027 .028
      141.303 .312 .316
         .324
Magnitudes
   Stars
      113.000
Main-Sequence Stars
   Calcium Emission
      116.008
Main-Sequence Stars
   Convection
      065.028
Main-Sequence Stars
   Evolution
      065.005 .124
Main-Sequence Stars
   Kinematics
      155.028
Main-Sequence Stars
   Magnetic Fields
      116.008
Main-Sequence Stars
   Opacities
      065.005
Main-Sequence Stars
   Stellar Envelopes
      064.008
Markarian Galaxies
      158.022 .036 .051
         .052 .061

Mars
  012.003
  097.000
Mars
  Albedo
  097.075
Mars
  Channels
  094.159
Mars
  Climates
  097.026 .077
Mars
  Craters
  094.185
  097.005 .021 .043
Mars
  Ionosphere
  093.037
  097.056 .057
Mars
  Magnetic Field
  097.020
Mars
  Magnetosphere
  097.048
Mars
  Maps
  097.003
Mars
  Meteorite Impact
  094.118
  097.078
Mars
  Oblateness
  097.019
Mars
  Photographs
  097.005
Mars
  Photometry
  091.023
  097.015
Mars
  Pictures
  097.054 .069 .080
Mars
  Positions
  097.039
Mars
  Surface
  097.002 .018 .028
    .041 .050 .065
Mars
  Surface Pressure
  097.036
Mars
  Surface Structures
  097.005 .021 .035
    .037 .040 .062
    .070 .071
Mars
  UV Spectra
  093.013
Mars
  Volcanism
  097.004 .055 .080
Mars Atmosphere
  097.011 .046 .068
    .074

Mars Atmosphere
  Carbon Dioxide
  097.016 .017 .066
Mars Atmosphere
  Clouds
  097.024 .038
Mars Atmosphere
  Dust
  097.006 .069
Mars Atmosphere
  Dust Clouds
  097.031 .032
Mars Atmosphere
  Lyman Alpha
  097.049 .052
Mars Atmosphere
  Ozone
  097.072
Mars Atmosphere
  Photolysis
  097.042
Mars Atmosphere
  Radiative Transfer
  093.012
  097.045
Mars Atmosphere
  UV Radiation
  097.007
Mars Satellites
  097.200
Mars Satellites
  Infrared Radiation
  097.203
Mars Satellites
  Positions
  097.039
Mascons
  Moon
  094.166 .172
Masers
  Interstellar
  131.031 .032 .033
    .077 .100 .103
    .128 .134 .509
Masers
  Interstellar Clouds
  131.104
Masers
  Interstellar Matter
  131.060
Masers
  Late Type Stars
  114.048 .132
Masers
  Line Profiles
  131.088
Masers
  OH Sources
  131.070
Mass
  Andromeda Nebula
  158.101
Mass
  Earth-Moon System
  091.010
Mass
  Galaxy
  155.006
Mass
  Mercury
  092.040

Mass
  Neptune
  101.007
Mass
  Venus
  093.022
Mass Loss
  064.000
Mass Loss
  Binaries
  117.007
  121.005
Mass Loss
  Close Binaries
  117.006
Mass Loss
  Late Type Stars
  064.032
Mass Loss
  Quasi-Stellar Objects
  141.001
Mass Loss
  Shell Stars
  064.001
Mass Loss
  Stellar Evolution
  065.021 .036
Mass Loss
  Wolf-Rayet Stars
  064.023
Mass Loss
  X-Ray Sources
  142.002
Mass-Lumin Relation
  Galaxies
  158.008
Masses
  Eclipsing Variables
  121.041
Masses
  Galaxies
  158.035 .090
Masses
  Globular Clusters
  154.011
Masses
  Minor Planets
  098.001
Masses
  Planets
  091.010
Masses
  Quasars
  141.068
Masses
  Spectroscopic Binaries
  118.017
Masses
  Stars
  115.000
Masses
  Visual Binaries
  118.008 .017 .025
Massive Stars
  Atmospheres
  065.021
Massive Stars
  Nucleosynthesis
  065.087
Mathematics
  021.000

Meetings Proceedings
    012.000
Meetings Reports
    011.000
Mercury
    092.000
Mercury
  Albedo
    092.004
Mercury
  Atmosphere
    092.001 .013 .015
      .018 .038 .040
Mercury
  Brightness Temperature
    093.001
Mercury
  Infrared Radiation
    092.034
Mercury
  Infrared Spectra
    092.001
Mercury
  Magnetic Field
    092.036 .037
Mercury
  Mass
    092.040
Mercury
  Obliquity
    092.031
Mercury
  Pictures
    092.017 .039
Mercury
  Radio Radiation
    092.029
Mercury
  Radius
    092.040
Mercury
  Solar Wind
    092.035
Mercury
  Surface
    092.011 .017 .039
      .041
Mercury
  Transit
    092.005 .012 .013
      .019 .021
Meridian Circles
    032.024 .037 .039
      .056 .057 .059
      .061
Meridian Cirlces
    032.038
Metal Abundances
  Cepheids
    122.005
Metal Abundances
  Galactic Clusters
    153.015 .023
Metal Abundances
  Globular Clusters
    114.102
    154.006
Metal Abundances
  Late Type Stars
    114.105

Metal Abundances
  M Giants
    114.023
Metal Abundances
  Meteorites
    105.003
Metal-Deficient Stars
  Spectra
    114.033
Metal-Poor Stars
  Atmospheres
    114.105
Metallic-Line Stars
  Atmospheres
    114.131
Metallic-Line Stars
  Catalogues
    114.097
Metallic-Line Stars
  Element Abundances
    114.051 .076
Metallic-Line Stars
  Models
    114.076
Metallic-Line Stars
  Photometry
    113.043
Meteor Streams
    104.000
Meteor Streams
  Radiants
    104.071
Meteor Trails
    104.033 .034 .035
      .037
Meteorite Craters
    105.000
Meteorites
    105.000
Meteorites
  Element Abundances
    094.395
    105.001 .045 .087
      .093 .151
Meteorites
  Iron Meteorites
    105.094 .121 .138
Meteorites
  Metal Abundances
    105.003
Meteorites
  Stone Meteorites
    105.001 .109 .149
      .150 .161
Meteorites
  Terrestrial Ages
    105.092 .101
Meteorites
  Thermal Evolution
    105.089
Meteors
    104.000
Meteors
  Fragmentation
    104.033
Meteors
  Heating
    104.001
Meteors
  Light Curves
    104.021

Meteors
  Orbits
    104.015 .022
Meteors
  Radar Echoes
    104.011 .050
Meteors
  Spectra
    104.041 .042 .044
      .045
Meteors
  Spectrophotometry
    104.040
Methane
  Jupiter Atmosphere
    099.007
Methane
  Saturn Satellites
    100.205
Methods of Observation
    031.000
Methods of Reduction
    031.000
Micrometeorites
    105.146
Micrometeors
    104.056 .057 .058
Microspherules
    105.088
Microturbulence
    063.012
Microturbulence
  Photosphere
    071.001
Microturbulence
  Stellar Atmospheres
    064.019 .028
Microwave Background
    066.042
    162.029
Minor Planets
    098.000
Minor Planets
  Brightnesses
    098.010
Minor Planets
  Colors
    098.010
Minor Planets
  Ephemerides
    098.043
Minor Planets
  Infrared Spectra
    099.207
Minor Planets
  Light Curves
    098.010 .012
Minor Planets
  Masses
    098.001
Minor Planets
  Motion
    098.009
Minor Planets
  Observations
    098.015 .042
Minor Planets
  Orbits
    042.043

Minor Planets
  Photometry
    098.012 .017
Minor Planets
  Polarization
    098.010
Minor Planets
  Resonances
    098.004
Mira Variables
    114.042 .075 .167
    115.009
    122.003 .008 .134
MK Types
  A Stars
    155.021
MK Types
  F Stars
    114.152
MK Types
  G Stars
    114.152
Molecules
  Interstellar Clouds
    131.010 .079 .089
        .090 .091 .124
Molecules
  Interstellar Matter
    022.049 .059
    131.004 .007 .009
        .011 .024 .025
        .026 .027 .030
        .034 .049 .050
        .060 .061 .062
        .063 .064 .068
        .069 .071 .093
        .095 .126 .129
        .133 .134 .139
    132.001 .003 .025
        .029 .034
    155.004
Molecules
  Spectra
    022.008 .046 .076
        .077 .078 .079
        .080
    141.602
Moon
  Albedo
    094.140 .146 .314
Moon
  Atmosphere
    094.149 .150
Moon
  Chemical Composition
    094.175 .177
Moon
  Craters
    094.143 .147 .185
        .321 .363 .369
        .370 .377 .385
        .405
Moon
  Density
    094.117
Moon
  Electric Conductivity
    094.113 .165
Moon
  Element Abundances
    094.375

Moon
  Evolution
    094.002 .008 .148
Moon
  Globules
    094.307
Moon
  Gravity
    094.115 .160 .162
        .166 .172
Moon
  Infrared Maps
    094.370
Moon
  Infrared Radiation
    094.114
Moon
  Interior
    094.102 .108 .111
        .126 .129 .155
        .156
Moon
  Landing Sites
    094.308 .372
Moon
  Laser Ranging Stations
    094.158
Moon
  Magnetic Field
    094.109 .112 .148
Moon
  Mare Origin
    094.147
Moon
  Mascons
    094.166 .172
Moon
  Meteorite Impact
    094.118 .161
Moon
  Models
    094.117 .182
Moon
  Orbit
    094.005
Moon
  Origin
    094.003 .010
Moon
  Radar Maps
    094.157 .367 .368
        .369 .370
Moon
  Radio Radiation
    094.164 .171
Moon
  Red Spots
    094.316
Moon
  Reference Points
    094.137
Moon
  Regolith
    094.308 .387
Moon
  Rilles
    094.159
Moon
  Rocks
    094.306 .310 .312
        .315 .317 .322

Moon
  Rocks
    094.364 .365 .373
        .379 .393 .394
        .406
    105.123
Moon
  Rotation
    094.138
Moon
  Samples
    094.174 .303 .304
        .305 .309 .320
        .325 .326 .360
        .361 .376 .382
        .383 .390 .391
        .392 .395 .398
        .400 .402 .403
        .404
Moon
  Seismicity
    094.108 .111 .318
Moon
  Soil
    094.311 .319 .322
        .324 .371 .380
        .381
Moon
  Solar Wind
    074.032 .079
    094.103 .150 .167
Moon
  Surface
    094.106 .107 .127
        .138 .144 .154
        .157 .168 .170
        .173 .183 .184
        .368 .369
Moon
  Surface Structures
    094.139 .180 .362
        .378
Moon
  Temperatures
    094.113
Moon
  Thermal History
    094.002
Moon
  UV Spectra
    094.140
Moon
  Volcanism
    094.154 .389
Moon Dynamics
    094.000
Moon Global Properties
    094.100
Moon Local Properties
    094.300
Moon Surface
  Helium Densities
    094.152
Moon Surface
  Hydrogen Densities
    094.152
Moon Surface
  Radon Emission
    094.116 .313

Moon Surface
   Reference Points
      094.141
Moonquakes
   094.108 .151
Moustaches
   073.025
Multiple Galaxies
   158.000
Multiple Stars
   117.000
M51
   158.131 .132 .133
M87
   142.114

N-Body Problem
   151.021 .022 .023
      .024 .025 .027
      .028 .053
Navigation
   046.000
Navigation
   Space Vehicles
      052.000
Nearby Dwarfs
   Parallaxes
      111.010
Nearby Dwarfs
   Photometry
      113.013 .044
Nearby Dwarfs
   Proper Motions
      111.010
Nearby Galaxies
   158.068 .072 .117
Nearby Stars
   Ages
      155.053
Nearby Stars
   Kinematics
      155.051 .053
Nearby Stars
   Space Distribution
      155.053
Nebulae
   Cometary Nebulae
      132.023
Neptune
   101.000
Neptune
   Albedo
      099.001
Neptune
   Atmosphere
      101.009
Neptune
   Interior
      099.004
Neptune
   Mass
      101.007
Neptune
   Satellites
      101.007
Neptune
   Spectra
      101.006

Neutral Hydrogen
   Galactic Plane
      155.012
Neutrino Astronomy
   061.000
Neutron Stars
   065.037 .038 .039
      .063 .102 .103
      .117 .118 .120
      .122 .123
   066.025 .130
   142.072
Neutron Stars
   Accretion
      065.045 .075 .099
Neutron Stars
   Binaries
      065.062
Neutron Stars
   Close Binaries
      117.004
Neutron Stars
   Envelopes
      061.004
Neutron Stars
   Formation
      065.064
Neutron Stars
   Magnetic Fields
      062.066
      065.084
Neutron Stars
   Models
      065.035 .062 .085
Neutron Stars
   Opacities
      065.092
Neutron Stars
   Polarization
      065.092
Neutron Stars
   Pulsars
      065.040 .054
      141.310 .320 .327
         .342
Neutron Stars
   Pulsations
      065.058
Neutron Stars
   Rotation
      062.066
      065.091
Neutron Stars
   Stability
      065.058
Neutron Stars
   Supernova Remnants
      125.006
Neutron Stars
   Supernovae
      125.011
Neutron Stars
   X-Ray Sources
      142.014 .046
News Notes
   002.000
Night-Sky Radiation
   082.023 .044 .094
Noctilucent Clouds
   082.072

Nova IV Cephei 1971
   124.104
Nova Delphini 1967
   124.103
Nova DQ Herculis
   124.100
Nova Herculis 1963
   124.102
Nova VW Hydri
   124.101
Nova V1017 Sagittarii
   124.106
Nova FH Serpentis
   124.107
Nova RR Telescopii
   124.105
Novae
   124.000
Novae
   Andromeda Nebula
      124.005
Novae
   Envelopes
      124.001 .009
Novae
   Explosions
      124.002
Novae
   Gamma Rays
      124.004
Novae
   Infrared Photometry
      124.003
Novae
   UV Photometry
      124.010 .107
Novae
   X-Ray Sources
      114.047
Nuclear Reactions
   061.040
Nuclear Reactions
   Interplanetary Matter
      091.016
Nuclear Reactions
   Solar Atmosphere
      073.015
Nuclear Reactions
   Solar Interior
      080.014
Nuclear Reactions
   Stellar Interiors
      065.001 .009 .023
         .047 .119
Nuclear Reactions
   Supermassive Stars
      065.121
Nuclear Reactions
   Supernovae
      125.001
Nucleosynthesis
   Massive Stars
      065.087
Nucleosynthesis
   Stellar
      065.008 .049 .095
      114.106
Nucleosynthesis
   Stellar Interiors
      065.026 .089 .094
         .104

Nucleosynthesis
　Supernovae
　　125.027

O Stars
　Distribution
　　155.033
O Stars
　H Alpha
　　114.016
O Stars
　Spectra
　　114.037
O Stars
　Spectrophotometry
　　114.050
OB Associations
　　151.048
OB Associations
　Formation
　　065.003
OB Stars
　　158.104
OB Stars
　Atmospheres
　　064.007
OB Stars
　Galactic Clusters
　　153.010
OB Stars
　Magellanic Clouds
　　159.003
OB Stars
　Photometry
　　113.026
Obituaries
　　007.000
Oblateness
　Mars
　　097.019
Obliquity
　Mercury
　　092.031
Observatories
　　008.000
Occultations
　Lunar
　　096.000
OH
　Interstellar Matter
　　131.078 .081 .092
OH Absorption
　Interstellar Matter
　　131.020
OH Emission
　Infrared Stars
　　141.609
OH Sources
　　114.018
　　131.013 .019 .023
　　　.046 .092 .502
　　　.523
OH Sources
　Masers
　　131.070
OH Sources
　Positions
　　131.001
Oort's Constants
　　155.006

Opacities
　Interstellar Matter
　　131.040
　　142.001
Opacities
　Main-Sequence Stars
　　065.005
Opacities
　Neutron Stars
　　065.092
Opacities
　Stellar Atmospheres
　　064.003 .024
Opacities
　Stellar Envelopes
　　064.022
Open Clusters
　　153.000
Optics
　Astronomical
　　031.000
Orbit
　Moon
　　094.005
Orbits
　Earth Satellites
　　052.009 .012 .014
　　　.015 .020 .021
　　　.024 .031 .035
　　　.036 .037 .038
　　　.041 .045 .046
Orbits
　Eclipsing Variables
　　121.003 .038 .047
　　　.087
Orbits
　Meteors
　　104.015 .022
Orbits
　Minor Planets
　　042.043
Orbits
　Periodic
　　042.002
Orbits
　Resonances
　　042.015
Orbits
　Spectroscopic Binaries
　　119.006 .008 .012
　　　.015
Orbits
　Visual Binaries
　　118.001 .002 .003
　　　.005 .010 .019
　　　.022 .027 .028
　　　.030 .031 .033
Organizations
　　010.000
Orion Nebula
　　131.060 .134 .531
　　　.533 .535
　　132.004 .014 .029
　　　.030 .034 .035
　　133.010
Orion Nebula
　Flare Stars
　　122.070
Orion Nebula
　Helium Abundance
　　132.012 .019

Orion Nebula
　Infrared Photometry
　　132.017
Orion Nebula
　Magnetic Field
　　132.025
Orion Nebula
　Radio Radiation
　　131.068
Orion Nebula
　Spectra
　　132.013 .022
Oscillations
　Chromosphere
　　077.025
Oscillations
　Eclipsing Variables
　　121.085
Oscillations
　Gaseous Spheres
　　065.011
Oscillations
　Photosphere
　　071.027 .047 .048
　　　.049
Oscillations
　Polytropes
　　065.057
Oscillations
　Solar Atmosphere
　　080.001 .004
Oscillations
　Solar Corona
　　077.056
Oscillations
　Solar Interior
　　080.013 .058
Oscillations
　Stellar Evolution
　　065.020
Oscillations
　Stellar Models
　　066.104
Oscillator Strengths
　　022.020 .028 .033
　　　.038 .039 .056
　　　.062

Parallaxes
　Nearby Dwarfs
　　111.010
Parallaxes
　Red Dwarfs
　　111.001
Parallaxes
　Spectroscopic Binaries
　　119.006
Parallaxes
　Stars
　　111.000
Parallaxes
　Visual Binaries
　　118.008 .010 .025
Peculiar A Stars
　　064.008
　　113.031
　　114.025 .043 .083
　　　.156
　　116.001 .002 .012

Peculiar A Stars
  Absorption Lines
    114.171
Peculiar A Stars
  Atmospheres
    064.041
Peculiar A Stars
  Catalogues
    114.097
Peculiar A Stars
  Element Abundances
    114.051
Peculiar A Stars
  Helium Abundance
    065.050
Peculiar A Stars
  Line Identifications
    114.078 .111
Peculiar A Stars
  Line Profiles
    114.026
Peculiar A Stars
  Magnetic Accretion
    116.010
Peculiar A Stars
  Magnetic Fields
    116.005
Peculiar A Stars
  Photometry
    114.006
Peculiar A Stars
  Rotation
    114.067
Peculiar A Stars
  Spectra
    114.068 .082
    116.013
Peculiar A Stars
  Spectroscopic Binaries
    119.009 .011
Peculiar A Stars
  UV Spectra
    114.078
Peculiar B Stars
  Helium Abundance
    065.050
Peculiar B Stars
  Luminosities
    115.016
Peculiar B Stars
  Magnetic Fields
    116.007
Peculiar G Stars
  Fine Analyses
    114.053
Peculiar G Stars
  Spectra
    114.053
Peculiar Objects
    158.300
Peculiar Stars
  Magellanic Clouds
    122.117
Peculiar Stars
  Spectra
    114.036
Periodic Orbits
  Stability
    042.002
Periodicals
    001.000

Perseids
    104.030 .032
Personal Notes
    006.000
Perturbation Theory
    042.042 .044 .054
    .068 .069 .071
    .075
    052.003 .007
Photography
    036.000
Photometers
    034.051 .052 .058
    .085
Photometric Systems
    113.003 .004 .005
    .031
Photometry
    113.000
Photometry
  A Stars
    155.021
Photometry
  B Stars
    113.026
Photometry
  Beta Cephei Stars
    122.043
Photometry
  Beta Lyrae Stars
    121.097
Photometry
  Carbon Stars
    113.028 .029 .032
    114.159
Photometry
  Cepheids
    122.002
Photometry
  Comets
    102.022
Photometry
  Delta Scuti Stars
    122.018
Photometry
  Dwarf Novae
    122.042
Photometry
  Early Type Stars
    113.025 .027 .039
Photometry
  Eclipsing Variables
    121.007 .009 .025
    .027 .039 .048
    .049 .050 .051
    .058 .061 .068
    .071
Photometry
  Emission-Line Stars
    114.121
Photometry
  Flare Stars
    113.023
    122.097
Photometry
  G Stars
    114.088
Photometry
  Galactic Clusters
    113.014
    153.002 .008 .009

Photometry
  Galactic Clusters
    153.013 .019 .020
    .022 .023 .024
    .029 .032 .033
Photometry
  Galactic Nebulae
    113.045
    132.002
Photometry
  Galaxies
    158.015 .019 .039
    .042 .056 .091
    .116
Photometry
  Gaseous Nebulae
    132.016
Photometry
  Globular Clusters
    113.014
    154.002 .005 .007
    .013 .014 .015
    .017 .019 .028
    .029
Photometry
  Helium-Weak Stars
    114.060
Photometry
  High-Speed
    031.011
Photometry
  Infrared Sources
    141.612
Photometry
  Jupiter
    099.023
Photometry
  K Stars
    114.088 .173
Photometry
  Late Type Stars
    114.070 .075
Photometry
  Magnetic Stars
    113.018
Photometry
  Mars
    091.023
    097.015
Photometry
  Metallic-Line Stars
    113.043
Photometry
  Minor Planets
    098.012 .017
Photometry
  Multicolor
    113.013 .019 .035
    .043 .044
Photometry
  Nearby Dwarfs
    113.013 .044
Photometry
  OB Stars
    113.026
Photometry
  Peculiar A Stars
    114.006
Photometry
  Photosphere
    071.040

Photometry
  Planetary Nebulae
    141.612
Photometry
  Planets
    091.023
Photometry
  Pluto
    101.005
Photometry
  Pulsating Stars
    122.094
Photometry
  Saturn
    100.007
Photometry
  Seyfert Galaxies
    158.062
Photometry
  Shell Stars
    113.021
Photometry
  Solar Corona
    074.050
Photometry
  Southern Stars
    113.035
Photometry
  Star Clusters
    153.025
Photometry
  Stellar Associations
    152.003
Photometry
  Sunspots
    072.034
Photometry
  Supernovae
    125.037
Photometry
  T Tauri Stars
    122.010
Photometry
  UBV
    113.012 .024
Photometry
  Variables
    122.098
Photometry
  Visual Binaries
    118.034
Photometry
  Wolf-Rayet Stars
    114.005
Photometry
  X-Ray Sources
    113.008
    117.030
    142.063 .067
Photometry
  Zodiacal Light
    106.006
Photomultipliers
    034.030 .049 .086
Photosphere
  Element Abundances
    071.004
Photosphere
  Faculae
    072.021

Photosphere
  Fine Structure
    071.015
Photosphere
  Magnetic Fields
    071.008 .012 .022
          .026 .041 .042
    073.038
    077.032
Photosphere
  Microturbulence
    071.001
Photosphere
  Models
    071.003 .016 .023
          .045
Photosphere
  Motions
    080.062
Photosphere
  Oscillations
    071.027 .047 .048
          .049
Photosphere
  Photometry
    071.040
Photosphere
  Polarization
    071.009
Photosphere
  Solar
    071.000
Photosphere
  Spectra
    071.016 .017 .046
Photosphere
  Temperatures
    071.024
Photosphere
  Velocity Fields
    071.013
Photospheres
  Stellar
    064.006
Physical Variables
    122.000
Physics
    022.000
Planetaria
    009.000
Planetary Atmospheres
    091.009 .026
    097.001
Planetary Atmospheres
  Carbon Dioxide
    091.017
Planetary Atmospheres
  Organic Molecules
    091.027
Planetary Atmospheres
  Radiative Transfer
    091.024
Planetary Atmospheres
  Reflection
    091.022
Planetary Atmospheres
  Scattering
    063.015
Planetary Atmospheres
  Solar Cosmic Rays
    091.008

Planetary Atmospheres
  Solar Wind
    074.022
Planetary Atmospheres
  Temperatures
    091.021
Planetary Atmospheres
  Water
    091.017
Planetary Nebulae
    131.511
    133.000
Planetary Nebulae
  Central Stars
    065.101
    133.004 .019
Planetary Nebulae
  Emission-Line Stars
    133.012
Planetary Nebulae
  Evolution
    133.020
Planetary Nebulae
  Filaments
    133.003
Planetary Nebulae
  Forbidden Lines
    133.010
Planetary Nebulae
  Heating
    133.015
Planetary Nebulae
  Helium Abundance
    133.005
Planetary Nebulae
  Infrared Radiation
    133.006 .007
Planetary Nebulae
  Line Profiles
    133.001
Planetary Nebulae
  Models
    133.002 .015
Planetary Nebulae
  Morphology
    133.002
Planetary Nebulae
  Nuclei
    126.014
    133.013
Planetary Nebulae
  Photometry
    141.612
Planetary Nebulae
  Positions
    133.011
Planetary Nebulae
  Radio Radiation
    133.008 .011
Planetary Nebulae
  Radio Sources
    141.018
Planetary Nebulae
  Shells
    133.017
Planetary Nebulae
  Structure
    133.014
Planetary Probes
    053.000

Planetary System
  Cosmogony
    107.000
Planetary System Physics
    091.000
Planetary Systems
  Extrasolar
    117.008
Planetary Theory
    042.001 .020 .037
    .042 .086
Planets
  Brightness Temperature
    091.011
Planets
  Evolution
    091.003
    107.005
Planets
  Figure
    091.000
Planets
  Giant Planets
    091.007
    099.062
Planets
  Gravity
    091.007
Planets
  Infrared Spectra
    114.101
Planets
  Interiors
    091.020
Planets
  Magnetospheres
    091.013
Planets
  Masses
    091.010
Planets
  Mean Elements
    091.010
Planets
  Minor
    098.000
Planets
  Models
    099.062
Planets
  Outer Planets
    042.066
    091.013
Planets
  Photometry
    091.023
Planets
  Surfaces
    094.183
Planets
  Terrestrial Planets
    091.003
Planets
  Thermal History
    091.003
Plasma
    062.000
Plasma
  Alfven Waves
    062.046 .051

Plasma
  Heating
    062.049
Plasma
  Hydromagnetic Waves
    062.047
Plasma
  Ionization
    062.049
Plasma
  Magnetic Fields
    062.012
Plasma
  Polarization
    062.056
Plasma
  Radiation Mechanisms
    062.002
Plasma
  Stability
    062.012
Plasma
  Wave Propagation
    062.034
Pleiades
    153.009 .012
Pleiades
  Flare Stars
    122.006 .020
Pleiades
  Spectrophotometry
    114.046
Pluto
    101.000
Pluto
  Atmosphere
    101.004
Pluto
  Motion
    101.001
Pluto
  Photometry
    101.005
Pluto
  Rotation
    101.005
Polar Motion
    045.000
Polarimeters
    034.036 .085
Polarization
  Binaries
    117.018
Polarization
  Cool Stars
    113.036
Polarization
  Crab Nebula
    134.001 .003
Polarization
  Early Type Stars
    131.082
Polarization
  Emission-Line Stars
    114.029 .161
Polarization
  Galactic Clusters
    131.052
Polarization
  Galaxies
    158.088

Polarization
  Interstellar Matter
    131.003 .039 .042
    .055 .083 .084
    .121
Polarization
  Magnetic Stars
    116.003 .009
Polarization
  Minor Planets
    098.010
Polarization
  Neutron Stars
    065.092
Polarization
  Photosphere
    071.009
Polarization
  Plasma
    062.056
Polarization
  Pulsars
    141.315 .318 .333
    .341
Polarization
  Quasars
    141.050
Polarization
  Radio Sources
    141.014 .019 .054
Polarization
  Solar Corona
    074.006 .011 .049
    .071 .112
Polarization
  Solar Limb
    071.034
Polarization
  Stellar Atmospheres
    064.012
Polarization
  Stellar Envelopes
    064.046
Polarization
  Supergiants
    131.085
Polarization
  Supernova Remnants
    125.015 .024
Polarization
  Variables
    122.098 .099
Polarization
  White Dwarfs
    063.002
    126.007 .008 .010
    .014 .015 .016
    .021 .023
Polarization
  X-Ray Sources
    142.066
Polarization of
  Starlight
    131.000
Polytropes
    061.060
Polytropes
  Oscillations
    065.057

Population II Cepheids
  HR Diagrams
    122.095
Population II Stars
  Evolution
    065.060
Positional Astronomy
  041.000
Positions
  Mars
    097.039
Positions
  Mars Satellites
    097.039
Positions
  OH Sources
    131.001
Positions
  Planetary Nebulae
    133.011
Positions
  Radio Sources
    141.010 .051 .073
Positions
  Star Catalogues
    041.024 .025
Positions
  X-Ray Sources
    142.062
Pre-Main-Sequence Stars
  113.002
  114.029
  141.601
Precession
  155.075
Prominences
  073.000
Prominences
  Electron Densities
    073.048
Prominences
  Electron Temperatures
    073.027
Prominences
  Formation
    073.010 .041
Prominences
  Line Intensities
    073.029
Prominences
  Line Profiles
    073.047
Prominences
  Magnetic Fields
    073.010 .020
Prominences
  Models
    073.023
Prominences
  Physical Conditions
    073.063
Prominences
  Spectra
    073.004 .011 .033 .049
Prominences
  Spectrophotometry
    073.012
Proper Motion Surveys
  112.013 .019

Proper Motions
  112.000
Proper Motions
  Catalogues
    041.000
    112.006 .007 .008 .014 .015
Proper Motions
  Galactic Clusters
    112.009 .010 .011
    153.019
Proper Motions
  Nearby Dwarfs
    111.010
Proper Motions
  Pulsars
    141.330
Proper Motions
  Red Dwarfs
    111.001
Protogalaxies
  Models
    151.009
Protoplanets
  099.055
  107.009 .013
Protostars
  065.044
Protostars
  Envelopes
    065.073
Protostars
  Evolution
    065.014
Protostars
  Models
    065.013
Pulsar
  Crab Nebula
    134.002 .006
    141.306 .326 .333
Pulsar
  Vela
    125.010
Pulsars
  141.300
Pulsars
  Atmospheres
    141.331
Pulsars
  Binaries
    141.313
Pulsars
  Birthrates
    065.055
Pulsars
  Cosmic Rays
    143.066
Pulsars
  Distances
    141.325
Pulsars
  Gamma Rays
    061.034
    141.332 .340
Pulsars
  Gravitation
    141.302
Pulsars
  Interstellar Matter
    131.022

Pulsars
  Interstellar Matter
    141.343
Pulsars
  Magnetic Fields
    141.327
Pulsars
  Magnetospheres
    062.027 .028
    141.303 .312 .316 .324
Pulsars
  Models
    141.338 .342
Pulsars
  Neutron Stars
    065.040 .054
    141.310 .320 .327 .342
Pulsars
  OH Lines
    141.305
Pulsars
  Periods
    141.317
Pulsars
  Polarization
    141.315 .318 .333 .341
Pulsars
  Proper Motions
    141.330
Pulsars
  Pulse Intensities
    141.314
Pulsars
  Pulse Structure
    141.332 .339 .341
Pulsars
  Radio Radiation
    141.098 .304 .323 .341
Pulsars
  Radio Spectra
    141.319
Pulsars
  Rotation
    141.301 .327
    156.002
Pulsars
  Supernova Remnants
    125.006
Pulsars
  Variations
    141.308
Pulsars
  X Rays
    141.326 .328 .329
Pulsating Stars
  122.125
Pulsating Stars
  Line Profiles
    065.086
    122.088
Pulsating Stars
  Photometry
    122.094
Pulsating Stars
  Stability
    065.090

Pulsation Theory
　122.000
Pulsations
　Neutron Stars
　065.058
Pulsations
　Stellar Envelopes
　065.127
Pulsations
　Stellar Evolution
　065.020 .025
Pulsations
　Stellar Interiors
　065.059 .093
Pulsations
　Supermassive Stars
　065.027
Pulsations
　White Dwarfs
　122.064
Pulsations
　Wolf-Rayet Stars
　114.093
Pulsations
　X-Ray Sources
　142.027 .029 .042
　　.061 .111 .116
　　.117

Quadrantids
　104.027
Quasars
　141.000
Quasars
　Absolute Magnitudes
　141.082
Quasars
　Brightnesses
　158.032
Quasars
　Chains
　141.044
Quasars
　Close Pairs
　141.124
Quasars
　Clusters of Galaxies
　141.053
Quasars
　Counts
　141.036
Quasars
　Identifications
　141.125
Quasars
　Intergalactic Matter
　141.057
Quasars
　Luminosities
　141.026 .089
Quasars
　Masses
　141.068
Quasars
　Models
　141.036
Quasars
　Nuclei
　141.040

Quasars
　Polarization
　141.050
Quasars
　Radio Radiation
　141.043
Quasars
　Radio Structure
　141.005
Quasars
　Redshifts
　141.021 .026 .059
　　.082 .089 .119
　　.123 .128
　162.018
Quasars
　Samples
　141.013
Quasars
　Spectra
　141.076
Quasars
　Spectrophotometry
　141.063
Quasars
　Variations
　141.050 .060
Quasi-Stellar Objects
　141.029 .058 .088
　　.111 .120 .121
　　.122
Quasi-Stellar Objects
　Atmospheres
　064.025
Quasi-Stellar Objects
　Binaries
　141.024
Quasi-Stellar Objects
　Close Pairs
　141.067
Quasi-Stellar Objects
　Colors
　141.084
Quasi-Stellar Objects
　Mass Loss
　141.001
Quasi-Stellar Objects
　Models
　141.062
Quasi-Stellar Objects
　Radio Radiation
　141.025 .080
Quasi-Stellar Objects
　Redshifts
　141.003 .083 .084
　　.097
Quasi-Stellar Objects
　Variations
　141.002 .060

R CrB Variables
　065.047
　122.013
Radar Echoes
　Meteors
　104.011 .050
Radar Echoes
　Venus
　093.028

Radar Maps
　Moon
　094.157 .367 .368
　　.369 .370
Radar Maps
　Venus
　093.018
Radial Velocities
　112.000
Radial Velocities
　Carbon Stars
　112.001
Radial Velocities
　Catalogues
　112.002 .017
Radial Velocities
　Clusters of Galaxies
　160.014
Radial Velocities
　Early Type Stars
　155.045
Radial Velocities
　Galaxies
　158.122
Radial Velocities
　Spectroscopic Binaries
　119.002 .016
Radial Velocities
　Visual Binaries
　118.004
Radial Velocities
　White Dwarfs
　126.001
Radiation Belts
　084.000
Radiation Belts
　Jupiter
　099.019 .044 .057
　　.058
Radiative Transfer
　063.000
Radiative Transfer
　Circumstellar Matter
　131.028
Radiative Transfer
　Circumstellar Shells
　064.061
Radiative Transfer
　Galactic Nebulae
　132.006 .026
Radiative Transfer
　Mars Atmosphere
　093.012
　097.045
Radiative Transfer
　Planetary Atmospheres
　091.024
Radiative Transfer
　Reflection Nebulae
　132.024
Radiative Transfer
　Solar Atmosphere
　071.043
Radiative Transfer
　Stellar Atmospheres
　064.036
Radiative Transfer
　Stellar Envelopes
　064.040

Radiative Transfer
  Venus Atmosphere
    093.012
    097.045
Radio Equipment
  033.000
Radio Galaxies
  158.050 .074
Radio Galaxies
  Clusters of Galaxies
    160.012
Radio Galaxies
  H I Regions
    131.516
Radio Galaxies
  Redshifts
    141.119
Radio Lines
  Interstellar Matter
    131.034
Radio Radiation
  Clusters of Galaxies
    160.001 .004 .005
    .016 .019 .025
Radio Radiation
  Flare Stars
    122.127
Radio Radiation
  Galactic
    157.000
Radio Radiation
  Galactic Center
    155.003 .008 .027
    .040
    157.003
Radio Radiation
  Galaxies
    158.040 .060
Radio Radiation
  H II Regions
    131.504 .511 .521
    .525 .536
    158.082
Radio Radiation
  Jupiter
    099.054 .059
Radio Radiation
  Mercury
    092.029
Radio Radiation
  Moon
    094.164 .171
Radio Radiation
  Orion Nebula
    131.068
Radio Radiation
  Planetary Nebulae
    133.008 .011
Radio Radiation
  Pulsars
    141.098 .304 .323
    .341
Radio Radiation
  Quasars
    141.043
Radio Radiation
  Quasi-Stellar Objects
    141.025 .080
Radio Radiation
  Seyfert Galaxies
    158.303

Radio Radiation
  Solar Corona
    074.042 .047
Radio Radiation
  Solar Flares
    073.008
    077.068
Radio Radiation
  Supernova Remnants
    125.058
    141.030
Radio Radiation
  Supernovae
    125.034
Radio Radiation
  T Tauri Stars
    114.073
Radio Radiation
  X-Ray Sources
    142.002 .036
Radio Sources
    141.000
Radio Sources
  Angular Diameters
    141.055 .056
Radio Sources
  Binaries
    141.016 .035
Radio Sources
  Catalogues
    141.086
Radio Sources
  Clusters of Galaxies
    141.012 .028
Radio Sources
  Counts
    141.011 .015
Radio Sources
  Distribution
    141.015
Radio Sources
  Extragalactic
    141.008 .009 .012
    .016 .023 .027
    .052 .073 .074
    .090 .109
    151.050
Radio Sources
  Fine Structure
    141.007
Radio Sources
  Flux Densities
    141.009 .038 .041
    .046 .071 .078
    .086 .109
Radio Sources
  Galaxies
    141.045
    158.088
Radio Sources
  Identifications
    141.119 .121
Radio Sources
  Lunar Occultations
    141.051 .087
Radio Sources
  Magnetic Fields
    141.037
Radio Sources
  Models
    141.014 .027 .079

Radio Sources
  Optical Identification
    141.010 .042 .065
    .066 .069 .070
    .072
    158.089
Radio Sources
  Planetary Nebulae
    141.018
Radio Sources
  Polarization
    141.014 .019 .054
Radio Sources
  Positions
    141.010 .051 .070
Radio Sources
  Redshifts
    141.099
Radio Sources
  Scintillations
    141.007 .064 .127
Radio Sources
  Spectra
    141.004 .008 .022
    .037
Radio Sources
  Spectroscopic Binaries
    119.010
Radio Sources
  Surveys
    141.007 .017 .047
    .086 .109
Radio Sources
  Synchrotron Radiation
    062.025
    141.037
Radio Sources
  Variations
    141.030 .039 .060
    .075 .088 .123
Radio Sources
  X Rays
    142.068
Radio Spectra
  Jupiter
    099.026
Radio Spectra
  Pulsars
    141.319
Radio Structure
  Galactic Disk
    155.055
Radio Surveys
  Interacting Galaxies
    158.059
Radio Telescopes
  033.000
Radiometers
  034.043
Radius
  Mercury
    092.040
Recombination Lines
  H I Regions
    131.501 .535 .544
Recombination Lines
  H II Regions
    131.540
Recombination Lines
  Interstellar Matter
    131.053 .081

Recombination Lines
Radio Frequencies
131.059 .501 .540
.544
132.004 .019
Recombination Lines
Supernovae
131.059
Red Dwarfs
Hyades
111.001
Red Dwarfs
Parallaxes
111.001
Red Dwarfs
Proper Motions
111.001
Red Giants
Envelopes
064.017
Red Giants
Globular Clusters
113.007
154.013
Red Giants
Infrared Photometry
113.007
114.008
Red Giants
Reflection Nebulae
114.008
Redshifts
Clusters of Galaxies
160.007 .010 .014
.015
Redshifts
Galaxies
131.553
141.097
158.004 .006 .047
162.023
Redshifts
Quasars
141.021 .026 .059
.082 .089 .119
.123 .128
162.018
Redshifts
Quasi-Stellar Objects
141.003 .083 .084
.097
Redshifts
Radio Galaxies
141.119
Redshifts
Radio Sources
141.099
Redshifts
Wolf-Rayet Binaries
117.017
Reflection Nebulae
132.000
Reflection Nebulae
Radiative Transfer
132.024
Reflection Nebulae
Red Giants
114.008
Refraction
012.023
031.052 .053

Refraction
Interstellar Matter
041.001 .058
082.000
Refractors
032.036
Relativistic
Astrophysics
066.000
Relativistic Stars
Rotation
065.033
Relativistic Stars
Stability
065.078 .079
066.052
Relativity Theory
Tests
066.129
Resonances
Minor Planets
098.004
Resonances
Orbits
042.015
Resonances
Saturn Satellites
100.201 .203
Rotating Fluids
Equilibrium Figures
062.020 .021
Rotating Masses
Structure
065.034
Rotating Stars
Line Profiles
114.071
Rotating Stars
Models
065.016
Rotating Stars
Stability
065.051
066.023
Rotation
Be Stars
114.077
Rotation
Binaries
117.026
Rotation
Deformable Bodies
042.018 .019
Rotation
Earth
044.000
Rotation
Emission-Line Stars
114.163
Rotation
Galaxies
158.008 .026 .035
.080
Rotation
Magnetic Stars
062.026
Rotation
Moon
094.138

Rotation
Neutron Stars
062.066
065.091
Rotation
Peculiar A Stars
114.067
Rotation
Pluto
101.005
Rotation
Pulsars
141.301 .327
156.002
Rotation
Relativistic Stars
065.033
Rotation
Solar Corona
074.009
Rotation
Stars
116.000
Rotation
Stellar Atmospheres
126.017
Rotation
Stellar Envelopes
064.015
Rotation
Stellar Evolution
065.064
Rotation
Sun
080.000
Rotation
Supermassive Stars
065.096
Rotation
Venus
093.003
Rotation
Visual Binaries
118.014
Rotation
White Dwarfs
126.012
RR Lyrae Stars
122.046 .048 .050
.051 .123
RR Lyrae Stars
Absolute Magnitudes
122.100
RR Lyrae Stars
Globular Clusters
122.027 .060 .093
RR Lyrae Stars
Light Curves
122.066
RR Lyrae Stars
Magellanic Clouds
122.130
RV Tauri Stars
Spectra
122.071
RW Aurigae Stars
Infrared Photometry
122.062

Sampson's Theory
    099.206 .217 .221
Satellite Geodesy
    046.000
Satellites
    Neptune
    101.007
Satellites
    Retrograde
    042.041
Saturn
    012.003
    100.000
Saturn
    Albedo
    091.015
Saturn
    Brightness Temperature
    093.001
Saturn
    Chemical Composition
    099.076
Saturn
    Formation
    099.043
Saturn
    Interior
    099.004 .029
Saturn
    Models
    099.075
Saturn
    Photometry
    100.007
Saturn
    Rings
    091.004
    100.003 .006 .009
        .010 .011 .013
        .014 .015
Saturn
    Spectra
    099.005
Saturn
    Temperatures
    099.038
Saturn Atmosphere
    099.030
    100.002
Saturn Atmosphere
    Chemical Composition
    099.045
Saturn Atmosphere
    Infrared Spectra
    100.001
Saturn Atmosphere
    Spectra
    099.063
Saturn Atmosphere
    Spectrophotometry
    100.004
Saturn Satellites
    100.200
Saturn Satellites
    Albedo
    100.205 .208
Saturn Satellites
    Light Variations
    100.210

Saturn Satellites
    Methane
    100.205
Saturn Satellites
    Resonances
    100.201 .203
Scattering
    063.000 .009 .014
Scattering
    Jupiter Atmosphere
    099.003
Scattering
    Planetary Atmospheres
    063.015
Scattering
    Solar Corona
    077.034 .035
Schmidt Telescopes
    032.028 .048 .049
Scintillation
    082.000
Scintillations
    Interplanetary Matter
    106.018 .038
    143.021 .022
Scintillations
    Radio Sources
    141.007 .064 .127
Secular Perturbations
    042.001
Seeing
    031.050
Seyfert Galaxies
    Colors
    158.303
Seyfert Galaxies
    Infrared Spectra
    158.034
Seyfert Galaxies
    Nuclei
    141.058 .061
    158.031 .032 .033
        .062 .073 .089
        .092 .093
Seyfert Galaxies
    Optical Variations
    158.029
Seyfert Galaxies
    Photometry
    158.062
Seyfert Galaxies
    Radio Radiation
    158.303
Seyfert Galaxies
    Spectra
    158.030 .063
Seyfert Galaxies
    Spectrophotometry
    158.093
Shell Stars
    Mass Loss
    064.001
Shell Stars
    Photometry
    113.021
Shell Stars
    Spectra
    114.035
    119.016

Shock Waves
    Interplanetary Matter
    106.031 .036
Shock Waves
    Interplanetary Space
    106.004
Shock Waves
    Interstellar Gas
    131.101
Shock Waves
    Solar Corona
    074.063 .064 .083
Shock Waves
    Solar Wind
    074.055
Shock Waves
    Stellar Atmospheres
    064.060
Shock Waves
    Stellar Models
    065.115
Shock Waves
    Supernova Remnants
    125.013
Shock Waves
    Supernovae
    125.001 .027
Silicon Stars
    114.022
Site Testing
    082.000
Societies
    010.000
Solar Activity
    072.000
Solar Activity
    Cycles
    072.012 .024 .025
        .044 .045
Solar Activity
    Distribution
    072.006
Solar Atmosphere
    Absorption Coefficient
    080.063
Solar Atmosphere
    Alfven Waves
    080.038
Solar Atmosphere
    Models
    071.051
    080.028
Solar Atmosphere
    Neutrons
    076.013
    080.032
Solar Atmosphere
    Nuclear Reactions
    073.015
Solar Atmosphere
    Oscillations
    080.001 .004
Solar Atmosphere
    Radiative Transfer
    071.043
Solar Chromosphere
    073.000
Solar Constant
    080.030
Solar Corona
    074.000

Solar Corona
  Active Regions
    074.075
Solar Corona
  Brightness
    074.070
Solar Corona
  Cosmic Rays
    074.026
Solar Corona
  Cycles
    074.030
Solar Corona
  Disturbances
    074.043
Solar Corona
  Electron Densities
    022.010
    074.068 .070
Solar Corona
  Electron Temperatures
    074.031
Solar Corona
  Element Abundances
    074.025
Solar Corona
  Emission Lines
    074.061 .066
Solar Corona
  Fe XIII Lines
    022.010
Solar Corona
  Heating
    074.052 .083
Solar Corona
  Line Identifications
    074.025
Solar Corona
  Line Profiles
    074.011 .046
Solar Corona
  Magnetic Fields
    074.029 .044 .048
    .061 .078 .087
Solar Corona
  Models
    066.002
    074.018
Solar Corona
  Oscillations
    077.056
Solar Corona
  Particles
    074.003
Solar Corona
  Photometry
    074.050
Solar Corona
  Polar Rays
    074.051
Solar Corona
  Polarization
    074.006 .011 .049
    .071 .112
Solar Corona
  Radio Radiation
    074.042 .047
Solar Corona
  Rotation
    074.009

Solar Corona
  Scattering
    077.034 .035
Solar Corona
  Shock Waves
    074.063 .064 .083
Solar Corona
  Spectra
    074.002 .012
Solar Corona
  Structure
    074.062 .089 .114
Solar Corona
  Temperatures
    074.013 .041
Solar Corona
  Thermal Emission
    074.008
Solar Corona
  Transients
    074.005 .027 .028
    077.018
Solar Corona
  X Rays
    074.025 .074
    076.002 .006
Solar Cosmic Radiation
    078.000
Solar Cosmic Rays
  Acceleration
    078.001 .014
Solar Cosmic Rays
  Electrons
    078.002 .007 .021
    .022 .029 .032
    .042
Solar Cosmic Rays
  Element Abundances
    078.042
Solar Cosmic Rays
  Particles
    078.008 .009 .010
    .011 .019 .023
    .025 .026 .030
    .032 .033 .034
    .035
Solar Cosmic Rays
  Planetary Atmospheres
    091.008
Solar Cosmic Rays
  Propagation
    106.014
Solar Cosmic Rays
  Protons
    077.063
    078.003 .005 .012
    .015 .017 .018
    .027 .028 .030
    .031 .032 .042
    .043
Solar Eclipse
  1966 November 12
    079.104
Solar Eclipse
  1971 February 25
    079.102
Solar Eclipse
  1972 July 10
    079.103

Solar Eclipse
  1973 June 30
    079.100
Solar Eclipse
  1973 December 24
    079.101
Solar Eclipse
  1974 June 20
    079.105
Solar Eclipses
    079.000
Solar Flares
    073.000
Solar Flares
  Cooling
    073.053
Solar Flares
  Cosmic Rays
    078.005 .045
Solar Flares
  Disturbances
    073.054 .055
    077.019
Solar Flares
  Electrons
    073.009 .032
Solar Flares
  Extreme UV
    076.005
Solar Flares
  Fine Structure
    073.008
Solar Flares
  H Alpha
    073.013
Solar Flares
  Line Identifications
    073.001
Solar Flares
  Lyman Alpha
    073.018
Solar Flares
  Magnetic Fields
    073.042 .057
    077.026
Solar Flares
  Models
    073.006 .007 .030
    .031 .062
Solar Flares
  Particles
    073.058
Solar Flares
  Protons
    073.009 .019 .024
    .032
    078.037
Solar Flares
  Radio Radiation
    073.008
    077.068
Solar Flares
  Spectra
    022.027
    073.031 .035 .052
    .060
Solar Flares
  X Rays
    062.004
    076.005 .020 .021

Solar Gamma Rays
076.000
Solar Interior
 Convection
 062.011
 080.015 .035 .036
 .037
Solar Interior
 Nuclear Reactions
 080.014
Solar Interior
 Oscillations
 080.013 .058
Solar Interior
 Temperatures
 080.037
Solar Limb
 Anomalous Redshifts
 080.002
Solar Limb
 Light Deflection
 066.087
Solar Limb
 Line Profiles
 080.031 .033
Solar Limb
 Polarization
 071.034
Solar Magnetic Fields
 080.010 .011 .021
 .026 .040 .041
 .042 .061 .062
Solar Motion
 041.026
 155.028 .031
Solar Neighborhood
 Flare Stars
 122.115
Solar Neighborhood
 Stellar Birthrate
 065.052
Solar Neighborhood
 Stellar Population
 155.002
Solar Neutrinos
 066.089
 080.012 .014 .016
 .018 .022 .029
 .039 .055 .057
 .059 .060
Solar Oblateness
 072.035
 080.023 .024
Solar Patrol
 075.000
Solar Photosphere
 071.000
Solar Radio Bursts
 073.055
 074.003 .004 .013
 .026 .044
 077.003 .006 .008
 .009 .013 .014
 .015 .016 .017
 .018 .019 .020
 .024 .026 .027
 .031 .033 .034
 .035 .036 .039
 .040 .041 .051
 .052 .053 .054
 .055 .056 .057

Solar Radio Bursts
 077.058 .063 .064
 .065 .066 .080
Solar Radio Radiation
 077.000
Solar Radio Radiation
 Fine Structure
 077.001 .021
Solar Radio Radiation
 Variations
 077.023 .032 .062
Solar Radio Spectra
 077.030
Solar Rotation
 080.000
Solar Seeing
 031.020
 092.005
Solar Spectrum
 071.000
Solar Spectrum
 Absorption Lines
 073.051
Solar Spectrum
 Beryllium Abundance
 071.029
Solar Spectrum
 Calcium Abundance
 071.031
Solar Spectrum
 Element Abundances
 071.036
Solar Spectrum
 Extreme UV
 074.046
Solar Spectrum
 Fraunhofer Lines
 071.006 .007 .021
 .044 .050
Solar Spectrum
 Germanium Abundance
 071.025
Solar Spectrum
 Line Broadening
 071.006
Solar Spectrum
 Line Formation
 080.006
Solar Spectrum
 Line Profiles
 071.007 .020 .066
 080.008 .009
Solar Spectrum
 Oxygen Abundance
 071.002
Solar Spectrum
 Thulium Abundance
 071.030
Solar Spectrum
 UV
 022.004
 073.016
 076.014
Solar System
 Chemical Composition
 107.014
Solar System
 Early History
 107.002

Solar System
 Origin
 107.010
Solar UV Radiation
 076.000
Solar Wind
 074.000
Solar Wind
 Alfven Waves
 074.007 .056 .058
Solar Wind
 Comets
 102.015
Solar Wind
 Cooling
 074.038
Solar Wind
 Dynamics
 074.088
Solar Wind
 Electron Temperatures
 074.031
Solar Wind
 Element Abundances
 074.110
Solar Wind
 Expansion
 074.014 .016
Solar Wind
 Heating
 062.064
 074.020
Solar Wind
 Helium Content
 074.073
Solar Wind
 Hydromagnetic Waves
 062.064
Solar Wind
 Interplanetary Matter
 074.037
 082.029
 106.037
Solar Wind
 Interstellar Matter
 143.030
Solar Wind
 Landau Damping
 077.027
Solar Wind
 Magnetic Fields
 074.014 .017 .021
 .084
Solar Wind
 Magnetosphere
 074.076
Solar Wind
 Mercury
 092.035
Solar Wind
 Models
 062.063
 074.053
Solar Wind
 Moon
 074.032 .079
 094.103 .150 .167
Solar Wind
 Planetary Atmospheres
 074.022

Solar Wind
  Shock Waves
    074.055
Solar Wind
  Structure
    072.006
    074.040
Solar Wind
  Turbulence
    074.001
Solar Wind
  Variations
    074.019
Solar Wind
  Velocities
    074.015 .045 .054
      .057 .082
    077.027
Solar Wind
  Venus Atmosphere
    093.021 .023
Solar X Rays
  076.000
Solar X Rays
  Bright Points
    076.009
Solar X Rays
  Bursts
    076.003 .008 .011
      .012 .039
Solar X Rays
  Fluxes
    076.015
Solar-Terrestrial
  Relations
    085.000
Southern Stars
  Photometry
    113.035
Southern Stars
  Surveys
    041.023
Space Motions
  112.000
Space Motions
  Cepheids
    155.015
Space Motions
  White Dwarfs
    126.001
Space Probes
  Observations
    055.000
Space Vehicles
  Navigation
    052.000
Space Velocities
  Cepheids
    122.017
Spaceflight
  051.000
Spectra
  Molecular
    022.057 .058
Spectral Classification
  114.063 .079 .097
    .103 .122 .152
  121.036
Spectral Types
  Carbon Stars
    114.098

Spectral Types
  F Stars
    114.063
Spectrographs
  032.047
  033.024
  034.024 .029 .039
    .048 .057 .083
Spectroheliograms
  073.026
Spectroheliograms
  Extreme UV
    071.008
Spectroheliograms
  Lyman Alpha
    076.001
Spectroheliographs
  034.035
Spectrometers
  034.002 .018 .038
    .041 .044 .051
    .053 .054 .078
Spectrophotometers
  034.001 .004
Spectrophotometry
  A Stars
    114.123 .124
Spectrophotometry
  Chromosphere
    073.012
Spectrophotometry
  Cygnus Loop
    132.015
Spectrophotometry
  Galaxies
    158.104
Spectrophotometry
  H II Regions
    131.529
Spectrophotometry
  Infrared Sources
    114.032
    141.608
Spectrophotometry
  Interstellar Matter
    131.002
Spectrophotometry
  Meteors
    104.040
Spectrophotometry
  O Stars
    114.050
Spectrophotometry
  Pleiades
    114.046
Spectrophotometry
  Prominences
    073.012
Spectrophotometry
  Quasars
    141.063
Spectrophotometry
  Saturn Atmosphere
    100.004
Spectrophotometry
  Seyfert Galaxies
    158.093
Spectrophotometry
  Supergiants
    114.012

Spectrophotometry
  UV
    114.007
Spectrophotometry
  White Dwarfs
    126.005 .019
Spectroscopic Binaries
  119.000
Spectroscopic Binaries
  Early Type
    119.005
Spectroscopic Binaries
  Hyades
    119.007
Spectroscopic Binaries
  Light Curves
    119.004
Spectroscopic Binaries
  Masses
    118.017
Spectroscopic Binaries
  Orbits
    119.006 .008 .012
      .015
Spectroscopic Binaries
  Parallaxes
    119.006
Spectroscopic Binaries
  Peculiar A Stars
    119.009 .011
Spectroscopic Binaries
  Radial Velocities
    119.002 .016
Spectroscopic Binaries
  Radio Sources
    119.010
Spectroscopic Binaries
  Spectra
    119.013
Spectroscopy
  114.000
Spectrum Variables
  114.005 .019 .026
    .049 .068 .084
    .094 .163 .165
  122.101
Spicules
  073.046
Spiral Arms
  155.043
Spiral Structure
  012.018
  131.073
  151.017 .018 .030
    .035 .036 .040
    .041 .048 .049
    .060
  155.002 .012 .055
  158.009 .133
Stability
  Neutron Stars
    065.058
Stability
  Periodic Orbits
    042.002
Stability
  Plasma
    062.012
Stability
  Pulsating Stars
    065.090

Stability
  Relativistic Stars
    065.078 .079
    066.052
Stability
  Rotating Stars
    065.051
    066.023
Stability
  Stellar Interiors
    065.090
Stability
  Stellar Models
    062.062
Stability
  Stellar Systems
    151.001 .003 .032
Stability
  Super-Massive Stars
    065.022
Star Catalogues
  Apparent Magnitudes
    113.020
Star Catalogues
  Comparisons
    041.030
Star Catalogues
  Declinations
    041.010 .011 .012
    .075
Star Catalogues
  Errors
    041.014
Star Catalogues
  Positions
    041.000 .024 .025
Star Catalogues
  Right Ascensions
    041.008
Star Clusters
  C-M Diagrams
    159.001
Star Clusters
  Dynamics
    151.020 .029 .038
Star Clusters
  Magellanic Clouds
    153.025
    159.001
Star Clusters
  Photometry
    153.025
Star Clusters
  Relativistic
    151.038
Star Formation
    065.003 .019 .116
    151.060
    155.002 .025
Star Formation
  Galaxies
    065.010
    158.006
Stars
  Diameters
    115.000
Stars
  Figure
    116.000

Stars
  Luminosities
    115.000
Stars
  Magnetic Field
    116.000
Stars
  Magnitudes
    113.000
Stars
  Masses
    115.000
Stars
  Parallaxes
    111.000
Stars
  Rotation
    116.000
Stars
  Temperatures
    114.000
Stellar Associations
    152.000
Stellar Associations
  Photometry
    152.003
Stellar Atmospheres
    064.000
Stellar Atmospheres
  Color Indices
    113.017
Stellar Atmospheres
  Convection
    064.018 .038 .043
Stellar Atmospheres
  Element Abundances
    064.042 .051 .059
Stellar Atmospheres
  Gravity-Darkening
    064.020
Stellar Atmospheres
  Instability
    064.045
Stellar Atmospheres
  Iron Abundance
    064.027
Stellar Atmospheres
  Line Profiles
    064.004 .013 .025
Stellar Atmospheres
  LTE Models
    064.044
Stellar Atmospheres
  Microturbulence
    064.019 .028
Stellar Atmospheres
  Models
    064.011 .024 .035
    .037 .049
    113.017
    114.074 .174
Stellar Atmospheres
  Opacities
    064.003 .024
Stellar Atmospheres
  Polarization
    064.012
Stellar Atmospheres
  Radiative Transfer
    064.036

Stellar Atmospheres
  Rotation
    126.017
Stellar Atmospheres
  Shock Waves
    064.060
Stellar Atmospheres
  Velocities
    064.019
Stellar Atmospheres
  X Rays
    064.029
Stellar Envelopes
    064.000
Stellar Envelopes
  Magnetic Fields
    064.015
Stellar Envelopes
  Main-Sequence Stars
    064.008
Stellar Envelopes
  Models
    064.001 .010
Stellar Envelopes
  Opacities
    064.022
Stellar Envelopes
  Polarization
    064.046
Stellar Envelopes
  Pulsations
    065.127
Stellar Envelopes
  Radiative Transfer
    064.040
Stellar Envelopes
  Rotation
    064.015
Stellar Envelopes
  Spectra
    064.033
Stellar Evolution
    065.000
Stellar Evolution
  Carbon Burning
    065.032
Stellar Evolution
  Galactic Clusters
    153.017
Stellar Evolution
  Galaxies
    158.100 .121
Stellar Evolution
  Helium Burning
    065.065
Stellar Evolution
  Magellanic Clouds
    065.097
Stellar Evolution
  Mass Loss
    065.021 .036
Stellar Evolution
  Oscillations
    065.020
Stellar Evolution
  Pulsations
    065.020 .025
Stellar Evolution
  Rotation
    065.064

Stellar Evolution
  Secular Stability
    065.006
Stellar Flares
  Gamma Rays
    061.035
Stellar Groups
    155.005 .021 .076
Stellar Groups
  Moving Groups
    155.022
Stellar Interiors
  Convection
    065.030 .060 .088
Stellar Interiors
  Cooling
    065.041
    142.009
Stellar Interiors
  Evolution
    065.101
Stellar Interiors
  Instabilities
    065.046
Stellar Interiors
  Magnetic Fields
    065.080
Stellar Interiors
  Nuclear Reactions
    065.001 .009 .023
    .047 .119
Stellar Interiors
  Nucleosynthesis
    065.026 .089 .094
    .104
Stellar Interiors
  Pulsations
    065.059 .093
Stellar Interiors
  Secular Stability
    065.007
Stellar Interiors
  Stability
    065.090
Stellar Models
  Core-Helium-Burning
    065.098
Stellar Models
  Differential Rotation
    062.062
Stellar Models
  Evolution
    065.015 .113
Stellar Models
  Line Formation
    064.027
Stellar Models
  Oscillations
    066.104
Stellar Models
  Secular Stability
    065.012 .074
Stellar Models
  Shock Waves
    065.115
Stellar Models
  Stability
    062.062
Stellar Occultations
  Jupiter
    099.037 .049 .050

Stellar Occultations
  Venus
    093.010
Stellar Orbits
  Galaxy
    155.032
Stellar Populations
    155.022
Stellar Populations
  Andromeda Nebula
    158.105
Stellar Rings
    113.010
Stellar Spectra
    114.000
Stellar Spectra
  Analyses
    114.009
Stellar Spectra
  Carbon Isotopes
    114.089
Stellar Spectra
  Chemical Composition
    114.074
Stellar Spectra
  Element Abundances
    114.066
Stellar Spectra
  Iron Abundance
    114.010
Stellar Spectra
  Line Identifications
    114.133
Stellar Spectra
  Line Profiles
    114.065
Stellar Spectra
  Silicon Lines
    114.021
Stellar Spectra
  Technetium Lines
    114.064
    122.044
Stellar Spectra
  UV
    114.130
Stellar Structure
    065.000
Stellar Systems
  Density Waves
    151.006 .036 .057
Stellar Systems
  Dynamics
    042.050
    151.000
Stellar Systems
  Kinematics
    151.000
Stellar Systems
  Models
    151.007 .010 .019
    .035 .044 .045
Stellar Systems
  Relativistic
    151.037
Stellar Systems
  Relaxation Mechanisms
    151.034
Stellar Systems
  Stability
    151.001 .003 .032

Stellar Systems
  Star Encounters
    151.004
Stellar Systems
  Velocity Distribution
    151.007
Stellar Winds
    064.001 .009 .025
    .058
    142.030 .059
Stephan's Quintet
    141.017
    158.001 .064
Stone Meteorites
    105.001 .109 .149
    .150 .161
Stonehenge
    004.030
Subdwarf B Stars
  Evolution
    065.002
Subdwarf O Stars
  Evolution
    065.002
Subdwarfs
    126.000
Sun
  Active Regions
    073.005
    077.002
    080.027
Sun
  Diameter
    080.031
Sun
  Differential Rotation
    071.033
    080.007 .036 .042
Sun
  Figure
    080.000
Sun
  Interior
    080.000
Sun
  Limb Darkening
    080.005
Sun
  Models
    080.016
Sun
  Rotation
    080.000
Sundials
    003.033 .114 .130
    035.003
Sunspot Groups
    071.035
    072.016 .020 .026
    .042 .043
Sunspots
    072.000
Sunspots
  Cooling
    072.023
Sunspots
  Cycles
    072.009
Sunspots
  Densities
    072.005

Sunspots
Distances
072.043
Sunspots
Distribution
072.036
Sunspots
Heat Transport
072.037
Sunspots
Light Bridges
072.011 .040
Sunspots
Magnetic Fields
072.004 .007 .027
.030 .033 .037
.040
Sunspots
Models
072.014 .023 .039
.041
Sunspots
Morphology
072.022
Sunspots
Motions
072.019
080.061
Sunspots
Penumbrae
072.022 .031
Sunspots
Photometry
072.034
Sunspots
Spectra
071.017
072.008 .017
Sunspots
Temperatures
072.005
Sunspots
Umbrae
072.001 .004 .022
.028 .032 .042
Super-Massive Stars
Stability
065.022
Supergiants
Atmospheres
064.039
Supergiants
Chemical Composition
065.083
Supergiants
Companion
114.014
Supergiants
Element Abundances
114.013
Supergiants
Evolution
065.094
Supergiants
F Type
114.039
Supergiants
Infrared Radiation
114.014

Supergiants
Magellanic Clouds
114.162
122.077
Supergiants
Oxygen-Rich
114.041
Supergiants
Polarization
131.085
Supergiants
Spectra
065.061
114.106 .169
Supergiants
Spectrophotometry
114.012
Supermassive Stars
Collapse
065.068
Supermassive Stars
Evolution
065.096
Supermassive Stars
Nuclear Reactions
065.121
Supermassive Stars
Pulsations
065.027
Supermassive Stars
Rotation
065.096
Supernova in IC 43
125.106
Supernova in NGC 493
125.104
Supernova in NGC 1058
125.103
Supernova in NGC 3310
125.109
Supernova in NGC 3627
125.105
Supernova in NGC 3916
125.110
Supernova in NGC 4038-39
125.111
Supernova in NGC 4156
125.107
Supernova in NGC 4414
125.112
Supernova in NGC 5161
125.108
Supernova in NGC 5236
125.101
Supernova in NGC 5253
125.102
Supernova in NGC 5457
125.100
Supernova Remnants
012.015
125.000
Supernova Remnants
Classification
125.025
Supernova Remnants
Envelopes
125.061
Supernova Remnants
Evolution
125.016 .026 .050
.059

Supernova Remnants
Galactic Center
155.003
Supernova Remnants
Galactic Distribution
125.072
Supernova Remnants
Galactic Interactions
125.028
Supernova Remnants
Gamma Rays
125.054
Supernova Remnants
Interstellar Matter
125.005 .028
Supernova Remnants
Magnetic Fields
125.060
Supernova Remnants
Models
125.064
Supernova Remnants
Neutron Stars
125.006
Supernova Remnants
Polarization
125.015 .024
Supernova Remnants
Pulsars
125.006
Supernova Remnants
Radio Radiation
125.058
141.030
Supernova Remnants
Shock Waves
125.013
Supernova Remnants
Spectra
125.051
132.015
Supernova Remnants
X Rays
125.013 .048 .049
.050 .064
Supernova Remnants
X-Ray Sources
125.010
Supernova Remnants
21 cm Radiation
125.019 .052
Supernovae
012.015
125.000
Supernovae
Binaries
117.038
125.007 .014
Supernovae
Distances
125.023
158.099
Supernovae
Explosions
061.004
Supernovae
Frequency Distribution
125.022
Supernovae
Gamma Rays
125.002 .009 .054

Supernovae
　Light Curves
　　125.038
Supernovae
　Models
　　125.008
Supernovae
　Neutron Stars
　　125.011
Supernovae
　Nuclear Reactions
　　125.001
Supernovae
　Nucleosynthesis
　　125.027
Supernovae
　Photometry
　　125.037
Supernovae
　Radio Radiation
　　125.034
Supernovae
　Recombination Lines
　　131.059
Supernovae
　Search
　　125.031 .032
Supernovae
　Shells
　　125.017
Supernovae
　Shock Waves
　　125.001 .027
Supernovae
　Spectra
　　125.040 .041
Supernovae
　Surveys
　　125.033
Supernovae
　White Dwarfs
　　125.003
Surges
　073.003
Symbiotic Stars
　064.017
　114.052 .061 .121
　　.165
Symposia Proceedings
　012.000
Symposia Reports
　011.000
Synchrotron Radiation
　063.036
Synchrotron Radiation
　Galactic Disk
　　155.023
Synchrotron Radiation
　Radio Sources
　　062.025
　　141.037

T Tauri Stars
　114.028
　131.066
T Tauri Stars
　Luminosities
　　115.015

T Tauri Stars
　Photometry
　　122.010
T Tauri Stars
　Radio Radiation
　　114.073
Taurids
　104.051
Teaching In Astronomy
　014.000
Tektites
　105.135 .163
Telescopes
　032.000
Telescopes
　Cassegrain Telescopes
　　032.035
Telescopes
　Schmidt Telescopes
　　032.028 .048 .049
Telescopes
　X-Ray Telescopes
　　032.054
Telescopes
　Zenith Telescopes
　　032.026
Television Cameras
　034.084
Television Photometry
　031.024
Temperatures
　B Stars
　　114.092 .117
Temperatures
　Chromosphere
　　072.005
Temperatures
　Gaseous Nebulae
　　132.016
Temperatures
　Interstellar Gas
　　131.057
Temperatures
　Jupiter
　　099.038
Temperatures
　Jupiter Atmosphere
　　099.050
Temperatures
　Moon
　　094.113
Temperatures
　Photosphere
　　071.024
Temperatures
　Planetary Atmospheres
　　091.021
Temperatures
　Saturn
　　099.038
Temperatures
　Solar Corona
　　074.013 .041
Temperatures
　Solar Interior
　　080.037
Temperatures
　Stars
　　114.000

Temperatures
　Sunspots
　　072.005
Temperatures
　Uranus
　　099.038
Temperatures
　Venus Atmosphere
　　093.014
Three-Body Problem
　042.009 .010 .029
　　.031 .038 .040
　　.050 .057 .059
　　.061
Three-Body Problem
　Restricted
　　042.002 .030 .032
　　.033 .039 .045
　　.046 .047 .051
　　.055 .056 .062
　　.065 .070 .073
　　.080
Time
　044.000
Titan
　003.065
　091.026
　100.202 .203 .205
　　.206 .207 .209
　　.215
Transit Circles
　032.024 .037 .039
　　.056 .057 .059
　　.061
Transit Cirlces
　032.038
Transition Probabilities
　022.004 .009 .015
　　.036 .037 .048
　　.056 .063 .074
　073.001
Transplutonian Planets
　101.000
Turbulence
　Earth Atmosphere
　　031.050
Turbulence
　Solar Wind
　　074.001
Turbulence
　Universe
　　162.004 .009 .030
Twilight
　082.044
Two-Body Problem
　042.048 .063

U Geminorum Stars
　122.096
　142.050
U Geminorum Stars
　X Rays
　　121.004
Universe
　Expansion
　　066.143
Universe
　Friedmann Universe
　　066.031

SUBJECT INDEX - VOL. 11

Universe
  Turbulence
    162.004 .009 .030
Universe Evolution
  162.000
Universe Structure
  162.000
Uranus
  101.000
Uranus
  Albedo
    091.015
    099.001
Uranus
  Atmosphere
    099.063
Uranus
  Interior
    099.004
Uranus
  Spectra
    101.006 .013
Uranus
  Temperatures
    099.038
UV Background
  106.005
UV Excess Stars
  113.010
UV Excess Stars
  Globular Clusters
    154.004
UV Photometry
  Early Type Stars
    113.006
UV Photometry
  Eclipsing Variables
    121.034
UV Photometry
  Novae
    124.010 .107
UV Radiation
  Galaxies
    158.020
UV Radiation
  Mars Atmosphere
    097.007
UV Sources
  064.014
UV Spectra
  A Stars
    114.114 .123 .124
UV Spectra
  B Stars
    114.116
UV Spectra
  Be Stars
    114.004
UV Spectra
  Beta Lyrae Stars
    121.059
UV Spectra
  Early Type Stars
    114.015 .110 .125
    .133
UV Spectra
  Emission-Line Stars
    121.030
UV Spectra
  Interstellar Matter
    113.001

UV Spectra
  Jupiter Atmosphere
    099.064
UV Spectra
  Late Type Stars
    114.072
UV Spectra
  Mars
    093.013
UV Spectra
  Moon
    094.140
UV Spectra
  Peculiar A Stars
    114.078
UV Spectra
  Venus
    093.013
UV Spectra
  Venus Atmosphere
    093.025
UV Spectrophotometry
  Eclipsing Variables
    121.067
UV Stars
  Globular Clusters
    065.089
UV Stars
  Interstellar Matter
    131.018

Variables
  Irregular
    122.035 .041 .079
    .084
Variables
  Long Period
    065.127
    114.075 .132
    122.008 .011 .019
    .079 .132
Variables
  Magellanic Clouds
    122.089 .117
Variables
  Photometry
    122.098
Variables
  Polarization
    122.098 .099
Variables
  Semiregular
    122.044 .054
Variables Catalogues
  120.000
Variables Eclipsing
  121.000
Variables Ephemerides
  120.000
Variables Observations
  123.000
Variables Physical
  122.000
Vela
  Pulsar
    125.010
Velocities
  H II Regions
    131.513

Velocities
  Solar Wind
    074.015 .045 .054
    .057 .082
    077.027
Velocities
  Stellar Atmospheres
    064.019
Velocities
  Venus Atmosphere
    093.034
Velocity Distribution
  Stellar Systems
    151.007
Velocity Fields
  H II Regions
    131.554
Velocity Fields
  Photosphere
    071.013
Velocity of Light
  022.050
Venus
  012.003
  093.000
Venus
  Albedo
    091.015
Venus
  Brightness Temperature
    093.001
Venus
  Gravity
    093.022
Venus
  Infrared Radiation
    093.020
Venus
  Internal Structure
    093.027
Venus
  Ionosphere
    093.004 .022 .036
    .037 .043
    097.056
Venus
  Magnetic Field
    093.023
    097.020
Venus
  Magnetosphere
    093.026
Venus
  Mass
    093.022
Venus
  Radar Echoes
    093.028
Venus
  Radar Maps
    093.018
Venus
  Rotation
    093.003
Venus
  Spectra
    093.008
Venus
  Stellar Occultations
    093.010

SUBJECT INDEX - VOL. 11

Venus
  UV Spectra
    093.013
Venus Atmosphere
    093.009 .010 .015
    .022 .033
Venus Atmosphere
  Ammonia
    093.035
Venus Atmosphere
  Carbon Dioxide
    093.041
Venus Atmosphere
  Circulation
    093.011 .024
Venus Atmosphere
  Clouds
    093.005 .042
    099.028
Venus Atmosphere
  Dust
    093.011
Venus Atmosphere
  Models
    093.031
Venus Atmosphere
  Radiative Transfer
    093.012
    097.045
Venus Atmosphere
  Solar Wind
    093.021 .023
Venus Atmosphere
  Structure
    093.024
Venus Atmosphere
  Temperatures
    093.014
Venus Atmosphere
  UV Spectra
    093.025
Venus Atmosphere
  Velocities
    093.034
Venus Atmosphere
  Water
    093.017 .035
Virgo
  Clusters of Galaxies
    113.009
Visual Binaries
  117.020
  118.000
Visual Binaries
  Absolute Magnitudes
    118.025
Visual Binaries
  Lists
    118.011
Visual Binaries
  Long Period
    118.013
Visual Binaries
  Lunar Occultations
    118.018
Visual Binaries
  Masses
    118.008 .017 .025
Visual Binaries
  Orbits
    118.001 .002 .003

Visual Binaries
  Orbits
    118.005 .010 .019
    .022 .027 .028
    .030 .031 .033
Visual Binaries
  Parallaxes
    118.008 .019 .025
Visual Binaries
  Photometry
    118.034
Visual Binaries
  Radial Velocities
    118.004
Visual Binaries
  Rotation
    118.014
Visual Multiple Stars
    118.000
VV Cephei Stars
    121.036

W UMa Stars
  117.031
  121.006 .010 .046
W UMa Stars
  Evolution
    121.069
W Virginis Stars
  065.007
  122.001 .029
W Virginis Stars
  Luminosities
    122.049
Water
  Interstellar Matter
    131.058
Water
  Planetary Atmospheres
    091.017
Water
  Venus Atmosphere
    093.017 .035
White Dwarfs
  065.025 .064
  126.000
White Dwarfs
  Absolute Magnitudes
    065.089
White Dwarfs
  Accretion
    065.099
White Dwarfs
  Binaries
    121.005
    122.064
    126.018
White Dwarfs
  Chemical Composition
    126.004
White Dwarfs
  Collapse
    125.003
White Dwarfs
  Cooling
    126.002
White Dwarfs
  Envelopes
    126.020

White Dwarfs
  Evolution
    126.012
White Dwarfs
  Magnetic Fields
    063.002
    126.006 .007
White Dwarfs
  Models
    126.002
White Dwarfs
  Polarization
    063.002
    126.007 .008 .010
    .014 .015 .016
    .021 .023
White Dwarfs
  Pulsations
    122.064
White Dwarfs
  Radial Velocities
    126.001
White Dwarfs
  Rotation
    126.012
White Dwarfs
  Space Motions
    126.001
White Dwarfs
  Spectra
    126.003 .017
White Dwarfs
  Spectrophotometry
    126.005 .019
White Dwarfs
  Supernovae
    125.003
White Dwarfs
  X Rays
    126.018
Wolf-Rayet Binaries
  Redshifts
    117.017
Wolf-Rayet Stars
    114.017 .024 .079
Wolf-Rayet Stars
  Atmospheres
    064.030 .037
Wolf-Rayet Stars
  Binaries
    114.093
    117.017
    119.001
Wolf-Rayet Stars
  Envelopes
    114.056
Wolf-Rayet Stars
  Infrared Spectra
    114.001
Wolf-Rayet Stars
  Mass Loss
    064.023
Wolf-Rayet Stars
  Models
    065.048
Wolf-Rayet Stars
  Photometry
    114.005
Wolf-Rayet Stars
  Pulsations
    114.093

SUBJECT INDEX - VOL. 11

Wolf-Rayet Stars
  Spectra
    064.030
    114.054 .055

X Rays
  Andromeda Nebula
    158.102
X Rays
  Cataclysmic Variables
    122.063
X Rays
  Clusters of Galaxies
    160.004 .006 .016
       .017
X Rays
  Cygnus Loop
    132.011
X Rays
  Galactic Center
    155.027
X Rays
  Galaxies
    158.096
X Rays
  Interstellar Matter
    131.040 .074
    142.001
X Rays
  Pulsars
    141.326 .328 .329
X Rays
  Radio Sources
    142.068
X Rays
  Solar
    076.000
X Rays
  Solar Corona
    074.025 .074
    076.002 .006
X Rays
  Solar Flares
    062.004
    076.005 .020 .021
X Rays
  Spectra
    143.034
X Rays
  Stellar Atmospheres
    064.029
X Rays
  Supernova Remnants
    125.013 .048 .049
       .050 .064
X Rays
  U Geminorum Stars
    121.004
X Rays
  White Dwarfs
    126.018
X-Ray Astronomy
  061.000
X-Ray Background
  061.003 .011
  142.045 .053 .113
X-Ray Sources
  142.000

X-Ray Sources
  Accretion
    065.075
X-Ray Sources
  Binaries
    114.081
    117.034
    121.001 .101
    122.102
    125.007
    141.098
    142.003 .006 .007
       .009 .010 .016
       .017 .018 .019
       .022 .025 .028
       .030 .034 .049
       .050 .054 .059
       .063 .069 .071
       .072 .073 .115
       .116 .117
X-Ray Sources
  Black Holes
    113.008
    142.010 .020 .025
       .046 .048 .049
       .069 .101
X-Ray Sources
  Bursts
    142.047
X-Ray Sources
  Catalogues
    142.035 .043
X-Ray Sources
  Clusters of Galaxies
    142.031 .114
    160.005
X-Ray Sources
  Compact
    022.011
X-Ray Sources
  Cosmic Rays
    142.008
X-Ray Sources
  Distances
    142.037 .112
X-Ray Sources
  Extragalactic
    141.610
    142.043 .096 .097
X-Ray Sources
  Galactic
    142.023 .032 .038
       .095 .098
X-Ray Sources
  Galactic Distribution
    142.053
X-Ray Sources
  Galactic Plane
    142.033
X-Ray Sources
  Gamma Rays
    142.021 .047
    155.001
X-Ray Sources
  Infrared Radiation
    142.015
X-Ray Sources
  Luminosities
    142.040

X-Ray Sources
  Magnetic Accretion
    142.016
X-Ray Sources
  Magnetic Fields
    142.027
X-Ray Sources
  Mass Loss
    142.002
X-Ray Sources
  Models
    142.012 .046 .070
       .110 .111
X-Ray Sources
  Neutron Stars
    142.014 .046
X-Ray Sources
  Novae
    114.047
X-Ray Sources
  Optical Identification
    113.011
    114.044
    142.032 .105 .106
X-Ray Sources
  Optical Observations
    142.065
X-Ray Sources
  Photometry
    113.008
    117.030
    142.063 .067
X-Ray Sources
  Polarization
    142.066
X-Ray Sources
  Positions
    142.062
X-Ray Sources
  Pulsations
    142.027 .029 .042
       .061 .111 .116
       .117
X-Ray Sources
  Radio Counterparts
    141.006
X-Ray Sources
  Radio Outbursts
    141.039
X-Ray Sources
  Radio Radiation
    142.002 .036
X-Ray Sources
  Soft X Rays
    142.044
X-Ray Sources
  Spectra
    142.051
X-Ray Sources
  Structure
    142.060
X-Ray Sources
  Supernova Remnants
    125.010
X-Ray Sources
  Transient
    142.004
X-Ray Sources
  Variations
    142.055 .060

X-Ray Sources
    21 cm Radiation
        131.532
X-Ray Telescopes
    032.054

Zenith Cameras
    032.029
Zenith Telescopes
    032.026 .030
Zeta Aurigae
    121.042
Zodiacal Light
    106.000

Zodiacal Light
    Photometry
        106.006
Zodiacal Light
    Spectra
        106.003
Zodiacal Light
    Variations
        106.010 .020

ASTRONOMY AND ASTROPHYSICS ABSTRACTS

A Publication of the
Astronomisches Rechen-Institut Heidelberg
Member of the Abstracting Board
of the International Council of Scientific Unions

Editors:
S. Böhme, W. Fricke, U. Güntzel-Lingner, F. Henn,
D. Krahn, U. Scheffer, G. Zech

Vol. 1 Literature 1969, Part 1, X + 435 pp. (1969)
Vol. 2 Literature 1969, Part 2, X + 516 pp. (1970)
Vol. 3 Literature 1970, Part 1, X + 490 pp. (1970)
Vol. 4 Literature 1970, Part 2, X + 562 pp. (1971)
Vol. 5 Literature 1971, Part 1, X + 505 pp. (1971)
Vol. 6 Literature 1971, Part 2, X + 560 pp. (1972)
Vol. 7 Literature 1972, Part 1, X + 526 pp. (1972)

Price: Vols. 1-7 each Cloth DM 72,-; US $ 29.40
Subscription price: each Cloth DM 57,60; US $ 23.50

Vol. 8 Literature 1972, Part 2, X + 594 pp. (1973)
Vol. 9 Literature 1973, Part 1, X + 610 pp. (1973)

Price: Vols. 8-9 each Cloth DM 78,-; US $ 31.90
Subscription price: each Cloth DM 62,40; US $ 25.50

Vol. 10 Literature 1973, Part 2, X + 661 pp. (1974)
Vol. 11 Literature 1974, Part 1, X + 579 pp. (1974)

Price: Vols. 10-11 each Cloth DM 86,-; US $ 35.10
Subscription price: each Cloth DM 68,80; US $ 28.10

Prices are subject to change without notice

Proceedings of the First European Astronomical Meeting

Held Under the Auspices of the International Astronomical Union in Athens, September 4-9, 1972 (in 3 volumes)

Vol. 1   **Solar Activity and Related Interplanetary and Terrestrial Phenomena**

Edited by J. Xanthakis. 78 figs. XV, 195 pages. 1973
Cloth DM 94,–; US $38.40. ISBN 3-540-06314-5

Vol. 2   **Stars and the Milky Way System**

Edited by L. N. Mavridis. 169 figs. XI, 368 pages. 1974
Cloth DM 138,–; US $56.30. ISBN 3-540-06383-8

Vol. 3   **Galaxies and Relativistic Astrophysics**

Edited by B. Barbanis and J.D. Hadjidemetriou.
61 figs. XII, 247 pages. 1974
Cloth DM 126,–; US $51.50 ISBN 3-540-06416-8

The First European Astronomical Meeting was attended by over 330 astronomers from 34 countries, many of them eminent in various fields of contemporary astronomy and astrophysics. The 24 general and invited papers and over 70 contributed papers contained in these three volumes range over such topics as solar activity, infrared astronomy, interstellar molecules, optical and radio work on nearby galaxies, pulsars, and high-energy astrophysics. Discussions are fully reported. There was in addition a special session of reports on the European Joint Activities (CESRA, EPS, ESO, ESRO, JOSO, and INTERCOSMOS) and the major National Projects (British Projects, INAG, Italian Projects, Max-Planck-Institut). Another session was devoted to plans for observing the 1973 solar eclipse.

**Springer-Verlag
Berlin Heidelberg New York**

GPSR Compliance

The European Union's (EU) General Product Safety Regulation (GPSR) is a set of rules that requires consumer products to be safe and our obligations to ensure this.

If you have any concerns about our products, you can contact us on

ProductSafety@springernature.com

In case Publisher is established outside the EU, the EU authorized representative is:

Springer Nature Customer Service Center GmbH
Europaplatz 3
69115 Heidelberg, Germany